Formulas from Algebra

Exponents

If all bases are nonzero:

$$u^m u^n = u^{m+n}$$

$$\frac{u^m}{u^n} = u^{m-n}$$

$$u^0 = 1$$

$$u^{-n} = \frac{1}{u^n}$$

$$(uv)^m = u^m v^m$$

$$(u^m)^n = u^{mn}$$

$$\left(\frac{u}{v}\right)^m = \frac{u^m}{v^m}$$

Radicals and Rational Exponents

If all roots are real numbers:

$$\sqrt[n]{uv} = \sqrt[n]{u} \cdot \sqrt[n]{v}$$

$$\sqrt[n]{\frac{u}{v}} = \frac{\sqrt[n]{u}}{\sqrt[n]{v}} \quad (v \neq 0)$$

$$\sqrt[m]{\sqrt[n]{u}} = \sqrt[mn]{u}$$

$$(\sqrt[n]{u})^n = u$$

$$\sqrt[n]{u^m} = (\sqrt[n]{u})^m$$

$$\sqrt[n]{u^n} = \begin{cases} |u| & n \text{ even} \\ u & n \text{ odd} \end{cases}$$

$$u^{1/n} = \sqrt[n]{u}$$

$$u^{m/n} = (u^{1/n})^m = (\sqrt[n]{u})^m$$

$$u^{m/n} = (u^m)^{1/n} = \sqrt[n]{u^m}$$

Special Products

$$(u + v)(u - v) = u^2 - v^2$$

$$(u + v)^2 = u^2 + 2uv + v^2$$

$$(u - v)^2 = u^2 - 2uv + v^2$$

$$(u + v)^3 = u^3 + 3u^2v + 3uv^2 + v^3$$

$$(u - v)^3 = u^3 - 3u^2v + 3uv^2 - v^3$$

Factoring Polynomials

$$u^2 - v^2 = (u + v)(u - v)$$

$$u^2 + 2uv + v^2 = (u + v)^2$$

$$u^2 - 2uv + v^2 = (u - v)^2$$

$$u^3 + v^3 = (u + v)(u^2 - uv + v^2)$$

$$u^3 - v^3 = (u - v)(u^2 + uv + v^2)$$

Inequalities

If $u < v$ and $v < w$, then $u < w$.

If $u < v$, then $u + w < v + w$.

If $u < v$ and $c > 0$, then $uc < vc$.

If $u < v$ and $c < 0$, then $uc > vc$.

If $c > 0$, $|u| < c$ is equivalent to $-c < u < c$.

If $c > 0$, $|u| > c$ is equivalent to $u < -c$ or $u > c$.

Quadratic Formula

If $a \neq 0$, the solutions of the equation $ax^2 + bx + c = 0$ are given by

$$x = \frac{-b \pm \sqrt{b^2 - 4ac}}{2a}.$$

Logarithms

If $0 < b \neq 1, 0 < a \neq 1, x, R, S, > 0$

$y = \log_b x$ if and only if $b^y = x$

$$\log_b 1 = 0$$

$$\log_b b = 1$$

$$\log_b b^y = y$$

$$b^{\log_b x} = x$$

$$\log_b RS = \log_b R + \log_b S$$

$$\log_b \frac{R}{S} = \log_b R - \log_b S$$

$$\log_b R^c = c \log_b R$$

$$\log_b x = \frac{\log_a x}{\log_a b}$$

Determinants

$$\begin{vmatrix} a & b \\ c & d \end{vmatrix} = ad - bc$$

Arithmetic Sequences and Series

$$a_n = a_1 + (n - 1)d$$

$$S_n = n\left(\frac{a_1 + a_n}{2}\right) \text{ or } S_n = \frac{n}{2}[2a_1 + (n - 1)d]$$

Geometric Sequences and Series

$$a_n = a_1 \cdot r^{n-1}$$

$$S_n = \frac{a_1(1 - r^n)}{1 - r} \quad (r \neq 1)$$

$$S = \frac{a_1}{1 - r} \quad (|r| < 1) \text{ infinite geometric series}$$

Factorial

$$n! = n \cdot (n - 1) \cdot (n - 2) \cdot \ldots \cdot 3 \cdot 2 \cdot 1$$

$$n \cdot (n - 1)! = n!, \; 0! = 1$$

Binomial Coefficient

$$\binom{n}{r} = \frac{n!}{r!(n - r)!} \text{ (integer } n \text{ and } r, n \geq r \geq 0)$$

Binomial Theorem

If n is a positive integer

$$(a + b)^n = \binom{n}{0}a^n + \binom{n}{1}a^{n-1}b$$

$$+ \cdots + \binom{n}{r}a^{n-r}b^r + \cdots + \binom{n}{n}b^n$$

Formulas from Geometry

Triangle

$h = a \sin \theta$

$\text{Area} = \dfrac{1}{2}bh$

Trapezoid

$\text{Area} = \dfrac{h}{2}(a + b)$

Circle

$\text{Area} = \pi r^2$

$\text{Circumference} = 2\pi r$

Sector of Circle

$\text{Area} = \dfrac{\theta r^2}{2}$ (θ in radians)

$s = r\theta$ (θ in radians)

Right Circular Cone

$\text{Volume} = \dfrac{\pi r^2 h}{3}$

$\text{Lateral surface area} = \pi r\sqrt{r^2 + h^2}$

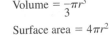

Right Circular Cylinder

$\text{Volume} = \pi r^2 h$

$\text{Lateral surface area} = 2\pi rh$

Right Triangle

Pythagorean Theorem:

$c^2 = a^2 + b^2$

Parallelogram

$\text{Area} = bh$

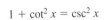

Circular Ring

$\text{Area} = \pi(R^2 - r^2)$

Ellipse

$\text{Area} = \pi ab$

Cone

$\text{Volume} = \dfrac{Ah}{3}$ ($A = \text{Area of base}$)

Sphere

$\text{Volume} = \dfrac{4}{3}\pi r^3$

$\text{Surface area} = 4\pi r^2$

Formulas from Trigonometry

Angular Measure

$\pi \text{ radians} = 180°$

So, 1 radian $= \dfrac{180}{\pi}$ degrees,

and 1 degree $= \dfrac{\pi}{180}$ radians.

Reciprocal Identities

$\sin x = \dfrac{1}{\csc x}$ 　　　　 $\csc x = \dfrac{1}{\sin x}$

$\cos x = \dfrac{1}{\sec x}$ 　　　　 $\sec x = \dfrac{1}{\cos x}$

$\tan x = \dfrac{1}{\cot x}$ 　　　　 $\cot x = \dfrac{1}{\tan x}$

Quotient Identities

$\tan x = \dfrac{\sin x}{\cos x}$ 　　　　 $\cot x = \dfrac{\cos x}{\sin x}$

Pythagorean Identities

$\sin^2 x + \cos^2 x = 1$

$\tan^2 x + 1 = \sec^2 x$

$1 + \cot^2 x = \csc^2 x$

Precalculus
Graphical, Numerical, Algebraic

SEVENTH EDITION

Franklin D. Demana   The Ohio State University

Bert K. Waits   The Ohio State University

Gregory D. Foley   Liberal Arts and Science Academy
of Austin

Daniel Kennedy   Baylor School

*AP is a registered trademark of the College Board, which was not involved in the production of, and does not endorse, this product.

Addison
Wesley

Boston San Francisco New York
London Toronto Sydney Tokyo Singapore Madrid
Mexico City Munich Paris Cape Town Hong Kong Montreal

Publisher	Greg Tobin
Executive Editor	Anne Kelly
Project Editor	Joanne Ha
Editorial Assistant	Ashley O'Shaughnessy
Managing Editor	Karen Wernholm
Senior Production Supervisor	Jeffrey Holcomb
Senior Designer	Barbara T. Atkinson
Photo Researcher	Beth Anderson
Supplements Coordinator	Emily Portwood
Media Producer	Michelle Murray
Software Development	John O'Brien and Mary Durnwald
Senior Marketing Manager	Becky Anderson
Marketing Assistant	Maureen McLaughlin
Senior Author Support/ Technology Specialist	Joe Vetere
Senior Prepress Supervisor	Caroline Fell
Senior Manufacturing Buyer	Evelyn Beaton
Developmental Editor	Elka Block
Cover Design	Suzanne Heiser
Text Design	Leslie Haimes
Project Management	Kathy Smith
Production Coordination	Harry Druding, Nesbitt Graphics, Inc.
Composition and Illustrations	Nesbitt Graphics, Inc.
Cover photo	© Royalty-Free/Corbis. Ferris wheel in Odaiba, Tokyo.

For permission to use copyrighted material, grateful acknowledgment is made to the copyright holders listed on page 1032, which is hereby made part of this copyright page.

Many of the designations used by manufacturers and sellers to distinguish their products are claimed as trademarks. Where those designations appear in this book, and Addison-Wesley was aware of a trademark claim, the designations have been printed in initial caps or all caps.

*AP is a registered trademark of the College Board, which was not involved in the production of, and does not endorse, this product.

Library of Congress Cataloging-in-Publication Data

Precalculus : graphical, numerical, algebraic / Franklin Demana … [et al.].—7th ed.
    p. cm.
  ISBN 0-13-227650-X (student edition)
   1. Algebra--Textbook. 2. Trigonometry--Textbooks. I. Demana, Franklin D., 1938-
QA154.3.P74 2006
512'.13--dc22
                    2005051543

9 10—CRK—10 09

Contents

CHAPTER 2

Polynomial, Power, and Rational Functions 169

CHAPTER 3

CHAPTER 4

CHAPTER 5

CHAPTER 8

Analytic Geometry in Two and Three Dimensions   631

APPENDIX A

Algebra Review

APPENDIX B

Key Formulas

APPENDIX C

Logic

About the Authors

Franklin D. Demana

Frank Demana received his master's degree in mathematics and his Ph.D. from Michigan State University. Currently, he is Professor Emeritus of Mathematics at The Ohio State University. As an active supporter of the use of technology to teach and learn mathematics, he is cofounder of the national Teachers Teaching with Technology (T³) professional development program. He has been the director and codirector of more than $10 million of National Science Foundation (NSF) and foundational grant activities. He is currently a co–principal investigator on a $3 million grant from the U.S. Department of Education Mathematics and Science Educational Research program awarded to The Ohio State University. Along with frequent presentations at professional meetings, he has published a variety of articles in the areas of computer- and calculator-enhanced mathematics instruction. Dr. Demana is also cofounder (with Bert Waits) of the annual International Conference on Technology in Collegiate Mathematics (ICTCM). He is co-recipient of the 1997 Glenn Gilbert National Leadership Award presented by the National Council of Supervisors of Mathematics, and co-recipient of the 1998 Christofferson-Fawcett Mathematics Education Award presented by the Ohio Council of Teachers of Mathematics.

Dr. Demana coauthored *Calculus: Graphical, Numerical, Algebraic; Essential Algebra: A Calculator Approach; Transition to College Mathematics; College Algebra and Trigonometry: A Graphing Approach; College Algebra: A Graphing Approach; Precalculus: Functions and Graphs;* and *Intermediate Algebra: A Graphing Approach.*

Bert K. Waits

Bert Waits received his Ph.D. from The Ohio State University and is currently Professor Emeritus of Mathematics there. Dr. Waits is cofounder of the national Teachers Teaching with Technology (T³) professional development program, and has been codirector or principal investigator on several large National Science Foundation projects. Dr. Waits has published articles in more than 50 nationally recognized professional journals. He frequently gives invited lectures, workshops, and minicourses at national meetings of the MAA and the National Council of Teachers of Mathematics (NCTM) on how to use computer technology to enhance the teaching and learning of mathematics. He has given invited presentations at the International Congress on Mathematical Education (ICME-6, -7, and -8) in Budapest (1988), Quebec (1992), and Seville (1996). Dr. Waits is co-recipient of the 1997 Glenn Gilbert National Leadership Award presented by the National Council of Supervisors of Mathematics, and is the cofounder (with Frank Demana) of the ICTCM. He is also co-recipient of the 1998 Christofferson-Fawcett Mathematics Education Award presented by the Ohio Council of Teachers of Mathematics.

Dr. Waits coauthored *Calculus: Graphical, Numerical, Algebraic; College Algebra and Trigonometry: A Graphing Approach; College Algebra: A Graphing Approach; Precalculus: Functions and Graphs;* and *Intermediate Algebra: A Graphing Approach.*

Gregory D. Foley

Greg Foley received B.A. and M.A. degrees in mathematics and a Ph.D. in mathematics education from The University of Texas at Austin. He is Director of the Liberal Arts and Science Academy of Austin, the advanced academic magnet high school program of the Austin Independent School District in Texas. Dr. Foley has taught elementary arithmetic through graduate-level mathematics, as well as upper division and graduate-level mathematics education classes. From 1977 to 2004, he held full time faculty positions at North Harris County College, Austin Community College, The Ohio State University, Sam Houston State University, and Appalachian State University, where he was Distinguished Professor of Mathematics Education in the Department of Mathematical Sciences and directed the Mathematics Education Leadership Training (MELT) program. Dr. Foley has presented over 200 lectures, workshops, and institutes throughout the United States and internationally, has directed a variety of funded projects, and has published articles in several professional journals. Active in various learned societies, he is a member of the Committee on the Mathematical Education of Teachers of the Mathematical Association of America (MAA). In 1998, Dr. Foley received the biennial American Mathematical Association of Two-Year Colleges (AMATYC) Award for Mathematics Excellence and in 2005, he received the annual Teachers Teaching with Technology (T³) Leadership Award.

Daniel Kennedy

Dan Kennedy received his undergraduate degree from the College of the Holy Cross and his master's degree and Ph.D. in mathematics from the University of North Carolina at Chapel Hill. Since 1973 he has taught mathematics at the Baylor School in Chattanooga, Tennessee, where he holds the Cartter Lupton Distinguished Professorship. Dr. Kennedy became an Advanced Placement Calculus reader in 1978, which led to an increasing level of involvement with the program as workshop consultant, table leader, and exam leader. He joined the Advanced Placement Calculus Test Development Committee in 1986, then in 1990 became the first high school teacher in 35 years to chair that committee. It was during his tenure as chair that the program moved to require graphing calculators and laid the early groundwork for the 1998 reform of the Advanced Placement Calculus curriculum. The author of the 1997 *Teacher's Guide—AP* Calculus,* Dr. Kennedy has conducted more than 50 workshops and institutes for high school calculus teachers. His articles on mathematics teaching have appeared in the *Mathematics Teacher* and the *American Mathematical Monthly,* and he is a frequent speaker on education reform at professional and civic meetings. Dr. Kennedy was named a Tandy Technology Scholar in 1992 and a Presidential Award winner in 1995.

Dr. Kennedy coauthored *Calculus: Graphical, Numerical, Algebraic; Prentice Hall Algebra I; Prentice Hall Geometry;* and *Prentice Hall Algebra 2.*

Preface

Although much attention has been paid since 1990 to reforming calculus courses, precalculus textbooks have remained surprisingly traditional. Now that the College Board's AP* Calculus curriculum has been accepted as a model fro a twenty-first-century calculus course, the path has been cleared for a new precalculus course to match the AP* goals and objectives. With this edition of *Precalculus: Graphical, Numerical, Algebraic* the authors of *Calculus: Graphical, Numerical, Algebraic,* the best-selling textbook in the AP* Calculus market, have designed such a precalculus course. For those students continuing on in a calculus course, this precalculus textbook concludes with a chapter that prepares students for the two central themes of calculus: instantaneous rate of change and continuous accumulation. This intuitively appealing preview of calculus is both more useful and more reasonable than the traditional, unmotivated foray into the computation of limits, and it is more in keeping with the stated goals and objectives of the AP* courses.

Recognizing that precalculus could be a capstone course for many students, the authors also include *quantitative literacy* topics such as probability, statistics, and the mathematics of finance. Their goal is to provide students with the good critical-thinking skills needed to succeed in any endeavor.

Continuing in the spirit of two earlier editions, the authors have integrated graphing technology throughout the course, not as an additional topic but as an essential tool for both mathematical discovery and effective problem solving. Graphing technology enables students to study a full catalog of basic functions at the beginning of the course, thereby giving them insights into function properties that are not seen in many books until later chapters. By connecting the algebra of functions to the visualization of their graphs, the authors are even able to introduce students to parametric equations, piecewise-defined functions, limit notation, and an intuitive understanding of continuity as early as Chapter 1.

Once students are comfortable with the language of functions, the authors guide them through a more traditional exploration of twelve basic functions and their algebraic properties, always reinforcing the connections among their algebraic, graphical, and numerical representations. The book uses a consistent approach to modeling, emphasizing in every chapter the use of particular types of functions to model behavior in the real world.

Finally, this textbook has faithfully incorporated not only the teaching strategies that have made *Calculus: Graphical, Numerical, Algebraic* so popular, but also some of the strategies from the popular Prentice Hall high-school algebra series, and thus has produced a seamless pedagogical transition from prealgebra through calculus for students. Although this book can certainly be appreciated on its own merits, teachers who seek continuity in their mathematics sequence might consider this deliberate alignment of pedagogy to be an additional asset of *Precalculus: Graphical, Numerical, Algebraic.*

Our Approach

The Rule of Four—A Balanced Approach

A principal feature of this edition is the balance among the algebraic, numerical, graphical, and verbal methods of representing problems: the rule of four. For instance, we obtain solutions algebraically when that is the most appropriate technique to use, and we obtain solutions graphically or numerically when algebra is difficult to use. We urge students to solve problems by one method and then support or confirm their solutions by using another method. We believe that students must learn the value of each of these methods or represen-

---

*AP is a registered trademark of the College Board, which was not involved in the production of, and does not endorse, this product.

tations and must learn to choose the one most appropriate for solving the particular problem under consideration. This approach reinforces the idea that to understand a problem fully, students need to understand it algebraically as well as graphically and numerically.

Problem-Solving Approach

Systematic problem solving is emphasized in the examples throughout the text, using the following variation of Polya's problem-solving process:

- *understand* the problem,
- *develop* a mathematical model,
- *solve* the mathematical model and support or confirm the solutions, and
- *interpret* the solution.

Students are encouraged to use this process throughout the text.

Twelve Basic Functions

Twelve Basic Functions are emphasized throughout the book as a major theme and focus. These functions are:

- The Identity Function
- The Squaring Function
- The Cubing Function
- The Reciprocal Function
- The Square Root Function
- The Exponential Function
- The Natural Logarithm Function
- The Sine Function
- The Cosine Function
- The Absolute Value Function
- The Greatest Integer Function
- The Logistic Function

One of the most distinctive features of this textbook is that it introduces students to the full vocabulary of functions early in the course. Students meet the twelve basic functions graphically in Chapter 1 and are able to compare and contrast them as they learn about concepts like domain, range, symmetry, continuity, end behavior, asymptotes, extrema, and even periodicity—concepts that are difficult to appreciate when the only examples a teacher can refer to are polynomials. With this book, students are able to characterize functions by their behavior within the first month of classes. (For example, thanks to graphing technology, it is no longer nec-

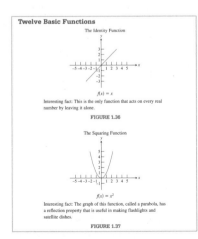

essary to understand radians before one can learn that the sine function is bounded, periodic, odd, and continuous, with domain $(-\infty, \infty)$ and range $[-1, 1]$.) Once students have a comfortable understanding of functions in general, the rest of the course consists of studying the various types of functions in greater depth, particularly with respect to their algebraic properties and modeling applications.

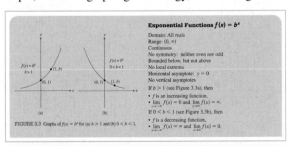

These functions are used to develop the fundamental analysis skills that are needed in calculus and advanced mathematics courses. Section 1.2 provides an overview of these functions by examining their graphs. A complete gallery of basic functions is included in Appendix B for easy reference.

Each basic function is revisited later in the book with a deeper analysis that includes investigation of the algebraic properties.

General characteristics of families of functions are also summarized.

Applications and Real Data

The majority of the applications in the text are based on real data from cited sources, and their presentations are self-contained; students will not need any experience in the fields from which the applications are drawn.

As they work through the applications, students are exposed to functions as mechanisms for modeling data and are motivated to learn about how various functions can help model real-life problems. They learn to analyze and model data, represent data graphically, interpret from graphs, and fit curves. Additionally, the tabular representation of data presented in this text highlights the concept that a function is a correspondence between numerical variables. This helps students build the connection between the numbers and graphs and recognize the importance of a full graphical, numerical, and algebraic understanding of a problem. For a complete listing of applications, please see the Applications Index on page 1015.

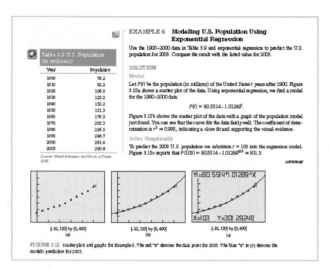

EXAMPLE 6 Modeling U.S. Population Using Exponential Regression

Use the 1900–2000 data in Table 3.9 and exponential regression to predict the U.S. population for 2003. Compare the result with the listed value for 2003.

SOLUTION

Model

Let $P(t)$ be the population (in millions) of the United States t years after 1900. Figure 3.15a shows a scatter plot of the data. Using exponential regression, we find a model for the 1900–2000 data

$$P(t) = 80.5514 \cdot 1.01289^t.$$

Figure 3.15b shows the scatter plot of the data with a graph of the population model just found. You can see that the curve fits the data fairly well. The coefficient of determination is $r^2 \approx 0.995$, indicating a close fit and supporting the visual evidence.

Solve Graphically

To predict the 2003 U.S. population we substitute $t = 103$ into the regression model. Figure 3.15c reports that $P(103) = 80.5514 \cdot 1.01289^{103} \approx 301.3$.

continued

Table 3.9 U.S. Population (in millions)

Year	Population
1900	76.2
1910	92.2
1920	106.0
1930	123.2
1940	132.2
1950	151.3
1960	179.3
1970	203.3
1980	226.5
1990	248.7
2000	281.4
2003	290.8

Source: World Almanac and Book of Facts 2005.

FIGURE 3.15 Scatter plots and graphs for Example 6. The red "x" denotes the data point for 2003. The blue "x" in (c) denotes the model's prediction for 2003.

Interpret

The model predicts the U.S. population was 301.3 million in 2003. The actual population was 290.8 million. We overestimated by 10.5 million, less than a 4% error.

Now try Exercise 43.

Content Changes to This Edition

For instructors, we have added additional coverage on topics that students usually find challenging, especially in Chapters 1, 2, and 9. In addition we have updated all the data in examples and exercises wherever appropriate. We have also trimmed back certain sections to better accommodate the length of the instructional periods, and we've added extensive resources for both new and experienced teachers. As a result, we believe that the changes made in this edition make this the most effective text available to students.

Chapter P

Complex numbers are now introduced in Section P.6, which is earlier than their previous placement in Chapter 2.

Chapter 1

Section 1.4 from the previous edition has been split into two sections to give more practice at function composition and to give inverse functions their own section. Graphical representations of absolute value compositions have been added.

Chapter 2

The section on complex numbers has been moved to Chapter P to make the length of this chapter more teachable. Subsections titled "Applications of Quadratic Functions" and "Monomial Functions and Their Graphs" have been included to highlight these topics.

Chapter 4

Exploration exercises have been added to introduce the arcsecant and arccosecant functions and the domain options associated with them.

Chapter 6

The material of this chapter is now unified under the title "Applications of Trigonometry." The section on vectors has been simplified, and there is a new subsection connecting the topics of polar curves and parametric curves. Geometric representation of complex numbers has been moved from Chapter 2 to Section 6.6.

Chapter 8

The updated Chapter Project titled "Ellipses as Models of Pendulum Motion" addresses the application of ellipses.

Chapter 9

There are now separate sections for sequences and series, with more examples and exercises involving each, as well as expanded treatment of sequence convergence.

Chapter 10

This preview of calculus first provides an historical perspective to calculus by presenting the classical studies of motion through the tangent line and area problems. Limits are then investigated further, and the chapter concludes with graphical and numerical examinations of derivatives and integrals.

New or Enhanced Features

Several features have been enhanced in this revision to assist students in achieving mastery of the skills and concepts of the course. We are pleased to offer the following new or enhanced features.

Chapter Openers include a motivating photograph and a general description of an application that can be solved with the topics in the chapter. The application is revisited later in the chapter with a specific problem that is solved. These problems enable students to explore realistic situations using graphical, numerical, and algebraic methods. Students are also asked to model problem situations using the functions studied in the chapter. In addition, the chapter sections are listed here.

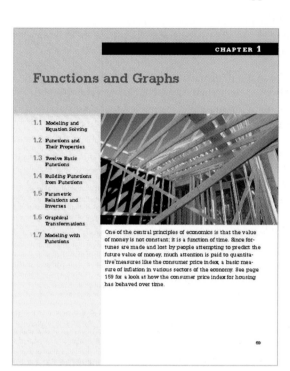

A **Chapter Overview** begins each chapter to give you a sense of what you are going to learn. This overview provides a roadmap of the chapter, as well as tells how the different topics in the chapter are connected under one big idea. It is always helpful to remember that mathematics isn't modular, but interconnected, and that the skills and concepts you are learning throughout the course build on one another to help you understand more complicated processes and relationships.

Similarly, the **What you'll learn about ... and why** feature gives you the big ideas in each section and explains their purpose. You should read this as you begin the

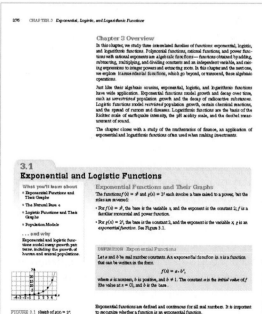

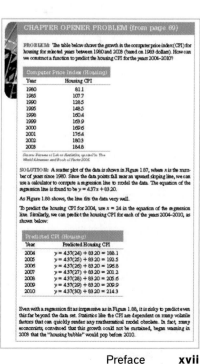

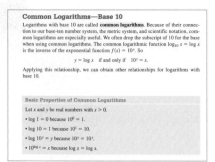

Common Logarithms—Base 10

Logarithms with base 10 are called **common logarithms**. Because of their connection to our base-ten number system, the metric system, and scientific notation, common logarithms are especially useful. We often drop the subscript of 10 for the base when using common logarithms. The common logarithmic function $\log_{10} x = \log x$ is the inverse of the exponential function $f(x) = 10^x$. So

$$y = \log x \quad \text{if and only if} \quad 10^y = x.$$

Applying this relationship, we can obtain other relationships for logarithms with base 10.

Basic Properties of Common Logarithms

Let x and y be real numbers with $x > 0$.
- $\log 1 = 0$ because $10^0 = 1$.
- $\log 10 = 1$ because $10^1 = 10$.
- $\log 10^y = y$ because $10^y = 10^y$.
- $10^{\log x} = x$ because $\log x = \log x$.

---

section and always review it after you have completed the section to make sure you understand all of the key topics that you have just studied.

Vocabulary is highlighted in yellow for easy reference. **Properties** are boxed in green so that you can easily find them.

Each example ends with a suggestion to **Now Try** a related exercise. Working the suggested exercise is an easy way for you to check your comprehension of the material while reading each section, instead of waiting to the end of each section or chapter to see if you "got it." In the *Annotated Teacher's Edition*, various examples are marked for the instructor with the 🖐 icon. Alternates are provided for these examples in the *Acetates and Transparencies* package.

Explorations appear throughout the text and provide you with the perfect opportunity to become an active learner and discover mathematics on your own. This will help hone your critical thinking and problem-solving skills. Some are technology-based and others involve exploring mathematical ideas and connections.

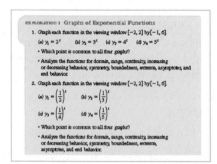

EXPLORATION 1 Graphs of Exponential Functions

1. Graph each function in the viewing window $[-2, 2]$ by $[-1, 6]$.
   (a) $y_1 = 2^x$ (b) $y_2 = 3^x$ (c) $y_3 = 4^x$ (d) $y_4 = 5^x$
   • Which point is common to all four graphs?
   • Analyze the functions for domain, range, continuity, increasing or decreasing behavior, symmetry, boundedness, extrema, asymptotes, and end behavior.
2. Graph each function in the viewing window $[-2, 2]$ by $[-1, 6]$.
   (a) $y_1 = \left(\frac{1}{2}\right)^x$ (b) $y_2 = \left(\frac{1}{3}\right)^x$
   (c) $y_3 = \left(\frac{1}{4}\right)^x$ (d) $y_4 = \left(\frac{1}{5}\right)^x$
   • Which point is common to all four graphs?
   • Analyze the functions for domain, range, continuity, increasing or decreasing behavior, symmetry, boundedness, extrema, asymptotes, and end behavior.

Margin Notes on various topics appear throughout the text. *Tips* offer practical advice to you on using your grapher to obtain the best, most accurate results. *Margin notes* include historical information, hints about examples, and provide additional insight to help you avoid common pitfalls and errors.

A BIT OF HISTORY

Logarithmic functions were developed around 1594 as computational tools by Scottish mathematician John Napier (1550–1617). He originally called them "artificial numbers," but changed the name to logarithms, which means "reckoning numbers."

The **Looking Ahead to Calculus** 📖 icon is found throughout the text next to many examples and topics to point out concepts that students will encounter again in calculus. Ideas that foreshadow calculus are highlighted, such as limits, maximum and minimum, asymptotes, and continuity. Early in

Graphs of Logarithmic Functions with Base b

Using the change-of-base formula we can rewrite any logarithmic function $g(x) = \log_b x$ as

$$g(x) = \frac{\ln x}{\ln b} = \frac{1}{\ln b}\ln x.$$

So every logarithmic function is a constant multiple of the natural logarithmic function $f(x) = \ln x$. If the base is $b > 1$, the graph of $g(x) = \log_b x$ is a vertical stretch or shrink of the graph of $f(x) = \ln x$ by the factor $1/\ln b$. If $0 < b < 1$, a reflection across the x-axis is required as well.

the text, the idea of the limit using an intuitive and conceptual approach is introduced. Some calculus notation and language is introduced in the early chapters and used throughout the text to establish familiarity.

The **Web/Real Data** 🌐 icon is used to mark the examples and exercises that use real cited data.

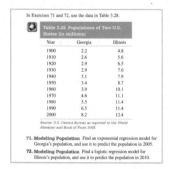

In Exercises 71 and 72, use the data in Table 3.28.

Table 3.28 Populations of Two U.S. States (in millions)

Year	Georgia	Illinois
1900	2.2	4.8
1910	2.6	5.6
1920	2.9	6.5
1930	2.9	7.6
1940	3.1	7.9
1950	3.4	8.7
1960	3.9	10.1
1970	4.6	11.1
1980	5.5	11.4
1990	6.5	11.4
2000	8.2	12.4

Source: U.S. Census Bureau as reported in the World Almanac and Book of Facts 2005.

71. **Modeling Population** Find an exponential regression model for Georgia's population, and use it to predict the population in 2005.
72. **Modeling Population** Find a logistic regression model for Illinois's population, and use it to predict the population in 2010.

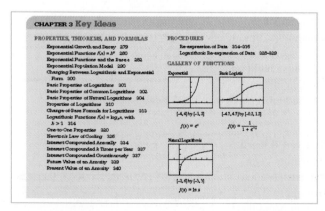

CHAPTER 3 Key Ideas

PROPERTIES, THEOREMS, AND FORMULAS
Exponential Growth and Decay   279
Exponential Functions $f(x) = b^x$ 280
Exponential Functions and the Base e 282
Exponential Population Model   290
Changing Between Logarithmic and Exponential Form   300
Basic Properties of Logarithms   301
Basic Properties of Common Logarithms   302
Basic Properties of Natural Logarithms   304
Properties of Logarithms   310
Change-of-Base Formula for Logarithms   313
Logarithmic Functions $f(x) = \log_b x$, with $b > 1$ 314
One-to-One Properties   320
Newton's Law of Cooling   326
Interest Compounded Annually   334
Interest Compounded k Times per Year   337
Interest Compounded Continuously   337
Future Value of an Annuity   339
Present Value of an Annuity   340

PROCEDURES
Re-expression of Data   314–316
Logarithmic Re-expression of Data   325–329

GALLERY OF FUNCTIONS

Exponential

$[-4, 4]$ by $[-1, 3]$
$f(x) = e^x$

Basic Logistic

$[-4.7, 4.7]$ by $[-0.5, 1.5]$
$f(x) = \dfrac{1}{1 + e^{-x}}$

Natural Logarithmic

$[-2, 6]$ by $[-3, 3]$
$f(x) = \ln x$

The **Chapter Review** material at the end of each chapter are sections dedicated to helping students review the chapter concepts. **Key Ideas** has three parts: Properties, Theorems, and Formulas; Procedures; and Gallery of Functions. The **Review Exercises** represent the full range of exercises covered in the chapter and give additional practice with the ideas developed in the chapter. The exercises with red numbers indicate problems that would make up a good practice test. **Chapter Projects** conclude each chapter and require students to analyze data. They can be assigned as either individual or group work. Each project expands upon concepts and ideas taught in the chapter, and many projects refer to the Web for further investigation of real data.

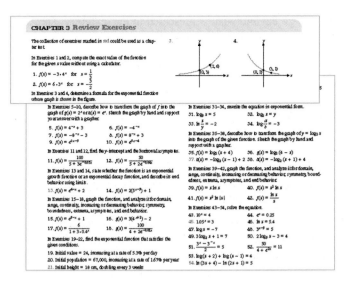

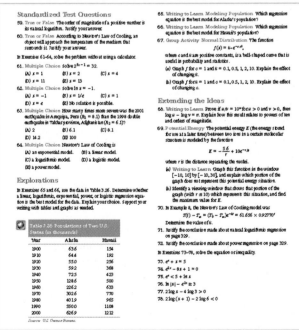

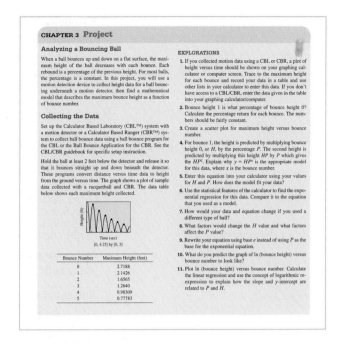

Exercise Sets

Each exercise set begins with a **Quick Review** to help you review skills needed in the exercise set, thus reminding you again that mathematics is not modular. There are also directions that give a *section to go to for help* so that students are prepared to do the Section Exercises.

There are over 6,000 exercises, including 680 Quick Review Exercises. Following the Quick Review are exercises that allow you to practice the algebraic skills learned in that section. These exercises have been carefully graded from routine to challenging. The following types of skills are tested in each exercise set:

- Algebraic and analytic manipulation
- Connecting algebra to geometry
- Interpretation of graphs
- Graphical and numerical representations of functions
- Data analysis

Also included in the exercise sets are thought-provoking exercises:

- **Standardized Test Questions** include two true-false problems with justifications and four multiple-choice questions.

- **Explorations** are opportunities for students to discover mathematics on their own or in groups. These exercises often require the use of critical thinking to explore the ideas.

- **Writing to Learn** exercises give you practice at communicating about mathematics and opportunities to demonstrate your understanding of important ideas.

- **Group Activity** exercises ask you to work on the problems in groups or solve them as individual or group projects.

- **Extending the Ideas** exercises go beyond what is presented in the textbook. These exercises are challenging extensions of the book's material.

This variety of exercises provides sufficient flexibility to emphasize the skills most needed for each student or class.

Supplements and Resources

For the Student

Student Edition, ISBN 0-13-227650-X

Student's Solutions Manual, ISBN 0-321-36994-7

- Contains detailed, worked-out solutions to odd-numbered exercises

Student Practice Workbook, ISBN 0-13-198580-9

- New examples that parallel key examples from each section in the book are provided, along with a detailed solution

- Related practice problems follow each example

Graphing Calculator Manual, ISBN 0-321-37000-7

- Grapher Workshop provides detailed instruction on important grapher features

- Features TI-83 Plus, Silver, TI-84, and TI-89 Titanium

For the Teacher

Annotated Teacher's Edition, ISBN 0-321-37423-1

- Answers included on the same page for most exercises

- Various examples are marked with the ◣ icon. Alternates are provided for these examples in the *Acetates and Transparencies* package.

- The *Annotated Teacher's Edition* also includes notes written specifically for the teacher. These notes include chapter and section objectives, suggested assignments, lesson guides, and teaching tips.

Resource Manual, ISBN 0-321-36995-5

- Major Concepts Review, Group Activity Worksheets, Sample Chapter Tests, Standardized Test Preparation Questions, Contest Problems

Solutions Manual, ISBN 0-321-36993-9

- Complete solutions to all exercises, including Quick Reviews, Exercises, Explorations, and Chapter Reviews

Tests and Quizzes, ISBN 0-321-36992-0

- Two parallel tests per chapter, two quizzes for every 3–4 sections, two parallel midterm tests covering Chapters P–5, two parallel end-of-year tests, covering Chapters 6–10

Acetates and Transparencies, ISBN 0-321-36997-1

- Full color transparencies of 10–15 useful general transparencies along with black and white transparency masters of all Quick Review Exercises

- One Alternate Example per lesson

Technology Resources

MathXL®, www.mathxl.com

MathXL® is a powerful online homework, tutorial, and assessment system that accompanies our textbooks in mathematics or statistics. With MathXL, instructors can create, edit, and assign online homework and tests using algorithmically generated exercises correlated at the objective level to the textbook. They can also create and assign their own online exercises and import TestGen tests for added flexibility. All student work is tracked in MathXL's online gradebook. Students can take chapter tests in MathXL and receive personalized study plans based on their test results. The study plan diagnoses weaknesses and links students directly to tutorial exercises for the objectives they need to study and retest. Students can also access video clips from selected exercises. For more information, visit our Web site at www.mathxl.com, or contact your local sales representative.

MathXL® Tutorials on CD

This interactive tutorial CD-ROM provides algorithmically generated practice exercises that are correlated at the objective level to the exercises in the textbook. Every practice exercise is accompanied by an example and a guided solution designed to involve students in the solution process. Selected exercises may also include a video clip to help students visualize concepts. The software provides helpful feedback for incorrect answers and can generate printed summaries of students' progress.

InterAct Math Tutorial Web site, www.interactmath.com

Get practice and tutorial help online! This interactive tutorial Web site provides algorithmically generated practice exercises that correlate directly to the exercises in the textbook. Students can retry an exercise as many times as they like with new values each time for unlimited practice and mastery. Every exercise is accompanied by an interactive guided solution that provides helpful feedback for incorrect answers, and students can also view a worked-out sample problem that steps them through an exercise similar to the one they're working on.

Video Lectures on CD

The video lectures for this text are also available on CD-ROM, making it easy and convenient for students to watch the videos from a computer at home or on campus. The complete digitized video set, affordable and portable for students, is ideal for distance learning or supplemental instruction. These videotaped lectures feature an engaging team of mathematics instructors who present comprehensive coverage of topics in the text.

TestGen®

TestGen® enables instructors to build, edit, print, and administer tests using a computerized bank of questions developed to cover all the objectives of the text. TestGen is algorithmically based, allowing instructors to create multiple but equivalent versions of the same question or test with the click of a button. Instructors can also modify test bank questions or add new questions. Tests can be printed or administered online. The software is available on a dual-platform Windows/Macintosh CD-ROM.

Presentation Express CD-ROM (PowerPoint® slides)

This time saving component includes classroom presentation slides that correlate to the topic sequence of the textbook. In addition, all transparencies are included in *PowerPoint®* format making it easier for you to teach and to customize based on your teaching preferences. All slides can be customized and edited.

Teacher Express CD-ROM (with LessonView)

- This is a new suite of instructional tools on CD-ROM to help teachers plan, teach, and assess at the click of a mouse. Powerful lesson planning, resource management, testing, and an interactive *Annotated Teacher's Edition* all in one place make class preparation quick and easy!

- Contents include: Planning Express, *Annotated Teacher's Edition*, Program Teaching Resources, Correlations, Links to Other Resources.

- Online resources require an Internet Connection.

Student Express CD-ROM (with PDF Text)

The perfect tool for test review or studying, this CD provides the complete student textbook in an electronic format.

Web Site

Our Web site, www.awl.com/demana, provides dynamic resources for instructors and students. Some of the resources include TI graphing calculator downloads, online quizzing, teaching tips, study tips, Explorations, and end-of-chapter projects.

Acknowledgments

We wish to express our gratitude to the reviewers of this and previous editions who provided such invaluable insight and comment. Special thanks is due to our consultant, Cynthia Schimek, *Secondary Mathematics Curriculum Specialist, Katy Independent School District, Texas,* for her guidance and invaluable insight on this revision.

Judy Ackerman
Montgomery College

Ignacio Alarcon
Santa Barbara City College

Ray Barton
Olympus High School

Nicholas G. Belloit
Florida Community College at Jacksonville

Margaret A. Blumberg
University of Southwestern Louisiana

Ray Cannon
Baylor University

Marilyn P. Carlson
Arizona State University

Edward Champy
Northern Essex Community College

Janis M. Cimperman
Saint Cloud State University

Wil Clarke
La Sierra University

Marilyn Cobb
Lake Travis High School

Donna Costello
Plano Senior High School

Gerry Cox
Lake Michigan College

Deborah A. Crocker
Appalachian State University

Marian J. Ellison
University of Wisconsin—Stout

Donna H. Foss
University of Central Arkansas

Betty Givan
Eastern Kentucky University

Brian Gray
Howard Community College

Daniel Harned
Michigan State University

Vahack Haroutunian
Fresno City College

Celeste Hernandez
Richland College

Rich Hoelter
Raritan Valley Community College

Dwight H. Horan
Wentworth Institute of Technology

Margaret Hovde
Grossmont College

Miles Hubbard
Saint Cloud State University

Sally Jackman
Richland College

T. J. Johnson
Hendrickson High School

Stephen C. King
University of South Carolina—Aiken

Jeanne Kirk
William Howard Taft High School

Georgianna Klein
Grand Valley State University

Deborah L. Kruschwitz-List
University of Wisconsin—Stout

Carlton A. Lane
Hillsborough Community College

James Larson
Lake Michigan University

Edward D. Laughbaum
Columbus State Community College

Ron Marshall
Western Carolina University

Janet Martin
Lubbock High School

Beverly K. Michael
University of Pittsburgh

Mary Margaret Shoaf-Grubbs
College of New Rochelle

Paul Mlakar
St. Mark's School of Texas

Malcolm Soule
California State University, Northridge

John W. Petro
Western Michigan University

Sandy Spears
Jefferson Community College

Cynthia M. Piez
University of Idaho

Shirley R. Stavros
Saint Cloud State University

Debra Poese
Montgomery College

Stuart Thomas
University of Oregon

Jack Porter
University of Kansas

Janina Udrys
Schoolcraft College

Antonio R. Quesada
The University of Akron

Mary Voxman
University of Idaho

Hilary Risser
Plano West Senior High

Eddie Warren
University of Texas at Arlington

Thomas H. Rousseau
Siena College

Steven J. Wilson
Johnson County Community College

David K. Ruch
Sam Houston State University

Gordon Woodward
University of Nebraska

Sid Saks
Cuyahoga Community College

Cathleen Zucco-Teveloff
Trinity College

We express special thanks to Chris Brueningsen, Linda Antinone, and Bill Bower for their work on the Chapter Projects. We would also like to thank Perian Herring, Frank Purcell, and Tom Wegleitner for their meticulous accuracy checking of the text. We are grateful to Nesbitt Graphics, who pulled off an amazing job on composition and proofreading, and specifically to Kathy Smith and Harry Druding for expertly managing the entire production process. Finally, our thanks as well are extended to the professional and remarkable staff at Addison-Wesley, for their advice and support in revising this text, particularly Anne Kelly, Becky Anderson, Greg Tobin, Rich Williams, Neil Heyden, Gary Schwartz, Marnie Greenhut, Joanne Ha, Karen Wernholm, Jeffrey Holcomb, Barbara Atkinson, Evelyn Beaton, Beth Anderson, Maureen McLaughlin, and Michelle Murray. Particular recognition is due Elka Block, who tirelessly helped us through the development and production of this book.

—F. D. D.
—B. K. W.
—G. D. F.
—D. K.

Prerequisites

Large distances are measured in *light years*, the distance light travels in one year. Astronomers use the speed of light, approximately 186,000 miles per second, to approximate distances between planets. See page 39 for examples.

Chapter P Overview

Historically, algebra was used to represent problems with symbols (algebraic models) and solve them by reducing the solution to algebraic manipulation of symbols. This technique is still important today. Graphing calculators are used today to approach problems by representing them with graphs (graphical models) and solve them with numerical and graphical techniques of the technology.

We begin with basic properties of real numbers and introduce absolute value, distance formulas, midpoint formulas, and write equations of circles. Slope of a line is used to write standard equations for lines and applications involving linear equations are discussed. Equations and inequalities are solved using both algebraic and graphical techniques.

P.1
Real Numbers

What you'll learn about

- Representing Real Numbers
- Order and Interval Notation
- Basic Properties of Algebra
- Integer Exponents
- Scientific Notation

. . . and why

These topics are fundamental in the study of mathematics and science.

Representing Real Numbers

A **real number** is any number that can be written as a decimal. Real numbers are represented by symbols such as $-8, 0, 1.75, 2.333..., 0.\overline{36}, 8/5, \sqrt{3}, \sqrt[3]{16}, e,$ and π.

The set of real numbers contains several important subsets:

The **natural (or counting) numbers**: $\{1, 2, 3, \ldots\}$

The **whole numbers**: $\{0, 1, 2, 3, \ldots\}$

The **integers**: $\{\ldots, -3, -2, -1, 0, 1, 2, 3, \ldots\}$

The braces $\{\ \}$ are used to enclose the **elements**, or **objects**, of the set. The rational numbers are another important subset of the real numbers. A **rational number** is any number that can be written as a ratio a/b of two integers, where $b \neq 0$. We can use **set-builder notation** to describe the rational numbers:

$$\left\{ \frac{a}{b} \,\middle|\, a, b \text{ are integers, and } b \neq 0 \right\}$$

The vertical bar that follows a/b is read "such that."

The decimal form of a rational number either **terminates** like $7/4 = 1.75$, or is **infinitely repeating** like $4/11 = 0.363636... = 0.\overline{36}$. The bar over the 36 indicates the block of digits that repeats. A real number is **irrational** if it is *not* rational. The decimal form of an irrational number is infinitely nonrepeating. For example, $\sqrt{3} = 1.7320508...$ and $\pi = 3.14159265....$

Real numbers are approximated with calculators by giving a few of its digits. Sometimes we can find the decimal form of rational numbers with calculators, but not very often.

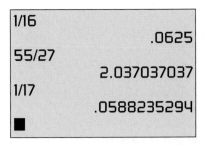

FIGURE P.1 Calculator decimal representations of 1/16, 55/27, and 1/17 with the calculator set in floating decimal mode. (Example 1)

EXAMPLE 1   Examining Decimal Forms of Rational Numbers

Determine the decimal form of 1/16, 55/27, and 1/17.

SOLUTION Figure P.1 suggests that the decimal form of 1/16 terminates and that of 55/27 repeats in blocks of 037.

$$\frac{1}{16} = 0.0625 \quad \text{and} \quad \frac{55}{27} = 2.\overline{037}$$

We cannot predict the *exact* decimal form of 1/17 from Figure P.1, however we can say that $1/17 \approx 0.0588235294$. The symbol $\approx$ is read "*is approximately equal to*." We can use long division (see Exercise 66) to show that

$$\frac{1}{17} = 0.\overline{0588235294117647}.$$

Now try Exercise 3.

The real numbers and the points of a line can be matched *one-to-one* to form a **real number line**. We start with a horizontal line and match the real number zero with a point O, the **origin**. **Positive numbers** are assigned to the right of the origin, and **negative numbers** to the left, as shown in Figure P.2.

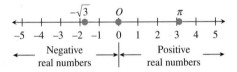

FIGURE P.2 The real number line.

Every real number corresponds to one and only one point on the real number line, and every point on the real number line corresponds to one and only one real number. Between every pair of real numbers on the number line there are infinitely many more real numbers.

The number associated with a point is **the coordinate of the point**. As long as the context is clear, we will follow the standard convention of using the real number for both the name of the point and its coordinate.

Order and Interval Notation

The set of real numbers is **ordered**. This means that we can compare any two real numbers that are not equal using inequalities and say that one is "less than" or "greater than" the other.

UNORDERED SYSTEMS

Not all number systems are ordered. For example, the complex number system, to be introduced in Section P.6, has no natural ordering.

OPPOSITES AND NUMBER LINE

$a < 0 \Rightarrow -a > 0$

If $a < 0$, then a is to the left of 0 on the real number line, and its opposite, $-a$, is to the right of 0. Thus, $-a > 0$.

Order of Real Numbers

Let a and b be any two real numbers.

Symbol	Definition	Read
$a > b$	$a - b$ is positive	a is greater than b
$a < b$	$a - b$ is negative	a is less than b
$a \geq b$	$a - b$ is positive or zero	a is greater than or equal to b
$a \leq b$	$a - b$ is negative or zero	a is less than or equal to b

The symbols $>$, $<$, $\geq$, and $\leq$ are **inequality symbols**.

Geometrically, $a > b$ means that a is to the right of b (equivalently b is to the left of a) on the real number line. For example, since $6 > 3$, 6 is to the right of 3 on the real number line. Note also that $a > 0$ means that $a - 0$, or simply a, is positive and $a < 0$ means that a is negative.

We are able to compare any two real numbers because of the following important property of the real numbers.

Trichotomy Property

Let a and b be any two real numbers. Exactly one of the following is true:
$$a < b, \quad a = b, \quad \text{or} \quad a > b.$$

Inequalities can be used to describe **intervals** of real numbers, as illustrated in Example 2.

EXAMPLE 2   Interpreting Inequalities

Describe and graph the interval of real numbers for the inequality.

(a) $x < 3$ **(b)** $-1 < x \leq 4$

SOLUTION

(a) The inequality $x < 3$ describes all real numbers less than 3 (Figure P.3a).

(b) The *double inequality* $-1 < x \leq 4$ represents all real numbers between -1 and 4, excluding -1 and including 4 (Figure P.3b).   *Now try Exercise 5.*

EXAMPLE 3   Writing Inequalities

Write an interval of real numbers using an inequality and draw its graph.

(a) The real numbers between -4 and -0.5

(b) The real numbers greater than or equal to zero

SOLUTION

(a) $-4 < x < -0.5$ (Figure P.3c)

(b) $x \geq 0$ (Figure P.3d)   *Now try Exercise 15.*

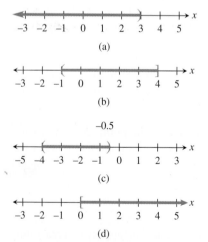

FIGURE P.3 In graphs of inequalities, parentheses correspond to $<$ and $>$ and brackets to $\leq$ and $\geq$. (Examples 2 and 3)

As shown in Example 2, inequalities define *intervals* on the real number line. We often use [2, 5] to describe the *bounded interval* determined by $2 \le x \le 5$. This interval is **closed** because it contains its *endpoints* 2 and 5. There are four types of **bounded intervals**.

Bounded Intervals of Real Numbers

Let a and b be real numbers with $a < b$.

Interval Notation	Interval Type	Inequality Notation	Graph
$[a, b]$	Closed	$a \le x \le b$	
(a, b)	Open	$a < x < b$	
$[a, b)$	Half-open	$a \le x < b$	
$(a, b]$	Half-open	$a < x \le b$	

The numbers a and b are the **endpoints** of each interval.

INTERVAL NOTATION AT $\pm\infty$

Because $-\infty$ is *not* a real number, we use $(-\infty, 2)$ instead of $[-\infty, 2)$ to describe $x < 2$. Similarly, we use $[-1, \infty)$ instead of $[-1, \infty]$ to describe $x \ge -1$.

The interval of real numbers determined by the inequality $x < 2$ can be described by the *unbounded interval* $(-\infty, 2)$. This interval is **open** because it does *not* contain its endpoint 2.

We use the interval notation $(-\infty, \infty)$ to represent the entire set of real numbers. The symbols $-\infty$ (*negative infinity*) and ∞ (*positive infinity*) allow us to use interval notation for unbounded intervals and are *not* real numbers. There are four types of **unbounded intervals**.

Unbounded Intervals of Real Numbers

Let a and b be real numbers.

Interval Notation	Interval Type	Inequality Notation	Graph
$[a, \infty)$	Closed	$x \ge a$	
(a, ∞)	Open	$x > a$	
$(-\infty, b]$	Closed	$x \le b$	
$(-\infty, b)$	Open	$x < b$	

Each of these intervals has exactly one endpoint, namely a or b.

EXAMPLE 4   Converting Between Intervals and Inequalities

Convert interval notation to inequality notation or vice versa. Find the endpoints and state whether the interval is bounded, its type, and graph the interval.

(a) $[-6, 3)$ **(b)** $(-\infty, -1)$ **(c)** $-2 \le x \le 3$

SOLUTION

(a) The interval $[-6, 3)$ corresponds to $-6 \le x < 3$ and is bounded and half-open (see Figure P.4a). The endpoints are -6 and 3.

(b) The interval $(-\infty, -1)$ corresponds to $x < -1$ and is unbounded and open (see Figure P.4b). The only endpoint is -1.

(c) The inequality $-2 \le x \le 3$ corresponds to the closed, bounded interval $[-2, 3]$ (see Figure P.4c). The endpoints are -2 and 3.

Now try Exercise 29.

(a)
(b)
(c)

FIGURE P.4 Graphs of the intervals of real numbers in Example 4.

Basic Properties of Algebra

Algebra involves the use of letters and other symbols to represent real numbers. A **variable** is a letter or symbol (for example, x, y, t, θ) that represents an unspecified real number. A **constant** is a letter or symbol (for example, -2, 0, $\sqrt{3}$, π) that represents a specific real number. An **algebraic expression** is a combination of variables and constants involving addition, subtraction, multiplication, division, powers, and roots.

We state some of the properties of the arithmetic operations of addition, subtraction, multiplication, and division, represented by the symbols $+$, $-$, $\times$ (or $\cdot$) and $\div$ (or $/$), respectively. Addition and multiplication are the primary operations. Subtraction and division are defined in terms of addition and multiplication.

Subtraction: $a - b = a + (-b)$

Division: $\dfrac{a}{b} = a\left(\dfrac{1}{b}\right), \; b \ne 0$

In the above definitions, $-b$ is the **additive inverse** or **opposite** of b, and $1/b$ is the **multiplicative inverse** or **reciprocal** of b. Perhaps surprisingly, additive inverses are not always negative numbers. The additive inverse of 5 is the negative number -5. However, the additive inverse of -3 is the positive number 3.

SUBTRACTION VS. NEGATIVE NUMBERS

On many calculators, there are two "$-$" keys, one for subtraction and one for negative numbers or opposites. Be sure you know how to use both keys correctly. Misuse can lead to incorrect results.

The following properties hold for real numbers, variables, and algebraic expressions.

Properties of Algebra

Let u, v, and w be real numbers, variables, or algebraic expressions.

1. Commutative property

Addition: $u + v = v + u$

Multiplication: $uv = vu$

2. Associative property

Addition:

$(u + v) + w = u + (v + w)$

Multiplication: $(uv)w = u(vw)$

3. Identity property

Addition: $u + 0 = u$

Multiplication: $u \cdot 1 = u$

4. Inverse property

Addition: $u + (-u) = 0$

Multiplication: $u \cdot \dfrac{1}{u} = 1, \, u \neq 0$

5. Distributive property

Multiplication over addition:

$u(v + w) = uv + uw$

$(u + v)w = uw + vw$

Multiplication over subtraction:

$u(v - w) = uv - uw$

$(u - v)w = uw - vw$

The left-hand sides of the equations for the distributive property show the **factored form** of the algebraic expressions, and the right-hand sides show the **expanded form**.

EXAMPLE 5 Using the Distributive Property

(a) Write the expanded form of $(a + 2)x$.

(b) Write the factored form of $3y - by$.

SOLUTION

(a) $(a + 2)x = ax + 2x$

(b) $3y - by = (3 - b)y$ *Now try Exercise 37.*

Here are some properties of the additive inverse together with examples that help illustrate their meanings.

Properties of the Additive Inverse

Let u and v be real numbers, variables, or algebraic expressions.

Property	Example
1. $-(-u) = u$	$-(-3) = 3$
2. $(-u)v = u(-v) = -(uv)$	$(-4)3 = 4(-3) = -(4 \cdot 3) = -12$
3. $(-u)(-v) = uv$	$(-6)(-7) = 6 \cdot 7 = 42$
4. $(-1)u = -u$	$(-1)5 = -5$
5. $-(u + v) = (-u) + (-v)$	$-(7 + 9) = (-7) + (-9) = -16$

Integer Exponents

Exponential notation is used to shorten products of factors that repeat. For example,

$$(-3)(-3)(-3)(-3) = (-3)^4 \quad \text{and} \quad (2x+1)(2x+1) = (2x+1)^2.$$

Exponential Notation

Let a be a real number, variable, or algebraic expression and n a positive integer. Then

$$a^n = \underbrace{a \cdot a \cdot \ldots \cdot a}_{n \text{ factors}},$$

where n is the **exponent**, a is the **base**, and a^n is the **nth power of a**, read as "a to the nth power."

The two exponential expressions in Example 6 have the same value but have different bases. Be sure you understand the difference.

UNDERSTANDING NOTATION

$(-3)^2 = 9$
$-3^2 = -9$
Be careful!

EXAMPLE 6   Identifying the Base

(a) In $(-3)^5$, the base is -3.

(b) In -3^5, the base is 3.   *Now try Exercise 43.*

Here are the basic properties of exponents together with examples that help illustrate their meanings.

Properties of Exponents

Let u and v be real numbers, variables, or algebraic expressions and m and n be integers. All bases are assumed to be nonzero.

Property	Example
1. $u^m u^n = u^{m+n}$	$5^3 \cdot 5^4 = 5^{3+4} = 5^7$
2. $\dfrac{u^m}{u^n} = u^{m-n}$	$\dfrac{x^9}{x^4} = x^{9-4} = x^5$
3. $u^0 = 1$	$8^0 = 1$
4. $u^{-n} = \dfrac{1}{u^n}$	$y^{-3} = \dfrac{1}{y^3}$
5. $(uv)^m = u^m v^m$	$(2z)^5 = 2^5 z^5 = 32z^5$
6. $(u^m)^n = u^{mn}$	$(x^2)^3 = x^{2 \cdot 3} = x^6$
7. $\left(\dfrac{u}{v}\right)^m = \dfrac{u^m}{v^m}$	$\left(\dfrac{a}{b}\right)^7 = \dfrac{a^7}{b^7}$

To simplify an expression involving powers means to rewrite it so that each factor appears only once, all exponents are positive, and exponents and constants are combined as much as possible.

MOVING FACTORS

Be sure you understand how exponent Property 4 permits us to move factors from the numerator to the denominator and vice versa:

$$\frac{v^{-m}}{u^{-n}} = \frac{u^n}{v^m}$$

EXAMPLE 7 Simplifying Expressions Involving Powers

(a) $(2ab^3)(5a^2b^5) = 10(aa^2)(b^3b^5) = 10a^3b^8$

(b) $\dfrac{u^2v^{-2}}{u^{-1}v^3} = \dfrac{u^2u^1}{v^2v^3} = \dfrac{u^3}{v^5}$

(c) $\left(\dfrac{x^2}{2}\right)^{-3} = \dfrac{(x^2)^{-3}}{2^{-3}} = \dfrac{x^{-6}}{2^{-3}} = \dfrac{2^3}{x^6} = \dfrac{8}{x^6}$

Now try Exercise 47.

Scientific Notation

Any positive number can be written in **scientific notation,**

$$c \times 10^m, \text{ where } 1 \le c < 10 \text{ and } m \text{ is an integer.}$$

This notation provides a way to work with very large and very small numbers. For example, the distance between the Earth and the Sun is about 93,000,000 miles. In scientific notation,

$$93,000,000 \text{ mi} = 9.3 \times 10^7 \text{ mi.}$$

The *positive exponent* 7 indicates that moving the decimal point in 9.3 to the right 7 places produces the decimal form of the number.

The mass of an oxygen molecule is about

$$0.000\ 000\ 000\ 000\ 000\ 000\ 000\ 053 \text{ grams.}$$

In scientific notation,

$$0.000\ 000\ 000\ 000\ 000\ 000\ 000\ 053 \text{ g} = 5.3 \times 10^{-23} \text{ g.}$$

The *negative exponent* -23 indicates that moving the decimal point in 5.3 to the left 23 places produces the decimal form of the number.

EXAMPLE 8 Converting to and from Scientific Notation

(a) $2.375 \times 10^8 = 237,500,000$

(b) $0.000000349 = 3.49 \times 10^{-7}$

Now try Exercises 57 and 59.

EXAMPLE 9   Using Scientific Notation

Simplify $\dfrac{(370{,}000)(4{,}500{,}000{,}000)}{18{,}000}$.

SOLUTION

Using Mental Arithmetic

$$\frac{(370{,}000)(4{,}500{,}000{,}000)}{18{,}000} = \frac{(3.7 \times 10^5)(4.5 \times 10^9)}{1.8 \times 10^4}$$

$$= \frac{(3.7)(4.5)}{1.8} \times 10^{5+9-4}$$

$$= 9.25 \times 10^{10}$$

$$= 92{,}500{,}000{,}000$$

Using a Calculator

Figure P.5 shows two ways to perform the computation. In the first, the numbers are entered in decimal form. In the second, the numbers are entered in scientific notation. The calculator uses "9.25E10" to stand for 9.25×10^{10}.

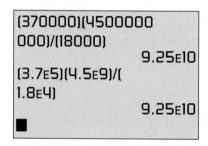

FIGURE P.5 Be sure you understand how your calculator displays scientific notation. (Example 9)

Now try Exercise 63.

QUICK REVIEW P.1

1. List the positive integers between -3 and 7.

2. List the integers between -3 and 7.

3. List all negative integers greater than -4.

4. List all positive integers less than 5.

In Exercises 5 and 6, use a calculator to evaluate the expression. Round the value to two decimal places.

5. (a) $4(-3.1)^3 - (-4.2)^5$ **(b)** $\dfrac{2(-5.5) - 6}{7.4 - 3.8}$

6. (a) $5[3(-1.1)^2 - 4(-0.5)^3]$ **(b)** $5^{-2} + 2^{-4}$

In Exercises 7 and 8, evaluate the algebraic expression for the given values of the variables.

7. $x^3 - 2x + 1, \; x = -2, \, 1.5$

8. $a^2 + ab + b^2, \; a = -3, \, b = 2$

In Exercises 9 and 10, list the possible remainders.

9. When the positive integer n is divided by 7.

10. When the positive integer n is divided by 13.

SECTION P.1 EXERCISES

In Exercises 1–4, find the decimal form for the rational number. State whether it repeats or terminates.

1. $-37/8$

2. $15/99$

3. $-13/6$

4. $5/37$

In Exercises 5–10, describe and graph the interval of real numbers.

5. $x \le 2$

6. $-2 \le x < 5$

7. $(-\infty, 7)$

8. $[-3, 3]$

9. x is negative

10. x is greater than or equal to 2 and less than or equal to 6.

In Exercises 11–16, use an inequality to describe the interval of real numbers.

11. $[-1, 1)$

12. $(-\infty, 4]$

13.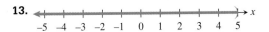

14.

15. x is between -1 and 2.

16. x is greater than or equal to 5.

In Exercises 17–22, use interval notation to describe the interval of real numbers.

17. $x > -3$

18. $-7 < x < -2$

19.

20.

21. x is greater than -3 and less than or equal to 4.

22. x is positive.

In Exercises 23–28, use words to describe the interval of real numbers.

23. $4 < x \le 9$

24. $x \ge -1$

25. $[-3, \infty)$

26. $(-5, 7)$

27.

28.

In Exercises 29–32, convert to inequality notation. Find the endpoints and state whether the interval is bounded or unbounded and its type.

29. $(-3, 4]$

30. $(-3, -1)$

31. $(-\infty, 5)$

32. $[-6, \infty)$

In Exercises 33–36, use both inequality and interval notation to describe the set of numbers. State the meaning of any variables you use.

33. Writing to Learn Bill is at least 29 years old.

34. Writing to Learn No item at Sarah's Variety Store costs more than $2.00.

35. Writing to Learn The price of a gallon of gasoline varies from $1.099 to $1.399.

36. Writing to Learn Salary raises at the State University of California at Chico will average between 2% and 6.5%.

In Exercises 37–40, use the distributive property to write the factored form or the expanded form of the given expression.

37. $a(x^2 + b)$

38. $(y - z^3)c$

39. $ax^2 + dx^2$

40. $a^3z + a^3w$

In Exercises 41 and 42, find the additive inverse of the number.

41. $6 - \pi$

42. -7

In Exercises 43 and 44, identify the base of the exponential expression.

43. -5^2

44. $(-2)^7$

45. Group Activity Discuss which algebraic property or properties are illustrated by the equation. Try to reach a consensus.

(a) $(3x)y = 3(xy)$

(b) $a^2b = ba^2$

(c) $a^2b + (-a^2b) = 0$

(d) $(x + 3)^2 + 0 = (x + 3)^2$

(e) $a(x + y) = ax + ay$

46. Group Activity Discuss which algebraic property or properties are illustrated by the equation. Try to reach a consensus.

(a) $(x + 2)\dfrac{1}{x + 2} = 1$

(b) $1 \cdot (x + y) = x + y$

(c) $2(x - y) = 2x - 2y$

(d) $2x + (y - z) = 2x + (y + (-z))$
$= (2x + y) + (-z) =$
$(2x + y) - z$

(e) $\dfrac{1}{a}(ab) = \left(\dfrac{1}{a}a\right)b = 1 \cdot b = b$

In Exercises 47–52, simplify the expression. Assume that the variables in the denominators are nonzero.

47. $\dfrac{x^4 y^3}{x^2 y^5}$ **48.** $\dfrac{(3x^2)^2 y^4}{3y^2}$

49. $\left(\dfrac{4}{x^2}\right)^2$ **50.** $\left(\dfrac{2}{xy}\right)^{-3}$

51. $\dfrac{(x^{-3} y^2)^{-4}}{(y^6 x^{-4})^{-2}}$ **52.** $\left(\dfrac{4a^3 b}{a^2 b^3}\right)\left(\dfrac{3b^2}{2a^2 b^4}\right)$

The data in Table P.1 gives the revenues in thousands of dollars for public elementary and secondary schools for the 2003–04 school year.

Table P.1 Department of Education	
Source	Amount (in $1000)
Federal	36,930,338
State	221,802,107
Local and Intermediate	193,175,805
Total	451,908,251

Source: National Education Association, as reported in The World Almanac and Book of Facts, 2005.

In Exercises 53–56, write the amount of revenue in dollars obtained from the source in scientific notation.

53. Federal

54. State

55. Local and Intermediate

56. Total

In Exercises 57 and 58, write the number in scientific notation.

57. The mean distance from Jupiter to the Sun is about 483,900,000 miles.

58. The electric charge, in coulombs, of an electron is about $-0.000\,000\,000\,000\,000\,000\,16$.

In Exercises 59–62, write the number in decimal form.

59. 3.33×10^{-8} **60.** 6.73×10^{11}

61. The distance that light travels in 1 year (*one light-year*) is about 5.87×10^{12} mi.

62. The mass of a neutron is about 1.6747×10^{-24} g.

In Exercises 63 and 64, use scientific notation to simplify.

63. $\dfrac{(1.35 \times 10^{-7})(2.41 \times 10^8)}{1.25 \times 10^9}$ **64.** $\dfrac{(3.7 \times 10^{-7})(4.3 \times 10^6)}{2.5 \times 10^7}$

$$\frac{(1.35 \times 2.41)(10^{-7} \times 10^8)}{1.25 \times 10^9}$$

Explorations

65. Investigating Exponents For positive integers m and n, we can use the definition to show that $a^m a^n = a^{m+n}$.

 (a) Examine the equation $a^m a^n = a^{m+n}$ for $n = 0$ and explain why it is reasonable to define $a^0 = 1$ for $a \neq 0$.

 (b) Examine the equation $a^m a^n = a^{m+n}$ for $n = -m$ and explain why it is reasonable to define $a^{-m} = 1/a^m$ for $a \neq 0$.

66. Decimal Forms of Rational Numbers Here is the third step when we divide 1 by 17. (The first two steps are not shown, because the quotient is 0 in both cases.)

$$
\begin{array}{r}
0.05 \\
17\overline{)1.00} \\
\underline{85} \\
15
\end{array}
$$

By convention we say that 1 is the first remainder in the long division process, 10 is the second, and 15 is the third remainder.

 (a) Continue this long division process until a remainder is repeated, and complete the following table:

Step	Quotient	Remainder
1	0	1
2	0	10
3	5	15
⋮	⋮	⋮

 (b) Explain why the digits that occur in the quotient between the pair of repeating remainders determine the infinitely repeating portion of the decimal representation. In this case

$$\frac{1}{17} = 0.\overline{0588235294117647}.$$

 (c) Explain why this procedure will always determine the infinitely repeating portion of a rational number whose decimal representation does not terminate.

Standardized Test Questions

67. True or False The additive inverse of a real number must be negative. Justify your answer.

68. True or False The reciprocal of a positive real number must be less than 1. Justify your answer.

In Exercises 69–72, solve these problems without using a calculator.

69. Multiple Choice Which of the following inequalities corresponds to the interval $[-2, 1)$?

(A) $x \leq -2$ **(B)** $-2 \leq x \leq 1$

(C) $-2 < x < 1$ **(D)** $-2 < x \leq 1$

(E) $-2 \leq x < 1$

70. Multiple Choice What is the value of $(-2)^4$?

(A) 16   **(B)** 8

(C) 6   **(D)** -8

(E) -16

71. Multiple Choice What is the base of the exponential expression -7^2?

(A) -7 **(B)** 7

(C) -2 **(D)** 2

(E) 1

72. Multiple Choice Which of the following is the simplified form of $\dfrac{x^6}{x^2}$, $x \neq 0$?

(A) x^{-4} **(B)** x^2

(C) x^3 **(D)** x^4

(E) x^8

Extending the Ideas

The **magnitude** of a real number is its distance from the origin.

73. List the whole numbers whose magnitudes are less than 7.

74. List the natural numbers whose magnitudes are less than 7.

75. List the integers whose magnitudes are less than 7.

P.2
Cartesian Coordinate System

Cartesian Plane

The points in a plane correspond to ordered pairs of real numbers, just as the points on a line can be associated with individual real numbers. This correspondence creates the **Cartesian plane**, or the **rectangular coordinate system** in the plane.

To construct a rectangular coordinate system, or a Cartesian plane, draw a pair of perpendicular real number lines, one horizontal and the other vertical, with the lines intersecting at their respective 0-points (Figure P.6). The horizontal line is usually the **x-axis** and the vertical line is usually the **y-axis**. The positive direction on the x-axis is to the right, and the positive direction on the y-axis is up. Their point of intersection, *O*, is the **origin of the Cartesian plane**.

Each point *P* of the plane is associated with an **ordered pair (x, y)** of real numbers, the **(Cartesian) coordinates of the point**. The **x-coordinate** represents the intersection of the x-axis with the perpendicular from *P*, and the **y-coordinate** represents the intersection of the y-axis with the perpendicular from *P*. Figure P.6 shows the points *P* and *Q* with coordinates (4, 2) and (−6, −4), respectively. As with real numbers and a number line, we use the ordered pair (*a*, *b*) for both the name of the point and its coordinates.

The coordinate axes divide the Cartesian plane into four **quadrants**, as shown in Figure P.7.

EXAMPLE 1   Plotting Data on U.S. Exports to Mexico

The value in billions of dollars of U.S. exports to Mexico from 1996 to 2003 is given in Table P.2. Plot the (year, export value) ordered pairs in a rectangular coordinate system.

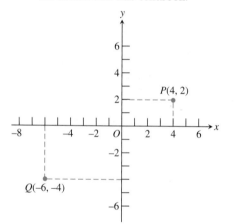

FIGURE P.6 The Cartesian coordinate plane.

Table P.2 U.S. Exports to Mexico	
Year	U.S. Exports (billions of dollars)
1996	56.8
1997	71.4
1998	78.8
1999	86.9
2000	111.3
2001	101.3
2002	97.5
2003	97.4

Source: U.S. Census Bureau, Statistical Abstract of the United States, 2001, 2004–2005.

SOLUTION

The points are plotted in Figure P.8 on page 15.

Now try Exercise 31.

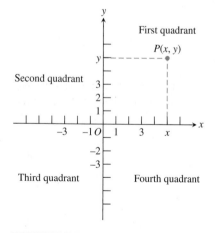

FIGURE P.7 The four quadrants. Points on the x- or y-axis are not in any quadrant.

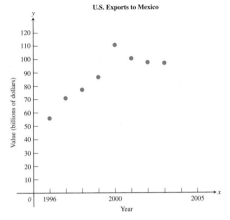

FIGURE P.8 The graph for Example 1.

A **scatter plot** is a plotting of the (x, y) data pairs on a Cartesian plane. Figure P.8 shows a scatter plot of the data from Table P.2.

Absolute Value of a Real Number

The *absolute value of a real number* suggests its **magnitude** (size). For example, the absolute value of 3 is 3 and the absolute value of -5 is 5.

DEFINITION Absolute Value of a Real Number

The **absolute value of a real number** a is

$$|a| = \begin{cases} a, & \text{if } a > 0 \\ -a, & \text{if } a < 0. \\ 0, & \text{if } a = 0 \end{cases}$$

EXAMPLE 2   Using the Definition of Absolute Value

Evaluate:

(a) $|-4|$ **(b)** $|\pi - 6|$

SOLUTION

(a) Because $-4 < 0$, $|-4| = -(-4) = 4$.

(b) Because $\pi \approx 3.14$, $\pi - 6$ is negative, so $\pi - 6 < 0$. Thus,
$|\pi - 6| = -(\pi - 6) = 6 - \pi \approx 2.858$.

Now try Exercise 9.

Here is a summary of some important properties of absolute value.

Properties of Absolute Value

Let a and b be real numbers.

1. $|a| \geq 0$ **2.** $|-a| = |a|$

3. $|ab| = |a||b|$ **4.** $\left|\dfrac{a}{b}\right| = \dfrac{|a|}{|b|}, b \neq 0$

Distance Formulas

The *distance* between -1 and 4 on the number line is 5 (see Figure P.9). This distance may be found by subtracting the smaller number from the larger: $4 - (-1) = 5$. If we use absolute value, the order of subtraction does not matter: $|4 - (-1)| = |-1 - 4| = 5$.

$$|4 - (-1)| = |-1 - 4| = 5$$

-3 -2 -1   0   1   2   3   4   5

FIGURE P.9 Finding the distance between -1 and 4.

ABSOLUTE VALUE AND DISTANCE

If we let $b = 0$ in the distance formula we see that the distance between a and 0 is $|a|$. Thus, the absolute value of a number is its distance from zero.

Distance Formula (Number Line)

Let a and b be real numbers. The **distance between a and b** is

$$|a - b|.$$

Note that $|a - b| = |b - a|$.

To find the *distance* between two points that lie on the same horizontal or vertical line in the Cartesian plane, we use the distance formula for points on a number line. For example, the distance between points x_1 and x_2 on the x-axis is $|x_1 - x_2| = |x_2 - x_1|$ and the distance between the points y_1 and y_2 on the y-axis is $|y_1 - y_2| = |y_2 - y_1|$.

To find the distance between two points $P(x_1, y_1)$ and $Q(x_2, y_2)$ that do not lie on the same horizontal or vertical line we form the right triangle determined by P, Q, and $R(x_2, y_1)$ (Figure P.10).

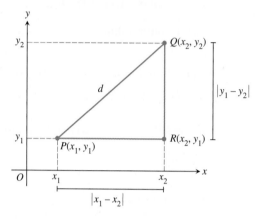

FIGURE P.10 Forming a right triangle with hypotenuse $\overline{PQ}$.

The distance from P to R is $|x_1 - x_2|$, and the distance from R to Q is $|y_1 - y_2|$. By the **Pythagorean Theorem** (see Figure P.11), the distance d between P and Q is

$$d = \sqrt{|x_1 - x_2|^2 + |y_1 - y_2|^2}.$$

Because $|x_1 - x_2|^2 = (x_1 - x_2)^2$ and $|y_1 - y_2|^2 = (y_1 - y_2)^2$, we obtain the following formula.

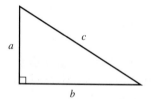

FIGURE P.11 The Pythagorean theorem: $c^2 = a^2 + b^2$.

Distance Formula (Coordinate Plane)

The **distance d between points $P(x_1, y_1)$ and $Q(x_2, y_2)$** in the coordinate plane is

$$d = \sqrt{(x_1 - x_2)^2 + (y_1 - y_2)^2}.$$

EXAMPLE 3   Finding the Distance Between Two Points

Find the distance d between the points $(1, 5)$ and $(6, 2)$.

SOLUTION

$$d = \sqrt{(1 - 6)^2 + (5 - 2)^2} \quad \text{The distance formula}$$
$$= \sqrt{(-5)^2 + 3^2}$$
$$= \sqrt{25 + 9}$$
$$= \sqrt{34} \approx 5.83 \quad \text{Using a calculator}$$

Now try Exercise 11.

Midpoint Formulas

When the endpoints of a segment in a number line are known, we take the average of their coordinates to find the midpoint of the segment.

Midpoint Formula (Number Line)

The **midpoint of the line segment with endpoints a and b** is

$$\frac{a + b}{2}.$$

EXAMPLE 4   Finding the Midpoint of a Line Segment

The midpoint of the line segment with endpoints -9 and 3 on a number line is

$$\frac{(-9) + 3}{2} = \frac{-6}{2} = -3.$$

See Figure P.12.   *Now try Exercise 23.*

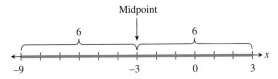

FIGURE P.12 Notice that the distance from the midpoint, -3, to 3 or to -9 is 6. (Example 4)

Just as with number lines, the midpoint of a line segment in the coordinate plane is determined by its endpoints. Each coordinate of the midpoint is the average of the corresponding coordinates of its endpoints.

Midpoint Formula (Coordinate Plane)

The **midpoint of the line segment with endpoints (a, b) and (c, d)** is

$$\left(\frac{a + c}{2}, \frac{b + d}{2} \right).$$

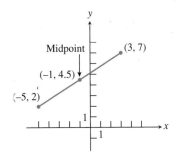

FIGURE P.13 (Example 5.)

EXAMPLE 5   Finding the Midpoint of a Line Segment

The midpoint of the line segment with endpoints $(-5, 2)$ and $(3, 7)$ is

$$(x, y) = \left(\frac{-5 + 3}{2}, \frac{2 + 7}{2} \right) = (-1, 4.5).$$

See Figure P.13.   *Now try Exercise 25.*

Equations of Circles

A **circle** is the set of points in a plane at a fixed distance (**radius**) from a fixed point (**center**). Figure P.14 shows the circle with center (h, k) and radius r. If (x, y) is any point on the circle, the distance formula gives

$$\sqrt{(x - h)^2 + (y - k)^2} = r.$$

Squaring both sides, we obtain the following equation for a circle.

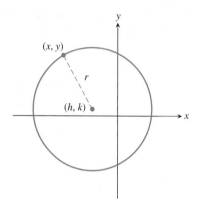

FIGURE P.14 The circle with center (h, k) and radius r.

DEFINITION Standard Form Equation of a Circle

The **standard form equation of a circle** with center (h, k) and radius r is

$$(x - h)^2 + (y - k)^2 = r^2.$$

EXAMPLE 6   Finding Standard Form Equations of Circles

Find the standard form equation of the circle.

(a) Center $(-4, 1)$, radius 8   **(b)** Center $(0, 0)$, radius 5

SOLUTION

(a) $(x - h)^2 + (y - k)^2 = r^2$ Standard form equation

$(x - (-4))^2 + (y - 1)^2 = 8^2$ Substitute $h = -4, k = 1, r = 8$.

$(x + 4)^2 + (y - 1)^2 = 64$

(b) $(x - h)^2 + (y - k)^2 = r^2$ Standard form equation

$(x - 0)^2 + (y - 0)^2 = 5^2$ Substitute $h = 0, k = 0, r = 5$.

$x^2 + y^2 = 25$

Now try Exercise 41.

Applications

EXAMPLE 7   Using an Inequality to Express Distance

We can state that "the distance between x and -3 is less than 9" using the inequality

$$|x - (-3)| < 9 \quad \text{or} \quad |x + 3| < 9.$$

Now try Exercise 51.

The converse of the Pythagorean theorem is true. That is, if the sum of squares of the lengths of the two sides of a triangle equals the square of the length of the third side, then the triangle is a right triangle.

EXAMPLE 8 Verifying Right Triangles

Use the converse of the Pythagorean theorem and the distance formula to show that the points $(-3, 4)$, $(1, 0)$, and $(5, 4)$ determine a right triangle.

SOLUTION The three points are plotted in Figure P.15. We need to show that the lengths of the sides of the triangle satisfy the Pythagorean relationship $a^2 + b^2 = c^2$. Applying the distance formula we find that

$$a = \sqrt{(-3 - 1)^2 + (4 - 0)^2} = \sqrt{32},$$
$$b = \sqrt{(1 - 5)^2 + (0 - 4)^2} = \sqrt{32},$$
$$c = \sqrt{(-3 - 5)^2 + (4 - 4)^2} = \sqrt{64}.$$

The triangle is a right triangle because

$$a^2 + b^2 = (\sqrt{32})^2 + (\sqrt{32})^2 = 32 + 32 = 64 = c^2.$$

Now try Exercise 39.

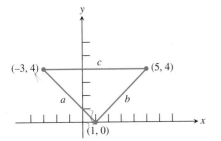

FIGURE P.15 The triangle in Example 8.

Properties of geometric figures can sometimes be confirmed using analytic methods such as the midpoint formulas.

EXAMPLE 9 Using the Midpoint Formula

It is a fact from geometry that the diagonals of a parallelogram bisect each other. Prove this with a midpoint formula.

SOLUTION We can position a parallelogram in the rectangular coordinate plane as shown in Figure P.16. Applying the midpoint formula for the coordinate plane to segments OB and AC, we find that

$$\text{midpoint of segment } OB = \left(\frac{0 + a + c}{2}, \frac{0 + b}{2}\right) = \left(\frac{a + c}{2}, \frac{b}{2}\right),$$

$$\text{midpoint of segment } AC = \left(\frac{a + c}{2}, \frac{b + 0}{2}\right) = \left(\frac{a + c}{2}, \frac{b}{2}\right).$$

The midpoints of segments OA and AC are the same, so the diagonals of the parallelogram $OABC$ meet at their midpoints and thus bisect each other.

Now try Exercise 37.

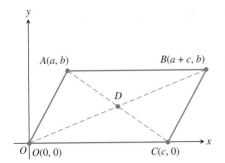

FIGURE P.16 The coordinates of B must be $(a + c, b)$ in order for CB to be parallel to OA. (Example 9)

QUICK REVIEW P.2

In Exercises 1 and 2, plot the two numbers on a number line. Then find the distance between them.

1. $\sqrt{7}, \sqrt{2}$

2. $-\dfrac{5}{3}, -\dfrac{9}{5}$

In Exercises 3 and 4, plot the real numbers on a number line.

3. $-3, 4, 2.5, 0, -1.5$

4. $-\dfrac{5}{2}, -\dfrac{1}{2}, \dfrac{2}{3}, 0, -1$

In Exercises 5 and 6, plot the points.

5. $A(3, 5), B(-2, 4), C(3, 0), D(0, -3)$

6. $A(-3, -5), B(2, -4), C(0, 5), D(-4, 0)$

In Exercises 7–10, use a calculator to evaluate the expression. Round your answer to two decimal places.

7. $\dfrac{-17 + 28}{2}$

8. $\sqrt{13^2 + 17^2}$

9. $\sqrt{6^2 + 8^2}$

10. $\sqrt{(17 - 3)^2 + (-4 - 8)^2}$

SECTION P.2 EXERCISES

In Exercises 1 and 2, estimate the coordinates of the points.

1.

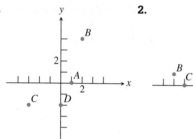

2.

In Exercises 3 and 4, name the quadrants containing the points.

3. (a) $(2, 4)$ **(b)** $(0, 3)$ **(c)** $(-2, 3)$ **(d)** $(-1, -4)$

4. (a) $\left(\dfrac{1}{2}, \dfrac{3}{2}\right)$ **(b)** $(-2, 0)$ **(c)** $(-1, -2)$ **(d)** $\left(-\dfrac{3}{2}, -\dfrac{7}{3}\right)$

In Exercises 5–8, evaluate the expression.

5. $3 + |-3|$

6. $2 - |-2|$

7. $|(-2)3|$

8. $\dfrac{-2}{|-2|}$

In Exercises 9 and 10, rewrite the expression without using absolute value symbols.

9. $|\pi - 4|$

10. $|\sqrt{5} - 5/2|$

In Exercises 11–18, find the distance between the points.

11. $-9.3, 10.6$

12. $-5, -17$

13. $(-3, -1), (5, -1)$

14. $(-4, -3), (1, 1)$

15. $(0, 0), (3, 4)$

16. $(-1, 2), (2, -3)$

17. $(-2, 0), (5, 0)$

18. $(0, -8), (0, -1)$

In Exercises 19–22, find the area and perimeter of the figure determined by the points.

19. $(-5, 3), (0, -1), (4, 4)$

20. $(-2, -2), (-2, 2), (2, 2), (2, -2)$

21. $(-3, -1), (-1, 3), (7, 3), (5, -1)$

22. $(-2, 1), (-2, 6), (4, 6), (4, 1)$

In Exercises 23–28, find the midpoint of the line segment with the given endpoints.

23. $-9.3, 10.6$

24. $-5, -17$

25. $(-1, 3), (5, 9)$

26. $(3, \sqrt{2}), (6, 2)$

27. $(-7/3, 3/4), (5/3, -9/4)$

28. $(5, -2), (-1, -4)$

In Exercises 29–34, draw a scatter plot of the data given in the table.

29. U.S. Aluminum Imports The total value y in billions of dollars of aluminum imported by the United States each year from 1997 to 2003 is given in the table. (*Source: U.S Census Bureau, Statistical Abstract of the United States, 2001, 2004–2005.*)

x	1997	1998	1999	2000	2001	2002	2003
y	5.6	6.0	6.3	6.9	6.4	6.6	7.2

30. U.S. Aluminum Exports The total value y in billions of dollars of aluminum exported by the United States each year from 1997 to 2003 is given in the table. (*Source: U.S. Census Bureau, Statistical Abstract of the United States, 2001, 2004–2005.*)

x	1997	1998	1999	2000	2001	2002	2003
y	3.8	3.6	3.6	3.8	3.3	2.9	2.9

31. U.S. Imports from Mexico The total in billions of dollars of U.S. imports from Mexico from 1996 to 2003 is given in Table P.3.

Table P.3 U.S. Imports from Mexico	
Year	U.S. Imports (billions of dollars)
1996	74.3
1997	85.9
1998	94.6
1999	109.7
2000	135.9
2001	131.3
2002	134.6
2003	138.1

Source: U.S. Census Bureau, Statistical Abstract of the United States, 2001, 2004–2005.

32. U.S. Agricultural Exports The total in billions of dollars of U.S. agricultural exports from 1996 to 2003 is given in Table P.4.

Table P.4 U.S. Agricultural Exports	
Year	U.S. Agricultural Exports (billions of dollars)
1996	60.6
1997	57.1
1998	52.0
1999	48.2
2000	53.0
2001	55.2
2002	54.8
2003	61.5

Source: U.S. Census Bureau, Statistical Abstract of the United States, 2004–2005.

33. U.S. Agricultural Trade Surplus The total in billions of dollars of U.S. agricultural trade surplus from 1996 to 2003 is given in Table P.5.

Table P.5 U.S. Agricultural Trade Surplus	
Year	U.S. Agricultural Trade Surplus (billions of dollars)
1996	28.1
1997	21.9
1998	16.3
1999	11.5
2000	13.8
2001	15.7
2002	12.8
2003	8.3

Source: U.S. Census Bureau, Statistical Abstract of the United States, 2004–2005.

34. U.S. Exports to Armenia The total in millions of dollars of U.S. exports to Armenia from 1996 to 2003 is given in Table P.6.

Table P.6 U.S. Exports to Armenia	
Year	U.S. Exports (millions of dollars)
1996	57.4
1997	62.1
1998	51.4
1999	51.2
2000	55.6
2001	49.9
2002	111.8
2003	102.8

Source: U.S. Census Bureau, Statistical Abstract of the United States, 2001, 2004–2005.

In Exercises 35 and 36, use the graph of the investment value of a $10,000 investment made in 1978 in Fundamental Investors™ of the American Funds™. The value as of January is shown for a few recent years in the graph below. (*Source: Annual report of Fundamental Investors for the year ending December 31, 2004.*)

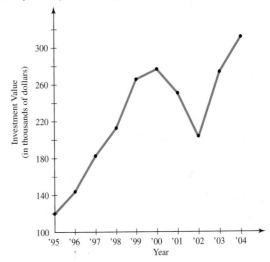

35. Reading from Graphs Use the graph to estimate the value of the investment as of

(a) January 1997 and (b) January 2000.

36. Percent Increase Estimate the percent increase in the value of the $10,000 investment from

(a) January 1996 to January 1997.

(b) January 2000 to January 2001.

(c) January 1995 to January 2004.

37. Prove that the figure determined by the points is an isosceles triangle: (1, 3), (4, 7), (8, 4)

38. Group Activity Prove that the diagonals of the figure determined by the points bisect each other.

  (a) Square $(-7, -1)$, $(-2, 4)$, $(3, -1)$, $(-2, -6)$

  (b) Parallelogram $(-2, -3)$, $(0, 1)$, $(6, 7)$, $(4, 3)$

39. (a) Find the lengths of the sides of the triangle in the figure.

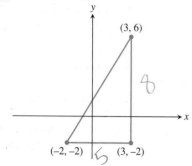

  (b) Writing to Learn Show that the triangle is a right triangle.

40. (a) Find the lengths of the sides of the triangle in the figure.

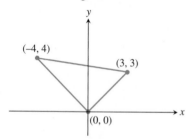

  (b) Writing to Learn Show that the triangle is a right triangle.

In Exercises 41–44, find the standard form equation for the circle.

41. Center (1, 2), radius 5

42. Center $(-3, 2)$, radius 1

43. Center $(-1, -4)$, radius 3

44. Center (0, 0), radius $\sqrt{3}$

In Exercises 45–48, find the center and radius of the circle.

45. $(x - 3)^2 + (y - 1)^2 = 36$

46. $(x + 4)^2 + (y - 2)^2 = 121$

47. $x^2 + y^2 = 5$

48. $(x - 2)^2 + (y + 6)^2 = 25$

In Exercises 49–52, write the statement using absolute value notation.

49. The distance between x and 4 is 3.

50. The distance between y and -2 is greater than or equal to 4.

51. The distance between x and c is less than d units.

52. y is more than d units from c.

53. Determining a Line Segment with Given Midpoint
Let (4, 4) be the midpoint of the line segment determined by the points (1, 2) and (a, b). Determine a and b.

54. Writing to Learn Isosceles but Not Equilateral Triangle
Prove that the triangle determined by the points (3, 0), $(-1, 2)$, and (5, 4) is isosceles but not equilateral.

55. Writing to Learn Equidistant Point from Vertices of a Right Triangle Prove that the midpoint of the hypotenuse of the right triangle with vertices (0, 0), (5, 0), and (0, 7) is equidistant from the three vertices.

56. Writing to Learn Describe the set of real numbers that satisfy $|x - 2| < 3$.

57. Writing to Learn Describe the set of real numbers that satisfy $|x + 3| \geq 5$.

Standardized Test Questions

58. True or False If a is a real number, then $|a| \geq 0$. Justify your answer.

59. True or False Consider the right triangle ABC shown at the right. If M is the midpoint of the segment AB, then M' is the midpoint of the segment AC. Justify your answer.

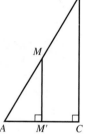

In Exercises 60–63, solve these problems without using a calculator.

60. Multiple Choice Which of the following is equal to $|1 - \sqrt{3}|$?

  (A) $1 - \sqrt{3}$ **(B)** $\sqrt{3} - 1$

  (C) $(1 - \sqrt{3})^2$ **(D)** $\sqrt{2}$

  (E) $\sqrt{1/3}$

61. Multiple Choice Which of the following is the midpoint of the line segment with endpoints -3 and 2?

  (A) 5/2   **(B)** 1

  (C) $-1/2$ **(D)** -1

  (E) $-5/2$

62. Multiple Choice Which of the following is the center of the circle $(x - 3)^2 + (y + 4)^2 = 2$?

(A) $(3, -4)$ (B) $(-3, 4)$

(C) $(4, -3)$ (D) $(-4, 3)$

(E) $(3/2, -2)$

63. Multiple Choice Which of the following points is in the third quadrant?

(A) $(0, -3)$ (B) $(-1, 0)$

(C) $(2, -1)$ (D) $(-1, 2)$

(E) $(-2, -3)$

Explorations

64. Dividing a Line Segment Into Thirds

(a) Find the coordinates of the points one-third and two-thirds of the way from $a = 2$ to $b = 8$ on a number line.

(b) Repeat (a) for $a = -3$ and $b = 7$.

(c) Find the coordinates of the points one-third and two-thirds of the way from a to b on a number line.

(d) Find the coordinates of the points one-third and two-thirds of the way from the point $(1, 2)$ to the point $(7, 11)$ in the coordinate plane.

(e) Find the coordinates of the points one-third and two-thirds of the way from the point (a, b) to the point (c, d) in the coordinate plane.

Extending the Ideas

65. Writing to Learn Equidistant Point from Vertices of a Right Triangle Prove that the midpoint of the hypotenuse of any right triangle is equidistant from the three vertices.

66. Comparing Areas Consider the four points $A(0, 0)$, $B(0, a)$, $C(a, a)$, and $D(a, 0)$. Let P be the midpoint of the line segment CD and Q the point one-fourth of the way from A to D on segment AD.

(a) Find the area of triangle BPQ.

(b) Compare the area of triangle BPQ with the area of square $ABCD$.

In Exercises 67–69, let $P(a, b)$ be a point in the first quadrant.

67. Find the coordinates of the point Q in the fourth quadrant so that PQ is perpendicular to the x-axis.

68. Find the coordinates of the point Q in the second quadrant so that PQ is perpendicular to the y-axis.

69. Find the coordinates of the point Q in the third quadrant so that the origin is the midpoint of the segment PQ.

70. Writing to Learn Prove that the distance formula for the number line is a special case of the distance formula for the Cartesian plane.

P.3
Linear Equations and Inequalities

Equations

An **equation** is a statement of equality between two expressions. Here are some properties of equality that we use to solve equations algebraically.

Properties of Equality	
Let u, v, w, and z be real numbers, variables, or algebraic expressions.	
1. Reflexive	$u = u$
2. Symmetric	If $u = v$, then $v = u$.
3. Transitive	If $u = v$, and $v = w$, then $u = w$.
4. Addition	If $u = v$ and $w = z$, then $u + w = v + z$.
5. Multiplication	If $u = v$ and $w = z$, then $uw = vz$.

Solving Equations

A **solution of an equation in x** is a value of x for which the equation is true. To **solve an equation in x** means to find all values of x for which the equation is true, that is, to find all solutions of the equation.

> **EXAMPLE 1   Confirming a Solution**
>
> Prove that $x = -2$ is a solution of the equation $x^3 - x + 6 = 0$.
>
> **SOLUTION**
>
> $$(-2)^3 - (-2) + 6 \stackrel{?}{=} 0$$
> $$-8 + 2 + 6 \stackrel{?}{=} 0$$
> $$0 = 0$$ *Now try Exercise 1.*

Linear Equations in One Variable

The most basic equation in algebra is a *linear equation*.

DEFINITION Linear Equation in x
A **linear equation in x** is one that can be written in the form
$$ax + b = 0,$$
where a and b are real numbers with $a \neq 0$.

The equation $2z - 4 = 0$ is linear in the variable z. The equation $3u^2 - 12 = 0$ is *not* linear in the variable u. A linear equation in one variable has exactly one solution. We solve such an equation by transforming it into an *equivalent equation* whose solution is obvious. Two or more equations are **equivalent** if they have the same solutions. For example, the equations $2z - 4 = 0$, $2z = 4$, and $z = 2$ are all equivalent. Here are operations that produce equivalent equations.

Operations for Equivalent Equations

An equivalent equation is obtained if one or more of the following operations are performed.

Operation	Given Equation	Equivalent Equation
1. Combine like terms, reduce fractions, and remove grouping symbols	$2x + x = \dfrac{3}{9}$	$3x = \dfrac{1}{3}$
2. Perform the same operation on both sides.		
(a) Add (-3).	$x + 3 = 7$	$x = 4$
(b) Subtract $(2x)$.	$5x = 2x + 4$	$3x = 4$
(c) Multiply by a nonzero constant $(1/3)$.	$3x = 12$	$x = 4$
(d) Divide by a nonzero constant (3).	$3x = 12$	$x = 4$

The next two examples illustrate how to use equivalent equations to solve linear equations.

EXAMPLE 2 Solving a Linear Equation

Solve $2(2x - 3) + 3(x + 1) = 5x + 2$. Support the result with a calculator.

SOLUTION

$$2(2x - 3) + 3(x + 1) = 5x + 2$$

$$4x - 6 + 3x + 3 = 5x + 2 \quad \text{Distributive property}$$

$$7x - 3 = 5x + 2 \qquad\qquad \text{Combine like terms.}$$

$$2x = 5 \qquad\qquad\qquad\quad \text{Add 3, and subtract } 5x.$$

$$x = 2.5 \qquad\qquad\qquad\quad \text{Divide by 2.}$$

To support our algebraic work we can use a calculator to evaluate the original equation for $x = 2.5$. Figure P.17 shows that each side of the original equation is equal to 14.5 if $x = 2.5$. ***Now try Exercise 23.***

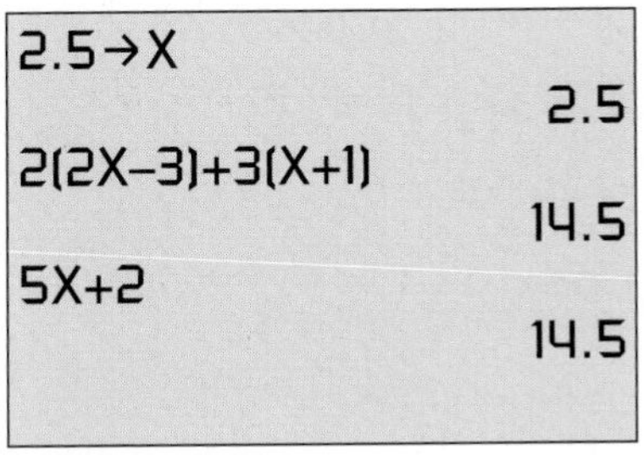

```
2.5→X
                    2.5
2(2X−3)+3(X+1)
                   14.5
5X+2
                   14.5
```

FIGURE P.17 The top line stores the number 2.5 into the variable x. (Example 2)

If an equation involves fractions, find the least common denominator (LCD) of the fractions and multiply both sides by the LCD. This is sometimes referred to as *clearing the equation of fractions*. Example 3 illustrates.

Notice in Example 2 that $2 = \dfrac{2}{1}$.

EXAMPLE 3   Solving a Linear Equation Involving Fractions

Solve

$$\frac{5y - 2}{8} = 2 + \frac{y}{4}$$

SOLUTION The denominators are 8, 1, and 4. The LCD of the fractions is 8. (See Appendix A.3 if necessary.)

$$\frac{5y - 2}{8} = 2 + \frac{y}{4}$$

$$8\left(\frac{5y - 2}{8}\right) = 8\left(2 + \frac{y}{4}\right) \qquad \text{Multiply by the LCD 8.}$$

$$8 \cdot \frac{5y - 2}{8} = 8 \cdot 2 + 8 \cdot \frac{y}{4} \qquad \text{Distributive property}$$

$$5y - 2 = 16 + 2y \qquad \text{Simplify.}$$

$$5y = 18 + 2y \qquad \text{Add 2.}$$

$$3y = 18 \qquad \text{Subtract } 2y.$$

$$y = 6 \qquad \text{Divide by 3.}$$

We leave it to you to check the solution either using paper and pencil or a calculator.

Now try Exercise 25.

Linear Inequalities in One Variable

We used inequalities to describe order on the number line in Section P.1. For example, if x is to the left of 2 on the number line, or if x is any real number less than 2, we write $x < 2$. The most basic inequality in algebra is a *linear inequality*.

DEFINITION Linear Inequality in *x*

A **linear inequality in *x*** is one that can be written in the form

$$ax + b < 0, \ ax + b \leq 0, \ ax + b > 0, \ \text{or} \ ax + b \geq 0,$$

where a and b are real numbers with $a \neq 0$.

To **solve an inequality in *x*** means to find all values of x for which the inequality is true. A **solution of an inequality in *x*** is a value of x for which the inequality is true. The set of all solutions of an inequality is the **solution set** of the inequality. We **solve an inequality** by finding its solution set. Here is a list of properties we use to solve inequalities.

DIRECTION OF AN INEQUALITY

Multiplying (or dividing) an inequality by a positive number preserves the direction of the inequality. Multiplying (or dividing) an inequality by a negative number reverses the direction.

Properties of Inequalities

Let u, v, w, and z be real numbers, variables, or algebraic expressions, and c a real number.

1. Transitive   If $u < v$ and $v < w$, then $u < w$.

2. Addition   If $u < v$, then $u + w < v + w$.
     If $u < v$ and $w < z$, then $u + w < v + z$.

3. Multiplication   If $u < v$ and $c > 0$, then $uc < vc$.
     If $u < v$ and $c < 0$, then $uc > vc$.

The above properties are true if $<$ is replaced by $\leq$. There are similar properties for $>$ and $\geq$.

The set of solutions of a linear inequality in one variable form an interval of real numbers. Just as with linear equations, we solve a linear inequality by transforming it into an *equivalent inequality* whose solutions are obvious. Two or more inequalities are **equivalent** if they have the same set of solutions. The properties of inequalities listed above describe operations that transform an inequality into an equivalent one.

EXAMPLE 4   Solving a Linear Inequality

Solve $3(x - 1) + 2 \leq 5x + 6$.

SOLUTION

$$3(x - 1) + 2 \leq 5x + 6$$
$$3x - 3 + 2 \leq 5x + 6 \qquad \text{Distributive property}$$
$$3x - 1 \leq 5x + 6 \qquad \text{Simplify.}$$
$$3x \leq 5x + 7 \qquad \text{Add 1.}$$
$$-2x \leq 7 \qquad \text{Subtract } 5x.$$
$$\left(-\frac{1}{2}\right) \cdot -2x \geq \left(-\frac{1}{2}\right) \cdot 7 \qquad \text{Multiply by } -1/2. \text{ (The inequality reverses.)}$$
$$x \geq -3.5$$

The solution set of the inequality is the set of all real numbers greater than or equal to -3.5. In interval notation, the solution set is $[-3.5, \infty)$.

Now try Exercise 41.

Because the solution set of a linear inequality is an interval of real numbers, we can display the solution set with a number line graph as illustrated in Example 5.

EXAMPLE 5   Solving a Linear Inequality Involving Fractions

Solve the inequality and graph its solution set.

$$\frac{x}{3} + \frac{1}{2} > \frac{x}{4} + \frac{1}{3}$$

continued

SOLUTION The LCD of the fractions is 12.

$$\frac{x}{3} + \frac{1}{2} > \frac{x}{4} + \frac{1}{3} \qquad \text{The original inequality}$$

$$12 \cdot \left(\frac{x}{3} + \frac{1}{2} \right) > 12 \cdot \left(\frac{x}{4} + \frac{1}{3} \right) \qquad \text{Multiply by the LCD 12.}$$

$$4x + 6 > 3x + 4 \qquad \text{Simplify.}$$

$$x + 6 > 4 \qquad \text{Subtract } 3x.$$

$$x > -2 \qquad \text{Subtract 6.}$$

The solution set is the interval $(-2, \infty)$. Its graph is shown in Figure P.18.

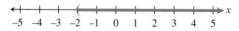

FIGURE P.18 The graph of the solution set of the inequality in Example 5.

Now try Exercise 43.

Sometimes two inequalities are combined in a **double inequality**, whose solution set is a double inequality with x isolated as the middle term. Example 6 illustrates.

EXAMPLE 6   Solving a Double Inequality

Solve the inequality and graph its solution set.

$$-3 < \frac{2x + 5}{3} \le 5$$

SOLUTION

$$-3 < \frac{2x + 5}{3} \le 5$$

$$-9 < 2x + 5 \le 15 \qquad \text{Multiply by 3.}$$

$$-14 < 2x \le 10 \qquad \text{Subtract 5.}$$

$$-7 < x \le 5 \qquad \text{Divide by 2.}$$

The solution set is the set of all real numbers greater than -7 and less than or equal to 5. In interval notation, the solution is set $(-7, 5]$. Its graph is shown in Figure P.19.

Now try Exercise 47.

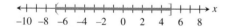

FIGURE P.19 The graph of the solution set of the double inequality in Example 6.

QUICK REVIEW P.3

In Exercises 1 and 2, simplify the expression by combining like terms.

1. $2x + 5x + 7 + y - 3x + 4y + 2$

2. $4 + 2x - 3z + 5y - x + 2y - z - 2$

In Exercises 3 and 4, use the distributive property to expand the products. Simplify the resulting expression by combining like terms.

3. $3(2x - y) + 4(y - x) + x + y$

4. $5(2x + y - 1) + 4(y - 3x + 2) + 1$

In Exercises 5–10, use the LCD to combine the fractions. Simplify the resulting fraction.

5. $\dfrac{2}{y} + \dfrac{3}{y}$

6. $\dfrac{1}{y - 1} + \dfrac{3}{y - 2}$

7. $2 + \dfrac{1}{x}$

8. $\dfrac{1}{x} + \dfrac{1}{y} - x$

9. $\dfrac{x + 4}{2} + \dfrac{3x - 1}{5}$

10. $\dfrac{x}{3} + \dfrac{x}{4}$

SECTION P.3 EXERCISES

In Exercises 1–4, find which values of x are solutions of the equation.

1. $2x^2 + 5x = 3$

    (a) $x = -3$ **(b)** $x = -\dfrac{1}{2}$ **(c)** $x = \dfrac{1}{2}$

2. $\dfrac{x}{2} + \dfrac{1}{6} = \dfrac{x}{3}$

    (a) $x = -1$ **(b)** $x = 0$ **(c)** $x = 1$

3. $\sqrt{1 - x^2} + 2 = 3$

    (a) $x = -2$ **(b)** $x = 0$ **(c)** $x = 2$

4. $(x - 2)^{1/3} = 2$

    (a) $x = -6$ **(b)** $x = 8$ **(c)** $x = 10$

In Exercises 5–10, determine whether the equation is linear in x.

5. $5 - 3x = 0$ **6.** $5 = 10/2$

7. $x + 3 = x - 5$ **8.** $x - 3 = x^2$

9. $2\sqrt{x} + 5 = 10$ **10.** $x + \dfrac{1}{x} = 1$

In Exercises 11–24, solve the equation.

11. $3x = 24$ **12.** $4x = -16$

13. $3t - 4 = 8$ **14.** $2t - 9 = 3$

15. $2x - 3 = 4x - 5$ **16.** $4 - 2x = 3x - 6$

17. $4 - 3y = 2(y + 4)$ **18.** $4(y - 2) = 5y$

19. $\dfrac{1}{2}x = \dfrac{7}{8}$ **20.** $\dfrac{2}{3}x = \dfrac{4}{5}$

21. $\dfrac{1}{2}x + \dfrac{1}{3} = 1$ **22.** $\dfrac{1}{3}x + \dfrac{1}{4} = 1$

23. $2(3 - 4z) - 5(2z + 3) = z - 17$

24. $3(5z - 3) - 4(2z + 1) = 5z - 2$

In Exercises 25–28, solve the equation. Support your answer with a calculator.

25. $\dfrac{2x - 3}{4} + 5 = 3x$ **26.** $2x - 4 = \dfrac{4x - 5}{3}$

27. $\dfrac{t + 5}{8} - \dfrac{t - 2}{2} = \dfrac{1}{3}$ **28.** $\dfrac{t - 1}{3} + \dfrac{t + 5}{4} = \dfrac{1}{2}$

29. Writing to Learn Write a statement about solutions of equations suggested by the computations in the figure.

(a)
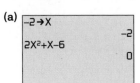

(b)
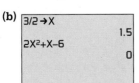

30. Writing to Learn Write a statement about solutions of equations suggested by the computations in the figure.

(a)
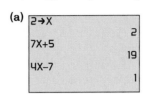

(b)
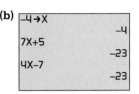

In Exercises 31–34, find which values of x are solutions of the inequality.

31. $2x - 3 < 7$

    (a) $x = 0$ **(b)** $x = 5$ **(c)** $x = 6$

32. $3x - 4 \geq 5$

    (a) $x = 0$ **(b)** $x = 3$ **(c)** $x = 4$

33. $-1 < 4x - 1 \leq 11$

    (a) $x = 0$ **(b)** $x = 2$ **(c)** $x = 3$

34. $-3 \leq 1 - 2x \leq 3$

    (a) $x = -1$ **(b)** $x = 0$ **(c)** $x = 2$

In Exercises 35–42, solve the inequality, and draw a number line graph of the solution set.

35. $x - 4 < 2$ **36.** $x + 3 > 5$

37. $2x - 1 \leq 4x + 3$ **38.** $3x - 1 \geq 6x + 8$

39. $2 \leq x + 6 < 9$ **40.** $-1 \leq 3x - 2 < 7$

41. $2(5 - 3x) + 3(2x - 1) \leq 2x + 1$

42. $4(1 - x) + 5(1 + x) > 3x - 1$

In Exercises 43–54, solve the inequality.

43. $\dfrac{5x + 7}{4} \leq -3$ **44.** $\dfrac{3x - 2}{5} > -1$

45. $4 \geq \dfrac{2y - 5}{3} \geq -2$ **46.** $1 > \dfrac{3y - 1}{4} > -1$

47. $0 \leq 2z + 5 < 8$ **48.** $-6 < 5t - 1 < 0$

49. $\dfrac{x - 5}{4} + \dfrac{3 - 2x}{3} < -2$ **50.** $\dfrac{3 - x}{2} + \dfrac{5x - 2}{3} < -1$

51. $\dfrac{2y - 3}{2} + \dfrac{3y - 1}{5} < y - 1$

52. $\dfrac{3 - 4y}{6} - \dfrac{2y - 3}{8} \geq 2 - y$

53. $\dfrac{1}{2}(x - 4) - 2x \leq 5(3 - x)$

54. $\dfrac{1}{2}(x + 3) + 2(x - 4) < \dfrac{1}{3}(x - 3)$

In Exercises 55–58, find the solutions of the equation or inequality displayed in Figure P.20.

55. $x^2 - 2x < 0$

56. $x^2 - 2x = 0$

57. $x^2 - 2x > 0$

58. $x^2 - 2x \leq 0$

X	Y1	
0	0	
1	-1	
2	0	
3	3	
4	8	
5	15	
6	24	

Y1 ◼ X²–2X

FIGURE P.20 The second column gives values of $y_1 = x^2 - 2x$ for $x = 0, 1, 2, 3, 4, 5$, and 6.

59. Writing to Learn Explain how the second equation was obtained from the first.

$$x - 3 = 2x + 3, \quad 2x - 6 = 4x + 6$$

60. Writing to Learn Explain how the second equation was obtained from the first.

$$2x - 1 = 2x - 4, \quad x - \frac{1}{2} = x - 2$$

61. Group Activity Determine whether the two equations are equivalent.

(a) $3x = 6x + 9, \quad x = 2x + 9$

(b) $6x + 2 = 4x + 10, \quad 3x + 1 = 2x + 5$

62. Group Activity Determine whether the two equations are equivalent.

(a) $3x + 2 = 5x - 7, \quad -2x + 2 = -7$

(b) $2x + 5 = x - 7, \quad 2x = x - 7$

Standardized Test Questions

63. True or False $-6 > -2$. Justify your answer.

64. True or False $2 \leq \frac{6}{3}$. Justify your answer.

In Exercises 65–68, you may use a graphing calculator to solve these problems.

65. Multiple Choice Which of the following equations is equivalent to the equation $3x + 5 = 2x + 1$?

(A) $3x = 2x$

(B) $3x = 2x + 4$

(C) $\frac{3}{2}x + \frac{5}{2} = x + 1$

(D) $3x + 6 = 2x$

(E) $3x = 2x - 4$

66. Multiple Choice Which of the following inequalities is equivalent to the inequality $-3x < 6$?

(A) $3x < -6$

(B) $x < 10$

(C) $x > -2$

(D) $x > 2$

(E) $x > 3$

67. Multiple Choice Which of the following is the solution to the equation $x(x + 1) = 0$?

(A) $x = 0$ or $x = -1$

(B) $x = 0$ or $x = 1$

(C) only $x = -1$

(D) only $x = 0$

(E) only $x = 1$

68. Multiple Choice Which of the following represents an equation equivalent to the equation

$$\frac{2x}{3} + \frac{1}{2} = \frac{x}{4} - \frac{1}{3}$$

that is cleared of fractions?

(A) $2x + 1 = x - 1$

(B) $8x + 6 = 3x - 4$

(C) $4x + 3 = \frac{3}{2}x - 2$

(D) $4x + 3 = 3x - 4$

(E) $4x + 6 = 3x - 4$

Explorations

69. Testing Inequalities on a Calculator

(a) The calculator we use indicates that the statement $2 < 3$ is true by returning the value 1 (for true) when $2 < 3$ is entered. Try it with your calculator.

(b) The calculator we use indicates that the statement $2 < 1$ is false by returning the value 0 (for false) when $2 < 1$ is entered. Try it with your calculator.

(c) Use your calculator to test which of these two numbers is larger: 799/800, 800/801.

(d) Use your calculator to test which of these two numbers is larger: $-102/101$, $-103/102$.

(e) If your calculator returns 0 when you enter $2x + 1 < 4$, what can you conclude about the value stored in x?

Extending the Ideas

70. Perimeter of a Rectangle The formula for the perimeter P of a rectangle is

$$P = 2(L + W).$$

Solve this equation for W.

71. Area of a Trapezoid The formula for the area A of a trapezoid is

$$A = \frac{1}{2}h(b_1 + b_2).$$

Solve this equation for b_1.

72. Volume of a Sphere The formula for the volume V of a sphere is

$$V = \frac{4}{3}\pi r^3.$$

Solve this equation for r.

73. Celsius and Fahrenheit The formula for Celsius temperature in terms of Fahrenheit temperature is

$$C = \frac{5}{9}(F - 32).$$

Solve the equation for F.

P.4
Lines in the Plane

What you'll learn about

- Slope of a Line

- Point-Slope Form Equation of a Line

- Slope-Intercept Form Equation of a Line

- Graphing Linear Equations in Two Variables

- Parallel and Perpendicular Lines

- Applying Linear Equations in Two Variables

. . . and why

Linear equations are used extensively in applications involving business and behavioral science.

Slope of a Line

The slope of a nonvertical line is the ratio of the amount of vertical change to the amount of horizontal change between two points. For the points (x_1, y_1) and (x_2, y_2), the vertical change is $\triangle y = y_2 - y_1$ and the horizontal change is $\triangle x = x_2 - x_1$. $\triangle y$ is read "delta" y. See Figure P.21.

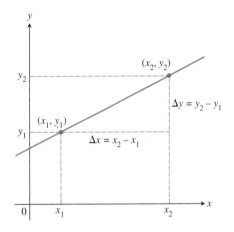

FIGURE P.21 The slope of a nonvertical line can be found from the coordinates of any two points of the line.

DEFINITION Slope of a Line

The **slope** of the nonvertical line through the points (x_1, y_1) and (x_2, y_2) is

$$m = \frac{\triangle y}{\triangle x} = \frac{y_2 - y_1}{x_2 - x_1}.$$

If the line is vertical, then $x_1 = x_2$ and the slope is undefined.

EXAMPLE 1   Finding the Slope of a Line

Find the slope of the line through the two points. Sketch a graph of the line.

(a) $(-1, 2)$ and $(4, -2)$ **(b)** $(1, 1)$, and $(3, 4)$

SOLUTION

(a) The two points are $(x_1, y_1) = (-1, 2)$ and $(x_2, y_2) = (4, -2)$. Thus,

$$m = \frac{y_2 - y_1}{x_2 - x_1} = \frac{(-2) - 2}{4 - (-1)} = -\frac{4}{5}.$$

continued

SLOPE FORMULA

The slope does not depend on the order of the points. We could use $(x_1, y_1) = (4, -2)$ and $(x_2, y_2) = (-1, 2)$ in Example 1a. Check it out.

EXAMPLE 5   Finding an Equation of a Parallel Line

Find an equation of the line through $P(1, -2)$ that is parallel to the line L with equation $3x - 2y = 1$.

SOLUTION We find the slope of L by writing its equation in slope-intercept form.

$$3x - 2y = 1 \qquad \text{Equation for } L$$

$$-2y = -3x + 1 \qquad \text{Subtract } 3x.$$

$$y = \frac{3}{2}x - \frac{1}{2} \qquad \text{Divide by } -2.$$

The slope of L is $3/2$.

The line whose equation we seek has slope $3/2$ and contains the point $(x_1, y_1) = (1, -2)$. Thus, the point-slope form equation for the line we seek is

$$y + 2 = \frac{3}{2}(x - 1)$$

$$y + 2 = \frac{3}{2}x - \frac{3}{2} \qquad \text{Distributive property}$$

$$y = \frac{3}{2}x - \frac{7}{2}$$

Now try Exercise 41(a).

EXAMPLE 6   Finding an Equation of a Perpendicular Line

Find an equation of the line through $P(2, -3)$ that is perpendicular to the line L with equation $4x + y = 3$. Support the result with a grapher.

SOLUTION We find the slope of L by writing its equation in slope-intercept form.

$$4x + y = 3 \qquad \text{Equation for } L$$

$$y = -4x + 3 \qquad \text{Subtract } 4x.$$

The slope of L is -4.

The line whose equation we seek has slope $-1/(-4) = 1/4$ and passes through the point $(x_1, y_1) = (2, -3)$. Thus, the point-slope form equation for the line we seek is

$$y - (-3) = \frac{1}{4}(x - 2)$$

$$y + 3 = \frac{1}{4}x - \frac{2}{4} \qquad \text{Distributive property}$$

$$y = \frac{1}{4}x - \frac{7}{2}$$

Figure P.28 shows the graphs of the two equations in a square viewing window and suggests that the graphs are perpendicular.

Now try Exercise 43(b).

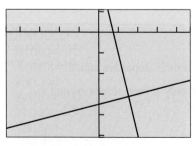

[-4.7, 4.7] by [-5.1, 1.1]

FIGURE P.28 The graphs of $y = -4x + 3$ and $y = (1/4)x - 7/2$ in this square viewing window appear to intersect at a right angle. (Example 6)

Applying Linear Equations in Two Variables

Linear equations and their graphs occur frequently in applications. Algebraic solutions to these application problems often require finding an equation of a line and solving a linear equation in one variable. Grapher techniques complement algebraic ones.

EXAMPLE 7 Finding the Depreciation of Real Estate

Camelot Apartments purchased a $50,000 building and depreciates it $2000 per year over a 25-year period.

(a) Write a linear equation giving the value y of the building in terms of the years x after the purchase.

(b) In how many years will the value of the building be $24,500?

SOLUTION

(a) We need to determine the value of m and b so that $y = mx + b$, where $0 \le x \le 25$. We know that $y = 50,000$ when $x = 0$, so the line has y-intercept $(0, 50,000)$ and $b = 50,000$. One year after purchase $(x = 1)$, the value of the building is $50,000 - 2,000 = 48,000$. So when $x = 1$, $y = 48,000$. Using algebra, we find

$$y = mx + b$$
$$48,000 = m \cdot 1 + 50,000 \qquad y = 48,000 \text{ when } x = 1$$
$$-2000 = m$$

The value y of the building x years after its purchase is

$$y = -2000x + 50,000.$$

(b) We need to find the value of x when $y = 24,500$.

$$y = -2000x + 50,000$$

Again, using algebra we find

$$24,500 = -2000x + 50,000 \qquad \text{Set } y = 24,500.$$
$$-25,500 = -2000x \qquad \text{Subtract } 50,000.$$
$$12.75 = x$$

The depreciated value of the building will be $24,500 exactly 12.75 years, or 12 years 9 months, after purchase by Camelot Apartments. We can support our algebraic work both graphically and numerically. The trace coordinates in Figure P.29a show graphically that $(12.75, 24,500)$ is a solution of $y = -2000x + 50,000$. This means that $y = 24,500$ when $x = 12.75$.

Figure P.29b is a table of values for $y = -2000x + 50,000$ for a few values of x. The fourth line of the table shows numerically that $y = 24,500$ when $x = 12.75$.

Now try Exercise 45.

Figure P.30 on page 38 shows Americans' income from 1998 to 2003 in trillions of dollars and a corresponding scatter plot of the data. In Example 8 we model the data in Figure P.30 with a linear equation.

X=12.75   Y=24500

[0, 23.5] by [0, 60000]

(a)

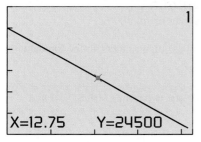

X	Y1	
12	26000	
12.25	25500	
12.5	25000	
12.75	24500	
13	24000	
13.25	23500	
13.5	23000	

Y1■ −2000X+50000

(b)

FIGURE P.29 A (a) graph and (b) table of values for $y = -2000x + 50,000$. (Example 7)

Year	Amount (trillions of dollars)
1998	7.4
1999	7.8
2000	8.4
2001	8.7
2002	8.9
2003	9.2

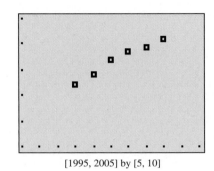

[1995, 2005] by [5, 10]

FIGURE P.30 Americans' Personal Income.
Source: U.S. Census Bureau, *Statistical Abstract of the United States, 2004–2005.* (Example 8)

EXAMPLE 8   Finding a Linear Model for Americans' Personal Income

American's personal income in trillions of dollars is given in Figure P.30.

(a) Write a linear equation for Americans' income y in terms of the year x using the points (1998, 7.4) and (1999, 7.8).

(b) Use the equation in (a) to estimate Americans' income in 2001.

(c) Use the equation in (a) to predict Americans' income in 2006.

(d) Superimpose a graph of the linear equation in (a) on a scatter plot of the data.

SOLUTION

(a) Let $y = mx + b$. The slope of the line through the two points (1998, 7.4) and (1999, 7.8) is

$$m = \frac{7.8 - 7.4}{1999 - 1998} = 0.4.$$

The value of 7.4 trillion dollars in 1998 gives $y = 7.4$ when $x = 1998$.

$$y = mx + b$$
$$y = 0.4x + b \qquad \text{\small $m = 0.4$}$$
$$7.4 = 0.4(1998) + b \qquad \text{\small $y = 7.4$ when $x = 1998$}$$
$$b = 7.4 - 0.4(1998)$$
$$b = -791.8$$

The linear equation we seek is $y = 0.4x - 791.8$.

(b) We need to find the value of y when $x = 2001$.

$$y = 0.4x - 791.8$$
$$y = 0.4(2001) - 791.8 \qquad \text{\small Set $x = 2001$.}$$
$$y = 8.6$$

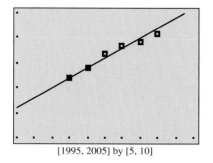

[1995, 2005] by [5, 10]

FIGURE P.31 Linear model for Americans' personal income. (Example 8)

Using the linear model we found in (a) we estimate Americans' income in 2001 to be 8.6 trillion dollars, a little less than the actual amount 8.7 trillion.

(c) We need to find the value of y when $x = 2006$.

$$y = 0.4x - 791.8$$

$$y = 0.4(2006) - 791.8 \quad \text{Set } x = 2006.$$

$$y = 10.6$$

Using the linear model we found in (a) we predict Americans' income in 2006 to be 10.6 trillion dollars.

(d) The graph and scatter plot are shown in Figure P.31.

Now try Exercise 51.

CHAPTER OPENER PROBLEM (from page 1)

PROBLEM: Assume that the speed of light is approximately 186,000 miles per second. (It took a long time to arrive at this number. See the note below about the speed of light.)

(a) If the distance from the Moon to the Earth is approximately 237,000 miles, find the length of time required for light to travel from the Earth to the Moon.

(b) If light travels from the Earth to the Sun in 8.32 minutes, approximate the distance from the Earth to the Sun.

(c) If it takes 5 hours and 29 seconds for light to travel from the Sun to Pluto, approximate the distance from the Sun to Pluto.

SOLUTION: We use the linear equation $d = r \times t$ (distance = rate × time) to make the calculations with $r = 186,000$ miles/second.

(a) Here $d = 237,000$ miles, so

$$t = \frac{d}{r} = \frac{237,000 \text{ miles}}{186,000 \text{ miles/second}} \approx 1.27 \text{ seconds}.$$

The length of time required for light to travel from the Earth to the Moon is about 1.27 seconds.

(b) Here $t = 8.32$ minutes $= 499.2$ seconds, so

$$d = r \times t = 186,000 \, \frac{\text{miles}}{\text{second}} \times 499.2 \text{ seconds} = 92,851,200 \text{ miles}.$$

The distance from the Earth to the Sun is about 93 million miles.

(c) Here $t = 5$ hours and 29 minutes $= 329$ minutes $= 19,740$ seconds, so

$$d = r \times t = 186,000 \, \frac{\text{miles}}{\text{second}} \times 19,740 \text{ seconds}$$

$$= 3,671,640,000 \text{ miles}.$$

The distance from the Sun to Pluto is about 3.7×10^9 miles.

THE SPEED OF LIGHT

Many scientists have tried to measure the speed of light. For example, Galileo Galilei (1564–1642) attempted to measure the speed of light without much success. Visit the following web site for some interesting information about this topic: *http://www.what-is-the-speed-of-light.com/*

QUICK REVIEW P.4

In Exercises 1–4, solve for x.

1. $-75x + 25 = 200$

2. $400 - 50x = 150$

3. $3(1 - 2x) + 4(2x - 5) = 7$

4. $2(7x + 1) = 5(1 - 3x)$

In Exercises 5–8, solve for y.

5. $2x - 5y = 21$

6. $\frac{1}{3}x + \frac{1}{4}y = 2$

7. $2x + y = 17 + 2(x - 2y)$

8. $x^2 + y = 3x - 2y$

In Exercises 9 and 10, simplify the fraction.

9. $\dfrac{9 - 5}{-2 - (-8)}$

10. $\dfrac{-4 - 6}{-14 - (-2)}$

SECTION P.4 EXERCISES

In Exercises 1 and 2, estimate the slope of the line.

1.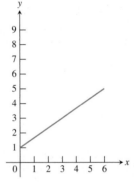

2.

In Exercises 3–6, find the slope of the line through the pair of points.

3. $(-3, 5)$ and $(4, 9)$

4. $(-2, 1)$ and $(5, -3)$

5. $(-2, -5)$ and $(-1, 3)$

6. $(5, -3)$ and $(-4, 12)$

In Exercises 7–10, find the value of x or y so that the line through the pair of points has the given slope.

Points	Slope
7. $(x, 3)$ and $(5, 9)$	$m = 2$
8. $(-2, 3)$ and $(4, y)$	$m = -3$
9. $(-3, -5)$ and $(4, y)$	$m = 3$
10. $(-8, -2)$ and $(x, 2)$	$m = 1/2$

In Exercises 11–14, find a *point-slope form* equation for the line through the point with given slope.

Point	Slope	Point	Slope
11. $(1, 4)$	$m = 2$	**12.** $(-4, 3)$	$m = -2/3$
13. $(5, -4)$	$m = -2$	**14.** $(-3, 4)$	$m = 3$

In Exercises 15–20, find a *general form equation* for the line through the pair of points.

15. $(-7, -2)$ and $(1, 6)$

16. $(-3, -8)$ and $(4, -1)$

17. $(1, -3)$ and $(5, -3)$

18. $(-1, -5)$ and $(-4, -2)$

19. $(-1, 2)$ and $(2, 5)$

20. $(4, -1)$ and $(4, 5)$

In Exercises 21–26, find a slope-intercept form equation for the line.

21. The line through $(0, 5)$ with slope $m = -3$

22. The line through $(1, 2)$ with slope $m = 1/2$

23. The line through the points $(-4, 5)$ and $(4, 3)$

24. The line through the points $(4, 2)$ and $(-3, 1)$

25. The line $2x + 5y = 12$

26. The line $7x - 12y = 96$

In Exercises 27–30, graph the linear equation on a grapher. Choose a viewing window that shows the line intersecting both the x- and y-axes.

27. $8x + y = 49$

28. $2x + y = 35$

29. $123x + 7y = 429$

30. $2100x + 12y = 3540$

In Exercises 31 and 32, the line contains the origin and the point in the upper right corner of the grapher screen.

31. Writing to Learn Which line shown here has the greater slope? Explain.

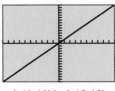 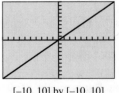

$[-10, 10]$ by $[-15, 15]$ $[-10, 10]$ by $[-10, 10]$

(a)   (b)

32. Writing to Learn Which line shown here has the greater slope? Explain.

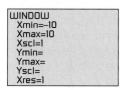 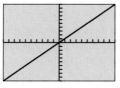

[−20, 20] by [−35, 35] [−5, 5] by [−20, 20]

(a) (b)

In Exercises 33–36, find the value of x and the value of y for which $(x, 14)$ and $(18, y)$ are points on the graph.

33. $y = 0.5x + 12$ **34.** $y = -2x + 18$

35. $3x + 4y = 26$ **36.** $3x - 2y = 14$

In Exercises 37–40, find the values for Ymin, Ymax, and Yscl that will make the graph of the line appear in the viewing window as shown here.

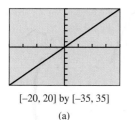 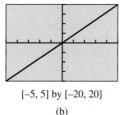

37. $y = 3x$ **38.** $y = 5x$

39. $y = \dfrac{2}{3}x$ **40.** $y = \dfrac{5}{4}x$

In Exercises 41–44, (a) find an equation for the line passing through the point and parallel to the given line, and (b) find an equation for the line passing through the point and perpendicular to the given line. Support your work graphically.

Point	Line
41. $(1, 2)$	$y = 3x - 2$
42. $(-2, 3)$	$y = -2x + 4$
43. $(3, 1)$	$2x + 3y = 12$
44. $(6, 1)$	$3x - 5y = 15$

45. Real Estate Appreciation Bob Michaels purchased a house 8 years ago for $42,000. This year it was appraised at $67,500.

(a) A linear equation $V = mt + b$, $0 \le t \le 15$, represents the value V of the house for 15 years after it was purchased. Determine m and b.

(b) Graph the equation and trace to estimate in how many years after purchase this house will be worth $72,500.

(c) Write and solve an equation algebraically to determine how many years after purchase this house will be worth $74,000.

(d) Determine how many years after purchase this house will be worth $80,250.

46. Investment Planning Mary Ellen plans to invest $18,000, putting part of the money x into a savings that pays 5% annually and the rest into an account that pays 8% annually.

(a) What are the possible values of x in this situation?

(b) If Mary Ellen invests x dollars at 5%, write an equation that describes the total interest I received from both accounts at the end of one year.

(c) Graph and trace to estimate how much Mary Ellen invested at 5% if she earned $1020 in total interest at the end of the first year.

(d) Use your grapher to generate a table of values for I to find out how much Mary Ellen should invest at 5% to earn $1185 in total interest in one year.

47. Navigation A commercial jet airplane climbs at takeoff with slope $m = 3/8$. How far in the horizontal direction will the airplane fly to reach an altitude of 12,000 ft above the takeoff point?

48. Grade of a Highway Interstate 70 west of Denver, Colorado, has a section posted as a 6% grade. This means that for a horizontal change of 100 ft there is a 6-ft vertical change.

6% grade

(a) Find the slope of this section of the highway.

(b) On a highway with a 6% grade what is the horizontal distance required to climb 250 ft?

(c) A sign along the highway says 6% grade for the next 7 mi. Estimate how many feet of vertical change there are along those next 7 mi. (There are 5280 ft in 1 mile.)

49. Writing to Learn Building Specifications Asphalt shingles do not meet code specifications on a roof that has less than a 4-12 pitch. A 4-12 pitch means there are 4 ft of vertical change in 12 ft of horizontal change. A certain roof has slope $m = 3/8$. Could asphalt shingles be used on that roof? Explain.

50. Revisiting Example 8 Use the linear equation found in Example 8 to estimate Americans' income in 2000, 2002, 2003 displayed in Figure P.30.

51. Americans' Spending Americans personal consumption expenditures from 1998 to 2003 in trillions of dollars is shown in the table.

x	1998	1999	2000	2001	2002	2003
y	5.9	6.3	6.7	7.0	7.4	7.8

(a) Write a linear equation for Americans' spending y in terms of the year x using the points (1998, 5.9) and (1999, 6.3).

(b) Use the equation in (a) to estimate Americans' expenditures in 2002.

(c) Use the equation in (a) to predict Americans' expenditures in 2006.

(d) Superimpose a graph of the linear equation in (a) on a scatter plot of the data.

52. U.S. Imports from Mexico The total y in billions of dollars of U.S. imports from Mexico for each year x from 1996 to 2003 is given in the table. *(Source: U.S. Census Bureau, Statistical Abstract of the United States, 2001, 2004–2005.)*

x	1996	1997	1998	1999	2000	2001	2002	2003
y	74.3	85.9	94.6	109.7	135.9	131.3	134.6	138.1

(a) Use the pairs (1997, 85.9) and (2001, 131.3) to write a linear equation for x and y.

(b) Superimpose the graph of the linear equation in (a) on a scatter plot of the data.

(c) Use the equation in (a) to predict the total U.S. imports from Mexico in 2006.

53. World Population The midyear world population for the years 1997 to 2004 (in millions) is shown in Table P.7.

Table P.7 World Population	
Year	Population (millions)
1997	5852
1998	5930
1999	6006
2000	6082
2001	6156
2002	6230
2003	6303
2004	6377

Source: http://www.census.gov/ipc/www/worldpop.html

STAT
EDIT

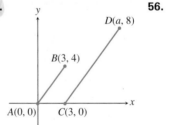

(a) Let $x = 0$ represent 1990, $x = 1$ represent 1991, and so forth. Draw a scatter plot of the data.

(b) Use the 1997 and 2004 data to write a linear equation for the population y in terms of the year x. Superimpose the graph of the linear equation on the scatter plot of the data.

(c) Use the equation in (b) to predict the midyear world population in 2006. Compare it with the Census Bureau estimate of 6525.

54. U.S. Exports to Japan The total in billions of dollars of U.S. exports to Japan from 1996 to 2003 is given in Table P.8.

Table P.8 U.S. Exports to Japan	
Year	U.S. Exports (billions of dollars)
1996	67.6
1997	65.5
1998	57.8
1999	57.5
2000	64.9
2001	57.4
2002	51.4
2003	52.1

Source: U.S. Census Bureau, Statistical Abstract of the United States, 2001, 2004–2005.

(a) Let $x = 0$ represent 1990, $x = 1$ represent 1991, and so forth. Draw a scatter plot of the data.

(b) Use the 1996 and 2003 data to write a linear equation for the U.S. exports to Japan y in terms of the year x. Superimpose the graph of the linear equation on the scatter plot in (a).

(c) Use the equation in (b) to predict the U.S. exports to Japan in 2006.

In Exercises 55 and 56, determine a so that the line segments AB and CD are parallel.

55.
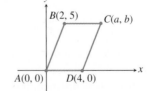

56.
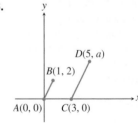

In Exercises 57 and 58, determine a and b so that figure $ABCD$ is a parallelogram.

57.

58.
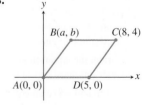

59. Writing to Learn Perpendicular Lines

(a) Is it possible for two lines with positive slopes to be perpendicular? Explain.

(b) Is it possible for two lines with negative slopes to be perpendicular? Explain.

60. Group Activity Parallel and Perpendicular Lines

(a) Assume that $c \neq d$ and a and b are not both zero. Show that $ax + by = c$ and $ax + by = d$ are parallel lines. Explain why the restrictions on a, b, c, and d are necessary.

(b) Assume that a and b are not both zero. Show that $ax + by = c$ and $bx - ay = d$ are perpendicular lines. Explain why the restrictions on a and b are necessary.

Standardized Test Questions

61. True or False The slope of a vertical line is zero. Justify your answer.

62. True or False The graph of any equation of the form $ax + by = c$, where a and b are not both zero, is always a line. Justify your answer.

In Exercises 63–66, you may use a graphing calculator to solve these problems.

63. Multiple Choice Which of the following is an equation of the line through the point $(-2, 3)$ with slope 4?

(A) $y - 3 = 4(x + 2)$ **(B)** $y + 3 = 4(x - 2)$

(C) $x - 3 = 4(y + 2)$ **(D)** $x + 3 = 4(y - 2)$

(E) $y + 2 = 4(x - 3)$

64. Multiple Choice Which of the following is an equation of the line with slope 3 and y-intercept -2?

(A) $y = 3x + 2$ **(B)** $y = 3x - 2$

(C) $y = -2x + 3$ **(D)** $x = 3y - 2$

(E) $x = 3y + 2$

65. Multiple Choice Which of the following lines is perpendicular to the line $y = -2x + 5$?

(A) $y = 2x + 1$ **(B)** $y = -2x - \dfrac{1}{5}$

(C) $y = -\dfrac{1}{2}x + \dfrac{1}{3}$ **(D)** $y = -\dfrac{1}{2}x + 3$

(E) $y = \dfrac{1}{2}x - 3$

66. Multiple Choice Which of the following is the slope of the line through the two points $(-2, 1)$ and $(1, -4)$?

(A) $-\dfrac{3}{5}$ **(B)** $\dfrac{3}{5}$

(C) $-\dfrac{5}{3}$ **(D)** $\dfrac{5}{3}$

(E) -3

Explorations

67. Exploring the Graph of $\dfrac{x}{a} + \dfrac{y}{b} = c, a \neq 0, b \neq 0$

Let $c = 1$.

(a) Draw the graph for $a = 3$, $b = -2$.

(b) Draw the graph for $a = -2$, $b = -3$.

(c) Draw the graph for $a = 5$, $b = 3$.

(d) Use your graphs in (a), (b), (c) to conjecture what a and b represent when $c = 1$. Prove your conjecture.

(e) Repeat (a)–(d) for $c = 2$.

(f) If $c = -1$, what do a and b represent?

68. Investigating Graphs of Linear Equations

(a) Graph $y = mx$ for $m = -3, -2, -1, 1, 2, 3$ in the window $[-8, 8]$ by $[-5, 5]$. What do these graphs have in common? How are they different?

(b) If $m > 0$, what do the graphs of $y = mx$ and $y = -mx$ have in common? How are they different?

(c) Graph $y = 0.3x + b$ for $b = -3, -2, -1, 0, 1, 2, 3$ in $[-8, 8]$ by $[-5, 5]$. What do these graphs have in common? How are they different?

Extending the Ideas

69. Connecting Algebra and Geometry Show that if the midpoints of consecutive sides of any quadrilateral (see figure) are connected, the result is a parallelogram.

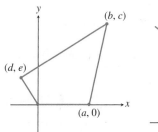

Art for Exercise 69

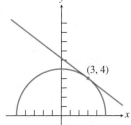

Art for Exercise 70

70. Connecting Algebra and Geometry Consider the semicircle of radius 5 centered at $(0, 0)$ as shown in the figure. Find an equation of the line tangent to the semicircle at the point $(3, 4)$. (*Hint:* A line tangent to a circle is perpendicular to the radius at the point of tangency.)

71. Connecting Algebra and Geometry Show that in any triangle (see figure), the line segment joining the midpoints of two sides is parallel to the third side and half as long.

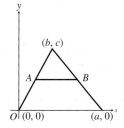

P.5
Solving Equations Graphically, Numerically, and Algebraically

. . . and why

These basic techniques are involved in using a graphing utility to solve equations in this textbook.

Solving Equations Graphically

The graph of the equation $y = 2x - 5$ (in x and y) can be used to solve the equation $2x - 5 = 0$ (in x). Using the techniques of Section P.3, we can show algebraically that $x = 5/2$ is a solution of $2x - 5 = 0$. Therefore, the ordered pair $(5/2, 0)$ is a solution of $y = 2x - 5$. Figure P.32 suggests that the x-intercept of the graph of the line $y = 2x - 5$ is the point $(5/2, 0)$ as it should be.

One way to solve an equation graphically is to find all its x-intercepts. There are many graphical techniques that can be used to find x-intercepts.

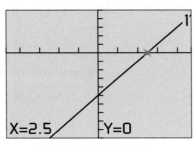

[−4.7, 4.7] by [−10, 5]

FIGURE P.32 Using the Trace feature of a grapher, we see that $(2.5, 0)$ is an x-intercept of the graph of $y = 2x - 5$ and, therefore, $x = 2.5$ is a solution of the equation $2x - 5 = 0$.

EXAMPLE 1   Solving by Finding *x*-Intercepts

Solve the equation $2x^2 - 3x - 2 = 0$ graphically.

SOLUTION

Solve Graphically Find the x-intercepts of the graph of $y = 2x^2 - 3x - 2$ (Figure P.33). We use Trace to see that $(-0.5, 0)$ and $(2, 0)$ are x-intercepts of this graph. Thus, the solutions of this equation are $x = -0.5$ and $x = 2$. Answers obtained graphically are really approximations, although in general they are very good approximations.

Solve Algebraically In this case, we can use factoring to find exact values.

$$2x^2 - 3x - 2 = 0$$

$$(2x + 1)(x - 2) = 0 \quad \text{Factor.}$$

We can conclude that

$$2x + 1 = 0 \quad \text{or} \quad x - 2 = 0,$$

$$x = -1/2 \quad \text{or} \quad x = 2.$$

So, $x = -1/2$ and $x = 2$ are the exact solutions of the original equation.

Now try Exercise 1.

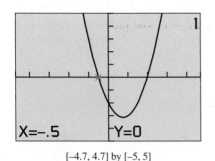

[−4.7, 4.7] by [−5, 5]

FIGURE P.33 It appears that $(-0.5, 0)$ and $(2, 0)$ are x-intercepts of the graph of $y = 2x^2 - 3x - 2$. (Example 1)

The algebraic solution procedure used in Example 1 is a special case of the following important property.

> **Zero Factor Property**
>
> Let a and b be real numbers.
>
> $$\text{If } ab = 0, \text{ then } a = 0 \text{ or } b = 0.$$

Solving Quadratic Equations

Linear equations ($ax + b = 0$) and *quadratic equations* are two members of the family of *polynomial equations*, which will be studied in more detail in Chapter 2.

> **DEFINITION Quadratic Equation in x**
>
> A **quadratic equation in x** is one that can be written in the form
>
> $$ax^2 + bx + c = 0,$$
>
> where a, b, and c are real numbers with $a \neq 0$.

We review some of the basic algebraic techniques for solving quadratic equations. One algebraic technique that we have already used in Example 1 is *factoring*.

Quadratic equations of the form $(ax + b)^2 = c$ are fairly easy to solve as illustrated in Example 2.

SQUARE ROOT PRINCIPLE

If $t^2 = K > 0$, then $t = \sqrt{K}$ or $t = -\sqrt{K}$.

EXAMPLE 2   Solving by Extracting Square Roots

Solve $(2x - 1)^2 = 9$ algebraically.

SOLUTION

$$(2x - 1)^2 = 9$$

$$2x - 1 = \pm 3 \qquad \text{Extract square roots.}$$

$$2x = 4 \quad \text{or} \quad 2x = -2$$

$$x = 2 \quad \text{or} \quad x = -1$$

Now try Exercise 9.

The technique of Example 2 is more general than you might think because every quadratic equation can be written in the form $(x + b)^2 = c$. The procedure we need to accomplish this is *completing the square*.

> **Completing the Square**
>
> To solve $x^2 + bx = c$ by **completing the square**, add $(b/2)^2$ to both sides of the equation and factor the left side of the new equation.
>
> $$x^2 + bx + \left(\frac{b}{2}\right)^2 = c + \left(\frac{b}{2}\right)^2$$
>
> $$\left(x + \frac{b}{2}\right)^2 = c + \frac{b^2}{4}$$

To solve a quadratic equation by completing the square, we simply divide both sides of the equation by the coefficient of x^2 and then complete the square as illustrated in Example 3.

EXAMPLE 3   Solving by Completing the Square

Solve $4x^2 - 20x + 17 = 0$ by completing the square.

SOLUTION

$$4x^2 - 20x + 17 = 0$$

$$x^2 - 5x + \frac{17}{4} = 0 \qquad \text{Divide by 4.}$$

$$x^2 - 5x = -\frac{17}{4} \qquad \text{Subtract } \left(\frac{17}{4}\right).$$

Completing the square on the equation above we obtain

$$x^2 - 5x + \left(-\frac{5}{2}\right)^2 = -\frac{17}{4} + \left(-\frac{5}{2}\right)^2 \qquad \text{Add } \left(-\frac{5}{2}\right)^2.$$

$$\left(x - \frac{5}{2}\right)^2 = 2 \qquad\qquad \text{Factor and simplify.}$$

$$x - \frac{5}{2} = \pm\sqrt{2} \qquad\qquad \text{Extract square roots.}$$

$$x = \frac{5}{2} \pm \sqrt{2}$$

$$x = \frac{5}{2} + \sqrt{2} \approx 3.91 \text{ or } x = \frac{5}{2} - \sqrt{2} \approx 1.09 \qquad \textit{Now try Exercise 13.}$$

The procedure of Example 3 can be applied to the general quadratic equation $ax^2 + bx + c = 0$ to produce the following formula for its solutions (see Exercise 68).

Quadratic Formula

The solutions of the quadratic equation $ax^2 + bx + c = 0$, where $a \neq 0$, are given by the **quadratic formula**

$$x = \frac{-b \pm \sqrt{b^2 - 4ac}}{2a}.$$

EXAMPLE 4   Solving Using the Quadratic Formula

Solve the equation $3x^2 - 6x = 5$.

SOLUTION First we subtract 5 from both sides of the equation to put it in the form $ax^2 + bx + c = 0$: $3x^2 - 6x - 5 = 0$. We can see that $a = 3$, $b = -6$, and $c = -5$.

$$x = \frac{-b \pm \sqrt{b^2 - 4ac}}{2a} \qquad\qquad \text{Quadratic formula}$$

$$x = \frac{-(-6) \pm \sqrt{(-6)^2 - 4(3)(-5)}}{2(3)} \qquad a = 3, b = -6, c = -5$$

$$x = \frac{6 \pm \sqrt{96}}{6} \qquad\qquad\qquad \text{Simplify.}$$

continued

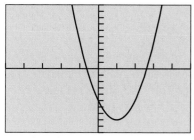

[–5, 5] by [–10, 10]

FIGURE P.34 The graph of
$y = 3x^2 - 6x - 5$. (Example 4)

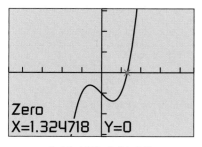

[–4.7, 4.7] by [–3.1, 3.1]

(a)

```
1.324718→X
                1.324718
X³–X–1
              1.823355E–7
```

(b)

FIGURE P.35 The graph of $y = x^3 - x - 1$. (a) shows that $(1.324718, 0)$ is an approximation to the x-intercept of the graph. (b) supports this conclusion. (Example 5)

$$x = \frac{6 + \sqrt{96}}{6} \approx 2.63 \quad \text{or} \quad x = \frac{6 - \sqrt{96}}{6} \approx -0.63$$

The graph of $y = 3x^2 - 6x - 5$ in Figure P.34 supports that the x-intercepts are approximately -0.63 and 2.63.   ***Now try Exercise 19.***

Solving Quadratic Equations Algebraically

There are four basic ways to solve quadratic equations algebraically.

1. **Factoring** (see Example 1)

2. **Extracting Square Roots** (see Example 2)

3. **Completing the Square** (see Example 3)

4. **Using the Quadratic Formula** (see Example 4)

Approximating Solutions of Equations Graphically

A solution of the equation $x^3 - x - 1 = 0$ is a value of x that makes the value of $y = x^3 - x - 1$ equal to zero. Example 5 illustrates a built-in procedure on graphing calculators to find such values of x.

EXAMPLE 5   Solving Graphically

Solve the equation $x^3 - x - 1 = 0$ graphically.

SOLUTION Figure P.35a suggests that $x = 1.324718$ is the solution we seek. Figure P.35b provides numerical support that $x = 1.324718$ is a close approximation to the solution because, when $x = 1.324718$, $x^3 - x - 1 \approx 1.82 \times 10^{-7}$, which is nearly zero.   ***Now try Exercise 31.***

When solving equations graphically we usually get approximate solutions and not exact solutions. We will use the following agreement about accuracy in this book.

Agreement about Approximate Solutions

For applications, round to a value that is reasonable for the context of the problem. For all others round to two decimal places unless directed otherwise.

With this accuracy agreement, we would report the solution found in Example 5 as 1.32.

Approximating Solutions of Equations Numerically with Tables

The table feature on graphing calculators provides a numerical *zoom-in procedure* that we can use to find accurate solutions of equations. We illustrate this procedure in Example 6 using the same equation of Example 5.

EXAMPLE 6   Solving Using Tables

Solve the equation $x^3 - x - 1 = 0$ using grapher tables.

SOLUTION From Figure P.35a, we know that the solution we seek is between $x = 1$ and $x = 2$. Figure P.36a sets the starting point of the table (TblStart = 1) at $x = 1$ and increments the numbers in the table ($\triangle$Tbl = 0.1) by 0.1. Figure P.36b shows that the zero of $x^3 - x - 1$ is between $x = 1.3$ and $x = 1.4$.

TABLE SETUP		
TblStart=1		
$\triangle$Tbl=.1		
Indpnt: **Auto** Ask		
Depend: **Auto** Ask		

X	Y₁	
1	–1	
1.1	–.769	
1.2	–.472	
1.3	–.103	
1.4	.344	
1.5	.875	
1.6	1.496	

Y₁ ▤ X³–X–1

(a)   (b)

FIGURE P.36 (a) gives the setup that produces the table in (b). (Example 6)

The next two steps in this process are shown in Figure P.37.

X	Y₁	
1.3	–.103	
1.31	–.0619	
1.32	–.02	
1.33	.02264	
1.34	.0661	
1.35	.11038	
1.36	.15546	

Y₁ ▤ X³–X–1

X	Y₁	
1.32	–.02	
1.321	–.0158	
1.322	–.0116	
1.323	–.0073	
1.324	–.0031	
1.325	.0012	
1.326	.00547	

Y₁ ▤ X³–X–1

(a)   (b)

FIGURE P.37 In (a) TblStart = 1.3 and $\triangle$Tbl = 0.01, and in (b) TblStart = 1.32 and $\triangle$Tbl = 0.001. (Example 6)

From Figure P.37a, we can read that the zero is between $x = 1.32$ and $x = 1.33$; from Figure P.37b, we can read that the zero is between $x = 1.324$ and $x = 1.325$. Because all such numbers round to 1.32, we can report the zero as 1.32 with our accuracy agreement.   *Now try Exercise 37.*

EXPLORATION 1 Finding Real Zeros of Equations

Consider the equation $4x^2 - 12x + 7 = 0$.

1. Use a graph to show that this equation has two real solutions, one between 0 and 1 and the other between 2 and 3.

2. Use the numerical zoom-in procedure illustrated in Example 6 to find each zero accurate to two decimal places.

3. Use the built-in zero finder (see Example 5) to find the two solutions. Then round them to two decimal places.

4. If you are familiar with the graphical zoom-in process, use it to find each solution accurate to two decimal places.

5. Compare the numbers obtained in parts 2, 3, and 4.

6. Support the results obtained in parts 2, 3, and 4 numerically.

7. Use the numerical zoom-in procedure illustrated in Example 6 to find each zero accurate to six decimal places. Compare with the answer found in part 3 with the zero finder.

Solving Equations by Finding Intersections

Sometimes we can rewrite an equation and solve it graphically by finding the *points of intersection* of two graphs. A point (a, b) is a **point of intersection** of two graphs if it lies on both graphs.

We illustrate this procedure with the absolute value equation in Example 7.

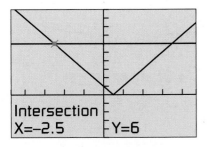

[−4.7, 4.7] by [−5, 10]

FIGURE P.38 The graphs of $y = |2x - 1|$ and $y = 6$ intersect at $(-2.5, 6)$ and $(3.5, 6)$. (Example 7)

EXAMPLE 7 Solving by Finding Intersections

Solve the equation $|2x - 1| = 6$.

SOLUTION Figure P.38 suggests that the V-shaped graph of $y = |2x - 1|$ intersects the graph of the horizontal line $y = 6$ twice. We can use Trace or the intersection feature of our grapher to see that the two points of intersection have coordinates $(-2.5, 6)$ and $(3.5, 6)$. This means that the original equation has two solutions: -2.5 and 3.5.

We can use algebra to find the exact solutions. The only two real numbers with absolute value 6 are 6 itself and -6. So, if $|2x - 1| = 6$, then

$$2x - 1 = 6 \quad \text{or} \quad 2x - 1 = -6$$

$$x = \frac{7}{2} = 3.5 \quad \text{or} \quad x = -\frac{5}{2} = -2.5$$

Now try Exercise 39.

QUICK REVIEW P.5

In Exercises 1–4, expand the product.

1. $(3x - 4)^2$

2. $(2x + 3)^2$

3. $(2x + 1)(3x - 5)$

4. $(3y - 1)(5y + 4)$

In Exercises 5–8, factor completely.

5. $25x^2 - 20x + 4$

6. $15x^3 - 22x^2 + 8x$

7. $3x^3 + x^2 - 15x - 5$

8. $y^4 - 13y^2 + 36$

In Exercises 9 and 10, combine the fractions and reduce the resulting fraction to lowest terms.

9. $\dfrac{x}{2x + 1} - \dfrac{2}{x + 3}$

10. $\dfrac{x + 1}{x^2 - 5x + 6} - \dfrac{3x + 11}{x^2 - x - 6}$

SECTION P.5 EXERCISES

In Exercises 1–6, solve the equation graphically by finding x-intercepts. Confirm by using factoring to solve the equation.

1. $x^2 - x - 20 = 0$

2. $2x^2 + 5x - 3 = 0$

3. $4x^2 - 8x + 3 = 0$

4. $x^2 - 8x = -15$

5. $x(3x - 7) = 6$

6. $x(3x + 11) = 20$

In Exercises 7–12, solve the equation by extracting square roots.

7. $4x^2 = 25$

8. $2(x - 5)^2 = 17$

9. $3(x + 4)^2 = 8$

10. $4(u + 1)^2 = 18$

11. $2y^2 - 8 = 6 - 2y^2$

12. $(2x + 3)^2 = 169$

In Exercises 13–18, solve the equation by completing the square.

13. $x^2 + 6x = 7$

14. $x^2 + 5x - 9 = 0$

15. $x^2 - 7x + \dfrac{5}{4} = 0$

16. $4 - 6x = x^2$

17. $2x^2 - 7x + 9 = (x - 3)(x + 1) + 3x$

18. $3x^2 - 6x - 7 = x^2 + 3x - x(x + 1) + 3$

In Exercises 19–24, solve the equation using the quadratic formula.

19. $x^2 + 8x - 2 = 0$

20. $2x^2 - 3x + 1 = 0$

21. $3x + 4 = x^2$

22. $x^2 - 5 = \sqrt{3}x$

23. $x(x + 5) = 12$

24. $x^2 - 2x + 6 = 2x^2 - 6x - 26$

In Exercises 25–28, estimate any x- and y-intercepts that are shown in the graph.

25.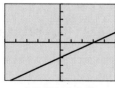

$[-5, 5]$ by $[-5, 5]$

26.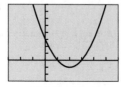

$[-3, 6]$ by $[-3, 8]$

27.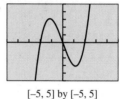

$[-5, 5]$ by $[-5, 5]$

28.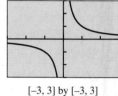

$[-3, 3]$ by $[-3, 3]$

In Exercises 29–34, solve the equation graphically by finding x-intercepts.

29. $x^2 + x - 1 = 0$

30. $4x^2 + 20x + 23 = 0$

31. $x^3 + x^2 + 2x - 3 = 0$

32. $x^3 - 4x + 2 = 0$

33. $x^2 + 4 = 4x$

34. $x^2 + 2x = -2$

In Exercises 35 and 36, the table permits you to estimate a zero of an expression. State the expression and give the zero as accurately as can be read from the table.

35.

X	Y1	
.4	-.04	
.41	-.0119	
.42	.0164	
.43	.0449	
.44	.0736	
.45	.1025	
.46	.1316	
Y1 ▤ X²+2X−1		

36.

X	Y1	
-1.735	-.0177	
-1.734	-.0117	
-1.733	-.0057	
-1.732	3E-4	
-1.731	.0063	
-1.73	.01228	
-1.729	.01826	
Y1 ▤ X³−3X		

In Exercises 37 and 38, use tables to find the indicated number of solutions of the equation accurate to two decimal places.

37. Two solutions of $x^2 - x - 1 = 0$

38. One solution of $-x^3 + x + 1 = 0$

In Exercises 39–44, solve the equation graphically by finding intersections. Confirm your answer algebraically.

39. $|t - 8| = 2$

40. $|x + 1| = 4$

41. $|2x + 5| = 7$

42. $|3 - 5x| = 4$

43. $|2x - 3| = x^2$

44. $|x + 1| = 2x - 3$

45. Interpreting Graphs The graphs in the two viewing windows shown here can be used to solve the equation $3\sqrt{x+4} = x^2 - 1$ graphically.

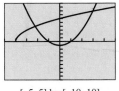
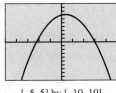

[-5, 5] by [-10, 10]	[-5, 5] by [-10, 10]
(a)	(b)

(a) The viewing window in (a) illustrates the intersection method for solving. Identify the two equations that are graphed.

(b) The viewing window in (b) illustrates the x-intercept method for solving. Identify the equation that is graphed.

(c) **Writing to Learn** How are the intersection points in (a) related to the x-intercepts in (b)?

46. Writing to Learn Revisiting Example 6 Explain why all real numbers x that satisfy $1.324 < x < 1.325$ round to 1.32.

In Exercises 47–56, use a method of your choice to solve the equation.

47. $x^2 + x - 2 = 0$

48. $x^2 - 3x = 12 - 3(x - 2)$

49. $|2x - 1| = 5$

50. $x + 2 - 2\sqrt{x+3} = 0$

51. $x^3 + 4x^2 - 3x - 2 = 0$

52. $x^3 - 4x + 2 = 0$

53. $|x^2 + 4x - 1| = 7$

54. $|x + 5| = |x - 3|$

55. $|0.5x + 3| = x^2 - 4$

56. $\sqrt{x+7} = -x^2 + 5$

57. Group Activity Discriminant of a Quadratic The radicand $b^2 - 4ac$ in the quadratic formula is called the **discriminant** of the quadratic polynomial $ax^2 + bx + c$ because it can be used to describe the nature of its zeros.

(a) **Writing to Learn** If $b^2 - 4ac > 0$, what can you say about the zeros of the quadratic polynomial $ax^2 + bx + c$? Explain your answer.

(b) **Writing to Learn** If $b^2 - 4ac = 0$, what can you say about the zeros of the quadratic polynomial $ax^2 + bx + c$? Explain your answer.

(c) **Writing to Learn** If $b^2 - 4ac < 0$, what can you say about the zeros of the quadratic polynomial $ax^2 + bx + c$? Explain your answer.

58. Group Activity Discriminant of a Quadratic Use the information learned in Exercise 57 to create a quadratic polynomial with the following numbers of real zeros. Support your answers graphically.

(a) Two real zeros

(b) Exactly one real zero

(c) No real zeros

59. Size of a Soccer Field Several of the World Cup '94 soccer matches were played in Stanford University's stadium in Menlo Park, California. The field is 30 yd longer than it is wide, and the area of the field is 8800 yd^2. What are the dimensions of this soccer field?

60. Height of a Ladder John's paint crew knows from experience that its 18-ft ladder is particularly stable when the distance from the ground to the top of the ladder is 5 ft more than the distance from the building to the base of the ladder as shown in the figure. In this position, how far up the building does the ladder reach?

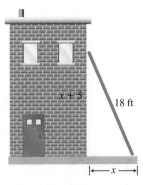

61. Finding the Dimensions of a Norman Window A Norman window has the shape of a square with a semicircle mounted on it. Find the width of the window if the total area of the square and the semicircle is to be 200 ft^2.

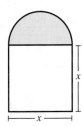

Standardized Test Questions

62. True or False If 2 is an x-intercept of the graph of $y = ax^2 + bx + c$, then 2 is a solution of the equation $ax^2 + bx + c = 0$. Justify your answer.

63. True or False If $2x^2 = 18$, then x must be equal to 3. Justify your answer.

In Exercises 64–67, you may use a graphing calculator to solve these problems.

64. Multiple Choice Which of the following are the solutions of the equation $x(x - 3) = 0$?

(A) Only $x = 3$ (B) Only $x = -3$

(C) $x = 0$ and $x = -3$ (D) $x = 0$ and $x = 3$

(E) There are no solutions.

65. Multiple Choice Which of the following replacements for ? make $x^2 - 5x + ?$ a perfect square?

(A) $-\dfrac{5}{2}$ **(B)** $\left(-\dfrac{5}{2}\right)^2$

(C) $(-5)^2$ **(D)** $\left(-\dfrac{2}{5}\right)^2$

(E) -6

66. Multiple Choice Which of the following are the solutions of the equation $2x^2 - 3x - 1 = 0$?

(A) $\dfrac{3}{4} \pm \sqrt{17}$ **(B)** $\dfrac{3 \pm \sqrt{17}}{4}$

(C) $\dfrac{3 \pm \sqrt{17}}{2}$ **(D)** $\dfrac{-3 \pm \sqrt{17}}{4}$

(E) $\dfrac{3 \pm 1}{4}$

67. Multiple Choice Which of the following are the solutions of the equation $|x - 1| = -3$?

(A) Only $x = 4$ **(B)** Only $x = -2$

(C) Only $x = 2$ **(D)** $x = 4$ and $x = -2$

(E) There are no solutions.

Explorations

68. Deriving the Quadratic Formula Follow these steps to use completing the square to solve $ax^2 + bx + c = 0$, $a \neq 0$.

(a) Subtract c from both sides of the original equation and divide both sides of the resulting equation by a to obtain

$$x^2 + \frac{b}{a}x = -\frac{c}{a}.$$

(b) Add the square of one-half of the coefficient of x in (a) to both sides and simplify to obtain

$$\left(x + \frac{b}{2a}\right)^2 = \frac{b^2 - 4ac}{4a^2}.$$

(c) Extract square roots in (b) and solve for x to obtain the quadratic formula

$$x = \frac{-b \pm \sqrt{b^2 - 4ac}}{2a}.$$

Extending the Ideas

69. Finding Number of Solutions Consider the equation $|x^2 - 4| = c$

(a) Find a value of c for which this equation has four solutions. (There are many such values.)

(b) Find a value of c for which this equation has three solutions. (There is only one such value.)

(c) Find a value of c for which this equation has two solutions. (There are many such values.)

(d) Find a value of c for which this equation has no solutions. (There are many such values.)

(e) Writing to Learn Are there any other possible numbers of solutions of this equation? Explain.

70. Sums and Products of Solutions of $ax^2 + bx + c = 0$, $a \neq 0$ Suppose that $b^2 - 4ac > 0$.

(a) Show that the sum of the two solutions of this equation is $-(b/a)$.

(b) Show that the product of the two solutions of this equation is c/a.

71. Exercise 70 Continued The equation $2x^2 + bx + c = 0$ has two solutions x_1 and x_2. If $x_1 + x_2 = 5$ and $x_1 \cdot x_2 = 3$, find the two solutions.

P.6
Complex Numbers

What you'll learn about

- Complex Numbers
- Operations with Complex Numbers
- Complex Conjugates and Division
- Complex Solutions of Quadratic Equations

. . . and why

The zeros of polynomials are complex numbers.

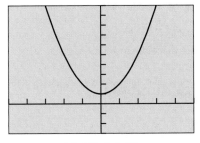

[–5, 5] by [–3, 10]

FIGURE P.39 The graph of $f(x) = x^2 + 1$ has no x-intercepts.

HISTORICAL NOTE

René Descartes (1596–1650) coined the term imaginary in a time when negative solutions to equations were considered *false*. Carl Friedrich Gauss (1777–1855) gave us the term complex number and the symbol i for $\sqrt{-1}$. Today practical applications of complex numbers abound.

Complex Numbers

Figure P.39 shows that the function $f(x) = x^2 + 1$ has no real zeros, so $x^2 + 1 = 0$ has no real-number solutions. To remedy this situation, mathematicians in the 17th century extended the definition of $\sqrt{a}$ to include negative real numbers a. First the number $i = \sqrt{-1}$ is defined as a solution of the equation $i^2 + 1 = 0$ and is the **imaginary unit**. Then for any negative real number $\sqrt{a} = \sqrt{|a|} \cdot i$.

The extended system of numbers, called the *complex numbers*, consists of all real numbers and sums of real numbers and real number multiples of i. The following are all examples of complex numbers:

$$-6, \quad 5i, \quad \sqrt{5}, \quad -7i, \quad \frac{5}{2}i + \frac{2}{3}, \quad -2 + 3i, \quad 5 - 3i, \quad \frac{1}{3} + \frac{4}{5}i.$$

DEFINITION Complex Number

A **complex number** is any number that can be written in the form

$$a + bi,$$

where a and b are real numbers. The real number a is the **real part**, the real number b is the **imaginary part**, and $a + bi$ is the **standard form**.

A real number a is the complex number $a + 0i$, so *all real numbers are also complex numbers*. If $a = 0$ and $b \neq 0$, then $a + bi$ becomes bi, and is an **imaginary number**. For instance, $5i$ and $-7i$ are imaginary numbers.

Two complex numbers are **equal** if and only if their real and imaginary parts are equal. For example,

$$x + yi = 2 + 5i \quad \text{if and only if } x = 2 \text{ and } y = 5.$$

Operations with Complex Numbers

Adding complex numbers is done by adding their real and imaginary parts separately. Subtracting complex numbers is also done using the same parts.

DEFINITION Addition and Subtraction of Complex Numbers

If $a + bi$ and $c + di$ are two complex numbers, then

Sum: $\qquad (a + bi) + (c + di) = (a + c) + (b + d)i,$

Difference: $\qquad (a + bi) - (c + di) = (a - c) + (b - d)i.$

EXAMPLE 1 Adding and Subtracting Complex Numbers

(a) $(7 - 3i) + (4 + 5i) = (7 + 4) + (-3 + 5)i = 11 + 2i$

(b) $(2 - i) - (8 + 3i) = (2 - 8) + (-1 - 3)i = -6 - 4i$

Now try Exercise 3.

The **additive identity** for the complex numbers is $0 = 0 + 0i$. The **additive inverse** of $a + bi$ is $-(a + bi) = -a - bi$ because

$$(a + bi) + (-a - bi) = 0 + 0i = 0.$$

Many of the properties of real numbers also hold for complex numbers. These include:

• Commutative properties of addition and multiplication,

• Associative properties of addition and multiplication, and

• Distributive properties of multiplication over addition and subtraction.

Using these properties and the fact that $i^2 = -1$, complex numbers can be multiplied by treating them as algebraic expressions.

EXAMPLE 2 Multiplying Complex Numbers

$$(2 + 3i) \cdot (5 - i) = 2(5 - i) + 3i(5 - i)$$
$$= 10 - 2i + 15i - 3i^2$$
$$= 10 + 13i - 3(-1)$$
$$= 13 + 13i$$

Now try Exercise 9.

```
(7−3i)+(4+5i)
                   11+2i
(2−i)−(8+3i)
                   −6−4i
(2+3i)*(5−i)
                   13+13i
■
```

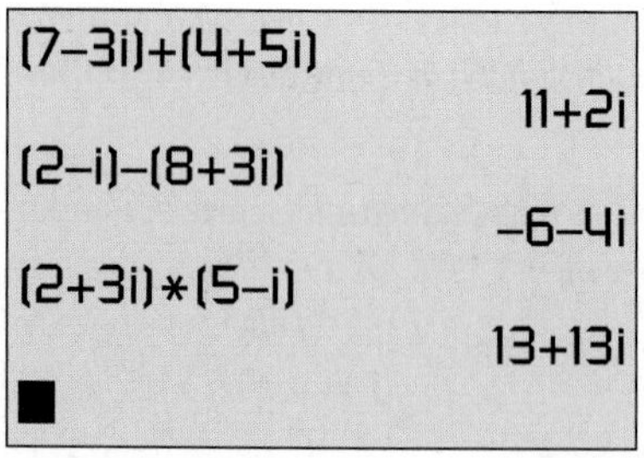

FIGURE P.40 Complex number operations on a grapher. (Examples 1 and 2)

We can generalize Example 2 as follows:

$$(a + bi)(c + di) = ac + adi + bci + bdi^2$$
$$= (ac - bd) + (ad + bc)i$$

Many graphers can perform basic calculations on complex numbers. Figure P.40 shows how the operations of Examples 1 and 2 look on some graphers.

We compute positive integer powers of complex numbers by treating them as algebraic expressions.

EXAMPLE 3 Raising a Complex Number to a Power

If $z = \dfrac{1}{2} + \dfrac{\sqrt{3}}{2} i$, find z^2 and z^3.

continued

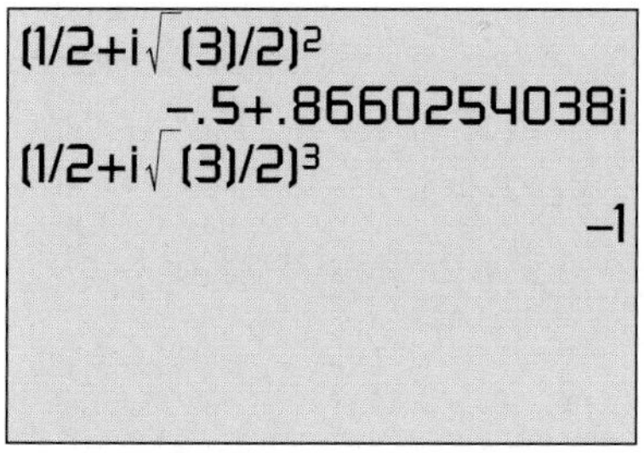

SOLUTION

$$z^2 = \left(\frac{1}{2} + \frac{\sqrt{3}}{2}\,i\right)\left(\frac{1}{2} + \frac{\sqrt{3}}{2}\,i\right)$$

$$= \frac{1}{4} + \frac{\sqrt{3}}{4}\,i + \frac{\sqrt{3}}{4}\,i + \frac{3}{4}\,i^2$$

$$= \frac{1}{4} + \frac{2\sqrt{3}}{4}\,i + \frac{3}{4}\,(-1)$$

$$= -\frac{1}{2} + \frac{\sqrt{3}}{2}\,i$$

$$z^3 = z^2 \cdot z = \left(-\frac{1}{2} + \frac{\sqrt{3}}{2}\,i\right)\left(\frac{1}{2} + \frac{\sqrt{3}}{2}\,i\right)$$

$$= -\frac{1}{4} - \frac{\sqrt{3}}{4}\,i + \frac{\sqrt{3}}{4}\,i + \frac{3}{4}\,i^2$$

$$= -\frac{1}{4} + 0i + \frac{3}{4}\,(-1)$$

$$= -1$$

Figure P.41 supports these results numerically.   ***Now try Exercise 27.***

Example 3 demonstrates that $1/2 + (\sqrt{3}/2)i$ is a cube root of -1 and a solution of $x^3 + 1 = 0$. In Section 2.5 complex zeros of polynomial functions will be explored in depth.

Complex Conjugates and Division

The product of the complex numbers $a + bi$ and $a - bi$ is a positive real number:

$$(a + bi) \cdot (a - bi) = a^2 - (bi)^2 = a^2 + b^2.$$

We introduce the following definition to describe this special relationship.

DEFINITION Complex Conjugate

The **complex conjugate** of the complex number $z = a + bi$ is

$$\bar{z} = \overline{a + bi} = a - bi.$$

The **multiplicative identity** for the complex numbers is $1 = 1 + 0i$. The **multiplicative inverse**, or **reciprocal**, of $z = a + bi$ is

$$z^{-1} = \frac{1}{z} = \frac{1}{a + bi} = \frac{1}{a + bi} \cdot \frac{a - bi}{a - bi} = \frac{a}{a^2 + b^2} - \frac{b}{a^2 + b^2}\,i.$$

In general, a quotient of two complex numbers, written in fraction form, can be simplified as we just simplified $1/z$—by multiplying the numerator and denominator of the fraction by the complex conjugate of the denominator.

(1/2+i√(3)/2)²
       −.5+.8660254038i
(1/2+i√(3)/2)³
                     −1

FIGURE P.41 The square and cube of a complex number. (Example 3)

EXAMPLE 4   Dividing Complex Numbers

Write the complex number in standard form.

(a) $\dfrac{2}{3-i}$

(b) $\dfrac{5+i}{2-3i}$

SOLUTION Multiply the numerator and denominator by the complex conjugate of the denominator.

(a) $\dfrac{2}{3-i}=\dfrac{2}{3-i}\cdot\dfrac{3+i}{3+i}$

$=\dfrac{6+2i}{3^2+1^2}$

$=\dfrac{6}{10}+\dfrac{2}{10}i$

$=\dfrac{3}{5}+\dfrac{1}{5}i$

(b) $\dfrac{5+i}{2-3i}=\dfrac{5+i}{2-3i}\cdot\dfrac{2+3i}{2+3i}$

$=\dfrac{10+15i+2i+3i^2}{2^2+3^2}$

$=\dfrac{7+17i}{13}$

$=\dfrac{7}{13}+\dfrac{17}{13}i$

Now try Exercise 33.

Complex Solutions of Quadratic Equations

Recall that the solutions of the quadratic equation $ax^2+bx+c=0$, where a, b, and c are real numbers and $a\neq 0$, are given by the quadratic formula

$$x=\dfrac{-b\pm\sqrt{b^2-4ac}}{2a}.$$

The radicand b^2-4ac is the **discriminant,** and tells us whether the solutions are real numbers. In particular, if $b^2-4ac<0$, the solutions involve the square root of a negative number and so lead to complex-number solutions. In all, there are three cases, which we now summarize:

Discriminant of a Quadratic Equation

For a quadratic equation $ax^2+bx+c=0$, where a, b, and c are real numbers and $a\neq 0$,

• If $b^2-4ac>0$, there are two distinct real solutions.

• If $b^2-4ac=0$, there is one repeated real solution.

• If $b^2-4ac<0$, there is a complex conjugate pair of solutions.

EXAMPLE 5 Solving a Quadratic Equation

Solve $x^2 + x + 1 = 0$.

SOLUTION

Solve Algebraically

Using the quadratic formula with $a = b = c = 1$, we obtain

$$x = \frac{-(1) \pm \sqrt{(1)^2 - 4(1)(1)}}{2(1)} = \frac{-1 \pm \sqrt{-3}}{2} = -\frac{1}{2} \pm \frac{\sqrt{3}}{2}i$$

So the solutions are $-1/2 + (\sqrt{3}/2)i$ and $-1/2 - (\sqrt{3}/2)i$, a complex conjugate pair.

Confirm Numerically

Substituting $-1/2 + (\sqrt{3}/2)i$ into the original equation, we obtain

$$\left(-\frac{1}{2} + \frac{\sqrt{3}}{2}i\right)^2 + \left(-\frac{1}{2} + \frac{\sqrt{3}}{2}i\right) + 1$$

$$= \left(-\frac{1}{2} - \frac{\sqrt{3}}{2}i\right) + \left(-\frac{1}{2} + \frac{\sqrt{3}}{2}i\right) + 1 = 0$$

By a similar computation we can confirm the second solution.

Now try Exercise 41.

QUICK REVIEW P.6

In Exercises 1–4, add or subtract, and simplify.

1. $(2x + 3) + (-x + 6)$ **2.** $(3y - x) + (2x - y)$

3. $(2a + 4d) - (a + 2d)$ **4.** $(6z - 1) - (z + 3)$

In Exercises 5–10, multiply and simplify.

5. $(x - 3)(x + 2)$

6. $(2x - 1)(x + 3)$

7. $(x - \sqrt{2})(x + \sqrt{2})$

8. $(x + 2\sqrt{3})(x - 2\sqrt{3})$

9. $[x - (1 + \sqrt{2})][x - (1 - \sqrt{2})]$

10. $[x - (2 + \sqrt{3})][x - (2 - \sqrt{3})]$

SECTION P.6 EXERCISES

In Exercises 1–8, write the sum or difference in the standard form $a + bi$.

1. $(2 - 3i) + (6 + 5i)$ **2.** $(2 - 3i) + (3 - 4i)$

3. $(7 - 3i) + (6 - i)$ **4.** $(2 + i) - (9i - 3)$

5. $(2 - i) + (3 - \sqrt{-3})$

6. $(\sqrt{5} - 3i) + (-2 + \sqrt{-9})$

7. $(i^2 + 3) - (7 + i^3)$

8. $(\sqrt{7} + i^2) - (6 - \sqrt{-81})$

In Exercises 9–16, write the product in standard form.

9. $(2 + 3i)(2 - i)$ **10.** $(2 - i)(1 + 3i)$

11. $(1 - 4i)(3 - 2i)$ **12.** $(5i - 3)(2i + 1)$

13. $(7i - 3)(2 + 6i)$ **14.** $(\sqrt{-4} + i)(6 - 5i)$

15. $(-3 - 4i)(1 + 2i)$ **16.** $(\sqrt{-2} + 2i)(6 + 5i)$

In Exercises 17–20, write the expression in the form bi, where b is a real number.

17. $\sqrt{-16}$ **18.** $\sqrt{-25}$

19. $\sqrt{-3}$ **20.** $\sqrt{-5}$

In Exercises 21–24, find the real numbers x and y that make the equation true.

21. $2 + 3i = x + yi$ **22.** $3 + yi = x - 7i$

23. $(5 - 2i) - 7 = x - (3 + yi)$

24. $(x + 6i) = (3 - i) + (4 - 2yi)$

In Exercises 25–28, write the complex number in standard form.

25. $(3 + 2i)^2$ **26.** $(1 - i)^3$

27. $\left(\dfrac{\sqrt{2}}{2} + \dfrac{\sqrt{2}}{2}i\right)^4$ **28.** $\left(\dfrac{\sqrt{3}}{2} + \dfrac{1}{2}i\right)^3$

In Exercises 29–32, find the product of the complex number and its conjugate.

29. $2 - 3i$ **30.** $5 - 6i$

31. $-3 + 4i$ **32.** $-1 - \sqrt{2}i$

In Exercises 33–40, write the expression in standard form.

33. $\dfrac{1}{2 + i}$ **34.** $\dfrac{i}{2 - i}$

35. $\dfrac{2 + i}{2 - i}$ **36.** $\dfrac{2 + i}{3i}$

37. $\dfrac{(2 + i)^2(-i)}{1 + i}$ **38.** $\dfrac{(2 - i)(1 + 2i)}{5 + 2i}$

39. $\dfrac{(1 - i)(2 - i)}{1 - 2i}$ **40.** $\dfrac{(1 - \sqrt{2}i)(1 + i)}{(1 + \sqrt{2}i)}$

In Exercises 41–44, solve the equation.

41. $x^2 + 2x + 5 = 0$ **42.** $3x^2 + x + 2 = 0$

43. $4x^2 - 6x + 5 = x + 1$ **44.** $x^2 + x + 11 = 5x - 8$

Standardized Test Questions

45. True or False There are no complex numbers z satisfying $z = -\bar{z}$. Justify your answer.

46. True or False For the complex number i, $i + i^2 + i^3 + i^4 = 0$. Justify your answer.

In Exercises 47–50, solve the problem without using a calculator.

47. Multiple Choice Which of the following is the standard form for the product $(2 + 3i)(2 - 3i)$?

(A) $-5 + 12i$ (B) $4 - 9i$ (C) $13 - 3i$ (D) -5 (E) 13

48. Multiple Choice Which of the following is the standard form for the quotient $\dfrac{1}{i}$?

(A) 1 (B) -1 (C) i (D) $-i$ (E) $-1 + i$

49. Multiple Choice Assume that $2 - 3i$ is a solution of $ax^2 + bx + c = 0$, where a, b, c are real numbers. Which of the following is also a solution of the equation?

(A) $2 + 3i$ (B) $-2 - 3i$ (C) $-2 + 3i$

(D) $3 + 2i$ (E) $\dfrac{1}{2 - 3i}$

50. Multiple Choice Which of the following is the standard form for the power $(1 - i)^3$?

(A) $-4i$ (B) $-2 + 2i$ (C) $-2 - 2i$ (D) $2 + 2i$ (E) $2 - 2i$

Explorations

51. Group Activity The Powers of i

(a) Simplify the complex numbers $i, i^2, \ldots, i^8$ by evaluating each one.

(b) Simplify the complex numbers $i^{-1}, i^{-2}, \ldots, i^{-8}$ by evaluating each one.

(c) Evaluate i^0.

(d) **Writing to Learn** Discuss your results from (a)–(c) with the members of your group, and write a summary statement about the integer powers of i.

52. Writing to Learn Describe the nature of the graph of $f(x) = ax^2 + bx + c$ when $a, b,$ and c are real numbers and the equation $ax^2 + bx + c = 0$ has nonreal complex solutions.

Extending the Ideas

53. Prove that the difference between a complex number and its conjugate is a complex number whose real part is 0.

54. Prove that the product of a complex number and its complex conjugate is a complex number whose imaginary part is zero.

55. Prove that the complex conjugate of a product of two complex numbers is the product of their complex conjugates.

56. Prove that the complex conjugate of a sum of two complex numbers is the sum of their complex conjugates.

57. Writing to Learn Explain why $-i$ is a solution of $x^2 - ix + 2 = 0$ but i is not.

P.7
Solving Inequalities Algebraically and Graphically

What you'll learn about

- Solving Absolute Value Inequalities
- Solving Quadratic Inequalities
- Approximating Solutions to Inequalities
- Projectile Motion

. . . and why

These techniques are involved in using a graphing utility to solve inequalities in this textbook.

Solving Absolute Value Inequalities

The methods for solving inequalities parallel the methods for solving equations. Here are two basic rules we apply to solve absolute value inequalities.

Solving Absolute Value Inequalities

Let u be an algebraic expression in x and let a be a real number with $a \geq 0$.

1. If $|u| < a$, then u is in the interval $(-a, a)$. That is,

$$|u| < a \quad \text{if and only if} \quad -a < u < a.$$

2. If $|u| > a$, then u is in the interval $(-\infty, -a)$ or (a, ∞), that is,

$$|u| > a \quad \text{if and only if} \quad u < -a \text{ or } u > a.$$

The inequalities $<$ and $>$ can be replaced with $\leq$ and $\geq$, respectively. See Figure P.42.

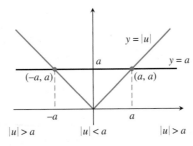

FIGURE P.42 The solution of $|u| < a$ is represented by the portion of the number line where the graph of $y = |u|$ is below the graph of $y = a$. The solution of $|u| > a$ is represented by the portion of the number line where the graph of $y = |a|$ is above the graph of $y = a$.

EXAMPLE 1   Solving an Absolute Value Inequality

Solve $|x - 4| < 8$.

SOLUTION

$$|x - 4| < 8 \qquad \text{Original inequality}$$

$$-8 < x - 4 < 8 \qquad \text{Equivalent double inequality}$$

$$-4 < x < 12 \qquad \text{Add 4.}$$

As an interval the solution is $(-4, 12)$.

Figure P.43 shows that points on the graph of $y = |x - 4|$ are below the points on the graph of $y = 8$ for values of x between -4 and 12.

Now try Exercise 3.

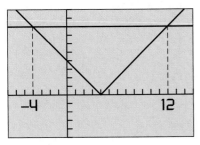

[−7, 15] by [−5, 10]

FIGURE P.43 The graphs of $y = |x - 4|$ and $y = 8$. (Example 1)

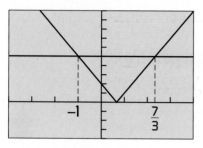

[−4, 4] by [−4, 10]

FIGURE P.44 The graphs of $y = |3x - 2|$ and $y = 5$. (Example 2)

UNION OF TWO SETS

The **union of two sets** *A* **and** *B*, denoted by $A \cup B$, is the set of all objects that belong to *A* or *B* or both.

EXAMPLE 2   Solving Another Absolute Value Inequality

Solve $|3x - 2| \geq 5$.

SOLUTION The solution of this absolute value inequality consists of the solutions of both of these inequalities.

$$3x - 2 \leq -5 \quad \text{or} \quad 3x - 2 \geq 5$$

$$3x \leq -3 \quad \text{or} \quad 3x \geq 7 \qquad \text{Add 2.}$$

$$x \leq -1 \quad \text{or} \quad x \geq \frac{7}{3} \qquad \text{Divide by 3.}$$

The solution consists of all numbers that are in either one of the two intervals $(-\infty, -1]$ or $[7/3, \infty)$, which may be written as $(-\infty, -1] \cup [7/3, \infty)$. The notation "$\cup$" is read as "union."

Figure P.44 shows that points on the graph of $y = |3x - 2|$ are above or on the points on the graph of $y = 5$ for values of x to the left of and including -1 and to the right of and including 7/3.   ***Now try Exercise 7.***

Solving Quadratic Inequalities

To solve a quadratic inequality such as $x^2 - x - 12 > 0$ we begin by solving the corresponding quadratic equation $x^2 - x - 12 = 0$. Then we determine the values of x for which the graph of $y = x^2 - x - 12$ lies above the x-axis.

EXAMPLE 3   Solving a Quadratic Inequality

Solve $x^2 - x - 12 > 0$.

SOLUTION First we solve the corresponding equation $x^2 - x - 12 = 0$.

$$x^2 - x - 12 = 0$$

$$(x - 4)(x + 3) = 0 \qquad \text{Factor.}$$

$$x - 4 = 0 \quad \text{or} \quad x + 3 = 0 \qquad ab = 0 \Rightarrow a = 0 \text{ or } b = 0$$

$$x = 4 \quad \text{or} \quad x = -3 \qquad \text{Solve for } x.$$

The solutions of the corresponding quadratic equation are -3 and 4, and they are not solutions of the original inequality because $0 > 0$ is false. Figure P.45 shows that the points on the graph of $y = x^2 - x - 12$ are above the x-axis for values of x to the left of -3 and to the right of 4.

The solution of the original inequality is $(-\infty, -3) \cup (4, \infty)$.

Now try Exercise 11.

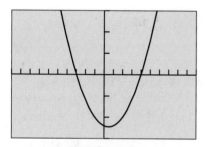

[−10, 10] by [−15, 15]

FIGURE P.45 The graph of $y = x^2 - x - 12$ appears to cross the x-axis at $x = -3$ and $x = 4$. (Example 3)

In Example 4, the quadratic inequality involves the symbol $\leq$. In this case, the solutions of the corresponding quadratic equation are also solutions of the inequality.

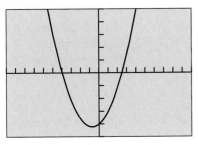

[−10, 10] by [−25, 25]

FIGURE P.46 The graph of $y = 2x^2 + 3x - 20$ appears to be below the x-axis for $-4 < x < 2.5$. (Example 4)

EXAMPLE 4   Solving Another Quadratic Inequality

Solve $2x^2 + 3x \le 20$.

SOLUTION First we subtract 20 from both sides of the inequality to obtain $2x^2 + 3x - 20 \le 0$. Next, we solve the corresponding quadratic equation $2x^2 + 3x - 20 = 0$.

$$2x^2 + 3x - 20 = 0$$

$$(x + 4)(2x - 5) = 0 \qquad \text{Factor.}$$

$$x + 4 = 0 \quad \text{or} \quad 2x - 5 = 0 \qquad ab = 0 \Rightarrow a = 0 \text{ or } b = 0$$

$$x = -4 \quad \text{or} \qquad x = \frac{5}{2} \qquad \text{Solve for } x.$$

The solutions of the corresponding quadratic equation are -4 and $5/2 = 2.5$. You can check that they are also solutions of the inequality.

Figure P.46 shows that the points on the graph of $y = 2x^2 + 3x - 20$ are below the x-axis for values of x between -4 and 2.5. The solution of the original inequality is $[-4, 2.5]$. We use square brackets because the numbers -4 and 2.5 are also solutions of the inequality.

Now try Exercise 9.

In Examples 3 and 4 the corresponding quadratic equation factored. If this doesn't happen we will need to approximate the zeros of the quadratic equation if it has any. Then we use our accuracy agreement from Section P.5 and write the endpoints of any intervals accurate to two decimal places as illustrated in Example 5.

EXAMPLE 5   Solving a Quadratic Inequality Graphically

Solve $x^2 - 4x + 1 \ge 0$ graphically.

SOLUTION We can use the graph of $y = x^2 - 4x + 1$ in Figure P.47 to determine that the solutions of the equation $x^2 - 4x + 1 = 0$ are about 0.27 and 3.73. Thus, the solution of the original inequality is $(-\infty, 0.27] \cup [3.73, \infty)$. We use square brackets because the zeros of the quadratic equation are solutions of the inequality even though we only have approximations to their values.

Now try Exercise 21.

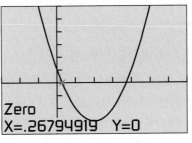

[−3, 7] by [−4, 6]

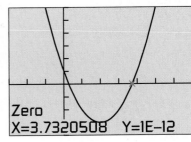

[−3, 7] by [−4, 6]

FIGURE P.47 This figure suggests that $y = x^2 - 4x + 1$ is zero for $x \approx 0.27$ and $x \approx 3.73$. (Example 5)

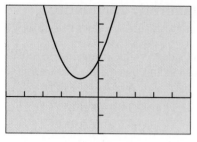

[–5, 5] by [–2, 5]

FIGURE P.48 The values of $y = x^2 + 2x + 2$ are never negative. (Example 6)

EXAMPLE 6   Showing that a Quadratic Inequality has no Solution

Solve $x^2 + 2x + 2 < 0$.

SOLUTION Figure P.48 shows that the graph of $y = x^2 + 2x + 2$ lies above the x-axis for all values for x. Thus, the inequality $x^2 + 2x + 2 < 0$ has *no* solution.

Now try Exercise 25.

Figure P.48 also shows that the solutions of the inequality $x^2 + 2x + 2 > 0$ is the set of all real numbers or, in interval notation, $(-\infty, \infty)$. A quadratic inequality can also have exactly one solution (see Exercise 31).

Approximating Solutions to Inequalities

To solve the inequality in Example 7 we approximate the zeros of the corresponding graph. Then we determine the values of x for which the corresponding graph is above or on the x-axis.

EXAMPLE 7   Solving a Cubic Inequality

Solve $x^3 + 2x^2 - 1 \geq 0$ graphically.

SOLUTION

We can use the graph of $y = x^3 + 2x^2 - 1$ in Figure P.49 to show that the solutions of the corresponding equation $x^3 + 2x^2 - 1 = 0$ are approximately -1.62, -1, and 0.62. The points on the graph of $y = x^3 + 2x^2 - 1$ are above the x-axis for values of x between -1.62 and -1, and for values of x to the right of 0.62.

The solution of the inequality is $[-1.62, -1] \cup [0.62, \infty)$. We use square brackets because the zeros of $y = x^3 + 2x^2 - 1$ are also solutions of the inequality.

Now try Exercise 27.

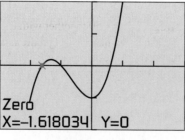

[–3, 3] by [–2, 2]

FIGURE P.49 The graph of $y = x^3 + 2x^2 - 1$ appears to be above the x-axis between the two negative x-intercepts and to the right of the positive x-intercept. (Example 7)

Projectile Motion

The movement of an object that is propelled vertically, but then subject only to the force of gravity, is an example of **projectile motion**.

Projectile Motion

Suppose an object is launched vertically from a point s_0 feet above the ground with an initial velocity of v_0 feet per second. The vertical position s (in feet) of the object t seconds after it is launched is

$$s = -16t^2 + v_0t + s_0.$$

EXAMPLE 8 Finding Height of a Projectile

A projectile is launched straight up from ground level with an initial velocity of 288 ft/sec.

(a) When will the projectile's height above ground be 1152 ft?

(b) When will the projectile's height above ground be at least 1152 ft?

SOLUTION

Here $s_0 = 0$ and $v_0 = 288$. So, the projectile's height is $s = -16t^2 + 288t$.

(a) We need to determine when $s = 1152$.

$$\begin{aligned}
s &= -16t^2 + 288t \\
1152 &= -16t^2 + 288t && \text{Substitute } s = 1152. \\
16t^2 - 288t + 1152 &= 0 && \text{Add } 16t^2 - 288t. \\
t^2 - 18t + 72 &= 0 && \text{Divide by 16.} \\
(t - 6)(t - 12) &= 0 && \text{Factor.} \\
t = 6 \quad \text{or} \quad t &= 12 && \text{Solve for } t.
\end{aligned}$$

The projectile is 1152 ft above ground twice; the first time at $t = 6$ sec on the way up, and the second time at $t = 12$ sec on the way down (Figure P.50).

(b) The projectile will be at least 1152 ft above ground when $s \geq 1152$. We can see from Figure P.50 together with the algebraic work in (a) that the solution is [6, 12]. This means that the projectile is at least 1152 ft above ground for times between $t = 6$ sec and $t = 12$ sec, including 6 and 12 sec.

In Exercise 32 we ask you to use algebra to solve the inequality $s = -16t^2 + 288t \geq 1152$.   ***Now try Exercise 33.***

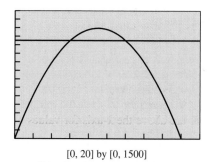

[0, 20] by [0, 1500]

FIGURE P.50 The graphs of $s = -16t^2 + 288t$ and $s = 1152$. We know from Example 8a that the two graphs intersect at (6, 1152) and (12, 1152).

QUICK REVIEW P.7

In Exercises 1–3, solve for x.

1. $-7 < 2x - 3 < 7$ **2.** $5x - 2 \geq 7x + 4$

3. $|x + 2| = 3$

In Exercises 4–6, factor the expression completely.

4. $4x^2 - 9$ **5.** $x^3 - 4x$

6. $9x^2 - 16y^2$

In Exercises 7 and 8, reduce the fraction to lowest terms.

7. $\dfrac{z^2 - 25}{z^2 - 5z}$ **8.** $\dfrac{x^2 + 2x - 35}{x^2 - 10x + 25}$

In Exercises 9 and 10, add the fractions and simplify.

9. $\dfrac{x}{x - 1} + \dfrac{x + 1}{3x - 4}$ **10.** $\dfrac{2x - 1}{x^2 - x - 2} + \dfrac{x - 3}{x^2 - 3x + 2}$

SECTION P.7 EXERCISES

In Exercises 1–8, solve the inequality algebraically. Write the solution in interval notation and draw its number line graph.

1. $|x + 4| \geq 5$

2. $|2x - 1| > 3.6$

3. $|x - 3| < 2$

4. $|x + 3| \leq 5$

5. $|4 - 3x| - 2 < 4$

6. $|3 - 2x| + 2 > 5$

7. $\left|\dfrac{x + 2}{3}\right| \geq 3$

8. $\left|\dfrac{x - 5}{4}\right| \leq 6$

In Exercises 9–16, solve the inequality. Use algebra to solve the corresponding equation.

9. $2x^2 + 17x + 21 \leq 0$

10. $6x^2 - 13x + 6 \geq 0$

11. $2x^2 + 7x > 15$

12. $4x^2 + 2 < 9x$

13. $2 - 5x - 3x^2 < 0$

14. $21 + 4x - x^2 > 0$

15. $x^3 - x \geq 0$

16. $x^3 - x^2 - 30x \leq 0$

In Exercises 17–26, solve the inequality graphically.

17. $x^2 - 4x < 1$

18. $12x^2 - 25x + 12 \geq 0$

19. $6x^2 - 5x - 4 > 0$

20. $4x^2 - 1 \leq 0$

21. $9x^2 + 12x - 1 \geq 0$

22. $4x^2 - 12x + 7 < 0$

23. $4x^2 + 1 > 4x$

24. $x^2 + 9 \leq 6x$

25. $x^2 - 8x + 16 < 0$

26. $9x^2 + 12x + 4 \geq 0$

In Exercises 27–30, solve the cubic inequality graphically.

27. $3x^3 - 12x + 2 \geq 0$

28. $8x - 2x^3 - 1 < 0$

29. $2x^3 + 2x > 5$

30. $4 \leq 2x^3 + 8x$

31. Group Activity Give an example of a quadratic inequality with the indicated solution.

(a) All real numbers

(b) No solution

(c) Exactly one solution

(d) $[-2, 5]$

(e) $(-\infty, -1) \cup (4, \infty)$

(f) $(-\infty, 0] \cup [4, \infty)$

32. Revisiting Example 8 Solve the inequality $-16t^2 + 288t \geq 1152$ algebraically and compare your answer with the result obtained in Example 10.

33. Projectile Motion A projectile is launched straight up from ground level with an initial velocity of 256 ft/sec.

(a) When will the projectile's height above ground be 768 ft?

(b) When will the projectile's height above ground be at least 768 ft?

(c) When will the projectile's height above ground be less than or equal to 768 ft?

34. Projectile Motion A projectile is launched straight up from ground level with an initial velocity of 272 ft/sec.

(a) When will the projectile's height above ground be 960 ft?

(b) When will the projectile's height above ground be more than 960 ft?

(c) When will the projectile's height above ground be less than or equal to 960 ft?

35. Writing to Learn Explain the role of equation solving in the process of solving an inequality. Give an example.

36. Travel Planning Barb wants to drive to a city 105 mi from her home in no more than 2 h. What is the lowest average speed she must maintain on the drive?

37. Connecting Algebra and Geometry Consider the collection of all rectangles that have length 2 in. less than twice their width.

(a) Find the possible widths (in inches) of these rectangles if their perimeters are less than 200 in.

(b) Find the possible widths (in inches) of these rectangles if their areas are less than or equal to 1200 in.^{2}.

38. Boyle's Law For a certain gas, $P = 400/V$, where P is pressure and V is volume. If $20 \leq V \leq 40$, what is the corresponding range for P?

39. Cash-Flow Planning A company has current assets (cash, property, inventory, and accounts receivable) of $200,000 and current liabilities (taxes, loans, and accounts payable) of $50,000. How much can it borrow if it wants its ratio of assets to liabilities to be no less than 2? Assume the amount borrowed is added to both current assets and current liabilities.

Standardized Test Questions

40. True or False The absolute value inequality $|x - a| < b$, where a and b are real numbers, always has at least one solution. Justify your answer.

41. True or False Every real number is a solution of the absolute value inequality $|x - a| \geq 0$, where a is a real number. Justify your answer.

In Exercises 42–45, solve these problems without using a calculator.

42. Multiple Choice Which of the following is the solution to $|x - 2| < 3$?

(A) $x = -1$ or $x = 5$

(B) $[-1, 5)$

(C) $[-1, 5]$

(D) $(-\infty, -1) \cup (5, \infty)$

(E) $(-1, 5)$

43. Multiple Choice Which of the following is the solution to $x^2 - 2x + 2 \geq 0$?

(A) $[0, 2]$

(B) $(-\infty, 0) \cup (2, \infty)$

(C) $(-\infty, 0] \cup [2, \infty)$

(D) All real numbers

(E) There is no solution.

44. Multiple Choice Which of the following is the solution to $x^2 > x$?

(A) $(-\infty, 0) \cup (1, \infty)$ **(B)** $(-\infty, 0] \cup [1, \infty)$

(C) $(1, \infty)$ **(D)** $(0, \infty)$

(E) There is no solution.

45. Multiple Choice Which of the following is the solution to $x^2 \leq 1$?

(A) $(-\infty, 1]$ **(B)** $(-1, 1)$

(C) $[1, \infty)$ **(D)** $[-1, 1]$

(E) There is no solution.

Explorations

46. Constructing a Box with No Top An open box is formed by cutting squares from the corners of a regular piece of cardboard (see figure) and folding up the flaps.

(a) What size corner squares should be cut to yield a box with a volume of 125 in.^{3}?

(b) What size corner squares should be cut to yield a box with a volume more than 125 in.^{3}?

(c) What size corner squares should be cut to yield a box with a volume of at most 125 in.^{3}?

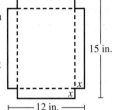

Extending the Ideas

In Exercises 47 and 48, use a combination of algebraic and graphical techniques to solve the inequalities.

47. $|2x^2 + 7x - 15| < 10$ **48.** $|2x^2 + 3x - 20| \geq 10$

CHAPTER P **Key Ideas**

PROPERTIES, THEOREMS, AND FORMULAS

PROCEDURES

CHAPTER P Review Exercises

The collection of exercises marked in red could be used as a chapter test.

In Exercises 1 and 2, find the endpoints and state whether the interval is bounded or unbounded.

1. $[0, 5]$ **2.** $(2, \infty)$

3. Distributive Property Use the distributive property to write the expanded form of $2(x^2 - x)$.

4. Distributive Property Use the distributive property to write the factored form of $2x^3 + 4x^2$.

In Exercises 5 and 6, simplify the expression. Assume that denominators are not zero.

5. $\dfrac{(uv^2)^3}{v^2 u^3}$ **6.** $(3x^2 y^3)^{-2}$

In Exercises 7 and 8, write the number in scientific notation.

7. The mean distance from Pluto to the sun is about 3,680,000,000 miles.

8. The diameter of a red blood corpuscle is about 0.000007 meter.

In Exercises 9 and 10, write the number in decimal form.

9. Our solar system is about 5×10^9 years old.

10. The mass of an electron is about 9.1094×10^{-28} g (gram).

11. The data in Table P.9 gives the spending on collegiate student financial aid from several sources. Write the spending amount in dollars by the given source in scientific notation.

Table P.9 Student Financial Aid	
Spending Source	Amount (dollars)
Federal Aid	50,711,000,000
State Grant Programs	4,630,000,000
State-sponsored Loans	500,000,000
Private-sector Loans	3,995,000,000
Institution and other Grants	14,497,000,000

Source: The College Board, as reported in The Chronicle of Higher Education, Almanac, 2002–3, August 30, 2002.

(a) Federal Aid

(b) State Grant Programs

(c) State-sponsored Loans

(d) Private-sector Loans

(e) Institution and other Grants

12. Decimal Form Find the decimal form for $-5/11$. State whether it repeats or terminates.

In Exercises 13 and 14, find **(a)** the distance between the points and **(b)** the midpoint of the line segment determined by the points.

13. -5 and 14   **14.** $(-4, 3)$ and $(5, -1)$

In Exercises 15 and 16, show that the figure determined by the points is the indicated type.

15. Right triangle: $(-2, 1)$, $(3, 11)$, $(7, 9)$

16. Equilateral triangle: $(0, 1)$, $(4, 1)$, $(2, 1 - 2\sqrt{3})$

In Exercises 17 and 18, find the standard form equation for the circle,

17. Center $(0, 0)$, radius 2

18. Center $(5, -3)$, radius 4

In Exercises 19 and 20, find the center and radius of the circle.

19. $(x + 5)^2 + (y + 4)^2 = 9$

20. $x^2 + y^2 = 1$

21. (a) Find the length of the sides of the triangle in the figure.

(b) Writing to Learn Show that the triangle is a right triangle.

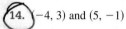

22. Distance and Absolute Value Use absolute value notation to write the statement that the distance between z and -3 is less than or equal to 1.

23. Finding a Line Segment with Given Midpoint Let $(3, 5)$ be the midpoint of the line segment with endpoints $(-1, 1)$ and (a, b). Determine a and b.

24. Finding Slope Find the slope of the line through the points $(-1, -2)$ and $(4, -5)$.

25. Finding Point-Slope Form Equation Find an equation in point-slope form for the line through the point $(2, -1)$ with slope $m = -2/3$.

26. Find an equation of the line through the points $(-5, 4)$ and $(2, -5)$ in the general form $Ax + By + C = 0$.

In Exercises 27–32, find an equation in slope-intercept form for the line.

27. The line through $(3, -2)$ with slope $m = 4/5$

28. The line through the points $(-1, -4)$ and $(3, 2)$

29. The line through $(-2, 4)$ with slope $m = 0$

30. The line $3x - 4y = 7$

31. The line through $(2, -3)$ and parallel to the line $2x + 5y = 3$.

32. The line through $(2, -3)$ and perpendicular to the line $2x + 5y = 3$.

33. SAT Math Scores The SAT scores are measured on an 800-point scale. The data in Table P.10 shows the average SAT math score for several years.

Table P.10 Average SAT Math Scores	
Year	SAT Math Score
1995	506
1997	511
1998	512
1999	511
2000	514
2001	514
2002	516
2003	519
2004	518

Source: The World Almanac and Book of Facts, The New York Times June, 2005.

(a) Let $x = 0$ represent 1990, $x = 1$ represent 1991, and so forth. Draw a scatter plot of the data.

(b) Use the 1995 and 2000 data to write a linear equation for the average SAT math score y in terms of the year x. Superimpose the graph of the linear equation on the scatter plot in (a).

(c) Use the equation in (b) to estimate the average SAT math score in 1996. Compare with the actual value of 508.

(d) Use the equation in (b) to predict the average SAT math score in 2006.

34. Consider the point $(-6, 3)$ and Line L: $4x - 3y = 5$. Write an equation **(a)** for the line passing through this point and parallel to L, and **(b)** for the line passing through this point and perpendicular to L. Support your work graphically.

In Exercises 35 and 36, assume that each graph contains the origin and the upper right-hand corner of the viewing window.

35. Find the slope of the line in the figure.

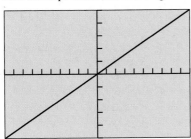

[−10, 10] by [−25, 25]

36. Writing to Learn Which line has the greater slope? Explain.

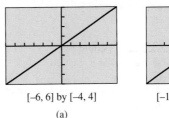 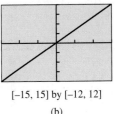

[−6, 6] by [−4, 4]   [−15, 15] by [−12, 12]

(a)   (b)

In Exercises 37–50, solve the equation algebraically.

37. $3x - 4 = 6x + 5$
38. $\dfrac{x - 2}{3} + \dfrac{x + 5}{2} = \dfrac{1}{3}$

39. $2(5 - 2y) - 3(1 - y) = y + 1$

40. $3(3x - 1)^2 = 21$
41. $x^2 - 4x - 3 = 0$

42. $16x^2 - 24x + 7 = 0$
43. $6x^2 + 7x = 3$

44. $2x^2 + 8x = 0$
45. $x(2x + 5) = 4(x + 7)$

46. $|4x + 1| = 3$
47. $4x^2 - 20x + 25 = 0$

48. $-9x^2 + 12x - 4 = 0$
49. $x^2 = 3x$

50. $4x^2 - 4x + 2 = 0$
51. $x^2 - 6x + 13 = 0$

52. $x^2 - 2x + 4 = 0$

53. Completing the Square Use completing the square to solve the equation $2x^2 - 3x - 1 = 0$.

54. Quadratic Formula Use the quadratic formula to solve the equation $3x^2 + 4x - 1 = 0$.

In Exercises 55–58, solve the equation graphically.

55. $3x^3 - 19x^2 - 14x = 0$
56. $x^3 + 2x^2 - 4x - 8 = 0$

57. $x^3 - 2x^2 - 2 = 0$
58. $|2x - 1| = 4 - x^2$

In Exercises 59 and 60, solve the inequality and draw a number line graph of the solution.

59. $-2 < x + 4 \le 7$
60. $5x + 1 \ge 2x - 4$

In Exercises 61–72, solve the inequality.

61. $\dfrac{3x - 5}{4} \le -1$

62. $|2x - 5| < 7$

63. $|3x + 4| \ge 2$

64. $4x^2 + 3x > 10$

65. $2x^2 - 2x - 1 > 0$

66. $9x^2 - 12x - 1 \le 0$

67. $x^3 - 9x \le 3$

68. $4x^3 - 9x + 2 > 0$

69. $\left| \dfrac{x + 7}{5} \right| > 2$

70. $2x^2 + 3x - 35 < 0$

71. $4x^2 + 12x + 9 \ge 0$

72. $x^2 - 6x + 9 < 0$

In Exercises 73–80, perform the indicated operation, and write the result in the standard form $a + bi$.

73. $(3 - 2i) + (-2 + 5i)$

74. $(5 - 7i) - (3 - 2i)$

75. $(1 + 2i)(3 - 2i)$

76. $(1 + i)^3$

77. $(1 + 2i)^2(1 - 2i)^2$

78. i^{29}

79. $\sqrt{-16}$

80. $\dfrac{2 + 3i}{1 - 5i}$

81. Projectile Motion A projectile is launched straight up from ground level with an initial velocity of 320 ft/sec.

(a) When will the projectile's height above ground be 1538 ft?

(b) When will the projectile's height above ground be at most 1538 ft?

(c) When will the projectile's height above ground be greater than or equal to 1538 ft?

82. Navigation A commercial jet airplane climbs at takeoff with slope $m = 4/9$. How far in the horizontal direction will the airplane fly to reach an altitude of 20,000 ft above the takeoff point?

83. Connecting Algebra and Geometry Consider the collection of all rectangles that have length 1 cm more than three times their width w.

(a) Find the possible widths (in cm) of these rectangles if their perimeters are less than or equal to 150 cm.

(b) Find the possible widths (in cm) of these rectangles if their areas are greater than 1500 cm^2.

Functions and Graphs

One of the central principles of economics is that the value of money is not constant; it is a function of time. Since fortunes are made and lost by people attempting to predict the future value of money, much attention is paid to quantitative measures like the consumer price index, a basic measure of inflation in various sectors of the economy. See page 159 for a look at how the consumer price index for housing has behaved over time.

Chapter 1 Overview

In this chapter we begin the study of functions that will continue throughout the book. Your previous courses have introduced you to some basic functions. These functions can be visualized using a graphing calculator, and their properties can be described using the notation and terminology that will be introduced in this chapter. A familiarity with this terminology will serve you well in later chapters when we explore properties of functions in greater depth.

1.1
Modeling and Equation Solving

What you'll learn about

- Numerical Models
- Algebraic Models
- Graphical Models
- The Zero Factor Property
- Problem Solving
- Grapher Failure and Hidden Behavior
- A Word About Proof

... and why

Numerical, algebraic, and graphical models provide different methods to visualize, analyze, and understand data.

Numerical Models

Scientists and engineers have always used mathematics to model the real world and thereby to unravel its mysteries. A **mathematical model** is a mathematical structure that approximates phenomena for the purpose of studying or predicting their behavior. Thanks to advances in computer technology, the process of devising mathematical models is now a rich field of study itself, **mathematical modeling**.

We will be concerned primarily with three types of mathematical models in this book: *numerical models*, *algebraic models*, and *graphical models*. Each type of model gives insight into real-world problems, but the best insights are often gained by switching from one kind of model to another. Developing the ability to do that will be one of the goals of this course.

Perhaps the most basic kind of mathematical model is the **numerical model**, in which numbers (or *data*) are analyzed to gain insights into phenomena. A numerical model can be as simple as the major league baseball standings or as complicated as the network of interrelated numbers that measure the global economy.

Table 1.1 The Minimum Hourly Wage

Year	Minimum Hourly Wage (MHW)	Purchasing Power in 1996 Dollars
1955	0.75	4.39
1960	1.00	5.30
1965	1.25	6.23
1970	1.60	6.47
1975	2.10	6.12
1980	3.10	5.90
1985	3.35	4.88
1990	3.80	4.56
1995	4.25	4.38
2000	5.15	4.69
2005	5.15	4.15

Source: www.infoplease.com

EXAMPLE 1   Tracking the Minimum Wage

The numbers in Table 1.1 show the growth of the minimum hourly wage (MHW) from 1955 to 2005. The table also shows the MHW adjusted to the purchasing power of 1996 dollars (using the CPI-U, the Consumer Price Index for all Urban Consumers). Answer the following questions using only the data in the table.

(a) In what five-year period did the actual MHW increase the most?

(b) In what year did a worker earning the MHW enjoy the greatest purchasing power?

(c) A worker on minimum wage in 1980 was earning nearly twice as much as a worker on minimum wage in 1970, and yet there was great pressure to raise the minimum wage again. Why?

SOLUTION

(a) In the period 1975 to 1980 it increased by $1.00. Notice that the minimum wage never goes down, so we can tell that there were no other increases of this magnitude even though the table does not give data from every year.

(b) In 1970.

(c) Although the MHW increased from $1.60 to $3.10 in that period, the purchasing power actually dropped by $0.57 (in 1996 dollars). This is one way inflation can affect the economy.   *Now try Exercise 11.*

The numbers in Table 1.1 provide a numerical model for one aspect of the U.S. economy by using another numerical model, the urban consumer price index (CPI-U), to adjust the data. Working with large numerical models is standard operating procedure in business and industry, where computers are relied upon to provide fast and accurate data processing.

EXAMPLE 2 Analyzing Prison Populations

Table 1.2 shows the growth in the number of prisoners incarcerated in state and federal prisons from 1980 to 2000. Is the proportion of female prisoners over the years increasing?

SOLUTION The *number* of female prisoners over the years is certainly increasing, but so is the total number of prisoners, so it is difficult to discern from the data whether the *proportion* of female prisoners is increasing. What we need is another column of numbers showing the ratio of female prisoners to total prisoners.

We could compute all the ratios separately, but it is easier to do this kind of repetitive calculation with a single command on a computer spreadsheet. You can also do this on a graphing calculator by manipulating lists (see Exercise 19). Table 1.3 shows the percentage of the total population each year that consists of female prisoners. With this data to extend our numerical model, it is clear that the proportion of female prisoners is increasing. ***Now try Exercise 19.***

Table 1.2 U.S. Prison Population (Thousands)

Year	Total	Male	Female
1980	316	304	12
1985	480	459	21
1990	740	699	41
1995	1085	1021	64
2000	1382	1290	92

Source: U.S. Justice Department.

Table 1.3 Female Percentage of U.S. Prison Population

Year	% Female
1980	3.8
1985	4.4
1990	5.5
1995	5.9
2000	6.7

Source: U.S. Justice Department.

Algebraic Models

An **algebraic model** uses formulas to relate variable quantities associated with the phenomena being studied. The added power of an algebraic model over a numerical model is that it can be used to generate numerical values of unknown quantities by relating them to known quantities.

EXAMPLE 3 Comparing Pizzas

A pizzeria sells a rectangular 18″ by 24″ pizza for the same price as its large round pizza (24″ diameter). If both pizzas are of the same thickness, which option gives the most pizza for the money?

SOLUTION We need to compare the *areas* of the pizzas. Fortunately, geometry has provided algebraic models that allow us to compute the areas from the given information.

For the rectangular pizza:

$$Area = l \times w = 18 \times 24 = 432 \text{ square inches.}$$

For the circular pizza:

$$Area = \pi r^2 = \pi \left(\frac{24}{2}\right)^2 = 144\pi \approx 452.4 \text{ square inches.}$$

The round pizza is larger and therefore gives more for the money.

Now try Exercise 21.

The algebraic models in Example 3 come from geometry, but you have probably encountered algebraic models from many other sources in your algebra and science courses.

EXPLORATION EXTENSIONS

Suppose that after the sale, the merchandise prices are increased by 25%. If m represents the marked price before the sale, find an algebraic model for the post-sale price, including tax.

EXPLORATION 1   Designing an Algebraic Model

A department store is having a sale in which everything is discounted 25% off the marked price. The discount is taken at the sales counter, and then a state sales tax of 6.5% and a local sales tax of 0.5% are added on.

1. The discount price d is related to the marked price m by the formula $d = km$, where k is a certain constant. What is k?

2. The actual sale price s is related to the discount price d by the formula $s = d + td$, where t is a constant related to the total sales tax. What is t?

3. Using the answers from steps 1 and 2 you can find a constant p that relates s directly to m by the formula $s = pm$. What is p?

4. If you only have \$30, can you afford to buy a shirt marked \$36.99?

5. If you have a credit card but are determined to spend no more than \$100, what is the maximum total value of your marked purchases before you present them at the sales counter?

The ability to generate numbers from formulas makes an algebraic model far more useful as a predictor of behavior than a numerical model. Indeed, one optimistic goal of scientists and mathematicians when modeling phenomena is to fit an algebraic model to numerical data and then (even more optimistically) to analyze why it works. Not all models can be used to make accurate predictions. For example, nobody has ever devised a successful formula for predicting the ups and downs of the stock market as a function of time, although that does not stop investors from trying.

If numerical data do behave reasonably enough to suggest that an algebraic model might be found, it is often helpful to look at a picture first. That brings us to graphical models.

Graphical Models

A **graphical model** is a visible representation of a numerical model or an algebraic model that gives insight into the relationships between variable quantities. Learning to interpret and use graphs is a major goal of this book.

EXAMPLE 4   Visualizing Galileo's Gravity Experiments

Galileo Galilei (1564–1642) spent a good deal of time rolling balls down inclined planes carefully recording the distance they traveled as a function of elapsed time. His experiments are commonly repeated in physics classes today, so it is easy to reproduce a typical table of Galilean data.

Elapsed time (seconds)	0	1	2	3	4	5	6	7	8
Distance traveled (inches)	0	0.75	3	6.75	12	18.75	27	36.75	48

What graphical model fits the data? Can you find an algebraic model that fits?

continued

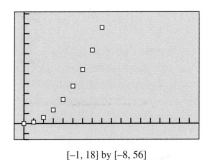

[–1, 18] by [–8, 56]

FIGURE 1.1 A scatter plot of the data from a Galileo gravity experiment. (Example 4)

SOLUTION A scatter plot of the data is shown in Figure 1.1.

Galileo's experience with quadratic functions suggested to him that this figure was a parabola with its vertex at the origin; he therefore modeled the effect of gravity as a quadratic function:

$$d = kt^2.$$

Because the ordered pair (1, 0.75) must satisfy the equation, it follows that $k = 0.75$, yielding the equation

$$d = 0.75t^2.$$

You can verify numerically that this algebraic model correctly predicts the rest of the data points. We will have much more to say about parabolas in Chapter 2.

Now try Exercise 23.

This insight led Galileo to discover several basic laws of motion that would eventually be named after Isaac Newton. While Galileo had found the algebraic model to describe the path of the ball, it would take Newton's calculus to explain why it worked.

EXAMPLE 5 Fitting a Curve to Data

We showed in Example 2 that the percentage of females in the U.S. prison population has been steadily growing over the years. Model this growth graphically and use the graphical model to suggest an algebraic model.

SOLUTION Let t be the number of years after 1980, and let F be the percentage of females in the prison population from year 0 to year 20. From the data in Table 1.3 we get the corresponding data in Table 1.4:

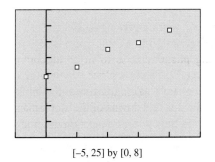

[–5, 25] by [0, 8]

FIGURE 1.2 A scatter plot of the data in Table 1.4. (Example 5)

Table 1.4 Percentage (F) of Females in the Prison Population t years after 1980				
t	0	5	10	15	20
F	3.8	4.4	5.5	5.9	6.7

Source: U.S. Justice Department.

A scatter plot of the data is shown in Figure 1.2.

This pattern looks linear. If we use a line as our graphical model, we can find an algebraic model by finding the equation of the line. We will describe in Chapter 2 how a statistician would find the best line to fit the data, but we can get a pretty good fit for now by finding the line through the points (0, 3.8) and (20, 6.7).

The slope is $(6.7 - 3.8)/(20 - 0) = 0.145$ and the y-intercept is 3.8. Therefore, the line has equation $y = 0.145x + 3.8$. You can see from Figure 1.3 that this line does a very nice job of modeling the data.

Now try Exercises 13 and 15.

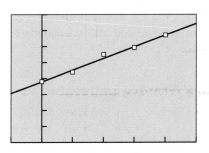

[–5, 25] by [0, 8]

FIGURE 1.3 The line with equation $y = 0.145x + 3.8$ is a good model for the data in Table 1.4. (Example 5)

EXPLORATION 2 **Interpreting the Model**

The parabola in Example 4 arose from a law of physics that governs falling objects, which should inspire more confidence than the linear model in Example 5. We can repeat Galileo's experiment many times with differently-sloped ramps, different units of measurement, and even on different planets, and a quadratic model will fit it every time. The purpose of this Exploration is to think more deeply about the linear model in the prison example.

1. The linear model we found will not continue to predict the percentage of female prisoners in the U.S. indefinitely. Why must it eventually fail?

2. Do you think that our linear model will give an accurate estimate of the percentage of female prisoners in the U.S. in 2009? Why or why not?

3. The linear model is such a good fit that it actually calls our attention to the unusual jump in the percentage of female prisoners in 1990. Statisticians would look for some unusual "confounding" factor in 1990 that might explain the jump. What sort of factors do you think might explain it?

4. Does Table 1.1 suggest a possible factor that might influence female crime statistics?

There are other ways of graphing numerical data that are particularly useful for statistical studies. We will treat some of them in Chapter 9. The scatter plot will be our choice of data graph for the time being, as it provides the closest connection to graphs of functions in the Cartesian plane.

The Zero Factor Property

The main reason for studying algebra through the ages has been to solve equations. We develop algebraic models for phenomena so that we can solve problems, and the solutions to the problems usually come down to finding solutions of algebraic equations.

If we are fortunate enough to be solving an equation in a single variable, we might proceed as in the following example.

EXAMPLE 6 Solving an Equation Algebraically

Find all real numbers x for which $6x^3 = 11x^2 + 10x$.

SOLUTION We begin by changing the form of the equation to $6x^3 - 11x^2 - 10x = 0$.

We can then solve this equation algebraically by factoring:

$$6x^3 - 11x^2 - 10x = 0$$

$$x(6x^2 - 11x - 10) = 0$$

$$x(2x - 5)(3x + 2) = 0$$

continued

$$x = 0 \quad \text{or} \quad 2x - 5 = 0 \quad \text{or} \quad 3x + 2 = 0$$

$$x = 0 \quad \text{or} \qquad x = \frac{5}{2} \quad \text{or} \qquad x = -\frac{2}{3}$$

Now try Exercise 31.

In Example 6, we used the important Zero Factor Property of real numbers.

The Zero Factor Property

A product of real numbers is zero if and only if at least one of the factors in the product is zero.

It is this property that algebra students use to solve equations in which an expression is set equal to zero. Modern problem solvers are fortunate to have an alternative way to find such solutions.

If we graph the expression, then the x-intercepts of the graph of the expression will be the values for which the expression equals 0.

EXAMPLE 7   Solving an Equation: Comparing Methods

Solve the equation $x^2 = 10 - 4x$.

SOLUTION

Solve Algebraically

The given equation is equivalent to $x^2 + 4x - 10 = 0$.

This quadratic equation has irrational solutions that can be found by the quadratic formula.

$$x = \frac{-4 + \sqrt{16 + 40}}{2} \approx 1.7416574$$

and

$$x = \frac{-4 - \sqrt{16 + 40}}{2} \approx -5.7416574$$

While the decimal answers are certainly accurate enough for all practical purposes, it is important to note that only the expressions found by the quadratic formula give the *exact* real number answers. The tidiness of exact answers is a worthy mathematical goal. Realistically, however, exact answers are often impossible to obtain, even with the most sophisticated mathematical tools.

Solve Graphically

We first find an equivalent equation with 0 on the right-hand side: $x^2 + 4x - 10 = 0$. We then graph the equation $y = x^2 + 4x - 10$, as shown in Figure 1.4.

We then use the grapher to locate the x-intercepts of the graph:

$$x \approx 1.7416574 \quad \text{and} \quad x \approx -5.741657.$$

Now try Exercise 35.

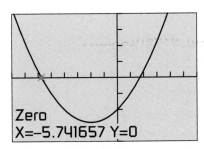

Zero
X=-5.741657 Y=0

[-8, 6] by [-20, 20]

FIGURE 1.4 The graph of $y = x^2 + 4x - 10$. (Example 7)

SOLVING EQUATIONS WITH TECHNOLOGY

Example 7 shows one method of solving an equation with technology. Some graphers could also solve the equation in Example 7 by finding the *intersection* of the graphs of $y = x^2$ and $y = 10 - 4x$. Some graphers have built-in equation solvers. Each method has its advantages and disadvantages, but we recommend the "finding the x-intercepts" technique for now because it most closely parallels the classical algebraic techniques for finding roots of equations, and makes the connection between the algebraic and graphical models easier to follow and appreciate.

We used the graphing utility of the calculator to **solve graphically** in Example 7. Most calculators also have solvers that would enable us to **solve numerically** for the same decimal approximations without considering the graph. Some calculators have computer algebra systems that will solve numerically to produce exact answers in certain cases. In this book we will distinguish between these two technological methods and the traditional pencil-and-paper methods used to **solve algebraically**.

Every method of solving an equation usually comes down to finding where an expression equals zero. If we use $f(x)$ to denote an algebraic expression in the variable x, the connections are as follows:

Fundamental Connection

If a is a real number that solves the equation $f(x) = 0$, then these three statements are equivalent:

1. The number a is a **root** (or **solution**) of the **equation** $f(x) = 0$.

2. The number a is a **zero** of $y = f(x)$.

3. The number a is an **x-intercept** of the **graph** of $y = f(x)$. (Sometimes the point $(a, 0)$ is referred to as an x-intercept.)

Problem Solving

George Pólya (1887–1985) is sometimes called the father of modern problem solving, not only because he was good at it (as he certainly was) but also because he published the most famous analysis of the problem-solving process: *How to Solve It: A New Aspect of Mathematical Method*. His "four steps" are well known to most mathematicians:

Pólya's Four Problem-Solving Steps

1. Understand the problem.
2. Devise a plan.
3. Carry out the plan.
4. Look back.

The problem-solving process that we recommend you use throughout this course will be the following version of Pólya's four steps.

A Problem-Solving Process

Step 1—Understand the problem.

- Read the problem as stated, several times if necessary.
- Be sure you understand the meaning of each term used.
- Restate the problem in your own words. Discuss the problem with others if you can.
- Identify clearly the information that you need to solve the problem.
- Find the information you need from the given data.

Step 2—Develop a mathematical model of the problem.

• Draw a picture to visualize the problem situation. It usually helps.

• Introduce a variable to represent the quantity you seek. (In some cases there may be more than one.)

• Use the statement of the problem to find an equation or inequality that relates the variables you seek to quantities that you know.

Step 3—Solve the mathematical model and support or confirm the solution.

• **Solve algebraically** using traditional algebraic methods and **support graphically or support numerically** using a graphing utility.

• **Solve graphically or numerically** using a graphing utility and **confirm algebraically** using traditional algebraic methods.

• **Solve graphically or numerically** because there is no other way possible.

Step 4—Interpret the solution in the problem setting.

• Translate your mathematical result into the problem setting and decide whether the result makes sense.

EXAMPLE 8 Applying the Problem-Solving Process

The engineers at an auto manufacturer pay students $0.08 per mile plus $25 per day to road test their new vehicles.

(a) How much did the auto manufacturer pay Sally to drive 440 miles in one day?

(b) John earned $93 test-driving a new car in one day. How far did he drive?

SOLUTION

Model A picture of a car or of Sally or John would not be helpful, so we go directly to designing the model. Both John and Sally earned $25 for one day, plus $0.08 per mile. Multiply dollars/mile by miles to get dollars.

So if p represents the pay for driving x miles in one day, our algebraic model is

$$p = 25 + 0.08x.$$

Solve Algebraically

(a) To get Sally's pay we let $x = 440$ and solve for p:

$$p = 25 + 0.08(440)$$
$$= 60.20$$

(b) To get John's mileage we let $p = 93$ and solve for x:

$$93 = 25 + 0.08x$$
$$68 = 0.08x$$
$$x = \frac{68}{0.08}$$
$$x = 850$$

continued

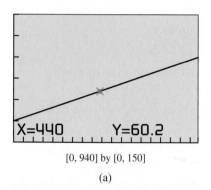

[0, 940] by [0, 150]

(a)

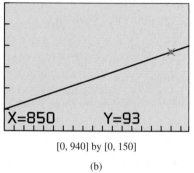

[0, 940] by [0, 150]

(b)

FIGURE 1.5 Graphical support for the algebraic solutions in Example 8.

Support Graphically

Figure 1.5a shows that the point (440, 60.20) is on the graph of $y = 25 + 0.08x$, supporting our answer to (a). Figure 1.5b shows that the point (850, 93) is on the graph of $y = 25 + 0.08x$, supporting our answer to (b). (We could also have **supported** our answer **numerically** by simply substituting in for each *x* and confirming the value of *p*.)

Interpret

Sally earned $60.20 for driving 440 miles in one day. John drove 850 miles in one day to earn $93.00.   *Now try Exercise 47.*

It is not really necessary to *show* written support as part of an algebraic solution, but it is good practice to support answers wherever possible simply to reduce the chance for error. We will often show written support of our solutions in this book in order to highlight the connections among the algebraic, graphical, and numerical models.

Grapher Failure and Hidden Behavior

While the graphs produced by computers and graphing calculators are wonderful tools for understanding algebraic models and their behavior, it is important to keep in mind that machines have limitations. Occasionally they can produce graphical models that misrepresent the phenomena we wish to study, a problem we call **grapher failure**. Sometimes the viewing window will be too large, obscuring details of the graph which we call **hidden behavior**. We will give an example of each just to illustrate what can happen, but rest assured that these difficulties rarely occur with graphical models that arise from real-world problems.

EXAMPLE 9   Seeing Grapher Failure

Look at the graph of $y = 3/(2x - 5)$ on a graphing calculator. Is there an *x*-intercept?

SOLUTION A graph is shown in Figure 1.6a.

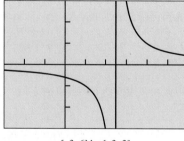

[−3, 6] by [−3, 3]

(a)

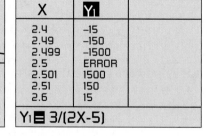

(b)

FIGURE 1.6 (a) A graph with a mysterious *x*-intercept. (b) As *x* approaches 2.5, the value of $3/(2x - 5)$ approaches $\pm\infty$. (Example 9)

continued

The graph seems to show an x-intercept about halfway between 2 and 3. To confirm this algebraically, we would set $y = 0$ and solve for x:

$$0 = \frac{3}{2x - 5}$$
$$0(2x - 5) = 3$$
$$0 = 3$$

The statement $0 = 3$ is false for all x, so there can be no value that makes $y = 0$, and hence there can be no x-intercept for the graph. What went wrong?

The answer is a simple form of grapher failure: The vertical line should not be there! As suggested by the table in Figure 1.6b, the actual graph of $y = 3/(2x - 5)$ approaches $-\infty$ to the left of $x = 2.5$, and comes down from $+\infty$ to the right of $x = 2.5$ (more on this later). The expression $3/(2x - 5)$ is undefined at $x = 2.5$, but the graph in Figure 1.6a does not reflect this. The grapher plots points at regular increments from left to right, *connecting the points* as it goes. It hits some low point off the screen to the left of 2.5, followed immediately by some high point off the screen to the right of 2.5, and it connects them with that unwanted line. ***Now try Exercise 49.***

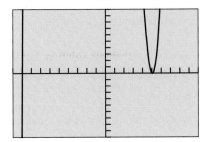

[–10, 10] by [–10, 10]
(a)

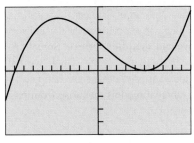

[–10, 10] by [–500, 500]
(b)

FIGURE 1.7 The graph of $y = x^3 - 1.1x^2 - 65.4x + 229.5$ in two viewing windows. (Example 10)

EXAMPLE 10 Not Seeing Hidden Behavior

Solve graphically: $x^3 - 1.1x^2 - 65.4x + 229.5 = 0$.

SOLUTION Figure 1.7a shows the graph in the standard $[-10, 10]$ by $[-10, 10]$ window, an inadequate choice because too much of the graph is off the screen. Our horizontal dimensions look fine, so we adjust our vertical dimensions to $[-500, 500]$, yielding the graph in Figure 1.7b.

We use the grapher to locate an x-intercept near -9 (which we find to be -9) and then an x-intercept near 5 (which we find to be 5). The graph leads us to believe that we are done. However, if we zoom in closer to observe the behavior near $x = 5$, the graph tells a new story (Figure 1.8).

In this graph we see that there are actually *two* x-intercepts near 5 (which we find to be 5 and 5.1). There are therefore three roots (or zeros) of the equation $x^3 - 1.1x^2 - 65.4x + 229.5 = 0$: $x = -9$, $x = 5$, and $x = 5.1$. ***Now try Exercise 51.***

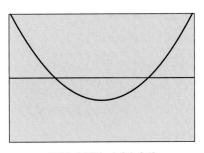

[4.95, 5.15] by [–0.1, 0.1]

FIGURE 1.8 A closer look at the graph of $y = x^3 - 1.1x^2 - 65.4x + 229.5$. (Example 10)

You might wonder if there could be still *more* hidden x-intercepts in Example 10! We will learn in Chapter 2 how the *Fundamental Theorem of Algebra* guarantees that there are not.

A Word About Proof

While Example 10 is still fresh on our minds, let us point out a subtle, but very important, consideration about our solution.

We *solved graphically* to find two solutions, then eventually three solutions, to the given equation. Although we did not show the steps, it is easy to *confirm numerically* that the three numbers found are actually solutions by substituting them into the equation. But the problem asked us to find *all* solutions. While we could explore that equation

TEACHER NOTE

Sometimes it is impossible to show all of the details of a graph in a single window. For example, in Example 10 the graph in Figure 1.8 reveals minute details of the graph, but it hides the overall shape of the graph.

graphically in a hundred more viewing windows and never find another solution, our failure to find them would not *prove* that they are not out there somewhere. That is why the Fundamental Theorem of Algebra is so important. It tells us that there can be at most three real solutions to *any* cubic equation, so we know for a fact that there are no more.

Exploration is encouraged throughout this book because it is how mathematical progress is made. Mathematicians are never satisfied, however, until they have *proved* their results. We will show you proofs in later chapters and we will ask you to produce proofs occasionally in the exercises. That will be a time for you to set the technology aside, get out a pencil, and show in a logical sequence of algebraic steps that something is undeniably and universally true. This process is called **deductive reasoning**.

EXAMPLE 11   Proving a Peculiar Number Fact

Prove that 6 is a factor of $n^3 - n$ for every positive integer n.

SOLUTION You can explore this expression for various values of n on your calculator. Table 1.5 shows it for the first 12 values of n.

Table 1.5 The first 12 values of $n^3 - n$											
n	1	2	3	4	5	6	7	8	9	10	11	12
$n^3 - n$	0	6	24	60	120	210	336	504	720	990	1320	1716

All of these numbers are divisible by 6, but that does not prove that they will continue to be divisible by 6 for all values of n. In fact, a table with a billion values, all divisible by 6, would not constitute a proof. Here is a proof:

Let n be *any* positive integer.

- We can factor $n^3 - n$ as the product of three numbers:
  $(n - 1)(n)(n + 1)$.

- The factorization shows that $n^3 - n$ is always the product of three consecutive integers.

- Every set of three consecutive integers must contain a multiple of 3.

- Since 3 divides a factor of $n^3 - n$, it follows that 3 is a factor of $n^3 - n$ itself.

- Every set of three consecutive integers must contain a multiple of 2.

- Since 2 divides a factor of $n^3 - n$, it follows that 2 is a factor of $n^3 - n$ itself.

- Since both 2 and 3 are factors of $n^3 - n$, we know that 6 is a factor of $n^3 - n$.

End of proof!   *Now try Exercise 53.*

QUICK REVIEW 1.1 *(For help, go to Section A.2.)*

Factor the following expressions completely over the real numbers.

1. $x^2 - 16$

2. $x^2 + 10x + 25$

3. $81y^2 - 4$

4. $3x^3 - 15x^2 + 18x$

5. $16h^4 - 81$

6. $x^2 + 2xh + h^2$

7. $x^2 + 3x - 4$

8. $x^2 - 3x + 4$

9. $2x^2 - 11x + 5$

10. $x^4 + x^2 - 20$

SECTION 1.1 EXERCISES

In Exercises 1–10, match the numerical model to the corresponding graphical model (*a–j*) and algebraic model (*k–t*).

1.

x	3	5	7	9	12	15
y	6	10	14	18	24	30

2.

x	0	1	2	3	4	5
y	2	3	6	11	18	27

3.

x	2	4	6	8	10	12
y	4	10	16	22	28	34

4.

x	5	10	15	20	25	30
y	90	80	70	60	50	40

5.

x	1	2	3	4	5	6
y	39	36	31	24	15	4

6.

x	1	2	3	4	5	6
y	5	7	9	11	13	15

7.

x	5	7	9	11	13	15
y	1	2	3	4	5	6

8.

x	4	8	12	14	18	24
y	20	72	156	210	342	600

9.

x	3	4	5	6	7	8
y	8	15	24	35	48	63

10.

x	4	7	12	19	28	39
y	1	2	3	4	5	6

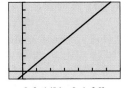

[–2, 14] by [–4, 36]
(a)

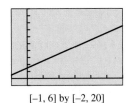

[–1, 6] by [–2, 20]
(b)

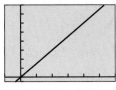

[–4, 40] by [–1, 7]
(c)

[–3, 18] by [–2, 32]
(d)

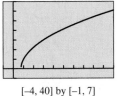

[–1, 7] by [–4, 40]
(e)

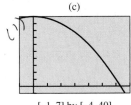

[–1, 7] by [–4, 40]
(f)

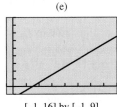

[–1, 16] by [–1, 9]
(g)

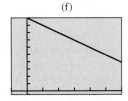

[–5, 30] by [–5, 100]
(h)

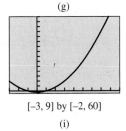

[–3, 9] by [–2, 60]
(i)

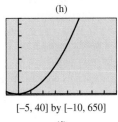

[–5, 40] by [–10, 650]
(j)

(k) $y = x^2 + x$

(l) $y = 40 - x^2$

(m) $y = (x + 1)(x - 1)$

(n) $y = \sqrt{x - 3}$

(o) $y = 100 - 2x$

(p) $y = 3x - 2$

(q) $y = 2x$

(r) $y = x^2 + 2$

(s) $y = 2x + 3$

(t) $y = \dfrac{x - 3}{2}$

Exercises 11–18 refer to the data in Table 1.6 below showing the percentage of the female and male populations in the United States employed in the civilian work force in selected years from 1954 to 2004.

Table 1.6 Employment Statistics

Year	Female	Male
1954	32.3	83.5
1959	35.1	82.3
1964	36.9	80.9
1969	41.1	81.1
1974	42.8	77.9
1979	47.7	76.5
1984	50.1	73.2
1989	54.9	74.5
1994	56.2	72.6
1999	58.5	74.0
2004	57.4	71.9

Source: www.bls.gov

11. (a) According to the numerical model, what has been the trend in females joining the work force since 1954?

(b) In what 5-year interval did the percentage of women who were employed change the most?

12. (a) According to the numerical model, what has been the trend in males joining the work force since 1954?

(b) In what 5-year interval did the percentage of men who were employed change the most?

13. Model the data graphically with two scatter plots on the same graph, one showing the percentage of women employed as a function of time and the other showing the same for men. Measure time in years since 1954.

14. Are the male percentages falling faster than the female percentages are rising, or vice versa?

15. Model the data algebraically with linear equations of the form $y = mx + b$. Write one equation for the women's data and another equation for the men's data. Use the 1954 and 1999 ordered pairs to compute the slopes.

16. If the percentages continue to follow the linear models you found in Exercise 15, what will the employment percentages for women and men be in the year 2009?

17. If the percentages continue to follow the linear models you found in Exercise 15, when will the percentages of women and men in the civilian work force be the same? What percentage will that be?

18. Writing to Learn Explain why the percentages cannot continue indefinitely to follow the linear models that you wrote in Exercise 15.

19. Doing Arithmetic with Lists Enter the data from the "Total" column of Table 1.2 of Example 2 into list L_1 in your calculator. Enter the data from the "Female" column into list L_2. Check a few computations to see that the procedures in (a) and (b) cause the cal-culator to divide each element of L_2 by the corresponding entry in L_1, multiply it by 100, and store the resulting list of percentages in L_3.

(a) On the home screen, enter the command: $100 \times L_2 / L_1 \rightarrow L_3$.

(b) Go to the top of list L_3 and enter $L_3 = 100(L_2/L_1)$.

20. Comparing Cakes A bakery sells a 9″ by 13″ cake for the same price as an 8″ diameter round cake. If the round cake is twice the height of the rectangular cake, which option gives the most cake for the money?

21. Stepping Stones A garden shop sells 12″ by 12″ square step-ping stones for the same price as 13″ round stones. If all of the stepping stones are the same thickness, which option gives the most rock for the money?

22. Free Fall of a Smoke Bomb At the Oshkosh, WI, air show, Jake Trouper drops a smoke bomb to signal the official beginning of the show. Ignoring air resistance, an object in free fall will fall d feet in t seconds, where d and t are related by the algebraic model $d = 16t^2$.

(a) How long will it take the bomb to fall 180 feet?

(b) If the smoke bomb is in free fall for 12.5 seconds after it is dropped, how high was the airplane when the smoke bomb was dropped?

23. Physics Equipment A physics student obtains the following data involving a ball rolling down an inclined plane, where t is the elapsed time in seconds and y is the distance traveled in inches.

t	0	1	2	3	4	5
y	0	1.2	4.8	10.8	19.2	30

Find an algebraic model that fits the data.

24. U.S. Air Travel The number of revenue passengers enplaned in the U.S. over the 14-year period from 1991 to 2004 is shown in Table 1.7.

Table 1.7 U.S. Air Travel

Year	Passengers (millions)	Year	Passengers (millions)
1991	452.3	1998	612.9
1992	475.1	1999	636.0
1993	488.5	2000	666.1
1994	528.8	2001	622.1
1995	547.8	2002	612.9
1996	581.2	2003	646.3
1997	594.7	2004	697.8

Source: www.airlines.org

(a) Graph a scatter plot of the data. Let x be the number of years since 1991.

(b) From 1991 to 2000 the data seem to follow a linear model. Use the 1991 and 2000 points to find an equation of the line and superimpose the line on the scatter plot.

(c) According to the linear model, in what year did the number of passengers seem destined to reach 900 million?

(d) What happened to disrupt the linear model?

Exercises 25–28 refer to the graph below, which shows the *minimum* salaries in major league baseball over a recent 18-year period and the *average* salaries in major league baseball over the same period. Salaries are measured in dollars and time is measured after the starting year (year 0).

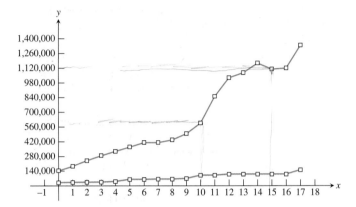

Source: Major League Baseball Players Association.

25. Which line is which, and how do you know?

26. After Peter Ueberroth's resignation as baseball commissioner in 1988 and his successor's untimely death in 1989, the team owners broke free of previous restrictions and began an era of competitive spending on player salaries. Identify where the 1990 salaries appear in the graph and explain how you can spot them.

27. The owners attempted to halt the uncontrolled spending by proposing a salary cap, which prompted a players' strike in 1994. The strike caused the 1995 season to be shortened and left many fans angry. Identify where the 1995 salaries appear in the graph and explain how you can spot them.

28. Writing to Learn Analyze the general patterns in the graphical model and give your thoughts about what the long-term implications might be for
   (a) the players;
   (b) the team owners;
   (c) the baseball fans.

In Exercises 29–38, solve the equation algebraically and graphically.

29. $v^2 - 5 = 8 - 2v^2$

30. $(x + 11)^2 = 121$

31. $2x^2 - 5x + 2 = (x - 3)(x - 2) + 3x$

32. $x^2 - 7x - \dfrac{3}{4} = 0$

33. $x(2x - 5) = 12$ **34.** $x(2x - 1) = 10$

35. $x(x + 7) = 14$

36. $x^2 - 3x + 4 = 2x^2 - 7x - 8$

37. $x + 1 - 2\sqrt{x + 4} = 0$ **38.** $\sqrt{x} + x = 1$

In Exercises 39–46, solve the equation graphically by converting it to an equivalent equation with 0 on the right-hand side and then finding the *x*-intercepts.

39. $2x - 5 = \sqrt{x + 4}$ **40.** $|3x - 2| = 2\sqrt{x + 8}$

41. $|2x - 5| = 4 - |x - 3|$ **42.** $\sqrt{x + 6} = 6 - 2\sqrt{5 - x}$

43. $2x - 3 = x^3 - 5$ **44.** $x + 1 = x^3 - 2x - 5$

45. $(x + 1)^{-1} = x^{-1} + x$ **46.** $x^2 = |x|$

47. Swan Auto Rental charges \$32 per day plus \$0.18 per mile for an automobile rental.
   (a) Elaine rented a car for one day and she drove 83 miles. How much did she pay?
   (b) Ramon paid \$69.80 to rent a car for one day. How far did he drive?

48. Connecting Graphs and Equations The curves on the graph below are the graphs of the three curves given by
$$y_1 = 4x + 5$$
$$y_2 = x^3 + 2x^2 - x + 3$$
$$y_3 = -x^3 - 2x^2 + 5x + 2.$$

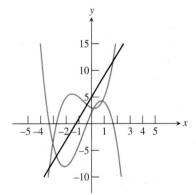

   (a) Write an equation that can be solved to find the points of intersection of the graphs of y_1 and y_2.
   (b) Write an equation that can be solved to find the *x*-intercepts of the graph of y_3.
   (c) Writing to Learn How does the graphical model reflect the fact that the answers to (a) and (b) are equivalent algebraically?
   (d) Confirm numerically that the *x*-intercepts of y_3 give the same values when substituted into the expressions for y_1 and y_2.

49. Exploring Grapher Failure Let $y = (x^{200})^{1/200}$.

(a) Explain algebraically why $y = x$ for all $x \geq 0$.

(b) Graph the equation $y = (x^{200})^{1/200}$ in the window [0, 1] by [0, 1].

(c) Is the graph different from the graph of $y = x$?

(d) Can you explain why the grapher failed?

50. Connecting Algebra and Geometry Explain how the algebraic equation $(x + b)^2 = x^2 + 2bx + b^2$ models the areas of the regions in the geometric figure shown below on the left:

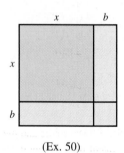 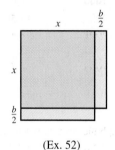

(Ex. 50)   (Ex. 52)

51. Exploring Hidden Behavior Solving graphically, find all real solutions to the following equations. Watch out for hidden behavior.

(a) $y = 10x^3 + 7.5x^2 - 54.85x + 37.95$

(b) $y = x^3 + x^2 - 4.99x + 3.03$

52. Connecting Algebra and Geometry The geometric figure shown on the right above is a large square with a small square missing.

(a) Find the area of the figure.

(b) What area must be added to complete the large square?

(c) Explain how the algebraic formula for completing the square models the completing of the square in (b).

53. Proving a Theorem Prove that if n is a positive integer, then $n^2 + 2n$ is either odd or a multiple of 4. Compare your proof with those of your classmates.

54. Writing to Learn The graph below shows the distance from home against time for a jogger. Using information from the graph, write a paragraph describing the jogger's workout.

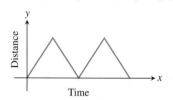

Standardized Test Questions

55. True or False A product of real numbers is zero if and only if every factor in the product is zero. Justify your answer.

56. True or False An algebraic model can always be used to make accurate predictions.

In Exercises 57–60, you may use a graphing calculator to decide which algebraic model corresponds to the given graphical or numerical model.

(A) $y = 2x + 3$ **(B)** $y = x^2 + 5$

(C) $y = 12 - 3x$ **(D)** $y = 4x + 3$

(E) $y = \sqrt{8 - x}$

57. Multiple Choice

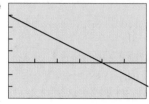

[0, 6] by [−9, 15]

58. Multiple Choice

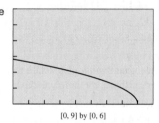

[0, 9] by [0, 6]

59. Multiple Choice

x	1	2	3	4	5	6
y	6	9	14	21	30	41

60. Multiple Choice

x	0	2	4	6	8	10
y	3	7	11	15	19	23

Explorations

61. Analyzing the Market Both Ahmad and LaToya watch the stock market throughout the year for stocks that make significant jumps from one month to another. When they spot one, each buys 100 shares. Ahmad's rule is to sell the stock if it fails to perform well for three months in a row. LaToya's rule is to sell in December if the stock has failed to perform well since its purchase.

The graph below shows the monthly performance in dollars (Jan–Dec) of a stock that both Ahmad and LaToya have been watching.

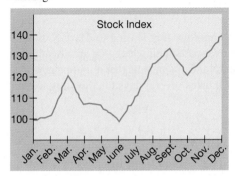

(a) Both Ahmad and LaToya bought the stock early in the year. In which month?

(b) At approximately what price did they buy the stock?

(c) When did Ahmad sell the stock?

(d) How much did Ahmad lose on the stock?

(e) Writing to Learn Explain why LaToya's strategy was better than Ahmad's for this particular stock in this particular year.

(f) Sketch a 12-month graph of a stock's performance that would favor Ahmad's strategy over LaToya's.

62. Group Activity Creating Hidden Behavior
You can create your own graphs with hidden behavior. Working in groups of two or three, try this exploration.

(a) Graph the equation $y = (x + 2)(x^2 - 4x + 4)$ in the window $[-4, 4]$ by $[-10, 10]$.

(b) Confirm algebraically that this function has zeros only at $x = -2$ and $x = 2$.

(c) Graph the equation $y = (x + 2)(x^2 - 4x + 4.01)$ in the window $[-4, 4]$ by $[-10, 10]$

(d) Confirm algebraically that this function has only one zero, at $x = -2$. (Use the discriminant.)

(e) Graph the equation $(x + 2)(x^2 - 4x + 3.99)$ in the window $[-4, 4]$ by $[-10, 10]$.

(f) Confirm algebraically that this function has three zeros. (Use the discriminant.)

Extending the Ideas

63. The Proliferation of Cell Phones Table 1.8 shows the number of cellular phone subscribers in the U.S. and their average monthly bill in the years from 1998 to 2004.

	Table 1.8 Cellular Phone Subscribers	
Year	Subscribers (millions)	Average Local Monthly Bill ($)
1998	69.2	39.43
1999	86.0	41.24
2000	109.5	45.27
2001	128.4	47.37
2002	140.8	48.40
2003	158.7	49.91
2004	180.4	50.64

Source: Cellular Telecommunication & Internet Association.

(a) Graph the scatter plots of the number of subscribers and the average local monthly bill as functions of time, letting time t = the number of years after 1990.

(b) One of the scatter plots clearly suggests a linear model in the form $y = mx + b$. Use the points at $t = 8$ and $t = 14$ to find a linear model.

(c) Superimpose the graph of the linear model onto the scatter plot. Does the fit appear to be good?

(d) Why does a linear model seem inappropriate for the other scatter plot? Can you think of a function that might fit the data better?

(e) In 1995 there were 33.8 million subscribers with an average local monthly bill of $51.00. Add these points to the scatter plots.

(f) Writing to Learn The 1995 points do not seem to fit the models used to represent the 1998–2004 data. Give a possible explanation for this.

64. Group Activity (Continuation of Exercise 63) Discuss the economic forces suggested by the two models in Exercise 63 and speculate about the future by analyzing the graphs.

1.2
Functions and Their Properties

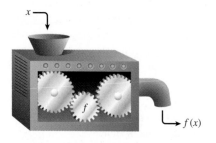

FIGURE 1.9 A "machine" diagram for a function.

A BIT OF HISTORY

The word function in its mathematical sense is generally attributed to Gottfried Leibniz (1646–1716), one of the pioneers in the methods of calculus. His attention to clarity of notation is one of his greatest contributions to scientific progress, which is why we still use his notation in calculus courses today. Ironically, it was not Leibniz but Leonhard Euler (1707–1783) who introduced the familiar notation $f(x)$.

In this section we will introduce the terminology that is used to describe functions throughout this book. Feel free to skim over parts with which you are already familiar, but take the time to become comfortable with concepts that might be new to you (like continuity and symmetry). Even if it takes several days to cover this section, it will be precalculus time well spent.

Function Definition and Notation

Mathematics and its applications abound with examples of formulas by which quantitative variables are related to each other. The language and notation of functions is ideal for that purpose. A function is actually a simple concept; if it were not, history would have replaced it with a simpler one by now. Here is the definition.

> **DEFINITION   Function, Domain, and Range**
>
> A function from a set D to a set R is a rule that assigns to every element in D a unique element in R. The set D of all input values is the **domain** of the function, and the set R of all output values is the **range** of the function.

There are many ways to look at functions. One of the most intuitively helpful is the "machine" concept (Figure 1.9), in which values of the domain (x) are fed into the machine (the function f) to produce range values (y). To indicate that y comes from the function acting on x, we use Euler's elegant **function notation** $y = f(x)$ (which we read as "**y equals f of x**" or "**the value of f at x**"). Here x is the **independent variable** and y is the **dependent variable**.

A function can also be viewed as a **mapping** of the elements of the domain onto the elements of the range. Figure 1.10a shows a function that maps elements from the domain X onto elements of the range Y. Figure 1.10b shows another such mapping, but *this one is not a function*, since the rule does not assign the element x_1 to a *unique* element of Y.

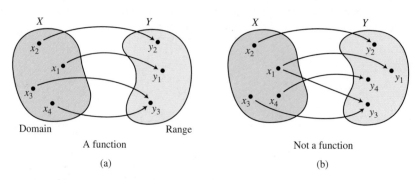

A function

(a)

Not a function

(b)

FIGURE 1.10 The diagram in (a) depicts a mapping from X to Y that is a function. The diagram in (b) depicts a mapping from X to Y that is not a function.

This uniqueness of the range value is very important to us as we study function behavior. Knowing that $f(2) = 8$ tells us something about f, and that understanding would be contradicted if we were to discover later that $f(2) = 4$. That is why you will never see a function defined by an ambiguous formula like $f(x) = 3x \pm 2$.

EXAMPLE 1 Defining a Function

Does the formula $y = x^2$ define y as a function of x?

SOLUTION Yes, y is a function of x. In fact, we can write the formula in function notation: $f(x) = x^2$. When a number x is substituted into the function, the square of x will be the output, and there is no ambiguity about what the square of x is.   *Now try Exercise 3.*

Another useful way to look at functions is graphically. The **graph of the function $y = f(x)$** is the set of all points $(x, f(x))$, x in the domain of f. We match domain values along the x-axis with their range values along the y-axis to get the ordered pairs that yield the graph of $y = f(x)$.

EXAMPLE 2 Seeing a Function Graphically

Of the three graphs shown in Figure 1.11, which is *not* the graph of a function? How can you tell?

SOLUTION The graph in (c) is not the graph of a function. There are three points on the graph with x-coordinate 0, so the graph does not assign a *unique* value to 0. (Indeed, we can see that there are plenty of numbers between -2 and 2 to which the graph assigns multiple values.) The other two graphs do not have a comparable problem because no vertical line intersects either of the other graphs in more than one point. Graphs that pass this *vertical line test* are the graphs of functions.

Now try Exercise 5.

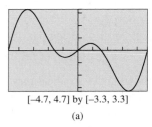
[-4.7, 4.7] by [-3.3, 3.3]
(a)

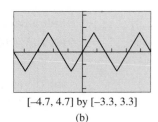
[-4.7, 4.7] by [-3.3, 3.3]
(b)

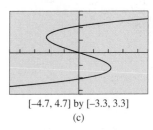
[-4.7, 4.7] by [-3.3, 3.3]
(c)

FIGURE 1.11 One of these is not the graph of a function. (Example 2)

Vertical Line Test

A graph (set of points (x, y)) in the xy-plane defines y as a function of x if and only if no vertical line intersects the graph in more than one point.

When moving from a numerical model to an algebraic model we will often use a function to approximate data pairs that by themselves violate our definition. In Figure 1.12 we can see that several pairs of data points fail the vertical line test, and yet the linear function approximates the data quite well.

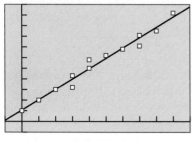

[–1, 10] by [–1, 11]

FIGURE 1.12 The data points fail the vertical line test but are nicely approximated by a linear function.

Domain and Range

We will usually define functions algebraically, giving the rule explicitly in terms of the domain variable. The rule, however, does not tell the complete story without some consideration of what the domain actually is.

For example, we can define the volume of a sphere as a function of its radius by the formula

$$V(r) = \frac{4}{3}\pi r^3 \text{ (Note that this is "V of r"—not "V \cdot r").}$$

This *formula* is defined for all real numbers, but the volume *function* is not defined for negative r values. So, if our intention were to study the volume function, we would restrict the domain to be all $r \geq 0$.

Agreement

Unless we are dealing with a model (like volume) that necessitates a restricted domain, we will assume that the domain of a function defined by an algebraic expression is the same as the domain of the algebraic expression, the **implied domain**. For models, we will use a domain that fits the situation, the **relevant domain**.

EXAMPLE 3   Finding the Domain of a Function

Find the domain of each of these functions:

(a) $f(x) = \sqrt{x + 3}$

(b) $g(x) = \dfrac{\sqrt{x}}{x - 5}$

(c) $A(s) = (\sqrt{3}/4)s^2$, where $A(s)$ is the area of an equilateral triangle with sides of length s.

SOLUTION

Solve Algebraically

(a) The expression under a radical may not be negative. We set $x + 3 \geq 0$ and solve to find $x \geq -3$. The domain of f is the interval $[-3, \infty)$.

(b) The expression under a radical may not be negative; therefore $x \geq 0$. Also, the denominator of a fraction may not be zero; therefore $x \neq 5$. The domain of g is the interval $[0, \infty)$ with the number 5 removed, which we can write as the *union* of two intervals: $[0, 5) \cup (5, \infty)$.

NOTE

The symbol "$\cup$" is read "union." It means that the elements of the two sets are combined to form one set.

(c) The algebraic expression has domain all real numbers, but the behavior being modeled restricts s from being negative. The domain of A is the interval $[0, \infty)$.

Support Graphically

We can support our answers in (a) and (b) graphically, as the calculator should not plot points where the function is undefined.

(a) Notice that the graph of $y = \sqrt{x + 3}$ (Figure 1.13a) shows points only for $x \geq -3$, as expected.

continued

(b) The graph of $y = \sqrt{x}/(x - 5)$ (Figure 1.13b) shows points only for $x \geq 0$, as expected, but shows an unexpected line through the *x*-axis at $x = 5$. This line, a form of grapher failure described in the previous section, should not be there. Ignoring it, we see that 5, as expected, is not in the domain.

(c) The graph of $y = (\sqrt{3}/4)s^2$ (Figure 1.13c) shows the unrestricted domain of the algebraic expression: all real numbers. The calculator has no way of knowing that *s* is the length of a side of a triangle. ***Now try Exercise 11.***

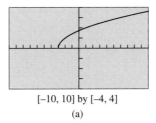
$[-10, 10]$ by $[-4, 4]$
(a)

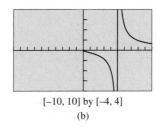
$[-10, 10]$ by $[-4, 4]$
(b)

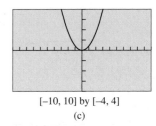
$[-10, 10]$ by $[-4, 4]$
(c)

FIGURE 1.13 Graphical support of the algebraic solutions in Example 3. The vertical line in (b) should be ignored because it results from grapher failure. The points in (c) with negative *x*-coordinates should be ignored because the calculator does not know that *x* is a length (but we do).

Finding the range of a function algebraically is often much harder than finding the domain, although graphically the things we look for are similar: To find the *domain* we look for all *x-coordinates* that correspond to points on the graph, and to find the *range* we look for all *y-coordinates* that correspond to points on the graph. A good approach is to use graphical and algebraic approaches simultaneously, as we show in Example 4.

EXAMPLE 4   Finding the Range of a Function

Find the range of the function $f(x) = \dfrac{2}{x}$.

SOLUTION

Solve Graphically The graph of $y = \dfrac{2}{x}$ is shown in Figure 1.14.

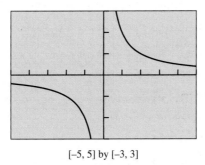
$[-5, 5]$ by $[-3, 3]$

FIGURE 1.14 The graph of $y = 2/x$. Is $y = 0$ in the range?

FUNCTION NOTATION

A grapher typically does not use function notation. So the function $f(x) = x^2 + 1$ is entered as $y_1 = x^2 + 1$. On some graphers you can evaluate *f* at $x = 3$ by entering $y_1(3)$ on the home screen. On the other hand, on other graphers $y_1(3)$ means $y_1 * 3$.

It appears that $x = 0$ is not in the domain (as expected, because a denominator cannot be zero). It also appears that the range consists of all real numbers except 0.

continued

Confirm Algebraically

We confirm that 0 is not in the range by trying to solve $2/x = 0$:

$$\frac{2}{x} = 0$$

$$2 = 0 \cdot x$$

$$2 = 0$$

Since the equation $2 = 0$ is never true, $2/x = 0$ has no solutions, and so $y = 0$ is not in the range. But how do we know that all other real numbers are in the range? We let k be any other real number and try to solve $2/x = k$:

$$\frac{2}{x} = k$$

$$2 = k \cdot x$$

$$x = \frac{2}{k}$$

As you can see, there was no problem finding an x this time, so 0 is the only number not in the range of f. We write the range $(-\infty, 0) \cup (0, \infty)$.

Now try Exercise 17.

You can see that this is considerably more involved than finding a domain, but we are hampered at this point by not having many tools with which to analyze function behavior. We will revisit the problem of finding ranges in Exercise 86, after having developed the tools that will simplify the analysis.

Continuity

One of the most important properties of the majority of functions that model real-world behavior is that they are *continuous*. Graphically speaking, a function is continuous at a point if the graph does not come apart at that point. We can illustrate the concept with a few graphs (Figure 1.15):

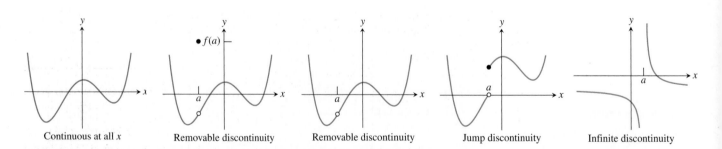

Continuous at all x Removable discontinuity   Removable discontinuity   Jump discontinuity   Infinite discontinuity

FIGURE 1.15 Some points of discontinuity.

Let's look at these cases individually.

This graph is continuous everywhere. Notice that the graph has no breaks. This means that if we are studying the behavior of the function f for x values close to any particular real number a, we can be assured that the $f(x)$ values will be close to $f(a)$.

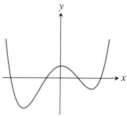

Continuous at all x

This graph is continuous everywhere except for the "hole" at $x = a$. If we are studying the behavior of this function f for x values close to a, we can *not* be assured that the $f(x)$ values will be close to $f(a)$. In this case, $f(x)$ is smaller than $f(a)$ for x near a. This is called a **removable discontinuity** because it can be patched by redefining $f(a)$ so as to plug the hole.

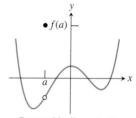

Removable discontinuity

This graph also has a **removable discontinuity** at $x = a$. If we are studying the behavior of this function f for x values close to a, we are still not assured that the $f(x)$ values will be close to $f(a)$, because in this case $f(a)$ doesn't even exist. It is removable because we could define $f(a)$ in such a way as to plug the hole and make f continuous at a.

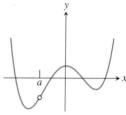

Removable discontinuity

Here is a discontinuity that is not removable. It is a **jump discontinuity** because there is more than just a hole at $x = a$; there is a *jump* in function values that makes the gap impossible to plug with a single point $(a, f(a))$, no matter how we try to redefine $f(a)$.

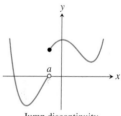

Jump discontinuity

This is a function with an **infinite discontinuity** at $x = a$. It is definitely not removable.

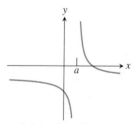

Infinite discontinuity

The simple geometric concept of an unbroken graph at a point is one of those visual notions that is extremely difficult to communicate accurately in the language of algebra. The key concept from the pictures seems to be that we want the point $(x, f(x))$ to slide smoothly onto the point $(a, f(a))$ without missing it, from either direction. This merely requires that $f(x)$ approach $f(a)$ as a *limit* as x approaches a. A function f is **continuous at $x = a$** if $\lim\limits_{x \to a} f(x) = f(a)$. A function f is **discontinuous at $x = a$** if it is not continuous at $x = a$.

CHOOSING VIEWING WINDOWS

Most viewing windows will show a vertical line for the function in Figure 1.16. It is sometimes possible to choose a viewing window in which the vertical line does not appear, as we did in Figure 1.16.

EXAMPLE 5 Identifying Points of Discontinuity

Judging from the graphs, which of the following figures shows functions that are discontinuous at $x = 2$? Are any of the discontinuities removable?

SOLUTION Figure 1.16 shows a function that is undefined at $x = 2$ and hence not continuous there. The discontinuity at $x = 2$ is not removable.

The function graphed in Figure 1.17 is a quadratic polynomial whose graph is a parabola, a graph that has no breaks because its domain includes all real numbers. It is continuous for all x.

The function graphed in Figure 1.18 is not defined at $x = 2$ and so can not be continuous there. The graph looks like the graph of the line $y = x + 2$, except that there is a hole where the point $(2, 4)$ should be. This is a removable discontinuity.

Now try Exercise 21.

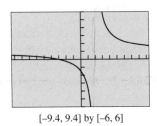

[−9.4, 9.4] by [−6, 6]

FIGURE 1.16 $f(x) = \dfrac{x + 3}{x - 2}$

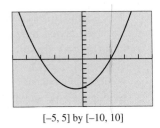

[−5, 5] by [−10, 10]

FIGURE 1.17 $g(x) = (x + 3)(x - 2)$

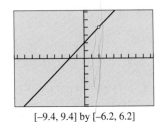

[−9.4, 9.4] by [−6.2, 6.2]

FIGURE 1.18 $h(x) = \dfrac{x^2 - 4}{x - 2}$

Increasing and Decreasing Functions

Another function concept that is easy to understand graphically is the property of being increasing, decreasing, or constant on an interval. We illustrate the concept with a few graphs (Figure 1.19):

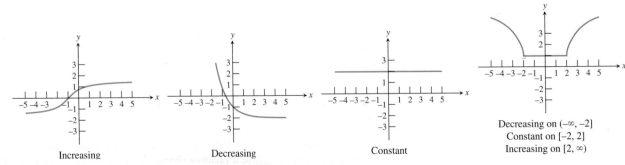

Increasing   Decreasing   Constant

Decreasing on $(-\infty, -2]$
Constant on $[-2, 2]$
Increasing on $[2, \infty)$

FIGURE 1.19 Examples of increasing, decreasing, or constant on an interval.

Once again the idea is easy to communicate graphically, but how can we identify these properties of functions algebraically? Exploration 1 will help to set the stage for the algebraic definition.

> **EXPLORATION 1 Increasing, Decreasing, and Constant Data**
>
> **1.** Of the three tables of numerical data below, which would be modeled by a function that is (a) increasing, (b) decreasing, (c) constant?
>
> | X | Y1 | | X | Y2 | | X | Y3 |
> |---|-----|---|---|-----|---|---|-----|
> | -2 | 12 | | -2 | 3 | | -2 | -5 |
> | -1 | 12 | | -1 | 1 | | -1 | -3 |
> | 0 | 12 | | 0 | 0 | | 0 | -1 |
> | 1 | 12 | | 1 | -2 | | 1 | 1 |
> | 3 | 12 | | 3 | -6 | | 3 | 4 |
> | 7 | 12 | | 7 | -12 | | 7 | 10 |
>
> **2.** Make a list of $\triangle Y1$, the *change* in Y1 values as you move down the list. As you move from $Y1 = a$ to $Y1 = b$, the change is $\triangle Y1 = b - a$. Do the same for the values of Y2 and Y3.
>
> | X moves from | $\triangle X$ | $\triangle Y1$ | X moves from | $\triangle X$ | $\triangle Y2$ | X moves from | $\triangle X$ | $\triangle Y3$ |
> |---|---|---|---|---|---|---|---|---|
> | -2 to -1 | 1 | | -2 to -1 | 1 | | -2 to -1 | 1 | |
> | -1 to 0 | 1 | | -1 to 0 | 1 | | -1 to 0 | 1 | |
> | 0 to 1 | 1 | | 0 to 1 | 1 | | 0 to 1 | 1 | |
> | 1 to 3 | 2 | | 1 to 3 | 2 | | 1 to 3 | 2 | |
> | 3 to 7 | 4 | | 3 to 7 | 4 | | 3 to 7 | 4 | |
>
> **3.** What is true about the quotients $\triangle Y/\triangle X$ for an increasing function? For a decreasing function? For a constant function?
>
> **4.** Where else have you seen the quotient $\triangle Y/\triangle X$? Does this reinforce your answers in part 3?

△LIST ON A CALCULATOR

Your calculator might be able to help you with the numbers in Exploration 1. Some calculators have a "△List" operation that will calculate the changes as you move down a list. For example, the command "△List (L1) → L3" will store the differences from L1 into L3. Note that △List (L1) is always one entry shorter than L1 itself.

Your analysis of the quotients $\triangle Y/\triangle X$ in the exploration should help you to understand the following definition.

> **DEFINITION Increasing, Decreasing, and Constant Function on an Interval**
>
> A function f is **increasing** on an interval if, for any two points in the interval, a positive change in x results in a positive change in $f(x)$.
>
> A function f is **decreasing** on an interval if, for any two points in the interval, a positive change in x results in a negative change in $f(x)$.
>
> A function f is **constant** on an interval if, for any two points in the interval, a positive change in x results in a zero change in $f(x)$.

EXAMPLE 6   Analyzing a Function for Increasing-Decreasing Behavior

For each function, tell the intervals on which it is increasing and the intervals on which it is decreasing.

(a) $f(x) = (x + 2)^2$ **(b)** $g(x) = \dfrac{x^2}{x^2 - 1}$

SOLUTION

Solve Graphically

(a) We see from the graph in Figure 1.20 that f is decreasing on $(-\infty, -2]$ and increasing on $[-2, \infty)$. (Notice that we include -2 in both intervals. Don't worry that this sets up some contradiction about what happens *at* -2, because we only talk about functions increasing or decreasing on *intervals*, and -2 is not an interval.)

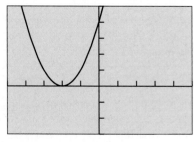

[–5, 5] by [–3, 5]

FIGURE 1.20 The function $f(x) = (x + 2)^2$ decreases on $(-\infty, -2]$ and increases on $[-2, \infty)$. (Example 6)

(b) We see from the graph in Figure 1.21 that g is increasing on $(-\infty, -1)$, increasing again on $(-1, 0]$, decreasing on $[0, 1)$, and decreasing again on $(1, \infty)$.

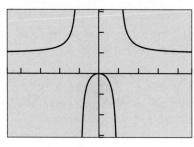

[–4.7, 4.7] by [–3.1, 3.1]

FIGURE 1.21 The function $g(x) = x^2/(x^2 - 1)$ increases on $(-\infty, -1)$ and $(-1, 0]$; the function decreases on $[0, 1)$ and $(1, \infty)$. (Example 6)

Now try Exercise 33.

You may have noticed that we are making some assumptions about the graphs. How do we know that they don't turn around somewhere off the screen? We will develop some ways to answer that question later in the book, but the most powerful methods will await you when you study calculus.

Boundedness

The concept of *boundedness* is fairly simple to understand both graphically and algebraically. We will move directly to the algebraic definition after motivating the concept with some typical graphs (Figure 1.22).

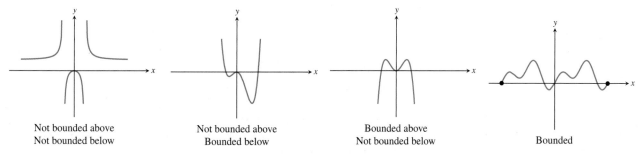

Not bounded above
Not bounded below

Not bounded above
Bounded below

Bounded above
Not bounded below

Bounded

FIGURE 1.22 Some examples of graphs bounded and not bounded above and below.

DEFINITION Lower Bound, Upper Bound, and Bounded

A function f is **bounded below** if there is some number b that is less than or equal to every number in the range of f. Any such number b is called a **lower bound** of f.

A function f is **bounded above** if there is some number B that is greater than or equal to every number in the range of f. Any such number B is called an **upper bound** of f.

A function f is **bounded** if it is bounded both above and below.

We can extend the above definition to the idea of **bounded on an interval** by restricting the domain of consideration in each part of the definition to the interval we wish to consider. For example, the function $f(x) = 1/x$ is bounded above on the interval $(-\infty, 0)$ and bounded below on the interval $(0, \infty)$.

EXAMPLE 7 Checking Boundedness

Identify each of these functions as bounded below, bounded above, or bounded.

(a) $w(x) = 3x^2 - 4$ **(b)** $p(x) = \dfrac{x}{1 + x^2}$

SOLUTION

Solve Graphically

The two graphs are shown in Figure 1.23. It appears that w is bounded below, and p is bounded.

Confirm Graphically

We can confirm that w is bounded below by finding a lower bound, as follows:

$$x^2 \geq 0 \qquad \text{\small x^2 is nonnegative}$$
$$3x^2 \geq 0 \qquad \text{\small Multiply by 3.}$$
$$3x^2 - 4 \geq 0 - 4 \qquad \text{\small Subtract 4.}$$
$$3x^2 - 4 \geq -4$$

Thus, -4 is a lower bound for $w(x) = 3x^2 - 4$.

We leave the verification that p is bounded as an exercise (Exercise 77).

Now try Exercise 37.

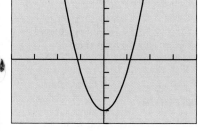

[–4, 4] by [–5, 5]
(a)

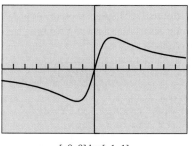

[–8, 8] by [–1, 1]
(b)

FIGURE 1.23 The graphs for Example 7. Which are bounded where?

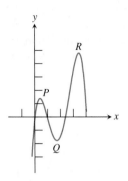

FIGURE 1.24 The graph suggests that f has a local maximum at P, a local minimum at Q, and a local maximum at R.

Local and Absolute Extrema

Many graphs are characterized by peaks and valleys where they change from increasing to decreasing and vice versa. The extreme values of the function (or *local extrema*) can be characterized as either *local maxima* or *local minima*. The distinction can be easily seen graphically. Figure 1.24 shows a graph with three local extrema: local maxima at points P and R and a local minimum at Q.

This is another function concept that is easier to see graphically than to describe algebraically. Notice that a local maximum does not have to be *the* maximum value of a function; it only needs to be the maximum value of the function on *some* tiny interval.

We have already mentioned that the best method for analyzing increasing and decreasing behavior involves calculus. Not surprisingly, the same is true for local extrema. We will generally be satisfied in this course with approximating local extrema using a graphing calculator, although sometimes an algebraic confirmation will be possible when we learn more about specific functions.

> **DEFINITION Local and Absolute Extrema**
>
> A **local maximum** of a function f is a value $f(c)$ that is greater than or equal to all range values of f on some open interval containing c. If $f(c)$ is greater than or equal to all range values of f, then $f(c)$ is the **maximum** (or **absolute maximum**) value of f.
>
> A **local minimum** of a function f is a value $f(c)$ that is less than or equal to all range values of f on some open interval containing c. If $f(c)$ is less than or equal to all range values of f, then $f(c)$ is the **minimum** (or **absolute minimum**) value of f.
>
> Local extrema are also called **relative extrema** .

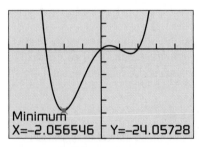

[–5, 5] by [–35, 15]

FIGURE 1.25 A graph of $y = x^4 - 7x^2 + 6x$. (Example 8)

USING A GRAPHER TO FIND LOCAL EXTREMA

Most modern graphers have built in "maximum" and "minimum" finders that identify local extrema by looking for sign changes in $\triangle y$. It is not easy to find local extrema by zooming in on them, as the graphs tend to flatten out and hide the very behavior you are looking at. If you use this method, keep narrowing the vertical window to maintain some curve in the graph.

EXAMPLE 8   Identifying Local Extrema

Decide whether $f(x) = x^4 - 7x^2 + 6x$ has any local maxima or local minima. If so, find each local maximum value or minimum value and the value of x at which each occurs.

SOLUTION The graph of $y = x^4 - 7x^2 + 6x$ (Figure 1.25) suggests that there are two local minimum values and one local maximum value. We use the graphing calculator to approximate local minima as -24.06 (which occurs at $x \approx -2.06$) and -1.77 (which occurs at $x \approx 1.60$). Similarly, we identify the (approximate) local maximum as 1.32 (which occurs at $x \approx 0.46$).   *Now try Exercise 41.*

Symmetry

In the graphical sense, the word "symmetry" in mathematics carries essentially the same meaning as it does in art: The picture (in this case, the graph) "looks the same" when viewed in more than one way. The interesting thing about mathematical symmetry is that it can be characterized numerically and algebraically as well. We will be looking at three particular types of symmetry, each of which can be spotted easily from a graph, a table of values, or an algebraic formula, once you know what to look for. Since it is the connections among the three models (graphical, numerical, and algebraic) that we need to emphasize in this section, we will illustrate the various symmetries in all three ways, side-by-side.

Symmetry with respect to the y-axis
Example: $f(x) = x^2$

Graphically

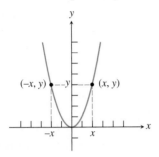

Numerically

x	$f(x)$
-3	9
-2	4
-1	1
1	1
2	4
3	9

Algebraically

For all x in the domain of f,

$$f(-x) = f(x)$$

Functions with this property (for example, x^n, n even) are **even** functions.

FIGURE 1.26 The graph looks the same to the left of the y-axis as it does to the right of it.

Symmetry with respect to the x-axis
Example: $x = y^2$

Graphically

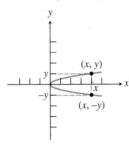

Numerically

x	y
9	-3
4	-2
1	-1
1	1
4	2
9	3

Algebraically

Graphs with this kind of symmetry are not functions (except the zero function), but we can say that $(x, -y)$ is on the graph whenever (x, y) is on the graph.

FIGURE 1.27 The graph looks the same above the x-axis as it does below it.

Symmetry with respect to the origin
Example: $f(x) = x^3$

| Graphically | Numerically | Algebraically |

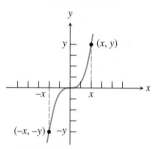

x	y
-3	-27
-2	-8
-1	-1
1	1
2	8
3	27

For all x in the domain of f,

$$f(-x) = -f(x).$$

Functions with this property (for example, x^n, n odd) are **odd** functions.

FIGURE 1.28 The graph looks the same upside-down as it does rightside-up.

EXAMPLE 9   Checking Functions for Symmetry

Tell whether each of the following functions is odd, even, or neither.

(a) $f(x) = x^2 - 3$ **(b)** $g(x) = x^2 - 2x - 2$ **(c)** $h(x) = \dfrac{x^3}{4 - x^2}$

SOLUTION

(a) Solve Graphically

The graphical solution is shown in Figure 1.29.

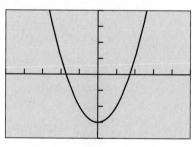

[−5, 5] by [−4, 4]

FIGURE 1.29 This graph appears to be symmetric with respect to the y-axis, so we conjecture that f is an even function.

Confirm Algebraically

We need to verify that

$$f(-x) = f(x)$$

for all x in the domain of f.

$$f(-x) = (-x)^2 - 3 = x^2 - 3$$
$$= f(x)$$

Since this identity is true for all x, the function f is indeed even.

continued

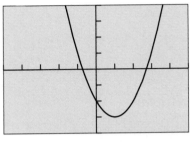

[−5, 5] by [−4, 4]

FIGURE 1.30 This graph does not appear to be symmetric with respect to either the y-axis or the origin, so we conjecture that g is neither even nor odd.

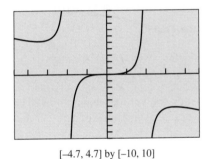

[−4.7, 4.7] by [−10, 10]

FIGURE 1.31 This graph appears to be symmetric with respect to the origin, so we conjecture that h is an odd function.

(b) Solve Graphically

The graphical solution is shown in Figure 1.30.

Confirm Algebraically

We need to verify that

$$g(-x) \neq g(x) \text{ and } g(-x) \neq -g(x).$$
$$g(-x) = (-x)^2 - 2(-x) - 2$$
$$= x^2 + 2x - 2$$

$$g(x) = x^2 - 2x - 2$$
$$-g(x) = -x^2 + 2x + 2$$

So $g(-x) \neq g(x)$ and $g(-x) \neq -g(x)$.

We conclude that g is neither odd nor even.

(c) Solve Graphically

The graphical solution is shown in Figure 1.31.

Confirm Algebraically

We need to verify that

$$h(-x) = -h(x)$$

for all x in the domain of h.

$$h(-x) = \frac{(-x)^3}{4 - (-x)^2} = \frac{-x^3}{4 - x^2}$$
$$= -h(x)$$

Since this identity is true for all x except ± 2 (which are not in the domain of h), the function h is odd. ***Now try Exercise 49.***

Asymptotes

Consider the graph of the function $f(x) = \dfrac{2x^2}{4 - x^2}$ in Figure 1.32.

The graph appears to flatten out to the right and to the left, getting closer and closer to the horizontal line $y = -2$. We call this line a *horizontal asymptote*. Similarly, the graph appears to flatten out as it goes off the top and bottom of the screen, getting closer and closer to the vertical lines $x = -2$ and $x = 2$. We call these lines *vertical asymptotes*. If we superimpose the asymptotes onto Figure 1.32 as dashed lines, you can see that they form a kind of template that describes the limiting behavior of the graph (Figure 1.33 on the next page).

Since asymptotes describe the behavior of the graph at its horizontal or vertical extremities, the definition of an asymptote can best be stated with limit notation. In this definition, note that $x \to a^-$ means "x approaches a from the left," while $x \to a^+$ means "x approaches a from the right."

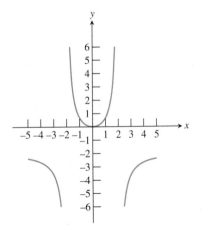

FIGURE 1.32 The graph of $f(x) = 2x^2/(4 - x^2)$ has two vertical asymptotes and one horizontal asymptote.

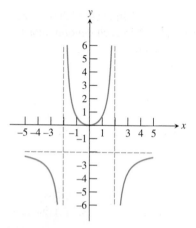

FIGURE 1.33 The graph of $f(x) = 2x^2/(4 - x^2)$ with the asymptotes shown as dashed lines.

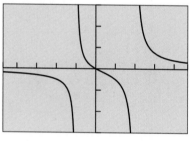

[−4.7, 4.7] by [−3, 3]

FIGURE 1.34 The graph of $y = x/(x^2 - x - 2)$ has vertical asymptotes of $x = -1$ and $x = 2$ and a horizontal asymptote of $y = 0$. (Example 10)

DEFINITION Horizontal and Vertical Asymptotes

The line $y = b$ is a **horizontal asymptote** of the graph of a function $y = f(x)$ if $f(x)$ approaches a limit of b as x approaches $+\infty$ or $-\infty$.

In limit notation:

$$\lim_{x \to -\infty} f(x) = b \quad \text{or} \quad \lim_{x \to +\infty} f(x) = b.$$

The line $x = a$ is a **vertical asymptote** of the graph of a function $y = f(x)$ if $f(x)$ approaches a limit of $+\infty$ or $-\infty$ as x approaches a from either direction.

In limit notation:

$$\lim_{x \to a^-} f(x) = \pm\infty \quad \text{or} \quad \lim_{x \to a^+} f(x) = \pm\infty.$$

EXAMPLE 10   Identifying the Asymptotes of a Graph

Identify any horizontal or vertical asymptotes of the graph of

$$y = \frac{x}{x^2 - x - 2}.$$

SOLUTION The quotient $x/(x^2 - x - 2) = x/((x + 1)(x - 2))$ is undefined at $x = -1$ and $x = 2$, which makes them likely sites for vertical asymptotes. The graph (Figure 1.34) provides support, showing vertical asymptotes of $x = -1$ and $x = 2$.

For large values of x, the numerator (a large number) is dwarfed by the denominator (a *product* of *two* large numbers), suggesting that $\lim_{x \to \infty} x/((x + 1)(x - 2)) = 0$. This would indicate a horizontal asymptote of $y = 0$. The graph (Figure 1.34) provides support, showing a horizontal asymptote of $y = 0$ as $x \to \infty$. Similar logic suggests that $\lim_{x \to -\infty} x/((x + 1)(x - 2)) = -0 = 0$, indicating the same horizontal asymptote as $x \to -\infty$. Again, the graph provides support for this.   ***Now try Exercise 57.***

End Behavior

A horizontal asymptote gives one kind of end behavior for a function because it shows how the function behaves as it goes off toward either "end" of the *x*-axis. Not all graphs approach lines, but it is helpful to consider what *does* happen "out there." We illustrate with a few examples.

EXAMPLE 11   Matching Functions Using End Behavior

Match the functions with the graphs in Figure 1.35 by considering end behavior. All graphs are shown in the same viewing window.

(a) $y = \dfrac{3x}{x^2 + 1}$ **(b)** $y = \dfrac{3x^2}{x^2 + 1}$ **(c)** $y = \dfrac{3x^3}{x^2 + 1}$ **(d)** $y = \dfrac{3x^4}{x^2 + 1}$

continued

Zooming-out is often a good way to investigate end behavior with a graphing calculator. Here are some useful zooming tips:

- Start with a "square" window.
- Set Xscl and Yscl to zero to avoid fuzzy axes.
- Be sure the zoom factors are both the same. (They will be unless you change them.)

SOLUTION When x is very large, the denominator $x^2 + 1$ in each of these functions is almost the same number as x^2. If we replace $x^2 + 1$ in each denominator by x^2 and then reduce the fractions, we get the simpler functions

(a) $y = \dfrac{3}{x}$ (close to $y = 0$ for large x)   **(b)** $y = 3$

(c) $y = 3x$ **(d)** $y = 3x^2$.

So, we look for functions that have end behavior resembling, respectively, the functions

(a) $y = 0$ **(b)** $y = 3$ **(c)** $y = 3x$ **(d)** $y = 3x^2$.

Graph (iv) approaches the line $y = 0$. Graph (iii) approaches the line $y = 3$. Graph (ii) approaches the line $y = 3x$. Graph (i) approaches the parabola $y = 3x^2$. So, (a) matches (iv), (b) matches (iii), (c) matches (ii), and (d) matches (i).

Now try Exercise 65.

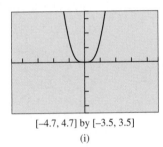
[–4.7, 4.7] by [–3.5, 3.5]
(i)

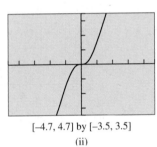
[–4.7, 4.7] by [–3.5, 3.5]
(ii)

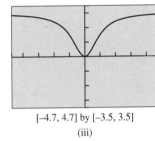
[–4.7, 4.7] by [–3.5, 3.5]
(iii)

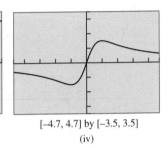
[–4.7, 4.7] by [–3.5, 3.5]
(iv)

FIGURE 1.35 Match the graphs with the functions in Example 11.

For more complicated functions we are often content with knowing whether the end behavior is bounded or unbounded in either direction.

QUICK REVIEW 1.2   *(For help, go to Sections A.3, P.3, and P.5.)*

In Exercises 1–4, solve the equation or inequality.

1. $x^2 - 16 = 0$ **2.** $9 - x^2 = 0$

3. $x - 10 < 0$ **4.** $5 - x \le 0$

In Exercises 5–10, find all values of x algebraically for which the algebraic expression is *not* defined. Support your answer graphically.

5. $\dfrac{x}{x - 16}$ **6.** $\dfrac{x}{x^2 - 16}$

7. $\sqrt{x - 16}$

8. $\dfrac{\sqrt{x^2 + 1}}{x^2 - 1}$

9. $\dfrac{\sqrt{x + 2}}{\sqrt{3 - x}}$

10. $\dfrac{x^2 - 2x}{x^2 - 4}$

SECTION 1.2 EXERCISES

In Exercises 1–4, determine whether the formula determines y as a function of x. If not, explain why not.

1. $y = \sqrt{x - 4}$ **2.** $y = x^2 \pm 3$

3. $x = 2y^2$ **4.** $x = 12 - y$

In Exercises 5–8, use the vertical line test to determine whether the curve is the graph of a function.

5.

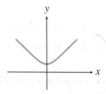

6.

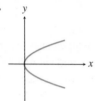

7.

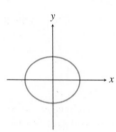

8.

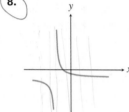

In Exercises 9–16, find the domain of the function algebraically and support your answer graphically.

9. $f(x) = x^2 + 4$ **10.** $h(x) = \dfrac{5}{x - 3}$

11. $f(x) = \dfrac{3x - 1}{(x + 3)(x - 1)}$ **12.** $f(x) = \dfrac{1}{x} + \dfrac{5}{x - 3}$

13. $g(x) = \dfrac{x}{x^2 - 5x}$ **14.** $h(x) = \dfrac{\sqrt{4 - x^2}}{x - 3}$

15. $h(x) = \dfrac{\sqrt{4 - x}}{(x + 1)(x^2 + 1)}$ **16.** $f(x) = \sqrt{x^4 - 16x^2}$

In Exercises 17–20, find the range of the function.

17. $f(x) = 10 - x^2$ **18.** $g(x) = 5 + \sqrt{4 - x}$

19. $f(x) = \dfrac{x^2}{1 - x^2}$ **20.** $g(x) = \dfrac{3 + x^2}{4 - x^2}$

In Exercises 21–24, graph the function and tell whether or not it has a point of discontinuity at $x = 0$. If there is a discontinuity, tell whether it is removable or nonremovable.

21. $g(x) = \dfrac{3}{x}$ **22.** $h(x) = \dfrac{x^3 + x}{x}$

23. $f(x) = \dfrac{|x|}{x}$ **24.** $g(x) = \dfrac{x}{x - 2}$

In Exercises 25–28, state whether each labeled point identifies a local minimum, a local maximum, or neither. Identify intervals on which the function is decreasing and increasing.

25.

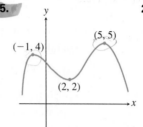

26.

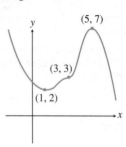

27.

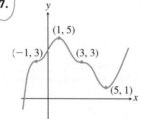

28.

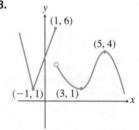

In Exercises 29–34, graph the function and identify intervals on which the function is increasing, decreasing, or constant.

29 $f(x) = |x + 2| - 1$

30. $f(x) = |x + 1| + |x - 1| - 3$

31. $g(x) = |x + 2| + |x - 1| - 2$

32. $h(x) = 0.5(x + 2)^2 - 1$

33. $g(x) = 3 - (x - 1)^2$

34. $f(x) = x^3 - x^2 - 2x$

In Exercises 35–40, determine whether the function is bounded above, bounded below, or bounded on its domain.

35. $y = 32$ **36.** $y = 2 - x^2$

37. $y = 2^x$ **38.** $y = 2^{-x}$

39. $y = \sqrt{1 - x^2}$ **40.** $y = x - x^3$

In Exercises 41–46, use a grapher to find all local maxima and minima and the values of x where they occur. Give values rounded to two decimal places.

41. $f(x) = 4 - x + x^2$ **42.** $g(x) = x^3 - 4x + 1$

43. $h(x) = -x^3 + 2x - 3$ **44.** $f(x) = (x + 3)(x - 1)^2$

45. $h(x) = x^2\sqrt{x + 4}$ **46.** $g(x) = x|2x + 5|$

In Exercises 47–54, state whether the function is odd, even, or neither. Support graphically and confirm algebraically.

47. $f(x) = 2x^4$

48. $g(x) = x^3$

49. $f(x) = \sqrt{x^2 + 2}$

50. $g(x) = \dfrac{3}{1 + x^2}$

51. $f(x) = -x^2 + 0.03x + 5$

52. $f(x) = x^3 + 0.04x^2 + 3$

53. $g(x) = 2x^3 - 3x$

54. $h(x) = \dfrac{1}{x}$

In Exercises 55–62, use a method of your choice to find all horizontal and vertical asymptotes of the function.

55. $f(x) = \dfrac{x}{x - 1}$

56. $q(x) = \dfrac{x - 1}{x}$

57. $g(x) = \dfrac{x + 2}{3 - x}$

58. $q(x) = 1.5^x$

59. $f(x) = \dfrac{x^2 + 2}{x^2 - 1}$

60. $p(x) = \dfrac{4}{x^2 + 1}$

61. $g(x) = \dfrac{4x - 4}{x^3 - 8}$

62. $h(x) = \dfrac{2x - 4}{x^2 - 4}$

In Exercises 63–66, match the function with the corresponding graph by considering end behavior and asymptotes. All graphs are shown in the same viewing window.

63. $y = \dfrac{x + 2}{2x + 1}$

64. $y = \dfrac{x^2 + 2}{2x + 1}$

65. $y = \dfrac{x + 2}{2x^2 + 1}$

66. $y = \dfrac{x^3 + 2}{2x^2 + 1}$

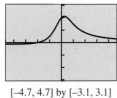

[−4.7, 4.7] by [−3.1, 3.1]
(a)

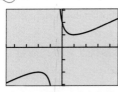

[−4.7, 4.7] by [−3.1, 3.1]
(c)

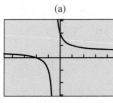

[−4.7, 4.7] by [−3.1, 3.1]
(b)

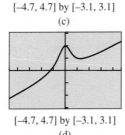

[−4.7, 4.7] by [−3.1, 3.1]
(d)

67. Can a graph cross its own asymptote? The Greek roots of the word "asymptote" mean "not meeting," since graphs tend to approach, but not meet, their asymptotes. Which of the following functions have graphs that *do* intersect their horizontal asymptotes?

(a) $f(x) = \dfrac{x}{x^2 - 1}$

(b) $g(x) = \dfrac{x}{x^2 + 1}$

(c) $h(x) = \dfrac{x^2}{x^3 + 1}$

68. Can a graph have two horizontal asymptotes? Although most graphs have at most one horizontal asymptote, it is possible for a graph to have more than one. Which of the following functions have graphs with more than one horizontal asymptote?

(a) $f(x) = \dfrac{|x^3 + 1|}{8 - x^3}$

(b) $g(x) = \dfrac{|x - 1|}{x^2 - 4}$

(c) $h(x) = \dfrac{x}{\sqrt{x^2 - 4}}$

69. Can a graph intersect its own vertical asymptote?

Graph the function $f(x) = \dfrac{x - |x|}{x^2} + 1$.

(a) The graph of this function does not intersect its vertical asymptote. Explain why it does not.

(b) Show how you can add a single point to the graph of f and get a graph that *does* intersect its vertical asymptote.

(c) Is the graph in (b) the graph of a function?

70. Writing to Learn Explain why a graph cannot have more than two horizontal asymptotes.

Standardized Test Questions

71. True or False The graph of function f is defined as the set of all points $(x, f(x))$ where x is in the domain of f. Justify your answer.

72. True or False A relation that is symmetric with respect to the x-axis cannot be a function. Justify your answer.

In Exercises 73–76, answer the question without using a calculator.

73. Multiple Choice Which function is continuous?

(A) Number of children enrolled in a particular school as a function of time

(B) Outdoor temperature as a function of time

(C) Cost of U.S. postage as a function of the weight of the letter

(D) Price of a stock as a function of time

(E) Number of soft drinks sold at a ballpark as a function of outdoor temperature

74. Multiple Choice Which function is *not* continuous?

(A) Your altitude as a function of time while flying from Reno to Dallas

(B) Time of travel from Miami to Pensacola as a function of driving speed

(C) Number of balls that can fit completely inside a particular box as a function of the radius of the balls

(D) Area of a circle as a function of radius

(E) Weight of a particular baby as a function of time after birth

75. Decreasing Function Which function is decreasing?

(A) Outdoor temperature as a function of time

(B) The Dow Jones Industrial Average as a function of time

(C) Air pressure in the Earth's atmosphere as a function of altitude

(D) World population since 1900 as a function of time

(E) Water pressure in the ocean as a function of depth

76. Increasing or Decreasing Which function cannot be classified as either increasing or decreasing?

(A) Weight of a lead brick as a function of volume

(B) Height of a ball that has been tossed upward as a function of time

(C) Time of travel from Buffalo to Syracuse as a function of driving speed

(D) Area of a square as a function of side length

(E) Height of a swinging pendulum as a function of time

Explorations

77. Bounded Functions As promised in Example 7 of this section, we will give you a chance to prove algebraically that $p(x) = x/(1 + x^2)$ is bounded.

(a) Graph the function and find the smallest integer k that appears to be an upper bound.

(b) Verify that $x/(1 + x^2) < k$ by proving the equivalent inequality $kx^2 - x + k > 0$. (Use the quadratic formula to show that the quadratic has no real zeros.)

(c) From the graph, find the greatest integer k that appears to be a lower bound.

(d) Verify that $x/(1 + x^2) > k$ by proving the equivalent inequality $kx^2 - x + k < 0$.

78. Baylor School Grade Point Averages Baylor School uses a sliding scale to convert the percentage grades on its transcripts to grade point averages (GPAs). Table 1.9 shows the GPA equivalents for selected grades:

Table 1.9 Converting Grades	
Grade (x)	GPA (y)
60	0.00
65	1.00
70	2.05
75	2.57
80	3.00
85	3.36
90	3.69
95	4.00
100	4.28

Source: Baylor School College Counselor.

(a) Considering GPA (y) as a function of percentage grade (x), is it increasing, decreasing, constant, or none of these?

(b) Make a table showing the *change* ($\triangle y$) in GPA as you move down the list. (See Exploration 1.)

(c) Make a table showing the change in $\triangle y$ as you move down the list. (This is $\triangle\triangle y$.) Considering the *change* ($\triangle y$) in GPA as a function of percentage grade (x), is it increasing, decreasing, constant, or none of these?

(d) In general, what can you say about the shape of the graph if y is an increasing function of x and $\triangle y$ is a decreasing function of x?

(e) Sketch the graph of a function y of x such that y is a decreasing function of x and $\triangle y$ is an increasing function of x.

79. Group Activity Sketch (freehand) a graph of a function f with domain all real numbers that satisfies all of the following conditions:

(a) f is continuous for all x;

(b) f is increasing on $(-\infty, 0]$ and on $[3, 5]$;

(c) f is decreasing on $[0, 3]$ and on $[5, \infty)$;

(d) $f(0) = f(5) = 2$;

(e) $f(3) = 0$.

80. Group Activity Sketch (freehand) a graph of a function f with domain all real numbers that satisfies all of the following conditions:

(a) f is decreasing on $(-\infty, 0)$ and decreasing on $(0, \infty)$;

(b) f has a nonremovable point of discontinuity at $x = 0$;

(c) f has a horizontal asymptote at $y = 1$;

(d) $f(0) = 0$;

(e) f has a vertical asymptote at $x = 0$.

81. Group Activity Sketch (freehand) a graph of a function f with domain all real numbers that satisfies all of the following conditions:

(a) f is continuous for all x;

(b) f is an even function;

(c) f is increasing on $[0, 2]$ and decreasing on $[2, \infty)$;

(d) $f(2) = 3$.

82. Group Activity Get together with your classmates in groups of two or three. Sketch a graph of a function, but do not show it to the other members of your group. Using the language of functions (as in Exercises 79–81), describe your function as completely as you can. Exchange descriptions with the others in your group and see if you can reproduce each other's graphs.

Extending the Ideas

83. A function that is bounded above has an infinite number of upper bounds, but there is always a *least upper bound*, i.e., an upper bound that is less than all the others. This least upper bound may or may not be in the range of f. For each of the following functions, find the least upper bound and tell whether or not it is in the range of the function.

(a) $f(x) = 2 - 0.8x^2$

(b) $g(x) = \dfrac{3x^2}{3 + x^2}$

(c) $h(x) = \dfrac{1 - x}{x^2}$

(d) $p(x) = 2 \sin (x)$

(e) $q(x) = \dfrac{4x}{x^2 + 2x + 1}$

84. Writing to Learn A continuous function f has domain all real numbers. If $f(-1) = 5$ and $f(1) = -5$, explain why f must have at least one zero in the interval $[-1, 1]$. (This generalizes to a property of continuous functions known as the Intermediate Value Theorem.)

85. Proving a Theorem Prove that the graph of every odd function with domain all real numbers must pass through the origin.

86. Finding the Range Graph the function $f(x) = \dfrac{3x^2 - 1}{2x^2 + 1}$ in the window $[-6, 6]$ by $[-2, 2]$.

(a) What is the apparent horizontal asymptote of the graph?

(b) Based on your graph, determine the apparent range of f.

(c) Show algebraically that $-1 \le \dfrac{3x^2 - 1}{2x^2 + 1} < 1.5$ for all x, thus confirming your conjecture in part (b).

1.3
Twelve Basic Functions

What you'll learn about

- What Graphs Can Tell Us
- Twelve Basic Functions
- Analyzing Functions Graphically

. . . and why

As you continue to study mathematics, you will find that the twelve basic functions presented here will come up again and again. By knowing their basic properties, you will recognize them when you see them.

What Graphs Can Tell Us

The preceding section has given us a vocabulary for talking about functions and their properties. We have an entire book ahead of us to study these functions in depth, but in this section we want to set the scene by just *looking* at the graphs of twelve "basic" functions that are available on your graphing calculator.

You will find that function attributes such as domain, range, continuity, asymptotes, extrema, increasingness, decreasingness, and end behavior are every bit as graphical as they are algebraic. Moreover, the visual cues are often much easier to spot than the algebraic ones.

In future chapters you will learn more about the algebraic properties that make these functions behave as they do. Only then will you able to *prove* what is visually apparent in these graphs.

Twelve Basic Functions

The Identity Function

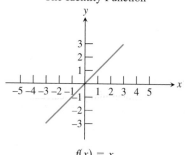

$$f(x) = x$$

Interesting fact: This is the only function that acts on every real number by leaving it alone.

FIGURE 1.36

The Squaring Function

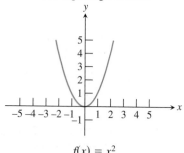

$$f(x) = x^2$$

Interesting fact: The graph of this function, called a parabola, has a reflection property that is useful in making flashlights and satellite dishes.

FIGURE 1.37

The Cubing Function

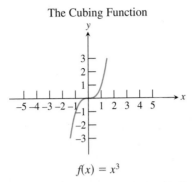

$$f(x) = x^3$$

Interesting fact: The origin is called a "point of inflection" for this curve because the graph changes curvature at that point.

FIGURE 1.38

The Reciprocal Function

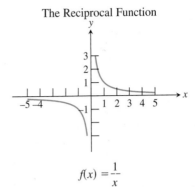

$$f(x) = \frac{1}{x}$$

Interesting fact: This curve, called a hyperbola, also has a reflection property that is useful in satellite dishes.

FIGURE 1.39

The Square Root Function

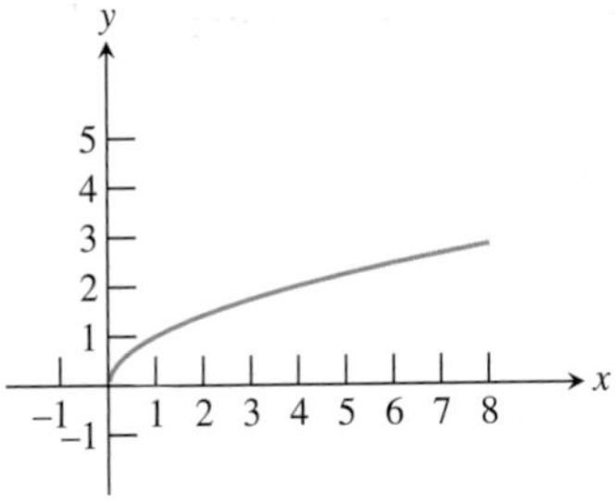

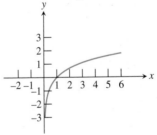

$$f(x) = \sqrt{x}$$

Interesting fact: Put any positive number into your calculator. Take the square root. Then take the square root again. Then take the square root again, and so on. Eventually you will always get 1.

FIGURE 1.40

The Exponential Function

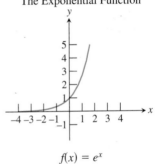

$$f(x) = e^x$$

Interesting fact: The number e is an irrational number (like π) that shows up in a variety of applications. The symbols e and π were both brought into popular use by the great Swiss mathematician Leonhard Euler (1707–1783).

FIGURE 1.41

The Natural Logarithm Function

$$f(x) = \ln x$$

Interesting fact: This function increases very slowly. If the x-axis and y-axis were both scaled with unit lengths of one inch, you would have to travel more than two and a half miles along the curve just to get a foot above the x-axis.

FIGURE 1.42

The Sine Function

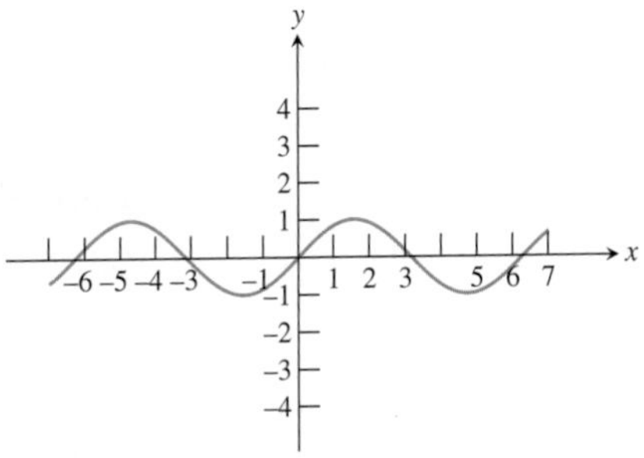

$$f(x) = \sin x$$

Interesting fact: This function and the sinus cavities in your head derive their names from a common root: the Latin word for "bay." This is due to a 12th-century mistake made by Robert of Chester, who translated a word incorrectly from an Arabic manuscript.

FIGURE 1.43

The Cosine Function

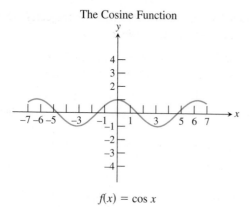

$$f(x) = \cos x$$

Interesting fact: The local extrema of the cosine function occur exactly at the zeros of the sine function, and vice versa.

FIGURE 1.44

The Absolute Value Function

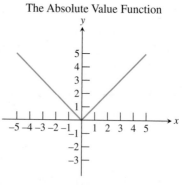

$$f(x) = |x| = \text{abs}\,(x)$$

Interesting fact: This function has an abrupt change of direction (a "corner") at the origin, while our other functions are all "smooth" on their domains.

FIGURE 1.45

The Greatest Integer Function

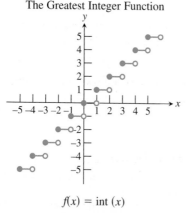

$$f(x) = \text{int}\,(x)$$

Interesting fact: This function has a jump discontinuity at every integer value of x. Similar-looking functions are called *step functions*.

FIGURE 1.46

The Logistic Function

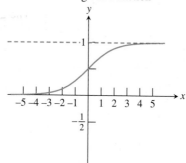

$$f(x) = \frac{1}{1 + e^{-x}}$$

Interesting fact: There are two horizontal asymptotes, the x-axis and the line $y = 1$. This function provides a model for many applications in biology and business.

FIGURE 1.47

EXAMPLE 1   Looking for Domains

(a) Nine of the functions have domain the set of all real numbers. Which three do not?

(b) One of the functions has domain the set of all reals except 0. Which function is it, and why isn't zero in its domain?

(c) Which two functions have no negative numbers in their domains? Of these two, which one is defined at zero?

SOLUTION

(a) Imagine dragging a vertical line along the x-axis. If the function has domain the set of all real numbers, then the line will always intersect the graph. The intersection might occur off screen, but the TRACE function on the calculator will show the y-coordinate if there is one. Looking at the graphs in Figures 1.39, 1.40, and 1.42, we conjecture that there are vertical lines that do not intersect the curve.

continued

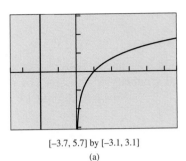

[−3.7, 5.7] by [−3.1, 3.1]

(a)

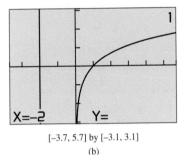

[−3.7, 5.7] by [−3.1, 3.1]

(b)

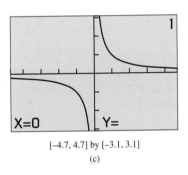

[−4.7, 4.7] by [−3.1, 3.1]

(c)

FIGURE 1.48 (a) A vertical line through −2 on the *x*-axis appears to miss the graph of $y = \ln x$. (b) A TRACE confirms that −2 is not in the domain. (c) A TRACE at $x = 0$ confirms that 0 is not in the domain of $y = 1/x$. (Example 1)

A TRACE at the suspected *x*-coordinates confirms our conjecture (Figure 1.48). The functions are $y = 1/x$, $y = \sqrt{x}$ and $y = \ln x$.

(b) The function $y = 1/x$, with a vertical asymptote at $x = 0$, is defined for all real numbers except 0. This is explained algebraically by the fact that division by zero is undefined.

(c) The functions $y = \sqrt{x}$ and $y = \ln x$ have no negative numbers in their domains. (We already knew that about the square root function.) While 0 is in the domain of $y = \sqrt{x}$, we can see by tracing that it is not in the domain of $y = \ln x$. We will see the algebraic reason for this in Chapter 3.

Now try Exercise 13.

EXAMPLE 2 Looking for Continuity

Only two of twelve functions have points of discontinuity. Are these points in the domain of the function?

SOLUTION All of the functions have continuous, unbroken graphs except for $y = 1/x$, and $y = \text{int}(x)$.

The graph of $y = 1/x$ clearly has an infinite discontinuity at $x = 0$ (Figure 1.39). We saw in Example 1 that 0 is not in the domain of the function. Since $y = 1/x$ is continuous for every point *in its domain*, it is called a **continuous function**.

The graph of $y = \text{int}(x)$ has a discontinuity at every integer value of *x* (Figure 1.46). Since this function has discontinuities at points in its domain, it is *not* a continuous function.

Now try Exercise 15.

EXAMPLE 3 Looking for Boundedness

Only three of the twelve basic functions are bounded (above and below). Which three?

SOLUTION A function that is bounded must have a graph that lies entirely between two horizontal lines. The sine, cosine, and logistic functions have this property (Figure 1.49). It looks like the graph of $y = \sqrt{x}$ might also have this property, but we know that the end behavior of the square root function is unbounded: $\lim_{x \to \infty} \sqrt{x} = \infty$, so it is really only bounded below. You will learn in Chapter 4 why the sine and cosine functions are bounded.

Now try Exercise 17.

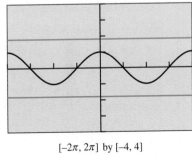

[−2π, 2π] by [−4, 4]

(a)

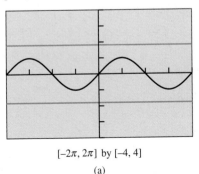

[−2π, 2π] by [−4, 4]

(b)

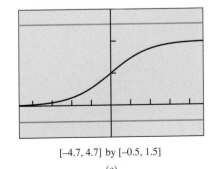

[−4.7, 4.7] by [−0.5, 1.5]

(c)

FIGURE 1.49 The graphs of $y = \sin x$, $y = \cos x$, and $y = 1/(1 + e^{-x})$ lie entirely between two horizontal lines and are therefore graphs of bounded functions. (Example 3)

EXAMPLE 4   Looking for Symmetry

Three of the twelve basic functions are even. Which are they?

SOLUTION Recall that the graph of an even function is symmetric with respect to the y-axis. Three of the functions exhibit the required symmetry: $y = x^2$, $y = \cos x$, and $y = |x|$ (Figure 1.50).   ***Now try Exercise 19.***

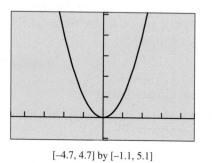
[−4.7, 4.7] by [−1.1, 5.1]
(a)

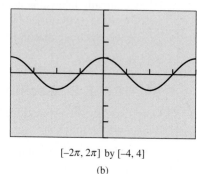
[−2π, 2π] by [−4, 4]
(b)

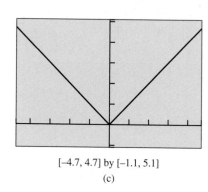
[−4.7, 4.7] by [−1.1, 5.1]
(c)

FIGURE 1.50 The graphs of $y = x^2$, $y = \cos x$, and $y = |x|$ are symmetric with respect to the y-axis, indicating that the functions are even. (Example 4)

 ## Analyzing Functions Graphically

We could continue to explore the twelve basic functions as in the first four examples, but we also want to make the point that there is no need to restrict ourselves to the basic twelve. We can alter the basic functions slightly and see what happens to their graphs, thereby gaining further visual insights into how functions behave.

EXAMPLE 5   Analyzing a Function Graphically

Graph the function $y = (x - 2)^2$. Then answer the following questions:

(a) On what interval is the function increasing? On what interval is it decreasing?

(b) Is the function odd, even, or neither?

(c) Does the function have any extrema?

(d) How does the graph relate to the graph of the basic function $y = x^2$?

SOLUTION The graph is shown in Figure 1.51.

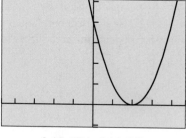
[−4.7, 4.7] by [−1.1, 5.1]

FIGURE 1.51 The graph of $y = (x - 2)^2$. (Example 5)

continued

(a) The function is increasing if its graph is headed upward as it moves from left to right. We see that it is increasing on the interval $[2, \infty)$. The function is decreasing if its graph is headed downward as it moves from left to right. We see that it is decreasing on the interval $(-\infty, 2]$.

(b) The graph is not symmetric with respect to the y-axis, nor is it symmetric with respect to the origin. The function is neither.

(c) Yes, we see that the function has a minimum value of 0 at $x = 2$. (This is easily confirmed by the algebraic fact that $(x - 2)^2 \geq 0$ for all x.)

(d) We see that the graph of $y = (x - 2)^2$ is just the graph of $y = x^2$ moved two units to the right. *Now try Exercise 35.*

EXPLORATION 1 Looking for Asymptotes

1. Two of the basic functions have vertical asymptotes at $x = 0$. Which two?

2. Form a new function by adding these functions together. Does the new function have a vertical asymptote at $x = 0$?

3. Three of the basic functions have horizontal asymptotes at $y = 0$. Which three?

4. Form a new function by adding these functions together. Does the new function have a horizontal asymptote at $y = 0$?

5. Graph $f(x) = 1/x$, $g(x) = 1/(2x^2 - x)$, and $h(x) = f(x) + g(x)$. Does $h(x)$ have a vertical asymptote at $x = 0$?

EXAMPLE 6 Identifying a Piecewise-Defined Function

Which of the twelve basic functions has the following **piecewise** definition over separate intervals of its domain?

$$f(x) = \begin{cases} x & \text{if } x \geq 0 \\ -x & \text{if } x < 0 \end{cases}$$

SOLUTION You may recognize this as the definition of the absolute value function (Chapter P). Or, you can reason that the graph of this function must look just like the line $y = x$ to the right of the y-axis, but just like the graph of the line $y = -x$ to the left of the y-axis. That is a perfect description of the absolute value graph in Figure 1.45. Either way, we recognize this as a piecewise definition of $f(x) = |x|$.

Now try Exercise 45.

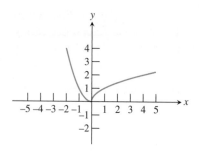

FIGURE 1.52 A piecewise-defined function. (Example 7)

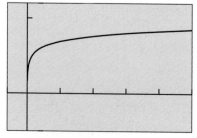

[–600, 5000] by [–5, 12]

FIGURE 1.53 The graph of $y = \ln x$ still appears to have a horizontal asymptote, despite the much larger window than in Figure 1.42. (Example 8)

EXAMPLE 7   Defining a Function Piecewise

Using basic functions from this section, construct a piecewise definition for the function whose graph is shown in Figure 1.52. Is your function continuous?

SOLUTION This appears to be the graph of $y = x^2$ to the left of $x = 0$ and the graph of $y = \sqrt{x}$ to the right of $x = 0$. We can therefore define it piecewise as

$$f(x) = \begin{cases} x^2 & \text{if } x \le 0 \\ \sqrt{x} & \text{if } x > 0 \end{cases}$$

The function is continuous.   *Now try Exercise 47.*

You can go a long way toward understanding a function's behavior by looking at its graph. We will continue that theme in the exercises and then revisit it throughout the book. However, you can't go *all* the way toward understanding a function by looking at its graph, as Example 8 shows.

EXAMPLE 8   Looking for a Horizontal Asymptote

Does the graph of $y = \ln x$ (Figure 1.42) have a horizontal asymptote?

SOLUTION In Figure 1.42 it certainly *looks* like there is a horizontal asymptote that the graph is approaching from below. If we choose a much larger window (Figure 1.53), it still looks that way. In fact, we could zoom out on this function all day long and it would *always* look like it is approaching some horizontal asymptote—but it is not. We will show algebraically in Chapter 3 that the end behavior of this function is $\lim_{x \to \infty} \ln x = \infty$, so its graph must eventually rise above the level of any horizontal line. That rules out any horizontal asymptote, even though there is no *visual* evidence of that fact that we can see by looking at its graph.

Now try Exercise 55.

EXAMPLE 9   Analyzing a Function

Give a complete analysis of the basic function $f(x) = |x|$.

SOLUTION

BASIC FUNCTION The Absolute Value Function

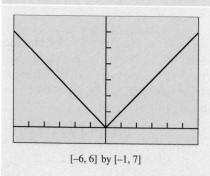

[–6, 6] by [–1, 7]

FIGURE 1.54 The graph of $f(x) = |x|$.

$f(x) = |x|$
Domain: All reals
Range: $[0, \infty)$
Continuous
Decreasing on $(\infty, 0]$; increasing on $[0, \infty)$
Symmetric with respect to the y-axis (an even function)
Bounded below
Local minimum at $(0, 0)$
No horizontal asymptotes
No vertical asymptotes
End behavior: $\lim_{x \to -\infty} |x| = \infty$ and $\lim_{x \to \infty} |x| = \infty$

Now try Exercise 67.

QUICK REVIEW 1.3 *(For help, go to Sections P.1, P.2, 3.1, and 3.3.)*

In Exercises 1–10, evaluate the expression without using a calculator.

1. $|-59.34|$

2. $|5 - \pi|$

3. $|\pi - 7|$

4. $\sqrt{(-3)^2}$

5. $\ln{(1)}$

6. e^0

7. $(\sqrt[3]{3})^3$

8. $\sqrt[3]{(-15)^3}$

9. $\sqrt[3]{-8^2}$

10. $|1 - \pi| - \pi$

SECTION 1.3 EXERCISES

In Exercises 1–12, each graph is a slight variation on the graph of one of the twelve basic functions described in this section. Match the graph to one of the twelve functions (a)–(l) and then support your answer by checking the graph on your calculator. (All graphs are shown in the window $[-4.7, 4.7]$ by $[-3.1, 3.1]$.)

(a) $y = -\sin x$ **(b)** $y = \cos x + 1$ **(c)** $y = e^x - 2$

(d) $y = (x + 2)^3$ **(e)** $y = x^3 + 1$ **(f)** $y = (x - 1)^2$

(g) $y = |x| - 2$ **(h)** $y = -1/x$ **(i)** $y = -x$

(j) $y = -\sqrt{x}$ **(k)** $y = \text{int}(x + 1)$ **(l)** $y = 2 - 4/(1 + e^{-x})$

1.

2.
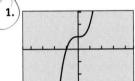

3.
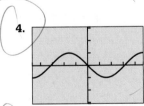

4.
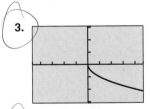

5.
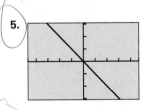

6.
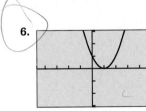

7.
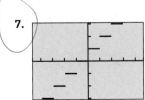

8.
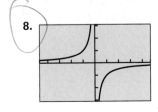

9.
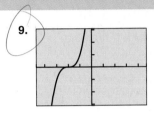

10.
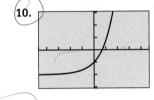

11.
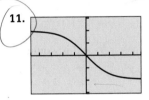

12.
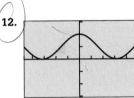

In Exercises 13–18, identify which of Exercises 1–12 display functions that fit the description given.

13. The function whose domain excludes zero.

14. The function whose domain consists of all nonnegative real numbers.

15. The two functions that have at least one point of discontinuity.

16. The function that is not a *continuous function.*

17. The six functions that are bounded below.

18. The four functions that are bounded above.

In Exercises 19–28, identify which of the twelve basic functions fit the description given.

19. The four functions that are odd.

20. The six functions that are increasing on their entire domains.

21. The three functions that are decreasing on the interval $(-\infty, 0)$.

22. The three functions with infinitely many local extrema.

23. The three functions with no zeros.

24. The three functions with range {all real numbers}.

25. The four functions that do *not* have end behavior
$\lim\limits_{x\to+\infty} f(x) = +\infty$.

26. The three functions with end behavior $\lim\limits_{x\to-\infty} f(x) = -\infty$.

27. The four functions whose graphs look the same when turned upside-down and flipped about the y-axis.

28. The two functions whose graphs are identical except for a horizontal shift.

In Exercises 29–34, use your graphing calculator to produce a graph of the function. Then determine the domain and range of the function by looking at its graph.

29. $f(x) = x^2 - 5$ **30.** $g(x) = |x - 4|$

31. $h(x) = \ln(x + 6)$ **32.** $k(x) = 1/x + 3$

33. $s(x) = \text{int}(x/2)$ **34.** $p(x) = (x + 3)^2$

In Exercises 35–42, graph the function. Then answer the following questions:

(a) On what interval, if any, is the function increasing? Decreasing?

(b) Is the function odd, even, or neither?

(c) Give the function's extrema, if any.

(d) How does the graph relate to a graph of one of the twelve basic functions?

35. $r(x) = \sqrt{x - 10}$ **36.** $f(x) = \sin(x) + 5$

37. $f(x) = 3/(1+e^{-x})$ **38.** $q(x) = e^x + 2$

39. $h(x) = |x| - 10$ **40.** $g(x) = 4\cos(x)$

41. $s(x) = |x - 2|$ **42.** $f(x) = 5 - \text{abs}(x)$

43. Find the horizontal asymptotes for the graph shown in Exercise 11.

44. Find the horizontal asymptotes for the graph of $f(x)$ in Exercise 37.

In Exercises 45–52, sketch the graph of the piecewise-defined function. (Try doing it without a calculator.) In each case, give any points of discontinuity.

45. $f(x) = \begin{cases} x & \text{if } x \le 0 \\ x^2 & \text{if } x > 0 \end{cases}$ **46.** $g(x) = \begin{cases} x^3 & \text{if } x \le 0 \\ e^x & \text{if } x > 0 \end{cases}$

47. $h(x) = \begin{cases} |x| & \text{if } x < 0 \\ \sin x & \text{if } x \ge 0 \end{cases}$ **48.** $w(x) = \begin{cases} 1/x & \text{if } x < 0 \\ \sqrt{x} & \text{if } x \ge 0 \end{cases}$

49. $f(x) = \begin{cases} \cos x & \text{if } x \le 0 \\ e^x & \text{if } x > 0 \end{cases}$ **50.** $g(x) = \begin{cases} |x| & \text{if } x < 0 \\ x^2 & \text{if } x \ge 0 \end{cases}$

51. $f(x) = \begin{cases} -3 - x & \text{if } x \le 0 \\ 1 & \text{if } 0 < x < 1 \\ x^2 & \text{if } x \ge 1 \end{cases}$

52. $f(x) = \begin{cases} x^2 & \text{if } x < -1 \\ |x| & \text{if } -1 \le x < 1 \\ \text{int}(x) & \text{if } x \ge 1 \end{cases}$

53. Writing to Learn The function $f(x) = \sqrt{x^2}$ is one of our twelve basic functions written in another form.

(a) Graph the function and identify which basic function it is.

(b) Explain algebraically why the two functions are equal.

54. Uncovering Hidden Behavior The function
$g(x) = \sqrt{x^2 + 0.0001} - 0.01$ is *not* one of our twelve basic functions written in another form.

(a) Graph the function and identify which basic function it appears to be.

(b) Verify numerically that it is not the basic function that it appears to be.

55. Writing to Learn The function $f(x) = \ln(e^x)$ is one of our twelve basic functions written in another form.

(a) Graph the function and identify which basic function it is.

(b) Explain how the equivalence of the two functions in (a) shows that the natural logarithm function is *not* bounded above (even though it *appears* to be bounded above in Figure 1.42).

56. Writing to Learn Let $f(x)$ be the function that gives the cost, in cents, to mail a letter that weighs x ounces. As of June 2002, the cost is 37 cents for a letter that weighs up to one ounce, plus 23 cents for each additional ounce or portion thereof.

(a) Sketch a graph of $f(x)$.

(b) How is this function similar to the greatest integer function? How is it different?

57. Analyzing a Function Set your calculator to DOT mode and graph the greatest integer function, $y = \text{int}(x)$, in the window $[-4.7, 4.7]$ by $[-3.1, 3.1]$. Then complete the following analysis.

BASIC FUNCTION
The Greatest Integer Function

$f(x) = \text{int } x$

Domain:
Range:
Continuity:
Increasing/decreasing behavior:
Symmetry:
Boundedness:
Local extrema:
Horizontal asymptotes:
Vertical asymptotes:
End behavior:

Standardized Test Questions

58. True or False The greatest integer function has an inverse function. Justify your answer.

59. True or False The logistic function has two horizontal asymptotes. Justify your answer.

In Exercises 60–63, you may use a graphing calculator to answer the question.

60. Multiple Choice Which function has range {all real numbers}?

(A) $f(x) = 4 + \ln x$

(B) $f(x) = 3 - 1/x$

(C) $f(x) = 5/(1 + e^{-x})$

(D) $f(x) = \text{int}(x - 2)$

(E) $f(x) = 4 \cos x$

61. Multiple Choice Which function is bounded both above and below?

(A) $f(x) = x^2 - 4$

(B) $f(x) = (x - 3)^3$

(C) $f(x) = 3e^x$

(D) $f(x) = 3 + 1/(1 + e^{-x})$

(E) $f(x) = 4 - |x|$

62. Multiple Choice Which of the following is the same as the restricted-domain function $f(x) = \text{int}(x), 0 \le x < 2$?

(A) $f(x) = \begin{cases} 0 & \text{if } 0 \le x < 1 \\ 1 & \text{if } x = 1 \\ 2 & \text{if } 1 < x < 2 \end{cases}$

(B) $f(x) = \begin{cases} 0 & \text{if } x = 0 \\ 1 & \text{if } 0 < x \le 1 \\ 2 & \text{if } 1 < x < 2 \end{cases}$

(C) $f(x) = \begin{cases} 0 & \text{if } 0 \le x < 1 \\ 1 & \text{if } 1 \le x < 2 \end{cases}$

(D) $f(x) = \begin{cases} 1 & \text{if } 0 \le x < 1 \\ 2 & \text{if } 1 \le x < 2 \end{cases}$

(E) $f(x) = \begin{cases} x & \text{if } 0 \le x < 1 \\ 1 + x & \text{if } 1 \le x < 2 \end{cases}$

63. Multiple Choice Increasing Functions Which function is increasing on the interval $(-\infty, \infty)$?

(A) $f(x) = \sqrt{3 + x}$

(B) $f(x) = \text{int}(x)$

(C) $f(x) = 2x^2$

(D) $f(x) = \sin x$

(E) $f(x) = 3/(1 + e^{-x})$

Explorations

64. Which is Bigger? For positive values of x, we wish to compare the values of the basic functions x^2, x, and $\sqrt{x}$.

(a) How would you order them from least to greatest?

(b) Graph the three functions in the viewing window [0, 30] by [0, 20]. Does the graph confirm your response in (a)?

(c) Now graph the three functions in the viewing window [0, 2] by [0, 1.5].

(d) Write a careful response to the question in (a) that accounts for all positive values of x.

65. Odds and Evens There are four odd functions and three even functions in the gallery of twelve basic functions. After multiplying these functions together pairwise in different combinations and exploring the graphs of the products, make a conjecture about the symmetry of:

(a) a product of two odd functions.

(b) a product of two even functions.

(c) a product of an odd function and an even function.

66. Group Activity Assign to each student in the class the name of one of the twelve basic functions, but secretly so that no student knows the "name" of another. (The same function name could be given to several students, but all the functions should be used at least once.) Let each student make a one-sentence self-introduction to the class that reveals something personal "about who I am that really identifies me." The rest of the students then write down their guess as to the function's identity. Hints should be subtle and cleverly anthropomorphic. (For example, the absolute value function saying "I have a very sharp smile" is subtle and clever, while "I am absolutely valuable" is not very subtle at all.)

67. Pepperoni Pizzas For a statistics project, a student counted the number of pepperoni slices on pizzas of various sizes at a local pizzeria, compiling the following table:

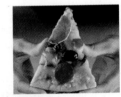

Table 1.10		
Type of Pizza	Radius	Pepperoni count
Personal	4"	12
Medium	6"	27
Large	7"	37
Extra Large	8"	48

(a) Explain why the pepperoni count (P) ought to be proportional to the square of the radius (r).

(b) Assuming that $P = k \cdot r^2$, use the data pair (4, 12) to find the value of k.

(c) Does the algebraic model fit the rest of the data well?

(d) Some pizza places have charts showing their kitchen staff how much of each topping should be put on each size of pizza. Do you think this pizzeria uses such a chart? Explain.

Extending the Ideas

68. Inverse Functions Two functions are said to be *inverses* of each other if the graph of one can be obtained from the graph of the other by reflecting it across the line $y = x$.
For example, the functions with the graphs shown below are inverses of each other:

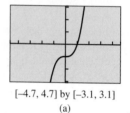 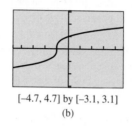

[–4.7, 4.7] by [–3.1, 3.1]   [–4.7, 4.7] by [–3.1, 3.1]
(a)   (b)

(a) Two of the twelve basic functions in this section are inverses of each other. Which are they?

(b) Two of the twelve basic functions in this section are their own inverses. Which are they?

(c) If you restrict the domain of one of the twelve basic functions to $[0, \infty)$, it becomes the inverse of another one. Which are they?

69. Identifying a Function by Its Properties

(a) Seven of the twelve basic functions have the property that $f(0) = 0$. Which five do not?

(b) Only one of the twelve basic functions has the property that $f(x + y) = f(x) + f(y)$ for all x and y in its domain. Which one is it?

(c) One of the twelve basic functions has the property that $f(x + y) = f(x)f(y)$ for all x and y in its domain. Which one is it?

(d) One of the twelve basic functions has the property that $f(xy) = f(x) + f(y)$ for all x and y in its domain. Which one is it?

(e) Four of the twelve basic functions have the property that $f(x) + f(-x) = 0$ for all x in their domains. Which four are they?

1.4
Building Functions from Functions

. . . and why

Most of the functions that you will encounter in calculus and in real life can be created by combining or modifying other functions.

Combining Functions Algebraically

Knowing how a function is "put together" is an important first step when applying the tools of calculus. Functions have their own algebra based on the same operations we apply to real numbers (addition, subtraction, multiplication, and division). One way to build new functions is to apply these operations, using the following definitions.

DEFINITION Sum, Difference, Product, and Quotient of Functions

Let f and g be two functions with intersecting domains. Then for all values of x in the intersection, the algebraic combinations of f and g are defined by the following rules:

Sum: $\qquad (f + g)(x) = f(x) + g(x)$

Difference: $\quad (f - g)(x) = f(x) - g(x)$

Product: $\qquad (fg)(x) = f(x)g(x)$

Quotient: $\qquad \left(\dfrac{f}{g}\right)(x) = \dfrac{f(x)}{g(x)}$, provided $g(x) \neq 0$

In each case, the domain of the new function consists of all numbers that belong to both the domain of f and the domain of g. As noted, the zeros of the denominator are excluded from the domain of the quotient.

Euler's function notation works so well in the above definitions that it almost obscures what is really going on. The "+" in the expression "$(f + g)(x)$" stands for a brand new operation called *function addition*. It builds a new function, $f + g$, from the given functions f and g. Like any function, $f + g$ is defined by what it does: It takes a domain value x and returns a range value $f(x) + g(x)$. Note that the "+" sign in "$f(x) + g(x)$" *does* stand for the familiar operation of real number addition. So, with the same symbol taking on different roles on either side of the equal sign, there is more to the above definitions than first meets the eye.

Fortunately, the definitions are easy to apply.

EXAMPLE 1   Defining New Functions Algebraically

Let $f(x) = x^2$ and $g(x) = \sqrt{x+1}$.

Find formulas for the functions $f + g, f - g, fg, f/g$, and gg. Give the domain of each.

SOLUTION We first determine that f has domain all real numbers and that g has domain $[-1, \infty)$. These domains overlap, the intersection being the interval $[-1, \infty)$. So:

$$(f + g)(x) = f(x) + g(x) = x^2 + \sqrt{x+1} \quad \text{with domain } [-1, \infty).$$

$$(f - g)(x) = f(x) - g(x) = x^2 - \sqrt{x+1} \quad \text{with domain } [-1, \infty).$$

$$(fg)(x) = f(x)g(x) = x^2\sqrt{x+1} \quad \text{with domain } [-1, \infty).$$

$$\left(\frac{f}{g}\right)(x) = \frac{f(x)}{g(x)} = \frac{x^2}{\sqrt{x+1}} \quad \text{with domain } (-1, \infty).$$

$$(gg)(x) = g(x)g(x) = (\sqrt{x+1})^2 \quad \text{with domain } [-1, \infty).$$

Note that we could express $(gg)(x)$ more simply as $x + 1$. That would be fine, but the simplification would not change the fact that the domain of gg is (by definition) the interval $[-1, \infty)$. Under other circumstances the function $h(x) = x + 1$ would have domain all real numbers, but under these circumstances it cannot; it is a product of two functions with restricted domains.   ***Now try Exercise 3.***

Composition of Functions

It is not hard to see that the function $\sin(x^2)$ is built from the basic functions $\sin x$ and x^2, but the functions are not put together by addition, subtraction, multiplication, or division. Instead, the two functions are combined by simply applying them in order—first the squaring function, then the sine function. This operation for combining functions, which has no counterpart in the algebra of real numbers, is called *function composition.*

DEFINITION Composition of Functions

Let f and g be two functions such that the domain of f intersects the range of g. The **composition f of g**, denoted $f \circ g$, is defined by the rule

$$(f \circ g)(x) = f(g(x)).$$

The domain of $f \circ g$ consists of all x-values in the domain of g that map to $g(x)$-values in the domain of f. (See Figure 1.55.)

The composition g of f, denoted $g \circ f$, is defined similarly. In most cases $g \circ f$ and $f \circ g$ are different functions. (In the language of algebra, "function composition is not commutative.")

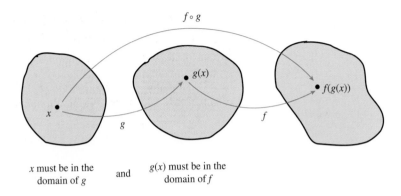

$$x \text{ must be in the domain of } g \quad \text{and} \quad g(x) \text{ must be in the domain of } f$$

FIGURE 1.55 In the composition $f \circ g$, the function g is applied first and then f. This is the reverse of the order in which we read the symbols.

EXAMPLE 2 Composing Functions

Let $f(x) = e^x$ and $g(x) = \sqrt{x}$. Find $(f \circ g)(x)$ and $(g \circ f)(x)$ and verify numerically that the functions $f \circ g$ and $g \circ f$ are not the same.

SOLUTION

$$(f \circ g)(x) = f(g(x)) = f(\sqrt{x}) = e^{\sqrt{x}}$$
$$(g \circ f)(x) = g(f(x)) = g(e^x) = \sqrt{e^x}$$

One verification that these functions are not the same is that they have different domains: $f \circ g$ is defined only for $x \geq 0$, while $g \circ f$ is defined for all real numbers. We could also consider their graphs (Figure 1.56), which agree only at $x = 0$ and $x = 4$.

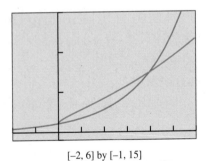

[−2, 6] by [−1, 15]

FIGURE 1.56 The graphs of $y = e^{\sqrt{x}}$ and $y = \sqrt{e^x}$ are not the same. (Example 2)

Finally, the graphs suggest a numerical verification: Find a single value of x for which $f(g(x))$ and $g(f(x))$ give different values. For example, $f(g(1)) = e$ and $g(f(1)) = \sqrt{e}$. The graph helps us to make a judicious choice of x. You do not want to check the functions at $x = 0$ and $x = 4$ and conclude that they are the same!

Now try Exercise 15.

EXPLORATION 1   Composition Calisthenics

One of the *f* functions in Column B can be composed with one of the *g* functions in Column C to yield each of the basic *f* ∘ *g* functions in Column A. Can you match the columns successfully without a graphing calculator? If you are having trouble, try it with a graphing calculator.

A	B	C
f ∘ *g*	*f*	*g*
x	$x - 3$	$x^{0.6}$
x^2	$2x - 3$	x^2
$\lvert x \rvert$	$\sqrt{x}$	$\dfrac{(x-2)(x+2)}{2}$
x^3	x^5	$\ln(e^3 x)$
$\ln x$	$\lvert 2x + 4 \rvert$	$\dfrac{x}{2}$
$\sin x$	$1 - 2x^2$	$\dfrac{x+3}{2}$
$\cos x$	$2 \sin x \cos x$	$\sin\left(\dfrac{x}{2}\right)$

EXAMPLE 3   Finding the Domain of a Composition

Let $f(x) = x^2 - 1$ and let $g(x) = \sqrt{x}$. Find the domains of the composite functions

(a) $g \circ f$ **(b)** $f \circ g$

SOLUTION

(a) We compose the functions in the order specified:

$$(g \circ f)(x) = g(f(x))$$
$$= \sqrt{x^2 - 1}$$

For x to be in the domain of $g \circ f$, we must first find $f(x) = x^2 - 1$, which we can do for all real x. Then we must take the square root of the result, which we can only do for nonnegative values of $x^2 - 1$.

Therefore, the domain of $g \circ f$ consists of all real numbers for which $x^2 - 1 \geq 0$, namely the union $(-\infty, -1] \cup [1, \infty)$.

(b) Again, we compose the functions in the order specified:

$$(f \circ g)(x) = f(g(x))$$
$$= (\sqrt{x})^2 - 1$$

CAUTION

We might choose to express ($f \circ g$) more simply as $x - 1$. However, you must remember that the composition is restricted to the domain of $g(x) = \sqrt{x}$, or $[0, \infty]$. The domain of $x - 1$ is all real numbers. It is a good idea to work out the domain of a composition before you simplify the expression for $f(g(x))$. One way to simplify and maintain the restriction on the domain in Example 3 is to write ($f \circ g$) $(x) = x - 1$, $x \geq 0$.

continued

For x to be in the domain of $f \circ g$, we must first be able to find $g(x) = \sqrt{x}$, which we can only do for nonnegative values of x. Then we must be able to square the result and subtract 1, which we can do for all real numbers.

Therefore, the domain of $f \circ g$ consists of the interval $[0, \infty)$.

Support Graphically

We can graph the composition functions to see if the grapher respects the domain restrictions. The screen to the left of each graph shows the set up in the "Y=" editor. Figure 1.57b shows the graph of $y = (g \circ f)(x)$, while Figure 1.57d shows the graph of $y = (f \circ g)(x)$. The graphs support our algebraic work quite nicely. ***Now try Exercise 17.***

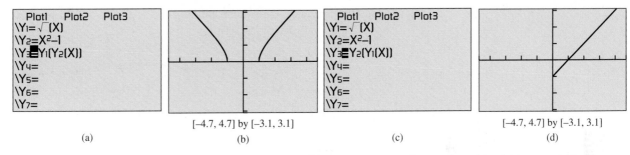

| (a) | (b) | (c) | (d) |

[-4.7, 4.7] by [-3.1, 3.1]   (b)   (c)   [-4.7, 4.7] by [-3.1, 3.1]   (d)

FIGURE 1.57 The functions Y1 and Y2 are composed to get the graphs of $y = (g \circ f)(x)$ and $y = (f \circ g)(x)$, respectively. The graphs support our conclusions about the domains of the two composite functions. (Example 3)

In Examples 2 and 3 two functions were *composed* to form new functions. There are times in calculus when we need to reverse the process. That is, we may begin with a function h and *decompose* it by finding functions whose composition is h.

EXAMPLE 4 Decomposing Functions

For each function h, find functions f and g such that $h(x) = f(g(x))$.

(a) $h(x) = (x + 1)^2 - 3(x + 1) + 4$

(b) $h(x) = \sqrt{x^3 + 1}$

SOLUTION

(a) We can see that h is quadratic in $x + 1$. Let $f(x) = x^2 - 3x + 4$ and let $g(x) = x + 1$. Then

$$h(x) = f(g(x)) = f(x + 1) = (x + 1)^2 - 3(x + 1) + 4.$$

(b) We can see that h is the square root of the function $x^3 + 1$. Let $f(x) = \sqrt{x}$ and let $g(x) = x^3 + 1$. Then

$$h(x) = f(g(x)) = f(x^3 + 1) = \sqrt{x^3 + 1}.$$

Now try Exercise 25.

There is often more than one way to decompose a function. For example, an alternate way to decompose $h(x) = \sqrt{x^3 + 1}$ in Example 4b is to let $f(x) = \sqrt{x + 1}$ and let $g(x) = x^3$. Then $h(x) = f(g(x)) = f(x^3) = \sqrt{x^3 + 1}$.

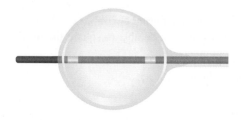

FIGURE 1.58 (Example 5)

EXAMPLE 5   Modeling with Function Composition

In the medical procedure known as angioplasty, doctors insert a catheter into a heart vein (through a large peripheral vein) and inflate a small, spherical balloon on the tip of the catheter. Suppose the balloon is inflated at a constant rate of 44 cubic millimeters per second. (See Figure 1.58.)

(a) Find the volume after t seconds

(b) When the volume is V, what is the radius r?

(c) Write an equation that gives the radius r as a function of the time. What is the radius after 5 seconds?

SOLUTION

(a) After t seconds, the volume will be $44t$.

(b) Solve Algebraically

$$\frac{4}{3}\pi r^3 = v$$

$$r^3 = \frac{3v}{4\pi}$$

$$r = \sqrt[3]{\frac{3v}{4\pi}}$$

(c) Substituting $44t$ for V gives $r = \sqrt[3]{\frac{3 \cdot 44t}{4\pi}}$ or $r = \sqrt[3]{\frac{33t}{\pi}}$. After 5 seconds, the

radius will be $r = \sqrt[3]{\frac{33 \cdot 5}{\pi}} \approx 3.74$ mm.

Now try Exercise 31.

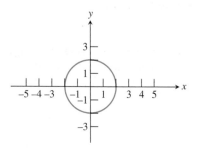

FIGURE 1.59 A circle of radius 2 centered at the origin. This set of ordered pairs (x, y) defines a *relation* that is not a function, because the graph fails the vertical line test.

GRAPHING RELATIONS

Relations that are not functions are often not easy to graph. We will study some special cases later in the course (circles, ellipses, etc.), but some simple-looking relations like Example 6 are difficult to graph. Nor do our calculators help much, because the equation can not be put into "Y1=" form. Interestingly, we *do* know that the graph of the relation in Example 6, whatever it looks like, fails the vertical line test.

Relations and Implicitly Defined Functions

There are many useful curves in mathematics that fail the vertical line test and therefore are not graphs of functions. One such curve is the circle in Figure 1.59. While y is not related to x as a function in this instance, there is certainly some sort of relationship going on. In fact, not only does the shape of the graph show a significant *geometric* relationship among the points, but the ordered pairs (x, y) exhibit a significant *algebraic* relationship as well: They consist exactly of the solutions to the equation $x^2 + y^2 = 4$.

The general term for a set of ordered pairs (x, y) is a **relation**. If the relation happens to relate a *single* value of y to each value of x, then the relation is also a function and its graph will pass the vertical line test. In the case of the circle with equation $x^2 + y^2 = 4$, both $(0, 2)$ and $(0, -2)$ are in the relation, so y is not a function of x.

EXAMPLE 6 Verifying Pairs in a Relation

Determine which of the ordered pairs $(2, -5)$, $(1, 3)$, and $(2, 1)$ are in the relation defined by $x^2y + y^2 = 5$. Is the relation a function?

SOLUTION We simply substitute the x- and y-coordinates of the ordered pairs into $x^2y + y^2$ and see if we get 5.

continued

(2, −5): $(2)^2(−5) + (−5)^2 = 5$ Substitute $x = 2, y = −5$.

(1, 3): $(1)^2(3) + (3)^2 = 12 \neq 5$ Substitute $x = 1, y = 3$.

(2, 1): $(2)^2(1) + (1)^2 = 5$ Substitute $x = 2, y = 1$.

So, $(2, −5)$ and $(2, 1)$ are in the relation, but $(1, 3)$ is not.

Since the equation relates two different y-values ($−5$ and 1) to the same x-value (2), the relation cannot be a function.

Now try Exercise 35.

Let us revisit the circle $x^2 + y^2 = 4$. While it is not a function itself, we can split it into two equations that *do* define functions, as follows:

$$x^2 + y^2 = 4$$
$$y^2 = 4 − x^2$$
$$y = +\sqrt{4 − x^2} \text{ or } y = −\sqrt{4 − x^2}$$

The graphs of these two functions are, respectively, the upper and lower semicircles of the circle in Figure 1.59. They are shown in Figure 1.60. Since all the ordered pairs in either of these functions satisfy the equation $x^2 + y^2 = 4$, we say that the relation given by the equation defines the two functions **implicitly**.

EXAMPLE 7 Using Implicitly Defined Functions

Describe the graph of the relation $x^2 + 2xy + y^2 = 1$.

SOLUTION This looks like a difficult task at first, but notice that the expression on the left of the equal sign is a factorable trinomial. This enables us to split the relation into two implicitly defined functions as follows:

$$x^2 + 2xy + y^2 = 1$$
$$(x + y)^2 = 1 \qquad \text{Factor.}$$
$$x + y = \pm 1 \qquad \text{Extract square roots.}$$
$$x + y = 1 \text{ or } x + y = −1$$
$$y = −x + 1 \text{ or } y = −x − 1 \qquad \text{Solve for } y.$$

The graph consists of two parallel lines (Figure 1.61), each the graph of one of the implicitly defined functions.

Now try Exercise 37.

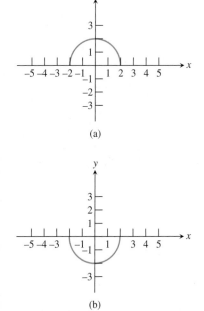

(a)

(b)

FIGURE 1.60 The graphs of
(a) $y = +\sqrt{4 − x^2}$ and
(b) $y = −\sqrt{4 − x^2}$. In each case, y is
defined as a function of x. These two
functions are defined *implicitly* by the relation
$x^2 + y^2 = 4$.

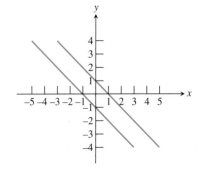

FIGURE 1.61 The graph of the relation $x^2 + 2xy + y^2 = 1$ (Example 7)

QUICK REVIEW 1.4   *(For help, go to Sections P.1, 1.2, and 1.3.)*

In Exercises 1–10, find the domain of the function and express it in interval notation.

1. $f(x) = \dfrac{x-2}{x+3}$ **2.** $g(x) = \ln(x-1)$

3. $f(t) = \sqrt{5-t}$ **4.** $g(x) = \dfrac{3}{\sqrt{2x-1}}$

5. $f(x) = \sqrt{\ln(x)}$ **6.** $h(x) = \sqrt{1-x^2}$

7. $f(t) = \dfrac{t+5}{t^2+1}$ **8.** $g(t) = \ln(|t|)$

9. $f(x) = \dfrac{1}{\sqrt{1-x^2}}$ **10.** $g(x) = 2$

SECTION 1.4 EXERCISES

In Exercises 1–4, find formulas for the functions $f+g, f-g$, and fg. Give the domain of each.

1. $f(x) = 2x-1; g(x) = x^2$ **2.** $f(x) = (x-1)^2; g(x) = 3-x$

3. $f(x) = \sqrt{x}; g(x) = \sin x$ **4.** $f(x) = \sqrt{x+5}; g(x) = |x+3|$

In Exercises 5–8, find formulas for f/g and g/f. Give the domain of each.

5. $f(x) = \sqrt{x+3}; g(x) = x^2$

6. $f(x) = \sqrt{x-2}; g(x) = \sqrt{x+4}$

7. $f(x) = x^2; g(x) = \sqrt{1-x^2}$

8. $f(x) = x^3; g(x) = \sqrt[3]{1-x^3}$

9. $f(x) = x^2$ and $g(x) = 1/x$ are shown below in the viewing window $[0, 5]$ by $[0, 5]$. Sketch the graph of the sum $(f+g)(x)$ by adding the y-coordinates directly from the graphs. Then graph the sum on your calculator and see how close you came.

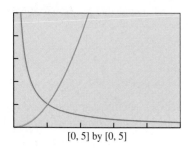

[0, 5] by [0, 5]

10. The graphs of $f(x) = x^2$ and $g(x) = 4-3x$ are shown in the viewing window $[-5, 5]$ by $[-10, 25]$. Sketch the graph of the difference $(f-g)(x)$ by subtracting the y-coordinates directly from the graphs. Then graph the difference on your calculator and see how close you came.

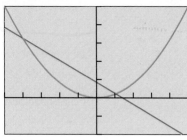

[–5, 5] by [–10, 25]

In Exercises 11–14, find $(f \circ g)(3)$ and $(g \circ f)(-2)$

11. $f(x) = 2x-3; g(x) = x+1$

12. $f(x) = x^2-1; g(x) = 2x-3$

13. $f(x) = x^2+4; g(x) = \sqrt{x+1}$

14. $f(x) = \dfrac{x}{x+1}; g(x) = 9-x^2$

In Exercises 15–22, find $f(g(x))$ and $g(f(x))$. State the domain of each.

15. $f(x) = 3x+2; g(x) = x-1$

16. $f(x) = x^2-1; g(x) = \dfrac{1}{x-1}$

17. $f(x) = x^2-2; g(x) = \sqrt{x+1}$

18. $f(x) = \dfrac{1}{x-1}; g(x) = \sqrt{x}$

19. $f(x) = x^2; g(x) = \sqrt{1-x^2}$

20. $f(x) = x^3; g(x) = \sqrt[3]{1-x^3}$

21. $f(x) = \dfrac{1}{2x}; g(x) = \dfrac{1}{3x}$

22. $f(x) = \dfrac{1}{x+1}; g(x) = \dfrac{1}{x-1}$

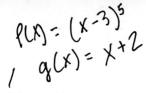

$$P(x) = (x-3)^5$$
$$g(x) = x+2$$

In Exercises 23–30, find $f(x)$ and $g(x)$ so that the function can be described as $y = f(g(x))$. (There may be more than one possible decomposition.)

23. $y = \sqrt{x^2 - 5x}$

24. $y = (x^3 + 1)^2$

25. $y = |3x - 2|$

26. $y = \dfrac{1}{x^3 - 5x + 3}$

27. $y = (x - 3)^5 + 2$

28. $y = e^{\sin x}$

29. $y = \cos(\sqrt{x})$

30. $y = (\tan x)^2 + 1$

31. Weather Balloons A high-altitude spherical weather balloon expands as it rises due to the drop in atmospheric pressure. Suppose that the radius r increases at the rate of 0.03 inches per second and that $r = 48$ inches at time $t = 0$. Determine an equation that models the volume V of the balloon at time t and find the volume when $t = 300$ seconds.

32. A Snowball's Chance Jake stores a small cache of 4-inch diameter snowballs in the basement freezer, unaware that the freezer's self-defrosting feature will cause each snowball to lose about 1 cubic inch of volume every 40 days. He remembers them a year later (call it 360 days) and goes to retrieve them. What is their diameter then?

33. Satellite Photography A satellite camera takes a rectangular-shaped picture. The smallest region that can be photographed is a 5-km by 7-km rectangle. As the camera zooms out, the length l and width w of the rectangle increase at a rate of 2 km/sec. How long does it take for the area A to be at least 5 times its original size?

34. Computer Imaging New Age Special Effects, Inc., prepares computer software based on specifications prepared by film directors. To simulate an approaching vehicle, they begin with a computer image of a 5-cm by 7-cm by 3-cm box. The program increases each dimension at a rate of 2 cm/sec. How long does it take for the volume V of the box to be at least 5 times its initial size?

35. Which of the ordered pairs $(1, 1)$, $(4, -2)$, and $(3, -1)$ are in the relation given by $3x + 4y = 5$?

36. Which of the ordered pairs $(5, 1)$, $(3, 4)$, and $(0, -5)$ are in the relation given by $x^2 + y^2 = 25$?

In Exercises 37–44, find two functions defined implicitly by the given relation.

37. $x^2 + y^2 = 25$

38. $x + y^2 = 25$

39. $x^2 - y^2 = 25$

40. $3x^2 - y^2 = 25$

41. $x + |y| = 1$

42. $x - |y| = 1$

43. $y^2 = x^2$

44. $y^2 = x$

Standardized Test Questions

45. True or False The domain of the quotient function $(f/g)(x)$ consists of all numbers that belong to both the domain of f and the domain of g. Justify your answer.

46. True or False The domain of the product function $(fg)(x)$ consists of all numbers that belong to either the domain of f or the domain of g. Justify your answer.

You may use a graphing calculator when solving Exercises 47–50.

47. Multiple Choice Suppose f and g are functions with domain all real numbers. Which of the following statements is *not* necessarily true?

(A) $(f + g)(x) = (g + f)(x)$ **(B)** $(fg)(x) = (gf)(x)$

(C) $f(g(x)) = g(f(x))$ **(D)** $(f - g)(x) = -(g - f)(x)$

(E) $(f \circ g)(x) = f(g(x))$

48. Multiple Choice If $f(x) = x - 7$ and $g(x) = \sqrt{4 - x}$, what is the domain of the function f/g?

(A) $(-\infty, 4)$ **(B)** $(-\infty, 4]$ **(C)** $(4, \infty)$

(D) $[4, \infty)$ **(E)** $(4, 7) \cup (7, \infty)$

49. Multiple Choice If $f(x) = x^2 + 1$, then $(f \circ f)(x) =$

(A) $2x^2 + 2$ **(B)** $2x^2 + 1$ **(C)** $x^4 + 1$

(D) $x^4 + 2x^2 + 1$ **(E)** $x^4 + 2x^2 + 2$

50. Multiple Choice Which of the following relations defines the function $y = |x|$ implicitly?

(A) $y = x$ **(B)** $y^2 = x^2$ **(C)** $y^3 = x^3$

(D) $x^2 + y^2 = 1$ **(E)** $x = |y|$

Explorations

51. Three on a Match Match each function f with a function g and a domain D so that $(f \circ g)(x) = x^2$ with domain D.

f	g	D
e^x	$\sqrt{2 - x}$	$x \neq 0$
$(x^2 + 2)^2$	$x + 1$	$x \neq 1$
$(x^2 - 2)^2$	$2 \ln x$	$(0, \infty)$
$\dfrac{1}{(x - 1)^2}$	$\dfrac{1}{x - 1}$	$[2, \infty)$
$x^2 - 2x + 1$	$\sqrt{x - 2}$	$(-\infty, 2]$
$\left(\dfrac{x + 1}{x}\right)^2$	$\dfrac{x + 1}{x}$	$(-\infty, \infty)$

52. Be a g Whiz Let $f(x) = x^2 + 1$. Find a function g so that

(a) $(fg)(x) = x^4 - 1$

(b) $(f + g)(x) = 3x^2$

(c) $(f/g)(x) = 1$

(d) $f(g(x)) = 9x^4 + 1$

(e) $g(f(x)) = 9x^4 + 1$

Extending the Ideas

53. Identifying Identities An *identity* for a function operation is a function that combines with a given function f to return the same function f. Find the identity functions for the following operations:

(a) Function addition. That is, find a function g such that
$(f + g)(x)) = (g + f)(x) = f(x)$.

(b) Function multiplication. That is, find a function g such that
$(fg)(x) = (gf)(x) = f(x)$.

(c) Function composition. That is, find a function g such that
$(f \circ g)(x) = (g \circ f)(x) = f(x)$.

54. Is Function Composition Associative? You already know that function composition is not commutative; that is, $(f \circ g)(x) \neq (g \circ f)(x)$. But is function composition associative? That is, does $(f \circ (g \circ h))(x) = ((f \circ g) \circ h))(x)$? Explain your answer.

55. Revisiting Example 6 Solve $x^2 y + y^2 = 5$ for y using the quadratic formula and graph the pair of implicit functions.

1.5
Parametric Relations and Inverses

What you'll learn about

- Defined Relations Parametrically

- Inverse Relations and Inverse Functions

. . . and why

Some functions and graphs can best be defined parametrically, while some others can be best understood as inverses of functions we already know.

Relations Defined Parametrically

Another natural way to define functions or, more generally, relations, is to define *both* elements of the ordered pair (x, y) in terms of another variable t, called a **parameter**. We illustrate with an example.

EXAMPLE 1 Defining a Function Parametrically

Consider the set of all ordered pairs (x, y) defined by the equations

$$x = t + 1$$
$$y = t^2 + 2t$$

where t is any real number.

(a) Find the points determined by $t = -3, -2, -1, 0, 1, 2,$ and 3.

(b) Find an algebraic relationship between x and y. (This is often called "eliminating the parameter.") Is y a function of x?

(c) Graph the relation in the (x, y) plane.

SOLUTION

(a) Substitute each value of t into the formulas for x and y to find the point that it determines parametrically:

t	$x = t + 1$	$y = t^2 + 2t$	(x, y)
-3	-2	3	$(-2, 3)$
-2	-1	0	$(-1, 0)$
-1	0	-1	$(0, -1)$
0	1	0	$(1, 0)$
1	2	3	$(2, 3)$
2	3	8	$(3, 8)$
3	4	15	$(4, 15)$

(b) We can find the relationship between x and y algebraically by the method of substitution. First solve for t in terms of x to obtain $t = x - 1$.

$$y = t^2 + 2t \qquad \text{Given}$$
$$y = (x - 1)^2 + 2(x - 1) \qquad t = x - 1$$
$$= x^2 - 2x + 1 + 2x - 2 \qquad \text{Expand.}$$
$$= x^2 - 1 \qquad \text{Simplify.}$$

This is consistent with the ordered pairs we had found in the table. As t varies over all real numbers, we will get all the ordered pairs in the relation $y = x^2 - 1$, which does indeed define y as a function of x.

continued

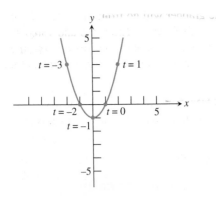

FIGURE 1.62 (Example 1)

(c) Since the parametrically defined relation consists of all ordered pairs in the relation $y = x^2 - 1$, we can get the graph by simply graphing the parabola $y = x^2 - 1$. See Figure 1.62.   ***Now try Exercise 5.***

EXAMPLE 2   Using a Graphing Calculator in Parametric Mode

Consider the set of all ordered pairs (x, y) defined by the equations

$$x = t^2 + 2t$$

$$y = t + 1$$

where t is any real number.

(a) Use a graphing calculator to find the points determined by $t = -3, 2, -1, 0, 1, 2,$ and 3.

(b) Use a graphing calculator to graph the relation in the (x, y) plane.

(c) Is y a function of x?

(d) Find an algebraic relationship between x and y.

t	(x, y)
-3	$(3, -2)$
-2	$(0, -1)$
-1	$(-1, 0)$
0	$(0, 1)$
1	$(3, 2)$
2	$(8, 3)$
3	$(15, 4)$

SOLUTION

(a) When the calculator is in *parametric mode*, the "Y =" screen provides a space to enter both X and Y as functions of the parameter T (Figure 1.63a). After entering the functions, use the table setup in Figure 1.63b to obtain the table shown in Figure 1.63c. The table shows, for example, that when T = −3 we have X1T = 3 and Y1T = −2, so the ordered pair corresponding to $t = -3$ is $(3, -2)$.

(b) In parametric mode, the "WINDOW" screen contains the usual x-axis information, as well as "Tmin," "Tmax," and "Tstep" (Figure 1.64a). To include most of the points listed in part (a), we set Xmin = −5, Xmax = 5, Ymin = −3, and Ymax = 3. Since $t = y - 1$, we set Tmin and Tmax to values one less than those for Ymin and Ymax.

continued

Plot1 Plot2 Plot3	TABLE SETUP	
\X₁ᴛ◼T²+2T	TblStart=−3	
Y₁ᴛ◼T+1	ΔTbl=1	
\X₂ᴛ=	Indpnt: **Auto** Ask	
Y₂ᴛ=	Depend: **Auto** Ask	
\X₃ᴛ=		
Y₃ᴛ=		
\X₄ᴛ=		

T	X₁ᴛ	**Y₁ᴛ**
−3	3	−2
−2	0	−1
−1	−1	0
0	0	1
1	3	2
2	8	3
3	15	4

Y₁ᴛ ◼ T+1

(a)   (b)   (c)

FIGURE 1.63 Using the table feature of a grapher set in parametric mode. (Example 2)

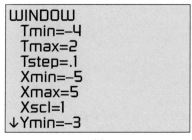

(a)

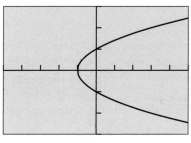

[–5, 5] by [–3, 3]

(b)

FIGURE 1.64 The graph of a parabola in parametric mode on a graphing calculator. (Example 2)

TIME FOR T

Functions defined by parametric equations are frequently encountered in problems of motion, where the x- and y-coordinates of a moving object are computed as functions of time. This makes time the parameter, which is why you almost always see parameters given as "t" in parametric equations.

The value of Tstep determines how far the grapher will go from one value of t to the next as it computes the ordered pairs. With Tmax − Tmin = 6 and Tstep = 0.1, the grapher will compute 60 points, which is sufficient. (The more points, the smoother the graph. See Exploration 1.) The graph is shown in Figure 1.64b. Use trace to find some of the points found in (a).

(c) No, y is not a function of x. We can see this from part (a) because $(0, -1)$ and $(0, 1)$ have the same x-value but different y-values. Alternatively, notice that the graph in (b) fails the vertical line test.

(d) We can use the same algebraic steps as in Example 1 to get the relation in terms of x and y: $x = y^2 - 1$. *Now try Exercise 7.*

EXPLORATION 1 Watching your Tstep

1. Graph the parabola in Example 2 in parametric mode as described in the solution. Press TRACE and observe the values of T, X, and Y. At what value of T does the calculator begin tracing? What point on the parabola results? (It's off the screen.) At what value of T does it stop tracing? What point on the parabola results? How many points are computed as you trace from start to finish?

2. Leave everything else the same and change the Tstep to 0.01. Do you get a smoother graph? Why or why not?

3. Leave everything else the same and change the Tstep to 1. Do you get a smoother graph? Why or why not?

4. What effect does the Tstep have on the speed of the grapher? Is this easily explained?

5. Now change the Tstep to 2. Why does the left portion of the parabola disappear? (It may help to TRACE along the curve.)

6. Change the Tstep back to 0.1 and change the Tmin to −1. Why does the bottom side of the parabola disappear? (Again, it may help to TRACE.)

7. Make a change to the window that will cause the grapher to show the bottom side of the parabola but not the top.

Inverse Relations and Inverse Functions

What happens when we reverse the coordinates of all the ordered pairs in a relation? We obviously get another relation, as it is another set of ordered pairs, but does it bear any resemblance to the original relation? If the original relation happens to be a function, will the new relation also be a function?

We can get some idea of what happens by examining Examples 1 and 2. The ordered pairs in Example 2 can be obtained by simply reversing the coordinates of the ordered pairs in Example 1. This is because we set up Example 2 by switching the parametric equations for x and y that we used in Example 1. We say that the relation in Example 2 is the *inverse relation* of the relation in Example 1.

> **DEFINITION** Inverse Relation
>
> The ordered pair (a, b) is in a relation if and only if the ordered pair (b, a) is in the **inverse relation**.

We will study the connection between a relation and its inverse. We will be most interested in inverse relations that happen to be *functions*. Notice that the graph of the inverse relation in Example 2 fails the vertical line test and is therefore not the graph of a function. Can we predict this failure by considering the graph of the original relation in Example 1? Figure 1.65 suggests that we can.

The inverse graph in Figure 1.65b fails the vertical line test because two different y-values have been paired with the same x-value. This is a direct consequence of the fact that the original relation in Figure 1.65a paired two different x-values with the same y-value. The inverse graph fails the *vertical* line test precisely because the original graph fails the *horizontal* line test. This gives us a test for relations whose inverses are functions.

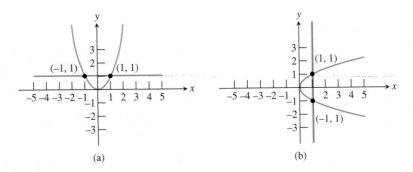

(a)   (b)

FIGURE 1.65 The inverse relation in (b) fails the vertical line test because the original relation in (a) fails the horizontal line test.

> Horizontal Line Test
>
> The inverse of a relation is a function if and only if each horizontal line intersects the graph of the original relation in at most one point.

EXAMPLE 3   Applying the Horizontal Line Test

Which of the graphs (1)–(4) in Figure 1.66 are graphs of

(a) relations that are functions?

(b) relations that have inverses that are functions?

SOLUTION

(a) Graphs (1) and (4) are graphs of functions because these graphs pass the vertical line test. Graphs (2) and (3) are not graphs of functions because these graphs fail the vertical line test.

continued

(b) Graphs (1) and (2) are graphs of relations whose inverses are functions because these graphs pass the horizontal line test. Graphs (3) and (4) fail the horizontal line test so their inverse relations are not functions.

Now try Exercise 9.

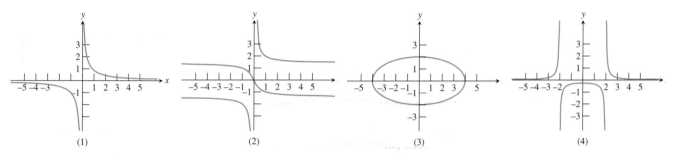

FIGURE 1.66 (Example 3)

A *function* whose inverse is a function has a graph that passes both the horizontal and vertical line tests (such as graph (1) in Example 3). Such a function is **one-to-one**, since every x is paired with a unique y and every y is paired with a unique x.

CAUTION ABOUT FUNCTION NOTATION

The symbol f^{-1} is read "f inverse" and should never be confused with the reciprocal of f. If f is a function, the symbol f^{-1}, can *only* mean f inverse. The reciprocal of f must be written as $1/f$.

DEFINITION Inverse Function

If f is a one-to-one function with domain D and range R, then the **inverse function of f**, denoted f^{-1}, is the function with domain R and range D defined by

$$f^{-1}(b) = a \quad \text{if and only if} \quad f(a) = b.$$

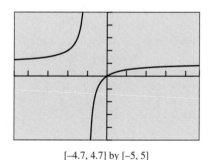

[−4.7, 4.7] by [−5, 5]

FIGURE 1.67 The graph of $f(x) = x/(x + 1)$. (Example 4)

EXAMPLE 4 Finding an Inverse Function Algebraically

Find an equation for $f^{-1}(x)$ if $f(x) = x/(x + 1)$.

SOLUTION The graph of f in Figure 1.67 suggests that f is one-to-one. The original function satisfies the equation $y = x/(x + 1)$. If f truly is one-to-one, the inverse function f^{-1} will satisfy the equation $x = y/(y + 1)$. (Note that we just switch the x and the y.)

If we solve this new equation for y we will have a formula for $f^{-1}(x)$:

$$x = \frac{y}{y + 1}$$

$$x(y + 1) = y \qquad \text{Multiply by } y + 1.$$

$$xy + x = y \qquad \text{Distributive property}$$

$$xy - y = -x \qquad \text{Isolate the } y \text{ terms.}$$

$$y(x - 1) = -x \qquad \text{Factor out } y.$$

$$y = \frac{-x}{x - 1} \qquad \text{Divide by } x - 1.$$

$$y = \frac{x}{1 - x} \qquad \text{Multiply numerator and denominator by } -1.$$

Therefore $f^{-1}(x) = x/(1 - x)$.

Now try Exercise 15.

Let us candidly admit two things regarding Example 4 before moving on to a graphical model for finding inverses. First, many functions are not one-to-one and so do not have inverse functions. Second, the algebra involved in finding an inverse function in the manner of Example 4 can be extremely difficult. We will actually find very few inverses this way. As you will learn in future chapters, we will usually rely on our understanding of how f maps x to y to understand how f^{-1} maps y to x.

It is possible to use the graph of f to produce a graph of f^{-1} without doing any algebra at all, thanks to the following geometric reflection property:

The Inverse Reflection Principle

The points (a, b) and (b, a) in the coordinate plane are symmetric with respect to the line $y = x$. The points (a, b) and (b, a) are **reflections** of each other across the line $y = x$.

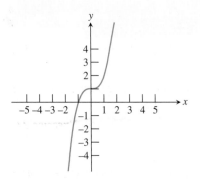

FIGURE 1.68 The graph of a one-to-one function. (Example 5)

EXAMPLE 5   Finding an Inverse Function Graphically

The graph of a function $y = f(x)$ is shown in Figure 1.68. Sketch a graph of the function $y = f^{-1}(x)$. Is f a one-to-one function?

SOLUTION We need not find a formula for $f^{-1}(x)$. All we need to do is to find the reflection of the given graph across the line $y = x$. This can be done geometrically.

Imagine a mirror along the line $y = x$ and draw the reflection of the given graph in the mirror. (See Figure 1.69.)

Another way to visualize this process is to imagine the graph to be drawn on a large pane of glass. Imagine the glass rotating around the line $y = x$ so that the *positive* x-axis switches places with the *positive* y-axis. (The back of the glass must be rotated to the front for this to occur.) The graph of f will then become the graph of f^{-1}.

Since the inverse of f has a graph that passes the horizontal and vertical line test, f is a one-to-one function.   ***Now try Exercise 23.***

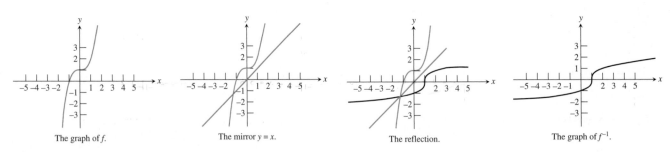

The graph of f.   The mirror $y = x$.   The reflection.   The graph of f^{-1}.

FIGURE 1.69 The Mirror Method. The graph of f is reflected in an imaginary mirror along the line $y = x$ to produce the graph of f^{-1}. (Example 5)

There is a natural connection between inverses and function composition that gives further insight into what an inverse actually does: It "undoes" the action of the original function. This leads to the following rule:

The Inverse Composition Rule

A function f is one-to-one with inverse function g if and only if

$$f(g(x)) = x \text{ for every } x \text{ in the domain of } g, \text{ and}$$

$$g(f(x)) = x \text{ for every } x \text{ in the domain of } f.$$

EXAMPLE 6   Verifying Inverse Functions

Show algebraically that $f(x) = x^3 + 1$ and $g(x) = \sqrt[3]{x - 1}$ are inverse functions.

SOLUTION   We use the Inverse Composition Rule.

$$f(g(x)) = f(\sqrt[3]{x - 1}) = (\sqrt[3]{x - 1})^3 + 1 = x - 1 + 1 = x$$

$$g(f(x)) = g(x^3 + 1) = \sqrt[3]{(x^3 + 1) - 1} = \sqrt[3]{x^3} = x$$

Since these equations are true for all x, the Inverse Composition Rule guarantees that f and g are inverses.

You do not have far to go to find graphical support of this algebraic verification, since these are the functions whose graphs are shown in Example 5!

Now try Exercise 27.

Some functions are so important that we need to study their inverses even though they are not one-to-one. A good example is the square root function, which is the "inverse" of the square function. It is not the inverse of the *entire* squaring function, because the full parabola fails the horizontal line test. Figure 1.70 shows that the function $y = \sqrt{x}$ is really the inverse of a "restricted-domain" version of $y = x^2$ defined only for $x \geq 0$.

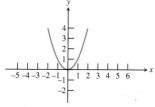

The graph of $y = x^2$ (not one-to-one).

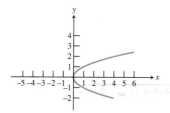

The inverse relation of $y = x^2$ (not a function).

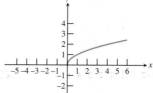

The graph of $y = \sqrt{x}$ (a function).

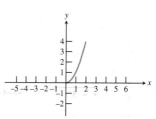

The graph of the function whose inverse is $y = \sqrt{x}$.

FIGURE 1.70 The function $y = x^2$ has no inverse function, but $y = \sqrt{x}$ is the inverse function of $y = x^2$ on the restricted domain $[0, \infty)$.

The consideration of domains adds a refinement to the algebraic inverse-finding method of Example 4, which we now summarize:

How to Find an Inverse Function Algebraically

Given a formula for a function f, proceed as follows to find a formula for f^{-1}.

1. Determine that there is a function f^{-1} by checking that f is one-to-one. State any restrictions on the domain of f. (Note that it might be necessary to impose some to get a one-to-one version of f.)

2. Switch x and y in the formula $y = f(x)$.

3. Solve for y to get the formula $y = f^{-1}(x)$. State any restrictions on the domain of f^{-1}.

EXAMPLE 7   Finding an Inverse Function

Show that $f(x) = \sqrt{x + 3}$ has an inverse function and find a rule for $f^{-1}(x)$. State any restrictions on the domains of f and f^{-1}.

SOLUTION

Solve Algebraically

The graph of f passes the horizontal line test, so f has an inverse function (Figure 1.71). Note that f has domain $[-3, \infty)$ and range $[0, \infty)$.

To find f^{-1} we write

$y = \sqrt{x + 3}$	where $x \geq -3,\ y \geq 0$	
$x = \sqrt{y + 3}$	where $y \geq -3,\ x \geq 0$	Interchange x and y.
$x^2 = y + 3$	where $y \geq -3,\ x \geq 0$	Square.
$y = x^2 - 3$	where $y \geq -3,\ x \geq 0$	Solve for y.

Thus $f^{-1}(x) = x^2 - 3$, with an "inherited" domain restriction of $x \geq 0$. Figure 1.71 shows the two functions. Note the domain restriction of $x \geq 0$ imposed on the parabola $y = x^2 - 3$.

Support Graphically

Use a grapher in parametric mode and compare the graphs of the two sets of parametric equations with Figure 1.71:

$$x = t \qquad \text{and} \qquad x = \sqrt{t + 3}$$
$$y = \sqrt{t + 3} \qquad\qquad\qquad y = t$$

Now try Exercise 17.

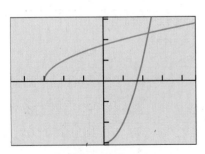

[−4.7, 4.7] by [−3.1, 3.1]

FIGURE 1.71 The graph of $f(x) = \sqrt{x + 3}$ and its inverse, a restricted version of $y = x^2 - 3$. (Example 7)

QUICK REVIEW 1.5 *(For help, go to Section P.3 and P.4.)*

In Exercises 1–10, solve the equation for y.

1. $x = 3y - 6$

2. $x = 0.5y + 1$

3. $x = y^2 + 4$

4. $x = y^2 - 6$

5. $x = \dfrac{y - 2}{y + 3}$

6. $x = \dfrac{3y - 1}{y + 2}$

7. $x = \dfrac{2y + 1}{y - 4}$

8. $x = \dfrac{4y + 3}{3y - 1}$

9. $x = \sqrt{y + 3}, y \geq -3$

10. $x = \sqrt{y - 2}, y \geq 2$

SECTION 1.5 EXERCISES

In Exercises 1–4, find the (x, y) pair for the value of the parameter.

1. $x = 3t$ and $y = t^2 + 5$ for $t = 2$

2. $x = 5t - 7$ and $y = 17 - 3t$ for $t = -2$

3. $x = t^3 - 4t$ and $y = \sqrt{t + 1}$ for $t = 3$

4. $x = |t + 3|$ and $y = 1/t$ for $t = -8$

In Exercises 5–8, complete the following. **(a)** Find the points determined by $t = -3, -2, -1, 0, 1, 2,$ and 3. **(b)** Find a direct algebraic relationship between x and y and determine whether the parametric equations determine y as a function of x. **(c)** Graph the relationship in the xy-plane.

5. $x = 2t$ and $y = 3t - 1$

6. $x = t + 1$ and $y = t^2 - 2t$

7. $x = t^2$ and $y = t - 2$

8. $x = \sqrt{t}$ and $y = 2t - 5$

In Exercises 9–12, the graph of a relation is shown. **(a)** Is the relation a function? **(b)** Does the relation have an inverse that is a function?

9.

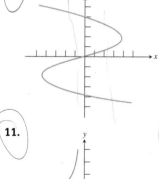

10.

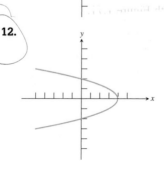

11.

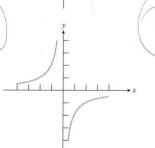

12.

In Exercises 13–22, find a formula for $f^{-1}(x)$. Give the domain of f^{-1}, including any restrictions "inherited" from f.

13. $f(x) = 3x - 6$

14. $f(x) = 2x + 5$

15. $f(x) = \dfrac{2x - 3}{x + 1}$

16. $f(x) = \dfrac{x + 3}{x - 2}$

17. $f(x) = \sqrt{x - 3}$

18. $f(x) = \sqrt{x + 2}$

19. $f(x) = x^3$

20. $f(x) = x^3 + 5$

21. $f(x) = \sqrt[3]{x + 5}$

22. $f(x) = \sqrt[3]{x - 2}$

In Exercises 23–26, determine whether the function is one-to-one. If it is one-to-one, sketch the graph of the inverse.

23. *don't sketch graph*

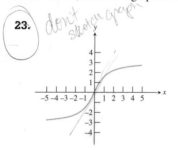

24.

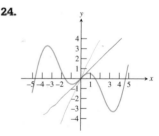

25.

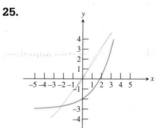

26.

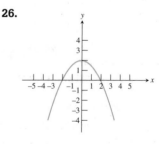

In Exercises 27–32, confirm that f and g are inverses by showing that $f(g(x)) = x$ and $g(f(x)) = x$.

27. $f(x) = 3x - 2$ and $g(x) = \dfrac{x + 2}{3}$

28. $f(x) = \dfrac{x + 3}{4}$ and $g(x) = 4x - 3$

29. $f(x) = x^3 + 1$ and $g(x) = \sqrt[3]{x - 1}$

30. $f(x) = \dfrac{7}{x}$ and $g(x) = \dfrac{7}{x}$

31. $f(x) = \dfrac{x + 1}{x}$ and $g(x) = \dfrac{1}{x - 1}$

32. $f(x) = \dfrac{x + 3}{x - 2}$ and $g(x) = \dfrac{2x + 3}{x - 1}$

33. Currency Conversion In May of 2002 the exchange rate for converting U.S. dollars (x) to euros (y) was $y = 1.08x$.

   (a) How many euros could you get for $100 U.S.?

   (b) What is the inverse function, and what conversion does it represent?

   (c) In the spring of 2002, a tourist had an elegant lunch in Provence, France ordering from a "fixed price" 48-euro menu. How much was that in U.S. dollars?

34. Temperature Conversion The formula for converting Celsius temperature (x) to Kelvin temperature is $k(x) = x + 273.16$. The formula for converting Fahrenheit temperature (x) to Celsius temperature is $c(x) = (5/9)(x - 32)$.

   (a) Find a formula for $c^{-1}(x)$. What is this formula used for?

   (b) Find $(k \circ c)(x)$. What is this formula used for?

35. Which pairs of basic functions (Section 1.3) are inverses of each other?

36. Which basic functions (Section 1.3) are their own inverses?

37. Which basic function can be defined parametrically as follows?
$$x = t^3 \text{ and } y = \sqrt{t^6} \text{ for } -\infty < t < \infty$$

38. Which basic function can be defined parametrically as follows?
$$x = 8t^3 \text{ and } y = (2t)^3 \text{ for } -\infty < t < \infty$$

Standardized Test Questions

39. True or False If f is a one-to-one function with domain D and range R, then f^{-1} is a one-to-one function with domain R and range D. Justify your answer.

40. True or False The set of points $(t + 1, 2t + 3)$ for all real numbers t form a line with slope 2. Justify your answer.

41. Multiple Choice Which ordered pair is in the *inverse* of the relation given by $x^2 y + 5y = 9$?

   (A) $(2, 1)$ **(B)** $(-2, 1)$ **(C)** $(-1, 2)$ **(D)** $(2, -1)$
   (E) $(1, -2)$

42. Multiple Choice Which ordered pair is not in the *inverse* of the relation given by $xy^2 - 3x = 12$?

   (A) $(0, -4)$ **(B)** $(4, 1)$ **(C)** $(3, 2)$ **(D)** $(2, 12)$
   (E) $(1, -6)$

43. Multiple Choice Which function is the *inverse* of the function $f(x) = 3x - 2$?

   (A) $g(x) = \dfrac{x}{3} + 2$

   (B) $g(x) = 2 - 3x$

   (C) $g(x) = \dfrac{x + 2}{3}$

   (D) $g(x) = \dfrac{x - 3}{2}$

   (E) $g(x) = \dfrac{x - 2}{3}$

44. Multiple Choice Which function is the *inverse* of the function $f(x) = x^3 + 1$?

   (A) $g(x) = \sqrt[3]{x - 1}$
   (B) $g(x) = \sqrt[3]{x} - 1$
   (C) $g(x) = x^3 - 1$
   (D) $g(x) = \sqrt[3]{x + 1}$
   (E) $g(x) = 1 - x^3$

Explorations

45. Function Properties Inherited by Inverses There are some properties of functions that are automatically shared by inverse functions (when they exist) and some that are not. Suppose that f has an inverse function f^{-1}. Give an algebraic or graphical argument (not a rigorous formal proof) to show that each of these properties of f must necessarily be shared by f^{-1}.

   (a) f is continuous.

   (b) f is one-to-one.

   (c) f is odd (graphically, symmetric with respect to the origin).

   (d) f is increasing.

46. Function Properties Not Inherited by Inverses There are some properties of functions that are not necessarily shared by inverse functions, even if the inverses exist. Suppose that f has an inverse function f^{-1}. For each of the following properties, give an example to show that f can have the property while f^{-1} does not.

   (a) f has a graph with a horizontal asymptote.

   (b) f has domain all real numbers.

   (c) f has a graph that is bounded above.

   (d) f has a removable discontinuity at $x = 5$.

47. Scaling Algebra Grades A teacher gives a challenging algebra test to her class. The lowest score is 52, which she decides to scale to 70. The highest score is 88, which she decides to scale to 97.

 (a) Using the points (52, 70) and (88, 97), find a linear equation that can be used to convert raw scores to scaled grades.

 (b) Find the inverse of the function defined by this linear equation. What does the inverse function do?

48. Writing to Learn (Continuation of Exercise 47) Explain why it is important for fairness that the scaling function used by the teacher be an *increasing* function. (Caution: It is *not* because "everyone's grade must go up." What would the scaling function in Exercise 47 do for a student who does enough "extra credit" problems to get a raw score of 136?)

Extending the Ideas

49. Modeling a Fly Ball Parametrically A baseball that leaves the bat at an angle of 60° from horizontal traveling 110 feet per second follows a path that can be modeled by the following pair of parametric equations. (You might enjoy verifying this if you have studied motion in physics.)

$$x = 110(t)\cos(60°)$$

$$y = 110(t)\sin(60°) - 16t^2$$

You can simulate the flight of the ball on a grapher. Set your grapher to parametric mode and put the functions above in for X2T and Y2T. Set X1T = 325 and Y1T = 5T to draw a 30-foot fence 325 feet from home plate. Set Tmin = 0, Tmax = 6, Tstep = 0.1, Xmin = 0, Xmax = 350, Xscl = 0, Ymin = 0, Ymax = 300, and Yscl = 0.

 (a) Now graph the function. Does the fly ball clear the fence?

 (b) Change the angle to 30° and run the simulation again. Does the ball clear the fence?

 (c) What angle is optimal for hitting the ball? Does it clear the fence when hit at that angle?

50. The Baylor GPA Scale Revisited (*See Problem 78 in Section 1.2.*) The function used to convert Baylor School percentage grades to GPAs on a 4-point scale is

$$y = \left(\frac{3^{1.7}}{30}(x - 65) \right)^{\frac{1}{1.7}} + 1.$$

The function has domain [65, 100]. Anything below 65 is a failure and automatically converts to a GPA of 0.

 (a) Find the inverse function algebraically. What can the inverse function be used for?

 (b) Does the inverse function have any domain restrictions?

 (c) Verify with a graphing calculator that the function found in (a) and the given function are really inverses.

51. Group Activity (Continuation of Exercise 50) The number 1.7 that appears in two places in the GPA scaling formula is called the scaling factor (k). The value of k can be changed to alter the curvature of the graph while keeping the points (65, 1) and (95, 4) fixed. It was felt that the lowest D (65) needed to be scaled to 1.0, while the middle A (95) needed to be scaled to 4.0. The faculty's Academic Council considered several values of k before settling on 1.7 as the number that gives the "fairest" GPAs for the other percentage grades.

Try changing k to other values between 1 and 2. What kind of scaling curve do you get when $k = 1$? Do you agree with the Baylor decision that $k = 1.7$ gives the fairest GPAs?

1.6
Graphical Transformations

What you'll learn about

- Transformations
- Vertical and Horizontal Translations
- Reflections Across Axes
- Vertical and Horizontal Stretches and Shrinks
- Combining Transformations

. . . and why

Studying transformations will help you to understand the relationships between graphs that have similarities but are not the same.

Transformations

The following functions are all different:

$$y = x^2$$
$$y = (x - 3)^2$$
$$y = 1 - x^2$$
$$y = x^2 - 4x + 5$$

However, a look at their graphs shows that, while no two are exactly the same, all four have the same identical *shape* and *size*. Understanding how algebraic alterations change the shapes, sizes, positions, and orientations of graphs is helpful for understanding the connection between algebraic and graphical models of functions.

In this section we relate graphs using **transformations**, which are functions that map real numbers to real numbers. By acting on the x-coordinates and y-coordinates of points, transformations change graphs in predictable ways. **Rigid transformations**, which leave the size and shape of a graph unchanged, include horizontal translations, vertical translations, reflections, or any combination of these. **Non-rigid transformations**, which generally distort the shape of a graph, include horizontal or vertical stretches and shrinks.

Vertical and Horizontal Translations

A **vertical translation** of the graph of $y = f(x)$ is a shift of the graph up or down in the coordinate plane. A **horizontal translation** is a shift of the graph to the left or the right. The following exploration will give you a good feel for what translations are and how they occur.

EXPLORATION 1   Introducing Translations

Set your viewing window to $[-5, 5]$ by $[-5, 15]$ and your graphing mode to sequential as opposed to simultaneous.

1. Graph the functions

$$y_1 = x^2 \qquad\qquad y_4 = y_1(x) - 2 = x^2 - 2$$
$$y_2 = y_1(x) + 3 = x^2 + 3 \qquad y_5 = y_1(x) - 4 = x^2 - 4$$
$$y_3 = y_1(x) + 1 = x^2 + 1$$

on the same screen. What effect do the $+3$, $+1$, -2, and -4 seem to have?

2. Graph the functions

$$y_1 = x^2 \qquad\qquad y_4 = y_1(x - 2) = (x - 2)^2$$
$$y_2 = y_1(x + 3) = (x + 3)^2 \qquad y_5 = y_1(x - 4) = (x - 4)^2$$
$$y_3 = y_1(x + 1) = (x + 1)^2$$

on the same screen. What effect do the $+3$, $+1$, -2, and -4 seem to have?

3. Repeat steps 1 and 2 for the functions $y_1 = x^3$, $y_1 = |x|$, and $y_1 = \sqrt{x}$. Do your observations agree with those you made after steps 1 and 2?

TECHNOLOGY ALERT

In Exploration 1, the notation $y_1(x + 3)$ means the function y_1, evaluated at $x + 3$. It does not mean multiplication.

In general, *replacing x by x − c* shifts the graph horizontally c units. Similarly, *replacing y by y − c* shifts the graph vertically c units. If c is positive the shift is to the right or up; if c is negative the shift is to the left or down.

This is a nice, consistent rule that unfortunately gets complicated by the fact that the c for a vertical shift rarely shows up being subtracted from y. Instead, it usually shows up on the other side of the equal sign being *added* to f(x). That leads us to the following rule, which only *appears* to be different for horizontal and vertical shifts:

Translations

Let c be a positive real number. Then the following transformations result in translations of the graph of $y = f(x)$:

Horizontal Translations

$$y = f(x - c)$$ a translation to the right by c units

$$y = f(x + c)$$ a translation to the left by c units

Vertical Translations

$$y = f(x) + c$$ a translation up by c units

$$y = f(x) - c$$ a translation down by c units

EXAMPLE 1   Vertical Translations

Describe how the graph of $y = |x|$ can be transformed to the graph of the given equation.

(a) $y = |x| - 4$ **(b)** $y = |x + 2|$

SOLUTION

(a) The equation is in the form $y = f(x) - 4$, a translation down by 4 units. See Figure 1.72.

(b) The equation is in the form $y = f(x + 2)$, a translation left by 2 units. See Figure 1.73 on the next page.   *Now try Exercise 3.*

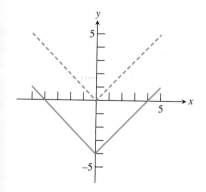

FIGURE 1.72 $y = |x| - 4$ (Example 1)

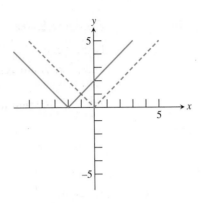

FIGURE 1.73 $y = |x + 2|$ (Example 1)

EXAMPLE 2   Finding Equations for Translations

Each view in Figure 1.74 shows the graph of $y_1 = x^3$ and a vertical or horizontal translation y_2. Write an equation for y_2 as shown in each graph.

SOLUTION

(a) $y_2 = x^3 - 3 = y_1(x) - 3$ (a vertical translation down by 3 units)

(b) $y_2 = (x + 2)^3 = y_1(x + 2)$ (a horizontal translation left by 2 units)

(c) $y_2 = (x - 3)^3 = y_1(x - 3)$ (a horizontal translation right by 3 units)

Now try Exercise 25.

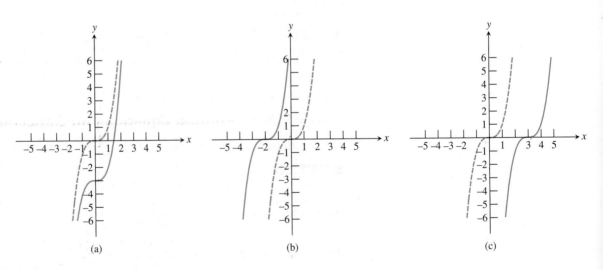

(a)   (b)   (c)

FIGURE 1.74 Translations of $y_1 = x^3$. (Example 2)

Reflections Across Axes

Points (x, y) and $(x, -y)$ are **reflections of each other across the x-axis**. Points (x, y) and $(-x, y)$ are **reflections of each other across the y-axis**. (See Figure 1.75.) Two points (or graphs) that are symmetric with respect to a line are **reflections of each other across that line**.

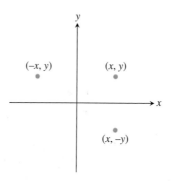

FIGURE 1.75 The point (x, y) and its reflections across the x- and y-axes.

Figure 1.75 suggests that a reflection across the x-axis results when y is replaced by $-y$, and a reflection across the y-axis results when x is replaced by $-x$.

Reflections

The following transformations result in reflections of the graph of $y = f(x)$:

Across the x-axis

$$y = -f(x)$$

Across the y-axis

$$y = f(-x)$$

EXAMPLE 3 Finding Equations for Reflections

Find an equation for the reflection of $f(x) = \dfrac{5x - 9}{x^2 + 3}$ across each axis.

SOLUTION

Solve Algebraically

Across the x-axis: $y = -f(x) = -\dfrac{5x - 9}{x^2 + 3} = \dfrac{9 - 5x}{x^2 + 3}$

Across the y-axis: $y = f(-x) = \dfrac{5(-x) - 9}{(-x)^2 + 3} = \dfrac{-5x - 9}{x^2 + 3}$

Support Graphically

The graphs in Figure 1.76 support our algebraic work.

continued

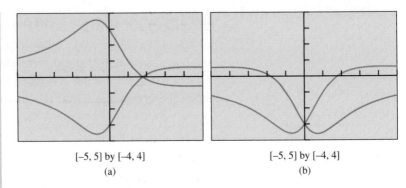

[–5, 5] by [–4, 4]   [–5, 5] by [–4, 4]

(a)   (b)

FIGURE 1.76 Reflections of $f(x) = (5x - 9)/(x^2 + 3)$ across (a) the x-axis and (b) the y-axis. (Example 3)   *Now try Exercise 29.*

You might expect that odd and even functions, whose graphs already possess special symmetries, would exhibit special behavior when reflected across the axes. They do, as shown by Example 4 and Exercises 33 and 34.

EXAMPLE 4   Reflecting Even Functions

Prove that the graph of an even function remains unchanged when it is reflected across the y-axis.

SOLUTION

Note that we can get plenty of graphical support for these statements by reflecting the graphs of various even functions, but what is called for here is **proof**, which will require algebra.

Let f be an even function; that is, $f(-x) = f(x)$ for all x in the domain of f. To reflect the graph of $y = f(x)$ across the y-axis, we make the transformation $y = f(-x)$. But $f(-x) = f(x)$ for all x in the domain of f, so this transformation results in $y = f(x)$. The graph of f therefore remains unchanged.

Now try Exercise 33.

Graphing Absolute Value Compositions

Given the graph of $y = f(x)$,

the graph of $y = |f(x)|$ can be obtained by reflecting the portion of the graph below the x-axis across the x-axis, leaving the portion above the x-axis unchanged;

the graph of $y = f(|x|)$ can be obtained by *replacing* the portion of the graph to the left of the y-axis by a reflection of the portion to the right of the y-axis across the y-axis, leaving the portion to the right of the y-axis unchanged. (The result will show even symmetry.)

Function compositions with absolute value can be realized graphically by reflecting portions of graphs, as you will see in the following Exploration.

EXPLORATION 2 Compositions with Absolute Value

The graph of $y = f(x)$ is shown at the right.
Match each of the graphs below with one of the
following equations and use the language of
function reflection to defend your match. Note
that two of the graphs will not be used.

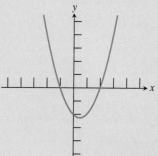

1. $y = |f(x)|$
2. $y = f(|x|)$
3. $y = -|f(x)|$
4. $y = |f(|x|)|$

(A)

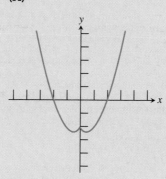

(B)

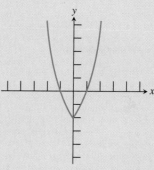

(C)

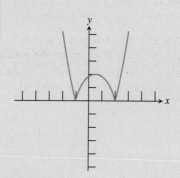

(D)

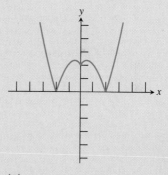

(E)

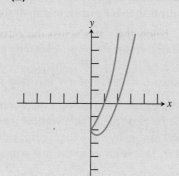

(F)

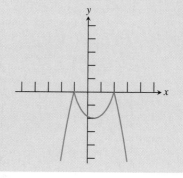

Vertical and Horizontal Stretches and Shrinks

We now investigate what happens when we multiply all the *y*-coordinates (or all the *x*-coordinates) of a graph by a fixed real number.

EXPLORATION 3 **Introducing Stretches and Shrinks**

Set your viewing window to $[-4.7, 4.7]$ by $[-1.1, 5.1]$ and your graphing mode to sequential as opposed to simultaneous.

1. Graph the functions

$$y_1 = \sqrt{4 - x^2}$$
$$y_2 = 1.5y_1(x) = 1.5\sqrt{4 - x^2}$$
$$y_3 = 2y_1(x) = 2\sqrt{4 - x^2}$$
$$y_4 = 0.5y_1(x) = 0.5\sqrt{4 - x^2}$$
$$y_5 = 0.25y_1(x) = 0.25\sqrt{4 - x^2}$$

on the same screen. What effect do the 1.5, 2, 0.5, and 0.25 seem to have?

2. Graph the functions

$$y_1 = \sqrt{4 - x^2}$$
$$y_2 = y_1(1.5x) = \sqrt{4 - (1.5x)^2}$$
$$y_3 = y_1(2x) = \sqrt{4 - (2x)^2}$$
$$y_4 = y_1(0.5x) = \sqrt{4 - (0.5x)^2}$$
$$y_5 = y_1(0.25x) = \sqrt{4 - (0.25x)^2}$$

on the same screen. What effect do the 1.5, 2, 0.5, and 0.25 seem to have?

Exploration 3 suggests that multiplication of *x* or *y* by a constant results in a horizontal or vertical stretching or shrinking of the graph.

In general, *replacing x by x/c* distorts the graph horizontally by a factor of *c*. Similarly, *replacing y by y/c* distorts the graph vertically by a factor of *c*. If *c* is greater than 1 the distortion is a stretch; if *c* is less than 1 the distortion is a shrink.

As with translations, this is a nice, consistent rule that unfortunately gets complicated by the fact that the *c* for a vertical stretch or shrink rarely shows up as a divisor of *y*. Instead, it usually shows up on the other side of the equal sign as a *factor* multiplied by $f(x)$. That leads us to the following rule:

EXAMPLE 5 Finding Equations for Stretches and Shrinks

Let C_1 be the curve defined by $y_1 = f(x) = x^3 - 16x$. Find equations for the following non-rigid transformations of C_1:

(a) C_2 is a vertical stretch of C_1 by a factor of 3.

(b) C_3 is a horizontal shrink of C_1 by a factor of 1/2.

SOLUTION

Solve Algebraically

(a) Denote the equation for C_2 by y_2. Then

$$y_2 = 3 \cdot f(x)$$
$$= 3(x^3 - 16x)$$
$$= 3x^3 - 48x$$

(b) Denote the equation for C_3 by y_3. Then

$$y_3 = f\left(\frac{x}{1/2}\right)$$
$$= f(2x)$$
$$= (2x)^3 - 16(2x)$$
$$= 8x^3 - 32x$$

Support Graphically

The graphs in Figure 1.77 support our algebraic work.

Now try Exercise 39.

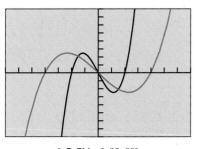

[–7, 7] by [–80, 80]

(a)

[–7, 7] by [–80, 80]

(b)

FIGURE 1.77 The graph of $y_1 = f(x) = x^3 - 16x$, shown with (a) a vertical stretch and (b) a horizontal shrink. (Example 5)

Combining Transformations

Transformations may be performed in succession—one after another. If the transformations include stretches, shrinks, or reflections, the order in which the transformations are performed may make a difference. In those cases, be sure to pay particular attention to order.

EXAMPLE 6   Combining Transformations in Order

(a) The graph of $y = x^2$ undergoes the following transformations, in order. Find the equation of the graph that results.

- a horizontal shift 2 units to the right

- a vertical stretch by a factor of 3

- a vertical translation 5 units up

(b) Apply the transformations in (a) in the opposite order and find the equation of the graph that results.

SOLUTION

(a) Applying the transformations in order, we have

$$x^2 \Rightarrow (x - 2)^2 \Rightarrow 3(x - 2)^2 \Rightarrow 3(x - 2)^2 + 5$$

Expanding the final expression, we get the function $y = 3x^2 - 12x + 17$.

(b) Applying the transformations in the opposite order, we have

$$x^2 \Rightarrow x^2 + 5 \Rightarrow 3(x^2 + 5) \Rightarrow 3((x - 2)^2 + 5)$$

Expanding the final expression, we get the function $y = 3x^2 - 12x + 27$.

The second graph is ten units higher than the first graph because the vertical stretch lengthens the vertical translation when the translation occurs first. Order often matters when stretches, shrinks, or reflections are involved.

Now try Exercise 47.

EXAMPLE 7   Transforming a Graph Geometrically

The graph of $y = f(x)$ is shown in Figure 1.78. Determine the graph of the composite function $y = 2f(x + 1) - 3$ by showing the effect of a sequence of transformations on the graph of $y = f(x)$.

SOLUTION

The graph of $y = 2f(x + 1) - 3$ can be obtained from the graph of $y = f(x)$ by the following sequence of transformations:

(a) a vertical stretch by a factor of 2 to get $y = 2f(x)$ (Figure 1.79a)

(b) a horizontal translation 1 unit to the left to get $y = 2f(x + 1)$ (Figure 1.79b)

(c) a vertical translation 3 units down to get $y = 2f(x + 1) - 3$ (Figure 1.79c)

(The order of the first two transformations can be reversed without changing the final graph.)

Now try Exercise 51.

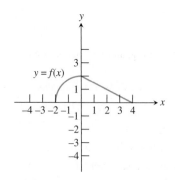

FIGURE 1.78 The graph of the function $y = f(x)$ in Example 7.

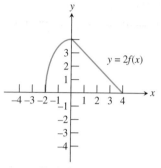
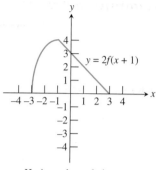
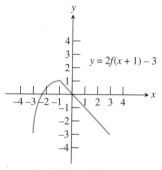

Vertical stretch
of factor 2

(a)

Horizontal translation
left 1 unit

(b)

Vertical translation
down 3 units

(c)

FIGURE 1.79 Transforming the graph of $y = f(x)$ in Figure 1.78 to get the graph of $y = 2f(x + 1) - 3$. (Example 7)

QUICK REVIEW 1.6 *(For help, go to Section A.2.)*

In Exercises 1–6, write the expression as a binomial squared.

1. $x^2 + 2x + 1$

2. $x^2 - 6x + 9$

3. $x^2 + 12x + 36$

4. $4x^2 + 4x + 1$

5. $x^2 - 5x + \dfrac{25}{4}$

6. $4x^2 - 20x + 25$

In Exercises 7–10, perform the indicated operations and simplify.

7. $(x - 2)^2 + 3(x - 2) + 4$

8. $2(x + 3)^2 - 5(x + 3) - 2$

9. $(x - 1)^3 + 3(x - 1)^2 - 3(x - 1)$

10. $2(x + 1)^3 - 6(x + 1)^2 + 6(x + 1) - 2$

SECTION 1.6 EXERCISES

In Exercises 1–8, describe how the graph of $y = x^2$ can be transformed to the graph of the given equation.

1. $y = x^2 - 3$

2. $y = x^2 + 5.2$

3. $y = (x + 4)^2$

4. $y = (x - 3)^2$

5. $y = (100 - x)^2$

6. $y = x^2 - 100$

7. $y = (x - 1)^2 + 3$

8. $y = (x + 50)^2 - 279$

In Exercises 9–12, describe how the graph of $y = \sqrt{x}$ can be transformed to the graph of the given equation.

9. $y = -\sqrt{x}$

10. $y = \sqrt{x - 5}$

11. $y = \sqrt{-x}$

12. $y = \sqrt{3 - x}$

In Exercises 13–16, describe how the graph of $y = x^3$ can be transformed to the graph of the given equation.

13. $y = 2x^3$

14. $y = (2x)^3$

15. $y = (0.2x)^3$

16. $y = 0.3x^3$

In Exercises 17–20, describe how to transform the graph of f into the graph of g.

17. $f(x) = \sqrt{x + 2}$ and $g(x) = \sqrt{x - 4}$

18. $f(x) = (x - 1)^2$ and $g(x) = -(x + 3)^2$

19. $f(x) = (x - 2)^3$ and $g(x) = -(x + 2)^3$

20. $f(x) = |2x|$ and $g(x) = 4|x|$

In Exercises 21–24, sketch the graphs of f, g, and h by hand. Support your answers with a grapher.

21. $f(x) = (x + 2)^2$
$g(x) = 3x^2 - 2$
$h(x) = -2(x - 3)^2$

22. $f(x) = x^3 - 2$
$g(x) = (x + 4)^3 - 1$
$h(x) = 2(x - 1)^3$

23. $f(x) = \sqrt[3]{x + 1}$
$g(x) = 2\sqrt[3]{x} - 2$
$h(x) = -\sqrt[3]{x} - 3$

24. $f(x) = -2|x| - 3$
$g(x) = 3|x + 5| + 4$
$h(x) = |3x|$

cises 25–28, the graph is that of a function $y = f(x)$ that can
obtained by transforming the graph of $y = \sqrt{x}$. Write a formula
the function f.

25.
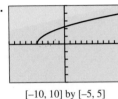
[−10, 10] by [−5, 5]

26.
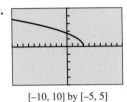
[−10, 10] by [−5, 5]

27.
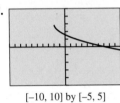
[−10, 10] by [−5, 5]

28.
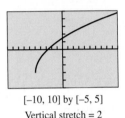
[−10, 10] by [−5, 5]
Vertical stretch = 2

In Exercises 29–32, find the equation of the reflection of f across **(a)**
the x-axis and **(b)** the y-axis.

29. $f(x) = x^3 - 5x^2 - 3x + 2$ **30.** $f(x) = 2\sqrt{x + 3} - 4$

31. $f(x) = \sqrt[3]{8x}$ **32.** $f(x) = 3|x + 5|$

33. Reflecting Odd Functions Prove that the graph of an odd
function is the same when reflected across the x-axis as it is when
reflected across the y-axis.

34. Reflecting Odd Functions Prove that if an odd function is
reflected about the y-axis and then reflected again about the
x-axis, the result is the original function.

Exercises 35–38 refer to the graph of $y = f(x)$ shown below. In each
case, sketch a graph of the new function.

35. $y = |f(x)|$ **36.** $y = f(|x|)$

37. $y = -f(|x|)$ **38.** $y = |f(|x|)|$

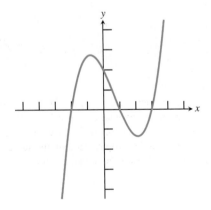

In Exercises 39–42, transform the given function by **(a)** a vertical
stretch by a factor of 2, and **(b)** a horizontal shrink by a factor of 1/3.

39. $f(x) = x^3 - 4x$ **40.** $f(x) = |x + 2|$

41. $f(x) = x^2 + x - 2$ **42.** $f(x) = \dfrac{1}{x + 2}$

In Exercises 43–46, describe a basic graph and a sequence of transformations that can be used to produce a graph of the given function.

43. $y = 2(x - 3)^2 - 4$ **44.** $y = -3\sqrt{x + 1}$

45. $y = (3x)^2 - 4$ **46.** $y = -2|x + 4| + 1$

In Exercises 47–50, a graph G is obtained from a graph of y by the
sequence of transformations indicated. Write an equation whose graph
is G.

47. $y = x^2$: a vertical stretch by a factor of 3, then a shift right
4 units.

48. $y = x^2$: a shift right 4 units, then a vertical stretch by a
factor of 3.

49. $y = |x|$: a shift left 2 units, then a vertical stretch by a factor of 2,
and finally a shift down 4 units.

50. $y = |x|$: a shift left 2 units, then a horizontal shrink by a
factor of 1/2, and finally a shift down 4 units.

Exercises 51–54 refer to the function f whose graph is shown below.

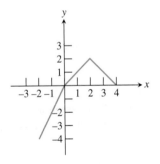

51. Sketch the graph of $y = 2 + 3f(x + 1)$.

52. Sketch the graph of $y = -f(x + 1) + 1$.

53. Sketch the graph of $y = f(2x)$.

54. Sketch the graph of $y = 2f(x - 1) + 2$.

55. Writing to Learn Graph some examples to convince yourself
that a reflection and a translation can have a different effect when
combined in one order than when combined in the opposite order.
Then explain in your own words why this can happen.

56. Writing to Learn Graph some examples to convince yourself
that vertical stretches and shrinks do not affect a graph's x-intercepts.
Then explain in your own words why this is so.

57. Celsius vs. Fahrenheit The graph shows the temperature in degrees Celsius in Windsor, Ontario, for one 24-hour period. Describe the transformations that convert this graph to one showing degrees Fahrenheit. [*Hint*: $F(t) = (9/5)C(t) + 32$.]

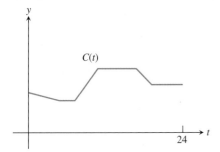

58. Fahrenheit vs. Celsius The graph shows the temperature in degrees Fahrenheit in Mt. Clemens, Michigan, for one 24-hour period. Describe the transformations that convert this graph to one showing degrees Celsius.
[*Hint*: $F(t) = (9/5)C(t) + 32$.]

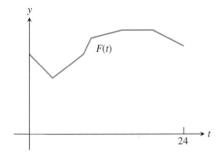

Standardized Test Questions

59. True or False The function $y = f(x + 3)$ represents a translation to the right by 3 units of the graph of $y = f(x)$. Justify your answer.

60. True or False The function $y = f(x) - 4$ represents a translation down 4 units of the graph of $y = f(x)$. Justify your answer.

In Exercises 61–64, you may use a graphing calculator to answer the question.

61. Multiple Choice Given a function f, which of the following represents a vertical stretch by a factor of 3?

(A) $y = f(3x)$ (B) $y = f(x/3)$

(C) $y = 3f(x)$ (D) $y = f(x)/3$

(E) $y = f(x) + 3$

62. Multiple Choice Given a function f, which of the following represents a horizontal translation of 4 units to the right?

(A) $y = f(x) + 4$ (B) $y = f(x) - 4$

(C) $y = f(x + 4)$ (D) $y = f(x - 4)$

(E) $y = 4f(x)$

63. Multiple Choice Given a function f, which of the following represents a vertical translation of 2 units upward, followed by a reflection across the y-axis?

(A) $y = f(-x) + 2$ (B) $y = 2 - f(x)$

(C) $y = f(2 - x)$ (D) $y = -f(x - 2)$

(E) $y = f(x) - 2$

64. Multiple Choice Given a function f, which of the following represents reflection across the x-axis, followed by a horizontal shrink by a factor of 1/2?

(A) $y = -2f(x)$ (B) $y = -f(x)/2$

(C) $y = f(-2x)$ (D) $y = -f(x/2)$

(E) $y = -f(2x)$

Explorations

65. International Finance Table 1.11 shows the price of a share of stock in Dell Computer for the first eight months of 2004:

Table 1.11 Dell Computer	
Month	Price ($)
1	33.44
2	32.65
3	33.62
4	34.78
5	35.24
6	35.82
7	35.47
8	34.84

Source: Yahoo! Finance

(a) Graph price (y) as a function of month (x) as a line graph, connecting the points to make a continuous graph.

(b) Explain what transformation you would apply to this graph to produce a graph showing the price of the stock in Japanese yen.

66. Group Activity Get with a friend and graph the function $y = x^2$ on both your graphers. Apply a horizontal or vertical stretch or shrink to the function on one of the graphers. Then change the *window* of that grapher to make the two graphs look the same. Can you formulate a general rule for how to find the window?

Extending the Ideas

67. The Absolute Value Transformation Graph the function
$f(x) = x^4 - 5x^3 + 4x^2 + 3x + 2$ in the viewing window $[-5, 5]$
by $[-10, 10]$. (Put the equation in Y1.)

(a) Study the graph and try to predict what the graph of
$y = |f(x)|$ will look like. Then turn Y1 off and graph
Y2 = abs (Y1). Did you predict correctly?

(b) Study the original graph again and try to predict what the
graph of $y = f(|x|)$ will look like. Then turn Y1 off and graph
Y2 = Y1(abs (X)). Did you predict correctly?

(c) Given the graph of $y = g(x)$ shown below, sketch a graph of
$y = |g(x)|$.

(d) Given the graph of $y = g(x)$ shown below, sketch a graph of
$y = g(|x|)$.

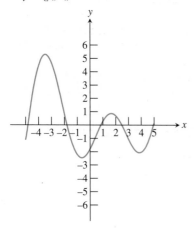

68. Parametric Circles and Ellipses Set your grapher to parametric and radian mode and your window as follows:

Tmin $= 0$, Tmax $= 7$, Tstep $= 0.1$

Xmin $= -4.7$, Xmax $= 4.7$, Xscl $= 1$

Ymin $= -3.1$, Ymax $= 3.1$, Yscl $= 1$

(a) Graph the parametric equations $x = \cos t$ and $y = \sin t$. You
should get a circle of radius 1.

(b) Use a transformation of the parametric function of x to produce
the graph of an ellipse that is 4 units wide and 2 units tall.

(c) Use a transformation of both parametric functions to produce a
circle of radius 3.

(d) Use a transformation of both functions to produce an ellipse
that is 8 units wide and 4 units tall.

(You will learn more about ellipses in Chapter 8.)

1.7
Modeling with Functions

... and why

Using a function to model a variable under observation in terms of another variable often allows one to make predictions in practical situations, such as predicting the future growth of a business based on known data.

Functions from Formulas

Now that you have learned more about what functions are and how they behave, we want to return to the modeling theme of Section 1.1. In that section we stressed that one of the goals of this course was to become adept at using numerical, algebraic, and graphical models of the real world in order to solve problems. We now want to focus your attention more precisely on modeling with *functions.*

You have already seen quite a few formulas in the course of your education. Formulas involving two variable quantities always relate those variables implicitly, and quite often the formulas can be solved to give one variable explicitly as a function of the other. In this book we will use a variety of formulas to pose and solve problems algebraically, although we will not assume prior familiarity with those formulas that we borrow from other subject areas (like physics or economics). We *will* assume familiarity with certain key formulas from mathematics.

EXAMPLE 1   Forming Functions from Formulas

Write the area A of a circle as a function of its

(a) radius r.

(b) diameter d.

(c) circumference C.

SOLUTION

(a) The familiar area formula from geometry gives A as a function of r:

$$A = \pi r^2.$$

(b) This formula is not so familiar. However, we know that $r = d/2$, so we can substitute that expression for r in the area formula:

$$A = \pi r^2 = \pi (d/2)^2 = (\pi/4)d^2.$$

(c) Since $C = 2\pi r$, we can solve for r to get $r = C/(2\pi)$. Then substitute to get A:
$A = \pi r^2 = \pi(C/(2\pi))^2 = \pi C^2/(4\pi^2) = C^2/(4\pi)$.

Now try Exercise 19.

EXAMPLE 2   A Maximum Value Problem

A square of side x inches is cut out of each corner of an 8 in. by 15 in. piece of cardboard and the sides are folded up to form an open-topped box (Figure 1.80).

(a) Write the volume V of the box as a function of x.

(b) Find the domain of V as a function of x. (Note that the model imposes restrictions on x.)

(c) Graph V as a function of x over the domain found in part (b) and use the maximum finder on your grapher to determine the maximum volume such a box can hold.

(d) How big should the cut-out squares be in order to produce the box of maximum volume?

SOLUTION

(a) The box will have a base with sides of width $8 - 2x$ and length $15 - 2x$. The depth of the box will be x when the sides are folded up. Therefore $V = x(8 - 2x)(15 - 2x)$.

(b) The formula for V is a polynomial with domain all reals. However, the depth x must be nonnegative, as must the width of the base, $8 - 2x$. Together, these two restrictions yield a domain of $[0, 4]$. (The endpoints give a box with no volume, which is as mathematically feasible as other zero concepts.)

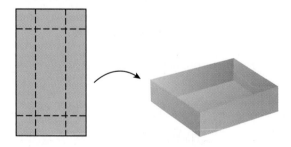

FIGURE 1.80 An open-topped box made by cutting the corners from a piece of cardboard and folding up the sides. (Example 2)

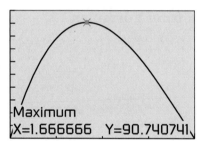

[0, 4] by [0, 100]

(a)

Maximum
X=1.666666   Y=90.740741

[0, 4] by [0, 100]

(b)

FIGURE 1.81 The graph of the volume of the box in Example 2.

(c) The graph is shown in Figure 1.81. The maximum finder shows that the maximum occurs at the point (5/3, 90.74). The maximum volume is about 90.74 in.³.

(d) Each square should have sides of one-and-two-thirds inches.

Now try Exercise 33.

Functions from Graphs

When "thinking graphically" becomes a genuine part of your problem-solving strategy, it is sometimes actually easier to start with the graphical model than it is to go straight to the algebraic formula. The graph provides valuable information about the function.

EXAMPLE 3   Protecting an Antenna

A small satellite dish is packaged with a cardboard cylinder for protection. The parabolic dish is 24 in. in diameter and 6 in. deep, and the diameter of the cardboard cylinder is 12 in. How tall must the cylinder be to fit in the middle of the dish and be flush with the top of the dish? (See Figure 1.82.)

SOLUTION

Solve Algebraically

The diagram in Figure 1.82a showing the cross section of this 3-dimensional problem is also a 2-dimensional graph of a quadratic function. We can transform our basic function $y = x^2$ with a vertical shrink so that it goes through the points $(12, 6)$ and $(-12, 6)$, thereby producing a graph of the parabola in the coordinate plane (Figure 1.82b).

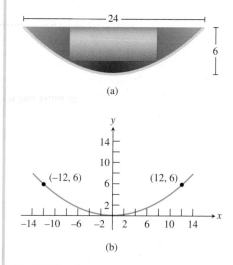

FIGURE 1.82 (a) A parabolic satellite dish with a protective cardboard cylinder in the middle for packaging. (b) The parabola in the coordinate plane. (Example 3)

$$y = kx^2 \qquad \text{Vertical shrink}$$

$$6 = k(\pm 12)^2 \qquad \text{Substitute } x = \pm 12, y = 6.$$

$$k = \frac{6}{144} = \frac{1}{24} \qquad \text{Solve for } k.$$

Thus, $y = \dfrac{1}{24}x^2$.

To find the height of the cardboard cylinder, we first find the y-coordinate of the parabola 6 inches from the center, that is, when $x = 6$:

$$y = \frac{1}{24}(6)^2 = 1.5$$

From that point to the top of the dish is $6 - 1.5 = 4.5$ in.

Now try Exercise 35.

Although Example 3 serves nicely as a "functions from graphs" example, it is also an example of a function that must be constructed by gathering relevant information from a verbal description and putting it together in the right way. People who do mathematics for a living are accustomed to confronting that challenge regularly as a necessary first step in modeling the real world. In honor of its importance, we have saved it until last to close out this chapter in style.

Functions from Verbal Descriptions

There is no fail-safe way to form a function from a verbal description. It can be hard work, frequently a good deal harder than the mathematics required to solve the problem once the function has been found. The 4-step problem-solving process in Section 1.1 gives you several valuable tips, perhaps the most important of which is to *read* the problem carefully. Understanding what the words say is critical if you hope to model the situation they describe.

EXAMPLE 4   Finding the Model and Solving

Grain is leaking through a hole in a storage bin at a constant rate of 8 cubic inches per minute. The grain forms a cone-shaped pile on the ground below. As it grows, the height of the cone always remains equal to its radius. If the cone is one foot tall now, how tall will it be in one hour?

FIGURE 1.83 A cone with equal height and radius. (Example 4)

SOLUTION

Reading the problem carefully, we realize that the formula for the volume of the cone is needed (Figure 1.83). From memory or by looking it up, we get the formula $V = (1/3)\pi r^2 h$. A careful reading also reveals that the height and the radius are always equal, so we can get volume directly as a function of height: $V = (1/3)\pi h^3$.

When $h = 12$ in., the volume is $V = (\pi/3)(12)^3 = 576\pi$ in.^{3}.

One hour later, the volume will have grown by $(60 \text{ min})(8 \text{ in.}^3/\text{min}) = 480$ in.^{3}. The total volume of the pile at that point will be $(576\pi + 480)$ in.^{3}. Finally, we use the volume formula once again to solve for h:

$$\frac{1}{3}\pi h^3 = 576\pi + 480$$

$$h^3 = \frac{3(576\pi + 480)}{\pi}$$

$$h = \sqrt[3]{\frac{3(576\pi + 480)}{\pi}}$$

$$h \approx 12.98 \text{ inches.}$$

Now try Exercise 37.

EXAMPLE 5 Letting Units Work for You

How many rotations does a 15 in. (radius) tire make per second on a sport utility vehicle traveling 70 mph?

SOLUTION It is the perimeter of the tire that comes in contact with the road, so we first find the circumference of the tire:

$$C = 2\pi r = 2\pi(15) = 30\pi \text{ in.}$$

That means that 1 rotation $= 30\pi$ in. From this point we proceed by converting "miles per hour" to "rotations per second" by a series of **conversion factors** that are really factors of 1:

$$\frac{70 \text{ miles}}{1 \text{ hour}} \times \frac{1 \text{ hour}}{60 \text{ min}} \times \frac{1 \text{ min}}{60 \text{ sec}} \times \frac{5280 \text{ feet}}{1 \text{ mile}} \times \frac{12 \text{ inches}}{1 \text{ foot}} \times \frac{1 \text{ rotation}}{30\pi \text{ inches}}$$

$$= \frac{70 \times 5280 \times 12 \text{ rotations}}{60 \times 60 \times 30\pi \text{ sec}} \approx 13.07 \text{ rotations per second}$$

Now try Exercise 39.

Functions from Data

In this course we will use the following 3-step strategy to construct functions from data.

Constructing a Function from Data

Given a set of data points of the form (x, y), to construct a formula that approximates y as a function of x:

1. Make a scatter plot of the data points. The points do not need to pass the vertical line test.
2. Determine from the shape of the plot whether the points seem to follow the graph of a familiar type of function (line, parabola, cubic, sine curve, etc.).
3. Transform a basic function of that type to fit the points as closely as possible.

Step 3 might seem like a lot of work, and for earlier generations it certainly was; it required all of the tricks of Section 1.6 and then some. We, however, will gratefully use technology to do this "curve-fitting" step for us, as shown in Example 6.

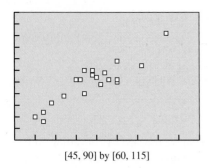

[45, 90] by [60, 115]

FIGURE 1.84 The scatter plot of the temperature data in Example 6.

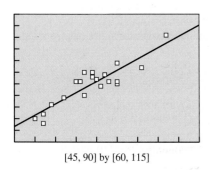

[45, 90] by [60, 115]

FIGURE 1.85 The temperature scatter plot with the regression line shown. (Example 6)

EXAMPLE 6 Curve-Fitting with Technology

Table 1.12 records the low and high daily temperatures observed on 9/9/1999 in 20 major American cities. Find a function that approximates the high temperature (y) as a function of the low temperature (x). Use this function to predict the high temperature that day for Madison, WI, given that the low was 46.

SOLUTION The scatter plot is shown in Figure 1.84.

Table 1.12 Temperature on 9/9/99					
City	Low	High	City	Low	High
New York, NY	70	86	Miami, FL	76	92
Los Angeles, CA	62	80	Honolulu, HI	70	85
Chicago, IL	52	72	Seattle, WA	50	70
Houston, TX	70	94	Jacksonville, FL	67	89
Philadelphia, PA	68	86	Baltimore, MD	64	88
Albuquerque, NM	61	86	St. Louis, MO	57	79
Phoenix, AZ	82	106	El Paso, TX	62	90
Atlanta, GA	64	90	Memphis, TN	60	86
Dallas, TX	65	87	Milwaukee, WI	52	68
Detroit, MI	54	76	Wilmington, DE	66	84

Source: AccuWeather, Inc.

Notice that the points do not fall neatly along a well-known curve, but they do seem to fall *near* an upwardly sloping line. We therefore choose to model the data with a function whose graph is a line. We could fit the line by sight (as we did in Example 5 in Section 1.1), but this time we will use the calculator to find the line of "best fit," called the **regression line**. (See your grapher's owner's manual for how to do this.) The regression line is found to be approximately $y = 0.97x + 23$. As Figure 1.85 shows, the line fits the data as well as can be expected.

If we use this function to predict the high temperature for the day in Madison, WI, we get $y = 0.97(46) + 23 = 67.62$. (For the record, the high that day was 67.)

Now try Exercise 47, parts (a) and (b).

Professional statisticians would be quick to point out that this function should not be trusted as a model for all cities, despite the fairly successful prediction for Madison. (For example, the prediction for San Francisco, with a low of 54 and a high of 64, is off by more than 11 degrees.) *The effectiveness of a data-based model is highly dependent on the number of data points and on the way they were selected.* The functions we construct from data in this book should be analyzed for how well they model the data, not for how well they model the larger population from which the data came.

In addition to lines, we can model scatter plots with several other curves by choosing the appropriate regression option on a calculator or computer. The options to which we will refer in this book (and the chapters in which we will study them) are shown in the following table:

Regression Type	Equation	Graph	Applications
Linear (Chaper 2)	$y = ax + b$		Fixed cost plus variable cost, linear growth, free-fall velocity, simple interest, linear depreciation, many others
Quadratic (Chapter 2)	$y = ax^2 + bx + c$ (requires at least 3 points)		Position during free fall, projectile motion, parabolic reflectors, area as a function of linear dimension, quadratic growth, etc.
Cubic (Chapter 2)	$y = ax^3 + bx^2 + cx + d$ (requires at least 4 points)		Volume as a function of linear dimension, cubic growth, miscellaneous applications where quadratic regression does not give a good fit
Quartic (Chapter 2)	$y = ax^4 + bx^3 + cx^2 + dx + e$ (requires at least 5 points)		Quartic growth, miscellaneous applications where quadratic and cubic regression do not give a good fit
Natural logarithmic (ln) (Chapter 3)	$y = a + b \ln x$ (requires $x > 0$)		Logarithmic growth, decibels (sound), Richter scale (earthquakes), inverse exponential models
Exponential ($b > 1$) (Chapter 3)	$y = a \cdot b^x$		Exponential growth, compound interest, population models
Exponential ($0 < b < 1$) (Chapter 3)	$y = a \cdot b^x$		Exponential decay, depreciation, temperature loss of a cooling body, etc.
Power (requires $x, y > 0$) (Chapter 2)	$y = a \cdot x^b$		Inverse-square laws, Kepler's third law
Logistic (Chapter 3)	$y = \dfrac{c}{1 + a \cdot e^{-bx}}$		Logistic growth: spread of a rumor, population models
Sinusoidal (Chapter 4)	$y = a \sin (bx + c) + d$		Periodic behavior: harmonic motion, waves, circular motion, etc.

DISPLAYING DIAGNOSTICS

If your calculator is giving regression formulas but not displaying the values of r or r^2 or R^2, you may be able to fix that. Go to the CATALOG menu and choose a command called "DiagnosticOn." Enter the command on the home screen and see the reply "Done." Your next regression should display the diagnostic values.

These graphs are only examples, as they can vary in shape and orientation. (For example, any of the curves could appear upside-down.) The grapher uses various strategies to fit these curves to the data, most of them based on combining function composition with linear regression. Depending on the regression type, the grapher may display a number r called the **correlation coefficient** or a number r^2 or R^2 called the **coefficient of determination**. In either case, a useful "rule of thumb" is: *the closer the absolute value of this number is to 1, the better the curve fits the data.*

We can use this fact to help choose a regression type, as in Exploration 1.

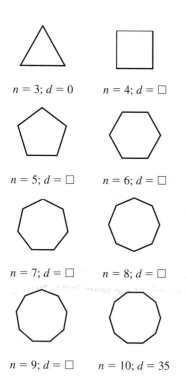

$n = 3; d = 0$ $n = 4; d = \square$

$n = 5; d = \square$ $n = 6; d = \square$

$n = 7; d = \square$ $n = 8; d = \square$

$n = 9; d = \square$ $n = 10; d = 35$

FIGURE 1.86 Some Polygons.
(Exploration 1)

EXPLORATION 1   Diagonals of a Regular Polygon

How many diagonals does a regular polygon have? Can the number be expressed as a function of the number of sides? Try this Exploration.

1. Draw in all the diagonals (i.e., segments connecting nonadjacent points) in each of the regular polygons shown and fill in the number (d) of diagonals in the space below the figure. The values of d for the triangle ($n = 3$) and the decagon ($n = 10$) are filled in for you.

2. Put the values of n in list L1, *beginning with $n = 4$.* (We want to avoid that $y = 0$ value for some of our regressions later.) Put the corresponding values of d in list L2. Display a scatter plot of the ordered pairs.

3. The graph shows an increasing function with some curvature, but it is not clear which kind of growth would fit it best. Try these regressions (preferably in the given order) and record the value of r^2 or R^2 for each: linear, power, quadratic, cubic, quartic. (Note that the curvature is not right for logarithmic, logistic, or sinusoidal curve-fitting, so it is not worth it to try those.)

4. What kind of curve is the best fit? (It might appear at first that there is a tie, but look more closely at the functions you get.) How good *is* the fit?

5. Looking back, could you have predicted the results of the cubic and quartic regressions after seeing the result of the quadratic regression?

6. The "best-fit" curve gives the actual formula for d as a function of n. (In Chapter 9 you will learn how to derive this formula for yourself.) Use the formula to find the number of diagonals of a 128-gon.

We will have more to say about curve fitting as we study the various function types in later chapters.

CHAPTER OPENER PROBLEM (from page 69)

PROBLEM: The table below shows the growth in the computer price index (CPI) for housing for selected years between 1980 and 2003 (based on 1983 dollars). How can we construct a function to predict the housing CPI for the years 2004–2010?

Computer Price Index (Housing)

Year	Housing CPI
1980	81.1
1985	107.7
1990	128.5
1995	148.5
1998	160.4
1999	163.9
2000	169.6
2001	176.4
2002	180.3
2003	184.8

Source: Bureau of Labor Statistics, quoted in The World Almanac and Book of Facts 2005.

SOLUTION: A scatter plot of the data is shown in Figure 1.87, where x is the number of years since 1980. Since the data points fall near an upward sloping line, we can use a calculator to compute a regression line to model the data. The equation of the regression line is found to be $y = 4.37x + 83.20$.

As Figure 1.88 shows, the line fits the data very well.

To predict the housing CPI for 2004, use $x = 24$ in the equation of the regression line. Similarly, we can predict the housing CPI for each of the years 2004–2010, as shown below:

Predicted CPI (Housing)

Year	Predicted Housing CPI
2004	$y = 4.37(24) + 83.20 = 188.1$
2005	$y = 4.37(25) + 83.20 = 192.5$
2006	$y = 4.37(26) + 83.20 = 196.8$
2007	$y = 4.37(27) + 83.20 = 201.2$
2008	$y = 4.37(28) + 83.20 = 205.6$
2009	$y = 4.37(29) + 83.20 = 209.9$
2010	$y = 4.37(30) + 83.20 = 214.3$

Even with a regression fit as impressive as in Figure 1.88, it is risky to predict even this far beyond the data set. Statistics like the CPI are dependent on many volatile factors that can quickly render any mathematical model obsolete. In fact, many economists, convinced that this growth could not be sustained, began warning in 2003 that the "housing bubble" would pop before 2010.

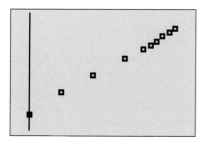

[–2, 25] by [63, 202]

FIGURE 1.87 Scatter plot of the data for the housing Consumer Price Index.

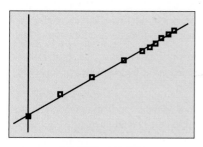

[–2, 25] by [63, 202]

FIGURE 1.88 Scatter plot with the regression line shown.

QUICK REVIEW 1.7 *(For help, go to Section P.3 and P.4.)*

In Exercises 1–10, solve the given formula for the given variable.

1. Area of a Triangle Solve for h: $A = \dfrac{1}{2}bh$

2. Area of a Trapezoid Solve for h: $A = \dfrac{1}{2}(b_1 + b_2)h$

3. Volume of a Right Circular Cylinder Solve for h:
$V = \pi r^2 h$

4. Volume of a Right Circular Cone Solve for h:
$V = \dfrac{1}{3}\pi r^2 h$

5. Volume of a Sphere Solve for r: $V = \dfrac{4}{3}\pi r^3$

6. Surface Area of a Sphere Solve for r: $A = 4\pi r^2$

7. Surface Area of a Right Circular Cylinder Solve for h:
$A = 2\pi rh + 2\pi r^2$

8. Simple Interest Solve for t: $I = Prt$

9. Compound Interest Solve for P: $A = P\left(1 + \dfrac{r}{n}\right)^{nt}$

10. Free-Fall from Height H Solve for t: $s = H - \dfrac{1}{2}gt^2$

SECTION 1.7 EXERCISES

In Exercises 1–10, write a mathematical expression for the quantity described verbally:

1. Five more than three times a number x.

2. A number x increased by 5 and then tripled.

3. Seventeen percent of a number x.

4. Four more than 5% of a number x.

5. Area of a Rectangle The area of a rectangle whose length is 12 more than its width x.

6. Area of a Triangle The area of a triangle whose altitude is 2 more than its base length x.

7. Salary Increase A salary after a 4.5% increase, if the original salary is x dollars.

8. Income Loss Income after a 3% drop in the current income of x dollars.

9. Sale Price Sale price of an item marked x dollars, if 40% is discounted from the marked price.

10. Including Tax Actual cost of an item selling for x dollars if the sales tax rate is 8.75%.

In Exercises 11–14, choose a variable and write a mathematical expression for the quantity described verbally.

11. Total Cost The total cost is $34,500 plus $5.75 for each item produced.

12. Total Cost The total cost is $28,000 increased by 9% plus $19.85 for each item produced.

13. Revenue The revenue when each item sells for $3.75.

14. Profit The profit consists of a franchise fee of $200,000 plus 12% of all sales.

In Exercises 15–20, write the specified quantity as a function of the specified variable. It will help in each case to draw a picture.

15. The height of a right circular cylinder equals its diameter. Write the volume of the cylinder as a function of its radius.

16. One leg of a right triangle is twice as long as the other. Write the length of the hypotenuse as a function of the length of the shorter leg.

17. The base of an isosceles triangle is half as long as the two equal sides. Write the area of the triangle as a function of the length of the base.

18. A square is inscribed in a circle. Write the area of the square as a function of the radius of the circle.

19. A sphere is contained in a cube, tangent to all six faces. Find the surface area of the cube as a function of the radius of the sphere.

20. An isosceles triangle has its base along the x-axis with one base vertex at the origin and its vertex in the first quadrant on the graph of $y = 6 - x^2$. Write the area of the triangle as a function of the length of the base.

In Exercises 21–36, write an equation for the problem and solve the problem.

21. One positive number is 4 times another positive number. The sum of the two numbers is 620. Find the two numbers.

22. When a number is added to its double and its triple, the sum is 714. Find the three numbers.

23. Salary Increase Mark received a 3.5% salary increase. His salary after the raise was $36,432. What was his salary before the raise?

24. Consumer Price Index The consumer price index for food and beverages in 2003 was 184.0 after a hefty 2.3% increase from the previous year. What was the consumer price index for food and beverages in 2002? (Source: *U.S. Bureau of Labor Statistics*)

25. Travel Time A traveler averaged 52 miles per hour on a 182-mile trip. How many hours were spent on the trip?

26. Travel Time On their 560-mile trip, the Bruins basketball team spent two more hours on the interstate highway than they did on local highways. They averaged 45 mph on local highways and 55 mph on the interstate highways. How many hours did they spend driving on local highways?

27. Sale Prices At a shirt sale, Jackson sees two shirts that he likes equally well. Which is the better bargain, and why?

40% off

25% off

28. Job Offers Ruth is weighing two job offers from the sales departments of two competing companies. One offers a base salary of $25,000 plus 5% of gross sales; the other offers a base salary of $20,000 plus 7% of gross sales. What would Ruth's gross sales total need to be to make the second job offer more attractive than the first?

29. Personal Computers From 1996 to 1997, the worldwide shipments of personal computers grew from 71,065,000 to 82,400,000. What was the percentage increase in worldwide personal computer shipments? (Source: *Dataquest*)

30. Personal Computers From 1996 to 1997, the U.S. shipments of personal computers grew from 26,650,000 to 30,989,000. What was the percentage increase in U.S. personal computer shipments? (Source: *Dataquest*)

31. Mixing Solutions How much 10% solution and how much 45% solution should be mixed together to make 100 gallons of 25% solution?

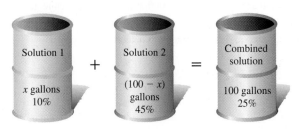

(a) Write an equation that models this problem.

(b) Solve the equation graphically.

32. Mixing Solutions The chemistry lab at the University of Hardwoods keeps two acid solutions on hand. One is 20% acid and the other is 35% acid. How much 20% acid solution and how much 35% acid solution should be used to prepare 25 liters of a 26% acid solution?

33. Maximum Value Problem A square of side x inches is cut out of each corner of a 10 in. by 18 in. piece of cardboard and the sides are folded up to form an open-topped box.

(a) Write the volume V of the box as a function of x.

(b) Find the domain of your function, taking into account the restrictions that the model imposes in x.

(c) Use your graphing calculator to determine the dimensions of the cut-out squares that will produce the box of maximum volume.

34. Residential Construction DDL Construction is building a rectangular house that is 16 feet longer than it is wide. A rain gutter is to be installed in four sections around the 136-foot perimeter of the house. What lengths should be cut for the four sections?

35. Protecting an Antenna In Example 3, suppose the parabolic dish has a 32 in. diameter and is 8 in. deep, and the radius of the cardboard cylinder is 8 in. Now how tall must the cylinder be to fit in the middle of the dish and be flush with the top of the dish?

36. Interior Design Renée's Decorating Service recommends putting a border around the top of the four walls in a dining room that is 3 feet longer than it is wide. Find the dimensions of the room if the total length of the border is 54 feet.

37. Finding the Model and Solving Water is stored in a conical tank with a faucet at the bottom. The tank has depth 24 inches and radius 9 in., and it is filled to the brim. If the faucet is opened to allow the water to flow at a rate of 5 cubic inches per second, what will the depth of the water be after 2 minutes?

38. Investment Returns Reggie invests $12,000, part at 7% annual interest and part at 8.5% annual interest. How much is invested at each rate if Reggie's total annual interest is $900?

39. Unit Conversion A tire of a moving bicycle has radius 16 in. If the tire is making 2 rotations per second, find the bicycle's speed in miles per hour.

40. Investment Returns Jackie invests $25,000, part at 5.5% annual interest and the balance at 8.3% annual interest. How much is invested at each rate if Jackie receives a 1-year interest payment of $1571?

Standardized Test Questions

41. True or False A correlation coefficient gives an indication of how closely a regression line or curve fits a set of data. Justify your answer.

42. True or False Linear regression is useful for modeling the position of an object in free fall. Justify your answer.

In Exercises 43–46, tell which type of regression is likely to give the most accurate model for the scatter plot shown without using a calculator.

(A) Linear regression

(B) Quadratic regression

(C) Cubic regression

(D) Exponential regression

(E) Sinusoidal regression

43. Multiple Choice

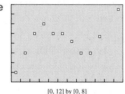

[0, 12] by [0, 8]

44. Multiple Choice

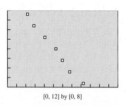

[0, 12] by [0, 8]

45. Multiple Choice

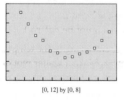

[0, 12] by [0, 8]

46. Multiple Choice

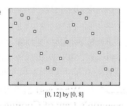

[0, 12] by [0, 8]

Exploration

47. Manufacturing The Buster Green Shoe Company determines that the annual cost C of making x pairs of one type of shoe is $30 per pair plus $100,000 in fixed overhead costs. Each pair of shoes that is manufactured is sold wholesale for $50.

(a) Find an equation that models the cost of producing x pairs of shoes.

(b) Find an equation that models the revenue produced from selling x pairs of shoes.

(c) Find how many pairs of shoes must be made and sold in order to break even.

(d) Graph the equations in (a) and (b). What is the graphical interpretation of the answer in (c)?

48. Employee Benefits John's company issues employees a contract that identifies salary and the company's contributions to pension, health insurance premiums, and disability insurance. The company uses the following formulas to calculate these values.

Salary	x (dollars)
Pension	12% of salary
Health Insurance	3% of salary
Disability Insurance	0.4% of salary

If John's total contract with benefits is worth $48,814.20, what is his salary?

49. Manufacturing Queen, Inc., a tennis racket manufacturer, determines that the annual cost C of making x rackets is $23 per racket plus $125,000 in fixed overhead costs. It costs the company $8 to string a racket.

(a) Find a function $y_1 = u(x)$ that models the cost of producing x unstrung rackets.

(b) Find a function $y_2 = s(x)$ that models the cost of producing x strung rackets.

(c) Find a function $y_3 = R_u(x)$ that models the revenue generated by selling x unstrung rackets.

(d) Find a function $y_4 = R_s(x)$ that models the revenue generated by selling x rackets.

(e) Graph y_1, y_2, y_3, and y_4 simultaneously in the window [0, 10,000] by [0, 500,000].

(f) Writing to Learn Write a report to the company recommending how they should manufacture their rackets, strung or unstrung. Assume that you can include the viewing window in (e) as a graph in the report, and use it to support your recommendation.

50. Hourly Earnings of U.S. Production Workers The average hourly earnings of U.S. production workers for 1990–2003 are shown in Table 1.13.

Table 1.13 Average Hourly Earnings

Year	Average Hourly Earnings
1990	10.19
1991	10.50
1992	10.76
1993	11.03
1994	11.32
1995	11.64
1996	12.03
1997	12.49
1998	13.00
1999	13.47
2000	14.00
2001	14.53
2002	14.95
2003	15.35

Source: Bureau of Labor Statistics, U.S. Dept. of Labor, as reported in The World Almanac and Book of Facts 2005.

(a) Produce a scatter plot of the hourly earnings (y) as a function of years since 1990 (x).

(b) Find the linear regression equation. Round the coefficients to the nearest 0.001.

(c) Does the value of r suggest that the linear model is appropriate?

(d) Find the quadratic regression equation. (Round the coefficients to the nearest 0.01.)

(e) Does the value of R^2 suggest that a quadratic model is appropriate?

(f) Use both curves to predict the housing CPI for the year 2010. How different are the estimates?

(g) Writing to Learn Use the results of parts (a)–(f) to explain why it is risky to predict y-values for x-values that are not very close to the data points, even when the regression plots fit the data points very well.

Extending the Ideas

51. Newton's Law of Cooling A 190° cup of coffee is placed on a desk in a 72° room. According to Newton's Law of Cooling, the temperature T of the coffee after t minutes will be $T = (190 - 72)b^t + 72$, where b is a constant that depends on how easily the cooling substance loses heat. The data in Table 1.14 are from a simulated experiment of gathering temperature readings from a cup of coffee in a 72° room at 20 one-minute intervals:

Table 1.14 Cooling a Cup of Coffee

Time	Temp	Time	Temp
1	184.3	11	140.0
2	178.5	12	136.1
3	173.5	13	133.5
4	168.6	14	130.5
5	164.0	15	127.9
6	159.2	16	125.0
7	155.1	17	122.8
8	151.8	18	119.9
9	147.0	19	117.2
10	143.7	20	115.2

(a) Make a scatter plot of the data, with the times in list L1 and the temperatures in list L2.

(b) Store L2 − 72 in list L3. The values in L3 should now be an exponential function ($y = a \times b^x$) of the values in L1.

(c) Find the exponential regression equation for L3 as a function of L1. How well does it fit the data?

52. Group Activity Newton's Law of Cooling If you have access to laboratory equipment (such as a CBL or CBR unit for your grapher), gather experimental data such as in Exercise 51 from a cooling cup of coffee. Proceed as follows:

(a) First, use the temperature probe to record the temperature of the room. It is a good idea to turn off fans and air conditioners that might affect the temperature of the room during the experiment. It should be a constant.

(b) Heat the coffee. It need not be boiling, but it should be at least 160°. (It also need not be coffee.)

(c) Make a new list consisting of the temperature values minus the room temperature. Make a scatter plot of this list (y) against the time values (x). It should appear to approach the x-axis as an asymptote.

(d) Find the equation of the exponential regression curve. How well does it fit the data?

(e) What is the equation predicted by Newton's Law of Cooling? (Substitute your initial coffee temperature and the temperature of your room for the 190 and 72 in the equation in Exercise 51.)

(f) Group Discussion What sort of factors would affect the value of b in Newton's Law of Cooling? Discuss your ideas with the group.

Math at Work

When I was a kid, I always liked seeing how things work. I read books to see how things worked and I liked to tinker with things like bicycles, roller skates, or whatever. So it was only natural that I would become a mechanical engineer.

I design methods of controlling turbine speed, so a turbine can be used to generate electric power. The turbine is pushed by steam. A valve is opened to let the steam in, and the speed of the turbine is dictated by how much steam is allowed into the system. Once the turbine is spinning it generates electricity, which runs the pumps, which determine the position of the valves. In this way, the turbine actually controls its own speed.

We use oil pressure to make all of this happen. Oil cannot be compressed, so when we push oil inside a tube, the oil opens the valve, which determines how much steam is pushing the turbine. We need the pumps to push the oil, which controls the opening or closing of the valve. The turbine needs to be able to get to speed in a given amount of time, as well as shut down fast enough in case of emergency, and the opening and closing of the valve controls this speed.

While computers do much of the math these days, we use calculus for stress and motion analysis, as well as vibration analysis, fluid flow, and heat analysis. One of the equations we use is:

John Jay

$$\gamma_1 + \frac{v_1{}^2}{2} + gh_1 = \gamma_2 + \frac{v_2{}^2}{2} + gh_2$$

This is an *energy balance equation.* The γ's represent density of the oil, the v's represent flow velocity of the oil, the g's represent the force of gravity, and the h's represent the height of the oil above the point where it exerts pressure. The subscript 1's indicate these quantities at one point in the system, and the subscript 2's indicate these quantities at another point in the system.

What this equation means is that the total energy at one point in the system must equal the total energy at another point in the system. Thus, we know how much energy is exerted against the valve controls, wherever they are located in the system.

CHAPTER 1 Key Ideas

PROPERTIES, THEOREMS, AND FORMULAS

PROCEDURES

CHAPTER 1 **Review Exercises**

The collection of exercises marked in red could be used as a chapter test.

In Exercises 1–10, match the graph with the corresponding function (a)–(j) from the list below. Use your knowledge of function behavior, *not* your grapher.

(a) $f(x) = x^2 - 1$

(b) $f(x) = x^2 + 1$

(c) $f(x) = (x - 2)^2$

(d) $f(x) = (x + 2)^2$

(e) $f(x) = \dfrac{x - 1}{2}$

(f) $f(x) = |x - 2|$

(g) $f(x) = |x + 2|$

(h) $f(x) = -\sin x$

(i) $f(x) = e^x - 1$

(j) $f(x) = 1 + \cos x$

1.

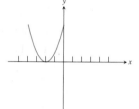

2.

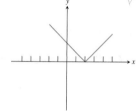

3.

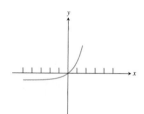

4.

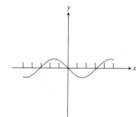

5.

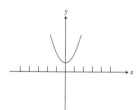

6.

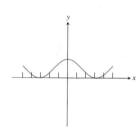

7.

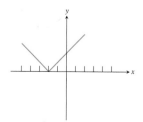

8.

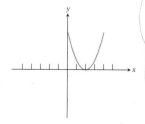

9.

10.

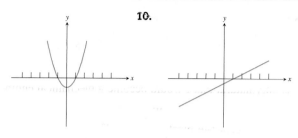

In Exercises 11–18, find (a) the domain and (b) the range of the function.

11. $g(x) = x^3$

12. $f(x) = 35x - 602$

13. $g(x) = x^2 + 2x + 1$

14. $h(x) = (x - 2)^2 + 5$

15. $g(x) = 3|x| + 8$

16. $k(x) = \sqrt{4 - x^2} - 2$

17. $f(x) = \dfrac{x}{x^2 - 2x}$

18. $k(x) = \dfrac{1}{\sqrt{9 - x^2}}$

In Exercises 19 and 20, graph the function, and state whether the function is continuous at $x = 0$. If it is discontinuous, state whether the discontinuity is removable or nonremovable.

19. $f(x) = \dfrac{x^2 - 3}{x + 2}$

20. $k(x) = \begin{cases} 2x + 3 & \text{if } x > 0 \\ 3 - x^2 & \text{if } x \le 0 \end{cases}$

In Exercises 21–24, find all (a) vertical asymptotes and (b) horizontal asymptotes of the graph of the function. Be sure to state your answers as equations of lines.

21. $y = \dfrac{5}{x^2 - 5x}$

22. $y = \dfrac{3x}{x - 4}$

23. $y = \dfrac{7x}{\sqrt{x^2 + 10}}$

24. $y = \dfrac{|x|}{x + 1}$

In Exercises 25–28, graph the function and state the intervals on which the function is *increasing*.

25. $y = \dfrac{x^3}{6}$

26. $y = 2 + |x - 1|$

27. $y = \dfrac{x}{1 - x^2}$

28. $y = \dfrac{x^2 - 1}{x^2 - 4}$

In Exercises 29–32, graph the function and tell whether the function is bounded above, bounded below, or bounded.

29. $f(x) = x + \sin x$

30. $g(x) = \dfrac{6x}{x^2 + 1}$

31. $h(x) = 5 - e^x$

32. $k(x) = 1000 + \dfrac{x}{1000}$

In Exercises 33–36, use a grapher to find all (a) relative maximum values and (b) relative minimum values of the function. Also state the value of x at which each relative extremum occurs.

33. $y = (x + 1)^2 - 7$

34. $y = x^3 - 3x$

35. $y = \dfrac{x^2 + 4}{x^2 - 4}$

36. $y = \dfrac{4x}{x^2 + 4}$

In Exercises 37–40, graph the function and state whether the function is odd, even, or neither.

37. $y = 3x^2 - 4|x|$

38. $y = \sin x - x^3$

39. $y = \dfrac{x}{e^x}$

40. $y = x \cos (x)$

In Exercises 41–44, find a formula for $f^{-1}(x)$.

41. $f(x) = 2x + 3$

42. $f(x) = \sqrt[3]{x - 8}$

43. $f(x) = \dfrac{2}{x}$

44. $f(x) = \dfrac{6}{x + 4}$

Exercises 45–52 refer to the function $y = f(x)$ whose graph is given below.

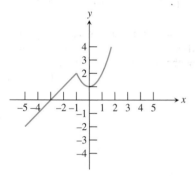

45. Sketch the graph of $y = f(x) - 1$.

46. Sketch the graph of $y = f(x - 1)$.

47. Sketch the graph of $y = f(-x)$.

48. Sketch the graph of $y = -f(x)$.

49. Sketch a graph of the inverse relation.

50. Does the inverse relation define y as a function of x?

51. Sketch a graph of $y = f(|x|)$.

52. Define f algebraically as a piecewise function. [*Hint:* the pieces are translations of two of our "basic" functions.]

In Exercises 53–58, let $f(x) = \sqrt{x}$ and let $g(x) = x^2 - 4$.

53. Find an expression for $(f \circ g)(x)$ and give its domain.

54. Find an expression for $(g \circ f)(x)$ and give its domain.

55. Find an expression for $(fg)(x)$ and give its domain.

56. Find an expression for $\left(\dfrac{f}{g}\right)(x)$ and give its domain.

57. Describe the end behavior of the graph of $y = f(x)$.

58. Describe the end behavior of the graph of $y = f(g(x))$.

In Exercises 59–64, write the specified quantity as a function of the specified variable. Remember that drawing a picture will help.

59. Square Inscribed in a Circle A square of side s is inscribed in a circle. Write the area of the circle as a function of s.

60. Circle Inscribed in a Square A circle is inscribed in a square of side s. Write the area of the circle as a function of s.

61. Volume of a Cylindrical Tank A cylindrical tank with diameter 20 feet is partially filled with oil to a depth of h feet. Write the volume of oil in the tank as a function of h.

62. Draining a Cylindrical Tank A cylindrical tank with diameter 20 feet is filled with oil to a depth of 40 feet. The oil begins draining at a constant rate of 2 cubic feet per second. Write the volume of the oil remaining in the tank t seconds later as a function of t.

63. Draining a Cylindrical Tank A cylindrical tank with diameter 20 feet is filled with oil to a depth of 40 feet. The oil begins draining at a constant rate of 2 cubic feet per second. Write the depth of the oil remaining in the tank t seconds later as a function of t.

64. Draining a Cylindrical Tank A cylindrical tank with diameter 20 feet is filled with oil to a depth of 40 feet. The oil begins draining so that the depth of oil in the tank decreases at a constant rate of 2 feet per hour. Write the volume of oil remaining in the tank t hours later as a function of t.

65. U.S. Crude Oil Imports The imports of crude oil to the U.S. from Canada in the years 1995–2004 (in thousands of barrels per day) are given in Table 1.15.

Table 1.15 Crude Oil Imports from Canada	
Year	Barrels/day × 1000
1995	1,040
1996	1,075
1997	1,198
1998	1,266
1999	1,178
2000	1,348
2001	1,356
2002	1,445
2003	1,549
2004	1,606

Source: Energy Information Administration, Petroleum Supply Monthly, as reported in The World Almanac and Book of Facts 2005.

(a) Sketch a scatter plot of import numbers in the right-hand column (y) as a function of years since 1990 (x).

(b) Find the equation of the regression line and superimpose it on the scatter plot.

(c) Based on the regression line, approximately how many thousands of barrels of oil would the U.S. import from Canada in 2010?

66. The winning times in the women's 100-meter freestyle event at the Summer Olympic Games since 1952 are shown in Table 1.16:

Table 1.16 Women's 100-Meter Freestyle			
Year	Time	Year	Time
1952	66.8	1980	54.79
1956	62.0	1984	55.92
1960	61.2	1988	54.93
1964	59.5	1992	54.64
1968	60.0	1996	54.50
1972	58.59	2000	53.83
1976	55.65	2004	53.84

Source: The World Almanac and Book of Facts 2005.

(a) Sketch a scatter plot of the times (y) as a function of the years (x) beyond 1900. (The values of x will run from 52 to 104.)

(b) Explain why a linear model cannot be appropriate for these times over the long term.

(c) The points appear to be approaching a horizontal asymptote of $y = 52$. What would this mean about the times in this Olympic event?

(d) Subtract 52 from all the times so that they will approach an asymptote of $y = 0$. Redo the scatter plot with the new y-values. Now find the *exponential* regression curve and superimpose its graph on the vertically-shifted scatter plot.

(e) According to the regression curve, what will be the winning time in the women's 100-meter freestyle event at the 2008 Olympics?

67. Inscribing a Cylinder Inside a Sphere A right circular cylinder of radius r and height h is inscribed inside a sphere of radius $\sqrt{3}$ inches.

(a) Use the Pythagorean Theorem to write h as a function of r.

(b) Write the volume V of the cylinder as a function of r.

(c) What values of r are in the domain of V?

(d) Sketch a graph of $V(r)$ over the domain $[0, \sqrt{3}]$.

(e) Use your grapher to find the maximum volume that such a cylinder can have.

68. Inscribing a Rectangle Under a Parabola A rectangle is inscribed between the x-axis and the parabola $y = 36 - x^2$ with one side along the x-axis, as shown in the figure below.

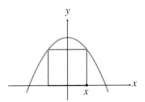

(a) Let x denote the x-coordinate of the point highlighted in the figure. Write the area A of the rectangle as a function of x.

(b) What values of x are in the domain of A?

(c) Sketch a graph of $A(x)$ over the domain.

(d) Use your grapher to find the maximum area that such a rectangle can have.

CHAPTER 1 Project

Modeling the Growth of a Business

In 1971, Starbucks Coffee opened its first location in Pike Place Market—Seattle's legendary open-air farmer's market. By 1987, the number of Starbucks stores had grown to 17 and by 1999 there were 2498 locations. The data in the table below (obtained from Starbucks Coffee's web site, www.starbucks. com) summarizes the growth of this company from 1987 through 1999.

Year	Number of locations
1987	17
1988	33
1989	55
1990	84
1991	116
1992	165
1993	272
1994	425
1995	676
1996	1015
1997	1412
1998	1886
1999	2498

Explorations

1. Enter the data in the table into your grapher or computer. (Let $t = 0$ represent 1987.) Draw a scatter plot for the data.

2. Refer to page 157 in this chapter. Look at the types of graphs displayed and the associated regression types. Notice that the Exponential Regression model with $b > 1$ seems to most closely match the plotted data. Use your grapher or computer to find an exponential regression equation to model this data set (see your grapher's guidebook for instructions on how to do this).

3. Use the exponential model you just found to predict the total number of Starbucks locations for 2000 and 2001.

4. There were 2498 Starbucks locations in 2000 and 4709 locations in 2001. (You can verify these numbers and find more up-to-date information in the investors's section of the Starbucks web site.) Why is there such a big difference between your predicted values and the actual number of Starbucks locations? What real-world feature of business growth was not accounted for in the exponential growth model?

5. You need to model the data set with an equation that takes into account the fact that a business's growth eventually levels out or reaches a carrying capacity. Refer to page xxx again. Notice that the Logistic Regression modeling graph appears to show exponential growth at first, but eventually levels out. Use your grapher or computer to find the logistic regression equation to model this data set (see your grapher's guidebook for instructions on how to do this).

6. Use the logistic model you just found to predict the total number of Starbucks locations for 2000 and 2001. How do your predictions compare with the actual number of locations for 2000 and 2001? How many locations do you think there will be in the year 2020?

Polynomial, Power, and Rational Functions

Humidity and relative humidity are measures used by weather forecasters. Humidity affects our comfort and our health. If humidity is too low, our skin can become dry and cracked, and viruses can live longer. If humidity is too high, it can make warm temperatures feel even warmer, and mold, fungi, and dust mites can live longer. See page 244 to learn how relative humidity is modeled as a rational function.

Chapter 2 Overview

Chapter 1 laid a foundation of the general characteristics of functions, equations, and graphs. In this chapter and the next two, we will explore the theory and applications of specific families of functions. We begin this exploration by studying three interrelated families of functions: polynomial, power, and rational functions. These three families of functions are used in the social, behavioral, and natural sciences.

This chapter includes a thorough study of the theory of polynomial equations. We investigate algebraic methods for finding both real- and complex-number solutions of such equations and relate these methods to the graphical behavior of polynomial and rational functions. The chapter closes by extending these methods to inequalities in one variable.

2.1
Linear and Quadratic Functions and Modeling

What you'll learn about

- Polynomial Functions

- Linear Functions and Their Graphs

- Average Rate of Change

- Linear Correlation and Modeling

- Quadratic Functions and Their Graphs

- Applications of Quadratic Functions

. . . and why

Many business and economic problems are modeled by linear functions. Quadratic and higher degree polynomial functions are used to model some manufacturing applications.

Polynomial Functions

Polynomial functions are among the most familiar of all functions.

DEFINITION Polynomial Function

Let n be a nonnegative integer and let $a_0, a_1, a_2, \ldots, a_{n-1}, a_n$ be real numbers with $a_n \neq 0$. The function given by

$$f(x) = a_n x^n + a_{n-1} x^{n-1} + \cdots + a_2 x^2 + a_1 x + a_0$$

is a **polynomial function of degree n**. The **leading coefficient** is a_n.

The zero function $f(x) = 0$ is a polynomial function. It has no degree and no leading coefficient.

Polynomial functions are defined and continuous on all real numbers. It is important to recognize whether a function is a polynomial function.

EXAMPLE 1   Identifying Polynomial Functions

Which of the following are polynomial functions? For those that are polynomial functions, state the degree and leading coefficient. For those that are not, explain why not.

(a) $f(x) = 4x^3 - 5x - \dfrac{1}{2}$

(b) $g(x) = 6x^{-4} + 7$

(c) $h(x) = \sqrt{9x^4 + 16x^2}$

(d) $k(x) = 15x - 2x^4$

continued

SOLUTION

(a) f is a polynomial function of degree 3 with leading coefficient 4.

(b) g is not a polynomial function because of the exponent -4.

(c) h is not a polynomial function because it cannot be simplified into polynomial form. Notice that $\sqrt{9x^4 + 16x^2} \neq 3x^2 + 4x$.

(d) k is a polynomial function of degree 4 with leading coefficient -2.

Now try Exercise 1.

The zero function and all constant functions are polynomial functions. Some other familiar functions are also polynomial functions, as shown below.

Polynomial Functions of No and Low Degree

Name	Form	Degree
Zero function	$f(x) = 0$	Undefined
Constant function	$f(x) = a \ (a \neq 0)$	0
Linear function	$f(x) = ax + b \ (a \neq 0)$	1
Quadratic function	$f(x) = ax^2 + bx + c \ (a \neq 0)$	2

We study polynomial functions of degree 3 and higher in Section 2.3. For the remainder of this section, we turn our attention to the nature and uses of linear and quadratic polynomial functions.

Linear Functions and Their Graphs

Linear equations and graphs of lines were reviewed in Sections P.3 and P.4, and some of the examples in Chapter 1 involved linear functions. We now take a closer look at the properties of linear functions.

A **linear function** is a polynomial function of degree 1 and so has the form

$$f(x) = ax + b, \text{ where } a \text{ and } b \text{ are constants and } a \neq 0.$$

If we use m for the leading coefficient instead of a and let $y = f(x)$, then this equation becomes the familiar slope-intercept form of a line:

$$y = mx + b.$$

SURRPRISING FACT

Not all lines in the Cartesian plane are graphs of linear functions.

Vertical lines are not graphs of functions because they fail the vertical line test, and horizontal lines are graphs of constant functions. A line in the Cartesian plane is the graph of a linear function if and only if it is a **slant line**, that is, neither horizontal nor vertical. We can apply the formulas and methods of Section P.4 to problems involving linear functions.

EXAMPLE 2   Finding an Equation of a Linear Function

Write an equation for the linear function f such that $f(-1) = 2$ and $f(3) = -2$.

SOLUTION

Solve Algebraically

We seek a line through the points $(-1, 2)$ and $(3, -2)$. The slope is

$$m = \frac{y_2 - y_1}{x_2 - x_1} = \frac{(-2) - 2}{3 - (-1)} = \frac{-4}{4} = -1.$$

Using this slope and the coordinates of $(-1, 2)$ with the point-slope formula, we have

$$y - y_1 = m(x - x_1)$$
$$y - 2 = -1(x - (-1))$$
$$y - 2 = -x - 1$$
$$y = -x + 1$$

Converting to function notation gives us the desired form:

$$f(x) = -x + 1.$$

Support Graphically

We can graph $y = -x + 1$ and see that it includes the points $(-1, 2)$ and $(3, -2)$. (See Figure 2.1.)

Confirm Numerically

Using $f(x) = -x + 1$ we prove that $f(-1) = 2$ and $f(3) = -2$:

$f(-1) = -(-1) + 1 = 1 + 1 = 2$, and $f(3) = -(3) + 1 = -3 + 1 = -2$.

Now try Exercise 7.

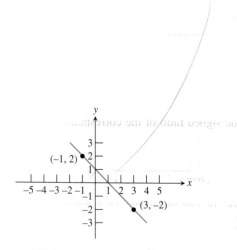

FIGURE 2.1 The graph of $y = -x + 1$ passes through $(-1, 2)$ and $(3, -2)$. (Example 2)

Average Rate of Change

Another property that characterizes a linear function is its *rate of change*. The **average rate of change** of a function $y = f(x)$ between $x = a$ and $x = b$, $a \neq b$, is

$$\frac{f(b) - f(a)}{b - a}.$$

You are asked to prove the following theorem in Exercise 85.

THEOREM Constant Rate of Change

A function defined on all real numbers is a linear function if and only if it has a constant nonzero average rate of change between any two points on its graph.

Because the average rate of change of a linear function is constant, it is called simply the **rate of change** of the linear function. The slope m in the formula $f(x) = mx + b$ is the rate of change of the linear function. In Exploration 1, we revisit Example 7 of Section P.4 in light of the rate of change concept.

EXPLORATION 1 Modeling Depreciation with a Linear Function

Camelot Apartments bought a $50,000 building and for tax purposes are depreciating it $2000 per year over a 25-yr period using straight-line depreciation.

1. What is the rate of change of the value of the building?
2. Write an equation for the value $v(t)$ of the building as a linear function of the time t since the building was placed in service.
3. Evaluate $v(0)$ and $v(16)$.
4. Solve $v(t) = 39,000$.

RATE AND RATIO

All rates are ratios, whether expressed as miles per hour, dollars per year, or even rise over run.

The rate of change of a linear function is the signed ratio of the corresponding line's rise over run. That is, for a linear function $f(x) = mx + b$,

$$\text{rate of change} = m = \frac{\text{rise}}{\text{run}} = \frac{\text{change in } y}{\text{change in } x} = \frac{\triangle y}{\triangle x}.$$

This formula allows us to interpret the slope, or rate of change, of a linear function numerically. For instance, in Exploration 1 the value of the apartment building fell from $50,000 to $0 over a 25-yr period. In Table 2.1 we compute $\triangle y/\triangle x$ for the apartment building's value (in dollars) as a function of time (in years). Because the average rate of change $\triangle y/\triangle x$ is the nonzero constant -2000, the building's value is a linear function of time decreasing at a rate of $2000/yr.

Table 2.1 Rate of Change of the Value of the Apartment Building in Exploration 1: $y = -2000x + 50{,}000$				
x (time)	y (value)	$\triangle x$	$\triangle y$	$\triangle y/\triangle x$
0	50,000			
		1	-2000	-2000
1	48,000			
		1	-2000	-2000
2	46,000			
		1	-2000	-2000
3	44,000			
		1	-2000	-2000
4	42,000			

In Exploration 1, as in other applications of linear functions, the constant term represents the value of the function for an input of 0. In general, for any function f, $f(0)$ is the **initial value of** f. So for a linear function $f(x) = mx + b$, the constant term b is the initial value of the function. For any polynomial function $f(x) = a_n x^n + \cdots + a_1 x + a_0$, the **constant term** $f(0) = a_0$ is the function's initial value. Finally, the initial value of any function—polynomial or otherwise—is the y-intercept of its graph.

We now summarize what we have learned about linear functions.

Characterizing the Nature of a Linear Function	
Point of View	**Characterization**
Verbal	polynomial of degree 1
Algebraic	$f(x) = mx + b \ (m \neq 0)$
Graphical	slant line with slope m and y-intercept b
Analytical	function with constant nonzero rate of change m: f is increasing if $m > 0$, decreasing if $m < 0$; initial value of the function $= f(0) = b$

Linear Correlation and Modeling

In Section 1.7 we approached modeling from several points of view. Along the way we learned how to use a grapher to create a scatter plot, compute a regression line for a data set, and overlay a regression line on a scatter plot. We touched on the notion of correlation coefficient. We now go deeper into these modeling and regression concepts.

Figure 2.2 shows five types of scatter plots. When the points of a scatter plot are clustered along a line, we say there is a **linear correlation** between the quantities represented by the data. When an oval is drawn around the points in the scatter plot, generally speaking, the narrower the oval, the stronger the linear correlation.

When the oval tilts like a line with positive slope (as in Figure 2.2a and b), the data have a **positive linear correlation**. On the other hand, when it tilts like a line with negative slope (as in Figure 2.2d and e), the data have a **negative linear correlation**. Some scatter plots exhibit little or no linear correlation (as in Figure 2.2c), or have non-linear patterns.

A number that measures the strength and direction of the linear correlation of a data set is the **(linear) correlation coefficient, r**.

Properties of the Correlation Coefficient, r
1. $-1 \leq r \leq 1$.
2. When $r > 0$, there is a positive linear correlation.
3. When $r < 0$, there is a negative linear correlation.
4. When $|r| \approx 1$, there is a strong linear correlation.
5. When $r \approx 0$, there is weak or no linear correlation.

Correlation informs the modeling process by giving us a measure of goodness of fit. Good modeling practice, however, demands that we have a theoretical reason for selecting a model. In business, for example, fixed cost is modeled by a constant function. (Otherwise, the cost would not be fixed.)

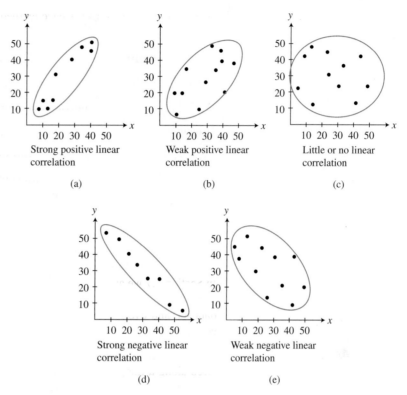

FIGURE 2.2 Five scatter plots and the types of linear correlation they suggest.

In economics, a linear model is often used for the demand for a product as a function of its price. For instance, suppose Twin Pixie, a large supermarket chain, conducts a market analysis on its store brand of doughnut-shaped oat breakfast cereal. The chain sets various prices for its 15-oz box at its different stores over a period of time. Then, using this data, the Twin Pixie researchers project the weekly sales at the entire chain of stores for each price and obtain the data shown in Table 2.2.

Table 2.2 Weekly Sales Data Based on Marketing Research	
Price per box	Boxes sold
$2.40	38,320
$2.60	33,710
$2.80	28,280
$3.00	26,550
$3.20	25,530
$3.40	22,170
$3.60	18,260

EXAMPLE 3   Modeling and Predicting Demand

Use the data in Table 2.2 on the preceding page to write a linear model for demand (in boxes sold per week) as a function of the price per box (in dollars). Describe the strength and direction of the linear correlation. Then use the model to predict weekly cereal sales if the price is dropped to $2.00 or raised to $4.00 per box.

SOLUTION

Model

We enter the data and obtain the scatter plot shown in Figure 2.3a. It appears that the data have a strong negative correlation.

We then find the linear regression model to be approximately

$$y = -15{,}358.93x + 73{,}622.50,$$

where x is the price per box of cereal and y the number of boxes sold.

Figure 2.3b shows the scatter plot for Table 2.2 together with a graph of the regression line. You can see that the line fits the data fairly well. The correlation coefficient of $r \approx -0.98$ supports this visual evidence.

Solve Graphically

Our goal is to predict the weekly sales for prices of $2.00 and $4.00 per box. Using the value feature of the grapher, as shown in Figure 2.3c, we see that y is about 42,900 when x is 2. In a similar manner we could find that $y \approx 12{,}190$ when x is 4.

Interpret

If Twin Pixie drops the price for its store brand of doughnut-shaped oat breakfast cereal to $2.00 per box, sales will rise to about 42,900 boxes per week. On the other hand, if they raise the price to $4.00 per box, sales will drop to around 12,190 boxes per week.   ***Now try Exercise 49.***

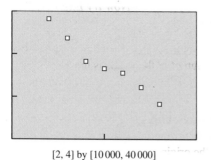

[2, 4] by [10 000, 40 000]

(a)

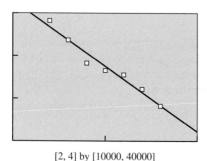

[2, 4] by [10000, 40000]

(b)

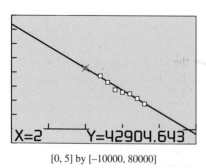

X=2   Y=42904.643

[0, 5] by [−10000, 80000]

(c)

FIGURE 2.3 Scatter plot and regression line graphs for Example 3.

We summarize for future reference the analysis used in Example 3.

Regression Analysis

1. Enter and plot the data (scatter plot).

2. Find the regression model that fits the problem situation.

3. Superimpose the graph of the regression model on the scatter plot, and observe the fit.

4. Use the regression model to make the predictions called for in the problem.

Quadratic Functions and Their Graphs

A **quadratic function** is a polynomial function of degree 2. Recall from Section 1.3 that the graph of the squaring function $f(x) = x^2$ is a parabola. We will see that the graph of every quadratic function is an upward or downward opening parabola. This is because the graph of any quadratic function can be obtained from the graph of the squaring function $f(x) = x^2$ by a sequence of translations, reflections, stretches, and shrinks.

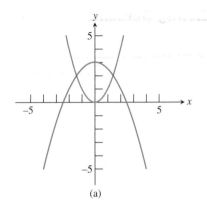

(a)

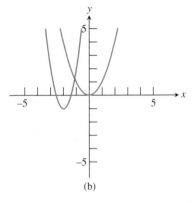

(b)

FIGURE 2.4 The graph of $f(x) = x^2$ (blue) shown with (a) $g(x) = -(1/2)x^2 + 3$ and (b) $h(x) = 3(x + 2)^2 - 1$. (Example 4)

EXAMPLE 4   Transforming the Squaring Function

Describe how to transform the graph of $f(x) = x^2$ into the graph of the given function. Sketch its graph by hand.

(a) $g(x) = -(1/2)x^2 + 3$ **(b)** $h(x) = 3(x + 2)^2 - 1$

SOLUTION

(a) The graph of $g(x) = -(1/2)x^2 + 3$ is obtained by vertically shrinking the graph of $f(x) = x^2$ by a factor of 1/2, reflecting the resulting graph across the x-axis, and translating the reflected graph up 3 units. See Figure 2.4a.

(b) The graph of $h(x) = 3(x + 2)^2 - 1$ is obtained by vertically stretching the graph of $f(x) = x^2$ by a factor of 3 and translating the resulting graph left 2 units and down 1 unit. See Figure 2.4b.   ***Now try Exercise 19.***

The graph of $f(x) = ax^2$, $a > 0$, is an upward-opening parabola. When $a < 0$, its graph is a downward-opening parabola. Regardless of the sign of a, the y-axis is the line of symmetry for the graph of $f(x) = ax^2$. The line of symmetry for a parabola is its **axis of symmetry**, or **axis** for short. The point on the parabola that intersects its axis is the **vertex** of the parabola. Because the graph of a quadratic function is always an upward- or downward-opening parabola, its vertex is always the lowest or highest point of the parabola. The vertex of $f(x) = ax^2$ is always the origin, as seen in Figure 2.5.

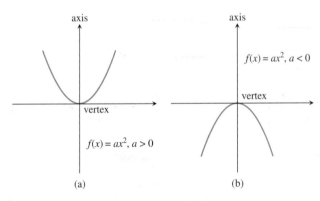

(a)   (b)

FIGURE 2.5 The graph $f(x) = ax^2$ for (a) $a > 0$ and (b) $a < 0$.

Expanding $f(x) = a(x - h)^2 + k$ and comparing the resulting coefficients with the **standard quadratic form** $ax^2 + bx + c$, where the powers of x are arranged in descending order, we can obtain formulas for h and k.

$$f(x) = a(x - h)^2 + k$$
$$= a(x^2 - 2hx + h^2) + k \quad\quad \text{Expand } (x - h)^2.$$
$$= ax^2 + (-2ah)x + (ah^2 + k) \quad \text{Distributive property}$$
$$= ax^2 + bx + c \quad\quad\quad\quad \text{Let } b = -2ah \text{ and } c = ah^2 + k.$$

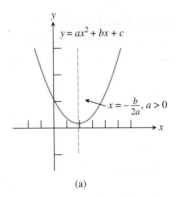

(a)

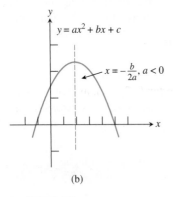

(b)

FIGURE 2.6 The vertex is at $x = -b/(2a)$, which therefore also describes the axis of symmetry. (a) When $a > 0$, the parabola opens upward. (b) When $a < 0$, the parabola opens downward.

Because $b = -2ah$ and $c = ah^2 + k$ in the last line above, $h = -b/(2a)$ and $k = c - ah^2$. Using these formulas, any quadratic function $f(x) = ax^2 + bx + c$ can be rewritten in the form

$$f(x) = a(x - h)^2 + k.$$

This *vertex form* for a quadratic function makes it easy to identify the vertex and axis of the graph of the function, and to sketch the graph.

Vertex Form of a Quadratic Function

Any quadratic function $f(x) = ax^2 + bx + c$, $a \neq 0$, can be written in the **vertex form**

$$f(x) = a(x - h)^2 + k.$$

The graph of f is a parabola with vertex (h, k) and axis $x = h$, where $h = -b/(2a)$ and $k = c - ah^2$. If $a > 0$, the parabola opens upward, and if $a < 0$, it opens downward. (See Figure 2.6.)

The formula $h = -b/(2a)$ is useful for locating the vertex and axis of the parabola associated with a quadratic function. To help you remember it, notice that $-b/(2a)$ is part of the quadratic formula

$$x = \frac{-b \pm \sqrt{b^2 - 4ac}}{2a}.$$

(Cover the radical term.) You need not remember $k = c - ah^2$ because you can use $k = f(h)$ instead.

EXAMPLE 5   Finding the Vertex and Axis of a Quadratic Function

Use the vertex form of a quadratic function to find the vertex and axis of the graph of $f(x) = 6x - 3x^2 - 5$. Rewrite the equation in vertex form.

SOLUTION

Solve Algebraically

The standard polynomial form of f is

$$f(x) = -3x^2 + 6x - 5.$$

So $a = -3$, $b = 6$, and $c = -5$, and the coordinates of the vertex are

$$h = -\frac{b}{2a} = -\frac{6}{2(-3)} = 1 \text{ and}$$

$$k = f(h) = f(1) = -3(1)^2 + 6(1) - 5 = -2.$$

The equation of the axis is $x = 1$, the vertex is $(1, -2)$, and the vertex form of f is

$$f(x) = -3(x - 1)^2 + (-2).$$

Now try Exercise 27.

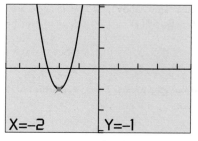

X=-2   Y=-1

[−4.7, 4.7] by [−3.1, 3.1]

FIGURE 2.7 The graphs of $f(x) = 3x^2 + 12x + 11$ and $y = 3(x + 2)^2 - 1$ appear to be identical. The vertex $(-2, -1)$ is highlighted. (Example 6)

EXAMPLE 6   Using Algebra to Describe the Graph of a Quadratic Function

Use completing the square to describe the graph of $f(x) = 3x^2 + 12x + 11$. Support your answer graphically.

SOLUTION

Solve Algebraically

$$f(x) = 3x^2 + 12x + 11$$

$$= 3(x^2 + 4x) + 11 \qquad \text{Factor 3 from the } x \text{ terms.}$$

$$= 3(x^2 + 4x + (\ ) - (\ )) + 11 \qquad \text{Prepare to complete the square.}$$

$$= 3(x^2 + 4x + (2^2) - (2^2)) + 11 \qquad \text{Complete the square.}$$

$$= 3(x^2 + 4x + 4) - 3(4) + 11 \qquad \text{Distribute the 3.}$$

$$= 3(x + 2)^2 - 1$$

The graph of f is an upward-opening parabola with vertex $(-2, -1)$, axis of symmetry $x = -2$, and intersects the x-axis at about -2.577 and -1.423. The exact values of the x-intercepts are $x = -2 \pm \sqrt{3}/3$.

Support Graphically

The graph in Figure 2.7 supports these results.   *Now try Exercise 33.*

We now summarize what we know about quadratic functions.

Characterizing the Nature of a Quadratic Function	
Point of View	**Characterization**
Verbal	polynomial of degree 2
Algebraic	$f(x) = ax^2 + bx + c$ or $f(x) = a(x - h)^2 + k \ (a \neq 0)$
Graphical	parabola with vertex (h, k) and axis $x = h$; opens upward if $a > 0$, opens downward if $a < 0$; initial value $= y$-intercept $= f(0) = c$; $x\text{-intercepts} = \dfrac{-b \pm \sqrt{b^2 - 4ac}}{2a}$

Applications of Quadratic Functions

In economics, when demand is linear, revenue is quadratic. Example 7 illustrates this by extending the Twin Pixie model of Example 3.

EXAMPLE 7   Predicting Maximum Revenue

Use the model $y = -15,358.93x + 73,622.50$ from Example 3 to develop a model for the weekly revenue generated by doughnut-shaped oat breakfast cereal sales. Determine the maximum revenue and how to achieve it.

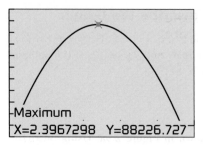

[0, 5] by [−10000, 100000]

FIGURE 2.8 The revenue model for Example 7.

SOLUTION

Model

Revenue can be found by multiplying the price per box, x, by the number of boxes sold, y. So the revenue is given by

$$R(x) = x \cdot y = -15{,}358.93x^2 + 73{,}622.50x,$$

a quadratic model.

Solve Graphically

In Figure 2.8, we find a maximum of about 88,227 occurs when x is about 2.40.

Interpret

To maximize revenue, Twin Pixie should set the price for its store brand of doughnut-shaped oat breakfast cereal at $2.40 per box. Based on the model, this will yield a weekly revenue of about $88,227.   *Now try Exercise 55.*

Recall that the average rate of change of a linear function is constant. In Exercise 78 you will see that the average rate of change of a quadratic function is not constant.

In calculus you will study not only average rate of change but also *instantaneous rate of change*. Such instantaneous rates include velocity and acceleration, which we now begin to investigate.

 Since the time of Galileo Galilei (1564–1642) and Isaac Newton (1642–1727), the vertical motion of a body in free fall has been well understood. The *vertical velocity* and *vertical position* (*height*) of a free-falling body (as functions of time) are classical applications of linear and quadratic functions.

Vertical Free-Fall Motion

The **height** s and **vertical velocity** v of an object in free fall are given by

$$s(t) = -\frac{1}{2}gt^2 + v_0t + s_0 \quad \text{and} \quad v(t) = -gt + v_0,$$

where t is time (in seconds), $g \approx 32 \text{ ft/sec}^2 \approx 9.8 \text{ m/sec}^2$ is the **acceleration due to gravity**, v_0 is the *initial vertical velocity* of the object, and s_0 is its *initial height*.

These formulas disregard air resistance, and the two values given for g are valid at sea level. We apply these formulas in Example 8, and will use them from time to time throughout the rest of the book.

The data in Table 2.3 were collected in Boone, North Carolina (about 1 km above sea level), using a Calculator-Based Ranger™ (CBR™) and a 15-cm rubber air-filled ball. The CBR™ was placed on the floor face up. The ball was thrown upward above the CBR™, and it landed directly on the face of the device.

Table 2.3 Rubber Ball Data from CBR™	
Time (sec)	Height (m)
0.0000	1.03754
0.1080	1.40205
0.2150	1.63806
0.3225	1.77412
0.4300	1.80392
0.5375	1.71522
0.6450	1.50942
0.7525	1.21410
0.8600	0.83173

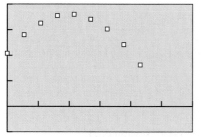

[0, 1.2] by [–0.5, 2.0]

(a)

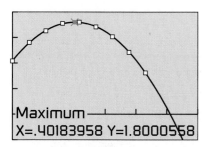

[0, 1.2] by [–0.5, 2.0]

(b)

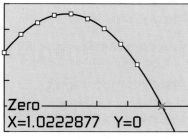

[0, 1.2] by [–0.5, 2.0]

(c)

FIGURE 2.9 Scatter plot and graph of height versus time for Example 8.

EXAMPLE 8   Modeling Vertical Free-Fall Motion

Use the data in Table 2.3 to write models for the height and vertical velocity of the rubber ball. Then use these models to predict the maximum height of the ball and its vertical velocity when it hits the face of the CBR™.

SOLUTION

Model

First we make a scatter plot of the data, as shown in Figure 2.9a. Using quadratic regression, we find the model for the height of the ball to be about

$$s(t) = -4.676t^2 + 3.758t + 1.045,$$

with $R^2 \approx 0.999$, indicating an excellent fit.

Our free-fall theory says the leading coefficient of -4.676 is $-g/2$, giving us a value for $g \approx 9.352$ m/sec^2, which is a bit less than the theoretical value of 9.8 m/sec^2. We also obtain $v_0 \approx 3.758$ m/sec. So the model for vertical velocity becomes

$$v(t) = -gt + v_0 \approx -9.352t + 3.758.$$

Solve Graphically and Numerically

The maximum height is the maximum value of $s(t)$, which occurs at the vertex of its graph. We can see from Figure 2.9b that the vertex has coordinates of about (0.402, 1.800).

In Figure 2.9c, to determine when the ball hits the face of the CBR™, we calculate the positive-valued zero of the height function, which is $t \approx 1.022$. We turn to our linear model to compute the vertical velocity at impact:

$$v(1.022) = -9.352(1.022) + 3.758 \approx -5.80 \text{ m/sec.}$$

Interpret

The maximum height the ball achieved was about 1.80 m above the face of the CBR™. The ball's downward rate is about 5.80 m/sec when it hits the CBR™.

The curve in Figure 2.9b appears to fit the data extremely well, and $R^2 \approx 0.999$. You may have noticed, however, that Table 2.3 contains the ordered pair (0.4300, 1.80392) and that $1.80392 > 1.800$, which is the maximum shown in Figure 2.9b. So, even though our model is theoretically based and an excellent fit to the data, it is not a perfect model. Despite its imperfections, the model provides accurate and reliable predictions about the CBR™ experiment. *Now try Exercise 63.*

REMINDER

Recall from Section 1.7 that R^2 is the coefficient of determination, which measures goodness of fit.

QUICK REVIEW 2.1 *(For help, go to Sections A.2. and P.4)*

In Exercises 1–2, write an equation in slope-intercept form for a line with the given slope m and y-intercept b.

1. $m = 8$, $b = 3.6$ **2.** $m = -1.8$, $b = -2$

In Exercises 3–4, write an equation for the line containing the given points. Graph the line and points.

3. $(-2, 4)$ and $(3, 1)$ **4.** $(1, 5)$ and $(-2, -3)$

In Exercises 5–8, expand the expression.

5. $(x + 3)^2$ **6.** $(x - 4)^2$

7. $3(x - 6)^2$ **8.** $-3(x + 7)^2$

In Exercises 9–10, factor the trinomial.

9. $2x^2 - 4x + 2$ **10.** $3x^2 + 12x + 12$

SECTION 2.1 EXERCISES

In Exercises 1–6, determine which are polynomial functions. For those that are, state the degree and leading coefficient. For those that are not, explain why not.

1. $f(x) = 3x^{-5} + 17$ **2.** $f(x) = -9 + 2x$

3. $f(x) = 2x^5 - \dfrac{1}{2}x + 9$ **4.** $f(x) = 13$

5. $h(x) = \sqrt[3]{27x^3 + 8x^6}$ **6.** $k(x) = 4x - 5x^2$

In Exercises 7–12, write an equation for the linear function f satisfying the given conditions. Graph $y = f(x)$.

7. $f(-5) = -1$ and $f(2) = 4$

8. $f(-3) = 5$ and $f(6) = -2$

9. $f(-4) = 6$ and $f(-1) = 2$

10. $f(1) = 2$ and $f(5) = 7$

11. $f(0) = 3$ and $f(3) = 0$

12. $f(-4) = 0$ and $f(0) = 2$

In Exercises 13–18, match a graph to the function. Explain your choice.

13. $f(x) = 2(x + 1)^2 - 3$ **14.** $f(x) = 3(x + 2)^2 - 7$

15. $f(x) = 4 - 3(x - 1)^2$ **16.** $f(x) = 12 - 2(x - 1)^2$

17. $f(x) = 2(x - 1)^2 - 3$ **18.** $f(x) = 12 - 2(x + 1)^2$

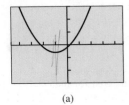

(a)

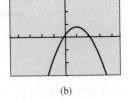

(b)

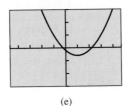

(c)

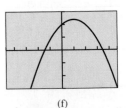

(d)

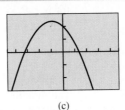

(e)

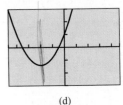

(f)

In Exercises 19–22, describe how to transform the graph of $f(x) = x^2$ into the graph of the given function. Sketch each graph by hand.

19. $g(x) = (x - 3)^2 - 2$ **20.** $h(x) = \dfrac{1}{4}x^2 - 1$

21. $g(x) = \dfrac{1}{2}(x + 2)^2 - 3$ **22.** $h(x) = -3x^2 + 2$

In Exercises 23–26, find the vertex and axis of the graph of the function.

23. $f(x) = 3(x - 1)^2 + 5$ **24.** $g(x) = -3(x + 2)^2 - 1$

25. $f(x) = 5(x - 1)^2 - 7$ **26.** $g(x) = 2(x - \sqrt{3})^2 + 4$

In Exercises 27–32, find the vertex and axis of the graph of the function. Rewrite the equation for the function in vertex form.

27. $f(x) = 3x^2 + 5x - 4$ **28.** $f(x) = -2x^2 + 7x - 3$

29. $f(x) = 8x - x^2 + 3$ **30.** $f(x) = 6 - 2x + 4x^2$

31. $g(x) = 5x^2 + 4 - 6x$ **32.** $h(x) = -2x^2 - 7x - 4$

In Exercises 33–38, use completing the square to describe the graph of each function. Support your answers graphically.

33. $f(x) = x^2 - 4x + 6$ **34.** $g(x) = x^2 - 6x + 12$

35. $f(x) = 10 - 16x - x^2$ **36.** $h(x) = 8 + 2x - x^2$

37. $f(x) = 2x^2 + 6x + 7$ **38.** $g(x) = 5x^2 - 25x + 12$

In Exercises 39–42, write an equation for the parabola shown, using the fact that one of the given points is the vertex.

39.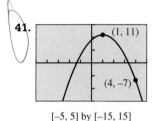

$(1, 5)$
$(-1, -3)$

[–5, 5] by [–15, 15]

40.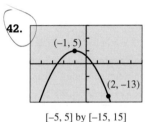

$(0, 5)$
$(2, -7)$

[–5, 5] by [–15, 15]

41.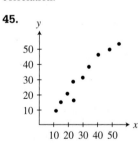

$(1, 11)$
$(4, -7)$

[–5, 5] by [–15, 15]

42.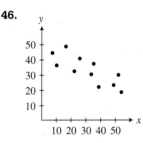

$(-1, 5)$
$(2, -13)$

[–5, 5] by [–15, 15]

In Exercises 43 and 44, write an equation for the quadratic function whose graph contains the given vertex and point.

43. Vertex $(1, 3)$, point $(0, 5)$

44. Vertex $(-2, -5)$, point $(-4, -27)$

In Exercises 45–48, describe the strength and direction of the linear correlation.

45.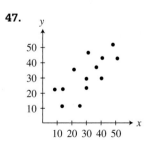

46.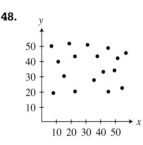

47.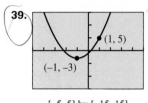

48.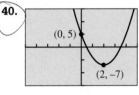

49. Comparing Age and Weight A group of male children were weighed. Their ages and weights are recorded in Table 2.4.

Table 2.4 Children's Age and Weight	
Age (months)	Weight (pounds)
18	23
20	25
24	24
26	32
27	33
29	29
34	35
39	39
42	44

(a) Draw a scatter plot of these data.

(b) Writing to Learn Describe the strength and direction of the correlation between age and weight.

50. Life Expectancy Table 2.5 shows the average number of additional years a U.S. citizen is expected to live for various ages.

Table 2.5 U.S. Life Expectancy	
Age (years)	Life Expectancy (years)
10	67.4
20	57.7
30	48.2
40	38.8
50	29.8
60	21.5
70	14.3
80	8.6

Source: U.S. National Center for Health Statistics, Vital Statistics of the United States.

(a) Draw a scatter plot of these data.

(b) Writing to Learn Describe the strength and direction of the correlation between age and life expectancy.

51. Straight-Line Depreciation Mai Lee bought a computer for her home office and depreciated it over 5 years using the straight-line method. If its initial value was $2350, what is its value 3 years later?

52. Costly Doll Making Patrick's doll-making business has weekly fixed costs of $350. If the cost for materials is $4.70 per doll and his total weekly costs average $500, about how many dolls does Patrick make each week?

53. Table 2.6 shows the average hourly compensation of production workers in manufacturing for several years. Let x be the number of years since 1970, so that $x = 5$ stands for 1975, and so forth.

Table 2.6 Production Worker Average	
Year	Hourly Compensation (dollars)
1975	6.36
1985	13.01
1995	17.19
2002	21.37

Source: U.S. Bureau of Labor Statistics as reported in The World Almanac and Book of Facts, 2005.

(a) Writing to Learn Find the linear regression model for the data. What does the slope in the regression model represent?

(b) Use the linear regression model to predict the production worker average hourly compensation in the year 2010.

54. Finding Maximum Area Among all the rectangles whose perimeters are 100 ft, find the dimensions of the one with maximum area.

55. Determining Revenue The per unit price p (in dollars) of a popular toy when x units (in thousands) are produced is modeled by the function

$$\text{price} = p = 12 - 0.025x.$$

The revenue (in thousands of dollars) is the product of the price per unit and the number of units (in thousands) produced. That is,

$$\text{revenue} = xp = x(12 - 0.025x).$$

(a) State the dimensions of a viewing window that shows a graph of the revenue model for producing 0 to 100,000 units.

(b) How many units should be produced if the total revenue is to be $1,000,000?

56. Finding the Dimensions of a Painting A large painting in the style of Rubens is 3 ft longer than it is wide. If the wooden frame is 12 in. wide, the area of the picture and frame is 208 ft^2, find the dimensions of the painting.

57. Using Algebra in Landscape Design Julie Stone designed a rectangular patio that is 25 ft by 40 ft. This patio is surrounded by a terraced strip of uniform width planted with small trees and shrubs. If the area A of this terraced strip is 504 ft^2, find the width x of the strip.

58. Management Planning The Welcome Home apartment rental company has 1600 units available, of which 800 are currently rented at $300 per month. A market survey indicates that each $5 decrease in monthly rent will result in 20 new leases.

(a) Determine a function $R(x)$ that models the total rental income realized by Welcome Home, where x is the number of $5 decreases in monthly rent.

(b) Find a graph of $R(x)$ for rent levels between $175 and $300 (that is, $0 \le x \le 25$) that clearly shows a maximum for $R(x)$.

(c) What rent will yield Welcome Home the maximum monthly income?

59. Group Activity Beverage Business The Sweet Drip Beverage Co. sells cans of soda pop in machines. It finds that sales average 26,000 cans per month when the cans sell for 50¢ each. For each nickel increase in the price, the sales per month drop by 1000 cans.

(a) Determine a function $R(x)$ that models the total revenue realized by Sweet Drip, where x is the number of $0.05 increases in the price of a can.

(b) Find a graph of $R(x)$ that clearly shows a maximum for $R(x)$.

(c) How much should Sweet Drip charge per can to realize the maximum revenue? What is the maximum revenue?

60. Group Activity Sales Manager Planning Jack was named District Manager of the Month at the Sylvania Wire Co. due to his hiring study. It shows that each of the 30 salespersons he supervises average $50,000 in sales each month, and that for each additional salesperson he would hire, the average sales would decrease $1000 per month. Jack concluded his study by suggesting a number of salespersons that he should hire to maximize sales. What was that number?

61. Free-Fall Motion As a promotion for the Houston Astros downtown ballpark, a competition is held to see who can throw a baseball the highest from the front row of the upper deck of seats, 83 ft above field level. The winner throws the ball with an initial vertical velocity of 92 ft/sec and it lands on the infield grass.

(a) Find the maximum height of the base ball.

(b) How much time is the ball in the air?

(c) Determine its vertical velocity when it hits the ground.

62. Baseball Throwing Machine The Sandusky Little League uses a baseball throwing machine to help train 10-year-old players to catch high pop-ups. It throws the baseball straight up with an initial velocity of 48 ft/sec from a height of 3.5 ft.

(a) Find an equation that models the height of the ball t seconds after it is thrown.

(b) What is the maximum height the baseball will reach? How many seconds will it take to reach that height?

63. Fireworks Planning At the Bakersville Fourth of July celebration, fireworks are shot by remote control into the air from a pit that is 10 ft below the earth's surface.

(a) Find an equation that models the height of an aerial bomb t seconds after it is shot upwards with an initial velocity of 80 ft/sec. Graph the equation.

(b) What is the maximum height above ground level that the aerial bomb will reach? How many seconds will it take to reach that height?

64. Landscape Engineering In her first project after being employed by Land Scapes International, Becky designs a decorative water fountain that will shoot water to a maximum height of 48 ft. What should be the initial velocity of each drop of water to achieve this maximum height? (*Hint:* Use a grapher and a guess-and-check strategy.)

65. Patent Applications Using quadratic regression on the data in Table 2.7, predict the year when the number of patent application will reach 450,000. Let $x = 0$ stand for 1980, $x = 10$ for 1990 and so forth.

Table 2.7 U.S. Patent Applications	
Year	Applications (thousands)
1980	113.0
1990	176.7
1995	228.8
1998	261.4
1999	289.5
2000	315.8
2001	346.6
2002	357.5
2003	367.0

Source: U.S. Census Bureau, Statistical Abstract of the United States, 2004–2005 (124th ed., Washington, D.C., 2004).

66. Highway Engineering Interstate 70 west of Denver, Colorado, has a section posted as a 6% grade. This means that for a horizontal change of 100 ft there is a 6-ft vertical change.

6% grade

(a) Find the slope of this section of the highway.

(b) On a highway with a 6% grade what is the horizontal distance required to climb 250 ft?

(c) A sign along the highway says 6% grade for the next 7 mi. Estimate how many feet of vertical change there are along those 7 mi. (There are 5280 ft in 1 mile.)

67. A group of female children were weighed. Their ages and weights are recorded in Table 2.8.

Table 2.8 Children's Ages and Weights	
Age (months)	Weight (pounds)
19	22
21	23
24	25
27	28
29	31
31	28
34	32
38	34
43	39

(a) Draw a scatter plot of the data.

(b) Find the linear regression model.

(c) Interpret the slope of the linear regression equation.

(d) Superimpose the regression line on the scatter plot.

(e) Use the regression model to predict the weight of a 30-month-old girl.

68. Table 2.9 shows the median U.S. family income (in 2003 dollars) for selected years. Let x be the number of years since 1940.

Table 2.9 Median Family Income in the U.S. (in 2003 dollars)	
Year	Median Family Income ($)
1947	21,201
1973	43,219
1979	45,989
1989	49,014
1995	48.679
2000	54,191
2003	52,680

Source: Economic Policy Institute, The State of Working America 2004/2005 (ILR Press, 2005).

(a) Find the linear regression model for the data.

(b) Use it to predict the median U.S. family income in 2010.

In Exercises 69–70, complete the analysis for the given Basic Function.

69. Analyzing a Function

> ### BASIC FUNCTION
> ### The Identity Function
>
> $f(x) = x$
> Domain:
> Range:
> Continuity:
> Increasing-decreasing behavior:
> Symmetry:
> Boundedness:
> Local extrema:
> Horizontal asymptotes:
> Vertical asymptotes:
> End behavior:

70. Analyzing a Function

> ### BASIC FUNCTION
> ### The Squaring Function
>
> $f(x) = x^2$
> Domain:
> Range:
> Continuity:
> Increasing-decreasing behavior:
> Symmetry:
> Boundedness:
> Local extrema:
> Horizontal asymptotes:
> Vertical asymptotes:
> End behavior:

Standardized Test Questions

71. True or False The initial value of $f(x) = 3x^2 + 2x - 3$ is 0. Justify your answer.

72. True or False The graph of the function $f(x) = x^2 - x + 1$ has no x-intercepts. Justify your answer.

In Exercises 73–76, you may use a graphing calculator to solve the problem.

In Exercises 73 and 74, $f(x) = mx + b$, $f(-2) = 3$, and $f(4) = 1$.

73. Multiple Choice What is the value of m?
   (A) 3   **(B)** -3 **(C)** -1 **(D)** 1/3   **(E)** $-1/3$

74. Multiple Choice What is the value of b?
   (A) 4   **(B)** 11/3   **(C)** 7/3   **(D)** 1   **(E)** $-1/3$

In Exercises 75 and 76, let $f(x) = 2(x + 3)^2 - 5$.

75. Multiple Choice What is the axis of symmetry of the graph of f?
   (A) $x = 3$ **(B)** $x = -3$ **(C)** $y = 5$ **(D)** $y = -5$ **(E)** $y = 0$

76. Multiple Choice What is the vertex of f?
   (A) $(0,0)$ **(B)** $(3,5)$ **(C)** $(3,-5)$ **(D)** $(-3,5)$ **(E)** $(-3,-5)$

Explorations

77. Writing to Learn Identifying Graphs of Linear Functions

   (a) Which of the lines graphed below are graphs of linear functions? Explain.

   (b) Which of the lines graphed below are graphs of functions? Explain.

   (c) Which of the lines graphed below are not graphs of functions? Explain.

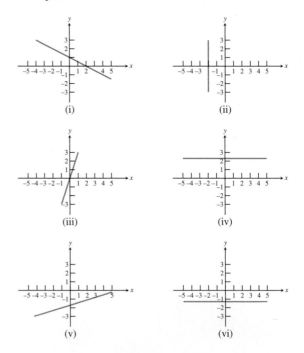

78. Average Rate of Change Let $f(x) = x^2$, $g(x) = 3x + 2$, $h(x) = 7x - 3$, $k(x) = mx + b$, and $l(x) = x^3$.

(a) Compute the average rate of change of f from $x = 1$ to $x = 3$.

(b) Compute the average rate of change of f from $x = 2$ to $x = 5$.

(c) Compute the average rate of change of f from $x = a$ to $x = c$.

(d) Compute the average rate of change of g from $x = 1$ to $x = 3$.

(e) Compute the average rate of change of g from $x = 1$ to $x = 4$.

(f) Compute the average rate of change of g from $x = a$ to $x = c$.

(g) Compute the average rate of change of h from $x = a$ to $x = c$.

(h) Compute the average rate of change of k from $x = a$ to $x = c$.

(i) Compute the average rate of change of l from $x = a$ to $x = c$.

Extending the Ideas

79. Minimizing Sums of Squares The linear regression line is often called the **least-square lines** because it minimizes the sum of the squares of the **residuals**, the differences between actual y values and predicted y values:

$$residual = y_i - (ax_i + b),$$

where (x_i, y_i) are the given data pairs and $y = ax + b$ is the regression equation, as shown in the figure.

Use these definitions to explain why the regression line obtained from reversing the ordered pairs in Table 2.2 is not the inverse of the function obtained in Example 3.

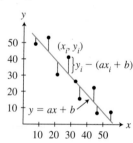

80. Median-Median Line Read about the median-median line by going to the Internet, your grapher owner's manual, or a library. Then use the following data set to complete this problem.

$\{(2, 8), (3, 6), (5, 9), (6, 8), (8, 11), (10, 13), (12, 14), (15, 4)\}$

(a) Draw a scatter plot of the data.

(b) Find the linear regression equation and graph it.

(c) Find the median-median line equation and graph it.

(d) Writing to Learn For this data, which of the two lines appears to be the line of better fit? Why?

81. Suppose $b^2 - 4ac > 0$ for the equation $ax^2 + bx + c = 0$.

(a) Prove that the sum of the two solutions of this equation is $-b/a$.

(b) Prove that the product of the two solutions of this equation is c/a.

82. Connecting Algebra and Geometry Prove that the axis of the graph of $f(x) = (x - a)(x - b)$ is $x = (a + b)/2$, where a and b are real numbers.

83. Connecting Algebra and Geometry Identify the vertex of the graph of $f(x) = (x - a)(x - b)$, where a and b are any real numbers.

84. Connecting Algebra and Geometry Prove that if x_1 and x_2 are real numbers and are zeros of the quadratic function $f(x) = ax^2 + bx + c$, then the axis of the graph of f is $x = (x_1 + x_2)/2$.

85. Prove the Constant Rate of Change Theorem, which is stated on page 172.

2.2

Power Functions with Modeling

Power Functions and Variation

Five of the basic functions introduced in Section 1.3 were power functions. Power functions are an important family of functions in their own right and are important building blocks for other functions.

DEFINITION Power Function

Any function that can be written in the form

$$f(x) = k \cdot x^a, \text{ where } k \text{ and } a \text{ are nonzero constants,}$$

is a **power function**. The constant a is the **power**, and k is the **constant of variation**, or **constant of proportion**. We say $f(x)$ **varies as** the ath power of x, or $f(x)$ **is proportional to** the ath power of x.

In general, if $y = f(x)$ varies as a constant power of x, then y is a power function of x. Many of the most common formulas from geometry and science are power functions.

Name	Formula	Power	Constant of Variation
Circumference	$C = 2\pi r$	1	2π
Area of a circle	$A = \pi r^2$	2	π
Force of gravity	$F = k/d^2$	-2	k
Boyle's Law	$V = k/P$	-1	k

These four power function models involve output-from-input relationships that can be expressed in the language of *variation and proportion*:

- The circumference of a circle varies directly as its radius.

- The area enclosed by a circle is directly proportional to the square of its radius.

- The force of gravity acting on an object is inversely proportional to the square of the distance from the object to the center of the Earth.

- Boyle's Law states that the volume of an enclosed gas (at a constant temperature) varies inversely as the applied pressure.

The power function formulas with positive powers are statements of **direct variation** and power function formulas with negative powers are statements of **inverse variation**. Unless the word *inversely* is included in a variation statement, the variation is assumed to be direct, as in Example 1.

EXAMPLE 1   Writing a Power Function Formula

From empirical evidence and the laws of physics it has been found that the period of time T for the full swing of a pendulum varies as the square root of the pendulum's length l, provided that the swing is small relative to the length of the pendulum. Express this relationship as a power function.

SOLUTION   Because it does not state otherwise, the variation is direct. So the power is positive. The wording tells us that T is a function of l. Using k as the constant of variation gives us

$$T(l) = k\sqrt{l} = k \cdot l^{1/2}.$$ ***Now try Exercise 17.***

Section 1.3 introduced five basic power functions:

$$x,\ x^2,\ x^3,\ x^{-1} = \frac{1}{x},\quad \text{and}\quad x^{1/2} = \sqrt{x}.$$

Example 2 describes two other power functions: the *cube root function* and the *inverse-square* function.

EXAMPLE 2   Analyzing Power Functions

State the power and constant of variation for the function, graph it, and analyze it.

(a) $f(x) = \sqrt[3]{x}$ **(b)** $g(x) = \dfrac{1}{x^2}$

SOLUTION

(a) Because $f(x) = \sqrt[3]{x} = x^{1/3} = 1 \cdot x^{1/3}$, its power is 1/3, and its constant of variation is 1. The graph of f is shown in Figure 2.10a.

Domain: All reals
Range: All reals
Continuous
Increasing for all x
Symmetric with respect to the origin (an odd function)
Not bounded above or below
No local extrema
No asymptotes
End behavior: $\displaystyle\lim_{x\to-\infty} \sqrt[3]{x} = -\infty$ and $\displaystyle\lim_{x\to\infty} \sqrt[3]{x} = \infty$

Interesting fact: The cube root function $f(x) = \sqrt[3]{x}$ is the inverse of the cubing function.

(b) Because $g(x) = 1/x^2 = x^{-2} = 1 \cdot x^{-2}$, its power is -2, and its constant of variation is 1. The graph of g is shown in Figure 2.10b.

Domain: $(-\infty, 0) \cup (0, \infty)$
Range: $(0, \infty)$
Continuous on its domain; discontinuous at $x = 0$
Increasing on $(-\infty, 0)$; decreasing on $(0, \infty)$
Symmetric with respect to the y-axis (an even function)
Bounded below, but not above
No local extrema

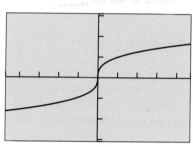

[−4.7, 4.7] by [−3.1, 3.1]
(a)

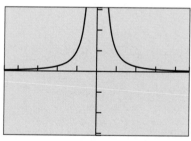

[−4.7, 4.7] by [−3.1, 3.1]
(b)

FIGURE 2.10   The graphs of
(a) $f(x) = \sqrt[3]{x} = x^{1/3}$ and
(b) $g(x) = 1/x^2 = x^{-2}$. (Example 2)

continued

Horizontal asymptote: $y = 0$; vertical asymptote: $x = 0$

End behavior: $\lim\limits_{x \to -\infty} (1/x^2) = 0$ and $\lim\limits_{x \to \infty}(1/x^2) = 0$

Interesting fact: $g(x) = 1/x^2$ is the basis of scientific *inverse-square laws*, for example, the inverse-square gravitational principle $F = k/d^2$ mentioned above.

So $g(x) = 1/x^2$ is sometimes called the inverse-square function, but it is *not* the inverse of the squaring function but rather its *multiplicative* inverse.

> **Now try Exercise 27.**

Monomial Functions and Their Graphs

A single-term polynomial function is a power function that is also called a *monomial function.*

DEFINITION Monomial Function

Any function that can be written as

$$f(x) = k \ \text{ or } \ f(x) = k \cdot x^n,$$

where k is a constant and n is a positive integer, is a **monomial function**.

So the zero function and constant functions are monomial functions, but the more typical monomial function is a power function with a positive integer power, which is the degree of the monomial. The basic functions x, x^2, and x^3 are typical monomial functions. It is important to understand the graphs of monomial functions because every polynomial function is either a monomial function or a sum of monomial functions.

In Exploration 1 we take a close look at six basic monomial functions. They have the form x^n for $n = 1, 2, \ldots, 6$. We group them by even and odd powers.

EXPLORATION 1 Comparing Graphs of Monomial Functions

Graph the triplets of functions in the stated windows and explain how the graphs are alike and how they are different. Consider the relevant aspects of analysis from Example 2. Which ordered pairs do all three graphs have in common?

1. $f(x) = x$, $g(x) = x^3$, and $h(x) = x^5$ in the window $[-2.35, 2.35]$ by $[-1.5, 1.5]$, then $[-5, 5]$ by $[-15, 15]$, and finally $[-20, 20]$ by $[-200, 200]$.

2. $f(x) = x^2$, $g(x) = x^4$, and $h(x) = x^6$ in the window $[-1.5, 1.5]$ by $[-0.5, 1.5]$, then $[-5, 5]$ by $[-5, 25]$, and finally $[-15, 15]$ by $[-50, 400]$.

From Exploration 1 we see that

$f(x) = x^n$ is an even function if n is even and an odd function if n is odd.

Because of this symmetry, it is enough to know the first quadrant behavior of $f(x) = x^n$. Figure 2.11 shows the graphs of $f(x) = x^n$ for $n = 1, 2, \ldots, 6$ in the first quadrant near the origin.

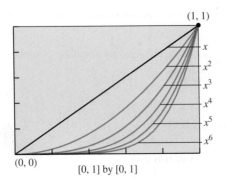

FIGURE 2.11 The graphs of $f(x) = x^n$, $0 \le x \le 1$, for $n = 1, 2, \ldots, 6$.

The following conclusions about the basic function $f(x) = x^3$ can be drawn from your investigations in Exploration 1.

BASIC FUNCTION The Cubing Function

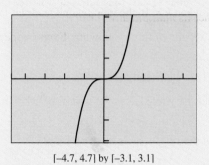

[−4.7, 4.7] by [−3.1, 3.1]

FIGURE 2.12 The graph of $f(x) = x^3$.

$f(x) = x^3$
Domain: All reals
Range: All reals
Continuous
Increasing for all x
Symmetric with respect to the origin (an odd function)
Not bounded above or below
No local extrema
No horizontal asymptotes
No vertical asymptotes
End behavior: $\lim\limits_{x \to -\infty} x^3 = -\infty$ and $\lim\limits_{x \to \infty} x^3 = \infty$

EXAMPLE 3 Graphing Monomial Functions

Describe how to obtain the graph of the given function from the graph of $g(x) = x^n$ with the same power n. Sketch the graph by hand and support your answer with a grapher.

(a) $f(x) = 2x^3$ 　　　　　　　　　　　**(b)** $f(x) = -\dfrac{2}{3}x^4$

SOLUTION

(a) We obtain the graph of $f(x) = 2x^3$ by vertically stretching the graph of $g(x) = x^3$ by a factor of 2. Both are odd functions. See Figure 2.13a.

(b) We obtain the graph of $f(x) = -(2/3)x^4$ by vertically shrinking the graph of $g(x) = x^4$ by a factor of 2/3 and then reflecting it across the x-axis. Both are even functions. See Figure 2.13b.   　　*Now try Exercise 31.*

We ask you to explore the graphical behavior of power functions of the form x^{-n} and $x^{1/n}$, where n is a positive integer, in Exercise 65.

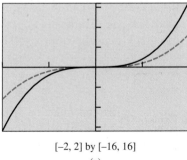

[−2, 2] by [−16, 16]
(a)

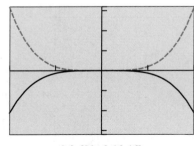

[−2, 2] by [−16, 16]
(b)

FIGURE 2.13 The graphs of (a) $f(x) = 2x^3$ with basic monomial $g(x) = x^3$, and (b) $f(x) = -(2/3)x^4$ with basic monomial $g(x) = x^4$. (Example 3)

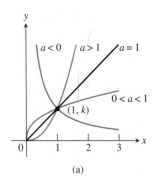

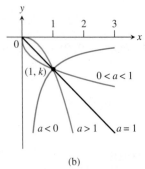

FIGURE 2.14 The graphs of $f(x) = k \cdot x^a$ for $x \geq 0$. (a) $k > 0$, (b) $k < 0$.

Graphs of Power Functions

The graphs in Figure 2.14 represent the four shapes that are possible for general power functions of the form $f(x) = kx^a$ for $x \geq 0$. In every case, the graph of f contains $(1, k)$. Those with positive powers also pass through $(0, 0)$. Those with negative exponents are asymptotic to both axes.

When $k > 0$, the graph lies in Quadrant I, but when $k < 0$, the graph is in Quadrant IV.

In general, for any power function $f(x) = k \cdot x^a$, one of three following things happens when $x < 0$.

- f is undefined for $x < 0$, as is the case for $f(x) = x^{1/2}$ and $f(x) = x^\pi$.

- f is an even function, so f is symmetric about the y-axis, as is the case for $f(x) = x^{-2}$ and $f(x) = x^{2/3}$.

- f is an odd function, so f is symmetric about the origin, as is the case for $f(x) = x^{-1}$ and $f(x) = x^{7/3}$.

Predicting the general shape of the graph of a power function is a two-step process as illustrated in Example 4.

EXAMPLE 4 Graphing Power Functions $f(x) = k \cdot x^a$

State the values of the constants k and a. Describe the portion of the curve that lies in Quadrant I or IV. Determine whether f is even, odd, or undefined for $x < 0$. Describe the rest of the curve if any. Graph the function to see whether it matches the description.

(a) $f(x) = 2x^{-3}$ **(b)** $f(x) = -0.4x^{1.5}$ **(c)** $f(x) = -x^{0.4}$

SOLUTION

(a) Because $k = 2$ is positive and $a = -3$ is negative, the graph passes through $(1, 2)$ and is asymptotic to both axes. The graph is decreasing in the first quadrant. The function f is odd because

$$f(-x) = 2(-x)^{-3} = \frac{2}{(-x)^3} = -\frac{2}{x^3} = -2x^{-3} = -f(x).$$

So its graph is symmetric about the origin. The graph in Figure 2.15a supports all aspects of the description.

(b) Because $k = -0.4$ is negative and $a = 1.5 > 1$, the graph contains $(0, 0)$ and passes through $(1, -0.4)$. In the fourth quadrant, it is decreasing. The function f is undefined for $x < 0$ because

$$f(x) = -0.4x^{1.5} = -\frac{2}{5}x^{3/2} = -\frac{2}{5}(\sqrt{x})^3,$$

and the square-root function is undefined for $x < 0$. So the graph of f has no points in Quadrants II or III. The graph in Figure 2.15b matches the description.

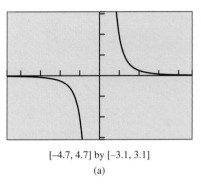

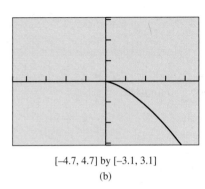

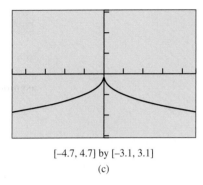

[−4.7, 4.7] by [−3.1, 3.1]
(a)

[−4.7, 4.7] by [−3.1, 3.1]
(b)

[−4.7, 4.7] by [−3.1, 3.1]
(c)

FIGURE 2.15 The graphs of (a) $f(x) = 2x^{-3}$, (b) $f(x) = -0.4x^{1.5}$, and (c) $f(x) = -x^{0.4}$. (Example 4)

(c) Because $k = -1$ is negative and $0 < a < 1$, the graph contains $(0, 0)$ and passes through $(1, -1)$. In the fourth quadrant, it is decreasing. The function f is even because

$$f(-x) = -(-x)^{0.4} = -(-x)^{2/5} = -(\sqrt[5]{-x})^2 = -(-\sqrt[5]{x})^2$$
$$= -(\sqrt[5]{x})^2 = -x^{0.4} = f(x).$$

So the graph of f is symmetric about the y-axis. The graph in Figure 2.15c fits the description. *Now try Exercise 43.*

The following information about the basic function $f(x) = \sqrt{x}$ follows from the investigation in Exercise 65.

BASIC FUNCTION The Square Root Function

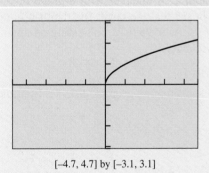

[−4.7, 4.7] by [−3.1, 3.1]

FIGURE 2.16 The graph of $f(x) = \sqrt{x}$.

$f(x) = \sqrt{x}$
Domain: $[0, \infty)$
Range: $[0, \infty)$
Continuous on $[0, \infty)$
Increasing on $[0, \infty)$
No symmetry
Bounded below but not above
Local minimum at $x = 0$
No horizontal asymptotes
No vertical asymptotes
End behavior: $\lim_{x \to \infty} \sqrt{x} = \infty$

Modeling with Power Functions

Noted astronomer Johannes Kepler (1571–1630) developed three laws of planetary motion that are used to this day. Kepler's Third Law states that the square of the period of orbit T (the time required for one full revolution around the Sun) for each planet is proportional to the cube of its average distance a from the Sun. Table 2.10 on the next page gives the relevant data for the six planets that were known in Kepler's time. The distances are given in millions of kilometers, or gigameters (Gm).

	Average Distance from Sun (Gm)	Period of Orbit (days)
Table 2.10 Average Distances and Orbit Periods for the Six Innermost Planets		
Planet		
Mercury	57.9	88
Venus	108.2	225
Earth	149.6	365.2
Mars	227.9	687
Jupiter	778.3	4332
Saturn	1427	10,760

Source: Shupe, Dorr, Payne, Hunsiker, et al., National Geographic Atlas of the World (rev. 6th ed.). Washington, DC: National Geographic Society, 1992, plate 116.

A BIT OF HISTORY

Example 5 shows the predictive power of a well founded model. Exercise 67 asks you to find Kepler's elegant form of the equation, $T^2 = a^3$, which he reported in *The Harmony of the World* in 1619.

EXAMPLE 5   Modeling Planetary Data with a Power Function

Use the data in Table 2.10 to obtain a power function model for orbital period as a function of average distance from the Sun. Then use the model to predict the orbital period for Neptune, which is 4497 Gm from the Sun on average.

SOLUTION

Model

First we make a scatter plot of the data, as shown in Figure 2.17a. Using power regression, we find the model for the orbital period to be about

$$T(a) \approx 0.20a^{1.5} = 0.20a^{3/2} = 0.20\sqrt{a^3}.$$

Figure 2.17b shows the scatter plot for Table 2.10 together with a graph of the power regression model just found. You can see that the curve fits the data quite well. The coefficient of determination is $r^2 \approx 0.999999912$, indicating an amazingly close fit and supporting the visual evidence.

Solve Numerically

To predict the orbit period for Neptune we substitute its average distance from the Sun in the power regression model:

$$T(4497) \approx 0.2(4497)^{1.5} \approx 60{,}313.$$

Interpret

It takes Neptune about 60,313 days to orbit the Sun, or about 165 years, which is the value given in the *National Geographic Atlas of the World*.

Figure 2.17c reports this result and gives some indication of the relative distances involved. Neptune is much farther from the Sun than the six innermost planets and especially the four closest to the Sun—Mercury, Venus, Earth, and Mars.

Now try Exercise 55.

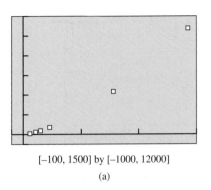

[−100, 1500] by [−1000, 12000]

(a)

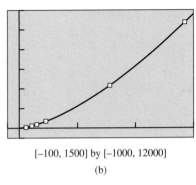

[−100, 1500] by [−1000, 12000]

(b)

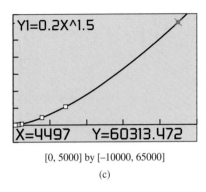

Y1=0.2X^1.5

X=4497   Y=60313.472

[0, 5000] by [−10000, 65000]

(c)

FIGURE 2.17 Scatter plot and graphs for Example 5.

In Example 6, we return to free-fall motion, with a new twist. The data in the table come from the same CBR™ experiment referenced in Example 8 of Section 2.1. This time we are looking at the downward distance (in meters) the ball has traveled since reaching its peak height and its downward speed (in meters per second). It can be shown (see Exercise 68) that free-fall speed is proportional to a power of the distance traveled.

EXAMPLE 6   Modeling Free-Fall Speed versus Distance

Use the data in Table 2.11 to obtain a power function model for speed p versus distance traveled d. Then use the model to predict the speed of the ball at impact given that impact occurs when $d \approx 1.80$ m.

Table 2.11 Rubber Ball Data from CBR™ Experiment	
Distance (m)	Speed (m/s)
0.00000	0.00000
0.04298	0.82372
0.16119	1.71163
0.35148	2.45860
0.59394	3.05209
0.89187	3.74200
1.25557	4.49558

SOLUTION

Model

Figure 2.18a on the next page is a scatter plot of the data. Using power regression, we find the model for speed p versus distance d to be about

$$p(d) \approx 4.03d^{0.5} = 4.03d^{1/2} = 4.03\sqrt{d}.$$

(See margin notes.) Figure 2.18b shows the scatter plot for Table 2.11 together with a graph of the power regression equation just found. You can see that the curve fits the data nicely. The coefficient of determination is $r^2 \approx 0.99770$, indicating a close fit and supporting the visual evidence.

continued

WHY p?

We use p for speed to distinguish it from velocity v. Recall that speed is the *absolute value* of velocity.

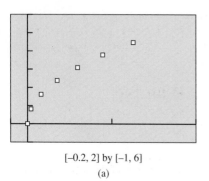

[–0.2, 2] by [–1, 6]

(a)

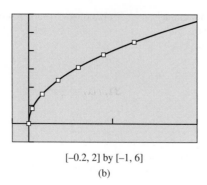

[–0.2, 2] by [–1, 6]

(b)

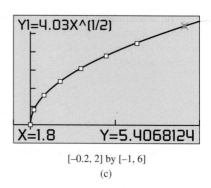

[–0.2, 2] by [–1, 6]

(c)

FIGURE 2.18 Scatter plot and graphs for Example 6.

A WORD OF WARNING

The regression routine traditionally used to compute power function models involves taking logarithms of the data, and therefore, all of the data must be strictly positive numbers. So we must leave out (0, 0) to compute the power regression equation.

Solve Numerically

To predict the speed at impact, we substitute $d \approx 1.80$ into the obtained power regression model:

$$p(1.80) \approx 5.4.$$

See Figure 2.18c.

Interpret

The speed at impact is about 5.4 m/sec. This is slightly less than the value obtained in Example 8 of Section 2.1, using a different modeling process for the same experiment.

Now try Exercise 57.

QUICK REVIEW 2.2 *(For help, go to Section A.1.)*

In Exercises 1–6, write the following expressions using only positive integer powers.

1. $x^{2/3}$

2. $p^{5/2}$

3. d^{-2}

4. x^{-7}

5. $q^{-4/5}$

6. $m^{-1.5}$

In Exercises 7–10, write the following expressions in the form $k \cdot x^a$ using a single rational number for the power a.

7. $\sqrt{9x^3}$

8. $\sqrt[3]{8x^5}$

9. $\sqrt[3]{\dfrac{5}{x^4}}$

10. $\dfrac{4x}{\sqrt{32x^3}}$

SECTION 2.2 EXERCISES

In Exercises 1–10, determine whether the function is a power function, given that c, g, k, and π represent constants. For those that are power functions, state the power and constant of variation.

1. $f(x) = -\dfrac{1}{2}x^5$

2. $f(x) = 9x^{5/3}$

3. $f(x) = 3 \cdot 2^x$

4. $f(x) = 13$

5. $E(m) = mc^2$

6. $KE(v) = \dfrac{1}{2}kv^5$

7. $d = \dfrac{1}{2}gt^2$

8. $V = \dfrac{4}{3}\pi r^3$

9. $I = \dfrac{k}{d^2}$

10. $F(a) = m \cdot a$

In Exercises 11–16, determine whether the function is a monomial function, given that l and π represent constants. For those that are monomial functions state the degree and leading coefficient. For those that are not, explain why not.

11. $f(x) = -4$

12. $f(x) = 3x^{-5}$

13. $y = -6x^7$

14. $y = -2 \cdot 5^x$

15. $S = 4\pi r^2$

16. $A = lw$

In Exercises 17–22, write the statement as a power function equation. Use k for the constant of variation if one is not given.

17. The area A of an equilateral triangle varies directly as the square of the length s of its sides.

18. The volume V of a circular cylinder with fixed height is proportional to the square of its radius r.

19. The current I in an electrical circuit is inversely proportional to the resistance R, with constant of variation V.

20. Charles's Law states the volume V of an enclosed ideal gas at a constant pressure varies directly as the absolute temperature T.

21. The energy E produced in a nuclear reaction is proportional to the mass m, with the constant of variation being c^2, the square of the speed of light.

22. The speed p of a free-falling object that has been dropped from rest varies as the square root of the distance traveled d, with a constant of variation $k = \sqrt{2g}$.

In Exercises 23–26, write a sentence that expresses the relationship in the formula, using the language of variation or proportion.

23. $w = mg$, where w and m are the weight and mass of an object and g is the constant acceleration due to gravity.

24. $C = \pi D$, where C and D are the circumference and diameter of a circle and π is the usual mathematical constant.

25. $n = c/v$, where n is the refractive index of a medium, v is the velocity of light in the medium, and c is the constant velocity of light in free space.

26. $d = p^2/(2g)$, where d is the distance traveled by a free-falling object dropped from rest, p is the speed of the object, and g is the constant acceleration due to gravity.

In Exercises 27–30, state the power and constant of variation for the function, graph it, and analyze it in the manner of Example 2 of this section.

27. $f(x) = 2x^4$

28. $f(x) = -3x^3$

29. $f(x) = \frac{1}{2}\sqrt[4]{x}$

30. $f(x) = -2x^{-3}$

In Exercises 31–36, describe how to obtain the graph of the given monomial function from the graph of $g(x) = x^n$ with the same power n. State whether f is even or odd. Sketch the graph by hand and support your answer with a grapher.

31. $f(x) = \frac{2}{3}x^4$

32. $f(x) = 5x^3$

33. $f(x) = -1.5x^5$

34. $f(x) = -2x^6$

35. $f(x) = \frac{1}{4}x^8$

36. $f(x) = \frac{1}{8}x^7$

In Exercises 37–42, match the equation to one of the curves labeled in the figure.

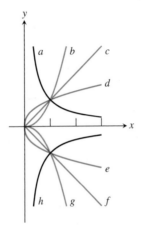

37. $f(x) = -\frac{2}{3}x^4$

38. $f(x) = \frac{1}{2}x^{-5}$

39. $f(x) = 2x^{1/4}$

40. $f(x) = -x^{5/3}$

41. $f(x) = -2x^{-2}$

42. $f(x) = 1.7x^{2/3}$

In Exercises 43–48, state the values of the constants k and a for the function $f(x) = k \cdot x^a$. Describe the portion of the curve that lies in Quadrant I or IV. Determine whether f is even, odd, or undefined for $x < 0$. Describe the rest of the curve if any. Graph the function to see whether it matches the description.

43. $f(x) = 3x^{1/4}$

44. $f(x) = -4x^{2/3}$

45. $f(x) = -2x^{4/3}$

46. $f(x) = \frac{2}{5}x^{5/2}$

47. $f(x) = \frac{1}{2}x^{-3}$

48. $f(x) = -x^{-4}$

In Exercises 49 and 50, data are given for y as a power function of x. Write an equation for the power function, and state its power and constant of variation.

49.

x	2	4	6	8	10
y	2	0.5	0.222...	0.125	0.08

50.

x	1	4	9	16	25
y	-2	-4	-6	-8	-10

51. Boyle's Law The volume of an enclosed gas (at a constant temperature) varies inversely as the pressure. If the pressure of a 3.46-L sample of neon gas at 302°K is 0.926 atm, what would the volume be at a pressure of 1.452 atm if the temperature does not change?

52. Charles's Law The volume of an enclosed gas (at a constant pressure) varies directly as the absolute temperature. If the pressure of a 3.46-L sample of neon gas at 302°K is 0.926 atm, what would the volume be at a temperature of 338°K if the pressure does not change?

53. Diamond Refraction Diamonds have the extremely high refraction index of $n = 2.42$ on average over the range of visible light. Use the formula from Exercise 25 and the fact that $c \approx 3.00 \times 10^8$ m/sec to determine the speed of light through a diamond.

54. Windmill Power The power P (in watts) produced by a windmill is proportional to the cube of the wind speed v (in mph). If a wind of 10 mph generates 15 watts of power, how much power is generated by winds of 20, 40, and 80 mph? Make a table and explain the pattern.

55. Keeping Warm For mammals and other warm-blooded animals to stay warm requires quite a bit of energy. Temperature loss is related to surface area, which is related to body weight, and temperature gain is related to circulation, which is related to pulse rate. In the final analysis, scientists have concluded that the pulse rate r of mammals is a power function of their body weight w.

(a) Draw a scatter plot of the data in Table 2.12.

(b) Find the power regression model.

(c) Superimpose the regression curve on the scatter plot.

(d) Use the regression model to predict the pulse rate for a 450-kg horse. Is the result close to the 38 beats/min reported by A. J. Clark in 1927?

Table 2.12 Weight and Pulse Rate of Selected Mammals

Mammal	Body weight (kg)	Pulse rate (beats/min)
Rat	0.2	420
Guinea pig	0.3	300
Rabbit	2	205
Small dog	5	120
Large dog	30	85
Sheep	50	70
Human	70	72

Source: A. J. Clark, Comparative Physiology of the Heart. New York: Macmillan, 1927.

56. Even and Odd Functions If n is an integer, $n \geq 1$, prove that $f(x) = x^n$ is an odd function if n is odd and is an even function if n is even.

57. Light Intensity Velma and Reggie gathered the data in Table 2.13 using a 100-watt light bulb and a Calculator-Based Laboratory™ (CBL™) with a light-intensity probe.

(a) Draw a scatter plot of the data in Table 2.13

(b) Find the power regression model. Is the power close to the theoretical value of $a = -2$?

(c) Superimpose the regression curve on the scatter plot.

(d) Use the regression model to predict the light intensity at distances of 1.7m and 3.4 m.

Table 2.13 Light Intensity Data for a 100-W Light Bulb

Distance (m)	Intensity (W/m²)
1.0	7.95
1.5	3.53
2.0	2.01
2.5	1.27
3.0	0.90

Standardized Test Questions

58. True or False The function $f(x) = x^{-2/3}$ is even. Justify your answer.

59. True or False The graph $f(x) = x^{1/3}$ is symmetric about the y-axis. Justify your answer.

In Exercises 60–63, solve the problem without using a calculator.

60. Multiple Choice Let $f(x) = 2x^{-1/2}$. What is the value of $f(4)$?

(A) 1   **(B)** -1 **(C)** $2\sqrt{2}$ **(D)** $\dfrac{1}{2\sqrt{2}}$ **(E)** 4

61. Multiple Choice Let $f(x) = -3x^{-1/3}$. Which of the following statements is true?

(A) $f(0) = 0$ **(B)** $f(-1) = -3$ **(C)** $f(1) = 1$
(D) $f(3) = 3$ **(E)** $f(0)$ is undefined

62. Multiple Choice Let $f(x) = x^{2/3}$. Which of the following statements is true?

(A) f is an odd function.

(B) f is an even function.

(C) f is neither an even nor an odd function.

(D) The graph f is symmetric with respect to the x-axis.

(E) The graph f is symmetric with respect to the origin.

63. Multiple Choice Which of the following is the domain of the function $f(x) = x^{3/2}$?

(A) All reals   **(B)** $[0,\infty)$ **(C)** $(0,\infty)$
(D) $(-\infty,0)$ **(E)** $(-\infty,0) \cup (0,\infty)$

Explorations

64. Group Activity Rational Powers Working in a group of three students, investigate the behavior of power functions of the form $f(x) = k \cdot x^{m/n}$, where m and n are positive with no factors in common. Have one group member investigate each of the following cases:

• n is even

• n is odd and m is even

• n is odd and m is odd

For each case, decide whether f is even, f is odd, or f is undefined for $x < 0$. Solve graphically and confirm algebraically in a way to convince the rest of your group and your entire class.

65. Comparing the Graphs of Power Functions Graph the functions in the stated windows and explain how the graphs are alike and how they are different. Consider the relevant aspects of analysis from Example 2. Which ordered pairs do all four graphs have in common?

(a) $f(x) = x^{-1}$, $g(x) = x^{-2}$, $h(x) = x^{-3}$, and $k(x) = x^{-4}$ in the windows [0, 1] by [0, 5], [0, 3] by [0, 3], and [−2, 2] by [−2, 2].

(b) $f(x) = x^{1/2}$, $g(x) = x^{1/3}$, $h(x) = x^{1/4}$, and $k(x) = x^{1/5}$ in the windows [0, 1] by [0, 1], [0, 3] by [0, 2], and [−3, 3] by [−2, 2].

Extending the Ideas

66. Writing to Learn Irrational Powers A negative number to an irrational power is undefined. Analyze the graphs of $f(x) = x^{\pi}, x^{1/\pi}, x^{-\pi}, -x^{\pi}, -x^{1/\pi}$, and $-x^{-\pi}$. Prepare a sketch of all six graphs on one set of axes, labeling each of the curves. Write an explanation for why each graph is positioned and shaped as it is.

67. Planetary Motion Revisited Convert the time and distance units in Table 2.10 to the Earth-based units of years and astronomical units using

$$1 \text{ yr} = 365.2 \text{ days} \quad \text{and} \quad 1 \text{ AU} = 149.6 \text{ Gm}.$$

Use this "re-expressed" data to obtain a power function model. Show algebraically that this model closely approximates Kepler's equation $T^2 = a^3$.

68. Free Fall Revisited The **speed** p of an object is the absolute value of its velocity v. The distance traveled d by an object dropped from an initial height s_0 with a current height s is given by

$$d = s_0 - s$$

until it hits the ground. Use this information and the free-fall motion formulas from Section 2.1 to prove that

$$d = \frac{1}{2} gt^2, p = gt, \text{ and therefore } p = \sqrt{2gd}.$$

Do the results of Example 6 approximate this last formula?

69. Prove that $g(x) = 1/f(x)$ is even if and only if $f(x)$ is even and that $g(x) = 1/f(x)$ is odd if and only if $f(x)$ is odd.

70. Use the results in Exercise 69 to prove that $g(x) = x^{-a}$ is even if and only if $f(x) = x^a$ is even and that $g(x) = x^{-a}$ is odd if and only if $f(x) = x^a$ is odd.

71. Joint Variation If a variable z varies as the product of the variables x and y, we say z **varies jointly** as x and y, and we write $z = k \cdot x \cdot y$, where k is the constant of variation. Write a sentence that expresses the relationship in each of the following formulas, using the language of joint variation.

(a) $F = m \cdot a$, where F and a are the force and acceleration acting on an object of mass m.

(b) $KE = (1/2)m \cdot v^2$, where KE and v are the kinetic energy and velocity of an object of mass m.

(c) $F = G \cdot m_1 \cdot m_2/r^2$, where F is the force of gravity acting on objects of masses m_1 and m_2 with a distance r between their centers and G is the universal gravitational constant.

Just as any two points in the Cartesian plane with different x values and different y values determine a unique slant line and its related linear function, any three noncollinear points with different x values determine a quadratic function. In general, $(n + 1)$ points positioned with sufficient generality determine a polynomial function of degree n. The process of fitting a polynomial of degree n to $(n + 1)$ points is **polynomial interpolation**. Exploration 2 involves two polynomial interpolation problems.

EXPLORATION 2 **Interpolating Points with a Polynomial**

1. Use cubic regression to fit a curve through the four points given in the table.

x	-2	1	3	8
y	2	0.5	-0.2	1.25

2. Use quartic regression to fit a curve through the five points given in the table.

x	3	4	5	6	8
y	-2	-4	-1	8	3

How good is the fit in each case? Why?

Generally we want a reason beyond "it fits well" to choose a model for genuine data. However, when no theoretical basis exists for picking a model, a balance between goodness of fit and simplicity of model is sought. For polynomials, we try to pick a model with the lowest possible degree that has a reasonably good fit.

QUICK REVIEW 2.3 *(For help, go to Sections A.2. and P.5.)*

In Exercises 1–6, factor the polynomial into linear factors.

1. $x^2 - x - 12$

2. $x^2 - 11x + 28$

3. $3x^2 - 11x + 6$

4. $6x^2 - 5x + 1$

5. $3x^3 - 5x^2 + 2x$

6. $6x^3 - 22x^2 + 12x$

In Exercises 7–10, solve the equation mentally.

7. $x(x - 1) = 0$

8. $x(x + 2)(x - 5) = 0$

9. $(x + 6)^3(x + 3)(x - 1.5) = 0$

10. $(x + 6)^2(x + 4)^4(x - 5)^3 = 0$

SECTION 2.3 EXERCISES

In Exercises 1–6, describe how to transform the graph of an appropriate monomial function $f(x) = x^n$ into the graph of the given polynomial function. Sketch the transformed graph by hand and support your answer with a grapher. Compute the location of the y-intercept as a check on the transformed graph.

1. $g(x) = 2(x - 3)^3$

2. $g(x) = -(x + 5)^3$

3. $g(x) = -\dfrac{1}{2}(x + 1)^3 + 2$

4. $g(x) = \dfrac{2}{3}(x - 3)^3 + 1$

5. $g(x) = -2(x + 2)^4 - 3$

6. $g(x) = 3(x - 1)^4 - 2$

In Exercises 7 and 8, graph the polynomial function, locate its extrema and zeros, and explain how it is related to the monomials from which it is built.

7. $f(x) = -x^4 + 2x$

8. $g(x) = 2x^4 - 5x^2$

In Exercises 9–12, match the polynomial function with its graph. Explain your choice. Do not use a graphing calculator.

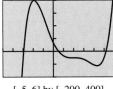

[–5, 6] by [–200, 400]

(a)

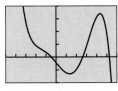

[–5, 6] by [–200, 400]

(b)

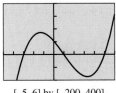

[–5, 6] by [–200, 400]

(c)

[–5, 6] by [–200, 400]

(d)

9. $f(x) = 7x^3 - 21x^2 - 91x + 104$

10. $f(x) = -9x^3 + 27x^2 + 54x - 73$

11. $f(x) = x^5 - 8x^4 + 9x^3 + 58x^2 - 164x + 69$

12. $f(x) = -x^5 + 3x^4 + 16x^3 - 2x^2 - 95x - 44$

In Exercises 13–16, graph the function pairs in the same series of viewing windows. Zoom out until the two graphs look nearly identical and state your final viewing window.

13. $f(x) = x^3 - 4x^2 - 5x - 3$ and $g(x) = x^3$

14. $f(x) = x^3 + 2x^2 - x + 5$ and $g(x) = x^3$

15. $f(x) = 2x^3 + 3x^2 - 6x - 15$ and $g(x) = 2x^3$

16. $f(x) = 3x^3 - 12x + 17$ and $g(x) = 3x^3$

In Exercises 17–24, graph the function in a viewing window that shows all of its extrema and x-intercepts. Describe the end behavior using limits.

17. $f(x) = (x - 1)(x + 2)(x + 3)$

18. $f(x) = (2x - 3)(4 - x)(x + 1)$

19. $f(x) = -x^3 + 4x^2 + 31x - 70$

20. $f(x) = x^3 - 2x^2 - 41x + 42$

21. $f(x) = (x - 2)^2(x + 1)(x - 3)$

22. $f(x) = (2x + 1)(x - 4)^3$

23. $f(x) = 2x^4 - 5x^3 - 17x^2 + 14x + 41$

24. $f(x) = -3x^4 - 5x^3 + 15x^2 - 5x + 19$

In Exercises 25–28, describe the end behavior of the polynomial function using $\lim\limits_{x \to \infty} f(x)$ and $\lim\limits_{x \to -\infty} f(x)$.

25. $f(x) = 3x^4 - 5x^2 + 3$

26. $f(x) = -x^3 + 7x^2 - 4x + 3$

27. $f(x) = 7x^2 - x^3 + 3x - 4$

28. $f(x) = x^3 - x^4 + 3x^2 - 2x + 7$

In Exercises 29–32, match the polynomial function with its graph. Approximate all of the real zeros of the function.

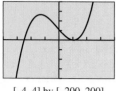

[−4, 4] by [−200, 200]

(a)

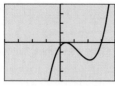

[−4, 4] by [−200, 200]

(b)

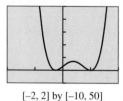

[−2, 2] by [−10, 50]

(c)

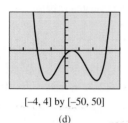

[−4, 4] by [−50, 50]

(d)

29. $f(x) = 20x^3 + 8x^2 - 83x + 55$

30. $f(x) = 35x^3 - 134x^2 + 93x - 18$

31. $f(x) = 44x^4 - 65x^3 + x^2 + 17x + 3$

32. $f(x) = 4x^4 - 8x^3 - 19x^2 + 23x - 6$

In Exercises 33–38, find the zeros of the function algebraically.

33. $f(x) = x^2 + 2x - 8$ **34.** $f(x) = 3x^2 + 4x - 4$

35. $f(x) = 9x^2 - 3x - 2$ **36.** $f(x) = x^3 - 25x$

37. $f(x) = 3x^3 - x^2 - 2x$ **38.** $f(x) = 5x^3 - 5x^2 - 10x$

In Exercises 39–42, state the degree and list the zeros of the polynomial function. State the multiplicity of each zero and whether the graph crosses the x-axis at the corresponding x-intercept. Then sketch the graph of the polynomial function by hand.

39. $f(x) = x(x - 3)^2$ **40.** $f(x) = -x^3(x - 2)$

41. $f(x) = (x - 1)^3(x + 2)^2$ **42.** $f(x) = 7(x - 3)^2(x + 5)^4$

In Exercises 43–48, graph the function in a viewing window that shows all of its x-intercepts and approximate all of its zeros.

43. $f(x) = 2x^3 + 3x^2 - 7x - 6$

44. $f(x) = -x^3 + 3x^2 + 7x - 2$

45. $f(x) = x^3 + 2x^2 - 4x - 7$

46. $f(x) = -x^4 - 3x^3 + 7x^2 + 2x + 8$

47. $f(x) = x^4 + 3x^3 - 9x^2 + 2x + 3$

48. $f(x) = 2x^5 - 11x^4 + 4x^3 + 47x^2 - 42x - 8$

In Exercises 49–52, find the zeros of the function algebraically or graphically.

49. $f(x) = x^3 - 36x$

50. $f(x) = x^3 + 2x^2 - 109x - 110$

51. $f(x) = x^3 - 7x^2 - 49x + 55$

52. $f(x) = x^3 - 4x^2 - 44x + 96$

In Exercises 53–56, using only algebra, find a cubic function with the given zeros. Support by graphing your answer.

53. $3, -4, 6$ **54.** $-2, 3, -5$

55. $\sqrt{3}, -\sqrt{3}, 4$ **56.** $1, 1 + \sqrt{2}, 1 - \sqrt{2}$

57. Use cubic regression to fit a curve through the four points given in the table.

x	−3	−1	1	3
y	22	25	12	−5

58. Use cubic regression to fit a curve through the four points given in the table.

x	−2	1	4	7
y	2	5	9	26

59. Use quartic regression to fit a curve through the five points given in the table.

x	3	4	5	6	8
y	−7	−4	−11	8	3

60. Use quartic regression to fit a curve through the five points given in the table.

x	0	4	5	7	13
y	−21	−19	−12	8	3

In Exercises 61–62, explain why the function has at least one real zero.

61. Writing to Learn $f(x) = x^7 + x + 100$

62. Writing to Learn $f(x) = x^9 - x + 50$

63. Stopping Distance A state highway patrol safety division collected the data on stopping distances in Table 2.14 on the next page.

(a) Draw a scatter plot of the data.

(b) Find the quadratic regression model.

(c) Superimpose the regression curve on the scatter plot.

(d) Use the regression model to predict the stopping distance for a vehicle traveling at 25 mph.

(e) Use the regression model to predict the speed of a car if the stopping distance is 300 ft.

Table 2.14 Highway Safety Division

Speed (mph)	Stopping Distance (ft)
10	15.1
20	39.9
30	75.2
40	120.5
50	175.9

64. Analyzing Profit Economists for Smith Brothers, Inc., find the company profit P by using the formula $P = R - C$, where R is the total revenue generated by the business and C is the total cost of operating the business.

(a) Using data from past years, the economists determined that $R(x) = 0.0125x^2 + 412x$ models total revenue, and $C(x) = 12{,}225 + 0.00135x^3$ models the total cost of doing business, where x is the number of customers patronizing the business. How many customers must Smith Bros. have to be profitable each year?

(b) How many customers must there be for Smith Bros. to realize an annual profit of $60,000?

65. Circulation of Blood Research conducted at a national health research project shows that the speed at which a blood cell travels in an artery depends on its distance from the center of the artery. The function $v = 1.19 - 1.87r^2$ models the velocity (in centimeters per second) of a cell that is r centimeters from the center of an artery.

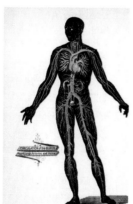

(a) Find a graph of v that reflects values of v appropriate for this problem. Record the viewing-window dimensions.

(b) If a blood cell is traveling at 0.975 cm/sec, estimate the distance the blood cell is from the center of the artery.

66. Volume of a Box Dixie Packaging Co. has contracted to manufacture a box with no top that is to be made by removing squares of width x from the corners of a 15-in. by 60-in. piece of cardboard.

(a) Show that the volume of the box is modeled by $V(x) = x(60 - 2x)(15 - 2x)$.

(b) Determine x so that the volume of the box is at least 450 in.³

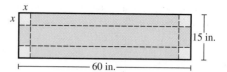

67. Volume of a Box Squares of width x are removed from a 10-cm by 25-cm piece of cardboard, and the resulting edges are folded up to form a box with no top. Determine all values of x so that the volume of the resulting box is at most 175 cm³.

68. Volume of a Box The function $V = 2666x - 210x^2 + 4x^3$ represents the volume of a box that has been made by removing squares of width x from each corner of a rectangular sheet of material and then folding up the sides. What values are possible for x?

Standardized Test Questions

69. True or False The graph of $f(x) = x^3 - x^2 - 2$ crosses the x-axis between $x = 1$ and $x = 2$. Justify your answer.

70. True or False If the graph of $g(x) = (x + a)^2$ is obtained by translating the graph of $f(x) = x^2$ to the right, then a must be positive. Justify your answer.

In Exercises 71 and 72, solve the problem without using a calculator.

71. Multiple Choice What is the y-intercept of the graph of $f(x) = 2(x - 1)^3 + 5$?

(A) 7   (B) 5   (C) 3   (D) 2   (E) 1

72. Multiple Choice What is the multiplicity of the zero $x = 2$ in $f(x) = (x - 2)^2(x + 2)^3(x + 3)^7$?

(A) 1   (B) 2   (C) 3   (D) 5   (E) 7

In Exercises 73 and 74, which of the specified functions might have the given graph?

73. Multiple Choice

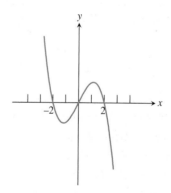

(A) $f(x) = -x(x + 2)(2 - x)$

(B) $f(x) = -x(x + 2)(x - 2)$

(C) $f(x) = -x^2(x + 2)(x - 2)$

(D) $f(x) = -x(x + 2)^2(x - 2)$

(E) $f(x) = -x(x + 2)(x - 2)^2$

74. Multiple Choice

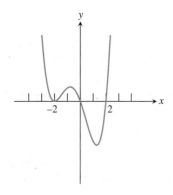

(A) $f(x) = x(x + 2)^2(x - 2)$

(B) $f(x) = x(x + 2)^2(2 - x)$

(C) $f(x) = x^2(x + 2)(x - 2)$

(D) $f(x) = x(x + 2)(x - 2)^2$

(E) $f(x) = x^2(x + 2)(x - 2)^2$

Explorations

In Exercises 75 and 76, two views of the function are given.

75. Writing to Learn Describe why each view of the function

$$f(x) = x^5 - 10x^4 + 2x^3 + 64x^2 - 3x - 55,$$

by itself, may be considered inadequate.

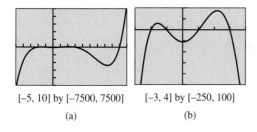

[-5, 10] by [-7500, 7500]   [-3, 4] by [-250, 100]

(a)   (b)

76. Writing to Learn Describe why each view of the function

$$f(x) = 10x^4 + 19x^3 - 121x^2 + 143x - 51,$$

by itself, may be considered inadequate.

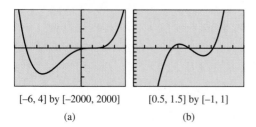

[-6, 4] by [-2000, 2000]   [0.5, 1.5] by [-1, 1]

(a)   (b)

In Exercises 77–80, the function has hidden behavior when viewed in the window $[-10, 10]$ by $[-10, 10]$. Describe what behavior is hidden, and state the dimensions of a viewing window that reveals the hidden behavior.

77. $f(x) = 10x^3 - 40x^2 + 50x - 20$

78. $f(x) = 0.5(x^3 - 8x^2 + 12.99x - 5.94)$

79. $f(x) = 11x^3 - 10x^2 + 3x + 5$

80. $f(x) = 33x^3 - 100x^2 + 101x - 40$

Extending the Ideas

81. Graph the left side of the equation

$$3(x^3 - x) = a(x - b)^3 + c.$$

Then explain why there are no real numbers a, b, and c that make the equation true. (*Hint*: Use your knowledge of $y = x^3$ and transformations.)

82. Graph the left side of the equation

$$x^4 + 3x^3 - 2x - 3 = a(x - b)^4 + c.$$

Then explain why there are no real numbers a, b, and c that make the equation true.

83. Looking Ahead to Calculus The figure shows a graph of both $f(x) = -x^3 + 2x^2 + 9x - 11$ and the line L defined by $y = 5(x - 2) + 7$.

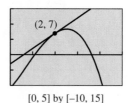

[0, 5] by [-10, 15]

(a) Confirm that the point $Q(2, 7)$ is a point of intersection of the two graphs.

(b) Zoom in at point Q to develop a visual understanding that $y = 5(x - 2) + 7$ is a *linear approximation* for $y = f(x)$ near $x = 2$.

(c) Recall that a line is *tangent* to a circle at a point P if it intersects the circle only at point P. View the two graphs in the window $[-5, 5]$ by $[-25, 25]$, and explain why that definition of tangent line is not valid for the graph of f.

84. Looking Ahead to Calculus Consider the function $f(x) = x^n$ where n is an odd integer.

(a) Suppose that a is a positive number. Show that the slope of the line through the points $P(a, f(a))$ and $Q(-a, f(-a))$ is a^{n-1}.

(b) Let $x_0 = a^{1/(n-1)}$. Find an equation of the line through point $(x_0, f(x_0))$ with the slope a^{n-1}.

(c) Consider the special case $n = 3$ and $a = 3$. Show both the graph of f and the line from part b in the window $[-5, 5]$ by $[-30, 30]$.

85. Derive an Algebraic Model of a Problem Show that the distance x in the figure is a solution of the equation $x^4 - 16x^3 + 500x^2 - 8000x + 32{,}000 = 0$ and find the value of D by following these steps.

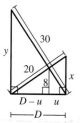

(a) Use the similar triangles in the diagram and the properties of proportions learned in geometry to show that

$$\frac{8}{x} = \frac{y - 8}{y}.$$

(b) Show that $y = \dfrac{8x}{x - 8}$.

(c) Show that $y^2 - x^2 = 500$. Then substitute for y, and simplify to obtain the desired degree 4 equation in x.

(d) Find the distance D.

86. Group Learning Activity Consider functions of the form $f(x) = x^3 + bx^2 + x + 1$ where b is a nonzero real number.

(a) Discuss as a group how the value of b affects the graph of the function.

(b) After completing (a), have each member of the group (individually) predict what the graphs of $f(x) = x^3 + 15x^2 + x + 1$ and $g(x) = x^3 - 15x^2 + x + 1$ will look like.

(c) Compare your predictions with each other. Confirm whether they are correct.

2.4
Real Zeros of Polynomial Functions

Long Division and the Division Algorithm

We have seen that factoring a polynomial reveals its zeros and much about its graph. Polynomial division gives us new and better ways to factor polynomials. First we observe that the division of polynomials closely resembles the division of integers:

$$
\begin{array}{r}
112 \\
32\overline{)3587} \\
\underline{32} \\
387 \\
\underline{32} \\
67 \\
\underline{64} \\
3
\end{array}
$$

$$
\begin{array}{rl}
& 1x^2 + 1x + 2 \qquad \leftarrow \text{Quotient} \\
3x + 2\overline{)3x^3 + 5x^2 + 8x + 7} & \leftarrow \text{Dividend} \\
\underline{3x^3 + 2x^2} & \leftarrow \text{Multiply: } 1x^2 \cdot (3x+2) \\
3x^2 + 8x + 7 & \leftarrow \text{Subtract} \\
\underline{3x^2 + 2x} & \leftarrow \text{Multiply: } 1x \cdot (3x+2) \\
6x + 7 & \leftarrow \text{Subtract} \\
\underline{6x + 4} & \leftarrow \text{Multiply: } 2 \cdot (3x+2) \\
3 & \leftarrow \text{Remainder}
\end{array}
$$

Division, whether integer or polynomial, involves a *dividend* divided by a *divisor* to obtain a *quotient* and a *remainder*. We can check and summarize our result with an equation of the form

$$(\text{Divisor})(\text{Quotient}) + \text{Remainder} = \text{Dividend}.$$

For instance, to check or summarize the long divisions shown above we could write

$$32 \times 112 + 3 = 3587 \qquad (3x + 2)(x^2 + x + 2) + 3 = 3x^3 + 5x^2 + 8x + 7.$$

The *division algorithm* contains such a summary *polynomial equation*, but with the dividend written on the left side of the equation.

Division Algorithm for Polynomials

Let $f(x)$ and $d(x)$ be polynomials with the degree of f greater than or equal to the degree of d, and $d(x) \neq 0$. Then there are unique polynomials $q(x)$ and $r(x)$, called the **quotient** and **remainder**, such that

$$f(x) = d(x) \cdot q(x) + r(x) \qquad (1)$$

where either $r(x) = 0$ or the degree of r is less than the degree of d.

The function $f(x)$ in the division algorithm is the **dividend**, and $d(x)$ is the **divisor**. If $r(x) = 0$, we say $d(x)$ **divides evenly** into $f(x)$.

The summary statement (1) is sometimes written in *fraction form* as follows:

$$\frac{f(x)}{d(x)} = q(x) + \frac{r(x)}{d(x)} \qquad (2)$$

For instance, to summarize the polynomial division example above we could write

$$\frac{3x^3 + 5x^2 + 8x + 7}{3x + 2} = x^2 + x + 2 + \frac{3}{3x + 2}.$$

EXAMPLE 1 Using Polynomial Long Division

Use long division to find the quotient and remainder when $2x^4 - x^3 - 2$ is divided by $2x^2 + x + 1$. Write a summary statement in both polynomial and fraction form.

SOLUTION

Solve Algebraically

$$
\begin{array}{r}
x^2 - x \qquad \leftarrow \text{Quotient} \\
2x^2 + x + 1{\overline{\smash{\big)}\,2x^4 - x^3 + 0x^2 + 0x - 2}} \\
\underline{2x^4 + x^3 + x^2 } \\
-2x^3 - x^2 + 0x - 2 \\
\underline{-2x^3 - x^2 - x } \\
x - 2 \qquad \leftarrow \text{Remainder}
\end{array}
$$

The division algorithm yields the polynomial form

$$2x^4 - x^3 - 2 = (2x^2 + x + 1)(x^2 - x) + (x - 2).$$

Using equation (2), we obtain the fraction form

$$\frac{2x^4 - x^3 - 2}{2x^2 + x + 1} = x^2 - x + \frac{x - 2}{2x^2 + x + 1}.$$

Support Graphically

Figure 2.34 supports the polynomial form of the summary statement.

Now try Exercise 1.

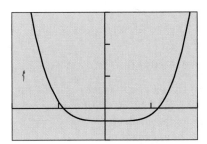

[–2, 2] by [–5, 15]

FIGURE 2.34 The graphs of $y_1 = 2x^4 - x^3 - 2$ and $y_2 = (2x^2 + x + 1)(x^2 - x) + (x - 2)$ are a perfect match. (Example 1)

Remainder and Factor Theorems

An important special case of the division algorithm occurs when the divisor is of the form $d(x) = x - k$, where k is a real number. Because the degree of $d(x) = x - k$ is 1, the remainder is a real number. So we obtain the following simplified summary statement for the division algorithm:

$$f(x) = (x - k)q(x) + r \qquad (3)$$

We use this special case of the division algorithm throughout the rest of the section.

Using Equation (3), we evaluate the polynomial $f(x)$ at $x = k$:

$$f(k) = (k - k)q(k) + r = 0 \cdot q(k) + r = 0 + r = r$$

So $f(k) = r$, which is the remainder. This reasoning yields the following theorem.

THEOREM Remainder Theorem

If a polynomial $f(x)$ is divided by $x - k$, then the remainder is $r = f(k)$.

Example 2 shows a clever use of the Remainder Theorem that gives information about the factors, zeros, and x-intercepts.

X	Y1	
-4	0	
-3	-14	
-2	-22	
-1	-24	
0	-20	
1	-10	
2	6	

Y1◼ 3X^2+7X–20

FIGURE 2.35 Table for $f(x) = 3x^2 + 7x - 20$ showing the remainders obtained when $f(x)$ is divided by $x - k$, for $k = -4, -3, \ldots, 1, 2$.

EXAMPLE 2   Using the Remainder Theorem

Find the remainder when $f(x) = 3x^2 + 7x - 20$ is divided by

(a) $x - 2$ **(b)** $x + 1$ **(c)** $x + 4$.

SOLUTION

Solve Numerically (by hand)

(a) We can find the remainder without doing long division! Using the Remainder Theorem with $k = 2$ we find that

$$r = f(2) = 3(2)^2 + 7(2) - 20 = 12 + 14 - 20 = 6.$$

(b) $r = f(-1) = 3(-1)^2 + 7(-1) - 20 = 3 - 7 - 20 = -24.$

(c) $r = f(-4) = 3(-4)^2 + 7(-4) - 20 = 48 - 28 - 20 = 0.$

Interpret Because the remainder in part (c) is 0, $x + 4$ *divides evenly* into $f(x) = 3x^2 + 7x - 20$. So, $x + 4$ is a factor of $f(x) = 3x^2 + 7x - 20$, -4 is a solution of $3x^2 + 7x - 20 = 0$, and -4 is an x-intercept of the graph of $y = 3x^2 + 7x - 20$. We know all of this without ever dividing, factoring, or graphing!

Support Numerically (using a grapher) We can find the remainders of several division problems at once using the table feature of a grapher. (See Figure 2.35.)

Now try Exercise 13.

Our interpretation of Example 2c leads us to the following theorem.

PROOF OF THE FACTOR THEOREM

If $f(x)$ has a factor $x - k$, there is a polynomial $g(x)$ such that

$f(x) = (x - k)g(x) = (x - k)g(x) + 0.$

By the uniqueness condition of the Division Algorithm $g(x)$ is the quotient and 0 is the remainder, and by the Remainder Theorem, $f(k) = 0$.

Conversely, if $f(k) = 0$, the remainder $r = 0$, $x - k$ divides evenly into $f(x)$, and $x - k$ is a factor of $f(x)$.

THEOREM Factor Theorem

A polynomial function $f(x)$ has a factor $x - k$ if and only if $f(k) = 0$.

Applying the ideas of the Factor Theorem to Example 2, we can factor $f(x) = 3x^2 + 7x - 20$ by dividing it by the known factor $x + 4$.

$$
\begin{array}{r}
3x - 5 \\
x + 4 \overline{)3x^2 + 7x - 20} \\
\underline{3x^2 + 12x} \\
-5x - 20 \\
\underline{-5x - 20} \\
0
\end{array}
$$

So, $f(x) = 3x^2 + 7x - 20 = (x + 4)(3x - 5)$. In this case, there really is no need to use long division or fancy theorems; traditional factoring methods can do the job. However, for polynomials of degree 3 and higher, these sophisticated methods can be quite helpful in solving equations and finding factors, zeros, and x-intercepts. Indeed, the Factor Theorem ties in nicely with earlier connections we have made in the following way.

Fundamental Connections for Polynomial Functions

For a polynomial function f and a real number k, the following statements are equivalent:

1. $x = k$ is a solution (or root) of the equation $f(x) = 0$.

2. k is a zero of the function f.

3. k is an x-intercept of the graph of $y = f(x)$.

4. $x - k$ is a factor of $f(x)$.

Synthetic Division

We continue with the important special case of polynomial division with the divisor $x - k$. The Remainder Theorem gave us a way to find remainders in this case without long division. We now learn a method for finding both quotients and remainders for division by $x - k$ without long division. This shortcut method for the division of a polynomial by a linear divisor $x - k$ is **synthetic division**.

We illustrate the evolution of this method below, progressing from long divsion through two intermediate stages to synthetic division.

Moving from stage to stage, focus on the coefficients and their relative positions. Moving from Stage 1 to Stage 2, we suppress the variable x and the powers of x, and then from Stage 2 to Stage 3, we eliminate unneeded duplications and collapse vertically.

Stage 1
Long Division

$$\begin{array}{r} 2x^2 + 3x + 4 \\ x - 3\overline{)2x^3 - 3x^2 - 5x - 12} \\ \underline{2x^3 - 6x^2} \\ 3x^2 - 5x - 12 \\ \underline{3x^2 - 9x} \\ 4x - 12 \\ \underline{4x - 12} \\ 0 \end{array}$$

Stage 2
Variables Suppressed

$$\begin{array}{r} 2 \quad 3 \quad 4 \\ -3\overline{)2 \; -3 \; -5 \; -12} \\ 2 \; -6 \\ 3 \; -5 \; -12 \\ 3 \; -9 \\ 4 \; -12 \\ 4 \; -12 \\ 0 \end{array}$$

Stage 3
Collapsed Vertically

$$\begin{array}{r} -3|\; 2 \; -3 \; -5 \; -12 \quad \text{Dividend}\\ -6 \; -9 \; -12 \\ \hline 2 \quad 3 \quad 4 \quad 0 \quad \text{Quotient, remainder} \end{array}$$

Finally, from Stage 3 to Stage 4, we change the sign of the number representing the divisor and the signs of the numbers on the second line of our division scheme. These sign changes yield two advantages:

• The number standing for the divisor $x - k$ is now k, its zero.

• Changing the signs in the second line allows us to add rather than subtract.

Stage 4
Synthetic Division

Zero of divisor $\rightarrow$
$$\begin{array}{r} 3|\; 2 \; -3 \; -5 \; -12 \quad \text{Dividend}\\ 6 \quad 9 \quad 12 \\ \hline 2 \quad 3 \quad 4 \quad 0 \quad \text{Quotient, remainder} \end{array}$$

With Stage 4 we have achieved our goal of synthetic division, a highly streamlined version of dividing a polynomial by $x - k$. How does this "bare bones" division work? Example 3 explains the steps.

EXAMPLE 3   Using Synthetic Division

Divide $2x^3 - 3x^2 - 5x - 12$ by $x - 3$ using synthetic division and write a summary statement in fraction form.

SOLUTION

Set Up

The zero of the divisor $x - 3$ is 3, which we put in the divisor position. Because the dividend is in standard form, we write its coefficients in order in the dividend position, *making sure to use a zero as a placeholder for any missing term.* We leave space for the line for products and draw a horizontal line below the space. (See below.)

Calculate

• Because the leading coefficient of the dividend must be the leading coefficient of the quotient, copy the 2 into the first quotient position.

$$\begin{array}{r|rrrr} \text{Zero of Divisor} & 3 & 2 & -3 & -5 & -12 \text{ Dividend} \\ \text{Line for products} & & & & & \\ \hline & & 2 & & & \end{array}$$

• Multiply the zero of the divisor (3) by the most recently determined coefficient of the quotient (2). Write the product above the line and one column to the right.

• Add the next coefficient of the dividend to the product just found and record the sum below the line in the same column.

• Repeat the "multiply" and "add" steps until the last row is completed.

$$\begin{array}{r|rrrr} \text{Zero of Divisor} & 3 & 2 & -3 & -5 & -12 \text{ Dividend} \\ \text{Line for products} & & & 6 & 9 & 12 \\ \text{Line for sums} & & 2 & 3 & 4 & 0 \text{ Remainder} \\ & & & \text{Quotient} \end{array}$$

Interpret

The last line of numbers are the coefficients of the quotient polynomial and the remainder. The quotient must be a quadratic function. (Why?) So the quotient is $2x^2 + 3x + 4$ and the remainder is 0. So we conclude that

$$\frac{2x^3 - 3x^2 - 5x - 12}{x - 3} = 2x^2 + 3x + 4, x \neq 3.$$

Now try Exercise 7.

Rational Zeros Theorem

Real zeros of polynomial functions are either **rational zeros**—zeros that are rational numbers—or **irrational zeros**—zeros that are irrational numbers. For example,

$$f(x) = 4x^2 - 9 = (2x + 3)(2x - 3)$$

has the rational zeros $-3/2$ and $3/2$, and

$$f(x) = x^2 - 2 = (x + \sqrt{2})(x - \sqrt{2})$$

has the irrational zeros $-\sqrt{2}$ and $\sqrt{2}$.

The Rational Zeros Theorem tells us how to make a list of all potential rational zeros for a polynomial function with integer coefficients.

> **THEOREM** **Rational Zeros Theorem**
>
> Suppose f is a polynomial function of degree $n \geq 1$ of the form
>
> $$f(x) = a_n x^n + a_{n-1} x^{n-1} + \cdots + a_0,$$
>
> with every coefficient an integer and $a_0 \neq 0$. If $x = p/q$ is a rational zero of f, where p and q have no common integer factors other than 1, then
>
> • p is an integer factor of the constant coefficient a_0, and
>
> • q is an integer factor of the leading coefficient a_n.

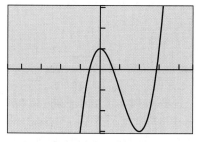

[−4.7, 4.7] by [−3.1, 3.1]

FIGURE 2.36 The function $f(x) = x^3 - 3x^2 + 1$ has three real zeros. (Example 4)

EXAMPLE 4 Finding the Rational Zeros

Find the rational zeros of $f(x) = x^3 - 3x^2 + 1$.

SOLUTION Because the leading and constant coefficients are both 1, according to the Rational Zeros Theorem, the only potential rational zeros of f are 1 and -1. So we check to see whether they are in fact zeros of f:

$$f(1) = (1)^3 - 3(1)^2 + 1 = -1 \neq 0$$

$$f(-1) = (-1)^3 - 3(-1)^2 + 1 = -3 \neq 0$$

So f has no rational zeros. Figure 2.36 shows that the graph of f has three x-intercepts. So f has three real zeros. All three must be irrational numbers.

Now try Exercise 33.

In Example 4 the Rational Zeros Theorem gave us only two candidates for rational zeros, neither of which "checked out." Often this theorem suggests many candidates, as we see in Example 5. In such a case, we use technology and a variety of algebraic methods to locate the rational zeros.

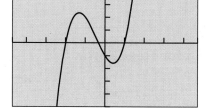

[−4.7, 4.7] by [−10, 10]

FIGURE 2.37 The function $f(x) = 3x^3 + 4x^2 - 5x - 2$ has three real zeros. (Example 5)

EXAMPLE 5 Finding the Rational Zeros

Find the rational zeros of $f(x) = 3x^3 + 4x^2 - 5x - 2$.

SOLUTION Because the leading coefficient is 3 and constant coefficient is -2, the Rational Zeros Theorem yields several potential rational zeros of f. So we take an organized approach to our solution.

Potential Rational Zeros:

$$\frac{\text{Factors of } -2}{\text{Factors of } 3} : \frac{\pm 1, \pm 2}{\pm 1, \pm 3} : \pm 1, \pm 2, \pm\frac{1}{3}, \pm\frac{2}{3}$$

Figure 2.37 suggests that, among our candidates, 1, -2, and possibly $-1/3$ or $-2/3$ are the most likely to be rational zeros. We use synthetic division because it tells us whether

continued

a number is a zero and, if so, how to factor the polynomial. To see whether 1 is a zero of *f*, we synthetically divide $f(x)$ by $x - 1$:

$$
\begin{array}{r|rrrr}
\text{Zero of Divisor} \quad 1 & 3 & 4 & -5 & -2 \quad \text{Dividend} \\
 & & 3 & 7 & 2 \\
\hline
 & 3 & 7 & 2 & 0 \quad \text{Remainder} \\
\end{array}
$$

Quotient

So because the remainder is 0, $x - 1$ is a factor of $f(x)$ and 1 is a zero of *f*. By the Division Algorithm and factoring, we conclude

$$
\begin{aligned}
f(x) &= 3x^3 + 4x^2 - 5x - 2 \\
&= (x - 1)(3x^2 + 7x + 2) \\
&= (x - 1)(3x + 1)(x + 2)
\end{aligned}
$$

So the rational zeros of *f* are 1, $-1/3$, and -2.   ***Now try Exercise 35.***

Upper and Lower Bounds

We narrow our search for real zeros by using a test that identifies upper and lower bounds for real zeros. A number *k* is an **upper bound for the real zeros** of *f* if $f(x)$ is never zero when *x* is greater than *k*. On the other hand, a number *k* is a **lower bound for the real zeros** of *f* if $f(x)$ is never zero when *x* is less than *k*. So if *c* is a lower bound and *d* is an upper bound for the real zeros of a function *f*, all of the real zeros of *f* must lie in the interval $[c, d]$. Figure 2.38 illustrates this situation.

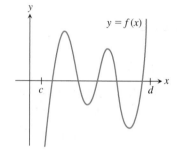

FIGURE 2.38 *c* is a lower bound and *d* is an upper bound for the real zeros of *f*.

Upper and Lower Bound Tests for Real Zeros

Let *f* be a polynomial function of degree $n \geq 1$ with a positive leading coefficient. Suppose $f(x)$ is divided by $x - k$ using synthetic division.

• If $k \geq 0$ and every number in the last line is nonnegative (positive or zero), then *k* is an *upper bound* for the real zeros of *f*.

• If $k \leq 0$ and the numbers in the last line are alternately nonnegative and nonpositive, then *k* is a *lower bound* for the real zeros of *f*.

EXAMPLE 6   Establishing Bounds for Real Zeros

Prove that all of the real zeros of $f(x) = 2x^4 - 7x^3 - 8x^2 + 14x + 8$ must lie in the interval $[-2, 5]$.

SOLUTION We must prove that 5 is an upper bound and -2 is a lower bound on the real zeros of *f*. The function *f* has a positive leading coefficient, so we employ the Upper and Lower Bound Tests, and use synthetic division:

continued

$$
\begin{array}{r|rrrrr}
5 & 2 & -7 & -8 & 14 & 8 \\
  &   & 10 & 15 & 35 & 245 \\
\hline
  & 2 & 3 & 7 & 49 & 253 \\
\end{array}
\quad \text{Last line}
$$

$$
\begin{array}{r|rrrrr}
-2 & 2 & -7 & -8 & 14 & 8 \\
   &   & -4 & 22 & -28 & 28 \\
\hline
   & 2 & -11 & 14 & -14 & 36 \\
\end{array}
\quad \text{Last line}
$$

Because the last line in the first division scheme consists of all positive numbers, 5 is an upper bound. Because the last line in the second division consists of numbers of alternating signs, -2 is a lower bound. All of the real zeros of f must therefore lie in the closed interval $[-2, 5]$.   ***Now try Exercise 37.***

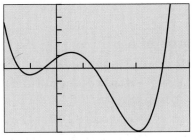

[−2, 5] by [−50, 50]

FIGURE 2.39 $f(x) = 2x^4 - 7x^3 - 8x^2 + 14x + 8$ has all of its real zeros in $[-2, 5]$. (Example 7)

EXAMPLE 7   Finding the Real Zeros of a Polynomial Function

Find all of the real zeros of $f(x) = 2x^4 - 7x^3 - 8x^2 + 14x + 8$.

SOLUTION From Example 6 we know that all of the real zeros of f must lie in the closed interval $[-2, 5]$. So in Figure 2.39 we set our Xmin and Xmax accordingly.

Next we use the Rational Zeros Theorem.

Potential Rational Zeros:

$$
\frac{\text{Factors of } 8}{\text{Factors of } 2} : \frac{\pm 1, \pm 2, \pm 4, \pm 8}{\pm 1, \pm 2} : \pm 1, \pm 2, \pm 4, \pm 8, \pm \frac{1}{2}
$$

We compare the x-intercepts of the graph in Figure 2.39 and our list of candidates, and decide 4 and $-1/2$ are the only potential rational zeros worth pursuing.

$$
\begin{array}{r|rrrrr}
4 & 2 & -7 & -8 & 14 & 8 \\
  &   & 8 & 4 & -16 & -8 \\
\hline
  & 2 & 1 & -4 & -2 & 0 \\
\end{array}
$$

From this first synthetic division we conclude

$$
f(x) = 2x^4 - 7x^3 - 8x^2 + 14x + 8
$$
$$
= (x - 4)(2x^3 + x^2 - 4x - 2)
$$

and we now divide the cubic factor $2x^3 + x^2 - 4x - 2$ by $x + 1/2$:

$$
\begin{array}{r|rrrr}
-1/2 & 2 & 1 & -4 & -2 \\
     &   & -1 & 0 & 2 \\
\hline
     & 2 & 0 & -4 & 0 \\
\end{array}
$$

continued

This second synthetic division allows to complete the factoring of $f(x)$.

$$f(x) = (x - 4)(2x^3 + x^2 - 4x - 2)$$

$$= (x - 4)\left(x + \frac{1}{2}\right)(2x^2 - 4)$$

$$= 2(x - 4)\left(x + \frac{1}{2}\right)(x^2 - 2)$$

$$= (x - 4)(2x + 1)(x + \sqrt{2})(x - \sqrt{2})$$

So the zeros of f are the rational numbers 4 and $-1/2$ and the irrational numbers $-\sqrt{2}$ and $\sqrt{2}$.

Now try Exercise 49.

A polynomial function cannot have more real zeros than its degree, but it can have fewer. When a polynomial has fewer real zeros than its degree, the Upper and Lower Bound Tests help us know that we have found them all, as illustrated by Example 8.

EXAMPLE 8   Finding the Real Zeros of a Polynomial Function

Prove that all of the real zeros of $f(x) = 10x^5 - 3x^2 + x - 6$ lie in the interval $[0, 1]$, and find them.

SOLUTION We first prove that 1 is an upper bound and 0 is a lower bound for the real zeros of f. The function f has a positive leading coefficient, so we use synthetic division and the Upper and Lower Bound Tests:

1	10	0	0	−3	1	−6	
		10	10	10	7	8	
	10	10	10	7	8	2	Last line
0	10	0	0	−3	1	−6	
		0	0	0	0	0	
	10	0	0	−3	1	−6 .	Last line

Because the last line in the first division scheme consists of all nonnegative numbers, 1 is an upper bound. Because the last line in the second division consists of numbers that are alternately nonnegative and nonpositive, 0 is a lower bound. All of the real zeros of f must therefore lie in the closed interval $[0, 1]$. So in Figure 2.40 we set our Xmin and Xmax accordingly.

Next we use the Rational Zeros Theorem.

Potential Rational Zeros:

$$\frac{\text{Factors of } -6}{\text{Factors of } 10} \cdot \frac{\pm 1, \pm 2, \pm 3, \pm 6}{\pm 1, \pm 2, \pm 5, \pm 10}:$$

$$\pm 1, \pm 2, \pm 3, \pm 6, \pm \frac{1}{2}, \pm \frac{3}{2}, \pm \frac{1}{5}, \pm \frac{2}{5}, \pm \frac{3}{5}, \pm \frac{6}{5}, \pm \frac{1}{10}, \pm \frac{3}{10}.$$

continued

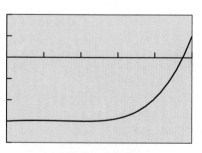

[0, 1] by [−8, 4]

FIGURE 2.40 $y = 10x^5 - 3x^2 + x - 6$. (Example 8)

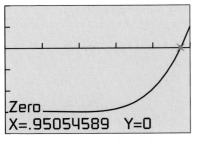

[0, 1] by [−8, 4]

FIGURE 2.41 An approximation for the irrational zero of $f(x) = 10x^5 - 3x^2 + x - 6$. (Example 8)

We compare the x-intercepts of the graph in Figure 2.40 and our list of candidates, and decide f has no rational zeros. From Figure 2.40 we see that f changes sign on the interval $[0.8, 1]$, so by the Intermediate Value Theorem must have a real zero on this interval. Because it is not rational we conclude that it is irrational. Figure 2.41 shows that this lone real zero of f is approximately 0.95.

Now try Exercise 55.

QUICK REVIEW 2.4 *(For help, go to Sections A.2. and A.3.)*

In Exercises 1–4, rewrite the expression as a polynomial in standard form.

1. $\dfrac{x^3 - 4x^2 + 7x}{x}$

2. $\dfrac{2x^3 - 5x^2 - 6x}{2x}$

3. $\dfrac{x^4 - 3x^2 + 7x^5}{x^2}$

4. $\dfrac{6x^4 - 2x^3 + 7x^2}{3x^2}$

In Exercises 5–10, factor the polynomial into linear factors.

5. $x^3 - 4x$

6. $6x^2 - 54$

7. $4x^2 + 8x - 60$

8. $15x^3 - 22x^2 + 8x$

9. $x^3 + 2x^2 - x - 2$

10. $x^4 + x^3 - 9x^2 - 9x$

SECTION 2.4 EXERCISES

In Exercises 1–6, divide $f(x)$ by $d(x)$, and write a summary statement in polynomial form and fraction form.

1. $f(x) = x^2 - 2x + 3;\ d(x) = x - 1$

2. $f(x) = x^3 - 1;\ d(x) = x + 1$

3. $f(x) = x^3 + 4x^2 + 7x - 9;\ d(x) = x + 3$

4. $f(x) = 4x^3 - 8x^2 + 2x - 1;\ d(x) = 2x + 1$

5. $f(x) = x^4 - 2x^3 + 3x^2 - 4x + 6;\ d(x) = x^2 + 2x - 1$

6. $f(x) = x^4 - 3x^3 + 6x^2 - 3x + 5;\ d(x) = x^2 + 1$

In Exercises 7–12, divide using synthetic division, and write a summary statement in fraction form.

7. $\dfrac{x^3 - 5x^2 + 3x - 2}{x + 1}$

8. $\dfrac{2x^4 - 5x^3 + 7x^2 - 3x + 1}{x - 3}$

9. $\dfrac{9x^3 + 7x^2 - 3x}{x - 10}$

10. $\dfrac{3x^4 + x^3 - 4x^2 + 9x - 3}{x + 5}$

11. $\dfrac{5x^4 - 3x + 1}{4 - x}$

12. $\dfrac{x^8 - 1}{x + 2}$

In Exercises 13–18, use the Remainder Theorem to find the remainder when $f(x)$ is divided by $x - k$.

13. $f(x) = 2x^2 - 3x + 1;\ k = 2$

14. $f(x) = x^4 - 5;\ k = 1$

15. $f(x) = x^3 - x^2 + 2x - 1;\ k = -3$

16. $f(x) = x^3 - 3x + 4;\ k = -2$

17. $f(x) = 2x^3 - 3x^2 + 4x - 7;\ k = 2$

18. $f(x) = x^5 - 2x^4 + 3x^2 - 20x + 3;\ k = -1$

In Exercises 19–24, use the Factor Theorem to determine whether the first polynomial is a factor of the second polynomial.

19. $x - 1$; $x^3 - x^2 + x - 1$ **20.** $x - 3$; $x^3 - x^2 - x - 15$

21. $x - 2$; $x^3 + 3x - 4$ **22.** $x - 2$; $x^3 - 3x - 2$

23. $x + 2$; $4x^3 + 9x^2 - 3x - 10$

24. $x + 1$; $2x^{10} - x^9 + x^8 + x^7 + 2x^6 - 3$

In Exercises 25 and 26, use the graph to guess possible linear factors of $f(x)$. Then completely factor $f(x)$ with the aid of synthetic division.

25. $f(x) = 5x^3 - 7x^2 - 49x + 51$

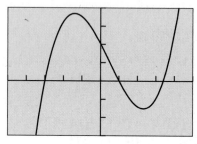

[–5, 5] by [–75, 100]

26. $f(x) = 5x^3 - 12x^2 - 23x + 42$

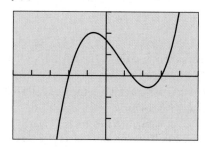

[–5, 5] by [–75, 75]

In Exercises 27–30, find the polynomial function with leading coefficient 2 that has the given degree and zeros.

27. Degree 3, with -2, 1, and 4 as zeros

28. Degree 3, with -1, 3, and -5 as zeros

29. Degree 3, with 2, $\frac{1}{2}$, and $\frac{3}{2}$ as zeros

30. Degree 4, with -3, -1, 0, and $\frac{5}{2}$ as zeros

In Exercises 31 and 32, using only algebraic methods, find the cubic function with the given table of values. Check with a grapher.

31.

x	-4	0	3	5
$f(x)$	0	180	0	0

32.

x	-2	-1	1	5
$f(x)$	0	24	0	0

In Exercises 33–36, use the Rational Zeros Theorem to write a list of all potential rational zeros. Then determine which ones, if any, are zeros.

33. $f(x) = 6x^3 - 5x - 1$ **34.** $f(x) = 3x^3 - 7x^2 + 6x - 14$

35. $f(x) = 2x^3 - x^2 - 9x + 9$

36. $f(x) = 6x^4 - x^3 - 6x^2 - x - 12$

In Exercises 37–40, use synthetic division to prove that the number k is an upper bound for the real zeros of the function f.

37. $k = 3$; $f(x) = 2x^3 - 4x^2 + x - 2$

38. $k = 5$; $f(x) = 2x^3 - 5x^2 - 5x - 1$

39. $k = 2$; $f(x) = x^4 - x^3 + x^2 + x - 12$

40. $k = 3$; $f(x) = 4x^4 - 6x^3 - 7x^2 + 9x + 2$

In Exercises 41–44, use synthetic division to prove that the number k is a lower bound for the real zeros of the function f.

41. $k = -1$; $f(x) = 3x^3 - 4x^2 + x + 3$

42. $k = -3$; $f(x) = x^3 + 2x^2 + 2x + 5$

43. $k = 0$; $f(x) = x^3 - 4x^2 + 7x - 2$

44. $k = -4$; $f(x) = 3x^3 - x^2 - 5x - 3$

In Exercises 45–48, use the Upper and Lower Bound Tests to decide whether there could be real zeros for the function outside the window shown. If so, check for additional zeros.

45. $f(x) = 6x^4 - 11x^3 - 7x^2 + 8x - 34$

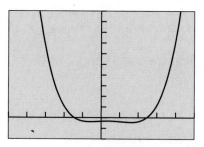

[–5, 5] by [–200, 1000]

46. $f(x) = x^5 - x^4 + 21x^2 + 19x - 3$

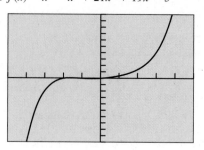

[–5, 5] by [–1000, 1000]

47. $f(x) = x^5 - 4x^4 - 129x^3 + 396x^2 - 8x + 3$

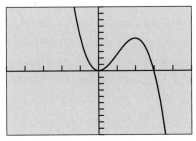

[−5, 5] by [−1000, 1000]

48. $f(x) = 2x^5 - 5x^4 - 141x^3 + 216x^2 - 91x + 25$

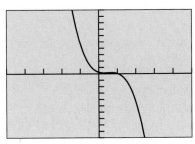

[−5, 5] by [−1000, 1000]

In Exercises 49–56, find all of the real zeros of the function, finding exact values whenever possible. Identify each zero as rational or irrational.

49. $f(x) = 2x^3 - 3x^2 - 4x + 6$

50. $f(x) = x^3 + 3x^2 - 3x - 9$

51. $f(x) = x^3 + x^2 - 8x - 6$

52. $f(x) = x^3 - 6x^2 + 7x + 4$

53. $f(x) = x^4 - 3x^3 - 6x^2 + 6x + 8$

54. $f(x) = x^4 - x^3 - 7x^2 + 5x + 10$

55. $f(x) = 2x^4 - 7x^3 - 2x^2 - 7x - 4$

56. $f(x) = 3x^4 - 2x^3 + 3x^2 + x - 2$

57. Setting Production Schedules The Sunspot Small Appliance Co. determines that the supply function for their EverCurl hair dryer is $S(p) = 6 + 0.001p^3$ and that its demand function is $D(p) = 80 - 0.02p^2$, where p is the price. Determine the price for which the supply equals the demand and the number of hair dryers corresponding to this equilibrium price.

58. Setting Production Schedules The Pentkon Camera Co. determines that the supply and demand functions for their 35 mm−70 mm zoom lens are $S(p) = 200 - p + 0.000007p^4$ and $D(p) = 1500 - 0.0004p^3$, where p is the price. Determine the price for which the supply equals the demand and the number of zoom lenses corresponding to this equilibrium price.

59. Find the remainder when $x^{40} - 3$ is divided by $x + 1$.

60. Find the remainder when $x^{63} - 17$ is divided by $x - 1$.

61. Let $f(x) = x^4 + 2x^3 - 11x^2 - 13x + 38$

   (a) Use the upper and lower bound tests to prove that all of the real zeros of f lie on the interval $[-5, 4]$.

   (b) Find all of the rational zeros of f.

   (c) Factor $f(x)$ using the rational zero(s) found in (b).

   (d) Approximate all of the irrational zeros of f.

   (e) Use synthetic division and the irrational zero(s) found in (d) to continue the factorization of $f(x)$ begun in (c).

62. Lewis's distance D from a motion detector is given by the data in Table 2.15.

Table 2.15 Motion Detector Data			
t (sec)	D (m)	t (sec)	D (m)
0.0	1.00	4.5	0.99
0.5	1.46	5.0	0.84
1.0	1.99	5.5	1.28
1.5	2.57	6.0	1.87
2.0	3.02	6.5	2.58
2.5	3.34	7.0	3.23
3.0	2.91	7.5	3.78
3.5	2.31	8.0	4.40
4.0	1.57		

   (a) Find a cubic regression model, and graph it together with a scatter plot of the data.

   (b) Use the cubic regression model to estimate how far Lewis is from the motion detector initially.

   (c) Use the cubic regression model to estimate when Lewis changes direction. How far from the motion detector is he when he changes direction?

Standardized Test Questions

63. True or False The polynomial function $f(x)$ has a factor $x + 2$ if and only if $f(2) = 0$. Justify your answer.

64. True or False If $f(x) = (x - 1)(2x^2 - x + 1) + 3$, then the remainder when $f(x)$ is divided by $x - 1$ is 3. Justify your answer.

In Exercises 65–68, you may use a graphing calculator to solve the problem.

65. Multiple Choice Let f be a polynomial function with $f(3) = 0$. Which of the following statements is not true?

   (A) $x + 3$ is a factor of $f(x)$.

   (B) $x - 3$ is a factor of $f(x)$.

   (C) $x = 3$ is a zero of $f(x)$.

   (D) 3 is an x-intercept of $f(x)$.

   (E) The remainder when $f(x)$ is divided by $x - 3$ is zero.

66. Multiple Choice Let $f(x) = 2x^3 + 7x^2 + 2x - 3$. Which of the following is not a possible rational root of f?

(A) -3 **(B)** -1 **(C)** 1 **(D)** $1/2$ **(E)** $2/3$

67. Multiple Choice Let $f(x) = (x + 2)(x^2 + x - 1) - 3$. Which of the following statements is not true?

(A) The remainder when $f(x)$ is divided by $x + 2$ is -3.

(B) The remainder when $f(x)$ is divided by $x - 2$ is -3.

(C) The remainder when $f(x)$ is divided by $x^2 + x - 1$ is -3.

(D) $x + 2$ is not a factor of $f(x)$.

(E) $f(x)$ is not evenly divisible by $x + 2$.

68. Multiple Choice Let $f(x) = (x^2 + 1)(x - 2) + 7$. Which of the following statements is not true?

(A) The remainder when $f(x)$ is divided by $x^2 + 1$ is 7.

(B) The remainder when $f(x)$ is divided by $x - 2$ is 7.

(C) $f(2) = 7$ **(D)** $f(0) = 5$

(E) f does not have a real root.

Explorations

69. Archimedes' Principle A spherical buoy has a radius of 1 m and a density one-fourth that of seawater. By Archimedes' Principle, the weight of the displaced water will equal the weight of the buoy.

- Let x = the depth to which the buoy sinks.
- Let d = the density of seawater.
- Let r = the radius of the circle formed where buoy, air, and water meet. See the figure below.

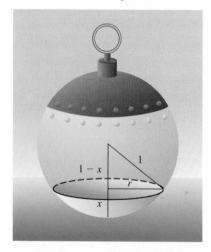

Notice that $0 < x < 1$ and that

$(1 - x)^2 + r^2 = 1$

$$r^2 = 1 - (1 - x)^2$$

$$= 2x - x^2$$

(a) Verify that the volume of the buoy is $4\pi/3$.

(b) Use your result from (a) to establish the weight of the buoy as $\pi d/3$.

(c) Prove the weight of the displaced water is $\pi d \cdot x(3r^2 + x^2)/6$.

(d) Approximate the depth to which the buoy will sink.

70. Archimedes' Principle Using the scenario of Exercise 69, find the depth to which the buoy will sink if its density is one-fifth that of seawater.

71. Biological Research Stephanie, a biologist who does research for the poultry industry, models the population P of wild turkeys, t days after being left to reproduce, with the function

$$P(t) = -0.00001t^3 + 0.002t^2 + 1.5t + 100.$$

(a) Graph the function $y = P(t)$ for appropriate values of t.

(b) Find what the maximum turkey population is and when it occurs.

(c) Assuming that this model continues to be accurate, when will this turkey population become extinct?

(d) Writing to Learn Create a scenario that could explain the growth exhibited by this turkey population.

72. Architectural Engineering Dave, an engineer at the Trumbauer Group, Inc., an architectural firm, completes structural specifications for a 172-ft-long steel beam, anchored at one end to a piling 20 ft above the ground. He knows that when a 200-lb object is placed d feet from the anchored end, the beam bends s feet where

$$s = (3 \times 10^{-7})d^2(550 - d).$$

(a) What is the independent variable in this polynomial function?

(b) What are the dimensions of a viewing window that shows a graph for the values that make sense in this problem situation?

(c) How far is the 200-lb object from the anchored end if the vertical deflection is 1.25 ft?

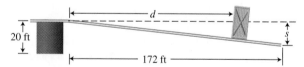

73. A classic theorem, **Descartes' Rule of Signs**, tells us about the number of positive and negative real zeros of a polynomial function, by looking at the polynomial's variations in sign. A *variation in sign* occurs when consecutive coefficients (in standard form) have opposite signs.

If $f(x) = a_n x^n + \cdots + a_0$ is a polynomial of degree n, then

- The number of positive real zeros of f is equal to the number of variations in sign of $f(x)$, or that number less some even number.

- The number of negative real zeros of f is equal to the number of variations in sign of $f(-x)$, or that number less some even number.

Use Descartes' Rule of Signs to determine the possible numbers of positive and negative real zeros of the function.

(a) $f(x) = x^3 + x^2 - x + 1$

(b) $f(x) = x^3 + x^2 + x + 1$

(c) $f(x) = 2x^3 + x - 3$

(d) $g(x) = 5x^4 + x^2 - 3x - 2$

Extending the Ideas

74. Writing to Learn Graph each side of the Example 3 summary equation:

$$f(x) = \frac{2x^3 - 3x^2 - 5x - 12}{x - 3} \quad \text{and}$$

$$g(x) = 2x^2 - 3x + 4, \quad x \neq 3$$

How are these functions related? Include a discussion of the domain and continuity of each function.

75. Writing to Learn Explain how to carry out the following division using synthetic division. Work through the steps with complete explanations. Interpret and check your result.

$$\frac{4x^3 - 5x^2 + 3x + 1}{2x - 1}$$

76. Writing to Learn The figure shows a graph of $f(x) = x^4 + 0.1x^3 - 6.5x^2 + 7.9x - 2.4$. Explain how to use a grapher to justify the statement.

$$f(x) = x^4 + 0.1x^3 - 6.5x^2 + 7.9x - 2.4$$

$$\approx (x + 3.10)(x - 0.5)(x - 1.13)(x - 1.37)$$

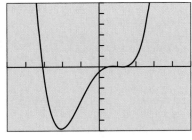

[-5, 5] by [-30, 30]

77. (a) Writing to Learn Write a paragraph that describes how the zeros of $f(x) = (1/3)x^3 + x^2 + 2x - 3$ are related to the zeros of $g(x) = x^3 + 3x^2 + 6x - 9$. In what ways does this example illustrate how the Rational Zeros Theorem can be applied to find the zeros of a polynomial with *rational* number coefficients?

(b) Find the rational zeros of $f(x) = x^3 - \dfrac{7}{6}x^2 - \dfrac{20}{3}x + \dfrac{7}{2}$.

(c) Find the rational zeros of $f(x) = x^3 - \dfrac{5}{2}x^2 - \dfrac{37}{12}x + \dfrac{5}{2}$.

78. Use the Rational Zeros Theorem to prove $\sqrt{2}$ is irrational.

79. Group Activity *Work in groups of three.* Graph $f(x) = x^4 + x^3 - 8x^2 - 2x + 7$.

(a) Use grapher methods to find approximate real number zeros.

(b) Identify a list of four linear factors whose product could be called an *approximate factorization of f(x)*.

(c) Discuss what graphical and numerical methods you could use to show that the factorization from part (b) is reasonable.

2.5
Complex Zeros and the Fundamental Theorem of Algebra

What you'll learn about

■ Two Major Theorems

■ Complex Conjugate Zeros

■ Factoring with Real Number Coefficients

. . . and why

These topics provide the complete story about the zeros and factors of polynomials with real number coefficients.

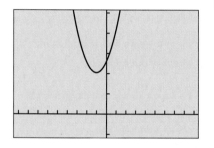

[−9.4, 9.4] by [−2, 10]

FIGURE 2.42 The graph of $f(x) = x^2 + 2x + 5$ has no x-intercepts, so f has no real zeros.

Two Major Theorems

In Section 2.3 we learned that a polynomial function of degree n has at most n real zeros. Figure 2.42 shows that the polynomial function $f(x) = x^2 + 2x + 5$ of degree 2 has no real zeros. (Why?) A little arithmetic, however, shows that the complex number $-1 + 2i$ is a zero of f:

$$f(-1 + 2i) = (-1 + 2i)^2 + 2(-1 + 2i) + 5$$
$$= (-3 - 4i) + (-2 + 4i) + 5$$
$$= 0 + 0i$$
$$= 0$$

The quadratic formula shows that $-1 \pm 2i$ are the two zeros of f and can be used to find the complex zeros for any polynomial function of degree 2. In this section we will learn about complex zeros of polynomial functions of higher degree and how to use these zeros to factor polynomial expressions.

THEOREM **Fundamental Theorem of Algebra**

A polynomial function of degree n has n complex zeros (real and nonreal). Some of these zeros may be repeated.

The Factor Theorem extends to the complex zeros of a polynomial function. Thus, k is a complex zero of a polynomial if and only if $x - k$ is a factor of the polynomial, even if k is not a real number. We combine this fact with the Fundamental Theorem of Algebra to obtain the following theorem.

THEOREM **Linear Factorization Theorem**

If $f(x)$ is a polynomial function of degree $n > 0$, then $f(x)$ has precisely n linear factors and

$$f(x) = a(x - z_1)(x - z_2) \cdots (x - z_n)$$

where a is the leading coefficient of $f(x)$ and $z_1, z_2, \ldots, z_n$ are the complex zeros of $f(x)$. The z_i are not necessarily distinct numbers; some may be repeated.

The Fundamental Theorem of Algebra and the Linear Factorization Theorem are *existence theorems*. They tell us of the existence of zeros and linear factors, but not how to find them.

One connection is lost going from real zeros to complex zeros. If k is a *nonreal* complex zero of a polynomial function $f(x)$, then k is *not* an x-intercept of the graph of f. The other connections hold whether k is real or nonreal:

Fundamental Polynomial Connections in the Complex Case

The following statements about a polynomial function f are equivalent if k is a complex number:

1. $x = k$ is a solution (or root) of the equation $f(x) = 0$.

2. k is a zero of the function f.

3. $x - k$ is a factor of $f(x)$.

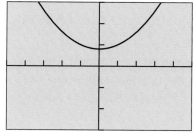

[–5, 5] by [–15, 15]

(a)

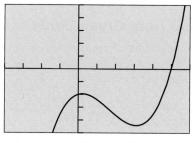

[–4, 6] by [–25, 25]

(b)

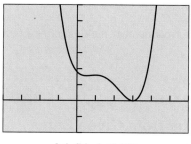

[–4, 6] by [–10, 30]

(c)

FIGURE 2.43 The graphs of (a) $y = x^2 + 4$, (b) $y = x^3 - 5x^2 + 2x - 10$, and (c) $y = x^4 - 6x^3 + 10x^2 - 6x + 9$. (Example 1)

EXAMPLE 1 Exploring Fundamental Polynomial Connections

Write the polynomial function in standard form, and identify the zeros of the function and the x-intercepts of its graph.

(a) $f(x) = (x - 2i)(x + 2i)$

(b) $f(x) = (x - 5)(x - \sqrt{2}i)(x + \sqrt{2}i)$

(c) $f(x) = (x - 3)(x - 3)(x - i)(x + i)$

SOLUTION

(a) The quadratic function $f(x) = (x - 2i)(x + 2i) = x^2 + 4$ has two zeros: $x = 2i$ and $x = -2i$. Because the zeros are not real, the graph of f has no x-intercepts.

(b) The cubic function

$$f(x) = (x - 5)(x - \sqrt{2}i)(x + \sqrt{2}i)$$
$$= (x - 5)(x^2 + 2)$$
$$= x^3 - 5x^2 + 2x - 10$$

has three zeros: $x = 5$, $x = \sqrt{2}i$, and $x = -\sqrt{2}i$. Of the three, only $x = 5$ is an x-intercept.

(c) The quartic function

$$f(x) = (x - 3)(x - 3)(x - i)(x + i)$$
$$= (x^2 - 6x + 9)(x^2 + 1)$$
$$= x^4 - 6x^3 + 10x^2 - 6x + 9$$

has four zeros: $x = 3$, $x = 3$, $x = i$, and $x = -i$. There are only three distinct zeros. The real zero $x = 3$ is a repeated zero of multiplicity two. Due to this even multiplicity, the graph of f touches but does not cross the x-axis at $x = 3$, the only x-intercept.

Figure 2.43 supports our conclusions regarding x-intercepts.

Now try Exercise 1.

Complex Conjugate Zeros

In Section P.6 we saw that, for quadratic equations $ax^2 + bx + c = 0$ with real coefficients, if the discriminant $b^2 - 4ac$ is negative, the solutions are a conjugate pair of complex numbers. This relationship generalizes to polynomial functions of higher degree in the following way:

THEOREM Complex Conjugate Zeros

Suppose that $f(x)$ is a polynomial function with *real coefficients*. If a and b are real numbers with $b \neq 0$ and $a + bi$ is a zero of $f(x)$, then its complex conjugate $a - bi$ is also a zero of $f(x)$.

EXPLORATION 1   What Can Happen If the Coefficients Are Not Real?

1. Use substitution to verify that $x = 2i$ and $x = -i$ are zeros of $f(x) = x^2 - ix + 2$. Are the conjugates of $2i$ and $-i$ also zeros of $f(x)$?

2. Use substitution to verify that $x = i$ and $x = 1 - i$ are zeros of $g(x) = x^2 - x + (1 + i)$. Are the conjugates of i and $1 - i$ also zeros of $g(x)$?

3. What conclusions can you draw from parts 1 and 2? Do your results contradict the theorem about complex conjugate zeros?

EXAMPLE 2   Finding a Polynomial from Given Zeros

Write a polynomial function of minimum degree in standard form with real coefficients whose zeros include -3, 4, and $2 - i$.

SOLUTION Because -3 and 4 are real zeros, $x + 3$ and $x - 4$ must be factors. Because the coefficients are real and $2 - i$ is a zero, $2 + i$ must also be a zero. Therefore, $x - (2 - i)$ and $x - (2 + i)$ must both be factors of $f(x)$. Thus,

$$f(x) = (x + 3)(x - 4)[x - (2 - i)][x - (2 + i)]$$
$$= (x^2 - x - 12)(x^2 - 4x + 5)$$
$$= x^4 - 5x^3 - 3x^2 + 43x - 60$$

is a polynomial of the type we seek. Any nonzero real number multiple of $f(x)$ will also be such a polynomial.   *Now try Exercise 7.*

EXAMPLE 3   Finding a Polynomial from Given Zeros

Write a polynomial function of minimum degree in standard form with real coefficients whose zeros include $x = 1$, $x = 1 + 2i$, $x = 1 - i$.

SOLUTION Because the coefficients are real and $1 + 2i$ is a zero, $1 - 2i$ must also be a zero. Therefore, $x - (1 + 2i)$ and $x - (1 - 2i)$ are both factors of $f(x)$.

continued

Likewise, because $1 - i$ is a zero, $1 + i$ must be a zero. It follows that $x - (1 - i)$ and $x - (1 + i)$ are both factors of $f(x)$. Therefore,

$$f(x) = (x - 1)[x - (1 + 2i)][x - (1 - 2i)][x - (1 + i)][x - (1 - i)]$$

$$= (x - 1)(x^2 - 2x + 5)(x^2 - 2x + 2)$$

$$= (x^3 - 3x^2 + 7x - 5)(x^2 - 2x + 2)$$

$$= x^5 - 5x^4 + 15x^3 - 25x^2 + 24x - 10$$

is a polynomial of the type we seek. Any nonzero real number multiple of $f(x)$ will also be such a polynomial. ***Now try Exercise 13.***

EXAMPLE 4 Factoring a Polynomial with Complex Zeros

Find all zeros of $f(x) = x^5 - 3x^4 - 5x^3 + 5x^2 - 6x + 8$, and write $f(x)$ in its linear factorization.

SOLUTION Figure 2.44 suggests that the real zeros of f are $x = -2$, $x = 1$, and $x = 4$.

Using synthetic division we can verify these zeros and show that $x^2 + 1$ is a factor of f. So $x = i$ and $x = -i$ are also zeros. Therefore,

$$f(x) = x^5 - 3x^4 - 5x^3 + 5x^2 - 6x + 8$$

$$= (x + 2)(x - 1)(x - 4)(x^2 + 1)$$

$$= (x + 2)(x - 1)(x - 4)(x - i)(x + i).$$

Now try Exercise 29.

[–4.7, 4.7] by [–125, 125]

FIGURE 2.44 $f(x) = x^5 - 3x^4 - 5x^3 + 5x^2 - 6x + 8$ has three real zeros. (Example 4)

Synthetic division can be used with complex number divisors in the same way it is used with real number divisors.

EXAMPLE 5 Finding Complex Zeros

The complex number $z = 1 - 2i$ is a zero of $f(x) = 4x^4 + 17x^2 + 14x + 65$. Find the remaining zeros of $f(x)$, and write it in its linear factorization.

SOLUTION We use a synthetic division to show that $f(1 - 2i) = 0$:

$$
\begin{array}{r|ccccc}
1 - 2i & 4 & 0 & 17 & 14 & 65 \\
& & 4 - 8i & -12 - 16i & -27 - 26i & -65 \\
\hline
& 4 & 4 - 8i & 5 - 16i & -13 - 26i & 0
\end{array}
$$

Thus $1 - 2i$ is a zero of $f(x)$. The conjugate $1 + 2i$ must also be a zero. We use synthetic division on the quotient found above to find the remaining quadratic factor:

$$
\begin{array}{r|cccc}
1 + 2i & 4 & 4 - 8i & 5 - 16i & -13 - 26i \\
& & 4 + 8i & 8 + 16i & 13 + 26i \\
\hline
& 4 & 8 & 13 & 0
\end{array}
$$

continued

Finally, we use the quadratic formula to find the two zeros of $4x^2 + 8x + 13$:

$$x = \frac{-8 \pm \sqrt{64 - 208}}{8}$$

$$= \frac{-8 \pm \sqrt{-144}}{8}$$

$$= \frac{-8 \pm 12i}{8}$$

$$= -1 \pm \frac{3}{2}i$$

Thus the four zeros of $f(x)$ are $1 - 2i$, $1 + 2i$, $-1 + (3/2)i$, and $-1 - (3/2)i$. Because the leading coefficient of $f(x)$ is 4, we obtain

$$f(x) = 4[x - (1 - 2i)]\,[x - (1 + 2i)]\left[x - \left(-1 + \tfrac{3}{2}i\right)\right]\left[x - \left(-1 - \tfrac{3}{2}i\right)\right].$$

If we wish to remove fractions in the factors, we can distribute the 4 to get

$$f(x) = [x - (1 - 2i)][x - (1 + 2i)][2x - (-2 + 3i)][2x - (-2 - 3i)].$$

Now try Exercise 33.

Factoring with Real Number Coefficients

Let $f(x)$ be a polynomial function with real coefficients. The Linear Factorization Theorem tells us that $f(x)$ can be factored into the form

$$f(x) = a(x - z_1)\,(x - z_2) \cdots (x - z_n),$$

where z_i are complex numbers. Recall, however, that nonreal complex zeros occur in conjugate pairs. The product of $x - (a + bi)$ and $x - (a - bi)$ is

$$[x - (a + bi)]\,[x - (a - bi)] = x^2 - (a - bi)x - (a + bi)x + (a + bi)\,(a - bi)$$
$$= x^2 - 2ax + (a^2 + b^2).$$

So the quadratic expression $x^2 - 2ax + (a^2 + b^2)$, whose coefficients are real numbers, is a factor of $f(x)$. Such a quadratic expression with real coefficients but no real zeros is **irreducible over the reals**. In other words, if we require that the factors of a polynomial have real coefficients, the factorization can be accomplished with linear factors and irreducible quadratic factors.

Factors of a Polynomial with Real Coefficients

Every polynomial function with real coefficients can be written as a product of linear factors and irreducible quadratic factors, each with real coefficients.

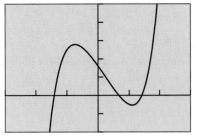

[–3, 3] by [–20, 50]

FIGURE 2.45 $f(x) = 3x^5 - 2x^4 + 6x^3 - 4x^2 - 24x + 16$ has three real zeros. (Example 6)

EXAMPLE 6 Factoring a Polynomial

Write $f(x) = 3x^5 - 2x^4 + 6x^3 - 4x^2 - 24x + 16$ as a product of linear and irreducible quadratic factors, each with real coefficients.

SOLUTION The Rational Zeros Theorem provides the candidates for the rational zeros of f. The graph of f in Figure 2.45 suggests which candidates to try first. Using synthetic division, we find that $x = 2/3$ is a zero. Thus,

$$f(x) = \left(x - \frac{2}{3}\right)(3x^4 + 6x^2 - 24)$$

$$= \left(x - \frac{2}{3}\right)(3)(x^4 + 2x^2 - 8)$$

$$= (3x - 2)(x^2 - 2)(x^2 + 4)$$

$$= (3x - 2)(x - \sqrt{2})(x + \sqrt{2})(x^2 + 4)$$

Because the zeros of $x^2 + 4$ are complex, any further factorization would introduce nonreal complex coefficients. We have taken the factorization of f as far as possible, subject to the condition that *each factor has real coefficients*.

Now try Exercise 37.

We have seen that if a polynomial function has real coefficients, then its nonreal complex zeros occur in conjugate pairs. *Because a polynomial of odd degree has an odd number of zeros, it must have at least one zero that is real.* This confirms Example 7 of Section 2.3 in light of complex numbers.

> **Polynomial Function of Odd Degree**
>
> Every polynomial function of odd degree with real coefficients has at least one real zero.

The function $f(x) = 3x^5 - 2x^4 + 6x^3 - 4x^2 - 24x + 16$ in Example 6 fits the conditions of this theorem, so we know immediately that we are on the right track in searching for at least one real zero.

QUICK REVIEW 2.5 *(For help, go to Sections P.5, P.6, and 2.4.)*

In Exercises 1–4, perform the indicated operation, and write the result in the form $a + bi$.

1. $(3 - 2i) + (-2 + 5i)$

2. $(5 - 7i) - (3 - 2i)$

3. $(1 + 2i)(3 - 2i)$ **4.** $\dfrac{2 + 3i}{1 - 5i}$

In Exercises 5 and 6, factor the quadratic expression.

5. $2x^2 - x - 3$ **6.** $6x^2 - 13x - 5$

In Exercises 7 and 8, solve the quadratic equation.

7. $x^2 - 5x + 11 = 0$ **8.** $2x^2 + 3x + 7 = 0$

In Exercises 9 and 10, list all potential rational zeros.

9. $3x^4 - 5x^3 + 3x^2 - 7x + 2$

10. $4x^5 - 7x^2 + x^3 + 13x - 3$

SECTION 2.5 EXERCISES

In Exercises 1–4, write the polynomial in standard form, and identify the zeros of the function and the x-intercepts of its graph.

1. $f(x) = (x - 3i)(x + 3i)$

2. $f(x) = (x + 2)(x - \sqrt{3}\,i)(x + \sqrt{3}\,i)$

3. $f(x) = (x - 1)(x - 1)(x + 2i)(x - 2i)$

4. $f(x) = x(x - 1)(x - 1 - i)(x - 1 + i)$

In Exercises 5–12, write a polynomial function of minimum degree in standard form with real coefficients whose zeros include those listed.

5. i and $-i$ **6.** $1 - 2i$ and $1 + 2i$

7. 1, $3i$, and $-3i$ **8.** $-4, 1 - i$, and $1 + i$

9. 2, 3, and i **10.** -1, 2, and $1 - i$

11. 5 and $3 + 2i$ **12.** -2 and $1 + 2i$

In Exercises 13–16, write a polynomial function of minimum degree in standard form with real coefficients whose zeros and their multiplicities include those listed.

13. 1 (multiplicity 2), -2 (multiplicity 3)

14. -1 (multiplicity 3), 3 (multiplicity 1)

15. 2 (multiplicity 2), $3 + i$ (multiplicity 1)

16. -1 (multiplicity 2), $-2 - i$ (multiplicity 1)

In Exercises 17–20, match the polynomial function graph to the given zeros and multiplicities.

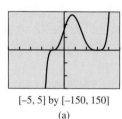

[−5, 5] by [−150, 150]

(a)

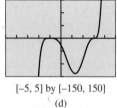

[−5, 5] by [−150, 150]

(b)

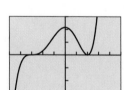

[−5, 5] by [−150, 150]

(c)

[−5, 5] by [−150, 150]

(d)

17. -3 (multiplicity 2), 2 (multiplicity 3)

18. -3 (multiplicity 3), 2 (multiplicity 2)

19. -1 (multiplicity 4), 3 (multiplicity 3)

20. -1 (multiplicity 3), 3 (multiplicity 4)

In Exercises 21–26, state how many complex and real zeros the function has.

21. $f(x) = x^2 - 2x + 7$

22. $f(x) = x^3 - 3x^2 + x + 1$

23. $f(x) = x^3 - x + 3$

24. $f(x) = x^4 - 2x^2 + 3x - 4$

25. $f(x) = x^4 - 5x^3 + x^2 - 3x + 6$

26. $f(x) = x^5 - 2x^2 - 3x + 6$

In Exercises 27–32, find all of the zeros and write a linear factorization of the function.

27. $f(x) = x^3 + 4x - 5$

28. $f(x) = x^3 - 10x^2 + 44x - 69$

29. $f(x) = x^4 + x^3 + 5x^2 - x - 6$

30. $f(x) = 3x^4 + 8x^3 + 6x^2 + 3x - 2$

31. $f(x) = 6x^4 - 7x^3 - x^2 + 67x - 105$

32. $f(x) = 20x^4 - 148x^3 + 269x^2 - 106x - 195$

In Exercises 33–36, using the given zero, find all of the zeros and write a linear factorization of $f(x)$.

33. $1 + i$ is a zero of $f(x) = x^4 - 2x^3 - x^2 + 6x - 6$.

34. $4i$ is a zero of $f(x) = x^4 + 13x^2 - 48$.

35. $3 - 2i$ is a zero of $f(x) = x^4 - 6x^3 + 11x^2 + 12x - 26$.

36. $1 + 3i$ is a zero of $f(x) = x^4 - 2x^3 + 5x^2 + 10x - 50$.

In Exercises 37–42, write the function as a product of linear and irreducible quadratic factors all with real coefficients.

37. $f(x) = x^3 - x^2 - x - 2$

38. $f(x) = x^3 - x^2 + x - 6$

39. $f(x) = 2x^3 - x^2 + 2x - 3$

40. $f(x) = 3x^3 - 2x^2 + x - 2$

41. $f(x) = x^4 + 3x^3 - 3x^2 + 3x - 4$

42. $f(x) = x^4 - 2x^3 + x^2 - 8x - 12$

In Exercises 43 and 44, use *Archimedes' principle*, which states that when a sphere of radius r with density d_S is placed in a liquid of density $d_L = 62.5$ lb/ft^3, it will sink to a depth h where

$$\frac{\pi}{3}(3rh^2 - h^3)d_L = \frac{4}{3}\pi r^3 d_S.$$

Find an approximate value for h if:

43. $r = 5$ ft and $d_S = 20$ lb/ft^3.

44. $r = 5$ ft and $d_S = 45$ lb/ft^3.

In Exercises 45–48, answer yes or no. If yes, include an example. If no, give a reason.

45. Writing to Learn Is it possible to find a polynomial of degree 3 with real number coefficients that has -2 as its only real zero?

46. Writing to Learn Is it possible to find a polynomial of degree 3 with real coefficients that has $2i$ as its only nonreal zero?

47. Writing to Learn Is it possible to find a polynomial $f(x)$ of degree 4 with real coefficients that has zeros -3, $1 + 2i$, and $1 - i$?

48. Writing to Learn Is it possible to find a polynomial $f(x)$ of degree 4 with real coefficients that has zeros $1 + 3i$ and $1 - i$?

In Exercises 49 and 50, find the unique polynomial with real coefficients that meets these conditions.

49. Degree 4; zeros at $x = 3$, $x = -1$, and $x = 2 - i$; $f(0) = 30$

50. Degree 4; zeros at $x = 1 - 2i$ and $x = 1 + i$; $f(0) = 20$

51. Sally's distance D from a motion detector is given by the data in Table 2.16.

(a) Find a cubic regression model, and graph it together with a scatter plot of the data.

(b) Describe Sally's motion.

(c) Use the cubic regression model to estimate when Sally changes direction. How far is she from the motion detector when she changes direction?

Table 2.16 Motion Detector Data

t (sec)	D (m)	t (sec)	D (m)
0.0	3.36	4.5	3.59
0.5	2.61	5.0	4.15
1.0	1.86	5.5	3.99
1.5	1.27	6.0	3.37
2.0	0.91	6.5	2.58
2.5	1.14	7.0	1.93
3.0	1.69	7.5	1.25
3.5	2.37	8.0	0.67
4.0	3.01		

52. Jacob's distance D from a motion detector is given by the data in Table 2.17.

(a) Find a quadratic regression model, and graph it together with a scatter plot of the data.

(b) Describe Jacob's motion.

(c) Use the quadratic regression model to estimate when Jacob changes direction. How far is he from the motion detector when he changes direction?

Table 2.17 Motion Detector Data

t (sec)	D (m)	t (sec)	D (m)
0.0	4.59	4.5	1.70
0.5	3.92	5.0	2.25
1.0	3.14	5.5	2.84
1.5	2.41	6.0	3.39
2.0	1.73	6.5	4.02
2.5	1.21	7.0	4.54
3.0	0.90	7.5	5.04
3.5	0.99	8.0	5.59
4.0	1.31		

Standardized Test Questions

53. True or False There is at least one polynomial with real coefficients with $1 - 2i$ as its only nonreal zero. Justify your answer.

54. True or False A polynomial of degree 3 with real coefficients must have two nonreal zeros. Justify your answer.

In Exercises 55–58, you may use a graphing calculator to solve the problem.

55. Multiple Choice Let z be a nonreal complex number and $\bar{z}$ its complex conjugate. Which of the following is not a real number?

(A) $z + \bar{z}$ **(B)** $z\bar{z}$ **(C)** $(z + \bar{z})^2$ **(D)** $(z\bar{z})^2$ **(E)** z^2

56. Multiple Choice Which of the following cannot be the number of real zeros of a polynomial of degree 5 with real coefficients?

(A) 0   **(B)** 1   **(C)** 2   **(D)** 3   **(E)** 4

57. Multiple Choice Which of the following cannot be the number of nonreal zeros of a polynomial of degree 5 with real coefficients?

(A) 0   **(B)** 2   **(C)** 3   **(D)** 4

(E) None of the above.

58. Multiple Choice Assume that $1 + 2i$ is a zero of the polynomial f with real coefficients. Which of the following statements is not true?

(A) $x - (1 + 2i)$ is a factor of $f(x)$.

(B) $x^2 - 2x + 5$ is a factor of $f(x)$.

(C) $x - (1 - 2i)$ is a factor of $f(x)$.

(D) $1 - 2i$ is a zero of f.

(E) The number of nonreal complex zeros of f could be 1.

Explorations

59. Group Activity The Powers of $1 + i$

  (a) Selected powers of $1 + i$ are displayed in Table 2.18. Find a pattern in the data, and use it to extend the table to power 7, 8, 9, and 10.

  (b) Compute $(1 + i)^7$, $(1 + i)^8$, $(1 + i)^9$, and $(1 + i)^{10}$ using the fact that $(1 + i)^6 = -8i$.

  (c) Compare your results from parts (a) and (b) and reconcile, if needed.

Table 2.18 Powers of $1 + i$

Power	Real Part	Imaginary Part
0	1	0
1	1	1
2	0	2
3	-2	2
4	-4	0
5	-4	-4
6	0	-8

60. Group Activity The Square Roots of i Let a and b be real numbers such that $(a + bi)^2 = i$.

  (a) Expand the left-hand side of the given equation.

  (b) Think of the right-hand side of the equation as $0 + 1i$, and separate the real and imaginary parts of the equation to obtain two equations.

  (c) Solve for a and b.

  (d) Check your answer by substituting them in the original equation.

  (e) What are the two square roots of i?

61. Verify that the complex number i is a zero of the polynomial $f(x) = x^3 - ix^2 + 2ix + 2$.

62. Verify that the complex number $-2i$ is a zero of the polynomial $f(x) = x^3 - (2 - i)x^2 + (2 - 2i)x - 4$.

Extending the Ideas

In Exercises 63 and 64, verify that $g(x)$ is a factor of $f(x)$. Then find $h(x)$ so that $f = g \cdot h$.

63. $g(x) = x - i;\quad f(x) = x^3 + (3 - i)x^2 - 4ix - 1$

64. $g(x) = x - 1 - i;\quad f(x) = x^3 - (1 + i)x^2 + x - 1 - i$

65. Find the three cube roots of 8 by solving $x^3 = 8$.

66. Find the three cube roots of -64 by solving $x^3 = -64$.

2.6
Graphs of Rational Functions

What you'll learn about

- Rational Functions
- Transformations of the Reciprocal Function
- Limits and Asymptotes
- Analyzing Graphs of Rational Functions
- Exploring Relative Humidity

. . . and why

Rational functions are used in calculus and in scientific applications such as inverse proportions.

Rational Functions

Rational functions are ratios (or quotients) of polynomial functions.

DEFINITION   Rational Functions

Let f and g be polynomial functions with $g(x) \neq 0$. Then the function given by

$$r(x) = \frac{f(x)}{g(x)}$$

is a **rational function**.

The domain of a rational function is the set of all real numbers except the zeros of its denominator. Every rational function is continuous on its domain.

EXAMPLE 1   Finding the Domain of a Rational Function

Find the domain of f and use limits to describe its behavior at value(s) of x not in its domain.

$$f(x) = \frac{1}{x - 2}$$

SOLUTION The domain of f is all real numbers $x \neq 2$. The graph in Figure 2.46 strongly suggests that f has a vertical asymptote at $x = 2$. As x approaches 2 from the left, the values of f decrease without bound. As x approaches 2 from the right, the values of f increase without bound. Using the notation introduced in Section 1.2 on page 99 we write

$$\lim_{x \to 2^-} f(x) = -\infty \quad \text{and} \quad \lim_{x \to 2^+} f(x) = \infty.$$

The tables in Figure 2.47 support this visual evidence numerically.

Now try Exercise 1.

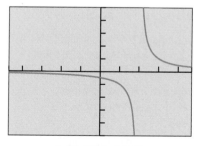

[−4.7, 4.7] by [−5, 5]

FIGURE 2.46 The graph of $f(x) = 1/(x - 2)$. (Example 1)

X	Y1		
2	ERROR		
2.01	100		
2.02	50		
2.03	33.333		
2.04	25		
2.05	20		
2.06	16.667		

Y1 ▤ 1/(X−2)

(a)

X	Y1		
2	ERROR		
1.99	−100		
1.98	−50		
1.97	−33.33		
1.96	−25		
1.95	−20		
1.94	−16.67		

Y1 ▤ 1/(X−2)

(b)

FIGURE 2.47 Table of values for $f(x) = 1/(x - 2)$ for values of x (a) to the right of 2, and (b) to the left of 2. (Example 1)

In Chapter 1 we defined horizontal and vertical asymptotes of the graph of a function $y = f(x)$. The line $y = b$ is a *horizontal asymptote* of the graph of f if

$$\lim_{x \to -\infty} f(x) = b \quad \text{or} \quad \lim_{x \to \infty} f(x) = b.$$

The line $x = a$ is a *vertical asymptote* of the graph of f if

$$\lim_{x \to a^-} f(x) = \pm\infty \quad \text{or} \quad \lim_{x \to a^+} f(x) = \pm\infty.$$

We can see from Figure 2.46 that $\lim_{x \to -\infty} 1/(x-2) = \lim_{x \to \infty} 1/(x-2) = 0$, so the line $y = 0$ is a horizontal asymptote of the graph of $f(x) = 1/(x-2)$. Because $\lim_{x \to 2^-} f(x) = -\infty$ and $\lim_{x \to 2^+} f(x) = \infty$, the line $x = 2$ is a vertical asymptote of $f(x) = 1/(x-2)$.

Transformations of the Reciprocal Function

One of the simplest rational functions is the reciprocal function

$$f(x) = \frac{1}{x},$$

which is one of the basic functions introduced in Chapter 1. It can be used to generate many other rational functions.

Here is what we know about the reciprocal function.

BASIC FUNCTION The Reciprocal Function

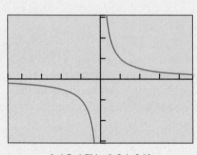

[–4.7, 4.7] by [–3.1, 3.1]

FIGURE 2.48 The graph of $f(x) = 1/x$.

$f(x) = \dfrac{1}{x}$

Domain: $(-\infty, 0) \cup (0, \infty)$

Range: $(-\infty, 0) \cup (0, \infty)$

Continuity: All $x \neq 0$

Decreasing on $(-\infty, 0)$ and $(0, \infty)$

Symmetric with respect to origin (an odd function)

Unbounded

No local extrema

Horizontal asymptote: $y = 0$

Vertical asymptote: $x = 0$

End behavior: $\lim\limits_{x \to -\infty} f(x) = \lim\limits_{x \to \infty} f(x) = 0$

EXPLORATION 1 Comparing Graphs of Rational Functions

1. Sketch the graph and find an equation for the function g whose graph is obtained from the reciprocal function $f(x) = 1/x$ by a translation of 2 units to the right.

2. Sketch the graph and find an equation for the function h whose graph is obtained from the reciprocal function $f(x) = 1/x$ by a translation of 5 units to the right, followed by a reflection across the x-axis.

3. Sketch the graph and find an equation for the function k whose graph is obtained from the reciprocal function $f(x) = 1/x$ by a translation of 4 units to the left, followed by a vertical stretch by a factor of 3, and finally a translation 2 units down.

The graph of any nonzero rational function of the form

$$g(x) = \frac{ax + b}{cx + d}, \quad c \neq 0$$

can be obtained through transformations of the graph of the reciprocal function. If the degree of the numerator is greater than or equal to the degree of the denominator, we can use polynomial division to rewrite the rational function.

EXAMPLE 2   Transforming the Reciprocal Function

Describe how the graph of the given function can be obtained by transforming the graph of the reciprocal function $f(x) = 1/x$. Identify the horizontal and vertical asymptotes and use limits to describe the corresponding behavior. Sketch the graph of the function.

(a) $g(x) = \dfrac{2}{x + 3}$ **(b)** $h(x) = \dfrac{3x - 7}{x - 2}$

SOLUTION

(a) $g(x) = \dfrac{2}{x + 3} = 2\left(\dfrac{1}{x + 3}\right) = 2f(x + 3)$

The graph of g is the graph of the reciprocal function shifted left 3 units and then stretched vertically by a factor of 2. So the lines $x = -3$ and $y = 0$ are vertical and horizontal asymptotes, respectively. Using limits we have $\lim_{x \to \infty} g(x) = \lim_{x \to -\infty} g(x) = 0$, $\lim_{x \to -3^+} g(x) = \infty$, and $\lim_{x \to -3^-} g(x) = -\infty$. The graph is shown in Figure 2.49a.

(b) We begin with polynomial division:

$$
\begin{array}{r}
3 \\
x - 2\overline{)3x - 7} \\
\underline{3x - 6} \\
-1
\end{array}
$$

So, $h(x) = \dfrac{3x - 7}{x - 2} = 3 - \dfrac{1}{x - 2} = -f(x - 2) + 3.$

Thus the graph of h is the graph of the reciprocal function translated 2 units to the right, followed by a reflection across the x-axis, and then translated 3 units up. (Note that the reflection must be executed before the vertical translation.) So the lines $x = 2$ and $y = 3$ are vertical and horizontal asymptotes, respectively. Using limits we have $\lim_{x \to \infty} h(x) = \lim_{x \to -\infty} h(x) = 3$, $\lim_{x \to 2^+} g(x) = -\infty$, and $\lim_{x \to 2^-} g(x) = \infty$. The graph is shown in Figure 2.49b.

Now try Exercise 5.

Limits and Asymptotes

In Example 2 we found asymptotes by translating the known asymptotes of the reciprocal function. In Example 3, we use graphing and algebra to find an asymptote.

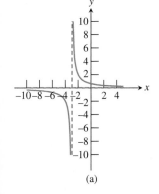

(a)

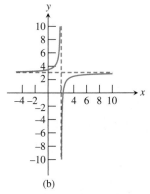

(b)

FIGURE 2.49 The graphs of (a) $g(x) = 2/(x + 3)$ and (b) $h(x) = (3x - 7)/(x - 2)$, with asymptotes shown in red.

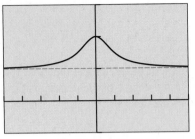

[–5, 5] by [–1, 3]

FIGURE 2.50 The graph of $f(x) = (x^2 + 2)/(x^2 + 1)$ with its horizontal asymptote $y = 1$.

EXAMPLE 3   Finding Asymptotes

Find the horizontal and vertical asymptotes of $f(x) = (x^2 + 2)/(x^2 + 1)$. Use limits to describe the corresponding behavior of f.

SOLUTION

Solve Graphically

The graph of f in Figure 2.50 suggests that

$$\lim_{x \to \infty} f(x) = \lim_{x \to -\infty} f(x) = 1$$

and that there are no vertical asymptotes. The horizontal asymptote is $y = 1$.

Solve Algebraically

The domain of f is all real numbers. So there are no vertical asymptotes. Using polynomial long division, we find that

$$f(x) = \frac{x^2 + 2}{x^2 + 1} = 1 + \frac{1}{x^2 + 1}.$$

When the value of $|x|$ is large, the denominator $x^2 + 1$ is a large positive number, and $1/(x^2 + 1)$ is a small positive number, getting closer to zero as $|x|$ increases. Therefore,

$$\lim_{x \to \infty} f(x) = \lim_{x \to -\infty} f(x) = 1,$$

so $y = 1$ is indeed a horizontal asymptote.   *Now try Exercise 19.*

Example 3 shows the connection between the end behavior of a rational function and its horizontal asymptote. We now generalize this relationship and summarize other features of the graph of a rational function:

Graph of a Rational Function

The graph of $y = f(x)/g(x) = (a_n x^n + \cdots)/(b_m x^m + \cdots)$ has the following characteristics:

1. **End behavior asymptote:**

   If $n < m$, the end behavior asymptote is the horizontal asymptote $y = 0$.

   If $n = m$, the end behavior asymptote is the horizontal asymptote $y = a_n/b_m$.

   If $n > m$, the end behavior asymptote is the quotient polynomial function $y = q(x)$, where $f(x) = g(x)q(x) + r(x)$. There is no horizontal asymptote.

2. **Vertical asymptotes:** These occur at the zeros of the denominator, provided that the zeros are not also zeros of the numerator of equal or greater multiplicity.

3. **x-intercepts:** These occur at the zeros of the numerator, which are not also zeros of the denominator.

4. **y-intercept:** This is the value of $f(0)$, if defined.

It is a good idea to find all of the asymptotes and intercepts when graphing a rational function. If the end behavior asymptote of a rational function is a slant line, we call it a **slant asymptote**, as illustrated in Example 4.

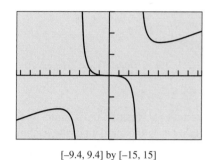

[–9.4, 9.4] by [–15, 15]

(a)

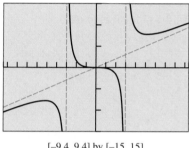

[–9.4, 9.4] by [–15, 15]

(b)

FIGURE 2.51 The graph of $f(x) = x^3/(x^2 - 9)$ (a) by itself and (b) with its asymptotes.

EXAMPLE 4   Graphing a Rational Function

Find the asymptotes and intercepts of the function $f(x) = x^3/(x^2 - 9)$ and graph the function.

SOLUTION The degree of the numerator is greater than the degree of the denominator, so the end behavior asymptote is the quotient polynomial. Using polynomial long division, we obtain

$$f(x) = \frac{x^3}{x^2 - 9} = x + \frac{9x}{x^2 - 9}.$$

So the quotient polynomial is $q(x) = x$, a slant asymptote. Factoring the denominator,

$$x^2 - 9 = (x - 3)(x + 3),$$

shows that the zeros of the denominator are $x = 3$ and $x = -3$. Consequently, $x = 3$ and $x = -3$ are the vertical asymptotes of f. The only zero of the numerator is 0, so $f(0) = 0$, and thus we see that the point $(0, 0)$ is the only x-intercept and the y-intercept of the graph of f.

The graph of f in Figure 2.51a passes through $(0, 0)$ and suggests the vertical asymptotes $x = 3$ and $x = -3$ and the slant asymptote $y = q(x) = x$. Figure 2.51b shows the graph of f with its asymptotes overlaid.

Now try Exercise 29.

Analyzing Graphs of Rational Functions

Because the degree of the numerator of the rational function in Example 5 is less than the degree of the denominator, we know that the graph of the function has $y = 0$ as a horizontal asymptote.

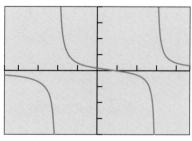

[–4.7, 4.7] by [–4, 4]

FIGURE 2.52 The graph of $f(x) = (x - 1)/(x^2 - x - 6)$. (Example 5)

EXAMPLE 5   Analyzing the Graph of a Rational Function

Find the intercepts, asymptotes, use limits to describe the behavior at the vertical asymptotes, and analyze and draw the graph of the rational function

$$f(x) = \frac{x - 1}{x^2 - x - 6}.$$

SOLUTION

The numerator is zero when $x = 1$, so the x-intercept is 1. Because $f(0) = 1/6$, the y-intercept is 1/6. The denominator factors as

$$x^2 - x - 6 = (x - 3)(x + 2),$$

so there are vertical asymptotes at $x = -2$ and $x = 3$. From the comment preceding this example we know that the horizontal asymptote is $y = 0$. Figure 2.52 supports this information and allows us to conclude that

$$\lim_{x \to -2^-} f(x) = -\infty, \ \lim_{x \to -2^+} f(x) = \infty, \ \lim_{x \to 3^-} f(x) = -\infty, \text{ and } \lim_{x \to 3^+} f(x) = \infty.$$

continued

Domain: $(-\infty, -2) \cup (-2, 3) \cup (3, \infty)$
Range: All reals
Continuity: All $x \neq -2, 3$
Decreasing on $(-\infty, -2)$, $(-2, 3)$, and $(3, \infty)$
Not symmetric
Unbounded
No local extrema
Horizontal asymptotes: $y = 0$
Vertical asymptotes: $x = -2$, $x = 3$
End behavior: $\lim\limits_{x \to -\infty} f(x) = \lim\limits_{x \to \infty} f(x) = 0$

Now try Exercise 39.

The degrees of the numerator and denominator of the rational function in Example 6 are equal. Thus, we know that the graph of the function has $y = 2$, the quotient of the leading coefficients, as its end-behavior asymptote.

EXAMPLE 6   Analyzing the Graph of a Rational Function

Find the intercepts, analyze, and draw the graph of the rational function

$$f(x) = \frac{2x^2 - 2}{x^2 - 4}.$$

SOLUTION The numerator factors as

$$2x^2 - 2 = 2(x^2 - 1) = 2(x + 1)(x - 1),$$

so the x-intercepts are -1 and 1. The y-intercept is $f(0) = 1/2$. The denominator factors as

$$x^2 - 4 = (x + 2)(x - 2),$$

so the vertical asymptotes are $x = -2$ and $x = 2$. From the comments preceding Example 6 we know that $y = 2$ is the horizontal asymptote. Figure 2.53 supports this information and allows us to conclude that

$$\lim_{x \to -2^-} f(x) = \infty, \lim_{x \to -2^+} f(x) = -\infty, \lim_{x \to 2^-} f(x) = -\infty, \lim_{x \to 2^+} f(x) = \infty.$$

Domain: $(-\infty, -2) \cup (-2, 2) \cup (2, \infty)$
Range: $(-\infty, 1/2] \cup (2, \infty)$
Continuity: All $x \neq -2, 2$
Increasing on $(-\infty, -2)$ and $(-2, 0]$; decreasing on $[0, 2)$ and $(2, \infty)$
Symmetric with respect to the y-axis (an even function)
Unbounded
Local maximum of $1/2$ at $x = 0$
Horizontal asymptotes: $y = 2$
Vertical asymptotes: $x = -2, x = 2$
End behavior: $\lim\limits_{x \to -\infty} f(x) = \lim\limits_{x \to \infty} f(x) = 2$

Now try Exercise 41.

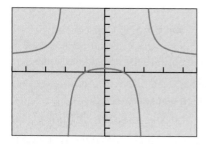

[−4.7, 4.7] by [−8, 8]

FIGURE 2.53 The graph of $f(x) = (2x^2 - 2)/(x^2 - 4)$. It can be shown that f takes on no value between $1/2$, the y-intercept, and 2, the horizontal asymptote. (Example 6)

In Examples 7 and 8 we will investigate the rational function

$$f(x) = \frac{x^3 - 3x^2 + 3x + 1}{x - 1}.$$

The degree of the numerator of f exceeds the degree of the denominator by 2. Thus, there is no horizontal asymptote. We will see that the end behavior asymptote is a polynomial of degree 2.

EXAMPLE 7 Finding an End-Behavior Asymptote

Find the end-behavior asymptote of

$$f(x) = \frac{x^3 - 3x^2 + 3x + 1}{x - 1}$$

and graph it together with f in two windows:

(a) one showing the details around the vertical asymptote of f,

(b) one showing a graph of f that resembles its end-behavior asymptote.

SOLUTION The graph of f has a vertical asymptote at $x = 1$. Divide $x^3 - 3x^2 + 3x + 1$ by $x - 1$ to show that

$$f(x) = \frac{x^3 - 3x^2 + 3x + 1}{x - 1} = x^2 - 2x + 1 + \frac{2}{x - 1}.$$

The end behavior asymptote of f is $y = x^2 - 2x + 1$.

(a) The graph of f in Figure 2.54 shows the details around the vertical asymptote. We have also overlaid the graph of its end-behavior asymptote as a dashed line.

(b) If we draw the graph of $f(x) = (x^3 - 3x^2 + 3x + 1)/(x - 1)$ and its end-behavior asymptote $y = x^2 - 2x + 1$ in a large enough viewing window the two graphs will appear to be identical. See Figure 2.55.

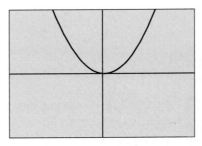

[–40, 40] by [–500, 500]

FIGURE 2.55 The graphs of $f(x) = (x^3 - 3x^2 + 3x + 1)/(x - 1)$ and its end-behavior asymptote $y = x^2 - 2x + 1$ appear to be identical. (Example 7)

Now try Exercise 47.

EXAMPLE 8 Analyzing the Graph of a Rational Function

Find the intercepts, analyze, and draw the graph of the rational function

$$f(x) = \frac{x^3 - 3x^2 + 3x + 1}{x - 1}.$$

continued

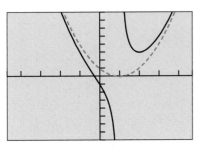

[–4.7, 4.7] by [–8, 8]

FIGURE 2.54 The graph of $f(x) = (x^3 - 3x^2 + 3x + 1)/(x - 1)$ as a solid black line and its end-behavior asymptote $y = x^2 - 2x + 1$ as a dashed blue line. (Example 7)

SOLUTION

f has only one *x*-intercept and we can use the graph of *f* in Figure 2.54 to show that it is about -0.26. The *y*-intercept is $f(0) = -1$. The vertical asymptote is $x = 1$ as we have seen. We know that the graph of *f* does not have a horizontal asymptote, and from Example 7 we know that the end-behavior asymptote is $y = x^2 - 2x + 1$. We can also use Figure 2.54 to show that *f* has a local minimum of 3 at $x = 2$. Figure 2.54 supports this information and allows us to conclude that

$$\lim_{x \to 1^-} f(x) = -\infty \text{ and } \lim_{x \to 1^+} f(x) = \infty.$$

Domain: All $x \neq 1$
Range: All reals
Continuity: All $x \neq 1$
Decreasing on $(-\infty, 1)$ and $(1, 2]$; increasing on $[2, \infty)$
Not symmetric
Unbounded
Local minimum of 3 at $x = 2$
No horizontal asymptotes; end-behavior asymptote: $y = x^2 - 2x + 1$
Vertical asymptote: $x = 1$
End behavior: $\lim_{x \to -\infty} f(x) = \lim_{x \to \infty} f(x) = \infty$ ***Now try Exercise 55.***

Exploring Relative Humidity

The phrase *relative humidity* is familiar from everyday weather reports. **Relative humidity** is the ratio of constant vapor pressure to saturated vapor pressure. So, relative humidity is inversely proportional to saturated vapor pressure.

CHAPTER OPENER PROBLEM (from page 169)

PROBLEM: Determine the relative humidity values that correspond to the saturated vapor pressures of 12, 24, 36, 48, and 60 millibars, at a constant vapor pressure of 12 millibars. (In practice, saturated vapor pressure increases as the temperature increases.)

SOLUTION: Relative humidity (RH) is found by dividing constant vapor pressure (CVP) by saturated vapor pressure (SVP). So, for example, for SVP = 24 millibars and CVP = 12 millibars, RH = 12/24 = 1/2 = 0.5 = 50%. See the table below, which is based on CVP = 12 millibars with increasing temperature.

SVP (millibars)	RH (%)
12	100
24	50
36	$33.\overline{3}$
48	25
60	20

QUICK REVIEW 2.6 *(For help, go to Section 2.4.)*

In Exercises 1–6, use factoring to find the real zeros of the function.

1. $f(x) = 2x^2 + 5x - 3$ **2.** $f(x) = 3x^2 - 2x - 8$

3. $g(x) = x^2 - 4$ **4.** $g(x) = x^2 - 1$

5. $h(x) = x^3 - 1$ **6.** $h(x) = x^2 + 1$

In Exercises 7–10, find the quotient and remainder when $f(x)$ is divided by $d(x)$.

7. $f(x) = 2x + 1, d(x) = x - 3$

8. $f(x) = 4x + 3, d(x) = 2x - 1$

9. $f(x) = 3x - 5, d(x) = x$

10. $f(x) = 5x - 1, d(x) = 2x$

SECTION 2.6 EXERCISES

In Exercises 1–4, find the domain of the function f. Use limits to describe the behavior of f at value(s) of x not in its domain.

1. $f(x) = \dfrac{1}{x + 3}$ **2.** $f(x) = \dfrac{-3}{x - 1}$

3. $f(x) = \dfrac{-1}{x^2 - 4}$ **4.** $f(x) = \dfrac{2}{x^2 - 1}$

In Exercises 5–10, describe how the graph of the given function can be obtained by transforming the graph of the reciprocal function $g(x) = 1/x$. Identify the horizontal and vertical asymptotes and use limits to describe the corresponding behavior. Sketch the graph of the function.

5. $f(x) = \dfrac{1}{x - 3}$ **6.** $f(x) = -\dfrac{2}{x + 5}$

7. $f(x) = \dfrac{2x - 1}{x + 3}$ **8.** $f(x) = \dfrac{3x - 2}{x - 1}$

9. $f(x) = \dfrac{5 - 2x}{x + 4}$ **10.** $f(x) = \dfrac{4 - 3x}{x - 5}$

In Exercises 11–14, evaluate the limit based on the graph of f shown.

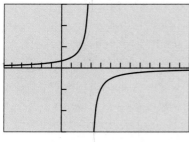

[–5.8, 13] by [–3, 3]

11. $\lim_{x \to 3^-} f(x)$ **12.** $\lim_{x \to 3^+} f(x)$

13. $\lim_{x \to \infty} f(x)$ **14.** $\lim_{x \to -\infty} f(x)$

In Exercises 15–18, evaluate the limit based on the graph of f shown.

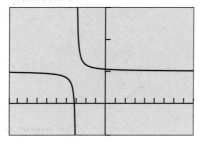

[–9.8, 9] by [–5, 15]

15. $\lim_{x \to -3^+} f(x)$ **16.** $\lim_{x \to -3^-} f(x)$

17. $\lim_{x \to -\infty} f(x)$ **18.** $\lim_{x \to \infty} f(x)$

In Exercises 19–22, find the horizontal and vertical asymptotes of $f(x)$. Use limits to describe the corresponding behavior.

19. $f(x) = \dfrac{2x^2 - 1}{x^2 + 3}$ **20.** $f(x) = \dfrac{3x^2}{x^2 + 1}$

21. $f(x) = \dfrac{2x + 1}{x^2 - x}$ **22.** $f(x) = \dfrac{x - 3}{x^2 + 3x}$

In Exercises 23–30, find the asymptotes and intercepts of the function, and graph the function.

23. $g(x) = \dfrac{x - 2}{x^2 - 2x - 3}$ **24.** $g(x) = \dfrac{x + 2}{x^2 + 2x - 3}$

25. $h(x) = \dfrac{2}{x^3 - x}$ **26.** $h(x) = \dfrac{3}{x^3 - 4x}$

27. $f(x) = \dfrac{2x^2 + x - 2}{x^2 - 1}$ **28.** $g(x) = \dfrac{-3x^2 + x + 12}{x^2 - 4}$

29. $f(x) = \dfrac{x^2 - 2x + 3}{x + 2}$ **30.** $g(x) = \dfrac{x^2 - 3x - 7}{x + 3}$

In Exercises 31–36, match the rational function with its graph. Identify the viewing window and the scale used on each axis.

31. $f(x) = \dfrac{1}{x - 4}$ **32.** $f(x) = -\dfrac{1}{x + 3}$

33. $f(x) = 2 + \dfrac{3}{x - 1}$ **34.** $f(x) = 1 + \dfrac{1}{x + 3}$

35. $f(x) = -1 + \dfrac{1}{4 - x}$ **36.** $f(x) = 3 - \dfrac{2}{x - 1}$

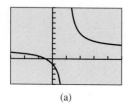

(a)

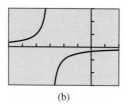

(b)

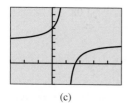

(c)

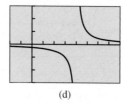

(d)

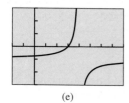

(e)

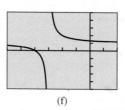

(f)

In Exercises 37–44, find the intercepts, asymptotes, use limits to describe the behavior at the vertical asymptotes, and analyze and draw the graph of the given rational function.

37. $f(x) = \dfrac{2}{2x^2 - x - 3}$ **38.** $g(x) = \dfrac{2}{x^2 + 4x + 3}$

39. $h(x) = \dfrac{x - 1}{x^2 - x - 12}$ **40.** $k(x) = \dfrac{x + 1}{x^2 - 3x - 10}$

41. $f(x) = \dfrac{x^2 + x - 2}{x^2 - 9}$ **42.** $g(x) = \dfrac{x^2 - x - 2}{x^2 - 2x - 8}$

43. $h(x) = \dfrac{x^2 + 2x - 3}{x + 2}$ **44.** $k(x) = \dfrac{x^2 - x - 2}{x - 3}$

In Exercises 45–50, find the end-behavior asymptote of the given rational function f and graph it together with f in two windows:

(a) One showing the details around the vertical asymptote(s) of f.

(b) One showing a graph of f that resembles its end-behavior asymptote.

45. $f(x) = \dfrac{x^2 - 2x + 3}{x - 5}$ **46.** $f(x) = \dfrac{2x^2 + 2x - 3}{x + 3}$

47. $f(x) = \dfrac{x^3 - x^2 + 1}{x + 2}$ **48.** $f(x) = \dfrac{x^3 + 1}{x - 1}$

49. $f(x) = \dfrac{x^4 - 2x + 1}{x - 2}$ **50.** $f(x) = \dfrac{x^5 + 1}{x^2 + 1}$

In Exercises 51–56, find the intercepts, analyze, and draw the graph of the given rational funciton.

51. $f(x) = \dfrac{3x^2 - 2x + 4}{x^2 - 4x + 5}$ **52.** $g(x) = \dfrac{4x^2 + 2x}{x^2 - 4x + 8}$

53. $h(x) = \dfrac{x^3 - 1}{x - 2}$ **54.** $k(x) = \dfrac{x^3 - 2}{x + 2}$

55. $f(x) = \dfrac{x^3 - 2x^2 + x - 1}{2x - 1}$ **56.** $g(x) = \dfrac{2x^3 - 2x^2 - x + 5}{x - 2}$

In Exercises 57–62, find the intercepts, asymptotes, end-behavior asymptote, and graph the function together with its end-behavior asymptote.

57. $h(x) = \dfrac{x^4 + 1}{x + 1}$ **58.** $k(x) = \dfrac{2x^5 + x^2 - x + 1}{x^2 - 1}$

59. $f(x) = \dfrac{x^5 - 1}{x + 2}$ **60.** $g(x) = \dfrac{x^5 + 1}{x - 1}$

61. $h(x) = \dfrac{2x^3 - 3x + 2}{x^3 - 1}$ **62.** $k(x) = \dfrac{3x^3 + x - 4}{x^3 + 1}$

Standardized Test Questions

63. True or False A rational function must have a vertical asymptote. Justify your answer.

64. True or False $f(x) = \dfrac{x^2 - x}{\sqrt{x^2 + 4}}$ is a rational function. Justify your answer.

In Exercises 65–68, you may use a graphing calculator to solve the problem.

65. Multiple Choice Let $f(x) = \dfrac{-2}{x^2 + 3x}$. What values of x have to be excluded from the domain of f.

(A) only 0   **(B)** only 3   **(C)** only −3

(D) only 0, 3   **(E)** only 0, −3

66. Multiple Choice Let $g(x) = \dfrac{2}{x + 3}$. Which of the transformations of $f(x) = \dfrac{2}{x}$ produce the graph of g?

(A) Translate the graph of f left 3 units.

(B) Translate the graph of f right 3 units.

(C) Translate the graph of f down 3 units.

(D) Translate the graph of f up 3 units.

(E) Vertically stretch the graph of f by a factor of 2.

67. Multiple Choice Let $f(x) = \dfrac{x^2}{x+5}$. Which of the following statements is true about the graph of f?

(A) There is no vertical asymptote.

(B) There is a horizontal asymptote but no vertical asymptote.

(C) There is a slant asymptote but no vertical asymptote.

(D) There is a vertical asymptote and a slant asymptote.

(E) There is a vertical and horizontal asymptote.

68. Multiple Choice What is the degree of the end-behavior asymptote of $f(x) = \dfrac{x^8+1}{x^4+1}$?

(A) 0   (B) 1   (C) 2   (D) 3   (E) 4

Explorations

69. Group Activity *Work in groups of two.* Compare the functions

$f(x) = \dfrac{x^2-9}{x-3}$ and $g(x) = x + 3$.

(a) Are the domains equal?

(b) Does f have a vertical asymptote? Explain.

(c) Explain why the graphs appear to be identical.

(d) Are the functions identical?

70. Group Activity Explain why the functions are identical or not. Include the graphs and a comparison of the functions' asymptotes, intercepts, and domain.

(a) $f(x) = \dfrac{x^2+x-2}{x-1}$ and $g(x) = x + 2$

(b) $f(x) = \dfrac{x^2-1}{x+1}$ and $g(x) = x - 1$

(c) $f(x) = \dfrac{x^2-1}{x^3-x^2-x+1}$ and $g(x) = \dfrac{1}{x-1}$

(d) $f(x) = \dfrac{x-1}{x^2+x-2}$ and $g(x) = \dfrac{1}{x+2}$

71. Boyle's Law This ideal gas law states that the volume of an enclosed gas at a fixed temperature varies inversely as the pressure.

(a) **Writing to Learn** Explain why Boyle's Law yields both a rational function model and a power function model.

(b) Which power functions are also rational functions?

(c) If the pressure of a 2.59-L sample of nitrogen gas at 291°K is 0.866 atm, what would the volume be at a pressure of 0.532 atm if the temperature does not change?

72. Light Intensity Aileen and Malachy gathered the data in Table 2.19 using a 75-watt light bulb and a Calculator-Based Laboratory™ (CBL™) with a light-intensity probe.

(a) Draw a scatter plot of the data in Table 2.19.

(b) Find an equation for the data assuming it has the form $f(x) = k/x^2$ for some constant k. Explain your method for choosing k.
    5.81

(c) Superimpose the regression curve on the scatter plot.

(d) Use the regression model to predict the light intensity at distances of 2.2 m and 4.4 m.

Table 2.19 Light Intensity Data for a 75-W Light Bulb	
Distance (m)	Intensity (W/m²)
1.0	6.09
1.5	2.51
2.0	1.56
2.5	1.08
3.0	0.74

Extending the Ideas

In Exercises 73–76, graph the function. Express the function as a piecewise-defined function without absolute value, and use the result to confirm the graph's asymptotes and intercepts algebraically.

73. $h(x) = \dfrac{2x-3}{|x|+2}$ **74.** $h(x) = \dfrac{3x+5}{|x|+3}$

75. $f(x) = \dfrac{5-3x}{|x|+4}$ **76.** $f(x) = \dfrac{2-2x}{|x|+1}$

77. Describe how the graph of a nonzero rational function

$$f(x) = \dfrac{ax+b}{cx+d}, c \neq 0$$

can be obtained from the graph of $y = 1/x$. (*Hint:* Use long division.)

78. Writing to Learn Let $f(x) = 1 + 1/(x - 1/x)$ and $g(x) = (x^3 + x^2 - x)/(x^3 - x)$. Does $f = g$? Support your answer by making a comparative analysis of all of the features of f and g, including asymptotes, intercepts, and domain.

2.7
Solving Equations in One Variable

What you'll learn about

- Solving Rational Equations
- Extraneous Solutions
- Applications

... and why

Applications involving rational functions as models often require that an equation involving fractions be solved.

Solving Rational Equations

Equations involving rational expressions or fractions (see Appendix A.3) are **rational equations**. Every rational equation can be written in the form

$$\frac{f(x)}{g(x)} = 0.$$

If $f(x)$ and $g(x)$ are polynomial functions with no common factors, then the zeros of $f(x)$ are the solutions of the equation.

Usually it is not necessary to put a rational equation in the form of $f(x)/g(x)$. To solve an equation involving fractions we begin by finding the LCD (least common denominator) of all the terms of the equation. Then we clear the equation of fractions by multiplying each side of the equation by the LCD. Sometimes the LCD contains variables.

When we multiply or divide an equation by an expression containing variables, the resulting equation may have solutions that are *not* solutions of the original equation. These are **extraneous solutions**. For this reason we must check each solution of the resulting equation in the original equation.

EXAMPLE 1   Solving by Clearing Fractions

Solve $x + \dfrac{3}{x} = 4$.

SOLUTION

Solve Algebraically

The LCD is x.

$$x + \frac{3}{x} = 4.$$

$$x^2 + 3 = 4x \qquad \text{Multiply by } x.$$

$$x^2 - 4x + 3 = 0 \qquad \text{Subtract } 4x.$$

$$(x - 1)(x - 3) = 0 \qquad \text{Factor.}$$

$$x - 1 = 0 \quad \text{or} \quad x - 3 = 0 \qquad \text{Zero factor property}$$

$$x = 1 \quad \text{or} \quad x = 3$$

Confirm Algebraically

For $x = 1$, $x + \dfrac{3}{x} = 1 + \dfrac{3}{1} = 4$, and for $x = 3$, $x + \dfrac{3}{x} = 3 + \dfrac{3}{3} = 4$.

Each value is a solution of the original equation.   *Now try Exercise 1.*

GRAPHER NOTE

The graph in Figure 2.56 contains a pseudo-vertical asymptote, which we should ignore.

When the fractions in Example 2 are cleared, we obtain a quadratic equation.

EXAMPLE 2   Solving a Rational Equation

Solve $x + \dfrac{1}{x - 4} = 0$.

SOLUTION

Solve Algebraically The LCD is $x - 4$.

$$x + \frac{1}{x - 4} = 0.$$

$$x(x - 4) + \frac{x - 4}{x - 4} = 0 \qquad \text{Multiply by } x - 4.$$

$$x^2 - 4x + 1 = 0 \qquad \text{Distributive property}$$

$$x = \frac{4 \pm \sqrt{(-4)^2 - 4(1)(1)}}{2(1)} \qquad \text{Quadratic formula}$$

$$x = \frac{4 \pm 2\sqrt{3}}{2} \qquad \text{Simplify.}$$

$$x = 2 \pm \sqrt{3} \qquad \text{Simplify.}$$

$$x \approx 0.268, 3.732$$

Solve Graphically

The graph in Figure 2.56 suggests that the function $y = x + 1/(x - 4)$ has two zeros. We can use the graph to find that the zeros are about 0.268 and 3.732, agreeing with the values found algebraically.   ***Now try Exercise 7.***

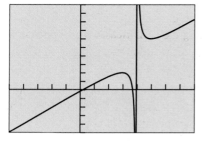

[-5, 8] by [-5, 10]

FIGURE 2.56 The graph of $y = x + 1/(x - 4)$. (Example 2)

Extraneous Solutions

We will find extraneous solutions in Examples 3 and 4.

EXAMPLE 3   Eliminating Extraneous Solutions

Solve the equation

$$\frac{2x}{x - 1} + \frac{1}{x - 3} = \frac{2}{x^2 - 4x + 3}.$$

SOLUTION

Solve Algebraically

The denominator of the right-hand side, $x^2 - 4x + 3$, factors into $(x - 1)(x - 3)$. So the least common denominator (LCD) of the equation is $(x - 1)(x - 3)$, and we multiply both sides of the equation by this LCD:

$$(x - 1)(x - 3)\left(\frac{2x}{x - 1} + \frac{1}{x - 3}\right) = (x - 1)(x - 3)\left(\frac{2}{x^2 - 4x + 3}\right)$$

$$2x(x - 3) + (x - 1) = 2 \qquad \text{Distributive property}$$

$$2x^2 - 5x - 3 = 0 \qquad \text{Distributive property}$$

$$(2x + 1)(x - 3) = 0 \qquad \text{Factor.}$$

$$x = -\frac{1}{2} \quad \text{or} \quad x = 3$$

continued

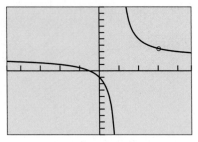

[–4.7, 4.7] by [–10, 10]

FIGURE 2.57 The graph of $f(x) =$ $2x/(x - 1) + 1/(x - 3) - 2/(x^2 - 4x + 3)$. (Example 3)

Confirm Numerically

We replace x by $-1/2$ in the original equation:

$$\frac{2(-1/2)}{(-1/2) - 1} + \frac{1}{(-1/2) - 3} \overset{?}{=} \frac{2}{(-1/2)^2 - 4(-1/2) + 3}$$

$$\frac{2}{3} - \frac{2}{7} \overset{?}{=} \frac{8}{21}$$

The equation is true, so $x = -1/2$ is a valid solution. The original equation is not defined for $x = 3$, so $x = 3$ is an extraneous solution.

Support Graphically

The graph of

$$f(x) = \frac{2x}{x - 1} + \frac{1}{x - 3} - \frac{2}{x^2 - 4x + 3}$$

in Figure 2.57 suggests that $x = -1/2$ is an x-intercept and $x = 3$ is not.

Interpret

Only $x = -1/2$ is a solution.   *Now try Exercise 13.*

We will see that Example 4 has no solutions.

EXAMPLE 4   Eliminating Extraneous Solutions

Solve

$$\frac{x - 3}{x} + \frac{3}{x + 2} + \frac{6}{x^2 + 2x} = 0.$$

SOLUTION   The LCD is $x(x + 2)$.

$$\frac{x - 3}{x} + \frac{3}{x + 2} + \frac{6}{x^2 + 2x} = 0$$

$$(x - 3)(x + 2) + 3x + 6 = 0 \qquad \text{Multiply by } x(x+2).$$

$$x^2 - x - 6 + 3x + 6 = 0 \qquad \text{Expand.}$$

$$x^2 + 2x = 0 \qquad \text{Simplify.}$$

$$x(x + 2) = 0 \qquad \text{Factor.}$$

$$x = 0 \quad \text{or} \quad x = -2$$

Substituting $x = 0$ or $x = -2$ into the original equation results in division by zero. So both of these numbers are extraneous solutions and the original equation has no solution.
                                                           Now try Exercise 17.

Applications

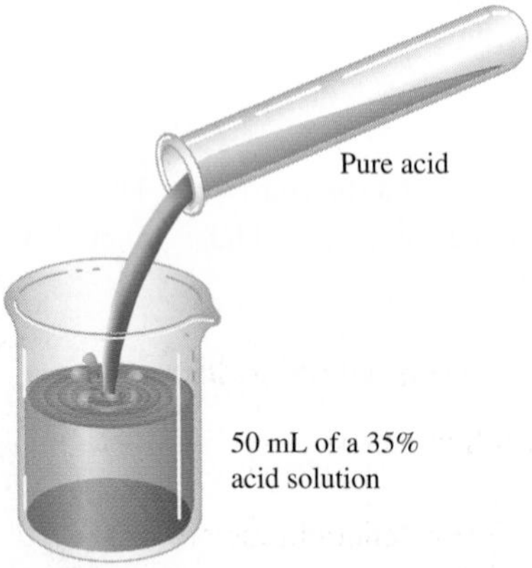

EXAMPLE 5   Calculating Acid Mixtures

How much pure acid must be added to 50 mL of a 35% acid solution to produce a mixture that is 75% acid? (See Figure 2.58.)   *continued*

Pure acid

50 mL of a 35% acid solution

FIGURE 2.58 Mixing solutions. (Example 5)

SOLUTION

Model

Word statement: $\dfrac{\text{mL of pure acid}}{\text{mL of mixture}}$ = concentration of acid

0.35×50 or 17.5 = mL of pure acid in 35% solution

x = mL of acid added

$x + 17.5$ = mL of pure acid in resulting mixture

$x + 50$ = mL of the resulting mixture

$\dfrac{x + 17.5}{x + 50}$ = concentration of acid

Solve Graphically

$$\dfrac{x + 17.5}{x + 50} = 0.75 \qquad \text{\small Equation to be solved}$$

Figure 2.59 shows graphs of $f(x) = (x + 17.5)/(x + 50)$ and $g(x) = 0.75$. The point of intersection is $(80, 0.75)$.

Interpret

We need to add 80 mL of pure acid to the 35% acid solution to make a solution that is 75% acid. ***Now try Exercise 31.***

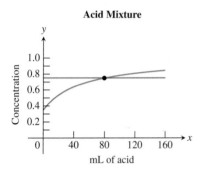

Acid Mixture

Intersection: $x = 80$; $y = .75$

FIGURE 2.59 The graphs of $f(x) = (x + 17.5)/(x + 50)$ and $g(x) = 0.75$ (Example 5)

EXAMPLE 6 Finding a Minimum Perimeter

Find the dimensions of the rectangle with minimum perimeter if its area is 200 square meters. Find this least perimeter.

SOLUTION

Model

$$A = 200 = x\left(\dfrac{200}{x}\right)$$

FIGURE 2.60 A rectangle with area 200 m². (Example 6)

Word Statement: Perimeter $= 2 \times$ length $+ 2 \times$ width

x = width in meters

$\dfrac{200}{x} = \dfrac{\text{area}}{\text{width}}$ = length in meters

Function to be minimized: $P(x) = 2x + 2\left(\dfrac{200}{x}\right) = 2x + \dfrac{400}{x}$

Minimum
X=14.142136   Y=56.568542

[0, 50] by [0, 150]

FIGURE 2.61 A graph of $P(x) = 2x + 400/x$. (Example 6)

continued

FIGURE 2.62 A tomato juice can. (Example 7)

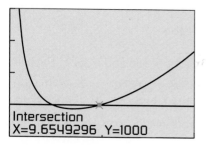

Intersection
X=9.6549296 Y=1000

[0, 20] by [0, 4000]

FIGURE 2.63 (Example 7)

The graph of P in Figure 2.61 shows a minimum of approximately 56.57, occurring when $x \approx 14.14$.

Interpret

The width is about 14.14 m, and the minimum perimeter is about 56.57 m. Because $200/14.14 \approx 14.14$, the dimensions of the rectangle with minimum perimeter are 14.14 m by 14.14 m, a square.   *Now try Exercise 35.*

EXAMPLE 7   Designing a Juice Can

Stewart Cannery will package tomato juice in 2-liter cylindrical cans. Find the radius and height of the cans if the cans have a surface area of 1000 cm². (See Figure 2.62.)

SOLUTION

Model

$$S = \text{surface area of can in cm}^2$$
$$r = \text{radius of can in centimeters}$$
$$h = \text{height of can in centimeters}$$

Using volume (V) and surface area (S) formulas and the fact that $1 \text{ L} = 1000 \text{ cm}^3$, we conclude that

$$V = \pi r^2 h = 2000 \quad \text{and} \quad S = 2\pi r^2 + 2\pi rh = 1000.$$

So

$$2\pi r^2 + 2\pi rh = 1000$$

$$2\pi r^2 + 2\pi r \left(\frac{2000}{\pi r^2} \right) = 1000 \qquad \text{Substitute } h = 2000/(\pi r^2).$$

$$2\pi r^2 + \frac{4000}{r} = 1000 \qquad \text{Equation to be solved}$$

Solve Graphically

Figure 2.63 shows the graphs of $f(x) = 2\pi r^2 + 4000/r$ and $g(x) = 1000$. One point of intersection occurs when r is approximately 9.65. A second point of intersection occurs when r is approximately 4.62.

Because $h = 2000/(\pi r^2)$, the corresponding values for h are

$$h = \frac{2000}{\pi (4.619\ldots)^2} \approx 29.83 \text{ and } h = \frac{2000}{\pi (9.654\ldots)^2} \approx 6.83.$$

Interpret

With a surface area of 1000 cm², the cans either have a radius of 4.62 cm and a height of 29.83 cm or have a radius of 9.65 cm and a height of 6.83 cm.

Now try Exercise 37.

QUICK REVIEW 2.7 *(For help, go to Sections A.3. and P.5.)*

In Exercises 1 and 2, find the missing numerator or denominator.

1. $\dfrac{2x}{x-3} = \dfrac{?}{x^2 + x - 12}$ **2.** $\dfrac{x-1}{x+1} = \dfrac{x^2 - 1}{?}$

In Exercises 3–6, find the LCD and rewrite the expression as a single fraction reduced to lowest terms.

3. $\dfrac{5}{12} + \dfrac{7}{18} - \dfrac{5}{6}$ **4.** $\dfrac{3}{x-1} - \dfrac{1}{x}$

5. $\dfrac{x}{2x+1} - \dfrac{2}{x-3}$

6. $\dfrac{x+1}{x^2 - 5x + 6} - \dfrac{3x+11}{x^2 - x - 6}$

In Exercises 7–10, use the quadratic formula to find the zeros of the quadratic polynomials.

7. $2x^2 - 3x - 1$ **8.** $2x^2 - 5x - 1$

9. $3x^2 + 2x - 2$ **10.** $x^2 - 3x - 9$

SECTION 2.7 EXERCISES

In Exercises 1–6, solve the equation algebraically. Support your answer numerically and identify any extraneous solutions.

1. $\dfrac{x-2}{3} + \dfrac{x+5}{3} = \dfrac{1}{3}$ **2.** $x + 2 = \dfrac{15}{x}$

3. $x + 5 = \dfrac{14}{x}$ **4.** $\dfrac{1}{x} - \dfrac{2}{x-3} = 4$

5. $x + \dfrac{4x}{x-3} = \dfrac{12}{x-3}$ **6.** $\dfrac{3}{x-1} + \dfrac{2}{x} = 8$

In Exercises 7–12, solve the equation algebraically and graphically. Check for extraneous solutions.

7. $x + \dfrac{10}{x} = 7$ **8.** $x + 2 = \dfrac{15}{x}$

9. $x + \dfrac{12}{x} = 7$ **10.** $x + \dfrac{6}{x} = -7$

11. $2 - \dfrac{1}{x+1} = \dfrac{1}{x^2 + x}$ **12.** $2 - \dfrac{3}{x+4} = \dfrac{12}{x^2 + 4x}$

In Exercises 13–18, solve the equation algebraically. Check for extraneous solutions. Support your answer graphically.

13. $\dfrac{3x}{x+5} + \dfrac{1}{x-2} = \dfrac{7}{x^2 + 3x - 10}$

14. $\dfrac{4x}{x+4} + \dfrac{3}{x-1} = \dfrac{15}{x^2 + 3x - 4}$

15. $\dfrac{x-3}{x} - \dfrac{3}{x+1} + \dfrac{3}{x^2 + x} = 0$

16. $\dfrac{x+2}{x} - \dfrac{4}{x-1} + \dfrac{2}{x^2 - x} = 0$

17. $\dfrac{3}{x+2} + \dfrac{6}{x^2 + 2x} = \dfrac{3-x}{x}$

18. $\dfrac{x+3}{x} - \dfrac{2}{x+3} = \dfrac{6}{x^2 + 3x}$

In Exercises 19–22, two possible solutions to the equation $f(x) = 0$ are listed. Use the given graph of $y = f(x)$ to decide which, if any, are extraneous.

19. $x = -5$ or $x = -2$

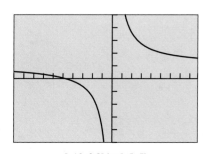

[−10, 8.8] by [−5, 5]

20. $x = -2$ or $x = 3$

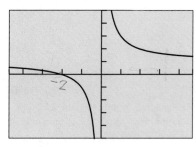

[−4.7, 4.7] by [−5, 5]

21. $x = -2$ or $x = 2$

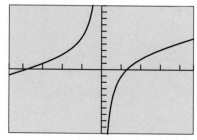

[–4.7, 4.7] by [–10, 10]

22. $x = 0$ or $x = 3$

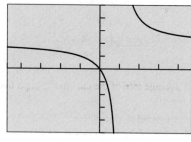

[–4.7, 4.7] by [–5, 5]

In Exercises 23–30, solve the equation.

23. $\dfrac{2}{x-1} + x = 5$

24. $\dfrac{x^2 - 6x + 5}{x^2 - 2} = 3$

25. $\dfrac{x^2 - 2x + 1}{x + 5} = 0$

26. $\dfrac{3x}{x+2} + \dfrac{2}{x-1} = \dfrac{5}{x^2 + x - 2}$

27. $\dfrac{4x}{x+4} + \dfrac{5}{x-1} = \dfrac{15}{x^2 + 3x - 4}$

28. $\dfrac{3x}{x+1} + \dfrac{5}{x-2} = \dfrac{15}{x^2 - x - 2}$

29. $x^2 + \dfrac{5}{x} = 8$

30. $x^2 - \dfrac{3}{x} = 7$

31. Acid Mixture Suppose that x mL of pure acid are added to 125 mL of a 60% acid solution. How many mL of pure acid must be added to obtain a solution of 83% acid?

(a) Explain why the concentration $C(x)$ of the new mixture is

$$C(x) = \dfrac{x + 0.6(125)}{x + 125}.$$

(b) Suppose the viewing window in the figure is used to find a solution to the problem. What is the equation of the horizontal line?

(c) Writing to Learn Write and solve an equation that answers the question of this problem. Explain your answer.

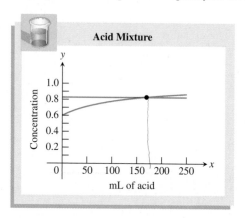

32. Acid Mixture Suppose that x mL of pure acid are added to 100 mL of a 35% acid solution.

(a) Express the concentration $C(x)$ of the new mixture as a function of x.

(b) Use a graph to determine how much pure acid should be added to the 35% solution to produce a new solution that is 75% acid.

(c) Solve (b) algebraically.

33. Breaking Even Mid Town Sports Apparel, Inc., has found that it needs to sell golf hats for $2.75 each in order to be competitive. It costs $2.12 to produce each hat, and it has weekly overhead costs of $3000.

(a) Let x be the number of hats produced each week. Express the average cost (including overhead costs) of producing one hat as a function of x.

(b) Solve algebraically to find the number of golf hats that must be sold each week to make a profit. Support your answer graphically.

(c) Writing to Learn How many golf hats must be sold to make a profit of $1000 in 1 week? Explain your answer.

34. Bear Population The number of bears at any time t (in years) in a federal game reserve is given by

$$P(t) = \dfrac{500 + 250t}{10 + 0.5t}.$$

(a) Find the population of bears when the value of t is 10, 40, and 100.

(b) Does the graph of the bear population have a horizontal asymptote? If so, what is it? If not, why not?

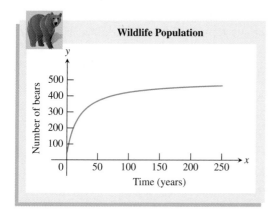

Wildlife Population

(c) **Writing to Learn** According to this model, what is the largest the bear population can become? Explain your answer.

35. **Minimizing Perimeter** Consider all rectangles with an area of 182 ft². Let x be the length of one side of such a rectangle.

(a) Express the perimeter P as a function of x.

(b) Find the dimensions of the rectangle that has the least perimeter. What is the least perimeter?

36. **Group Activity Page Design** Hendrix Publishing Co. wants to design a page that has a 0.75-in. left border, a 1.5-in. top border, and borders on the right and bottom of 1-in. They are to surround 40 in.² of print material. Let x be the width of the print material.

(a) Express the area of the page as a function of x.

(b) Find the dimensions of the page that has the least area. What is the least area?

37. **Industrial Design** Drake Cannery will pack peaches in 0.5-L cylindrical cans. Let x be the radius of the can in cm.

(a) Express the surface area S of the can as a function of x.

(b) Find the radius and height of the can if the surface area is 900 cm².

38. **Group Activity Designing a Swimming Pool** Thompson Recreation, Inc., wants to build a rectangular swimming pool with the top of the pool having surface area 1000 ft². The pool is required to have a walk of uniform width 2 ft surrounding it. Let x be the length of one side of the pool.

(a) Express the area of the plot of land needed for the pool and surrounding sidewalk as a function of x.

(b) Find the dimensions of the plot of land that has the least area. What is the least area?

39. **Resistors** The total electrical resistance R of two resistors connected in parallel with resistances R_1 and R_2 is given by

$$\frac{1}{R} = \frac{1}{R_1} + \frac{1}{R_2}$$

One resistor has a resistance of 2.3 ohms. Let x be the resistance of the second resistor.

(a) Express the total resistance R as a function of x.

(b) Find the resistance of the second resistor if the total resistance of the pair is 1.7 ohms.

40. **Designing Rectangles** Consider all rectangles with an area of 200 m². Let x be the length of one side of such a rectangle.

(a) Express the perimeter P as a function of x.

(b) Find the dimensions of a rectangle whose perimeter is 70 m.

41. **Swimming Pool Drainage** Drains A and B are used to empty a swimming pool. Drain A alone can empty the pool in 4.75 h. Let t be the time it takes for drain B alone to empty the pool.

(a) Express as a function of t the part D of the drainage that can be done in 1 h with both drains open at the same time.

(b) Find graphically the time it takes for drain B alone to empty the pool if both drains, when open at the same time, can empty the pool in 2.6 h. Confirm algebraically.

42. **Time-Rate Problem** Josh rode his bike 17 mi from his home to Columbus, and then traveled 53 mi by car from Columbus to Dayton. Assume that the average rate of the car was 43 mph faster than the average rate of the bike.

(a) Express the total time required to complete the 70-mi trip (bike and car) as a function of the rate x of the bike.

(b) Find graphically the rate of the bike if the total time of the trip was 1 h 40 min. Confirm algebraically.

43. **Fast Food Sales** The total amount in sales in billions of dollars by fast food business for several years is given in Table 2.20. Let $x = 0$ represent 1990, $x = 1$ represent 1991, and so forth. A model for the data is given by

$$y = 120 - \frac{500}{x + 8}$$

(a) Graph the model together with a scatter plot of the data.

(b) Use the model to estimate the amount of sales by fast food business in 2005.

Table 2.20 Fast Food Sales	
Year	Amount (in billions)
1992	70.6
1993	74.9
1994	78.5
1995	82.5
1996	85.9
1997	88.8
1998	92.5
1999	97.5
2000	101.4
2001	105.5

Source: Technomic, as reported in USA Today July 3–4, 2002.

44. Number of Wineries The number of wineries for several years is given in Table 2.21. Let $x = 0$ represent 1970, $x = 1$, represent 1971, and so forth. A model for this data is given by

$$y = 3000 - \frac{39,500}{x + 9}$$

(a) Graph the model together with a scatter plot of the data.

(b) Use the model to estimate the number of wineries in 2005.

Table 2.21 Number of Wineries

Year	Number
1975	579
1980	912
1985	1375
1990	1625
1995	1813
2000	2188

Source: American Vintners Association as reported in USA TODAY on June 28, 2002.

Standardized Test Questions

45. True or False An extraneous solution of a rational equation is also a solution of the equation. Justify your answer.

46. True or False The equation $1/(x^2 - 4) = 0$ has no solution. Justify your answer.

In Exercises 47–50, solve the problem without using a calculator.

47. Multiple Choice Which of the following are the solutions of the equation $x - \dfrac{3x}{x + 2} = \dfrac{6}{x + 2}$?

(A) $x = -2$ or $x = 3$

(B) $x = -1$ or $x = 3$

(C) only $x = -2$

(D) only $x = 3$

(E) There are no solutions.

48. Multiple Choice Which of the following are the solutions of the equation $1 - \dfrac{3}{x} = \dfrac{6}{x^2 + 2x}$?

(A) $x = -2$ or $x = 4$

(B) $x = -3$ or $x = 0$

(C) $x = -3$ or $x = 4$

(D) only $x = -3$

(E) There are no solutions.

49. Multiple Choice Which of the following are the solutions of the equation $\dfrac{x}{x + 2} + \dfrac{2}{x - 5} = \dfrac{14}{x^2 - 3x - 10}$?

(A) $x = -5$ or $x = 3$

(B) $x = -2$ or $x = 5$

(C) only $x = 3$

(D) only $x = -5$

(E) There are no solutions.

50. Multiple Choice Ten liters of a 20% acid solution are mixed with 30 liters of a 30% acid solution. Which of the following is the percent of acid in the final mixture?

(A) 21% **(B)** 22.5% **(C)** 25%   **(D)** 27.5 **(E)** 28%

Explorations

51. Revisit Example 4 Consider the following equation, which we solved in Example 4.

$$\frac{x - 3}{x} + \frac{3}{x + 2} + \frac{6}{x^2 + 2x} = 0$$

Let $f(x) = \dfrac{x - 3}{x} + \dfrac{3}{x + 2} + \dfrac{6}{x^2 + 2x}$.

(a) Combine the fractions in $f(x)$ but do *not* reduce to lowest terms.

(b) What is the domain of f?

(c) Write f as a piecewise-defined function.

(d) **Writing to Learn** Graph f and explain how the graph supports your answers in (b) and (c).

Extending the Ideas

In Exercises 52–55, solve for x.

52. $y = 1 + \dfrac{1}{1 + x}$

53. $y = 1 - \dfrac{1}{1 - x}$

54. $y = 1 + \dfrac{1}{1 + \dfrac{1}{x}}$

55. $y = 1 + \dfrac{1}{1 + \dfrac{1}{1 - x}}$

2.8
Solving Inequalities in One Variable

. . . and why

Designing containers as well as other types of applications often require that an inequality be solved.

Polynomial Inequalities

A polynomial inequality takes the form $f(x) > 0$, $f(x) \geq 0$, $f(x) < 0$, $f(x) \leq 0$ or $f(x) \neq 0$, where $f(x)$ is a polynomial. There is a fundamental connection between inequalities and the positive or negative sign of the corresponding expression $f(x)$:

- To solve $f(x) > 0$ is to find the values of x that make $f(x)$ *positive*.

- To solve $f(x) < 0$ is to find the values of x that make $f(x)$ *negative*.

If the expression $f(x)$ is a product, we can determine its sign by determining the sign of each of its factors. Example 1 illustrates that a polynomial function changes sign only at its real zeros of odd multiplicity.

EXAMPLE 1 Finding where a Polynomial is Zero, Positive, or Negative

Let $f(x) = (x + 3)(x^2 + 1)(x - 4)^2$. Determine the real number values of x that cause $f(x)$ to be **(a)** zero, **(b)** positive, **(c)** negative.

SOLUTION We begin by verbalizing our analysis of the problem:

(a) The real zeros of $f(x)$ are -3 (with multiplicity 1) and 4 (with multiplicity 2). So $f(x)$ is zero if $x = -3$ or $x = 4$.

(b) The factor $x^2 + 1$ is positive for all real numbers x. The factor $(x - 4)^2$ is positive for all real numbers x except $x = 4$, which makes $(x - 4)^2 = 0$. The factor $x + 3$ is positive if and only if $x > -3$. So $f(x)$ is positive if $x > -3$ and $x \neq 4$.

(c) By process of elimination, $f(x)$ is negative if $x < -3$.

This verbal reasoning process is aided by the following **sign chart**, which shows the x-axis as a number line with the real zeros displayed as the locations of potential sign change and the factors displayed with their sign value in the corresponding interval:

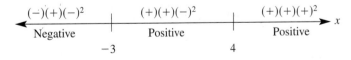

Figure 2.64 supports our findings because the graph of f is above the x-axis for x in $(-3, 4)$ or $(4, \infty)$, is on the x-axis for $x = -3$ or $x = 4$, and is below the x-axis for x in $(-\infty, -3)$. ***Now try Exercise 1.***

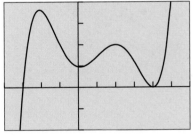

[−4, 6] by [−100, 200]

FIGURE 2.64 The graph of $f(x) = (x + 3)(x^2 + 1)(x - 4)^2$. (Example 1)

Our work in Example 1 allows us to report the solutions of four polynomial inequalities:

- The solution of $(x + 3)(x^2 + 1)(x - 4)^2 > 0$ is $(-3, 4) \cup (4, \infty)$.

- The solution of $(x + 3)(x^2 + 1)(x - 4)^2 \geq 0$ is $[-3, \infty)$.

• The solution of $(x + 3)(x^2 + 1)(x - 4)^2 < 0$ is $(-\infty, -3)$.

• The solution of $(x + 3)(x^2 + 1)(x - 4)^2 \leq 0$ is $(-\infty, -3] \cup \{4\}$.

Example 1 illustrates some important general characteristics of polynomial functions and polynomial inequalities. The polynomial function $f(x) = (x + 3)(x^2 + 1)(x - 4)^2$ in Example 1 and Figure 2.64

• changes sign at its real zero of odd multiplicity ($x = -3$);

• touches the x-axis but does not change sign at its real zero of even multiplicity ($x = 4$);

• has no x-intercepts or sign changes at its nonreal complex zeros associated with the irreducible quadratic factor ($x^2 + 1$).

This is consistent with what we learned about the relationships between zeros and graphs of polynomial functions in Sections 2.3–2.5. The real zeros and their multiplicity together with the end behavior of a polynomial function give us sufficient information about the polynomial to sketch its graph well enough to obtain a correct sign chart without considering the signs of the factors. See Figure 2.65.

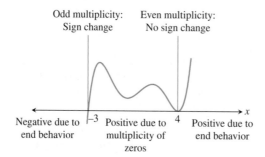

FIGURE 2.65 The sign chart and graph of $f(x) = (x + 3)(x^2 + 1)(x - 4)^2$ overlaid.

EXPLORATION 1   Sketching a Graph of a Polynomial from Its Sign Chart

Use your knowledge of end behavior and multiplicity of real zeros to create a sign chart and sketch the graph of the function. Check your sign chart using the factor method of Example 1. Then check your sketch using a grapher.

1. $f(x) = 2(x - 2)^3(x + 3)^2$

2. $f(x) = -(x + 2)^4(x + 1)(2x^2 + x + 1)$

3. $f(x) = 3(x - 2)^2(x + 4)^3(-x^2 - 2)$

So far in this section all of the polynomials have been presented in factored form and all of the inequalities have had zero on one of the sides. Examples 2 and 3 show us how to solve polynomial inequalities when they are not given in such a convenient form.

EXAMPLE 2 Solving a Polynomial Inequality Analytically

Solve $2x^3 - 7x^2 - 10x + 24 > 0$ analytically.

SOLUTION Let $f(x) = 2x^3 - 7x^2 - 10x + 24$. The Rational Zeros Theorem yields several possible rational zeros of f for factoring purposes:

$$\pm 1, \pm 2, \pm 3, \pm 4, \pm 6, \pm 8, \pm 12, \pm 24, \pm \frac{1}{2}, \pm \frac{3}{2}.$$

A table or graph of f can suggest which of these candidates to try. In this case, $x = 4$ is a rational zero of f, as the following synthetic division shows:

$$\begin{array}{r|rrrr} 4 & 2 & -7 & -10 & 24 \\ & & 8 & 4 & -24 \\ \hline & 2 & 1 & -6 & 0 \end{array}$$

The synthetic division lets us start the factoring process, which can then be completed using basic factoring methods:

$$f(x) = 2x^3 - 7x^2 - 10x + 24$$
$$= (x - 4)(2x^2 + x - 6)$$
$$= (x - 4)(2x - 3)(x + 2)$$

So the zeros of f are 4, 3/2, and −2. They are all real and all of multiplicity 1, so each will yield a sign change in $f(x)$. Because the degree of f is odd and its leading coefficient is positive, the end behavior of f is given by

$$\lim_{x \to \infty} f(x) = \infty \quad \text{and} \quad \lim_{x \to -\infty} f(x) = -\infty.$$

Our analysis yields the following sign chart:

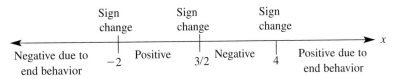

The solution of $2x^3 - 7x^2 - 10x + 24 > 0$ is $(-2, 3/2) \cup (4, \infty)$.

Now try Exercise 11.

EXAMPLE 3 Solving a Polynomial Inequality Graphically

Solve $x^3 - 6x^2 \le 2 - 8x$ graphically.

SOLUTION First we rewrite the inequality as $x^3 - 6x^2 + 8x - 2 \le 0$. Then we let $f(x) = x^3 - 6x^2 + 8x - 2$ and find the real zeros of f graphically as shown in Figure 2.66. The three real zeros are approximately 0.32, 1.46, and 4.21. The solution consists of the x values for which the graph is on or below the x-axis. So the solution of $x^3 - 6x^2 \le 2 - 8x$ is approximately $(-\infty, 0.32] \cup [1.46, 4.21]$.

The end points of these intervals are accurate to two decimal places. We use square brackets because the zeros of the polynomial are solutions of the inequality even though we only have approximations of their values. *Now try Exercise 13.*

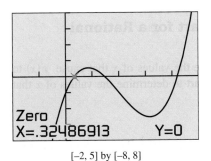

[−2, 5] by [−8, 8]

FIGURE 2.66 The graph of $f(x) = x^3 - 6x^2 + 8x - 2$, with one of three real zeros highlighted. (Example 3)

$$\pi \cdot 4.62^2 < \quad \pi r^2 \quad < \pi \cdot 9.65^2 \qquad \text{Multiply by } \pi.$$

$$\frac{1}{\pi \cdot 4.62^2} > \frac{1}{\pi r^2} > \frac{1}{\pi \cdot 9.65^2} \qquad 0 < a < b \Rightarrow \frac{1}{a} > \frac{1}{b}.$$

$$\frac{2000}{\pi \cdot 4.62^2} > \frac{2000}{\pi r^2} > \frac{2000}{\pi \cdot 9.65^2} \qquad \text{Multiply by } 2000.$$

$$\frac{2000}{\pi(4.619\ldots)^2} > \quad h \quad > \frac{2000}{\pi(9.654\ldots)^2} \qquad \text{Use the extra decimal places now.}$$

$$29.83 > \quad h \quad > 6.83 \qquad \text{Compute.}$$

Interpret

The surface area of the can will be less than 1000 cm^3 if its radius is between 4.62 cm and 9.65 cm and its height is between 6.83 cm and 29.83 cm. For any particular can, h must equal $2000/(\pi r^2)$.   ***Now try Exercise 61.***

QUICK REVIEW 2.8 *(For help, go to Sections A.2, A.3, and 2.3.)*

In Exercises 1–4, use limits to state the end behavior of the function.

1. $f(x) = 2x^3 + 3x^2 - 2x + 1$

2. $f(x) = -3x^4 - 3x^3 + x^2 - 1$

3. $g(x) = \dfrac{x^3 - 2x^2 + 1}{x - 2}$

4. $g(x) = \dfrac{2x^2 - 3x + 1}{x + 1}$

In Exercises 5–8, combine the fractions and reduce your answer to lowest terms.

5. $x^2 + \dfrac{5}{x}$

6. $x^2 - \dfrac{3}{x}$

7. $\dfrac{x}{2x + 1} - \dfrac{2}{x - 3}$

8. $\dfrac{x}{x - 1} + \dfrac{x + 1}{3x - 4}$

In Exercises 9 and 10, (a) list all the possible rational zeros of the polynomial and (b) factor the polynomial completely.

9. $2x^3 + x^2 - 4x - 3$ **10.** $3x^3 - x^2 - 10x + 8$

SECTION 2.8 EXERCISES

In Exercises 1–6, determine the x values that cause the polynomial function to be **(a)** zero, **(b)** positive, and **(c)** negative.

1. $f(x) = (x + 2)(x + 1)(x - 5)$

2. $f(x) = (x - 7)(3x + 1)(x + 4)$

3. $f(x) = (x + 7)(x + 4)(x - 6)^2$

4. $f(x) = (5x + 3)(x^2 + 6)(x - 1)$

5. $f(x) = (2x^2 + 5)(x - 8)^2(x + 1)^3$

6. $f(x) = (x + 2)^3(4x^2 + 1)(x - 9)^4$

In Exercises 7–12, complete the factoring if needed, and solve the polynomial inequality using a sign chart. Support graphically.

7. $(x + 1)(x - 3)^2 > 0$

8. $(2x + 1)(x - 2)(3x - 4) \le 0$

9. $(x + 1)(x^2 - 3x + 2) < 0$

10. $(2x - 7)(x^2 - 4x + 4) > 0$

11. $2x^3 - 3x^2 - 11x + 6 \ge 0$

12. $x^3 - 4x^2 + x + 6 \le 0$

In Exercises 13–20, solve the polynomial inequality graphically.

13. $x^3 - x^2 - 2x \ge 0$

14. $2x^3 - 5x^2 + 3x < 0$

15. $2x^3 - 5x^2 - x + 6 > 0$

16. $x^3 - 4x^2 - x + 4 \le 0$

17. $3x^3 - 2x^2 - x + 6 \ge 0$

18. $-x^3 - 3x^2 - 9x + 4 < 0$

19. $2x^4 - 3x^3 - 6x^2 + 5x + 6 < 0$

20. $3x^4 - 5x^3 - 12x^2 + 12x + 16 \ge 0$

In Exercises 21–24, solve the following inequalities for the given function $f(x)$.

(a) $f(x) > 0$ **(b)** $f(x) \geq 0$ **(c)** $f(x) < 0$ **(d)** $f(x) \leq 0$

21. $f(x) = (x^2 + 4)(2x^2 + 3)$

22. $f(x) = (x^2 + 1)(-2 - 3x^2)$

23. $f(x) = (2x^2 - 2x + 5)(3x - 4)^2$

24. $f(x) = (x^2 + 4)(3 - 2x)^2$

In Exercises 25–32, determine the x values that cause the function to be **(a)** zero, **(b)** undefined, **(c)** positive, and **(d)** negative.

25. $f(x) = \dfrac{x - 1}{(2x + 3)(x - 4)}$ **26.** $f(x) = \dfrac{(2x - 7)(x + 1)}{x + 5}$

27. $f(x) = x\sqrt{x + 3}$ **28.** $f(x) = x^2|2x + 9|$

29. $f(x) = \dfrac{\sqrt{x + 5}}{(2x + 1)(x - 1)}$ **30.** $f(x) = \dfrac{x - 1}{(x - 4)\sqrt{x + 2}}$

31. $f(x) = \dfrac{(2x + 5)\sqrt{x - 3}}{(x - 4)^2}$ **32.** $f(x) = \dfrac{3x - 1}{(x + 3)\sqrt{x - 5}}$

In Exercises 33–44, solve the inequality using a sign chart. Support graphically.

33. $\dfrac{x - 1}{x^2 - 4} < 0$ **34.** $\dfrac{x + 2}{x^2 - 9} < 0$

35. $\dfrac{x^2 - 1}{x^2 + 1} \leq 0$ **36.** $\dfrac{x^2 - 4}{x^2 + 4} > 0$

37. $\dfrac{x^2 + x - 12}{x^2 - 4x + 4} > 0$ **38.** $\dfrac{x^2 + 3x - 10}{x^2 - 6x + 9} < 0$

39. $\dfrac{x^3 - x}{x^2 + 1} \geq 0$ **40.** $\dfrac{x^3 - 4x}{x^2 + 2} \leq 0$

41. $x|x - 2| > 0$ **42.** $\dfrac{x - 3}{|x + 2|} < 0$

43. $(2x - 1)\sqrt{x + 4} < 0$ **44.** $(3x - 4)\sqrt{2x + 1} \geq 0$

In Exercises 45–54, solve the inequality.

45. $\dfrac{x^3(x - 2)}{(x + 3)^2} < 0$ **46.** $\dfrac{(x - 5)^4}{x(x + 3)} \geq 0$

47. $x^2 - \dfrac{2}{x} > 0$ **48.** $x^2 + \dfrac{4}{x} \geq 0$

49. $\dfrac{1}{x + 1} + \dfrac{1}{x - 3} \leq 0$ **50.** $\dfrac{1}{x + 2} - \dfrac{2}{x - 1} > 0$

51. $(x + 3)|x - 1| \geq 0$ **52.** $(3x + 5)^2|x - 2| < 0$

53. $\dfrac{(x - 5)|x - 2|}{\sqrt{2x - 3}} \geq 0$ **54.** $\dfrac{x^2(x - 4)^3}{\sqrt{x + 1}} < 0$

55. Writing to Learn Write a paragraph that explains two ways to solve the inequality $3(x - 1) + 2 \leq 5x + 6$.

56. Company Wages Pederson Electric Co. charges $25 per service call plus $18 per hour for home repair work. How long did an electrician work if the charge was less than $100? Assume the electrician rounds off the time to the nearest quarter hour.

57. Connecting Algebra and Geometry Consider the collection of all rectangles that have lengths 2 in. less than twice their widths. Find the possible widths (in inches) of these rectangles if their perimeters are less than 200 in.

58. Planning for Profit The Grovenor Candy Co. finds that the cost of making a certain candy bar is $0.13 per bar. Fixed costs amount to $2000 per week. If each bar sells for $0.35, find the minimum number of candy bars that will earn the company a profit.

59. Designing a Cardboard Box Picaro's Packaging Plant wishes to design boxes with a volume of *not more than* 100 in.^{3}. Squares are to be cut from the corners of a 12-in. by 15-in. piece of cardboard (see figure), with the flaps folded up to make an open box. What size squares should be cut from the cardboard?

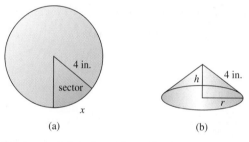

60. Cone Problem Beginning with a circular piece of paper with a 4-inch radius, as shown in (a), cut out a sector with an arc of length x. Join the two radial edges of the remaining portion of the paper to form a cone with radius r and height h, as shown in (b). What length of arc will produce a cone with a volume greater than 21 in.^{3}?

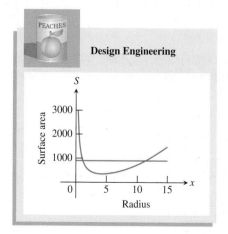

(a)   (b)

61. Design a Juice Can Flannery Cannery packs peaches in 0.5-L cylindrical cans.

(a) Express the surface area S of the can as a function of the radius x (in cm).

(b) Find the dimensions of the can if the surface is less than 900 cm^2.

(c) Find the least possible surface area of the can.

62. Resistors The total electrical resistance R of two resistors connected in parallel with resistances R_1 and R_2 is given by

$$\frac{1}{R} = \frac{1}{R_1} + \frac{1}{R_2}.$$

One resistor has a resistance of 2.3 ohms. Let x be the resistance of the second resistor.

(a) Express the total resistance R as a function of x.

(b) Find the resistance in the second resistor if the total resistance of the pair is at least 1.7 ohms.

63. Per Capita Income The U.S. average per capita income for several years from 1990 to 2003 is given in Table 2.22 in 2003 dollars. Let $x = 0$ represent 1990, $x = 1$ represent 1991, and so forth.

(a) Find the linear regression model for the data.

(b) Use the model to predict when the per capita income will exceed $40,000.

Table 2.22 Personal Income Per Capita	
Year	Income (dollars)
1990	19,477
1992	20,958
1994	22,435
1996	24,488
1998	27,126
2000	29,847
2002	30,906
2003	31,632

Source: U.S. Census Bureau, Statistical Abstract of the United States, 2001, 2004–2005.

64. Single Family House Cost The median sales prices of new, privately owned one-family houses sold in the U.S. are given for selected years in Table 2.23. Let x be the number of years since 1980.

(a) Find the cubic regression model for the data.

(b) Use the model to predict when the median cost of a new home will exceed $250,000.

Table 2.23 Median Sales Price of a New House	
Year	Price (dollars)
1980	64,600
1985	84,300
1990	122,900
1995	133,900
2000	169,000
2003	195,000

Source: U.S. Census Bureau, Statistical Abstract of the United States, 2004–2005.

Standardized Test Questions

65. True or False The graph of $f(x) = x^4 (x + 3)^2(x - 1)^3$ changes sign at $x = 0$. Justify your answer.

66. True or False The graph $r(x) = \dfrac{2x - 1}{(x + 2)(x - 1)}$ changes sign at $x = -2$. Justify your answer.

In Exercises 67–70, solve the problem without using a calculator.

67. Multiple Choice Which of the following is the solution to $x^2 < x$?

(A) $(0, \infty)$ **(B)** $(1, \infty)$ **(C)** $(0, 1)$

(D) $(-\infty, 1)$ **(E)** $(-\infty, 0) \cup (1, \infty)$

68. Multiple Choice Which of the following is the solution to $\dfrac{1}{(x + 2)^2} \geq 0$?

(A) $(-2, \infty)$ **(B)** All $x \neq -2$ **(C)** All $x \neq 2$

(D) All real numbers   **(E)** There are no solutions.

69. Multiple Choice Which of the following is the solution to $\dfrac{x^2}{x - 3} < 0$?

(A) $(-\infty, 3)$ **(B)** $(-\infty, 3]$ **(C)** $(-\infty, 0] \cup (0, 3)$

(D) $(-\infty, 0) \cup (0, 3)$ **(E)** There are no solutions.

70. Multiple Choice Which of the following is the solution to $(x^2 - 1)^2 \leq 0$?

(A) $\{-1, 1\}$ **(B)** $\{1\}$ **(C)** $[-1, 1]$

(D) $[0, 1]$ **(E)** There are no solutions.

Explorations

In Exercises 71 and 72, find the vertical asymptotes and intercepts of the rational function. Then use a sign chart and a table of values to sketch the function by hand. Support your result using a grapher. (Hint: You may need to graph the function in more than one window to see different parts of the overall graph.)

71. $f(x) = \dfrac{(x - 1)(x + 2)^2}{(x - 3)(x + 1)}$ **72.** $g(x) = \dfrac{(x - 3)^4}{x^2 + 4x}$

Extending the Ideas

73. Group Activity Looking Ahead to Calculus Let $f(x) = 3x - 5$.

(a) Assume x is in the interval defined by $|x - 3| < 1/3$. Give a convincing argument that $|f(x) - 4| < 1$.

(b) Writing to Learn Explain how (a) is modeled by the figure on the next page.

(c) Show how the algebra used in (a) can be modified to show that if $|x - 3| < 0.01$, then $|f(x) - 4| < 0.03$. How would the figure on the next page change to reflect these inequalities?

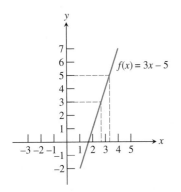

f(x) = 3x − 5

74. Writing to Learn Boolean Operators The Test menu of many graphers contains inequality symbols that can be used to construct inequality statements, as shown in (a). An answer of 1 indicates the statement is true, and 0 indicates the statement is false. In (b), the graph of $Y_1 = (x^2 - 4 \geq 0)$ is shown using Dot

mode and the window $[-4.7, 4.7]$ by $[-3.1, 3.1]$. Experiment with the Test menu, and then write a paragraph explaining how to interpret the graph in (b).

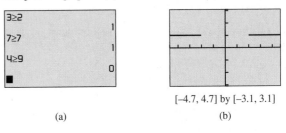

$[-4.7, 4.7]$ by $[-3.1, 3.1]$

(a)   (b)

In Exercises 75 and 76, use the properties of inequality from Chapter P to prove the statement.

75. If $0 < a < b$, then $a^2 < b^2$.

76. If $0 < a < b$, then $\dfrac{1}{a} > \dfrac{1}{b}$.

Math at Work

I have a Ph.D. in computational neuroscience. I became interested in this field after working as a software engineer. Now, I'm here, a back-to-school mom at an engineering school in Switzerland! After years of software programming, I found I had forgotten most of my mathematics, but I've found it's like any other skill—it just takes practice.

In this field we use computer simulations to explore ways of understanding what neurons do in the brain and the rest of our nervous systems. The brain contains so many neurons—and so many different kinds of them—that if we try to make a really "realistic" model of a part of the brain, not only do the simulations run very slowly on the computer, but also they can be too complicated to understand!

What we do is use very simple models for the individual neurons. We use mathematics to analytically calculate quantities such as the response of a group of neurons to a specific input signal or how

much information the neurons can transmit under various conditions. Then we use computer simulations to test whether our theoretical prediction really works. When the theory matches the experiment, it's really exciting!

An equation that is commonly used in my field is that for the membrane potential of a neuron. This is the amount of current the neuron is able to store before it "fires" an action poten-

Alix Kamakaokalani Herrmann

tial, which is how it communicates to other neurons, or even muscles, telling them to move. If we keep some of the quantities in the equation constant, we can solve for the others. For instance, if the input current and resistance are kept constant, we can solve for the membrane potential.

CHAPTER 2 Key Ideas

PROPERTIES, THEOREMS, AND FORMULAS

PROCEDURES

GALLERY OF FUNCTIONS

Identity

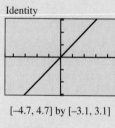

[−4.7, 4.7] by [−3.1, 3.1]

$$f(x) = x$$

Squaring

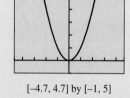

[−4.7, 4.7] by [−1, 5]

$$f(x) = x^2$$

Cubing

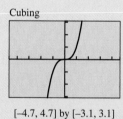

[−4.7, 4.7] by [−3.1, 3.1]

$$f(x) = x^3$$

Reciprocal

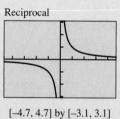

[−4.7, 4.7] by [−3.1, 3.1]

$$f(x) = 1/x = x^{-1}$$

Square Root

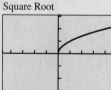

[−4.7, 4.7] by [−3.1, 3.1]

$$f(x) = \sqrt{x} = x^{1/2}$$

Cube Root

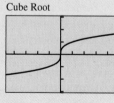

[−4.7, 4.7] by [−3.1, 3.1]

$$f(x) = \sqrt[3]{x} = x^{1/3}$$

Inverse-Square

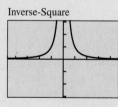

[−4.7, 4.7] by [−3.1, 3.1]

$$f(x) = 1/x^2 = x^{-2}$$

CHAPTER 2 **Review Exercises**

The collection of exercises marked in red could be used as a chapter test.

In Exercises 1 and 2, write an equation for the linear function f satisfying the given conditions. Graph $y = f(x)$.

1. $f(-3) = -2$ and $f(4) = -9$

2. $f(-3) = 6$ and $f(1) = -2$

In Exercises 3 and 4, describe how to transform the graph of $f(x) = x^2$ into the graph of the given function. Sketch the graph by hand and support your answer with a grapher.

3. $h(x) = 3(x - 2)^2 + 4$

4. $g(x) = -(x + 3)^2 + 1$

In Exercises 5–8, find the vertex and axis of the graph of the function. Support your answer graphically.

5. $f(x) = -2(x + 3)^2 + 5$ **6.** $g(x) = 4(x - 5)^2 - 7$

7. $f(x) = -2x^2 - 16x - 31$ **8.** $g(x) = 3x^2 - 6x + 2$

In Exercises 9 and 10, write an equation for the quadratic function whose graph contains the given vertex and point.

9. Vertex $(-2, -3)$, point $(1, 2)$

10. Vertex $(-1, 1)$, point $(3, -2)$

In Exercises 11 and 12, write an equation for the quadratic function with graph shown, given one of the labeled points is the vertex of the parabola.

11.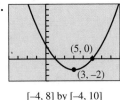

[–4, 8] by [–4, 10]

12.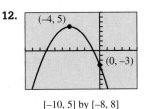

[–10, 5] by [–8, 8]

In Exercises 13–16, graph the function in a viewing window that shows all of its extrema and x-intercepts.

13. $f(x) = x^2 + 3x - 40$ **14.** $f(x) = -8x^2 + 16x - 19$

15. $f(x) = x^3 + x^2 + x + 5$ **16.** $f(x) = x^3 - x^2 - 20x - 2$

In Exercises 17 and 18, write the statement as a power function equation. Let k be the constant of variation.

17. The surface area S of a sphere varies directly as the square of the radius r.

18. The force of gravity F acting on an object is inversely proportional to the square of the distance d from the object to the center of the earth.

In Exercises 19 and 20, write a sentence that expresses the relationship in the formula, using the language of variation or proportion.

19. $F = kx$, where F is the force it takes to stretch a spring x units from its unstressed length and k is the spring's force constant.

20. $A = \pi \cdot r^2$, where A and r are the area and radius of a circle and π is the usual mathematical constant.

In Exercises 21–24, state the values of the constants k and a for the function $f(x) = k \cdot x^a$. Describe the portion of the curve that lies in Quadrant I or IV. Determine whether f is even, odd, or undefined for $x < 0$. Describe the rest of the curve if any. Graph the function to see whether it matches the description.

21. $f(x) = 4x^{1/3}$ **22.** $f(x) = -2x^{3/4}$

23. $f(x) = -2x^{-3}$ **24.** $f(x) = (2/3)x^{-4}$

In Exercises 25–28, divide $f(x)$ by $d(x)$, and write a summary statement in polynomial form.

25. $f(x) = 2x^3 - 7x^2 + 4x - 5$; $d(x) = x - 3$

26. $f(x) = x^4 + 3x^3 + x^2 - 3x + 3$; $d(x) = x + 2$

27. $f(x) = 2x^4 - 3x^3 + 9x^2 - 14x + 7$; $d(x) = x^2 + 4$

28. $f(x) = 3x^4 - 5x^3 - 2x^2 + 3x - 6$; $d(x) = 3x + 1$

In Exercises 29 and 30, use the Remainder Theorem to find the remainder when $f(x)$ is divided by $x - k$. Check by using synthetic division.

29. $f(x) = 3x^3 - 2x^2 + x - 5$; $k = -2$

30. $f(x) = -x^2 + 4x - 5$; $k = 3$

In Exercises 31 and 32, use the Factor Theorem to determine whether the first polynomial is a factor of the second polynomial.

31. $x - 2$; $x^3 - 4x^2 + 8x - 8$

32. $x + 3$; $x^3 + 2x^2 - 4x - 2$

In Exercises 33 and 34, use synthetic division to prove that the number k is an upper bound for the real zeros of the function f.

33. $k = 5$; $f(x) = x^3 - 5x^2 + 3x + 4$

34. $k = 4$; $f(x) = 4x^4 - 16x^3 + 8x^2 + 16x - 12$

In Exercises 35 and 36, use synthetic division to prove that the number k is a lower bound for the real zeros of the function f.

35. $k = -3$; $f(x) = 4x^4 + 4x^3 - 15x^2 - 17x - 2$

36. $k = -3$; $f(x) = 2x^3 + 6x^2 + x - 6$

In Exercises 37 and 38, use the Rational Zeros Theorem to write a list of all potential rational zeros. Then determine which ones, if any, are zeros.

37. $f(x) = 2x^4 - x^3 - 4x^2 - x - 6$

38. $f(x) = 6x^3 - 20x^2 + 11x + 7$

In Exercises 39–42, perform the indicated operation, and write the result in the form $a + bi$.

39. $(1 + i)^3$

40. $(1 + 2i)^2(1 - 2i)^2$

41. i^{29}

42. $\sqrt{-16}$

In Exercises 43 and 44, solve the equation.

43. $x^2 - 6x + 13 = 0$

44. $x^2 - 2x + 4 = 0$

In Exercises 45–48, match the polynomial function with its graph. Explain your choice.

45. $f(x) = (x - 2)^2$

46. $f(x) = (x - 2)^3$

47. $f(x) = (x - 2)^4$

48. $f(x) = (x - 2)^5$

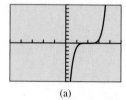

(a)

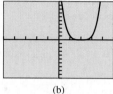

(b)

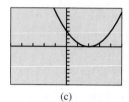

(c)

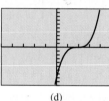

(d)

In Exercises 49–52, find all of the real zeros of the function, finding exact values whenever possible. Identify each zero as rational or irrational. State the number of nonreal complex zeros.

49. $f(x) = x^4 - 10x^3 + 23x^2$

50. $k(t) = t^4 - 7t^2 + 12$

51. $h(x) = x^3 - 2x^2 - 8x + 5$

52. $k(x) = x^4 - x^3 - 14x^2 + 24x + 5$

In Exercises 53–56, find all of the zeros and write a linear factorization of the function.

53. $f(x) = 2x^3 - 9x^2 + 2x + 30$

54. $f(x) = 5x^3 - 24x^2 + x + 12$

55. $f(x) = 6x^4 + 11x^3 - 16x^2 - 11x + 10$

56. $f(x) = x^4 - 8x^3 + 27x^2 - 50x + 50$, given that $1 + 2i$ is a zero.

In Exercises 57–60, write the function as a product of linear and irreducible quadratic factors all with real coefficients.

57. $f(x) = x^3 - x^2 - x - 2$

58. $f(x) = 9x^3 - 3x^2 - 13x - 1$

59. $f(x) = 2x^4 - 9x^3 + 23x^2 - 31x + 15$

60. $f(x) = 3x^4 - 7x^3 - 3x^2 + 17x + 10$

In Exercises 61–66, write a polynomial function with real coefficients whose zeros and their multiplicities include those listed.

61. Degree 3; zeros: $\sqrt{5}, -\sqrt{5}, 3$

62. Degree 2; -3 only real zero

63. Degree 4; zeros: $3, -2, 1/3, -1/2$

64. Degree 3; zeros: $1 + i, 2$

65. Degree 4; zeros: -2(multiplicity 2), 4(multiplicity 2)

66. Degree 3; zeros: $2 - i, -1$, and $f(2) = 6$

In Exercises 67 and 68, describe how the graph of the given function can be obtained by transforming the graph of the reciprocal function $f(x) = 1/x$. Identify the horizontal and vertical asymptotes.

67. $f(x) = \dfrac{-x + 7}{x - 5}$

68. $f(x) = \dfrac{3x + 5}{x + 2}$

In Exercises 69–72, find the asymptotes and intercepts of the function, and graph it.

69. $f(x) = \dfrac{x^2 + x + 1}{x^2 - 1}$

70. $f(x) = \dfrac{2x^2 + 7}{x^2 + x - 6}$

71. $f(x) = \dfrac{x^2 - 4x + 5}{x + 3}$

72. $g(x) = \dfrac{x^2 - 3x - 7}{x + 3}$

In Exercises 73–74, find the intercepts and analyze and draw the graph of the given rational function.

73. $f(x) = \dfrac{x^3 + x^2 - 2x + 5}{x + 2}$

74. $f(x) = \dfrac{-x^4 + x^2 + 1}{x - 1}$

In Exercises 75–82, solve the equation or inequality algebraically, and support graphically.

75. $2x + \dfrac{12}{x} = 11$

76. $\dfrac{x}{x + 2} + \dfrac{5}{x - 3} = \dfrac{25}{x^2 - x - 6}$

77. $2x^3 + 3x^2 - 17x - 30 < 0$

78. $3x^4 + x^3 - 36x^2 + 36x + 16 \geq 0$

79. $\dfrac{x + 3}{x^2 - 4} \geq 0$

80. $\dfrac{x^2 - 7}{x^2 - x - 6} < 1$

81. $(2x - 1)^2|x + 3| \leq 0$

82. $\dfrac{(x - 1)|x - 4|}{\sqrt{x + 3}} > 0$

83. Writing to Learn Determine whether
$$f(x) = x^5 - 10x^4 - 3x^3 + 28x^2 + 20x - 2$$
has a zero outside the viewing window. Explain. (See graph on next page.)

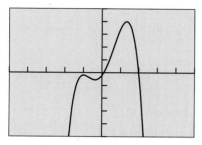

[–5, 5] by [–50, 50]

84. Launching a Rock Larry uses a slingshot to launch a rock straight up from a point 6 ft above level ground with an initial velocity of 170 ft/sec.

(a) Find an equation that models the height of the rock t seconds after it is launched and graph the equation. (See Example 8 in Section 2.1.)

(b) What is the maximum height of the rock? When will it reach that height?

(c) When will the rock hit the ground?

85. Volume of a Box Villareal Paper Co. has contracted to manufacture a box with no top that is to be made by removing squares of width x from the corners of a 30-in. by 70-in. piece of cardboard.

(a) Find an equation that models the volume of the box.

(b) Determine x so that the box has a volume of 5800 in.^{3}.

86. Architectural Engineering DeShanna, an engineer at J. P. Cook, Inc., completes structural specifications for a 255-ft-long steel beam anchored between two pilings 50 ft above ground, as shown in the figure. She knows that when a 250-lb object is placed d feet from the west piling, the beam bends s feet where

$$s = (8.5 \times 10^{-7})d^2(255 - d).$$

(a) Graph the function s.

(b) What are the dimensions of a viewing window that shows a graph for the values that make sense in this problem situation?

(c) What is the greatest amount of vertical deflection s, and where does it occur?

(d) Writing to Learn Give a possible scenario explaining why the solution to part (c) does not occur at the halfway point.

West   East

87. Storage Container A liquid storage container on a truck is in the shape of a cylinder with hemispheres on each end as shown in the figure. The cylinder and hemispheres have the same radius. The total length of the container is 140 ft.

(a) Determine the volume V of the container as a function of the radius x.

(b) Graph the function $y = V(x)$.

(c) What is the radius of the container with the largest possible volume? What is the volume?

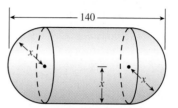

88. Pell Grants The maximum loan permitted under the federal student-aid program is given in Table 2.24 for several years. Let $x = 0$ represent 1990, $x = 1$ represent 1991, and so forth.

Table 2.24 Maximum Pell Grant

Year	Amount (dollars)
1990	2300
1991	2400
1992	2400
1993	2300
1994	2300
1995	2340
1996	2470
1997	2700
1998	3000
1999	3125
2000	3300
2001	3750
2002	4000

Source: U.S. Education Department, as reported in The Chronicle of Higher Education, February 15, 2002.

(a) Find a quadratic regression model for the Pell Grant amounts, and graph it together with a scatter plot of the data.

(b) Find a quartic regression model for the Pell Grant amounts, and graph it together with a scatter plot of the data.

(c) Use each regression equation to predict the amount of a Pell Grant in 2006.

(d) Writing to Learn Determine the end behavior of the two regression models. What does the end behavior say about future Pell Grant amounts?

89. National Institute of Health Spending Table 2.25 shows the spending at the National Institute of Health for several years. Let $x = 0$ represent 1990, $x = 1$ represent 1991, and so forth.

Table 2.25 Spending at the National Institute of Health	
Year	Amount (billion)
1993	10.3
1994	11.0
1995	11.3
1996	11.9
1997	12.7
1998	13.6
1999	15.6
2000	17.9
2001	20.5
2002	23.6

Source: National Institute of Health as reported in The Chronicle of Higher Education November 26, 1999 and February 15, 2002.

(a) Find a linear regression model, and graph it together with a scatter plot of the data.

(b) Find a quadratic regression model, and graph it together with a scatter plot of the data.

(c) Use the linear and quadratic regression models to estimate when the amount of spending will exceed $30 billion.

90. Breaking Even Midtown Sporting Goods has determined that it needs to sell its soccer shinguards for $5.25 a pair in order to be competitive. It costs $4.32 to produce each pair of shinguards, and the weekly overhead cost is $4000.

(a) Express the average cost that includes the overhead of producing one shinguard as a function of the number x of shinguards produced each week.

(b) Solve algebraically to find the number of shinguards that must be sold each week to make $8000 in profit. Support your work graphically.

91. Deer Population The number of deer P at any time t (in years) in a federal game reserve is given by

$$P(t) = \frac{800 + 640t}{20 + 0.8t}.$$

(a) Find the number of deer when t is 15, 70, and 100.

(b) Find the horizontal asymptote of the graph of $y = P(t)$.

(c) According to the model, what is the largest possible deer population?

92. Resistors The total electrical resistance R of two resistors connected in parallel with resistances R_1 and R_2 is given by

$$\frac{1}{R} = \frac{1}{R_1} + \frac{1}{R_2}.$$

The total resistance is 1.2 ohms. Let $x = R_1$.

(a) Express the second resistance R_2 as a function of x.

(b) Find R_2 if x_1 is 3 ohms.

93. Acid Mixture Suppose that x ounces of distilled water are added to 50 oz of pure acid.

(a) Express the concentration $C(x)$ of the new mixture as a function of x.

(b) Use a graph to determine how much distilled water should be added to the pure acid to produce a new solution that is less than 60% acid.

(c) Solve (b) algebraically.

94. Industrial Design Johnson Cannery will pack peaches in 1-L cylindrical cans. Let x be the radius of the base of the can in centimeters.

(a) Express the surface area S of the can as a function of x.

(b) Find the radius and height of the can if the surface area is 900 cm^2.

(c) What dimensions are possible for the can if the surface area is to be less than 900 cm^2?

95. Industrial Design Gilman Construction is hired to build a rectangular tank with a square base and no top. The tank is to hold 1000 ft^3 of water. Let x be a length of the base.

(a) Express the outside surface area S of the tank as a function of x.

(b) Find the length, width, and height of the tank if the outside surface area is 600 ft^2.

(c) What dimensions are possible for the tank if the outside surface area is to be less than 600 ft^2?

CHAPTER 2 **Project**

Modeling the Height of a Bouncing Ball

When a ball is bouncing up and down on a flat surface, its height with respect to time can be modeled using a quadratic function. One form of a quadratic function is the vertex form:

$$y = a(x - h)^2 + k$$

In this equation, y represents the height of the ball and x represents the elapsed time. For this project, you will use a motion detection device to collect distance and time data for a bouncing ball, then find a mathematical model that describes the position of the ball.

Explorations

Total elapsed time (seconds)	Height of the ball (meters)	Total elapsed time (seconds)	Height of the ball (meters)
0.688	0	1.118	0.828
0.731	0.155	1.161	0.811
0.774	0.309	1.204	0.776
0.817	0.441	1.247	0.721
0.860	0.553	1.290	0.650
0.903	0.643	1.333	0.563
0.946	0.716	1.376	0.452
0.989	0.773	1.419	0.322
1.032	0.809	1.462	0.169
1.075	0.828		

1. If you collected motion data using a CBL2 or CBR™, a plot of height versus time or distance versus time should be shown on your grapher or computer screen. Either plot will work for this project. If you do not have access to a CBL2/CBR™, enter the data from the table above into your grapher or computer. Create a scatter plot for the data.

2. Find values for a, h, and k so that the equation $y = a(x - h)^2 + k$ fits one of the bounces contained in the data plot. Approximate the vertex (h, k) from your data plot and solve for the value of a algebraically.

3. Change the values of a, h, and k in the model found above and observe how the graph of the function is affected on your grapher or computer. Generalize how each of these changes affects the graph.

4. Expand the equation you found in #2 above so that it is in the standard quadratic form: $y = ax^2 + bx + c$.

5. Use your grapher or computer to select the data from the bounce you modeled above and then use quadratic regression to find a model for this data set. (See your grapher's guidebook for instructions on how to do this.) How does this model compare with the standard quadratic form found in #4?

6. Complete the square to transform the regression model to the vertex form of a quadratic and compare it to the original vertex model found in #2. (Round the values of a, b, and c to the nearest 0.001 before completing the square if desired.)

Exponential, Logistic, and Logarithmic Functions

The loudness of a sound we hear is based on the intensity of the associated sound wave. This sound intensity is the energy per unit time of the wave over a given area, measured in watts per square meter (W/m^2). The intensity is greatest near the source and decreases as you move away, whether the sound is rustling leaves or rock music. Because of the wide range of audible sound intensities, they are generally converted into *decibels*, which are based on logarithms. See page 307.

Chapter 3 Overview

In this chapter, we study three interrelated families of functions: exponential, logistic, and logarithmic functions. Polynomial functions, rational functions, and power functions with rational exponents are **algebraic functions**— functions obtained by adding, subtracting, multiplying, and dividing constants and an independent variable, and raising expressions to integer powers and extracting roots. In this chapter and the next one, we explore **transcendental functions**, which go beyond, or transcend, these algebraic operations.

Just like their algebraic cousins, exponential, logistic, and logarithmic functions have wide application. Exponential functions model growth and decay over time, such as *unrestricted* population growth and the decay of radioactive substances. Logistic functions model *restricted* population growth, certain chemical reactions, and the spread of rumors and diseases. Logarithmic functions are the basis of the Richter scale of earthquake intensity, the pH acidity scale, and the decibel measurement of sound.

The chapter closes with a study of the mathematics of finance, an application of exponential and logarithmic functions often used when making investments.

3.1
Exponential and Logistic Functions

What you'll learn about

- Exponential Functions and Their Graphs
- The Natural Base *e*
- Logistic Functions and Their Graphs
- Population Models

. . . and why

Exponential and logistic functions model many growth patterns, including the growth of human and animal populations.

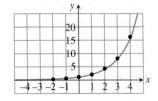

FIGURE 3.1 Sketch of $g(x) = 2^x$.

Exponential Functions and Their Graphs

The functions $f(x) = x^2$ and $g(x) = 2^x$ each involve a base raised to a power, but the roles are reversed:

- For $f(x) = x^2$, the base is the variable x, and the exponent is the constant 2; f is a familiar monomial and power function.

- For $g(x) = 2^x$, the base is the constant 2, and the exponent is the variable x; g is an *exponential function*. See Figure 3.1.

DEFINITION Exponential Functions

Let a and b be real number constants. An **exponential function** in x is a function that can be written in the form

$$f(x) = a \cdot b^x,$$

where a is nonzero, b is positive, and $b \neq 1$. The constant a is the *initial value* of f (the value at $x = 0$), and b is the **base**.

Exponential functions are defined and continuous for all real numbers. It is important to recognize whether a function is an exponential function.

EXAMPLE 1 Identifying Exponential Functions

(a) $f(x) = 3^x$ is an exponential function, with an initial value of 1 and base of 3.

(b) $g(x) = 6x^{-4}$ is *not* an exponential function because the base x is a variable and the exponent is a constant; g is a power function.

(c) $h(x) = -2 \cdot 1.5^x$ is an exponential function, with an initial value of -2 and base of 1.5.

(d) $k(x) = 7 \cdot 2^{-x}$ is an exponential function, with an initial value of 7 and base of 1/2 because $2^{-x} = (2^{-1})^x = (1/2)^x$.

(e) $q(x) = 5 \cdot 6^{\pi}$ is *not* an exponential function because the exponent π is a constant; q is a constant function. **Now try Exercise 1.**

One way to evaluate an exponential function, when the inputs are rational numbers, is to use the properties of exponents.

EXAMPLE 2 Computing Exponential Function Values for Rational Number Inputs

For $f(x) = 2^x$,

(a) $f(4) = 2^4 = 2 \cdot 2 \cdot 2 \cdot 2 = 16.$

(b) $f(0) = 2^0 = 1$

(c) $f(-3) = 2^{-3} = \dfrac{1}{2^3} = \dfrac{1}{8} = 0.125$

(d) $f\left(\dfrac{1}{2}\right) = 2^{1/2} = \sqrt{2} = 1.4142\ldots$

(e) $f\left(-\dfrac{3}{2}\right) = 2^{-3/2} = \dfrac{1}{2^{3/2}} = \dfrac{1}{\sqrt{2^3}} = \dfrac{1}{\sqrt{8}} = 0.35355\ldots$ **Now try Exercise 7.**

There is no way to use properties of exponents to express an exponential function's value for *irrational* inputs. For example, if $f(x) = 2^x$, $f(\pi) = 2^{\pi}$, but what does 2^{π} mean? Using properties of exponents, $2^3 = 2 \cdot 2 \cdot 2$, $2^{3.1} = 2^{31/10} = \sqrt[10]{2^{31}}$. So we can find meaning for 2^{π} by using successively closer *rational* approximations to π as shown in Table 3.1.

Table 3.1 Values of $f(x) = 2^x$ for Rational Numbers x Approaching $\pi = 3.14159265\ldots$						
x	3	3.1	3.14	3.141	3.1415	3.14159
2^x	8	8.5...	8.81...	8.821...	8.8244...	8.82496...

We can conclude that $f(\pi) = 2^{\pi} \approx 8.82$, which could be found directly using a grapher. The methods of calculus permit a more rigorous definition of exponential functions than we give here, a definition that allows for both rational and irrational inputs.

The way exponential functions change makes them useful in applications. This pattern of change can best be observed in tabular form.

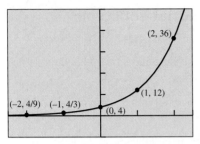

[−2.5, 2.5] by [−10, 50]

(a)

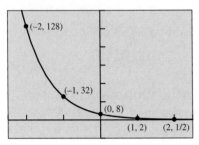

[−2.5, 2.5] by [−25, 150]

(b)

FIGURE 3.2 Graphs of (a) $g(x) = 4 \cdot 3^x$ and (b) $h(x) = 8 \cdot (1/4)^x$. (Example 3)

EXAMPLE 3   Finding an Exponential Function from its Table of Values

Determine formulas for the exponential functions g and h whose values are given in Table 3.2.

Table 3.2 Values for Two Exponential Functions		
x	$g(x)$	$h(x)$
−2	4/9 $\rangle \times 3$	128 $\rangle \times \frac{1}{4}$
−1	4/3 $\rangle \times 3$	32 $\rangle \times \frac{1}{4}$
0	4 $\rangle \times 3$	8 $\rangle \times \frac{1}{4}$
1	12 $\rangle \times 3$	2 $\rangle \times \frac{1}{4}$
2	36	1/2

SOLUTION Because g is exponential, $g(x) = a \cdot b^x$. Because $g(0) = 4$, the initial value a is 4. Because $g(1) = 4 \cdot b^1 = 12$, the base b is 3. So,

$$g(x) = 4 \cdot 3^x.$$

Because h is exponential, $h(x) = a \cdot b^x$. Because $h(0) = 8$, the initial value a is 8. Because $h(1) = 8 \cdot b^1 = 2$, the base b is 1/4. So,

$$h(x) = 8 \cdot \left(\frac{1}{4}\right)^x.$$

Figure 3.2 shows the graphs of these functions pass through the points whose coordinates are given in Table 3.2.   ***Now try Exercise 11.***

Observe the patterns in the $g(x)$ and $h(x)$ columns of Table 3.2. The $g(x)$ values increase by a factor of 3 and the $h(x)$ values decrease by a factor of 1/4, as we add 1 to x moving from one row of the table to the next. In each case, the change factor is the base of the exponential function. This pattern generalizes to all exponential functions as illustrated in Table 3.3.

Table 3.3 Values for a General Exponential Function $f(x) = a \cdot b^x$	
x	$a \cdot b^x$
−2	ab^{-2} $\rangle \times b$
−1	ab^{-1} $\rangle \times b$
0	a $\rangle \times b$
1	ab $\rangle \times b$
2	ab^2

In Table 3.3, as x increases by 1, the function value is multiplied by the base b. This relationship leads to the following *recursive formula*.

Exponential Growth and Decay

For any exponential function $f(x) = a \cdot b^x$ and any real number x,

$$f(x + 1) = b \cdot f(x).$$

If $a > 0$ and $b > 1$, the function f is increasing and is an **exponential growth function**. The base b is its **growth factor**.

If $a > 0$ and $b < 1$, f is decreasing and is an **exponential decay function**. The base b is its **decay factor**.

In Example 3, g is an exponential growth function, and h is an exponential decay function. As x increases by 1, $g(x) = 4 \cdot 3^x$ grows by a factor of 3, and $h(x) = 8 \cdot (1/4)^x$ decays by a factor of 1/4. The base of an exponential function, like the slope of a linear function, tells us whether the function is increasing or decreasing and by how much.

So far, we have focused most of our attention on the algebraic and numerical aspects of exponential functions. We now turn our attention to the graphs of these functions.

EXPLORATION 1 Graphs of Exponential Functions

1. Graph each function in the viewing window $[-2, 2]$ by $[-1, 6]$.

   (a) $y_1 = 2^x$ **(b)** $y_2 = 3^x$ **(c)** $y_3 = 4^x$ **(d)** $y_4 = 5^x$

   • Which point is common to all four graphs?

   • Analyze the functions for domain, range, continuity, increasing or decreasing behavior, symmetry, boundedness, extrema, asymptotes, and end behavior.

2. Graph each function in the viewing window $[-2, 2]$ by $[-1, 6]$.

   (a) $y_1 = \left(\dfrac{1}{2}\right)^x$ **(b)** $y_2 = \left(\dfrac{1}{3}\right)^x$

   (c) $y_3 = \left(\dfrac{1}{4}\right)^x$ **(d)** $y_4 = \left(\dfrac{1}{5}\right)^x$

   • Which point is common to all four graphs?

   • Analyze the functions for domain, range, continuity, increasing or decreasing behavior, symmetry, boundedness, extrema, asymptotes, and end behavior.

We summarize what we have learned about exponential functions with an initial value of 1.

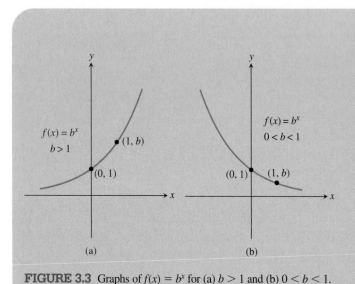

(a)

(b)

FIGURE 3.3 Graphs of $f(x) = b^x$ for (a) $b > 1$ and (b) $0 < b < 1$.

Exponential Functions $f(x) = b^x$

Domain: All reals
Range: $(0, \infty)$
Continuous
No symmetry: neither even nor odd
Bounded below, but not above
No local extrema
Horizontal asymptote: $y = 0$
No vertical asymptotes
If $b > 1$ (see Figure 3.3a), then

- f is an increasing function,
- $\lim\limits_{x \to -\infty} f(x) = 0$ and $\lim\limits_{x \to \infty} f(x) = \infty$.

If $0 < b < 1$ (see Figure 3.3b), then

- f is a decreasing function,
- $\lim\limits_{x \to -\infty} f(x) = \infty$ and $\lim\limits_{x \to \infty} f(x) = 0$.

The translations, reflections, stretches, and shrinks studied in Section 1.5 together with our knowledge of the graphs of basic exponential functions allow us to predict the graphs of the functions in Example 4.

EXAMPLE 4 Transforming Exponential Functions

Describe how to transform the graph of $f(x) = 2^x$ into the graph of the given function. Sketch the graphs by hand and support your answer with a grapher.

(a) $g(x) = 2^{x-1}$ **(b)** $h(x) = 2^{-x}$ **(c)** $k(x) = 3 \cdot 2^x$

SOLUTION

(a) The graph of $g(x) = 2^{x-1}$ is obtained by translating the graph of $f(x) = 2^x$ by 1 unit to the right (Figure 3.4a).

(b) We can obtain the graph of $h(x) = 2^{-x}$ by reflecting the graph of $f(x) = 2^x$ across the y-axis (Figure 3.4b). Because $2^{-x} = (2^{-1})^x = (1/2)^x$, we can also think of h as an exponential function with an initial value of 1 and a base of 1/2.

(c) We can obtain the graph of $k(x) = 3 \cdot 2^x$ by vertically stretching the graph of $f(x) = 2^x$ by a factor of 3 (Figure 3.4c).   *Now try Exercise 15.*

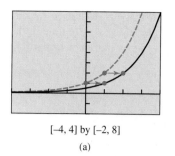
[−4, 4] by [−2, 8]
(a)

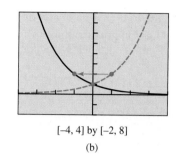
[−4, 4] by [−2, 8]
(b)

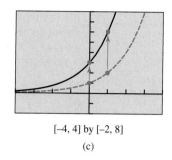
[−4, 4] by [−2, 8]
(c)

FIGURE 3.4 The graph of $f(x) = 2^x$ shown with (a) $g(x) = 2^{x-1}$, (b) $h(x) = 2^{-x}$, and (c) $k(x) = 3 \cdot 2^x$. (Example 4)

The Natural Base *e*

The function $f(x) = e^x$ is one of the basic functions introduced in Section 1.3, and is an exponential growth function.

BASIC FUNCTION **The Exponential Function**

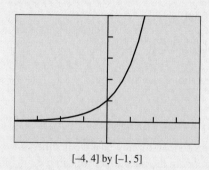
[−4, 4] by [−1, 5]

FIGURE 3.5 The graph of $f(x) = e^x$.

$f(x) = e^x$
Domain: All reals
Range: $(0, \infty)$
Continuous
Increasing for all x
No symmetry
Bounded below, but not above
No local extrema
Horizontal asymptote: $y = 0$
No vertical asymptotes
End behavior: $\lim\limits_{x \to -\infty} e^x = 0$ and $\lim\limits_{x \to \infty} e^x = \infty$

 Because $f(x) = e^x$ is increasing, it is an exponential growth function, so $e > 1$. But what is e, and what makes this exponential function *the* exponential function?

The letter e is the initial of the last name of Leonhard Euler (1707–1783), who introduced the notation. Because $f(x) = e^x$ has special calculus properties that simplify many calculations, e is the *natural base* of exponential functions for calculus purposes, and $f(x) = e^x$ is considered the *natural exponential function*.

DEFINITION **The Natural Base *e***

$$e = \lim_{x \to \infty} \left(1 + \frac{1}{x}\right)^x$$

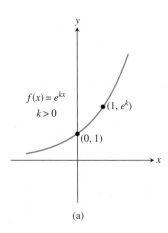

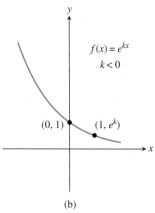

FIGURE 3.6 Graphs of $f(x) = e^{kx}$ for (a) $k > 0$ and (b) $k < 0$.

We cannot compute the irrational number e directly, but using this definition we can obtain successively closer approximations to e, as shown in Table 3.4. Continuing the process in Table 3.4 with a sufficiently accurate computer, it can be shown that $e \approx 2.718281828459$.

Table 3.4 Approximations Approaching the Natural Base e						
x	1	10	100	1000	10,000	100,000
$(1 + 1/x)^x$	2	2.5...	2.70...	2.716...	2.7181...	2.71826...

We are usually more interested in the exponential function $f(x) = e^x$ and variations of this function than in the irrational number e. In fact, *any* exponential function can be expressed in terms of the natural base e.

> **THEOREM Exponential Functions and the Base e**
>
> Any exponential function $f(x) = a \cdot b^x$ can be rewritten as
>
> $$f(x) = a \cdot e^{kx},$$
>
> for an appropriately chosen real number constant k.
>
> If $a > 0$ and $k > 0$, $f(x) = a \cdot e^{kx}$ is an exponential growth function. (See Figure 3.6a.)
>
> If $a > 0$ and $k < 0$, $f(x) = a \cdot e^{kx}$ is an exponential decay function. (See Figure 3.6b.)

In Section 3.3 we will develop some mathematics so that, for any positive number $b \neq 1$, we can easily find the value of k such that $e^{kx} = b^x$. In the meantime, we can use graphical and numerical methods to approximate k, as you will discover in Exploration 2.

> **EXPLORATION 2 Choosing k so that $e^{kx} = 2^x$**
>
> 1. Graph $f(x) = 2^x$ in the viewing window $[-4, 4]$ by $[-2, 8]$.
> 2. One at a time, overlay the graphs of $g(x) = e^{kx}$ for $k = 0.4, 0.5, 0.6, 0.7$, and 0.8. For which of these values of k does the graph of g most closely match the graph of f?
> 3. Using tables, find the 3-decimal-place value of k for which the values of g most closely approximate the values of f.

EXAMPLE 5   Transforming Exponential Functions

Describe how to transform the graph of $f(x) = e^x$ into the graph of the given function. Sketch the graphs by hand and support your answer with a grapher.

(a) $g(x) = e^{2x}$ **(b)** $h(x) = e^{-x}$ **(c)** $k(x) = 3e^x$

continued

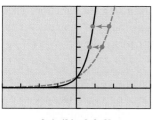

[-4, 4] by [-2, 8]

(a)

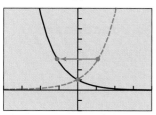

[-4, 4] by [-2, 8]

(b)

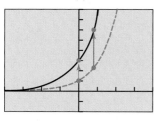

[-4, 4] by [-2, 8]

(c)

FIGURE 3.7 The graph of $f(x) = e^x$ shown with (a) $g(x) = e^{2x}$, (b) $h(x) = e^{-x}$, and (c) $k(x) = 3e^x$. (Example 5)

SOLUTION

(a) The graph of $g(x) = e^{2x}$ is obtained by horizontally shrinking the graph of $f(x) = e^x$ by a factor of 2 (Figure 3.7a).

(b) We can obtain the graph of $h(x) = e^{-x}$ by reflecting the graph of $f(x) = e^x$ across the y-axis (Figure 3.7b).

(c) We can obtain the graph of $k(x) = 3e^x$ by vertically stretching the graph of $f(x) = e^x$ by a factor of 3 (Figure 3.7c). ***Now try Exercise 21.***

Logistic Functions and Their Graphs

Exponential growth is *unrestricted*. An exponential growth function increases at an ever increasing rate and is not bounded above. In many growth situations, however, there is a limit to the possible growth. A plant can only grow so tall. The number of goldfish in an aquarium is limited by the size of the aquarium. In such situations the growth often begins in an exponential manner, but the growth eventually slows and the graph levels out. The associated growth function is bounded both below and above by horizontal asymptotes.

DEFINITION Logistic Growth Functions

Let a, b, c, and k be positive constants, with $b < 1$. A **logistic growth function** in x is a function that can be written in the form

$$f(x) = \frac{c}{1 + a \cdot b^x} \text{ or } f(x) = \frac{c}{1 + a \cdot e^{-kx}}$$

where the constant c is the **limit to growth**.

If $b > 1$ or $k < 0$, these formulas yield **logistic decay functions**. Unless otherwise stated, all *logistic functions* in this book will be logistic growth functions.

By setting $a = c = k = 1$, we obtain the **logistic function**

$$f(x) = \frac{1}{1 + e^{-x}}.$$

This function, though related to the exponential function e^x, *cannot* be obtained from e^x by translations, reflections, and horizontal and vertical stretches and shrinks. So we give the logistic function a formal introduction:

BASIC FUNCTION **The Logistic Function**

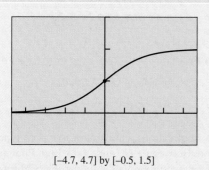

[-4.7, 4.7] by [-0.5, 1.5]

FIGURE 3.8 The graph of $f(x) = 1/(1 + e^{-x})$.

$$f(x) = \frac{1}{1 + e^{-x}}$$

Domain: All reals
Range: (0, 1)
Continuous
Increasing for all x
Symmetric about (0, 1/2), but neither even nor odd
Bounded below and above
No local extrema
Horizontal asymptotes: $y = 0$ and $y = 1$
No vertical asymptotes
End behavior: $\lim\limits_{x \to -\infty} f(x) = 0$ and $\lim\limits_{x \to \infty} f(x) = 1$

All logistic growth functions have graphs much like the basic logistic function. Their end behavior is always described by the equations

$$\lim_{x \to -\infty} f(x) = 0 \text{ and } \lim_{x \to \infty} f(x) = c,$$

where c is the limit to growth (see Exercise 73). All logistic functions are bounded by their horizontal asymptotes, $y = 0$ and $y = c$, and have a range of $(0, c)$. Although every logistic function is symmetric about the point of its graph with y-coordinate $c/2$, this point of symmetry is usually not the y-intercept, as we can see in Example 6.

EXAMPLE 6   Graphing Logistic Growth Functions

Graph the function. Find the y-intercept and the horizontal asymptotes.

(a) $f(x) = \dfrac{8}{1 + 3 \cdot 0.7^x}$ **(b)** $g(x) = \dfrac{20}{1 + 2e^{-3x}}$

SOLUTION

(a) The graph of $f(x) = 8/(1 + 3 \cdot 0.7^x)$ is shown in Figure 3.9a. The y-intercept is

$$f(0) = \frac{8}{1 + 3 \cdot 0.7^0} = \frac{8}{1 + 3} = 2.$$

Because the limit to growth is 8, the horizontal asymptotes are $y = 0$ and $y = 8$.

(b) The graph of $g(x) = 20/(1 + 2e^{-3x})$ is shown in Figure 3.9b. The y-intercept is

$$g(0) = \frac{20}{1 + 2e^{-3 \cdot 0}} = \frac{20}{1 + 2} = 20/3 \approx 6.67.$$

Because the limit to growth is 20, the horizontal asymptotes are $y = 0$ and $y = 20$.

Now try Exercise 41.

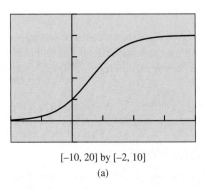
[−10, 20] by [−2, 10]
(a)

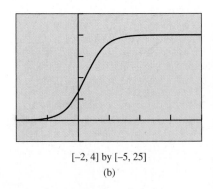
[−2, 4] by [−5, 25]
(b)

FIGURE 3.9 The graphs of (a) $f(x) = 8/(1 + 3 \cdot 0.7^x)$ and (b) $g(x) = 20/(1 + 2e^{-3x})$. (Example 6)

Population Models

Exponential and logistic functions have many applications. One area where both types of functions are used is in modeling population. Between 1990 and 2000, both Phoenix and San Antonio passed the 1 million mark. With its Silicon Valley industries, San Jose,

A NOTE ON POPULATION DATA

When the U.S. Census Bureau reports a population for a given year, it generally represents the population at the middle of the year, or July 1. We will assume this to be the case when interpreting our results to population problems.

Table 3.5 The Population of San Jose, California	
Year	Population
1990	782,248
2000	895,193

Source: *World Almanac and Book of Facts 2005.*

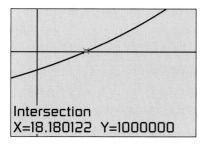

Intersection
X=18.180122 Y=1000000

[−10, 60] by [0, 1 500 000]

FIGURE 3.10 A population model for San Jose, California. (Example 7)

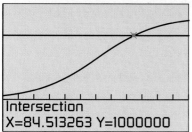

Intersection
X=84.513263 Y=1000000

[0, 120] by [−500 000, 1 500 000]

FIGURE 3.11 A population model for Dallas, Texas. (Example 8)

California appears to be the next U.S. city destined to surpass 1 million residents. When a city's population is growing rapidly, as in the case of San Jose, exponential growth is a reasonable model.

EXAMPLE 7 Modeling San Jose's Population

Using the data in Table 3.5 and assuming the growth is exponential, when will the population of San Jose surpass 1 million persons?

SOLUTION

Model Let $P(t)$ be the population of San Jose t years after 1990. Because P is exponential, $P(t) = P_0 \cdot b^t$, where P_0 is the initial (1990) population of 782,248. From Table 3.5 we see that $P(10) = 782,248 b^{10} = 895,193$. So,

$$b = \sqrt[10]{\frac{895,193}{782,248}} \approx 1.0136$$

and $P(t) = 782,248 \cdot 1.0136^t$.

Solve Graphically Figure 3.10 shows that this population model intersects $y = 1,000,000$ when the independent variable is about 18.18.

Interpret Because $1990 + 18 = 2008$, if the growth of its population is exponential, San Jose would surpass the 1 million mark in 2008. *Now try Exercise 51.*

While San Jose's population is soaring, in other major cities, such as Dallas, the population growth is slowing. The once sprawling Dallas is now *constrained* by its neighboring cities. *A logistic function is often an appropriate model for restricted growth*, such as the growth that Dallas is experiencing.

EXAMPLE 8 Modeling Dallas's Population

Based on recent census data, a logistic model for the population of Dallas, t years after 1900, is as follows:

$$P(t) = \frac{1,301,642}{1 + 21.602e^{-0.05054t}}$$

According to this model, when was the population 1 million?

SOLUTION

Figure 3.11 shows that the population model intersects $y = 1,000,000$ when the independent variable is about 84.51. Because $1900 + 85 = 1985$, if Dallas's population has followed this logistic model, its population was 1 million at the beginning of 1985. *Now try Exercise 55.*

QUICK REVIEW 3.1   *(For help, go to Sections A.1 and P.1.)*

In Exercises 1–4, evaluate the expression without using a calculator.

1. $\sqrt[3]{-216}$

2. $\sqrt[3]{\dfrac{125}{8}}$

3. $27^{2/3}$

4. $4^{5/2}$

In Exercises 5–8, rewrite the expression using a single positive exponent.

5. $(2^{-3})^4$ **6.** $(3^4)^{-2}$

7. $(a^{-2})^3$ **8.** $(b^{-3})^{-5}$

In Exercises 9–10, use a calculator to evaluate the expression.

9. $\sqrt[5]{-5.37824}$ **10.** $\sqrt[4]{92.3521}$

SECTION 3.1 EXERCISES

In Exercises 1–6, which of the following are exponential functions? For those that are exponential functions, state the initial value and the base. For those that are not, explain why not.

1. $y = x^8$

2. $y = 3^x$

3. $y = 5^x$

4. $y = 4^2$

5. $y = x^{\sqrt{x}}$

6. $y = x^{1.3}$

In Exercises 7–10, compute the exact value of the function for the given x-value without using a calculator.

7. $f(x) = 3 \cdot 5^x$ for $x = 0$

8. $f(x) = 6 \cdot 3^x$ for $x = -2$

9. $f(x) = -2 \cdot 3^x$ for $x = 1/3$

10. $f(x) = 8 \cdot 4^x$ for $x = -3/2$

In Exercises 11 and 12, determine a formula for the exponential function whose values are given in Table 3.6.

11. $f(x)$

12. $g(x)$

Table 3.6 Values for Two Exponential Functions		
x	$f(x)$	$g(x)$
-2	6	108
-1	3	36
0	3/2	12
1	3/4	4
2	3/8	4/3

In Exercises 13 and 14, determine a formula for the exponential function whose graph is shown in the figure.

13. $f(x)$ **14.** $g(x)$

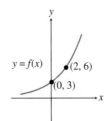
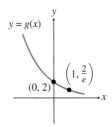

In Exercises 15–24, describe how to transform the graph of f into the graph of g. Sketch the graphs by hand and support your answer with a grapher.

15. $f(x) = 2^x,\ g(x) = 2^{x-3}$

16. $f(x) = 3^x,\ g(x) = 3^{x+4}$

17. $f(x) = 4^x,\ g(x) = 4^{-x}$

18. $f(x) = 2^x,\ g(x) = 2^{5-x}$

19. $f(x) = 0.5^x,\ g(x) = 3 \cdot 0.5^x + 4$

20. $f(x) = 0.6^x,\ g(x) = 2 \cdot 0.6^{3x}$

21. $f(x) = e^x,\ g(x) = e^{-2x}$

22. $f(x) = e^x,\ g(x) = -e^{-3x}$

23. $f(x) = e^x,\ g(x) = 2e^{3-3x}$

24. $f(x) = e^x,\ g(x) = 3e^{2x} - 1$

In Exercises 25–30, **(a)** match the given function with its graph. **(b) Writing to Learn** Explain how to make the choice without using a grapher.

25. $y = 3^x$

26. $y = 2^{-x}$

27. $y = -2^x$

28. $y = -0.5^x$

29. $y = 3^{-x} - 2$

30. $y = 1.5^x - 2$

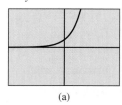

(a)

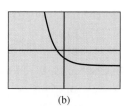

(b)

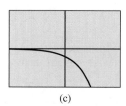

(c)

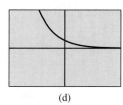

(d)

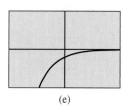

(e)

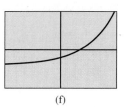

(f)

In Exercises 31–34, state whether the function is an exponential growth function or exponential decay function, and describe its end behavior using limits.

31. $f(x) = 3^{-2x}$

32. $f(x) = \left(\dfrac{1}{e}\right)^x$

33. $f(x) = 0.5^x$

34. $f(x) = 0.75^{-x}$

In Exercises 35–38, solve the inequality graphically.

35. $9^x < 4^x$

36. $6^{-x} > 8^{-x}$

37. $\left(\dfrac{1}{4}\right)^x > \left(\dfrac{1}{3}\right)^x$

38. $\left(\dfrac{1}{3}\right)^x < \left(\dfrac{1}{2}\right)^x$

Group Activity In Exercises 39 and 40, use the properties of exponents to prove that two of the given three exponential functions are identical. Support graphically.

39. (a) $y_1 = 3^{2x+4}$

   (b) $y_2 = 3^{2x} + 4$

   (c) $y_3 = 9^{x+2}$

40. (a) $y_1 = 4^{3x-2}$

   (b) $y_2 = 2(2^{3x-2})$

   (c) $y_3 = 2^{3x-1}$

In Exercises 41–44, use a grapher to graph the function. Find the y-intercept and the horizontal asymptotes.

41. $f(x) = \dfrac{12}{1 + 2 \cdot 0.8^x}$

42. $f(x) = \dfrac{18}{1 + 5 \cdot 0.2^x}$

43. $f(x) = \dfrac{16}{1 + 3e^{-2x}}$

44. $g(x) = \dfrac{9}{1 + 2e^{-x}}$

In Exercises 45–50, graph the function and analyze it for domain, range, continuity, increasing or decreasing behavior, symmetry, boundedness, extrema, asymptotes, and end behavior.

45. $f(x) = 3 \cdot 2^x$

46. $f(x) = 4 \cdot 0.5^x$

47. $f(x) = 4 \cdot e^{3x}$

48. $f(x) = 5 \cdot e^{-x}$

49. $f(x) = \dfrac{5}{1 + 4 \cdot e^{-2x}}$

50. $f(x) = \dfrac{6}{1 + 2 \cdot e^{-x}}$

51. Population Growth Using the data in Table 3.7 and assuming the growth is exponential, when would the population of Austin surpass 800,000 persons?

52. Population Growth Using the data in Table 3.7 and assuming the growth is exponential, when would the population of Columbus surpass 800,000 persons?

53. Population Growth Using the data in Table 3.7 and assuming the growth is exponential, when would the populations of Austin and Columbus be equal?

54. Population Growth Using the data in Table 3.7 and assuming the growth is exponential, which city—Austin or Columbus—would reach a population of 1 million first, and in what year?

Table 3.7 Populations of Two Major U.S. Cities

City	1990 Population	2000 Population
Austin, Texas	465,622	656,562
Columbus, Ohio	632,910	711,265

Source: World Almanac and Book of Facts 2005.

55. Population Growth Using 20th-century U.S. census data, the population of Ohio can be modeled by $P(t) = 12.79/(1 + 2.402e^{-0.0309x})$, where P is the population in millions and t is the number of years since 1900. Based on this model, when was the population of Ohio 10 million?

56. Population Growth Using 20th century U.S. census data, the population of New York state can be modeled by

$$P(t) = \frac{19.875}{1 + 57.993e^{-0.035005t}},$$

where P is the population in millions and t is the number of years since 1800. Based on this model,

(a) What was the population of New York in 1850?

(b) What will New York state's population be in 2010?

(c) What is New York's *maximum sustainable population* (limit to growth)?

57. Bacteria Growth The number B of bacteria in a petri dish culture after t hours is given by

$$B = 100e^{0.693t}.$$

(a) What was the initial number of bacteria present?

(b) How many bacteria are present after 6 hours?

58. Carbon Dating The amount C in grams of carbon-14 present in a certain substance after t years is given by

$$C = 20e^{-0.0001216t}.$$

(a) What was the initial amount of carbon-14 present?

(b) How much is left after 10,400 years? When will the amount left be 10 g?

Standardized Test Questions

59. True or False Every exponential function is strictly increasing. Justify your answer.

60. True or False Every logistic growth function has two horizontal asymptotes. Justify your answer.

In Exercises 61–64, solve the problem without using a calculator.

61. Multiple Choice Which of the following functions is exponential?

(A) $f(x) = a^2$

(B) $f(x) = x^3$

(C) $f(x) = x^{2/3}$

(D) $f(x) = \sqrt[3]{x}$

(E) $f(x) = 8^x$

62. Multiple Choice What point do all functions of the form $f(x) = b^x$ $(b > 0)$ have in common?

(A) $(1, 1)$

(B) $(1, 0)$

(C) $(0, 1)$

(D) $(0, 0)$

(E) $(-1, -1)$

63. Multiple Choice The growth factor for $f(x) = 4 \cdot 3^x$ is

(A) 3.   **(B)** 4.   **(C)** 12.

(D) 64.   **(E)** 81.

64. Multiple Choice For $x > 0$, which of the following is **true?**

(A) $3^x > 4^x$ **(B)** $7^x > 5^x$ **(C)** $(1/6)^x > (1/2)^x$

(D) $9^{-x} > 8^{-x}$ **(E)** $0.17^x > 0.32^x$

Explorations

65. Graph each function and analyze it for domain, range, increasing or decreasing behavior, boundedness, extrema, asymptotes, and end behavior.

(a) $f(x) = x \cdot e^x$ **(b)** $g(x) = \dfrac{e^{-x}}{x}$

66. Use the properties of exponents to solve each equation. Support graphically.

(a) $2^x = 4^2$ **(b)** $3^x = 27$

(c) $8^{x/2} = 4^{x+1}$ **(d)** $9^x = 3^{x+1}$

Extending the Ideas

67. Writing to Learn Table 3.8 gives function values for $y = f(x)$ and $y = g(x)$. Also, three different graphs are shown.

Table 3.8 Data for Two Functions

x	$f(x)$	$g(x)$
1.0	5.50	7.40
1.5	5.35	6.97
2.0	5.25	6.44
2.5	5.17	5.76
3.0	5.13	4.90
3.5	5.09	3.82
4.0	5.06	2.44
4.5	5.05	0.71

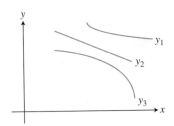

(a) Which curve of those shown in the graph most closely resembles the graph of $y = f(x)$? Explain your choice.

(b) Which curve most closely resembles the graph of $y = g(x)$? Explain your choice.

68. Writing to Learn Let $f(x) = 2^x$. Explain why the graph of $f(ax + b)$ can be obtained by applying one transformation to the graph of $y = c^x$ for an appropriate value of c. What is c?

Exercises 69–72 refer to the expression $f(a, b, c) = a \cdot b^c$. If $a = 2$, $b = 3$, and $c = x$, the expression is $f(2, 3, x) = 2 \cdot 3^x$, an exponential function.

69. If $b = x$, state conditions on a and c under which the expression $f(a, b, c)$ is a quadratic power function.

70. If $b = x$, state conditions on a and c under which the expression $f(a, b, c)$ is a decreasing linear function.

71. If $c = x$, state conditions on a and b under which the expression $f(a, b, c)$ is an increasing exponential function.

72. If $c = x$, state conditions on a and b under which the expression $f(a, b, c)$ is a decreasing exponential function.

73. Prove that $\lim\limits_{x \to -\infty} \dfrac{c}{1 + a \cdot b^x} = 0$ and $\lim\limits_{x \to \infty} \dfrac{c}{1 + a \cdot b^x} = c$, for constants a, b, and c, with $a > 0$, $0 < b < 1$, and $c > 0$.

3.2

Exponential and Logistic Modeling

. . . and why

Exponential functions model many types of unrestricted growth; logistic functions model restricted growth, including the spread of disease and the spread of rumors.

Constant Percentage Rate and Exponential Functions

Suppose that a population is changing at a **constant percentage rate r**, where r is the percent rate of change expressed in decimal form. Then the population follows the pattern shown.

Time in years	Population
0	$P(0) = P_0 =$ initial population
1	$P(1) = P_0 + P_0 r = P_0(1 + r)$
2	$P(2) = P(1) \cdot (1 + r) = P_0(1 + r)^2$
3	$P(3) = P(2) \cdot (1 + r) = P_0(1 + r)^3$
$\vdots$	$\vdots$
t	$P(t) = P_0(1 + r)^t$

So, in this case, the population is an exponential function of time.

Exponential Population Model

If a population P is changing at a constant percentage rate r each year, then

$$P(t) = P_0(1 + r)^t,$$

where P_0 is the initial population, r is expressed as a decimal, and t is time in years.

If $r > 0$, then $P(t)$ is an exponential growth function, and its *growth factor* is the base of the exponential function, $1 + r$.

On the other hand, if $r < 0$, the base $1 + r < 1$, $P(t)$ is an exponential decay function, and $1 + r$ is the *decay factor* for the population.

EXAMPLE 1   Finding Growth and Decay Rates

Tell whether the population model is an exponential growth function or exponential decay function, and find the constant percentage rate of growth or decay.

(a) San Jose: $P(t) = 782{,}248 \cdot 1.0136^t$

(b) Detroit: $P(t) = 1{,}203{,}368 \cdot 0.9858^t$

SOLUTION

(a) Because $1 + r = 1.0136$, $r = 0.0136 > 0$. So, P is an exponential growth function with a growth rate of 1.36%.

(b) Because $1 + r = 0.9858$, $r = -0.0142 < 0$. So, P is an exponential decay function with a decay rate of 1.42%.   *Now try Exercise 1.*

EXAMPLE 2   Finding an Exponential Function

Determine the exponential function with initial value $= 12$, increasing at a rate of 8% per year.

SOLUTION

Because $P_0 = 12$ and $r = 8\% = 0.08$, the function is $P(t) = 12(1 + 0.08)^t$ or $P(t) = 12 \cdot 1.08^t$. We could write this as $f(x) = 12 \cdot 1.08^x$, where x represents time.

Now try Exercise 7.

Exponential Growth and Decay Models

Exponential growth and decay models are used for populations of animals, bacteria, and even radioactive atoms. Exponential growth and decay apply to any situation where the growth is proportional to the current size of the quantity of interest. Such situations are frequently encountered in biology, chemistry, business, and the social sciences.

Exponential growth models can be developed in terms of the time it takes a quantity to double. On the flip side, exponential decay models can be developed in terms of the time it takes for a quantity to be halved. Examples 3–5 use these strategies.

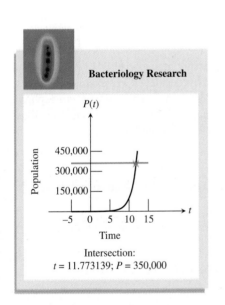

FIGURE 3.12 Rapid growth of a bacteria population. (Example 3)

EXAMPLE 3   Modeling Bacteria Growth

Suppose a culture of 100 bacteria is put into a petri dish and the culture doubles every hour. Predict when the number of bacteria will be 350,000.

SOLUTION

Model

$$200 = 100 \cdot 2 \qquad \text{Total bacteria after 1 hr}$$
$$400 = 100 \cdot 2^2 \qquad \text{Total bacteria after 2 hr}$$
$$800 = 100 \cdot 2^3 \qquad \text{Total bacteria after 3 hr}$$
$$\vdots$$
$$P(t) = 100 \cdot 2^t \qquad \text{Total bacteria after } t \text{ hr}$$

So the function $P(t) = 100 \cdot 2^t$ represents the bacteria population t hr after it is placed in the petri dish.

Solve Graphically Figure 3.12 shows that the population function intersects $y = 350,000$ when $t \approx 11.77$.

Interpret The population of the bacteria in the petri dish will be 350,000 in about 11 hr and 46 min.
Now try Exercise 15.

Exponential decay functions model the amount of a radioactive substance present in a sample. The number of atoms of a specific element that change from a radioactive state to a nonradioactive state is a fixed fraction per unit time. The process is called **radioactive decay**, and the time it takes for half of a sample to change its state is the **half-life** of the radioactive substance.

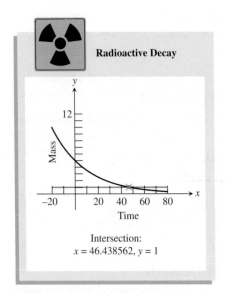

FIGURE 3.13 Radioactive decay.
(Example 4)

EXAMPLE 4   Modeling Radioactive Decay

Suppose the half-life of a certain radioactive substance is 20 days and there are 5 g (grams) present initially. Find the time when there will be 1 g of the substance remaining.

SOLUTION

Model If t is the time in days, the number of half-lives will be $t/20$.

$$\frac{5}{2} = 5\left(\frac{1}{2}\right)^{20/20} \qquad \text{Grams after 20 days}$$

$$\frac{5}{4} = 5\left(\frac{1}{2}\right)^{40/20} \qquad \text{Grams after } 2(20) = 40 \text{ days}$$

$$\vdots$$

$$f(t) = 5\left(\frac{1}{2}\right)^{t/20} \qquad \text{Grams after } t \text{ days}$$

Thus the function $f(t) = 5 \cdot 0.5^{t/20}$ models the mass in grams of the radioactive substance at time t.

Solve Graphically Figure 3.13 shows that the graph of $f(t) = 5 \cdot 0.5^{t/20}$ intersects $y = 1$ when $t \approx 46.44$.

Interpret There will be 1 g of the radioactive substance left after approximately 46.44 days, or about 46 days, 11 hr.   ***Now try Exercise 33.***

Scientists have established that atmospheric pressure at sea level is 14.7 lb/in.², and the pressure is reduced by half for each 3.6 mi above sea level. For example, the pressure 3.6 mi above sea level is $(1/2)(14.7) = 7.35$ lb/in.². This rule for atmospheric pressure holds for altitudes up to 50 mi above sea level. Though the context is different, the mathematics of atmospheric pressure closely resembles the mathematics of radioactive decay.

EXAMPLE 5   Determining Altitude from Atmospheric Pressure

Find the altitude above sea level at which the atmospheric pressure is 4 lb/in.².

SOLUTION
Model

$$7.35 = 14.7 \cdot 0.5^{3.6/3.6} \qquad \text{Pressure at 3.6 mi}$$

$$3.675 = 14.7 \cdot 0.5^{7.2/3.6} \qquad \text{Pressure at } 2(3.6) = 7.2 \text{ mi}$$

$$\vdots$$

$$P(h) = 14.7 \cdot 0.5^{h/3.6} \qquad \text{Pressure at } h \text{ mi}$$

So $P(h) = 14.7 \cdot 0.5^{h/3.6}$ models the atmospheric pressure P (in pounds per square inch) as a function of the height h (in miles above sea level). We must find the value of h that satisfies the equation

$$14.7 \cdot 0.5^{h/3.6} = 4.$$

continued

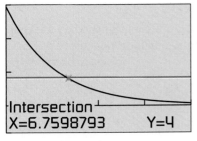

Intersection
X=6.7598793   Y=4

[0, 20] by [−4, 15]

FIGURE 3.14 A model for atmospheric pressure. (Example 5)

Solve Graphically Figure 3.14 shows that the graph of $P(h) = 14.7 \cdot 0.5^{h/3.6}$ intersects $y = 4$ when $h \approx 6.76$.

Interpret The atmospheric pressure is 4 lb/in.^{2} at an altitude of approximately 6.76 mi above sea level. **Now try Exercise 41.**

Using Regression to Model Population

So far, our models have been given to us or developed algebraically. We now use exponential and logistic regression to build models from population data.

Due to the post-World War II baby boom and other factors, exponential growth is not a perfect model for the U.S. population. It does, however, provide a means to make approximate predictions, as illustrated in Example 6.

Table 3.9 U.S. Population (in millions)

Year	Population
1900	76.2
1910	92.2
1920	106.0
1930	123.2
1940	132.2
1950	151.3
1960	179.3
1970	203.3
1980	226.5
1990	248.7
2000	281.4
2003	290.8

Source: World Almanac and Book of Facts 2005.

EXAMPLE 6   Modeling U.S. Population Using Exponential Regression

Use the 1900–2000 data in Table 3.9 and exponential regression to predict the U.S. population for 2003. Compare the result with the listed value for 2003.

SOLUTION

Model

Let $P(t)$ be the population (in millions) of the United States t years after 1900. Figure 3.15a shows a scatter plot of the data. Using exponential regression, we find a model for the 1990–2000 data

$$P(t) = 80.5514 \cdot 1.01289^t.$$

Figure 3.15b shows the scatter plot of the data with a graph of the population model just found. You can see that the curve fits the data fairly well. The coefficient of determination is $r^2 \approx 0.995$, indicating a close fit and supporting the visual evidence.

Solve Graphically

To predict the 2003 U.S. population we substitute $t = 103$ into the regression model. Figure 3.15c reports that $P(103) = 80.5514 \cdot 1.01289^{103} \approx 301.3$.

continued

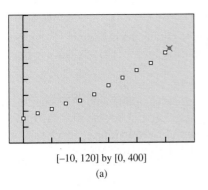

[−10, 120] by [0, 400]

(a)

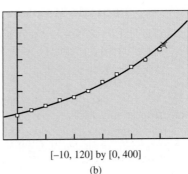

[−10, 120] by [0, 400]

(b)

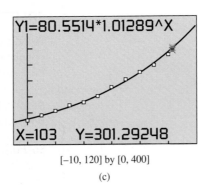

Y1=80.5514*1.01289^X

X=103   Y=301.29248

[−10, 120] by [0, 400]

(c)

FIGURE 3.15 Scatter plots and graphs for Example 6. The red "x" denotes the data point for 2003. The blue "x" in (c) denotes the model's prediction for 2003.

Interpret

The model predicts the U.S. population was 301.3 million in 2003. The actual population was 290.8 million. We overestimated by 10.5 million, less than a 4% error.

Now try Exercise 43.

Exponential growth is unrestricted, but population growth often is not. For many populations, the growth begins exponentially, but eventually slows and approaches a limit to growth called the **maximum sustainable population**.

In Section 3.1 we modeled Dallas's population with a logistic function. We now use logistic regression to do the same for the populations of Florida and Pennsylvania. As the data in Table 3.10 suggest, Florida had rapid growth in the second half of the 20th century, whereas Pennsylvania appears to be approaching its maximum sustainable population.

Table 3.10 Populations of Two U.S. States (in millions)

Year	Florida	Pennsylvania
1900	0.5	6.3
1910	0.8	7.7
1920	1.0	8.7
1930	1.5	9.6
1940	1.9	9.9
1950	2.8	10.5
1960	5.0	11.3
1970	6.8	11.8
1980	9.7	11.9
1990	12.9	11.9
2000	16.0	12.3

Source: U.S. Census Bureau.

EXAMPLE 7   Modeling Two States' Populations Using Logistic Regression

Use the data in Table 3.10 and logistic regression to predict the maximum sustainable populations for Florida and Pennsylvania. Graph the logistic models and interpret their significance.

SOLUTION Let $F(t)$ and $P(t)$ be the populations (in millions) of Florida and Pennsylvania, respectively, t years *after 1800*. Figure 3.16a shows a scatter plot of the data for both states; the data for Florida is shown in black, and for Pennsylvania, in red. Using logistic regression, we obtain the models for the two states:

$$F(t) = \frac{28.021}{1 + 9018.63e^{-0.047015t}} \quad \text{and} \quad P(t) = \frac{12.579}{1 + 29.0003e^{-0.034315t}}$$

continued

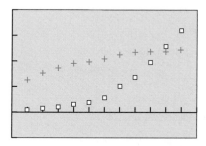

[90, 210] by [−5, 20]

(a)

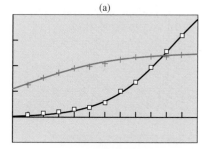

[90, 210] by [−5, 20]

(b)

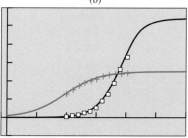

[−10, 300] by [−5, 30]

(c)

FIGURE 3.16 Scatter plots and graphs for Example 7.

Figure 3.16b shows the scatter plots of the data with graphs of the two population models. You can see that the curves fit the data fairly well. From the numerators of the models we see that

$$\lim_{t \to \infty} F(t) = 28.021 \quad \text{and} \quad \lim_{t \to \infty} P(t) = 12.579.$$

So the maximum sustainable population for Florida is about 28.0 million, and for Pennsylvania is about 12.6 million.

Figure 3.16c shows a three-century span for the two states. Pennsylvania had rapid growth in the 19th century and first half of the 20th century, and is now approaching its limit to growth. Florida, on the other hand, is currently experiencing extremely rapid growth but should be approaching its maximum sustainable population by the end of the 21st century.   ***Now try Exercise 50.***

Other Logistic Models

In Example 3, the bacteria cannot continue to grow exponentially forever because they cannot grow beyond the confines of the petri dish. In Example 7, though Florida's population is booming now, it will eventually level off, just as Pennsylvania's has done. Sunflowers and many other plants grow to a natural height following a logistic pattern. Chemical acid-base titration curves are logistic. Yeast cultures grow logistically. Contagious diseases and even rumors spread according to logistic models.

EXAMPLE 8   Modeling a Rumor

Watauga High School has 1200 students. Bob, Carol, Ted, and Alice start a rumor, which spreads logistically so that $S(t) = 1200/(1 + 39 \cdot e^{-0.9t})$ models the number of students who have heard the rumor by the end of t days, where $t = 0$ is the day the rumor begins to spread.

(a) How many students have heard the rumor by the end of Day 0?

(b) How long does it take for 1000 students to hear the rumor?

SOLUTION

(a) $S(0) = \dfrac{1200}{1 + 39 \cdot e^{-0.9 \cdot 0}} = \dfrac{1200}{1 + 39} = 30$. So, 30 students have heard the rumor by the end of Day 0.

(b) We need to solve $\dfrac{1200}{1 + 39e^{-0.9t}} = 1000$.

Figure 3.17 shows that the graph of $S(t) = 1200/(1 + 39 \cdot e^{-0.9t})$ intersects $y = 1000$ when $t \approx 5.86$. So toward the end of Day 6 the rumor has reached the ears of 1000 students.   ***Now try Exercise 45.***

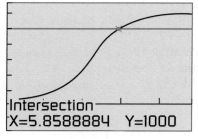

Intersection
X=5.8588884   Y=1000

[0, 10] by [−400, 1400]

FIGURE 3.17 The spread of a rumor. (Example 8)

QUICK REVIEW 3.2 *(For help, go to Section P.5.)*

In Exercises 1 and 2, convert the percent to decimal form or the decimal into a percent.

1. 15%

2. 0.04

3. Show how to increase 23 by 7% using a single multiplication.

4. Show how to decrease 52 by 4% using a single multiplication.

In Exercises 5 and 6, solve the equation algebraically.

5. $40 \cdot b^2 = 160$

6. $243 \cdot b^3 = 9$

In Exercises 7–10, solve the equation numerically.

7. $782b^6 = 838$ 　　　　**8.** $93b^5 = 521$

9. $672b^4 = 91$ 　　　　**10.** $127b^7 = 56$

SECTION 3.2 EXERCISES

In Exercises 1–6, tell whether the function is an exponential growth function or exponential decay function, and find the constant percentage rate of growth or decay.

1. $P(t) = 3.5 \cdot 1.09^t$ 　　　　**2.** $P(t) = 4.3 \cdot 1.018^t$

3. $f(x) = 78,963 \cdot 0.968^x$ 　　**4.** $f(x) = 5607 \cdot 0.9968^x$

5. $g(t) = 247 \cdot 2^t$ 　　　　**6.** $g(t) = 43 \cdot 0.05^t$

In Exercises 7–18, determine the exponential function that satisfies the given conditions.

7. Initial value = 5, increasing at a rate of 17% per year

8. Initial value = 52, increasing at a rate of 2.3% per day

9. Initial value = 16, decreasing at a rate of 50% per month

10. Initial value = 5, decreasing at a rate of 0.59% per week

11. Initial population = 28,900, decreasing at a rate of 2.6% per year

12. Initial population = 502,000, increasing at a rate of 1.7% per year

13. Initial height = 18 cm, growing at a rate of 5.2% per week

14. Initial mass = 15 g, decreasing at a rate of 4.6% per day

15. Initial mass = 0.6 g, doubling every 3 days

16. Initial population = 250, doubling every 7.5 hours

17. Initial mass = 592 g, halving once every 6 years

18. Initial mass = 17 g, halving once every 32 hours

In Exercises 19 and 20, determine a formula for the exponential function whose values are given in Table 3.11.

19. $f(x)$ 　　　　　　**20.** $g(x)$

Table 3.11 Values for Two Exponential Functions

x	$f(x)$	$g(x)$
-2	1.472	-9.0625
-1	1.84	-7.25
0	2.3	-5.8
1	2.875	-4.64
2	3.59375	-3.7123

In Exercises 21 and 22, determine a formula for the exponential function whose graph is shown in the figure.

21. 　　　　　　　　**22.**

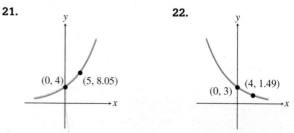

In Exercises 23–26, find the logistic function that satisfies the given conditions.

23. Initial value = 10, limit to growth = 40, passing through (1, 20).

24. Initial value = 12, limit to growth = 60, passing through (1, 24).

25. Initial population = 16, maximum sustainable population = 128, passing through (5, 32).

26. Initial height = 5, limit to growth = 30, passing through (3, 15).

In Exercises 27 and 28, determine a formula for the logistic function whose graph is shown in the figure.

27. **28.**

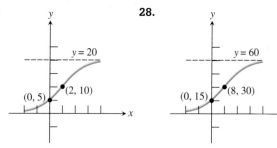

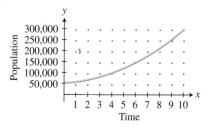

29. Exponential Growth The 2000 population of Jacksonville, Florida was 736,000 and was increasing at the rate of 1.49% each year. At that rate, when will the population be 1 million?

30. Exponential Growth The 2000 population of Las Vegas, Nevada was 478,000 and is increasing at the rate of 6.28% each year. At that rate, when will the population be 1 million?

31. Exponential Growth The population of Smallville in the year 1890 was 6250. Assume the population increased at a rate of 2.75% per year.

(a) Estimate the population in 1915 and 1940.

(b) Predict when the population reached 50,000.

32. Exponential Growth The population of River City in the year 1910 was 4200. Assume the population increased at a rate of 2.25% per year.

(a) Estimate the population in 1930 and 1945.

(b) Predict when the population reached 20,000.

33. Radioactive Decay The half-life of a certain radioactive substance is 14 days. There are 6.6 g present initially.

(a) Express the amount of substance remaining as a function of time t.

(b) When will there be less than 1 g remaining?

34. Radioactive Decay The half-life of a certain radioactive substance is 65 days. There are 3.5 g present initially.

(a) Express the amount of substance remaining as a function of time t.

(b) When will there be less than 1 g remaining?

35. Writing to Learn Without using formulas or graphs, compare and contrast exponential functions and linear functions.

36. Writing to Learn Without using formulas or graphs, compare and contrast exponential functions and logistic functions.

37. Writing to Learn Using the population model that is graphed, explain why the time it takes the population to double (doubling time) is independent of the population size.

38. Writing to Learn Explain why the half-life of a radioactive substance is independent of the initial amount of the substance that is present.

39. Bacteria Growth The number B of bacteria in a petri dish culture after t hours is given by

$$B = 100e^{0.693t}.$$

When will the number of bacteria be 200? Estimate the doubling time of the bacteria.

40. Radiocarbon Dating The amount C in grams of carbon-14 present in a certain substance after t years is given by

$$C = 20e^{-0.0001216t}.$$

Estimate the half-life of carbon-14.

41. Atmospheric Pressure Determine the atmospheric pressure outside an aircraft flying at 52,800 ft (10 mi above sea level).

42. Atmospheric Pressure Find the altitude above sea level at which the atmospheric pressure is 2.5 lb/in.^{2}.

43. Population Modeling Use the 1950–2000 data in Table 3.12 and exponential regression to predict Los Angeles's population for 2003. Compare the result with the listed value for 2003.

44. Population Modeling Use the 1950–2000 data in Table 3.12 and exponential regression to predict Phoenix's population for 2003. Compare the result with the listed value for 2003. Repeat these steps using 1960–2000 data to create the model.

Table 3.12 Populations of Two U.S. Cities (in thousands)

Year	Los Angeles	Phoenix
1950	1970	107
1960	2479	439
1970	2812	584
1980	2969	790
1990	3485	983
2000	3695	1321
2003	3820	1388

Source: World Almanac and Book of Facts, 2002, 2005.

45. Spread of Flu The number of students infected with flu at Springfield High School after t days is modeled by the function

$$P(t) = \frac{800}{1 + 49e^{-0.2t}}.$$

(a) What was the initial number of infected students?

(b) When will the number of infected students be 200?

(c) The school will close when 300 of the 800-student body are infected. When will the school close?

46. Population of Deer The population of deer after t years in Cedar State Park is modeled by the function

$$P(t) = \frac{1001}{1 + 90e^{-0.2t}}.$$

(a) What was the initial population of deer?

(b) When will the number of deer be 600?

(c) What is the maximum number of deer possible in the park?

47. Population Growth Using all of the data in Table 3.9, compute a logistic regression model, and use it to predict the U.S. population in 2010.

48. Population Growth Using the data in Table 3.13, confirm the model used in Example 8 of Section 3.1.

Table 3.13 Population of Dallas, Texas

Year	Population
1950	434,462
1960	679,684
1970	844,401
1980	904,599
1990	1,006,877
2000	1,188,589

Source: U.S. Census Bureau.

49. Population Growth Using the data in Table 3.14, confirm the model used in Exercise 56 of Section 3.1.

50. Population Growth Using the data in Table 3.14, compute a logistic regression model for Arizona's population for t years since 1800. Based on your model and the New York population model from Exercise 56 of Section 3.1, will the population of Arizona ever surpass that of New York? If so, when?

Table 3.14 Populations of Two U.S. States (in millions)

Year	Arizona	New York
1900	0.1	7.3
1910	0.2	9.1
1920	0.3	10.3
1930	0.4	12.6
1940	0.5	13.5
1950	0.7	14.8
1960	1.3	16.8
1970	1.8	18.2
1980	2.7	17.6
1990	3.7	18.0
2000	5.1	19.0

Source: U.S. Census Bureau.

Standardized Test Questions

51. True or False Exponential population growth is constrained with a maximum sustainable population. Justify your answer.

52. True or False If the constant percentage rate of an exponential function is negative, then the base of the function is negative. Justify your answer.

In Exercises 53–56, you may use a graphing calculator to solve the problem.

53. Multiple Choice What is the constant percentage growth rate of $P(t) = 1.23 \cdot 1.049^t$?

(A) 49%   **(B)** 23%   **(C)** 4.9%   **(D)** 2.3%   **(E)** 1.23%

54. Multiple Choice What is the constant percentage decay rate of $P(t) = 22.7 \cdot 0.834^t$?

(A) 22.7%   **(B)** 16.6%   **(C)** 8.34%

(D) 2.27%   **(E)** 0.834%

55. Multiple Choice A single cell amoeba doubles every 4 days. About how long will it take one amoeba to produce a population of 1000?

(A) 10 days   **(B)** 20 days   **(C)** 30 days

(D) 40 days   **(E)** 50 days

56. Multiple Choice A rumor spreads logistically so that $S(t) = 789/(1 + 16 \cdot e^{-0.8t})$ models the number of persons who have heard the rumor by the end of t days. Based on this model, which of the following is **true**?

(A) After 0 days, 16 people have heard the rumor.

(B) After 2 days, 439 people have heard the rumor.

(C) After 4 days, 590 people have heard the rumor.

(D) After 6 days, 612 people have heard the rumor.

(E) After 8 days, 769 people have heard the rumor.

Explorations

57. Population Growth **(a)** Use the 1900–1990 data in Table 3.9 and *logistic* regression to predict the U.S. population for 2000.

(b) Writing to Learn Compare the prediction with the value listed in the table for 2000.

(c) Noting the results of Example 6, which model—exponential or logistic—makes the better prediction in this case?

58. Population Growth Use the data in Tables 3.9 and 3.15.

(a) Based on exponential growth models, will Mexico's population surpass that of the United States, and if so, when?

(b) Based on logistic growth models, will Mexico's population surpass that of the United States, and if so, when?

(c) What are the maximum sustainable populations for the two countries?

(d) Writing to Learn Which model—exponential or logistic—is more valid in this case? Justify your choice.

Table 3.15 Population of Mexico (in millions)	
Year	Population
1900	13.6
1950	25.8
1960	34.9
1970	48.2
1980	66.8
1990	88.1
2001	101.9
2025	130.2
2050	154.0

Sources: 1992 Statesman's Yearbook and World Almanac and Book of Facts 2002.

Extending the Ideas

59. The **hyperbolic sine function** is defined by $\sinh(x) = (e^x - e^{-x})/2$. Prove that sinh is an odd function.

60. The **hyperbolic cosine function** is defined by $\cosh(x) = (e^x + e^{-x})/2$. Prove that cosh is an even function.

61. The **hyperbolic tangent function** is defined by $\tanh(x) = (e^x - e^{-x})/(e^x + e^{-x})$.

(a) Prove that $\tanh(x) = \sinh(x)/\cosh(x)$.

(b) Prove that tanh is an odd function.

(c) Prove that $f(x) = 1 + \tanh(x)$ is a logistic function.

3.3
Logarithmic Functions and Their Graphs

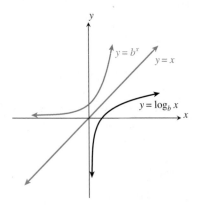

FIGURE 3.19 Because logarithmic functions are inverses of exponential functions, we can obtain the graph of a logarithmic function by the mirror or rotational methods discussed in Section 1.4.

A BIT OF HISTORY

Logarithmic functions were developed around 1594 as computational tools by Scottish mathematician John Napier (1550–1617). He originally called them "artificial numbers," but changed the name to logarithms, which means "reckoning numbers."

Inverses of Exponential Functions

In Section 1.4 we learned that, if a function passes the *horizontal line test*, then the inverse of the function is also a function. So an exponential function $f(x) = b^x$, has an inverse that is a function. See Figure 3.18. This inverse is the **logarithmic function with base b**, denoted $\log_b(x)$ or $\log_b x$. That is, if $f(x) = b^x$ with $b > 0$ and $b \neq 1$, then $f^{-1}(x) = \log_b x$. See Figure 3.19.

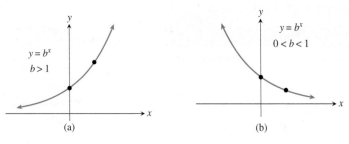

FIGURE 3.18 Exponential functions are either (a) increasing or (b) decreasing.

An immediate and useful consequence of this definition is the link between an exponential equation and its logarithmic counterpart.

Changing Between Logarithmic and Exponential Form

If $x > 0$ and $0 < b \neq 1$, then

$$y = \log_b(x) \quad \text{if and only if} \quad b^y = x.$$

This linking statement says that *a logarithm is an exponent*. Because logarithms are exponents, we can evaluate simple logarithmic expressions using our understanding of exponents.

EXAMPLE 1   Evaluating Logarithms

(a) $\log_2 8 = 3$ because $2^3 = 8$.

(b) $\log_3 \sqrt{3} = 1/2$ because $3^{1/2} = \sqrt{3}$.

(c) $\log_5 \dfrac{1}{25} = -2$ because $5^{-2} = \dfrac{1}{5^2} = \dfrac{1}{25}$.

(d) $\log_4 1 = 0$ because $4^0 = 1$.

(e) $\log_7 7 = 1$ because $7^1 = 7$.   *Now try Exercise 1.*

We can generalize the relationships observed in Example 1.

Basic Properties of Logarithms

For $0 < b \neq 1$, $x > 0$, and any real number y,

- $\log_b 1 = 0$ because $b^0 = 1$.
- $\log_b b = 1$ because $b^1 = b$.
- $\log_b b^y = y$ because $b^y = b^y$.
- $b^{\log_b x} = x$ because $\log_b x = \log_b x$.

GENERALLY $b > 1$

In practice, logarithmic bases are almost always greater than 1.

These properties give us efficient ways to evaluate simple logarithms and some exponential expressions. The first two parts of Example 2 are the same as the first two parts of Example 1.

EXAMPLE 2   Evaluating Logarithmic and Exponential Expressions

(a) $\log_2 8 = \log_2 2^3 = 3$.

(b) $\log_3 \sqrt{3} = \log_3 3^{1/2} = 1/2$.

(c) $6^{\log_6 11} = 11$.   *Now try Exercise 5.*

Logarithmic functions are inverses of exponential functions. So the inputs and outputs are switched. Table 3.16 illustrates this relationship for $f(x) = 2^x$ and $f^{-1}(x) = \log_2 x$.

Table 3.16 An Exponential Function and Its Inverse

x	$f(x) = 2^x$	x	$f^{-1}(x) = \log_2 x$
-3	$1/8$	$1/8$	-3
-2	$1/4$	$1/4$	-2
-1	$1/2$	$1/2$	-1
0	1	1	0
1	2	2	1
2	4	4	2
3	8	8	3

This relationship can be used to produce both tables and graphs for logarithmic functions, as you will discover in Exploration 1.

The geometric transformations studied in Section 1.5 together with our knowledge of the graphs of $y = \ln x$ and $y = \log x$ allow us to predict the graphs of the functions in Example 8.

EXAMPLE 8   Transforming Logarithmic Graphs

Describe how to transform the graph of $y = \ln x$ or $y = \log x$ into the graph of the given function.

(a) $g(x) = \ln (x + 2)$ **(b)** $h(x) = \ln (3 - x)$

(c) $g(x) = 3 \log x$ **(d)** $h(x) = 1 + \log x$

SOLUTION

(a) The graph of $g(x) = \ln (x + 2)$ is obtained by translating the graph of $y = \ln (x)$ 2 units to the left. See Figure 3.25a.

(b) $h(x) = \ln (3 - x) = \ln [-(x - 3)]$. So we obtain the graph of $h(x) = \ln (3 - x)$ from the graph of $y = \ln x$ by applying, in order, a reflection across the y-axis followed by a translation 3 units to the right. See Figure 3.25b.

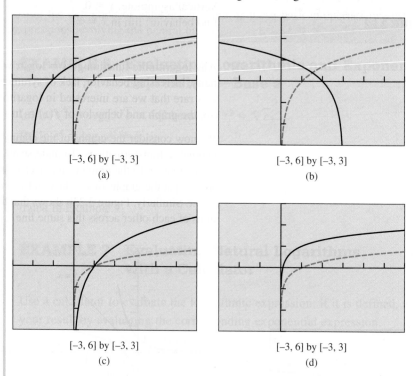

[−3, 6] by [−3, 3]

(a)

[−3, 6] by [−3, 3]

(b)

[−3, 6] by [−3, 3]

(c)

[−3, 6] by [−3, 3]

(d)

FIGURE 3.25 Transforming $y = \ln x$ to obtain (a) $g(x) = \ln(x + 2)$ and (b) $h(x) = \ln(3 - x)$; and $y = \log x$ to obtain (c) $g(x) = 3 \log x$ and (d) $h(x) = 1 + \log x$. (Example 8)

(c) The graph of $g(x) = 3 \log x$ is obtained by vertically stretching the graph of $f(x) = \log x$ by a factor of 3. See Figure 3.25c.

(d) We can obtain the graph of $h(x) = 1 + \log x$ from the graph of $f(x) = \log x$ by a translation 1 unit up. See Figure 3.25d.   ***Now try Exercise 41.***

Measuring Sound Using Decibels

Table 3.17 lists assorted sounds. Notice that a jet at takeoff is 100 trillion times as loud as a soft whisper. Because the range of audible sound intensities is so great, common logarithms (powers of 10) are used to compare how loud sounds are.

DEFINITION **Decibels**

The level of sound intensity in **decibels** (dB) is

$$\beta = 10 \log(I/I_0),$$

where β (beta) is the number of decibels, I is the sound intensity in W/m^2, and $I_0 = 10^{-12}$ W/m^2 is the threshold of human hearing (the quietest audible sound intensity).

BEL IS FOR BELL

The original unit for sound intensity level was the *bel* (B), which proved to be inconveniently large. So the decibel, one-tenth of a bel, has replaced it. The bel was named in honor of Scottish-born American Alexander Graham Bell (1847–1922), inventor of the telephone.

Table 3.17 Approximate Intensities of Selected Sounds

Sound	Intensity (W/m^2)
Hearing threshold	10^{-12}
Soft whisper at 5 m	10^{-11}
City traffic	10^{-5}
Subway train	10^{-2}
Pain threshold	10^0
Jet at takeoff	10^3

Source: Adapted from R. W. Reading, Exploring Physics: Concepts and Applications (Belmont, CA: Wadsworth, 1984).

CHAPTER OPENER PROBLEM (from page 275)

PROBLEM: How loud is a train inside a subway tunnel?

SOLUTION: Based on the data in Table 3.17,

$\beta = 10 \log(I/I_0)$

$\quad = 10 \log(10^{-2}/10^{-12})$

$\quad = 10 \log(10^{10})$

$\quad = 10 \cdot 10 = 100$

So the sound intensity level inside the subway tunnel is 100 dB.

QUICK REVIEW 3.3   *(For help, go to Section A.2.)*

In Exercises 1–6, evaluate the expression without using a calculator.

1. 5^{-2}

2. 10^{-3}

3. $\dfrac{4^0}{5}$

4. $\dfrac{1^0}{2}$

5. $\dfrac{8^{11}}{2^{28}}$

6. $\dfrac{9^{13}}{27^8}$

In Exercises 7–10, rewrite as a base raised to a rational number exponent.

7. $\sqrt{5}$

8. $\sqrt[3]{10}$

9. $\dfrac{1}{\sqrt{e}}$

10. $\dfrac{1}{\sqrt[3]{e^2}}$

SECTION 3.3 EXERCISES

In Exercises 1–18, evaluate the logarithmic expression without using a calculator.

1. $\log_4 4$

2. $\log_6 1$

3. $\log_2 32$

4. $\log_3 81$

5. $\log_5 \sqrt[3]{25}$

6. $\log_6 \dfrac{1}{\sqrt[5]{36}}$

7. $\log 10^3$

8. $\log 10{,}000$

9. $\log 100{,}000$

10. $\log 10^{-4}$

11. $\log \sqrt[3]{10}$

12. $\log \dfrac{1}{\sqrt{1000}}$

13. $\ln e^3$

14. $\ln e^{-4}$

15. $\ln \dfrac{1}{e}$

16. $\ln 1$

17. $\ln \sqrt[4]{e}$

18. $\ln \dfrac{1}{\sqrt{e^7}}$

In Exercises 19–24, evaluate the expression without using a calculator.

19. $7^{\log_7 3}$

20. $5^{\log_5 8}$

21. $10^{\log(0.5)}$

22. $10^{\log 14}$

23. $e^{\ln 6}$

24. $e^{\ln(1/5)}$

In Exercises 25–32, use a calculator to evaluate the logarithmic expression if it is defined, and check your result by evaluating the corresponding exponential expression.

25. $\log 9.43$

26. $\log 0.908$

27. $\log (-14)$

28. $\log (-5.14)$

29. $\ln 4.05$

30. $\ln 0.733$

31. $\ln (-0.49)$

32. $\ln (-3.3)$

In Exercises 33–36, solve the equation by changing it to exponential form.

33. $\log x = 2$

34. $\log x = 4$

35. $\log x = -1$

36. $\log x = -3$

In Exercises 37–40, match the function with its graph.

37. $f(x) = \log (1 - x)$

38. $f(x) = \log (x + 1)$

39. $f(x) = -\ln (x - 3)$

40. $f(x) = -\ln (4 - x)$

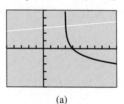

(a)

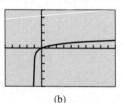

(b)

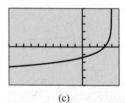

(c)

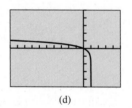

(d)

In Exercises 41–46, describe how to transform the graph of $y = \ln x$ into the graph of the given function. Sketch the graph by hand and support your sketch with a grapher.

41. $f(x) = \ln (x + 3)$

42. $f(x) = \ln (x) + 2$

43. $f(x) = \ln (-x) + 3$

44. $f(x) = \ln (-x) - 2$

45. $f(x) = \ln (2 - x)$

46. $f(x) = \ln (5 - x)$

In Exercises 47–52, describe how to transform the graph of $y = \log x$ into the graph of the given function. Sketch the graph by hand and support with a grapher.

47. $f(x) = -1 + \log (x)$

48. $f(x) = \log (x - 3)$

49. $f(x) = -2 \log (-x)$

50. $f(x) = -3 \log (-x)$

51. $f(x) = 2 \log (3 - x) - 1$

52. $f(x) = -3 \log (1 - x) + 1$

In Exercises 53–58, graph the function, and analyze it for domain, range, continuity, increasing or decreasing behavior, boundedness, extrema, symmetry, asymptotes, and end behavior.

53. $f(x) = \log (x - 2)$ **54.** $f(x) = \ln (x + 1)$

55. $f(x) = -\ln (x - 1)$ **56.** $f(x) = -\log (x + 2)$

57. $f(x) = 3 \log (x) - 1$ **58.** $f(x) = 5 \ln (2 - x) - 3$

59. Sound Intensity Use the data in Table 3.17 to compute the sound intensity in decibels for **(a)** a soft whisper, **(b)** city traffic, and **(c)** a jet at takeoff.

60. Light Absorption The Beer-Lambert law of absorption applied to Lake Erie states that the light intensity I (in lumens), at a depth of x feet, satisfies the equation

$$\log \frac{I}{12} = -0.00235x.$$

Find the intensity of the light at a depth of 30 ft.

61. Population Growth Using the data in Table 3.18, compute a logarithmic regression model, and use it to predict when the population of San Antonio will be 1,500,000.

Table 3.18 Population of San Antonio

Year	Population
1970	654,153
1980	785,940
1990	935,933
2000	1,151,305

Source: World Alamanac and Book of Facts 2005.

62. Population Decay Using the data in Table 3.19, compute a logarithmic regression model, and use it to predict when the population of Milwaukee will be 500,000.

Table 3.19 Population of Milwaukee

Year	Population
1970	717,372
1980	636,297
1990	628,088
2000	596,974

Source: World Alamanac and Book of Facts 2005.

Standardized Test Questions

63. True or False A logarithmic function is the inverse of an exponential function. Justify your answer.

64. True or False Common logarithms are logarithms with base 10. Justify your answer.

In Exercises 65–68, you may use a graphing calculator to solve the problem.

65. Multiple Choice What is the approximate value of the common log of 2?

(A) 0.10523 **(B)** 0.20000

(C) 0.30103 **(D)** 0.69315

(E) 3.32193

66. Multiple Choice Which statement is **false**?

(A) $\log 5 = 2.5 \log 2$ **(B)** $\log 5 = 1 - \log 2$

(C) $\log 5 > \log 2$ **(D)** $\log 5 < \log 10$

(E) $\log 5 = \log 10 - \log 2$

67. Multiple Choice Which statement is **false** about $f(x) = \ln x$?

(A) It is increasing on its domain.

(B) It is symmetric about the origin.

(C) It is continuous on its domain.

(D) It is unbounded.

(E) It has a vertical asymptote.

68. Multiple Choice Which of the following is the inverse of $f(x) = 2 \cdot 3^x$?

(A) $f^{-1}(x) = \log_3 (x/2)$ **(B)** $f^{-1}(x) = \log_2 (x/3)$

(C) $f^{-1}(x) = 2 \log_3 (x)$ **(D)** $f^{-1}(x) = 3 \log_2 (x)$

(E) $f^{-1}(x) = 0.5 \log_3 (x)$

Explorations

69. Writing to Learn Parametric Graphing In the manner of Exploration 1, make tables and graphs for $f(x) = 3^x$ and its inverse $f^{-1}(x) = \log_3 x$. Write a comparative analysis of the two functions regarding domain, range, intercepts, and asymptotes.

70. Writing to Learn Parametric Graphing In the manner of Exploration 1, make tables and graphs for $f(x) = 5^x$ and its inverse $f^{-1}(x) = \log_5 x$. Write a comparative analysis of the two functions regarding domain, range, intercepts, and asymptotes.

71. Group Activity Parametric Graphing In the manner of Exploration 1, find the number $b > 1$ such that the graphs of $f(x) = b^x$ and its inverse $f^{-1}(x) = \log_b x$ have exactly one point of intersection. What is the one point that is in common to the two graphs?

72. Writing to Learn Explain why zero is not in the domain of the logarithmic functions $y = \log_3 x$ and $y = \log_5 x$.

Extending the Ideas

73. Describe how to transform the graph of $f(x) = \ln x$ into the graph of $g(x) = \log_{1/e} x$.

74. Describe how to transform the graph of $f(x) = \log x$ into the graph of $g(x) = \log_{0.1} x$.

3.4
Properties of Logarithmic Functions

. . . and why

The applications of logarithms
are based on their many special
properties, so learn them well.

PROPERTIES OF EXPONENTS

Let b, x, and y be real numbers with
$b > 0$.

1. $b^x \cdot b^y = b^{x+y}$
2. $\dfrac{b^x}{b^y} = b^{x-y}$
3. $(b^x)^y = b^{xy}$

Properties of Logarithms

Logarithms have special algebraic traits that historically made them indispensable for calculations and that still make them valuable in many areas of applications and modeling. In Section 3.3 we learned about the inverse relationship between exponents and logarithms and how to apply some basic properties of logarithms. We now delve deeper into the nature of logarithms to prepare for equation solving and modeling.

Properties of Logarithms

Let b, R, and S be positive real numbers with $b \neq 1$, and c any real number.

• **Product rule**: $\log_b (RS) = \log_b R + \log_b S$

• **Quotient rule**: $\log_b \dfrac{R}{S} = \log_b R - \log_b S$

• **Power rule**: $\log_b R^c = c \log_b R$

The properties of exponents in the margin are the basis for these three properties of logarithms. For instance, the first exponent property listed in the margin is used to verify the product rule.

EXAMPLE 1   Proving the Product Rule for Logarithms

Prove $\log_b (RS) = \log_b R + \log_b S$.

SOLUTION Let $x = \log_b R$ and $y = \log_b S$. The corresponding exponential statements are $b^x = R$ and $b^y = S$. Therefore,

$$RS = b^x \cdot b^y$$

$$= b^{x+y} \qquad \text{First property of exponents}$$

$$\log_b (RS) = x + y \qquad \text{Change to logarithmic form.}$$

$$= \log_b R + \log_b S \qquad \text{Use the definitions of } x \text{ and } y.$$

Now try Exercise 37.

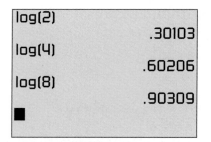

FIGURE 3.26 An arithmetic pattern of logarithms. (Exploration 1)

EXPLORATION 1 Exploring the Arithmetic of Logarithms

Use the 5-decimal place approximations shown in Figure 3.26 to support the properties of logarithms numerically.

1. Product $\log (2 \cdot 4) = \log 2 + \log 4$

2. Quotient $\log \left(\dfrac{8}{2}\right) = \log 8 - \log 2$

3. Power $\log 2^3 = 3 \log 2$

Now evaluate the common logs of other positive integers using the information given in Figure 3.26 and without using your calculator.

4. Use the fact that $5 = 10/2$ to evaluate log 5.

5. Use the fact that 16, 32, and 64 are powers of 2 to evaluate log 16, log 32, and log 64.

6. Evaluate log 25, log 40, and log 50.

List all of the positive integers less than 100 whose common logs can be evaluated knowing only log 2 and the properties of logarithms and without using a calculator.

When we solve equations algebraically that involve logarithms, we often have to rewrite expressions using properties of logarithms. Sometimes we need to expand as far as possible, and other times we condense as much as possible. The next three examples illustrate how properties of logarithms can be used to change the form of expressions involving logarithms.

EXAMPLE 2 Expanding the Logarithm of a Product

Assuming x and y are positive, use properties of logarithms to write $\log (8xy^4)$ as a sum of logarithms or multiples of logarithms.

SOLUTION $\log (8xy^4) = \log 8 + \log x + \log y^4$ Product rule

$= \log 2^3 + \log x + \log y^4$ $8 = 2^3$

$= 3 \log 2 + \log x + 4 \log y$ Power rule

Now try Exercise 1.

EXAMPLE 3 Expanding the Logarithm of a Quotient

Assuming x is positive, use properties of logarithms to write $\ln (\sqrt{x^2 + 5}/x)$ as a sum or difference of logarithms or multiples of logarithms.

SOLUTION $\ln \dfrac{\sqrt{x^2 + 5}}{x} = \ln \dfrac{(x^2 + 5)^{1/2}}{x}$

$= \ln (x^2 + 5)^{1/2} - \ln x$ Quotient rule

$= \dfrac{1}{2} \ln (x^2 + 5) - \ln x$ Power rule

Now try Exercise 9.

EXAMPLE 4   Condensing a Logarithmic Expression

Assuming x and y are positive, write $\ln x^5 - 2 \ln (xy)$ as a single logarithm.

SOLUTION $\ln x^5 - 2 \ln (xy) = \ln x^5 - \ln (xy)^2$ Power rule

$$= \ln x^5 - \ln (x^2 y^2)$$

$$= \ln \frac{x^5}{x^2 y^2}$$ Quotient rule

$$= \ln \frac{x^3}{y^2}$$ *Now try Exercise 13.*

As we have seen, logarithms have some surprising properties. It is easy to overgeneralize and fall into misconceptions about logarithms. Exploration 2 should help you discern what is true and false about logarithmic relationships.

EXPLORATION 2   Discovering Relationships and Nonrelationships

Of the eight relationships suggested here, four are *true* and four are *false* (using values of x within the domains of both sides of the equations). Thinking about the properties of logarithms, make a prediction about the truth of each statement. Then test each with some specific numerical values for x. Finally, compare the graphs of the two sides of the equation.

1. $\ln (x + 2) = \ln x + \ln 2$ **2.** $\log_3 (7x) = 7 \log_3 x$

3. $\log_2 (5x) = \log_2 5 + \log_2 x$ **4.** $\ln \dfrac{x}{5} = \ln x - \ln 5$

5. $\log \dfrac{x}{4} = \dfrac{\log x}{\log 4}$ **6.** $\log_4 x^3 = 3 \log_4 x$

7. $\log_5 x^2 = (\log_5 x)(\log_5 x)$ **8.** $\log |4x| = \log 4 + \log |x|$

Which four are true, and which four are false?

Change of Base

When working with a logarithmic expression with an undesirable base, it is possible to change the expression into a quotient of logarithms with a different base. For example, it is hard to evaluate $\log_4 7$ because 7 is not a simple power of 4 and there is no $\boxed{\log_4}$ key on a calculator or grapher.

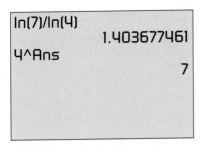

FIGURE 3.27 Evaluating and checking $\log_4 7$.

We can work around this problem with some algebraic trickery. First let $y = \log_4 7$. Then

$$4^y = 7 \qquad \text{Switch to exponential form.}$$

$$\ln 4^y = \ln 7 \qquad \text{Apply ln.}$$

$$y\ln 4 = \ln 7 \qquad \text{Power rule}$$

$$y = \frac{\ln 7}{\ln 4} \qquad \text{Divide by ln 4.}$$

So using our grapher (see Figure 3.27), we see that

$$\log_4 7 = \frac{\ln 7}{\ln 4} = 1.4036\ldots$$

We generalize this useful trickery as the change-of-base formula:

Change-of-Base Formula for Logarithms

For positive real numbers a, b, and x with $a \neq 1$ and $b \neq 1$,

$$\log_b x = \frac{\log_a x}{\log_a b}.$$

Calculators and graphers generally have two logarithm keys—$\boxed{\text{LOG}}$ and $\boxed{\text{LN}}$—which correspond to the bases 10 and e, respectively. So we often use the change-of-base formula in one of the following two forms:

$$\log_b x = \frac{\log x}{\log b} \quad \text{or} \quad \log_b x = \frac{\ln x}{\ln b}$$

These two forms are useful in evaluating logarithms and graphing logarithmic functions.

EXAMPLE 5   Evaluating Logarithms by Changing the Base

(a) $\log_3 16 = \dfrac{\ln 16}{\ln 3} = 2.523\ldots \approx 2.52$

(b) $\log_6 10 = \dfrac{\log 10}{\log 6} = \dfrac{1}{\log 6} = 1.285\ldots \approx 1.29$

(c) $\log_{1/2} 2 = \dfrac{\ln 2}{\ln (1/2)} = \dfrac{\ln 2}{\ln 1 - \ln 2} = \dfrac{\ln 2}{-\ln 2} = -1$ *Now try Exercise 23.*

Graphs of Logarithmic Functions with Base b

Using the change-of-base formula we can rewrite any logarithmic function $g(x) = \log_b x$ as

$$g(x) = \frac{\ln x}{\ln b} = \frac{1}{\ln b}\ln x.$$

So every logarithmic function is a constant multiple of the natural logarithmic function $f(x) = \ln x$. If the base is $b > 1$, the graph of $g(x) = \log_b x$ is a vertical stretch or shrink of the graph of $f(x) = \ln x$ by the factor $1/\ln b$. If $0 < b < 1$, a reflection across the x-axis is required as well.

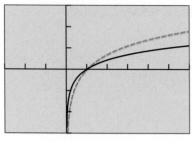

[–3, 6] by [–3, 3]

(a)

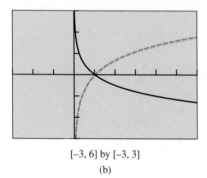

[–3, 6] by [–3, 3]

(b)

FIGURE 3.28 Transforming $f(x) = \ln x$ to obtain (a) $g(x) = \log_5 x$ and (b) $h(x) = \log_{1/4} x$. (Example 6)

EXAMPLE 6 Graphing logarithmic functions

Describe how to transform the graph of $f(x) = \ln x$ into the graph of the given function. Sketch the graph by hand and support your answer with a grapher.

(a) $g(x) = \log_5 x$

(b) $h(x) = \log_{1/4} x$

SOLUTION

(a) Because $g(x) = \log_5 x = \ln x/\ln 5$, its graph is obtained by vertically shrinking the graph of $f(x) = \ln x$ by a factor of $1/\ln 5 \approx 0.62$. See Figure 3.28a.

(b) $h(x) = \log_{1/4} x = \dfrac{\ln x}{\ln 1/4} = \dfrac{\ln x}{\ln 1 - \ln 4} = \dfrac{\ln x}{-\ln 4} = -\dfrac{1}{\ln 4} \ln x$. So we can obtain the graph of h from the graph of $f(x) = \ln x$ by applying, in either order, a reflection across the x-axis and a vertical shrink by a factor of $1/\ln 4 \approx 0.72$. See Figure 3.28b.

Now try Exercise 39.

We can generalize Example 6b in the following way: If $b > 1$, then $0 < 1/b < 1$ and

$$\log_{1/b} x = -\log_b x.$$

So when given a function like $h(x) = \log_{1/4} x$, with a base between 0 and 1, we can immediately rewrite it as $h(x) = -\log_4 x$. Because we can so readily change the base of logarithms with bases between 0 and 1, such logarithms are rarely encountered or used. Instead, we work with logarithms that have bases $b > 1$, which behave much like natural and common logarithms, as we now summarize.

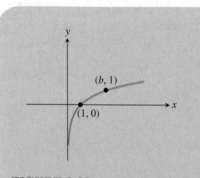

FIGURE 3.29 $f(x) = \log_b x,\ b > 1$

Logarithmic Functions $f(x) = \log_b x$, with $b > 1$

Domain: $(0, \infty)$
Range: All reals
Continuous
Increasing on its domain
No symmetry: neither even nor odd
Not bounded above or below
No local extrema
No horizontal asymptotes
Vertical asymptote: $x = 0$
End behavior: $\lim\limits_{x \to \infty} \log_b x = \infty$

ASTRONOMICALLY SPEAKING

An astronomical unit (AU) is the average distance between the Earth and the Sun, about 149.6 million kilometers (149.6 Gm).

Re-expressing Data

When seeking a model for a set of data it is often helpful to transform the data by applying a function to one or both of the variables in the data set. We did this already when we treated the years 1900–2000 as 0–100. Such a transformation of a data set is a **re-expression** of the data.

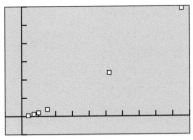

[−1, 10] by [−5, 30]

(a)

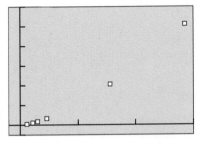

[−100, 1500] by [−1000, 12 000]

(b)

FIGURE 3.30 Scatter plots of the planetary data from (a) Table 3.20 and (b) Table 2.10.

Recall from Section 2.2 that Kepler's Third Law states that the square of the orbit period T for each planet is proportional to the cube of its average distance a from the Sun. If we re-express the Kepler planetary data in Table 2.10 using Earth-based units, the constant of proportion becomes 1 and the "is proportional to" in Kepler's Third Law becomes "equals." We can do this by dividing the "average distance" column by 149.6 Gm/AU and the "period of orbit" column by 365.2 days/yr. The re-expressed data are shown in Table 3.20.

Table 3.20 Average Distances and Orbit Periods for the Six Innermost Planets

Planet	Average Distance from Sun (AU)	Period of Orbit (yr)
Mercury	0.3870	0.2410
Venus	0.7233	0.6161
Earth	1.000	1.000
Mars	1.523	1.881
Jupiter	5.203	11.86
Saturn	9.539	29.46

Source: Re-expression of data from: Shupe, et al., *National Geographic Atlas of the World* (rev. 6th ed.), Washington, DC: National Geographic Society, 1992, plate 116.

Notice that the pattern in the scatter plot of these re-expressed data, shown in Figure 3.30a, is essentially the same as the pattern in the plot of the original data, shown in Figure 3.30b. What we have done is to make the numerical values of the data more convenient and to guarantee that our plot contains the ordered pair (1, 1) for Earth, which could potentially simplify our model. What we have *not* done and still wish to do is to clarify the relationship between the variables a (distance from the Sun) and T (orbit period).

Logarithms can be used to re-express data and help us clarify relationships and uncover hidden patterns. For the planetary data, if we plot (ln a, ln T) pairs instead of (a, T) pairs, the pattern is much clearer. In Example 7, we carry out this re-expression of the data and then use an algebraic *tour de force* to obtain Kepler's Third Law.

EXAMPLE 7 Establishing Kepler's Third Law Using Logarithmic Re-expression

Re-express the (a, T) data pairs in Table 3.20 as (ln a, ln T) pairs. Find a linear regression model for the (ln a, ln T) pairs. Rewrite the linear regression in terms of a and T, and rewrite the equation in a form with no logs or fractional exponents.

SOLUTION

Model

We use grapher list operations to obtain the (ln a, ln T) pairs (see Figure 3.31a on the next page). We make a scatter plot of the re-expressed data in Figure 3.31b on the next page. The (ln a, ln T) pairs appear to lie along a straight line.

continued

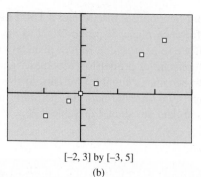

L2	L3	L4
.241	−.9493	−1.423
.6161	−.3239	−.4843
1	0	0
1.881	.42068	.6318
11.86	1.6492	2.4732
29.46	2.2554	3.383
------	------	------

L4 = ln(L2)

(a)

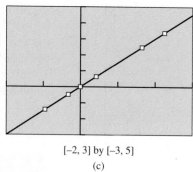

[−2, 3] by [−3, 5]

(b)

[−2, 3] by [−3, 5]

(c)

FIGURE 3.31 Scatter plot and graphs for Example 7.

We let $y = \ln T$ and $x = \ln a$. Then using linear regression, we obtain the following model:

$$y = 1.49950x + 0.00070 \approx 1.5x.$$

Figure 3.31c shows the scatter plot for the $(x, y) = (\ln a, \ln T)$ pairs together with a graph of $y = 1.5x$. You can see that the line fits the re-expressed data remarkably well.

Remodel

Returning to the original variables a and T, we obtain:

$$\ln T = 1.5 \cdot \ln a \qquad \text{$y = 1.5x$}$$

$$\frac{\ln T}{\ln a} = 1.5 \qquad \text{Divide by $\ln a$.}$$

$$\log_a T = \frac{3}{2} \qquad \text{Change of base}$$

$$T = a^{3/2} \qquad \text{Switch to exponential form.}$$

$$T^2 = a^3 \qquad \text{Square both sides.}$$

Interpret

This is Kepler's Third Law!

Now try Exercise 65.

QUICK REVIEW 3.4 (*For help, go to Sections A.1 and 3.3.*)

In Exercises 1–4, evaluate the expression without using a calculator.

1. $\log 10^2$

2. $\ln e^3$

3. $\ln e^{-2}$

4. $\log 10^{-3}$

In Exercises 5–10, simplify the expression.

5. $\dfrac{x^5 y^{-2}}{x^2 y^{-4}}$

6. $\dfrac{u^{-3} v^7}{u^{-2} v^2}$

7. $(x^6 y^{-2})^{1/2}$

8. $(x^{-8} y^{12})^{3/4}$

9. $\dfrac{(u^2 v^{-4})^{1/2}}{(27 u^6 v^{-6})^{1/3}}$

10. $\dfrac{(x^{-2} y^3)^{-2}}{(x^3 y^{-2})^{-3}}$

SECTION 3.4 EXERCISES

In Exercises 1–12, assuming x and y are positive, use properties of logarithms to write the expression as a sum or difference of logarithms or multiples of logarithms.

1. $\ln 8x$

2. $\ln 9y$

3. $\log \dfrac{3}{x}$

4. $\log \dfrac{2}{y}$

5. $\log_2 y^5$

6. $\log_2 x^{-2}$

7. $\log x^3 y^2$

8. $\log xy^3$

9. $\ln \dfrac{x^2}{y^3}$

10. $\log 1000x^4$

11. $\log \sqrt[4]{\dfrac{x}{y}}$

12. $\ln \dfrac{\sqrt[3]{x}}{\sqrt[3]{y}}$

In Exercises 13–22, assuming x, y, and z are positive, use properties of logarithms to write the expression as a single logarithm.

13. $\log x + \log y$

14. $\log x + \log 5$

15. $\ln y - \ln 3$

16. $\ln x - \ln y$

17. $\dfrac{1}{3} \log x$

18. $\dfrac{1}{5} \log z$

19. $2 \ln x + 3 \ln y$

20. $4 \log y - \log z$

21. $4 \log (xy) - 3 \log (yz)$

22. $3 \ln (x^3 y) + 2 \ln (yz^2)$

In Exercises 23–28, use the change-of-base formula and your calculator to evaluate the logarithm.

23. $\log_2 7$

24. $\log_5 19$

25. $\log_8 175$

26. $\log_{12} 259$

27. $\log_{0.5} 12$

28. $\log_{0.2} 29$

In Exercises 29–32, write the expression using only natural logarithms.

29. $\log_3 x$

30. $\log_7 x$

31. $\log_2 (a + b)$

32. $\log_5 (c - d)$

In Exercises 33–36, write the expression using only common logarithms.

33. $\log_2 x$

34. $\log_4 x$

35. $\log_{1/2} (x + y)$

36. $\log_{1/3} (x - y)$

37. Prove the quotient rule of logarithms.

38. Prove the power rule of logarithms.

In Exercises 39–42, describe how to transform the graph of $g(x) = \ln x$ into the graph of the given function. Sketch the graph by hand and support with a grapher.

39. $f(x) = \log_4 x$

40. $f(x) = \log_7 x$

41. $f(x) = \log_{1/3} x$

42. $f(x) = \log_{1/5} x$

In Exercises 43–46, match the function with its graph. Identify the window dimensions, Xscl, and Yscl of the graph.

43. $f(x) = \log_4 (2 - x)$

44. $f(x) = \log_6 (x - 3)$

45. $f(x) = \log_{0.5} (x - 2)$

46. $f(x) = \log_{0.7} (3 - x)$

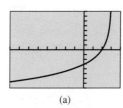
(a)

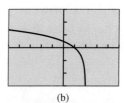
(b)

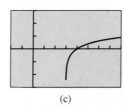
(c)

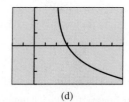
(d)

In Exercises 47–50, graph the function, and analyze it for domain, range, continuity, increasing or decreasing behavior, asymptotes, and end behavior.

47. $f(x) = \log_2 (8x)$

48. $f(x) = \log_{1/3} (9x)$

49. $f(x) = \log (x^2)$

50. $f(x) = \ln (x^3)$

51. Sound Intensity Compute the sound intensity level in decibels for each sound listed in Table 3.21.

Table 3.21 Approximate Intensities for Selected Sounds

Sound	Intensity (Watts/m^2)
(a) Hearing threshold	10^{-12}
(b) Rustling leaves	10^{-11}
(c) Conversation	10^{-6}
(d) School cafeteria	10^{-4}
(e) Jack hammer	10^{-2}
(f) Pain threshold	1

Sources: J. J. Dwyer, College Physics (Belmont, CA: Wadsworth, 1984), and E. Connally et al., Functions Modeling Change (New York: Wiley, 2000).

52. Earthquake Intensity The **Richter scale** magnitude R of an earthquake is based on the features of the associated seismic wave and is measured by

$$R = \log(a/T) + B,$$

where a is the amplitude in μm (micrometers), T is the period in seconds, and B accounts for the weakening of the seismic wave due to the distance from the epicenter. Compute the earthquake magnitude R for each set of values.

(a) $a = 250$, $T = 2$, and $B = 4.25$

(b) $a = 300$, $T = 4$, and $B = 3.5$

53. Light Intensity in Lake Erie The relationship between intensity I of light (in lumens) at a depth of x feet in Lake Erie is given by

$$\log \frac{I}{12} = -0.00235x.$$

What is the intensity at a depth of 40 ft?

54. Light Intensity in Lake Superior The relationship between intensity I of light (in lumens) at a depth of x feet in Lake Superior is given by

$$\log \frac{I}{12} = -0.0125x.$$

What is the intensity at a depth of 10 ft?

55. Writing to Learn Use the change-of-base formula to explain how we know that the graph of $f(x) = \log_3 x$ can be obtained by applying a transformation to the graph of $g(x) = \ln x$.

56. Writing to Learn Use the change-of-base formula to explain how the graph of $f(x) = \log_{0.8} x$ can be obtained by applying transformations to the graph of $g(x) = \log x$.

Standardized Test Questions

57. True or False The logarithm of the product of two positive numbers is the sum of the logarithms of the numbers. Justify your answer.

58. True or False The logarithm of a positive number is positive. Justify your answer.

In Exercises 59–62, solve the problem without using a calculator.

59. Multiple Choice $\log 12 =$

(A) $3 \log 4$ (B) $\log 3 + \log 4$

(C) $4 \log 3$ (D) $\log 3 \cdot \log 4$

(E) $2 \log 6$

60. Multiple Choice $\log_9 64 =$

(A) $5 \log_3 2$ (B) $(\log_3 8)^2$

(C) $(\ln 64)/(\ln 9)$ (D) $2 \log_9 32$

(E) $(\log 64)/9$

61. Multiple Choice $\ln x^5 =$

(A) $5 \ln x$ (B) $2 \ln x^3$

(C) $x \ln 5$ (d) $3 \ln x^2$

(E) $\ln x^2 \cdot \ln x^3$

62. Multiple Choice $\log_{1/2} x^2 =$

(A) $-2 \log_2 x$ (B) $2 \log_2 x$

(C) $-0.5 \log_2 x$ (D) $0.5 \log_2 x$

(E) $-2 \log_2 |x|$

Explorations

63. (a) Compute the power regression model for the following data.

x	4	6.5	8.5	10
y	2816	31,908	122,019	275,000

(b) Predict the y value associated with $x = 7.1$ using the power regression model.

(c) Re-express the data in terms of their natural logarithms and make a scatter plot of $(\ln x, \ln y)$.

(d) Compute the linear regression model $(\ln y) = a(\ln x) + b$ for $(\ln x, \ln y)$.

(e) Confirm that $y = e^b \cdot x^a$ is the power regression model found in (a).

64. (a) Compute the power regression model for the following data.

x	2	3	4.8	7.7
y	7.48	7.14	6.81	6.41

(b) Predict the y value associated with $x = 9.2$ using the power regression model.

(c) Re-express the data in terms of their natural logarithms and make a scatter plot of $(\ln x, \ln y)$.

(d) Compute the linear regression model $(\ln y) = a(\ln x) + b$ for $(\ln x, \ln y)$.

(e) Confirm that $y = e^b \cdot x^a$ is the power regression model found in (a).

65. Keeping Warm—Revisited Recall from Exercise 55 of Section 2.2 that scientists have found the pulse rate r of mammals to be a power function of their body weight w.

(a) Re-express the data in Table 3.22 in terms of their *common* logarithms and make a scatter plot of $(\log w, \log r)$.

(b) Compute the linear regression model $(\log r) = a(\log w) + b$ for $(\log w, \log r)$.

(c) Superimpose the regression curve on the scatter plot.

(d) Use the regression equation to predict the pulse rate for a 450-kg horse. Is the result close to the 38 beats/min reported by A. J. Clark in 1927?

(e) Writing to Learn Why can we use either common or natural logarithms to re-express data that fit a power regression model?

66. Let $a = \log 2$ and $b = \log 3$. Then, for example $\log 6 = a + b$. List the common logs of all the positive integers less than 100 that can be expressed in terms of a and b, writing equations such as $\log 6 = a + b$ for each case.

Extending the Ideas

67. Solve $\ln x > \sqrt[3]{x}$.

68. Solve $1.2^x \le \log_{1.2} x$.

69. Group Activity Work in groups of three. Have each group member graph and compare the domains for one pair of functions.

(a) $f(x) = 2 \ln x + \ln (x - 3)$ and $g(x) = \ln x^2(x - 3)$

(b) $f(x) = \ln (x + 5) - \ln (x - 5)$ and $g(x) = \ln \dfrac{x + 5}{x - 5}$

(c) $f(x) = \log (x + 3)^2$ and $g(x) = 2 \log (x + 3)$

Writing to Learn After discussing your findings, write a brief group report that includes your overall conclusions and insights.

70. Prove the change-of-base formula for logarithms.

71. Prove that $f(x) = \log x / \ln x$ is a constant function with restricted domain by finding the exact value of the constant $\log x / \ln x$ expressed as a common logarithm.

72. Graph $f(x) = \ln (\ln (x))$, and analyze it for domain, range, continuity, increasing or decreasing behavior, symmetry, asymptotes, end behavior, and invertibility.

Table 3.22 Weight and Pulse Rate of Selected Mammals

Mammal	Body weight (kg)	Pulse rate (beats/min)
Rat	0.2	420
Guinea pig	0.3	300
Rabbit	2	205
Small dog	5	120
Large dog	30	85
Sheep	50	70
Human	70	72

Source: A. J. Clark, *Comparative Physiology of the Heart* (New York: Macmillan, 1927).

3.5
Equation Solving and Modeling

What you'll learn about

- Solving Exponential Equations
- Solving Logarithmic Equations
- Orders of Magnitude and Logarithmic Models
- Newton's Law of Cooling
- Logarithmic Re-expression

. . . and why

The Richter scale, pH, and Newton's Law of Cooling are among the most important uses of logarithmic and exponential functions.

Solving Exponential Equations

Some logarithmic equations can be solved by changing to exponential form, as we saw in Example 5 of Section 3.3. For other equations, the properties of exponents or the properties of logarithms are used. A property of both exponential and logarithmic functions that is often helpful for solving equations is that they are one-to-one functions.

One-to-One Properties

For any exponential function $f(x) = b^x$,

- If $b^u = b^v$, then $u = v$.

For any logarithmic function $f(x) = \log_b x$,

- If $\log_b u = \log_b v$, then $u = v$.

Example 1 shows how the one-to-oneness of exponential functions can be used. Examples 3 and 4 use the one-to-one property of logarithms.

EXAMPLE 1   Solving an Exponential Equation Algebraically

Solve $20(1/2)^{x/3} = 5$.

SOLUTION

$$20\left(\frac{1}{2}\right)^{x/3} = 5$$

$$\left(\frac{1}{2}\right)^{x/3} = \frac{1}{4} \qquad \text{Divide by 20.}$$

$$\left(\frac{1}{2}\right)^{x/3} = \left(\frac{1}{2}\right)^2 \qquad \frac{1}{4} = \left(\frac{1}{2}\right)^2$$

$$\frac{x}{3} = 2 \qquad \text{One-to-one property}$$

$$x = 6 \qquad \text{Multiply by 3.}$$

Now try Exercise 1.

The equation in Example 2 involves a difference of two exponential functions, which makes it difficult to solve algebraically. So we start with a graphical approach.

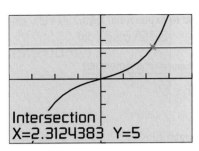

Intersection
X=2.3124383 Y=5

[–4, 4] by [–10, 10]

FIGURE 3.32 $y = (e^x - e^{-x})/2$ and $y = 5$. (Example 2)

A CINCH?

You may recognize the left-hand side of the equation in Example 2 as the *hyperbolic sine function* that was introduced in Exercise 59 of Section 3.2. This function is often used in calculus. We write $\sinh(x) = (e^x - e^{-x})/2$. "Sinh" is pronounced as if spelled "cinch."

EXAMPLE 2 Solving an Exponential Equation

Solve $(e^x - e^{-x})/2 = 5$.

SOLUTION

Solve Graphically Figure 3.32 shows that the graphs of $y = (e^x - e^{-x})/2$ and $y = 5$ intersect when $x \approx 2.31$.

Confirm Algebraically The algebraic approach involves some ingenuity. If we multiply each side of the original equation by $2e^x$ and rearrange the terms we can obtain a quadratic equation in e^x:

$$\frac{e^x - e^{-x}}{2} = 5$$

$$e^{2x} - e^0 = 10e^x \qquad \text{Multiply by } 2e^x.$$

$$(e^x)^2 - 10(e^x) - 1 = 0 \qquad \text{Subtract } 10e^x.$$

If we let $w = e^x$, this equation becomes $w^2 - 10w - 1 = 0$, and the quadratic formula gives

$$w = e^x = \frac{10 \pm \sqrt{104}}{2} = 5 \pm \sqrt{26}.$$

Because e^x is always positive, we reject the possibility that e^x has the negative value $5 - \sqrt{26}$. Therefore,

$$e^x = 5 + \sqrt{26}$$

$$x = \ln(5 + \sqrt{26}) \qquad \text{Convert to logarithmic form.}$$

$$x = 2.312\ldots \approx 2.31 \qquad \text{Approximate with a grapher.}$$

Now try Exercise 31.

Solving Logarithmic Equations

When logarithmic equations are solved algebraically, it is important to keep track of the domain of each expression in the equation as it is being solved. A particular algebraic method may introduce extraneous solutions or worse yet *lose* some valid solutions, as illustrated in Example 3.

EXAMPLE 3 Solving a Logarithmic Equation

Solve $\log x^2 = 2$.

SOLUTION

Method 1 Use the one-to-one property of logarithms.

$$\log x^2 = 2$$

$$\log x^2 = \log 10^2 \qquad y = \log 10^y$$

$$x^2 = 10^2 \qquad \text{One-to-one property}$$

$$x^2 = 100 \qquad 10^2 = 100$$

$$x = 10 \quad \text{or} \quad x = -10$$

continued

Logarithmic Re-expression

In Example 7 of Section 3.4 we learned that data pairs (x, y) that fit a power model have a linear relationship when re-expressed as $(\ln x, \ln y)$ pairs. We now illustrate that data pairs (x, y) that fit a logarithmic or exponential regression model can also be *linearized* through *logarithmic re-expression*.

Regression Models Related by Logarithmic Re-Expression	
• **Linear regression:**	$y = ax + b$
• **Natural logarithmic regression:**	$y = a + b \ln x$
• **Exponential regression:**	$y = a \cdot b^x$
• **Power regression:**	$y = a \cdot x^b$

When we examine a scatter plot of data pairs (x, y), we should ask whether one of these four regression models could be the best choice. If the data plot appears to be linear, a linear regression may be the best choice. But when it is visually evident that the data plot is not linear, the best choice may be a natural logarithmic, exponential, or power regression.

Knowing the shapes of logarithmic, exponential, and power function graphs helps us choose an appropriate model. In addition, it is often helpful to re-express the (x, y) data pairs as $(\ln x, y)$, $(x, \ln y)$, or $(\ln x, \ln y)$ and create scatter plots of the re-expressed data. If any of the scatter plots appear to be linear, then we have a likely choice for an appropriate model. See page 329.

The three regression models can be justified algebraically. We give the justification for exponential regression, and leave the other two justifications as exercises.

$$v = ax + b$$

$$\ln y = ax + b \qquad {\scriptstyle v = \ln y}$$

$$y = e^{ax+b} \qquad {\scriptstyle \text{Change to exponential form.}}$$

$$y = e^{ax} \cdot e^b \qquad {\scriptstyle \text{Use the laws of exponents.}}$$

$$y = e^b \cdot (e^a)^x$$

$$y = c \cdot d^x \qquad {\scriptstyle \text{Let } c = e^b \text{ and } d = e^a.}$$

Example 9 illustrates how a combination of knowledge about the shapes of logarithmic, exponential, and power function graphs is used in combination with logarithmic re-expression to choose a curve of best fit.

Three Types of Logarithmic Re-expression

1. Natural Logarithmic Regression Re-expressed: $(x, y) \rightarrow (\ln x, y)$

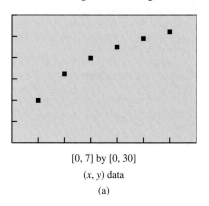

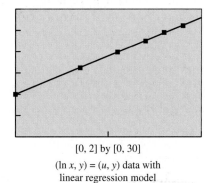

Conclusion:

$y = a \ln x + b$ is the logarithmic regression model for the (x, y) data.

[0, 7] by [0, 30]

(x, y) data

(a)

[0, 2] by [0, 30]

$(\ln x, y) = (u, y)$ data with linear regression model
$y = au + b$

(b)

2. Exponential Regression Re-expressed: $(x, y) \rightarrow (x, \ln y)$

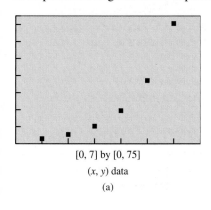

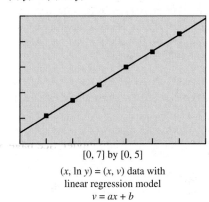

Conclusion:

$y = c(d^x)$, where $c = e^b$ and $d = e^a$, is the exponential regression model for the (x, y) data.

[0, 7] by [0, 75]

(x, y) data

(a)

[0, 7] by [0, 5]

$(x, \ln y) = (x, v)$ data with linear regression model
$v = ax + b$

(b)

3. Power Regression Re-expressed: $(x, y) \rightarrow (\ln x, \ln y)$

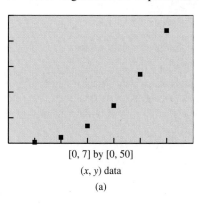

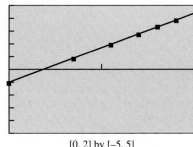

Conclusion:

$y = c(x^a)$, where $c = e^b$, is the power regression model for the (x, y) data.

[0, 7] by [0, 50]

(x, y) data

(a)

[0, 2] by [−5, 5]

$(\ln x, \ln y) = (u, v)$ data with linear regression model
$v = au + b$

(b)

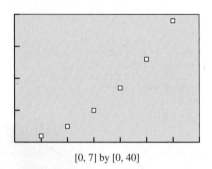

[0, 7] by [0, 40]

FIGURE 3.38 A scatter plot of the original data of Example 9.

EXAMPLE 9   Selecting a Regression Model

Decide whether these data can be best modeled by logarithmic, exponential, or power regression. Find the appropriate regression model.

x	1	2	3	4	5	6
y	2	5	10	17	26	38

SOLUTION The shape of the data plot in Figure 3.38 suggests that the data could be modeled by an exponential or power function.

Figure 3.39a shows the $(x, \ln y)$ plot, and Figure 3.39b shows the $(\ln x, \ln y)$ plot. Of these two plots, the $(\ln x, \ln y)$ plot appears to be more linear, so we find the power regression model for the original data.

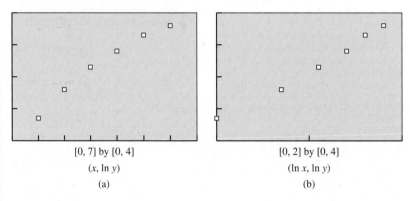

[0, 7] by [0, 4]	[0, 2] by [0, 4]
$(x, \ln y)$	$(\ln x, \ln y)$
(a)	(b)

FIGURE 3.39 Two logarithmic re-expressions of the data of Example 9.

Figure 3.40 shows the scatter plot of the original (x, y) data with the graph of the power regression model $y = 1.7910x^{1.6472}$ superimposed.

Now try Exercise 55.

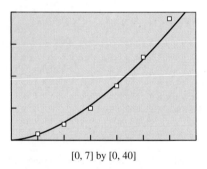

[0, 7] by [0, 40]

FIGURE 3.40 A power regression model fits the data of Example 9.

QUICK REVIEW 3.5   *(For help, go to Sections P.1 and 1.4.)*

In Exercises 1–4, prove that each function in the given pair is the inverse of the other.

1. $f(x) = e^{2x}$ and $g(x) = \ln(x^{1/2})$

2. $f(x) = 10^{x/2}$ and $g(x) = \log x^2, x > 0$

3. $f(x) = (1/3) \ln x$ and $g(x) = e^{3x}$

4. $f(x) = 3 \log x^2, x > 0$ and $g(x) = 10^{x/6}$

In Exercises 5 and 6, write the number in scientific notation.

5. The mean distance from Jupiter to the Sun is about 778,300,000 km.

6. An atomic nucleus has a diameter of about 0.000000000000001 m.

In Exercises 7 and 8, write the number in decimal form.

7. Avogadro's number is about 6.02×10^{23}.

8. The atomic mass unit is about 1.66×10^{-27} kg.

In Exercises 9 and 10, use scientific notation to simplify the expression; leave your answer in scientific notation.

9. $(186,000)(31,000,000)$ **10.** $\dfrac{0.0000008}{0.000005}$

SECTION 3.5 EXERCISES

In Exercises 1–10, find the exact solution algebraically, and check it by substituting into the original equation.

1. $36 \left(\frac{1}{3}\right)^{x/5} = 4$

2. $32 \left(\frac{1}{4}\right)^{x/3} = 2$

3. $2 \cdot 5^{x/4} = 250$

4. $3 \cdot 4^{x/2} = 96$

5. $2(10^{-x/3}) = 20$

6. $3(5^{-x/4}) = 15$

7. $\log x = 4$

8. $\log_2 x = 5$

9. $\log_4 (x - 5) = -1$

10. $\log_4 (1 - x) = 1$

In Exercises 11–18, solve each equation algebraically. Obtain a numerical approximation for your solution and check it by substituting into the original equation.

11. $1.06^x = 4.1$

12. $0.98^x = 1.6$

13. $50e^{0.035x} = 200$

14. $80e^{0.045x} = 240$

15. $3 + 2e^{-x} = 6$

16. $7 - 3e^{-x} = 2$

17. $3 \ln (x - 3) + 4 = 5$

18. $3 - \log (x + 2) = 5$

In Exercises 19–24, state the domain of each function. Then match the function with its graph. (Each graph shown has a window of $[-4.7, 4.7]$ by $[-3.1, 3.1]$.)

19. $f(x) = \log [x(x + 1)]$

20. $g(x) = \log x + \log (x + 1)$

21. $f(x) = \ln \dfrac{x}{x + 1}$

22. $g(x) = \ln x - \ln (x + 1)$

23. $f(x) = 2 \ln x$

24. $g(x) = \ln x^2$

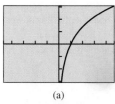

(a)

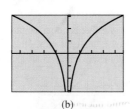

(b)

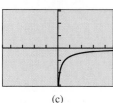

(c)

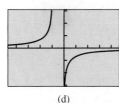

(d)

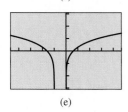

(e)

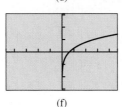

(f)

In Exercises 25–38, solve each equation by the method of your choice. Support your solution by a second method.

25. $\log x^2 = 6$

26. $\ln x^2 = 4$

27. $\log x^4 = 2$

28. $\ln x^6 = 12$

29. $\dfrac{2^x - 2^{-x}}{3} = 4$

30. $\dfrac{2^x + 2^{-x}}{2} = 3$

31. $\dfrac{e^x + e^{-x}}{2} = 4$

32. $2e^{2x} + 5e^x - 3 = 0$

33. $\dfrac{500}{1 + 25e^{0.3x}} = 200$

34. $\dfrac{400}{1 + 95e^{-0.6x}} = 150$

35. $\dfrac{1}{2} \ln (x + 3) - \ln x = 0$ **36.** $\log x - \dfrac{1}{2} \log (x + 4) = 1$

37. $\ln (x - 3) + \ln (x + 4) = 3 \ln 2$

38. $\log (x - 2) + \log (x + 5) = 2 \log 3$

In Exercises 39–44, determine how many orders of magnitude the quantities differ.

39. A \$100 bill and a dime

40. A canary weighing 20 g and a hen weighing 2 kg

41. An earthquake rated 7 on the Richter scale and one rated 5.5.

42. Lemon juice with pH = 2.3 and beer with pH = 4.1

43. The sound intensities of a riveter at 95 dB and ordinary conversation at 65 dB

44. The sound intensities of city traffic at 70 dB and rustling leaves at 10 dB

45. Comparing Earthquakes How many times more severe was the 1978 Mexico City earthquake ($R = 7.9$) than the 1994 Los Angeles earthquake ($R = 6.6$)?

46. Comparing Earthquakes How many times more severe was the 1995 Kobe, Japan, earthquake ($R = 7.2$) than the 1994 Los Angeles earthquake ($R = 6.6$)?

47. Chemical Acidity The pH of carbonated water is 3.9 and the pH of household ammonia is 11.9.

   (a) What are their hydrogen-ion concentrations?

   (b) How many times greater is the hydrogen-ion concentration of the carbonated water than that of the ammonia?

   (c) By how many orders of magnitude do the concentrations differ?

48. Chemical Acidity Stomach acid has a pH of about 2.0, and blood has a pH of 7.4.

   (a) What are their hydrogen-ion concentrations?

   (b) How many times greater is the hydrogen-ion concentration of the stomach acid than that of the blood?

   (c) By how many orders of magnitude do the concentrations differ?

49. Newton's Law of Cooling A cup of coffee has cooled from 92°C to 50°C after 12 min in a room at 22°C. How long will the cup take to cool to 30°C?

50. Newton's Law of Cooling A cake is removed from an oven at 350°F and cools to 120°F after 20 min in a room at 65°F. How long will the cake take to cool to 90°F?

51. Newton's Law of Cooling Experiment A thermometer is removed from a cup of coffee and placed in water with a temperature (T_m) of 10°C. The data in Table 3.24 were collected over the next 30 sec.

Table 3.24 Experimental Data		
Time t	Temp T	$T - T_m$
2	80.47	70.47
5	69.39	59.39
10	49.66	39.66
15	35.26	25.26
20	28.15	18.15
25	23.56	13.56
30	20.62	10.62

(a) Draw a scatter plot of the data $T - T_m$.

(b) Find an exponential regression equation for the $T - T_m$ data. Superimpose its graph on the scatter plot.

(c) Estimate the thermometer reading when it was removed from the coffee.

52. Newton's Law of Cooling Experiment A thermometer was removed from a cup of hot chocolate and placed in water with temperature $T_m = 0$°C. The data in Table 3.25 were collected over the next 30 sec.

(a) Draw a scatter plot of the data $T - T_m$.

(b) Find an exponential regression equation for the $T - T_m$ data. Superimpose its graph on the scatter plot.

(c) Estimate the thermometer reading when it was removed from the hot chocolate.

Table 3.25 Experimental Data		
Time t	Temp T	$T - T_m$
2	74.68	74.68
5	61.99	61.99
10	34.89	34.89
15	21.95	21.95
20	15.36	15.36
25	11.89	11.89
30	10.02	10.02

53. Penicillin Use The use of penicillin became so widespread in the 1980s in Hungary that it became practically useless against common sinus and ear infections. Now the use of more effective antibiotics has caused a decline in penicillin resistance. The bar graph shows the use of penicillin in Hungary for selected years.

(a) From the bar graph we read the data pairs to be approximately (1, 11), (8, 6), (15, 4.8), (16, 4), and (17, 2.5), using $t = 1$ for 1976, $t = 8$ for 1983, and so on. Complete a scatter plot for these data.

(b) Writing to Learn Discuss whether the bar graph shown or the scatter plot that you completed best represents the data and why.

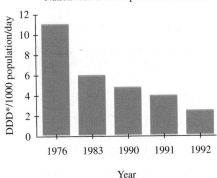

Nationwide Consumption of Penicillin

*Defined Daily Dose
Source: Science, vol. 264, April 15, 1994, American Association for the Advancement of Science.

54. Writing to Learn Which regression model would you use for the data in Exercise 53? Discuss various options, and explain why you chose the model you did. Support your writing with tables and graphs as needed.

Writing to Learn In Exercises 55–58, tables of (x, y) data pairs are given. Determine whether a linear, logarithmic, exponential, or power regression equation is the best model for the data. Explain your choice. Support your writing with tables and graphs as needed.

55.

x	1	2	3	4
y	3	4.4	5.2	5.8

56.

x	1	2	3	4
y	6	18	54	162

57.

x	1	2	3	4
y	3	6	12	24

58.

x	1	2	3	4
y	5	7	9	11

Standardized Test Questions

59. True or False The order of magnitude of a positive number is its natural logarithm. Justify your answer.

60. True or False According to Newton's Law of Cooling, an object will approach the temperature of the medium that surrounds it. Justify your answer.

In Exercises 61–64, solve the problem without using a calculator.

61. Multiple Choice Solve $2^{3x-1} = 32$.

   (A) $x = 1$ **(B)** $x = 2$ **(C)** $x = 4$

   (D) $x = 11$ **(E)** $x = 13$

62. Multiple Choice Solve $\ln x = -1$.

   (A) $x = -1$ **(B)** $x = 1/e$ **(C)** $x = 1$

   (D) $x = e$ **(E)** No solution is possible.

63. Multiple Choice How many times more severe was the 2001 earthquake in Arequipa, Peru ($R_1 = 8.1$) than the 1998 double earthquake in Takhar province, Afghanistan ($R_2 = 6.1$)?

   (A) 2   **(B)** 6.1   **(C)** 8.1

   (D) 14.2   **(E)** 100

64. Multiple Choice Newton's Law of Cooling is

   (A) an exponential model.   **(B)** a linear model.

   (C) a logarithmic model.   **(D)** a logistic model.

   (E) a power model.

Explorations

In Exercises 65 and 66, use the data in Table 3.26. Determine whether a linear, logarithmic, exponential, power, or logistic regression equation is the best model for the data. Explain your choice. Support your writing with tables and graphs as needed.

Table 3.26 Populations of Two U.S. States (in thousands)		
Year	Alaska	Hawaii
1900	63.6	154
1910	64.4	192
1920	55.0	256
1930	59.2	368
1940	72.5	423
1950	128.6	500
1960	226.2	633
1970	302.6	770
1980	401.9	965
1990	550.0	1108
2000	626.9	1212

Source: U.S. Census Bureau.

65. Writing to Learn Modeling Population Which regression equation is the best model for Alaska's population?

66. Writing to Learn Modeling Population Which regression equation is the best model for Hawaii's population?

67. Group Activity Normal Distribution The function

$$f(x) = k \cdot e^{-cx^2},$$

where c and k are positive constants, is a bell-shaped curve that is useful in probability and statistics.

   (a) Graph f for $c = 1$ and $k = 0.1, 0.5, 1, 2, 10$. Explain the effect of changing k.

   (b) Graph f for $k = 1$ and $c = 0.1, 0.5, 1, 2, 10$. Explain the effect of changing c.

Extending the Ideas

68. Writing to Learn Prove if $u/v = 10^n$ for $u > 0$ and $v > 0$, then $\log u - \log v = n$. Explain how this result relates to powers of ten and orders of magnitude.

69. Potential Energy The potential energy E (the energy stored for use at a later time) between two ions in a certain molecular structure is modeled by the function

$$E = -\frac{5.6}{r} + 10e^{-r/3}$$

where r is the distance separating the nuclei.

   (a) Writing to Learn Graph this function in the window $[-10, 10]$ by $[-10, 30]$, and explain which portion of the graph does not represent this potential energy situation.

   (b) Identify a viewing window that shows that portion of the graph (with $r \le 10$) which represents this situation, and find the maximum value for E.

70. In Example 8, the Newton's Law of Cooling model was

$$T(t) - T_m = (T_0 - T_m)e^{-kt} = 61.656 \times 0.92770^t$$

Determine the value of k.

71. Justify the conclusion made about natural logarithmic regression on page 329.

72. Justify the conclusion made about power regression on page 329.

In Exercises 73–78, solve the equation or inequality.

73. $e^x + x = 5$

74. $e^{2x} - 8x + 1 = 0$

75. $e^x < 5 + \ln x$

76. $\ln |x| - e^{2x} \ge 3$

77. $2 \log x - 4 \log 3 > 0$

78. $2 \log (x + 1) - 2 \log 6 < 0$

3.6

Mathematics of Finance

What you'll learn about

- Interest Compounded Annually

- Interest Compounded k Times per Year

- Interest Compounded Continuously

- Annual Percentage Yield

- Annuities—Future Value

- Loans and Mortgages— Present Value

. . . and why

The mathematics of finance is the science of letting your money work for you—valuable information indeed!

Interest Compounded Annually

In business, as the saying goes, "time is money." We must pay interest for the use of property or money over time. When we borrow money, we pay interest, and when we loan money, we receive interest. When we invest in a savings account, we are actually lending money to the bank.

Suppose a principal of P dollars is invested in an account bearing an interest rate r expressed in decimal form and calculated at the end of each year. If A_n represents the total amount in the account at the end of n years, then the value of the investment follows the growth pattern shown in Table 3.27.

Table 3.27 Interest Computed Annually	
Time in years	Amount in the account
0	$A_0 = P =$ principal
1	$A_1 = P + P \cdot r = P(1 + r)$
2	$A_2 = A_1 \cdot (1 + r) = P(1 + r)^2$
3	$A_3 = A_2 \cdot (1 + r) = P(1 + r)^3$
$\vdots$	$\vdots$
n	$A = A_n = P(1 + r)^n$

Notice that this is the constant percentage growth pattern studied in Section 3.2, and so the value of an investment is an exponential function of time. We call interest computed in this way **compound interest** because the interest becomes part of the investment, so that interest is earned on the interest itself.

Interest Compounded Annually

If a principal P is invested at a fixed annual interest rate r, calculated at the end of each year, then the value of the investment after n years is

$$A = P(1 + r)^n,$$

where r is expressed as a decimal.

EXAMPLE 1   Compounding Annually

Suppose Quan Li invests $500 at 7% interest compounded annually. Find the value of her investment 10 years later.

SOLUTION Letting $P = 500$, $r = 0.07$, and $n = 10$,

$A = 500(1 + 0.07)^{10} = 983.575. \ldots$ Rounding to the nearest cent, we see that the value of Quan Li's investment after 10 years is $983.58.

Now try Exercise 1.

Interest Compounded *k* Times per Year

Suppose a principal P is invested at an annual interest rate r compounded k times a year for t years. Then r/k is the interest rate per compounding period, and kt is the number of compounding periods. The amount A in the account after t years is

$$A = P\left(1 + \frac{r}{k}\right)^{kt}.$$

EXAMPLE 2 Compounding Monthly

Suppose Roberto invests \$500 at 9% annual interest *compounded monthly*, that is, compounded 12 times a year. Find the value of his investment 5 years later.

SOLUTION Letting $P = 500$, $r = 0.09$, $k = 12$, and $t = 5$,

$$A = 500\left(1 + \frac{0.09}{12}\right)^{12(5)} = 782.840.\ldots$$

So the value of Roberto's investment after 5 years is \$782.84.

Now try Exercise 5.

The problems in Examples 1 and 2 required that we calculate A. Examples 3 and 4 illustrate situations that require us to determine the values of other variables in the compound interest formula.

EXAMPLE 3 Finding the Time Period of an Investment

Judy has \$500 to invest at 9% annual interest compounded monthly. How long will it take for her investment to grow to \$3000?

SOLUTION

Model Let $P = 500$, $r = 0.09$, $k = 12$, and $A = 3000$ in the equation

$$A = P\left(1 + \frac{r}{k}\right)^{kt},$$

and solve for t.

Solve Graphically For

$$3000 = 500\left(1 + \frac{0.09}{12}\right)^{12t},$$

we let

$$f(t) = 500\left(1 + \frac{0.09}{12}\right)^{12t} \quad \text{and} \quad y = 3000,$$

and then find the point of intersection of the graphs. Figure 3.41 shows that this occurs at $t \approx 19.98$.

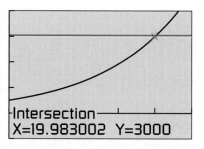

Intersection
X=19.983002 Y=3000

[0, 25] by [−1000, 4000]

FIGURE 3.41 Graph for Example 3.

continued

Confirm Algebraically

$$3000 = 500(1 + 0.09/12)^{12t}$$

$$6 = 1.0075^{12t} \qquad \text{Divide by 500.}$$

$$\ln 6 = \ln (1.0075^{12t}) \qquad \text{Apply ln to each side.}$$

$$\ln 6 = 12t \ln (1.0075) \qquad \text{Power rule}$$

$$t = \frac{\ln 6}{12 \ln 1.0075} \qquad \text{Divide by 12 ln 1.0075.}$$

$$= 19.983\ldots \qquad \text{Calculate.}$$

Interpret So it will take Judy 20 years for the value of the investment to reach (and slightly exceed) $3000. *Now try Exercise 21.*

EXAMPLE 4   Finding an Interest Rate

Stephen has $500 to invest. What annual interest rate *compounded quarterly* (four times per year) is required to double his money in 10 years?

SOLUTION

Model Letting $P = 500$, $k = 4$, $t = 10$, and $A = 1000$ yields the equation

$$1000 = 500\left(1 + \frac{r}{4}\right)^{4(10)}$$

that we solve for r.

Solve Graphically Figure 3.42 shows that $f(r) = 500(1 + r/4)^{40}$ and $y = 1000$ intersect at $r \approx 0.0699$, or $r = 6.99\%$.

Interpret Stephen's investment of $500 will double in 10 years at an annual interest rate of 6.99% compounded quarterly. *Now try Exercise 25.*

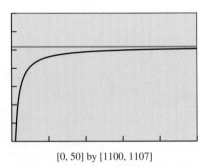

Intersection
X=.06991877 Y=1000

[0, 0.15] by [−500, 1500]

FIGURE 3.42 Graph for Example 4.

Interest Compounded Continuously

In Exploration 1, $1000 is invested for 1 year at a 10% interest rate. We investigate the value of the investment at the end of 1 year as the number of compounding periods k increases. In other words, we determine the "limiting" value of the expression $1000(1 + 0.1/k)^k$ as k assumes larger and larger integer values.

[0, 50] by [1100, 1107]

FIGURE 3.43 Graph for Exploration 1.

EXPLORATION 1   Increasing the Number of Compounding Periods Boundlessly

Let $A = 1000\left(1 + \dfrac{0.1}{k}\right)^k$.

1. Complete a table of values of A for $k = 10, 20, \ldots, 100$. What pattern do you observe?

2. Figure 3.43 shows the graphs of the function $A(k) = 1000(1 + 0.1/k)^k$ and the horizontal line $y = 1000e^{0.1}$. Interpret the meanings of these graphs.

Recall from Section 3.1 that $e = \lim\limits_{x \to \infty}(1 + 1/x)^x$. Therefore, for a fixed interest rate r, if we let $x = k/r$,

$$\lim_{k \to \infty}\left(1 + \frac{r}{k}\right)^{k/r} = e.$$

We do not know enough about limits yet, but with some calculus, it can be proved that $\lim_{k \to \infty} P(1 + r/k)^{kt} = Pe^{rt}$. So $A = Pe^{rt}$ is the formula used when interest is **compounded continuously**. In nearly any situation, one of the following two formulas can be used to compute compound interest:

Compound Interest—Value of an Investment

Suppose a principal P is invested at a fixed annual interest rate r. The value of the investment after t years is

- $A = P\left(1 + \dfrac{r}{k}\right)^{kt}$ when interest compounds k times per year,

- $A = Pe^{rt}$ when interest compounds continuously.

EXAMPLE 5   Compounding Continuously

Suppose LaTasha invests $100 at 8% annual interest compounded continuously. Find the value of her investment at the end of each of the years 1, 2, . . . , 7.

SOLUTION   Substituting into the formula for continuous compounding, we obtain $A(t) = 100e^{0.08t}$. Figure 3.44 shows the values of $y_1 = A(x) = 100e^{0.08x}$ for $x = 1$, 2, . . . , 7. For example, the value of her investment is $149.18 at the end of 5 years, and $175.07 at the end of 7 years.   *Now try Exercise 9.*

X	Y₁
1	108.33
2	117.35
3	127.12
4	137.71
5	149.18
6	161.61
7	175.07

Y₁ ■ 100e^(0.08X)

FIGURE 3.44 Table of values for Example 5.

Annual Percentage Yield

With so many different interest rates and methods of compounding it is sometimes difficult for a consumer to compare two different options. For example, would you prefer an investment earning 8.75% annual interest compounded quarterly or one earning 8.7% compounded monthly?

A common basis for comparing investments is the **annual percentage yield (APY)**—the percentage rate that, compounded annually, would yield the same return as the given interest rate with the given compounding period.

EXAMPLE 6   Computing Annual Percentage Yield (APY)

Ursula invests $2000 with Crab Key Bank at 5.15% annual interest compounded quarterly. What is the equivalent APY?

SOLUTION   Let $x =$ the equivalent APY. The value of the investment at the end of 1 year using this rate is $A = 2000(1 + x)$. Thus, we have

$$2000(1 + x) = 2000\left(1 + \frac{0.0515}{4}\right)^4$$

$$(1 + x) = \left(1 + \frac{0.0515}{4}\right)^4 \qquad \text{Divide by 2000.}$$

$$x = \left(1 + \frac{0.0515}{4}\right)^4 - 1 \qquad \text{Subtract 1.}$$

$$\approx 0.0525 \qquad \text{Calculate.}$$

continued

The annual percentage yield is 5.25%. In other words, Ursula's $2000 invested at 5.15% compounded quarterly for 1 year earns the same interest and yields the same value as $2000 invested elsewhere paying 5.25% interest once at the end of the year.

Now try Exercise 41.

Example 6 shows that the APY does not depend on the principal P because both sides of the equation were divided by $P = 2000$. So we can assume that $P = 1$ when comparing investments.

EXAMPLE 7   Comparing Annual Percentage Yields (APYs)

Which investment is more attractive, one that pays 8.75% compounded quarterly or another that pays 8.7% compounded monthly?

SOLUTION
Let

$$r_1 = \text{the APY for the 8.75\% rate,}$$
$$r_2 = \text{the APY for the 8.7\% rate.}$$

$$1 + r_1 = \left(1 + \frac{0.0875}{4}\right)^4 \qquad 1 + r_2 = \left(1 + \frac{0.087}{12}\right)^{12}$$

$$r_1 = \left(1 + \frac{0.0875}{4}\right)^4 - 1 \qquad r_2 = \left(1 + \frac{0.087}{12}\right)^{12} - 1$$

$$\approx 0.09041 \qquad\qquad \approx 0.09055$$

The 8.7% rate compounded monthly is more attractive because its APY is 9.055% compared with 9.041% for the 8.75% rate compounded quarterly.

Now try Exercise 45.

Annuities—Future Value

So far, in all of the investment situations we have considered, the investor has made a single *lump-sum* deposit. But suppose an investor makes regular deposits monthly, quarterly, or yearly—the same amount each time. This is an *annuity* situation.

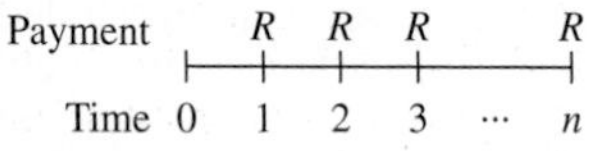

An **annuity** is a sequence of equal periodic payments. The annuity is **ordinary** if deposits are made at the end of each period at the same time the interest is posted in the account. Figure 3.45 represents this situation graphically. We will consider only ordinary annuities in this textbook.

FIGURE 3.45 Payments into an ordinary annuity.

Let's consider an example. Suppose Sarah makes quarterly $500 payments at the end of each quarter into a retirement account that pays 8% interest compounded quarterly. How much will be in Sarah's account at the end of the first year? Notice the pattern.

End of Quarter 1:

$500 = $500

End of Quarter 2:

$500 + $500(1.02) = $1010

End of Quarter 3:

$$\$500 + \$500(1.02) + \$500(1.02)^2 = \$1530.20$$

End of the year:

$$\$500 + \$500(1.02) + \$500(1.02)^2 + \$500(1.02)^3 \approx \$2060.80$$

Thus the total value of the investment returned from an annuity consists of all the periodic payments together with all the interest. This value is called the **future value** of the annuity because it is typically calculated when projecting into the future.

Future Value of an Annuity

The future value FV of an annuity consisting of n equal periodic payments of R dollars at an interest rate i per compounding period (payment interval) is

$$FV = R\frac{(1 + i)^n - 1}{i}.$$

EXAMPLE 8   Calculating the Value of an Annuity

At the end of each quarter year, Emily makes a $500 payment into the Lanaghan Mutual Fund. If her investments earn 7.88% annual interest compounded quarterly, what will be the value of Emily's annuity in 20 years?

SOLUTION Let $R = 500$, $i = 0.0788/4$, $n = 20(4) = 80$. Then,

$$FV = R\frac{(1 + i)^n - 1}{i}$$

$$FV = 500 \cdot \frac{(1 + 0.0788/4)^{80} - 1}{0.0788/4}$$

$$FV = 95{,}483.389\ldots$$

So the value of Emily's annuity in 20 years will be $95,483.39.

> *Now try Exercise 13.*

Loans and Mortgages—Present Value

An annuity is a sequence of equal period payments. The net amount of money put into an annuity is its **present value**. The net amount returned from the annuity is its future value. The periodic and equal payments on a loan or mortgage actually constitute an annuity.

How does the bank determine what the periodic payments should be? It considers what would happen to the present value of an investment with interest compounding over the term of the loan and compares the result to the future value of the loan repayment annuity.

We illustrate this reasoning by assuming that a bank lends you a present value $PV = \$50{,}000$ at 6% to purchase a house with the expectation that you will make a mortgage payment each month (at the monthly interest rate of $0.06/12 = 0.005$).

• The future value of an investment at 6% compounded monthly for n months is

$$PV(1 + i)^n = 50{,}000(1 + 0.005)^n.$$

• The future value of an annuity of R dollars (the loan payments) is

$$R\frac{(1+i)^n - 1}{i} = R\frac{(1+0.005)^n - 1}{0.005}.$$

To find R, we would solve the equation

$$50,000(1+0.005)^n = R\frac{(1+0.005)^n - 1}{0.005}.$$

In general, the monthly payments of R dollars for a loan of PV dollars must satisfy the equation

$$PV(1+i)^n = R\frac{(1+i)^n - 1}{i}.$$

Dividing both sides by $(1+i)^n$ leads to the following formula for the present value of an annuity.

Present Value of an Annuity

The present value PV of an annuity consisting of n equal payments of R dollars earning an interest rate i per period (payment interval) is

$$PV = R\frac{1-(1+i)^{-n}}{i}.$$

The annual interest rate charged on consumer loans is the **annual percentage rate (APR)**. The APY for the lender is higher than the APR. See Exercise 58.

EXAMPLE 9   Calculating Loan Payments

Carlos purchases a new pickup truck for $18,500. What are the monthly payments for a 4-year loan with a $2000 down payment if the annual interest rate (APR) is 2.9%?

SOLUTION

Model The down payment is $2000, so the amount borrowed is $16,500. Since APR $= 2.9\%$, $i = 0.029/12$ and the monthly payment is the solution to

$$16,500 = R\frac{1-(1+0.029/12)^{-4(12)}}{0.029/12}.$$

Solve Algebraically

$$R\left[1 - \left(1 + \frac{0.029}{12}\right)^{-4(12)}\right] = 16,500\left(\frac{0.029}{12}\right)$$

$$R = \frac{16,500(0.029/12)}{1-(1+0.029/12)^{-48}}$$

$$= 364.487\ldots$$

Interpret Carlos will have to pay $364.49 per month for 47 months, and slightly less the last month.   *Now try Exercise 19.*

QUICK REVIEW 3.6

1. Find 3.5% of 200.

2. Find 2.5% of 150.

3. What is one-fourth of 7.25%?

4. What is one-twelfth of 6.5%?

5. 78 is what percent of 120?

6. 28 is what percent of 80?

7. 48 is 32% of what number?

8. 176.4 is 84% of what number?

9. How much does Jane have at the end of 1 year if she invests $300 at 5% simple interest?

10. How much does Reggie have at the end of 1 year if he invests $500 at 4.5% simple interest?

SECTION 3.6 EXERCISES

In Exercises 1–4, find the amount A accumulated after investing a principal P for t years at an interest rate r compounded annually.

1. $P = \$1500$, $r = 7\%$, $t = 6$

2. $P = \$3200$, $r = 8\%$, $t = 4$

3. $P = \$12,000$, $r = 7.5\%$, $t = 7$

4. $P = \$15,500$, $r = 9.5\%$, $t = 12$

In Exercises 5–8, find the amount A accumulated after investing a principal P for t years at an interest rate r compounded k times per year.

5. $P = \$1500$, $r = 7\%$, $t = 5$, $k = 4$

6. $P = \$3500$, $r = 5\%$, $t = 10$, $k = 4$

7. $P = \$40,500$, $r = 3.8\%$, $t = 20$, $k = 12$

8. $P = \$25,300$, $r = 4.5\%$, $t = 25$, $k = 12$

In Exercises 9–12, find the amount A accumulated after investing a principal P for t years at interest rate r compounded continuously.

9. $P = \$1250$, $r = 5.4\%$, $t = 6$

10. $P = \$3350$, $r = 6.2\%$, $t = 8$

11. $P = \$21,000$, $r = 3.7\%$, $t = 10$

12. $P = \$8,875$, $r = 4.4\%$, $t = 25$

In Exercises 13–15, find the future value FV accumulated in an annuity after investing periodic payments R for t years at an annual interest rate r, with payments made and interest credited k times per year.

13. $R = \$500$, $r = 7\%$, $t = 6$, $k = 4$

14. $R = \$300$, $r = 6\%$, $t = 12$, $k = 4$

15. $R = \$450$, $r = 5.25\%$, $t = 10$, $k = 12$

16. $R = \$610$, $r = 6.5\%$, $t = 25$, $k = 12$

In Exercises 17 and 18, find the present value PV of a loan with an annual interest rate r and periodic payments R for a term of t years, with payments made and interest charged 12 times per year.

17. $r = 4.7\%$, $R = \$815.37$, $t = 5$

18. $r = 6.5\%$, $R = \$1856.82$, $t = 30$

In Exercises 19 and 20, find the periodic payment R of a loan with present value PV and an annual interest rate r for a term of t years, with payments made and interest charged 12 times per year.

19. $PV = \$18,000$, $r = 5.4\%$, $t = 6$

20. $PV = \$154,000$, $r = 7.2\%$, $t = 15$

21. **Finding Time** If John invests $2300 in a savings account with a 9% interest rate compounded quarterly, how long will it take until John's account has a balance of $4150?

22. **Finding Time** If Joelle invests $8000 into a retirement account with a 9% interest rate compounded monthly, how long will it take until this single payment has grown in her account to $16,000?

23. **Trust Officer** Megan is the trust officer for an estate. If she invests $15,000 into an account that carries an interest rate of 8% compounded monthly, how long will it be until the account has a value of $45,000 for Megan's client?

24. **Chief Financial Officer** Willis is the financial officer of a private university with the responsibility for managing an endowment. If he invests $1.5 million at an interest rate of 8% compounded quarterly, how long will it be until the account exceeds $3.75 million?

25. **Finding the Interest Rate** What interest rate compounded daily (365 days/year) is required for a $22,000 investment to grow to $36,500 in 5 years?

26. **Finding the Interest Rate** What interest rate compounded monthly is required for an $8500 investment to triple in 5 years?

27. **Pension Officer** Jack is an actuary working for a corporate pension fund. He needs to have $14.6 million grow to $22 million in 6 years. What interest rate compounded annually does he need for this investment?

28. **Bank President** The president of a bank has $18 million in his bank's investment portfolio that he wants to grow to $25 million in 8 years. What interest rate compounded annually does he need for this investment?

29. Doubling Your Money Determine how much time is required for an investment to double in value if interest is earned at the rate of 5.75% compounded quarterly.

30. Tripling Your Money Determine how much time is required for an investment to triple in value if interest is earned at the rate of 6.25% compounded monthly.

In Exercises 31–34, complete the table about continuous compounding.

	Initial Investment	APR	Time to Double	Amount in 15 years
31.	$12,500	9%	?	?
32.	$32,500	8%	?	?
33.	$ 9,500	?	4 years	?
34.	$16,800	?	6 years	?

In Exercises 35–40, complete the table about doubling time of an investment.

	APR	Compounding Periods	Time to Double
35.	4%	Quarterly	?
36.	8%	Quarterly	?
37.	7%	Annually	?
38.	7%	Quarterly	?
39.	7%	Monthly	?
40.	7%	Continuously	?

In Exercises 41–44, find the annual percentage yield (APY) for the investment.

41. $3000 at 6% compounded quarterly

42. $8000 at 5.75% compounded daily

43. *P* dollars at 6.3% compounded continuously

44. *P* dollars at 4.7% compounded monthly

45. Comparing Investments Which investment is more attractive, 5% compounded monthly or 5.1% compounded quarterly?

46. Comparing Investments Which investment is more attractive, $5\frac{1}{8}$% compounded annually or 5% compounded continuously?

In Exercises 47–50, payments are made and interest is credited at the end of each month.

47. An IRA Account Amy contributes $50 per month into the Lincoln National Bond Fund that earns 7.26% annual interest. What is the value of Amy's investment after 25 years?

48. An IRA Account Andrew contributes $50 per month into the Hoffbrau Fund that earns 15.5% annual interest. What is the value of his investment after 20 years?

49. An Investment Annuity Jolinda contributes to the Celebrity Retirement Fund that earns 12.4% annual interest. What should her monthly payments be if she wants to accumulate $250,000 in 20 years?

50. An Investment Annuity Diego contributes to a Commercial National money market account that earns 4.5% annual interest. What should his monthly payments be if he wants to accumulate $120,000 in 30 years?

51. Car Loan Payment What is Kim's monthly payment for a 4-year $9000 car loan with an APR of 7.95% from Century Bank?

52. Car Loan Payment What is Ericka's monthly payment for a 3-year $4500 car loan with an APR of 10.25% from County Savings Bank?

53. House Mortgage Payment Gendo obtains a 30-year $86,000 house loan with an APR of 8.75% from National City Bank. What is her monthly payment?

54. House Mortgage Payment Roberta obtains a 25-year $100,000 house loan with an APR of 9.25% from NBD Bank. What is her monthly payment?

55. Mortgage Payment Planning An $86,000 mortgage for 30 years at 12% APR requires monthly payments of $884.61. Suppose you decided to make monthly payments of $1050.00.

(a) When would the mortgage be completely paid?

(b) How much do you save with the greater payments compared with the original plan?

56. Mortgage Payment Planning Suppose you make payments of $884.61 for the $86,000 mortgage in Exercise 53 for 10 years and then make payments of $1050 until the loan is paid.

(a) When will the mortgage be completely paid under these circumstances?

(b) How much do you save with the greater payments compared with the original plan?

57. Writing to Learn Explain why computing the APY for an investment does not depend on the actual amount being invested. Give a formula for the APY on a $1 investment at annual rate *r* compounded *k* times a year. How do you extend the result to a $1000 investment?

58. Writing to Learn Give reasons why banks might not announce their APY on a loan they would make to you at a given APR. What is the bank's APY on a loan that they make at 4.5% APR?

59. Group Activity Work in groups of three or four. Consider population growth of humans or other animals, bacterial growth, radioactive decay, and compounded interest. Explain how these problem situations are similar and how they are different. Give examples to support your point of view.

60. Simple Interest versus Compounding Annually Steve purchases a $1000 certificate of deposit and will earn 6% each year. The interest will be mailed to him, so he will not earn interest on his interest.

(a) **Writing to Learn** Explain why after t years, the total amount of interest he receives from his investment plus the original $1000 is given by

$$f(t) = 1000(1 + 0.06t).$$

(b) Steve invests another $1000 at 6% compounded annually. Make a table that compares the value of the two investments for $t = 1, 2, \ldots, 10$ years.

Standardized Test Questions

61. True or False If $100 is invested at 5% annual interest for 1 year, there is no limit to the final value of the investment if it is compounded sufficiently often. Justify your answer.

62. True or False The total interest paid on a 15-year mortgage is less than half of the total interest paid on a 30-year mortgage with the same loan amount and APR. Justify your answer.

In Exercises 63–66, you may use a graphing calculator to solve the problem.

63. Multiple Choice What is the total value after 6 years of an initial investment of $2250 that earns 7% interest compounded quarterly?

(A) $3376.64   (B) $3412.00   (C) $3424.41

(D) $3472.16   (E) $3472.27

64. Multiple Choice The annual percentage yield of an account paying 6% compounded monthly is

(A) 6.03%.   (B) 6.12%.   (C) 6.17%.

(D) 6.20%.   (E) 6.24%.

65. Multiple Choice Mary Jo deposits $300 each month into her retirement account that pays 4.5% APR (0.375% per month). Use the formula $FV = R((1 + i)^n - 1)/i$ to find the value of her annuity after 20 years.

(A) $71,625.00

(B) $72,000.00

(C) $72,375.20

(D) $73,453.62

(E) $116,437.31

66. Multiple Choice To finance their home, Mr. and Mrs. Dass have agreed to a $120,000 mortgage loan at 7.25% APR. Use the formula $PV = R(1 - (1 + i)^{-n})/i$ to determine their monthly payments if the loan has a term of 15 years.

(A) $1095.44

(B) $1145.44

(C) $1195.44

(D) $1245.44

(E) $1295.44

Explorations

67. Loan Payoff Use the information about Carlos's truck loan in Example 9 to make a spreadsheet of the payment schedule. The first few lines of the spreadsheet should look like the following table:

Month No.	Payment	Interest	Principal	Balance
0				$16,500.00
1	$364.49	$39.88	$324.61	$16,175.39
2	$364.49	$39.09	$325.40	$15,849.99

To create the spreadsheet successfully, however, you need to use formulas for many of the cells, as shown in boldface type in the following sample:

Month No.	Payment	Interest	Principal	Balance
0				$16,500.00
=A2+1	$364.49	**=round(E2*2.9%/12,2)**	**=B3–C3**	**=E2–D3**
=A3+1	$364.49	**=round(E3*2.9%/12,2)**	**=B4–C4**	**=E3–D4**

Continue the spreadsheet using copy-and-paste techniques, and determine the amount of the 48th and final payment so that the final balance is $0.00.

68. Writing to Learn Loan Payoff Which of the following graphs is an accurate graph of the loan balance as a function of time, based on Carlos's truck loan in Example 9 and Exercise 67? Explain your choice based on increasing or decreasing behavior and other analytical characteristics. Would you expect the graph of loan balance versus time for a 30-year mortgage loan at twice the interest rate to have the same shape or a different shape as the one for the truck loan? Explain.

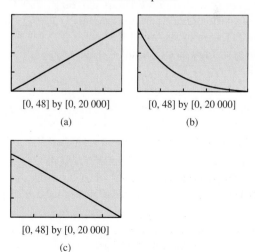

[0, 48] by [0, 20 000]   [0, 48] by [0, 20 000]
          (a)   (b)

[0, 48] by [0, 20 000]
          (c)

Extending the Ideas

69. The function

$$f(x) = 100 \frac{(1 + 0.08/12)^x - 1}{0.08/12}$$

describes the future value of a certain annuity.

(a) What is the annual interest rate?

(b) How many payments per year are there?

(c) What is the amount of each payment?

70. The function

$$f(x) = 200 \frac{1 - (1 + 0.08/12)^{-x}}{0.08/12}$$

describes the present value of a certain annuity.

(a) What is the annual interest rate?

(b) How many payments per year are there?

(c) What is the amount of each payment?

CHAPTER 3 **Key Ideas**

PROPERTIES, THEOREMS, AND FORMULAS

Exponential Growth and Decay   279
Exponential Functions $f(x) = b^x$ 280
Exponential Functions and the Base e 282
Exponential Population Model   290
Changing Between Logarithmic and Exponential
    Form   300
Basic Properties of Logarithms   301
Basic Properties of Common Logarithms   302
Basic Properties of Natural Logarithms   304
Properties of Logarithms   310
Change-of-Base Formula for Logarithms   313
Logarithmic Functions $f(x) = \log_b x$, with
    $b > 1$ 314
One-to-One Properties   320
Newton's Law of Cooling   326
Interest Compounded Annually   334
Interest Compounded k Times per Year   337
Interest Compounded Countinuously   337
Future Value of an Annuity   339
Present Value of an Annuity   340

PROCEDURES

Re-expression of Data   314–316
Logarithmic Re-expression of Data   328–329

GALLERY OF FUNCTIONS

Exponential

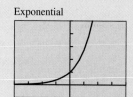

[−4, 4] by [−1, 5]

$$f(x) = e^x$$

Basic Logistic

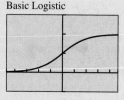

[−4.7, 4.7] by [−0.5, 1.5]

$$f(x) = \frac{1}{1 + e^{-x}}$$

Natural Logarithmic

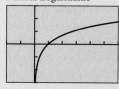

[−2, 6] by [−3, 3]

$$f(x) = \ln x$$

CHAPTER 3 **Review Exercises**

The collection of exercises marked in red could be used as a chapter test.

In Exercises 1 and 2, compute the exact value of the function for the given x value without using a calculator.

1. $f(x) = -3 \cdot 4^x$ for $x = \dfrac{1}{3}$

2. $f(x) = 6 \cdot 3^x$ for $x = -\dfrac{3}{2}$

In Exercises 3 and 4, determine a formula for the exponential function whose graph is shown in the figure.

 3.

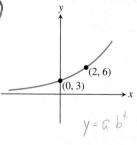

4.

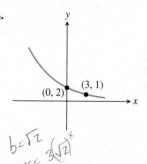

In Exercises 5–10, describe how to transform the graph of f into the graph of $g(x) = 2^x$ or $h(x) = e^x$. Sketch the graph by hand and support your answer with a grapher.

5. $f(x) = 4^{-x} + 3$ **6.** $f(x) = -4^{-x}$

7. $f(x) = -8^{-x} - 3$ **8.** $f(x) = 8^{-x} + 3$

9. $f(x) = e^{2x-3}$ **10.** $f(x) = e^{3x-4}$

In Exercises 11 and 12, find the y-intercept and the horizontal asymptotes.

11. $f(x) = \dfrac{100}{5 + 3e^{-0.05x}}$ **12.** $f(x) = \dfrac{50}{5 + 2e^{-0.04x}}$

In Exercises 13 and 14, state whether the function is an exponential growth function or an exponential decay function, and describe its end behavior using limits.

13. $f(x) = e^{4-x} + 2$ **14.** $f(x) = 2(5^{x-3}) + 1$

In Exercises 15–18, graph the function, and analyze it for domain, range, continuity, increasing or decreasing behavior, symmetry, boundedness, extrema, asymptotes, and end behavior.

15. $f(x) = e^{3-x} + 1$ **16.** $g(x) = 3(4^{x+1}) - 2$

17. $f(x) = \dfrac{6}{1 + 3 \cdot 0.4^x}$ **18.** $g(x) = \dfrac{100}{4 + 2e^{-0.01x}}$

In Exercises 19–22, find the exponential function that satisfies the given conditions.

19. Initial value = 24, increasing at a rate of 5.3% per day

20. Initial population = 67,000, increasing at a rate of 1.67% per year

21. Initial height = 18 cm, doubling every 3 weeks

22. Initial mass = 117 g, halving once every 262 hours

In Exercises 23 and 24, find the logistic function that satisfies the given conditions.

23. Initial value = 12, limit to growth = 30, passing through (2, 20).

24. Initial height = 6, limit to growth = 20, passing through (3, 15).

In Exercises 25 and 26, determine a formula for the logistic function whose graph is shown in the figure.

25.

26.

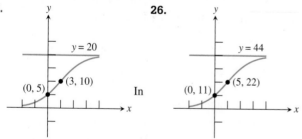

Exercises 27–30, evaluate the logarithmic expression without using a calculator.

27. $\log_2 32$ **28.** $\log_3 81$

29. $\log \sqrt[3]{10}$ **30.** $\ln \dfrac{1}{\sqrt{e^7}}$

In Exercises 31–34, rewrite the equation in exponential form.

31. $\log_3 x = 5$ **32.** $\log_2 x = y$

33. $\ln \dfrac{x}{y} = -2$ **34.** $\log \dfrac{a}{b} = -3$

In Exercises 35–38, describe how to transform the graph of $y = \log_2 x$ into the graph of the given function. Sketch the graph by hand and support with a grapher.

35. $f(x) = \log_2 (x + 4)$ **36.** $g(x) = \log_2 (4 - x)$

37. $h(x) = -\log_2 (x - 1) + 2$ **38.** $h(x) = -\log_2 (x + 1) + 4$

In Exercises 39–42, graph the function, and analyze it for domain, range, continuity, increasing or decreasing behavior, symmetry, boundedness, extrema, asymptotes, and end behavior.

39. $f(x) = x \ln x$ **40.** $f(x) = x^2 \ln x$

41. $f(x) = x^2 \ln |x|$ **42.** $f(x) = \dfrac{\ln x}{x}$

In Exercises 43–54, solve the equation.

43. $10^x = 4$ **44.** $e^x = 0.25$

45. $1.05^x = 3$ **46.** $\ln x = 5.4$

47. $\log x = -7$ **48.** $3^{x-3} = 5$

49. $3 \log_2 x + 1 = 7$ **50.** $2 \log_3 x - 3 = 4$

51. $\dfrac{3^x - 3^{-x}}{2} = 5$ **52.** $\dfrac{50}{4 + e^{2x}} = 11$

53. $\log (x + 2) + \log (x - 1) = 4$

54. $\ln (3x + 4) - \ln (2x + 1) = 5$

In Exercises 55 and 56, write the expression using only natural logarithms.

55. $\log_2 x$ **56.** $\log_{1/6} (6x^2)$

In Exercises 57 and 58, write the expression using only common logarithms.

57. $\log_5 x$ **58.** $\log_{1/2} (4x^3)$

In Exercises 59–62, match the function with its graph. All graphs are drawn in the window $[-4.7, 4.7]$ by $[-3.1, 3.1]$.

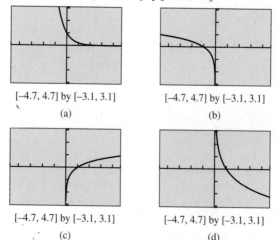

$[-4.7, 4.7]$ by $[-3.1, 3.1]$
(a)

$[-4.7, 4.7]$ by $[-3.1, 3.1]$
(b)

$[-4.7, 4.7]$ by $[-3.1, 3.1]$
(c)

$[-4.7, 4.7]$ by $[-3.1, 3.1]$
(d)

59. $f(x) = \log_5 x$ **60.** $f(x) = \log_{0.5} x$

61. $f(x) = \log_5 (-x)$ **62.** $f(x) = 5^{-x}$

63. **Compound Interest** Find the amount A accumulated after investing a principal $P = \$450$ for 3 years at an interest rate of 4.6% compounded annually.

64. **Compound Interest** Find the amount A accumulated after investing a principal $P = \$4800$ for 17 years at an interest rate 6.2% compounded quarterly.

65. **Compound Interest** Find the amount A accumulated after investing a principal P for t years at interest rate r compounded continuously.

66. **Future Value** Find the future value FV accumulated in an annuity after investing periodic payments R for t years at an annual interest rate r, with payments made and interest credited k times per year.

67. **Present Value** Find the present value PV of a loan with an annual interest rate $r = 5.5\%$ and periodic payments $R = \$550$ for a term of $t = 5$ years, with payments made and interest charged 12 times per year.

68. **Present Value** Find the present value PV of a loan with an annual interest rate $r = 7.25\%$ and periodic payments $R = \$953$ for a term of $t = 15$ years, with payments made and interest charged 26 times per year.

In Exercises 69 and 70, determine the value of k so that the graph of f passes through the given point.

69. $f(x) = 20e^{-kx}$, $(3, 50)$ **70.** $f(x) = 20e^{-kx}$, $(1, 30)$

In Exercises 71 and 72, use the data in Table 3.28.

Table 3.28 Populations of Two U.S. States (in millions)		
Year	Georgia	Illinois
1900	2.2	4.8
1910	2.6	5.6
1920	2.9	6.5
1930	2.9	7.6
1940	3.1	7.9
1950	3.4	8.7
1960	3.9	10.1
1970	4.6	11.1
1980	5.5	11.4
1990	6.5	11.4
2000	8.2	12.4

Source: U.S. Census Bureau as reported in the World Almanac and Book of Facts 2005.

71. **Modeling Population** Find an exponential regression model for Georgia's population, and use it to predict the population in 2005.

72. **Modeling Population** Find a logistic regression model for Illinois's population, and use it to predict the population in 2010.

73. **Drug Absorption** A drug is administered intravenously for pain. The function $f(t) = 90 - 52 \ln(1 + t)$, where $0 \le t \le 4$, gives the amount of the drug in the body after t hours.

(a) What was the initial ($t = 0$) number of units of drug administered?

(b) How much is present after 2 hr?

(c) Draw the graph of f.

74. **Population Decrease** The population of Metroville is 123,000 and is decreasing by 2.4% each year.

(a) Write a function that models the population as a function of time t.

(b) Predict when the population will be 90,000.

75. **Population Decrease** The population of Preston is 89,000 and is decreasing by 1.8% each year.

(a) Write a function that models the population as a function of time t.

(b) Predict when the population will be 50,000.

76. **Spread of Flu** The number P of students infected with flu at Northridge High School t days after exposure is modeled by

$$P(t) = \frac{300}{1 + e^{4-t}}.$$

(a) What was the initial ($t = 0$) number of students infected with the flu?

(b) How many students were infected after 3 days?

(c) When will 100 students be infected?

(d) What would be the maximum number of students infected?

77. **Rabbit Population** The number of rabbits in Elkgrove doubles every month. There are 20 rabbits present initially.

(a) Express the number of rabbits as a function of the time t.

(b) How many rabbits were present after 1 year? after 5 years?

(c) When will there be 10,000 rabbits?

78. **Guppy Population** The number of guppies in Susan's aquarium doubles every day. There are four guppies initially.

(a) Express the number of guppies as a function of time t.

(b) How many guppies were present after 4 days? after 1 week?

(c) When will there be 2000 guppies?

79. **Radioactive Decay** The half-life of a certain radioactive substance is 1.5 sec. The initial amount of substance is S_0 grams.

(a) Express the amount of substance S remaining as a function of time t.

(b) How much of the substance is left after 1.5 sec? after 3 sec?

(c) Determine S_0 if there was 1 g left after 1 min.

80. **Radioactive Decay** The half-life of a certain radioactive substance is 2.5 sec. The initial amount of substance is S_0 grams.

(a) Express the amount of substance S remaining as a function of time t.

(b) How much of the substance is left after 2.5 sec? after 7.5 sec?

(c) Determine S_0 if there was 1 g left after 1 min.

Trigo

81. Richter Scale Afghanistan suffered two major earthquakes in 1998. The one on February 4 had a Richter magnitude of 6.1, causing about 2300 deaths, and the one on May 30 measured 6.9 on the Richter scale, killing about 4700 people. How many times more powerful was the deadlier quake?

82. Chemical Acidity The pH of seawater is 7.6, and the pH of milk of magnesia is 10.5.

(a) What are their hydrogen-ion concentrations?

(b) How many times greater is the hydrogen-ion concentration of the seawater than that of milk of magnesia?

(c) By how many orders of magnitude do the concentrations differ?

83. Annuity Finding Time If Joenita invests $1500 into a retirement account with an 8% interest rate compounded quarterly, how long will it take this single payment to grow to $3750?

84. Annuity Finding Time If Juan invests $12,500 into a retirement account with a 9% interest rate compounded continuously, how long will it take this single payment to triple in value?

85. Monthly Payments The time t in months that it takes to pay off a $60,000 loan at 9% annual interest with monthly payments of x dollars is given by

$$t = 133.83 \ln\left(\frac{x}{x - 450}\right).$$

Estimate the length (term) of the $60,000 loan if the monthly payments are $700.

86. Monthly Payments Using the equation in Exercise 85, estimate the length (term) of the $60,000 loan if the monthly payments are $500.

87. Finding APY Find the annual percentage yield for an investment with an interest rate of 8.25% compounded monthly.

88. Finding APY Find the annual percentage yield that can be used to advertise an account that pays interest at 7.20% compounded countinuously.

89. Light Absorption The Beer-Lambert law of absorption applied to Lake Superior states that the light intensity I (in lumens) at a depth of x feet satisfies the equation

$$\log \frac{I}{12} = -0.0125x.$$

Find the light intensity at a depth of 25 ft.

90. For what values of b is $\log_b x$ a vertical stretch of $y = \ln x$? A vertical shrink of $y = \ln x$?

91. For what values of b is $\log_b x$ a vertical stretch of $y = \log x$? A vertical shrink of $y = \log x$?

92. If $f(x) = ab^x$, $a > 0$, $b > 0$, prove that $g(x) = \ln f(x)$ is a linear function. Find its slope and y-intercept.

93. Spread of Flu The number of students infected with flu after t days at Springfield High School is modeled by the function

$$P(t) = \frac{1600}{1 + 99e^{-0.4t}}.$$

(a) What was the initial number of infected students?

(b) When will 800 students be infected?

(c) The school will close when 400 of the 1600 student body are infected. When would the school close?

94. Population of Deer The population P of deer after t years in Briggs State Park is modeled by the function

$$P(t) = \frac{1200}{1 + 99e^{-0.4t}}.$$

(a) What was the inital population of deer?

(b) When will there be 1000 deer?

(c) What is the maximum number of deer planned for the park?

95. Newton's Law of Cooling A cup of coffee cooled from 96°C to 65°C after 8 min in a room at 20°C. When will it cool to 25°C?

96. Newton's Law of Cooling A cake is removed from an oven at 220°F and cools to 150°F after 35 min in a room at 75°F. When will it cool to 95°F?

97. The function

$$f(x) = 100 \frac{(1 + 0.09/4)^x - 1}{0.09/4}$$

describes the future value of a certain annuity.

(a) What is the annual interest rate?

(b) How many payments per year are there?

(c) What is the amount of each payment?

98. The function

$$g(x) = 200 \frac{1 - (1 + 0.11/4)^{-x}}{0.11/4}$$

describes the present value of a certain annuity.

(a) What is the annual interest rate?

(b) How many payments per year are there?

(c) What is the amount of each payment?

99. Simple Interest versus Compounding Continuously Grace purchases a $1000 certificate of deposit that will earn 5% each year. The interest will be mailed to her, so she will not earn interest on her interest.

(a) Show that after t years, the total amount of interest she receives from her investment plus the original $1000 is given by

$$f(t) = 1000(1 + 0.05t).$$

(b) Grace invests another $1000 at 5% compounded continuously. Make a table that compares the values of the two investments for $t = 1, 2, \ldots, 10$ years.

CHAPTER 3

Analyzing a B

When a ball bounce
mum height of the
rebound is a percenta
the percentage is a
motion detection dev
ing underneath a m
model that describes
of bounce number.

Collecting the

Set up the Calculato
a motion detector or
tem to collect ball b
the CBL or the Ball
CBL/CBR guideboc

Hold the ball at leas
that it bounces stra
These programs co
from the ground ver
data collected with
below shows each n

Bounce N
0
1
2
3
4
5

EXAMPLE 4   Designing a Running Track

The running lanes at the Emery Sears track at Bluffton College are 1 meter wide. The
inside radius of lane 1 is 33 meters and the inside radius of lane 2 is 34 meters. How
much longer is lane 2 than lane 1 around one turn? (See Figure 4.5.)

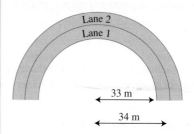

FIGURE 4.5 Two lanes of the track described in Example 4.

SOLUTION We think this solution through in radians. Each lane is a semicircle
with central angle $\theta = \pi$ and length $s = r\theta = r\pi$. The difference in their lengths,
therefore, is $34\pi - 33\pi = \pi$. Lane 2 is about 3.14 meters longer than lane 1.

Now try Exercise 37.

Angular and Linear Motion

In applications it is sometimes necessary to connect *angular speed* (measured in units
like revolutions per minute) to *linear speed* (measured in units like miles per hour). The
connection is usually provided by one of the arc length formulas or by a conversion factor that equates "1 radian" of angular measure to "1 radius" of arc length.

EXAMPLE 5   Using Angular Speed

Albert Juarez's truck has wheels 36 inches in diameter. If the wheels are rotating at
630 rpm (revolutions per minute), find the truck's speed in miles per hour.

SOLUTION We convert revolutions per minute to miles per hour by a series of unit
conversion factors. Note that the conversion factor $\dfrac{18 \text{ in.}}{1 \text{ radian}}$ works for this example
because the radius is 18 in.

$$\frac{630 \text{ rev}}{1 \text{ min}} \times \frac{60 \text{ min}}{1 \text{ hr}} \times \frac{2\pi \text{ radians}}{1 \text{ rev}} \times \frac{18 \text{ in.}}{1 \text{ radian}} \times \frac{1 \text{ ft}}{12 \text{ in.}} \times \frac{1 \text{ mi}}{5280 \text{ ft}}$$

$$\approx 67.47 \, \frac{\text{mi}}{\text{hr}}$$

Now try Exercise 45.

A **nautical mile** mile (naut mi) is the length of 1 minute of arc along Earth's equator.
Figure 4.6 shows, though not to scale, a central angle *AOB* of Earth that measures 1/60
of a degree. It intercepts an arc 1 naut mi long.

The arc length formula allows us to convert between nautical miles and **statute miles** (stat mi), the familiar "land mile" of 5280 feet.

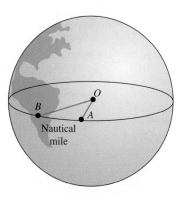

FIGURE 4.6 Although Earth is not a perfect sphere, its diameter is, on average, 7912.18 statute miles. A nautical mile is 1′ of Earth's circumference at the equator.

EXAMPLE 6 Converting to Nautical Miles

Megan McCarty, a pilot for Western Airlines, frequently pilots flights from Boston to San Francisco, a distance of 2698 stat mi. Captain McCarty's calculations of flight time are based on nautical miles. How many nautical miles is it from Boston to San Francisco?

SOLUTION The radius of the earth at the equator is approximately 3956 stat mi. Convert 1 minute to radians:

$$1' = \left(\frac{1}{60}\right)^{\circ} \times \frac{\pi \text{ rad}}{180^{\circ}} = \frac{\pi}{10{,}800} \text{ radians.}$$

Now we can apply the formula $s = r\theta$:

$$1 \text{ naut mi} = (3956)\left(\frac{\pi}{10{,}800}\right) \text{ stat mi}$$

$$\approx 1.15 \text{ stat mi}$$

$$1 \text{ stat mi} = \left(\frac{10{,}800}{3956\pi}\right) \text{ naut mi}$$

$$\approx 0.87 \text{ naut mi}$$

The distance from Boston to San Francisco is

$$2698 \text{ stat mi} = \frac{2698 \cdot 10{,}800}{3956\pi} \approx 2345 \text{ naut mi.}$$

Now try Exercise 51.

Distance Conversions

1 statute mile $\approx$ .87 nautical mile

1 nautical mile $\approx$ 1.15 statute mile

QUICK REVIEW 4.1 *(For help, go to Section 1.7.)*

In Exercises 1 and 2, find the circumference of the circle with the given radius r. State the correct unit.

1. $r = 2.5$ in.   **2.** $r = 4.6$ m

In Exercises 3 and 4, find the radius of the circle with the given circumference C.

3. $C = 12$ m   **4.** $C = 8$ ft

In Exercises 5 and 6, evaluate the expression for the given values of the variables. State the correct unit.

5. $s = r\theta$

    (a) $r = 9.9$ ft $\theta = 4.8$ rad

    (b) $r = 4.1$ km $\theta = 9.7$ rad

6. $v = r\omega$

    (a) $r = 8.7$ m $\omega = 3.0$ rad/sec

    (b) $r = 6.2$ ft $\omega = 1.3$ rad/sec

In Exercises 7–10, convert from miles per hour to feet per second or from feet per second to miles per hour.

7. 60 mph   **8.** 45 mph

9. 8.8 ft/sec   **10.** 132 ft/sec

  1-37 odd

BROOKE

SECTION 4.1 EXERCISES

In Exercises 1–4, convert from DMS to decimal form.

1. $23°12'$ **2.** $35°24'$

3. $118°44'15''$ **4.** $48°30'36''$

In Exercises 5–8, convert from decimal form to degrees, minutes, seconds (DMS).

5. $21.2°$ **6.** $49.7°$

7. $118.32°$ **8.** $99.37°$

In Exercises 9–16, convert from DMS to radians.

9. $60°$ **10.** $90°$

11. $120°$ **12.** $150°$

13. $71.72°$ **14.** $11.83°$

15. $61°24'$ **16.** $75°30'$

In Exercises 17–24, convert from radians to degrees.

17. $\pi/6$ **18.** $\pi/4$

19. $\pi/10$ **20.** $3\pi/5$

21. $7\pi/9$ **22.** $13\pi/20$

23. 2   **24.** 1.3

In Exercises 25–32, use the appropriate arc length formula to find the missing information.

s	r	θ
25. ?	2 in.	25 rad
26. ?	1 cm	70 rad
27. 1.5 ft	?	$\pi/4$ rad
28. 2.5 cm	?	$\pi/3$ rad
29. 3 m	1 m	?
30. 4 in.	7 in.	?
31. 40 cm	?	$20°$
32. ?	5 ft	$18°$

In Exercises 33 and 34, a central angle θ intercepts arcs s_1 and s_2 on two concentric circles with radii r_1 and r_2 respectively. Find the missing information.

θ	r_1	s_1	r_2	s_2
33. ?	11 cm	9 cm	44 cm	?
34. ?	8 km	36 km	?	72 km

35. To the nearest inch, find the perimeter of a 10-degree sector cut from a circular disc of radius 11 inches.

36. A 100-degree arc of a circle has a length of 7 cm. To the nearest centimeter, what is the radius of the circle?

37. It takes ten identical pieces to form a circular track for a pair of toy racing cars. If the inside arc of each piece is 3.4 inches shorter than the outside arc, what is the width of the track?

38. The concentric circles on an archery target are 6 inches apart. The inner circle (red) has perimeter of 37.7 inches. What is the perimeter of the next-largest (yellow) circle?

Exercises 39–42 refer to the 16 compass bearings shown. North corresponds to an angle of 0°, and other angles are measured clockwise from north.

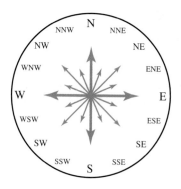

39. Compass Reading Find the angle in degrees that describes the compass bearing.

(a) NE (northeast)

(b) NNE (north-northeast)

(c) WSW (west-southwest)

40. Compass Reading Find the angle in degrees that describes the compass bearing.

(a) SSW (south-southwest)

(b) WNW (west-northwest)

(c) NNW (north-northwest)

41. Compass Reading Which compass direction is closest to a bearing of 121°?

42. Compass Reading Which compass direction is closest to a bearing of 219°?

43. Navigation Two Coast Guard patrol boats leave Cape May at the same time. One travels with a bearing of 42°30′ and the other with a bearing of 52°12′. If they travel at the same speed, approximately how far apart will they be when they are 25 statute miles from Cape May?

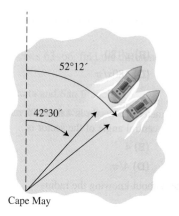

Cape May

44. Automobile Design Table 4.1 shows the size specifications for the tires that come as standard equipment on three different American vehicles.

Table 4.1 **Tire Sizes for Three Vehicles**		
Vehicle	Tire Type	Tire Diameter
Ford Taurus	215/60–16	26.16 inches
Dodge Charger RT	225/60–18	28.63 inches
Mercury Mariner	235/70–16	28.95 inches

(a) Find the speed of each vehicle in mph when the wheels are turning at 800 revolutions per minute.

(b) Compared to the Mercury Mariner, how many more revolutions must the tire of the Ford Taurus make in order to travel a mile?

(c) Writing to Learn It is unwise (and in some cases illegal) to equip a vehicle with wheels of a larger diameter than those for which it was designed. If a 2006 Ford Taurus were equipped with 28-inch tires, how would it affect the odometer (which measures mileage) and speedometer readings?

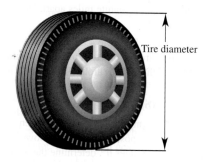

Tire diameter

45. Bicycle Racing Cathy Nguyen races on a bicycle with 13-inch radius wheels. When she is traveling at a speed of 44 ft/sec, how many revolutions per minute are her wheels making?

46. Tire Sizing The numbers in the "tire type" column in Exercise 44 give the size of the tire in the P-metric system. Each number is of the form $W/R-D$, where W is the width of the tire in millimeters, $R/100$ is the ratio of the sidewall (S) of the tire to its width W, and D is the diameter (in inches) of the wheel without the tire.

(a) Show that $S = WR/100$ millimeters $= WR/2540$ inches.

(b) The tire diameter is $D + 2S$. Derive a formula for the tire diameter that only involves the variables D, W, and R.

(c) Use the formula in part (b) to verify the tire diameters in Exercise 44. Then find the tire diameter for the 2006 Honda Ridgeline, which comes with 245/65–17 tires.

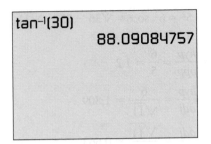

FIGURE 4.14 This is not cot (30°).

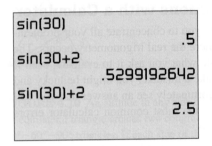

FIGURE 4.15 A correct and incorrect way to find sin (30°) + 2.

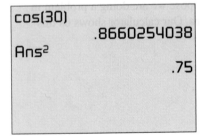

FIGURE 4.16 (Example 4)

There is also a key on the calculator for "TAN⁻¹"—but this is *not* the cotangent function! Remember that an exponent of −1 on a *function* is *never* used to denote a reciprocal; it is always used to denote the *inverse function*. We will study the inverse trigonometric functions in a later section, but meanwhile you can see that it is a bad way to evaluate cot (30) (Figure 4.14).

3. **Using Function Shorthand that the Calculator Does Not Recognize** This error is less dangerous because it usually results in an error message. We will often abbreviate powers of trig functions, writing (for example) "$\sin^3 \theta - \cos^3 \theta$" instead of the more cumbersome "$(\sin(\theta))^3 - (\cos(\theta))^3$." The calculator does not recognize the shorthand notation and interprets it as a syntax error.

4. **Not Closing Parentheses** This general algebraic error is easy to make on calculators that automatically open a parenthesis pair whenever you type a function key. Check your calculator by pressing the SIN key. If the screen displays "sin ("instead of just "sin" then you have such a calculator. The danger arises because the calculator will automatically *close* the parenthesis pair at the end of a command if you have forgotten to do so. That is fine if you *want* the parenthesis at the end of the command, but it is bad if you want it somewhere else. For example, if you want "sin (30)" and you type "sin (30", you will get away with it. But if you want "sin (30) + 2" and you type "sin (30 + 2", you will not (Figure 4.15).

It is usually impossible to find an "exact" answer on a calculator, especially when evaluating trigonometric functions. The actual values are usually irrational numbers with nonterminating, nonrepeating decimal expansions. However, you can find some exact answers if you know what you are looking for, as in Example 4.

EXAMPLE 4   Getting an "Exact Answer" on a Calculator

Find the exact value of cos 30° on a calculator.

SOLUTION As you see in Figure 4.16, the calculator gives the answer 0.8660254038. However, if we recognize 30° as one of our special angles (see Example 2 in this section), we might recall that the exact answer can be written in terms of a square root. We square our answer and get 0.75, which suggests that the exact value of cos 30° is $\sqrt{3/4} = \sqrt{3}/2$.

Now try Exercise 25.

Applications of Right Triangle Trigonometry

A triangle has six "parts", three angles and three sides, but you do not need to know all six parts to determine a triangle up to congruence. In fact, three parts are usually sufficient. The trigonometric functions take this observation a step further by giving us the means for actually *finding* the rest of the parts once we have enough parts to establish congruence. Using some of the parts of a triangle to solve for all the others is **solving a triangle**.

We will learn about solving general triangles in Sections 5.5 and 5.6, but we can already do some right triangle solving just by using the trigonometric ratios.

FIGURE 4.17 (Example 5)

EXAMPLE 5   Solving a Right Triangle

A right triangle with a hypotenuse of 8 includes a 37° angle (Figure 4.17). Find the measures of the other two angles and the lengths of the other two sides.

SOLUTION Since it is a right triangle, one of the other angles is 90°. That leaves $180° - 90° - 37° = 53°$ for the third angle.

Referring to the labels in Figure 4.17, we have

$$\sin 37° = \frac{a}{8} \qquad\qquad \cos 37° = \frac{b}{8}$$

$$a = 8 \sin 37° \qquad\qquad b = 8 \cos 37°$$

$$a \approx 4.81 \qquad\qquad b \approx 6.39$$

Now try Exercise 55.

The real-world applications of triangle-solving are many, reflecting the frequency with which one encounters triangular shapes in everyday life.

EXAMPLE 6   Finding the Height of a Building

From a point 340 feet away from the base of the Peachtree Center Plaza in Atlanta, Georgia, the angle of elevation to the top of the building is 65°. (See Figure 4.18.) Find the height h of the building.

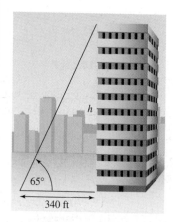

FIGURE 4.18 (Example 6)

A WORD ABOUT ROUNDING ANSWERS

Notice in Example 6 that we rounded the answer to the nearest integer. In applied problems it is illogical to give answers with more decimal places of accuracy than can be guaranteed for the input values. An answer of 729.132 feet implies razor-sharp accuracy, whereas the reported height of the building (340 feet) implies a much less precise measurement. (So does the angle of 65°.) Indeed, an engineer following specific rounding criteria based on "significant digits" would probably report the answer to Example 6 as 730 feet. We will not get too picky about rounding, but we will try to be sensible.

SOLUTION We need a ratio that will relate an angle to its opposite and adjacent sides. The tangent function is the appropriate choice.

$$\tan 65° = \frac{h}{340}$$

$$h = 340 \tan 65°$$

$$h \approx 729 \text{ feet}$$

Now try Exercise 61.

QUICK REVIEW 4.2 *(For help, go to Sections P.2 and 1.7.)*

In Exercises 1–4, use the Pythagorean theorem to solve for *x*.

1.

2.

3.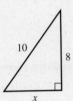

4.

In Exercises 5 and 6, convert units.

5. 8.4 ft to inches   **6.** 940 ft to miles

In Exercises 7–10, solve the equation. State the correct unit.

7. $0.388 = \dfrac{a}{20.4 \text{ km}}$ **8.** $1.72 = \dfrac{23.9 \text{ ft}}{b}$

9. $\dfrac{2.4 \text{ in.}}{31.6 \text{ in.}} = \dfrac{a}{13.3}$ **10.** $\dfrac{5.9}{\beta} = \dfrac{8.66 \text{ cm}}{6.15 \text{ cm}}$

SECTION 4.2 EXERCISES

In Exercises 1–8, find the values of all six trigonometric functions of the angle *θ*.

1.

2.

3.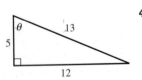

4.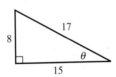

5.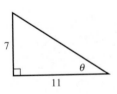

6.

7.

8.

In Exercises 9–18, assume that *θ* is an acute angle in a right triangle satisfying the given conditions. Evaluate the remaining trigonometric functions.

9. $\sin \theta = \dfrac{3}{7}$ **10.** $\sin \theta = \dfrac{2}{3}$

11. $\cos \theta = \dfrac{5}{11}$ **12.** $\cos \theta = \dfrac{5}{8}$

13. $\tan \theta = \dfrac{5}{9}$ **14.** $\tan \theta = \dfrac{12}{13}$

15. $\cot \theta = \dfrac{11}{3}$ **16.** $\csc \theta = \dfrac{12}{5}$

17. $\csc \theta = \dfrac{23}{9}$ **18.** $\sec \theta = \dfrac{17}{5}$

In Exercises 19–24, evaluate *without* using a calculator.

19. $\sin\left(\dfrac{\pi}{3}\right)$ **20.** $\tan\left(\dfrac{\pi}{4}\right)$

21. $\cot\left(\dfrac{\pi}{6}\right)$ **22.** $\sec\left(\dfrac{\pi}{3}\right)$

23. $\cos\left(\dfrac{\pi}{4}\right)$ **24.** $\csc\left(\dfrac{\pi}{3}\right)$

In Exercises 25–28, evaluate using a calculator. Give an exact value, not an approximate answer. (See Example 4.)

25. $\sec 45°$

26. $\sin 60°$

27. $\csc\left(\dfrac{\pi}{3}\right)$

28. $\tan\left(\dfrac{\pi}{3}\right)$

In Exercises 29–40, evaluate using a calculator. Be sure the calculator is in the correct mode. Give answers correct to three decimal places.

29. $\sin 74°$

30. $\tan 8°$

31. $\cos 19°23'$

32. $\tan 23°42'$

33. $\tan\left(\dfrac{\pi}{12}\right)$

34. $\sin\left(\dfrac{\pi}{15}\right)$

35. $\sec 49°$

36. $\csc 19°$

37. $\cot 0.89$

38. $\sec 1.24$

39. $\cot\left(\dfrac{\pi}{8}\right)$

40. $\csc\left(\dfrac{\pi}{10}\right)$

In Exercises 41–48, find the acute angle θ that satisfies the given equation. Give θ in both degrees and radians. You should do these problems without a calculator.

41. $\sin \theta = \dfrac{1}{2}$

42. $\sin \theta = \dfrac{\sqrt{3}}{2}$

43. $\cot \theta = \dfrac{1}{\sqrt{3}}$

44. $\cos \theta = \dfrac{\sqrt{2}}{2}$

45. $\sec \theta = 2$

46. $\cot \theta = 1$

47. $\tan \theta = \dfrac{\sqrt{3}}{3}$

48. $\cos \theta = \dfrac{\sqrt{3}}{2}$

In Exercises 49–54, solve for the variable shown.

49.

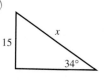

50.

51.

52.

53.

54.

In Exercises 55–58, solve the right $\triangle ABC$ for all of its unknown parts.

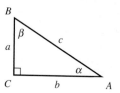

55. $\alpha = 20°$; $a = 12.3$

56. $\alpha = 41°$; $c = 10$

57. $\beta = 55°$; $a = 15.58$

58. $a = 5$; $\beta = 59°$

59. Writing to Learn What is $\lim_{\theta \to 0} \sin \theta$? Explain your answer in terms of right triangles in which θ gets smaller and smaller.

60. Writing to Learn What is $\lim_{\theta \to 0} \cos \theta$? Explain your answer in terms of right triangles in which θ gets smaller and smaller.

61. Height A guy wire from the top of the transmission tower at WJBC forms a $75°$ angle with the ground at a 55-foot distance from the base of the tower. How tall is the tower?

62. Height Kirsten places her surveyor's telescope on the top of a tripod 5 feet above the ground. She measures an $8°$ elevation above the horizontal to the top of a tree that is 120 feet away. How tall is the tree?

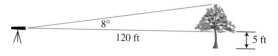

63. Group Activity Area For locations between 20° and 60° north latitude a solar collector panel should be mounted so that its angle with the horizontal is 20 greater than the local latitude. Consequently, the solar panel mounted on the roof of Solar Energy, Inc., in Atlanta (latitude 34°) forms a 54° angle with the horizontal. The bottom edge of the 12-ft long panel is resting on the roof, and the high edge is 5 ft above the roof. What is the total area of this rectangular collector panel?

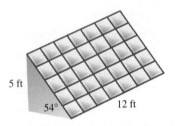

64. Height The Chrysler Building in New York City was the tallest building in the world at the time it was built. It casts a shadow approximately 130 feet long on the street when the sun's rays form an 82.9° angle with the earth. How tall is the building?

65. Distance DaShanda's team of surveyors had to find the distance *AC* across the lake at Montgomery County Park. Field assistants positioned themselves at points *A* and *C* while DaShanda set up an angle-measuring instrument at point *B,* 100 feet from *C* in a perpendicular direction. DaShanda measured ∠*ABC* as 75°12′42″. What is the distance *AC*?

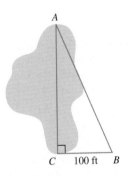

66. Group Activity Garden Design Allen's garden is in the shape of a quarter-circle with radius 10 ft. He wishes to plant his garden in four parallel strips, as shown in the diagram on the left below, so that the four arcs along the circular edge of the garden are all of equal length. After measuring four equal arcs, he carefully measures the widths of the four strips and records his data in the table shown at the right below.

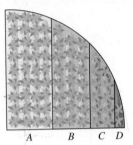

Strip	Width
A	3.827 ft
B	3.344 ft
C	2.068 ft
D	0.761 ft

Alicia sees Allen's data and realizes that he could have saved himself some work by figuring out the strip widths by trigonometry. By checking his data with a calculator she is able to correct two measurement errors he has made. Find Allen's two errors and correct them.

Standardized Test Questions

67. True or False If θ is an angle in any triangle, then $\tan \theta$ is the length of the side opposite θ divided by the length of the side adjacent to θ. Justify your answer.

68. True or False If *A* and *B* are angles of a triangle such that $A > B$, then $\cos A > \cos B$. Justify your answer.

You should answer these questions without using a calculator.

69. Multiple Choice Which of the following expressions does not represent a real number?

(A) $\sin 30°$ (B) $\tan 45°$ (C) $\cos 90°$

(D) $\csc 90°$ (E) $\sec 90°$

70. Multiple Choice If θ is the smallest angle in a 3–4–5 right triangle, then $\sin \theta =$

(A) $\dfrac{3}{5}$.   (B) $\dfrac{3}{4}$.   (C) $\dfrac{4}{5}$.

(D) $\dfrac{5}{4}$.   (E) $\dfrac{5}{3}$.

71. Multiple Choice If a nonhorizontal line has slope $\sin \theta$, it will be perpendicular to a line with slope

(A) $\cos \theta$.   (B) $-\cos \theta$.   (C) $\csc \theta$.

(D) $-\csc \theta$.   (E) $-\sin \theta$.

72. Multiple Choice Which of the following trigonometric ratios could *not* be π?

(A) $\tan\theta$ **(B)** $\cos\theta$ **(C)** $\cot\theta$

(D) $\sec\theta$ **(E)** $\csc\theta$

73. Trig Tables Before calculators became common classroom tools, students used trig tables to find trigonometric ratios. Below is a simplified trig table for angles between 40° and 50°. *Without using a calculator*, can you determine which column gives sine values, which gives cosine values, and which gives tangent values?

Trig Tables for Sine, Cosine, and Tangent			
Angle	?	?	?
40°	0.8391	0.6428	0.7660
42°	0.9004	0.6691	0.7431
44°	0.9657	0.6947	0.7193
46°	1.0355	0.7193	0.6947
48°	1.1106	0.7431	0.6691
50°	1.1917	0.7660	0.6428

74. Trig Tables Below is a simplified trig table for angles between 30° and 40°. *Without using a calculator*, can you determine which column gives cotangent values, which gives secant values, and which gives cosecant values?

Trig Tables for Cotangent, Secant, and Cosecant			
Angle	?	?	?
30°	1.1547	1.7321	2.0000
32°	1.1792	1.6003	1.8871
34°	1.2062	1.4826	1.7883
36°	1.2361	1.3764	1.7013
38°	1.2690	1.2799	1.6243
40°	1.3054	1.1918	1.5557

Explorations

75. Mirrors In the figure, a light ray shining from point A to point P on the mirror will bounce to point B in such a way that the *angle of incidence* α will equal the *angle of reflection* β. This is the *law of reflection* derived from physical experiments. Both angles are measured from the *normal line*, which is perpendicular to the mirror at the point of reflection P. If A is 2 m farther from the mirror than is B, and if $\alpha = 30°$ and $AP = 5$ m, what is the length PB?

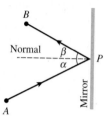

76. Pool On the pool table shown in the figure, where along the portion CD of the railing should you direct ball A so that it will bounce off CD and strike ball B? Assume that A obeys the law of reflection relative to rail CD.

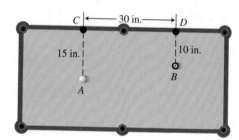

Extending the Ideas

77. Using the labeling of the triangle below, prove that if θ is an acute angle in any right triangle, $(\sin\theta)^2 + (\cos\theta)^2 = 1$

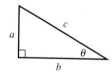

78. Using the labeling of the triangle below, prove that the area of the triangle is equal to $(1/2)\,ab\sin\theta$. [*Hint:* Start by drawing the altitude to side b and finding its length.]

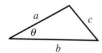

4.3
Trigonometry Extended: The Circular Functions

Trigonometric Functions of Any Angle

We now extend the definitions of the six basic trigonometric functions beyond triangles so that we do not have to restrict our attention to acute angles, or even to positive angles.

In geometry we think of an angle as a union of two rays with a common vertex. Trigonometry takes a more dynamic view by thinking of an angle in terms of a rotating ray. The beginning position of the ray, the **initial side**, is rotated about its endpoint, called the **vertex**. The final position is called the **terminal side**. The **measure of an angle** is a number that describes the amount of rotation from the initial side to the terminal side of the angle. **Positive angles** are generated by counterclockwise rotations and **negative angles** are generated by clockwise rotations. Figure 4.19 shows an angle of measure α, where α is a positive number.

FIGURE 4.19 An angle with positive measure α.

To bring the power of coordinate geometry into the picture (literally), we usually place an angle in **standard position** in the Cartesian plane, with the vertex of the angle at the origin and its initial side lying along the positive *x*-axis. Figure 4.20 shows two angles in standard position, one with positive measure α and the other with negative measure β.

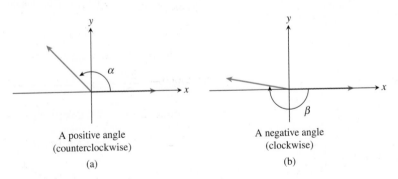

A positive angle
(counterclockwise)

(a)

A negative angle
(clockwise)

(b)

FIGURE 4.20 Two angles in standard position. In (a) the counterclockwise rotation generates an angle with positive measure α. In (b) the clockwise rotation generates an angle with negative measure β.

Two angles in this expanded angle-measurement system can have the same initial side and the same terminal side, yet have different measures. We call such angles **coterminal angles**. (See Figure 4.21 on the next page.) For example, angles of 90°,

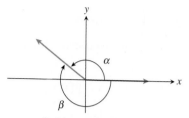

Positive and negative
coterminal angles

(a)

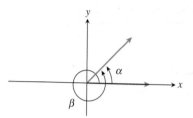

Two positive
coterminal angles

(b)

FIGURE 4.21 Coterminal angles. In (a) a positive angle and a negative angle are coterminal, while in (b) both coterminal angles are positive.

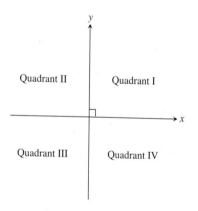

FIGURE 4.23 The four quadrants of the Cartesian plane. Both *x* and *y* are positive in QI (Quadrant I). Quadrants, like Super Bowls, are invariably designated by Roman numerals.

450°, and $-270°$ are all coterminal, as are angles of π radians, 3π radians, and -99π radians. In fact, angles are coterminal whenever they differ by an integer multiple of 360 degrees or by an integer multiple of 2π radians.

EXAMPLE 1   Finding Coterminal Angles

Find and draw a positive angle and a negative angle that are coterminal with the given angle.

(a) 30° **(b)** $-150°$ **(c)** $\dfrac{2\pi}{3}$ radians

SOLUTION There are infinitely many possible solutions; we will show two for each angle.

(a) Add 360°: $30° + 360° = 390°$

Subtract 360°: $30° - 360° = -330°$

Figure 4.22 shows these two angles, which are coterminal with the 30° angle.

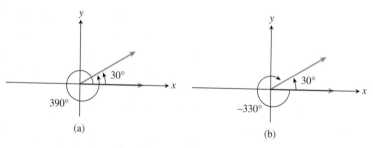

(a)   (b)

FIGURE 4.22 Two angles coterminal with 30°. (Example 1a)

(b) Add 360°: $-150° + 360° = 210°$

Subtract 720°: $-150° - 720° = -870°$

We leave it to you to draw the coterminal angles.

(c) Add 2π: $\dfrac{2\pi}{3} + 2\pi = \dfrac{2\pi}{3} + \dfrac{6\pi}{3} = \dfrac{8\pi}{3}$

Subtract 2π: $\dfrac{2\pi}{3} - 2\pi = \dfrac{2\pi}{3} - \dfrac{6\pi}{3} = -\dfrac{4\pi}{3}$

Again, we leave it to you to draw the coterminal angles.

Now try Exercise 1.

Extending the definitions of the six basic trigonometric functions so that they can apply to any angle is surprisingly easy, but first you need to see how our current definitions relate to the (x, y) coordinates in the Cartesian plane. We start in the first quadrant (see Figure 4.23), where the angles are all acute. Work through Exploration 1 before moving on.

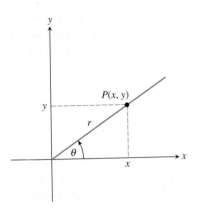

FIGURE 4.24 A point $P(x, y)$ in Quadrant I determines an acute angle θ. The number r denotes the distance from P to the origin. (Exploration 1)

EXPLORATION 1 **Investigating First Quadrant Trigonometry**

Let $P(x, y)$ be any point in the first quadrant (QI), and let r be the distance from P to the origin. (See Figure 4.24.)

1. Use the acute angle definition of the sine function (Section 4.2) to prove that $\sin \theta = y/r$.

2. Express $\cos \theta$ in terms of x and r.

3. Express $\tan \theta$ in terms of x and y.

4. Express the remaining three basic trigonometric functions in terms of x, y, and r.

If you have successfully completed Exploration 1, you should have no trouble verifying the solution to Example 2, which we show without the details.

EXAMPLE 2 **Evaluating Trig Functions Determined by a Point in QI**

Let θ be the acute angle in standard position whose terminal side contains the point $(5, 3)$. Find the six trigonometric functions of θ.

SOLUTION The distance from $(5, 3)$ to the origin is $\sqrt{34}$.

So
$$\sin \theta = \frac{3}{\sqrt{34}} \approx 0.514 \qquad \csc \theta = \frac{\sqrt{34}}{3} \approx 1.944$$

$$\cos \theta = \frac{5}{\sqrt{34}} \approx 0.857 \qquad \sec \theta = \frac{\sqrt{34}}{5} \approx 1.166$$

$$\tan \theta = \frac{3}{5} = 0.6 \qquad \cot \theta = \frac{5}{3} \approx 1.667$$

Now try Exercise 5.

Now we have an easy way to extend the trigonometric functions to any angle: Use the same definitions in terms of x, y, and r—*whether or not x and y are positive*. Compare Example 3 to Example 2.

EXAMPLE 3 **Evaluating Trig Functions Determined by a Point Not in QI**

Let θ be any angle in standard position whose terminal side contains the point $(-5, 3)$. Find the six trigonometric functions of θ.

SOLUTION The distance from $(-5, 3)$ to the origin is $\sqrt{34}$.

So
$$\sin \theta = \frac{3}{\sqrt{34}} \approx 0.514 \qquad \csc \theta = \frac{\sqrt{34}}{3} \approx 1.944$$

$$\cos \theta = \frac{-5}{\sqrt{34}} \approx -0.857 \qquad \sec \theta = \frac{\sqrt{34}}{-5} \approx -1.166$$

$$\tan \theta = \frac{3}{-5} = -0.6 \qquad \cot \theta = \frac{-5}{3} \approx -1.667$$

Now try Exercise 11.

Notice in Example 3 that θ is *any* angle in standard position whose terminal side contains the point $(-5, 3)$. There are infinitely many coterminal angles that could play the role of θ, some of them positive and some of them negative. The values of the six trigonometric functions would be the same for all of them.

We are now ready to state the formal definition.

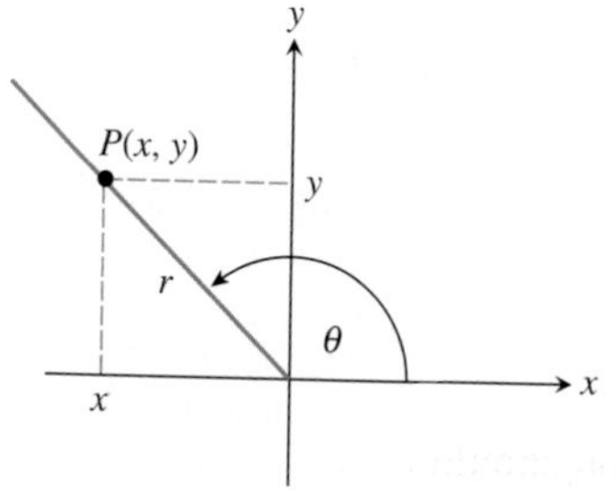

DEFINITION Trigonometric Functions of any Angle

Let θ be any angle in standard position and let $P(x, y)$ be any point on the terminal side of the angle (except the origin). Let r denote the distance from $P(x, y)$ to the origin, i.e., let $r = \sqrt{x^2 + y^2}$. (See Figure 4.25.) Then

$$\sin \theta = \frac{y}{r} \qquad\qquad \csc \theta = \frac{r}{y} \ (y \neq 0)$$

$$\cos \theta = \frac{x}{r} \qquad\qquad \sec \theta = \frac{r}{x} \ (x \neq 0)$$

$$\tan \theta = \frac{y}{x} \ (x \neq 0) \qquad\qquad \cot \theta = \frac{x}{y} \ (y \neq 0)$$

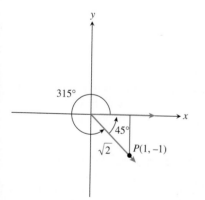

FIGURE 4.25 Defining the six trig functions of θ.

Examples 2 and 3 both began with a point $P(x, y)$ rather than an angle θ. Indeed, the point gave us so much information about the trigonometric ratios that we were able to compute them all without ever finding θ. So what do we do if we start with an angle θ in standard position and we want to evaluate the trigonometric functions? We try to find a point (x, y) on its terminal side. We illustrate this process with Example 4.

EXAMPLE 4 Evaluating the Trig Functions of 315°

Find the six trigonometric functions of 315°.

SOLUTION First we draw an angle of 315° in standard position. Without declaring a scale, pick a point P on the terminal side and connect it to the x-axis with a perpendicular segment. Notice that the triangle formed (called a **reference triangle**) is a 45°–45°–90° triangle. If we arbitrarily choose the horizontal and vertical sides of the reference triangle to be of length 1, then P has coordinates $(1, -1)$. (See Figure 4.26.)

We can now use the definitions with $x = 1$, $y = -1$, and $r = \sqrt{2}$.

$$\sin 315° = \frac{-1}{\sqrt{2}} = -\frac{\sqrt{2}}{2} \qquad \csc 315° = \frac{\sqrt{2}}{-1} = -\sqrt{2}$$

$$\cos 315° = \frac{1}{\sqrt{2}} = \frac{\sqrt{2}}{2} \qquad \sec 315° = \frac{\sqrt{2}}{1} = \sqrt{2}$$

$$\tan 315° = \frac{-1}{1} = -1 \qquad \cot 315° = \frac{1}{-1} = -1$$

Now try Exercise 25.

FIGURE 4.26 An angle of 315° in standard position determines a 45°–45°–90° *reference triangle.* (Example 4)

The happy fact that the reference triangle in Example 4 was a 45°–45°–90° triangle enabled us to label a point P on the terminal side of the 315° angle and then to find the trigonometric function values. We would also be able to find P if the given angle were to produce a 30°–60°–90° reference triangle.

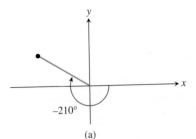

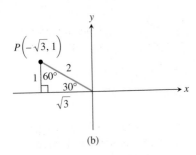

FIGURE 4.27 (Example 5a)

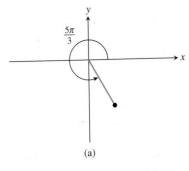

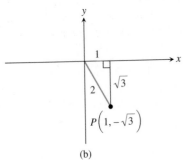

FIGURE 4.28 (Example 5b)

Evaluating Trig Functions of a Nonquadrantal Angle θ

1. Draw the angle θ in standard position, being careful to place the terminal side in the correct quadrant.

2. Without declaring a scale on either axis, label a point P (other than the origin) on the terminal side of θ.

3. Draw a perpendicular segment from P to the x-axis, determining the *reference triangle*. If this triangle is one of the triangles whose ratios you know, label the sides accordingly. If it is not, then you will need to use your calculator.

4. Use the sides of the triangle to determine the coordinates of point P, making them positive or negative according to the signs of x and y in that particular quadrant.

5. Use the coordinates of point P and the definitions to determine the six trig functions.

EXAMPLE 5   Evaluating More Trig Functions

Find the following without a calculator:

(a) sin (–210°)

(b) tan (5π/3)

(c) sec (–3π/4)

SOLUTION

(a) An angle of $-210°$ in standard position determines a 30°–60°–90° reference triangle in the second quadrant (Figure 4.27). We label the sides accordingly, then use the lengths of the sides to determine the point $P(-\sqrt{3}, 1)$. (Note that the x-coordinate is negative in the second quadrant.) The hypotenuse is $r = 2$. Therefore sin $(-210°) = y/r = 1/2$.

(b) An angle of $5\pi/3$ radians in standard position determines a 30°–60°–90° reference triangle in the fourth quadrant (Figure 4.28). We label the sides accordingly, then use the lengths of the sides to determine the point $P(1, -\sqrt{3})$. (Note that the y-coordinate is negative in the fourth quadrant.) The hypotenuse is $r = 2$. Therefore tan $(5\pi/3) = y/x = -\sqrt{3}/1 = -\sqrt{3}$.

(c) An angle of $-3\pi/4$ radians in standard position determines a 45°–45°–90° reference triangle in the third quadrant. (See Figure 4.29 on the next page.) We label the sides accordingly, then use the lengths of the sides to determine the point $P(-1, -1)$. (Note that both coordinates are negative in the third quadrant.) The hypotenuse is $r = \sqrt{2}$. Therefore sec $(-3\pi/4) = r/x = \sqrt{2}/-1 = -\sqrt{2}$.

Now try Exercise 29.

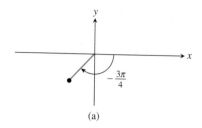

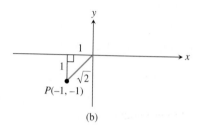

FIGURE 4.29 (Example 5c)

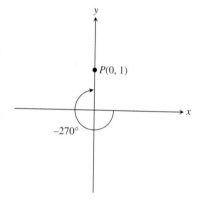

FIGURE 4.30 (Example 6a)

WHY NOT USE A CALCULATOR?

You might wonder why we would go through this procedure to produce values that could be found so easily with a calculator. The answer is to understand how trigonometry *works* in the coordinate plane. Ironically, technology has made these computational exercises more important than ever, since calculators have eliminated the need for the repetitive evaluations that once gave students their initial insights into the basic trig functions.

Angles whose terminal sides lie along one of the coordinate axes are called **quadrantal angles**, and although they do not produce reference triangles at all, it is easy to pick a point *P* along one of the axes.

EXAMPLE 6 Evaluating Trig Functions of Quadrantal Angles

Find each of the following, if it exists. If the value is undefined, write "undefined."

(a) $\sin(-270°)$

(b) $\tan 3\pi$

(c) $\sec \dfrac{11\pi}{2}$

SOLUTION

(a) In standard position, the terminal side of an angle of $-270°$ lies along the positive *y*-axis (Figure 4.30). A convenient point *P* along the positive *y*-axis is the point for which $r = 1$, namely $(0, 1)$. Therefore

$$\sin(-270°) = \frac{y}{r} = \frac{1}{1} = 1.$$

(b) In standard position, the terminal side of an angle of 3π lies along the negative *x*-axis. (See Figure 4.31 on the next page.) A convenient point *P* along the negative *x*-axis is the point for which $r = 1$, namely $(-1, 0)$. Therefore

$$\tan 3\pi = \frac{y}{x} = \frac{0}{-1} = 0.$$

(c) In standard position, the terminal side of an angle of $11\pi/2$ lies along the negative *y*-axis. (See Figure 4.32 on the next page.) A convenient point *P* along the negative *y*-axis is the point for which $r = 1$, namely $(0, -1)$. Therefore

$$\sec \frac{11\pi}{2} = \frac{r}{x} = \frac{1}{0}, \quad \text{undefined}$$

Now try Exercise 41.

Another good exercise is to use information from one trigonometric ratio to produce the other five. We do not need to know the angle θ, although we do need a hint as to the location of its terminal side so that we can sketch a reference triangle in the correct quadrant (or place a quadrantal angle on the correct side of the origin). Example 7 illustrates how this is done.

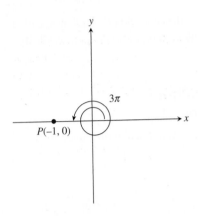

FIGURE 4.31 (Example 6b)

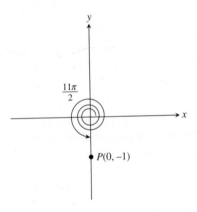

FIGURE 4.32 (Example 6c)

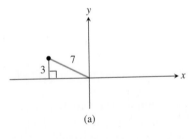

(a)

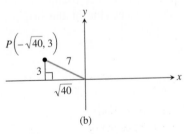

(b)

FIGURE 4.33 (Example 7a)

EXAMPLE 7   Using One Trig Ratio to Find the Others

Find $\cos \theta$ and $\tan \theta$ by using the given information to construct a reference triangle.

(a) $\sin \theta = \dfrac{3}{7}$ and $\tan \theta < 0$

(b) $\sec \theta = 3$ and $\sin \theta > 0$

(c) $\cot \theta$ is undefined and $\sec \theta$ is negative

SOLUTION

(a) Since $\sin \theta$ is positive, the terminal side is either in QI or QII. The added fact that $\tan \theta$ is negative means that the terminal side is in QII. We draw a reference triangle in QII with $r = 7$ and $y = 3$ (Figure 4.33); then we use the Pythagorean theorem to find that $x = -1\sqrt{7^2 - 3^2} = -\sqrt{40}$. (Note that x is negative in QII.)

We then use the definitions to get

$$\cos \theta = \frac{-\sqrt{40}}{7} \approx -0.904 \text{ and } \tan \theta = \frac{3}{-\sqrt{40}} \approx -0.474.$$

(b) Since $\sec \theta$ is positive, the terminal side is either in QI or QIV. The added fact that $\sin \theta$ is positive means that the terminal side is in QI. We draw a reference triangle in QI with $r = 3$ and $x = 1$ (Figure 4.34); then we use the Pythagorean theorem to find that $y = \sqrt{3^2 - 1^2} = \sqrt{8}$. (Note that y is positive in QI.)

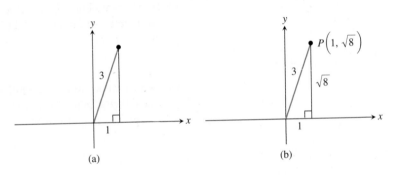

(a)   (b)

FIGURE 4.34 (Example 7b)

We then use the definitions to get

$$\cos \theta = \frac{1}{3} \approx 0.333 \text{ and } \tan \theta = \frac{\sqrt{8}}{1} \approx 2.828.$$

(We could also have found $\cos \theta$ directly as the reciprocal of $\sec \theta$.)

(c) Since $\cot \theta$ is undefined, we conclude that $y = 0$ and that θ is a quadrantal angle on the x-axis. The added fact that $\sec \theta$ is negative means that the terminal side is along the negative x-axis. We choose the point $(-1, 0)$ on the terminal side and use the definitions to get

$$\cos \theta = -1 \text{ and } \tan \theta = \frac{0}{-1} = 0.$$

Now try Exercise 43.

WHY NOT DEGREES?

One could actually develop a consistent theory of trigonometric functions based on a re-scaled *x*-axis with "degrees." For example, your graphing calculator will probably produce reasonable-looking graphs in degree mode. Calculus, however, uses rules that *depend* on radian measure for all trigonometric functions, so it is prudent for precalculus students to become accustomed to that now.

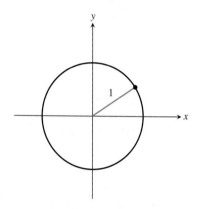

FIGURE 4.35 The Unit Circle.

Trigonometric Functions of Real Numbers

Now that we have extended the six basic trigonometric functions to apply to any angle, we are ready to appreciate them as functions of real numbers and to study their behavior. First, for reasons discussed in the first section of this chapter, we must agree to measure θ in radian mode so that the real number units of the input will match the real number units of the output.

When considering the trigonometric functions as functions of real numbers, the angles will be measured in radians.

DEFINITION Unit Circle

The unit circle is a circle of radius 1 centered at the origin (Figure 4.35).

The unit circle provides an ideal connection between triangle trigonometry and the trigonometric functions. Because arc length along the unit circle corresponds exactly to radian measure, we can use the circle itself as a sort of "number line" for the input values of our functions. This involves the **wrapping function**, which associates points on the number line with points on the circle.

Figure 4.36 shows how the wrapping function works. The real line is placed tangent to the unit circle at the point (1, 0), the point from which we measure angles in standard position. When the line is wrapped around the unit circle in both the positive (counterclockwise) and negative (clockwise) directions, each point t on the real line will fall on a point of the circle that lies on the terminal side of an angle of t radians in standard position. Using the coordinates (x, y) of this point, we can find the six trigonometric ratios for the angle t just as we did in Example 7—except even more easily, since $r = 1$.

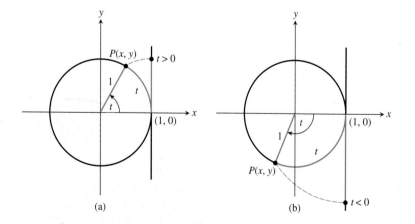

(a) (b)

FIGURE 4.36 How the number line is wrapped onto the unit circle. Note that each number t (positive or negative) is "wrapped" to a point P that lies on the terminal side of an angle of t radians in standard position.

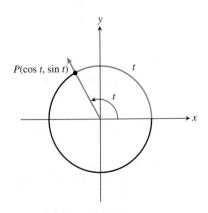

FIGURE 4.37 The real number t always wraps onto the point $(\cos t, \sin t)$ on the unit circle.

DEFINITION **Trigonometric Functions of Real Numbers**

Let t be any real number, and let $P(x, y)$ be the point corresponding to t when the number line is wrapped onto the unit circle as described above. Then

$$\sin t = y \qquad\qquad \csc t = \frac{1}{y}\ (y \neq 0)$$

$$\cos t = x \qquad\qquad \sec t = \frac{1}{x}\ (x \neq 0)$$

$$\tan t = \frac{y}{x}\ (x \neq 0) \qquad\qquad \cot t = \frac{x}{y}\ (y \neq 0)$$

Therefore, the number t on the number line always wraps onto the point $(\cos t, \sin t)$ on the unit circle (Figure 4.37).

Although it is still helpful to draw reference triangles inside the unit circle to see the ratios geometrically, this latest round of definitions does not invoke triangles at all. The real number t determines a point on the unit circle, and the (x, y) coordinates of the point determine the six trigonometric ratios. For this reason, the trigonometric functions when applied to real numbers are usually called the **circular functions**.

EXPLORATION 2 **Exploring the Unit Circle**

This works well as a group exploration. Get together in groups of two or three and explain to each other why these statements are true. Base your explanations on the unit circle (Figure 4.37). Remember that $-t$ wraps the same distance as t, but in the opposite direction.

1. For any t, the value of $\cos t$ lies between -1 and 1 inclusive.

2. For any t, the value of $\sin t$ lies between -1 and 1 inclusive.

3. The values of $\cos t$ and $\cos (-t)$ are always equal to each other. (Recall that this is the check for an *even* function.)

4. The values of $\sin t$ and $\sin (-t)$ are always opposites of each other. (Recall that this is the check for an *odd* function.)

5. The values of $\sin t$ and $\sin (t + 2\pi)$ are always equal to each other. In fact, that is true of all six trig functions on their domains, and for the same reason.

6. The values of $\sin t$ and $\sin (t + \pi)$ are always opposites of each other. The same is true of $\cos t$ and $\cos (t + \pi)$.

7. The values of $\tan t$ and $\tan (t + \pi)$ are always equal to each other (unless they are both undefined).

8. The sum $(\cos t)^2 + (\sin t)^2$ always equals 1.

9. (Challenge) Can you discover a similar relationship that is not mentioned in our list of eight? There are some to be found.

Periodic Functions

Statements 5 and 7 in Exploration 2 reveal an important property of the circular functions that we need to define for future reference.

DEFINITION Periodic Function

A function $y = f(t)$ is **periodic** if there is a positive number c such that

$f(t + c) = f(t)$ for all values of t in the domain of f. The smallest such

number c is called the **period** of the function.

Exploration 2 suggests that the sine and cosine functions have period 2π and that the tangent function has period π. We use this periodicity later to model predictably repetitive behavior in the real world, but meanwhile we can also use it to solve little non-calculator training problems like in some of the previous examples in this section.

EXAMPLE 8   Using Periodicity

Find each of the following numbers without a calculator.

(a) $\sin\left(\dfrac{57{,}801\pi}{2}\right)$

(b) $\cos(288.45\pi) - \cos(280.45\pi)$

(c) $\tan\left(\dfrac{\pi}{4} - 99{,}999\pi\right)$

SOLUTION

(a) $\sin\left(\dfrac{57{,}801\pi}{2}\right) = \sin\left(\dfrac{\pi}{2} + \dfrac{57{,}800\pi}{2}\right) = \sin\left(\dfrac{\pi}{2} + 28{,}900\pi\right)$

$= \sin\left(\dfrac{\pi}{2}\right) = 1$

Notice that $28{,}900\pi$ is just a large multiple of 2π, so $\pi/2$ and $((\pi/2) + 28{,}900\pi)$ wrap to the same point on the unit circle, namely $(0, 1)$.

(b) $\cos(288.45\pi) - \cos(280.45\pi) =$
$\cos(280.45\pi + 8\pi) - \cos(280.45\pi) = 0$

Notice that 280.45π and $(280.45\pi + 8\pi)$ wrap to the same point on the unit circle, so the cosine of one is the same as the cosine of the other.

(c) Since the period of the tangent function is π rather than 2π, $99{,}999\pi$ is a large multiple of the period of the tangent function. Therefore,

$$\tan\left(\frac{\pi}{4} - 99{,}999\pi\right) = \tan\left(\frac{\pi}{4}\right) = 1.$$

Now try Exercise 49.

We take a closer look at the properties of the six circular functions in the next two sections.

The 16-Point Unit Circle

At this point you should be able to use reference triangles and quadrantal angles to evaluate trigonometric functions for all integer multiples of 30° or 45° (equivalently, $\pi/6$ radians or $\pi/4$ radians). All of these special values wrap to the 16 special points shown on the unit circle below. Study this diagram until you are confident that you can find the coordinates of these points easily, but avoid using it as a reference when doing problems.

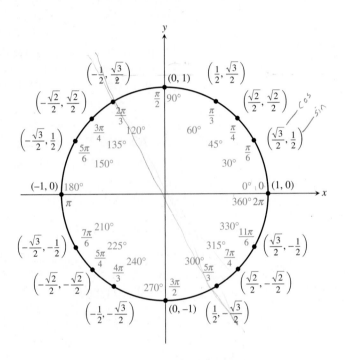

QUICK REVIEW 4.3   *(For help, go to Section 4.1.)*

In Exercises 1–4, give the value of the angle θ in degrees.

1. $\theta = -\dfrac{\pi}{6}$

2. $\theta = -\dfrac{5\pi}{6}$

3. $\theta = \dfrac{25\pi}{4}$

4. $\theta = \dfrac{16\pi}{3}$

In Exercises 5–8, use special triangles to evaluate:

5. $\tan \dfrac{\pi}{6}$

6. $\cot \dfrac{\pi}{4}$

7. $\csc \dfrac{\pi}{4}$

8. $\sec \dfrac{\pi}{3}$

In Exercises 9 and 10, use a right triangle to find the other five trigonometric functions of the *acute* angle θ.

9. $\sin \theta = \dfrac{5}{13}$

10. $\cos \theta = \dfrac{15}{17}$

SECTION 4.3 EXERCISES

In Exercises 1 and 2, identify the one angle that is not coterminal with all the others.

1. 150°, 510°, −210°, 450°, 870°

2. $\dfrac{5\pi}{3}, -\dfrac{5\pi}{3}, \dfrac{11\pi}{3}, -\dfrac{7\pi}{3}, \dfrac{365\pi}{3}$

In Exercises 3–6, evaluate the six trigonometric functions of the angle θ.

3.
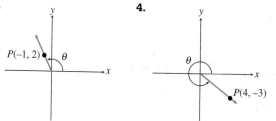

4.

5.
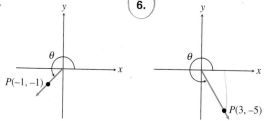

6.

In Exercises 7–12, point *P* is on the terminal side of angle θ. Evaluate the six trigonometric functions for θ. If the function is undefined, write "undefined."

7. $P(3, 4)$

8. $P(-4, -6)$

9. $P(0, 5)$

10. $P(-3, 0)$

11. $P(5, -2)$

12. $P(22, -22)$

In Exercises 13–16, state the sign (+ or −) of **(a)** sin *t*, **(b)** cos *t*, and **(c)** tan *t* for values of *t* in the interval given.

13. $\left(0, \dfrac{\pi}{2}\right)$

14. $\left(\dfrac{\pi}{2}, \pi\right)$

15. $\left(\pi, \dfrac{3\pi}{2}\right)$

16. $\left(\dfrac{3\pi}{2}, 2\pi\right)$

In Exercises 17–20, determine the sign (+ or −) of the given value without the use of a calculator.

17. cos 143°

18. tan 192°

19. $\cos \dfrac{7\pi}{8}$

20. $\tan \dfrac{4\pi}{5}$

In Exercises 21–24, choose the point on the terminal side of θ.

21. θ = 45°

   (a) (2, 2)   **(b)** $(1, \sqrt{3})$ **(c)** $(\sqrt{3}, 1)$

22. $\theta = \dfrac{2\pi}{3}$

   (a) $(-1, 1)$ **(b)** $(-1, \sqrt{3})$ **(c)** $(-\sqrt{3}, 1)$

23. $\theta = \dfrac{7\pi}{6}$

   (a) $(-\sqrt{3}, -1)$ **(b)** $(-1, \sqrt{3})$ **(c)** $(-\sqrt{3}, 1)$

24. θ = −60°

   (a) $(-1, -1)$ **(b)** $(1, -\sqrt{3})$ **(c)** $(-\sqrt{3}, 1)$

In Exercises 25–36, evaluate without using a calculator by using ratios in a reference triangle.

25. cos 120°

26. tan 300°

27. $\sec \dfrac{\pi}{3}$

28. $\csc \dfrac{3\pi}{4}$

29. $\sin \dfrac{13\pi}{6}$

30. $\cos \dfrac{7\pi}{3}$

31. $\tan -\dfrac{15\pi}{4}$

32. $\cot \dfrac{13\pi}{4}$

33. $\cos \dfrac{23\pi}{6}$

34. $\cos \dfrac{17\pi}{4}$

35. $\sin \dfrac{11\pi}{3}$

36. $\cot \dfrac{19\pi}{6}$

In Exercises 37–42, find **(a)** sin θ, **(b)** cos θ, and **(c)** tan θ for the given quadrantal angle. If the value is undefined, write "undefined."

37. −450°

38. −270°

39. 7π

40. $\dfrac{11\pi}{2}$

41. $-\dfrac{7\pi}{2}$

42. −4π

In Exercises 43–48, evaluate without using a calculator.

43. Find sin θ and tan θ if $\cos \theta = \dfrac{2}{3}$ and cot θ > 0.

44. Find cos θ and cot θ if $\sin \theta = \dfrac{1}{4}$ and tan θ < 0.

45. Find tan θ and sec θ if $\sin \theta = -\dfrac{2}{5}$ and cos θ > 0.

46. Find sin θ and cos θ if $\cot \theta = \dfrac{3}{7}$ and sec θ ≷ 0.

47. Find sec θ and csc θ if $\cot \theta = -\dfrac{4}{3}$ and cos θ < 0.

48. Find csc θ and cot θ if $\tan \theta = -\dfrac{4}{3}$ and sin θ > 0.

In Exercises 49–52, evaluate by using the period of the function.

49. $\sin\left(\dfrac{\pi}{6} + 49{,}000\pi\right)$

50. $\tan\left(1{,}234{,}567\pi\right) - \tan\left(7{,}654{,}321\pi\right)$

51. $\cos\left(\dfrac{5{,}555{,}555\pi}{2}\right)$

52. $\tan\left(\dfrac{3\pi - 70{,}000\pi}{2}\right)$

53. Group Activity Use a calculator to evaluate the expressions in Exercises 49–52. Does your calculator give the correct answers? Many calculators miss all four. Give a brief explanation of what probably goes wrong.

54. Writing to Learn Give a convincing argument that the period of $\sin t$ is 2π. That is, show that there is no smaller positive real number p such that $\sin(t + p) = \sin t$ for all real numbers t.

55. Refracted Light Light is *refracted* (bent) as it passes through glass. In the figure below θ_1 is the angle of incidence and θ_2 is the *angle of refraction*. The *index of refraction* is a constant μ that satisfies the equation

$$\sin\theta_1 = \mu \sin\theta_2.$$

If $\theta_1 = 83°$ and $\theta_2 = 36°$ for a certain piece of flint glass, find the index of refraction.

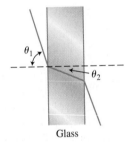

Glass

56. Refracted Light A certain piece of crown glass has an index of refraction of 1.52. If a light ray enters the glass at an angle $\theta_1 = 42°$, what is $\sin\theta_2$?

57. Damped Harmonic Motion A weight suspended from a spring is set into motion. Its dispacement d from equilibrium is modeled by the equation

$$d = 0.4e^{-0.2t}\cos 4t.$$

where d is the displacement in inches and t is the time in seconds. Find the displacement at the given time. Use radian mode.

(a) $t = 0$

(b) $t = 3$

58. Swinging Pendulum The Columbus Museum of Science and Industry exhibits a Foucault pendulum 32 ft long that swings back and forth on a cable once in approximately 6 sec. The angle θ (in radians) between the cable and an imaginary vertical line is modeled by the equation

$$\theta = 0.25 \cos t.$$

Find the measure of angle θ when $t = 0$ and $t = 2.5$.

59. Too Close for Comfort An F-15 aircraft flying at an altitude of 8000 ft passes directly over a group of vacationers hiking at 7400 ft. If θ is the angle of elevation from the hikers to the F-15, find the distance d from the group to the jet for the given angle.

(a) $\theta = 45°$ **(b)** $\theta = 90°$ **(c)** $\theta = 140°$

60. Manufacturing Swimwear Get Wet, Inc. manufactures swimwear, a seasonal product. The monthly sales x (in thousands) for Get Wet swimsuits are modeled by the equation

$$x = 72.4 + 61.7 \sin\frac{\pi t}{6},$$

where $t = 1$ represents January, $t = 2$ February, and so on. Estimate the number of Get Wet swimsuits sold in January, April, June, October, and December. For which two of these months are sales the same? Explain why this might be so.

Standardized Test Questions

61. True or False If θ is an angle of a triangle such that $\cos\theta < 0$, then θ is obtuse. Justify your answer.

62. True or False If θ is an angle in standard position determined by the point $(8, -6)$, then $\sin\theta = -0.6$. Justify your answer.

You should answer these questions without using a calculator.

63. Multiple Choice If $\sin\theta = 0.4$, then $\sin(-\theta) + \csc\theta =$

(A) -0.15 **(B)** 0 **(C)** 0.15 **(D)** 0.65 **(E)** 2.1

64. Multiple Choice If $\cos\theta = 0.4$, then $\cos(\theta + \pi) =$

(A) -0.6 **(B)** -0.4 **(C)** 0.4 **(D)** 0.6 **(E)** 3.54

65. Multiple Choice The range of the function $f(t) = (\sin t)^2 + (\cos t)^2$ is

(A) $\{1\}$ **(B)** $[-1, 1]$ **(C)** $[0, 1]$

(D) $[0, 2]$ **(E)** $[0, \infty)$

66. Multiple Choice If $\cos\theta = -\dfrac{5}{13}$ and $\tan\theta > 0$, then $\sin\theta =$

(A) $-\dfrac{12}{13}$ **(B)** $-\dfrac{5}{12}$ **(C)** $\dfrac{5}{13}$ **(D)** $\dfrac{5}{12}$ **(E)** $\dfrac{12}{13}$

Explorations

In Exercises 67–70, find the value of the unique real number θ between 0 and 2π that satisfies the two given conditions.

67. $\sin\theta = \dfrac{1}{2}$ and $\tan\theta < 0$.

68. $\cos\theta = \dfrac{\sqrt{3}}{2}$ and $\sin\theta < 0$.

69. $\tan\theta = -1$ and $\sin\theta < 0$.

70. $\sin\theta = -\dfrac{\sqrt{2}}{2}$ and $\tan\theta > 0$.

Exercises 71–74 refer to the unit circle in this figure. Point P is on the terminal side of an angle t and point Q is on the terminal side of an angle $t + \pi/2$.

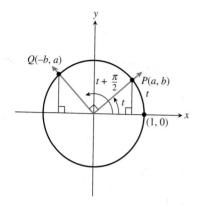

71. Using Geometry in Trigonometry Drop perpendiculars from points P and Q to the x-axis to form two right triangles. Explain how the right triangles are related.

72. Using Geometry in Trigonometry If the coordinates of point P are (a, b), explain why the coordinates of point Q are $(-b, a)$.

73. Explain why $\sin\left(t + \dfrac{\pi}{2}\right) = \cos t$.

74. Explain why $\cos\left(t + \dfrac{\pi}{2}\right) = -\sin t$.

75. Writing to Learn In the figure for Exercises 71–74, t is an angle with radian measure $0 < t < \pi/2$. Draw a similar figure for an angle with radian measure $\pi/2 < t < \pi$ and use it to explain why $\sin(t + \pi/2) = \cos t$.

76. Writing to Learn Use the accompanying figure to explain each of the following.

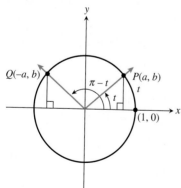

(a) $\sin(\pi - t) = \sin t$ **(b)** $\cos(\pi - t) = -\cos t$

Extending the Ideas

77. Approximation and Error Analysis Use your grapher to complete the table to show that $\sin\theta \approx \theta$ (in radians) when $|\theta|$ is small. Physicists often use the approximation $\sin\theta \approx \theta$ for small values of θ. For what values of θ is the *magnitude of the error* in approximating $\sin\theta$ by θ less than 1% of $\sin\theta$? That is, solve the relation

$$|\sin\theta - \theta| < 0.01 \, |\sin\theta|.$$

(*Hint:* Extend the table to include a column for values of
$$\dfrac{|\sin\theta - \theta|}{|\sin\theta|}.)$$

θ	$\sin\theta$	$\sin\theta - \theta$
-0.03		
-0.02		
-0.01		
0		
0.01		
0.02		
0.03		

78. Proving a Theorem If t is any real number, prove that $1 + (\tan t)^2 = (\sec t)^2$.

Taylor Polynomials Radian measure allows the trigonometric functions to be approximated by simple polynomial functions. For example, in Exercises 79 and 80, sine and cosine are approximated by Taylor polynomials, named after the English mathematician Brook Taylor (1685–1731). Complete each table showing a Taylor polynomial in the third column. Describe the patterns in the table.

79.

θ	$\sin\theta$	$\theta - \dfrac{\theta^3}{6}$	$\sin\theta - \left(\theta - \dfrac{\theta^3}{6}\right)$
-0.3	$-0.295\ldots$		
-0.2	$-0.198\ldots$		
-0.1	$-0.099\ldots$		
0	0		
0.1	$0.099\ldots$		
0.2	$0.198\ldots$		
0.3	$0.295\ldots$		

80.

θ	$\cos\theta$	$1 - \dfrac{\theta^2}{2} + \dfrac{\theta^4}{24}$	$\cos\theta - \left(1 - \dfrac{\theta^2}{2} + \dfrac{\theta^4}{24}\right)$
-0.3	$0.955\ldots$		
-0.2	$0.980\ldots$		
-0.1	$0.995\ldots$		
0	1		
0.1	$0.995\ldots$		
0.2	$0.980\ldots$		
0.3	$0.955\ldots$		

4.4
Graphs of Sine and Cosine: Sinusoids

What you'll learn about

- The Basic Waves Revisited

- Sinusoids and Transformations

- Modeling Periodic Behavior with Sinusoids

... and why

Sine and cosine gain added significance when used to model waves and periodic behavior

The Basic Waves Revisited

In the first three sections of this chapter you saw how the trigonometric functions are rooted in the geometry of triangles and circles. It is these connections with geometry that give trigonometric functions their mathematical power and make them widely applicable in many fields.

The unit circle in Section 4.3 was the key to defining the trigonometric functions as functions of real numbers. This makes them available for the same kind of analysis as the other functions introduced in Chapter 1. (Indeed, two of our "Twelve Basic Functions" are trigonometric.) We now take a closer look at the algebraic, graphical, and numerical properties of the trigonometric functions, beginning with sine and cosine.

Recall that we can learn quite a bit about the sine function by looking at its graph. Here is a summary of sine facts:

BASIC FUNCTION The Sine Function

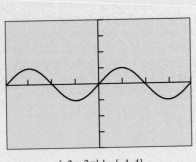

$[-2\pi, 2\pi]$ by $[-4, 4]$

FIGURE 4.37A

$f(x) = \sin x$
Domain: All reals
Range: $[-1, 1]$
Continuous
Alternately increasing and decreasing in periodic waves
Symmetric with respect to the origin (odd)
Bounded
Absolute maximum of 1
Absolute minimum of -1
No horizontal asymptotes
No vertical asymptotes
End behavior: $\lim\limits_{x\to-\infty} \sin x$ and $\lim\limits_{x\to\infty} \sin x$ do not exist. (The function values continually oscillate between -1 and 1 and approach no limit.)

We can add to this list that $y = \sin x$ is *periodic*, with period 2π. We can also add understanding of where the sine function comes from: by definition, $\sin t$ is the y-coordinate of the point P on the unit circle to which the real number t gets wrapped (or, equivalently, the point P on the unit circle determined by an angle of t radians in standard position). In fact, now we can see where the wavy graph comes from. Try Exploration 1.

EXPLORATION 1 Graphing sin *t* as a Function of *t*

Set your grapher to radian mode, parametric, and "simultaneous" graphing modes.

Set Tmin = 0, Tmax = 6.3, Tstep = $\pi/24$.

Set the (x, y) window to $[-1.2, 6.3]$ by $[-2.5, 2.5]$.

Set $X_{1T} = \cos(T)$ and $Y_{1T} = \sin(T)$. This will graph the unit circle. Set $X_{2T} = T$ and $Y_{2T} = \sin(T)$. This will graph sin (T) as a function of T.

Now start the graph and watch the point go counterclockwise around the unit circle as *t* goes from 0 to 2π in the positive direction. You will simultaneously see the *y*-coordinate of the point being graphed as a function of *t* along the horizontal *t*-axis. You can clear the drawing and watch the graph as many times as you need to in order to answer the following questions.

1. Where is the point on the unit circle when the wave is at its highest?

2. Where is the point on the unit circle when the wave is at its lowest?

3. Why do both graphs cross the *x*-axis at the same time?

4. Double the value of Tmax and change the window to $[-2.4, 12.6]$ by $[-5, 5]$. If your grapher can change "style" to show a moving point, choose that style for the unit circle graph. Run the graph and watch how the sine curve tracks the *y*-coordinate of the point as it moves around the unit circle.

5. Explain from what you have seen why the period of the sine function is 2π.

6. Challenge: Can you modify the grapher settings to show dynamically how the cosine function tracks the *x*-coordinate as the point moves around the unit circle?

Although a static picture does not do the dynamic simulation justice, Figure 4.38 shows the final screens for the two graphs in Exploration 1.

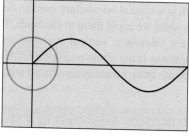 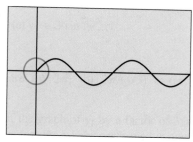

$[-1.2, 6.3]$ by $[-2.5, 2.5]$
(a)

$[-2.4, 12.6]$ by $[-5, 5]$
(b)

FIGURE 4.38 The graph of $y = \sin t$ tracks the *y*-coordinate of the point determined by *t* as it moves around the unit circle.

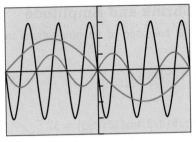

$[-3\pi, 3\pi]$ by $[-4, 4]$

FIGURE 4.40 Sinusoids (in this case, sine curves) of different amplitudes and periods. (Example 2)

Support Graphically The graphs of the three functions are shown in Figure 4.40. You should be able to tell which is which quite easily by checking the periods or the amplitudes.   *Now try Exercise 9.*

In some applications, the *frequency* of a sinusoid is an important consideration. The frequency is simply the reciprocal of the period.

Frequency of a Sinusoid

The **frequency** of the sinusoid $f(x) = a \sin(bx + c) + d$ is $|b|/2\pi$. Similarly, the frequency of $f(x) = a \cos(bx + c) + d$ is $|b|/2\pi$.

Graphically, the frequency is the number of complete cycles the wave completes in a unit interval.

EXAMPLE 3   Finding the Frequency of a Sinusoid

Find the frequency of the function $f(x) = 4 \sin(2x/3)$ and interpret its meaning graphically.

Sketch the graph in the window $[-3\pi, 3\pi]$ by $[-4, 4]$.

SOLUTION The frequency is $(2/3) \div 2\pi = 1/(3\pi)$. This is the reciprocal of the period, which is 3π. The graphical interpretation is that the graph completes 1 full cycle per interval of length 3π. (That, of course, is what having a period of 3π is all about.) The graph is shown in Figure 4.41.   *Now try Exercise 17.*

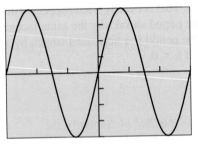

$[-3\pi, 3\pi]$ by $[-4, 4]$

FIGURE 4.41 The graph of the function $f(x) = 4 \sin(2x/3)$. It has frequency $1/(3\pi)$, so it completes 1 full cycle per interval of length 3π. (Example 3)

Recall from Section 1.5 that the graph of $y = f(x + c)$ is a translation of the graph of $y = f(x)$ by c units to the left when $c > 0$. That is exactly what happens with sinusoids, but using terminology with its roots in electrical engineering, we say that the wave undergoes a **phase shift** of $-c$.

EXAMPLE 4   Getting one Sinusoid from Another by a Phase Shift

(a) Write the cosine function as a phase shift of the sine function.

(b) Write the sine function as a phase shift of the cosine function.

SOLUTION

(a) The function $y = \sin x$ has a maximum at $x = \pi/2$, while the function $y = \cos x$ has a maximum at $x = 0$. Therefore, we need to shift the sine curve $\pi/2$ units to the *left* to get the cosine curve:

$$\cos x = \sin(x + \pi/2).$$

(b) It follows from the work in (a) that we need to shift the cosine curve $\pi/2$ units to the right to get the sine curve:

$$\sin x = \cos(x - \pi/2).$$

You can support with your grapher that these statements are true. Incidentally, there are many other translations that would have worked just as well. Adding any integral multiple of 2π to the phase shift would result in the same graph.

Now try Exercise 41.

One note of caution applies when combining these transformations. A horizontal stretch or shrink affects the variable along the horizontal axis, so it *also affects the phase shift.* Consider the transformation in Example 5.

EXAMPLE 5 Combining a Phase Shift with a Period Change

Construct a sinusoid with period $\pi/5$ and amplitude 6 that goes through $(2, 0)$.

SOLUTION

To find the coefficient of x, we set $2\pi/|b| = \pi/5$ and solve to find that $b = \pm 10$. We arbitrarily choose $b = 10$. (Either will satisfy the specified conditions.)

For amplitude 6, we have $|a| = 6$. Again, we arbitrarily choose the positive value. The graph of $y = 6 \sin(10x)$ has the required amplitude and period, but it does not go through the point $(2, 0)$. It does, however, go through the point $(0, 0)$, so all that is needed is a phase shift of $+2$ to finish our function. Replacing x by $x - 2$, we get

$$y = 6 \sin(10(x - 2)) = 6 \sin(10x - 20).$$

Notice that we did *not* get the function $y = 6 \sin(10x - 2)$. That function would represent a phase shift of $y = \sin(10x)$, but only by $2/10$, not 2. Parentheses are important when combining phase shifts with horizontal stretches and shrinks.

Now try Exercise 59.

Graphs of Sinusoids

The graphs of $y = a \sin(b(x - h)) + k$ and $y = a \cos(b(x - h)) + k$ (where $a \neq 0$ and $b \neq 0$) have the following characteristics:

$$\text{amplitude} = |a|;$$

$$\text{period} = \frac{2\pi}{|b|};$$

$$\text{frequency} = \frac{|b|}{2\pi}.$$

When compared to the graphs of $y = a \sin bx$ and $y = a \cos bx$, respectively, they also have the following characteristics:

a phase shift of h;

a vertical translation of k.

EXAMPLE 6 Constructing a Sinusoid by Transformations

Construct a sinusoid $y = f(x)$ that rises from a minimum value of $y = 5$ at $x = 0$ to a maximum value of $y = 25$ at $x = 32$. (See Figure 4.42.)

SOLUTION

Solve Algebraically The amplitude of this sinusoid is half the height of the graph: $(25 - 5)/2 = 10$. So $|a| = 10$. The period is 64 (since a full period goes from minimum to maximum and back down to the minimum). So set $2\pi/|b| = 64$. Solving, we get $|b| = \pi/32$.

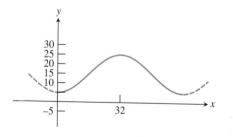

FIGURE 4.42 A sinusoid with specifications. (Example 6)

continued

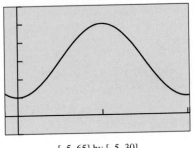

[-5, 65] by [-5, 30]

FIGURE 4.43 The graph of the function $y = -10 \cos((\pi/32)x) + 15$. (Example 8)

We need a sinusoid that takes on its minimum value at $x = 0$. We could shift the graph of sine or cosine horizontally, but it is easier to take the cosine curve (which assumes its *maximum* value at $x = 0$) and turn it upside down. This reflection can be obtained by letting $a = -10$ rather than 10.

So far we have:

$$y = -10 \cos\left(\pm\frac{\pi}{32}x\right)$$

$$= -10 \cos\left(\frac{\pi}{32}x\right) \qquad \text{(Since cos is an even function)}$$

Finally, we note that this function ranges from a minimum of -10 to a maximum of 10. We shift the graph vertically by 15 to obtain a function that ranges from a minimum of 5 to a maximum of 25, as required. Thus

$$y = -10 \cos\left(\frac{\pi}{32}x\right) + 15.$$

Support Graphically We support our answer graphically by graphing the function (Figure 4.43).   ***Now try Exercise 69.***

Modeling Periodic Behavior with Sinusoids

Example 6 was intended as more than just a review of the graphical transformations. Constructing a sinusoid with specific properties is often the key step in modeling physical situations that exhibit periodic behavior over time. The procedure we followed in Example 6 can be summarized as follows:

Constructing a Sinusoidal Model Using Time

1. Determine the maximum value M and minimum value m. The amplitude A of the sinusoid will be $A = \dfrac{M - m}{2}$, and the vertical shift will be $C = \dfrac{M + m}{2}$.

2. Determine the period p, the time interval of a single cycle of the periodic function. The horizontal shrink (or stretch) will be $B = \dfrac{2\pi}{p}$.

3. Choose an appropriate sinusoid based on behavior at some given time T. For example, at time T:

    $f(t) = A \cos(B(t - T)) + C$ attains a maximum value;

    $f(t) = -A \cos(B(t - T)) + C$ attains a minimum value;

    $f(t) = A \sin(B(t - T)) + C$ is halfway between a minimum and a maximum value;

    $f(t) = -A \sin(B(t - T)) + C$ is halfway between a maximum and a minimum value.

We apply the procedure in Example 7 to model the ebb and flow of a tid...

EXAMPLE 7 Calculating the Ebb and Flow of Tides

One particular July 4th in Galveston, TX, high tide occurred at 9:36 A.M. At that time the water at the end of the 61st Street Pier was 2.7 meters deep. Low tide occurred at 3:48 P.M., at which time the water was only 2.1 meters deep. Assume that the depth of the water is a sinusoidal function of time with a period of half a lunar day (about 12 hours 24 minutes).

(a) At what time on the 4th of July did the first low tide occur?

(b) What was the approximate depth of the water at 6:00 A.M. and at 3:00 P.M. that day?

(c) What was the first time on July 4th when the water was 2.4 meters deep?

SOLUTION

Model

We want to model the depth D as a sinusoidal function of time t. The depth varies from a maximum of 2.7 meters to a minimum of 2.1 meters, so the amplitude $A = \dfrac{2.7 - 2.1}{2} = 0.3$, and the vertical shift will be $C = \dfrac{2.7 + 2.1}{2} = 2.4$. The period is 12 hours 24 minutes, which converts to 12.4 hours, so $B = \dfrac{2\pi}{12.4} = \dfrac{\pi}{6.2}$.

We need a sinusoid that assumes its maximum value at 9:36 A.M. (which converts to 9.6 hours after midnight, a convenient time 0). We choose the cosine model. Thus,

$$D(t) = 0.3 \cos\left(\frac{\pi}{6.2}(t - 9.6)\right) + 2.4.$$

Solve Graphically The graph over the 24-hour period of July 4th is shown in Figure 4.44.

We now use the graph to answer the questions posed.

(a) The first low tide corresponds to the first local minimum on the graph. We find graphically that this occurs at $t = 3.4$. This translates to $3 + (0.4)(60) = 3{:}24$ A.M.

(b) The depth at 6:00 A.M. is $D(6) \approx 2.32$ meters. The depth at 3:00 P.M. is $D(12 + 3) = D(15) \approx 2.12$ meters.

(c) The first time the water is 2.4 meters deep corresponds to the leftmost intersection of the sinusoid with the line $y = 2.4$. We use the grapher to find that $t = 0.3$. This translates to $0 + (0.3)(60) = 00{:}18$ A.M., which we write as 12:18 A.M.

Now try Exercise 75.

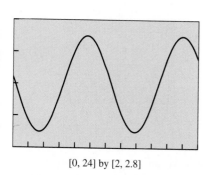

[0, 24] by [2, 2.8]

FIGURE 4.44 The Galveston tide graph. (Example 7)

We will see more applications of this kind when we look at *simple harmonic motion* in Section 4.8.

QUICK REVIEW 4.4 *(For help, go to Sections 1.6, 4.1, and 4.2.)*

In Exercises 1–3, state the sign (positive or negative) of the function in each quadrant.

1. $\sin x$ **2.** $\cos x$

3. $\tan x$

In Exercises 4–6, give the radian measure of the angle.

4. $135°$ **5.** $-150°$ **6.** $450°$

In Exercises 7–10, find a transformation that will transform the graph of y_1 to the graph of y_2.

7. $y_1 = \sqrt{x}$ and $y_2 = 3\sqrt{x}$

8. $y_1 = e^x$ and $y_2 = e^{-x}$

9. $y_1 = \ln x$ and $y_2 = 0.5 \ln x$

10. $y_1 = x^3$ and $y_2 = x^3 - 2$

SECTION 4.4 EXERCISES

In Exercises 1–6, find the amplitude of the function and use the language of transformations to describe how the graph of the function is related to the graph of $y = \sin x$.

1. $y = 2 \sin x$ **2.** $y = \dfrac{2}{3} \sin x$

3. $y = -4 \sin x$ **4.** $y = -\dfrac{7}{4} \sin x$

5. $y = 0.73 \sin x$ **6.** $y = -2.34 \sin x$

In Exercises 7–12, find the period of the function and use the language of transformations to describe how the graph of the function is related to the graph of $y = \cos x$.

7. $y = \cos 3x$ **8.** $y = \cos x/5$

9. $y = \cos (-7x)$ **10.** $y = \cos (-0.4x)$

11. $y = 3 \cos 2x$ **12.** $y = \dfrac{1}{4} \cos \dfrac{2x}{3}$

In Exercises 13–16, find the amplitude, period, and frequency of the function and use this information (not your calculator) to sketch a graph of the function in the window $[-3\pi, 3\pi]$ by $[-4, 4]$.

13. $y = 3 \sin \dfrac{x}{2}$ **14.** $y = 2 \cos \dfrac{x}{3}$

15. $y = -\dfrac{3}{2} \sin 2x$ **16.** $y = -4 \sin \dfrac{2x}{3}$

In Exercises 17–22, graph one period of the function. Use your understanding of transformations, not your graphing calculators. Be sure to show the scale on both axes.

17. $y = 2 \sin x$ **18.** $y = 2.5 \sin x$

19. $y = 3 \cos x$ **20.** $y = -2 \cos x$

21. $y = -0.5 \sin x$ **22.** $y = 4 \cos x$

In Exercises 23–28, graph three periods of the function. Use your understanding of transformations, not your graphing calculators. Be sure to show the scale on both axes.

23. $y = 5 \sin 2x$ **24.** $y = 3 \cos \dfrac{x}{2}$

25. $y = 0.5 \cos 3x$ **26.** $y = 20 \sin 4x$

27. $y = 4 \sin \dfrac{x}{4}$ **28.** $y = 8 \cos 5x$

In Exercises 29–34, specify the period and amplitude of each function. Then give the viewing window in which the graph is shown. Use your understanding of transformations, not your graphing calculators.

29. $y = 1.5 \sin 2x$ **30.** $y = 2 \cos 3x$

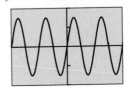 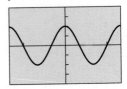

31. $y = -3 \cos 2x$ **32.** $y = 5 \sin \dfrac{x}{2}$

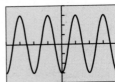 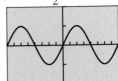

33. $y = -4 \sin \dfrac{\pi}{3}x$ **34.** $y = 3 \cos \pi x$

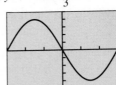 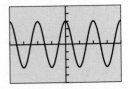

In Exercises 35–40, identify the maximum and minimum values and the zeros of the function in the interval $[-2\pi, 2\pi]$. Use your understanding of transformations, not your graphing calculators.

35. $y = 2 \sin x$ **36.** $y = 3 \cos \dfrac{x}{2}$

37. $y = \cos 2x$ **38.** $y = \dfrac{1}{2} \sin x$

39. $y = -\cos 2x$ **40.** $y = -2 \sin x$

41. Write the function $y = -\sin x$ as a phase shift of $y = \sin x$.

42. Write the function $y = -\cos x$ as a phase shift of $y = \sin x$.

In Exercises 43–48, describe the transformations required to obtain the graph of the given function from a basic trigonometric graph.

43. $y = 0.5 \sin 3x$

44. $y = 1.5 \cos 4x$

45. $y = -\dfrac{2}{3} \cos \dfrac{x}{3}$

46. $y = \dfrac{3}{4} \sin \dfrac{x}{5}$

47. $y = 3 \cos \dfrac{2\pi x}{3}$

48. $y = -2 \sin \dfrac{\pi x}{4}$

In Exercises 49–52, describe the transformations required to obtain the graph of y_2 from the graph of y_1.

49. $y_1 = \cos 2x$ and $y_2 = \dfrac{5}{3} \cos 2x$

50. $y_1 = 2 \cos \left(x + \dfrac{\pi}{3}\right)$ and $y_2 = \cos \left(x + \dfrac{\pi}{4}\right)$

51. $y_1 = 2 \cos \pi x$ and $y_2 = 2 \cos 2\pi x$

52. $y_1 = 3 \sin \dfrac{2\pi x}{3}$ and $y_2 = 2 \sin \dfrac{\pi x}{3}$

In Exercises 53–56, select the pair of functions that have identical graphs.

53. (a) $y = \cos x$

(b) $y = \sin \left(x + \dfrac{\pi}{2}\right)$

(c) $y = \cos \left(x + \dfrac{\pi}{2}\right)$

54. (a) $y = \sin x$

(b) $y = \cos \left(x - \dfrac{\pi}{2}\right)$

(c) $y = \cos x$

55. (a) $y = \sin \left(x + \dfrac{\pi}{2}\right)$

(b) $y = -\cos (x - \pi)$

(c) $y = \cos \left(x - \dfrac{\pi}{2}\right)$

56. (a) $y = \sin \left(2x + \dfrac{\pi}{4}\right)$

(b) $y = \cos \left(2x - \dfrac{\pi}{2}\right)$

(c) $y = \cos \left(2x - \dfrac{\pi}{4}\right)$

In Exercises 57–60, construct a sinusoid with the given amplitude and period that goes through the given point.

57. Amplitude 3, period π, point $(0, 0)$

58. Amplitude 2, period 3π, point $(0, 0)$

59. Amplitude 1.5, period $\pi/6$, point $(1, 0)$

60. Amplitude 3.2, period $\pi/7$, point $(5, 0)$

In Exercises 61–68, state the amplitude and period of the sinusoid, and (relative to the basic function) the phase shift and vertical translation.

61. $y = -2 \sin \left(x - \dfrac{\pi}{4}\right) + 1$

62. $y = -3.5 \sin \left(2x - \dfrac{\pi}{2}\right) - 1$

63. $y = 5 \cos \left(3x - \dfrac{\pi}{6}\right) + 0.5$

64. $y = 3 \cos (x + 3) - 2$

65. $y = 2 \cos 2\pi x + 1$

66. $y = 4 \cos 3\pi x - 2$

67. $y = \dfrac{7}{3} \sin \left(x + \dfrac{5}{2}\right) - 1$

68. $y = \dfrac{2}{3} \cos \left(\dfrac{x - 3}{4}\right) + 1$

In Exercises 69 and 70, find values a, b, h, and k so that the graph of the function $y = a \sin (b(x + h)) + k$ is the curve shown.

69.

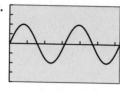

[0, 6.28] by [–4, 4]

70.

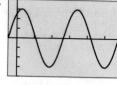

[–0.5, 5.78] by [–4, 4]

71. Points of Intersection Graph $y = 1.3^{-x}$ and $y = 1.3^{-x} \cos x$ for x in the interval $[-1, 8]$.

(a) How many points of intersection do there appear to be?

(b) Find the coordinates of each point of intersection.

72. Motion of a Buoy A signal buoy in the Chesapeake Bay bobs up and down with the height h of its transmitter (in feet) above sea level modeled by $h = a \sin bt + 5$. During a small squall its height varies from 1 ft to 9 ft and there are 3.5 sec from one 9-ft height to the next. What are the values of the constants a and b?

73. Ferris Wheel A Ferris wheel 50 ft in diameter makes one revolution every 40 sec. If the center of the wheel is 30 ft above the ground, how long after reaching the low point is a rider 50 ft above the ground?

74. Tsunami Wave An earthquake occurred at 9:40 A.M. on Nov. 1, 1755, at Lisbon, Portugal, and started a *tsunami* (often called a tidal wave) in the ocean. It produced waves that traveled more than 540 ft/sec (370 mph) and reached a height of 60 ft. If the period of the waves was 30 min or 1800 sec, estimate the length L between the crests.

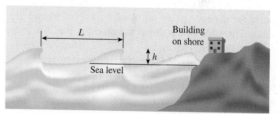

75. Ebb and Flow On a particular Labor Day, the high tide in southern California occurs at 7:12 A.M. At that time you measure the water at the end of the Santa Monica Pier to be 11 ft deep. At 1:24 P.M. it is low tide, and you measure the water to be only 7 ft deep. Assume the depth of the water is a sinusoidal function of time with a period of 1/2 a lunar day, which is about 12 hr 24 min.

(a) At what time on that Labor Day does the first low tide occur?

(b) What was the approximate depth of the water at 4:00 A.M. and at 9:00 P.M.?

(c) What is the first time on that Labor Day that the water is 9 ft deep?

76. Blood Pressure The function
$$P = 120 + 30 \sin 2\pi t$$
models the blood pressure (in millimeters of mercury) for a person who has a (high) blood pressure of 150/90; t represents seconds.

(a) What is the period of this function?

(b) How many heartbeats are there each minute?

(c) Graph this function to model a 10-sec time interval.

77. Bouncing Block A block mounted on a spring is set into motion directly above a motion detector, which registers the distance to the block at intervals of 0.1 seconds. When the block is released, it is 7.2 cm above the motion detector. The table below shows the data collected by the motion detector during the first two seconds, with distance d measured in centimeters:

(a) Make a scatter plot of d as a function of t and estimate the maximum d visually. Use this number and the given minimum (7.2) to compute the amplitude of the block's motion.

(b) Estimate the period of the block's motion visually from the scatter plot.

(c) Model the motion of the block as a sinusoidal function $d(t)$.

(d) Graph your function with the scatter plot to support your model graphically.

t	0.1	0.2	0.3	0.4	0.5	0.6	0.7	0.8	0.9	1.0
d	9.2	13.9	18.8	21.4	20.0	15.6	10.5	7.4	8.1	12.1

t	1.1	1.2	1.3	1.4	1.5	1.6	1.7	1.8	1.9	2.0
d	17.3	20.8	20.8	17.2	12.0	8.1	7.5	10.5	15.6	19.9

78. LP Turntable A suction-cup-tipped arrow is secured vertically to the outer edge of a turntable designed for playing LP phonograph records (ask your parents). A motion detector is situated 60 cm away. The turntable is switched on and a motion detector measures the distance to the arrow as it revolves around the turntable. The table below shows the distance d as a function of time during the first 4 seconds.

t	0.2	0.4	0.6	0.8	1.0	1.2	1.4	1.6	1.8	2.0
d	63.5	71.6	79.8	84.7	84.7	79.8	71.6	63.5	60.0	63.5

t	2.2	2.4	2.6	2.8	3.0	3.2	3.4	3.6	3.8	4.0
d	71.6	79.8	84.7	84.7	79.8	71.6	63.5	60.0	63.5	71.6

(a) If the turntable is 25.4 cm in diameter, find the amplitude of the arrow's motion.

(b) Find the period of the arrow's motion by analyzing the data.

(c) Model the motion of the arrow as a sinusoidal function $d(t)$.

(d) Graph your function with a scatter plot to support your model graphically.

79. Temperature Data The normal monthly Fahrenheit temperatures in Albuquerque, NM are shown in the table below (month 1 = Jan, month 2 = Feb, etc.):

Month	1	2	3	4	5	6	7	8	9	10	11	12
Temp	36	41	48	56	65	75	79	76	69	57	44	36

Source: National Climatic Data Center, as reported in The World Almanac and Book of Facts, *2005.*

Model the temperature T as a sinusoidal function of time, using 36 as the minimum value and 79 as the maximum value. Support your answer graphically by graphing your function with a scatter plot.

80. Temperature Data The normal monthly Fahrenheit temperatures in Helena, MT are shown in the table below (month 1 = Jan, month 2 = Feb, etc.):

Month	1	2	3	4	5	6	7	8	9	10	11	12
Temp	20	26	35	44	53	61	68	67	56	45	31	21

Source: National Climatic Data Center, as reported in The World Almanac and Book of Facts, *2005.*

Model the temperature T as a sinusoidal function of time, using 20 as the minimum value and 68 as the maximum value. Support your answer graphically by graphing your function with a scatter plot.

Standardized Test Questions

81. True or False The graph of $y = \sin 2x$ has half the period of the graph of $y = \sin 4x$. Justify your answer.

82. True or False Every sinusoid can be written as $y = A \cos (Bx + C)$ for some real numbers A, B, and C. Justify your answer.

You may use a graphing calculator when answering these questions.

83. Multiple Choice A sinusoid with amplitude 4 has a minimum value of 5. Its maximum value is

(A) 7.   **(B)** 9.   **(C)** 11.

(D) 13.   **(E)** 15.

84. Multiple Choice The graph of $y = f(x)$ is a sinusoid with period 45 passing through the point $(6, 0)$. Which of the following can be determined from the given information?

I. $f(0)$ **II.** $f(6)$ **III.** $f(96)$

(A) I only   **(B)** II only

(C) I and III only   **(D)** II and III only

(E) I, II, and III only

85. Multiple Choice The period of the function $f(x) = 210 \sin (420x + 840)$ is

(A) $\pi/840$.   **(B)** $\pi/420$.   **(C)** $\pi/210$.

(D) $210/\pi$.   **(E)** $420/\pi$.

86. Multiple Choice The number of solutions to the equation $\sin (2000x) = 3/7$ in the interval $[0, 2\pi]$ is

(A) 1000.   **(B)** 2000.   **(C)** 4000.

(D) 6000.   **(E)** 8000.

Explorations

87. Approximating Cosine

(a) Draw a scatter plot $(x, \cos x)$ for the 17 special angles x, where $-\pi \le x \le \pi$.

(b) Find a quartic regression for the data.

(c) Compare the approximation to the cosine function given by the quartic regression with the Taylor polynomial approximations given in Exercise 80 of Section 4.3.

88. Approximating Sine

(a) Draw a scatter plot $(x, \sin x)$ for the 17 special angles x, where $-\pi \le x \le \pi$.

(b) Find a cubic regression for the data.

(c) Compare the approximation to the sine function given by the cubic regression with the Taylor polynomial approximations given in Exercise 79 of Section 4.3.

89. Visualizing a Musical Note A piano tuner strikes a tuning fork for the note middle C and creates a sound wave that can be modeled by

$$y = 1.5 \sin 524\pi t,$$

where t is the time in seconds.

(a) What is the period p of this function?

(b) What is the frequency $f = 1/p$ of this note?

(c) Graph the function.

90. Writing to Learn In a certain video game a cursor bounces back and forth horizontally across the screen at a constant rate. Its distance d from the center of the screen varies with time t and hence can be described as a function of t. Explain why this horizontal distance d from the center of the screen *does not vary* according to an equation $d = a \sin bt$, where t represents seconds. You may find it helpful to include a graph in your explanation.

91. Group Activity Using only integer values of a and b between 1 and 9 inclusive, look at graphs of functions of the form

$$y = \sin (ax) \cos (bx) - \cos (ax) \sin (bx)$$

for various values of a and b. (A group can look at more graphs at a time than one person can.)

(a) Some values of a and b result in the graph of $y = \sin x$. Find a general rule for such values of a and b.

(b) Some values of a and b result in the graph of $y = \sin 2x$. Find a general rule for such values of a and b.

(c) Can you guess which values of a and b will result in the graph of $y = \sin kx$ for an arbitrary integer k?

92. Group Activity Using only integer values of a and b between 1 and 9 inclusive, look at graphs of functions of the form

$$y = \cos (ax) \cos (bx) + \sin (ax) \sin (bx)$$

for various values of a and b. (A group can look at more graphs at a time than one person can.)

(a) Some values of a and b result in the graph of $y = \cos x$. Find a general rule for such values of a and b.

(b) Some values of a and b result in the graph of $y = \cos 2x$. Find a general rule for such values of a and b.

(c) Can you guess which values of a and b will result in the graph of $y = \cos kx$ for an arbitrary integer k?

Extending the Ideas

In Exercises 93–96, the graphs of the sine and cosine functions are waveforms like the figure below. By correctly labeling the coordinates of points A, B, and C, you will get the graph of the function given.

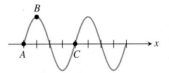

93. $y = 3 \cos 2x$ and $A = \left(-\dfrac{\pi}{4}, 0\right)$. Find B and C.

94. $y = 4.5 \sin \left(x - \dfrac{\pi}{4}\right)$ and $A = \left(\dfrac{\pi}{4}, 0\right)$. Find B and C.

95. $y = 2 \sin \left(3x - \dfrac{\pi}{4}\right)$ and $A = \left(\dfrac{\pi}{12}, 0\right)$. Find B and C.

96. $y = 3 \sin (2x - \pi)$, and A is the first x-intercept on the right of the y-axis. Find A, B, and C.

97. The Ultimate Sinusoidal Equation It is an interesting fact that any sinusoid can be written in the form

$$y = a \sin [b(x - H)] + k,$$

where both a and b are positive numbers.

(a) Explain why you can assume b is positive. (*Hint:* Replace b by $-B$ and simplify.)

(b) Use one of the horizontal translation identities to prove that the equation

$$y = a \cos [b(x - h)] + k$$

has the same graph as

$$y = a \sin [b(x - H)] + k$$

for a correctly chosen value of H. Explain how to choose H.

(c) Give a unit-circle argument for the identity $\sin (\theta + \pi) = -\sin \theta$. Support your unit-circle argument graphically.

(d) Use the identity from (c) to prove that

$$y = -a \sin [b(x - h)] + k,\ a > 0,$$

has the same graph as

$$y = a \sin [b(x - H)] + k,\ a > 0$$

for a correctly chosen value of H. Explain how to choose H.

(e) Combine your results from (a)–(d) to prove that any sinusoid can be represented by the equation

$$y = a \sin [b(x - H)] + k$$

where a and b are both positive.

4.5
Graphs of Tangent, Cotangent, Secant, and Cosecant

What you'll learn about

- The Tangent Function
- The Cotangent Function
- The Secant Function
- The Cosecant Function

. . . and why

This will give us functions for the remaining trigonometric ratios.

The Tangent Function

The graph of the tangent function is shown below. As with the sine and cosine graphs, this graph tells us quite a bit about the function's properties. Here is a summary of tangent facts:

THE TANGENT FUNCTION

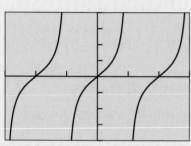

[−3π/2, 3π/2] by [−4, 4]

FIGURE 4.44A

$f(x) = \tan x$
Domain: All reals except odd multiples of $\pi/2$
Range: All reals
Continuous (i.e., continuous on its domain)
Increasing on each interval in its domain
Symmetric with respect to the origin (odd).
Not bounded above or below
No local extrema
No horizontal asymptotes
Vertical asymptotes: $x = k \cdot (\pi/2)$ for all odd integers k
End behavior: $\lim\limits_{x \to -\infty} \tan x$ and $\lim\limits_{x \to \infty} \tan x$ do not exist. (The function values continually oscillate between $-\infty$ and ∞ and approach no limit.)

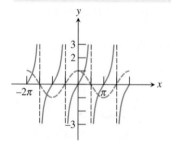

FIGURE 4.45 The tangent function has asymptotes at the zeros of cosine.

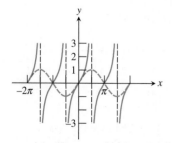

FIGURE 4.46 The tangent function has zeros at the zeros of sine.

We now analyze why the graph of $f(x) = \tan x$ behaves the way it does. It follows from the definitions of the trigonometric functions (Section 4.2) that

$$\tan x = \frac{\sin x}{\cos x}.$$

Unlike the sinusoids, the tangent function has a denominator that might be zero, which makes the function undefined. Not only does this actually happen, but it happens an infinite number of times: at all the values of x for which $\cos x = 0$. That is why the tangent function has vertical asymptotes at those values (Figure 4.45). The zeros of the tangent function are the same as the zeros of the sine function: all the integer multiples of π (Figure 4.46).

Because $\sin x$ and $\cos x$ are both periodic with period 2π, you might expect the period of the tangent function to be 2π also. The graph shows, however, that it is π.

The constants $a, b, h,$ and k influence the behavior of $y = a \tan(b(x - h)) + k$ in much the same way that they do for the graph of $y = a \sin(b(x - h)) + k$. The constant a yields a vertical stretch or shrink, b affects the period, h causes a horizontal translation, and k causes a vertical translation. The terms *amplitude* and *phase shift*, however, are not used, as they apply only to sinusoids.

EXAMPLE 1 Graphing a Tangent Function

Describe the graph of the function $y = -\tan 2x$ in terms of a basic trigonometric function. Locate the vertical asymptotes and graph four periods of the function.

SOLUTION The effect of the 2 is a horizontal shrink of the graph of $y = \tan x$ by a factor of 1/2, while the effect of the -1 is a reflection across the x-axis. Since the vertical asymptotes of $y = \tan x$ are all odd multiples of $\pi/2$, the shrink factor causes the vertical asymptotes of $y = \tan 2x$ to be all odd multiples of $\pi/4$ (Figure 4.47a). The reflection across the x-axis (Figure 4.47b) does not change the asymptotes.

Since the period of the function $y = \tan x$ is π, the period of the function $y = -\tan 2x$ is (thanks again to the shrink factor) $\pi/2$. Thus, any window of horizontal length 2π will show four periods of the graph. Figure 4.47b uses the window $[-\pi, \pi]$ by $[-4, 4]$.

Now try Exercise 5.

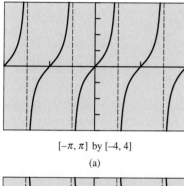

$[-\pi, \pi]$ by $[-4, 4]$

(a)

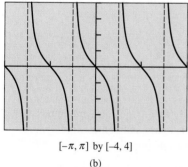

$[-\pi, \pi]$ by $[-4, 4]$

(b)

FIGURE 4.47 The graph of (a) $y = \tan 2x$ is reflected across the x-axis to produce the graph of (b) $y = -\tan 2x$. (Example 1)

The other three trigonometric functions (cotangent, secant, and cosecant) are reciprocals of tangent, cosine, and sine, respectively. (This is the reason that you probably do not have buttons for them on your calculators.) As functions they are certainly interesting, but as basic functions they are unnecessary—we can do our trigonometric modeling and equation-solving with the other three. Nonetheless, we give each of them a brief section of its own in this book.

The Cotangent Function

The cotangent function is the reciprocal of the tangent function. Thus,

$$\cot x = \frac{\cos x}{\sin x}.$$

The graph of $y = \cot x$ will have asymptotes at the zeros of the sine function (Figure 4.48) and zeros at the zeros of the cosine function (Figure 4.49).

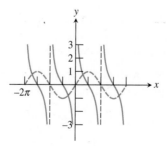

FIGURE 4.48 The cotangent has asymptotes at the zeros of the sine function.

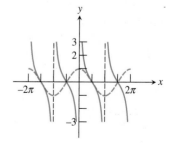

FIGURE 4.49 The cotangent has zeros at the zeros of the cosine function.

If your calculator does not have a "cotan" button, you can use the fact that cotangent and tangent are reciprocals. For example, the function in Example 2 can be entered in a calculator as $y = 3/\tan(x/2) + 1$ or as $y = 3(\tan(x/2))^{-1} + 1$. Remember that it can *not* be entered as $y = 3\tan^{-1}(x/2) + 1$. (The -1 exponent in that position represents a function *inverse*, not a *reciprocal*.)

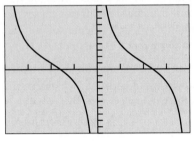

$[-2\pi, 2\pi]$ by $[-10, 10]$

FIGURE 4.50 Two periods of $f(x) = 3 \cot(x/2) + 1$. (Example 2)

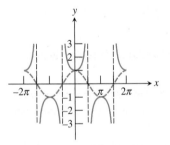

FIGURE 4.51 Characteristics of the secant function are inferred from the fact that it is the reciprocal of the cosine function.

EXAMPLE 2 Graphing a Cotangent Function

Describe the graph of $f(x) = 3 \cot(x/2) + 1$ in terms of a basic trigonometric function. Locate the vertical asymptotes and graph two periods.

SOLUTION The graph is obtained from the graph of $y = \cot x$ by effecting a horizontal stretch by a factor of 2, a vertical stretch by a factor of 3, and a vertical translation up 1 unit. The horizontal stretch makes the period of the function 2π (twice the period of $y = \cot x$), and the asymptotes are at the even multiples of π. Figure 4.50 shows two periods of the graph of f.

Now try Exercise 9.

The Secant Function

Important characteristics of the secant function can be inferred from the fact that it is the reciprocal of the cosine function.

Whenever $\cos x = 1$, its reciprocal $\sec x$ is also 1. The graph of the secant function has asymptotes at the zeros of the cosine function. The period of the secant function is 2π, the same as its reciprocal, the cosine function.

The graph of $y = \sec x$ is shown with the graph of $y = \cos x$ in Figure 4.51. A local maximum of $y = \cos x$ corresponds to a local minimum of $y = \sec x$, while a local minimum of $y = \cos x$ corresponds to a local maximum of $y = \sec x$.

EXPLORATION 1 **Proving a Graphical Hunch**

Figure 4.52 shows that the graphs of $y = \sec x$ and $y = -2 \cos x$ never seem to intersect.

If we stretch the reflected cosine graph vertically by a large enough number, will it continue to miss the secant graph? Or is there a large enough (positive) value of k so that the graph of $y = \sec x$ *does* intersect the graph of $y = -k \cos x$?

1. Try a few other values of k in your calculator. Do the graphs intersect?

2. Your exploration should lead you to conjecture that the graphs of $y = \sec x$ and $y = -k \cos x$ will *never* intersect for any positive value of k. Verify this conjecture by proving **algebraically** that the equation

$$-k \cos x = \sec x$$

has no real solutions when k is a positive number.

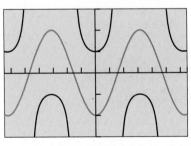

$[-6.5, 6.5]$ by $[-3, 3]$

FIGURE 4.52 The graphs of $y = \sec x$ and $y = -2 \cos x$. (Exploration 1)

**EXAMPLE 3 Solving a Trigonometric Equation
              Algebraically**

Find the value of x between π and $3\pi/2$ that solves $\sec x = -2$.

SOLUTION We construct a reference triangle in the third quadrant that has the appropriate ratio, *hyp/adj*, equal to -2. This is most easily accomplished by choosing an x-coordinate of -1 and a hypotenuse of 2 (Figure 4.53a). We recognize this as a $30°$–$60°$–$90°$ triangle that determines an angle of $240°$, which converts to $4\pi/3$ radians (Figure 4.53b).

Therefore the answer is $4\pi/3$.

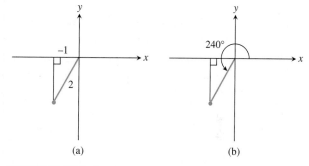

(a)   (b)

FIGURE 4.53 A reference triangle in the third quadrant (a) with *hyp/adj* $= -2$ determines an angle (b) of 240 degrees, which converts to $4\pi/3$ radians (Example 3).

Now try Exercise 29.

The Cosecant Function

Important characteristics of the cosecant function can be inferred from the fact that it is the reciprocal of the sine function.

Whenever $\sin x = 1$, its reciprocal $\csc x$ is also 1. The graph of the cosecant function has asymptotes at the zeros of the sine function. The period of the cosecant function is 2π, the same as its reciprocal, the sine function.

The graph of $y = \csc x$ is shown with the graph of $y = \sin x$ in Figure 4.54. A local maximum of $y = \sin x$ corresponds to a local minimum of $y = \csc x$, while a local minimum of $y = \sin x$ corresponds to a local maximum of $y = \csc x$.

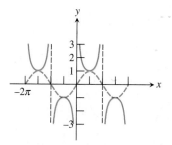

FIGURE 4.54 Characteristics of the cosecant function are inferred from the fact that it is the reciprocal of the sine function.

EXAMPLE 4 Solving a Trigonometric Equation Graphically

Find the smallest positive number x such that $x^2 = \csc x$.

SOLUTION There is no algebraic attack that looks hopeful, so we solve this equation graphically. The intersection point of the graphs of $y = x^2$ and $y = \csc x$ that has the smallest positive x-coordinate is shown in Figure 4.55. We use the grapher to determine that $x \approx 1.068$.

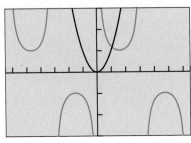

[−6.5, 6.5] by [−3, 3]

FIGURE 4.55 A graphical solution of a trigonometric equation. (Example 3)

Now try Exercise 39.

ARE COSECANT CURVES PARABOLAS?

Figure 4.55 shows a parabola intersecting one of the infinite number of U-shaped curves that make up the graph of the cosecant function. In fact, the parabola intersects *all* of those curves that lie above the *x*-axis, since the parabola must spread out to cover the entire domain of $y = x^2$, which is all real numbers! The cosecant curves do not keep spreading out, as they are hemmed in by asymptotes. That means that the U-shaped curves in the cosecant function are not parabolas.

To close this section, we summarize the properties of the six basic trigonometric functions in tabular form. The "n" that appears in several places should be understood as taking on all possible integer values: $0, \pm 1, \pm 2, \pm 3, \ldots$.

Summary: Basic Trigonometric Functions

Function	Period	Domain	Range	Asymptotes	Zeros	Even/Odd
$\sin x$	2π	All reals	$[-1, 1]$	None	$n\pi$	Odd
$\cos x$	2π	All reals	$[-1, 1]$	None	$\pi/2 + n\pi$	Even
$\tan x$	π	$x \neq \pi/2 + n\pi$	All reals	$x = \pi/2 + n\pi$	$n\pi$	Odd
$\cot x$	π	$x \neq n\pi$	All reals	$x = n\pi$	$\pi/2 + n\pi$	Odd
$\sec x$	2π	$x \neq \pi/2 + n\pi$	$(-\infty, -1] \cup [1, \infty)$	$x = \pi/2 + n\pi$	None	Even
$\csc x$	2π	$x \neq n\pi$	$(-\infty, -1] \cup [1, \infty)$	$x = n\pi$	None	Odd

---

QUICK REVIEW 4.5 *(For help, go to Sections 1.2, 2.6, and 4.3.)*

In Exercises 1–4, state the period of the function.

1. $y = \cos 2x$

2. $y = \sin 3x$

3. $y = \sin \dfrac{1}{3}x$

4. $y = \cos \dfrac{1}{2}x$

In Exercises 5–8, find the zeros and vertical asymptotes of the function.

5. $y = \dfrac{x - 3}{x + 4}$

6. $y = \dfrac{x + 5}{x - 1}$

7. $y = \dfrac{x + 1}{(x - 2)(x + 2)}$

8. $y = \dfrac{x + 2}{x(x - 3)}$

In Exercises 9 and 10, tell whether the function is odd, even, or neither.

9. $y = x^2 + 4$

10. $y = \dfrac{1}{x}$

SECTION 4.5 EXERCISES

In Exercises 1–4, identify the graph of each function. Use your understanding of transformations, not your graphing calculator.

1. Graphs of one period of csc x and 2 csc x are shown.

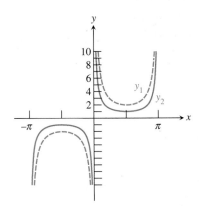

2. Graphs of two periods of 0.5 tan x and 5 tan x are shown.

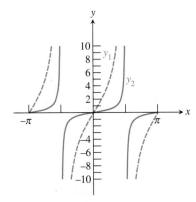

3. Graphs of csc x and 3 csc $2x$ are shown.

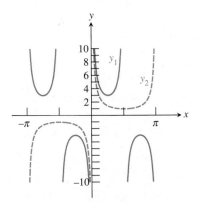

4. Graphs of cot x and cot $(x - 0.5) + 3$ are shown.

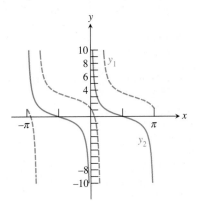

In Exercises 5–12, describe the graph of the function in terms of a basic trigonometric function. Locate the vertical asymptotes and graph two periods of the function.

5. $y = \tan 2x$ **6.** $y = -\cot 3x$

7. $y = \sec 3x$ **8.** $y = \csc 2x$

9. $y = 2 \cot 2x$ **10.** $y = 3 \tan (x/2)$

11. $y = \csc (x/2)$ **12.** $y = 3 \sec 4x$

In Exercises 13–16, match the trigonometric function with its graph. Then give the Xmin and Xmax values for the viewing window in which the graph is shown. Use your understanding of transformations, not your graphing calculator.

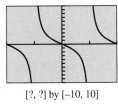

[?, ?] by [–10, 10]
(a)

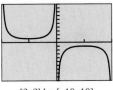

[?, ?] by [–10, 10]
(b)

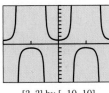

[?, ?] by [–10, 10]
(c)

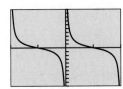

[?, ?] by [–10, 10]
(d)

13. $y = -2 \tan x$ **14.** $y = \cot x$

15. $y = \sec 2x$ **16.** $y = -\csc x$

In Exercises 17–20, analyze each function for domain, range, continuity, increasing or decreasing behavior, symmetry, boundedness, extrema, asymptotes, and end behavior.

17. $f(x) = \cot x$

18. $f(x) = \sec x$

19. $f(x) = \csc x$

20. $f(x) = \tan(x/2)$

In Exercises 21–28, describe the transformations required to obtain the graph of the given function from a basic trigonometric graph.

21. $y = 3 \tan x$

22. $y = -\tan x$

23. $y = 3 \csc x$

24. $y = 2 \tan x$

25. $y = -3 \cot \dfrac{1}{2}x$

26. $y = -2 \sec \dfrac{1}{2}x$

27. $y = -\tan \dfrac{\pi}{2}x + 2$

28. $y = 2 \tan \pi x - 2$

In Exercises 29–34, solve for x in the given interval. You should be able to find these numbers without a calculator, using reference triangles in the proper quadrants.

29. $\sec x = 2$, $0 \le x \le \pi/2$

30. $\csc x = 2$, $\pi/2 \le x \le \pi$

31. $\cot x = -\sqrt{3}$, $\pi/2 \le x \le \pi$

32. $\sec x = -\sqrt{2}$, $\pi \le x \le 3\pi/2$

33. $\csc x = 1$, $2\pi \le x \le 5\pi/2$

34. $\cot x = 1$, $-\pi \le x \le -\pi/2$

In Exercises 35–40, use a calculator to solve for x in the given interval.

35. $\tan x = 1.3$, $0 \le x \le \dfrac{\pi}{2}$

36. $\sec x = 2.4$, $0 \le x \le \dfrac{\pi}{2}$

37. $\cot x = -0.6$, $\dfrac{3\pi}{2} \le x \le 2\pi$

38. $\csc x = -1.5$, $\pi \le x \le \dfrac{3\pi}{2}$

39. $\csc x = 2$, $0 \le x \le 2\pi$

40. $\tan x = 0.3$, $0 \le x \le 2\pi$

41. Writing to Learn The figure shows a unit circle and an angle t whose terminal side is in quadrant III.

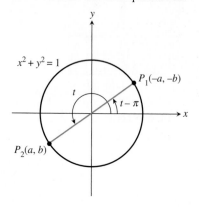

(a) If the coordinates of point P_2 are (a, b), explain why the coordinates of point P_1 on the circle and the terminal side of angle $t - \pi$ are $(-a, -b)$.

(b) Explain why $\tan t = \dfrac{b}{a}$.

(c) Find $\tan(t - \pi)$, and show that $\tan t = \tan(t - \pi)$.

(d) Explain why the period of the tangent function is π.

(e) Explain why the period of the cotangent function is π.

42. Writing to Learn Explain why it is correct to say $y = \tan x$ is the slope of the terminal side of angle x in standard position. P is on the unit circle.

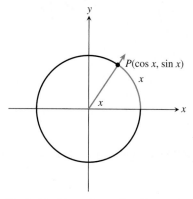

43. Periodic Functions Let f be a periodic function with period p. That is, p is the smallest positive number such that

$$f(x + p) = f(x)$$

for any value of x in the domain of f. Show that the reciprocal $1/f$ is periodic with period p.

44. Identities Use the unit circle to give a convincing argument for the identity.

(a) $\sin(t + \pi) = -\sin t$

(b) $\cos(t + \pi) = -\cos t$

(c) Use (a) and (b) to show that $\tan(t + \pi) = \tan t$. Explain why this is *not* enough to conclude that the period of tangent is π.

45. Lighthouse Coverage The Bolivar Lighthouse is located on a small island 350 ft from the shore of the mainland as shown in the figure.

(a) Express the distance d as a function of the angle x.

(b) If x is 1.55 rad, what is d?

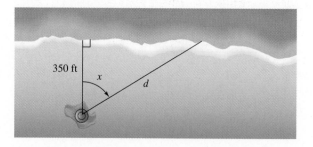

46. Hot-Air Balloon A hot-air balloon over Albuquerque, New Mexico, is being blown due east from point P and traveling at a constant height of 800 ft. The angle y is formed by the ground and the line of vision from P to the balloon. This angle changes as the balloon travels.

(a) Express the horizontal distance x as a function of the angle y.

(b) When the angle is $\pi/20$ rad, what is its horizontal distance from P?

(c) An angle of $\pi/20$ rad is equivalent to how many degrees?

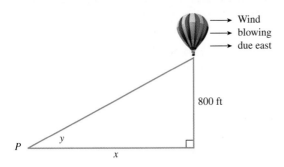

In Exercises 47–50, find approximate solutions for the equation in the interval $-\pi < x < \pi$.

47. $\tan x = \csc x$ **48.** $\sec x = \cot x$

49. $\sec x = 5 \cos x$ **50.** $4 \cos x = \tan x$

Standardized Test Questions

51. True or False The function $f(x) = \tan x$ is increasing on the interval $(-\infty, \infty)$. Justify your answer.

52. True or False If $x = a$ is an asymptote of the secant function, then $\cot a = 0$. Justify your answer.

You should answer these questions without using a calculator.

53. Multiple Choice The graph of $y = \cot x$ can be obtained by a horizontal shift of the graph of $y =$

(A) $-\tan x$. (B) $-\cot x$. (C) $\sec x$.

(D) $\tan x$. (E) $\csc x$.

54. Multiple Choice The graph of $y = \sec x$ never intersects the graph of $y =$

(A) x. (B) x^2. (C) $\csc x$.

(D) $\cos x$. (E) $\sin x$.

55. Multiple Choice If $k \neq 0$, what is the range of the function $y = k \csc x$?

(A) $[-k, k]$ (B) $(-k, k)$

(C) $(-\infty, -k) \cup (k, \infty)$ (D) $(-\infty, -k] \cup [k, \infty)$

(E) $(-\infty, -1/k] \cup [1/k, \infty)$

56. Multiple Choice The graph of $y = \csc x$ has the same set of asymptotes as the graph of $y =$

(A) $\sin x$. (B) $\tan x$. (C) $\cot x$.

(D) $\sec x$. (E) $\csc 2x$.

Explorations

In Exercises 57 and 58, graph both f and g in the $[-\pi, \pi]$ by $[-10, 10]$ viewing window. Estimate values in the interval $[-\pi, \pi]$ for which $f > g$.

57. $f(x) = 5 \sin x$ and $g(x) = \cot x$

58. $f(x) = -\tan x$ and $g(x) = \csc x$

59. Writing to Learn Graph the function $f(x) = -\cot x$ on the interval $(-\pi, \pi)$. Explain why it is correct to say that f is increasing on the interval $(0, \pi)$, but it is not correct to say that f is increasing on the interval $(-\pi, \pi)$.

60. Writing to Learn Graph functions $f(x) = -\sec x$ and

$$g(x) = \frac{1}{x - (\pi/2)}$$

simultaneously in the viewing window $[0, \pi]$ by $[-10, 10]$. Discuss whether you think functions f and g are equivalent.

61. Write $\csc x$ as a horizontal translation of $\sec x$.

62. Write $\cot x$ as the reflection about the x-axis of a horizontal translation of $\tan x$.

Extending the Ideas

63. Group Activity Television Coverage A television camera is on a platform 30 m from the point on High Street where the Worthington Memorial Day Parade will pass. Express the distance d from the camera to a particular parade float as a function of the angle x, and graph the function over the interval $-\pi/2 < x < \pi/2$.

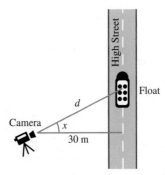

64. What's In a Name?   The word *sine* comes from the Latin word *sinus*, which means "bay" or "cove." It entered the language through a mistake (variously attributed to Gerardo of Cremona or Robert of Chester) in translating the Arabic word "jiba" (chord) as if it were "jaib" (bay). This was due to the fact that the Arabs abbreviated their technical terms, much as we do today. Imagine someone unfamiliar with the technical term "cosecant" trying to reconstruct the English word that is abbreviated by "csc." It might well enter their language as their word for "cascade."

The names for the other trigonometric functions can all be explained.

(a) *Cosine* means "*sine* of the *complement.*" Explain why this is a logical name for cosine.

(b) In the figure below, *BC* is perpendicular to *OC*, which is a radius of the unit circle. By a familiar geometry theorem, *BC* is tangent to the circle. *OB* is part of a secant that intersects the unit circle at *A*. It lies along the terminal side of an angle of *t* radians in standard position. Write the coordinates of *A* as functions of *t*.

(c) Use similar triangles to find length *BC* as a trig function of *t*.

(d) Use similar triangles to find length *OB* as a trig function of *t*.

(e) Use the results from parts (a), (c), and (d) to explain where the names "tangent, cotangent, secant," and "cosecant" came from.

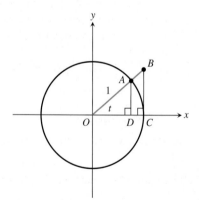

65. Capillary Action   A film of liquid in a thin (capillary) tube has surface tension γ (gamma) given by

$$\gamma = \frac{1}{2}h\rho gr \sec \phi,$$

where *h* is the height of the liquid in the tube, ρ (rho) is the density of the liquid, $g = 9.8$ m/sec^2 is the acceleration due to gravity, *r* is the radius of the tube, and φ (phi) is the angle of contact between the tube and the liquid's surface. Whole blood has a surface tension of 0.058 N/m (newton per meter) and a density of 1050 kg/m^3. Suppose that blood rises to a height of 1.5 m in a capillary blood vessel of radius 4.7×10^{-6} m. What is the contact angle between the capillary vessel and the blood surface? (1 N = 1 (kg · m)/sec^2)

66. Advanced Curve Fitting   A researcher has reason to believe that the data in the table below can best be described by an algebraic model involving the secant function:

$$y = a \sec (bx).$$

Unfortunately, her calculator will only do sine regression. She realizes that the following two facts will help her:

$$\frac{1}{y} = \frac{1}{a \sec (bx)} = \frac{1}{a}\cos (bx)$$

and

$$\cos (bx) = \sin\left(bx + \frac{\pi}{2}\right).$$

(a) Use these two facts to show that

$$\frac{1}{y} = \frac{1}{a}\sin\left(bx + \frac{\pi}{2}\right).$$

(b) Store the *x* values in the table in L1 in your calculator and the *y* values in L2. Store the *reciprocals* of the *y* values in L3. Then do a sine regression for L3 (1/*y*) as a function of L1 (*x*). Write the regression equation.

(c) Use the regression equation in (b) to determine the values of *a* and *b*.

(d) Write the secant model: $y = a \sec (bx)$. Does the curve fit the (L1, L2) scatter plot?

x	1	2	3	4
y	5.0703	5.2912	5.6975	6.3622

x	5	6	7	8
y	7.4359	9.2541	12.716	21.255

4.6
Graphs of Composite Trigonometric Functions

What you'll learn about

- Combining Trigonometric and Algebraic Functions
- Sums and Differences of Sinusoids
- Damped Oscillation

... and why

Function composition extends our ability to model periodic phenomena like heartbeats and sound waves.

Combining Trigonometric and Algebraic Functions

A theme of this text has been "families of functions." We have studied polynomial functions, exponential functions, logarithmic functions, and rational functions (to name a few), and in this chapter we have studied trigonometric functions. Now we consider adding, multiplying, or composing trigonometric functions with functions from these other families.

The notable property that distinguishes the trigonometric function from others we have studied is periodicity. Example 1 shows that when a trigonometric function is combined with a polynomial, the resulting function may or may not be periodic.

EXAMPLE 1 Combining the Sine Function With x^2

Graph each of the following functions for $-2\pi \le x \le 2\pi$, adjusting the vertical window as needed. Which of the functions appear to be periodic?

(a) $y = \sin x + x^2$

(b) $y = x^2 \sin x$

(c) $y = (\sin x)^2$

(d) $y = \sin(x^2)$

SOLUTION We show the graphs and their windows in Figure 4.56 on the next page. Only the graph of $y = (\sin x)^2$ exhibits periodic behavior in the interval $-2\pi \le x \le 2\pi$. (You can widen the window to see further graphical evidence that this is indeed the only periodic function among the four.)

Now try Exercise 5.

EXAMPLE 2 Verifying Periodicity Algebraically

Verify algebraically that $f(x) = (\sin x)^2$ is periodic and determine its period graphically.

SOLUTION We use the fact that the period of the basic sine function is 2π, that is, $\sin(x + 2\pi) = \sin(x)$ for all x. It follows that

$$f(x + 2\pi) = (\sin(x + 2\pi))^2$$

$$= (\sin(x))^2 \quad \text{By periodicity of sine}$$

$$= f(x)$$

So $f(x)$ is also periodic, with some period that divides 2π. The graph in Figure 4.56c on the next page shows that the period is actually π.

Now try Exercise 9.

EXPONENT NOTATION

Example 3 introduces a shorthand notation for powers of trigonometric functions: $(\sin \theta)^n$ can be written as $\sin^n \theta$. (Caution: This shorthand notation will probably not be recognized by your calculator.)

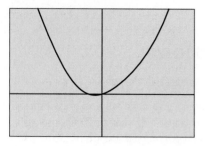

$[-2\pi, 2\pi]$ by $[-10, 20]$

(a)

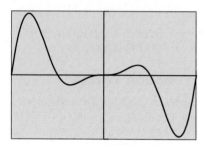

$[-2\pi, 2\pi]$ by $[-25, 25]$

(b)

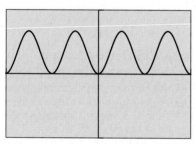

$[-2\pi, 2\pi]$ by $[-1.5, 1.5]$

(c)

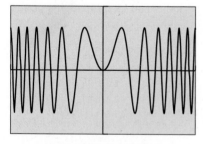

$[-2\pi, 2\pi]$ by $[-1.5, 1.5]$

(d)

FIGURE 4.56 The graphs of the four functions in Example 1. Only graph (c) exhibits periodic behavior over the interval $-2\pi \le x \le 2\pi$.

EXAMPLE 3   Composing $y = \sin x$ and $y = x^3$

Prove algebraically that $f(x) = \sin^3 x$ is periodic and find the period graphically. State the domain and range and sketch a graph showing two periods.

SOLUTION To prove that $f(x) = \sin^3 x$ is periodic, we show that $f(x + 2\pi) = f(x)$ for all x.

$$f(x + 2\pi) = \sin^3(x + 2\pi)$$

$$= (\sin(x + 2\pi))^3 \qquad \text{Changing notation}$$

$$= (\sin(x))^3 \qquad \text{By periodicity of sine}$$

$$= \sin^3(x) \qquad \text{Changing notation}$$

$$= f(x)$$

Thus $f(x)$ is periodic with a period that divides 2π. Graphing the function over the interval $-2\pi \le x \le 2\pi$ (Figure 4.57), we see that the period must be 2π.

Since both functions being composed have domain $(-\infty, \infty)$, the domain of f is also $(-\infty, \infty)$. Since cubing all numbers in the interval $[-1, 1]$ gives all numbers in the interval $[-1, 1]$, the range is $[-1, 1]$ (as supported by the graph).

Now try Exercise 13.

Comparing the graphs of $y = \sin^3 x$ and $y = \sin x$ over a single period (Figure 4.58), we see that the two functions have the same zeros and extreme points, but otherwise the graph of $y = \sin^3 x$ is closer to the x-axis than the graph of $y = \sin x$. This is because $|y^3| < |y|$ whenever y is between -1 and 1. In fact, higher odd powers of $\sin x$ yield graphs that are "sucked in" more and more, but always with the same zeros and extreme points.

The absolute value of a periodic function is also a periodic function. We consider two such functions in Example 4.

EXAMPLE 4   Analyzing Nonnegative Periodic Functions

Find the domain, range, and period of each of the following functions. Sketch a graph showing four periods.

(a) $f(x) = |\tan x|$

(b) $g(x) = |\sin x|$

SOLUTION

(a) Whenever $\tan x$ is defined, so is $|\tan x|$. Therefore, the domain of f is the same as the domain of the tangent function, that is, all real numbers except $\pi/2 + n\pi$, $n = 0, \pm 1, \ldots$. Because $f(x) = |\tan x| \ge 0$ and the range of $\tan x$ is $(-\infty, \infty)$, the range of f is $[0, \infty)$. The period of f, like that of $y = \tan x$, is π. The graph of $y = f(x)$ is shown in Figure 4.59.

(b) Whenever $\sin x$ is defined, so is $|\sin x|$. Therefore, the domain of g is the same as the domain of the sine function, that is, all real numbers. Because $g(x) = |\sin x| \ge 0$ and the range of $\sin x$ is $[-1, 1]$ the range of g is $[0, 1]$.

continued

The period of g is only half the period of $y = \sin x$, for reasons that are apparent from viewing the graph. The negative sections of the sine curve below the x-axis are reflected above the x-axis, where they become repetitions of the positive sections. The graph of $y = g(x)$ is shown in Figure 4.60 on the next page.

Now try Exercise 15.

When a sinusoid is added to a (nonconstant) linear function, the result is *not* periodic, The graph repeats its *shape* at regular intervals, but the function takes on different values over those intervals. The graph will show a curve oscillating between two parallel lines, as in Example 5.

EXAMPLE 5   Adding a Sinusoid to a Linear Function

The graph of $f(x) = 0.5x + \sin x$ oscillates between two parallel lines (Figure 4.61). What are the equations of the two lines?

SOLUTION As $\sin x$ oscillates between -1 and 1, $f(x)$ oscillates between $0.5x - 1$ and $0.5x + 1$. Therefore, the two lines are $y = 0.5x - 1$ and $y = 0.5x + 1$. Graphing the two lines and $f(x)$ in the same window provides graphical support. Of course, the graph should resemble Figure 4.61 on the next page if your lines are correct.

Now try Exercise 19.

Sums and Differences of Sinusoids

Section 4.4 introduced you to sinusoids, functions that can be written in the form

$$y = a \sin (b(x - h)) + k$$

and therefore have the shape of a sine curve.

Sinusoids model a variety of physical and social phenomena—such as sound waves, voltage in alternating electrical current, the velocity of air flow during the human respiratory cycle, and many others. Sometimes these phenomena interact in an additive fashion. For example, if y_1 models the sound of one tuning fork and y_2 models the sound of a second tuning fork, then $y_1 + y_2$ models the sound when they are both struck simultaneously. So we are interested in whether the sums and differences of sinusoids are again sinusoids.

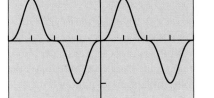

[$-2\pi, 2\pi$] by [$-1.5, 1.5$]

FIGURE 4.57 The graph of $f(x) = \sin^3 x$. (Example 3)

[$0, 2\pi$] by [$-1.5, 1.5$]

FIGURE 4.58 The graph suggests that $|\sin^3 x| \leq |\sin x|$.

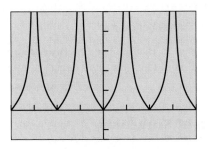

[$-2\pi, 2\pi$] by [$-1.5, 5$]

FIGURE 4.59 $f(x) = |\tan x|$ has the same period as $y = \tan x$. (Example 4a)

EXPLORATION 1   Investigating Sinusoids

Graph these functions, one at a time, in the viewing window $[-2\pi, 2\pi]$ by $[-6, 6]$. Which ones appear to be sinusoids?

$$y = 3 \sin x + 2 \cos x \qquad\qquad y = 2 \sin x - 3 \cos x$$
$$y = 2 \sin 3x - 4 \cos 2x \qquad\qquad y = 2 \sin (5x + 1) - 5 \cos 5x$$
$$y = \cos \left(\frac{7x - 2}{5}\right) + \sin \left(\frac{7x}{5}\right) \qquad y = 3 \cos 2x + 2 \sin 7x$$

What relationship between the sine and cosine functions ensures that their sum or difference will again be a sinusoid? Check your guess on a graphing calculator by constructing your own examples.

A MINOR EXCEPTION

A sinusoid and its exact opposite have the same period and sum to the zero function, which is not considered to be a sinusoid.

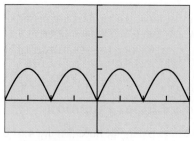

[−2π, 2π] by [−1, 3]

FIGURE 4.60 $g(x) = |\sin x|$ has half the period of $y = \sin x$. (Example 4b)

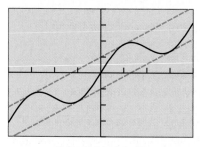

[−2π, 2π] by [−4, 4]

FIGURE 4.61 The graph of $f(x) = 0.5x + \sin x$ oscillates between the lines $y = 0.5x + 1$ and $y = 0.5x - 1$. Although the wave repeats its shape, it is not periodic. (Example 5)

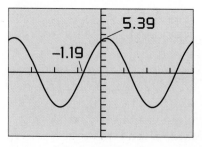

[−2π, 2π] by [−10, 10]

FIGURE 4.62 The sum of two sinusoids: $f(x) = 2 \sin x + 5 \cos x$. (Example 7)

The rule turns out to be fairly simple: Sums and differences of sinusoids with the same period are again sinusoids. We state this rule more explicitly as follows.

Sums That Are Sinusoid Functions

If $y_1 = a_1 \sin (b(x - h_1))$ and $y_2 = a_2 \cos (b(x - h_2))$, then

$$y_1 + y_2 = a_1 \sin (b(x - h_1)) + a_2 \cos (b(x - h_2))$$

is a sinusoid with period $2\pi/|b|$.

For the sum to be a sinusoid, the two sinusoids being added together must have the same period, and the sum has the same period as both of them. Also, although the rule is stated in terms of a sine function being added to a cosine function, the fact that every cosine function is a translation of a sine function (and vice versa) makes the rule equally applicable to the sum of two sine functions or the sum of two cosine functions. If they have the same period, their sum is a sinusoid.

EXAMPLE 6   Identifying a Sinusoid

Determine whether each of the following functions is or is not a sinusoid.

(a) $f(x) = 5 \cos x + 3 \sin x$

(b) $f(x) = \cos 5x + \sin 3x$

(c) $f(x) = 2 \cos 3x - 3 \cos 2x$

(d) $f(x) = a \cos\left(\dfrac{3x}{7}\right) - b \cos\left(\dfrac{3x}{7}\right) + c \sin\left(\dfrac{3x}{7}\right)$

SOLUTION

(a) Yes, since both functions in the sum have period 2π.

(b) No, since $\cos 5x$ has period $2\pi/5$ and $\sin 3x$ has period $2\pi/3$.

(c) No, since $2 \cos 3x$ has period $2\pi/3$ and $3 \cos 2x$ has period π.

(d) Yes, since all three functions in the sum have period $14\pi/3$. (The first two sum to a sinusoid with the same period as the third, so adding the third function still yields a sinusoid.)   ***Now try Exercise 25.***

EXAMPLE 7   Expressing the Sum of Sinusoids as a Sinusoid

Let $f(x) = 2 \sin x + 5 \cos x$. From the discussion above, you should conclude that $f(x)$ is a sinusoid.

(a) Find the period of f.

(b) Estimate the amplitude and phase shift graphically (to the nearest hundredth).

(c) Give a sinusoid $a \sin (b(x - h))$ that approximates $f(x)$.

continued

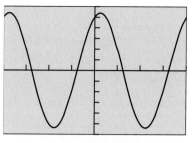

[−2π, 2π] by [−6, 6]

FIGURE 4.63 The graphs of
$y = 2 \sin x + 5 \cos x$ and
$y = 5.39 \sin (x + 1.19)$ appear to be
identical. (Example 7)

SOLUTION

(a) The period of f is the same as the period of $\sin x$ and $\cos x$, namely 2π.

Solve Graphically

(b) We will learn an algebraic way to find the amplitude and phase shift in the next chapter, but for now we will find this information graphically. Figure 4.62 suggests that f is indeed a sinusoid. That is, for some a and b,

$$2 \sin x + 5 \cos x = a \sin (x - h).$$

The *maximum value*, rounded to the nearest hundredth, is 5.39, so the amplitude of f is about 5.39. The *x-intercept* closest to $x = 0$, rounded to the nearest hundredth, is −1.19, so the phase shift of the sine function is about −1.19. We conclude that

$$f(x) = a \sin (x + h) \approx 5.39 \sin (x + 1.19).$$

(c) We support our answer graphically by showing that the graphs of $y = 2 \sin x + 5 \cos x$ and $y = 5.39 \sin (x + 1.19)$ are virtually identical (Figure 4.63).

Now try Exercise 29.

The sum of two sinusoids with different periods, while not a sinusoid, will often be a periodic function. Finding the period of a sum of periodic functions can be tricky. Here is a useful fact to keep in mind. If f is periodic, and if $f(x + s) = f(x)$ for all x in the domain of f, then the period of f divides s exactly. In other words, s is either the period or a multiple of the period.

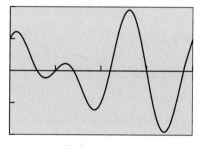

[0, 2π] by [−2, 2]

FIGURE 4.64 One period of
$f(x) = \sin 2x + \cos 3x$. (Example 8)

EXAMPLE 8 Showing a Function is Periodic but Not a Sinusoid

Show that $f(x) = \sin 2x + \cos 3x$ is periodic but not a sinusoid. Graph one period.

SOLUTION Since $\sin 2x$ and $\cos 3x$ have different periods, the sum is not a sinusoid. Next we show that 2π is a candidate for the period of f, that is, $f(x + 2\pi) = f(x)$ for all x.

$$f(x + 2\pi) = \sin (2(x + 2\pi)) + \cos (3(x + 2\pi))$$
$$= \sin (2x + 4\pi) + \cos (3x + 6\pi)$$
$$= \sin 2x + \cos 3x$$
$$= f(x)$$

This means either that 2π is the period of f or that the period is an exact divisor of 2π. Figure 4.64 suggests that the period is not smaller than 2π, so it must be 2π.

The graph shows that indeed f is not a sinusoid. *Now try Exercise 35.*

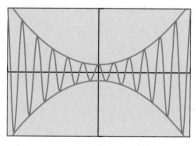

[−2π, 2π] by [−40, 40]

FIGURE 4.65 The graph of
$y = (x^2 + 5)\cos 6x$ shows **damped**
oscillation.

Damped Oscillation

Because the values of $\sin bt$ and $\cos bt$ oscillate between −1 and 1, something interesting happens when either of these functions is multiplied by another function. For example, consider the function $y = (x^2 + 5) \cos 6x$, graphed in Figure 4.65. The (blue) graph of the function oscillates between the (red) graphs of $y = x^2 + 5$ and $y = -(x^2 + 5)$. The "squeezing" effect that can be seen near the origin is called **damping**.

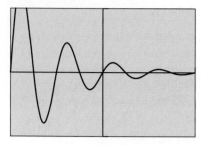

$[-\pi, \pi]$ by $[-5, 5]$

(a)

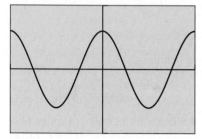

$[-\pi, \pi]$ by $[-5, 5]$

(b)

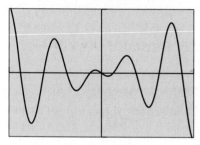

$[-2\pi, 2\pi]$ by $[-12, 12]$

(c)

FIGURE 4.66 The graphs of functions (a), (b), and (c) in Example 9. The wave in graph (b) does not exhibit damped oscillation.

Damped Oscillation

The graph of $y = f(x) \cos bx$ (or $y = f(x) \sin bx$) oscillates between the graphs of $y = f(x)$ and $y = -f(x)$. When this reduces the amplitude of the wave, it is called **damped oscillation**. The factor $f(x)$ is called the **damping factor**.

EXAMPLE 9   Identifying Damped Oscillation

For each of the following functions, determine if the graph shows damped oscillation. If it does, identify the damping factor and tell where the damping occurs.

(a) $f(x) = 2^{-x} \sin 4x$

(b) $f(x) = 3 \cos 2x$

(c) $f(x) = -2x \cos 2x$

SOLUTION The graphs are shown in Figure 4.66.

(a) This is damped oscillation. The damping factor is 2^{-x} and the damping occurs as $x \to \infty$.

(b) This wave has a constant amplitude of 3. No damping occurs.

(c) This is damped oscillation. The damping factor is $-2x$. The damping occurs as $x \to 0$.   ***Now try Exercise 43.***

EXAMPLE 10   A Damped Oscillating Spring

Dr. Sanchez's physics class collected data for an air table glider that oscillates between two springs. The class determined from the data that the equation

$$y = 0.22e^{-0.065t} \cos 2.4t$$

modeled the displacement y of the spring from its original position as a function of time t.

(a) Identify the damping factor and tell where the damping occurs.

(b) Approximately how long does it take for the spring to be damped so that $-0.1 \le y \le 0.1$?

SOLUTION The graph is shown in Figure 4.67.

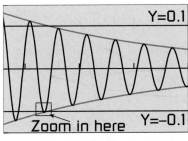

[8, 25] by [–0.15, 0.15]

(a)

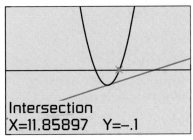

[11, 12.4] by [–0.11, –0.09]

(b)

FIGURE 4.68 The damped oscillation in Example 10 eventually gets to be less than 0.1 in either direction.

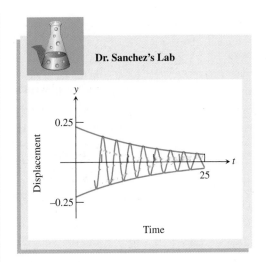

FIGURE 4.67 Damped oscillation in the physics lab. (Example 10)

(a) The damping factor is $0.22e^{-0.065t}$. The damping occurs as $t \to \infty$.

(b) We want to find how soon the curve $y = 0.22e^{-0.065t} \cos 2.4t$ falls entirely between the lines $y = -0.1$ and $y = 0.1$. By zooming in on the region indicated in Figure 4.68a and using grapher methods, we find that it takes approximately 11.86 seconds until the graph of $y = 0.22e^{-0.065t} \cos 2.4t$ lies entirely between $y = -0.1$ and $y = 0.1$ (Figure 4.68b).

Now try Exercise 71.

QUICK REVIEW 4.6 *(For help, go to Sections 1.2 and 1.4.)*

In Exercises 1–6, state the domain and range of the function.

1. $f(x) = 3 \sin 2x$ **2.** $f(x) = -2 \cos 3x$

3. $f(x) = \sqrt{x - 1}$ **4.** $f(x) = \sqrt{x}$

5. $f(x) = |x| - 2$ **6.** $f(x) = |x + 2| + 1$

In Exercises 7 and 8, describe the end behavior of the function, that is, the behavior as $|x| \to \infty$.

7. $f(x) = 5e^{-2x}$ **8.** $f(x) = -0.2(5^{-0.1x})$

In Exercises 9 and 10, form the compositions $f \circ g$ and $g \circ f$. State the domain of each function.

9. $f(x) = x^2 - 4$ and $g(x) = \sqrt{x}$

10. $f(x) = x^2$ and $g(x) = \cos x$

SECTION 4.6 EXERCISES

In Exercises 1–8, graph the function for $-2\pi \leq x \leq 2\pi$, adjusting the vertical window as needed. State whether or not the function appears to be periodic.

1. $f(x) = (\sin x)^2$ **2.** $f(x) = (1.5 \cos x)^2$

3. $f(x) = x^2 + 2 \sin x$ **4.** $f(x) = x^2 - 2 \cos x$

5. $f(x) = x \cos x$ **6.** $f(x) = x^2 \cos x$

7. $f(x) = (\sin x + 1)^3$ **8.** $f(x) = (2 \cos x - 4)^2$

In Exercises 9–12, verify algebraically that the function is periodic and determine its period graphically. Sketch a graph showing two periods.

9. $f(x) = \cos^2 x$ **10.** $f(x) = \cos^3 x$

11. $f(x) = \sqrt{\cos^2 x}$ **12.** $f(x) = |\cos^3 x|$

In Exercises 13–18, state the domain and range of the function and sketch a graph showing four periods.

13. $y = \cos^2 x$

14. $y = |\cos x|$

15. $y = |\cot x|$

16. $y = \cos|x|$

17. $y = -\tan^2 x$

18. $y = -\sin^2 x$

The graph of each function in Exercises 19–22 oscillates between two parallel lines, as in Example 5. Find the equations of the two lines and graph the lines and the function in the same viewing window.

19. $y = 2x + \cos x$

20. $y = 1 - 0.5x + \cos 2x$

21. $y = 2 - 0.3x + \cos x$

22. $y = 1 + x + \cos 3x$

In Exercises 23–28, determine whether $f(x)$ is a sinusoid.

23. $f(x) = \sin x - 3 \cos x$

24. $f(x) = 4 \cos x + 2 \sin x$

25. $f(x) = 2 \cos \pi x + \sin \pi x$

26. $f(x) = 2 \sin x - \tan x$

27. $f(x) = 3 \sin 2x - 5 \cos x$

28. $f(x) = \pi \sin 3x - 4\pi \sin 2x$

In Exercises 29–34, find a, b, and h so that $f(x) \approx a \sin(b(x - h))$.

29. $f(x) = 2 \sin 2x - 3 \cos 2x$

30. $f(x) = \cos 3x + 2 \sin 3x$

31. $f(x) = \sin \pi x - 2 \cos \pi x$

32. $f(x) = \cos 2\pi x + 3 \sin 2\pi x$

33. $f(x) = 2 \cos x + \sin x$

34. $f(x) = 3 \sin 2x - \cos 2x$

In Exercises 35–38, the function is periodic but not a sinusoid. Find the period graphically and sketch a graph showing one period.

35. $y = 2 \cos x + \cos 3x$

36. $y = 2 \sin 2x + \cos 3x$

37. $y = \cos 3x - 4 \sin 2x$

38. $y = \sin 2x + \sin 5x$

In Exercises 39–42, match the function with its graph.

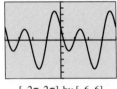

[$-2\pi, 2\pi$] by [$-6, 6$]
(a)

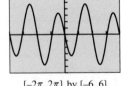

[$-2\pi, 2\pi$] by [$-6, 6$]
(b)

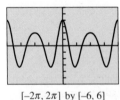

[$-2\pi, 2\pi$] by [$-6, 6$]
(c)

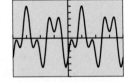

[$-2\pi, 2\pi$] by [$-6, 6$]
(d)

39. $y = 2 \cos x - 3 \sin 2x$

40. $y = 2 \sin 5x - 3 \cos 2x$

41. $y = 3 \cos 2x + \cos 3x$

42. $y = \sin x - 4 \sin 2x$

In Exercises 43–48, tell whether the function exhibits damped oscillation. If so, identify the damping factor and tell whether the damping occurs as $x \to 0$ or as $x \to \infty$.

43. $f(x) = e^{-x} \sin 3x$

44. $f(x) = x \sin 4x$

45. $f(x) = \sqrt{5} \cos 1.2x$

46. $f(x) = \pi^2 \cos \pi x$

47. $f(x) = x^3 \sin 5x$

48. $f(x) = \left(\dfrac{2}{3}\right)^x \sin\left(\dfrac{2x}{3}\right)$

In Exercises 49–52, graph both f and plus or minus its damping factor in the same viewing window. Describe the behavior of the function f for $x > 0$. What is the end behavior of f?

49. $f(x) = 1.2^{-x} \cos 2x$

50. $f(x) = 2^{-x} \sin 4x$

51. $f(x) = x^{-1} \sin 3x$

52. $f(x) = e^{-x} \cos 3x$

In Exercises 53–56, find the period and graph the function over two periods.

53. $y = \sin 3x + 2 \cos 2x$

54. $y = 4 \cos 2x - 2 \cos (3x - 1)$

55. $y = 2 \sin (3x + 1) - \cos (5x - 1)$

56. $y = 3 \cos (2x - 1) - 4 \sin (3x - 2)$

In Exercises 57–62, graph f over the interval $[-4\pi, 4\pi]$. Determine whether the function is periodic and, if it is, state the period.

57. $f(x) = \left|\sin \dfrac{1}{2}x\right| + 2$

58. $f(x) = 3x + 4 \sin 2x$

59. $f(x) = x - \cos x$

60. $f(x) = x + \sin 2x$

61. $f(x) = \dfrac{1}{2}x + \cos 2x$

62. $f(x) = 3 - x + \sin 3x$

In Exercises 63–70, find the domain and range of the function.

63. $f(x) = 2x + \cos x$

64. $f(x) = 2 - x + \sin x$

65. $f(x) = |x| + \cos x$

66. $f(x) = -2x + |3 \sin x|$

67. $f(x) = \sqrt{\sin x}$

68. $f(x) = \sin |x|$

69. $f(x) = \sqrt{|\sin x|}$

70. $f(x) = \sqrt{\cos x}$

71. Oscillating Spring The oscillations of a spring subject to friction are modeled by the equation $y = 0.43e^{-0.55t} \cos 1.8t$.

(a) Graph y and its two damping curves in the same viewing window for $0 \le t \le 12$.

(b) Approximately how long does it take for the spring to be damped so that $-0.2 \le y \le 0.2$?

72. Predicting Economic Growth The business manager of a small manufacturing company finds that she can model the company's annual growth as roughly exponential, but with cyclical fluctuations. She uses the function $S(t) = 75(1.04)^t + 4 \sin (\pi t/3)$ to estimate sales (in millions of dollars), t years after 2005.

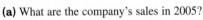

(a) What are the company's sales in 2005?

(b) What is the approximate annual growth rate?

(c) What does the model predict for sales in 2013?

(d) How many years are in each economic cycle for this company?

73. Writing to Learn Example 3 shows that the function $y = \sin^3 x$ is periodic. Explain whether you think that $y = \sin x^3$ is periodic and why.

74. Writing to Learn Example 4 shows that $y = |\tan x|$ is periodic. Write a convincing argument that $y = \tan |x|$ is not a periodic function.

In Exercises 75 and 76, select the one correct inequality, **(a)** or **(b)**. Give a convincing argument.

75. (a) $x - 1 \leq x + \sin x \leq x + 1$ for all x.

(b) $x - \sin x \leq x + \sin x$ for all x.

76. (a) $-x \leq x \sin x \leq x$ for all x.

(b) $-|x| \leq x \sin x \leq |x|$ for all x.

In Exercises 77–80, match the function with its graph. In each case state the viewing window.

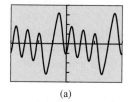

(a)

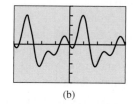

(b)

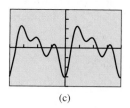

(c)

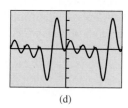

(d)

77. $y = \cos x - \sin 2x - \cos 3x + \sin 4x$

78. $y = \cos x - \sin 2x - \cos 3x + \sin 4x - \cos 5x$

79. $y = \sin x + \cos x - \cos 2x - \sin 3x$

80. $y = \sin x - \cos x - \cos 2x - \cos 3x$

Standardized Test Questions

81. True or False The function $f(x) = \sin |x|$ is periodic. Justify your answer.

82. True or False The sum of two sinusoids is a sinusoid. Justify your answer.

You may use a graphing calculator when answering these questions.

83. Multiple Choice What is the period of the function $f(x) = |\sin x|$?

(A) $\pi/2$ **(B)** π **(C)** 2π

(D) 3π **(E)** None; the function is not periodic.

84. Multiple Choice The function $f(x) = x \sin x$ is

(A) discontinuous. **(B)** bounded. **(C)** even.

(D) one-to-one. **(E)** periodic.

85. Multiple Choice The function $f(x) = x + \sin x$ is

(A) discontinuous. **(B)** bounded. **(C)** even.

(D) odd. **(E)** periodic.

86. Multiple Choice Which of the following functions is *not* a sinusoid?

(A) $2 \cos (2x)$ **(B)** $3 \sin (2x)$ **(C)** $3 \sin (2x) + 2 \cos (2x)$

(D) $3 \sin (3x) + 2 \cos (2x)$ **(E)** $\sin (3x + 3) + \cos (3x + 2)$

Explorations

87. Group Activity Inaccurate or Misleading Graphs

(a) Set Xmin = 0 and Xmax = 2π. Move the cursor along the *x*-axis. What is the distance between one pixel and the next (to the nearest hundreth)?

(b) What is the period of $f(x) = \sin 250x$? Consider that the period is the length of one full cycle of the graph. Approximately how many cycles should there be between two adjacent pixels? Can your grapher produce an accurate graph of this function between 0 and 2π?

88. Group Activity Length of Days The graph shows the number of hours of daylight in Boston as a function of the day of the year, from September 21, 1983, to December 15, 1984. Key points are labeled and other critical information is provided. Write a formula for the sinusoidal function and check it by graphing.

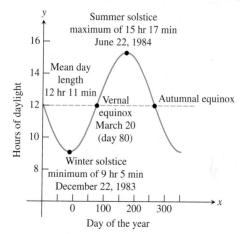

Extending the Ideas

In Exercises 89–96, first try to predict what the graph will look like (without too much effort, that is, just for fun). Then graph the function in one or more viewing windows to determine the main features of the graph, and draw a summary sketch. Where applicable, name the period, amplitude, domain, range, asymptotes, and zeros.

89. $f(x) = \cos e^x$

90. $g(x) = e^{\tan x}$

91. $f(x) = \sqrt{x} \sin x$

92. $g(x) = \sin \pi x + \sqrt{4 - x^2}$

93. $f(x) = \dfrac{\sin x}{x}$

94. $g(x) = \dfrac{\sin x}{x^2}$

95. $f(x) = x \sin \dfrac{1}{x}$

96. $g(x) = x^2 \sin \dfrac{1}{x}$

4.7
Inverse Trigonometric Functions

What you'll learn about

- Inverse Sine Function
- Inverse Cosine and Tangent Functions
- Composing Trigonometric and Inverse Trigonometric Functions
- Applications of Inverse Trigonometric Functions

. . . and why

Inverse trig functions can be used to solve trigonometric equations.

Inverse Sine Function

You learned in Section 1.4 that each function has an inverse relation, and that this inverse relation is a function only if the original function is one-to-one. The six basic trigonometric functions, being periodic, fail the horizontal line test for one-to-oneness rather spectacularly. However, you also learned in Section 1.4 that some functions are important enough that we want to study their inverse behavior despite the fact that they are not one-to-one. We do this by restricting the domain of the original function to an interval on which it *is* one-to-one, then finding the inverse of the restricted function. (We did this when defining the square root function, which is the inverse of the function $y = x^2$ restricted to a nonnegative domain.)

If you restrict the domain of $y = \sin x$ to the interval $[-\pi/2, \pi/2]$, as shown in Figure 4.69a, the restricted function is one-to-one. The **inverse sine function** $y = \sin^{-1} x$ is the inverse of this restricted portion of the sine function (Figure 4.69b).

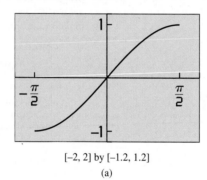
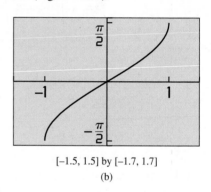

$[-2, 2]$ by $[-1.2, 1.2]$

(a)

$[-1.5, 1.5]$ by $[-1.7, 1.7]$

(b)

FIGURE 4.69 The (a) restriction of $y = \sin x$ is one-to-one and (b) has an inverse, $y = \sin^{-1} x$.

By the usual inverse relationship, the statements

$$y = \sin^{-1} x \qquad \text{and} \qquad x = \sin y$$

are equivalent for y-values in the restricted domain $[-\pi/2, \pi/2]$ and x-values in $[-1, 1]$. This means that $\sin^{-1} x$ can be thought of as *the angle between $-\pi/2$ and $\pi/2$ whose sine is x*. Since angles and directed arcs on the unit circle have the same measure, the angle $\sin^{-1} x$ is also called the **arcsine of x**.

Inverse Sine Function (Arcsine Function)

The unique angle y in the interval $[-\pi/2, \pi/2]$ such that $\sin y = x$ is the **inverse sine** (or **arcsine**) of x, denoted $\mathbf{sin^{-1} x}$ or $\mathbf{arcsin\, x}$.

The domain of $y = \sin^{-1} x$ is $[-1, 1]$ and the range is $[-\pi/2, \pi/2]$.

FIGURE 4.70 The values of $y = \sin^{-1} x$ will always be found on the right-hand side of the unit circle, between $-\pi/2$ and $\pi/2$.

It helps to think of the range of $y = \sin^{-1} x$ as being along the right-hand side of the unit circle, which is traced out as angles range from $-\pi/2$ to $\pi/2$ (Figure 4.70).

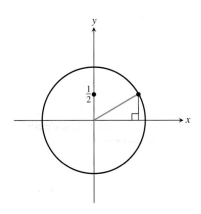

FIGURE 4.71 $\sin^{-1}(1/2) = \pi/6$.
(Example 1a)

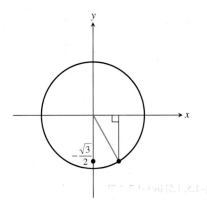

FIGURE 4.72 $\sin^{-1}(-\sqrt{3}/2) = -\pi/3$.
(Example 1b)

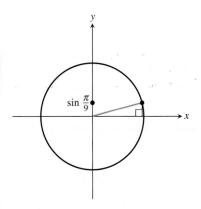

FIGURE 4.73 $\sin^{-1}(\sin(\pi/9)) = \pi/9$.
(Example 1d)

EXAMPLE 1 Evaluating $\sin^{-1} x$ without a Calculator

Find the exact value of each expression without a calculator.

(a) $\sin^{-1}\left(\dfrac{1}{2}\right)$ **(b)** $\sin^{-1}\left(-\dfrac{\sqrt{3}}{2}\right)$ **(c)** $\sin^{-1}\left(\dfrac{\pi}{2}\right)$

(d) $\sin^{-1}\left(\sin\left(\dfrac{\pi}{9}\right)\right)$ **(e)** $\sin^{-1}\left(\sin\left(\dfrac{5\pi}{6}\right)\right)$

SOLUTION

(a) Find the point on the right half of the unit circle whose y-coordinate is $1/2$ and draw a reference triangle (Figure 4.71). We recognize this as one of our special ratios, and the angle in the interval $[-\pi/2, \pi/2]$ whose sine is $1/2$ is $\pi/6$. Therefore

$$\sin^{-1}\left(\frac{1}{2}\right) = \frac{\pi}{6}.$$

(b) Find the point on the right half of the unit circle whose y-coordinate is $-\sqrt{3}/2$ and draw a reference triangle (Figure 4.72). We recognize this as one of our special ratios, and the angle in the interval $[-\pi/2, \pi/2]$ whose sine is $-\sqrt{3}/2$ is $-\pi/3$. Therefore

$$\sin^{-1}\left(-\frac{\sqrt{3}}{2}\right) = -\frac{\pi}{3}.$$

(c) $\sin^{-1}(\pi/2)$ does not exist, because the domain of $\sin^{-1}$ is $[-1, 1]$ and $\pi/2 > 1$.

(d) Draw an angle of $\pi/9$ in standard position and mark its y-coordinate on the y-axis (Figure 4.73). The angle in the interval $[-\pi/2, \pi/2]$ whose sine is this number is $\pi/9$. Therefore

$$\sin^{-1}\left(\sin\left(\frac{\pi}{9}\right)\right) = \frac{\pi}{9}.$$

(e) Draw an angle of $5\pi/6$ in standard position (notice that this angle is *not* in the interval $[-\pi/2, \pi/2]$) and mark its y-coordinate on the y-axis. (See Figure 4.74 on the next page.) The angle in the interval $[-\pi/2, \pi/2]$ whose sine is this number is $\pi - 5\pi/6 = \pi/6$. Therefore

$$\sin^{-1}\left(\sin\left(\frac{5\pi}{6}\right)\right) = \frac{\pi}{6}.$$

Now try Exercise 1.

EXAMPLE 2 Evaluating $\sin^{-1} x$ with a Calculator

Use a calculator in radian mode to evaluate these inverse sine values:

(a) $\sin^{-1}(-0.81)$

(b) $\sin^{-1}(\sin(3.49\pi))$

SOLUTION

(a) $\sin^{-1}(-0.81) = -0.9441521\ldots \approx -0.944$

(b) $\sin^{-1}(\sin(3.49\pi)) = -1.5393804\ldots \approx -1.539$

continued

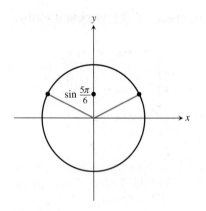

FIGURE 4.74 $\sin^{-1}(\sin(5\pi/6)) = \pi/6$. (Example 1e)

WHAT ABOUT THE INVERSE COMPOSITION RULE?

Does Example 1e violate the Inverse Composition Rule of Section 1.4? That rule guarantees that $f^{-1}(f(x)) = x$ for every x in the domain of f. Keep in mind, however, that the domain of f might need to be restricted in order for f^{-1} to exist. That is certainly the case with the sine function. So Example 1e does not violate the Inverse Composition Rule, because that rule *does not apply* at $x = 5\pi/6$. It lies outside the (restricted) domain of sine.

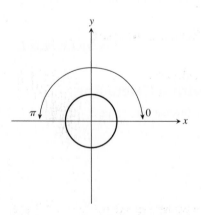

FIGURE 4.76 The values of $y = \cos^{-1} x$ will always be found on the top half of the unit circle, between 0 and π.

Although this is a calculator answer, we can use it to get an exact answer if we are alert enough to expect a multiple of π. Divide the answer by π:

$$\text{Ans}/\pi = -0.49.$$

Therefore, we conclude that $\sin^{-1}(\sin(3.49\pi)) = -0.49\pi$.

You should also try to work Example 2b without a calculator. It is possible!

Now try Exercise 19.

Inverse Cosine and Tangent Functions

If you restrict the domain of $y = \cos x$ to the interval $[0, \pi]$, as shown in Figure 4.75a, the restricted function is one-to-one. The **inverse cosine function** $y = \cos^{-1} x$ is the inverse of this restricted portion of the cosine function (Figure 4.75b).

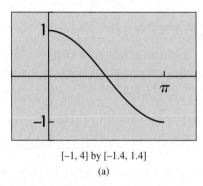
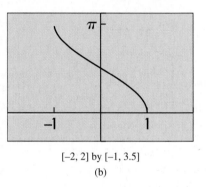

$[-1, 4]$ by $[-1.4, 1.4]$
(a)

$[-2, 2]$ by $[-1, 3.5]$
(b)

FIGURE 4.75 The (a) restriction of $y = \cos x$ is one-to-one and (b) has an inverse, $y = \cos^{-1} x$.

By the usual inverse relationship, the statements

$$y = \cos^{-1} x \quad \text{and} \quad x = \cos y$$

are equivalent for y-values in the restricted domain $[0, \pi]$ and x-values in $[-1, 1]$. This means that $\cos^{-1} x$ can be thought of as *the angle between 0 and π whose cosine is x.* The angle $\cos^{-1} x$ is also the **arccosine of x**.

Inverse Cosine Function (Arccosine Function)

The unique angle y in the interval $[0, \pi]$ such that $\cos y = x$ is the **inverse cosine** (or **arccosine**) of x, denoted **$\cos^{-1} x$** or **arccos x**. The domain of $y = \cos^{-1} x$ is $[-1, 1]$ and the range is $[0, \pi]$.

It helps to think of the range of $y = \cos^{-1} x$ as being along the top half of the unit circle, which is traced out as angles range from 0 to π (Figure 4.76).

If you restrict the domain of $y = \tan x$ to the interval $(-\pi/2, \pi/2)$, as shown in Figure 4.77a, the restricted function is one-to-one. The **inverse tangent function**

$y = \tan^{-1} x$ is the inverse of this restricted portion of the tangent function (Figure 4.77b).

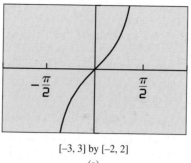
[-3, 3] by [-2, 2]
(a)

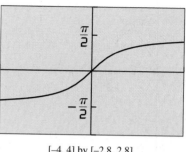
[-4, 4] by [-2.8, 2.8]
(b)

FIGURE 4.77 The (a) restriction of $y = \tan x$ is one-to-one and (b) has an inverse, $y = \tan^{-1} x$.

By the usual inverse relationship, the statements

$$y = \tan^{-1} x \quad \text{and} \quad x = \tan y$$

are equivalent for y-values in the restricted domain $(-\pi/2, \pi/2)$ and x-values in $(-\infty, \infty)$. This means that $\tan^{-1} x$ can be thought of as *the angle between $-\pi/2$ and $\pi/2$ whose tangent is x*. The angle $\tan^{-1} x$ is also the **arctangent of x**.

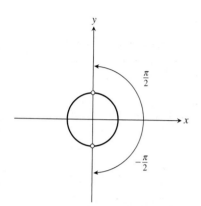

FIGURE 4.78 The values of $y = \tan^{-1} x$ will always be found on the right-hand side of the unit circle, between (but not including) $-\pi/2$ and $\pi/2$.

Inverse Tangent Function (Arctangent Function)

The unique angle y in the interval $(-\pi/2, \pi/2)$ such that $\tan y = x$ is the **inverse tangent** (or **arctangent**) of x, denoted **$\tan^{-1} x$** or **arctan x**.
The domain of $y = \tan^{-1} x$ is $(-\infty, \infty)$ and the range is $(-\pi/2, \pi/2)$.

It helps to think of the range of $y = \tan^{-1} x$ as being along the right-hand side of the unit circle (minus the top and bottom points), which is traced out as angles range from $-\pi/2$ to $\pi/2$ (noninclusive) (Figure 4.78).

EXAMPLE 3 Evaluating Inverse Trig Functions without a Calculator

Find the exact value of the expression without a calculator.

(a) $\cos^{-1}\left(-\dfrac{\sqrt{2}}{2}\right)$

(b) $\tan^{-1}\sqrt{3}$

(c) $\cos^{-1}(\cos(-1.1))$

SOLUTION

(a) Find the point on the top half of the unit circle whose x-coordinate is $-\sqrt{2}/2$ and draw a reference triangle (Figure 4.79). We recognize this as one of our special ratios, and the angle in the interval $[0, \pi]$ whose cosine is $-\sqrt{2}/2$ is $3\pi/4$. Therefore

$$\cos^{-1}\left(-\frac{\sqrt{2}}{2}\right) = \frac{3\pi}{4}.$$

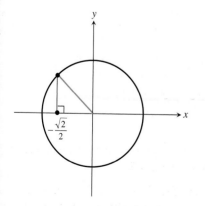

FIGURE 4.79 $\cos^{-1}(-\sqrt{2}/2) = 3\pi/4$.
(Example 3a)

continued

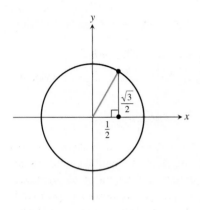

FIGURE 4.80 $\tan^{-1}\sqrt{3} = \pi/3$.
(Example 3b)

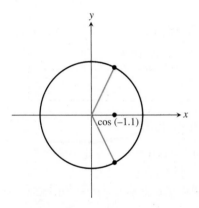

FIGURE 4.81 $\cos^{-1}(\cos(-1.1)) = 1.1$.
(Example 3c)

WHAT ABOUT ARCCOT, ARCSEC, AND ARCCSC?

Because we already have inverse functions for their reciprocals, we do not really need inverse functions for cot, sec, and csc for computational purposes. Moreover, the decision of how to choose the range of arcsec and arccsc is not as straightforward as with the other functions. See Exercises 63, 71, and 72.

(b) Find the point on the right side of the unit circle whose y-coordinate is $\sqrt{3}$ times its x-coordinate and draw a reference triangle (Figure 4.80). We recognize this as one of our special ratios, and the angle in the interval $(-\pi/2, \pi/2)$ whose tangent is $\sqrt{3}$ is $\pi/3$. Therefore

$$\tan^{-1}\sqrt{3} = \frac{\pi}{3}.$$

(c) Draw an angle of -1.1 in standard position (notice that this angle is *not* in the interval $[0, \pi]$) and mark its x-coordinate on the x-axis (Figure 4.81). The angle in the interval $[0, \pi]$ whose cosine is this number is 1.1. Therefore

$$\cos^{-1}(\cos(-1.1)) = 1.1.$$

Now try Exercises 5 and 7.

EXAMPLE 4   Describing End Behavior

Describe the end behavior of the function $y = \tan^{-1} x$.

SOLUTION We can get this information most easily by considering the graph of $y = \tan^{-1} x$, remembering how it relates to the restricted graph of $y = \tan x$. (See Figure 4.82.)

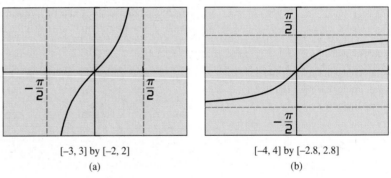

$[-3, 3]$ by $[-2, 2]$ $[-4, 4]$ by $[-2.8, 2.8]$
(a)   (b)

FIGURE 4.82 The graphs of (a) $y = \tan x$ (restricted) and (b) $y = \tan^{-1} x$. The vertical asymptotes of $y = \tan x$ are reflected to become the horizontal asymptotes of $y = \tan^{-1} x$. (Example 4)

When we reflect the graph of $y = \tan x$ about the line $y = x$ to get the graph of $y = \tan^{-1} x$, the vertical asymptotes $x = \pm\pi/2$ become horizontal asymptotes $y = \pm\pi/2$. We can state the end behavior accordingly:

$$\lim_{x \to -\infty} \tan^{-1} x = -\frac{\pi}{2} \quad \text{and} \quad \lim_{x \to +\infty} \tan^{-1} x = \frac{\pi}{2}.$$

Now try Exercise 21.

Composing Trigonometric and Inverse Trigonometric Functions

We have already seen the need for caution when applying the Inverse Composition Rule to the trigonometric functions and their inverses (Examples 1e and 3c). The following equations are *always* true whenever they are defined:

$$\sin{(\sin^{-1}(x))} = x \qquad \cos{(\cos^{-1}(x))} = x \qquad \tan{(\tan^{-1}(x))} = x$$

On the other hand, the following equations are only true for x values in the "restricted" domains of sin, cos, and tan:

$$\sin^{-1}(\sin{(x)}) = x \qquad \cos^{-1}(\cos{(x)}) = x \qquad \tan^{-1}(\tan{(x)}) = x$$

 An even more interesting phenomenon occurs when we compose inverse trigonometric functions of one kind with trigonometric functions of another kind, as in $\sin{(\tan^{-1}{x})}$. Surprisingly, these trigonometric compositions reduce to algebraic functions that involve no trigonometry at all! This curious situation has profound implications in calculus, where it is sometimes useful to decompose nontrigonometric functions into trigonometric components that seem to come out of nowhere. Try Exploration 1.

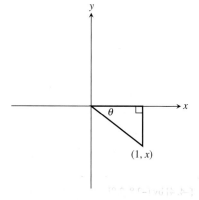

FIGURE 4.83 If $x < 0$, then $\theta = \tan^{-1}{x}$ is an angle in the fourth quadrant. (Exploration 1)

EXPLORATION 1 Finding Inverse Trig Functions of Trig Functions

In the right triangle shown to the right, the angle θ is measured in radians.

1. Find $\tan{\theta}$.
2. Find $\tan^{-1}{x}$.
3. Find the hypotenuse of the triangle as a function of x.
4. Find $\sin{(\tan^{-1}(x))}$ as a ratio involving no trig functions.
5. Find $\sec{(\tan^{-1}(x))}$ as a ratio involving no trig functions.
6. If $x < 0$, then $\tan^{-1}{x}$ is a negative angle in the fourth quadrant (Figure 4.83). Verify that your answers to parts (4) and (5) are still valid in this case.

 EXAMPLE 5 Composing Trig Functions with Arcsine

Compose each of the six basic trig functions with $\sin^{-1}{x}$ and reduce the composite function to an algebraic expression involving no trig functions.

SOLUTION This time we begin with the triangle shown in Figure 4.84, in which $\theta = \sin^{-1}{x}$. (This triangle could appear in the fourth quadrant if x were negative, but the trig ratios would be the same.)

The remaining side of the triangle (which is $\cos{\theta}$) can be found by the Pythagorean theorem. If we denote the unknown side by s, we have

$$s^2 + x^2 = 1$$
$$s^2 = 1 - x^2$$
$$s = \pm\sqrt{1 - x^2}$$

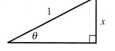

FIGURE 4.84 A triangle in which $\theta = \sin^{-1}{x}$. (Example 5)

continued

Note the ambiguous sign, which requires a further look. Since $\sin^{-1} x$ is always in Quadrant I or IV, the horizontal side of the triangle can only be positive. Therefore, we can actually write s unambiguously as $\sqrt{1 - x^2}$, giving us the triangle in Figure 4.85.

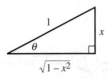

FIGURE 4.85 If $\theta = \sin^{-1} x$, then $\cos \theta = \sqrt{1 - x^2}$. Note that $\cos \theta$ will be positive because $\sin^{-1} x$ can only be in Quadrant I or IV. (Example 5)

We can now read all the required ratios straight from the triangle:

$$\sin\left(\sin^{-1}(x)\right) = x \qquad\qquad \csc\left(\sin^{-1}(x)\right) = \frac{1}{x}$$

$$\cos\left(\sin^{-1}(x)\right) = \sqrt{1 - x^2} \qquad\qquad \sec\left(\sin^{-1}(x)\right) = \frac{1}{\sqrt{1 - x^2}}$$

$$\tan\left(\sin^{-1}(x)\right) = \frac{x}{\sqrt{1 - x^2}} \qquad\qquad \cot\left(\sin^{-1}(x)\right) = \frac{\sqrt{1 - x^2}}{x}$$

Now try Exercise 47.

Applications of Inverse Trigonometric Functions

When an application involves an angle as a dependent variable, as in $\theta = f(x)$, then to solve for x, it is natural to use an inverse trigonometric function and find $x = f^{-1}(\theta)$.

EXAMPLE 6   Calculating a Viewing Angle

The bottom of a 20-foot replay screen at Dodger Stadium is 45 feet above the playing field. As you move away from the wall, the angle formed by the screen at your eye changes. There is a distance from the wall at which the angle is the greatest. What is that distance?

SOLUTION

Model

The angle subtended by the screen is represented in Figure 4.86 by θ, and $\theta = \theta_1 - \theta_2$. Since $\tan \theta_1 = 65/x$, it follows that $\theta_1 = \tan^{-1}(65/x)$. Similarly, $\theta_2 = \tan^{-1}(45/x)$. Thus,

$$\theta = \tan^{-1}\frac{65}{x} - \tan^{-1}\frac{45}{x}.$$

Solve Graphically

Figure 4.87 shows a graph of θ that reflects degree mode. The question about distance for maximum viewing angle can be answered by finding the x-coordinate of the maximum point of this graph. Using grapher methods we see that this maximum occurs when $x \approx 54$ feet.

continued

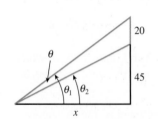

FIGURE 4.86 The diagram for the stadium screen. (Example 6)

Replay Screen Viewing Angle
(Dodger Stadium)

Distance (feet)

FIGURE 4.87 Viewing angle θ as a function of distance x from the wall. (Example 6)

Therefore the maximum angle subtended by the replay screen occurs about 54 feet from the wall.

Now try Exercise 55.

QUICK REVIEW 4.7 *(For help, go to Section 4.3.)*

In Exercises 1–4, state the sign (positive or negative) of the sine, cosine, and tangent in the quadrant.

1. Quadrant I

2. Quadrant II

3. Quadrant III

4. Quadrant IV

In Exercises 5–10, find the exact value.

5. $\sin(\pi/6)$

6. $\tan(\pi/4)$

7. $\cos(2\pi/3)$

8. $\sin(2\pi/3)$

9. $\sin(-\pi/6)$

10. $\cos(-\pi/3)$

SECTION 4.7 EXERCISES

In Exercises 1–12, find the exact value.

1. $\sin^{-1}\left(\dfrac{\sqrt{3}}{2}\right)$

2. $\sin^{-1}\left(-\dfrac{1}{2}\right)$

3. $\tan^{-1} 0$

4. $\cos^{-1} 1$

5. $\cos^{-1}\left(\dfrac{1}{2}\right)$

6. $\tan^{-1} 1$

7. $\tan^{-1}(-1)$

8. $\cos^{-1}\left(-\dfrac{\sqrt{3}}{2}\right)$

9. $\sin^{-1}\left(-\dfrac{1}{\sqrt{2}}\right)$

10. $\tan^{-1}(-\sqrt{3})$

11. $\cos^{-1} 0$

12. $\sin^{-1} 1$

In Exercises 13–16, use a calculator to find the approximate value. Express your answer in degrees.

13. $\sin^{-1}(0.362)$

14. $\arcsin 0.67$

15. $\tan^{-1}(-12.5)$

16. $\cos^{-1}(-0.23)$

In Exercises 17–20, use a calculator to find the approximate value. Express your result in radians.

17. $\tan^{-1}(2.37)$

18. $\tan^{-1}(22.8)$

19. $\sin^{-1}(-0.46)$

20. $\cos^{-1}(-0.853)$

In Exercises 21 and 22, describe the end behavior of the function.

21. $y = \tan^{-1}(x^2)$

22. $y = (\tan^{-1} x)^2$

In Exercises 23–32, find the exact value without a calculator.

23. $\cos(\sin^{-1}(1/2))$

24. $\sin(\tan^{-1} 1)$

25. $\sin^{-1}(\cos(\pi/4))$

26. $\cos^{-1}(\cos(7\pi/4))$

27. $\cos(2\sin^{-1}(1/2))$

28. $\sin(\tan^{-1}(-1))$

29. $\arcsin(\cos(\pi/3))$

30. $\arccos(\tan(\pi/4))$

31. $\cos(\tan^{-1}\sqrt{3})$

32. $\tan^{-1}(\cos\pi)$

3–36, analyze each function for domain, range, continu-
ng or decreasing behavior, symmetry, boundedness,
asymptotes, and end behavior.

$f(x) = \sin^{-1}x$ **34.** $f(x) = \cos^{-1}x$

5. $f(x) = \tan^{-1}x$

36. $f(x) = \cot^{-1}x$ (See graph in Exercise 67.)

In Exercises 37–40, use transformations to describe how the graph of
the function is related to a basic inverse trigonometric graph. State the
domain and range.

37. $f(x) = \sin^{-1}(2x)$ **38.** $g(x) = 3\cos^{-1}(2x)$

39. $h(x) = 5\tan^{-1}(x/2)$ **40.** $g(x) = 3\arccos(x/2)$

In Exercises 41–46, find an exact solution to the equation without a
calculator.

41. $\sin(\sin^{-1}x) = 1$ **42.** $\cos^{-1}(\cos x) = 1$

43. $2\sin^{-1}x = 1$ **44.** $\tan^{-1}x = -1$

45. $\cos(\cos^{-1}x) = 1/3$ **46.** $\sin^{-1}(\sin x) = \pi/10$

In Exercises 47–52, find an algebraic expression equivalent to the
given expression. (*Hint:* Form a right triangle as done in Example 5.)

47. $\sin(\tan^{-1}x)$ **48.** $\cos(\tan^{-1}x)$

49. $\tan(\arcsin x)$ **50.** $\cot(\arccos x)$

51. $\cos(\arctan 2x)$ **52.** $\sin(\arccos 3x)$

53. Group Activity Viewing Angle You are standing in an art
museum viewing a picture. The bottom of the picture is 2 ft above
your eye level, and the picture is 12 ft tall. Angle θ is formed by
the lines of vision to the bottom and to the top of the picture.

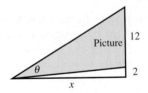

(a) Show that $\theta = \tan^{-1}\left(\dfrac{14}{x}\right) - \tan^{-1}\left(\dfrac{2}{x}\right)$.

(b) Graph θ in the [0, 25] by [0, 55] viewing window using
degree mode. Use your grapher to show that the maximum
value of θ occurs approximately 5.3 ft from the picture.

(c) How far (to the nearest foot) are you standing from the wall if
$\theta = 35°$?

54. Group Activity Analysis of a Lighthouse A rotating
beacon L stands 3 m across the harbor from the nearest point P
along a straight shoreline. As the light rotates, it forms an angle θ
as shown in the figure, and illuminates a point Q on the same
shoreline as P.

(a) Show that $\theta = \tan^{-1}\left(\dfrac{x}{3}\right)$.

(b) Graph θ in the viewing window $[-20, 20]$ by $[-90, 90]$
using degree mode. What do negative values of x represent
in the problem? What does a positive angle represent?
A negative angle?

(c) Find θ when $x = 15$.

55. Rising Hot-Air Balloon The hot-air balloon festival held
each year in Phoenix, Arizona, is a popular event for
photographers. Jo Silver, an award-winning photographer at the
event, watches a balloon rising from ground level from a point
500 ft away on level ground.

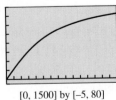

(a) Write θ as a function of the height s of the balloon.

(b) Is the change in θ greater as s changes from 10 ft to 20 ft, or
as s changes from 200 ft to 210 ft? Explain.

(c) Writing to Learn In the graph of
this relationship shown here, do you
think that the x-axis represents the
height s and the y-axis angle θ, or
does the x-axis represent angle θ
and the y-axis height s? Explain.

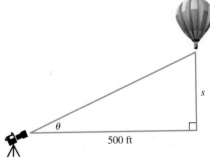

[0, 1500] by [–5, 80]

56. Find the domain and range of each of the following functions.

(a) $f(x) = \sin(\sin^{-1} x)$ (b) $g(x) = \sin^{-1}(x) + \cos^{-1}(x)$

(c) $h(x) = \sin^{-1}(\sin x)$ (d) $k(x) = \sin(\cos^{-1} x)$

(e) $q(x) = \cos^{-1}(\sin x)$

Standardized Test Questions

57. True or False $\sin(\sin^{-1}x) = x$ for all real numbers x. Justify your answer.

58. True or False The graph of $y = \arctan x$ has two horizontal asymptotes. Justify your answer.

You should answer these questions without using a calculator.

59. Multiple Choice $\cos^{-1}\left(-\dfrac{\sqrt{3}}{2}\right) =$

(A) $-\dfrac{7\pi}{6}$ (B) $-\dfrac{\pi}{3}$ (C) $-\dfrac{\pi}{6}$

(D) $\dfrac{2\pi}{3}$ (E) $\dfrac{5\pi}{6}$

60. Multiple Choice $\sin^{-1}(\sin \pi) =$

(A) -2π (B) $-\pi$ (C) 0

(D) π (E) 2π

61. Multiple Choice $\sec(\tan^{-1}x) =$

(A) x (B) $\csc x$ (C) $\sqrt{1 + x^2}$

(D) $\sqrt{1 - x^2}$ (E) $\dfrac{\sin x}{(\cos x)^2}$

62. Multiple Choice The range of the function $f(x) = \arcsin x$ is

(A) $(-\infty, \infty)$.   (B) $(-1, 1)$.   (C) $[-1, 1]$.

(D) $[0, \pi]$.   (E) $[-\pi/2, \pi/2]$.

Explorations

63. Writing to Learn Using the format demonstrated in this section for the inverse sine, cosine, and tangent functions, give a careful definition of the inverse cotangent function. (*Hint:* The range of $y = \cot^{-1} x$ is $(0, \pi)$.)

64. Writing to Learn Use an appropriately labeled triangle to explain why $\sin^{-1} x + \cos^{-1} x = \pi/2$. For what values of x is the left-hand side of this equation defined?

65. Graph each of the following functions and interpret the graph to find the domain, range, and period of each function. Which of the three functions has points of discontinuity? Are the discontinuities removable or nonremovable?

(a) $y = \sin^{-1}(\sin x)$

(b) $y = \cos^{-1}(\cos x)$

(c) $y = \tan^{-1}(\tan x)$

Extending the Ideas

66. Practicing for Calculus Express each of the following functions as an algebraic expression involving no trig functions.

(a) $\cos(\sin^{-1} 2x)$ (b) $\sec^2(\tan^{-1} x)$

(c) $\sin(\cos^{-1} \sqrt{x})$ (d) $-\csc^2(\cot^{-1} x)$

(e) $\tan(\sec^{-1} x^2)$

67. Arccotangent on the Calculator Most graphing calculators do not have a button for the inverse cotangent. The graph is shown below. Find an expression that you can put into your calculator to produce a graph of $y = \cot^{-1} x$.

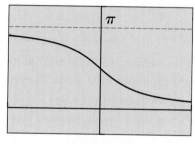

[–3, 3] by [–1, 4]

68. Advanced Decomposition Decompose each of the following algebraic functions by writing it as a trig function of an arctrig function.

(a) $\sqrt{1 - x^2}$ (b) $\dfrac{x}{\sqrt{1 + x^2}}$ (c) $\dfrac{x}{\sqrt{1 - x^2}}$

69. Use elementary transformations and the arctangent function to construct a function with domain all real numbers that has horizontal asymptotes at $y = 24$ and $y = 42$.

70. Avoiding Ambiguities When choosing the right triangle in Example 5, we used a hypotenuse of 1. It is sometimes necessary to use a variable quantity for the hypotenuse, in which case it is a good idea to use x^2 rather than x, just in case x is negative. (All of our definitions of the trig functions have involved triangles in which the hypotenuse is assumed to be positive.)

(a) If we use the triangle below to represent $\theta = \sin^{-1}(1/x)$, explain why side s must be positive regardless of the sign of x.

(b) Use the triangle in part (a) to find $\tan(\sin^{-1}(1/x))$.

(c) Using an appropriate triangle, find $\sin(\cos^{-1}(1/x))$.

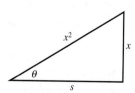

71. Defining Arcsecant The range of the secant function is $(-\infty, -1] \cup [1, \infty)$, which must become the domain of the arcsecant function. The graph of $y = \text{arcsec } x$ must therefore be the union of two unbroken curves. Two possible graphs with the correct domain are shown below.

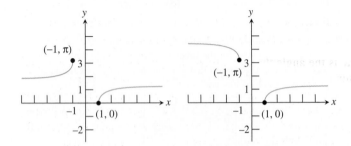

(a) The graph on the left has one horizontal asymptote. What is it?

(b) The graph on the right has two horizontal asymptotes. What are they?

(c) Which of these graphs is also the graph of $y = \cos^{-1}(1/x)$?

(d) Which of these graphs is increasing on both connected intervals?

72. Defining Arccosecant The range of the cosecant function is, $(-\infty, -1] \cup [1, \infty)$ which must become the domain of the arccosecant function. The graph of $y = \text{arccsc } x$ must therefore be the union of two unbroken curves. Two possible graphs with the correct domain are shown below.

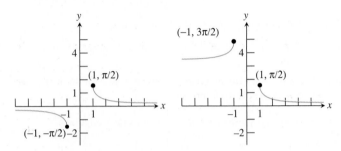

(a) The graph on the left has one horizontal asymptote. What is it?

(b) The graph on the right has two horizontal asymptotes. What are they?

(c) Which of these graphs is also the graph of $y = \sin^{-1}(1/x)$?

(d) Which of these graphs is decreasing on both connected intervals?

4.8
Solving Problems with Trigonometry

What you'll learn about

■ More Right Triangle Problems

■ Simple Harmonic Motion

. . . and why

These problems illustrate some of the better-known applications of trigonometry.

More Right Triangle Problems

We close this first of two trigonometry chapters by revisiting some of the applications of Section 4.2 (right triangle trigonometry) and Section 4.4 (sinusoids).

An **angle of elevation** is the angle through which the eye moves up from horizontal to look at something above, and an **angle of depression** is the angle through which the eye moves down from horizontal to look at something below. For two observers at different elevations looking at each other, the angle of elevation for one equals the angle of depression for the other. The concepts are illustrated in Figure 4.88 as they might apply to observers at Mount Rushmore or the Grand Canyon.

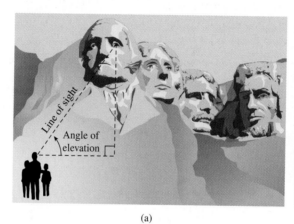

(a)   (b)

FIGURE 4.88 (a) Angle of elevation at Mount Rushmore. (b) Angle of depression at the Grand Canyon.

EXAMPLE 1   Using Angle of Depression

The angle of depression of a buoy from the top of the Barnegat Bay lighthouse 130 feet above the surface of the water is 6°. Find the distance x from the base of the lighthouse to the buoy.

SOLUTION Figure 4.89 models the situation.

In the diagram, $\theta = 6°$ because the angle of elevation from the buoy equals the angle of depression from the lighthouse. We **solve algebraically** using the tangent function:

$$\tan \theta = \tan 6° = \frac{130}{x}$$

$$x = \frac{130}{\tan 6°} \approx 1236.9$$

Interpreting We find that the buoy is about 1237 feet from the base of the lighthouse.

Now try Exercise 3.

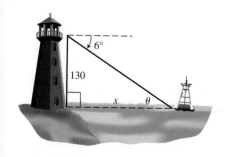

FIGURE 4.89 A big lighthouse and a little buoy. (Example 1)

EXAMPLE 2   Making Indirect Measurements

From the top of the 100-ft-tall Altgelt Hall a man observes a car moving toward the building. If the angle of depression of the car changes from 22° to 46° during the period of observation, how far does the car travel?

SOLUTION

Solve Algebraically Figure 4.90 models the situation. Notice that we have labeled the acute angles at the car's two positions as 22° and 46° (because the angle of elevation from the car equals the angle of depression from the building). Denote the distance the car moves as x. Denote its distance from the building at the second observation as d.

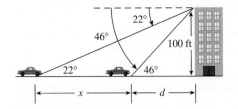

FIGURE 4.90 A car approaches Altgelt Hall. (Example 2)

From the smaller right triangle we conclude:

$$\tan 46° = \frac{100}{d}$$

$$d = \frac{100}{\tan 46°}$$

From the larger right triangle we conclude:

$$\tan 22° = \frac{100}{x + d}$$

$$x + d = \frac{100}{\tan 22°}$$

$$x = \frac{100}{\tan 22°} - d$$

$$x = \frac{100}{\tan 22°} - \frac{100}{\tan 46°}$$

$$x \approx 150.9$$

Interpreting our answer, we find that the car travels about 151 feet.

> *Now try Exercise 7.*

EXAMPLE 3   Finding Height Above Ground

A large, helium-filled penguin is moored at the beginning of a parade route awaiting the start of the parade. Two cables attached to the underside of the penguin make angles of 48° and 40° with the ground and are in the same plane as a perpendicular line from the penguin to the ground. (See Figure 4.91.) If the cables are attached to the ground 10 feet from each other, how high above the ground is the penguin?

SOLUTION We can simplify the drawing to the two right triangles in Figure 4.92 that share the common side h.

Model

By the definition of the tangent function,

$$\frac{h}{x} = \tan 48° \quad \text{and} \quad \frac{h}{x + 10} = \tan 40°.$$

FIGURE 4.91 A large, helium-filled penguin. (Example 3)

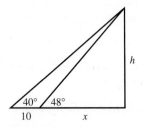

FIGURE 4.92 (Example 3)

continued

Solve Algebraically

Solving for h,

$$h = x \tan 48° \quad \text{and} \quad h = (x + 10) \tan 40°.$$

Set these two expressions for h equal to each other and solve the equation for x:

$$x \tan 48° = (x + 10) \tan 40° \qquad \text{Both equal } h.$$

$$x \tan 48° = x \tan 40° + 10 \tan 40°$$

$$x \tan 48° - x \tan 40° = 10 \tan 40° \qquad \text{Isolate } x \text{ terms}$$

$$x(\tan 48° - \tan 40°) = 10 \tan 40° \qquad \text{Factor out } x$$

$$x = \frac{10 \tan 40°}{\tan 48° - \tan 40°} \approx 30.90459723$$

We retain the full display for x because we are not done yet; we need to solve for h:

$$h = x \tan 48° = (30.90459723) \tan 48° \approx 34.32$$

The penguin is approximately 34 feet above ground level.

Now try Exercise 15.

EXAMPLE 4 Using Trigonometry in Navigation

A U.S. Coast Guard patrol boat leaves Port Cleveland and averages 35 **knots** (nautical mph) traveling for 2 hours on a course of 53° and then 3 hours on a course of 143°. What is the boat's bearing and distance from Port Cleveland?

SOLUTION Figure 4.93 models the situation.

Solve Algebraically In the diagram, line AB is a transversal that cuts a pair of parallel lines. Thus, $\beta = 53°$ because they are alternate interior angles. Angle α, as the supplement of a 143° angle, is 37°. Consequently, $\angle ABC = 90°$ and AC is the hypotenuse of right $\triangle ABC$.

Use distance = rate × time to determine distances AB and BC.

$$AB = (35 \text{ knots})(2 \text{ hours}) = 70 \text{ nautical miles}$$

$$BC = (35 \text{ knots})(3 \text{ hours}) = 105 \text{ nautical miles}$$

Solve the right triangle for AC and θ.

$$AC = \sqrt{70^2 + 105^2} \qquad \text{Pythagorean theorem}$$

$$AC \approx 126.2$$

$$\theta = \tan^{-1}\left(\frac{105}{70}\right)$$

$$\theta \approx 56.3°$$

Interpreting We find that the boat's bearing from Port Cleveland is $53° + \theta$, or approximately 109.3°. They are about 126 nautical miles out.

Now try Exercise 17.

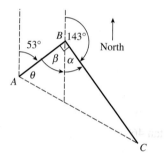

FIGURE 4.93 Path of travel for a Coast Guard boat that corners well at 35 knots. (Example 4)

Simple Harmonic Motion

Because of their periodic nature, the sine and cosine functions are helpful in describing the motion of objects that oscillate, vibrate, or rotate. For example, the linkage in Figure 4.94 converts the rotary motion of a motor to the back-and-forth motion needed for some machines. When the wheel rotates, the piston moves back and forth.

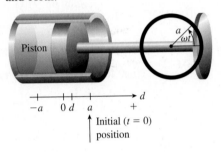

FIGURE 4.94 A piston operated by a wheel rotating at a constant rate demonstrates simple harmonic motion.

If the wheel rotates at a constant rate ω radians per second, the back-and-forth motion of the piston is an example of *simple harmonic motion* and can be modeled by an equation of the form

$$d = a \cos \omega t, \quad \omega > 0,$$

where a is the radius of the wheel and d is the directed distance of the piston from its center of oscillation.

For the sake of simplicity, we will define simple harmonic motion in terms of a point moving along a number line.

FREQUENCY AND PERIOD

Notice that harmonic motion is sinusoidal, with amplitude $|a|$ and period $2\pi/\omega$. The frequency is the reciprocal of the period.

Simple Harmonic Motion

A point moving on a number line is in **simple harmonic motion** if its directed distance d from the origin is given by either

$$d = a \sin \omega t \quad \text{or} \quad d = a \cos \omega t,$$

where a and ω are real numbers and $\omega > 0$. The motion has **frequency** $\omega/2\pi$, which is the number of oscillations per unit of time.

EXPLORATION 1 Watching Harmonic Motion

You can watch harmonic motion on your graphing calculator. Set your grapher to parametric mode and set $X_{1T} = \cos(T)$ and $Y_{1T} = \sin(T)$. Set Tmin = 0, Tmax = 25, Tstep = 0.2, Xmin = −1.5, Xmax = 1.5, Xscl = 1, Ymin = −100, Ymax = 100, Yscl = 0.

If your calculator allows you to change style to graph a moving ball, choose that style. When you graph the function, you will see the ball moving along the x-axis between −1 and 1 in simple harmonic motion. If your grapher does not have the moving ball option, wait for the grapher to finish graphing, then press TRACE and keep your finger pressed on the right arrow key to see the tracer move in simple harmonic motion.

1. For each value of T, the parametrization gives the point $(\cos(T), \sin(T))$. What well-known curve should this parametrization produce?
2. Why does the point seem to go back and forth on the x-axis when it should be following the curve identified in part (1)? (*Hint:* Check that viewing window again!)
3. Why does the point slow down at the extremes and speed up in the middle? (*Hint:* Remember that the grapher is really following the curve identified in part (1).)
4. How can you tell that this point moves in simple harmonic motion?

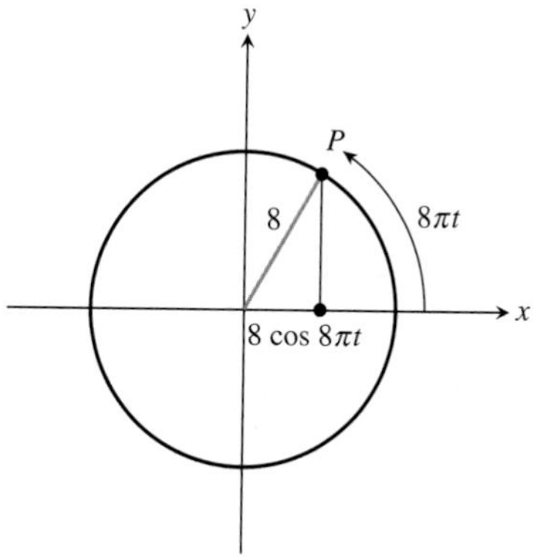

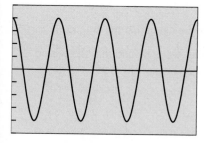

FIGURE 4.95 Modeling the path of a piston by a sinusoid. (Example 5)

[0, 1] by [−10, 10]

FIGURE 4.96 A sinusoid with frequency 4 models the motion of the piston in Example 5.

EXAMPLE 5 Calculating Harmonic Motion

In a mechanical linkage like the one shown in Figure 4.94, a wheel with an 8-cm radius turns with an angular velocity of 8π radians/sec.

(a) What is the frequency of the piston?

(b) What is the distance from the starting position ($t = 0$) exactly 3.45 seconds after starting?

SOLUTION Imagine the wheel to be centered at the origin and let $P(x, y)$ be a point on its perimeter (Figure 4.95). As the wheel rotates and P goes around, the motion of the piston follows the path of the x-coordinate of P along the x-axis. The angle determined by P at any time t is $8\pi t$, so its x-coordinate is $8 \cos 8\pi t$. Therefore, the sinusoid $d = 8 \cos 8\pi t$ models the motion of the piston.

(a) The frequency of $d = 8 \cos 8\pi t$ is $8\pi/2\pi$, or 4. The piston makes four complete back-and-forth strokes per second. The graph of d as a function of t is shown in Figure 4.96. The four cycles of the sinusoidal graph in the interval [0, 1] model the four cycles of the motor or the four strokes of the piston. Note that the sinusoid has a period of 1/4, the reciprocal of the frequency.

continued

(b) We must find the distance between the positions at $t = 0$ and $t = 3.45$.

The initial position at $t = 0$ is

$$d(0) = 8.$$

The position at $t = 3.45$ is

$$d(3.45) = 8 \cos (8\pi \cdot 3.45) \approx 2.47.$$

The distance between the two positions is approximately $8 - 2.47 = 5.53$.

Interpreting our answer, we conclude that the piston is approximately 5.53 cm from its starting position after 3.45 seconds.   ***Now try Exercise 27.***

EXAMPLE 6   Calculating Harmonic Motion

A mass oscillating up and down on the bottom of a spring (assuming perfect elasticity and no friction or air resistance) can be modeled as harmonic motion. If the weight is displaced a maximum of 5 cm, find the modeling equation if it takes 2 seconds to complete one cycle. (See Figure 4.97.)

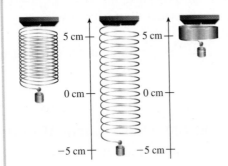

FIGURE 4.97 The mass and spring in Example 6.

SOLUTION We have our choice between the two equations $d = a \sin \omega t$ or $d = a \cos \omega t$. Assuming that the spring is at the origin of the coordinate system when $t = 0$, we choose the equation $d = a \sin \omega t$.

Because the maximum displacement is 5 cm, we conclude that the amplitude $a = 5$.

Because it takes 2 seconds to complete one cycle, we conclude that the period is 2 and the frequency is 1/2. Therefore,

$$\frac{\omega}{2\pi} = \frac{1}{2},$$

$$\omega = \pi.$$

Putting it all together, our modeling equation is $d = 5 \sin \pi t$.

Now try Exercise 29.

CHAPTER OPENER PROBLEM (from page 349)

PROBLEM: If we know that the musical note A above middle C has a pitch of 440 Hertz, how can we model the sound produced by it at 80 decibels?

SOLUTION: Sound is modeled by simple harmonic motion, with frequency perceived as pitch and measured in cycles per second, and amplitude perceived as loudness and measured in decibels. So for the musical note A with a pitch of 440 hertz, we have frequency $= \omega/2\pi = 440$ and therefore $\omega = 2\pi440 = 880\pi$.

If this note is played at a loudness of 80 decibels, we have $|a| = 80$. Using the simple harmonic motion model $d = a \sin \omega t$ we have

$$d = 80 \sin 880\pi t.$$

QUICK REVIEW 4.8 *(For help, go to Sections 4.1, 4.2, and 4.3.)*

In Exercises 1–4, find the lengths a, b, and c.

1.
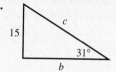

2.
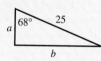

3.
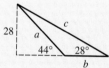

4.
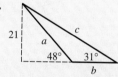

In Exercises 5 and 6, find the complement and supplement of the angle.

5. 32°   **6.** 73°

In Exercises 7 and 8, state the bearing that describes the direction.

7. NE (northeast)   **8.** SSW (south-southwest)

In Exercises 9 and 10, state the amplitude and period of the sinusoid.

9. $-3 \sin 2(x - 1)$ **10.** $4 \cos 4(x + 2)$

SECTION 4.8 EXERCISES

In Exercises 1–43, solve the problem using your knowledge of geometry and the techniques of this section. Sketch a figure if one is not provided.

1. Finding a Cathedral Height The angle of elevation of the top of the Ulm Cathedral from a point 300 ft away from the base of its steeple on level ground is 60°. Find the height of the cathedral.

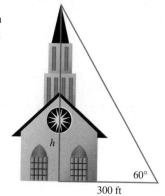

2. Finding a Monument Height From a point 100 ft from its base, the angle of elevation of the top of the Arch of Septimus Severus, in Rome, Italy, is 34°13′12″. How tall is this monument?

3. Finding a Distance The angle of depression from the top of the Smoketown Lighthouse 120 ft above the surface of the water to a buoy is 10°. How far is the buoy from the lighthouse?

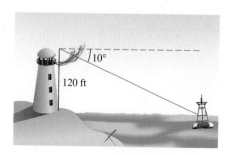

4. Finding a Baseball Stadium Dimension The top row of the red seats behind home plate at Cincinnati's Riverfront Stadium is 90 ft above the level of the playing field. The angle of depression to the base of the left field wall is 14°. How far is the base of the left field wall from a point on level ground directly below the top row?

5. Finding a Guy-Wire Length A guy wire connects the top of an antenna to a point on level ground 5 ft from the base of the antenna. The angle of elevation formed by this wire is 80°. What are the length of the wire and the height of the antenna?

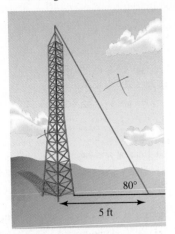

6. Finding a Length A wire stretches from the top of a vertical pole to a point on level ground 16 ft from the base of the pole. If the wire makes an angle of 62° with the ground, find the height of the pole and the length of the wire.

7. Height of Eiffel Tower The angle of elevation of the top of the TV antenna mounted on top of the Eiffel Tower in Paris is measured to be 80°1′12″ at a point 185 ft from the base of the tower. How tall is the tower plus TV antenna?

8. Finding the Height of Tallest Chimney The world's tallest smokestack at the International Nickel Co., Sudbury, Ontario, casts a shadow that is approximately 1580 ft long when the sun's angle of elevation (measured from the horizon) is 38°. How tall is the smokestack?

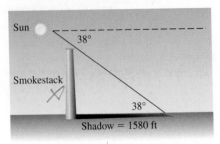

9. Cloud Height To measure the height of a cloud, you place a bright searchlight directly below the cloud and shine the beam straight up. From a point 100 ft away from the searchlight, you measure the angle of elevation of the cloud to be 83°12′. How high is the cloud?

10. Ramping Up A ramp leading to a freeway overpass is 470 ft long and rises 32 ft. What is the average angle of inclination of the ramp to the nearest tenth of a degree?

11. Antenna Height A guy wire attached to the top of the KSAM radio antenna is anchored at a point on the ground 10 m from the antenna's base. If the wire makes an angle of 55° with level ground, how high is the KSAM antenna?

12. Building Height To determine the height of the Louisiana-Pacific (LP) Tower, the tallest building in Conroe, Texas, a surveyor stands at a point on the ground, level with the base of the LP building. He measures the point to be 125 ft from the building's base and the angle of elevation to the top of the building to be 29°48′. Find the height of the building.

13. Navigation The *Paz Verde*, a whalewatch boat, is located at point P, and L is the nearest point on the Baja California shore. Point Q is located 4.25 mi down the shoreline from L and $\overline{PL} \perp \overline{LQ}$. Determine the distance that the *Paz Verde* is from the shore if $\angle PQL = 35°$.

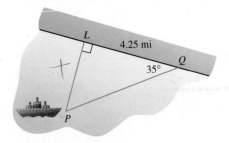

14. Recreational Hiking While hiking on a level path toward Colorado's front range, Otis Evans determines that the angle of elevation to the top of Long's Peak is 30°. Moving 1000 ft closer to the mountain, Otis determines the angle of elevation to be 35°. How much higher is the top of Long's Peak than Otis's elevation?

15. Civil Engineering The angle of elevation from an observer to the bottom edge of the Delaware River drawbridge observation deck located 200 ft from the observer is 30°. The angle of elevation from the observer to the top of the observation deck is 40°. What is the height of the observation deck?

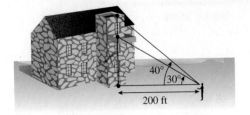

16. Traveling Car From the top of a 100-ft building a man observes a car moving toward him. If the angle of depression of the car changes from 15° to 33° during the period of observation, how far does the car travel?

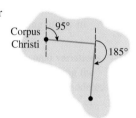

17. Navigation The Coast Guard cutter *Angelica* travels at 30 knots from its home port of Corpus Christi on a course of 95° for 2 hr and then changes to a course of 185° for 2 hr. Find the distance and the bearing from the Corpus Christi port to the boat.

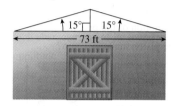

18. Navigation The *Cerrito Lindo* travels at a speed of 40 knots from Fort Lauderdale on a course of 65° for 2 hr and then changes to a course of 155° for 4 hr. Determine the distance and the bearing from Fort Lauderdale to the boat.

19. Land Measure The angle of depression is 19° from a point 7256 ft above sea level on the north rim of the Grand Canyon level to a point 6159 ft above sea level on the south rim. How wide is the canyon at that point?

20. Ranger Fire Watch A ranger spots a fire from a 73-ft tower in Yellowstone National Park. She measures the angle of depression to be 1°20′. How far is the fire from the tower?

21. Civil Engineering The bearing of the line of sight to the east end of the Royal Gorge footbridge from a point 325 ft due north of the west end of the footbridge across the Royal Gorge is 117°. What is the length *l* of the bridge?

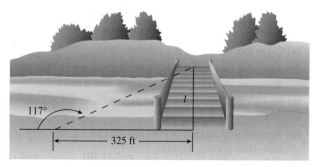

22. Space Flight The angle of elevation of a space shuttle from Cape Canaveral is 17° when the shuttle is directly over a ship 12 mi downrange. What is the altitude of the shuttle when it is directly over the ship?

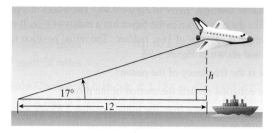

23. Architectural Design A barn roof is constructed as shown in the figure. What is the height of the vertical center span?

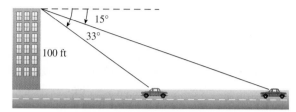

24. Recreational Flying A hot-air balloon over Park City, Utah, is 760 ft above the ground. The angle of depression from the balloon to an observer is 5.25°. Assuming the ground is relatively flat, how far is the observer from a point on the ground directly under the balloon?

25. Navigation A shoreline runs north-south, and a boat is due east of the shoreline. The bearings of the boat from two points on the shore are 110° and 100°. Assume the two points are 550 ft apart. How far is the boat from the shore?

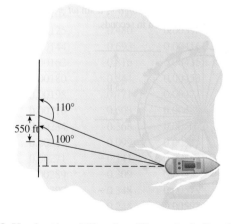

26. Navigation Milwaukee, Wisconsin, is directly west of Grand Haven, Michigan, on opposite sides of Lake Michigan. On a foggy night, a law enforcement boat leaves from Milwaukee on a course of 105° at the same time that a small smuggling craft steers a course of 195° from Grand Haven. The law enforcement boat averages 23 knots and collides with the smuggling craft. What was the smuggling boat's average speed?

44. Writing to Learn Human sleep-awake cycles at three different ages are described by the accompanying graphs. The portions of the graphs above the horizontal lines represent times awake, and the portions below represent times asleep.

Newborn

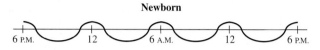

Four years

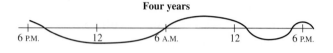

Adult

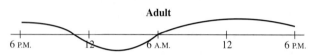

(a) What is the period of the sleep-awake cycle of a new-born? Of a four year old? Of an adult?

(b) Which of these three sleep-awake cycles is the closest to being modeled by a function $y = a \sin bx$?

Using Trigonometry in Geometry In a *regular polygon* all sides have equal length and all angles have equal measure. In Exercises 45 and 46, consider the regular seven-sided polygon whose sides are 5 cm.

45. Find the length of the *apothem*, the segment from the center of the seven-sided polygon to the midpoint of a side.

46. Find the radius of the circumscribed circle of the regular seven-sided polygon.

47. A *rhombus* is a quadrilateral with all sides equal in length. Recall that a rhombus is also a parallelogram. Find length AC and length BD in the rhombus shown here.

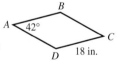

Extending the Ideas

48. A roof has two sections, one with a 50° elevation and the other with a 20° elevation, as shown in the figure.

(a) Find the height BE.

(b) Find the height CD.

(c) Find the length $AE + ED$, and double it to find the length of the roofline.

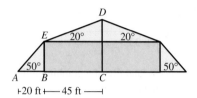

49. Steep Trucking The *percentage grade* of a road is its slope expressed as a percentage. A tractor-trailer rig passes a sign that reads, "6% grade next 7 miles." What is the average angle of inclination of the road?

50. Television Coverage Many satellites travel in *geosynchronous orbits*, which means that the satellite stays over the same point on the Earth. A satellite that broadcasts cable television is in geosynchronous orbit 100 mi above the Earth. Assume that the Earth is a sphere with radius 4000 mi, and find the arc length of coverage area for the cable television satellite on the Earth's surface.

51. Group Activity A musical note like that produced with a tuning fork or pitch meter is a pressure wave. Typically, frequency is measured in hertz (1 Hz = 1 cycle per second). Table 4.5 gives frequency (in Hz) of several musical notes. The time-vs.-pressure tuning fork data in Table 4.6 was collected using a CBL and a microphone.

Table 4.5 Tuning Fork Data	
Note	Frequency (Hz)
C	262
C♯ or D♭	277
D	294
D♯ or E♭	311
E	330
F	349
F♯ or G♭	370
G	392
G♯ or A♭	415
A	440
A♯ or B♭	466
B	494
C (next octave)	524

Table 4.6 Tuning Fork Data			
Time (sec)	Pressure	Time (sec)	Pressure
0.0002368	1.29021	0.0049024	−1.06632
0.0005664	1.50851	0.0051520	0.09235
0.0008256	1.51971	0.0054112	1.44694
0.0010752	1.51411	0.0056608	1.51411
0.0013344	1.47493	0.0059200	1.51971
0.0015840	0.45619	0.0061696	1.51411
0.0018432	−0.89280	0.0064288	1.43015
0.0020928	−1.51412	0.0066784	0.19871
0.0023520	−1.15588	0.0069408	−1.06072
0.0026016	−0.04758	0.0071904	−1.51412
0.0028640	1.36858	0.0074496	−0.97116
0.0031136	1.50851	0.0076992	0.23229
0.0033728	1.51971	0.0079584	1.46933
0.0036224	1.51411	0.0082080	1.51411
0.0038816	1.45813	0.0084672	1.51971
0.0041312	0.32185	0.0087168	1.50851
0.0043904	−0.97676	0.0089792	1.36298
0.0046400	−1.51971		

(a) Graph a scatter plot of the data.

(b) Determine a, b, and h so that the equation
$y = a \sin(b(t - h))$ is a model for the data.

(c) Determine the frequency of the sinusoid in part (b), and u.
Table 4.5 to identify the musical note produced by the tuning
fork.

(d) Identify the musical note produced by the tuning fork used in
Exercise 43.

...**Ideas**

...**EMS, AND FORMULAS**

PROCEDURES

GALLERY OF FUNCTIONS

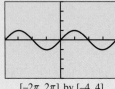

$[-2\pi, 2\pi]$ by $[-4, 4]$
$f(x) = \sin x$

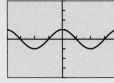

$[-2\pi, 2\pi]$ by $[-4, 4]$
$f(x) = \cos x$

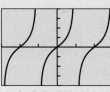

$[-3\pi/2, 3\pi/2]$ by $[-4, 4]$
$f(x) = \tan x$

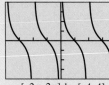

$[-2\pi, 2\pi]$ by $[-4, 4]$
$f(x) = \cot x$

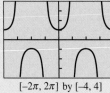

$[-2\pi, 2\pi]$ by $[-4, 4]$
$f(x) = \sec x$

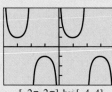

$[-2\pi, 2\pi]$ by $[-4, 4]$
$f(x) = \csc x$

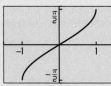

$[-1.5, 1.5]$ by $[-1.7, 1.7]$
$f(x) = \sin^{-1} x$

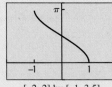

$[-2, 2]$ by $[-1, 3.5]$
$f(x) = \cos^{-1} x$

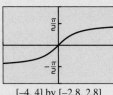

$[-4, 4]$ by $[-2.8, 2.8]$
$f(x) = \tan^{-1} x$

CHAPTER 4 Review Exercises

The collection of exercises marked in red could be used as a chapter test.

In Exercises 1–8, determine the quadrant of the terminal side of the angle in standard position. Convert degree measures to radians and radian measures to degrees.

1. $\dfrac{5\pi}{2}$

2. $\dfrac{3\pi}{4}$

3. $-135°$

4. $-45°$

5. $78°$

6. $112°$

7. $\dfrac{\pi}{12}$

8. $\dfrac{7\pi}{10}$

In Exercises 9 and 10, determine the angle measure in both degrees and radians. Draw the angle in standard position if its terminal side is obtained as described.

9. A three-quarters counterclockwise rotation

10. Two and one-half counterclockwise rotations

In Exercises 11–16, the point is on the terminal side of an angle in standard position. Give the smallest positive angle measure in both degrees and radians.

11. $(\sqrt{3}, 1)$

12. $(-1, 1)$

13. $(-1, \sqrt{3})$

14. $(-3, -3)$

15. $(6, -12)$

16. $(2, 4)$

In Exercises 17–28, evaluate the expression exactly without a calculator.

17. $\sin 30°$

18. $\cos 330°$

19. $\tan(-135°)$

20. $\sec(-135°)$

21. $\sin\dfrac{5\pi}{6}$

22. $\csc\dfrac{2\pi}{3}$

23. $\sec\left(-\dfrac{\pi}{3}\right)$

24. $\tan\left(-\dfrac{2\pi}{3}\right)$

25. $\csc 270°$

26. $\sec 180°$

27. $\cot(-90°)$

28. $\tan 360°$

In Exercises 29–32, evaluate exactly all six trigonometric functions of the angle. Use reference triangles and not your calculator.

29. $-\dfrac{\pi}{6}$

30. $\dfrac{19\pi}{4}$

31. $-135°$

32. $420°$

33. Find all six trigonometric functions of α in $\triangle ABC$.

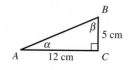

In Exercises 45–48, x is an angle in standard position with $0 \le x \le 2\pi$. Determine the quadrant of x.

34. Use a right triangle to determine the values of all trigonometric functions of θ, where $\cos\theta = 5/7$.

35. Use a right triangle to determine the values of all trigonometric functions of θ, where $\tan\theta = 15/8$.

36. Use a calculator in degree mode to solve $\cos\theta = 3/7$ if $0° \le \theta \le 90°$.

37. Use a calculator in radian mode to solve $\tan x = 1.35$ if $\pi \le x \le 3\pi/2$.

38. Use a calculator in radian mode to solve $\sin x = 0.218$ if $0 \le x \le 2\pi$.

In Exercises 39–44, solve the right $\triangle ABC$.

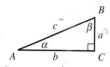

39. $\alpha = 35°$, $c = 15$

40. $b = 8$, $c = 10$

41. $\beta = 48°$, $a = 7$

42. $\alpha = 28°$, $c = 8$

43. $b = 5$, $c = 7$

44. $a = 2.5$, $b = 7.3$

In Exercises 45–48, x is an angle in standard position with $0 \le x \le 2\pi$. Determine the quadrant of x.

45. $\sin x < 0$ and $\tan x > 0$

46. $\cos x < 0$ and $\csc x > 0$

47. $\tan x < 0$ and $\sin x > 0$

48. $\sec x < 0$ and $\csc x > 0$

In Exercises 49–52, point P is on the terminal side of angle θ. Evaluate the six trigonometric functions for θ.

49. $(-3, 6)$

50. $(12, 7)$

51. $(-5, -3)$

52. $(4, 9)$

In Exercises 53–60, use transformations to describe how the graph of the function is related to a basic trigonometric graph. Graph two periods.

53. $y = \sin(x + \pi)$

54. $y = 3 + 2\cos x$

55. $y = -\cos(x + \pi/2) + 4$

56. $y = -2 - 3\sin(x - \pi)$

57. $y = \tan 2x$

58. $y = -2\cot 3x$

59. $y = -2\sec\dfrac{x}{2}$

60. $y = \csc \pi x$

In Exercises 61–66, state the amplitude, period, phase shift, domain, and range for the sinusoid.

61. $f(x) = 2\sin 3x$

62. $g(x) = 3\cos 4x$

63. $f(x) = 1.5\sin(2x - \pi/4)$

64. $g(x) = -2\sin(3x - \pi/3)$

65. $y = 4\cos(2x - 1)$

66. $g(x) = -2\cos(3x + 1)$

In Exercises 67 and 68, graph the function. Then estimate the values of a, b, and h so that $f(x) \approx a\sin(b(x - h))$.

67. $f(x) = 2\sin x - 4\cos x$

68. $f(x) = 3\cos 2x - 2\sin 2x$

In Exercises 69–72, use a calculator to evaluate the expression. Express your answer in both degrees and radians.

69. $\sin^{-1}(0.766)$

70. $\cos^{-1}(0.479)$

71. $\tan^{-1} 1$

72. $\sin^{-1}\left(\dfrac{\sqrt{3}}{2}\right)$

In Exercises 73–76, use transformations to describe how the graph of the function is related to a basic inverse trigonometric graph. State the domain and range.

73. $y = \sin^{-1} 3x$

74. $y = \tan^{-1} 2x$

75. $y = \sin^{-1}(3x - 1) + 2$

76. $y = \cos^{-1}(2x + 1) - 3$

In Exercises 77–82, find the exact value of x without using a calculator.

77. $\sin x = 0.5, \quad \pi/2 \le x \le \pi$

78. $\cos x = \sqrt{3}/2, \quad 0 \le x \le \pi$

79. $\tan x = -1, \quad 0 \le x \le \pi$

80. $\sec x = 2, \quad \pi \le x \le 2\pi$

81. $\csc x = -1, \quad 0 \le x \le 2\pi$

82. $\cot x = -\sqrt{3}, \quad 0 \le x \le \pi$

In Exercises 83 and 84, describe the end behavior of the function.

83. $\dfrac{\sin x}{x^2}$

84. $\dfrac{3}{5} e^{-x/12} \sin(2x - 3)$

In Exercises 85–88, evaluate the expression without a calculator.

85. $\tan(\tan^{-1} 1)$

86. $\cos^{-1}(\cos \pi/3)$

87. $\tan(\sin^{-1} 3/5)$

88. $\cos^{-1}(\cos -\pi/7)$

In Exercises 89–92, determine whether the function is periodic. State the period (if applicable), the domain, and the range.

89. $f(x) = |\sec x|$

90. $g(x) = \sin |x|$

91. $f(x) = 2x + \tan x$

92. $g(x) = 2 \cos 2x + 3 \sin 5x$

93. Arc Length Find the length of the arc intercepted by a central angle of $2\pi/3$ rad in a circle with radius 2.

94. Algebraic Expression Find an algebraic expression equivalent to $\tan(\cos^{-1} x)$.

95. Height of Building The angle of elevation of the top of a building from a point 100 m away from the building on level ground is 78°. Find the height of the building.

96. Height of Tree A tree casts a shadow 51 ft long when the angle of elevation of the sun (measured with the horizon) is 25°. How tall is the tree?

97. Traveling Car From the top of a 150-ft building Flora observes a car moving toward her. If the angle of depression of the car changes from 18° to 42° during the observation, how far does the car travel?

98. Finding Distance A lighthouse L stands 4 mi from the closest point P along a straight shore (see figure). Find the distance from P to a point Q along the shore if $\angle PLQ = 22°$.

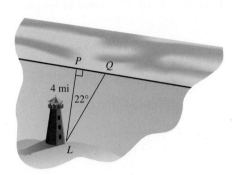

99. Navigation An airplane is flying due east between two signal towers. One tower is due north of the other. The bearing from the plane to the north tower is 23°, and to the south tower is 128°. Use a drawing to show the exact location of the plane.

100. Finding Distance The bearings of two points on the shore from a boat are 115° and 123°. Assume the two points are 855 ft apart. How far is the boat from the nearest point on shore if the shore is straight and runs north-south?

101. Height of Tree Dr. Thom Lawson standing on flat ground 62 ft from the base of a Douglas fir measures the angle of elevation to the top of the tree as 72°24′. What is the height of the tree?

102. Storing Hay A 75-ft-long conveyor is used at the Lovelady Farm to put hay bales up for winter storage. The conveyor is tilted to an angle of elevation of 22°.

(a) To what height can the hay be moved?

(b) If the conveyor is repositioned to an angle of 27°, to what height can the hay be moved?

103. Swinging Pendulum In the Hardy Boys Adventure *While the Clock Ticked*, the pendulum of the grandfather clock at the Purdy place is 44 in. long and swings through an arc of 6°. Find the length of the arc that the pendulum traces.

104. Finding Area A windshield wiper on a vintage 1994 Plymouth Acclaim is 20 in. long and has a blade 16 in. long. If the wiper sweeps through an angle of 110°, how large an area does the wiper blade clean? (See Exercise 71 in Section 4.1.)

105. Modeling Mean Temperature The average daily air temperature (°F) for Fairbanks, Alaska, from 1975 to 2004, can be modeled by the equation

$$T(x) = 37.3 \sin\left[\frac{2\pi}{365}(x - 114)\right] + 26,$$

where x is time in days with $x = 1$ representing January 1. On what days do you expect the average temperature to be 32°F?

Source: National Climatic Data Center, as reported in the World Almanac and Book of Facts 2005.

106. Taming The Beast The Beast is a featured roller coaster at the King Island's amusement park just north of Cincinnati. On its first and biggest hill, The Beast drops from a height of 52 ft above the ground along a sinusoidal path to a depth 18 ft underground as it enters a frightening tunnel. The mathematical model for this part of track is

$$h(x) = 35 \cos\left(\frac{x}{35}\right) + 17, \ 0 \le x \le 110,$$

where x is the horizontal distance from the top of the hill and $h(x)$ is the vertical position relative to ground level (both in feet). What is the horizontal distance from the top of the hill to the point where the track reaches ground level?

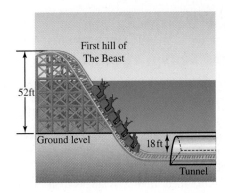

CHAPTER 4 Project

Modeling the Motion of a Pendulum

As a simple pendulum swings back and forth, its displacement can be modeled using a standard sinusoidal equation of the form:

$$y = a \cos (b(x - h)) + k$$

where y represents the pendulum's distance from a fixed point and x represents total elapsed time. In this project, you will use a motion detection device to collect distance and time data for a swinging pendulum, then find a mathematical model that describes the pendulum's motion.

COLLECTING THE DATA

To start, construct a simple pendulum by fastening about 1 meter of string to the end of a ball. Setup the Calculator Based Laboratory (CBL) system with a motion detector or a Calculator Based Ranger (CBR) system to collect time and distance readings for between 2 and 4 seconds (enough time to capture at least one complete swing of the pendulum). See the CBL/CBR guidebook for specific setup instructions. Start the pendulum swinging in front of the detector, then activate the system. The data table below shows a sample set of data collected as a pendulum swung back and forth in front of a CBR.

Total elapsed time (seconds)	Distance from the CBR (meters)
0	0.665
0.1	0.756
0.2	0.855
0.3	0.903
0.4	0.927
0.5	0.931
0.6	0.897
0.7	0.837
0.8	0.753
0.9	0.663
1.0	0.582
1.1	0.525
1.2	0.509
1.3	0.495
1.4	0.521
1.5	0.575
1.6	0.653
1.7	0.741
1.8	0.825
1.9	0.888
2.0	0.921

EXPLORATIONS

1. If you collected motion data using a CBL or CBR, a plot of distance versus time should be shown on your graphing calculator or computer screen. If you don't have access to a CBL/CBR, enter the data in the table above into your graphing calculator/computer. Create a scatter plot for the data.

2. Find values for a, b, h, and k so that the equation $y = a \cos (b(x - h)) + k$ fits the distance versus time data plot. Refer to the information box on page 390 in this chapter to review sinusoidal graph characteristics.

3. What are the physical meanings of the constants a and k in the modeling equation $y = a \cos (b(x - h)) + k$? (Hint: What distances do a and k measure?).

4. Which, if any, of the values of a, b, h, and/or k would change if you used the equation $y = a \sin (b(x - h)) + k$ to model the data set?

5. Use your calculator or computer to find a sinusoidal regression equation to model this data set (see your grapher's guidebook for instructions on how to do this). If your calculator/computer uses a different sinusoidal form, compare it to the modeling equation you found earlier, $y = a \cos (b(x - h)) + k$.

Analytic Trigonometry

It is no surprise that naturalists seeking to estimate wildlife populations must have an understanding of geometry (a word which literally means "earth measurement"). You will learn in this chapter that trigonometry, with its many connections to triangles and circles, enables us to extend the problem-solving tools of geometry significantly. On page 493 we will apply a result called Heron's Formula (which we prove trigonometrically) to estimate the local density of a deer population.

CHAPTER 5

Chapter 5 Overview

Although the title of this chapter suggests that we are now moving into the analytic phase of our study of trigonometric functions, the truth is that we have been in that phase for several sections already. Once the transition is made from triangle ratios to functions and their graphs, one is on analytic soil. But our primary applications of trigonometry so far have been computational; we have not made full use of the properties of the functions to study the connections among the trigonometric functions themselves. In this chapter we will shift our emphasis more toward theory and proof, exploring where the properties of these special functions lead us, often with no immediate concern for real-world relevance at all. We hope in the process to give you an appreciation for the rich and intricate tapestry of interlocking patterns that can be woven from the six basic trigonometric functions—patterns that will take on even greater beauty later on when you can view them through the lens of calculus.

5.1
Fundamental Identities

What you'll learn about

- Identities
- Basic Trigonometric Identities
- Pythagorean Identities
- Cofunction Identities
- Odd-Even Identities
- Simplifying Trigonometric Expressions
- Solving Trigonometric Equations

. . . and why

Identities are important when working with trigonometric functions in calculus.

Identities

As you probably realize by now, the symbol "=" means several different things in mathematics.

1. $1 + 1 = 2$ means *equality of real numbers*. It is a true sentence.

2. $2(x - 3) = 2x - 6$ signifies *equivalent expressions*. It is a true sentence.

3. $x^2 + 3 = 7$ is an *open sentence,* because it can be true or false, depending on whether x is a solution to the equation.

4. $(x^2 - 1)/(x + 1) = x - 1$ is an *identity.* It is a true sentence (very much like (2) above), but with the important qualification that *x must be in the domain of both expressions.* If either side of the equality is undefined, the sentence is meaningless. Substituting -1 into both sides of the equation in (3) gives a sentence that is mathematically false (i.e., $4 = 7$), whereas substituting -1 into both sides of the identity in (4) gives a sentence that is meaningless.

Statements like "$\tan \theta = \sin \theta/\cos \theta$" and "$\csc \theta = 1/\sin \theta$" are trigonometric **identities** because they are true for all values of the variable for which both sides of the equation are defined. The set of all such values is called the **domain of validity** of the identity. We will spend much of this chapter exploring trigonometric identities, their proofs, their implications, and their applications.

Basic Trigonometric Identities

Some trigonometric identities follow directly from the definitions of the six basic trigonometric functions. These *basic identities* consists of the *reciprocal identities* and the *quotient identities*.

Basic Trigonometric Identities

Reciprocal Identities

$$\csc\theta = \frac{1}{\sin\theta} \qquad \sec\theta = \frac{1}{\cos\theta} \qquad \cot\theta = \frac{1}{\tan\theta}$$

$$\sin\theta = \frac{1}{\csc\theta} \qquad \cos\theta = \frac{1}{\sec\theta} \qquad \tan\theta = \frac{1}{\cot\theta}$$

Quotient Identities

$$\tan\theta = \frac{\sin\theta}{\cos\theta} \qquad \cot\theta = \frac{\cos\theta}{\sin\theta}$$

EXPLORATION 1 **Making a Point about Domain of Validity**

1. $\theta = 0$ is in the domain of validity of exactly three of the basic identities. Which three?

2. For exactly two of the basic identities, one side of the equation is defined at $\theta = 0$ and the other side is not. Which two?

3. For exactly three of the basic identities, both sides of the equation are undefined at $\theta = 0$. Which three?

Pythagorean Identities

Exploration 2 in Section 4.3 introduced you to the fact that, for any real number t, the numbers $(\cos t)^2$ and $(\sin t)^2$ always sum to 1. This is clearly true for the quadrantal angles that wrap to the points $(\pm 1, 0)$ and $(0, \pm 1)$, and it is true for any other t because $\cos t$ and $\sin t$ are the (signed) lengths of legs of a reference triangle with hypotenuse 1 (Figure 5.1). No matter what quadrant the triangle lies in, the Pythagorean theorem guarantees the following identity: $(\cos t)^2 + (\sin t)^2 = 1$.

If we divide each term of the identity by $(\cos t)^2$, we get an identity that involves tangent and secant:

$$\frac{(\cos t)^2}{(\cos t)^2} + \frac{(\sin t)^2}{(\cos t)^2} = \frac{1}{(\cos t)^2}$$

$$1 + (\tan t)^2 = (\sec t)^2$$

If we divide each term of the identity by $(\sin t)^2$, we get an identity that involves cotangent and cosecant:

$$\frac{(\cos t)^2}{(\sin t)^2} + \frac{(\sin t)^2}{(\sin t)^2} = \frac{1}{(\sin t)^2}$$

$$(\cot t)^2 + 1 = (\csc t)^2$$

These three identities are called the *Pythagorean identities*, which we re-state using the shorthand notation for powers of trigonometric functions.

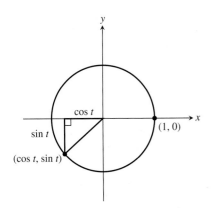

FIGURE 5.1 By the Pythagorean theorem, $(\cos t)^2 + (\sin t)^2 = 1$.

> **Pythagorean Identities**
>
> $$\cos^2 \theta + \sin^2 \theta = 1$$
>
> $$1 + \tan^2 \theta = \sec^2 \theta$$
>
> $$\cot^2 \theta + 1 = \csc^2 \theta$$

EXAMPLE 1 Using Identities

Find $\sin \theta$ and $\cos \theta$ if $\tan \theta = 5$ and $\cos \theta > 0$.

SOLUTION We could solve this problem by the reference triangle techniques of Section 4.3 (see Example 7 in that section), but we will show an alternate solution here using only identities.

First, we note that $\sec^2 \theta = 1 + \tan^2 \theta = 1 + 5^2 = 26$, so $\sec \theta = \pm\sqrt{26}$. Since $\sec \theta = \pm\sqrt{26}$, we have $\cos \theta = 1/\sec \theta = 1/\pm\sqrt{26}$. But $\cos \theta > 0$, so $\cos \theta = 1/\sqrt{26}$. Finally,

$$\tan \theta = 5$$

$$\frac{\sin \theta}{\cos \theta} = 5$$

$$\sin \theta = 5 \cos \theta = 5\left(\frac{1}{\sqrt{26}}\right) = \frac{5}{\sqrt{26}}.$$

Therefore, $\sin \theta = \dfrac{5}{\sqrt{26}}$ and $\cos \theta = \dfrac{1}{\sqrt{26}}$.   *Now try Exercise 1.*

If you find yourself preferring the reference triangle method, that's fine. Remember that combining the power of geometry and algebra to solve problems is one of the themes of this book, and the instinct to do so will serve you well in calculus.

Cofunction Identities

If C is the right angle in right $\triangle ABC$, then angles A and B are complements. Notice what happens if we use the usual triangle ratios to define the six trigonometric functions of angles A and B (Figure 5.2).

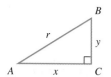

FIGURE 5.2 Angles A and B are complements in right $\triangle ABC$.

Angle A: $\sin A = \dfrac{y}{r}$ $\tan A = \dfrac{y}{x}$ $\sec A = \dfrac{r}{x}$

$\cos A = \dfrac{x}{r}$ $\cot A = \dfrac{x}{y}$ $\csc A = \dfrac{r}{y}$

Angle B: $\sin B = \dfrac{x}{r}$ $\tan B = \dfrac{x}{y}$ $\sec B = \dfrac{r}{y}$

$\cos B = \dfrac{y}{r}$ $\cot B = \dfrac{y}{x}$ $\csc B = \dfrac{r}{x}$

Do you see what happens? In every case, the value of a function at A is the same as the value of its cofunction at B. This always happens with complementary angles; in fact, it is this phenomenon that gives a "co"function its name. The "co" stands for "complement."

Cofunction Identities

$$\sin\left(\frac{\pi}{2} - \theta\right) = \cos\theta \qquad \cos\left(\frac{\pi}{2} - \theta\right) = \sin\theta$$

$$\tan\left(\frac{\pi}{2} - \theta\right) = \cot\theta \qquad \cot\left(\frac{\pi}{2} - \theta\right) = \tan\theta$$

$$\sec\left(\frac{\pi}{2} - \theta\right) = \csc\theta \qquad \csc\left(\frac{\pi}{2} - \theta\right) = \sec\theta$$

Although our argument on behalf of these equations was based on acute angles in a triangle, these equations are genuine identities, valid for all real numbers for which both sides of the equation are defined. We could extend our acute-angle argument to produce a general proof, but it will be easier to wait and use the identities of Section 5.3. We will revisit this particular set of fundamental identities in that section.

Odd-Even Identities

We have seen that every basic trigonometric function is either odd or even. Either way, the usual function relationship leads to another fundamental identity.

Odd-Even Identities

$$\sin(-x) = -\sin x \qquad \cos(-x) = \cos x \qquad \tan(-x) = -\tan x$$

$$\csc(-x) = -\csc x \qquad \sec(-x) = \sec x \qquad \cot(-x) = -\cot x$$

EXAMPLE 2   Using More Identities

If $\cos\theta = 0.34$, find $\sin(\theta - \pi/2)$.

SOLUTION This problem can best be solved using identities.

$$\sin\left(\theta - \frac{\pi}{2}\right) = -\sin\left(\frac{\pi}{2} - \theta\right) \quad \text{Sine is odd.}$$

$$= -\cos\theta \qquad \text{Cofunction identity}$$

$$= -0.34 \qquad \textit{Now try Exercise 7.}$$

Simplifying Trigonometric Expressions

In calculus it is often necessary to deal with expressions that involve trigonometric functions. Some of those expressions start out looking fairly complicated, but it is often possible to use identities along with algebraic techniques (e.g., factoring or combining fractions over a common denominator) to *simplify* the expressions before dealing with them. In some cases the simplifications can be dramatic.

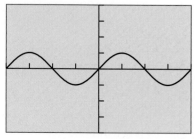

$[-2\pi, 2\pi]$ by $[-4, 4]$

(a)

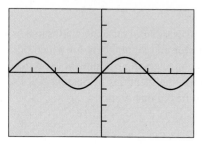

$[-2\pi, 2\pi]$ by $[-4, 4]$

(b)

FIGURE 5.3 Graphical support of the identity $\sin^3 x + \sin x \cos^2 x = \sin x$. (Example 3)

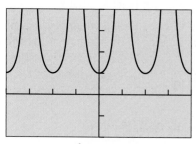

$[-2\pi, 2\pi]$ by $[-2, 4]$

(a)

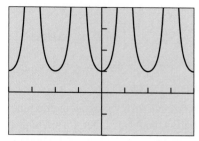

$[-2\pi, 2\pi]$ by $[-2, 4]$

(b)

FIGURE 5.4 Graphical support of the identity $(\sec x + 1)(\sec x - 1)/\sin^2 x = \sec^2 x$. (Example 4)

EXAMPLE 3   Simplifying by Factoring and Using Identities

Simplify the expression $\sin^3 x + \sin x \cos^2 x$.

SOLUTION

Solve Algebraically

$$\sin^3 x + \sin x \cos^2 x = \sin x \,(\sin^2 x + \cos^2 x)$$

$$= \sin x \,(1) \qquad\qquad \text{Pythagorean identity}$$

$$= \sin x$$

Support Graphically

We recognize the graph of $y = \sin^3 x + \sin x \cos^2 x$ (Figure 5.3a) as the same as the graph of $y = \sin x$. (Figure 5.3b)   ***Now try Exercise 13.***

EXAMPLE 4   Simplifying by Expanding and Using Identities

Simplify the expression $[(\sec x + 1)(\sec x - 1)]/\sin^2 x$.

SOLUTION

Solve Algebraically

$$\frac{(\sec x + 1)(\sec x - 1)}{\sin^2 x} = \frac{\sec^2 x - 1}{\sin^2 x} \qquad (a + b)(a - b) = a^2 - b^2$$

$$= \frac{\tan^2 x}{\sin^2 x} \qquad \text{Pythagorean identity}$$

$$= \frac{\sin^2 x}{\cos^2 x} \cdot \frac{1}{\sin^2 x} \qquad \tan x = \frac{\sin x}{\cos x}$$

$$= \frac{1}{\cos^2 x}$$

$$= \sec^2 x$$

Support Graphically

The graphs of $y = \dfrac{(\sec x + 1)(\sec x - 1)}{\sin^2 x}$ and $y = \sec^2 x$ appear to be identical, as expected. (Figure 5.4)   ***Now try Exercise 25.***

EXAMPLE 5   Simplifying by Combining Fractions and Using Identities

Simplify the expression $\dfrac{\cos x}{1 - \sin x} - \dfrac{\sin x}{\cos x}$.

continued

SOLUTION

$$\frac{\cos x}{1 - \sin x} - \frac{\sin x}{\cos x}$$

$$= \frac{\cos x}{1 - \sin x} \cdot \frac{\cos x}{\cos x} - \frac{\sin x}{\cos x} \cdot \frac{1 - \sin x}{1 - \sin x} \qquad \text{Rewrite using common denominator.}$$

$$= \frac{(\cos x)(\cos x) - (\sin x)(1 - \sin x)}{(1 - \sin x)(\cos x)}$$

$$= \frac{\cos^2 x - \sin x + \sin^2 x}{(1 - \sin x)(\cos x)}$$

$$= \frac{1 - \sin x}{(1 - \sin x)(\cos x)} \qquad \text{Pythagorean identity}$$

$$= \frac{1}{\cos x}$$

$$= \sec x$$

(We leave it to you to provide the graphical support.)

Now try Exercise 37.

We will use these same simplifying techniques to prove trigonometric identities in Section 5.2.

Solving Trigonometric Equations

The equation-solving capabilities of calculators have made it possible to solve trigonometric equations without understanding much trigonometry. This is fine, to the extent that solving equations is our goal. However, since understanding trigonometry is also a goal, we will occasionally pause in our development of identities to solve some trigonometric equations with paper and pencil, just to get some practice in using the identities.

EXAMPLE 6 Solving a Trigonometric Equation

Find all values of x in the interval $[0, 2\pi)$ that solve $\cos^3 x / \sin x = \cot x$.

SOLUTION

$$\frac{\cos^3 x}{\sin x} = \cot x$$

$$\frac{\cos^3 x}{\sin x} = \frac{\cos x}{\sin x}$$

$$\cos^3 x = \cos x \qquad \text{Multiply both sides by } \sin x.$$

$$\cos^3 x - \cos x = 0$$

$$(\cos x)(\cos^2 x - 1) = 0$$

$$(\cos x)(-\sin^2 x) = 0 \qquad \text{Pythagorean identity}$$

$$\cos x = 0 \quad \text{or} \quad \sin x = 0$$

continued

We reject the possibility that $\sin x = 0$ because it would make both sides of the original equation undefined.

The values in the interval $[0, 2\pi)$ that solve $\cos x = 0$ (and therefore $\cos^3 x / \sin x = \cot x$) are $\pi/2$ and $3\pi/2$.   ***Now try Exercise 51.***

EXAMPLE 7   Solving a Trigonometric Equation by Factoring

Find all solutions to the trigonometric equation $2 \sin^2 x + \sin x = 1$.

SOLUTION Let $y = \sin x$. The equation $2y^2 + y = 1$ can be solved by factoring:

$$2y^2 + y = 1$$

$$2y^2 + y - 1 = 0$$

$$(2y - 1)(y + 1) = 0$$

$$2y - 1 = 0 \quad \text{or} \quad y + 1 = 0$$

$$y = \frac{1}{2} \quad \text{or} \quad y = -1$$

So, in the original equation, $\sin x = 1/2$ or $\sin x = -1$. Figure 5.5 shows that the solutions in the interval $[0, 2\pi)$ are $\pi/6$, $5\pi/6$, and $3\pi/2$.

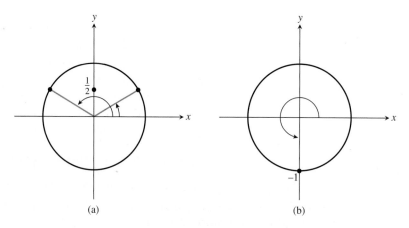

(a)   (b)

FIGURE 5.5 (a) $\sin x = 1/2$ has two solutions in $[0, 2\pi)$: $\pi/6$ and $5\pi/6$. (b) $\sin x = -1$ has one solution in $[0, 2\pi)$: $3\pi/2$. (Example 7)

To get *all* real solutions, we simply add integer multiples of the period, 2π, of the periodic function $\sin x$:

$$x = \frac{\pi}{6} + 2n\pi \quad \text{or} \quad x = \frac{5\pi}{6} + 2n\pi \quad \text{or} \quad x = \frac{3\pi}{2} + 2n\pi$$

$$(n = 0, \pm 1, \pm 2, \ldots)$$

Now try Exercise 57.

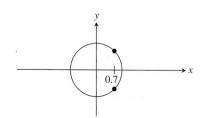

FIGURE 5.6 There are two points on the unit circle with x-coordinate 0.7. (Example 8)

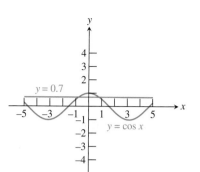

FIGURE 5.7 Intersecting the graphs of $y = \cos x$ and $y = 0.7$ gives two solutions to the equation $\cos t = 0.7$. (Example 8)

You might try solving the equation in Example 7 on your graph— parison. Finding *all* real solutions still requires an understand— finding *exact* solutions requires the savvy to divide your calcul— likely that anyone who knows that much trigonometry will actually ..... — solution to be easier!

EXAMPLE 8 Solving a Trig Equation with a Calculator

Find all solutions to the equation $\cos t = 0.7$, using a calculator where needed.

SOLUTION Figure 5.6 shows that there are two points on the unit circle with an x-coordinate of 0.7. We do not recognize this value as one of our special triangle ratios, but we can use a graphing calculator to find the smallest positive and negative values for which $\cos x = 0.7$ by intersecting the graphs of $y = \cos x$ and $y = 0.7$ (Figure 5.7).

The two values are predictably opposites of each other: $t \approx \pm 0.7954$. Using the period of cosine (which is 2π), we get the complete solution set: $\{\pm 0.7954 + 2n\pi \mid n = 0, \pm 1, \pm 2, \pm 3, \ldots\}$.

Now try Exercise 63.

QUICK REVIEW 5.1 *(For help, go to Sections A.2, A.3, and 4.7.)*

In Exercises 1–4, evaluate the expression.

1. $\sin^{-1}\left(\dfrac{12}{13}\right)$

2. $\cos^{-1}\left(\dfrac{3}{5}\right)$

3. $\cos^{-1}\left(-\dfrac{4}{5}\right)$

4. $\sin^{-1}\left(-\dfrac{5}{13}\right)$

In Exercises 5–8, factor the expression into a product of linear factors.

5. $a^2 - 2ab + b^2$ **6.** $4u^2 + 4u + 1$

7. $2x^2 - 3xy - 2y^2$ **8.** $2v^2 - 5v - 3$

In Exercises 9–12, simplify the expression.

9. $\dfrac{1}{x} - \dfrac{2}{y}$ **10.** $\dfrac{a}{x} + \dfrac{b}{y}$

11. $\dfrac{x + y}{(1/x) + (1/y)}$ **12.** $\dfrac{x}{x - y} - \dfrac{y}{x + y}$

SECTION 5.1 EXERCISES

In Exercises 1–4, evaluate without using a calculator. Use the Pythagorean identities rather than reference triangles. (See Example 1.)

1. Find $\sin \theta$ and $\cos \theta$ if $\tan \theta = 3/4$ and $\sin \theta > 0$.

2. Find $\sec \theta$ and $\csc \theta$ if $\tan \theta = 3$ and $\cos \theta > 0$.

3. Find $\tan \theta$ and $\cot \theta$ if $\sec \theta = 4$ and $\sin \theta < 0$.

4. Find $\sin \theta$ and $\tan \theta$ if $\cos \theta = 0.8$ and $\tan \theta < 0$.

In Exercises 5–8, use identities to find the value of the expression.

5. If $\sin \theta = 0.45$, find $\cos (\pi/2 - \theta)$.

6. If $\tan(\pi/2 - \theta) = -5.32$, find $\cot \theta$.

7. If $\sin(\theta - \pi/2) = 0.73$, find $\cos (-\theta)$.

8. If $\cot(-\theta) = 7.89$, find $\tan (\theta - \pi/2)$.

In Exercises 9–16, use basic identities to simplify the expression.

9. $\tan x \cos x$ **10.** $\cot x \tan x$

11. $\sec y \sin (\pi/2 - y)$ **12.** $\cot u \sin u$

13. $\dfrac{1 + \tan^2 x}{\csc^2 x}$ **14.** $\dfrac{1 - \cos^2 \theta}{\sin \theta}$

15. $\cos x - \cos^3 x$ **16.** $\dfrac{\sin^2 u + \tan^2 u + \cos^2 u}{\sec u}$

In Exercises 17–22, simplify the expression to either 1 or -1.

17. $\sin x \csc (-x)$

18. $\sec (-x) \cos (-x)$

19. $\cot (-x) \cot (\pi/2 - x)$

cot $(-x)$ tan $(-x)$

21. $\sin^2(-x) + \cos^2(-x)$

22. $\sec^2(-x) - \tan^2 x$

In Exercises 23–26, simplify the expression to either a constant or a basic trigonometric function. Support your result graphically.

23. $\dfrac{\tan(\pi/2 - x)\csc x}{\csc^2 x}$

24. $\dfrac{1 + \tan x}{1 + \cot x}$

25. $(\sec^2 x + \csc^2 x) - (\tan^2 x + \cot^2 x)$

26. $\dfrac{\sec^2 u - \tan^2 u}{\cos^2 v + \sin^2 v}$

In Exercises 27–32, use the basic identities to change the expression to one involving only sines and cosines. Then simplify to a basic trigonometric function.

27. $(\sin x)(\tan x + \cot x)$

28. $\sin \theta - \tan \theta \cos \theta + \cos(\pi/2 - \theta)$

29. $\sin x \cos x \tan x \sec x \csc x$

30. $\dfrac{(\sec y - \tan y)(\sec y + \tan y)}{\sec y}$

31. $\dfrac{\tan x}{\csc^2 x} + \dfrac{\tan x}{\sec^2 x}$

32. $\dfrac{\sec^2 x \csc x}{\sec^2 x + \csc^2 x}$

In Exercises 33–38, combine the fractions and simplify to a multiple of a power of a basic trigonometric function (e.g., $3 \tan^2 x$).

33. $\dfrac{1}{\sin^2 x} + \dfrac{\sec^2 x}{\tan^2 x}$

34. $\dfrac{1}{1 - \sin x} + \dfrac{1}{1 + \sin x}$

35. $\dfrac{\sin x}{\cot^2 x} - \dfrac{\sin x}{\cos^2 x}$

36. $\dfrac{1}{\sec x - 1} - \dfrac{1}{\sec x + 1}$

37. $\dfrac{\sec x}{\sin x} - \dfrac{\sin x}{\cos x}$

38. $\dfrac{\sin x}{1 - \cos x} + \dfrac{1 - \cos x}{\sin x}$

In Exercises 39–46, write each expression in factored form as an algebraic expression of a single trigonometric function (e.g., $(2 \sin x + 3)(\sin x - 1)$).

39. $\cos^2 x + 2 \cos x + 1$

40. $1 - 2 \sin x + \sin^2 x$

41. $1 - 2 \sin x + (1 - \cos^2 x)$

42. $\sin x - \cos^2 x - 1$

43. $\cos x - 2 \sin^2 x + 1$

44. $\sin^2 x + \dfrac{2}{\csc x} + 1$

45. $4 \tan^2 x - \dfrac{4}{\cot x} + \sin x \csc x$

46. $\sec^2 x - \sec x + \tan^2 x$

In Exercises 47–50, write each expression as an algebraic expression of a single trigonometric function (e.g., $2 \sin x + 3$).

47. $\dfrac{1 - \sin^2 x}{1 + \sin x}$

48. $\dfrac{\tan^2 \alpha - 1}{1 + \tan \alpha}$

49. $\dfrac{\sin^2 x}{1 + \cos x}$

50. $\dfrac{\tan^2 x}{\sec x + 1}$

In Exercises 51–56, find all solutions to the equation in the interval $[0, 2\pi)$. You do not need a calculator.

51. $2 \cos x \sin x - \cos x = 0$

52. $\sqrt{2} \tan x \cos x - \tan x = 0$

53. $\tan x \sin^2 x = \tan x$

54. $\sin x \tan^2 x = \sin x$

55. $\tan^2 x = 3$

56. $2 \sin^2 x = 1$

In Exercises 57–62, find all solutions to the equation. You do not need a calculator.

57. $4 \cos^2 x - 4 \cos x + 1 = 0$

58. $2 \sin^2 x + 3 \sin x + 1 = 0$

59. $\sin^2 \theta - 2 \sin \theta = 0$

60. $3 \sin t = 2 \cos^2 t$

61. $\cos(\sin x) = 1$

62. $2 \sin^2 x + 3 \sin x = 2$

In Exercises 63–68, find all solutions to the trigonometric equation, using a calculator where needed.

63. $\cos x = 0.37$

64. $\cos x = 0.75$

65. $\sin x = 0.30$

66. $\tan x = 5$

67. $\cos^2 x = 0.4$

68. $\sin^2 x = 0.4$

In Exercises 69–74, make the suggested trigonometric substitution, and then use Pythagorean identities to write the resulting function as a multiple of a basic trigonometric function.

69. $\sqrt{1 - x^2}, \quad x = \cos \theta$

70. $\sqrt{x^2 + 1}, \quad x = \tan \theta$

71. $\sqrt{x^2 - 9}, \quad x = 3 \sec \theta$

72. $\sqrt{36 - x^2}, \quad x = 6 \sin \theta$

73. $\sqrt{x^2 + 81}, \quad x = 9 \tan \theta$

74. $\sqrt{x^2 - 100}, \quad x = 10 \sec \theta$

Standardized Test Questions

75. True or False If $\sec(x - \pi/2) = 34$, then $\csc x = 34$. Justify your answer.

76. True or False The domain of validity for the identity $\sin \theta = \tan \theta \cos \theta$ is the set of all real numbers. Justify your answer.

You should answer these questions without using a calculator.

77. Multiple Choice Which of the following could *not* be set equal to sin x as an identity?

(A) $\cos\left(\dfrac{\pi}{2} - x\right)$ **(B)** $\cos\left(x - \dfrac{\pi}{2}\right)$

(C) $\sqrt{1 - \cos^2 x}$ **(D)** $\tan x \sec x$

(E) $-\sin(-x)$

78. Multiple Choice Exactly four of the six basic trigonometric functions are

(A) odd.   **(B)** even.

(C) periodic.   **(D)** continuous.

(E) bounded.

79. Multiple Choice A simpler expression for $(\sec\theta + 1)(\sec\theta - 1)$ is

(A) $\sin^2\theta$.   **(B)** $\cos^2\theta$.

(C) $\tan^2\theta$.   **(D)** $\cot^2\theta$.

(E) $\sec^2\theta$.

80. Multiple Choice How many numbers between 0 and 2π solve the equation $3\cos^2 x + \cos x = 2$?

(A) none   **(B)** one

(C) two   **(D)** three

(E) four

Explorations

81. Write all six basic trigonometric functions entirely in terms of sin x.

82. Write all six basic trigonometric functions entirely in terms of cos x.

83. Writing to Learn Graph the functions $y = \sin^2 x$ and $y = -\cos^2 x$ in the standard trigonometric viewing window. Describe the apparent relationship between these two graphs and verify it with a trigonometric identity.

84. Writing to Learn Graph the functions $y = \sec^2 x$ and $y = \tan^2 x$ in the standard trigonometric viewing window. Describe the apparent relationship between these two graphs and verify it with a trigonometric identity.

85. Orbit of the Moon Because its orbit is elliptical, the distance from the Moon to the Earth in miles (measured from the center of the Moon to the center of the Earth) varies periodically. On Monday, January 18, 2002, the Moon was at its apogee (farthest from the earth). The distance of the Moon from the Earth each Friday from January 23 to March 27 are recorded in Table 5.1.

Table 5.1 Distance from Earth to Moon

Date	Day	Distance
Jan 23	0	251,966
Jan 30	7	238,344
Feb 6	14	225,784
Feb 13	21	240,385
Feb 20	28	251,807
Feb 27	35	236,315
Mar 6	42	226,101
Mar 13	49	242,390
Mar 20	56	251,333
Mar 27	63	234,347

Source: The World Almanac and Book of Facts, 2005.

(a) Draw a scatter plot of the data, using "day" as x and "distance" as y.

(b) Use your calculator to do a sine regression of y on x. Find the equation of the best-fit sine curve and superimpose its graph on the scatter plot.

(c) What is the approximate number of days from apogee to apogee? Interpret this number in terms of the orbit of the moon.

(d) Approximately how far is the Moon from the Earth at perigee (closest distance)?

(e) Since the data begin at apogee, perhaps a cosine curve would be a more appropriate model. Use the sine curve in part (b) and a cofunction identity to write a cosine curve that fits the data.

86. Group Activity Divide your class into six groups, each assigned to one of the basic trigonometric functions. With your group, construct a list of five different expressions that can be simplified to your assigned function. When you are done, exchange lists with your "cofunction" group to check one another for accuracy.

Extending the Ideas

87. Prove that $\sin^4\theta - \cos^4\theta = \sin^2\theta - \cos^2\theta$.

88. Find all values of k that result in $\sin^2 x + 1 = k\sin x$ having an infinite solution set.

89. Use the cofunction identities and odd-even identities to prove that $\sin(\pi - x) = \sin x$.
[*Hint*: $\sin(\pi - x) = \sin(\pi/2 - (x - \pi/2))$.]

90. Use the cofunction identities and odd-even identities to prove that $\cos(\pi - x) = -\cos x$.
[*Hint*: $\cos(\pi - x) = \cos(\pi/2 - (x - \pi/2))$.]

91. Use the identity in Exercise 89 to prove that in any $\triangle ABC$, $\sin(A + B) = \sin C$.

92. Use the identities in Exercises 89 and 90 to find an identity for simplifying $\tan(\pi - x)$.

Proving Trigonometric Identities

. . . and why

Proving identities gives you excellent insights into the way mathematical proofs are constructed.

A Proof Strategy

We now arrive at the best opportunity in the precalculus curriculum for you to try your hand at constructing analytic proofs: trigonometric identities. Some are easy and some can be quite challenging, but in every case the *identity itself* frames your work with a beginning and ending. The proof consists of filling in what lies between.

The strategy for proving an identity is very different from the strategy for solving an equation, most notably in the very first step. Usually the first step in solving an equation is to write down the equation. If you do this with an identity, however, you will have a beginning and an ending—with no proof in between! With an identity, you begin by writing down *one function* and end by writing down *the other*. Example 1 will illustrate what we mean.

EXAMPLE 1   Proving an Algebraic Identity

Prove the algebraic identity $\dfrac{x^2 - 1}{x - 1} - \dfrac{x^2 - 1}{x + 1} = 2$.

SOLUTION We prove this identity by showing a sequence of expressions, each one easily seen to be equivalent to its preceding expression:

$$\frac{x^2 - 1}{x - 1} - \frac{x^2 - 1}{x + 1} = \frac{(x + 1)(x - 1)}{x - 1} - \frac{(x + 1)(x - 1)}{x + 1} \qquad \text{Factoring difference of squares}$$

$$= (x + 1)\left(\frac{x - 1}{x - 1}\right) - (x - 1)\left(\frac{x + 1}{x + 1}\right) \qquad \text{Algebraic manipulation}$$

$$= (x + 1)(1) - (x - 1)(1) \qquad \text{Reducing fractions}$$

$$= x + 1 - x + 1 \qquad \text{Algebraic manipulation}$$

$$= 2$$

Notice that the first thing we wrote down was the expression on the left-hand side (LHS), and the last thing we wrote down was the expression on the right-hand side (RHS). The proof would have been just as legitimate going from RHS to LHS, but it is more natural to move from the more complicated side to the less complicated side. Incidentally, the margin notes on the right, called "floaters," are included here for instructional purposes and are not really part of the proof. A good proof should consist of steps for which a knowledgeable reader could readily supply the floaters.

Now try Exercise 1.

These, then, are our first general strategies for proving an identity:

General Strategies I

1. The proof begins with the expression on one side of the identity.

2. The proof ends with the expression on the other side.

3. The proof in between consists of showing a sequence of expressions, each one easily seen to be equivalent to its preceding expression.

Proving Identities

Trigonometric identity proofs follow General Strategies I. We are told that two expressions are equal, and the object is to prove that they are equal. We do this by changing one expression into the other by a series of intermediate steps that follow the important rule that *every intermediate step yields an expression that is equivalent to the first.*

The changes at every step are accomplished by algebraic manipulations or identities, but the manipulations or identities should be sufficiently obvious as to require no additional justification. Since "obvious" is often in the eye of the beholder, it is usually safer to err on the side of including too many steps than too few.

By working through several examples, we try to give you a sense for what is appropriate as we illustrate some of the algebraic tools that you have at your disposal.

EXAMPLE 2   Proving an Identity

Prove the identity: $\tan x + \cot x = \sec x \csc x$.

SOLUTION We begin by deciding whether to start with the expression on the right or the left. It is usually best to start with the more complicated expression, as it is easier to proceed from the complex toward the simple than to go in the other direction. The expression on the left is slightly more complicated because it involves two terms.

$$\tan x + \cot x = \frac{\sin x}{\cos x} + \frac{\cos x}{\sin x} \qquad \text{Basic identities}$$

$$= \frac{\sin x}{\cos x} \cdot \frac{\sin x}{\sin x} + \frac{\cos x}{\sin x} \cdot \frac{\cos x}{\cos x} \qquad \begin{array}{l}\text{Setting up common} \\ \text{denominator}\end{array}$$

$$= \frac{\sin^2 x + \cos^2 x}{\cos x \cdot \sin x}$$

$$= \frac{1}{\cos x \cdot \sin x} \qquad \text{Pythagorean identity}$$

$$\qquad\qquad\qquad\qquad \begin{array}{l}\text{(A step you could} \\ \text{choose to omit)}\end{array}$$

$$= \frac{1}{\cos x} \cdot \frac{1}{\sin x}$$

$$= \sec x \csc x \qquad \text{Basic identities}$$

(Remember that the "floaters" are not really part of the proof.)

Now try Exercise 13.

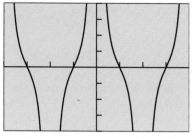

$[-2\pi, 2\pi]$ by $[-4, 4]$

$$f(x) = \frac{1}{\sec x - 1} + \frac{1}{\sec x + 1}$$

(a)

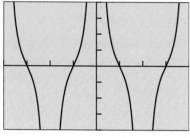

$[-2\pi, 2\pi]$ by $[-4, 4]$

$y = 2 \cot x \csc x$

(b)

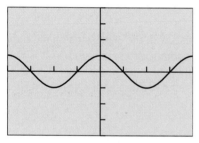

$[-2\pi, 2\pi]$ by $[-4, 4]$

$$y = \frac{1}{\sec x}$$

(c)

FIGURE 5.8 A grapher can be useful for identifying possible identities. (Example 3)

The preceding example illustrates three general strategies that are often useful in proving trigonometric identities.

General Strategies II

1. Begin with the more complicated expression and work toward the less complicated expression.

2. If no other move suggests itself, convert the entire expression to one involving sines and cosines.

3. Combine fractions by combining them over a common denominator.

EXAMPLE 3   Identifying and Proving an Identity

Match the function

$$f(x) = \frac{1}{\sec x - 1} + \frac{1}{\sec x + 1}$$

with one of the following. Then confirm the match with a proof.

(i) $2 \cot x \csc x$ **(ii)** $\dfrac{1}{\sec x}$

SOLUTION Figures 5.8a, b, and c show the graphs of the functions $y = f(x)$, $y = 2 \cot x \csc x$, and $y = 1/\sec x$, respectively. The graphs in (a) and (c) show that $f(x)$ is not equal to the expression in (ii). From the graphs in (a) and (b), it appears that $f(x)$ is equal to the expression in (i). To confirm, we begin with the expression for $f(x)$.

$$\frac{1}{\sec x - 1} + \frac{1}{\sec x + 1}$$

$$= \frac{\sec x + 1}{(\sec x - 1)(\sec x + 1)} + \frac{\sec x - 1}{(\sec x - 1)(\sec x + 1)} \qquad \text{Common denominator}$$

$$= \frac{\sec x + 1 + \sec x - 1}{\sec^2 x - 1}$$

$$= \frac{2 \sec x}{\tan^2 x} \qquad \text{Pythagorean identity}$$

$$= \frac{2}{\cos x} \cdot \frac{\cos^2 x}{\sin^2 x} \qquad \text{Basic identities}$$

$$= \frac{2 \cos x}{\sin x} \cdot \frac{1}{\sin x}$$

$$= 2 \cot x \csc x$$

Now try Exercise 55.

The next example illustrates how the algebraic identity $(a + b)(a - b) = a^2 - b^2$ can be used to set up a Pythagorean substitution.

EXAMPLE 4 Setting up a Difference of Squares

Prove the identity: $\cos t/(1 - \sin t) = (1 + \sin t)/\cos t$.

SOLUTION The left-hand expression is slightly more complicated, as we can handle extra terms in a numerator more easily than in a denominator. So we begin with the left.

$$\frac{\cos t}{1 - \sin t} = \frac{\cos t}{1 - \sin t} \cdot \frac{1 + \sin t}{1 + \sin t} \qquad \text{Setting up a difference of squares}$$

$$= \frac{(\cos t)(1 + \sin t)}{1 - \sin^2 t}$$

$$= \frac{(\cos t)(1 + \sin t)}{\cos^2 t} \qquad \text{Pythagorean identity}$$

$$= \frac{1 + \sin t}{\cos t}$$

Now try Exercise 39.

Notice that we kept $(\cos t)(1 + \sin t)$ in factored form in the hope that we could eventually eliminate the factor $\cos t$ and be left with the numerator we need. It is always a good idea to keep an eye on the "target" expression toward which your proof is aimed.

General Strategies III

1. Use the algebraic identity $(a + b)(a - b) = a^2 - b^2$ to set up applications of the Pythagorean identities.

2. Always be mindful of the "target" expression, and favor manipulations that bring you closer to your goal.

In more complicated identities (as in word ladders) it is sometimes helpful to see if both sides can be manipulated toward a common intermediate expression. The proof can then be reconstructed in a single path.

EXAMPLE 5 Working from Both Sides

Prove the identity: $\cot^2 u/(1 + \csc u) = (\cot u)(\sec u - \tan u)$.

SOLUTION Both sides are fairly complicated, but the left-hand side looks like it needs more work. We start on the left.

$$\frac{\cot^2 u}{1 + \csc u} = \frac{\csc^2 u - 1}{1 + \csc u} \qquad \text{Pythagorean identity}$$

$$= \frac{(\csc u + 1)(\csc u - 1)}{\csc u + 1} \qquad \text{Factor}$$

$$= \csc u - 1$$

At this point it is not clear how we can get from this expression to the one on the right-hand side of our identity. However, we now have reason to believe that the right-hand side must simplify to $\csc u - 1$, so we try simplifying the right-hand side.

continued

$$(\cot u)(\sec u - \tan u) = \left(\frac{\cos u}{\sin u}\right)\left(\frac{1}{\cos u} - \frac{\sin u}{\cos u}\right) \qquad \text{Basic identities}$$

$$= \frac{1}{\sin u} - 1 \qquad \text{Distribute the product}$$

$$= \csc u - 1$$

Now we can reconstruct the proof by going through $\csc u - 1$ as an intermediate step.

$$\frac{\cot^2 u}{1 + \csc u} = \frac{\csc^2 u - 1}{1 + \csc u}$$

$$= \frac{(\csc u + 1)(\csc u - 1)}{\csc u + 1}$$

$$= \csc u - 1 \qquad \text{Intermediate step}$$

$$= \frac{1}{\sin u} - 1$$

$$= \left(\frac{\cos u}{\sin u}\right)\left(\frac{1}{\cos u} - \frac{\sin u}{\cos u}\right)$$

$$= (\cot u)(\sec u - \tan u)$$

Now try Exercise 41.

Disproving Non-Identities

Obviously, not every equation involving trigonometric expressions is an identity. How can we spot a non-identity before embarking on a futile attempt at a proof? Try the following exploration.

EXPLORATION 1 Confirming a Non-Identity

Prove or disprove that this is an identity: $\cos 2x = 2 \cos x$.

1. Graph $y = \cos 2x$ and $y = 2 \cos x$ in the same window. Interpret the graphs to make a conclusion about whether or not the equation is an identity.

2. With the help of the graph, find a value of x for which $\cos 2x \neq 2 \cos x$.

3. Does the existence of the x value in part (2) *prove* that the equation is *not* an identity?

4. Graph $y = \cos 2x$ and $y = \cos^2 x - \sin^2 x$ in the same window. Interpret the graphs to make a conclusion about whether or not $\cos 2x = \cos^2 x - \sin^2 x$ is an identity.

5. Do the graphs in part (4) *prove* that $\cos 2x = \cos^2 x - \sin^2 x$ is an identity? Explain your answer.

Exploration 1 suggests that we can use graphers to help confirm a
only have to produce a single value of *x* for which the two com
defined but unequal. On the other hand, we cannot use graphers to
is an identity, since, for example, the graphers can never prove that two irrational numbers are equal. Also, graphers cannot show behavior over infinite domains.

Identities in Calculus

 In most calculus problems where identities play a role, the object is to make a complicated expression simpler for the sake of computational ease. Occasionally it is actually
necessary to make a *simple* expression *more complicated* for the sake of computational
ease. Each of the following identities (just a sampling of many) represents a useful substitution in calculus wherein the expression on the right is simpler to deal with (even
though it does not look that way). We prove one of these identities in Example 6 and
leave the rest for the exercises or for future sections.

1. $\cos^3 x = (1 - \sin^2 x)(\cos x)$

2. $\sec^4 x = (1 + \tan^2 x)(\sec^2 x)$

3. $\sin^2 x = \dfrac{1}{2} - \dfrac{1}{2}\cos 2x$

4. $\cos^2 x = \dfrac{1}{2} + \dfrac{1}{2}\cos 2x$

5. $\sin^5 x = (1 - 2\cos^2 x + \cos^4 x)(\sin x)$

6. $\sin^2 x \cos^5 x = (\sin^2 x - 2\sin^4 x + \sin^6 x)(\cos x)$

EXAMPLE 6 Proving an Identity Useful in Calculus

Prove the following identity:

$$\sin^2 x \cos^5 x = (\sin^2 x - 2\sin^4 x + \sin^6 x)(\cos x).$$

SOLUTION We begin with the expression on the left.

$$
\begin{aligned}
\sin^2 x \cos^5 x &= \sin^2 x \cos^4 x \cos x \\
&= (\sin^2 x)(\cos^2 x)^2(\cos x) \\
&= (\sin^2 x)(1 - \sin^2 x)^2(\cos x) \\
&= (\sin^2 x)(1 - 2\sin^2 x + \sin^4 x)(\cos x) \\
&= (\sin^2 x - 2\sin^4 x + \sin^6 x)(\cos x)
\end{aligned}
$$

Now try Exercise 51.

QUICK REVIEW 5.2   (For help, go to Section 5.1.)

In Exercises 1–6, write the expression in terms of sines and cosines only. Express your answer as a single fraction.

1. $\csc x + \sec x$

2. $\tan x + \cot x$

3. $\cos x \csc x + \sin x \sec x$

4. $\sin \theta \cot \theta - \cos \theta \tan \theta$

5. $\dfrac{\sin x}{\csc x} + \dfrac{\cos x}{\sec x}$

6. $\dfrac{\sec \alpha}{\cos \alpha} - \dfrac{\sin \alpha}{\csc \alpha \cos^2 \alpha}$

In Exercises 7–12, determine whether or not the equation is an identity. If not, find a single value of x for which the two expressions are not equal.

7. $\sqrt{x^2} = x$

8. $\sqrt[3]{x^3} = x$

9. $\sqrt{1 - \cos^2 x} = \sin x$

10. $\sqrt{\sec^2 x - 1} = \tan x$

11. $\ln \dfrac{1}{x} = -\ln x$

12. $\ln x^2 = 2 \ln x$

SECTION 5.2 EXERCISES

In Exercises 1–4, prove the algebraic identity by starting with the LHS expression and supplying a sequence of equivalent expressions that ends with the RHS expression.

1. $\dfrac{x^3 - x^2}{x} - (x - 1)(x + 1) = 1 - x$

2. $\dfrac{1}{x} - \dfrac{1}{2} = \dfrac{2 - x}{2x}$

3. $\dfrac{x^2 - 4}{x - 2} - \dfrac{x^2 - 9}{x + 3} = 5$

4. $(x - 1)(x + 2) - (x + 1)(x - 2) = 2x$

In Exercises 5–10, tell whether or not $f(x) = \sin x$ is an identity.

5. $f(x) = \dfrac{\sin^2 x + \cos^2 x}{\csc x}$

6. $f(x) = \dfrac{\tan x}{\sec x}$

7. $f(x) = \cos x \cdot \cot x$

8. $f(x) = \cos (x - \pi/2)$

9. $f(x) = (\sin^3 x)(1 + \cot^2 x)$

10. $f(x) = \dfrac{\sin 2x}{2}$

In Exercises 11–51, prove the identity.

11. $(\cos x)(\tan x + \sin x \cot x) = \sin x + \cos^2 x$

12. $(\sin x)(\cot x + \cos x \tan x) = \cos x + \sin^2 x$

13. $(1 - \tan x)^2 = \sec^2 x - 2 \tan x$

14. $(\cos x - \sin x)^2 = 1 - 2 \sin x \cos x$

15. $\dfrac{(1 - \cos u)(1 + \cos u)}{\cos^2 u} = \tan^2 u$

16. $\tan x + \sec x = \dfrac{\cos x}{1 - \sin x}$

17. $\dfrac{\cos^2 x - 1}{\cos x} = -\tan x \sin x$

18. $\dfrac{\sec^2 \theta - 1}{\sin \theta} = \dfrac{\sin \theta}{1 - \sin^2 \theta}$

19. $(1 - \sin \beta)(1 + \csc \beta) = 1 - \sin \beta + \csc \beta - \sin \beta \csc \beta$

20. $\dfrac{1}{1 - \cos x} + \dfrac{1}{1 + \cos x} = 2 \csc^2 x$

21. $(\cos t - \sin t)^2 + (\cos t + \sin t)^2 = 2$

22. $\sin^2 \alpha - \cos^2 \alpha = 1 - 2 \cos^2 \alpha$

23. $\dfrac{1 + \tan^2 x}{\sin^2 x + \cos^2 x} = \sec^2 x$

24. $\dfrac{1}{\tan \beta} + \tan \beta = \sec \beta \csc \beta$

25. $\dfrac{\cos \beta}{1 + \sin \beta} = \dfrac{1 - \sin \beta}{\cos \beta}$

26. $\dfrac{\sec x + 1}{\tan x} = \dfrac{\sin x}{1 - \cos x}$

27. $\dfrac{\tan^2 x}{\sec x + 1} = \dfrac{1 - \cos x}{\cos x}$

28. $\dfrac{\cot v - 1}{\cot v + 1} = \dfrac{1 - \tan v}{1 + \tan v}$

29. $\cot^2 x - \cos^2 x = \cos^2 x \cot^2 x$

30. $\tan^2 \theta - \sin^2 \theta = \tan^2 \theta \sin^2 \theta$

31. $\cos^4 x - \sin^4 x = \cos^2 x - \sin^2 x$

32. $\tan^4 t + \tan^2 t = \sec^4 t - \sec^2 t$

33. $(x \sin \alpha + y \cos \alpha)^2 + (x \cos \alpha - y \sin \alpha)^2 = x^2 + y^2$

34. $\dfrac{1 - \cos \theta}{\sin \theta} = \dfrac{\sin \theta}{1 + \cos \theta}$

35. $\dfrac{\tan x}{\sec x - 1} = \dfrac{\sec x + 1}{\tan x}$

36. $\dfrac{\sin t}{1 + \cos t} + \dfrac{1 + \cos t}{\sin t} = 2 \csc t$

37. $\dfrac{\sin x - \cos x}{\sin x + \cos x} = \dfrac{2 \sin^2 x - 1}{1 + 2 \sin x \cos x}$

38. $\dfrac{1 + \cos x}{1 - \cos x} = \dfrac{\sec x + 1}{\sec x - 1}$

39. $\dfrac{\sin t}{1 - \cos t} + \dfrac{1 + \cos t}{\sin t} = \dfrac{2(1 + \cos t)}{\sin t}$

40. $\dfrac{\sin A \cos B + \cos A \sin B}{\cos A \cos B - \sin A \sin B} = \dfrac{\tan A + \tan B}{1 - \tan A \tan B}$

41. $\sin^2 x \cos^3 x = (\sin^2 x - \sin^4 x)(\cos x)$

42. $\sin^5 x \cos^2 x = (\cos^2 x - 2 \cos^4 x + \cos^6 x)(\sin x)$

43. $\cos^5 x = (1 - 2 \sin^2 x + \sin^4 x)(\cos x)$

44. $\sin^3 x \cos^3 x = (\sin^3 x - \sin^5 x)(\cos x)$

45. $\dfrac{\tan x}{1 - \cot x} + \dfrac{\cot x}{1 - \tan x} = 1 + \sec x \csc x$

46. $\dfrac{\cos x}{1 + \sin x} + \dfrac{\cos x}{1 - \sin x} = 2 \sec x$

47. $\dfrac{2 \tan x}{1 - \tan^2 x} + \dfrac{1}{2 \cos^2 x - 1} = \dfrac{\cos x + \sin x}{\cos x - \sin x}$

48. $\dfrac{1 - 3 \cos x - 4 \cos^2 x}{\sin^2 x} = \dfrac{1 - 4 \cos x}{1 - \cos x}$

49. $\cos^3 x = (1 - \sin^2 x)(\cos x)$

50. $\sec^4 x = (1 + \tan^2 x)(\sec^2 x)$

51. $\sin^5 x = (1 - 2 \cos^2 x + \cos^4 x)(\sin x)$

In Exercises 52–57, match the function with an equivalent expression from the following list. Then confirm the match with a proof. (The matching is not one-to-one.)

(a) $\sec^2 x \csc^2 x$ **(b)** $\sec x + \tan x$ **(c)** $2 \sec^2 x$

(d) $\tan x \sin x$ **(e)** $\sin x \cos x$

52. $\dfrac{1 + \sin x}{\cos x}$

53. $(1 + \sec x)(1 - \cos x)$

54. $\sec^2 x + \csc^2 x$

55. $\dfrac{1}{1 + \sin x} + \dfrac{1}{1 - \sin x}$

56. $\dfrac{1}{\tan x + \cot x}$

57. $\dfrac{1}{\sec x - \tan x}$

Standardized Test Questions

58. True or False The equation $\sqrt{x^2} = x$ is an identity. Justify your answer.

59. True or False The equation $(\sqrt{x})^2 = x$ is an identity. Justify your answer.

You should answer these questions without using a calculator.

60. Multiple Choice If $f(x) = g(x)$ is an identity with domain of validity D, which of the following must be true?

I. For any x in D, $f(x)$ is defined.

II. For any x in D, $g(x)$ is defined.

III. For any x in D, $f(x) = g(x)$.

(A) None

(B) I and II only

(C) I and III only

(D) III only

(E) I, II, and III

61. Multiple Choice Which of these is an efficient first step in proving the identity $\dfrac{\sin x}{1 - \cos x} = \dfrac{1 + \cos x}{\sin x}$?

(A) $\dfrac{\sin x}{1 - \cos x} = \dfrac{\cos\left(\dfrac{\pi}{2} - x\right)}{1 - \cos x}$

(B) $\dfrac{\sin x}{1 - \cos x} = \dfrac{\sin x}{\sin^2 x + \cos^2 x - \cos x}$

(C) $\dfrac{\sin x}{1 - \cos x} = \dfrac{\sin x}{1 - \cos x} \cdot \dfrac{\csc x}{\csc x}$

(D) $\dfrac{\sin x}{1 - \cos x} = \dfrac{\sin x}{1 - \cos x} \cdot \dfrac{1 - \cos x}{1 - \cos x}$

(E) $\dfrac{\sin x}{1 - \cos x} = \dfrac{\sin x}{1 - \cos x} \cdot \dfrac{1 + \cos x}{1 + \cos x}$

62. Multiple Choice Which of the following could be an intermediate expression in a proof of the identity $\tan \theta + \sec \theta = \dfrac{\cos \theta}{1 - \sin \theta}$?

(A) $\sin \theta + \cos \theta$

(B) $\tan \theta + \csc \theta$

(C) $\dfrac{\sin \theta + 1}{\cos \theta}$

(D) $\dfrac{\cos \theta}{1 + \sin \theta}$

(E) $\cos \theta - \cot \theta$

63. Multiple Choice If $f(x) = g(x)$ is an identity and $\dfrac{f(x)}{g(x)} = k$, which of the following must be *false*?

(A) $g(x) \neq 0$

(B) $f(x) = 0$

(C) $k = 1$

(D) $f(x) - g(x) = 0$

(E) $f(x)g(x) > 0$

Explorations

In Exercises 64–69, identify a simple function that has the same graph. Then confirm your choice with a proof.

64. $\sin x \cot x$

65. $\cos x \tan x$

66. $\dfrac{\sin x}{\csc x} + \dfrac{\cos x}{\sec x}$

67. $\dfrac{\csc x}{\sin x} - \dfrac{\cot x \csc x}{\sec x}$

68. $\dfrac{\sin x}{\tan x}$

69. $(\sec^2 x)(1 - \sin^2 x)$

70. Writing to Learn Let θ be any number that is in the domain of all six trig functions. Explain why the natural logarithms of all six basic trig functions of θ sum to 0.

71. If A and B are complementary angles, prove that $\sin^2 A + \sin^2 B = 1$.

72. Group Activity If your class contains $2n$ students, write the two expressions from n different identities on separate pieces of paper. (If your class contains an odd number of students, invite your teacher to join you for this activity.) You can use the identities from Exercises 11–51 in this section or from other textbooks, but be sure to write them all in the variable x. Mix up the slips of paper and give one to each student in your class. Then see how long it takes you as a class, without looking at the book, to pair yourselves off as identities. (This activity takes on an added degree of difficulty if you try it without calculators.)

Extending the Ideas

In Exercises 73–78, confirm the identity.

73. $\sqrt{\dfrac{1 - \sin t}{1 + \sin t}} = \dfrac{1 - \sin t}{|\cos t|}$

74. $\sqrt{\dfrac{1 + \cos t}{1 - \cos t}} = \dfrac{1 + \cos t}{|\sin t|}$

75. $\sin^6 x + \cos^6 x = 1 - 3 \sin^2 x \cos^2 x$

76. $\cos^6 x - \sin^6 x = (\cos^2 x - \sin^2 x)(1 - \cos^2 x \sin^2 x)$

77. $\ln |\tan x| = \ln |\sin x| - \ln |\cos x|$

78. $\ln |\sec \theta + \tan \theta| + \ln |\sec \theta - \tan \theta| = 0$

79. Writing to Learn Let $y_1 = [\sin (x + 0.001) - \sin x]/0.001$ and $y_2 = \cos x$.

(a) Use graphs and tables to decide whether $y_1 = y_2$.

(b) Find a value for h so that the graph of $y_3 = y_1 - y_2$ in $[-2\pi, 2\pi]$ by $[-h, h]$ appears to be a sinusoid. Give a convincing argument that y_3 is a sinusoid.

80. Hyperbolic Functions The *hyperbolic trigonometric functions* are defined as follows:

$$\sinh x = \frac{e^x - e^{-x}}{2} \qquad \cosh x = \frac{e^x + e^{-x}}{2} \qquad \tanh x = \frac{\sinh x}{\cosh x}$$

$$\csch x = \frac{1}{\sinh x} \qquad \sech x = \frac{1}{\cosh x} \qquad \coth x = \frac{1}{\tanh x}$$

Confirm the identity.

(a) $\cosh^2 x - \sinh^2 x = 1$

(b) $1 - \tanh^2 x = \sech^2 x$

(c) $\coth^2 x - 1 = \csch^2 x$

81. Writing to Learn Write a paragraph to explain why

$$\cos x = \cos x + \sin (10\pi x)$$

appears to be an identity when the two sides are graphed in a decimal window. Give a convincing argument that it is not an identity.

5.3
Sum and Difference Identities

. . . and why

These identities provide clear examples of how different the algebra of functions can be from the algebra of real numbers.

Cosine of a Difference

There is a powerful instinct in all of us to believe that all functions obey the following law of additivity:

$$f(u + v) = f(u) + f(v).$$

In fact, very few do. If there were a Hall of Fame for algebraic blunders, the following would probably be the first two inductees:

$$(u + v)^2 = u^2 + v^2$$

$$\sqrt{u + v} = \sqrt{u} + \sqrt{v}$$

So, before we derive the true sum formulas for sine and cosine, let us clear the air with the following exploration.

EXPLORATION 1 Getting Past the Obvious but Incorrect Formulas

1. Let $u = \pi$ and $v = \pi/2$.
   Find $\sin(u + v)$. Find $\sin(u) + \sin(v)$.
   Does $\sin(u + v) = \sin(u) + \sin(v)$?

2. Let $u = 0$ and $v = 2\pi$.
   Find $\cos(u + v)$. Find $\cos(u) + \cos(v)$.
   Does $\cos(u + v) = \cos(u) + \cos(v)$?

3. Find your own values of u and v that will confirm that $\tan(u + v) \neq \tan(u) + \tan(v)$.

We could also show easily that

$$\cos(u - v) \neq \cos(u) - \cos(v) \quad \text{and} \quad \sin(u - v) \neq \sin(u) - \sin(v).$$

As you might expect, there *are* formulas for $\sin(u \pm v)$, $\cos(u \pm v)$, and $\tan(u \pm v)$, but Exploration 1 shows that they are not the ones our instincts would suggest. In a sense, that makes them all the more interesting. We will derive them all, beginning with the formula for $\cos(u - v)$.

Figure 5.9a on the next page shows angles u and v in standard position on the unit circle, determining points A and B with coordinates $(\cos u, \sin u)$ and $(\cos v, \sin v)$, respectively. Figure 5.9b shows the triangle ABO rotated so that the angle $\theta = u - v$ is in standard position. The angle θ determines point C with coordinates $(\cos \theta, \sin \theta)$.

The chord opposite angle θ has the same length in both circles, even though the coordinatization of the endpoints is different. Use the distance formula to find the length in each case, and set the formulas equal to each other:

$$AB = CD$$

$$\sqrt{(\cos v - \cos u)^2 + (\sin v - \sin u)^2} = \sqrt{(\cos \theta - 1)^2 + (\sin \theta - 0)^2}$$

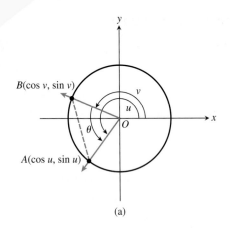

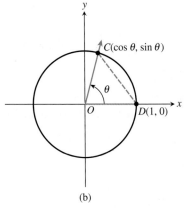

FIGURE 5.9 Angles u and v are in standard position in (a), while angle $\theta = u - v$ is in standard position in (b). The chords shown in the two circles are equal in length.

Square both sides to eliminate the radical and expand the binomials to get

$$\cos^2 u - 2 \cos u \cos v + \cos^2 v + \sin^2 u - 2 \sin u \sin v + \sin^2 v$$
$$= \cos^2 \theta - 2 \cos \theta + 1 + \sin^2 \theta$$

$$(\cos^2 u + \sin^2 u) + (\cos^2 v + \sin^2 v) - 2 \cos u \cos v - 2 \sin u \sin v$$
$$= (\cos^2 \theta + \sin^2 \theta) + 1 - 2 \cos \theta$$

$$2 - 2 \cos u \cos v - 2 \sin u \sin v = 2 - 2 \cos \theta$$

$$\cos u \cos v + \sin u \sin v = \cos \theta$$

Finally, since $\theta = u - v$, we can write

$$\cos (u - v) = \cos u \cos v + \sin u \sin v.$$

EXAMPLE 1 Using the Cosine-of-a-Difference Identity

Find the exact value of cos 15° without using a calculator.

SOLUTION The trick is to write cos 15° as cos (45° − 30°); then we can use our knowledge of the special angles.

$$\cos 15° = \cos (45° - 30°)$$
$$= \cos 45° \cos 30° + \sin 45° \sin 30° \qquad \text{Cosine difference identity}$$
$$= \left(\frac{\sqrt{2}}{2}\right)\left(\frac{\sqrt{3}}{2}\right) + \left(\frac{\sqrt{2}}{2}\right)\left(\frac{1}{2}\right)$$
$$= \frac{\sqrt{6} + \sqrt{2}}{4} \qquad\qquad\qquad \textbf{\textit{Now try Exercise 5.}}$$

Cosine of a Sum

Now that we have the formula for the cosine of a difference, we can get the formula for the cosine of a sum almost for free by using the odd-even identities.

$$\cos (u + v) = \cos (u - (-v))$$
$$= \cos u \cos (-v) + \sin u \sin (-v) \qquad \text{Cosine difference identity}$$
$$= \cos u \cos v + \sin u(-\sin v) \qquad \text{Odd-even identities}$$
$$= \cos u \cos v - \sin u \sin v$$

We can combine the sum and difference formulas for cosine as follows:

Cosine of a Sum or Difference

$$\cos (u \pm v) = \cos u \cos v \mp \sin u \sin v$$

(Note the sign switch in either case.)

We pointed out in Section 5.1 that the Cofunction Identities would be easier to prove with the results of Section 5.3. Here is what we mean.

EXAMPLE 2   Confirming Cofunction Identities

Prove the identities **(a)** $\cos\left((\pi/2) - x\right) = \sin x$ and
(b) $\sin\left((\pi/2) - x\right) = \cos x$.

SOLUTION

(a) $\cos\left(\dfrac{\pi}{2} - x\right) = \cos\left(\dfrac{\pi}{2}\right)\cos x + \sin\left(\dfrac{\pi}{2}\right)\sin x$ Cosine sum identity

$\phantom{\cos\left(\dfrac{\pi}{2} - x\right)} = 0 \cdot \cos x + 1 \cdot \sin x$

$\phantom{\cos\left(\dfrac{\pi}{2} - x\right)} = \sin x$

(b) $\sin\left(\dfrac{\pi}{2} - x\right) = \cos\left(\dfrac{\pi}{2} - \left(\dfrac{\pi}{2} - x\right)\right)$ $\sin\theta = \cos\left((\pi/2) - \theta\right)$ by previous proof

$\phantom{\sin\left(\dfrac{\pi}{2} - x\right)} = \cos\left(0 + x\right)$

$\phantom{\sin\left(\dfrac{\pi}{2} - x\right)} = \cos x$

Now try Exercise 41.

Sine of a Difference or Sum

We can use the cofunction identities in Example 2 to get the formula for the sine of a sum from the formula for the cosine of a difference.

$\sin\left(u + v\right) = \cos\left(\dfrac{\pi}{2} - \left(u + v\right)\right)$ Cofunction identity

$ = \cos\left(\left(\dfrac{\pi}{2} - u\right) - v\right)$ A little algebra

$ = \cos\left(\dfrac{\pi}{2} - u\right)\cos v + \sin\left(\dfrac{\pi}{2} - u\right)\sin v$ Cosine difference identity

$ = \sin u \cos v + \cos u \sin v$ Cofunction identities

Then we can use the odd-even identities to get the formula for the sine of a difference from the formula for the sine of a sum.

$\sin\left(u - v\right) = \sin\left(u + \left(-v\right)\right)$ A little algebra

$ = \sin u \cos\left(-v\right) + \cos u \sin\left(-v\right)$ Sine sum identity

$ = \sin u \cos v + \cos u \left(-\sin v\right)$ Odd-even identities

$ = \sin u \cos v - \cos u \sin v$

We can combine the sum and difference formulas for sine as follows:

Sine of a Sum or Difference

$$\sin\left(u \pm v\right) = \sin u \cos v \pm \cos u \sin v$$

(Note that the sign does *not* switch in either case.)

EXAMPLE 3   Using the Sum/Difference Formulas

Write each of the following expressions as the sine or cosine of an angle.

(a) $\sin 22° \cos 13° + \cos 22° \sin 13°$

(b) $\cos \dfrac{\pi}{3} \cos \dfrac{\pi}{4} + \sin \dfrac{\pi}{3} \sin \dfrac{\pi}{4}$

(c) $\sin x \sin 2x - \cos x \cos 2x$

SOLUTION The key in each case is recognizing which formula applies. (Indeed, the real purpose of such exercises is to help you remember the formulas.)

(a) $\sin 22° \cos 13° + \cos 22° \sin 13°$ Recognizing sine of sum formula

$\qquad = \sin (22° + 13°)$

$\qquad = \sin 35°$

(b) $\cos \dfrac{\pi}{3} \cos \dfrac{\pi}{4} + \sin \dfrac{\pi}{3} \sin \dfrac{\pi}{4}$ Recognizing cosine of difference formula

$\qquad = \cos \left(\dfrac{\pi}{3} - \dfrac{\pi}{4} \right)$

$\qquad = \cos \dfrac{\pi}{12}$

(c) $\sin x \sin 2x - \cos x \cos 2x$ Recognizing *opposite* of cos sum formula

$\qquad = -(\cos x \cos 2x - \sin x \sin 2x)$

$\qquad = -\cos (x + 2x)$ Applying formula

$\qquad = -\cos 3x$ ***Now try Exercise 19.***

If one of the angles in a sum or difference is a quadrantal angle (that is, a multiple of 90° or of $\pi/2$ radians), then the sum-difference identities yield single-termed expressions. Since the effect is to reduce the complexity, the resulting identity is called a **reduction formula**.

EXAMPLE 4   Proving Reduction Formulas

Prove the reduction formulas:

(a) $\sin (x + \pi) = -\sin x$

(b) $\cos \left(x + \dfrac{3\pi}{2} \right) = \sin x$

SOLUTION

(a) $\sin (x + \pi) = \sin x \cos \pi + \cos x \sin \pi$

$\qquad\qquad\qquad = \sin x \cdot (-1) + \cos x \cdot 0$

$\qquad\qquad\qquad = -\sin x$

(b) $\cos \left(x + \dfrac{3\pi}{2} \right) = \cos x \cos \dfrac{3\pi}{2} - \sin x \sin \dfrac{3\pi}{2}$

$\qquad\qquad\qquad = \cos x \cdot 0 - \sin x \cdot (-1)$

$\qquad\qquad\qquad = \sin x$ ***Now try Exercise 23.***

Tangent of a Difference or Sum

We can derive a formula for tan $(u \pm v)$ directly from the corresponding formulas for sine and cosine, as follows:

$$\tan (u \pm v) = \frac{\sin (u \pm v)}{\cos (u \pm v)} = \frac{\sin u \cos v \pm \cos u \sin v}{\cos u \cos v \mp \sin u \sin v}.$$

There is also a formula for tan $(u \pm v)$ that is written entirely in terms of tangent functions:

$$\tan (u \pm v) = \frac{\tan u \pm \tan v}{1 \mp \tan u \tan v}$$

We will leave the proof of the all-tangent formula to the Exercises.

EXAMPLE 5 Proving a Tangent Reduction Formula

Prove the reduction formula: tan $(\theta - (3\pi/2)) = -\cot \theta$.

SOLUTION We can't use the all-tangent formula (Do you see why?), so we convert to sines and cosines.

$$\tan \left(\theta - \frac{3\pi}{2} \right) = \frac{\sin (\theta - (3\pi/2))}{\cos (\theta - (3\pi/2))}$$

$$= \frac{\sin \theta \cos (3\pi/2) - \cos \theta \sin (3\pi/2)}{\cos \theta \cos (3\pi/2) + \sin \theta \sin (3\pi/2)}$$

$$= \frac{\sin \theta \cdot 0 - \cos \theta \cdot (-1)}{\cos \theta \cdot 0 + \sin \theta \cdot (-1)}$$

$$= -\cot \theta \qquad \qquad \textit{Now try Exercise 39.}$$

Verifying a Sinusoid Algebraically

Example 7 of Section 4.6 asked us to verify that the function $f(x) = 2 \sin x + 5 \cos x$ is a sinusoid. We solved graphically, concluding that $f(x) \approx 5.39 \sin (x + 1.19)$. We now have a way of solving this kind of problem algebraically, with exact values for the amplitude and phase shift. Example 6 illustrates the technique.

EXAMPLE 6 Expressing a Sum of Sinusoids as a Sinusoid

Express $f(x) = 2 \sin x + 5 \cos x$ as a sinusoid in the form $f(x) = a \sin (bx + c)$.

SOLUTION Since $a \sin (bx + c) = a (\sin bx \cos c + \cos bx \sin c)$, we have

$$2 \sin x + 5 \cos x = a (\sin bx \cos c + \cos bx \sin c)$$

$$= (a \cos c) \sin bx + (a \sin c) \cos bx.$$

Comparing coefficients, we see that $b = 1$ and that $a \cos c = 2$ and $a \sin c = 5$.

continued

We can solve for a as follows:

$$(a \cos c)^2 + (a \sin c)^2 = 2^2 + 5^2$$

$$a^2 \cos^2 c + a^2 \sin^2 c = 29$$

$$a^2(\cos^2 c + \sin^2 c) = 29$$

$$a^2 = 29 \qquad \text{Pythogorean identity}$$

$$a = \pm\sqrt{29}$$

If we choose a to be positive, then $\cos c = 2/\sqrt{29}$ and $\sin c = 5/\sqrt{29}$. We can identify an acute angle c with those specifications as either $\cos^{-1}(2/\sqrt{29})$ or $\sin^{-1}(5/\sqrt{29})$, which are equal. So, an exact sinusoid for f is

$$f(x) = 2 \sin x + 5 \cos x$$

$$= a \sin(bx + c)$$

$$= \sqrt{29} \sin(x + \cos^{-1}(2/\sqrt{29})) \text{ or } \sqrt{29} \sin(x + \sin^{-1}(5/\sqrt{29}))$$

Now try Exercise 43.

QUICK REVIEW 5.3   *(For help, go to Sections 4.2 and 5.1.)*

In Exercises 1–6, express the angle as a sum or difference of special angles (multiples of 30°, 45°, π/6, or π/4). Answers are not unique.

1. 15°

2. 75°

3. 165°

4. $\pi/12$

5. $5\pi/12$

6. $7\pi/12$

In Exercises 7–10, tell whether or not the identity $f(x + y) = f(x) + f(y)$ holds for the function f.

7. $f(x) = \ln x$

8. $f(x) = e^x$

9. $f(x) = 32x$

10. $f(x) = x + 10$

SECTION 5.3 EXERCISES

In Exercises 1–10, use a sum or difference identity to find an exact value.

1. $\sin 15°$

2. $\tan 15°$

3. $\sin 75°$

4. $\cos 75°$

5. $\cos \dfrac{\pi}{12}$

6. $\sin \dfrac{7\pi}{12}$

7. $\tan \dfrac{5\pi}{12}$

8. $\tan \dfrac{11\pi}{12}$

9. $\cos \dfrac{7\pi}{12}$

10. $\sin \dfrac{-\pi}{12}$

In Exercises 11–22, write the expression as the sine, cosine, or tangent of an angle.

11. $\sin 42° \cos 17° - \cos 42° \sin 17°$

12. $\cos 94° \cos 18° + \sin 94° \sin 18°$

13. $\sin \dfrac{\pi}{5} \cos \dfrac{\pi}{2} + \sin \dfrac{\pi}{2} \cos \dfrac{\pi}{5}$

14. $\sin \dfrac{\pi}{3} \cos \dfrac{\pi}{7} - \sin \dfrac{\pi}{7} \cos \dfrac{\pi}{3}$

15. $\dfrac{\tan 19° + \tan 47°}{1 - \tan 19° \tan 47°}$

16. $\dfrac{\tan(\pi/5) - \tan(\pi/3)}{1 + \tan(\pi/5) \tan(\pi/3)}$

17. $\cos \dfrac{\pi}{7} \cos x + \sin \dfrac{\pi}{7} \sin x$ **18.** $\cos x \cos \dfrac{\pi}{7} - \sin x \sin \dfrac{\pi}{7}$

19. $\sin 3x \cos x - \cos 3x \sin x$

20. $\cos 7y \cos 3y - \sin 7y \sin 3y$

21. $\dfrac{\tan 2y + \tan 3x}{1 - \tan 2y \tan 3x}$

22. $\dfrac{\tan 3\alpha - \tan 2\beta}{1 + \tan 3\alpha \tan 2\beta}$

In Exercises 23–30, prove the identity.

23. $\sin\left(x - \dfrac{\pi}{2}\right) = -\cos x$ **24.** $\tan\left(x - \dfrac{\pi}{2}\right) = -\cot x$

25. $\cos\left(x - \dfrac{\pi}{2}\right) = \sin x$

26. $\cos\left[\left(\dfrac{\pi}{2} - x\right) - y\right] = \sin (x + y)$

27. $\sin\left(x + \dfrac{\pi}{6}\right) = \dfrac{\sqrt{3}}{2}\sin x + \dfrac{1}{2}\cos x$

28. $\cos\left(x - \dfrac{\pi}{4}\right) = \dfrac{\sqrt{2}}{2}(\cos x + \sin x)$

29. $\tan\left(\theta + \dfrac{\pi}{4}\right) = \dfrac{1 + \tan \theta}{1 - \tan \theta}$

30. $\cos\left(\theta + \dfrac{\pi}{2}\right) = -\sin \theta$

In Exercises 31–34, match each graph with a pair of the following equations. Use your knowledge of identities and transformations, not your grapher.

(a) $y = \cos (3 - 2x)$

(b) $y = \sin x \cos 1 + \cos x \sin 1$

(c) $y = \cos (x - 3)$

(d) $y = \sin (2x - 5)$

(e) $y = \cos x \cos 3 + \sin x \sin 3$

(f) $y = \sin (x + 1)$

(g) $y = \cos 3 \cos 2x + \sin 3 \sin 2x$

(h) $y = \sin 2x \cos 5 - \cos 2x \sin 5$

31.   **32.**

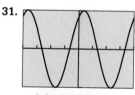 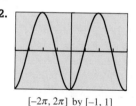

$[-2\pi, 2\pi]$ by $[-1, 1]$ $[-2\pi, 2\pi]$ by $[-1, 1]$

33.   **34.**

$[-2\pi, 2\pi]$ by $[-1, 1]$ $[-2\pi, 2\pi]$ by $[-1, 1]$

In Exercises 35 and 36, use sum or difference identities (and not your grapher) to solve the equation exactly.

35. $\sin 2x \cos x = \cos 2x \sin x$ **36.** $\cos 3x \cos x = \sin 3x \sin x$

In Exercises 37–42, prove the reduction formula.

37. $\sin\left(\dfrac{\pi}{2} - u\right) = \cos u$ **38.** $\tan\left(\dfrac{\pi}{2} - u\right) = \cot u$

39. $\cot\left(\dfrac{\pi}{2} - u\right) = \tan u$ **40.** $\sec\left(\dfrac{\pi}{2} - u\right) = \csc u$

41. $\csc\left(\dfrac{\pi}{2} - u\right) = \sec u$ **42.** $\cos\left(x + \dfrac{\pi}{2}\right) = -\sin x$

In Exercises 43–46, express the function as a sinusoid in the form $y = a \sin (bx + c)$.

43. $y = 3 \sin x + 4 \cos x$ **44.** $y = 5 \sin x - 12 \cos x$

45. $y = \cos 3x + 2 \sin 3x$ **46.** $y = 3 \cos 2x - 2 \sin 2x$

In Exercises 47–55, prove the identity.

47. $\sin (x - y) + \sin (x + y) = 2 \sin x \cos y$

48. $\cos (x - y) + \cos (x + y) = 2 \cos x \cos y$

49. $\cos 3x = \cos^3 x - 3 \sin^2 x \cos x$

50. $\sin 3u = 3 \cos^2 u \sin u - \sin^3 u$

51. $\cos 3x + \cos x = 2 \cos 2x \cos x$

52. $\sin 4x + \sin 2x = 2 \sin 3x \cos x$

53. $\tan (x + y) \tan (x - y) = \dfrac{\tan^2 x - \tan^2 y}{1 - \tan^2 x \tan^2 y}$

54. $\tan 5u \tan 3u = \dfrac{\tan^2 4u - \tan^2 u}{1 - \tan^2 4u \tan^2 u}$

55. $\dfrac{\sin (x + y)}{\sin (x - y)} = \dfrac{(\tan x + \tan y)}{(\tan x - \tan y)}$

Standardized Test Questions

56. True or False If A and B are supplementary angles, then $\cos A + \cos B = 0$. Justify your answer.

57. True or False If $\cos A + \cos B = 0$, then A and B are supplementary angles. Justify your answer.

You should answer these questions without using a calculator.

58. Multiple Choice If $\cos A \cos B = \sin A \sin B$, then $\cos (A + B) =$

(A) 0 **(B)** 1

(C) $\cos A + \cos B$ **(D)** $\cos B + \cos A$

(E) $\cos A \cos B + \sin A \sin B$

59. Multiple Choice The function $y = \sin x \cos 2x + \cos x \sin 2x$ has amplitude

(A) 1   (B) 1.5   (C) 2   (D) 3   (E) 6

60. Multiple Choice $\sin 15° =$

(A) $\dfrac{1}{4}$

(B) $\dfrac{\sqrt{3}}{4}$

(C) $\dfrac{\sqrt{3} + \sqrt{2}}{4}$

(D) $\dfrac{\sqrt{6} - \sqrt{2}}{4}$

(E) $\dfrac{\sqrt{6} + \sqrt{2}}{4}$

61. Multiple Choice A function with the property

$$f(1 + 2) = \frac{f(1) + f(2)}{1 - f(1)f(2)}$$ is

(A) $f(x) = \sin x$ (B) $f(x) = \tan x$

(C) $f(x) = \sec x$ (D) $f(x) = e^x$

(E) $f(x) = -1$

Explorations

62. Prove the identity $\tan (u + v) = \dfrac{\tan u + \tan v}{1 - \tan u \tan v}$.

63. Prove the identity $\tan (u - v) = \dfrac{\tan u - \tan v}{1 + \tan u \tan v}$.

64. Writing to Learn Explain why the identity in Exercise 62 cannot be used to prove the reduction formula $\tan (x + \pi/2) = -\cot x$. Then prove the reduction formula.

65. Writing to Learn Explain why the identity in Exercise 63 cannot be used to prove the reduction formula $\tan (x - 3\pi/2) = -\cot x$. Then prove the reduction formula.

66. An Identity for Calculus Prove the following identity, which is used in calculus to prove an important differentiation formula.

$$\frac{\sin (x + h) - \sin x}{h} = \sin x \left(\frac{\cos h - 1}{h} \right) + \cos x \frac{\sin h}{h}.$$

67. An Identity for Calculus Prove the following identity, which is used in calculus to prove another important differentiation formula.

$$\frac{\cos (x + h) - \cos x}{h} = \cos x \left(\frac{\cos h - 1}{h} \right) - \sin x \frac{\sin h}{h}.$$

68. Group Activity Place 24 points evenly spaced around the unit circle, starting with the point (1, 0). Using only your knowledge of the special angles and the sum and difference identities, work with your group to find the exact coordinates of all 24 points.

Extending the Ideas

In Exercises 69–72, assume that A, B, and C are the three angles of some $\triangle ABC$. (Note, then, that $A + B + C = \pi$.) Prove the following identities.

69. $\sin (A + B) = \sin C$

70. $\cos C = \sin A \sin B - \cos A \cos B$

71. $\tan A + \tan B + \tan C = \tan A \tan B \tan C$

72. $\cos A \cos B \cos C - \sin A \sin B \cos C - \sin A \cos B \sin C - \cos A \sin B \sin C = -1$

73. Writing to Learn The figure shows graphs of $y_1 = \cos 5x \cos 4x$ and $y_2 = -\sin 5x \sin 4x$ in one viewing window. Discuss the question, "How many solutions are there to the equation $\cos 5x \cos 4x = -\sin 5x \sin 4x$ in the interval $[-2\pi, 2\pi]$?" Give an algebraic argument that answers the question more convincingly than the graph does. Then support your argument with an *appropriate* graph.

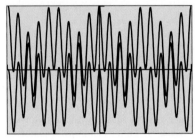

$[-2\pi, 2\pi]$ by $[-1, 1]$

74. Harmonic Motion Alternating electric current, an oscillating spring, or any other harmonic oscillator can be modeled by the equation

$$x = a \cos \left(\frac{2\pi}{T} t + \delta \right),$$

where T is the time for one period and δ is the phase constant. Show that this motion can also be modeled by the following sum of cosine and sine, each with zero phase constant:

$$a_1 \cos \left(\frac{2\pi}{T} \right) t + a_2 \sin \left(\frac{2\pi}{T} \right) t,$$

where $a_1 = a \cos \delta$ and $a_2 = -a \sin \delta$.

75. Magnetic Fields A magnetic field B can sometimes be modeled as the sum of an incident and a reflective field as

$$B = B_{in} + B_{ref},$$

where $\quad B_{in} = \dfrac{E_0}{c} \cos \left(\omega t - \dfrac{\omega x}{c} \right)$, and

$$B_{ref} = \frac{E_0}{c} \cos \left(\omega t + \frac{\omega x}{c} \right).$$

Show that $B = 2 \dfrac{E_0}{c} \cos \omega t \cos \dfrac{\omega x}{c}$.

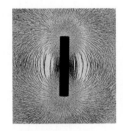

5.4
Multiple-Angle Identities

What you'll learn about

- Double-Angle Identities
- Power-Reducing Identities
- Half-Angle Identities
- Solving Trigonometric Equations

. . . and why

These identities are useful in calculus courses.

Double-Angle Identities

The formulas that result from letting $u = v$ in the angle sum identities are called the *double-angle identities*. We will state them all and prove one, leaving the rest of the proofs as exercises. (See Exercises 1–4.)

Double-Angle Identities

$$\sin 2u = 2 \sin u \cos u$$

$$\cos 2u = \begin{cases} \cos^2 u - \sin^2 u \\ 2 \cos^2 u - 1 \\ 1 - 2 \sin^2 u \end{cases}$$

$$\tan 2u = \frac{2 \tan u}{1 - \tan^2 u}$$

There are three identities for $\cos 2u$. This is not unusual; indeed, there are plenty of other identities one could supply for $\sin 2u$ as well, such as $2 \sin u \sin (\pi/2 - u)$. We list the three identities for $\cos 2u$ because they are all *useful* in various contexts and therefore worth memorizing.

EXAMPLE 1   Proving a Double-Angle Identity

Prove the identity: $\sin 2u = 2 \sin u \cos u$.

SOLUTION

$$\sin 2u = \sin (u + u)$$

$$= \sin u \cos u + \cos u \sin u \qquad \text{Sine of a sum } (v = u)$$

$$= 2 \sin u \cos u$$

Now try Exercise 1.

Power-Reducing Identities

One immediate use for two of the three formulas for $\cos 2u$ is to derive the *Power-Reducing Identities*. Some simple-looking functions like $y = \sin^2 u$ would be quite difficult to handle in certain calculus contexts if it were not for the existence of these identities.

Power-Reducing Identities

$$\sin^2 u = \frac{1 - \cos 2u}{2}$$

$$\cos^2 u = \frac{1 + \cos 2u}{2}$$

$$\tan^2 u = \frac{1 - \cos 2u}{1 + \cos 2u}$$

We will also leave the proofs of these identities as Exercises. (See Exercises 37 and 38.)

EXAMPLE 2 Proving an Identity

Prove the identity: $\cos^4 \theta - \sin^4 \theta = \cos 2\theta$.

SOLUTION

$$\cos^4 \theta - \sin^4 \theta = (\cos^2 \theta + \sin^2 \theta)(\cos^2 \theta - \sin^2 \theta)$$

$$= 1 \cdot (\cos^2 \theta - \sin^2 \theta) \qquad \text{Pythagorean identity}$$

$$= \cos 2\theta \qquad \text{Double-angle identity}$$

Now try Exercise 15.

EXAMPLE 3 Reducing a Power of 4

Rewrite $\cos^4 x$ in terms of trigonometric functions with no power greater than 1.

SOLUTION

$$\cos^4 x = (\cos^2 x)^2$$

$$= \left(\frac{1 + \cos 2x}{2}\right)^2 \qquad \text{Power-reducing identity}$$

$$= \left(\frac{1 + 2\cos 2x + \cos^2 2x}{4}\right)$$

$$= \frac{1}{4} + \frac{1}{2}\cos 2x + \frac{1}{4}\left(\frac{1 + \cos 4x}{2}\right) \qquad \text{Power-reducing identity}$$

$$= \frac{1}{4} + \frac{1}{2}\cos 2x + \frac{1}{8} + \frac{1}{8}\cos 4x$$

$$= \frac{1}{8}(3 + 4\cos 2x + \cos 4x)$$

Now try Exercise 39.

Half-Angle Identities

The power-reducing identities can be used to extend our stock of "special" angles whose trigonometric ratios can be found without a calculator. As usual, we are not suggesting that this algebraic procedure is any more practical than using a calculator, but we are suggesting that this sort of exercise helps you to understand how the functions behave. In Exploration 1, for example, we use a power-reduction formula to find the exact values of $\sin(\pi/8)$ and $\sin(9\pi/8)$ without a calculator.

EXPLORATION 1 Finding the Sine of Half an Angle

Recall the power-reducing formula $\sin^2 u = (1 - \cos 2u)/2$.

1. Use the power-reducing formula to show that $\sin^2(\pi/8) = (2 - \sqrt{2})/4$.

2. Solve for $\sin(\pi/8)$. Do you take the positive or negative square root? Why?

3. Use the power-reducing formula to show that $\sin^2(9\pi/8) = (2 - \sqrt{2})/4$.

4. Solve for $\sin(9\pi/8)$. Do you take the positive or negative square root? Why?

A little alteration of the power-reducing identities results in the *Half-Angle Identities*, which can be used directly to find trigonometric functions of $u/2$ in terms of trigonometric functions of u. As Exploration 1 suggests, there is an unavoidable ambiguity of sign involved with the square root that must be resolved in particular cases by checking the quadrant in which $u/2$ lies.

Half-Angle Identities

$$\sin\frac{u}{2} = \pm\sqrt{\frac{1 - \cos u}{2}}$$

$$\cos\frac{u}{2} = \pm\sqrt{\frac{1 + \cos u}{2}} \qquad \tan\frac{u}{2} = \begin{cases} \pm\sqrt{\dfrac{1 - \cos u}{1 + \cos u}} \\[2ex] \dfrac{1 - \cos u}{\sin u} \\[2ex] \dfrac{\sin u}{1 + \cos u} \end{cases}$$

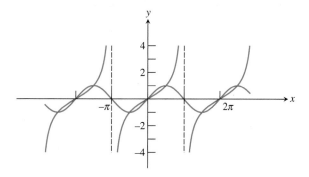

FIGURE 5.10 The functions $\sin u$ and $\tan(u/2)$ always have the same sign.

Solving Trigonometric Equations

New identities always provide new tools for solving trigonometric equations algebraically. Under the right conditions, they even lead to exact solutions. We assert again that we are not presenting these algebraic solutions for their practical value (as the calculator solutions are certainly sufficient for most applications and unquestionably much quicker to obtain), but rather as ways to observe the behavior of the trigonometric functions and their interwoven tapestry of identities.

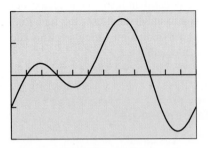

$[0, 2\pi]$ by $[-2, 2]$

FIGURE 5.11 The function $y = \sin 2x - \cos x$ for $0 \le x \le 2\pi$. The scale on the x-axis shows intervals of length $\pi/6$. This graph supports the solution found algebraically in Example 4.

EXAMPLE 4   Using a Double-Angle Identity

Solve algebraically in the interval $[0, 2\pi)$: $\sin 2x = \cos x$.

SOLUTION

$$\sin 2x = \cos x$$
$$2 \sin x \cos x = \cos x$$
$$2 \sin x \cos x - \cos x = 0$$
$$\cos x (2 \sin x - 1) = 0$$
$$\cos x = 0 \quad \text{or} \quad 2 \sin x - 1 = 0$$
$$\cos x = 0 \quad \text{or} \quad \sin x = \frac{1}{2}$$

The two solutions of $\cos x = 0$ are $x = \pi/2$ and $x = 3\pi/2$. The two solutions of $\sin x = 1/2$ are $x = \pi/6$ and $x = 5\pi/6$. Therefore, the solutions of $\sin 2x = \cos x$ are

$$\frac{\pi}{6}, \quad \frac{\pi}{2}, \quad \frac{5\pi}{6}, \quad \frac{3\pi}{2}.$$

We can **support** this result **graphically** by verifying the four x-intercepts of the function $y = \sin 2x - \cos x$ in the interval $[0, 2\pi)$ (Figure 5.11).

Now try Exercise 23.

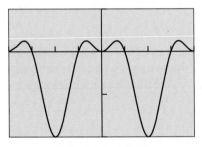

$[-2\pi, 2\pi]$ by $[-2, 1]$

FIGURE 5.12 The graph of $y = \sin^2 x - 2 \sin^2 (x/2)$ suggests that $\sin^2 x = 2 \sin^2 (x/2)$ has three solutions in $[0, 2\pi)$. (Example 5)

EXAMPLE 5   Using Half-Angle Identities

Solve $\sin^2 x = 2 \sin^2 (x/2)$.

SOLUTION The graph of $y = \sin^2 x - 2 \sin^2 (x/2)$ in Figure 5.12 suggests that this function is periodic with period 2π and that the equation $\sin^2 x = 2 \sin^2 (x/2)$ has three solutions in $[0, 2\pi)$.

Solve Algebraically

$$\sin^2 x = 2 \sin^2 \frac{x}{2}$$

$$\sin^2 x = 2 \left(\frac{1 - \cos x}{2} \right) \qquad \text{Half-angle identity}$$

$$1 - \cos^2 x = 1 - \cos x \qquad \text{Convert to all cosines.}$$

$$\cos x - \cos^2 x = 0$$

continued

$$\cos x\,(1 - \cos x) = 0$$

$$\cos x = 0 \quad \text{or} \quad \cos x = 1$$

$$x = \frac{\pi}{2} \quad \text{or} \quad \frac{3\pi}{2} \quad \text{or} \quad 0$$

The rest of the solutions are obtained by periodicity:

$$x = 2n\pi, \quad x = \frac{\pi}{2} + 2n\pi, \quad x = \frac{3\pi}{2} + 2n\pi, \quad n = 0, \pm1, \pm2, \dots$$

Now try Exercise 43.

QUICK REVIEW 5.4 *(For help, go to Section 5.1.)*

In Exercises 1–8, find the general solution of the equation.

1. $\tan x - 1 = 0$

2. $\tan x + 1 = 0$

3. $(\cos x)(1 - \sin x) = 0$

4. $(\sin x)(1 + \cos x) = 0$

5. $\sin x + \cos x = 0$

6. $\sin x - \cos x = 0$

7. $(2 \sin x - 1)(2 \cos x + 1) = 0$

8. $(\sin x + 1)(2 \cos x - \sqrt{2}) = 0$

9. Find the area of the trapezoid.

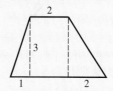

10. Find the height of the isosceles triangle.

SECTION 5.4 EXERCISES

In Exercises 1–4, use the appropriate sum or difference identity to prove the double-angle identity.

1. $\cos 2u = \cos^2 u - \sin^2 u$

2. $\cos 2u = 2 \cos^2 u - 1$

3. $\cos 2u = 1 - 2 \sin^2 u$

4. $\tan 2u = \dfrac{2 \tan u}{1 - \tan^2 u}$

In Exercises 5–10, find all solutions to the equation in the interval $[0, 2\pi)$.

5. $\sin 2x = 2 \sin x$

6. $\sin 2x = \sin x$

7. $\cos 2x = \sin x$

8. $\cos 2x = \cos x$

9. $\sin 2x - \tan x = 0$

10. $\cos^2 x + \cos x = \cos 2x$

In Exercises 11–14, write the expression as one involving only $\sin \theta$ and $\cos \theta$.

11. $\sin 2\theta + \cos \theta$

12. $\sin 2\theta + \cos 2\theta$

13. $\sin 2\theta + \cos 3\theta$

14. $\sin 3\theta + \cos 2\theta$

In Exercises 15–22, prove the identity.

15. $\sin 4x = 2 \sin 2x \cos 2x$

16. $\cos 6x = 2 \cos^2 3x - 1$

17. $2 \csc 2x = \csc^2 x \tan x$

18. $2 \cot 2x = \cot x - \tan x$

19. $\sin 3x = (\sin x)(4 \cos^2 x - 1)$

20. $\sin 3x = (\sin x)(3 - 4 \sin^2 x)$

21. $\cos 4x = 1 - 8 \sin^2 x \cos^2 x$

22. $\sin 4x = (4 \sin x \cos x)(2 \cos^2 x - 1)$

In Exercises 23–30, solve algebraically for exact solutions in the interval $[0, 2\pi)$. Use your grapher only to support your algebraic work.

23. $\cos 2x + \cos x = 0$

24. $\cos 2x + \sin x = 0$

25. $\cos x + \cos 3x = 0$

26. $\sin x + \sin 3x = 0$

27. $\sin 2x + \sin 4x = 0$

28. $\cos 2x + \cos 4x = 0$

29. $\sin 2x - \cos 3x = 0$

30. $\sin 3x + \cos 2x = 0$

In Exercises 31–36, use half-angle identities to find an exact value without a calculator.

31. $\sin 15°$

32. $\tan 195°$

33. $\cos 75°$

34. $\sin (5\pi/12)$

35. $\tan (7\pi/12)$

36. $\cos (\pi/8)$

37. Prove the power-reducing identities:

(a) $\sin^2 u = \dfrac{1 - \cos 2u}{2}$ (b) $\cos^2 u = \dfrac{1 + \cos 2u}{2}$

38. (a) Use the identities in Exercise 37 to prove the power-reducing identity $\tan^2 u = \dfrac{1 - \cos 2u}{1 + \cos 2u}$.

(b) **Writing to Learn** Explain why the identity in part (a) does not imply that $\tan u = \sqrt{\dfrac{1 - \cos 2u}{1 + \cos 2u}}$.

In Exercises 39–42, use the power-reducing identities to prove the identity.

39. $\sin^4 x = \dfrac{1}{8}(3 - 4\cos 2x + \cos 4x)$

40. $\cos^3 x = \left(\dfrac{1}{2}\cos x\right)(1 + \cos 2x)$

41. $\sin^3 2x = \left(\dfrac{1}{2}\sin 2x\right)(1 - \cos 4x)$

42. $\sin^5 x = \left(\dfrac{1}{8}\sin x\right)(3 - 4\cos 2x + \cos 4x)$

In Exercises 43–46, use the half-angle identities to find all solutions in the interval $[0, 2\pi)$. Then find the general solution.

43. $\cos^2 x = \sin^2\left(\dfrac{x}{2}\right)$ **44.** $\sin^2 x = \cos^2\left(\dfrac{x}{2}\right)$

45. $\tan\left(\dfrac{x}{2}\right) = \dfrac{1 - \cos x}{1 + \cos x}$ **46.** $\sin^2\left(\dfrac{x}{2}\right) = 2\cos^2 x - 1$

Standardized Test Questions

47. True or False The product of two functions with period 2π has period 2π. Justify your answer.

48. True or False The function $f(x) = \cos^2 x$ is a sinusoid. Justify your answer.

You should answer these questions without using a calculator.

49. Multiple Choice If $f(x) = \sin x$ and $g(x) = \cos x$, then $f(2x) =$

(A) $2 f(x)$ (B) $f(2) f(x)$ (C) $f(x) g(x)$

(D) $2 f(x) g(x)$ (E) $f(2) g(x) + g(2) f(x)$

50. Multiple Choice $\sin 22.5° =$

(A) $\dfrac{\sqrt{2}}{4}$ (B) $\dfrac{\sqrt{3}}{4}$ (C) $\dfrac{\sqrt{6} - \sqrt{2}}{4}$

(D) $\sqrt{\dfrac{2 - \sqrt{2}}{2}}$ (E) $\dfrac{\sqrt{2 - \sqrt{2}}}{2}$

51. Multiple Choice How many numbers between 0 and 2π satisfy the equation $\sin 2x = \cos x$?

(A) none   (B) one   (C) two   (D) three   (E) four

52. Multiple Choice The period of the function $\sin^2 x - \cos^2 x$ is

(A) $\dfrac{\pi}{4}$.   (B) $\dfrac{\pi}{2}$.   (C) π.   (D) 2π.   (E) 4π.

Explorations

53. Connecting Trigonometry and Geometry In a regular polygon all sides are the same length and all angles are equal in measure.

(a) If the perpendicular distance from the center of the polygon with n sides to the midpoint of a side is R, and if the length of the side of the polygon is x, show that

$$x = 2R\tan\dfrac{\theta}{2}$$

where $\theta = 2\pi/n$ is the central angle subtended by one side.

Regular polygon with n sides

(b) If the length of one side of a regular 11-sided polygon is approximately 5.87 and R is a whole number, what is the value of R?

54. Connecting Trigonometry and Geometry A rhombus is a quadrilateral with equal sides. The diagonals of a rhombus bisect the angles of the rhombus and are perpendicular bisectors of each other. Let $\angle ABC = \theta$, $d_1 = $ length of AC, and $d_2 = $ length of BD.

(a) Show that $\cos\dfrac{\theta}{2} = \dfrac{d_2}{2x}$ and $\sin\dfrac{\theta}{2} = \dfrac{d_1}{2x}$.

(b) Show that $\sin\theta = \dfrac{d_1 d_2}{2x^2}$.

55. Group Activity Maximizing Volume The ends of a 10-foot-long water trough are isosceles trapezoids as shown in the figure. Find the value of θ that maximizes the volume of the trough and the maximum volume.

56. Group Activity Tunnel Problem A rectangular tunnel is cut through a mountain to make a road. The upper vertices of the rectangle are on the circle $x^2 + y^2 = 400$, as illustrated in the figure.

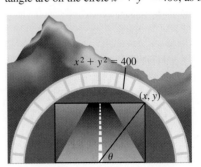

(a) Show that the cross-sectional area of the end of the tunnel is 400 sin 2θ.

(b) Find the dimensions of the rectangular end of the tunnel that maximizes its cross-sectional area.

Extending the Ideas

In Exercises 57–61, prove the double-angle formulas.

57. $\csc 2u = \dfrac{1}{2}\csc u \sec u$

58. $\cot 2u = \dfrac{\cot^2 u - 1}{2 \cot u}$

59. $\sec 2u = \dfrac{\csc^2 u}{\csc^2 u - 2}$

60. $\sec 2u = \dfrac{\sec^2 u}{2 - \sec^2 u}$

61. $\sec 2u = \dfrac{\sec^2 u \csc^2 u}{\csc^2 u - \sec^2 u}$

62. Writing to Learn Explain why

$$\sqrt{\frac{1 - \cos 2x}{2}} = |\sin x|$$

is an identity but

$$\sqrt{\frac{1 - \cos 2x}{2}} = \sin x$$

is not an identity.

63. Sahara Sunset Table 5.2 gives the time of day that astronomical twilight began in northeastern Mali on the first day of each month of 2005.

Table 5.2 Astronomical Twilight			
Date	Day	Time	18:00 +
Jan 1	1	17:32	−28
Feb 1	32	17:52	−8
Mar 1	60	18:05	5
Apr 1	91	18:14	14
May 1	121	18:24	24
Jun 1	152	18:36	36
Jul 1	182	18:43	43
Aug 1	213	18:37	37
Sep 1	244	18:15	15
Oct 1	274	17:48	−12
Nov 1	305	17:25	−35
Dec 1	335	17:19	−41

Source: The World Almanac 2005.

The second column gives the date as the day of the year, and the fourth column gives the time as number of minutes past 18:00.

(a) Enter the numbers in column 2 (day) in list L1 and the numbers in column 4 (minutes past 18:00) in list L2. Make a scatter plot with *x*-coordinates from L1 and *y*-coordinates from L2.

(b) Using sine regression, find the regression curve through the points and store its equation in Y1. Superimpose the graph of the curve on the scatter plot. Is it a good fit?

(c) Make a new column showing the *residuals* (the difference between the actual *y*-value at each data point and the *y*-value predicted by the regression curve) and store them in list L3. Your calculator might have a list called RESID, in which case the command RESID → L3 will perform this operation. You could also enter L2 − Y1(L1) → L3.

(d) Make a scatter plot with *x*-coordinates from L1 and *y*-coordinates from L3. Find the sine regression curve through *these* points and superimpose it on the scatter plot. Is it a good fit?

(e) Writing to Learn Interpret what the two regressions seem to indicate about the periodic behavior of astronomical twilight as a function of time. This is not an unusual phenomenon in astronomical data, and it kept ancient astronomers baffled for centuries.

The Law of Sines

What you'll learn about

- Deriving the Law of Sines

- Solving Triangles (AAS, ASA)

- The Ambiguous Case (SSA)

- Applications

. . . and why

The Law of Sines is a powerful extension of the triangle congruence theorems of Euclidean geometry.

Deriving the Law of Sines

Recall from geometry that a triangle has six parts (three sides (S), three angles (A)), but that its size and shape can be completely determined by fixing only three of those parts, provided they are the right three. These threesomes that determine *triangle congruence* are known by their acronyms: AAS, ASA, SAS, and SSS. The other two acronyms represent match-ups that don't quite work: AAA determines similarity only, while SSA does not even determine similarity.

With trigonometry we can find the other parts of the triangle once congruence is established. The tools we need are the Law of Sines and the Law of Cosines, the subjects of our last two trigonometric sections.

The **Law of Sines** states that the ratio of the sine of an angle to the length of its opposite side is the same for all three angles of any triangle.

Law of Sines

In any $\triangle ABC$ with angles A, B, and C opposite sides a, b, and c, respectively, the following equation is true:

$$\frac{\sin A}{a} = \frac{\sin B}{b} = \frac{\sin C}{c}.$$

The derivation of the Law of Sines refers to the two triangles in Figure 5.13, in each of which we have drawn an altitude to side c. Right triangle trigonometry applied to either of the triangles in Figure 5.13 tells us that

$$\sin A = \frac{h}{b}.$$

In the acute triangle on the top,

$$\sin B = \frac{h}{a},$$

while in the obtuse triangle on the bottom,

$$\sin (\pi - B) = \frac{h}{a}.$$

But $\sin (\pi - B) = \sin B$, so in either case

$$\sin B = \frac{h}{a}.$$

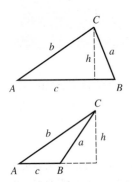

FIGURE 5.13 The Law of Sines.

Solving for h in both equations yields $h = b \sin A = a \sin B$. The equation $b \sin A = a \sin B$ is equivalent to

$$\frac{\sin A}{a} = \frac{\sin B}{b}.$$

If we were to draw an altitude to side a and repeat the same steps as above, we would reach the conclusion that

$$\frac{\sin B}{b} = \frac{\sin C}{c}.$$

Putting the results together,

$$\frac{\sin A}{a} = \frac{\sin B}{b} = \frac{\sin C}{c}.$$

Solving Triangles (AAS, ASA)

Two angles and a side of a triangle, in any order, determine the size and shape of a triangle completely. Of course, two angles of a triangle determine the third, so we really get one of the missing three parts for free. We solve for the remaining two parts (the unknown sides) with the Law of Sines.

EXAMPLE 1   Solving a Triangle Given Two Angles and a Side

Solve $\triangle ABC$ given that $\angle A = 36°$, $\angle B = 48°$, and $a = 8$. (See Figure 5.14.)

SOLUTION First, we note that $\angle C = 180° - 36° - 48° = 96°$. We then apply the Law of Sines:

$$\frac{\sin A}{a} = \frac{\sin B}{b} \quad \text{and} \quad \frac{\sin A}{a} = \frac{\sin C}{c}$$

$$\frac{\sin 36°}{8} = \frac{\sin 48°}{b} \qquad\qquad \frac{\sin 36°}{8} = \frac{\sin 96°}{c}$$

$$b = \frac{8 \sin 48°}{\sin 36°} \qquad\qquad c = \frac{8 \sin 96°}{\sin 36°}$$

$$b \approx 10.115 \qquad\qquad c \approx 13.536$$

The six parts of the triangle are:

$$\angle A = 36° \qquad\qquad a = 8$$

$$\angle B = 48° \qquad\qquad b \approx 10.115$$

$$\angle C = 96° \qquad\qquad c \approx 13.536$$

Now try Exercise 1.

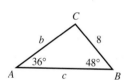

FIGURE 5.14 A triangle determined by AAS. (Example 1)

The Ambiguous Case (SSA)

While two angles and a side of a triangle are always sufficient to determine its size and shape, the same can not be said for two sides and an angle. Perhaps unexpectedly, it depends on where that angle is. If the angle is included between the two sides (the SAS case), then the triangle is uniquely determined up to congruence. If the angle is opposite one of the sides (the SSA case), then there might be one, two, or zero triangles determined.

Solving a triangle in the SAS case involves the Law of Cosines and will be handled in the next section. Solving a triangle in the SSA case is done with the Law of Sines, but with an eye toward the possibilities, as seen in the following Exploration.

FIGURE 5.15 The diagram for part 1. (Exploration 1)

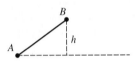

FIGURE 5.16 The diagram for part 2. (Exploration 1)

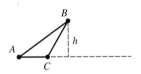

FIGURE 5.17 The diagram for parts 3–5. (Exploration 1)

> **EXPLORATION 1 Determining the Number of Triangles**
>
> We wish to construct △*ABC* given angle *A*, side *AB*, and side *BC*.
>
> 1. Suppose ∠*A* is obtuse and that side *AB* is as shown in Figure 5.15. To complete the triangle, side *BC* must determine a point on the dotted horizontal line (which extends infinitely to the left). Explain from the picture why a *unique* triangle △*ABC* is determined if *BC* > *AB*, but *no* triangle is determined if *BC* ≤ *AB*.
>
> 2. Suppose ∠*A* is acute and that side *AB* is as shown in Figure 5.16. To complete the triangle, side *BC* must determine a point on the dotted horizontal line (which extends infinitely to the right). Explain from the picture why a *unique* triangle △*ABC* is determined if *BC* = *h*, but *no* triangle is determined if *BC* < *h*.
>
> 3. Suppose ∠*A* is acute and that side *AB* is as shown in Figure 5.17. If *AB* > *BC* > *h*, then we can form a triangle as shown. Find a *second* point *C* on the dotted horizontal line that gives a side *BC* of the same length, but determines a different triangle. (This is the "ambiguous case.")
>
> 4. Explain why sin *C* is the same in both triangles in the ambiguous case. (This is why the Law of Sines is also ambiguous in this case.)
>
> 5. Explain from Figure 5.17 why a *unique* triangle is determined if *BC* ≥ *AB*.

Now that we know what can happen, let us try the algebra.

EXAMPLE 2   Solving a Triangle Given Two Sides and an Angle

Solve △*ABC* given that $a = 7$, $b = 6$, and ∠*A* = 26.3°. (See Figure 5.18.)

SOLUTION By drawing a reasonable sketch (Figure 5.18), we can assure ourselves that this is not the ambiguous case. (In fact, this is the case described in step 5 of Exploration 1.)

Begin by solving for the *acute* angle *B*, using the Law of Sines:

$$\frac{\sin A}{a} = \frac{\sin B}{b} \qquad \text{Law of Sines}$$

$$\frac{\sin 26.3°}{7} = \frac{\sin B}{6}$$

$$\sin B = \frac{6 \sin 26.3°}{7}$$

FIGURE 5.18 A triangle determined by SSA. (Example 2)

continued

$$B = \sin^{-1}\left(\frac{6 \sin 26.3°}{7}\right)$$

$$B = 22.3° \qquad \text{Round to match accuracy of given angle.}$$

Then, find the obtuse angle C by subtraction:

$$C = 180° - 26.3° - 22.3°$$

$$= 131.4°$$

Finally, find side c:

$$\frac{\sin A}{a} = \frac{\sin C}{c}$$

$$\frac{\sin 26.3°}{7} = \frac{\sin 131.4°}{c}$$

$$c = \frac{7 \sin 131.4°}{\sin 26.3°}$$

$$c \approx 11.9$$

The six parts of the triangle are:

$$\angle A = 26.3° \qquad\qquad a = 7$$

$$\angle B = 22.3° \qquad\qquad b = 6$$

$$\angle C = 131.4° \qquad\qquad c \approx 11.9$$

Now try Exercise 9.

EXAMPLE 3 Handling the Ambiguous Case

Solve $\triangle ABC$ given that $a = 6$, $b = 7$, and $\angle A = 30°$.

SOLUTION By drawing a reasonable sketch (Figure 5.19), we see that two triangles are possible with the given information. We keep this in mind as we proceed.

We begin by using the Law of Sines to find angle B.

$$\frac{\sin A}{a} = \frac{\sin B}{b} \qquad \text{Law of Sines}$$

$$\frac{\sin 30°}{6} = \frac{\sin B}{7}$$

$$\sin B = \frac{7 \sin 30°}{6}$$

$$B = \sin^{-1}\left(\frac{7 \sin 30°}{6}\right)$$

$$B = 35.7° \qquad \text{Round to match accuracy of given angle.}$$

Notice that the calculator gave us one value for B, not two. That is because we used the *function* $\sin^{-1}$, which cannot give two output values for the same input value. Indeed, the function $\sin^{-1}$ will *never give an obtuse angle*, which is why we chose to start with the acute angle in Example 2. In this case, the calculator has found the angle B shown in Figure 5.19a.

continued

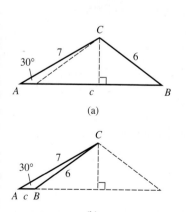

FIGURE 5.19 Two triangles determined by the same SSA values. (Example 3)

Find the obtuse angle C by subtraction:

$$C = 180° - 30.0° - 35.7° = 114.3°.$$

Finally, find side c:

$$\frac{\sin A}{a} = \frac{\sin C}{c}$$

$$\frac{\sin 30.0°}{6} = \frac{\sin 114.3°}{c}$$

$$c = \frac{6 \sin 114.3°}{\sin 30°}$$

$$c \approx 10.9$$

So, under the assumption that angle B is *acute* (see Figure 5.19a), the six parts of the triangle are:

$$\angle A = 30.0° \qquad a = 6$$

$$\angle B = 35.7° \qquad b = 7$$

$$\angle C = 114.3° \qquad c \approx 10.9$$

If angle B is *obtuse*, then we can see from Figure 5.19b that it has measure $180° - 35.7° = 144.3°$.

By subtraction, the acute angle $C = 180° - 30.0° - 144.3° = 5.7°$. We then recompute c:

$$c = \frac{6 \sin 5.7°}{\sin 30°} \approx 1.2 \quad \text{\small Substitute 5.7 for 114.3 in earlier computation.}$$

So, under the assumption that angle B is *obtuse* (see Figure 5.19b), the six parts of the triangle are:

$$\angle A = 30.0° \qquad\qquad a = 6$$

$$\angle B = 144.3° \qquad\qquad b = 7$$

$$\angle C = 5.7° \qquad\qquad c \approx 1.2 \qquad\qquad \textbf{\textit{Now try Exercise 19.}}$$

Applications

Many problems involving angles and distances can be solved by superimposing a triangle onto the situation and solving the triangle.

EXAMPLE 4   Locating a Fire

Forest Ranger Chris Johnson at ranger station A sights a fire in the direction $32°$ east of north. Ranger Rick Thorpe at ranger station B, 10 miles due east of A, sights the same fire on a line $48°$ west of north. Find the distance from each ranger station to the fire.

continued

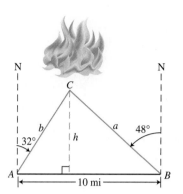

FIGURE 5.20 Determining the location of a fire. (Example 4)

SOLUTION Let *C* represent the location of the fire. A sketch (Fi
the superimposed triangle, △*ABC*, in which angles *A* and *B* and the
(*AB*) are known. This is a setup for the Law of Sines.

Note that $\angle A = 90° - 32° = 58°$ and $\angle B = 90° - 48° = 42°$. By subtraction, we find that $\angle C = 180° - 58° - 42° = 80°$.

$$\frac{\sin A}{a} = \frac{\sin C}{c} \quad \text{and} \quad \frac{\sin B}{b} = \frac{\sin C}{c} \qquad \text{Law of Sines}$$

$$\frac{\sin 58°}{a} = \frac{\sin 80°}{10} \qquad \qquad \frac{\sin 42°}{b} = \frac{\sin 80°}{10}$$

$$a = \frac{10 \sin 58°}{\sin 80°} \qquad \qquad b = \frac{10 \sin 42°}{\sin 80°}$$

$$a \approx 8.6 \qquad \qquad b \approx 6.8$$

The fire is about 6.8 miles from ranger station *A* and about 8.6 miles from ranger station *B*. ***Now try Exercise 45.***

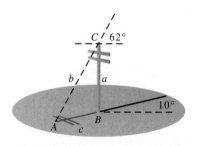

FIGURE 5.21 A telephone pole on a slope. (Example 5))

EXAMPLE 5 Finding the Height of a Pole

A road slopes 10° above the horizontal, and a vertical telephone pole stands beside the road. The angle of elevation of the Sun is 62°, and the pole casts a 14.5-foot shadow downhill along the road. Find the height of the telephone pole.

SOLUTION This is an interesting variation on a typical application of right-triangle trigonometry. The slope of the road eliminates the convenient right angle, but we can still solve the problem by solving a triangle.

Figure 5.21 shows the superimposed triangle, △*ABC*. A little preliminary geometry is required to find the measure of angles *A* and *C*. Due to the slope of the road, angle *A* is 10° less than the angle of elevation of the Sun and angle *B* is 10° more than a right angle. That is,

$$\angle A = 62° - 10° = 52°$$

$$\angle B = 90° + 10° = 100°$$

$$\angle C = 180° - 52° - 100° = 28°$$

Therefore,

$$\frac{\sin A}{a} = \frac{\sin C}{c} \qquad \text{Law of Sines}$$

$$\frac{\sin 52°}{a} = \frac{\sin 28°}{14.5}$$

$$a = \frac{14.5 \sin 52°}{\sin 28°}$$

$$a \approx 24.3 \qquad \text{Round to match accuracy of input.}$$

The pole is approximately 24.3 feet high. ***Now try Exercise 39.***

QUICK REVIEW 5.5 *(For help, go to Sections 4.2 and 4.7.)*

In Exercises 1–4, solve the equation $a/b = c/d$ for the given variable.

1. a **2.** b **3.** c **4.** d

In Exercises 5 and 6, evaluate the expression.

5. $\dfrac{7 \sin 48°}{\sin 23°}$ **6.** $\dfrac{9 \sin 121°}{\sin 14°}$

In Exercises 7–10, solve for the angle x.

7. $\sin x = 0.3, \quad 0° < x < 90°$

8. $\sin x = 0.3, \quad 90° < x < 180°$

9. $\sin x = -0.7, \quad 180° < x < 270°$

10. $\sin x = -0.7, \quad 270° < x < 360°$

SECTION 5.5 EXERCISES

In Exercises 1–4, solve the triangle.

1.

2.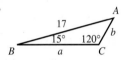

3.

4.

In Exercises 5–8, solve the triangle.

5. $A = 40°, \quad B = 30°, \quad b = 10$

6. $A = 50°, \quad B = 62°, \quad a = 4$

7. $A = 33°, \quad B = 70°, \quad b = 7$

8. $B = 16°, \quad C = 103°, \quad c = 12$

In Exercises 9–12, solve the triangle.

9. $A = 32°, \quad a = 17, \quad b = 11$

10. $A = 49°, \quad a = 32, \quad b = 28$

11. $B = 70°, \quad b = 14, \quad c = 9$

12. $C = 103°, \quad b = 46, \quad c = 61$

In Exercises 13–18, state whether the given measurements determine zero, one, or two triangles.

13. $A = 36°, \quad a = 2, \quad b = 7$

14. $B = 82°, \quad b = 17, \quad c = 15$

15. $C = 36°, \quad a = 17, \quad c = 16$

16. $A = 73°, \quad a = 24, \quad b = 28$

17. $C = 30°, \quad a = 18, \quad c = 9$

18. $B = 88°, \quad b = 14, \quad c = 62$

In Exercises 19–22, two triangles can be formed using the given measurements. Solve both triangles.

19. $A = 64°, \quad a = 16, \quad b = 17$

20. $B = 38°, \quad b = 21, \quad c = 25$

21. $C = 68°, \quad a = 19, \quad c = 18$

22. $B = 57°, \quad a = 11, \quad b = 10$

23. Determine the values of b that will produce the given number of triangles if $a = 10$ and $B = 42°$.

   (a) two triangles   **(b)** one triangle   **(c)** zero triangles

24. Determine the values of c that will produce the given number of triangles if $b = 12$ and $C = 53°$.

   (a) two triangles   **(b)** one triangle   **(c)** zero triangles

In Exercises 25 and 26, decide whether the triangle can be solved using the Law of Sines. If so, solve it. If not, explain why not.

25.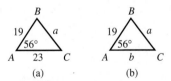

26.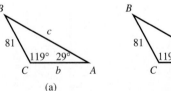

In Exercises 27–36, respond in one of the following ways:

(a) State, "Cannot be solved with the Law of Sines."

(b) State, "No triangle is formed."

(c) Solve the triangle.

27. $A = 61°, \quad a = 8, \quad b = 21$

28. $B = 47°, \quad a = 8, \quad b = 21$

29. $A = 136°$, $a = 15$, $b = 28$

30. $C = 115°$, $b = 12$, $c = 7$

31. $B = 42°$, $c = 18$, $C = 39°$

32. $A = 19°$, $b = 22$, $B = 47°$

33. $C = 75°$, $b = 49$, $c = 48$

34. $A = 54°$, $a = 13$, $b = 15$

35. $B = 31°$, $a = 8$, $c = 11$

36. $C = 65°$, $a = 19$, $b = 22$

37. Surveying a Canyon Two markers A and B on the same side of a canyon rim are 56 ft apart. A third marker C, located across the rim, is positioned so that $\angle BAC = 72°$ and $\angle ABC = 53°$.

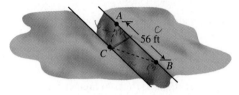

(a) Find the distance between C and A.

(b) Find the distance between the two canyon rims. (Assume they are parallel.)

38. Weather Forecasting Two meteorologists are 25 mi apart located on an east-west road. The meteorologist at point A sights a tornado 38° east of north. The meteorologist at point B sights the same tornado at 53° west of north. Find the distance from each meteorologist to the tornado. Also find the distance between the tornado and the road.

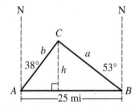

39. Engineering Design A vertical flagpole stands beside a road that slopes at an angle of 15° with the horizontal. When the angle of elevation of the Sun is 62°, the flagpole casts a 16-ft shadow downhill along the road. Find the height of the flagpole.

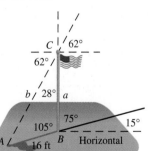

40. Altitude Observers 2.32 mi apart see a hot-air balloon directly between them but at the angles of elevation shown in the figure. Find the altitude of the balloon.

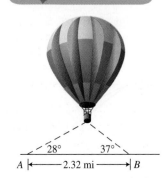

41. Reducing Air Resistance A 4-ft airfoil attached to the cab of a truck reduces wind resistance. If the angle between the airfoil and the cab top is 18° and angle B is 10°, find the length of a vertical brace positioned as shown in the figure.

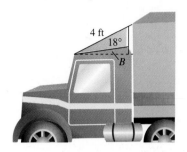

42. Group Activity Ferris Wheel Design A Ferris wheel has 16 evenly spaced cars. The distance between adjacent chairs is 15.5 ft. Find the radius of the wheel (to the nearest 0.1 ft).

43. Finding Height Two observers are 600 ft apart on opposite sides of a flagpole. The angles of elevation from the observers to the top of the pole are 19° and 21°. Find the height of the flagpole.

44. Finding Height Two observers are 400 ft apart on opposite sides of a tree. The angles of elevation from the observers to the top of the tree are 15° and 20°. Find the height of the tree.

45. Finding Distance Two lighthouses A and B are known to be exactly 20 mi apart on a north-south line. A ship's captain at S measures $\angle ASB$ to be 33°. A radio operator at B measures $\angle ABS$ to be 52°. Find the distance from the ship to each lighthouse.

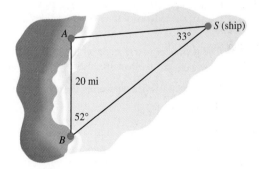

46. Using Measurement Data A geometry class is divided into ten teams, each of which is given a yardstick and a protractor to find the distance from a point A on the edge of a pond to a tree at a point C on the opposite shore. After they mark points A and B with stakes, each team uses a protractor to measure angles A and B and a yardstick to measure distance AB. Their measurements are given in the table on the next page.

A	B	AB
79°	84°	26′ 4″
81°	82°	25′ 5″
79°	83°	26′ 0″
80°	87°	26′ 1″
79°	87°	25′ 11″

A	B	AB
83°	84°	25′ 3″
82°	82°	26′ 5″
78°	85°	25′ 8″
77°	83°	26′ 4″
79°	82°	25′ 7″

Use the data to find the class's best estimate for the distance *AC*.

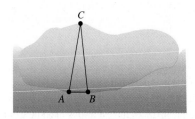

Standardized Test Questions

47. True or False The ratio of the sines of any two angles in a triangle equals the ratio of the lengths of their opposite sides. Justify your answer.

48. True or False The perimeter of a triangle with two 10-inch sides and two 40° angles is greater than 36. Justify your answer.

You may use a graphing calculator when answering these questions.

49. Multiple Choice The length *x* in the triangle shown at the right is

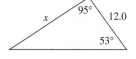

(A) 8.6.   (B) 15.0.   (C) 18.1.

(D) 19.2.   (E) 22.6.

50. Multiple Choice Which of the following three triangle parts do not necessarily determine the other three parts?

(A) AAS   (B) ASA   (C) SAS

(D) SSA   (E) SSS

51. Multiple Choice The shortest side of a triangle with angles 50°, 60°, and 70° has length 9.0. What is the length of the longest side?

(A) 11.0   (B) 11.5   (C) 12.0

(D) 12.5   (E) 13.0

52. Multiple Choice How many noncongruent triangles *ABC* can be formed if *AB* = 5, *A* = 60°, and *BC* = 8?

(A) none   (B) one   (C) two

(D) three   (E) infinitely many

Explorations

53. Writing to Learn

(a) Show that there are infinitely many triangles with AAA given if the sum of the three positive angles is 180°.

(b) Give three examples of triangles where *A* = 30°, *B* = 60°, and *C* = 90°.

(c) Give three examples where *A* = *B* = *C* = 60°.

54. Use the Law of Sines and the cofunction identities to derive the following formulas from right-triangle trigonometry:

(a) $\sin A = \dfrac{opp}{hyp}$ (b) $\cos A = \dfrac{adj}{hyp}$ (c) $\tan A = \dfrac{opp}{adj}$

55. Wrapping up Exploration 1 Refer to Figures 5.16 and 5.17 in Exploration 1 of this section.

(a) Express *h* in terms of angle *A* and length *AB*.

(b) In terms of the given angle *A* and the given length *AB*, state the conditions on length *BC* that will result in no triangle being formed.

(c) In terms of the given angle *A* and the given length *AB*, state the conditions on length *BC* that will result in a unique triangle being formed.

(d) In terms of the given angle *A* and the given length *AB*, state the conditions on length *BC* that will result in two possible triangles being formed.

Extending the Ideas

56. Solve this triangle assuming that ∠*B* is obtuse. (*Hint*: Draw a perpendicular from *A* to the line through *B* and *C*.)

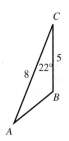

57. Pilot Calculations Towers *A* and *B* are known to be 4.1 mi apart on level ground. A pilot measures the angles of depression to the towers to be 36.5° and 25°, respectively, as shown in the figure. Find distances *AC* and *BC* and the height of the airplane.

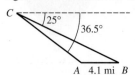

5.6
The Law of Cosines

What you'll learn about

- Deriving the Law of Cosines
- Solving Triangles (SAS, SSS)
- Triangle Area and Heron's Formula
- Applications

... and why

The Law of Cosines is an important extension of the Pythagorean theorem, with many applications.

Deriving the Law of Cosines

Having seen the Law of Sines, you will probably not be surprised to learn that there is a Law of Cosines. There are many such parallels in mathematics. What you might find surprising is that the Law of Cosines has absolutely no resemblance to the Law of Sines. Instead, it resembles the Pythagorean theorem. In fact, the Law of Cosines is often called the "generalized Pythagorean theorem" because it contains that classic theorem as a special case.

Law of Cosines

Let $\triangle ABC$ be any triangle with sides and angles labeled in the usual way (Figure 5.22).

Then

$$a^2 = b^2 + c^2 - 2bc \cos A$$

$$b^2 = a^2 + c^2 - 2ac \cos B$$

$$c^2 = a^2 + b^2 - 2ab \cos C$$

We derive only the first of the three equations, since the other two are derived in exactly the same way. Set the triangle in a coordinate plane so that the angle that appears in the formula (in this case, A) is at the origin in standard position, with side c along the positive x-axis. Depending on whether angle A is right (Figure 5.23a), acute (Figure 5.23b), or obtuse (Figure 5.23c), the point C will be on the y-axis, in QI, or in QII.

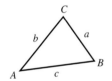

FIGURE 5.22 A triangle with the usual labeling (angles A, B, C; opposite sides a, b, c).

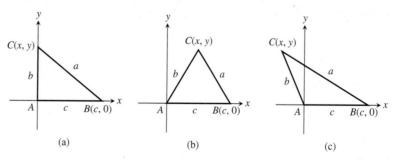

FIGURE 5.23 Three cases for proving the Law of Cosines.

In each of these three cases, C is a point on the terminal side of angle A in standard position, at distance b from the origin. Denote the coordinates of C by (x, y). By our definitions for trigonometric functions of any angle (Section 4.3), we can conclude that

$$\frac{x}{b} = \cos A \quad \text{and} \quad \frac{y}{b} = \sin A,$$

and therefore

$$x = b \cos A \quad \text{and} \quad y = b \sin A.$$

Now set a equal to the distance from C to B using the distance formula:

$$a = \sqrt{(x - c)^2 + (y - 0)^2} \qquad \text{Distance formula}$$

$$a^2 = (x - c)^2 + y^2 \qquad \text{Square both sides.}$$

$$= (b \cos A - c)^2 + (b \sin A)^2 \qquad \text{Substitution}$$

$$= b^2 \cos^2 A - 2bc \cos A + c^2 + b^2 \sin^2 A$$

$$= b^2(\cos^2 A + \sin^2 A) + c^2 - 2bc \cos A$$

$$= b^2 + c^2 - 2bc \cos A \qquad \text{Pythagorean identity}$$

Solving Triangles (SAS, SSS)

While the Law of Sines is the tool we use to solve triangles in the AAS and ASA cases, the Law of Cosines is the required tool for SAS and SSS. (Both methods can be used in the SSA case, but remember that there might be 0, 1, or 2 triangles.)

EXAMPLE 1 Solving a Triangle (SAS)

Solve $\triangle ABC$ given that $a = 11$, $b = 5$, and $C = 20°$. (See Figure 5.24.)

SOLUTION

$$c^2 = a^2 + b^2 - 2ab \cos C$$

$$= 11^2 + 5^2 - 2(11)(5) \cos 20°$$

$$= 42.6338\ldots$$

$$c = \sqrt{42.6338\ldots} \approx 6.5$$

We could now use either the Law of Cosines or the Law of Sines to find one of the two unknown angles. As a general rule, it is better to use the Law of Cosines to find angles, since the arccosine function will distinguish obtuse angles from acute angles.

$$a^2 = b^2 + c^2 - 2bc \cos A$$

$$11^2 = 5^2 + (6.529\ldots)^2 - 2(5)(6.529\ldots) \cos A$$

$$\cos A = \frac{5^2 + (6.529\ldots)^2 - 11^2}{2(5)(6.529\ldots)}$$

$$A = \cos^{-1}\left(\frac{5^2 + (6.529\ldots)^2 - 11^2}{2(5)(6.529\ldots)}\right)$$

$$\approx 144.8°$$

$$B = 180° - 144.8° - 20°$$

$$= 15.2°$$

So the six parts of the triangle are:

$$A = 144.8° \qquad a = 11$$

$$B = 15.2° \qquad b = 5$$

$$C = 20° \qquad c \approx 6.5$$

Now try Exercise 1.

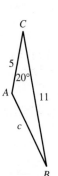

FIGURE 5.24 A triangle with two sides and an included angle known. (Example 1)

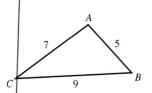

FIGURE 5.25 A triangle with three sides known. (Example 2)

EXAMPLE 2   Solving a Triangle (SSS)

Solve $\triangle ABC$ if $a = 9$, $b = 7$, and $c = 5$. (See Figure 5.25.)

SOLUTION We use the Law of Cosines to find two of the angles. The third angle can be found by subtraction from $180°$.

$$a^2 = b^2 + c^2 - 2bc \cos A \qquad\qquad b^2 = a^2 + c^2 - 2ac \cos B$$

$$9^2 = 7^2 + 5^2 - 2(7)(5) \cos A \qquad\qquad 7^2 = 9^2 + 5^2 - 2(9)(5) \cos B$$

$$70 \cos A = -7 \qquad\qquad\qquad\qquad 90 \cos B = 57$$

$$A = \cos^{-1}(-0.1) \qquad\qquad\qquad B = \cos^{-1}(57/90)$$

$$\approx 95.7° \qquad\qquad\qquad\qquad\quad \approx 50.7°$$

Then $C = 180° - 95.7° - 50.7° = 33.6°$.   *Now try Exercise 3.*

Triangle Area and Heron's Formula

The same parts that determine a triangle also determine its area. If the parts happen to be two sides and an included angle (SAS), we get a simple area formula in terms of those three parts that does not require finding an altitude.

Observe in Figure 5.23 (used in explaining the Law of Cosines) that each triangle has base c and altitude $y = b \sin A$. Applying the standard area formula, we have

$$\triangle \text{ Area} = \frac{1}{2}(base)(height) = \frac{1}{2}(c)(b \sin A) = \frac{1}{2}bc \sin A.$$

This is actually three formulas in one, as it does not matter which side we use as the base.

Area of a Triangle

$$\triangle \text{ Area} = \frac{1}{2}bc \sin A = \frac{1}{2}ac \sin B = \frac{1}{2}ab \sin C$$

EXAMPLE 3   Finding the Area of a Regular Polygon

Find the area of a regular octagon (8 equal sides, 8 equal angles) inscribed inside a circle of radius 9 inches.

SOLUTION Figure 5.26 shows that we can split the octagon into 8 congruent triangles. Each triangle has two 9-inch sides with an included angle of $\theta = 360°/8 = 45°$. The area of each triangle is

$$\triangle \text{ Area} = (1/2)(9)(9) \sin 45° = (81/2) \sin 45° = 81\sqrt{2}/4.$$

Therefore, the area of the octagon is

$$\triangle \text{ Area} = 8 \triangle \text{ Area} = 162\sqrt{2} \approx 229 \text{ square inches.}$$

Now try Exercise 31.

FIGURE 5.26 A regular octagon inscribed inside a circle of radius 9 inches. (Example 3)

There is also an area formula that can be used when the three sides of the triangle are known.

HERON'S FORMULA

The formula is named after Heron of Alexandria, whose proof of the formula is the oldest on record, but ancient Arabic scholars claimed to have known it from the works of Archimedes of Syracuse centuries earlier. Archimedes (c. 287–212 BC) is considered to be the greatest mathematician of all antiquity.

Although Heron proved this theorem using only classical geometric methods, we prove it, as most people do today, by using the tools of trigonometry

---

THEOREM Heron's Formula

Let a, b, and c be the sides of $\triangle ABC$, and let s denote the **semiperimeter**
$$(a + b + c)/2.$$
Then the area of $\triangle ABC$ is given by Area $= \sqrt{s(s - a)(s - b)(s - c)}$.

Proof

$$\text{Area} = \frac{1}{2}ab \sin C$$

$$4(\text{Area}) = 2ab \sin C$$

$$16(\text{Area})^2 = 4a^2b^2 \sin^2 C$$

$$= 4a^2b^2(1 - \cos^2 C) \qquad \text{Pythagorean identity}$$

$$= 4a^2b^2 - 4a^2b^2 \cos^2 C$$

$$= 4a^2b^2 - (2ab \cos C)^2$$

$$= 4a^2b^2 - (a^2 + b^2 - c^2)^2 \qquad \text{Law of Cosines}$$

$$= (2ab - (a^2 + b^2 - c^2))(2ab + (a^2 + b^2 - c^2)) \qquad \text{Difference of squares}$$

$$= (c^2 - (a^2 - 2ab + b^2))((a^2 + 2ab + b^2) - c^2)$$

$$= (c^2 - (a - b)^2)((a + b)^2 - c^2)$$

$$= (c - (a - b))(c + (a - b))((a + b) - c)((a + b) + c) \qquad \text{Difference of squares}$$

$$= (c - a + b)(c + a - b)(a + b - c)(a + b + c)$$

$$= (2s - 2a)(2s - 2b)(2s - 2c)(2s) \qquad 2s = a + b + c$$

$$16(\text{Area})^2 = 16s(s - a)(s - b)(s - c)$$

$$(\text{Area})^2 = s(s - a)(s - b)(s - c)$$

$$\text{Area} = \sqrt{s(s - a)(s - b)(s - c)}$$

EXAMPLE 4   Using Heron's Formula

Find the area of a triangle with sides 13, 15, 18.

SOLUTION First we compute the semiperimeter: $s = (13 + 15 + 18)/2 = 23$.

Then we use Heron's Formula

$$\text{Area} = \sqrt{23\,(23 - 13)(23 - 15)(23 - 18)}$$
$$= \sqrt{23 \cdot 10 \cdot 8 \cdot 5} = \sqrt{9200} = 20\sqrt{23}.$$

The approximate area is 96 square units.   ***Now try Exercise 21.***

Applications

We end this section with a few applications.

EXAMPLE 5 Measuring a Baseball Diamond

The bases on a baseball diamond are 90 feet apart, and the front edge of the pitcher's rubber is 60.5 feet from the back corner of home plate. Find the distance from the center of the front edge of the pitcher's rubber to the far corner of first base.

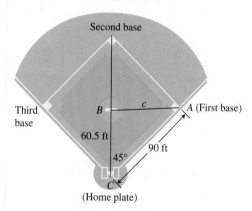

FIGURE 5.27 The diamond-shaped part of a baseball diamond. (Example 5)

SOLUTION Figure 5.27 shows first base as *A*, the pitcher's rubber as *B*, and home plate as *C*. The distance we seek is side *c* in $\triangle ABC$.

By the Law of Cosines,

$$c^2 = 60.5^2 + 90^2 - 2(60.5)(90) \cos 45°$$
$$c = \sqrt{60.5^2 + 90^2 - 2(60.5)(90) \cos 45°}$$
$$\approx 63.7$$

The distance from first base to the pitcher's rubber is about 63.7 feet.

Now try Exercise 37.

EXAMPLE 6 Measuring a Dihedral Angle (Solid Geometry)

A regular tetrahedron is a solid with four faces, each of which is an equilateral triangle. Find the measure of the *dihedral angle* formed along the common edge of two intersecting faces of a regular tetrahedron with edges of length 2.

continued

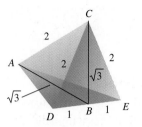

FIGURE 5.28 The measure of $\angle ABC$ is the same as the measure of any dihedral angle formed by two of the tetrahedron's faces. (Example 6)

SOLUTION Figure 5.28 shows the tetrahedron. Point B is the midpoint of edge DE, and A and C are the endpoints of the opposite edge. The measure of $\angle ABC$ is the same as the measure of the dihedral angle formed along edge DE, so we will find the measure of $\angle ABC$.

Because both $\triangle ADB$ and $\triangle CDB$ are 30°–60°–90° triangles, AB and BC both have length $\sqrt{3}$. If we apply the Law of Cosines to $\triangle ABC$, we obtain

$$2^2 = (\sqrt{3})^2 + (\sqrt{3})^2 - 2\sqrt{3}\sqrt{3}\cos(\angle ABC)$$

$$\cos(\angle ABC) = \frac{1}{3}$$

$$\angle ABC = \cos^{-1}\left(\frac{1}{3}\right) \approx 70.53°$$

The dihedral angle has the same measure as $\angle ABC$, approximately 70.53°. (We chose sides of length 2 for computational convenience, but in fact this is the measure of a dihedral angle in a regular tetrahedron of any size.)

Now try Exercise 43.

EXPLORATION 1   Estimating Acreage of a Plot of Land

Jim and Barbara are house-hunting and need to estimate the size of an irregular adjacent lot that is described by the owner as "a little more than an acre." With Barbara stationed at a corner of the plot, Jim starts at another corner and walks a straight line toward her, counting his paces. They then shift corners and Jim paces again, until they have recorded the dimensions of the lot (in paces) as in Figure 5.29. They later measure Jim's pace as 2.2 feet. What is the approximate acreage of the lot?

1. Use Heron's formula to find the area in square paces.
2. Convert the area to square feet, using the measure of Jim's pace.
3. There are 5280 feet in a mile. Convert the area to square miles.
4. There are 640 square acres in a square mile. Convert the area to acres.
5. Is there good reason to doubt the owner's estimate of the acreage of the lot?
6. Would Jim and Barbara be able to modify their system to estimate the area of an irregular lot with five straight sides?

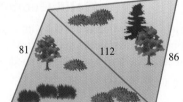

FIGURE 5.29 Dimensions (in paces) of an irregular plot of land. (Exploration 1)

♪ you suck anyway ♥ ♪♪

CHAPTER OPENER PROBL
(from page 443)

PROBLEM: Because deer require food, wa
weather and predators, and living space for
limits to the number of deer that a given pl
tions in national parks average 14 animals per square
region with sides of 3 kilometers, 4 kilometers, and 6 kilometers has a pop
tion of 50 deer, how close is the population on this land to the average national
park population?

SOLUTION: We can find the area of the land region

6 km

3 km   4 km

By using Heron's formula with

$$s = (3 + 4 + 6)/2 = 13/2$$

and

$$\text{Area} = \sqrt{s(s - a)(s - b)(s - c)}$$

$$= \sqrt{\frac{13}{2}\left(\frac{13}{2} - 3\right)\left(\frac{13}{2} - 4\right)\left(\frac{13}{2} - 6\right)}$$

$$= \sqrt{\frac{13}{2}\left(\frac{7}{2}\right)\left(\frac{5}{2}\right)\left(\frac{1}{2}\right)} \approx 5.3$$

so the area of the land region is 5.3 km^2.

If this land were to support 14 deer/km^2, it would have
$(5.3... \text{ km}^2)(14 \text{ deer/km}^2) = 74.7 \approx 75$ deer. Thus, the land supports
25 deer less than the average.

QUICK REVIEW 5.6 *(For help, go to Sections 2.4 and 4.7.)*

In Exercises 1–4, find an angle between $0°$ and $180°$ that is a solution to the equation.

1. $\cos A = 3/5$

2. $\cos C = -0.23$

3. $\cos A = -0.68$

4. $3 \cos C = 1.92$

In Exercises 5 and 6, solve the equation (in terms of x and y) for
(a) $\cos A$ and **(b)** A, $0 \le A \le 180°$.

5. $9^2 = x^2 + y^2 - 2xy \cos A$ **6.** $y^2 = x^2 + 25 - 10 \cos A$

In Exercises 7–10, find a quadratic polynomial with real coefficients
that satisfies the given condition.

7. Has two positive zeros

8. Has one positive and one negative zero

9. Has no real zeros

10. Has exactly one positive zero

N 5.6 EXERCISES

1–4, solve the triangle.

1.

2.

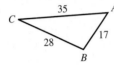

3.

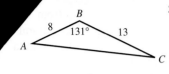

4.

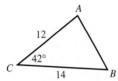

In Exercises 5–16, solve the triangle.

5. $A = 55°$, $b = 12$, $c = 7$

6. $B = 35°$, $a = 43$, $c = 19$

7. $a = 12$, $b = 21$, $C = 95°$

8. $b = 22$, $c = 31$, $A = 82°$

9. $a = 1$, $b = 5$, $c = 4$

10. $a = 1$, $b = 5$, $c = 8$

11. $a = 3.2$, $b = 7.6$, $c = 6.4$

12. $a = 9.8$, $b = 12$, $c = 23$

13. $A = 42°$, $a = 7$, $b = 10$

14. $A = 57°$, $a = 11$, $b = 10$

15. $A = 63°$, $a = 8.6$, $b = 11.1$

16. $A = 71°$, $a = 9.3$, $b = 8.5$

In Exercises 17–20, find the area of the triangle.

17. $A = 47°$, $b = 32$ ft, $c = 19$ ft

18. $A = 52°$, $b = 14$ m, $c = 21$ m

19. $B = 101°$, $a = 10$ cm, $c = 22$ cm

20. $C = 112°$, $a = 1.8$ in., $b = 5.1$ in.

In Exercises 21–28, decide whether a triangle can be formed with the given side lengths. If so, use Heron's formula to find the area of the triangle.

21. $a = 4$, $b = 5$, $c = 8$

22. $a = 5$, $b = 9$, $c = 7$

23. $a = 3$, $b = 5$, $c = 8$

24. $a = 23$, $b = 19$, $c = 12$

25. $a = 19.3$, $b = 22.5$, $c = 31$

26. $a = 8.2$, $b = 12.5$, $c = 28$

27. $a = 33.4$, $b = 28.5$, $c = 22.3$

28. $a = 18.2$, $b = 17.1$, $c = 12.3$

29. Find the radian measure of the largest angle in the triangle with sides of 4, 5, and 6.

30. A parallelogram has sides of 18 and 26 ft, and an angle of 39°. Find the shorter diagonal.

31. Find the area of a regular hexagon inscribed in a circle of radius 12 inches.

32. Find the area of a regular nonagon (9 sides) inscribed in a circle of radius 10 inches.

33. Find the area of a regular hexagon circumscribed about a circle of radius 12 inches. [*Hint*: start by finding the distance from a vertex of the hexagon to the center of the circle.]

34. Find the area of a regular nonagon (9 sides) circumscribed about a circle of radius 10 inches.

35. Measuring Distance Indirectly
Juan wants to find the distance between two points A and B on opposite sides of a building. He locates a point C that is 110 ft from A and 160 ft from B, as illustrated in the figure. If the angle at C is 54°, find distance AB.

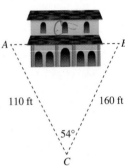

36. Designing a Baseball Field

(a) Find the distance from the center of the front edge of the pitcher's rubber to the far corner of second base. How does this distance compare with the distance from the pitcher's rubber to first base? (See Example 5.)

(b) Find $\angle B$ in $\triangle ABC$.

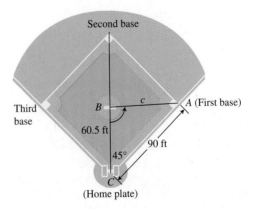

37. Designing a Softball Field In softball, adjacent bases are 60 ft apart. The distance from the center of the front edge of the pitcher's rubber to the far corner of home plate is 40 ft.

(a) Find the distance from the center of the pitcher's rubber to the far corner of first base.

(b) Find the distance from the center of the pitcher's rubber to the far corner of second base.

(c) Find $\angle B$ in $\triangle ABC$.

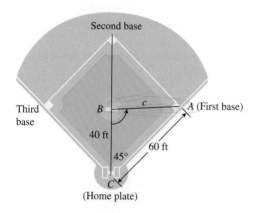

38. Surveyor's Calculations Tony must find the distance from A to B on opposite sides of a lake. He locates a point C that is 860 ft from A and 175 ft from B. He measures the angle at C to be $78°$. Find distance AB.

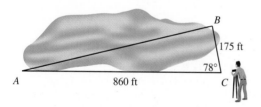

39. Construction Engineering A manufacturer is designing the roof truss that is modeled in the figure shown.

(a) Find the measure of $\angle CAE$.

(b) If $AF = 12$ ft, find the length DF.

(c) Find the length EF.

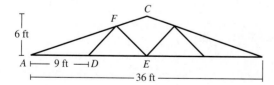

40. Navigation Two airplanes flying together in formation take off in different directions. One flies due east at 350 mph, and the other flies east-northeast at 380 mph. How far apart are the two airplanes 2 hr after they separate, assuming that they fly at the same altitude?

41. Football Kick The player waiting to receive a kickoff stands at the 5 yard line (point A) as the ball is being kicked 65 yd up the field from the opponent's 30 yard line. The kicked ball travels 73 yd at an angle of $8°$ to the right of the receiver, as shown in the figure (point B). Find the distance the receiver runs to catch the ball.

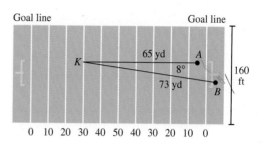

42. Group Activity Architectural Design Building Inspector Julie Wang checks a building in the shape of a regular octagon, each side 20 ft long. She checks that the contractor has located the corners of the foundation correctly by measuring several of the diagonals. Calculate what the lengths of diagonals HB, HC, and HD should be.

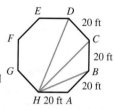

43. Connecting Trigonometry and Geometry $\angle CAB$ is inscribed in a rectangular box whose sides are 1, 2, and 3 ft long as shown. Find the measure of $\angle CAB$.

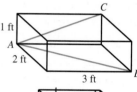

44. Group Activity Connecting Trigonometry and Geometry A cube has edges of length 2 ft. Point A is the midpoint of an edge. Find the measure of $\angle ABC$.

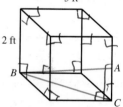

Standardized Test Questions

45. True or False If $\triangle ABC$ is any triangle with sides and angles labeled in the usual way, then $b^2 + c^2 > 2bc \cos A$. Justify your answer.

46. True or False If a, b, and θ are two sides and an included angle of a parallelogram, the area of the parallelogram is $ab \sin \theta$. Justify your answer.

You may use a graphing calculator when answering these questions.

47. Multiple Choice What is the area of a regular dodecagon (12-sided figure) inscribed in a circle of radius 12?

(A) 427 **(B)** 432 **(C)** 437 **(D)** 442 **(E)** 447

48. Multiple Choice The area of a triangle with sides 7, 8, and 9 is

(A) $6\sqrt{15}$. **(B)** $12\sqrt{5}$. **(C)** $16\sqrt{3}$. **(D)** $17\sqrt{3}$. **(E)** $18\sqrt{3}$.

49. Multiple Choice Two boats start at the same point and speed away along courses that form a 110° angle. If one boat travels at 24 miles per hour and the other boat travels at 32 miles per hour, how far apart are the boats after 30 minutes?

(A) 21 miles   (B) 22 miles   (C) 23 miles

(D) 24 miles   (E) 25 miles

50. Multiple Choice What is the measure of the smallest angle in a triangle with sides 12, 17, and 25?

(A) 21°   (B) 22°   (C) 23°   (D) 24°   (E) 25°

Explorations

51. Find the area of a regular polygon with n sides inscribed inside a circle of radius r. (Express your answer in terms of n and r.)

52. (a) Prove the identity: $\dfrac{\cos A}{a} = \dfrac{b^2 + c^2 - a^2}{2abc}$.

(b) Prove the (tougher) identity:

$$\frac{\cos A}{a} + \frac{\cos B}{b} + \frac{\cos C}{c} = \frac{a^2 + b^2 + c^2}{2abc}.$$

[*Hint*: use the identity in part (a), along with its other variations.]

53. Navigation Two ships leave a common port at 8:00 A.M. and travel at a constant rate of speed. Each ship keeps a log showing its distance from port and its distance from the other ship. Portions of the logs from later that morning for both ships are shown in the following tables.

Time	Naut mi from port	Naut mi from ship B
9:00	15.1	8.7
10:00	30.2	17.3

Time	Naut mi from port	Naut mi from ship A
9:00	12.4	8.7
11:00	37.2	26.0

(a) How fast is each ship traveling? (Express your answer in knots, which are nautical miles per hour.)

(b) What is the angle of intersection of the courses of the two ships? ○

(c) How far apart are the ships at 12:00 noon if they maintain the same courses and speeds?

Extending the Ideas

54. Prove that the area of a triangle can be found with the formula

$$\triangle \text{ Area} = \frac{a^2 \sin B \sin C}{2 \sin A}.$$

55. A **segment** of a circle is the region enclosed between a chord of a circle and the arc intercepted by the chord. Find the area of a segment intercepted by a 7-inch chord in a circle of radius 5 inches.

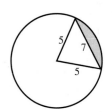

Math at Work

I got into medicine because I always liked a good challenge. Medicine can be like solving a puzzle, which I enjoy. I chose anesthesiology because it's even more challenging than other fields of medicine. Any surgical specialty tends to offer more difficult problems than, say, treating the sniffles.

What I enjoy most about my job is trying to gain people's confidence in a 5 minute interview. I can tell that some people will accept my judgments after our first interview, and others will question everything I do.

One good example of how we use math in medicine is when a patient goes into shock. Typically, the patient's blood pressure bottoms out, and his natural mechanisms are unable to raise it.

Ernest Newkirk, M.D.,

We give the patient dopamine to raise his blood pressure, but we need to make sure there is a consistent level of dopamine entering the bloodstream. For instance, if we have 400 mg of dopamine mixed into 250 cc of saline, we need to calculate how fast the intravenous drip should be so that there is 5 mg of dopamine in the bloodstream per kilo of body weight. Since all patients have different body weights, all patients require different rates of intravenous drip.

CHAPTER 5 Key Ideas

PROPERTIES, THEOREMS, AND FORMULAS

Reciprocal Identities   445
Quotient Identities   445
Pythagorean Identities   446
Cofunction Identities   447
Odd-Even Identities   447
Sum/Difference Identities   464–465
Double-Angle Identities   471
Power-Reducing Identities   472

Half-Angle Identities   473
Law of Sines   478
Law of Cosines   487
Triangle Area   489
Heron's Formula   490

PROCEDURES

Strategies for Proving an Identity   455–457

CHAPTER 5 Review Exercises

The collection of exercises marked in red could be used as a chapter test.

In Exercises 1 and 2, write the expression as the sine, cosine, or tangent of an angle.

1. $2 \sin 100° \cos 100°$

2. $\dfrac{2 \tan 40°}{1 - \tan^2 40°}$

In Exercises 3 and 4, simplify the expression to a single term. Support your answer graphically.

3. $(1 - 2 \sin^2 \theta)^2 + 4 \sin^2 \theta \cos^2 \theta$

4. $1 - 4 \sin^2 x \cos^2 x$

In Exercises 5–22, prove the identity.

5. $\cos 3x = 4 \cos^3 x - 3 \cos x$

6. $\cos^2 2x - \cos^2 x = \sin^2 x - \sin^2 2x$

7. $\tan^2 x - \sin^2 x = \sin^2 x \tan^2 x$

8. $2 \sin \theta \cos^3 \theta + 2 \sin^3 \theta \cos \theta = \sin 2\theta$

9. $\csc x - \cos x \cot x = \sin x$

10. $\dfrac{\tan \theta + \sin \theta}{2 \tan \theta} = \cos^2 \left(\dfrac{\theta}{2}\right)$

11. $\dfrac{1 + \tan \theta}{1 - \tan \theta} + \dfrac{1 + \cot \theta}{1 - \cot \theta} = 0$

12. $\sin 3\theta = 3 \cos^2 \theta \sin \theta - \sin^3 \theta$

13. $\cos^2 \left(\dfrac{t}{2}\right) = \dfrac{1 + \sec t}{2 \sec t}$

14. $\dfrac{\tan^3 \gamma - \cot^3 \gamma}{\tan^2 \gamma + \csc^2 \gamma} = \tan \gamma - \cot \gamma$

15. $\dfrac{\cos \phi}{1 - \tan \phi} + \dfrac{\sin \phi}{1 - \cot \phi} = \cos \phi + \sin \phi$

16. $\dfrac{\cos (-z)}{\sec (-z) + \tan (-z)} = 1 + \sin z$

17. $\sqrt{\dfrac{1 - \cos y}{1 + \cos y}} = \dfrac{1 - \cos y}{|\sin y|}$

18. $\sqrt{\dfrac{1 - \sin \gamma}{1 + \sin \gamma}} = \dfrac{|\cos \gamma|}{1 + \sin \gamma}$

19. $\tan \left(u + \dfrac{3\pi}{4}\right) = \dfrac{\tan u - 1}{1 + \tan u}$

20. $\dfrac{1}{4} \sin 4\gamma = \sin \gamma \cos^3 \gamma - \cos \gamma \sin^3 \gamma$

21. $\tan \dfrac{1}{2}\beta = \csc \beta - \cot \beta$

22. $\arctan t = \dfrac{1}{2} \arctan \dfrac{2t}{1 - t^2}, \quad -1 < t < 1$

In Exercises 23 and 24, use a grapher to conjecture whether the equation is likely to be an identity. Confirm your conjecture.

23. $\sec x - \sin x \tan x = \cos x$

24. $(\sin^2 \alpha - \cos^2 \alpha)(\tan^2 \alpha + 1) = \tan^2 \alpha - 1$

In Exercises 25–28, write the expression in terms of $\sin x$ and $\cos x$ only.

25. $\sin 3x + \cos 3x$

26. $\sin 2x + \cos 3x$

27. $\cos^2 2x - \sin 2x$

28. $\sin 3x - 3 \sin 2x$

In Exercises 29–34, find the general solution without using a calculator. Give exact answers.

29. $\sin 2x = 0.5$

30. $\cos x = \dfrac{\sqrt{3}}{2}$

31. $\tan x = -1$

32. $2 \sin^{-1} x = \sqrt{2}$

33. $\tan^{-1} x = 1$

34. $2 \cos 2x = 1$

In Exercises 35–38, solve the equation graphically. Find all solutions in the interval $[0, 2\pi)$.

35. $\sin^2 x - 3 \cos x = -0.5$

36. $\cos^3 x - 2 \sin x - 0.7 = 0$

37. $\sin^4 x + x^2 = 2$

38. $\sin 2x = x^3 - 5x^2 + 5x + 1$

In Exercises 39–44, find all solutions in the interval $[0, 2\pi)$ without using a calculator. Give exact answers.

39. $2 \cos x = 1$

40. $\sin 3x = \sin x$

41. $\sin^2 x - 2 \sin x - 3 = 0$

42. $\cos 2t = \cos t$

43. $\sin (\cos x) = 1$

44. $\cos 2x + 5 \cos x = 2$

In Exercises 45–48, solve the inequality. Use any method, but give exact answers.

45. $2 \cos 2x > 1$ for $0 \le x < 2\pi$

46. $\sin 2x > 2 \cos x$ for $0 < x \le 2\pi$

47. $2 \cos x < 1$ for $0 \le x < 2\pi$

48. $\tan x < \sin x$ for $-\dfrac{\pi}{2} < x < \dfrac{\pi}{2}$

In Exercises 49 and 50, find an equivalent equation of the form $y = a \sin (bx + c)$. Support your work graphically.

49. $y = 3 \sin 3x + 4 \cos 3x$ **50.** $y = 5 \sin 2x - 12 \cos 2x$

In Exercises 51–58, solve $\triangle ABC$.

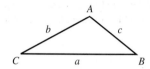

51. $A = 79°$, $B = 33°$, $a = 7$

52. $a = 5$, $b = 8$, $B = 110°$

53. $a = 8$, $b = 3$, $B = 30°$

54. $a = 14.7$, $A = 29.3°$, $C = 33°$

55. $A = 34°$, $B = 74°$, $c = 5$

56. $c = 41$, $A = 22.9°$, $C = 55.1°$

57. $a = 5$, $b = 7$, $c = 6$

58. $A = 85°$, $a = 6$, $b = 4$

In Exercises 59 and 60, find the area of $\triangle ABC$.

59. $a = 3$, $b = 5$, $c = 6$ **60.** $a = 10$, $b = 6$, $C = 50°$

61. If $a = 12$ and $B = 28°$, determine the values of b that will produce the indicated number of triangles:

   (a) Two   **(b)** One   **(c)** Zero

62. Surveying a Canyon Two markers A and B on the same side of a canyon rim are 80 ft apart, as shown in the figure. A hiker is located across the rim at point C. A surveyor determines that $\angle BAC = 70°$ and $\angle ABC = 65°$.

   (a) What is the distance between the hiker and point A?

   (b) What is the distance between the two canyon rims? (Assume they are parallel.)

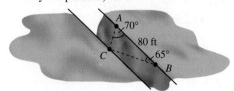

63. Altitude A hot-air balloon is seen over Tucson, Arizona, simultaneously by two observers at points A and B that are 1.75 mi apart on level ground and in line with the balloon. The angles of elevation are as shown here. How high above ground is the balloon?

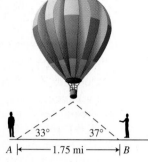

64. Finding Distance In order to determine the distance between two points A and B on opposite sides of a lake, a surveyor chooses a point C that is 900 ft from A and 225 ft from B, as shown in the figure. If the measure of the angle at C is 70°, find the distance between A and B.

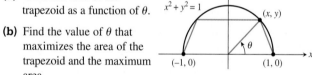

65. Finding Radian Measure Find the radian measure of the largest angle of the triangle whose sides have lengths 8, 9, and 10.

66. Finding a Parallelogram A parallelogram has sides of 15 and 24 ft, and an angle of 40°. Find the diagonals.

67. Maximizing Area A trapezoid is inscribed in the upper half of a unit circle, as shown in the figure.

   (a) Write the area of the trapezoid as a function of θ.

   (b) Find the value of θ that maximizes the area of the trapezoid and the maximum area.

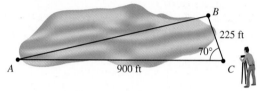

68. Beehive Cells A single cell in a beehive is a regular hexagonal prism open at the front with a trihedral cut at the back. Trihedral refers to a vertex formed by three faces of a polyhedron. It can be shown that the surface area of a cell is given by

$$S(\theta) = 6ab + \frac{3}{2}b^2 \left(-\cot \theta + \frac{\sqrt{3}}{\sin \theta} \right),$$

where θ is the angle between the axis of the prism and one of the back faces, a is the depth of the prism, and b is the length of the hexagonal front. Assume $a = 1.75$ in. and $b = 0.65$ in.

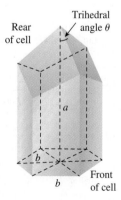

(a) Graph the function $y = S(\theta)$

(b) What value of θ gives the minimum surface area?
(*Note:* This answer is quite close to the observed angle in nature.)

(c) What is the minimum surface area?

69. Cable Television Coverage A cable broadcast satellite S orbits a planet at a height h (in miles) above the Earth's surface, as shown in the figure. The two lines from S are tangent to the Earth's surface. The part of the Earth's surface that is in the broadcast area of the satellite is determined by the central angle θ indicated in the figure.

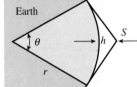

(a) Assuming that the Earth is spherical with a radius of 4000 mi, write h as a function of θ.

(b) Approximate θ for a satellite 200 mi above the surface of the Earth.

70. Finding Extremum Values The graph of

$$y = \cos x - \frac{1}{2}\cos 2x + \frac{1}{3}\cos 3x$$

is shown in the figure. The x-values that correspond to local maximum and minimum points are solutions of the equation $\sin x - \sin 2x + \sin 3x = 0$. Solve this equation algebraically, and support your solution using the graph of y.

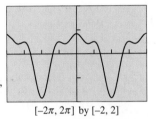

$[-2\pi, 2\pi]$ by $[-2, 2]$

71. Using Trigonometry in Geometry A regular hexagon whose sides are 16 cm is inscribed in a circle. Find the area inside the circle and outside the hexagon.

72. Using Trigonometry in Geometry A circle is inscribed in a regular pentagon whose sides are 12 cm. Find the area inside the pentagon and outside the circle.

73. Using Trigonometry in Geometry A wheel of cheese in the shape of a right circular cylinder is 18 cm in diameter and 5 cm thick. If a wedge of cheese with a central angle of 15° is cut from the wheel, find the volume of the cheese wedge.

74. Product-to-Sum Formulas Prove the following identities, which are called the **product-to-sum formulas**.

(a) $\sin u \sin v = \frac{1}{2}(\cos(u - v) - \cos(u + v))$

(b) $\cos u \cos v = \frac{1}{2}(\cos(u - v) + \cos(u + v))$

(c) $\sin u \cos v = \frac{1}{2}(\sin(u + v) + \sin(u - v))$

75. Sum-to-Product Formulas Use the product-to-sum formulas in Exercise 74 to prove the following identities, which are called the **sum-to-product formulas**.

(a) $\sin u + \sin v = 2\sin\dfrac{u + v}{2}\cos\dfrac{u - v}{2}$

(b) $\sin u - \sin v = 2\sin\dfrac{u - v}{2}\cos\dfrac{u + v}{2}$

(c) $\cos u + \cos v = 2\cos\dfrac{u + v}{2}\cos\dfrac{u - v}{2}$

(d) $\cos u - \cos v = -2\sin\dfrac{u + v}{2}\sin\dfrac{u - v}{2}$

76. Catching Students Faking Data
Carmen and Pat both need to make up a missed physics lab. They are to measure the total distance ($2x$) traveled by a beam of light from point A to point B and record it in 20° increments of θ as they adjust the mirror (C) upward vertically. They report

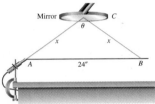

the following measurements. However, only one of the students actually did the lab; the other skipped it and faked the data. Who faked the data, and how can you tell?

	CARMEN		PAT
θ	$2x$	θ	$2x$
160°	24.4″	160°	24.5″
140°	25.6″	140°	25.2″
120°	28.0″	120°	26.4″
100°	31.2″	100°	30.4″
80°	37.6″	80°	35.2″
60°	48.0″	60°	48.0″
40°	70.4″	40°	84.0″
20°	138.4″	20°	138.4″

77. An Interesting Fact about (sin A)/a The ratio $(\sin A)/a$ that shows up in the Law of Sines shows up another way in the geometry of $\triangle ABC$: It is the reciprocal of the radius of the circumscribed circle.

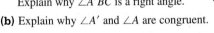

(a) Let $\triangle ABC$ be circumscribed as shown in the diagram, and construct diameter CA'. Explain why $\angle A'BC$ is a right angle.

(b) Explain why $\angle A'$ and $\angle A$ are congruent.

(c) If a, b, and c are the sides opposite angles A, B, and C as usual, explain why $\sin A' = a/d$, where d is the diameter of the circle.

(d) Finally, explain why $(\sin A)/a = 1/d$.

(e) Do $(\sin B)/b$ and $(\sin C)/c$ also equal $1/d$? Why?

31.

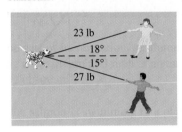

32.

In Exercises 33–38, find the magnitude and direction angle of the vector.

33. $\langle 3, 4 \rangle$

34. $\langle -1, 2 \rangle$

35. $3\mathbf{i} - 4\mathbf{j}$

36. $-3\mathbf{i} - 5\mathbf{j}$

37. $7(\cos 135° \, \mathbf{i} + \sin 135° \, \mathbf{j})$

38. $2(\cos 60° \, \mathbf{i} + \sin 60° \, \mathbf{j})$

In Exercises 39 and 40, find the vector $\mathbf{v}$ with the given magnitude and the same direction as $\mathbf{u}$.

39. $|\mathbf{v}| = 2, \mathbf{u} = \langle 3, -3 \rangle$

40. $|\mathbf{v}| = 5, \mathbf{u} = \langle -5, 7 \rangle$

41. Navigation An airplane is flying on a bearing of 335° at 530 mph. Find the component form of the velocity of the airplane.

42. Navigation An airplane is flying on a bearing of 170° at 460 mph. Find the component form of the velocity of the airplane.

43. Flight Engineering An airplane is flying on a compass heading (bearing) of 340° at 325 mph. A wind is blowing with the bearing 320° at 40 mph.

(a) Find the component form of the velocity of the airplane.

(b) Find the actual ground speed and direction of the plane.

44. Flight Engineering An airplane is flying on a compass heading (bearing) of 170° at 460 mph. A wind is blowing with the bearing 200° at 80 mph.

(a) Find the component form of the velocity of the airplane.

(b) Find the actual ground speed and direction of the airplane.

45. Shooting a Basketball A basketball is shot at a 70° angle with the horizontal direction with an initial speed of 10 m/sec.

(a) Find the component form of the initial velocity.

(b) Writing to Learn Give an interpretation of the horizontal and vertical components of the velocity.

46. Moving a Heavy Object In a warehouse a box is being pushed up a 15° inclined plane with a force of 2.5 lb, as shown in the figure.

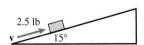

(a) Find the component form of the force.

(b) Writing to Learn Give an interpretation of the horizontal and vertical components of the force.

47. Moving a Heavy Object Suppose the box described in Exercise 46 is being towed up the inclined plane, as shown in the figure below. Find the force $\mathbf{w}$ needed in order for the component of the force parallel to the inclined plane to be 2.5 lb. Give the answer in component form.

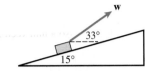

48. Combining Forces Juana and Diego Gonzales, ages six and four respectively, own a strong and stubborn puppy named Corporal. It is so hard to take Corporal for a walk that they devise a scheme to use two leashes. If Juana and Diego pull with forces of 23 lb and 27 lb at the angles shown in the figure, how hard is Corporal pulling if the puppy holds the children at a standstill?

In Exercises 49 and 50, find the direction and magnitude of the resultant force.

49. Combining Forces A force of 50 lb acts on an object at an angle of 45°. A second force of 75 lb acts on the object at an angle of −30°.

50. Combining Forces Three forces with magnitudes 100, 50, and 80 lb, act on an object at angles of 50°, 160°, and −20°, respectively.

51. Navigation A ship is heading due north at 12 mph. The current is flowing southwest at 4 mph. Find the actual bearing and speed of the ship.

52. Navigation A motor boat capable of 20 mph keeps the bow of the boat pointed straight across a mile-wide river. The current is flowing left to right at 8 mph. Find where the boat meets the opposite shore.

53. Group Activity A ship heads due south with the current flowing northwest. Two hours later the ship is 20 miles in the direction 30° west of south from the original starting point. Find the speed with no current of the ship and the rate of the current.

54. Group Activity Express each vector in component form and prove the following properties of vectors.

(a) $\mathbf{u} + \mathbf{v} = \mathbf{v} + \mathbf{u}$

(b) $(\mathbf{u} + \mathbf{v}) + \mathbf{w} = \mathbf{u} + (\mathbf{v} + \mathbf{w})$

(c) $\mathbf{u} + \mathbf{0} = \mathbf{u}$, where $\mathbf{0} = \langle 0, 0 \rangle$

(d) $\mathbf{u} + (-\mathbf{u}) = \mathbf{0}$, where $-\langle a, b \rangle = \langle -a, -b \rangle$

(e) $a(\mathbf{u} + \mathbf{v}) = a\mathbf{u} + a\mathbf{v}$ **(f)** $(a + b)\mathbf{u} = a\mathbf{u} + b\mathbf{u}$

(g) $(ab)\mathbf{u} = a(b\mathbf{u})$ **(h)** $a\mathbf{0} = \mathbf{0}, 0\mathbf{u} = \mathbf{0}$

(i) $(1)\mathbf{u} = \mathbf{u}, (-1)\mathbf{u} = -\mathbf{u}$ **(j)** $|a\mathbf{u}| = |a|\,|\mathbf{u}|$

Standardized Test Questions

55. True or False If $\mathbf{u}$ is a unit vector, then $-\mathbf{u}$ is also a unit vector. Justify your answer.

56. True or False If $\mathbf{u}$ is a unit vector, then $1/\mathbf{u}$ is also a unit vector. Justify your answer.

In Exercises 57–60, you may use a graphing calculator to solve the problem.

57. Multiple Choice Which of the following is the magnitude of the vector $\langle 2, -1 \rangle$?

(A) 1   **(B)** $\sqrt{3}$ **(C)** $\dfrac{\sqrt{5}}{5}$

(D) $\sqrt{5}$ **(E)** 5

58. Multiple Choice Let $\mathbf{u} = \langle -2, 3 \rangle$ and $\mathbf{v} = \langle 4, -1 \rangle$. Which of the following is equal to $\mathbf{u} - \mathbf{v}$?

(A) $\langle 6, -4 \rangle$ **(B)** $\langle 2, 2 \rangle$ **(C)** $\langle -2, 2 \rangle$

(D) $\langle -6, 2 \rangle$ **(E)** $\langle -6, 4 \rangle$

59. Multiple Choice Which of the following represents the vector $\mathbf{v}$ shown in the figure below?

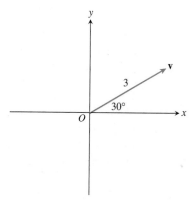

(A) $\langle 3 \cos 30°, 3 \sin 30° \rangle$ **(B)** $\langle 3 \sin 30°, 3 \cos 30° \rangle$

(C) $\langle 3 \cos 60°, 3 \sin 60° \rangle$ **(D)** $\langle \sqrt{3} \cos 30°, \sqrt{3} \sin 30° \rangle$

(E) $\langle \sqrt{3} \cos 30°, \sqrt{3} \sin 30° \rangle$

60. Multiple Choice Which of the following is a unit vector in the direction of $\mathbf{v} = -\mathbf{i} + 3\mathbf{j}$?

(A) $-\dfrac{1}{10}\mathbf{i} + \dfrac{3}{10}\mathbf{j}$ **(B)** $\dfrac{1}{10}\mathbf{i} - \dfrac{3}{10}\mathbf{j}$ **(C)** $-\dfrac{1}{\sqrt{10}}\mathbf{i} + \dfrac{3}{\sqrt{10}}\mathbf{j}$

(D) $\dfrac{1}{\sqrt{10}}\mathbf{i} - \dfrac{3}{\sqrt{10}}\mathbf{j}$ **(E)** $-\dfrac{1}{\sqrt{8}}\mathbf{i} + \dfrac{3}{\sqrt{8}}\mathbf{j}$

Explorations

61. Dividing a Line Segment in a Given Ratio Let A and B be two points in the plane, as shown in the figure.

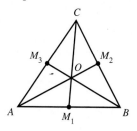

(a) Prove that $\overrightarrow{BA} = \overrightarrow{OA} - \overrightarrow{OB}$, where O is the origin.

(b) Let C be a point on the line segment BA which divides the segment in the ratio $x : y$ where $x + y = 1$. That is,

$$\frac{|\overrightarrow{BC}|}{|\overrightarrow{CA}|} = \frac{x}{y}.$$

Show that $\overrightarrow{OC} = x\overrightarrow{OA} + y\overrightarrow{OB}$.

62. Medians of a Triangle Perform the following steps to use vectors to prove that the medians of a triangle meet at a point O which divides each median in the ratio $1 : 2$. $M_1, M_2,$ and M_3 are midpoints of the sides of the triangle shown in the figure.

(a) Use Exercise 61 to prove that

$$\overrightarrow{OM_1} = \frac{1}{2}\overrightarrow{OA} + \frac{1}{2}\overrightarrow{OB}$$

$$\overrightarrow{OM_2} = \frac{1}{2}\overrightarrow{OC} + \frac{1}{2}\overrightarrow{OB}$$

$$\overrightarrow{OM_3} = \frac{1}{2}\overrightarrow{OA} + \frac{1}{2}\overrightarrow{OC}$$

(b) Prove that each of $2\overrightarrow{OM_1} + \overrightarrow{OC}, 2\overrightarrow{OM_2} + \overrightarrow{OA}, 2\overrightarrow{OM_3} + \overrightarrow{OB}$ is equal to $\overrightarrow{OA} + \overrightarrow{OB} + \overrightarrow{OC}$.

(c) Writing to Learn Explain why part (b) establishes the desired result.

Extending the Ideas

63. Vector Equation of a Line Let L be the line through the two points A and B. Prove that $C = (x, y)$ is on the line L if and only if $\overrightarrow{OC} = t\overrightarrow{OA} + (1 - t)\overrightarrow{OB}$, where t is a real number and O is the origin.

64. Connecting Vectors and Geometry Prove that the lines which join one vertex of a parallelogram to the midpoints of the opposite sides trisect the diagonal.

6.2
Dot Product of Vectors

What you'll learn about

- The Dot Product
- Angle Between Vectors
- Projecting One Vector onto Another
- Work

. . . and why

Vectors are used extensively in mathematics and science applications such as determining the net effect of several forces acting on an object and computing the work done by a force acting on an object.

The Dot Product

Vectors can be multiplied in two different ways, both of which are derived from their usefulness for solving problems in vector applications. The *cross product* (or *vector product* or *outer product*) results in a vector perpendicular to the plane of the two vectors being multiplied, which takes us into a third dimension and outside the scope of this chapter. The *dot product* (or *scalar product* or *inner product*) results in a scalar. In other words, the dot product of two vectors is not a vector but a real number! It is the important information conveyed by that number that makes the dot product so worthwhile, as you will see.

Now that you have some experience with vectors and arrows, we hope we won't confuse you if we occasionally resort to the common convention of using arrows to name the vectors they represent. For example, we might write "$\mathbf{u} = \overrightarrow{PQ}$" as a shorthand for "$\mathbf{u}$ is the vector represented by $\overrightarrow{PQ}$." This greatly simplifies the discussion of concepts like vector projection. Also, we will continue to use both vector notations, $\langle a, b \rangle$ and $a\mathbf{i} + b\mathbf{j}$, so you will get some practice with each.

DEFINITION Dot Product

The **dot product** or **inner product** of $\mathbf{u} = \langle u_1, u_2 \rangle$ and $\mathbf{v} = \langle v_1, v_2 \rangle$ is

$$\mathbf{u} \cdot \mathbf{v} = u_1 v_1 + u_2 v_2.$$

DOT PRODUCT AND STANDARD UNIT VECTORS

$(u_1\mathbf{i} + u_2\mathbf{j}) \cdot (v_1\mathbf{i} + v_2\mathbf{j}) = u_1 v_1 + u_2 v_2$

Dot products have many important properties that we make use of in this section. We prove the first two and leave the rest for the Exercises.

Properties of the Dot Product

Let $\mathbf{u}$, $\mathbf{v}$, and $\mathbf{w}$ be vectors and let c be a scalar.

1. $\mathbf{u} \cdot \mathbf{v} = \mathbf{v} \cdot \mathbf{u}$
2. $\mathbf{u} \cdot \mathbf{u} = |\mathbf{u}|^2$
3. $\mathbf{0} \cdot \mathbf{u} = 0$
4. $\mathbf{u} \cdot (\mathbf{v} + \mathbf{w}) = \mathbf{u} \cdot \mathbf{v} + \mathbf{u} \cdot \mathbf{w}$
   $(\mathbf{u} + \mathbf{v}) \cdot \mathbf{w} = \mathbf{u} \cdot \mathbf{w} + \mathbf{v} \cdot \mathbf{w}$
5. $(c\mathbf{u}) \cdot \mathbf{v} = \mathbf{u} \cdot (c\mathbf{v}) = c(\mathbf{u} \cdot \mathbf{v})$

Proof

Let $\mathbf{u} = \langle u_1, u_2 \rangle$ and $\mathbf{v} = \langle v_1, v_2 \rangle$.

Property 1

$$\mathbf{u} \cdot \mathbf{v} = u_1 v_1 + u_2 v_2 \quad \text{Definition of } \mathbf{u} \cdot \mathbf{v}$$
$$= v_1 u_1 + v_2 u_2 \quad \text{Commutative property of real numbers}$$
$$= \mathbf{v} \cdot \mathbf{u} \quad \text{Definition of } \mathbf{u} \cdot \mathbf{v}$$

Property 2

$$\mathbf{u} \cdot \mathbf{u} = u_1^2 + u_2^2 \quad \text{Definition of } \mathbf{u} \cdot \mathbf{u}$$
$$= \left(\sqrt{u_1^2 + u_2^2}\right)^2$$
$$= |\mathbf{u}|^2 \quad \text{Definition of } |\mathbf{u}|$$

EXAMPLE 1   Finding Dot Products

Find each dot product.

(a) $\langle 3, 4 \rangle \cdot \langle 5, 2 \rangle$

(b) $\langle 1, -2 \rangle \cdot \langle -4, 3 \rangle$

(c) $(2\mathbf{i} - \mathbf{j}) \cdot (3\mathbf{i} - 5\mathbf{j})$

SOLUTION

(a) $\langle 3, 4 \rangle \cdot \langle 5, 2 \rangle = (3)(5) + (4)(2) = 23$

(b) $\langle 1, -2 \rangle \cdot \langle -4, 3 \rangle = (1)(-4) + (-2)(3) = -10$

(c) $(2\mathbf{i} - \mathbf{j}) \cdot (3\mathbf{i} - 5\mathbf{j}) = (2)(3) + (-1)(-5) = 11$ *Now try Exercise 3.*

DOT PRODUCTS ON CALCULATORS

It is really a waste of time to compute a simple dot product of two-dimensional vectors using a calculator, but it can be done. Some calculators do vector operations outright, and others can do vector operations via matrices. If you have learned about matrix multiplication already, you will know why the matrix product $[u_1, u_2] \cdot \begin{bmatrix} v_1 \\ v_2 \end{bmatrix}$ yields the dot product $\langle u_1, u_2 \rangle \cdot \langle v_1, v_2 \rangle$ as a 1-by-1 matrix. (The same trick works with vectors of higher dimensions.) This book will cover matrix multiplication in Chapter 7.

Property 2 of the dot product gives us another way to find the length of a vector, as illustrated in Example 2.

EXAMPLE 2   Using Dot Product to Find Length

Use the dot product to find the length of the vector $\mathbf{u} = \langle 4, -3 \rangle$.

SOLUTION It follows from Property 2 that $|\mathbf{u}| = \sqrt{\mathbf{u} \cdot \mathbf{u}}$. Thus,

$$|\langle 4, -3 \rangle| = \sqrt{\langle 4, -3 \rangle \cdot \langle 4, -3 \rangle} = \sqrt{(4)(4) + (-3)(-3)} = \sqrt{25} = 5.$$

Now try Exercise 9.

Angle Between Vectors

Let $\mathbf{u}$ and $\mathbf{v}$ be two nonzero vectors in standard position as shown in Figure 6.16. The **angle between u and v** is the angle θ, $0 \le \theta \le \pi$ or $0° \le \theta \le 180°$. The angle between any two nonzero vectors is the corresponding angle between their respective standard position representatives.

We can use the dot product to find the angle between nonzero vectors, as we prove in the next theorem.

FIGURE 6.16 The angle θ between nonzero vectors $\mathbf{u}$ and $\mathbf{v}$.

THEOREM Angle Between Two Vectors

If θ is the angle between the nonzero vectors $\mathbf{u}$ and $\mathbf{v}$, then

$$\cos \theta = \frac{\mathbf{u} \cdot \mathbf{v}}{|\mathbf{u}||\mathbf{v}|}$$

$$\text{and } \theta = \cos^{-1}\left(\frac{\mathbf{u} \cdot \mathbf{v}}{|\mathbf{u}||\mathbf{v}|}\right)$$

Proof

We apply the Law of Cosines to the triangle determined by **u**, **v**, and **v** − **u** in Figure 6.16, and use the properties of the dot product.

$$|\mathbf{v} - \mathbf{u}|^2 = |\mathbf{u}|^2 + |\mathbf{v}|^2 - 2|\mathbf{u}||\mathbf{v}| \cos \theta$$

$$(\mathbf{v} - \mathbf{u}) \cdot (\mathbf{v} - \mathbf{u}) = |\mathbf{u}|^2 + |\mathbf{v}|^2 - 2|\mathbf{u}||\mathbf{v}| \cos \theta$$

$$\mathbf{v} \cdot \mathbf{v} - \mathbf{v} \cdot \mathbf{u} - \mathbf{u} \cdot \mathbf{v} + \mathbf{u} \cdot \mathbf{u} = |\mathbf{u}|^2 + |\mathbf{v}|^2 - 2|\mathbf{u}||\mathbf{v}| \cos \theta$$

$$|\mathbf{v}|^2 - 2\mathbf{u} \cdot \mathbf{v} + |\mathbf{u}|^2 = |\mathbf{u}|^2 + |\mathbf{v}|^2 - 2|\mathbf{u}||\mathbf{v}| \cos \theta$$

$$-2\mathbf{u} \cdot \mathbf{v} = -2|\mathbf{u}||\mathbf{v}| \cos \theta$$

$$\cos \theta = \frac{\mathbf{u} \cdot \mathbf{v}}{|\mathbf{u}||\mathbf{v}|}$$

$$\theta = \cos^{-1}\left(\frac{\mathbf{u} \cdot \mathbf{v}}{|\mathbf{u}||\mathbf{v}|}\right)$$

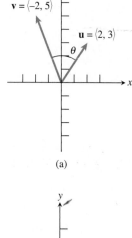

(a)

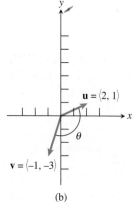

(b)

FIGURE 6.17 The vectors in (a) Example 3a and (b) Example 3b.

EXAMPLE 3   Finding the Angle Between Vectors

Find the angle between the vectors **u** and **v**.

(a) $\mathbf{u} = \langle 2, 3 \rangle$, $\mathbf{v} = \langle -2, 5 \rangle$ **(b)** $\mathbf{u} = \langle 2, 1 \rangle$, $\mathbf{v} = \langle -1, -3 \rangle$

SOLUTION

(a) See Figure 6.17a. Using the Angle Between Two Vectors Theorem, we have

$$\cos \theta = \frac{\mathbf{u} \cdot \mathbf{v}}{|\mathbf{u}||\mathbf{v}|} = \frac{\langle 2, 3 \rangle \cdot \langle -2, 5 \rangle}{|\langle 2, 3 \rangle||\langle -2, 5 \rangle|} = \frac{11}{\sqrt{13}\,\sqrt{29}}.$$

So,

$$\theta = \cos^{-1}\left(\frac{11}{\sqrt{13}\,\sqrt{29}}\right) \approx 55.5°.$$

(b) See Figure 6.17b. Again using the Angle Between Two Vectors Theorem, we have

$$\cos \theta = \frac{\mathbf{u} \cdot \mathbf{v}}{|\mathbf{u}||\mathbf{v}|} = \frac{\langle 2, 1 \rangle \cdot \langle -1, -3 \rangle}{|\langle 2, 1 \rangle||\langle -1, -3 \rangle|} = \frac{-5}{\sqrt{5}\,\sqrt{10}} = \frac{-1}{\sqrt{2}}.$$

So,

$$\theta = \cos^{-1}\left(\frac{-1}{\sqrt{2}}\right) = 135°.$$

Now try Exercise 13.

If vectors **u** and **v** are perpendicular, that is, if the angle between them is 90°, then

$$\mathbf{u} \cdot \mathbf{v} = |\mathbf{u}||\mathbf{v}| \cos 90° = 0$$

because cos 90° = 0.

DEFINITION Orthogonal Vectors

The vectors **u** and **v** are **orthogonal** if and only if $\mathbf{u} \cdot \mathbf{v} = 0$.

The terms "perpendicular" and "orthogonal" almost mean the same thing. The zero vector has no direction angle, so technically speaking, the zero vector is not perpendicular to any vector. However, the zero vector is orthogonal to every vector. Except for this special case, orthogonal and perpendicular are the same.

EXAMPLE 4   Proving Vectors are Orthogonal

Prove that the vectors $\mathbf{u} = \langle 2, 3 \rangle$ and $\mathbf{v} = \langle -6, 4 \rangle$ are orthogonal.

SOLUTION We must prove that their dot product is zero.

$$\mathbf{u} \cdot \mathbf{v} = \langle 2, 3 \rangle \cdot \langle -6, 4 \rangle = -12 + 12 = 0$$

The two vectors are orthogonal. *Now try Exercise 23.*

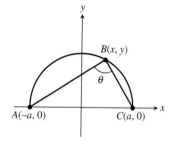

FIGURE 6.18 The angle $\angle ABC$ inscribed in the upper half of the circle $x^2 + y^2 = a^2$. (Exploration 1)

EXPLORATION 1   Angles Inscribed in Semicircles

Figure 6.18 shows $\angle ABC$ inscribed in the upper half of the circle $x^2 + y^2 = a^2$.

1. For $a = 2$, find the component form of the vectors $\mathbf{u} = \overrightarrow{BA}$ and $\mathbf{v} = \overrightarrow{BC}$.

2. Find $\mathbf{u} \cdot \mathbf{v}$. What can you conclude about the angle θ between these two vectors?

3. Repeat parts 1 and 2 for arbitrary a.

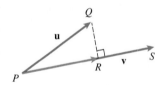

FIGURE 6.19 The vectors $\mathbf{u} = \overrightarrow{PQ}$, $\mathbf{v} = \overrightarrow{PS}$, and the vector projection of $\mathbf{u}$ onto $\mathbf{v}$, $\overrightarrow{PR} = \text{proj}_\mathbf{v}\mathbf{u}$.

Projecting One Vector onto Another

The **vector projection** of $\mathbf{u} = \overrightarrow{PQ}$ onto a nonzero vector $\mathbf{v} = \overrightarrow{PS}$ is the vector $\overrightarrow{PR}$ determined by dropping a perpendicular from Q to the line PS (Figure 6.19). We have resolved $\mathbf{u}$ into components $\overrightarrow{PR}$ and $\overrightarrow{RQ}$

$$\mathbf{u} = \overrightarrow{PR} + \overrightarrow{RQ}$$

with $\overrightarrow{PR}$ and $\overrightarrow{RQ}$ perpendicular.

The standard notation for $\overrightarrow{PR}$, the vector projection of $\mathbf{u}$ onto $\mathbf{v}$, is $\overrightarrow{PR} = \text{proj}_\mathbf{v}\mathbf{u}$. With this notation, $\overrightarrow{RQ} = \mathbf{u} - \text{proj}_\mathbf{v}\mathbf{u}$. We ask you to establish the following formula in the Exercises (see Exercise 58).

Projection of u onto v

If $\mathbf{u}$ and $\mathbf{v}$ are nonzero vectors, the projection of $\mathbf{u}$ onto $\mathbf{v}$ is

$$\text{proj}_\mathbf{v}\mathbf{u} = \left(\frac{\mathbf{u} \cdot \mathbf{v}}{|\mathbf{v}|^2} \right)\mathbf{v}.$$

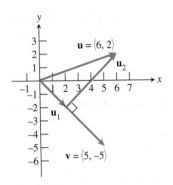

FIGURE 6.20 The vectors $\mathbf{u} = \langle 6, 2 \rangle$, $\mathbf{v} = \langle 5, -5 \rangle$, $\mathbf{u}_1 = \text{proj}_{\mathbf{v}}\mathbf{u}$, and $\mathbf{u}_2 = \mathbf{u} - \mathbf{u}_1$. (Example 5)

EXAMPLE 5 Decomposing a Vector into Perpendicular Components

Find the vector projection of $\mathbf{u} = \langle 6, 2 \rangle$ onto $\mathbf{v} = \langle 5, -5 \rangle$. Then write $\mathbf{u}$ as the sum of two orthogonal vectors, one of which is $\text{proj}_{\mathbf{v}}\mathbf{u}$.

SOLUTION We write $\mathbf{u} = \mathbf{u}_1 + \mathbf{u}_2$ where $\mathbf{u}_1 = \text{proj}_{\mathbf{v}}\mathbf{u}$ and $\mathbf{u}_2 = \mathbf{u} - \mathbf{u}_1$ (Figure 6.20).

$$\mathbf{u}_1 = \text{proj}_{\mathbf{v}}\mathbf{u} = \left(\frac{\mathbf{u} \cdot \mathbf{v}}{|\mathbf{v}|^2} \right)\mathbf{v} = \frac{20}{50}\langle 5, -5 \rangle = \langle 2, -2 \rangle$$

$$\mathbf{u}_2 = \mathbf{u} - \mathbf{u}_1 = \langle 6, 2 \rangle - \langle 2, -2 \rangle = \langle 4, 4 \rangle$$

Thus, $\mathbf{u}_1 + \mathbf{u}_2 = \langle 2, -2 \rangle + \langle 4, 4 \rangle = \langle 6, 2 \rangle = \mathbf{u}$.

Now try Exercise 25.

If $\mathbf{u}$ is a force, then $\text{proj}_{\mathbf{v}}\mathbf{u}$ represents the effective force in the direction of $\mathbf{v}$ (Figure 6.21).

We can use vector projections to determine the amount of force required in problem situations like Example 6.

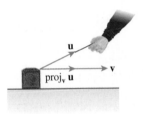

FIGURE 6.21 If we pull on a box with force $\mathbf{u}$, the effective force in the direction of $\mathbf{v}$ is $\text{proj}_{\mathbf{v}}\mathbf{u}$, the vector projection of $\mathbf{u}$ onto $\mathbf{v}$.

EXAMPLE 6 Finding a Force

Juan is sitting on a sled on the side of a hill inclined at $45°$. The combined weight of Juan and the sled is 140 pounds. What force is required for Rafaela to keep the sled from sliding down the hill? (See Figure 6.22.)

SOLUTION We can represent the force due to gravity as $\mathbf{F} = -140\mathbf{j}$ because gravity acts vertically downward. We can represent the side of the hill with the vector

$$\mathbf{v} = (\cos 45°)\mathbf{i} + (\sin 45°)\mathbf{j} = \frac{\sqrt{2}}{2}\mathbf{i} + \frac{\sqrt{2}}{2}\mathbf{j}.$$

The force required to keep the sled from sliding down the hill is

$$\mathbf{F}_1 = \text{proj}_{\mathbf{v}}\mathbf{F} = \left(\frac{\mathbf{F} \cdot \mathbf{v}}{|\mathbf{v}|^2} \right)\mathbf{v} = (\mathbf{F} \cdot \mathbf{v})\mathbf{v}$$

because $|\mathbf{v}| = 1$. So,

$$\mathbf{F}_1 = (\mathbf{F} \cdot \mathbf{v})\mathbf{v} = (-140)\left(\frac{\sqrt{2}}{2} \right)\mathbf{v} = -70(\mathbf{i} + \mathbf{j}).$$

The magnitude of the force that Rafaela must exert to keep the sled from sliding down the hill is $70\sqrt{2} \approx 99$ pounds.

Now try Exercise 45.

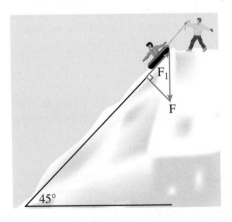

FIGURE 6.22 The sled in Example 6.

Work

If $\mathbf{F}$ is a constant force whose direction is the same as the direction of $\overrightarrow{AB}$, then the **work** W done by $\mathbf{F}$ in moving an object from A to B is

$$W = |\mathbf{F}|\,|\overrightarrow{AB}|.$$

If **F** is a constant force in any direction, then the **work** W done by **F** in moving an object from A to B is

$$W = \mathbf{F} \cdot \overrightarrow{AB}$$

$$= |\mathbf{F}||\overrightarrow{AB}| \cos \theta$$

where θ is the angle between **F** and $\overrightarrow{AB}$. Except for the sign, the work is the magnitude of the effective force in the direction of $\overrightarrow{AB}$ times $\overrightarrow{AB}$.

EXAMPLE 7   Finding Work

Find the work done by a 10 pound force acting in the direction $\langle 1, 2 \rangle$ in moving an object 3 feet from $(0, 0)$ to $(3, 0)$.

SOLUTION The force **F** has magnitude 10 and acts in the direction $\langle 1, 2 \rangle$, so

$$\mathbf{F} = 10 \frac{\langle 1, 2 \rangle}{|\langle 1, 2 \rangle|} = \frac{10}{\sqrt{5}} \langle 1, 2 \rangle.$$

UNITS FOR WORK

Work is usually measured in foot-pounds or Newton-meters. One Newton-meter is commonly referred to as one Joule.

The direction of motion is from $A = (0, 0)$ to $B = (3, 0)$, so $\overrightarrow{AB} = \langle 3, 0 \rangle$. Thus, the work done by the force is

$$\mathbf{F} \cdot \overrightarrow{AB} = \frac{10}{\sqrt{5}} \langle 1, 2 \rangle \cdot \langle 3, 0 \rangle = \frac{30}{\sqrt{5}} \approx 13.42 \text{ foot-pounds.}$$

Now try Exercise 53.

QUICK REVIEW 6.2 *(For help, go to Section 6.1.)*

In Exercises 1–4, find $|\mathbf{u}|$.

1. $\mathbf{u} = \langle 2, -3 \rangle$

2. $\mathbf{u} = -3\mathbf{i} - 4\mathbf{j}$

3. $\mathbf{u} = \cos 35° \, \mathbf{i} + \sin 35° \, \mathbf{j}$

4. $\mathbf{u} = 2(\cos 75° \mathbf{i} + \sin 75° \mathbf{j})$

In Exercises 5–8, the points A and B lie on the circle $x^2 + y^2 = 4$. Find the component form of the vector $\overrightarrow{AB}$.

5. $A = (-2, 0)$, $B = (1, \sqrt{3})$ **6.** $A = (2, 0)$, $B = (1, \sqrt{3})$

7. $A = (2, 0)$, $B = (1, -\sqrt{3})$

8. $A = (-2, 0)$, $B = (1, -\sqrt{3})$

In Exercises 9 and 10, find a vector **u** with the given magnitude in the direction of **v**.

9. $|\mathbf{u}| = 2$, $\mathbf{v} = \langle 2, 3 \rangle$ **10.** $|\mathbf{u}| = 3$, $\mathbf{v} = -4\mathbf{i} + 3\mathbf{j}$

SECTION 6.2 EXERCISES

In Exercises 1–8, find the dot product of **u** and **v**.

1. $\mathbf{u} = \langle 5, 3 \rangle$, $\mathbf{v} = \langle 12, 4 \rangle$

2. $\mathbf{u} = \langle -5, 2 \rangle$, $\mathbf{v} = \langle 8, 13 \rangle$

3. $\mathbf{u} = \langle 4, 5 \rangle$, $\mathbf{v} = \langle -3, -7 \rangle$

4. $\mathbf{u} = \langle -2, 7 \rangle$, $\mathbf{v} = \langle -5, -8 \rangle$

5. $\mathbf{u} = -4\mathbf{i} - 9\mathbf{j}$, $\mathbf{v} = -3\mathbf{i} - 2\mathbf{j}$

6. $\mathbf{u} = 2\mathbf{i} - 4\mathbf{j}$, $\mathbf{v} = -8\mathbf{i} + 7\mathbf{j}$

7. $\mathbf{u} = 7\mathbf{i}$, $\mathbf{v} = -2\mathbf{i} + 5\mathbf{j}$

8. $\mathbf{u} = 4\mathbf{i} - 11\mathbf{j}$, $\mathbf{v} = -3\mathbf{j}$

In Exercises 9–12, use the dot product to find $|\mathbf{u}|$.

9. $\mathbf{u} = \langle 5, -12 \rangle$

10. $\mathbf{u} = \langle -8, 15 \rangle$

11. $\mathbf{u} = -4\mathbf{i}$

12. $\mathbf{u} = 3\mathbf{j}$

In Exercises 13–22, find the angle θ between the vectors.

13. $\mathbf{u} = \langle -4, -3 \rangle$, $\mathbf{v} = \langle -1, 5 \rangle$

14. $\mathbf{u} = \langle 2, -2 \rangle$, $\mathbf{v} = \langle -3, -3 \rangle$

15. $\mathbf{u} = \langle 2, 3 \rangle$, $\mathbf{v} = \langle -3, 5 \rangle$　　**16.** $\mathbf{u} = \langle 5, 2 \rangle$, $\mathbf{v} = \langle -6, -1 \rangle$

17. $\mathbf{u} = 3\mathbf{i} - 3\mathbf{j}$, $\mathbf{v} = -2\mathbf{i} + 2\sqrt{3}\mathbf{j}$

18. $\mathbf{u} = -2\mathbf{i}$, $\mathbf{v} = 5\mathbf{j}$

19. $\mathbf{u} = \left(2 \cos \dfrac{\pi}{4} \right)\mathbf{i} + \left(2 \sin \dfrac{\pi}{4} \right)\mathbf{j}$, $\mathbf{v} = \left(\cos \dfrac{3\pi}{2} \right)\mathbf{i} + \left(\sin \dfrac{3\pi}{2} \right)\mathbf{j}$

20. $\mathbf{u} = \left(\cos \dfrac{\pi}{3} \right)\mathbf{i} + \left(\sin \dfrac{\pi}{3} \right)\mathbf{j}$, $\mathbf{v} = \left(3 \cos \dfrac{5\pi}{6} \right)\mathbf{i} + \left(3 \sin \dfrac{5\pi}{6} \right)\mathbf{j}$

21.

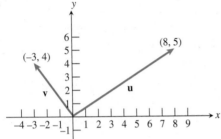

22.

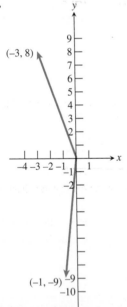

In Exercises 23–24, prove that the vectors $\mathbf{u}$ and $\mathbf{v}$ are orthogonal.

23. $\mathbf{u} = \langle 2, 3 \rangle$, $\mathbf{v} = \langle 3/2, -1 \rangle$

24. $\mathbf{u} = \langle -4, -1 \rangle$, $\mathbf{v} = \langle 1, -4 \rangle$

In Exercises 25–28, find the vector projection of $\mathbf{u}$ onto $\mathbf{v}$. Then write $\mathbf{u}$ as a sum of two orthogonal vectors, one of which is $\text{proj}_{\mathbf{v}}\mathbf{u}$.

25. $\mathbf{u} = \langle -8, 3 \rangle$, $\mathbf{v} = \langle -6, -2 \rangle$　　**26.** $\mathbf{u} = \langle 3, -7 \rangle$, $\mathbf{v} = \langle -2, -6 \rangle$

27. $\mathbf{u} = \langle 8, 5 \rangle$, $\mathbf{v} = \langle -9, -2 \rangle$　　**28.** $\mathbf{u} = \langle -2, 8 \rangle$, $\mathbf{v} = \langle 9, -3 \rangle$

In Exercises 29 and 30, find the interior angles of the triangle with given vertices.

29. $(-4, 5)$, $(1, 10)$, $(3, 1)$　　**30.** $(-4, 1)$, $(1, -6)$, $(5, -1)$

In Exercises 31 and 32, find $\mathbf{u} \cdot \mathbf{v}$ satisfying the given conditions where θ is the angle between $\mathbf{u}$ and $\mathbf{v}$.

31. $\theta = 150°$, $|\mathbf{u}| = 3$, $|\mathbf{v}| = 8$　　**32.** $\theta = \dfrac{\pi}{3}$, $|\mathbf{u}| = 12$, $|\mathbf{v}| = 40$

In Exercises 33–38, determine whether the vectors $\mathbf{u}$ and $\mathbf{v}$ are parallel, orthogonal, or neither.

33. $\mathbf{u} = \langle 5, 3 \rangle$, $\mathbf{v} = \left\langle -\dfrac{10}{4}, -\dfrac{3}{2} \right\rangle$

34. $\mathbf{u} = \langle 2, 5 \rangle$, $\mathbf{v} = \left\langle \dfrac{10}{3}, \dfrac{4}{3} \right\rangle$

35. $\mathbf{u} = \langle 15, -12 \rangle$, $\mathbf{v} = \langle -4, 5 \rangle$

36. $\mathbf{u} = \langle 5, -6 \rangle$, $\mathbf{v} = \langle -12, -10 \rangle$

37. $\mathbf{u} = \langle -3, 4 \rangle$, $\mathbf{v} = \langle 20, 15 \rangle$

38. $\mathbf{u} = \langle 2, -7 \rangle$, $\mathbf{v} = \langle -4, 14 \rangle$

In Exercises 39–42, find

(a) the x-intercept A and y-intercept B of the line.

(b) the coordinates of the point P so that $\overrightarrow{AP}$ is perpendicular to the line and $|\overrightarrow{AP}| = 1$. (There are two answers.)

39. $3x - 4y = 12$　　　　**40.** $-2x + 5y = 10$

41. $3x - 7y = 21$　　　　**42.** $x + 2y = 6$

In Exercises 43 and 44, find the vector(s) $\mathbf{v}$ satisfying the given conditions.

43. $\mathbf{u} = \langle 2, 3 \rangle$, $\mathbf{u} \cdot \mathbf{v} = 10$, $|\mathbf{v}|^2 = 17$

44. $\mathbf{u} = \langle -2, 5 \rangle$, $\mathbf{u} \cdot \mathbf{v} = -11$, $|\mathbf{v}|^2 = 10$

45. Sliding Down a Hill Ojemba is sitting on a sled on the side of a hill inclined at 60°. The combined weight of Ojemba and the sled is 160 pounds. What is the magnitude of the force required for Mandisa to keep the sled from sliding down the hill?

46. Revisiting Example 6 Suppose Juan and Rafaela switch positions. The combined weight of Rafaela and the sled is 125 pounds. What is the magnitude of the force required for Juan to keep the sled from sliding down the hill?

47. Braking Force A 2000 pound car is parked on a street that makes an angle of 12° with the horizontal (see figure).

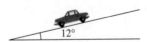

(a) Find the magnitude of the force required to keep the car from rolling down the hill.

(b) Find the force perpendicular to the street.

48. Effective Force A 60 pound force **F** that makes an angle of 25° with an inclined plane is pulling a box up the plane. The inclined plane makes an 18° angle with the horizontal (see figure). What is the magnitude of the effective force pulling the box up the plane?

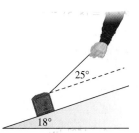

49. Work Find the work done lifting a 2600 pound car 5.5 feet.

50. Work Find the work done lifting a 100 pound bag of potatoes 3 feet.

51. Work Find the work done by a force **F** of 12 pounds acting in the direction $\langle 1, 2 \rangle$ in moving an object 4 feet from $(0, 0)$ to $(4, 0)$.

52. Work Find the work done by a force **F** of 24 pounds acting in the direction $\langle 4, 5 \rangle$ in moving an object 5 feet from $(0, 0)$ to $(5, 0)$.

53. Work Find the work done by a force **F** of 30 pounds acting in the direction $\langle 2, 2 \rangle$ in moving an object 3 feet from $(0, 0)$ to a point in the first quadrant along the line $y = (1/2)x$.

54. Work Find the work done by a force **F** of 50 pounds acting in the direction $\langle 2, 3 \rangle$ in moving an object 5 feet from $(0, 0)$ to a point in the first quadrant along the line $y = x$.

55. Work The angle between a 200 pound force **F** and $\overrightarrow{AB} = 2\mathbf{i} + 3\mathbf{j}$ is 30°. Find the work done by **F** in moving an object from A to B.

56. Work The angle between a 75 pound force **F** and $\overrightarrow{AB}$ is 60°, where $A = (-1, 1)$ and $B = (4, 3)$. Find the work done by **F** in moving an object from A to B.

57. Properties of the Dot Product Let **u**, **v**, and **w** be vectors and let c be a scalar. Use the component form of vectors to prove the following properties.

(a) $\mathbf{0} \cdot \mathbf{u} = 0$

(b) $\mathbf{u} \cdot (\mathbf{v} + \mathbf{w}) = \mathbf{u} \cdot \mathbf{v} + \mathbf{u} \cdot \mathbf{w}$

(c) $(\mathbf{u} + \mathbf{v}) \cdot \mathbf{w} = \mathbf{u} \cdot \mathbf{w} + \mathbf{v} \cdot \mathbf{w}$

(d) $(c\mathbf{u}) \cdot \mathbf{v} = \mathbf{u} \cdot (c\mathbf{v}) = c(\mathbf{u} \cdot \mathbf{v})$

58. Group Activity Projection of a Vector Let **u** and **v** be nonzero vectors. Prove that

(a) $\text{proj}_\mathbf{v}\mathbf{u} = \left(\dfrac{\mathbf{u} \cdot \mathbf{v}}{|\mathbf{v}|^2} \right)\mathbf{v}$

(b) $(\mathbf{u} - \text{proj}_\mathbf{v}\mathbf{u}) \cdot (\text{proj}_\mathbf{v}\mathbf{u}) = 0$

59. Group Activity Connecting Geometry and Vectors Prove that the sum of the squares of the diagonals of a parallelogram is equal to the sum of the squares of its sides.

60. If **u** is any vector, prove that we can write **u** as

$$\mathbf{u} = (\mathbf{u} \cdot \mathbf{i})\mathbf{i} + (\mathbf{u} \cdot \mathbf{j})\mathbf{j}.$$

Standardized Test Questions

61. True or False If $\mathbf{u} \cdot \mathbf{v} = 0$, then **u** and **v** are perpendicular. Justify your answer.

62. True or False If **u** is a unit vector, then $\mathbf{u} \cdot \mathbf{u} = 1$. Justify your answer.

In Exercises 63–66, you may use a graphing calculator to solve the problem.

63. Multiple Choice Let $\mathbf{u} = \langle 1, 1 \rangle$ and $\mathbf{v} = \langle -1, 1 \rangle$. Which of the following is the angle between **u** and **v**?

(A) 0°   **(B)** 45°   **(C)** 60°   **(D)** 90°   **(E)** 135°

64. Multiple Choice Let $\mathbf{u} = \langle 4, -5 \rangle$ and $\mathbf{v} = \langle -2, -3 \rangle$. Which of the following is equal to $\mathbf{u} \cdot \mathbf{v}$?

(A) -23 **(B)** -7 **(C)** 7   **(D)** 23   **(E)** $\sqrt{7}$

65. Multiple Choice Let $\mathbf{u} = \langle 3/2, -3/2 \rangle$ and $\mathbf{v} = \langle 2, 0 \rangle$. Which of the following is equal to $\text{proj}_\mathbf{v}\mathbf{u}$?

(A) $\langle 3/2, 0 \rangle$ **(B)** $\langle 3, 0 \rangle$ **(C)** $\langle -3/2, 0 \rangle$

(D) $\langle 3/2, 3/2 \rangle$ **(E)** $\langle -3/2, -3/2 \rangle$

66. Multiple Choice Which of the following vectors describes a 5 lb force acting in the direction of $\mathbf{u} = \langle -1, 1 \rangle$?

(A) $5 \langle -1, 1 \rangle$ **(B)** $\dfrac{5}{\sqrt{2}} \langle -1, 1 \rangle$ **(C)** $5 \langle 1, -1 \rangle$

(D) $\dfrac{5}{\sqrt{2}} \langle 1, -1 \rangle$ **(E)** $\dfrac{5}{2} \langle -1, 1 \rangle$

Explorations

67. Distance from a Point to a Line Consider the line L with equation $2x + 5y = 10$ and the point $P = (3, 7)$.

(a) Verify that $A = (0, 2)$ and $B = (5, 0)$ are the y- and x-intercepts of L.

(b) Find $\mathbf{w}_1 = \text{proj}_{\overrightarrow{AB}} \overrightarrow{AP}$ and $\mathbf{w}_2 = \overrightarrow{AP} - \text{proj}_{\overrightarrow{AB}} \overrightarrow{AP}$.

(c) Writing to Learn Explain why $|\mathbf{w}_2|$ is the distance from P to L. What is this distance?

(d) Find a formula for the distance of any point $P = (x_0, y_0)$ to L.

(e) Find a formula for the distance of any point $P = (x_0, y_0)$ to the line $ax + by = c$.

Extending the Ideas

68. Writing to Learn Let $\mathbf{w} = (\cos t) \mathbf{u} + (\sin t) \mathbf{v}$ where **u** and **v** are not parallel.

(a) Can the vector **w** be parallel to the vector **u**? Explain.

(b) Can the vector **w** be parallel to the vector **v**? Explain.

(c) Can the vector **w** be parallel to the vector $\mathbf{u} + \mathbf{v}$? Explain.

69. If the vectors **u** and **v** are not parallel, prove that

$$a\mathbf{u} + b\mathbf{v} = c\mathbf{u} + d\mathbf{v} \Rightarrow a = c, b = d.$$

6.3
Parametric Equations and Motion

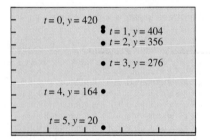

[0, 5] by [−10, 500]

FIGURE 6.23 The position of the rock at 0, 1, 2, 3, 4, and 5 seconds.

Parametric Equations

Imagine that a rock is dropped from a 420-ft tower. The rock's height y in feet above the ground t seconds later (ignoring air resistance) is modeled by $y = -16t^2 + 420$ as we saw in Section 2.1. Figure 6.23 shows a coordinate system imposed on the scene so that the line of the rock's fall is on the vertical line $x = 2.5$.

The rock's original position and its position after each of the first 5 seconds are the points

$$(2.5, 420), \quad (2.5, 404), \quad (2.5, 356), \quad (2.5, 276), \quad (2.5, 164), \quad (2.5, 20),$$

which are described by the pair of equations

$$x = 2.5, \qquad y = -16t^2 + 420,$$

when $t = 0, 1, 2, 3, 4, 5$. These two equations are an example of *parametric equations* with *parameter t*. As is often the case, the parameter t represents time.

Parametric Curves

In this section we study the graphs of *parametric equations* and investigate motion of objects that can be modeled with parametric equations.

> **DEFINITION Parametric Curve, Parametric Equations**
>
> The graph of the ordered pairs (x, y) where
>
> $$x = f(t), \quad y = g(t)$$
>
> are functions defined on an interval I of t-values is a **parametric curve**. The equations are **parametric equations** for the curve, the variable t is a **parameter**, and I is the **parameter interval**.

When we give parametric equations and a parameter interval for a curve, we have **parametrized** the curve. A **parametrization** of a curve consists of the parametric equations and the interval of t-values. Sometimes parametric equations are used by companies in their design plans. It is then easier for the company to make larger and smaller objects efficiently by just changing the parameter t.

Graphs of parametric equations can be obtained using parametric mode on a grapher.

EXAMPLE 1   Graphing Parametric Equations

For the given parameter interval, graph the parametric equations

$$x = t^2 - 2, \quad y = 3t.$$

(a) $-3 \leq t \leq 1$ **(b)** $-2 \leq t \leq 3$ **(c)** $-3 \leq t \leq 3$

continued

SOLUTION In each case, set Tmin equal to the left endpoint of the interval and Tmax equal to the right endpoint of the interval. Figure 6.24 shows a graph of the parametric equations for each parameter interval. The corresponding relations are different because the parameter intervals are different. *Now try Exercise 7.*

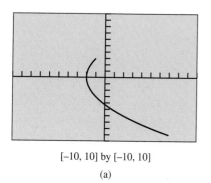

[−10, 10] by [−10, 10]

(a)

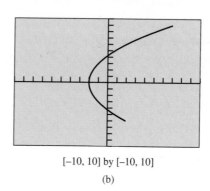

[−10, 10] by [−10, 10]

(b)

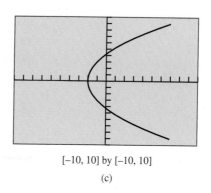

[−10, 10] by [−10, 10]

(c)

FIGURE 6.24 Three different relations defined parametrically. (Example 1)

Eliminating the Parameter

When a curve is defined parametrically it is sometimes possible to *eliminate the parameter* and obtain a rectangular equation in x and y that represents the curve. This often helps us identify the graph of the parametric curve as illustrated in Example 2.

EXAMPLE 2 Eliminating the Parameter

Eliminate the parameter and identify the graph of the parametric curve

$$x = 1 - 2t, \quad y = 2 - t, \quad -\infty < t < \infty.$$

SOLUTION We solve the first equation for t:

$$x = 1 - 2t$$

$$2t = 1 - x$$

$$t = \frac{1}{2}(1 - x)$$

Then we substitute this expression for t into the second equation:

$$y = 2 - t$$

$$y = 2 - \frac{1}{2}(1 - x)$$

$$y = 0.5x + 1.5$$

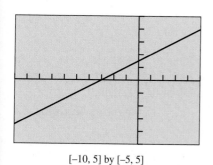

[−10, 5] by [−5, 5]

FIGURE 6.25 The graph of $y = 0.5x + 1.5$. (Example 2)

The graph of the equation $y = 0.5x + 1.5$ is a line with slope 0.5 and y-intercept 1.5 (Figure 6.25). *Now try Exercise 11.*

EXPLORATION 1 **Graphing the Curve of Example 2 Parametrically**

1. Use the parametric mode of your grapher to reproduce the graph in Figure 6.25. Use -2 for Tmin and 5.5 for Tmax.

2. Prove that the point $(17, 10)$ is on the graph of $y = 0.5x + 1.5$. Find the corresponding value of t that produces this point.

3. Repeat part 2 for the point $(-23, -10)$.

4. Assume that (a, b) is on the graph of $y = 0.5x + 1.5$. Find the corresponding value of t that produces this point.

5. How do you have to choose Tmin and Tmax so that the graph in Figure 6.25 fills the window?

If we do not specify a parameter interval for the parametric equations $x = f(t)$, $y = g(t)$, it is understood that the parameter t can take on all values which produce real numbers for x and y. We use this agreement in Example 3.

EXAMPLE 3   Eliminating the Parameter

Eliminate the parameter and identify the graph of the parametric curve

$$x = t^2 - 2, \quad y = 3t.$$

SOLUTION Here t can be any real number. We solve the second equation for t obtaining $t = y/3$ and substitute this value for y into the first equation.

$$x = t^2 - 2$$
$$x = \left(\frac{y}{3}\right)^2 - 2$$
$$x = \frac{y^2}{9} - 2$$
$$y^2 = 9(x + 2)$$

Figure 6.24c shows what the graph of these parametric equations looks like. In Chapter 8 we will call this a parabola that opens to the right. Interchanging x and y we can identify this graph as the inverse of the graph of the parabola $x^2 = 9(y + 2)$.

Now try Exercise 15.

PARABOLAS

The inverse of a parabola that opens up or down is a parabola that opens left or right. We will investigate these curves in more detail in Chapter 8.

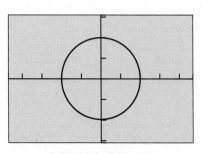

[–4.7, 4.7] by [–3.1, 3.1]

FIGURE 6.26 The graph of the circle of Example 4.

EXAMPLE 4   Eliminating the Parameter

Eliminate the parameter and identify the graph of the parametric curve

$$x = 2 \cos t, \quad y = 2 \sin t, \quad 0 \le t \le 2\pi.$$

SOLUTION The graph of the parametric equations in the square viewing window of Figure 6.26 suggests that the graph is a circle of radius 2 centered at the origin. We confirm this result algebraically.

continued

$$x^2 + y^2 = 4\cos^2 t + 4\sin^2 t$$
$$= 4(\cos^2 t + \sin^2 t)$$
$$= 4(1) \qquad \cos^2 t + \sin^2 t = 1$$
$$= 4$$

The graph of $x^2 + y^2 = 4$ is a circle of radius 2 centered at the origin. Increasing the length of the interval $0 \le t \le 2\pi$ will cause the grapher to trace all or part of the circle more than once. Decreasing the length of the interval will cause the grapher to only draw a portion of the complete circle. Try it! **Now try Exercise 23.**

In Exercise 65, you will find parametric equations for any circle in the plane.

Lines and Line Segments

We can use vectors to help us find parametric equations for a line as illustrated in Example 5.

EXAMPLE 5   Finding Parametric Equations for a Line

Find a parametrization of the line through the points $A = (-2, 3)$ and $B = (3, 6)$.

SOLUTION Let $P(x, y)$ be an arbitrary point on the line through A and B. As you can see from Figure 6.27, the vector $\overrightarrow{OP}$ is the tail-to-head vector sum of $\overrightarrow{OA}$ and $\overrightarrow{AP}$. You can also see that $\overrightarrow{AP}$ is a scalar multiple of $\overrightarrow{AB}$.

If we let the scalar be t, we have

$$\overrightarrow{OP} = \overrightarrow{OA} + \overrightarrow{AP}$$
$$\overrightarrow{OP} = \overrightarrow{OA} + t \cdot \overrightarrow{AB}$$
$$\langle x, y \rangle = \langle -2, 3 \rangle + t \langle 3-(-2), 6 - 3 \rangle$$
$$\langle x, y \rangle = \langle -2, 3 \rangle + t \langle 5, 3 \rangle$$
$$\langle x, y \rangle = \langle -2 + 5t, 3 + 3t \rangle$$

This vector equation is equivalent to the parametric equations $x = -2 + 5t$ and $y = 3 + 3t$. Together with the parameter interval $(-\infty, \infty)$, these equations define the line.

We can confirm our work numerically as follows: If $t = 0$, then $x = -2$ and $y = 3$, which gives the point A. Similarly, if $t = 1$, then $x = 3$ and $y = 6$, which gives the point B. **Now try Exercise 27.**

The fact that $t = 0$ yields point A and $t = 1$ yields point B in Example 5 is no accident, as a little reflection on Figure 6.27 and the vector equation $\overrightarrow{OP} = \overrightarrow{OA} + t \cdot \overrightarrow{AB}$ should suggest. We use this fact in Example 6.

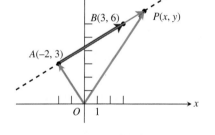

FIGURE 6.27 Example 5 uses vectors to construct a parametrization of the line through A and B.

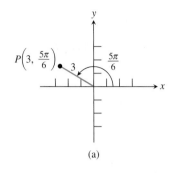

(a)

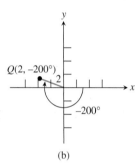

(b)

FIGURE 6.39 The points P and Q in Example 3.

SOLUTION

(a) For $P(3, 5\pi/6)$, $r = 3$ and $\theta = 5\pi/6$:

$$x = r \cos \theta \qquad\qquad\qquad y = r \sin \theta$$

$$x = 3 \cos \frac{5\pi}{6} \qquad \text{and} \qquad y = 3 \sin \frac{5\pi}{6}$$

$$x = 3\left(-\frac{\sqrt{3}}{2}\right) \approx -2.60 \qquad\qquad y = 3\left(\frac{1}{2}\right) = 1.5$$

The rectangular coordinates for P are $(-3\sqrt{3}/2, 1.5) \approx (-2.60, 1.5)$ (Figure 6.39a).

(b) For $Q(2, -200°)$, $r = 2$ and $\theta = -200°$:

$$x = r \cos \theta \qquad\qquad\qquad y = r \sin \theta$$

$$\qquad\qquad\qquad\qquad \text{and}$$

$$x = 2 \cos(-200°) \approx -1.88 \qquad\qquad y = 2 \sin(-200°) \approx 0.68$$

The rectangular coordinates for Q are approximately $(-1.88, 0.68)$ (Figure 6.39b).

Now try Exercise 15.

When converting rectangular coordinates to polar coordinates, we must remember that there are infinitely many possible polar coordinate pairs. In Example 4 we report two of the possibilities.

EXAMPLE 4   Converting from Rectangular to Polar Coordinates

Find two polar coordinate pairs for the points with given rectangular coordinates.

(a) $P(-1, 1)$ **(b)** $Q(-3, 0)$

SOLUTION

(a) For $P(-1, 1)$, $x = -1$ and $y = 1$:

$$r^2 = x^2 + y^2 \qquad\qquad \tan \theta = \frac{y}{x}$$

$$r^2 = (-1)^2 + (1)^2 \qquad\qquad \tan \theta = \frac{1}{-1} = -1$$

$$r = \pm\sqrt{2} \qquad\qquad \theta = \tan^{-1}(-1) + n\pi = -\frac{\pi}{4} + n\pi$$

We use the angles $-\pi/4$ and $-\pi/4 + \pi = 3\pi/4$. Because P is on the ray opposite the terminal side of $-\pi/4$, the value of r corresponding to this angle is negative (Figure 6.40). Because P is on the terminal side of $3\pi/4$, the value of r corresponding to this angle is positive. So two polar coordinate pairs of point P are

$$\left(-\sqrt{2}, -\frac{\pi}{4}\right) \quad \text{and} \quad \left(\sqrt{2}, \frac{3\pi}{4}\right).$$

(b) For $Q(-3, 0)$, $x = -3$ and $y = 0$. Thus, $r = \pm 3$ and $\theta = n\pi$. We use the angles 0 and π. So two polar coordinates pairs for point Q are

$$(-3, 0) \quad \text{and} \quad (3, \pi).$$

Now try Exercise 27.

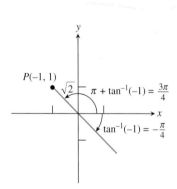

FIGURE 6.40 The point P in Example 4a.

EXPLORATION 1 **Using a Grapher to Convert Coordinates**

Most graphers have the capability to convert polar coordinates to rectangular coordinates and vice versa. Usually they give just one possible polar coordinate pair for a given rectangular coordinate pair.

1. Use your grapher to check the conversions in Examples 3 and 4.
2. Use your grapher to convert the polar coordinate pairs $(2, \pi/3)$, $(-1, \pi/2)$, $(2, \pi)$, $(-5, 3\pi/2)$, $(3, 2\pi)$, to rectangular coordinate pairs.
3. Use your grapher to convert the rectangular coordinate pairs $(-1, -\sqrt{3})$, $(0, 2)$, $(3, 0)$, $(-1, 0)$, $(0, -4)$ to polar coordinate pairs.

Equation Conversion

We can use the Coordinate Conversion Equations to convert polar form to rectangular form and vice versa. For example, the polar equation $r = 4 \cos \theta$ can be converted to rectangular form as follows:

$$r = 4 \cos \theta$$
$$r^2 = 4r \cos \theta$$
$$x^2 + y^2 = 4x \qquad \text{\footnotesize $r^2 = x^2 + y^2$, $r \cos \theta = x$}$$
$$x^2 - 4x + 4 + y^2 = 4 \qquad \text{\footnotesize Subtract $4x$ and add 4.}$$
$$(x - 2)^2 + y^2 = 4 \qquad \text{\footnotesize Factor.}$$

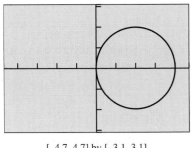

[−4.7, 4.7] by [−3.1, 3.1]

FIGURE 6.41 The graph of the polar equation $r = 4 \cos \theta$ in $0 \le \theta \le 2\pi$.

Thus the graph of $r = 4 \cos \theta$ is all or part of the circle with center $(2, 0)$ and radius 2.

Figure 6.41 shows the graph of $r = 4 \cos \theta$ for $0 \le \theta \le 2\pi$ obtained using the polar graphing mode of our grapher. So, the graph of $r = 4 \cos \theta$ is the entire circle.

Just as with parametric equations, the domain of a polar equation in r and θ is understood to be all values of θ for which the corresponding values of r are real numbers. You must also select a value for θ min and θ max to graph in polar mode.

You may be surprised by the polar form for a vertical line in Example 5.

EXAMPLE 5 **Converting from Polar Form to Rectangular Form**

Convert $r = 4 \sec \theta$ to rectangular form and identify the graph. Support your answer with a polar graphing utility.

SOLUTION

$$r = 4 \sec \theta$$
$$\frac{r}{\sec \theta} = 4 \qquad \text{\footnotesize Divide by $\sec \theta$.}$$
$$r \cos \theta = 4 \qquad \text{\footnotesize $\cos \theta = \frac{1}{\sec \theta}$.}$$
$$x = 4 \qquad \text{\footnotesize $r \cos \theta = x$}$$

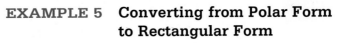

[−2, 8] by [−10, 10]

FIGURE 6.42 The graph of the vertical line $r = 4 \sec \theta$ ($x = 4$). (Example 5)

The graph is the vertical line $x = 4$ (Figure 6.42). *Now try Exercise 35.*

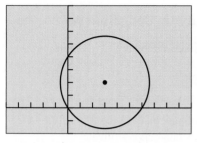

[–5, 10] by [–2, 8]

FIGURE 6.43 The graph of the circle $r = 6 \cos \theta + 4 \sin \theta$. (Example 6)

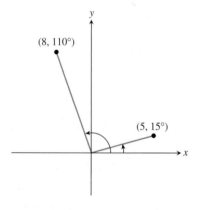

FIGURE 6.44 The distance and direction of two airplanes from a radar source. (Example 7)

EXAMPLE 6   Converting from Rectangular Form to Polar Form

Convert $(x - 3)^2 + (y - 2)^2 = 13$ to polar form.

SOLUTION

$$(x - 3)^2 + (y - 2)^2 = 13$$
$$x^2 - 6x + 9 + y^2 - 4y + 4 = 13$$
$$x^2 + y^2 - 6x - 4y = 0$$

Substituting r^2 for $x^2 + y^2$, $r \cos \theta$ for x, and $r \sin \theta$ for y gives the following:

$$r^2 - 6r \cos \theta - 4r \sin \theta = 0$$
$$r(r - 6 \cos \theta - 4 \sin \theta) = 0$$
$$r = 0 \quad \text{or} \quad r - 6 \cos \theta - 4 \sin \theta = 0$$

The graph of $r = 0$ consists of a single point, the origin, which is also on the graph of $r - 6 \cos \theta - 4 \sin \theta = 0$. Thus, the polar form is

$$r = 6 \cos \theta + 4 \sin \theta.$$

The graph of $r = 6 \cos \theta + 4 \sin \theta$ for $0 \le \theta \le 2\pi$ is shown in Figure 6.43 and appears to be a circle with center $(3, 2)$ and radius $\sqrt{13}$, as expected.

Now try Exercise 43.

Finding Distance Using Polar Coordinates

A radar tracking system sends out high-frequency radio waves and receives their reflection from an object. The distance and direction of the object from the radar is often given in polar coordinates.

EXAMPLE 7   Using a Radar Tracking System

Radar detects two airplanes at the same altitude. Their polar coordinates are (8 mi, 110°) and (5 mi, 15°). (See Figure 6.44.) How far apart are the airplanes?

SOLUTION By the Law of Cosines (Section 5.6),

$$d^2 = 8^2 + 5^2 - 2 \cdot 8 \cdot 5 \cos (110° - 15°)$$
$$d = \sqrt{8^2 + 5^2 - 2 \cdot 8 \cdot 5 \cos 95°}$$
$$d \approx 9.80$$

The airplanes are about 9.80 mi apart.

Now try Exercise 51.

We can also use the Law of Cosines to derive a formula for the distance between points in the polar coordinate system. See Exercise 61.

QUICK REVIEW 6.4 *(For help, go to Sections P.2, 4.3, and 5.6.)*

In Exercises 1 and 2, determine the quadrants containing the terminal side of the angles.

1. (a) $5\pi/6$ **(b)** $-3\pi/4$

2. (a) $-300°$ **(b)** $210°$

In Exercises 3–6, find a positive and a negative angle coterminal with the given angle.

3. $-\pi/4$ **4.** $\pi/3$

5. $160°$ **6.** $-120°$

In Exercises 7 and 8, write a standard form equation for the circle.

7. Center $(3, 0)$ and radius 2   **8.** Center $(0, -4)$ and radius 3

In Exercises 9 and 10, use The Law of Cosines to find the measure of the third side of the given triangle.

9.

10.

SECTION 6.4 EXERCISES

In Exercises 1–4, the polar coordinates of a point are given. Find its rectangular coordinates.

1.

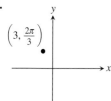

2.

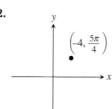

3.

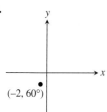

4.

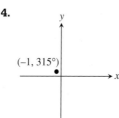

In Exercises 5 and 6, **(a)** complete the table for the polar equation and **(b)** plot the corresponding points.

5. $r = 3 \sin \theta$

θ	$\pi/4$	$\pi/2$	$5\pi/6$	π	$4\pi/3$	2π
r						

6. $r = 2 \csc \theta$

θ	$\pi/4$	$\pi/2$	$5\pi/6$	π	$4\pi/3$	2π
r						

In Exercises 7–14, plot the point with the given polar coordinates.

7. $(3, 4\pi/3)$ **8.** $(2, 5\pi/6)$ **9.** $(-1, 2\pi/5)$

10. $(-3, 17\pi/10)$ **11.** $(2, 30°)$ **12.** $(3, 210°)$

13. $(-2, 120°)$ **14.** $(-3, 135°)$

In Exercises 15–22, find the rectangular coordinates of the point with given polar coordinates.

15. $(1.5, 7\pi/3)$ **16.** $(2.5, 17\pi/4)$

17. $(-3, -29\pi/7)$ **18.** $(-2, -14\pi/5)$

19. $(-2, \pi)$ **20.** $(1, \pi/2)$

21. $(2, 270°)$ **22.** $(-3, 360°)$

In Exercises 23–26, polar coordinates of point P are given. Find all of its polar coordinates.

23. $P = (2, \pi/6)$ **24.** $P = (1, -\pi/4)$

25. $P = (1.5, -20°)$ **26.** $P = (-2.5, 50°)$

In Exercises 27–30, rectangular coordinates of point P are given. Find all polar coordinates of P that satisfy

(a) $0 \le \theta \le 2\pi$ **(b)** $-\pi \le \theta \le \pi$ **(c)** $0 \le \theta \le 4\pi$

27. $P = (1, 1)$ **28.** $P = (1, 3)$

29. $P = (-2, 5)$ **30.** $P = (-1, -2)$

In Exercises 31–34, use your grapher to match the polar equation with its graph.

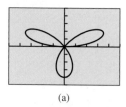

(a)

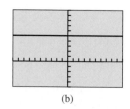

(b)

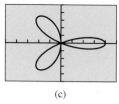

(c)

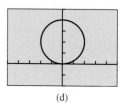

(d)

31. $r = 5 \csc \theta$ **32.** $r = 4 \sin \theta$

33. $r = 4 \cos 3\theta$ **34.** $r = 4 \sin 3\theta$

In Exercises 35–42, convert the polar equation to rectangular form and identify the graph. Support your answer by graphing the polar equation.

35. $r = 3 \sec \theta$ **36.** $r = -2 \csc \theta$ **37.** $r = -3 \sin \theta$

38. $r = -4 \cos \theta$ **39.** $r \csc \theta = 1$ **40.** $r \sec \theta = 3$

41. $r = 2 \sin \theta - 4 \cos \theta$ **42.** $r = 4 \cos \theta - 4 \sin \theta$

In Exercises 43–50, convert the rectangular equation to polar form. Graph the polar equation.

43. $x = 2$ **44.** $x = 5$

45. $2x - 3y = 5$ **46.** $3x + 4y = 2$

47. $(x - 3)^2 + y^2 = 9$ **48.** $x^2 + (y - 1)^2 = 1$

49. $(x + 3)^2 + (y + 3)^2 = 18$ **50.** $(x - 1)^2 + (y + 4)^2 = 17$

51. Tracking Airplanes The location, given in polar coordinates, of two planes approaching the Vicksburg airport are (4 mi, 12°) and (2 mi, 72°). Find the distance between the airplanes.

52. Tracking Ships The location of two ships from Mays Landing Lighthouse, given in polar coordinates, are (3 mi, 170°) and (5 mi, 150°). Find the distance between the ships.

53. Using Polar Coordinates in Geometry A square with sides of length a and center at the origin has two sides parallel to the x-axis. Find polar coordinates of the vertices.

54. Using Polar Coordinates in Geometry A regular pentagon whose center is at the origin has one vertex on the positive x-axis at a distance a from the center. Find polar coordinates of the vertices.

Standardized Test Questions

55. True or False Every point in the plane has exactly two polar coordinates. Justify your answer.

56. True or False If r_1 and r_2 are not 0, and if (r_1, θ) and $(r_2, \theta + \pi)$ represent the same point in the plane, then $r_1 = -r_2$. Justify your answer.

In Exercises 57–60, solve the problem without using a calculator.

57. Multiple Choice If $r \neq 0$, which of the following polar coordinate pairs represents the same point as the point with polar coordinates (r, θ)?

   (A) $(-r, \theta)$ **(B)** $(-r, \theta + 2\pi)$ **(C)** $(-r, \theta + 3\pi)$

   (D) $(r, \theta + \pi)$ **(E)** $(r, \theta + 3\pi)$

58. Multiple Choice Which of the following are the rectangular coordinates of the point with polar coordinate $(-2, -\pi/3)$?

   (A) $(-\sqrt{3}, 1)$ **(B)** $(-1, -\sqrt{3})$ **(C)** $(-1, \sqrt{3})$

   (D) $(1, -\sqrt{3})$ **(E)** $(1, \sqrt{3})$

59. Multiple Choice Which of the following polar coordinate pairs represent the same point as the point with polar coordinates $(2, 110°)$?

   (A) $(-2, -70°)$ **(B)** $(-2, 110°)$ **(C)** $(-2, -250°)$

   (D) $(2, -70°)$ **(E)** $(2, 290°)$

60. Multiple Choice Which of the following polar coordinate pairs does *not* represent the point with rectangular coordinates $(-2, -2)$?

   (A) $(2\sqrt{2}, -135°)$ **(B)** $(2\sqrt{2}, 225°)$ **(C)** $(-2\sqrt{2}, -315°)$

   (D) $(-2\sqrt{2}, 45°)$ **(E)** $(-2\sqrt{2}, 135°)$

Explorations

61. Polar Distance Formula Let P_1 and P_2 have polar coordinates (r_1, θ_1) and (r_2, θ_2), respectively.

   (a) If $\theta_1 - \theta_2$ is a multiple of π, write a formula for the distance between P_1 and P_2.

   (b) Use the Law of Cosines to prove that the distance between P_1 and P_2 is given by

$$d = \sqrt{r_1^2 + r_2^2 - 2r_1 r_2 \cos (\theta_1 - \theta_2)}$$

   (c) Writing to Learn Does the formula in part (b) agree with the formula(s) you found in part (a)? Explain.

62. Watching Your θ-Step Consider the polar curve $r = 4 \sin \theta$. Describe the graph for each of the following.

   (a) $0 \leq \theta \leq \pi/2$ **(b)** $0 \leq \theta \leq 3\pi/4$

   (c) $0 \leq \theta \leq 3\pi/2$ **(d)** $0 \leq \theta \leq 4\pi$

In Exercises 63–66, use the results of Exercise 61 to find the distance between the points with given polar coordinates.

63. $(2, 10°), (5, 130°)$ **64.** $(4, 20°), (6, 65°)$

65. $(-3, 25°), (-5, 160°)$ **66.** $(6, -35°), (8, -65°)$

Extending the Ideas

67. Graphing Polar Equations Parametrically Find parametric equations for the polar curve $r = f(\theta)$.

Group Activity In Exercises 68–71, use what you learned in Exercise 67 to write parametric equations for the given polar equation. Support your answers graphically.

68. $r = 2 \cos \theta$ **69.** $r = 5 \sin \theta$

70. $r = 2 \sec \theta$ **71.** $r = 4 \csc \theta$

6.5
Graphs of Polar Equations

Polar Curves and Parametric Curves

Polar curves are actually just special cases of parametric curves. Keep in mind that polar curves are graphed in the (x, y) plane, despite the fact that they are given in terms of r and θ. That is why the polar graph of $r = 4 \cos \theta$ is a circle (see Figure 6.41 in Section 6.4) rather than a cosine curve.

In function mode, points are determined by a vertical coordinate that changes as the horizontal coordinate moves left to right. In polar mode, points are determined by a directed distance from the pole that changes as the angle sweeps around the pole. The connection is provided by the Coordinate Conversion Equations from Section 6.4, which show that the graph of $r = f(\theta)$ is really just the graph of the parametric equations

$$x = f(\theta) \cos \theta$$
$$y = f(\theta) \sin \theta$$

for all values of θ in some parameter interval that suffices to produce a complete graph. (In many of our examples, $0 \le \theta < 2\pi$ will do.)

Since modern graphing calculators produce these graphs so easily in polar mode, we are frankly going to assume that you do not have to sketch them by hand. Instead we will concentrate on analyzing the properties of the curves. In later courses you can discover further properties of the curves using the tools of calculus.

Symmetry

You learned algebraic tests for symmetry for equations in rectangular form in Section 1.2. Algebraic tests also exist for polar form.

Figure 6.45 on the next page shows a rectangular coordinate system superimposed on a polar coordinate system, with the origin and the pole coinciding and the positive x-axis and the polar axis coinciding.

The three types of symmetry figures to be considered will have are:

1. The x-axis (polar axis) as a line of symmetry (Figure 6.45a).
2. The y-axis (the line $\theta = \pi/2$) as a line of symmetry (Figure 6.45b).
3. The origin (the pole) as a point of symmetry (Figure 6.45c).

All three algebraic tests for symmetry in polar forms require replacing the pair (r, θ), which satisfies the polar equation, with another coordinate pair and determining whether it also satisfies the polar equation.

Symmetry Tests for Polar Graphs

The graph of a polar equation has the indicated symmetry if either replacement produces an equivalent polar equation.

To Test for Symmetry	Replace	By
1. about the x-axis,	(r, θ)	$(r, -\theta)$ or $(-r, \pi - \theta)$.
2. about the y-axis,	(r, θ)	$(-r, -\theta)$ or $(r, \pi - \theta)$.
3. about the origin,	(r, θ)	$(-r, \theta)$ or $(r, \theta + \pi)$.

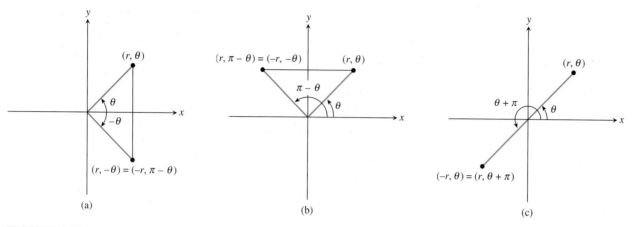

FIGURE 6.45 Symmetry with respect to (a) the *x*-axis (polar axis), (b) the *y*-axis (the line $\theta = \pi/2$), and (c) the origin (the pole).

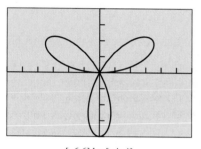

[–6,6] by [–4, 4]

FIGURE 6.46 The graph of $r = 4 \sin 3\theta$ is symmetric about the *y*-axis. (Example 1)

EXAMPLE 1   Testing for Symmetry

Use the symmetry tests to prove that the graph of $r = 4 \sin 3\theta$ is symmetric about the *y*-axis.

SOLUTION Figure 6.46 suggests that the graph of $r = 4 \sin 3\theta$ is symmetric about the *y*-axis and not symmetric about the *x*-axis or origin.

$$r = 4 \sin 3\theta$$

$$-r = 4 \sin 3(-\theta) \qquad \text{Replace } (r, \theta) \text{ by } (-r, -\theta).$$

$$-r = 4 \sin (-3\theta)$$

$$-r = -4 \sin 3\theta \qquad \text{sin } \theta \text{ is an odd function of } \theta.$$

$$r = 4 \sin 3\theta \qquad \text{(Same as original.)}$$

Because the equations $-r = 4 \sin 3(-\theta)$ and $r = 4 \sin 3\theta$ are equivalent, there is symmetry about the *y*-axis.   ***Now try Exercise 13.***

Analyzing Polar Graphs

We analyze graphs of polar equations in much the same way that we analyze the graphs of rectangular equations. For example, the function *r* of Example 1 is a continuous function of θ. Also $r = 0$ when $\theta = 0$ and when θ is any integer multiple of $\pi/3$. The domain of this function is the set of all real numbers.

Trace can be used to help determine the range of this polar function (Figure 6.47). It can be shown that $-4 \le r \le 4$.

Usually, we are more interested in the maximum value of $|r|$ rather than the range of *r* in polar equations. In this case, $|r| \le 4$ so we can conclude that the graph is bounded.

A maximum value for $|r|$ is a **maximum *r*-value** for a polar equation. A maximum *r*-value occurs at a point on the curve that is the maximum distance from the pole. In Figure 6.47, a maximum *r*-value occurs at $(4, \pi/6)$ and $(-4, \pi/2)$. In fact, we get a maximum *r*-value at every (r, θ) which represents the tip of one of the three petals.

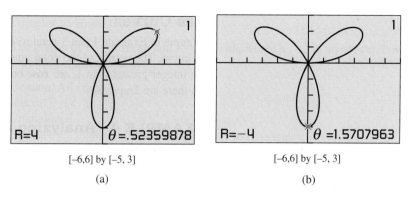

[−6,6] by [−5, 3]

(a)

[−6,6] by [−5, 3]

(b)

FIGURE 6.47 The values of r in $r = 4 \sin 3\theta$ vary from (a) 4 to (b) −4.

To find maximum r-values we must find maximum values of $|r|$ as opposed to the directed distance r. Example 2 shows one way to find maximum r-values graphically.

EXAMPLE 2   Finding Maximum *r*-Values

Find the maximum r-value of $r = 2 + 2 \cos \theta$.

SOLUTION Figure 6.48a shows the graph of $r = 2 + 2 \cos \theta$ for $0 \le \theta \le 2\pi$. Because we are only interested in the values of r, we use the graph of the rectangular equation $y = 2 + 2 \cos x$ in function graphing mode (Figure 6.48b). From this graph we can see that the maximum value of r, or y, is 4. It occurs when θ is any multiple of 2π.   ***Now try Exercise 21.***

EXAMPLE 3   Finding Maximum *r*-Values

Identify the points on the graph of $r = 3 \cos 2\theta$ for $0 \le \theta \le 2\pi$ that give maximum r-values.

SOLUTION Using trace in Figure 6.49 we can show that there are four points on the graph of $r = 3 \cos 2\theta$ in $0 \le \theta < 2\pi$ at maximum distance of 3 from the pole:

$$(3, 0), \quad (-3, \pi/2), \quad (3, \pi), \quad \text{and} \quad (-3, 3\pi/2).$$

Figure 6.50a shows the directed distances r as the y-values of $y_1 = 3 \cos 2x$, and Figure 6.50b shows the distances $|r|$ as the y-values of $y_2 = |3 \cos 2x|$. There are four maximum values of y_2 (i.e., $|r|$) in part (b) corresponding to the four extreme values of y_1 (i.e., r) in part (a).   ***Now try Exercise 23.***

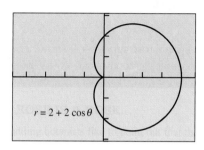

[−4.7, 4.7] by [−3.1, 3.1]

Polar coordinates

(a)

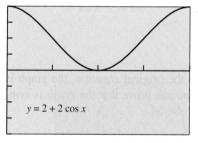

[0, 2π] by [−4, 4]

Rectangular coordinates

(b)

FIGURE 6.48 With $\theta = x$, the y-values in (b) are the same as the directed distance from the pole to (r, θ) in (a).

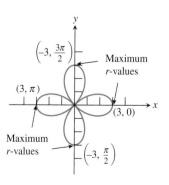

FIGURE 6.49 The graph of $r = 3 \cos 2\theta$. (Example 3)

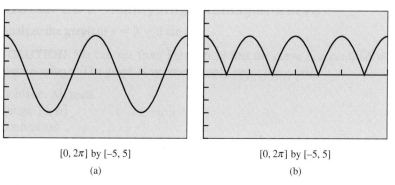

[0, 2π] by [−5, 5]

(a)

[0, 2π] by [−5, 5]

(b)

FIGURE 6.50 The graph of (a) $y_1 = 3 \cos 2x$ and (b) $y_2 = |3 \cos 2x|$ in function graphing mode. (Example 3)

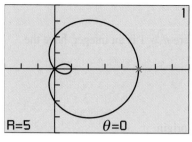

[−3, 8] by [−4, 4]

FIGURE 6.54 The graph of a limaçon with an inner loop. (Example 6)

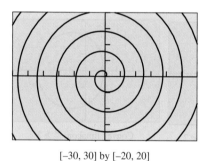

[−30, 30] by [−20, 20]

(a)

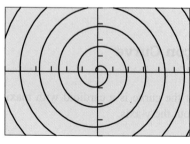

[−30, 30] by [−20, 20]

(b)

FIGURE 6.55 The graph of $r = \theta$ for (a) $\theta \geq 0$ (set θmin $= 0$, θmax $= 45$, θstep $= 0.1$) and (b) $\theta \leq 0$ (set θmin $= -45$, θmax $= 0$, θstep $= 0.1$). (Example 7)

EXAMPLE 6   Analyzing a Limaçon Curve

Analyze the graph of $r = 2 + 3 \cos \theta$.

SOLUTION We can see from Figure 6.54 that the curve is a limaçon with an inner loop and maximum r-value 5. The graph is symmetric only about the x-axis.

Domain: All reals.
Range: $[-1, 5]$
Continuous
Symmetric about the x-axis.
Bounded
Maximum r-value: 5
No asymptotes.

Now try Exercise 39.

Graphs of Limaçon Curves

The graphs of $r = a \pm b \sin \theta$ and $r = a \pm b \cos \theta$, where $a > 0$ and $b > 0$, have the following characteristics:

Domain: All reals
Range: $[a - b, a + b]$
Continuous
Symmetry: $r = a \pm b \sin \theta$, symmetric about y-axis
            $r = a \pm b \cos \theta$, symmetric about x-axis
Bounded
Maximum r-value: $a + b$
No asymptotes

EXPLORATION 1   Limaçon Curves

Try several values for a and b to convince yourself of the characteristics of limaçon curves listed above.

Other Polar Curves

All the polar curves we have graphed so far have been bounded. The spiral in Example 7 is unbounded.

EXAMPLE 7   Analyzing the Spiral of Archimedes

Analyze the graph of $r = \theta$.

SOLUTION We can see from Figure 6.55 that the curve has no maximum r-value and is symmetric about the y-axis.

Domain: All reals.
Range: All reals.
Continuous
Symmetric about the y-axis.
Unbounded
No maximum r-value.
No asymptotes.

Now try Exercise 41.

The **lemniscate curves** are graphs of polar equations of the form

$$r^2 = a^2 \sin 2\theta \quad \text{and} \quad r^2 = a^2 \cos 2\theta.$$

EXAMPLE 8 Analyzing a Lemniscate Curve

Analyze the graph of $r^2 = 4 \cos 2\theta$ for $[0, 2\pi]$.

SOLUTION It turns out that you can get the complete graph using $r = 2\sqrt{\cos 2\theta}$. You also need to choose a very small θ step to produce the graph in Figure 6.56.

Domain: $[0, \pi/4] \cup [3\pi/4, 5\pi/4] \cup [7\pi/4, 2\pi]$
Range: $[-2, 2]$
Symmetric about the x-axis, the y-axis, and the origin.
Continuous (on its domain)
Bounded
Maximum r-value: 2
No asymptotes.   *Now try Exercise 43.*

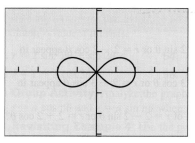

[-4.7, 4.7] by [-3.1, 3.1]

FIGURE 6.56 The graph of the lemniscate $r^2 = 4 \cos 2\theta$. (Example 8)

EXPLORATION 2 Revisiting Example 8

1. Prove that θ-values in the intervals $(\pi/4, 3\pi/4)$ and $(5\pi/4, 7\pi/4)$ are not in the domain of the polar equation $r^2 = 4 \cos 2\theta$.

2. Explain why $r = -2\sqrt{\cos 2\theta}$ produces the same graph as $r = 2\sqrt{\cos 2\theta}$ in the interval $[0, 2\pi]$.

3. Use the symmetry tests to show that the graph of $r^2 = 4 \cos 2\theta$ is symmetric about the x-axis.

4. Use the symmetry tests to show that the graph of $r^2 = 4 \cos 2\theta$ is symmetric about the y-axis.

5. Use the symmetry tests to show that the graph of $r^2 = 4 \cos 2\theta$ is symmetric about the origin.

QUICK REVIEW 6.5 *(For help, go to Sections 1.2 and 5.3.)*

In Exercises 1–4, find the absolute maximum value and absolute minimum value in $[0, 2\pi]$ and where they occur.

1. $y = 3 \cos 2x$ **2.** $y = 2 + 3 \cos x$

3. $y = 2\sqrt{\cos 2x}$ **4.** $y = 3 - 3 \sin x$

In Exercises 5 and 6, determine if the graph of the function is symmetric about the **(a)** x-axis, **(b)** y-axis, and **(c)** origin.

5. $y = \sin 2x$ **6.** $y = \cos 4x$

In Exercises 7–10, use trig identities to simplify the expression.

7. $\sin (\pi - \theta)$ **8.** $\cos (\pi - \theta)$

9. $\cos 2(\pi + \theta)$ **10.** $\sin 2(\pi + \theta)$

6.6
De Moivre's Theorem and *n*th Roots

What you'll learn about

- The Complex Plane

- Trigonometric Form of Complex Numbers

- Multiplication and Division of Complex Numbers

- Powers of Complex Numbers

- Roots of Complex Numbers

. . . and why

This material extends your equation-solving technique to include equations of the form $z^n = c$, *n* an integer and *c* a complex number.

The Complex Plane

You might be curious as to why we reviewed complex numbers in Section P.6, then proceeded to ignore them for the next six chapters. (Indeed, after this section we will pretty much ignore them again.) The reason is simply because the key to understanding calculus is the graphing of functions in the Cartesian plane, which consists of two perpendicular real (not complex) lines.

We are not saying that complex numbers are impossible to graph. Just as every real number is associated with a point of the real number line, every complex number can be associated with a point of the **complex plane**. This idea evolved through the work of Caspar Wessel (1745–1818), Jean-Robert Argand (1768–1822) and Carl Friedrich Gauss (1777–1855). Real numbers are placed along the horizontal axis (the **real axis**) and imaginary numbers along the vertical axis (the **imaginary axis**), thus associating the complex number $a + bi$ with the point (a, b). In Figure 6.57 we show the graph of $2 + 3i$ as an example.

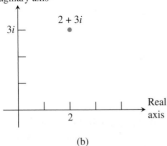

(a)

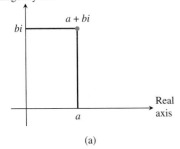

(b)

FIGURE 6.57 Plotting points in the complex plane.

IS THERE A CALCULUS OF COMPLEX FUNCTIONS?

There is a calculus of complex functions. If you study it someday, it should only be after acquiring a pretty firm algebraic and geometric understanding of the calculus of real functions.

> ### EXAMPLE 1   Plotting Complex Numbers
>
> Plot $u = 1 + 3i$, $v = 2 - i$, and $u + v$ in the complex plane. These three points and the origin determine a quadrilateral. Is it a parallelogram?
>
> **SOLUTION** First notice that $u + v = (1 + 3i) + (2 + i) = 3 + 2i$. The numbers u, v, and $u + v$ are plotted in Figure 6.58a. The quadrilateral is a parallelogram because the arithmetic is exactly the same as in vector addition (Figure 6.58b).
>
> *Now try Exercise 1.*

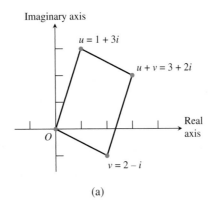

(a)

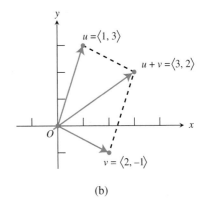

(b)

FIGURE 6.58 (a) Two numbers and their sum are plotted in the complex plane. (b) The arithmetic is the same as in vector addition. (Example 1)

Example 1 shows how the complex plane representation of complex number addition is virtually the same as the Cartesian plane representation of vector addition. Another similarity between complex numbers and two-dimensional vectors is the definition of absolute value.

> **DEFINITION Absolute Value (Modulus) of a Complex Number**
>
> The **absolute value** or **modulus** of a complex number $z = a + bi$ is
>
> $$|z| = |a + bi| = \sqrt{a^2 + b^2}.$$
>
> In the complex plane, $|a + bi|$ is the distance of $a + bi$ from the origin.

Trigonometric Form of Complex Numbers

Figure 6.59 shows the graph of $z = a + bi$ in the complex plane. The distance r from the origin is the modulus of z. If we define a direction angle θ for z just as we did with vectors, we see that $a = r \cos \theta$ and $b = r \sin \theta$. Substituting these expressions for a and b gives us the **trigonometric form** (or **polar form**) of the complex number z.

POLAR FORM

What's in a cis?

Trigonometric (or polar) form appears frequently enough in scientific texts to have an abbreviated form. The expression "$\cos \theta + i \sin \theta$" is often shortened to "cis θ" (pronounced "kiss θ"). Thus $z = r$ cis θ.

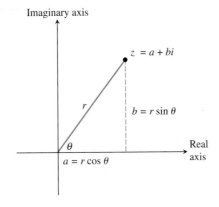

FIGURE 6.59 If r is the distance of $z = a + bi$ from the origin and θ is the directional angle shown, then $z = r (\cos \theta + i \sin \theta)$, which is the trigonometric form of z.

> **DEFINITION Trigonometric Form of a Complex Number**
>
> The **trigonometric form** of the complex number $z = a + bi$ is
>
> $$z = r(\cos \theta + i \sin \theta)$$
>
> where $a = r \cos \theta$, $b = r \sin \theta$, $r = \sqrt{a^2 + b^2}$, and $\tan \theta = b/a$. The number r is the *absolute value* or *modulus* of z, and θ is an **argument** of z.

An angle θ for the trigonometric form of z can always be chosen so that $0 \le \theta \le 2\pi$, although any angle coterminal with θ could be used. Consequently, the *angle θ* and *argument* of a complex number z are not unique. It follows that the trigonometric form of a complex number z is not unique.

Imaginary axis

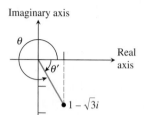

FIGURE 6.60 The complex number for Example 2a.

Imaginary axis

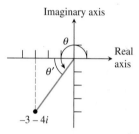

FIGURE 6.61 The complex number for Example 2b.

EXAMPLE 2 Finding Trigonometric Forms

Find the trigonometric form with $0 \le \theta < 2\pi$ for the complex number.

(a) $1 - \sqrt{3}i$ **(b)** $-3 - 4i$

SOLUTION

(a) For $1 - \sqrt{3}i$,

$$r = |1 - \sqrt{3}i| = \sqrt{(1)^2 + (\sqrt{3})^2} = 2.$$

Because the reference angle θ' for θ is $-\pi/3$ (Figure 6.60),

$$\theta = 2\pi + \left(-\frac{\pi}{3}\right) = \frac{5\pi}{3}.$$

Thus,

$$1 - \sqrt{3}i = 2 \cos \frac{5\pi}{3} + 2i \sin \frac{5\pi}{3}.$$

(b) For $-3 - 4i$,

$$|-3 - 4i| = \sqrt{(-3)^2 + (-4)^2} = 5.$$

The reference angle θ' for θ (Figure 6.61) satisfies the equation

$$\tan \theta' = \frac{4}{3}, \quad \text{so}$$

$$\theta' = \tan^{-1} \frac{4}{3} = 0.927.\ldots$$

Because the terminal side of θ is in the third quadrant, we conclude that

$$\theta = \pi + \theta' \approx 4.07.$$

Therefore,

$$-3 - 4i \approx 5(\cos 4.07 + i \sin 4.07).$$

Now try Exercise 5.

Multiplication and Division of Complex Numbers

The trigonometric form for complex numbers is particularly convenient for multiplying and dividing complex numbers. The product involves the product of the moduli and the sum of the arguments. (*Moduli* is the plural of *modulus*.) The quotient involves the quotient of the moduli and the difference of the arguments.

Product and Quotient of Complex Numbers

Let $z_1 = r_1 (\cos \theta_1 + i \sin \theta_1)$ and $z_2 = r_2 (\cos \theta_2 + i \sin \theta_2)$. Then

1. $z_1 \cdot z_2 = r_1 r_2 [\cos (\theta_1 + \theta_2) + i \sin (\theta_1 + \theta_2)].$

2. $\dfrac{z_1}{z_2} = \dfrac{r_1}{r_2} [\cos (\theta_1 - \theta_2) + i \sin (\theta_1 - \theta_2)], \quad r_2 \ne 0.$

Proof of the Product Formula

$$z_1 \cdot z_2 = r_1(\cos \theta_1 + i \sin \theta_1) \cdot r_2(\cos \theta_2 + i \sin \theta_2)$$

$$= r_1 r_2 [(\cos \theta_1 \cos \theta_2 - \sin \theta_1 \sin \theta_2) + i(\sin \theta_1 \cos \theta_2 + \cos \theta_1 \sin \theta_2)]$$

$$= r_1 r_2 [\cos(\theta_1 + \theta_2) + i \sin(\theta_1 + \theta_2)]$$

You will be asked to prove the quotient formula in Exercise 63.

EXAMPLE 3   Multiplying Complex Numbers

Express the product of z_1 and z_2 in standard form:

$$z_1 = 25\sqrt{2}\left(\cos \frac{-\pi}{4} + i \sin \frac{-\pi}{4}\right), \quad z_2 = 14\left(\cos \frac{\pi}{3} + i \sin \frac{\pi}{3}\right).$$

SOLUTION

$$z_1 \cdot z_2 = 25\sqrt{2}\left(\cos \frac{-\pi}{4} + i \sin \frac{-\pi}{4}\right) \cdot 14\left(\cos \frac{\pi}{3} + i \sin \frac{\pi}{3}\right)$$

$$= 25 \cdot 14\sqrt{2}\left[\cos\left(\frac{-\pi}{4} + \frac{\pi}{3}\right) + i \sin\left(\frac{-\pi}{4} + \frac{\pi}{3}\right)\right]$$

$$= 350\sqrt{2}\left(\cos \frac{\pi}{12} + i \sin \frac{\pi}{12}\right)$$

$$\approx 478.11 + 128.11i$$

Now try Exercise 19.

EXAMPLE 4   Dividing Complex Numbers

Express the quotient z_1/z_2 in standard form:

$$z_1 = 2\sqrt{2}(\cos 135° + i \sin 135°), \quad z_2 = 6(\cos 300° + i \sin 300°).$$

SOLUTION

$$\frac{z_1}{z_2} = \frac{2\sqrt{2}\,(\cos 135° + i \sin 135°)}{6(\cos 300° + i \sin 300°)}$$

$$= \frac{\sqrt{2}}{3}[\cos(135° - 300°) + i \sin(135° - 300°)]$$

$$= \frac{\sqrt{2}}{3}[\cos(-165°) + i \sin(-165°)]$$

$$\approx -0.46 - 0.12i$$

Now try Exercise 23.

Powers of Complex Numbers

We can use the product formula to raise a complex number to a power. For example, let $z = r (\cos \theta + i \sin \theta)$. Then

$$z^2 = z \cdot z$$

$$= r(\cos \theta + i \sin \theta) \cdot r(\cos \theta + i \sin \theta)$$

$$= r^2[\cos (\theta + \theta) + i \sin (\theta + \theta)]$$

$$= r^2(\cos 2\theta + i \sin 2\theta)$$

Figure 6.62 gives a geometric interpretation of squaring a complex number: its argument is doubled and its distance from the origin is multiplied by a factor of r, increased if $r > 1$ or decreased if $r < 1$.

We can find z^3 by multiplying z by z^2:

$$z^3 = z \cdot z^2$$

$$= r(\cos \theta + i \sin \theta) \cdot r^2(\cos 2\theta + i \sin 2\theta)$$

$$= r^3[\cos (\theta + 2\theta) + i \sin (\theta + 2\theta)]$$

$$= r^3(\cos 3\theta + i \sin 3\theta)$$

Similarly,

$$z^4 = r^4(\cos 4\theta + i \sin 4\theta)$$

$$z^5 = r^5(\cos 5\theta + i \sin 5\theta)$$

$$\vdots$$

This pattern can be generalized to the following theorem, named after the mathematician Abraham De Moivre (1667–1754), who also made major contributions to the field of probability.

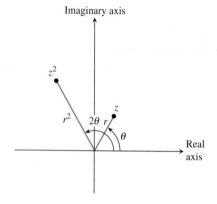

FIGURE 6.62 A geometric interpretation of z^2.

De Moivre's Theorem

Let $z = r(\cos \theta + i \sin \theta)$ and let n be a positive integer. Then

$$z^n = [r(\cos \theta + i \sin \theta)]^n = r^n(\cos n\theta + i \sin n\theta).$$

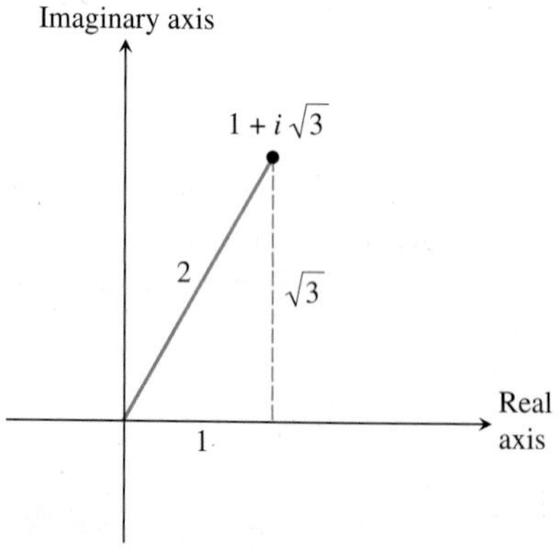

EXAMPLE 5   Using De Moivre's theorem

Find $(1 + i\sqrt{3})^3$ using De Moivre's theorem.

SOLUTION

Solve Algebraically   See Figure 6.63. The argument of $z = 1 + i\sqrt{3}$ is $\theta = \pi/3$, and its modulus is $|1 + i\sqrt{3}| = \sqrt{1 + 3} = 2$. Therefore,

$$z = 2\left(\cos \frac{\pi}{3} + i \sin \frac{\pi}{3}\right)$$

$$z^3 = 2^3\left[\cos \left(3 \cdot \frac{\pi}{3}\right) + i \sin \left(3 \cdot \frac{\pi}{3}\right)\right]$$

$$= 8(\cos \pi + i \sin \pi)$$

$$= 8(-1 + 0i) = -8$$

FIGURE 6.63 The complex number in Example 5.

continued

Support Numerically Figure 6.64a sets the graphing calculator we use in complex number mode. Figure 6.64b supports the result obtained algebraically.

Now try Exercise 31.

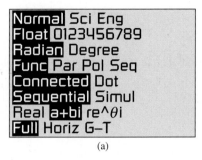
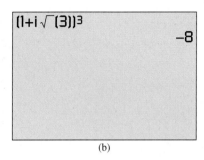

(a) (b)

FIGURE 6.64 (a) Setting a graphing calculator in complex number mode. (b) Computing $(1 + i\sqrt{3})^3$ with a graphing calculator.

EXAMPLE 6 Using De Moivre's Theorem

Find $[(-\sqrt{2}/2) + i(\sqrt{2}/2)]^8$ using De Moivre's theorem.

SOLUTION The argument of $z = (-\sqrt{2}/2) + i(\sqrt{2}/2)$ is $\theta = 3\pi/4$, and its modulus is

$$\left| \frac{-\sqrt{2}}{2} + i\frac{\sqrt{2}}{2} \right| = \sqrt{\frac{1}{2} + \frac{1}{2}} = 1.$$

Therefore,

$$z = \cos\frac{3\pi}{4} + i\sin\frac{3\pi}{4}$$

$$z^8 = \cos\left(8 \cdot \frac{3\pi}{4}\right) + i\sin\left(8 \cdot \frac{3\pi}{4}\right)$$

$$= \cos 6\pi + i\sin 6\pi$$

$$= 1 + i \cdot 0 = 1$$

Now try Exercise 35.

Roots of Complex Numbers

The complex number $1 + i\sqrt{3}$ in Example 5 is a solution of $z^3 = -8$, and the complex number $(-\sqrt{2}/2) + i(\sqrt{2}/2)$ in Example 6 is a solution of $z^8 = 1$. The complex number $1 + i\sqrt{3}$ is a third root of -8 and $(-\sqrt{2}/2) + i(\sqrt{2}/2)$ is an eighth root of 1.

nth Root of a Complex Number

A complex number $v = a + bi$ is an **nth root of z** if

$$v^n = z.$$

If $z = 1$, then v is an **nth root of unity**.

We use De Moivre's theorem to develop a general formula for finding the nth roots of a nonzero complex number. Suppose that $v = s(\cos \alpha + i \sin \alpha)$ is an nth root of $z = r(\cos \theta + i \sin \theta)$. Then

$$v^n = z$$

$$[s(\cos \alpha + i \sin \alpha)]^n = r(\cos \theta + i \sin \theta)$$

$$s^n(\cos n\alpha + i \sin n\alpha) = r(\cos \theta + i \sin \theta) \qquad\qquad (1)$$

Next, we take the absolute value of both sides:

$$|s^n(\cos n\alpha + i \sin na)| = |r(\cos \theta + i \sin \theta)|$$

$$\sqrt{s^{2n}(\cos^2 n\alpha + \sin^2 n\alpha)} = \sqrt{r^2(\cos^2 \theta + \sin^2 \theta)}$$

$$\sqrt{s^{2n}} = \sqrt{r^2}$$

$$s^n = r \qquad\qquad s > 0, r > 0$$

$$s = \sqrt[n]{r}$$

Substituting $s^n = r$ into Equation (1), we obtain

$$\cos n\alpha + i \sin n\alpha = \cos \theta + i \sin \theta.$$

Therefore, $n\alpha$ can be any angle coterminal with θ. Consequently, for any integer k, v is an nth root of z if $s = \sqrt[n]{r}$ and

$$n\alpha = \theta + 2\pi k$$

$$\alpha = \frac{\theta + 2\pi k}{n}.$$

The expression for v takes on n different values for $k = 0, 1, \ldots, n - 1$, and the values start to repeat for $k = n, n + 1, \ldots$.

We summarize this result.

Finding nth Roots of a Complex Number

If $z = r(\cos \theta + i \sin \theta)$, then the n distinct complex numbers

$$\sqrt[n]{r}\left(\cos \frac{\theta + 2\pi k}{n} + i \sin \frac{\theta + 2\pi k}{n}\right),$$

where $k = 0, 1, 2, \ldots, n - 1$, are the nth roots of the complex number z.

EXAMPLE 7   Finding Fourth Roots

Find the fourth roots of $z = 5(\cos (\pi/3) + i \sin (\pi/3))$.

SOLUTION The fourth roots of z are the complex numbers

$$\sqrt[4]{5}\left(\cos \frac{\pi/3 + 2\pi k}{4} + i \sin \frac{\pi/3 + 2\pi k}{4}\right)$$

for $k = 0, 1, 2, 3$.

continued

Taking into account that $(\pi/3 + 2\pi k)/4 = \pi/12 + \pi k/2$, the list becomes

$$z_1 = \sqrt[4]{5}\left[\cos\left(\frac{\pi}{12} + \frac{0}{2}\right) + i\sin\left(\frac{\pi}{12} + \frac{0}{2}\right)\right]$$

$$= \sqrt[4]{5}\left[\cos\frac{\pi}{12} + i\sin\frac{\pi}{12}\right]$$

$$z_2 = \sqrt[4]{5}\left[\cos\left(\frac{\pi}{12} + \frac{\pi}{2}\right) + i\sin\left(\frac{\pi}{12} + \frac{\pi}{2}\right)\right]$$

$$= \sqrt[4]{5}\left[\cos\frac{7\pi}{12} + i\sin\frac{7\pi}{12}\right]$$

$$z_3 = \sqrt[4]{5}\left[\cos\left(\frac{\pi}{12} + \frac{2\pi}{2}\right) + i\sin\left(\frac{\pi}{12} + \frac{2\pi}{2}\right)\right]$$

$$= \sqrt[4]{5}\left[\cos\frac{13\pi}{12} + i\sin\frac{13\pi}{12}\right]$$

$$z_4 = \sqrt[4]{5}\left[\cos\left(\frac{\pi}{12} + \frac{3\pi}{2}\right) + i\sin\left(\frac{\pi}{12} + \frac{3\pi}{2}\right)\right]$$

$$= \sqrt[4]{5}\left[\cos\frac{19\pi}{12} + i\sin\frac{19\pi}{12}\right]$$

Now try Exercise 45.

EXAMPLE 8 Finding Cube Roots

Find the cube roots of -1 and plot them.

SOLUTION First we write the complex number $z = -1$ in trigonometric form

$$z = -1 + 0i = \cos\pi + i\sin\pi.$$

The third roots of $z = -1 = \cos\pi + i\sin\pi$ are the complex numbers

$$\cos\frac{\pi + 2\pi k}{3} + i\sin\frac{\pi + 2\pi k}{3},$$

for $k = 0, 1, 2$. The three complex numbers are

$$z_1 = \cos\frac{\pi}{3} + i\sin\frac{\pi}{3} \qquad = \frac{1}{2} + \frac{\sqrt{3}}{2}i,$$

$$z_2 = \cos\frac{\pi + 2\pi}{3} + i\sin\frac{\pi + 2\pi}{3} = -1 + 0i,$$

$$z_3 = \cos\frac{\pi + 4\pi}{3} + i\sin\frac{\pi + 4\pi}{3} = \frac{1}{2} - \frac{\sqrt{3}}{2}i.$$

Figure 6.65 shows the graph of the three cube roots z_1, z_2, and z_3. They are evenly spaced (with distance of $2\pi/3$ radians) around the unit circle.

Now try Exercise 57.

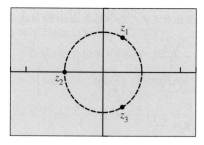

[–2.4, 2.4] by [–1.6, 1.6]

FIGURE 6.65 The three cube roots z_1, z_2, and z_3 of -1 displayed on the unit circle (dashed). (Example 8)

63. Quotient Formula Let $z_1 = r_1(\cos \theta_1 + i \sin \theta_1)$ and
$z_2 = r_2(\cos \theta_2 + i \sin \theta_2)$, $r_2 \neq 0$. Verify that
$z_1/z_2 = r_1/r_2[\cos(\theta_1 - \theta_2) + i \sin(\theta_1 - \theta_2)]$.

64. Group Activity *n*th **Roots** Show that the *n*th roots of the complex number $r(\cos \theta + i \sin \theta)$ are spaced $2\pi/n$ radians apart on a circle with radius $\sqrt[n]{r}$.

Standardized Test Questions

65. True or False The trigonometric form of a complex number is unique. Justify your answer.

66. True or False The complex number i is a cube root of $-i$. Justify your answer.

In Exercises 67–70, you may use a graphing calculator to solve the problem.

67. Multiple Choice Which of the following is a trigonometric form of the complex number $-1 + \sqrt{3}i$?

(A) $2\left(\cos \dfrac{\pi}{3} + i \sin \dfrac{\pi}{3}\right)$ **(B)** $2\left(\cos \dfrac{2\pi}{3} + i \sin \dfrac{2\pi}{3}\right)$

(C) $2\left(\cos \dfrac{4\pi}{3} + i \sin \dfrac{4\pi}{3}\right)$ **(D)** $2\left(\cos \dfrac{5\pi}{3} + i \sin \dfrac{5\pi}{3}\right)$

(E) $2\left(\cos \dfrac{7\pi}{3} + i \sin \dfrac{7\pi}{3}\right)$

68. Multiple Choice Which of the following is the number of distinct complex number solutions of $z^5 = 1 + i$?

(A) 0   **(B)** 1   **(C)** 3   **(D)** 4   **(E)** 5

69. Multiple Choice Which of the following is the standard form for the product

of $\sqrt{2}\left(\cos \dfrac{\pi}{4} + i \sin \dfrac{\pi}{4}\right)$ and $\sqrt{2}\left(\cos \dfrac{7\pi}{4} + i \sin \dfrac{7\pi}{4}\right)$?

(A) 2 **(B)** -2 **(C)** $-2i$ **(D)** $-1 + i$ **(E)** $1 - i$

70. Multiple Choice Which of the following is not a fourth root of 1?

(A) i^2 **(B)** $-i^2$ **(C)** $\sqrt{-1}$ **(D)** $-\sqrt{-1}$ **(E)** $\sqrt{i}$

Explorations

71. Complex Conjugates The complex conjugate of $z = a + bi$ is $\bar{z} = a - bi$. Let $z = r(\cos \theta + i \sin \theta)$.

(a) Prove that $\bar{z} = r[\cos(-\theta) + i \sin(-\theta)]$.

(b) Use the trigonometric form to find $z \cdot \bar{z}$.

(c) Use the trigonometric form to find $z/\bar{z}$, if $\bar{z} \neq 0$.

(d) Prove that $-z = r[\cos(\theta + \pi) + i \sin(\theta + \pi)]$.

72. Modulus of Complex Numbers Let $z = r(\cos \theta + i \sin \theta)$.

(a) Prove that $|z| = |r|$.

(b) Use the trigonometric form for the complex numbers z_1 and z_2 to prove that $|z_1 \cdot z_2| = |z_1| \cdot |z_2|$.

Extending the Ideas

73. Using Polar Form on a Graphing Calculator The complex number $r(\cos \theta + i \sin \theta)$ can be entered in polar form on some graphing calculators as $re^{i\theta}$.

(a) Support the result of Example 3 by entering the complex numbers z_1 and z_2 in polar form on your graphing calculator and computing the product with your graphing calculator.

(b) Support the result of Example 4 by entering the complex numbers z_1 and z_2 in polar form on your graphing calculator and computing the quotient with your graphing calculator.

(c) Support the result of Example 5 by entering the complex number in polar form on your graphing calculator and computing the power with your graphing calculator.

74. Visualizing Roots of Unity Set your graphing calculator in parametric mode with $0 \leq T \leq 8$, Tstep $= 1$, Xmin $= -2.4$, Xmax $= 2.4$, Ymin $= -1.6$, and Ymax $= 1.6$.

(a) Let $x = \cos((2\pi/8)t)$ and $y = \sin((2\pi/8)t)$. Use trace to visualize the eight eighth roots of unity. We say that $2\pi/8$ *generates* the eighth roots of unity. (Try both dot mode and connected mode.)

(b) Replace $2\pi/8$ in part (a) by the arguments of other eighth roots of unity. Do any others *generate* the eighth roots of unity?

(c) Repeat parts (a) and (b) for the fifth, sixth, and seventh roots of unity, using appropriate functions for x and y.

(d) What would you conjecture about an *n*th root of unity that generates all the *n*th roots of unity in the sense of part (a)?

75. Parametric Graphing Write parametric equations that represent $(\sqrt{2} + i)^n$ for $n = t$. Draw and label an *accurate* spiral representing $(\sqrt{2} + i)^n$ for $n = 0, 1, 2, 3, 4$.

76. Parametric Graphing Write parametric equations that represent $(-1 + i)^n$ for $n = t$. Draw and label an *accurate* spiral representing $(-1 + i)^n$ for $n = 0, 1, 2, 3, 4$.

77. Explain why the triangles formed by 0, 1, and z_1 and by 0, z_2 and z_1z_2 shown in the figure are similar triangles.

78. Compass and Straightedge Construction Using only a compass and straightedge, construct the location of z_1z_2 given the location of 0, 1, z_1, and z_2.

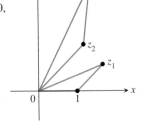

In Exercises 79–84, find all solutions of the equation (real and complex).

79. $x^3 - 1 = 0$ **80.** $x^4 - 1 = 0$

81. $x^3 + 1 = 0$ **82.** $x^4 + 1 = 0$

83. $x^5 + 1 = 0$ **84.** $x^5 - 1 = 0$

CHAPTER 6 **Key Ideas**

PROPERTIES, THEOREMS, AND FORMULAS

PROCEDURES

GALLERY OF FUNCTIONS

Rose Curves: $r = a \cos n\theta$ and $r = a \sin n\theta$

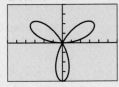

[–6,6] by [–4, 4]

$r = 4 \sin 3\theta$

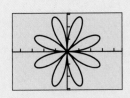

[–4.7, 4.7] by [–3.1, 3.1]

$r = 3 \sin 4\theta$

Limaçon Curves: $r = a \pm b \sin \theta$ and $r = a \pm b \cos \theta$ with $a > 0$ and $b > 0$

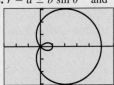

Limaçon with an inner loop: $\dfrac{a}{b} < 1$

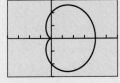

Cardioid: $\dfrac{a}{b} = 1$

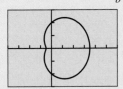

Dimpled limaçon: $1 < \dfrac{a}{b} < 2$

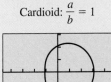

Convex limaçon: $\dfrac{a}{b} \geq 2$

Spiral of Archimedes:

[–30, 30] by [–20, 20]

$r = \theta,\, 0 \leq \theta \leq 45$

Lemniscate Curves: $r^2 = a^2 \sin 2\theta$ and $r^2 = a^2 \cos 2\theta$

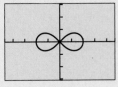

[–4.7, 4.7] by [–3.1, 3.1]

$r^2 = 4 \cos 2\theta$

CHAPTER 6 Review Exercises

The collection of exercises marked in red could be used as a chapter test.

In Exercises 1–6, let $\mathbf{u} = \langle 2, -1 \rangle$, $\mathbf{v} = \langle 4, 2 \rangle$, and $\mathbf{w} = \langle 1, -3 \rangle$ be vectors. Find the indicated expression.

1. $\mathbf{u} - \mathbf{v}$

2. $2\mathbf{u} - 3\mathbf{w}$

3. $|\mathbf{u} + \mathbf{v}|$

4. $|\mathbf{w} - 2\mathbf{u}|$

5. $\mathbf{u} \cdot \mathbf{v}$

6. $\mathbf{u} \cdot \mathbf{w}$

In Exercises 7–10, let $A = (2, -1)$, $B = (3, 1)$, $C = (-4, 2)$, and $D = (1, -5)$. Find the component form and magnitude of the vector.

7. $3\overrightarrow{AB}$

8. $\overrightarrow{AB} + \overrightarrow{CD}$

9. $\overrightarrow{AC} + \overrightarrow{BD}$

10. $\overrightarrow{CD} - \overrightarrow{AB}$

In Exercises 11 and 12, find **(a)** a unit vector in the direction of $\overrightarrow{AB}$ and **(b)** a vector of magnitude 3 in the opposite direction.

11. $A = (4, 0)$, $B = (2, 1)$

12. $A = (3, 1)$, $B = (5, 1)$

In Exercises 13 and 14, find **(a)** the direction angles of $\mathbf{u}$ and $\mathbf{v}$ and **(b)** the angle between $\mathbf{u}$ and $\mathbf{v}$.

13. $\mathbf{u} = \langle 4, 3 \rangle$, $\mathbf{v} = \langle 2, 5 \rangle$

14. $\mathbf{u} = \langle -2, 4 \rangle$, $\mathbf{v} = \langle 6, 4 \rangle$

In Exercises 15–18, convert the polar coordinates to rectangular coordinates.

15. $(-2.5, 25°)$

16. $(-3.1, 135°)$

17. $(2, -\pi/4)$

18. $(3.6, 3\pi/4)$

In Exercises 19 and 20, polar coordinates of point P are given. Find all of its polar coordinates.

19. $P = (-1, -2\pi/3)$

20. $P = (-2, 5\pi/6)$

In Exercises 21–24, rectangular coordinates of point P are given. Find polar coordinates of P that satisfy these conditions:

(a) $0 \le \theta \le 2\pi$ **(b)** $-\pi \le \theta \le \pi$ **(c)** $0 \le \theta \le 4\pi$

21. $P = (2, -3)$

22. $P = (-10, 0)$

23. $P = (5, 0)$

24. $P = (0, -2)$

In Exercises 25–30, eliminate the parameter t and identify the graph.

25. $x = 3 - 5t$, $y = 4 + 3t$

26. $x = 4 + t$, $y = -8 - 5t$, $-3 \le t \le 5$

27. $x = 2t^2 + 3$, $y = t - 1$

28. $x = 3 \cos t$, $y = 3 \sin t$

29. $x = e^{2t} - 1$, $y = e^t$

30. $x = t^3$, $y = \ln t$, $t > 0$

In Exercises 31 and 32, find a parametrization for the curve.

31. The line through the points $(-1, -2)$ and $(3, 4)$.

32. The line segment with endpoints $(-2, 3)$ and $(5, 1)$.

Exercises 33 and 34 refer to the complex number z_1 shown in the figure.

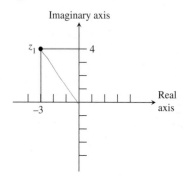

33. If $z_1 = a + bi$, find a, b, and $|z_1|$.

34. Find the trigonometric form of z_1.

In Exercises 35–38, write the complex number in standard form.

35. $6(\cos 30° + i \sin 30°)$

36. $3(\cos 150° + i \sin 150°)$

37. $2.5\left(\cos \dfrac{4\pi}{3} + i \sin \dfrac{4\pi}{3}\right)$

38. $4(\cos 2.5 + i \sin 2.5)$

In Exercises 39–42, write the complex number in trigonometric form where $0 \le \theta \le 2\pi$. Then write three other possible trigonometric forms for the number.

39. $3 - 3i$

40. $-1 + i\sqrt{2}$

41. $3 - 5i$

42. $-2 - 2i$

In Exercises 43 and 44, write the complex numbers $z_1 \cdot z_2$ and z_1/z_2 in trigonometric form.

43. $z_1 = 3(\cos 30° + i \sin 30°)$ and $z_2 = 4(\cos 60° + i \sin 60°)$

44. $z_1 = 5(\cos 20° + i \sin 20°)$ and $z_2 = -2(\cos 45° + i \sin 45°)$

In Exercises 45–48, use De Moivre's theorem to find the indicated power of the complex number. Write your answer in **(a)** trigonometric form and **(b)** standard form.

45. $\left[3\left(\cos \dfrac{\pi}{4} + i \sin \dfrac{\pi}{4}\right)\right]^5$

46. $\left[2\left(\cos \dfrac{\pi}{12} + i \sin \dfrac{\pi}{12}\right)\right]^8$

47. $\left[5\left(\cos \dfrac{5\pi}{3} + i \sin \dfrac{5\pi}{3}\right)\right]^3$

48. $\left[7\left(\cos \dfrac{\pi}{24} + i \sin \dfrac{\pi}{24}\right)\right]^6$

In Exercises 49–52, find and graph the nth roots of the complex number for the specified value of n.

49. $3 + 3i$, $n = 4$

50. 8, $n = 3$

51. 1, $n = 5$

52. -1, $n = 6$

In Exercises 53–60, decide whether the graph of the given polar equation appears among the four graphs shown.

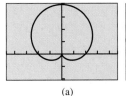

(a)

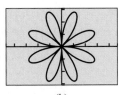

(b)

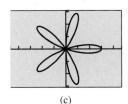

(c)

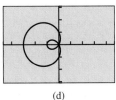

(d)

53. $r = 3 \sin 4\theta$

54. $r = 2 + \sin \theta$

55. $r = 2 + 2 \sin \theta$

56. $r = 3|\sin 3\theta|$

57. $r = 2 - 2 \sin \theta$

58. $r = 1 - 2 \cos \theta$

59. $r = 3 \cos 5\theta$

60. $r = 3 - 2 \tan \theta$

In Exercises 61–64, convert the polar equation to rectangular form and identify the graph.

61. $r = -2$

62. $r = -2 \sin \theta$

63. $r = -3 \cos \theta - 2 \sin \theta$

64. $r = 3 \sec \theta$

In Exercises 65–68, convert the rectangular equation to polar form. Graph the polar equation.

65. $y = -4$

66. $x = 5$

67. $(x - 3)^2 + (y + 1)^2 = 10$

68. $2x - 3y = 4$

In Exercises 69–72, analyze the graph of the polar curve.

69. $r = 2 - 5 \sin \theta$

70. $r = 4 - 4 \cos \theta$

71. $r = 2 \sin 3\theta$

72. $r^2 = 2 \sin 2\theta, 0 \le \theta \le 2\pi$

73. Graphing Lines Using Polar Equations

(a) Explain why $r = a \sec \theta$ is a polar form for the line $x = a$.

(b) Explain why $r = b \csc \theta$ is a polar form for the line $y = b$.

(c) Let $y = mx + b$. Prove that

$$r = \frac{b}{\sin \theta - m \cos \theta}$$

is a polar form for the line. What is the domain of r?

(d) Illustrate the result in part (c) by graphing the line $y = 2x + 3$ using the polar form from part (c).

74. Flight Engineering An airplane is flying on a bearing of 80° at 540 mph. A wind is blowing with the bearing 100° at 55 mph.

(a) Find the component form of the velocity of the airplane.

(b) Find the actual speed and direction of the airplane.

75. Flight Engineering An airplane is flying on a bearing of 285° at 480 mph. A wind is blowing with the bearing 265° at 30 mph.

(a) Find the component form of the velocity of the airplane.

(b) Find the actual speed and direction of the airplane.

76. Combining Forces A force of 120 lb acts on an object at an angle of 20°. A second force of 300 lb acts on the object at an angle of −5°. Find the direction and magnitude of the resultant force.

77. Braking Force A 3000 pound car is parked on a street that makes an angle of 16° with the horizontal (see figure).

(a) Find the force required to keep the car from rolling down the hill.

(b) Find the component of the force perpendicular to the street.

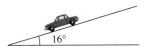

16°

78. Work Find the work done by a force **F** of 36 pounds acting in the direction given by the vector $\langle 3, 5 \rangle$ in moving an object 10 feet from $(0, 0)$ to $(10, 0)$.

79. Height of an Arrow Stewart shoots an arrow straight up from the top of a building with initial velocity of 245 ft/sec. The arrow leaves from a point 200 ft above level ground.

(a) Write an equation that models the height of the arrow as a function of time t.

(b) Use parametric equations to simulate the height of the arrow.

(c) Use parametric equations to graph height against time.

(d) How high is the arrow after 4 sec?

(e) What is the maximum height of the arrow? When does it reach its maximum height?

(f) How long will it be before the arrow hits the ground?

80. Ferris Wheel Problem Lucinda is on a Ferris wheel of radius 35 ft that turns at the rate of one revolution every 20 sec. The lowest point of the Ferris wheel (6 o'clock) is 15 ft above ground level at the point $(0, 15)$ of a rectangular coordinate system. Find parametric equations for the position of Lucinda as a function of time t in seconds if Lucinda starts ($t = 0$) at the point $(35, 50)$.

81. Ferris Wheel Problem The lowest point of a Ferris wheel (6 o'clock) of radius 40 ft is 10 ft above the ground, and the center is on the y-axis. Find parametric equations for Henry's position as a function of time t in seconds if his starting position ($t = 0$) is the point $(0, 10)$ and the wheel turns at the rate of one revolution every 15 sec.

82. Ferris Wheel Problem Sarah rides the Ferris wheel described in Exercise 81. Find parametric equations for Sarah's position as a function of time t in seconds if her starting position ($t = 0$) is the point (0, 90) and the wheel turns at the rate of one revolution every 18 sec.

83. Epicycloid The graph of the parametric equations

$$x = 4 \cos t - \cos 4t, \quad y = 4 \sin t - \sin 4t$$

is an *epicycloid*. The graph is the path of a point P on a circle of radius 1 rolling along the outside of a circle of radius 3, as suggested in the figure.

(a) Graph simultaneously this epicycloid and the circle of radius 3.

(b) Suppose the large circle has a radius of 4. Experiment! How do you think the equations in part (a) should be changed to obtain defining equations? What do you think the epicycloid would look like in this case? Check your guesses.

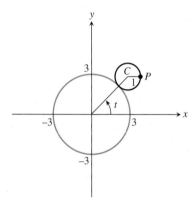

84. Throwing a Baseball Sharon releases a baseball 4 ft above the ground with an initial velocity of 66 ft/sec at an angle of 5° with the horizontal. How many seconds after the ball is thrown will it hit the ground? How far from Sharon will the ball be when it hits the ground?

85. Throwing a Baseball Diego releases a baseball 3.5 ft above the ground with an initial velocity of 66 ft/sec at an angle of 12° with the horizontal. How many seconds after the ball is thrown will it hit the ground? How far from Diego will the ball be when it hits the ground?

86. Field Goal Kicking Spencer practices kicking field goals 40 yd from a goal post with a crossbar 10 ft high. If he kicks the ball with an initial velocity of 70 ft/sec at a 45° angle with the horizontal (see figure), will Spencer make the field goal if the kick sails "true"?

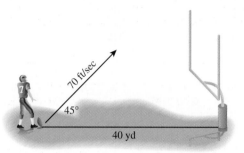

87. Hang Time An NFL place-kicker kicks a football downfield with an initial velocity of 85 ft/sec. The ball leaves his foot at the 15 yard line at an angle of 56° with the horizontal. Determine the following:

(a) The ball's maximum height above the field.

(b) The "hang time" (the total time the football is in the air).

88. Baseball Hitting Brian hits a baseball straight toward a 15-ft-high fence that is 400 ft from home plate. The ball is hit when it is 2.5 ft above the ground and leaves the bat at an angle of 30° with the horizontal. Find the initial velocity needed for the ball to clear the fence.

89. Throwing a Ball at a Ferris Wheel A 60-ft-radius Ferris wheel turns counterclockwise one revolution every 12 sec. Sam stands at a point 80 ft to the left of the bottom (6 o'clock) of the wheel. At the instant Kathy is at 3 o'clock, Sam throws a ball with an initial velocity of 100 ft/sec and an angle with the horizontal of 70°. He releases the ball from the same height as the bottom of the Ferris wheel. Find the minimum distance between the ball and Kathy.

90. Yard Darts Gretta and Lois are launching yard darts 20 ft from the front edge of a circular target of radius 18 in. If Gretta releases the dart 5 ft above the ground with an initial velocity of 20 ft/sec and at a 50° angle with the horizontal, will the dart hit the target?

CHAPTER 6 **Project**

Parametrizing Ellipses

As you discovered in the Chapter 4 Data Project, it is possible to model the displacement of a swinging pendulum using a sinusoidal equation of the form

$$x = a \sin (b(t - c)) + d$$

where x represents the pendulum's distance from a fixed point and t represents total elapsed time. In fact, a pendulum's velocity behaves sinusoidally as well: $y = ab \cos (b(t - c))$, where y represents the pendulum's velocity and a, b, and c are constants common to both the displacement and velocity equations.

Use a motion detection device to collect distance, velocity, and time data for a pendulum, then determine how a resulting plot of velocity versus displacement (called a phase-space plot) can be modeled using parametric equations.

COLLECTING THE DATA

Construct a simple pendulum by fastening about 1 meter of string to the end of a ball. Collect time, distance, and velocity readings for between 2 and 4 seconds (enough time to capture at least one complete swing of the pendulum). Start the pendulum swinging in front of the detector, then activate the system. The data table below shows a sample set of data collected as a pendulum swung back and forth in front of a CBR where t is total elapsed time in seconds, d = distance from the CBR in meters, v = velocity in meters/second.

t	d	v	t	d	v	t	d	v
0	1.021	0.325	0.7	0.621	−0.869	1.4	0.687	0.966
0.1	1.038	0.013	0.8	0.544	−0.654	1.5	0.785	1.013
0.2	1.023	−0.309	0.9	0.493	−0.359	1.6	0.880	0.826
0.3	0.977	−0.598	1.0	0.473	−0.044	1.7	0.954	0.678
0.4	0.903	−0.819	1.1	0.484	0.263	1.8	1.008	0.378
0.5	0.815	−0.996	1.2	0.526	0.573	1.9	1.030	0.049
0.6	0.715	−0.979	1.3	0.596	0.822	2.0	1.020	−0.260

EXPLORATIONS

1. Create a scatter plot for the data you collected or the data above.

2. With your calculator/computer in function mode, find values for a, b, c, and d so that the equation $y = a \sin (b(x - c)) + d$ (where y is distance and x is time) fits the distance versus time data plot.

3. Make a scatter plot of velocity versus time. Using the same a, b, and c values you found in (2), verify that the equation $y = ab \cos (b(x - c))$ (where y is velocity and x is time) fits the velocity versus time data plot.

4. What do you think a plot of velocity versus distance (with velocity on the vertical axis and distance on the horizontal axis) would look like? Make a rough sketch of your prediction, then create a scatter plot of velocity versus distance. How well did your predicted graph match the actual data plot?

5. With your calculator/computer in parametric mode, graph the parametric curve $x = a \sin (b(t - c)) + d$, $y = ab \cos (b(t - c))$, $0 \le t \le 2$ where x represents distance, y represents velocity, and t is the time parameter. How well does this curve match the scatter plot of velocity versus time?

Systems and Matrices

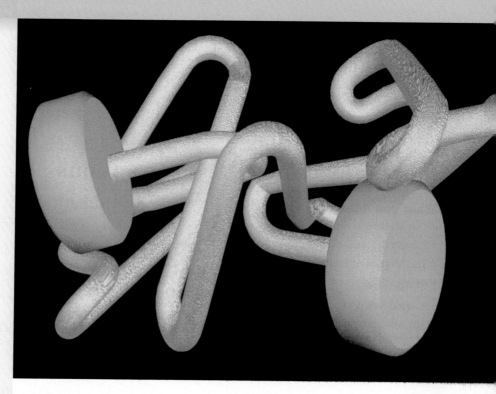

Scientists studying hemoglobin molecules, as represented in the photo, can make new discoveries by viewing the image on a computer. To see all the details, they may need to move the image up or down (translation), turn it around (rotation), or change the size (scaling). In computer graphics these operations are performed using matrix operations. See a related problem involving scaling a triangle on page 589.

Chapter 7 Overview

Many applications of mathematics in science, engineering, business, and other areas involve the use of systems of equations or inequalities in two or more variables as models for the corresponding problems. We investigate several techniques commonly used to solve such systems; and we investigate matrices, which play a central role in several of these techniques. The information age has made the use of matrices widespread because of their use in handling vast amounts of data.

We decompose a rational function into a sum of simpler rational functions using the method of partial fractions. This technique can be used to analyze a rational function, and is used in calculus to integrate rational functions analytically. Finally, we introduce linear programming, a method used to solve problems concerned with decision making in management science.

7.1
Solving Systems of Two Equations

What you'll learn about

- The Method of Substitution
- Solving Systems Graphically
- The Method of Elimination
- Applications

. . . and why

Many applications in business and science can be modeled using systems of equations.

The Method of Substitution

Here is an example of a system of two linear equations in the two variables x and y:

$$2x - y = 10$$
$$3x + 2y = 1.$$

A **solution of a system** of two equations in two variables is an ordered pair of real numbers that is a solution of each equation. For example, the ordered pair $(3, -4)$ is a solution to the above system. We can verify this by showing that $(3, -4)$ is a solution of each equation. Substituting $x = 3$ and $y = -4$ into each equation, we obtain

$$2x - y = 2(3) - (-4) = 6 + 4 = 10,$$
$$3x + 2y = 3(3) + 2(-4) = 9 - 8 = 1.$$

So, both equations are satisfied.

We have **solved the system of equations** when we have found all its solutions. In Example 1, we use the method of substitution to see that $(3, -4)$ is the only solution of this system.

EXAMPLE 1   Using the Substitution Method

Solve the system

$$2x - y = 10$$
$$3x + 2y = 1.$$

SOLUTION

Solve Algebraically Solving the first equation for y yields $y = 2x - 10$. Then substitute the expression for y into the second equation.   *continued*

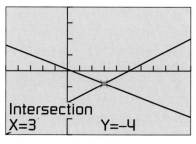

[–5, 10] by [–20, 20]

FIGURE 7.1 The two lines $y = 2x - 10$ and $y = -1.5x + 0.5$ intersect in the point $(3, -4)$. (Example 1)

$$3x + 2y = 1 \qquad \text{Second equation}$$
$$3x + 2(2x - 10) = 1 \qquad \text{Replace } y \text{ by } 2x - 10.$$
$$3x + 4x - 20 = 1 \qquad \text{Distributive property}$$
$$7x = 21 \qquad \text{Collect like terms.}$$
$$x = 3 \qquad \text{Divide by 7.}$$
$$y = -4 \qquad \text{Use } y = 2x - 10.$$

Support Graphically

The graph of each equation is a line. Figure 7.1 shows that the two lines intersect in the single point $(3, -4)$.

Interpret

The solution of the system is $x = 3$, $y = -4$, or the ordered pair $(3, -4)$.

Now try Exercise 5.

The method of substitution can sometimes be applied when the equations in the system are not linear, as illustrated in Example 2.

EXAMPLE 2 Solving a Nonlinear System by Substitution

Find the dimensions of a rectangular garden that has perimeter 100 ft and area 300 ft².

SOLUTION

Model

Let x and y be the lengths of adjacent sides of the garden (Figure 7.2). Then

$$2x + 2y = 100 \qquad \text{Perimeter is 100.}$$
$$xy = 300. \qquad \text{Area is 300.}$$

Solve Algebraically

Solving the first equation for y yields $y = 50 - x$. Then substitute the expression for y into the second equation.

$$xy = 300 \qquad \text{Second equation}$$
$$x(50 - x) = 300 \qquad \text{Replace } y \text{ by } 50 - x.$$
$$50x - x^2 = 300 \qquad \text{Distributive property}$$
$$x^2 - 50x + 300 = 0$$
$$x = \frac{50 \pm \sqrt{(-50)^2 - 4(300)}}{2} \qquad \text{Quadratic formula}$$
$$x = 6.972\ldots \quad \text{or} \quad x = 43.027\ldots \qquad \text{Evaluate.}$$
$$y = 43.027\ldots \quad \text{or} \quad y = 6.972\ldots \qquad \text{Use } y = 50 - x.$$

continued

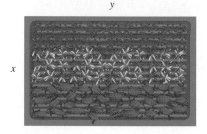

y

x

FIGURE 7.2 The rectangular garden in Example 2.

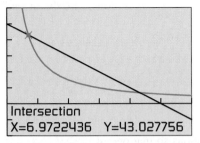

Intersection
X=6.9722436   Y=43.027756

[0, 60] by [–20, 60]

FIGURE 7.3 We can assume $x \geq 0$ and $y \geq 0$ because x and y are lengths. (Example 2)

ROUNDING AT THE END

In Example 2, we did *not* round the values found for x until we computed the values for y. For the sake of accuracy, do not round *intermediate results*. Carry all decimals on your calculator computations and then round the final answer(s).

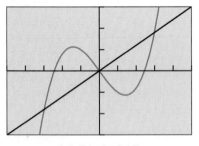

[–5, 5] by [–15, 15]

FIGURE 7.4 The graphs of $y = x^3 - 6x$ and $y = 3x$ have three points of intersection. (Example 3)

Support Graphically Figure 7.3 shows that the graphs of $y = 50 - x$ and $y = 300/x$ have two points of intersections.

Interpret The two ordered pairs $(6.972\ldots, 43.027\ldots)$ and $(43.027\ldots, 6.972\ldots)$ produce the same rectangle whose dimensions are approximately 7 ft by 43 ft.

Now try Exercise 11.

EXAMPLE 3   Solving a Nonlinear System Algebraically

Solve the system

$$y = x^3 - 6x$$
$$y = 3x.$$

Support your solution graphically.

SOLUTION

Substituting the value of y from the first equation into the second equation yields

$$x^3 - 6x = 3x$$
$$x^3 - 9x = 0$$
$$x(x - 3)(x + 3) = 0$$
$$x = 0, x = 3, x = -3 \quad \text{Zero factor property}$$
$$y = 0, y = 9, y = -9 \quad \text{Use } y = 3x.$$

The system of equations has three solutions: $(-3, -9)$, $(0, 0)$, and $(3, 9)$.

Support Graphically The graphs of the two equations in Figure 7.4 suggests that the three solutions found algebraically are correct. *Now try Exercise 13.*

Solving Systems Graphically

Sometimes the method of substitution leads to an equation in one variable that we are not able to solve using the standard algebraic techniques we have studied in this text. In these cases we can solve the system graphically by finding intersections as illustrated in Exploration 1.

EXPLORATION 1 **Solving a System Graphically**

Consider the system:

$$y = \ln x$$
$$y = x^2 - 4x + 2$$

1. Draw the graphs of the two equations in the $[0, 10]$ by $[-5, 5]$ viewing window.

2. Use the graph in part 1 to find the coordinates of the points of intersection shown in the viewing window.

3. Use your knowledge about the graphs of logarithmic and quadratic functions to explain why this system has exactly two solutions.

Substituting the expression for y of the first equation of Exploration 1 into the second equation yields

$$\ln x = x^2 - 4x + 2.$$

We have no standard algebraic technique to solve this equation.

The Method of Elimination

Consider a system of two linear equations in x and y. To **solve by elimination**, we rewrite the two equations as two equivalent equations so that one of the variables has opposite coefficients. Then we add the two equations to eliminate that variable.

EXAMPLE 4　Using the Elimination Method

Solve the system

$$2x + 3y = 5$$
$$-3x + 5y = 21.$$

SOLUTION

Solve Algebraically Multiply the first equation by 3 and the second equation by 2 to obtain

$$6x + 9y = 15$$
$$-6x + 10y = 42.$$

Then add the two equations to eliminate the variable x.

$$19y = 57$$

Next divide by 19 to solve for y.

$$y = 3$$

Finally, substitute $y = 3$ into either of the two original equations to determine that $x = -2$.

The solution of the original system is $(-2, 3)$.　　　　　*Now try Exercise 19.*

EXAMPLE 5　Finding No Solution

Solve the system

$$x - 3y = -2$$
$$2x - 6y = 4.$$

SOLUTION We use the elimination method.

Solve Algebraically

$$-2x + 6y = 4 \qquad \text{Multiply first equation by } -2.$$
$$2x - 6y = 4 \qquad \text{Second equation}$$
$$0 = 8 \qquad \text{Add.}$$

The last equation is true for *no* values of x and y. The system has no solution.

continued

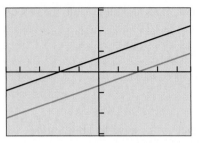

[−4.7, 4.7] by [−3.1, 3.1]

FIGURE 7.5 The graph of the two lines in Example 5 in this square viewing window appear to be parallel.

Support Graphically

Figure 7.5 suggests that the two lines that are the graphs of the two equations in the system are parallel. Solving for *y* in each equation yields

$$y = \frac{1}{3}x + \frac{2}{3}$$

$$y = \frac{1}{3}x - \frac{2}{3}.$$

The two lines have the same slope of 1/3 and are therefore parallel.

Now try Exercise 23.

An easy way to determine the *number of solutions* of a system of two linear equations in two variables is to look at the graphs of the two lines. There are three possibilities. The two lines can intersect in a single point, producing exactly *one* solution as in Examples 1 and 4. The two lines can be parallel, producing *no* solution as in Example 5. The two lines can be the same, producing infinitely many solutions as illustrated in Example 6.

EXAMPLE 6   Finding Infinitely Many Solutions

Solve the system

$$4x - 5y = 2$$

$$-12x + 15y = -6.$$

SOLUTION

$12x - 15y = 6$	Multiply first equation by 3.
$-12x + 15y = -6$	Second equation
$0 = 0$	Add.

The last equation is true for all values of *x* and *y*. Thus, every ordered pair that satisfies one equation satisfies the other equation. The system has infinitely many solutions.

Another way to see that there are infinitely many solutions is to solve each equation for *y*. Both equations yield

$$y = \frac{4}{5}x - \frac{2}{5}.$$

The two lines are the same.

Now try Exercise 25.

Applications

Table 7.1 shows the personal consumption expenditures (in billions of dollars) for dentists and health insurance in the U.S. for several years.

Table 7.1 U.S. Personal Consumption Expenditures		
Year	Dentists (billions of dollars)	Health Insurance (billions of dollars)
1995	45.4	60.7
1998	54.1	71.6
1999	57.4	76.1
2000	61.8	84.0
2001	66.8	89.4
2002	72.2	96.1
2003	75.0	106.0

Source: Bureau of Economic Analysis, U.S. Department of Commerce, as reported in The World Almanac and Book of Facts, 2005.

EXAMPLE 7   Estimating Personal Expenditures with Linear Models

(a) Find linear regression equations for the U.S. personal consumption expenditures for dentists and health insurance in Table 7.1. Superimpose their graphs on a scatter plot of the data.

(b) Use the models in part (a) to estimate when the U.S. personal consumption expenditures for dentists will be the same as that for health insurance and the corresponding amount.

SOLUTION

(a) Let $x = 0$ stand for 1990, $x = 1$ for 1991, and so forth. We use a graphing calculator to find linear regression equations for the amount of expenditures for dentists, y_D, and the amount of expenditures for health insurance, y_{HI}:

$$y_D \approx 3.8507x + 24.4079$$

$$y_{HI} \approx 5.6099x + 28.9184$$

Figure 7.6 shows the two regression equations together with a scatter plot of the two sets of data.

(b) Figure 7.6 shows that the graphs of y_D and y_{HI} intersect at approximately $(-2.56, 14.53)$. $x = -3$ stands for 1987, so Figure 7.6 suggests that the personal consumption expenditures for dentists and for health insurance were both about 14.5 billion sometime during 1987.

Now try Exercise 45.

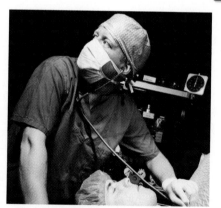

Intersection
X=-2.56395   Y=14.5349

[−10, 20] by [−40, 120]

FIGURE 7.6 The scatter plot and regression equations for the data in Table 7.1. Dentist (□), health insurance (+). (Example 7)

Suppliers will usually increase production, x, if they can get higher prices, p, for their products. So, as one variable increases, the other also increases. Normal mathematical practice would be to use p as the independent variable and x as the dependent variable. However, most economists put x on the horizontal axis and p on the vertical axis. In keeping with this practice, we write $p = f(x)$ for a **supply curve**. On one hand, as the price increases (vertical axis) so does the willingness for suppliers to increase production x (horizontal axis).

On the other hand, the demand, x, for a product by consumers will decrease as the price, p, goes up. So, as one variable increases, the other decreases. Again economists put x (demand) on the horizontal axis and p (price) on the vertical axis, even though it seems like p should be the dependent variable. In keeping with this practice, we write $p = g(x)$ for a **demand curve**.

Finally, a point where the supply curve and demand curve intersect is an **equilibrium point**. The corresponding price is the **equilibrium price**.

EXAMPLE 8 Determining the Equilibrium Price

Nibok Manufacturing has determined that production and price of a new tennis shoe should be geared to the equilibrium point for this system of equations.

$$p = 160 - 5x \quad \text{Demand curve}$$

$$p = 35 + 20x \quad \text{Supply curve}$$

The price, p, is in dollars and the number of shoes, x, is in millions of pairs. Find the equilibrium point.

SOLUTION

We use substitution to solve the system.

$$160 - 5x = 35 + 20x$$

$$25x = 125$$

$$x = 5$$

Substitute this value of x into the demand curve and solve for p.

$$p = 160 - 5x$$

$$p = 160 - 5(5) = 135$$

The equilibrium point is $(5, 135)$. The equilibrium price is \$135, the price for which supply and demand will be equal at 5 million pairs of tennis shoes.

Now try Exercise 43.

QUICK REVIEW 7.1 *(For help, go to Sections P.4 and P.5.)*

In Exercises 1 and 2, solve for y in terms of x.

1. $2x + 3y = 5$ **2.** $xy + x = 4$

In Exercises 3–6, solve the equation algebraically.

3. $3x^2 - x - 2 = 0$ **4.** $2x^2 + 5x - 10 = 0$

5. $x^3 = 4x$ **6.** $x^3 + x^2 = 6x$

7. Write an equation for the line through the point $(-1, 2)$ and parallel to the line $4x + 5y = 2$.

8. Write an equation for the line through the point $(-1, 2)$ and perpendicular to the line $4x + 5y = 2$.

9. Write an equation equivalent to $2x + 3y = 5$ with coefficient of x equal to -4.

10. Find the points of intersection of the graphs of $y = 3x$ and $y = x^3 - 6x$ graphically.

SECTION 7.1 EXERCISES

In Exercises 1 and 2, determine whether the ordered pair is a solution of the system.

1. $5x - 2y = 8$
$2x - 3y = 1$

   (a) $(0, 4)$ **(b)** $(2, 1)$

   (c) $(-2, -9)$

2. $y = x^2 - 6x + 5$
$y = 2x - 7$

   (a) $(2, -3)$ **(b)** $(1, -5)$

   (c) $(6, 5)$

In Exercises 3–12, solve the system by substitution.

3. $x + 2y = 5$
$\quad\quad y = -2$

4. $\quad x = 3$
$x - y = 20$

5. $\quad 3x + y = 20$
$\quad x - 2y = 10$

6. $2x - 3y = -23$
$\quad x + y = 0$

7. $2x - 3y = -7$
$4x + 5y = 8$

8. $3x + 2y = -5$
$2x - 5y = -16$

9. $\quad x - 3y = 6$
$-2x + 6y = 4$

10. $\quad 3x - y = -2$
$-9x + 3y = 6$

11. $\quad y = x^2$
$y - 9 = 0$

12. $\quad x = y + 3$
$x - y^2 = 3y$

In Exercises 13–18, solve the system algebraically. Support your answer graphically.

13. $\quad y = 6x^2$
$7x + y = 3$

14. $\quad y = 2x^2 + x$
$2x + y = 20$

15. $y = x^3 - x^2$
$y = 2x^2$

16. $y = x^3 + x^2$
$y = -x^2$

17. $x^2 + y^2 = 9$
$x - 3y = -1$

18. $\quad x^2 + y^2 = 16$
$4x + 7y = 13$

In Exercises 19–26, solve the system by elimination.

19. $x - y = 10$
$x + y = 6$

20. $\quad 2x + y = 10$
$\quad x - 2y = -5$

21. $3x - 2y = 8$
$5x + 4y = 28$

22. $4x - 5y = -23$
$3x + 4y = 6$

23. $\quad 2x - 4y = -10$
$-3x + 6y = -21$

24. $\quad 2x - 4y = 8$
$-x + 2y = -4$

25. $\quad 2x - 3y = 5$
$-6x + 9y = -15$

26. $\quad 2x - y = 3$
$-4x + 2y = 5$

In Exercises 27–30, use the graph to estimate any solutions of the system.

27. $y = 1 + 2x - x^2$
    $y = 1 - x$

28. $6x - 2y = 7$
    $2x + y = 4$

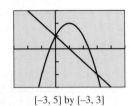

      [–3, 5] by [–3, 3]

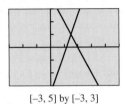

      [–3, 5] by [–3, 3]

29. $x + 2y = 0$
     $0.5x + y = 2$

30. $x^2 + y^2 = 16$
     $y + 4 = x^2$

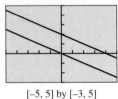

    [–5, 5] by [–3, 5]

    [–9.4, 9.4] by [–6.2, 6.2]

In Exercises 31–34, use graphs to determine the number of solutions the system has.

31. $3x + 5y = 7$
    $4x - 2y = -3$

32. $3x - 9y = 6$
    $2x - 6y = 1$

33. $2x - 4y = 6$
    $3x - 6y = 9$

34. $x - 7y = 9$
    $3x + 4y = 1$

In Exercises 35–42, solve the system graphically. Support your answer numerically.

35. $y = \ln x$
    $1 = 2x + y$

36. $y = 3 \cos x$
    $1 = 2x - y$

37. $y = x^3 - 4x$
    $4 = x - 2y$

38. $y = x^2 - 3x - 5$
    $1 = 2x - y$

39. $x^2 + y^2 = 4$
    $x + 2y = 2$

40. $x^2 + y^2 = 4$
    $x - 2y = 2$

41. $x^2 + y^2 = 9$
        $y = x^2 - 2$

42. $x^2 + y^2 = 9$
        $y = 2 - x^2$

In Exercises 43 and 44, find the equilibrium point for the given demand and supply curve.

43. $p = 200 - 15x$
    $p = 50 + 25x$

44. $p = 15 - \dfrac{7}{100}x$

    $p = 2 + \dfrac{3}{100}x$

45. Medicare Expenditure Table 7.2 shows expenditures (in billions of dollars) for benefits and administrative cost from federal hospital and medical insurance trust funds for several years. Let $x = 0$ stand for 1980, $x = 1$ for 1981, and so forth.

(a) Find the quadratic regression equation and superimpose its graph on a scatter plot of the data.

(b) Find the logistic regression equation and superimpose its graph on a scatter plot of the data.

(c) When will the two models predict expenditures of 300 billion dollars?

(d) Writing to Learn Explain the long range implications of using the quadratic regression equation to predict future expenditures.

(e) Writing to Learn Explain the long range implications of using the logistic regression equation to predict future expenditures.

 Table 7.2 Medicare National Health Expenditures

Year	Expenditures (billions)
1990	110.2
1995	183.2
1997	209.5
1998	210.2
1999	213.5
2000	225.1
2001	246.5
2002	267.1

Source: U.S. Health Care Financing Administration, Health Care Financing Review, Summer 2001, in Statistical Abstract of the U.S., 2004–2005.

46. Personal Income Table 7.3 on the next page gives the total personal income (in billions of dollars) for residents of the states of Iowa and Nevada for several years. Let $x = 0$ stand for 1990, $x = 1$ for 1991, and so forth.

(a) Find the linear regression equation for the Iowa data and superimpose its graph on a scatter plot of the Iowa data.

(b) Find the linear regression equation for the Nevada data and superimpose its graph on a scatter plot of the Nevada data.

(c) Using the models in parts (a) and (b), when will the personal income of the two states be the same?

Table 7.3 Total Personal Income		
Year	Iowa (billions)	Nevada (billions)
1990	48.4	24.8
2000	77.8	61.4
2001	80.2	63.6
2002	82.5	66.2

Source: U.S. Bureau of Economic Analysis, Survey of Current Business, May 1998, in Statistical Abstract of the U.S., 2004–2005.

47. Population Table 7.4 gives the population (in thousands) of the states of Arizona and Massachusetts for several years. Let $x = 0$ stand for 1980, $x = 1$ for 1981, and so forth.

(a) Find the linear regression equation for Arizona's data and superimpose its graph on a scatter plot of Arizona's data.

(b) Find the linear regression equation for Massachusetts's data and superimpose its graph on a scatter plot of Massachusetts data.

(c) Using the models in parts (a) and (b), when will the population of the two states be the same?

Table 7.4 Population		
Year	Arizona (thousands)	Massachusetts (thousands)
1980	2718	5737
1990	3665	6016
1995	4432	6141
1998	4883	6272
1999	5024	6317
2000	5131	6349
2001	5298	6400
2002	5441	6422
2003	5581	6433

Source: U.S. Bureau of the Census, in Statistical Abstract of the U.S., 2004–2005.

48. Group Activity Describe all possibilities for the number of solutions to a system of two equations in two variables if the graphs of the two equations are **(a)** a line and a circle, and **(b)** a circle and a parabola.

49. Garden Problem Find the dimensions of a rectangle with a perimeter of 200 m and an area of 500 m².

50. Cornfield Dimensions Find the dimensions of a rectangular cornfield with a perimeter of 220 yd and an area of 3000 yd².

51. Rowing Speed Hank can row a boat 1 mi upstream (against the current) in 24 min. He can row the same distance downstream in 13 min. If both the rowing speed and current speed are constant, find Hank's rowing speed and the speed of the current.

52. Airplane Speed An airplane flying with the wind from Los Angeles to New York City takes 3.75 hr. Flying against the wind, the airplane takes 4.4 hr for the return trip. If the air distance between Los Angeles and New York is 2500 mi and the airplane speed and wind speed are constant, find the airplane speed and the wind speed.

53. Food Prices At Philip's convenience store the total cost of one medium and one large soda is $1.74. The large soda costs $0.16 more than the medium soda. Find the cost of each soda.

54. Nut Mixture A 5-lb nut mixture is worth $2.80 per pound. The mixture contains peanuts worth $1.70 per pound and cashews worth $4.55 per pound. How many pounds of each type of nut are in the mixture?

55. Connecting Algebra and Functions Determine a and b so that the graph of $y = ax + b$ contains the two points $(-1, 4)$ and $(2, 6)$.

56. Connecting Algebra and Functions Determine a and b so that the graph of $ax + by = 8$ contains the two points $(2, -1)$ and $(-4, -6)$.

57. Rental Van Pedro has two plans to choose from to rent a van.

Company A: a flat fee of $40 plus 10 cents a mile.

Company B: a flat fee of $25 plus 15 cents a mile.

(a) How many miles can Pedro drive in order to be charged the same amount by the two companies?

(b) Writing to Learn Give reasons why Pedro might choose one plan over the other. Explain.

58. Salary Package Stephanie is offered two different salary options to sell major household appliances.

Plan A: a $300 weekly salary plus 5% of her sales.

Plan B: a $600 weekly salary plus 1% of her sales.

(a) What must Stephanie's sales be to earn the same amount on the two plans?

(b) Writing to Learn Give reasons why Stephanie might choose one plan over the other. Explain.

Standardized Test Questions

59. True or False Let a and b be real numbers. The following system of equations can have exactly two solutions:

$2x + 5y = a$

$3x - 4y = b.$

Justify your answer.

60. True or False If the resulting equation after using elimination correctly on a system of two linear equations in two variables is $7 = 0$, then the system has infinitely many solutions. Justify your answer.

In Exercises 61–64, solve the problem without using a calculator.

61. Multiple Choice Which of the following is a solution of the system $2x - 3y = 12$

$x + 2y = -1$?

(A) $(-3, 1)$ **(B)** $(-1, 0)$ **(C)** $(3, -2)$

(D) $(3, 2)$ **(E)** $(6, 0)$

62. Multiple Choice Which of the following cannot be the number of solutions of a system of two equations in two variables whose graphs are a circle and a parabola?

(A) 0   **(B)** 1   **(C)** 2   **(D)** 3   **(E)** 5

63. Multiple Choice Which of the following cannot be the number of solutions of a system of two equations in two variables whose graphs are parabolas?

(A) 1   **(B)** 2   **(C)** 4

(D) 5   **(E)** Infinitely many

64. Multiple Choice Which of the following is the number of solutions of a system of two linear equations in two variables if the resulting equation after using elimination correctly is $4 = 4$?

(A) 0   **(B)** 1   **(C)** 2

(D) 3   **(E)** Infinitely many

Explorations

65. An Ellipse and a Line Consider the system of equations

$$\frac{x^2}{4} + \frac{y^2}{9} = 1$$

$$x + y = 1.$$

(a) Solve the equation $x^2/4 + y^2/9 = 1$ for y in terms of x to determine the two implicit functions determined by the equation.

(b) Solve the system of equations graphically.

(c) Use substitution to confirm the solutions found in part (b).

66. A Hyperbola and a Line Consider the system of equations

$$\frac{x^2}{4} - \frac{y^2}{9} = 1$$

$$x - y = 0.$$

(a) Solve the equation $x^2/4 - y^2/9 = 1$ for y in terms of x to determine the two implicit functions determined by the equation.

(b) Solve the system of equations graphically.

(c) Use substitution to confirm the solutions found in part (b).

Extending the Ideas

In Exercises 67 and 68, use the elimination method to solve the system of equations.

67. $x^2 - 2y = -6$

$x^2 + y = 4$

68. $x^2 + y^2 = 1$

$x^2 - y^2 = 1$

In Exercises 69 and 70, $p(x)$ is the demand curve. The total revenue if x units are sold is $R = px$. Find the number of units sold that gives the maximum revenue.

69. $p = 100 - 4x$

70. $p = 80 - x^2$

7.2
Matrix Algebra

What you'll learn about

- Matrices
- Matrix Addition and Subtraction
- Matrix Multiplication
- Identity and Inverse Matrices
- Determinant of a Square Matrix
- Applications

. . . and why

Matrix algebra provides a powerful technique to manipulate large data sets and solve the related problems that are modeled by the matrices.

Matrices

A *matrix* is a rectangular array of numbers. Matrices provide an efficient way to solve systems of linear equations and to record data. The tables of data presented in this textbook are examples of matrices.

DEFINITION Matrix

Let m and n be positive integers. An **$m \times n$ matrix** (read "m by n matrix") is a rectangular array of m rows and n columns of real numbers.

$$\begin{bmatrix} a_{11} & a_{12} & \cdots & a_{1n} \\ a_{21} & a_{22} & \cdots & a_{2n} \\ \vdots & \vdots & & \vdots \\ a_{m1} & a_{m2} & \cdots & a_{mn} \end{bmatrix}$$

We also use the shorthand notation $[a_{ij}]$ for this matrix.

Each **element**, or **entry**, a_{ij}, of the matrix uses *double subscript* notation. The **row subscript** is the first subscript i, and the **column subscript** is j. The element a_{ij} is in the ith row and jth column. In general, the **order of an $m \times n$ matrix** is $m \times n$. If $m = n$, the matrix is a **square matrix**. Two matrices are **equal matrices** if they have the same order and their corresponding elements are equal.

EXAMPLE 1   Determining the Order of a Matrix

(a) The matrix $\begin{bmatrix} 1 & -2 & 3 \\ 2 & 0 & 4 \end{bmatrix}$ has order 2×3.

(b) The matrix $\begin{bmatrix} 1 & -1 \\ 0 & 4 \\ 2 & -1 \\ 3 & 2 \end{bmatrix}$ has order 4×2.

(c) The matrix $\begin{bmatrix} 1 & 2 & 3 \\ 4 & 5 & 6 \\ 7 & 8 & 9 \end{bmatrix}$ has order 3×3 and is a square matrix.

Now try Exercise 1.

Matrix Addition and Subtraction

We add or subtract two matrices of the same order by adding or subtracting their corresponding entries. Matrices of different orders can *not* be added or subtracted.

HISTORICAL NOTE

Methods used by the Chinese between 200 BC and 100 BC to solve problems involving several unknowns were similar to modern methods which use matrices. Matrices were formally developed in the 18th century by several mathematicians, including Leibniz, Cauchy, and Gauss.

DEFINITION Matrix Addition and Matrix Subtraction

Let $A = [a_{ij}]$ and $B = [b_{ij}]$ be matrices of order $m \times n$.

1. The **sum $A + B$** is the $m \times n$ matrix

$$A + B = [a_{ij} + b_{ij}].$$

2. The **difference $A - B$** is the $m \times n$ matrix

$$A - B = [a_{ij} - b_{ij}].$$

EXAMPLE 2   Using Matrix Addition

Matrix A gives the mean SAT verbal scores for the six New England states over the time period from 2001 to 2004. (*Source: The College Board, World Almanac and Book of Facts, 2005.*) Matrix B gives the mean SAT mathematics scores for the same 4-year period. Express the mean combined scores for the New England states from 2001 to 2004 as a single matrix.

$$
A = \begin{array}{c} \\ \text{CT} \\ \text{ME} \\ \text{MA} \\ \text{NH} \\ \text{RI} \\ \text{VT} \end{array}
\begin{array}{cccc} 01 & 02 & 03 & 04 \\ \left[\begin{array}{cccc} 509 & 509 & 512 & 515 \\ 506 & 503 & 503 & 505 \\ 511 & 512 & 516 & 518 \\ 520 & 519 & 522 & 522 \\ 501 & 504 & 502 & 503 \\ 511 & 512 & 515 & 516 \end{array}\right] \end{array}
\quad
B = \begin{array}{c} \\ \text{CT} \\ \text{ME} \\ \text{MA} \\ \text{NH} \\ \text{RI} \\ \text{VT} \end{array}
\begin{array}{cccc} 01 & 02 & 03 & 04 \\ \left[\begin{array}{cccc} 510 & 509 & 514 & 515 \\ 500 & 502 & 501 & 501 \\ 515 & 516 & 522 & 523 \\ 516 & 519 & 521 & 521 \\ 499 & 503 & 504 & 502 \\ 506 & 510 & 512 & 512 \end{array}\right] \end{array}
$$

SOLUTION The combined scores can be obtained by adding the two matrices:

$$
A + B = \begin{array}{c} \\ \text{CT} \\ \text{ME} \\ \text{MA} \\ \text{NH} \\ \text{RI} \\ \text{VT} \end{array}
\begin{array}{cccc} 01 & 02 & 03 & 04 \\ \left[\begin{array}{cccc} 1019 & 1018 & 1026 & 1030 \\ 1006 & 1005 & 1004 & 1006 \\ 1026 & 1028 & 1038 & 1041 \\ 1036 & 1038 & 1043 & 1043 \\ 1000 & 1007 & 1006 & 1005 \\ 1017 & 1022 & 1027 & 1028 \end{array}\right] \end{array}
$$

Now try Exercise 11.

POWER OF MATRIX ALGEBRA

The result in Example 2 is fairly simple, but it is significant that we found (essentially) 24 pieces of information with a single mathematical operation. That is the power of matrix algebra.

When we work with matrices, real numbers are **scalars**. The product of the real number k and the $m \times n$ matrix $A = [a_{ij}]$ is the $m \times n$ matrix

$$kA = [ka_{ij}].$$

The matrix $kA = [ka_{ij}]$ is a **scalar multiple of A**.

EXAMPLE 3   Using Scalar Multiplication

A consumer advocacy group has computed the mean retail prices for brand name products and generic products at three different stores in a major city. The prices are shown in the 3×2 matrix on the next page.

continued

$$
\begin{array}{cc}
& \text{Brand Generic} \\
\begin{array}{l}
\text{Store A} \\
\text{Store B} \\
\text{Store C}
\end{array}
&
\begin{bmatrix}
3.97 & 3.64 \\
3.78 & 3.69 \\
3.75 & 3.67
\end{bmatrix}
\end{array}
$$

The city has a combined sales tax of 7.25%. Construct a matrix showing the comparative prices with sales tax included.

SOLUTION Multiply the original matrix by the scalar 1.0725 to add the sales tax to every price.

$$
1.0725 \times
\begin{bmatrix}
3.97 & 3.64 \\
3.78 & 3.69 \\
3.75 & 3.67
\end{bmatrix}
\approx
\begin{array}{cc}
& \text{Brand Generic} \\
\begin{array}{l}
\text{Store A} \\
\text{Store B} \\
\text{Store C}
\end{array}
&
\begin{bmatrix}
4.26 & 3.90 \\
4.05 & 3.96 \\
4.02 & 3.94
\end{bmatrix}
\end{array}
$$

Now try Exercise 13.

Matrices inherit many properties possessed by the real numbers. Let $A = [a_{ij}]$ be any $m \times n$ matrix. The $m \times n$ matrix $O = [0]$ consisting entirely of zeros is the **zero matrix** because $A + O = A$. In other words, O is the **additive identity** for the set of all $m \times n$ matrices. The $m \times n$ matrix $B = [-a_{ij}]$ consisting of the *additive inverses* of the entries of A is the **additive inverse of A** because $A + B = O$. We also write $B = -A$. Just as with real numbers,

$$
A - B = [a_{ij} - b_{ij}] = [a_{ij} + (-b_{ij})] = [a_{ij}] + [-b_{ij}] = A + (-B).
$$

Thus, subtracting B from A is the same as adding the additive inverse of B to A.

EXPLORATION 1 Computing with Matrices

Let $A = [a_{ij}]$ and $B = [b_{ij}]$ be 2×2 matrices with $a_{ij} = 3i - j$ and $b_{ij} = i^2 + j^2 - 3$ for $i = 1, 2$ and $j = 1, 2$.

1. Determine A and B.

2. Determine the additive inverse $-A$ of A and verify that $A + (-A) = [0]$. What is the order of $[0]$?

3. Determine $3A - 2B$.

Matrix Multiplication

To form the *product AB* of two matrices, the number of columns of the matrix A on the left must be equal to the number of rows of the matrix B on the right. In this case, any row of A has the same number of entries as any column of B. Each entry of the product is obtained by summing the products of the entries of a row of A by the corresponding entries of a column of B.

DEFINITION Matrix Multiplication

Let $A = [a_{ij}]$ be an $m \times r$ matrix and $B = [b_{ij}]$ an $r \times n$ matrix.
The **product** $AB = [c_{ij}]$ is the $m \times n$ matrix where
$$c_{ij} = a_{i1}b_{1j} + a_{i2}b_{2j} + \cdots + a_{ir}b_{rj}.$$

The key to understanding how to form the product of any two matrices is to first consider the product of a $1 \times r$ matrix $A = [a_{1j}]$ with an $r \times 1$ matrix $B = [b_{j1}]$. According to the definition, $AB = [c_{11}]$ is the 1×1 matrix where $c_{11} = a_{11}b_{11} + a_{12}b_{21} + \cdots + a_{1r}b_{r1}$. For example, the product AB of the 1×3 matrix A and the 3×1 matrix B, where

$$A = [1 \quad 2 \quad 3] \quad \text{and} \quad B = \begin{bmatrix} 4 \\ 5 \\ 6 \end{bmatrix}$$

is

$$A \cdot B = [1 \quad 2 \quad 3] \cdot \begin{bmatrix} 4 \\ 5 \\ 6 \end{bmatrix} = [1 \cdot 4 + 2 \cdot 5 + 3 \cdot 6] = [32].$$

Then, the ij-entry of the product AB of an $m \times r$ matrix with an $r \times n$ matrix is the product of the ith row of A, considered as a $1 \times r$ matrix, with the jth column of B, considered as a $r \times 1$ matrix, as illustrated in Example 4.

EXAMPLE 4   Finding the Product of Two Matrices

Find the product AB if possible, where

(a) $A = \begin{bmatrix} 2 & 1 & -3 \\ 0 & 1 & 2 \end{bmatrix}$ and $B = \begin{bmatrix} 1 & -4 \\ 0 & 2 \\ 1 & 0 \end{bmatrix}$.

(b) $A = \begin{bmatrix} 2 & 1 & -3 \\ 0 & 1 & 2 \end{bmatrix}$ and $B = \begin{bmatrix} 3 & -4 \\ 2 & 1 \end{bmatrix}$.

SOLUTION

(a) The number of columns of A is 3 and the number of rows of B is 3, so the product AB is defined. The product $AB = [c_{ij}]$ is a 2×2 matrix where

$$c_{11} = [2 \quad 1 \quad -3] \begin{bmatrix} 1 \\ 0 \\ 1 \end{bmatrix} = 2 \cdot 1 + 1 \cdot 0 + (-3) \cdot 1 = -1,$$

$$c_{12} = [2 \quad 1 \quad -3] \begin{bmatrix} -4 \\ 2 \\ 0 \end{bmatrix} = 2 \cdot (-4) + 1 \cdot 2 + (-3) \cdot 0 = -6,$$

$$c_{21} = [0 \quad 1 \quad 2] \begin{bmatrix} 1 \\ 0 \\ 1 \end{bmatrix} = 0 \cdot 1 + 1 \cdot 0 + 2 \cdot 1 = 2,$$

$$c_{22} = [0 \quad 1 \quad 2] \begin{bmatrix} -4 \\ 2 \\ 0 \end{bmatrix} = 0 \cdot (-4) + 1 \cdot 2 + 2 \cdot 0 = 2.$$

continued

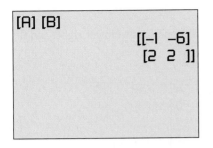

FIGURE 7.7 The matrix product AB of Example 4. Notice that the grapher displays the rows of the product as 1×2 matrices.

Thus, $AB = \begin{bmatrix} -1 & -6 \\ 2 & 2 \end{bmatrix}$. Figure 7.7 supports this computation.

(b) The number of columns of A is 3 and the number of rows of B is 2, so the product AB is *not* defined. *Now try Exercise 19.*

EXAMPLE 5 Using Matrix Multiplication

A florist makes three different cut flower arrangements for Mother's Day (I, II, and III), each involving roses, carnations, and lilies. Matrix A shows the number of each type of flower used in each arrangement.

$$\begin{array}{c} \begin{array}{ccc} \text{I} & \text{II} & \text{III} \end{array} \\ A = \begin{array}{r} \text{Roses} \\ \text{Carnations} \\ \text{Lilies} \end{array} \begin{bmatrix} 5 & 8 & 7 \\ 6 & 6 & 7 \\ 4 & 3 & 3 \end{bmatrix} \end{array}$$

The florist can buy his flowers from two different wholesalers (W1 and W2), but wants to give all his business to one or the other. The cost of the three flower types from the two wholesalers is shown in matrix B.

$$\begin{array}{c} \begin{array}{cc} \text{W1} & \text{W2} \end{array} \\ B = \begin{array}{r} \text{Roses} \\ \text{Carnations} \\ \text{Lilies} \end{array} \begin{bmatrix} 1.50 & 1.35 \\ 0.95 & 1.00 \\ 1.30 & 1.35 \end{bmatrix} \end{array}$$

Construct a matrix showing the cost of making each of the three flower arrangements from flowers supplied by the two different wholesalers.

SOLUTION

We can use the labeling of the matricies to help us. We want the columns of A to match up with the rows of B (since that's how the matrix multiplication works). We therefore switch the rows and columns of A to get the flowers along the columns. (The new matrix is called the **transpose** of A, denoted by A^T.) We then find the product $A^T B$:

$$\begin{array}{r} \begin{array}{ccc} \text{Rose} & \text{Carn} & \text{Lily} \end{array} \\ \begin{array}{r} \text{I} \\ \text{II} \\ \text{III} \end{array} \begin{bmatrix} 5 & 6 & 4 \\ 8 & 6 & 3 \\ 7 & 7 & 3 \end{bmatrix} \end{array} \times \begin{array}{r} \begin{array}{cc} \text{W1} & \text{W2} \end{array} \\ \begin{array}{r} \text{Rose} \\ \text{Carn} \\ \text{Lilly} \end{array} \begin{bmatrix} 1.50 & 1.35 \\ 0.95 & 1.00 \\ 1.30 & 1.35 \end{bmatrix} \end{array} = \begin{array}{r} \begin{array}{cc} \text{W1} & \text{W2} \end{array} \\ \begin{array}{r} \text{I} \\ \text{II} \\ \text{III} \end{array} \begin{bmatrix} 18.40 & 18.15 \\ 21.60 & 20.85 \\ 21.05 & 20.50 \end{bmatrix} \end{array}$$

Figure 7.8 shows the product $A^T B$ and supports our computation.

Now try Exercise 47.

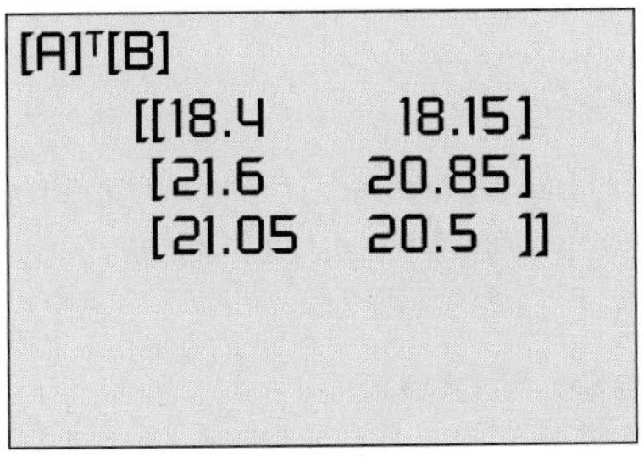

```
[A]ᵀ[B]
  [[18.4      18.15]
   [21.6      20.85]
   [21.05     20.5 ]]
```

FIGURE 7.8 The product $A^T B$ for the matrices A and B of Example 5.

Identity and Inverse Matrices

The $n \times n$ matrix I_n with 1's on the main diagonal (upper left to lower right) and 0's elsewhere is the **identity matrix of order $n \times n$**

$$I_n = \begin{bmatrix} 1 & 0 & 0 & \cdots & 0 \\ 0 & 1 & 0 & \cdots & 0 \\ 0 & 0 & 1 & \cdots & 0 \\ \vdots & \vdots & \vdots & & \vdots \\ 0 & 0 & 0 & \cdots & 1 \end{bmatrix}.$$

For example,

$$I_2 = \begin{bmatrix} 1 & 0 \\ 0 & 1 \end{bmatrix}, \quad I_3 = \begin{bmatrix} 1 & 0 & 0 \\ 0 & 1 & 0 \\ 0 & 0 & 1 \end{bmatrix}, \quad \text{and} \quad I_4 = \begin{bmatrix} 1 & 0 & 0 & 0 \\ 0 & 1 & 0 & 0 \\ 0 & 0 & 1 & 0 \\ 0 & 0 & 0 & 1 \end{bmatrix}.$$

If $A = [a_{ij}]$ is any $n \times n$ matrix, we can prove (see Exercise 56) that

$$AI_n = I_nA = A,$$

that is, I_n is the **multiplicative identity** for the set of $n \times n$ matrices.

If a is a nonzero real number, then $a^{-1} = 1/a$ is the multiplicative inverse of a, that is, $aa^{-1} = a(1/a) = 1$. The definition of the *multiplicative inverse* of a square matrix is similar.

DEFINITION Inverse of a Square Matrix

Let $A = [a_{ij}]$ be an $n \times n$ matrix. If there is a matrix B such that

$$AB = BA = I_n,$$

then B is the **inverse** of A. We write $B = A^{-1}$ (read "A inverse").

We will see that not every square matrix (Example 7) has an inverse. If a square matrix A has an inverse, then A is **nonsingular**. If A has no inverse, then A is **singular**.

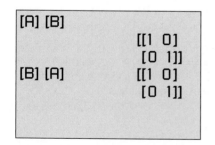

FIGURE 7.9 Showing A and B are inverse matrices. (Example 6)

EXAMPLE 6 Verifying an Inverse Matrix

Prove that

$$A = \begin{bmatrix} 3 & -2 \\ -1 & 1 \end{bmatrix} \quad \text{and} \quad B = \begin{bmatrix} 1 & 2 \\ 1 & 3 \end{bmatrix}$$

are inverse matrices.

SOLUTION Figure 7.9 shows that $AB = BA = I_2$. Thus, $B = A^{-1}$ and $A = B^{-1}$.

Now try Exercise 33.

EXAMPLE 7   Showing a Matrix has No Inverse

Prove that the matrix $A = \begin{bmatrix} 6 & 3 \\ 2 & 1 \end{bmatrix}$ is singular, that is, A has no inverse.

SOLUTION Suppose A has an inverse $B = \begin{bmatrix} x & y \\ z & w \end{bmatrix}$. Then, $AB = I_2$.

$$AB = \begin{bmatrix} 6 & 3 \\ 2 & 1 \end{bmatrix}\begin{bmatrix} x & y \\ z & w \end{bmatrix} = \begin{bmatrix} 1 & 0 \\ 0 & 1 \end{bmatrix}$$

$$= \begin{bmatrix} 6x + 3z & 6y + 3w \\ 2x + z & 2y + w \end{bmatrix} = \begin{bmatrix} 1 & 0 \\ 0 & 1 \end{bmatrix}$$

Using equality of matrices we obtain:

$$6x + 3z = 1 \qquad 6y + 3w = 0$$

$$2x + z = 0 \qquad 2y + w = 1$$

Multiplying both sides of the equation $2x + z = 0$ by 3 yields $6x + 3z = 0$. There are no values for x and z for which the value of $6x + 3z$ is both 0 and 1. Thus, A does not have an inverse.   ***Now try Exercise 37.***

Determinant of a Square Matrix

There is a simple test that determines if a 2×2 matrix has an inverse.

Inverse of a 2 × 2 Matrix

If $ad - bc \neq 0$, then

$$\begin{bmatrix} a & b \\ c & d \end{bmatrix}^{-1} = \frac{1}{ad - bc}\begin{bmatrix} d & -b \\ -c & a \end{bmatrix}.$$

The number $ad - bc$ is the **determinant** of the 2×2 matrix $A = \begin{bmatrix} a & b \\ c & d \end{bmatrix}$ and is denoted

$$\det A = \begin{vmatrix} a & b \\ c & d \end{vmatrix} = ad - bc.$$

To define the determinant of a higher order square matrix we need to introduce the *minors* and *cofactors* associated with the entries of a square matrix. Let $A = [a_{ij}]$ be an $n \times n$ matrix. The **minor** (short for "minor determinant") M_{ij} corresponding to the element a_{ij} is the determinant of the $(n-1) \times (n-1)$ matrix obtained by deleting the row and column containing a_{ij}. The **cofactor** corresponding to a_{ij} is $A_{ij} = (-1)^{i+j}M_{ij}$.

DEFINITION Determinant of a Square Matrix

Let $A = [a_{ij}]$ be a matrix of order $n \times n$ $(n > 2)$. The determinant of A, denoted by $\det A$ or $|A|$, is the sum of the entries in any row or any column multiplied by their respective cofactors. For example, expanding by the ith row gives

$$\det A = |A| = a_{i1}A_{i1} + a_{i2}A_{i2} + \cdots + a_{in}A_{in}.$$

If $A = [a_{ij}]$ is a 3×3 matrix, then, using the definition of determinant applied to the second row, we obtain

$$\begin{vmatrix} a_{11} & a_{12} & a_{13} \\ a_{21} & a_{22} & a_{23} \\ a_{31} & a_{32} & a_{33} \end{vmatrix} = a_{21}A_{21} + a_{22}A_{22} + a_{23}A_{23}$$

$$= a_{21}(-1)^3 \begin{vmatrix} a_{12} & a_{13} \\ a_{32} & a_{33} \end{vmatrix} + a_{22}(-1)^4 \begin{vmatrix} a_{11} & a_{13} \\ a_{31} & a_{33} \end{vmatrix}$$

$$+ a_{23}(-1)^5 \begin{vmatrix} a_{11} & a_{12} \\ a_{31} & a_{32} \end{vmatrix}$$

$$= -a_{21}(a_{12}a_{33} - a_{13}a_{32}) + a_{22}(a_{11}a_{33} - a_{13}a_{31})$$
$$- a_{23}(a_{11}a_{32} - a_{12}a_{31})$$

The determinant of a 3×3 matrix involves three determinants of 2×2 matrices, the determinant of a 4×4 matrix involves four determinants of 3×3 matrices, and so forth. This is a tedious definition to apply. Most of the time we use a grapher to evaluate determinants in this textbook.

EXPLORATION 2 **Investigating the Definition of Determinant**

1. Complete the expansion of the determinant of the 3×3 matrix $A = [a_{ij}]$ started above. Explain why each term in the expansion contains an element from each row and each column.

2. Use the first row of the 3×3 matrix to expand the determinant and compare to the expression in 1.

3. Prove that the determinant of a square matrix with a zero row or a zero column is zero.

We can now state the condition under which square matrices have inverses.

THEOREM **Inverses of $n \times n$ Matrices**

An $n \times n$ matrix A has an inverse if and only if $\det A \neq 0$.

There are complicated formulas for finding the inverses of nonsingular matrices of order 3×3 or higher. We will use a grapher instead of these formulas to find inverses of square matrices.

EXAMPLE 8 **Finding Inverse Matrices**

Determine whether the matrix has an inverse. If so, find its inverse matrix.

(a) $A = \begin{bmatrix} 3 & 1 \\ 4 & 2 \end{bmatrix}$ **(b)** $B = \begin{bmatrix} 1 & 2 & -1 \\ 2 & -1 & 3 \\ -1 & 0 & 1 \end{bmatrix}$

SOLUTION

(a) Since $\det A = ad - bc = 3 \cdot 2 - 1 \cdot 4 = 2 \neq 0$, we conclude that A has an inverse. Using the formula for the inverse of a 2×2 matrix, we obtain

continued

$$A^{-1} = \frac{1}{ad - bc}\begin{bmatrix} d & -b \\ -c & a \end{bmatrix} = \frac{1}{2}\begin{bmatrix} 2 & -1 \\ -4 & 3 \end{bmatrix}$$

$$= \begin{bmatrix} 1 & -0.5 \\ -2 & 1.5 \end{bmatrix}$$

You can check that $A^{-1}A = A^{-1}A = I_2$.

(b) Figure 7.10 shows that det $B = -10 \neq 0$ and

$$B^{-1} = \begin{bmatrix} 0.1 & 0.2 & -0.5 \\ 0.5 & 0 & 0.5 \\ 0.1 & 0.2 & 0.5 \end{bmatrix}.$$

You can use your grapher to check that $B^{-1}B = BB^{-1} = I_3$.

Now try Exercise 41.

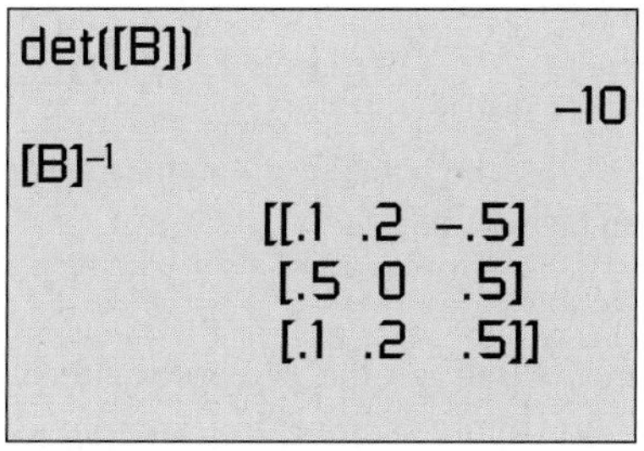

```
det([B])
                          -10
[B]⁻¹
        [[.1  .2  -.5]
         [.5  0    .5]
         [.1  .2   .5]]
```

FIGURE 7.10 The matrix B is nonsingular and so has an inverse. (Example 8b)

We list some of the important properties of matrices, some of which you will be asked to prove in the exercises.

Properties of Matrices

Let A, B, and C be matrices whose orders are such that the following sums, differences, and products are defined.

1. Commutative property
Addition:
$A + B = B + A$
Multiplication:
(Does not hold in general)

2. Associative property
Addition:
$(A + B) + C = A + (B + C)$
Multiplication:
$(AB)C = A(BC)$

3. Identity property
Addition: $A + O = A$
Multiplication: order of $A = n \times n$
$A \cdot I_n = I_n \cdot A = A$

4. Inverse property
Addition: $A + (-A) = O$
Multiplication: order of $A = n \times n$
$AA^{-1} = A^{-1}A = I_n, \quad |A| \neq 0$

5. Distributive property

Multiplication over addition
$A(B + C) = AB + AC$
$(A + B)C = AC + BC$

Multiplication over subtraction
$A(B - C) = AB - AC$
$(A - B)C = AC - BC$

Applications

Points in the Cartesian coordinate plane can be represented by 1×2 matrices. For example, the point $(2, -3)$ can be represented by the 1×2 matrix $[2 \;\; -3]$. We can calculate the images of points acted upon by some of the transformations studied in Section 1.5 using matrix multiplication as illustrated in Example 9.

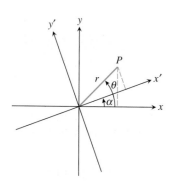

FIGURE 7.11 Rotating the xy-coordinate system through the angle α to obtain the $x'y'$-coordinate system. (Example 10)

EXAMPLE 9   Reflecting with Respect to the *x*-Axis as Matrix Multiplication

Prove that the image of a point under a reflection across the x-axis can be obtained by multiplying by $\begin{bmatrix} 1 & 0 \\ 0 & -1 \end{bmatrix}$.

SOLUTION The image of the point (x, y) under a reflection across the x-axis is $(x, -y)$. The product

$$[x \;\; y]\begin{bmatrix} 1 & 0 \\ 0 & -1 \end{bmatrix} = [x \;\; -y]$$

shows that the point (x, y) (in matrix form $[x \;\; y]$) is moved to the point $(x, -y)$ (in matrix form $[x \;\; -y]$).   ***Now try Exercise 57.***

Figure 7.11 shows the xy-coordinate system rotated through the angle α to obtain the $x'y'$-coordinate system. In Example 10, we see that the coordinates of a point in the $x'y'$-coordinate system can be obtained by multiplying the coordinates of the point in the xy-coordinate system by an appropriate 2×2 matrix. In Exercise 71, you will see that the reverse is also true.

EXAMPLE 10   Rotating a Coordinate System

Prove that the (x', y') coordinates of P in Figure 7.11 are related to the (x, y) coordinates of P by the equations

$$x' = x \cos \alpha + y \sin \alpha$$
$$y' = -x \sin \alpha + y \cos \alpha.$$

Then, prove that the coordinates (x', y') can be obtained from the (x, y) coordinates by matrix multiplication. We use this result in Section 8.4 when we study conic sections.

SOLUTION Using the right triangle formed by P and the $x'y'$-coordinate system, we obtain

$$x' = r \cos (\theta - \alpha) \quad \text{and} \quad y' = r \sin (\theta - \alpha).$$

Expanding the above expressions for x' and y' using trigonometric identities for $\cos (\theta - \alpha)$ and $\sin (\theta - \alpha)$ yields

$$x' = r \cos \theta \cos \alpha + r \sin \theta \sin \alpha, \quad \text{and}$$
$$y' = r \sin \theta \cos \alpha - r \cos \theta \sin \alpha.$$

continued

It follows from the right triangle formed by P and the xy-coordinate system that $x = r \cos \theta$ and $y = r \sin \theta$. Substituting these values for x and y into the above pair of equations yields

$$x' = x \cos \alpha + y \sin \alpha \quad \text{and} \quad y' = y \cos \alpha - x \sin \alpha$$
$$= -x \sin \alpha + y \cos \alpha,$$

which is what we were asked to prove. Finally, matrix multiplication shows that

$$[x' \quad y'] = [x \quad y] \begin{bmatrix} \cos \alpha & -\sin \alpha \\ \sin \alpha & \cos \alpha \end{bmatrix}.$$

Now try Exercise 71.

CHAPTER OPENER PROBLEM
(from page 567)

PROBLEM: If we have a triangle with vertices at $(0, 0)$, $(1, 1)$, and $(2, 0)$, and we want to double the lengths of the sides of the triangle, where would the vertices of the enlarged triangle be?

SOLUTION: Given a triangle with vertices at $(0, 0)$, $(1, 1)$, and $(2, 0)$, as in Figure 7.12, we can find the vertices of a new triangle whose sides are twice as long by multiplying by the scale matrix.

$$\begin{bmatrix} 2 & 0 \\ 0 & 2 \end{bmatrix}$$

For the point $(0, 0)$, we have

$$[x' \quad y'] = [0 \quad 0] \begin{bmatrix} 2 & 0 \\ 0 & 2 \end{bmatrix} = [0 \quad 0].$$

For the point $(1, 1)$, we have

$$[x' \quad y'] = [1 \quad 1] \begin{bmatrix} 2 & 0 \\ 0 & 2 \end{bmatrix} = [2 \quad 2].$$

And for the point $(2, 0)$, we have

$$[x' \quad y'] = [2 \quad 0] \begin{bmatrix} 2 & 0 \\ 0 & 2 \end{bmatrix} = [4 \quad 0].$$

So the new triangle has vertices $(0, 0)$, $(2, 2)$, and $(4, 0)$, as Figure 7.13 shows.

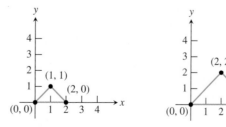

FIGURE 7.12   **FIGURE 7.13**

QUICK REVIEW 7.2 *(For help, go to Sections 1.5, 5.3, and 6.4.)*

In Exercises 1–4, the points **(a)** $(3, -2)$ and **(b)** (x, y) are reflected across the given line. Find the coordinates of the reflected points.

1. The *x*-axis

2. The *y*-axis

3. The line $y = x$

4. The line $y = -x$

In Exercises 5 and 6, express the coordinates of *P* in terms of θ.

5.
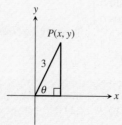

6.
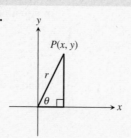

In Exercises 7–10, expand the expression.

7. $\sin(\alpha + \beta)$

8. $\sin(\alpha - \beta)$

9. $\cos(\alpha + \beta)$

10. $\cos(\alpha - \beta)$

SECTION 7.2 EXERCISES

In Exercises 1–6, determine the order of the matrix. Indicate whether the matrix is square.

1. $\begin{bmatrix} 2 & 3 & -1 \\ 1 & 0 & 5 \end{bmatrix}$ **2.** $\begin{bmatrix} 1 & 3 \\ -1 & 2 \end{bmatrix}$ **3.** $\begin{bmatrix} 5 & 6 \\ -1 & 2 \\ 0 & 0 \end{bmatrix}$

4. $\begin{bmatrix} -1 & 0 & 6 \end{bmatrix}$ **5.** $\begin{bmatrix} 2 \\ -1 \\ 0 \end{bmatrix}$ **6.** $\begin{bmatrix} 0 \end{bmatrix}$

In Exercises 7–10, identify the element specified for the following matrix.

$$\begin{bmatrix} -2 & 0 & 3 & 4 \\ 3 & 1 & 5 & -1 \\ 1 & 4 & -1 & 3 \end{bmatrix}$$

7. a_{13} **8.** a_{24} **9.** a_{32} **10.** a_{33}

In Exercises 11–16, find **(a)** $A + B$, **(b)** $A - B$, **(c)** $3A$, and **(d)** $2A - 3B$.

11. $A = \begin{bmatrix} 2 & 3 \\ -1 & 5 \end{bmatrix}$, $B = \begin{bmatrix} 1 & -3 \\ -2 & -4 \end{bmatrix}$

12. $A = \begin{bmatrix} -1 & 0 & 2 \\ 4 & 1 & -1 \\ 2 & 0 & 1 \end{bmatrix}$, $B = \begin{bmatrix} 2 & 1 & 0 \\ -1 & 0 & 2 \\ 4 & -3 & -1 \end{bmatrix}$

13. $A = \begin{bmatrix} -3 & 1 \\ 0 & -1 \\ 2 & 1 \end{bmatrix}$, $B = \begin{bmatrix} 4 & 0 \\ -2 & 1 \\ -3 & -1 \end{bmatrix}$

14. $A = \begin{bmatrix} 5 & -2 & 3 & 1 \\ -1 & 0 & 2 & 2 \end{bmatrix}$, $B = \begin{bmatrix} -2 & 3 & 1 & 0 \\ 4 & 0 & -1 & -2 \end{bmatrix}$

15. $A = \begin{bmatrix} -2 \\ 1 \\ 0 \end{bmatrix}$, $B = \begin{bmatrix} -1 \\ 0 \\ 4 \end{bmatrix}$

16. $A = \begin{bmatrix} -1 & -2 & 0 & 3 \end{bmatrix}$, and $B = \begin{bmatrix} 1 & 2 & -2 & 0 \end{bmatrix}$

In Exercises 17–22, use the definition of matrix multiplication to find **(a)** AB and **(b)** BA. Support your answer with the matrix feature of your grapher.

17. $A = \begin{bmatrix} 2 & 3 \\ -1 & 5 \end{bmatrix}$, $B = \begin{bmatrix} 1 & -3 \\ -2 & -4 \end{bmatrix}$

18. $A = \begin{bmatrix} 1 & -4 \\ 2 & 6 \end{bmatrix}$, $B = \begin{bmatrix} 5 & 1 \\ -2 & -3 \end{bmatrix}$

19. $A = \begin{bmatrix} 2 & 0 & 1 \\ 1 & 4 & -3 \end{bmatrix}$, $B = \begin{bmatrix} 1 & 2 \\ -3 & 1 \\ 0 & -2 \end{bmatrix}$

20. $A = \begin{bmatrix} 1 & 0 & -2 & 3 \\ 2 & 1 & 4 & -1 \end{bmatrix}$, $B = \begin{bmatrix} 5 & -1 \\ 0 & 2 \\ -1 & 3 \\ 4 & 2 \end{bmatrix}$

21. $A = \begin{bmatrix} -1 & 0 & 2 \\ 4 & 1 & -1 \\ 2 & 0 & 1 \end{bmatrix}$, $B = \begin{bmatrix} 2 & 1 & 0 \\ -1 & 0 & 2 \\ 4 & -3 & -1 \end{bmatrix}$

22. $A = \begin{bmatrix} -2 & 3 & 0 \\ 1 & -2 & 4 \\ 3 & 2 & 1 \end{bmatrix}$, $B = \begin{bmatrix} 4 & -1 & 2 \\ 0 & 2 & 3 \\ -1 & 3 & -1 \end{bmatrix}$

In Exercises 23–28, find **(a)** AB and **(b)** BA, or state that the product is not defined.

23. $A = \begin{bmatrix} 2 & -1 & 3 \end{bmatrix}, \quad B = \begin{bmatrix} -5 \\ 4 \\ 2 \end{bmatrix}$

24. $A = \begin{bmatrix} -2 \\ 3 \\ -4 \end{bmatrix}, \quad B = \begin{bmatrix} -1 & 2 & 4 \end{bmatrix}$

25. $A = \begin{bmatrix} -1 & 2 \\ 3 & 4 \end{bmatrix}, \quad B = \begin{bmatrix} -3 & 5 \end{bmatrix}$

26. $A = \begin{bmatrix} -1 & 3 \\ 0 & 1 \\ 1 & 0 \\ -3 & -1 \end{bmatrix}, \quad B = \begin{bmatrix} 5 & -6 \\ 2 & 3 \end{bmatrix}$

27. $A = \begin{bmatrix} 0 & 0 & 1 \\ 0 & 1 & 0 \\ 1 & 0 & 0 \end{bmatrix}, \quad B = \begin{bmatrix} 1 & 2 & 1 \\ 2 & 0 & 1 \\ -1 & 3 & 4 \end{bmatrix}$

28. $A = \begin{bmatrix} 0 & 0 & 1 & 0 \\ 0 & 1 & 0 & 0 \\ 1 & 0 & 0 & 0 \\ 0 & 0 & 0 & 1 \end{bmatrix}, \quad B = \begin{bmatrix} -1 & 2 & 3 & -4 \\ 2 & 1 & 0 & -1 \\ -3 & 2 & 1 & 3 \\ 4 & 0 & 2 & -1 \end{bmatrix}$

In Exercises 29–32, solve for a and b.

29. $\begin{bmatrix} a & -3 \\ 4 & 2 \end{bmatrix} = \begin{bmatrix} 5 & -3 \\ 4 & b \end{bmatrix}$

30. $\begin{bmatrix} 1 & -1 & 0 \\ a & -2 & 1 \end{bmatrix} = \begin{bmatrix} 1 & b & 0 \\ 3 & -2 & 1 \end{bmatrix}$

31. $\begin{bmatrix} 2 & a-1 \\ 2 & 3 \\ -1 & 2 \end{bmatrix} = \begin{bmatrix} 2 & -3 \\ b+2 & 3 \\ -1 & 2 \end{bmatrix}$

32. $\begin{bmatrix} a+3 & 2 \\ 0 & 5 \end{bmatrix} = \begin{bmatrix} 4 & 2 \\ 0 & b-1 \end{bmatrix}$

In Exercises 33 and 34, verify that the matrices are inverses of each other.

33. $A = \begin{bmatrix} 2 & 1 \\ 3 & 4 \end{bmatrix}, \quad B = \begin{bmatrix} 0.8 & -0.2 \\ -0.6 & 0.4 \end{bmatrix}$

34. $A = \begin{bmatrix} -2 & 1 & 3 \\ 1 & 2 & -2 \\ 0 & 1 & -1 \end{bmatrix}, \quad B = \begin{bmatrix} 0 & 1 & -2 \\ 0.25 & 0.5 & -0.25 \\ 0.25 & 0.5 & -1.25 \end{bmatrix}$

In Exercises 35–40, find the inverse of the matrix if it has one, or state that the inverse does not exist.

35. $\begin{bmatrix} 2 & 3 \\ 2 & 2 \end{bmatrix}$

36. $\begin{bmatrix} 6 & 3 \\ 10 & 5 \end{bmatrix}$

37. $\begin{bmatrix} 1 & 2 & -1 \\ 2 & -1 & 3 \\ 3 & 1 & 2 \end{bmatrix}$

38. $\begin{bmatrix} 2 & 3 & -1 \\ -1 & 0 & 4 \\ 0 & 1 & 1 \end{bmatrix}$

39. $A = [a_{ij}], a_{ij} = (-1)^{i+j}, 1 \le i \le 4, 1 \le j \le 4$

40. $B = [b_{ij}], b_{ij} = |i - j|, 1 \le i \le 3, 1 \le j \le 3$

In Exercises 41 and 42, use the definition to evaluate the determinant of the matrix.

41. $\begin{bmatrix} 2 & 1 & 1 \\ -1 & 0 & 2 \\ 1 & 3 & -1 \end{bmatrix}$

42. $\begin{bmatrix} 1 & 0 & 2 & 0 \\ 0 & 1 & 2 & 3 \\ 1 & -1 & 0 & 2 \\ 1 & 0 & 0 & 3 \end{bmatrix}$

In Exercises 43 and 44, solve for X.

43. $3X + A = B$, where $A = \begin{bmatrix} 1 \\ 3 \end{bmatrix}$ and $B = \begin{bmatrix} 4 \\ 2 \end{bmatrix}$.

44. $2X + A = B$, where $A = \begin{bmatrix} -1 & 2 \\ 0 & 3 \end{bmatrix}$ and $B = \begin{bmatrix} 1 & 4 \\ 1 & -1 \end{bmatrix}$.

45. Symmetric Matrix The matrix below gives the road mileage between Atlanta (A), Baltimore (B), Cleveland (C), and Denver (D). (*Source: AAA Road Atlas*)

(a) Writing to Learn Explain why the entry in the ith row and jth column is the same as the entry in the jth row and ith column. A matrix with this property is **symmetric**.

(b) Writing to Learn Why are the entries along the diagonal all 0's?

		A	B	C	D
A		0	689	774	1406
B		689	0	371	1685
C		774	371	0	1340
D		1406	1685	1340	0

46. Production Jordan Manufacturing has two factories, each of which manufactures three products. The number of units of product i produced at factory j in one week is represented by a_{ij} in the matrix

$$A = \begin{bmatrix} 120 & 70 \\ 150 & 110 \\ 80 & 160 \end{bmatrix}.$$

If production levels are increased by 10%, write the new production levels as a matrix B. How is B related to A?

47. Egg Production Happy Valley Farms produces three types of eggs: 1 (large), 2 (X-large), 3 (jumbo). The number of dozens of type i eggs sold to grocery store j is represented by a_{ij} in the matrix.

$$A = \begin{bmatrix} 100 & 60 \\ 120 & 70 \\ 200 & 120 \end{bmatrix}.$$

The per dozen price Happy Valley Farms charges for egg type i is represented by b_{i1} in the matrix

$$B = \begin{bmatrix} \$0.80 \\ \$0.85 \\ \$1.00 \end{bmatrix}.$$

(a) Find the product $B^T A$.

(b) Writing to Learn What does the matrix $B^T A$ represent?

48. Inventory A company sells four models of one name brand "all-in-one fax, printer, copier, and scanner machine" at three retail stores. The inventory at store i of model j is represented by s_{ij} in the matrix

$$S = \begin{bmatrix} 16 & 10 & 8 & 12 \\ 12 & 0 & 10 & 4 \\ 4 & 12 & 0 & 8 \end{bmatrix}.$$

The wholesale and retail prices of model i are represented by p_{i1} and p_{i2}, respectively, in the matrix

$$P = \begin{bmatrix} \$180 & \$269.99 \\ \$275 & \$399.99 \\ \$355 & \$499.99 \\ \$590 & \$799.99 \end{bmatrix}.$$

(a) Determine the product SP.

(b) **Writing to Learn** What does the matrix SP represent?

49. Profit A discount furniture store sells four types of 5-piece bedroom sets. The price charged for a bedroom set of type j is represented by a_{1j} in the matrix

$$A = [\$398 \quad \$598 \quad \$798 \quad \$998].$$

The number of sets of type j sold in one period is represented by b_{1j} in the matrix

$$B = [35 \quad 25 \quad 20 \quad 10].$$

The cost to the furniture store for a bedroom set of type j is given by c_{1j} in the matrix

$$C = [\$199 \quad \$268 \quad \$500 \quad \$670].$$

(a) Write a matrix product that gives the total revenue made from the sale of the bedroom sets in the one period.

(b) Write an expression using matrices that gives the profit produced by the sale of the bedroom sets in the one period.

50. Construction A building contractor has agreed to build six ranch-style houses, seven Cape Cod-style houses, and 14 colonial-style houses. The number of units of raw materials that go into each type of house are shown in the matrix

	Steel	Wood	Glass	Paint	Labor
Ranch	5	22	14	7	17
$R =$ Cape Cod	7	20	10	9	21
Colonial	6	27	8	5	13

Assume that steel costs $1600 a unit, wood $900 a unit, glass $500 a unit, paint $100 a unit, and labor $1000 a unit.

(a) Write a 1×3 matrix B that represents the number of each type of house to be built.

(b) Write a matrix product that gives the number of units of each raw material needed to build the houses.

(c) Write a 5×1 matrix C that represents the per unit cost of each type of raw material.

(d) Write a matrix product that gives the cost of each house.

(e) **Writing to Learn** Compute the product BRC. What does this matrix represent?

51. Rotating Coordinate Systems The xy-coordinate system is rotated through the angle $30°$ to obtain the $x'y'$-coordinate system.

(a) If the coordinates of a point in the xy-coordinate system are $(1, 1)$, what are the coordinates of the rotated point in the $x'y'$-coordinate system?

(b) If the coordinates of a point in the $x'y'$-coordinate system are $(1, 1)$, what are the coordinates of the point in the xy-coordinate system that was rotated to it?

52. Group Activity Let A, B, and C be matrices whose orders are such that the following expressions are defined. Prove that the following properties are true.

(a) $A + B = B + A$

(b) $(A + B) + C = A + (B + C)$

(c) $A(B + C) = AB + AC$

(d) $(A - B)C = AC - BC$

53. Group Activity Let A and B be $m \times n$ matrices and c and d scalars. Prove that the following properties are true.

(a) $c(A + B) = cA + cB$ (b) $(c + d)A = cA + dA$

(c) $c(dA) = (cd)A$ (d) $1 \cdot A = A$

54. Writing to Learn Explain why the definition given for the determinant of a square matrix agrees with the definition given for the determinant of a 2×2 matrix. (Assume that the determinant of a 1×1 matrix is the entry.)

55. Inverse of a 2 × 2 Matrix Prove that the inverse of the matrix

$$A = \begin{bmatrix} a & b \\ c & d \end{bmatrix} \quad \text{is} \quad A^{-1} = \frac{1}{ad - bc} \begin{bmatrix} d & -b \\ -c & a \end{bmatrix}$$

provided $ad - bc \neq 0$.

56. Identity Matrix Let $A = [a_{ij}]$ be an $n \times n$ matrix. Prove that $AI_n = I_nA = A$.

In Exercises 57–61, prove that the image of a point under the given transformation of the plane can be obtained by matrix multiplication.

57. A reflection across the y-axis

58. A reflection across the line $y = x$

59. A reflection across the line $y = -x$

60. A vertical stretch or shrink by a factor of a

61. A horizontal stretch or shrink by a factor of c

Standardized Test Questions

62. True or False Every square matrix has an inverse. Justify your answer.

63. True or False The determinant $|A|$ of the square matrix A is greater than or equal to 0. Justify your answer.

In Exercises 64–67, solve the problem without using a calculator.

64. Multiple Choice Which of the following is equal to the determinant of $A = \begin{bmatrix} 2 & 4 \\ -3 & -1 \end{bmatrix}$?

(A) 4   **(B)** -4 **(C)** 10   **(D)** -10 **(E)** -14

65. Multiple Choice Let A be a matrix of order 3×2 and B a matrix of order 2×4. Which of the following gives the order of the product AB?

(A) 2×2 **(B)** 3×4 **(C)** 4×3 **(D)** 6×8

(E) The product is not defined.

66. Multiple Choice Which of the following is the inverse of the matrix $\begin{bmatrix} 2 & 7 \\ 1 & 4 \end{bmatrix}$?

(A) $\begin{bmatrix} -4 & 7 \\ 1 & -2 \end{bmatrix}$ **(B)** $\begin{bmatrix} 2 & -7 \\ -1 & 4 \end{bmatrix}$ **(C)** $\begin{bmatrix} 2 & -1 \\ -7 & 4 \end{bmatrix}$

(D) $\begin{bmatrix} 4 & -1 \\ -7 & 2 \end{bmatrix}$ **(E)** $\begin{bmatrix} 4 & -7 \\ -1 & 2 \end{bmatrix}$

67. Multiple Choice Which of the following is the value of a_{13} in the matrix $[a_{ij}] = \begin{bmatrix} 1 & 2 & 3 \\ 4 & 5 & 6 \\ 7 & 8 & 9 \end{bmatrix}$?

(A) -7 **(B)** 7   **(C)** -3 **(D)** 3   **(E)** 10

Explorations

68. Continuation of Exploration 2 Let $A = [a_{ij}]$ be an $n \times n$ matrix.

(a) Prove that the determinant of A changes sign if two rows or two columns are interchanged. Start with a 3×3 matrix and compare the expansion by expanding by the same row (or column) before and after the interchange. [*Hint:* Compare without expanding the minors.] How can you generalize from the 3×3 case?

(b) Prove that the determinant of a square matrix with two identical rows or two identical columns is zero.

(c) Prove that if a scalar multiple of a row (or column) is added to another row (or column) the value of the determinant of a square matrix is unchanged. [*Hint:* Expand by the row (or column) being added to.]

69. Continuation of Exercise 68 Let $A = [a_{ij}]$ be an $n \times n$ matrix.

(a) Prove that if every element of a row or column of a matrix is multiplied by the real number c, then the determinant of the matrix is multiplied by c.

(b) Prove that if all the entries above the main diagonal (or all below it) of a matrix are zero, the determinant is the product of the elements on the main diagonal.

70. Writing Equations for Lines Using Determinants Consider the equation

$$\begin{vmatrix} 1 & x & y \\ 1 & x_1 & y_1 \\ 1 & x_2 & y_2 \end{vmatrix} = 0.$$

(a) Verify that the equation is linear in x and y.

(b) Verify that the two points (x_1, y_1) and (x_2, y_2) lie on the line in part (a).

(c) Use a determinant to state that the point (x_3, y_3) lies on the line in part (a).

(d) Use a determinant to state that the point (x_3, y_3) does not lie on the line in part (a).

71. Continuation of Example 10 The xy-coordinate system is rotated through the angle α to obtain the $x'y'$-coordinate system (see Figure 7.11).

(a) Show that the inverse of the matrix

$$A = \begin{bmatrix} \cos \alpha & -\sin \alpha \\ \sin \alpha & \cos \alpha \end{bmatrix}$$

of Example 10 is

$$A^{-1} = \begin{bmatrix} \cos \alpha & \sin \alpha \\ -\sin \alpha & \cos \alpha \end{bmatrix}.$$

(b) Prove that the (x, y) coordinates of P in Figure 7.11 are related to the (x', y') coordinates of P by the equations

$$x = x' \cos \alpha - y' \sin \alpha$$
$$y = x' \sin \alpha + y' \cos \alpha.$$

(c) Prove that the coordinates (x, y) can be obtained from the (x', y') coordinates by matrix multiplication. How is this matrix related to A?

Extending the Ideas

72. Characteristic Polynomial Let $A = [a_{ij}]$ be a 2×2 matrix and define $f(x) = \det (xI_2 - A)$.

(a) Expand the determinant to show that $f(x)$ is a polynomial of degree 2. (The **characteristic polynomial** of A.)

(b) How is the constant term of $f(x)$ related to $\det A$?

(c) How is the coefficient of x related to A?

(d) Prove that $f(A) = 0$.

73. Characteristic Polynomial Let $A = [a_{ij}]$ be a 3×3 matrix and define $f(x) = \det (xI_3 - A)$.

(a) Expand the determinant to show that $f(x)$ is a polynomial of degree 3. (The characteristic polynomial of A.)

(b) How is the constant term of $f(x)$ related to $\det A$?

(c) How is the coefficient of x^2 related to A?

(d) Prove that $f(A) = 0$.

7.3
Multivariate Linear Systems and Row Operations

Triangular Form for Linear Systems

The method of elimination used in Section 7.1 can be extended to systems of linear (first-degree) equations in more than two variables. The goal of the elimination method is to rewrite the system as an *equivalent system* of equations whose solution is obvious. Two systems of equations are **equivalent** if they have the same solution.

A *triangular form* of a system is an equivalent form from which the solution is easy to read. Here is an example of a system in triangular form.

$$x - 2y + z = 7$$
$$y - 2z = -7$$
$$z = 3$$

This convenient triangular form allows us to solve the system using substitution as illustrated in Example 1.

EXAMPLE 1   Solving by Substitution

Solve the system

$$x - 2y + z = 7$$
$$y - 2z = -7$$
$$z = 3.$$

SOLUTION The third equation determines z, namely $z = 3$. Substitute the value of z into the second equation to determine y.

$$y - 2z = -7 \qquad \text{Second equation}$$
$$y - 2(3) = -7 \qquad \text{Substitute } z = 3.$$
$$y = -1$$

Finally, substitute the values for y and z into the first equation to determine x.

$$x - 2y + z = 7 \qquad \text{First equation}$$
$$x - 2(-1) + 3 = 7 \qquad \text{Substitute } y = -1, z = 3.$$
$$x = 2$$

The solution of the system is $x = 2$, $y = -1$, $z = 3$, or the ordered triple $(2, -1, 3)$.

Now try Exercise 1.

BACK SUBSTITUTION

The method of solution used in Example 1 is sometimes referred to as *back substitution*.

Gaussian Elimination

Transforming a system to triangular form is **Gaussian elimination**, named after the famous German mathematician Carl Friedrich Gauss (1777–1855). Here are the operations needed to transform a system of linear equations into triangular form.

Equivalent Systems of Linear Equations

The following operations produce an equivalent system of linear equations.

1. Interchange any two equations of the system.

2. Multiply (or divide) one of the equations by any nonzero real number.

3. Add a multiple of one equation to any other equation in the system.

Watch how we use property 3 to bring the system in Example 2 to triangular form.

EXAMPLE 2 Using Gaussian Elimination

Solve the system

$$x - 2y + z = 7$$
$$3x - 5y + z = 14$$
$$2x - 2y - z = 3.$$

SOLUTION Each step in the following process leads to a system of equations equivalent to the original system.

Multiply the first equation by -3 and add the result to the second equation, replacing the second equation. (Leave the first and third equations unchanged.)

$$x - 2y + z = 7$$
$$y - 2z = -7 \qquad \begin{array}{l} -3x + 6y - 3z = -21 \\ 3x - 5y + z = 14 \end{array}$$
$$2x - 2y - z = 3$$

Multiply the first equation by -2 and add the result to the third equation, replacing the third equation.

$$x - 2y + z = 7$$
$$y - 2z = -7$$
$$2y - 3z = -11 \qquad \begin{array}{l} -2x + 4y - 2z = -14 \\ 2x - 2y - z = 3 \end{array}$$

Multiply the second equation by -2 and add the result to the third equation, replacing the third equation.

$$x - 2y + z = 7$$
$$y - 2z = -7$$
$$z = 3 \qquad \begin{array}{l} -2y + 4z = 14 \\ 2y - 3z = -11 \end{array}$$

This is the same system of Example 1 and is a triangular form of the original system. We know from Example 1 that the solution is $(2, -1, 3)$.   ***Now try Exercise 3.***

For a system of equations which has exactly one solution, the final system in Example 2 is in **triangular form**. In this case, the leading term of each equation has coefficient 1, the third equation has one variable (z), the second equation has at most two variables including one not in the third equation (y), and the first one has the remaining variable, x in this case.

EXAMPLE 3   Finding No Solution

Solve the system

$$x - 3y + z = 4$$
$$-x + 2y - 5z = 3$$
$$5x - 13y + 13z = 8.$$

SOLUTION Use Gaussian elimination.

$$x - 3y + z = 4$$
$$-y - 4z = 7 \qquad \text{Add 1st equation to 2nd equation.}$$
$$5x - 13y + 13z = 8$$

$$x - 3y + z = 4$$
$$-y - 4z = 7$$
$$2y + 8z = -12 \qquad \text{Multiply 1st equation by } -5 \text{ and add to 3rd equation.}$$

$$x - 3y + z = 4$$
$$-y - 4z = 7$$
$$0 = 2 \qquad \text{Multiply 2nd equation by 2 and add to 3rd equation.}$$

Since $0 = 2$ is never true, we conclude that this system has no solution.

Now try Exercise 5.

Elementary Row Operations and Row Echelon Form

When we solve a system of linear equations using Gaussian elimination, all the action is really on the coefficients of the variables. Matrices can be used to record the coefficients as we go through the steps of the Gaussian elimination process. We illustrate with the system of Example 2.

$$x - 2y + z = 7$$
$$3x - 5y + z = 14$$
$$2x - 2y - z = 3$$

The **augmented matrix** of this system of equations is

$$\begin{bmatrix} 1 & -2 & 1 & 7 \\ 3 & -5 & 1 & 14 \\ 2 & -2 & -1 & 3 \end{bmatrix}.$$

The entries in the last column are the numbers on the right-hand side of the equations. For the record, the **coefficient matrix** of this system is

$$\begin{bmatrix} 1 & -2 & 1 \\ 3 & -5 & 1 \\ 2 & -2 & -1 \end{bmatrix}.$$

Here the entries are the coefficients of the variables. We use this matrix to solve certain linear systems later in this section.

We repeat the Gaussian elimination process used in Example 2 and record the corresponding action on the augmented matrix.

System of Equations   **Augmented Matrix**

$$\begin{aligned} x - 2y + z &= 7 \\ y - 2z &= -7 \\ 2x - 2y - z &= 3 \end{aligned} \qquad \begin{bmatrix} 1 & -2 & 1 & 7 \\ 0 & 1 & -2 & -7 \\ 2 & -2 & -1 & 3 \end{bmatrix}$$

Multiply Eq. 1 (Row 1) by -3, add result to Eq. 2 (Row 2) replacing Eq. 2 (Row 2).

$$\begin{aligned} x - 2y + z &= 7 \\ y - 2z &= -7 \\ 2y - 3z &= -11 \end{aligned} \qquad \begin{bmatrix} 1 & -2 & 1 & 7 \\ 0 & 1 & -2 & -7 \\ 0 & 2 & -3 & -11 \end{bmatrix}$$

Multiply Eq. 1 (Row 1) by -2, add result to Eq. 3 (Row 3) replacing Eq. 3 (Row 3).

$$\begin{aligned} x - 2y + z &= 7 \\ y - 2z &= -7 \\ z &= 3 \end{aligned} \qquad \begin{bmatrix} 1 & -2 & 1 & 7 \\ 0 & 1 & -2 & -7 \\ 0 & 0 & 1 & 3 \end{bmatrix}$$

Multiply Eq. 2 (Row 2) by -2, add result to Eq. 3 (Row 3) replacing Eq. 3 (Row 3).

The augmented matrix above, corresponding to the triangular form of the original system of equations, is a *row echelon form* of the augmented matrix of the original system of equations. In general, the last few rows of a row echelon form of a matrix *can* consist of all 0's. We will see examples like this in a moment.

DEFINITION Row Echelon Form of a Matrix

A matrix is in **row echelon form** if the following conditions are satisfied.

1. Rows consisting entirely of 0's (if there are any) occur at the bottom of the matrix.

2. The first entry in any row with nonzero entries is 1.

3. The column subscript of the leading 1 entries increases as the row subscript increases.

Another way to phrase parts 2 and 3 of the above definition is to say that the leading 1's move to the right as we move down the rows.

Our goal is to take a system of equations, write the corresponding augmented matrix, and transform it to row echelon form without carrying along the equations. From there we can read off the solutions to the system fairly easily.

The operations that we use to transform a linear system to equivalent triangular form correspond to elementary row operations of the corresponding augmented matrix of the linear system.

Elementary Row Operations on a Matrix

A combination of the following operations will transform a matrix to row echelon form.

1. Interchange any two rows.

2. Multiply all elements of a row by a nonzero real number.

3. Add a multiple of one row to any other row.

Example 4 illustrates how we can transform the augmented matrix to row echelon form to solve a system of linear equations.

NOTATION

1. R_{ij} indicates interchanging the ith and jth row of a matrix.

2. kR_i indicates multiplying the ith row by the nonzero real number k.

3. $kR_i + R_j$ indicates adding k times the ith row to the jth row.

ROW ECHELON FORM

A word of caution! You can use your grapher to find a row echelon form of a matrix. However, row echelon form is *not* unique. Your grapher may produce a row echelon form different from the one you obtained by paper-and-pencil. Fortunately, all row echelon forms produce the same solution to the system of equations. (Correspondingly, a triangular form of a linear system is also not unique.)

EXAMPLE 4   Finding a Row Echelon Form

Solve the system

$$x - y + 2z = -3$$
$$2x + y - z = 0$$
$$-x + 2y - 3z = 7.$$

SOLUTION We apply elementary row operations to find a row echelon form of the augmented matrix. The elementary row operations used are recorded above the arrows using the notation in the margin.

$$\begin{bmatrix} 1 & -1 & 2 & -3 \\ 2 & 1 & -1 & 0 \\ -1 & 2 & -3 & 7 \end{bmatrix} \xrightarrow{(-2)R_1 + R_2} \begin{bmatrix} 1 & -1 & 2 & -3 \\ 0 & 3 & -5 & 6 \\ -1 & 2 & -3 & 7 \end{bmatrix} \xrightarrow{(1)R_1 + R_3}$$

$$\begin{bmatrix} 1 & -1 & 2 & -3 \\ 0 & 3 & -5 & 6 \\ 0 & 1 & -1 & 4 \end{bmatrix} \xrightarrow{R_{23}} \begin{bmatrix} 1 & -1 & 2 & -3 \\ 0 & 1 & -1 & 4 \\ 0 & 3 & -5 & 6 \end{bmatrix} \xrightarrow{(-3)R_2 + R_3}$$

$$\begin{bmatrix} 1 & -1 & 2 & -3 \\ 0 & 1 & -1 & 4 \\ 0 & 0 & -2 & -6 \end{bmatrix} \xrightarrow{(-1/2)R_3} \begin{bmatrix} 1 & -1 & 2 & -3 \\ 0 & 1 & -1 & 4 \\ 0 & 0 & 1 & 3 \end{bmatrix}$$

The last matrix is in row echelon form. Then we convert each row into equation form and complete the solution by substitution.

continued

$$y - z = 4 \qquad x - y + 2z = -3$$
$$z = 3 \qquad y - 3 = 4 \qquad x - 7 + 2(3) = -3$$
$$y = 7 \qquad x = -2$$

The solution of the original system of equations is $(-2, 7, 3)$.

Now try Exercise 33.

Reduced Row Echelon Form

If we continue to apply elementary row operations to a row echelon form of a matrix, we can obtain a matrix in which every column that has a leading 1 has 0's elsewhere. This is the **reduced row echelon form** of the matrix. It is usually easier to read the solution from the reduced row echelon form.

We apply elementary row operations to the row echelon form found in Example 4 until we find the reduced row echelon form.

$$\begin{bmatrix} 1 & -1 & 2 & -3 \\ 0 & 1 & -1 & 4 \\ 0 & 0 & 1 & 3 \end{bmatrix} \xrightarrow{(1)R_2 + R_1} \begin{bmatrix} 1 & 0 & 1 & 1 \\ 0 & 1 & -1 & 4 \\ 0 & 0 & 1 & 3 \end{bmatrix} \xrightarrow{(-1)R_3 + R_1}$$

$$\begin{bmatrix} 1 & 0 & 0 & -2 \\ 0 & 1 & -1 & 4 \\ 0 & 0 & 1 & 3 \end{bmatrix} \xrightarrow{(1)R_3 + R_2} \begin{bmatrix} 1 & 0 & 0 & -2 \\ 0 & 1 & 0 & 7 \\ 0 & 0 & 1 & 3 \end{bmatrix}$$

From this reduced row echelon form, we can immediately read the solution to the system of Example 4: $x = -2$, $y = 7$, $z = 3$. Figure 7.14 shows that the above final matrix is the reduced row echelon form of the augmented matrix of Example 4.

rref([A])

[[1 0 0 –2]
 [0 1 0 7]
 [0 0 1 3]]

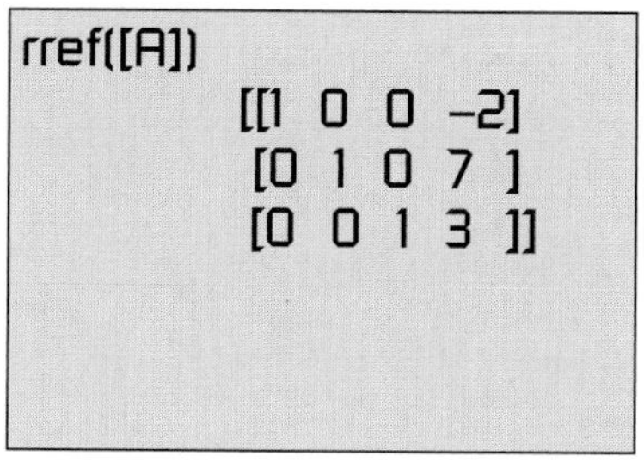

FIGURE 7.14 *A* is the augmented matrix of the system of linear equations in Example 4. "rref" stands for the grapher-produced reduced row echelon form of *A*.

EXAMPLE 5 Finding Infinitely Many Solutions

Solve the system

$$x + y + z = 3$$
$$2x + y + 4z = 8$$
$$x + 2y - z = 1.$$

SOLUTION Figure 7.15 shows the reduced row echelon form for the augmented matrix of the system. So, the following system of equations is equivalent to the original system.

$$x + 3z = 5$$
$$y - 2z = -2$$
$$0 = 0$$

Solving the first two equations for x and y in terms of z yields:

$$x = -3z + 5$$
$$y = 2z - 2.$$

rref([A])

[[1 0 3 5]
 [0 1 –2 –2]
 [0 0 0 0]]

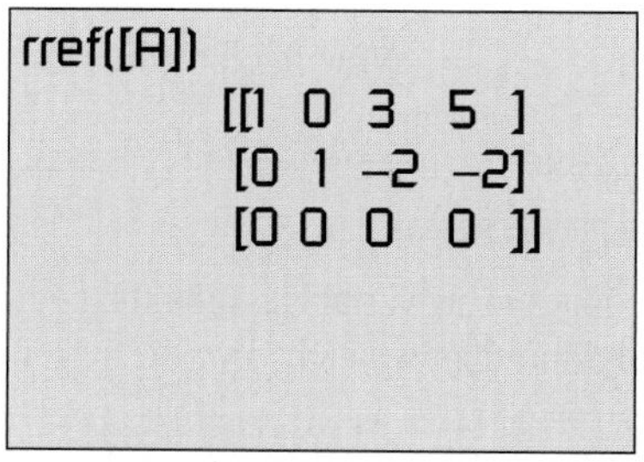

FIGURE 7.15 The reduced row echelon form for the augmented matrix of Example 5.

continued

This system has infinitely many solutions because for every value of z we can use these two equations to find corresponding values for x and y.

Interpret

The solution is the set of all ordered triples of the form $(-3z + 5, 2z - 2, z)$ where z is any real number. *Now try Exercise 39.*

We can also solve linear systems with more than three variables, or more than three equations, or both, by finding a row (or reduced row) echelon form. The solution set may become more complicated as illustrated in Example 6.

EXAMPLE 6 Finding Infinitely Many Solutions

Solve the system

$$\begin{aligned} x + 2y - 3z &= -1 \\ 2x + 3y - 4z + w &= -1 \\ 3x + 5y - 7z + w &= -2. \end{aligned}$$

SOLUTION The 3×5 augmented matrix is

$$\begin{bmatrix} 1 & 2 & -3 & 0 & -1 \\ 2 & 3 & -4 & 1 & -1 \\ 3 & 5 & -7 & 1 & -2 \end{bmatrix}.$$

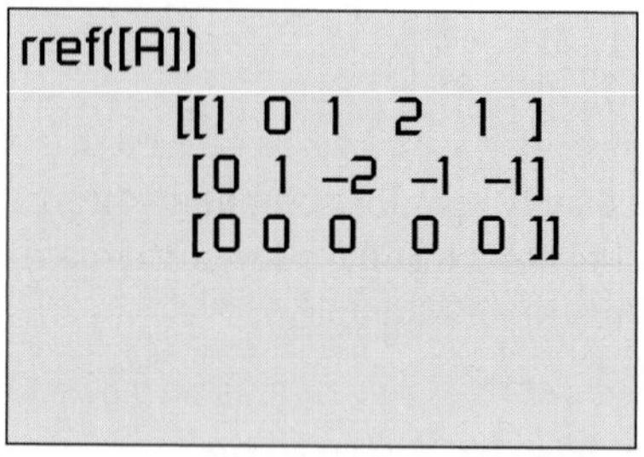
rref([A])
[[1 0 1 2 1]
[0 1 -2 -1 -1]
[0 0 0 0 0]]

FIGURE 7.16 The reduced row echelon form for the augmented matrix of Example 6.

Figure 7.16 shows the reduced row echelon form from which we can read that

$$\begin{aligned} x &= -z - 2w + 1 \\ y &= 2z + w - 1. \end{aligned}$$

This system has infinitely many solutions because for every pair of values for z and w we can use these two equations to find corresponding values for x and y.

Interpret

The solution is the set of all ordered 4-tuples of the form $(-z - 2w + 1, 2z + w - 1, z, w)$ where z and w are any real numbers.

Now try Exercise 43.

Solving Systems with Inverse Matrices

If a linear system consists of the same number of equations as variables, then the coefficient matrix is square. If this matrix is also nonsingular, then we can solve the system using the technique illustrated in Example 7.

EXAMPLE 7 Solving a System Using Inverse Matrices

Solve the system

$$\begin{aligned} 3x - 2y &= 0 \\ -x + y &= 5. \end{aligned}$$

continued

LINEAR EQUATIONS

If a and b are real numbers with $a \neq 0$, the linear equation $ax = b$ has a unique solution $x = a^{-1}b$. A similar statement holds for the linear matrix equation $AX = B$ when A is a nonsingular square matrix. (See the Invertible Square Linear Systems Theorem.)

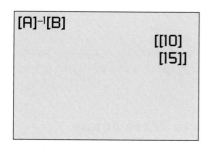

FIGURE 7.17 The solution of the matrix equation of Example 7.

SOLUTION First we write the system as a matrix equation. Let

$$A = \begin{bmatrix} 3 & -2 \\ -1 & 1 \end{bmatrix}, \qquad X = \begin{bmatrix} x \\ y \end{bmatrix}, \qquad \text{and} \qquad B = \begin{bmatrix} 0 \\ 5 \end{bmatrix}.$$

Then

$$A \cdot X = \begin{bmatrix} 3 & -2 \\ -1 & 1 \end{bmatrix} \cdot \begin{bmatrix} x \\ y \end{bmatrix} = \begin{bmatrix} 3x - 2y \\ -x + y \end{bmatrix}$$

so that

$$AX = B,$$

where A is the coefficient matrix of the system. You can easily check that $\det A = 1$, so A^{-1} exists. From Figure 7.17, we obtain

$$X = A^{-1}B = \begin{bmatrix} 10 \\ 15 \end{bmatrix}.$$

The solution of the system is $x = 10$, $y = 15$, or $(10, 15)$.

Now try Exercise 49.

Examples 7 and 8 are two instances of the following theorem.

THEOREM Invertible Square Linear Systems

Let A be the coefficient matrix of a system of n linear equations in n variables given by $AX = B$, where X is the $n \times 1$ matrix of variables and B is the $n \times 1$ matrix of numbers on the right-hand side of the equations. If A^{-1} exists, then the system of equations has the unique solution

$$X = A^{-1}B.$$

EXAMPLE 8   Solving a System Using Inverse Matrices

Solve the system

$$3x - 3y + 6z = 20$$
$$x - 3y + 10z = 40$$
$$-x + 3y - 5z = 30.$$

SOLUTION Let

$$A = \begin{bmatrix} 3 & -3 & 6 \\ 1 & -3 & 10 \\ -1 & 3 & -5 \end{bmatrix}, \qquad X = \begin{bmatrix} x \\ y \\ z \end{bmatrix}, \qquad \text{and} \qquad B = \begin{bmatrix} 20 \\ 40 \\ 30 \end{bmatrix}.$$

The system of equations can be written as

$$A \cdot X = B.$$

continued

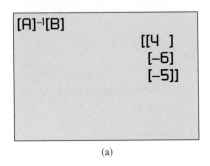

FIGURE 7.18 The solution of the system in Example 8.

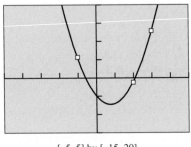

(a)

[−5, 5] by [−15, 20]
(b)

FIGURE 7.19 (a) The solution of the matrix equation of Example 9. (b) A graph of $f(x) = 4x^2 - 6x - 5$ superimposed on a scatter plot of the three points $(-1, 5)$, $(2, -1)$, and $(3, 13)$.

Figure 7.18 shows that det $A = -30 \neq 0$, so A^{-1} exists and

$$X = A^{-1}B = \begin{bmatrix} 18 \\ 39.\overline{3} \\ 14 \end{bmatrix}.$$

Interpret

The solution of the system of equations is $x = 18$, $y = 39\frac{1}{3}$, and $z = 14$, or $(18, 39\frac{1}{3}, 14)$.

Now try Exercise 51.

Applications

Any three noncollinear points with distinct x-coordinates determine exactly one second-degree polynomial as illustrated in Example 9. The graph of a second-degree polynomial is a parabola.

EXAMPLE 9   Fitting a Parabola to Three Points

Determine a, b, and c so that the points $(-1, 5)$, $(2, -1)$, and $(3, 13)$ are on the graph of $f(x) = ax^2 + bx + c$.

SOLUTION

Model

We must have $f(-1) = 5$, $f(2) = -1$, and $f(3) = 13$:

$$f(-1) = a - b + c = 5$$

$$f(2) = 4a + 2b + c = -1$$

$$f(3) = 9a + 3b + c = 13.$$

The above system of three linear equations in the three variables a, b, and c can be written in matrix form $AX = B$, where

$$A = \begin{bmatrix} 1 & -1 & 1 \\ 4 & 2 & 1 \\ 9 & 3 & 1 \end{bmatrix}, \qquad X = \begin{bmatrix} a \\ b \\ c \end{bmatrix}, \qquad \text{and} \qquad B = \begin{bmatrix} 5 \\ -1 \\ 13 \end{bmatrix}.$$

Solve Numerically

Figure 7.19a shows that

$$X = A^{-1}B = \begin{bmatrix} 4 \\ -6 \\ -5 \end{bmatrix}.$$

Thus, $a = 4$, $b = -6$, and $c = -5$. The second-degree polynomial $f(x) = 4x^2 - 6x - 5$ contains the points $(-1, 5)$, $(2, -1)$, and $(3, 13)$ (Figure 7.19b).

Now try Exercise 67.

EXPLORATION 1 **Mixing Solutions**

Aileen's Drugstore needs to prepare a 60-L mixture that is 40% acid using three concentrations of acid. The first concentration is 15% acid, the second is 35% acid, and the third is 55% acid. Because of the amounts of acid solution on hand, they need to use twice as much of the 35% solution as the 55% solution. How much of each solution should they use?

Let x = the number of liters of 15% solution used, y = the number of liters of 35% solution used, and z = the number of liters of 55% solution used.

1. Explain how the equation $x + y + z = 60$ is related to the problem.

2. Explain how the equation $0.15x + 0.35y + 0.55z = 24$ is related to the problem.

3. Explain how the equation $y = 2z$ is related to the problem.

4. Write the system of three equations obtained from parts 1–3 in the form $AX = B$, where A is the coefficient matrix of the system. What are A, B, and X?

5. Solve the matrix equation in part 4.

6. Interpret the solution in part 5 in terms of the problem situation.

QUICK REVIEW 7.3 *(For help, go to Sections 1.2 and 7.2.)*

In Exercises 1 and 2, find the amount of pure acid in the solution.

1. 40 L of a 32% acid solution

2. 60 mL of a 14% acid solution

In Exercises 3 and 4, find the amount of water in the solution.

3. 50 L of a 24% acid solution

4. 80 mL of a 70% acid solution

In Exercises 5 and 6, determine which points are on the graph of the function.

5. $f(x) = 2x^2 - 3x + 1$
   (a) $(-1, 6)$ (b) $(2, 1)$

6. $f(x) = x^3 - 4x - 1$
   (a) $(0, -1)$ (b) $(-2, -17)$

In Exercises 7 and 8, solve for x or y in terms of the other variables.

7. $y + z - w = 1$

8. $x - 2z + w = 3$

In Exercises 9 and 10, find the inverse of the matrix.

9. $\begin{bmatrix} 1 & 3 \\ -2 & -2 \end{bmatrix}$

10. $\begin{bmatrix} 0 & 0 & 2 \\ -2 & 1 & 3 \\ 0 & 2 & -2 \end{bmatrix}$

SECTION 7.3 EXERCISES

In Exercises 1 and 2, use substitution to solve the system of equations.

1. $x - 3y + z = 0$
$ 2y + 3z = 1$
$ z = -2$

2. $3x - y + 2z = -2$
$ y + 3z = 3$
$ 2z = 4$

In Exercises 3–8, use Gaussian elimination to solve the system of equations.

3. $x - y + z = 0$
$ 2x - 3z = -1$
$-x - y + 2z = -1$

4. $ 2x - y = 0$
$ x + 3y - z = -3$
$ 3y + z = 8$

5. $ x + y + z = -3$
$ 4x - y = -5$
$-3x + 2y + z = 4$

6. $ x + y - 3z = -1$
$ 2x - 3y + z = 4$
$ 3x - 7y + 5z = 4$

7. $x + y - z = 4$
$ y + w = -4$
$x - y = 1$
$x + z + w = 1$

8. $\dfrac{1}{2}x - y + z - w = 1$
$-x + y + z + 2w = -3$
$ x - z = 2$
$ y + w = 0$

In Exercises 9–12, perform the indicated elementary row operation on the matrix

$$\begin{bmatrix} 2 & -6 & 4 \\ 1 & 2 & -3 \\ -3 & 1 & -2 \end{bmatrix}.$$

9. $(3/2)R_1 + R_3$

10. $(1/2)R_1$

11. $(-2)R_2 + R_1$

12. $(1)R_1 + R_2$

In Exercises 13–16, what elementary row operations applied to

$$\begin{bmatrix} -2 & 1 & -1 & 2 \\ 1 & -2 & 3 & 0 \\ 3 & 1 & -1 & 2 \end{bmatrix}$$

will yield the given matrix?

13. $\begin{bmatrix} 1 & -2 & 3 & 0 \\ -2 & 1 & -1 & 2 \\ 3 & 1 & -1 & 2 \end{bmatrix}$

14. $\begin{bmatrix} 0 & -3 & 5 & 2 \\ 1 & -2 & 3 & 0 \\ 3 & 1 & -1 & 2 \end{bmatrix}$

15. $\begin{bmatrix} -2 & 1 & -1 & 2 \\ 1 & -2 & 3 & 0 \\ 0 & 7 & -10 & 2 \end{bmatrix}$

16. $\begin{bmatrix} -2 & 1 & -1 & 2 \\ 1 & -2 & 3 & 0 \\ 0.75 & 0.25 & -0.25 & 0.5 \end{bmatrix}$

In Exercises 17–20, find a row echelon form for the matrix.

17. $\begin{bmatrix} 1 & 3 & -1 \\ 2 & 1 & 4 \\ -3 & 0 & 1 \end{bmatrix}$

18. $\begin{bmatrix} 1 & 2 & -3 \\ -3 & -6 & 10 \\ -2 & -4 & 7 \end{bmatrix}$

19. $\begin{bmatrix} 1 & 2 & 3 & -4 \\ -2 & 6 & -6 & 2 \\ 3 & 12 & 6 & 12 \end{bmatrix}$

20. $\begin{bmatrix} 3 & 6 & 9 & -6 \\ 2 & 5 & 5 & -3 \end{bmatrix}$

In Exercises 21–24, find the reduced row echelon form for the matrix.

21. $\begin{bmatrix} 1 & 0 & 2 & 1 \\ 3 & 2 & 4 & 7 \\ 2 & 1 & 3 & 4 \end{bmatrix}$

22. $\begin{bmatrix} 1 & -2 & 2 & 1 & 1 \\ 3 & -5 & 6 & 3 & -1 \\ -2 & 4 & -3 & -2 & 5 \\ 3 & -5 & 6 & 4 & -3 \end{bmatrix}$

23. $\begin{bmatrix} 1 & 2 & 3 & 1 \\ -3 & -5 & -7 & -4 \end{bmatrix}$

24. $\begin{bmatrix} 3 & -6 & 3 & -3 \\ 2 & -4 & 2 & -2 \\ -3 & 6 & -3 & 3 \end{bmatrix}$

In Exercises 25–28, write the augmented matrix corresponding to the system of equations.

25. $2x - 3y + z = 1$
$-x + y - 4z = -3$
$ 3x - z = 2$

26. $3x - 4y + z - w = 1$
$ x + z - 2w = 4$

27. $2x - 5y + z - w = -3$
$ x - 2z + w = 4$
$ 2y - 3z - w = 5$

28. $3x - 2y = 5$
$-x + 5y = 7$

In Exercises 29–32, write the system of equations corresponding to the augmented matrix.

29. $\begin{bmatrix} 3 & 2 & -1 \\ -4 & 5 & 2 \end{bmatrix}$

30. $\begin{bmatrix} 1 & 0 & -1 & 2 & -3 \\ 2 & 1 & 0 & -1 & 4 \\ -1 & 1 & 2 & 0 & 0 \end{bmatrix}$

31. $\begin{bmatrix} 2 & 0 & 1 & 3 \\ -1 & 1 & 0 & 2 \\ 0 & 2 & -3 & -1 \end{bmatrix}$

32. $\begin{bmatrix} 2 & 1 & -2 & 4 \\ -3 & 0 & 2 & -1 \end{bmatrix}$

In Exercises 33–34, solve the system of equations by finding a row echelon form for the augmented matrix.

33. $ x - 2y + z = 8$
$ 2x + y - 3z = -9$
$-3x + y + 3z = 5$

34. $3x + 7y - 11z = 44$
$ x + 2y - 3z = 12$
$4x + 9y - 13z = 53$

In Exercises 35–44, solve the system of equations by finding the reduced row echelon form for the augmented matrix.

35.
$$x + 2y - z = 3$$
$$3x + 7y - 3z = 12$$
$$-2x - 4y + 3z = -5$$

36.
$$x - 2y + z = -2$$
$$2x - 3y + 2z = 2$$
$$4x - 8y + 5z = -5$$

37.
$$x + y + 3z = 2$$
$$3x + 4y + 10z = 5$$
$$x + 2y + 4z = 3$$

38.
$$x - z = 2$$
$$-2x + y + 3z = -5$$
$$2x + y - z = 3$$

39.
$$x + z = 2$$
$$2x + y + z = 5$$

40.
$$x + 2y - 3z = 1$$
$$-3x - 5y + 8z = -29$$

41.
$$x + 2y = 4$$
$$3x + 4y = 5$$
$$2x + 3y = 4$$

42.
$$x + y = 3$$
$$2x + 3y = 8$$
$$2x + 2y = 6$$

43.
$$x + y - 3z = 1$$
$$x - z - w = 2$$
$$2x + y - 4z - w = 3$$

44.
$$x - y - z + 2w = -3$$
$$2x - y - 2z + 3w = -3$$
$$x - 2y - z + 3w = -6$$

In Exercises 45 and 46, write the system of equations as a matrix equation $AX = B$, with A as the coefficient matrix of the system.

45.
$$2x + 5y = -3$$
$$x - 2y = 1$$

46.
$$5x - 7y + z = 2$$
$$2x - 3y - z = 3$$
$$x + y + z = -3$$

In Exercises 47 and 48, write the matrix equation as a system of equations.

47.
$$\begin{bmatrix} 3 & -1 \\ 2 & 4 \end{bmatrix} \begin{bmatrix} x \\ y \end{bmatrix} = \begin{bmatrix} -1 \\ 3 \end{bmatrix}$$

48.
$$\begin{bmatrix} 1 & 0 & -3 \\ 2 & -1 & 3 \\ -2 & 3 & -4 \end{bmatrix} \begin{bmatrix} x \\ y \\ z \end{bmatrix} = \begin{bmatrix} 3 \\ -1 \\ 2 \end{bmatrix}$$

In Exercises 49–54, solve the system of equations by using an inverse matrix.

49.
$$2x - 3y = -13$$
$$4x + y = -5$$

50.
$$x + 2y = -2$$
$$3x - 4y = 9$$

51.
$$2x - y + z = -6$$
$$x + 2y - 3z = 9$$
$$3x - 2y + z = -3$$

52.
$$x + 4y - 2z = 0$$
$$2x + y + z = 6$$
$$-3x + 3y - 5z = -13$$

53.
$$2x - y + z + w = -3$$
$$x + 2y - 3z + w = 12$$
$$3x - y - z + 2w = 3$$
$$-2x + 3y + z - 3w = -3$$

54.
$$2x + y + 2z = 8$$
$$3x + 2y - z - w = 10$$
$$-2x + y - 3w = -1$$
$$4x - 3y + 2z - 5w = 39$$

In Exercises 55–66, use a method of your choice to solve the system of equations.

55.
$$2x - y = 10$$
$$x - z = -1$$
$$y + z = -9$$

56.
$$1.25x + z = -2$$
$$y - 5.5z = -2.75$$
$$3x - 1.5y = -6$$

57.
$$x + 2y + 2z + w = 5$$
$$2x + y + 2z = 5$$
$$3x + 3y + 3z + 2w = 12$$
$$x + z + w = 1$$

58.
$$x - y + w = -4$$
$$-2x + y + z = 8$$
$$2x - 2y - z = -10$$
$$-2x + z + w = 5$$

59.
$$x - y + z = 6$$
$$x + y + 2z = -2$$

60.
$$x - 2y + z = 3$$
$$2x + y - z = -4$$

61.
$$2x + y + z + 4w = -1$$
$$x + 2y + z + w = 1$$
$$x + y + z + 2w = 0$$

62.
$$2x + 3y + 3z + 7w = 0$$
$$x + 2y + 2z + 5w = 0$$
$$x + y + 2z + 3w = -1$$

63.
$$2x + y + z + 2w = -3.5$$
$$x + y + z + w = -1.5$$

64.
$$2x + y + 4w = 6$$
$$x + y + z + w = 5$$

65.
$$x + y - z + 2w = 0$$
$$y - z + 2w = -1$$
$$x + y + 3w = 3$$
$$2x + 2y - z + 5w = 4$$

66.
$$x + y + w = 2$$
$$x + 4y + z - 2w = 3$$
$$x + 3y + z - 3w = 2$$
$$x + y + w = 2$$

In Exercises 67–70, determine f so that its graph contains the given points.

67. Curve Fitting $f(x) = ax^2 + bx + c$
$(-1, 3), (1, -3), (2, 0)$

68. Curve Fitting $f(x) = ax^3 + bx^2 + cx + d$.
$(-2, -37), (-1, -11), (0, -5), (2, 19)$

69. Family of Curves $f(x) = ax^2 + bx + c$
$(-1, -4), (1, -2)$

70. Family of Curves $f(x) = ax^3 + bx^2 + cx + d$
$(-1, -6), (0, -1), (1, 2)$

71. Population Table 7.5 gives the population (in thousands) for Corpus Christi, TX, and Garland, TX, for several years. Use $x = 0$ for 1980, $x = 1$ for 1981, and so forth.

(a) Find the linear regression equation for the Corpus Christi data and superimpose its graph on a scatter plot of the data.

(b) Find the linear regression equation for the Garland data and superimpose its graph on a scatter plot of the data.

(c) Estimate when the population of the two cities will be the same.

 Table 7.5 Population

Year	Corpus Christi (thousands)	Garland (thousands)
1980	232	139
1990	258	181
2000	277	216
2003	279	218

Source: U.S. Census Bureau, Statistical Abstract of the United States, 2004–2005.

72. Population Table 7.6 gives the population (in thousands) for Anaheim, CA, and Anchorage, AK, for several years. Use $x = 0$ for 1970, $x = 1$ for 1971, and so forth.

(a) Find the linear regression equation for the Anaheim data and superimpose its graph on a scatter plot of the data.

(b) Find the linear regression equation for the Anchorage data and superimpose its graph on a scatter plot of the data.

(c) Estimate when the population of the two cities will be the same.

Table 7.6 Population		
Year	Anaheim (thousands)	Anchorage (thousands)
1970	166	48
1980	219	174
1990	267	226
2000	328	260
2003	332	271

Source: U.S. Census Bureau, Statistical Abstract of the United States, 2001, 2004–2005.

73. Train Tickets At the Pittsburgh zoo, children ride a train for 25 cents, adults pay $1.00, and senior citizens 75 cents. On a given day, 1400 passengers paid a total of $740 for the rides. There were 250 more children riders than all other riders. Find the number of children, adult, and senior riders.

74. Manufacturing Stewart's Metals has three silver alloys on hand. One is 22% silver, another is 30% silver, and the third is 42% silver. How many grams of each alloy is required to produce 80 grams of a new alloy that is 34% silver if the amount of 30% alloy used is twice the amount of 22% alloy used.

75. Investment Monica receives an $80,000 inheritance. She invests part of it in CDs (certificates of deposit) earning 6.7% APY (annual percentage yield), part in bonds earning 9.3% APY, and the remainder in a growth fund earning 15.6% APY. She invests three times as much in the growth fund as in the other two combined. How much does she have in each investment if she receives $10,843 interest the first year?

76. Investments Oscar invests $20,000 in three investments earning 6% APY, 8% APY, and 10% APY. He invests $9000 more in the 10% investment than in the 6% investment. How much does he have invested at each rate if he receives $1780 interest the first year?

77. Investments Morgan has $50,000 to invest and wants to receive $5000 interest the first year. He puts part in CDs earning 5.75% APY, part in bonds earning 8.7% APY, and the rest in a growth fund earning 14.6% APY. How much should he invest at each rate if he puts the least amount possible in the growth fund?

78. Mixing Acid Solutions Simpson's Drugstore needs to prepare a 40-L mixture that is 32% acid from three solutions: a 10% acid solution, a 25% acid solution, and a 50% acid solution. How much of each solution should be used if Simpson's wants to use as little of the 50% solution as possible?

79. Loose Change Matthew has 74 coins consisting of nickels, dimes, and quarters in his coin box. The total value of the coins is $8.85. If the number of nickels and quarters is four more than the number of dimes, find how many of each coin Matthew has in his coin box.

80. Vacation Money Heather has saved $177 to take with her on the family vacation. She has 51 bills consisting of $1, $5, and $10 bills. If the number of $5 bills is three times the number of $10 bills, find how many of each bill she has.

In Exercises 81–82, use inverse matrices to find the equilibrium point for the demand and supply curves.

81. $p = 100 - 5x$ Demand curve

$p = 20 + 10x$ Supply curve

82. $p = 150 - 12x$ Demand curve

$p = 30 + 24x$ Supply curve

83. Writing to Learn Explain why adding one row to another row in a matrix is an elementary row operation.

84. Writing to Learn Explain why subtracting one row from another row in a matrix is an elementary row operation.

Standardized Test Questions

85. True or False Every nonzero square matrix has an inverse. Justify your answer.

86. True or False The reduced row echelon form of the augmented matrix of a system of three linear equations in three variables must be of the form

$$\begin{bmatrix} 1 & 0 & 0 & a \\ 0 & 1 & 0 & b \\ 0 & 0 & 1 & c \end{bmatrix},$$

where a, b, c, are real numbers. Justify your answer.

In Exercises 87–90, you may use a graphing calculator to solve the problem.

87. Multiple Choice Which of the following is the determinant of the matrix $\begin{bmatrix} 2 & 2 \\ -1 & 3 \end{bmatrix}$?

(A) 0   (B) 4

(C) -4 (D) 8

(E) -8

88. Multiple Choice Which of the following is the augmented matrix of the system of equations

$$x + 2y + z = -1$$
$$2x - y + 3z = -4?$$
$$3x + y - z = -2$$

(A) $\begin{bmatrix} 1 & 2 & 1 & -1 \\ 2 & -1 & 3 & -4 \\ 3 & 1 & -1 & -2 \end{bmatrix}$ (B) $\begin{bmatrix} 1 & 2 & 1 & 1 \\ 2 & -1 & 3 & 4 \\ 3 & 1 & -1 & 2 \end{bmatrix}$

(C) $\begin{bmatrix} 1 & 2 & 1 & 0 \\ 2 & -1 & 3 & 0 \\ 3 & 1 & -1 & 0 \end{bmatrix}$ (D) $\begin{bmatrix} 1 & 2 & 1 \\ 2 & -1 & 3 \\ 3 & 1 & -1 \end{bmatrix}$

(E) $\begin{bmatrix} 1 & 2 & -1 \\ 2 & -1 & -3 \\ 3 & 1 & 1 \end{bmatrix}$

89. Multiple Choice The matrix

$$\begin{bmatrix} 1 & 2 & 3 \\ 2 & 1 & 0 \\ 7 & 8 & 9 \end{bmatrix} \text{ was obtained from } \begin{bmatrix} 1 & 2 & 3 \\ 4 & 5 & 6 \\ 7 & 8 & 9 \end{bmatrix} \text{ by an}$$

elementary row operation. Which of the following describes the elementary row operation?

(A) $(-2)R_1$ (B) $(-2)R_1 + R_2$

(C) $(-2)R_2 + R_1$ (D) $(2)R_1 + R_2$

(E) $(2)R_2 + R_1$

90. Multiple Choice Which of the following is the reduced row echelon form for the augmented matrix of

$$x + 2y - z = 8$$
$$-x + 3y + 2z = 3 \qquad ?$$
$$2x - y + 3z = -19$$

(A) $\begin{bmatrix} 1 & 2 & 0 & 4 \\ 0 & 1 & 0 & 3 \\ 0 & 0 & 1 & -4 \end{bmatrix}$ (B) $\begin{bmatrix} 1 & 0 & 0 & 2 \\ 0 & 1 & 0 & -3 \\ 0 & 0 & 0 & 4 \end{bmatrix}$

(C) $\begin{bmatrix} 1 & 0 & 0 & -2 \\ 0 & 1 & 0 & 3 \\ 0 & 0 & 0 & -4 \end{bmatrix}$ (D) $\begin{bmatrix} 1 & 0 & 0 & 2 \\ 0 & 1 & 0 & -3 \\ 0 & 0 & 1 & 4 \end{bmatrix}$

(E) $\begin{bmatrix} 1 & 0 & 0 & -2 \\ 0 & 1 & 0 & 3 \\ 0 & 0 & 1 & -4 \end{bmatrix}$

Explorations

91. Group Activity Investigating the Solution of a System of 3 Linear Equations in 3 Variables Assume that the graph of a linear equation in three variables is a plane in 3-dimensional space. (You will study these in Chapter 8.)

(a) Explain geometrically how such a system can have a unique solution.

(b) Explain geometrically how such a system can have no solution. Describe several possibilities.

(c) Explain geometrically how such a system can have infinitely many solutions. Describe several possibilities. Construct physical models if you find that helpful.

Extending the Ideas

92. Writing to Learn Explain why a row echelon form of a matrix is not unique. That is, show that a matrix can have two unequal row echelon forms. Give an example.

The roots of the characteristic polynomial $C(x) = \det(xI_n - A)$ of the $n \times n$ matrix A are the **eigenvalues** of A (see Section 7.2, Exercises 72 and 73). Use this information in Exercises 93 and 94.

93. Let $A = \begin{bmatrix} 3 & 2 \\ 1 & 5 \end{bmatrix}$.

(a) Find the characteristic polynomial $C(x)$ of A.

(b) Find the graph of $y = C(x)$.

(c) Find the eigenvalues of A.

(d) Compare $\det A$ with the y-intercept of the graph of $y = C(x)$.

(e) Compare the sum of the main diagonal elements of A $(a_{11} + a_{22})$ with the sum of the eigenvalues.

94. Let $A = \begin{bmatrix} 2 & -1 \\ -5 & 2 \end{bmatrix}$.

(a) Find the characteristic polynomial $C(x)$ of A.

(b) Find the graph of $y = C(x)$.

(c) Find the eigenvalues of A.

(d) Compare $\det A$ with the y-intercept of the graph of $y = C(x)$.

(e) Compare the sum of the main diagonal elements of A $(a_{11} + a_{22})$ with the sum of the eigenvalues.

7.4
Partial Fractions

. . . and why

Partial fraction decompositions are used in calculus in integration and can be used to guide the sketch of the graph of a rational function.

Partial Fraction Decomposition

In Section 2.6 we saw that a polynomial with real coefficients could be factored into a product of factors with real coefficients, where each factor was either a linear factor or an irreducible quadratic factor. In this section we show that a rational function can be expressed as a sum of rational functions where each denominator is a power of a linear factor or a power of an irreducible quadratic factor.

For example,

$$\frac{3x - 4}{x^2 - 2x} = \frac{2}{x} + \frac{1}{x - 2}.$$

Each fraction in the sum is a **partial fraction**, and the sum is a **partial fraction decomposition** of the original rational function.

Partial Fraction Decomposition of $f(x)/d(x)$

1. Degree of $f \geq$ degree of d: Use the division algorithm to divide f by d to obtain the quotient q and remainder r and write

$$\frac{f(x)}{d(x)} = q(x) + \frac{r(x)}{d(x)}.$$

2. Factor $d(x)$ into a product of factors of the form $(mx + n)^u$ or $(ax^2 + bx + c)^v$, where $ax^2 + bx + c$ is irreducible.

3. For each factor $(mx + n)^u$: The partial fraction decomposition of $r(x)/d(x)$ must include the sum

$$\frac{A_1}{mx + n} + \frac{A_2}{(mx + n)^2} + \cdots + \frac{A_u}{(mx + n)^u},$$

where $A_1, A_2, \ldots, A_u$ are real numbers.

4. For each factor $(ax^2 + bx + c)^v$: The partial fraction decomposition of $r(x)/d(x)$ must include the sum

$$\frac{B_1 x + C_1}{ax^2 + bx + c} + \frac{B_2 x + C_2}{(ax^2 + bx + c)^2} + \cdots + \frac{B_v x + C_v}{(ax^2 + bx + c)^v},$$

where $B_1, B_2, \ldots, B_v$ and $C_1, C_2, \ldots, C_v$ are real numbers.

The partial fraction decomposition of the original rational function is the sum of $q(x)$ and the fractions in parts 3 and 4.

EXAMPLE 1   Writing the Decomposition Factors

Write the terms for the partial fraction decomposition of the rational function

$$\frac{5x - 1}{x^3(x + 3)(x^2 + 1)},$$

but do not solve for the corresponding constants.

continued

SOLUTION Applying part 3 to the factor x^3 of the denominator produces the expression

$$\frac{A_1}{x} + \frac{A_2}{x^2} + \frac{A_3}{x^3}.$$

Then, applying part 3 to the factor $(x + 3)$ of the denominator produces the expression

$$\frac{B_1}{x + 3}.$$

Finally, applying part 4 to the factor $(x^2 + 1)$ of the denominator produces the expression

$$\frac{C_1 x + D_1}{x^2 + 1}.$$

Adding these terms produces the partial fraction decomposition for the rational function

$$\frac{5x - 1}{x^3(x + 3)(x^2 + 1)} = \frac{A_1}{x} + \frac{A_2}{x^2} + \frac{A_3}{x^3} + \frac{B_1}{x + 3} + \frac{C_1 x + D_1}{x^2 + 1}.$$

Now try Exercise 1.

Denominators with Linear Factors

Examples 2 and 3 illustrate how the constants A_i in part 3 of the partial fraction decomposition procedure can be found.

EXAMPLE 2 Decomposing a Fraction with Distinct Linear Factors

Find the partial fraction decomposition of

$$\frac{5x - 1}{x^2 - 2x - 15}.$$

SOLUTION The denominator factors into $(x + 3)(x - 5)$. We write

$$\frac{5x - 1}{x^2 - 2x - 15} = \frac{A_1}{x + 3} + \frac{A_2}{x - 5}$$

and then "clear fractions" by multiplying both sides of the above equation by $x^2 - 2x - 15$ to obtain

$$5x - 1 = A_1(x - 5) + A_2(x + 3)$$

$$5x - 1 = (A_1 + A_2)x + (-5A_1 + 3A_2).$$

Comparing coefficients on the left and right side of the above equation, we obtain the following system of two equations in the two variables A_1 and A_2:

$$A_1 + A_2 = 5$$

$$-5A_1 + 3A_2 = -1.$$

continued

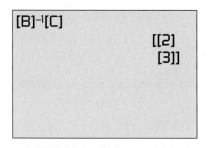

FIGURE 7.20 The solution of the system of equations in Example 2.

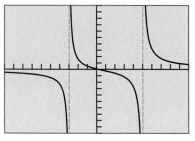

[−10, 10] by [−10, 10]

FIGURE 7.21 The graphs of $y = (5x − 1)/(x^2 − 2x − 15)$ and $y = 2/(x + 3) + 3/(x − 5)$ appear to be the same. (Example 2)

We can write this system in matrix form as $BX = C$ where

$$B = \begin{bmatrix} 1 & 1 \\ -5 & 3 \end{bmatrix}, \quad X = \begin{bmatrix} A_1 \\ A_2 \end{bmatrix}, \quad \text{and} \quad C = \begin{bmatrix} 5 \\ -1 \end{bmatrix},$$

and read from Figure 7.20 that

$$X = \begin{bmatrix} 2 \\ 3 \end{bmatrix}.$$

Thus, $A_1 = 2$, $A_2 = 3$, and

$$\frac{5x - 1}{x^2 - 2x - 15} = \frac{2}{x + 3} + \frac{3}{x - 5}.$$

Support Graphically

Figure 7.21 suggests that the following two functions are the same:

$$y = \frac{5x - 1}{x^2 - 2x - 15} \quad \text{and} \quad y = \frac{2}{x + 3} + \frac{3}{x - 5}.$$

Now try Exercise 17.

EXAMPLE 3   Decomposing a Fraction with a Repeated Linear Factor

Find the partial fraction decomposition of

$$\frac{-x^2 + 2x + 4}{x^3 - 4x^2 + 4x}.$$

SOLUTION The denominator factors into $x(x - 2)^2$. Because the factor $x - 2$ is squared, it contributes two terms to the decomposition:

$$\frac{-x^2 + 2x + 4}{x^3 - 4x^2 + 4x} = \frac{A_1}{x} + \frac{A_2}{x - 2} + \frac{A_3}{(x - 2)^2}.$$

We clear fractions by multiplying both sides of the above equation by $x^3 - 4x^2 + 4x$.

$$-x^2 + 2x + 4 = A_1(x - 2)^2 + A_2 x(x - 2) + A_3 x$$

Expanding and combining like terms in the above equation we obtain:

$$-x^2 + 2x + 4 = (A_1 + A_2)x^2 + (-4A_1 - 2A_2 + A_3)x + 4A_1.$$

Comparing coefficients of powers of x on the left and right side of the above equation, we obtain the following system of equations:

$$A_1 + A_2 = -1$$
$$-4A_1 - 2A_2 + A_3 = 2$$
$$4A_1 = 4.$$

The reduced row echelon form of the augmented matrix

$$\begin{bmatrix} 1 & 1 & 0 & -1 \\ -4 & -2 & 1 & 2 \\ 4 & 0 & 0 & 4 \end{bmatrix}$$

continued

of the preceding system of equations is

$$\begin{bmatrix} 1 & 0 & 0 & 1 \\ 0 & 1 & 0 & -2 \\ 0 & 0 & 1 & 2 \end{bmatrix}.$$

Thus $A_1 = 1, A_2 = -2, A_3 = 2$, and

$$\frac{-x^2 + 2x + 4}{x^3 - 4x^2 + 4x} = \frac{1}{x} + \frac{-2}{x - 2} + \frac{2}{(x - 2)^2}.$$

Now try Exercise 25.

Sometimes we can solve for the variables introduced in a partial fraction decomposition by substituting strategic values for x, as illustrated in Exploration 1.

EXPLORATION 1 Revisiting Examples 2 and 3

1. When we cleared fractions in Example 2 we obtained the equation
$5x - 1 = A_1(x - 5) + A_2(x + 3)$.

(a) Substitute $x = 5$ into this equation and solve for A_2.

(b) Substitute $x = -3$ into this equation and solve for A_1.

2. When we cleared fractions in Example 3 we obtained the equation
$-x^2 + 2x + 4 = A_1(x - 2)^2 + A_2x(x - 2) + A_3x$.

(a) Substitute $x = 2$ into this equation and solve for A_3.

(b) Substitute $x = 0$ into this equation and solve for A_1.

(c) Substitute any other value for x and use the values found for A_1 and A_3 to solve for A_2.

Denominators with Irreducible Quadratic Factors

Example 4 shows how to find the partial fraction decomposition for a rational function whose denominator has an irreducible quadratic factor.

EXAMPLE 4 Decomposing a Fraction with an Irreducible Quadratic Factor

Find the partial fraction decomposition of

$$\frac{x^2 + 4x + 1}{x^3 - x^2 + x - 1}.$$

SOLUTION We factor the denominator by grouping terms:

$$x^3 - x^2 + x - 1 = x^2(x - 1) + (x - 1)$$

$$= (x - 1)(x^2 + 1).$$

continued

Each factor occurs once, so each one leads to one term in the decomposition:

$$\frac{x^2 + 4x + 1}{x^3 - x^2 + x - 1} = \frac{A}{x - 1} + \frac{Bx + C}{x^2 + 1}.$$

We clear fractions by multiplying both sides of the above equation by $x^3 - x^2 + x - 1$:

$$x^2 + 4x + 1 = A(x^2 + 1) + (Bx + C)(x - 1).$$

Expanding and combining like terms in the above equation we obtain:

$$x^2 + 4x + 1 = (A + B)x^2 + (-B + C)x + (A - C).$$

Comparing coefficients of powers of x on the left and right side of the above equation, we obtain the following system of equations:

$$A + B = 1$$

$$-B + C = 4$$

$$A - C = 1.$$

Using any of the techniques of Section 7.3, we find $A = 3$, $B = -2$, and $C = 2$. Thus,

$$\frac{x^2 + 4x + 1}{x^3 - x^2 + x - 1} = \frac{3}{x - 1} + \frac{-2x + 2}{x^2 + 1}.$$

Now try Exercise 31.

EXAMPLE 5   Decomposing a Fraction with a Repeated Irreducible Quadratic Factor

Find the partial fraction decomposition of

$$\frac{2x^3 - x^2 + 5x}{(x^2 + 1)^2}.$$

SOLUTION The factor $(x^2 + 1)^2$ in the denominator leads to two terms in the partial fraction decomposition:

$$\frac{2x^3 - x^2 + 5x}{(x^2 + 1)^2} = \frac{B_1x + C_1}{x^2 + 1} + \frac{B_2x + C_2}{(x^2 + 1)^2}.$$

We clear fractions by multiplying both sides of the above equation by $(x^2 + 1)^2$:

$$2x^3 - x^2 + 5x = (B_1x + C_1)(x^2 + 1) + B_2x + C_2$$

$$= B_1x^3 + C_1x^2 + (B_1 + B_2)x + (C_1 + C_2).$$

Comparing coefficients of powers of x on the left and right side of the above equation, we see that $B_1 = 2$, $C_1 = -1$, $B_1 + B_2 = 5$, and $C_1 + C_2 = 0$. It follows that $B_2 = 3$ and $C_2 = 1$. Thus,

$$\frac{2x^3 - x^2 + 5x}{(x^2 + 1)^2} = \frac{2x - 1}{x^2 + 1} + \frac{3x + 1}{(x^2 + 1)^2}.$$

Now try Exercise 29.

Applications

Each part of the partial fraction decomposition of a rational function plays a central role in the analysis of its graph. One summand can be used to describe the end behavior of the graph. The other parts can be used to describe the behavior of the graph at one of its vertical asymptotes, as illustrated in Example 6.

EXAMPLE 6 Investigating the Graph of a Rational Function

Compare the graph of the rational function

$$f(x) = \frac{2x^2 + x - 14}{x^2 - 4}$$

with the graphs of the terms in its partial fraction decomposition.

SOLUTION We use division to rewrite $f(x)$ in the form

$$f(x) = 2 + \frac{x - 6}{x^2 - 4}.$$

Then we use the techniques of this section to find the partial fraction decomposition of $(x - 6)/(x^2 - 4)$, and, in turn, that of f:

$$f(x) = 2 + \frac{x - 6}{x^2 - 4} = 2 + \frac{2}{x + 2} + \frac{-1}{x - 2}.$$

Figure 7.22 shows the graph of f. You can see the relation of this graph to the graph of the end behavior asymptote $y = 2$, one of the terms of f. The graph of the term $y = 2/(x + 2)$ is very similar to the graph of f near $x = -2$ (Figure 7.23a). The graph of the term $y = -1/(x - 2)$ is very similar to the graph of f near $x = 2$ (Figure 7.23b).

Now try Exercise 33.

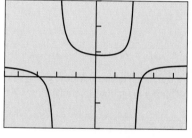

[–4.7, 4.7] by [–8, 12]

FIGURE 7.22 The graph of $f(x) = (2x^2 + x - 14)/(x^2 - 4)$. (Example 6)

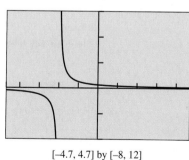

[–4.7, 4.7] by [–8, 12]

(a)

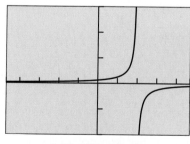

[–4.7, 4.7] by [–8, 12]

(b)

FIGURE 7.23 The graphs of (a) $y = 2/(x + 2)$ and (b) $y = -1/(x - 2)$. (Example 6)

QUICK REVIEW 7.4 *(For help, go to Sections A.2, A.3, and 2.4.)*

In Exercises 1–4, perform the indicated operations and write your answer as a single reduced fraction.

1. $\dfrac{1}{x-1} + \dfrac{2}{x-3}$

2. $\dfrac{5}{x+4} - \dfrac{2}{x+1}$

3. $\dfrac{1}{x} + \dfrac{3}{x+1} + \dfrac{1}{(x+1)^2}$

4. $\dfrac{3}{x^2+1} - \dfrac{x+1}{(x^2+1)^2}$

In Exercises 5 and 6, divide $f(x)$ by $d(x)$ to obtain a quotient $q(x)$ and remainder $r(x)$. Write a summary statement in fraction form: $q(x) + r(x)/d(x)$.

5. $f(x) = 3x^3 - 6x^2 - 2x + 7, \quad d(x) = x - 2$

6. $f(x) = 2x^3 + 3x^2 - 14x - 8, \quad d(x) = x^2 + x - 6$

In Exercises 7 and 8, write the polynomial as a product of linear and irreducible quadratic factors with real coefficients.

7. $x^4 - 2x^3 + x^2 - 8x - 12$ **8.** $x^4 - x^3 - 15x^2 - 23x - 10$

In Exercises 9 and 10, assume that $f(x) = g(x)$. What can you conclude about $A, B, C,$ and D?

9. $f(x) = Ax^2 + Bx + C + 1$

  $g(x) = 3x^2 - x + 2$

10. $f(x) = (A+1)x^3 + Bx^2 + Cx + D$

  $g(x) = -x^3 + 2x^2 - x - 5$

SECTION 7.4 EXERCISES

In Exercises 1–4, write the terms for the partial fraction decomposition of the rational function. Do not solve for the constants.

1. $\dfrac{x^2 - 7}{x(x^2 - 4)}$

2. $\dfrac{x^4 + 3x^2 - 1}{(x^2 + x + 1)^2(x^2 - x + 1)}$

3. $\dfrac{x^5 - 2x^4 + x - 1}{x^3(x-1)^2(x^2+9)}$

4. $\dfrac{x^2 + 3x + 2}{(x^3 - 1)^3}$

Exercises 5–8, use inverse matrices to find the partial fraction decomposition.

5. $\dfrac{x + 22}{(x+4)(x-2)} = \dfrac{A}{x+4} + \dfrac{B}{x-2}$

6. $\dfrac{x - 3}{x(x+3)} = \dfrac{A}{x+3} + \dfrac{B}{x}$

7. $\dfrac{3x^2 + 2x + 2}{(x^2+1)^2} = \dfrac{Ax + B}{x^2+1} + \dfrac{Cx + D}{(x^2+1)^2}$

8. $\dfrac{4x + 4}{x^2(x+2)} = \dfrac{A}{x} + \dfrac{B}{x^2} + \dfrac{C}{x+2}$

In Exercises 9–12, use the reduced row echelon form for the augmented matrix to find the partial fraction decomposition.

9. $\dfrac{x^2 - 2x + 1}{(x-2)^3} = \dfrac{A}{x-2} + \dfrac{B}{(x-2)^2} + \dfrac{C}{(x-2)^3}$

10. $\dfrac{5x^3 - 10x^2 + 5x - 5}{(x^2+4)(x^2+9)} = \dfrac{Ax + B}{x^2+4} + \dfrac{Cx + D}{x^2+9}$

11. $\dfrac{5x^5 + 22x^4 + 36x^3 + 53x^2 + 71x + 20}{(x+3)^2(x^2+2)^2}$

  $= \dfrac{A}{x+3} + \dfrac{B}{(x+3)^2} + \dfrac{Cx + D}{x^2+2} + \dfrac{Ex + F}{(x^2+2)^2}$

12. $\dfrac{-x^3 - 6x^2 - 5x + 87}{(x-1)^2(x+4)^2}$

  $= \dfrac{A}{x-1} + \dfrac{B}{(x-1)^2} + \dfrac{C}{x+4} + \dfrac{D}{(x+4)^2}$

In Exercises 13–16, find the partial fraction decomposition. Confirm your answer algebraically by combining the partial fractions.

13. $\dfrac{2}{(x-5)(x-3)}$

14. $\dfrac{4}{(x+3)(x+7)}$

15. $\dfrac{4}{x^2 - 1}$

16. $\dfrac{6}{x^2 - 9}$

In Exercises 17–20, find the partial fraction decomposition. Support your answer graphically.

17. $\dfrac{1}{x^2 + 2x}$

18. $\dfrac{-6}{x^2 - 3x}$

19. $\dfrac{-x + 10}{x^2 + x - 12}$

20. $\dfrac{7x - 7}{x^2 - 3x - 10}$

In Exercises 21–32, find the partial fraction decomposition.

21. $\dfrac{x + 17}{2x^2 + 5x - 3}$

22. $\dfrac{4x - 11}{2x^2 - x - 3}$

23. $\dfrac{2x^2 + 5}{(x^2 + 1)^2}$

24. $\dfrac{3x^2 + 4}{(x^2 + 1)^2}$

25. $\dfrac{x^2 - x + 2}{x^3 - 2x^2 + x}$

26. $\dfrac{-6x + 25}{x^3 - 6x^2 + 9x}$

27. $\dfrac{3x^2 - 4x + 3}{x^3 - 3x^2}$

28. $\dfrac{5x^2 + 7x - 4}{x^3 + 4x^2}$

29. $\dfrac{2x^3 + 4x - 1}{(x^2 + 2)^2}$

30. $\dfrac{3x^3 + 6x - 1}{(x^2 + 2)^2}$

31. $\dfrac{x^2 + 3x + 2}{x^3 - 1}$

32. $\dfrac{2x^2 - 4x + 3}{x^3 + 1}$

In Exercises 33–36, use division to write the rational function in the form $q(x) + r(x)/d(x)$, where the degree of $r(x)$ is less than the degree of $d(x)$. Then find the partial fraction decomposition of $r(x)/d(x)$. Compare the graphs of the rational function with the graphs of its terms in the partial fraction decomposition.

33. $\dfrac{2x^2 + x + 3}{x^2 - 1}$

34. $\dfrac{3x^2 + 2x}{x^2 - 4}$

35. $\dfrac{x^3 - 2}{x^2 + x}$

36. $\dfrac{x^3 + 2}{x^2 - x}$

In Exercises 37–42, match the function with its graph. Do this without using your grapher.

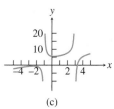

(a)

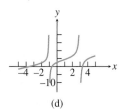

(b)

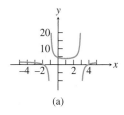

(c)

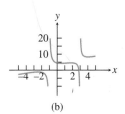

(d)

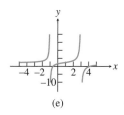

(e)

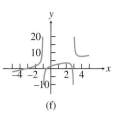

(f)

37. $y = x + 3 + \dfrac{2}{x + 1} - \dfrac{2}{x - 3}$

38. $y = x + 3 - \dfrac{1}{x + 1} + \dfrac{1}{x - 3}$

39. $y = x + 3 - \dfrac{1}{x + 1} - \dfrac{2}{x - 3}$

40. $y = x + 3 + \dfrac{2}{x + 1} + \dfrac{1}{x - 3}$

41. $y = 2 + \dfrac{2}{x + 1} - \dfrac{2}{x - 3}$

42. $y = 2 - \dfrac{1}{x + 1} - \dfrac{2}{x - 3}$

43. Group Activity Find the partial fraction decomposition of
$$\frac{1}{x(x - a)}.$$

44. Group Activity Find the partial fraction decomposition of
$$\frac{-1}{(x - 2)(x - b)}.$$

45. Group Activity Find the partial fraction decomposition of
$$\frac{3}{(x - a)(x - b)}.$$

46. Group Activity Find the partial fraction decomposition of
$$\frac{2}{x^2 - a^2}.$$

Standardized Test Questions

47. True or False If $f(x) = \dfrac{1}{x - 3} + \dfrac{1}{x^2 + 1}$,
then $\lim\limits_{x \to 3^-} f(x) = -\infty$. Justify your answer.

48. True or False If $f(x) = -1 + \dfrac{1}{x - 2} - \dfrac{1}{(x - 3)^2}$,
then $\lim\limits_{x \to \infty} f(x) = -1$. Justify your answer.

In Exercises 49–52, solve the problem without using a calculator.

49. Multiple Choice Which of the following gives the form of the partial fraction decomposition of $\dfrac{3x - 1}{x^2(x^2 + 2)}$?

(A) $\dfrac{A_1}{x} + \dfrac{B_1x + C_1}{x^2 + 2}$

(B) $\dfrac{B_1x + C_1}{x^2 + 2}$

(C) $\dfrac{A_1}{x} + \dfrac{A_2}{x^2} + \dfrac{B_1}{x^2 + 2}$

(D) $\dfrac{A_1}{x} + \dfrac{A_2}{x^2} + \dfrac{B_1x}{x^2 + 2}$

(E) $\dfrac{A_1}{x} + \dfrac{A_2}{x^2} + \dfrac{B_1x + C_1}{x^2 + 2}$

50. Multiple Choice Which of the following gives the form of the partial fraction decomposition of $\dfrac{2x^2 - x + 1}{(x + 3)^2(x^2 + 4)^2}$?

(A) $\dfrac{A_1}{x + 3} + \dfrac{B_1x + C_1}{x^2 + 4}$

(B) $\dfrac{A_1}{x + 3} + \dfrac{B_1x + C_1}{x^2 + 4} + \dfrac{B_2x + C_2}{(x^2 + 4)^2}$

(C) $\dfrac{A_1}{x + 3} + \dfrac{A_2}{(x + 3)^2} + \dfrac{B_1x + C_1}{x^2 + 4} + \dfrac{B_2x + C_2}{(x^2 + 4)^2}$

(D) $\dfrac{A_1}{x + 3} + \dfrac{A_2}{(x + 3)^2} + \dfrac{B_1x + C_1}{x^2 + 4} + \dfrac{(B_2x + C_2)^2}{x^2 + 4}$

(E) $\dfrac{A_1}{x + 3} + \dfrac{A_2}{(x + 3)^2} + \dfrac{B_1}{x^2 + 4} + \dfrac{B_2}{(x^2 + 4)^2}$

51. Multiple Choice Which of the following could be the

graph of $3 + \dfrac{2}{x-2} - \dfrac{3}{x+1}$?

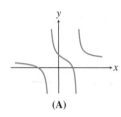

(A)

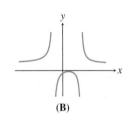

(B)

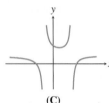

(C)

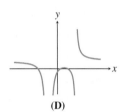

(D)

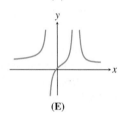

(E)

52. Multiple Choice Which of the following could be the graph

of $-2 - \dfrac{3}{x-1} - \dfrac{1}{(x+2)^2}$?

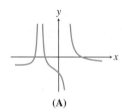

(A)

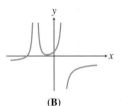

(B)

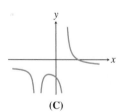

(C)

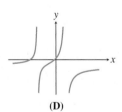

(D)

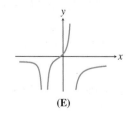

(E)

Explorations

53. Revisiting Example 3 When we cleared fractions in Example 3 we obtained the equation

$$x^2 + 4x + 1 = A(x^2 + 1) + (Bx + C)(x - 1).$$

(a) Substitute $x = 1$ into this equation and solve for A.

(b) Substitute $x = i$ and $x = -i$ into this equation to find a system of 2 equations to solve for B and C.

54. Writing to Learn Explain why it is valid in this section to obtain the systems of equations by equating coefficients of powers of x.

Extending the Ideas

55. Group Activity Examine the graph of

$$f(x) = \frac{a}{x-1} + \frac{b}{(x-1)^2} \quad \text{for}$$

(i) $a = b = 1$ **(ii)** $a = 1, b = -1$

(iii) $a = -1, b = 1$ **(iv)** $a = -1, b = -1$

Based on this examination, which of the two functions $y = a/(x - 1)$ or $y = b/(x - 1)^2$ has the greater effect on the graph of $f(x)$ near $x = 1$?

56. Writing to Learn Use partial fraction decomposition to explain why the graphs of

$$f(x) = \frac{2x - 3}{(x - 1)^2} \quad \text{and} \quad g(x) = \frac{2x + 3}{(x - 1)^2}$$

are so different near $x = 1$.

7.5
Systems of Inequalities in Two Variables

What you'll learn about
- Graph of an Inequality
- Systems of Inequalities
- Linear Programming

... and why
Linear programming is used in business and industry to maximize profits, minimize costs, and to help management make decisions.

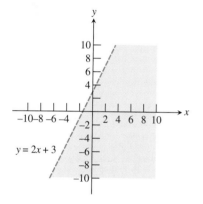

FIGURE 7.24 A graph of $y = 2x + 3$ (dashed line) and $y < 2x + 3$ (shaded area). The line is dashed to indicate it is *not* part of the solution of $y < 2x + 3$.

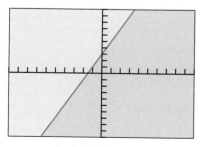

[–10, 10] by [–10, 10]

FIGURE 7.25 The graph of $y \geq 2x + 3$. (Example 1)

Graph of an Inequality

An ordered pair (a, b) of real numbers is a **solution of an inequality** in x and y if the substitution $x = a$ and $y = b$ satisfies the inequality. For example, the ordered pair $(2, 5)$ is a solution of $y < 2x + 3$ because

$$5 < 2(2) + 3 = 7.$$

However, the ordered pair $(2, 8)$ is *not* a solution because

$$8 \not< 2(2) + 3 = 7.$$

When we have found all the solutions we have **solved the inequality**.

The **graph of an inequality** in x and y consists of all pairs (x, y) that are solutions of the inequality. The graph of an inequality involving two variables typically is a region of the coordinate plane.

The point $(2, 7)$ is on the graph of the line $y = 2x + 3$ but is not a solution of $y < 2x + 3$. A point $(2, y)$ below the line $y = 2x + 3$ is on the graph of $y < 2x + 3$, and those above it are not. The graph of $y < 2x + 3$ is the set of all points below the line $y = 2x + 3$. The graph of the line $y = 2x + 3$ is the *boundary* of the region (Figure 7.24).

We can summarize our observations about the graph of an inequality in two variables with the following procedure.

Steps for Drawing the Graph of an Inequality in Two Variables

1. Draw the graph of the equation obtained by replacing the inequality sign by an equal sign. Use a dashed line if the inequality is $<$ or $>$. Use a solid line if the inequality is $\leq$ or $\geq$.

2. Check a point in each of the two regions of the plane determined by the graph of the equation. If the point satisfies the inequality, then shade the region containing the point.

EXAMPLE 1 Graphing a Linear Inequality

Draw the graph of $y \geq 2x + 3$. State the boundary of the region.

SOLUTION

Step 1. Because of "$\geq$," the graph of the line $y = 2x + 3$ is part of the graph of the inequality and should be drawn as a solid line.

Step 2. The point $(0, 4)$ is above the line and satisfies the inequality because

$$4 \geq 2(0) + 3 = 3.$$

Thus, the graph of $y \geq 2x + 3$ consists of all points on or above the line $y = 2x + 3$. The boundary is the graph of $y = 2x + 3$ (Figure 7.25).

Now try Exercise 9.

The graph of the linear inequality $y \geq ax + b$, $y > ax + b$, $y \leq ax + b$, or $y < ax + b$ is a **half-plane**. The graph of the line $y = ax + b$ is the **boundary** of the region.

EXAMPLE 2 Graphing Linear Inequalities

Draw the graph of the inequality. State the boundary of the region.

(a) $x \geq 2$ **(b)** $y < -3$

SOLUTION

(a) Step 1. Replacing "$\geq$" by "$=$" we obtain the equation $x = 2$ whose graph is a vertical line.

    Step 2. The graph of $x \geq 2$ is the set of all points on and to the right of the vertical line $x = 2$ (Figure 7.26a). The line $x = 2$ is the boundary of the region.

(b) Step 1. Replacing "$<$" by "$=$" we obtain the equation $y = -3$ whose graph is a horizontal line.

    Step 2. The graph of $y < -3$ is the set of all points below the horizontal line $y = -3$ (Figure 7.26b). The line $y = -3$ is the boundary of the region.

Now try Exercise 7.

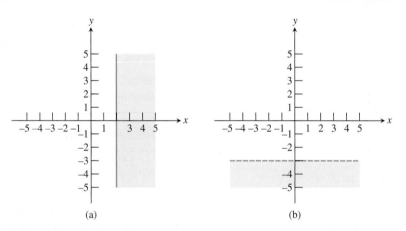

FIGURE 7.26 The graphs of (a) $x \geq 2$ and (b) $y < -3$. (Example 2)

EXAMPLE 3 Graphing a Quadratic Inequality

Draw the graph of $y \geq x^2 - 3$. State the boundary of the region.

SOLUTION

Step 1. Replacing "$\geq$" by "$=$" we obtain the equation $y = x^2 - 3$ whose graph is a parabola.

Step 2. The pair $(0, 1)$ is a solution of the inequality because

$$1 \geq (0)^2 - 3 = -3.$$

Thus, the graph of $y \geq x^2 - 3$ is the parabola together with the region inside the parabola (Figure 7.27). The parabola is the boundary of the region.

Now try Exercise 11.

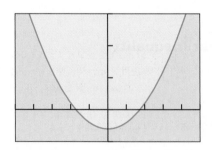

[–5, 5] by [–5, 15]

FIGURE 7.27 The graph of $y \geq x^2 - 3$. (Example 3)

Systems of Inequalities

A **solution** of a system of inequalities in x and y is an ordered pair (x, y) that satisfies each inequality in the system. When we have found all the common solutions we have **solved the system of inequalities**.

The technique for solving a system of inequalities graphically is similar to that for solving a system of equations graphically. We graph each inequality and determine the points common to the individual graphs.

EXAMPLE 4   Solving a System of Inequalities Graphically

Solve the system.

$$y > x^2$$

$$2x + 3y < 4$$

SOLUTION The graph of $y > x^2$ is shaded in Figure 7.28a. It does not include its boundary $y = x^2$. The graph of $2x + 3y < 4$ is shaded in Figure 7.28b. It does not include its boundary $2x + 3y = 4$. The solution to the system is the intersection of these two graphs, as shaded in Figure 7.28c.

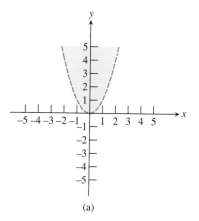

(a)

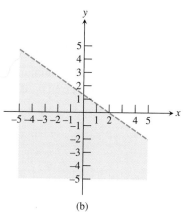

(b)

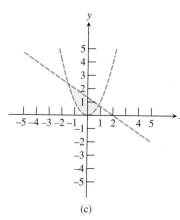

(c)

FIGURE 7.28 The graphs of (a) $y > x^2$, (b) $2x + 3y < 4$, and (c) the system of Example 4.

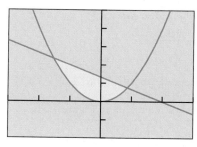

[–3, 3] by [–2, 5]

FIGURE 7.29 The solution of the system in Example 4. Most graphers cannot distinguish between dashed and solid boundaries.

Support with a Grapher

Figure 7.29 shows what our grapher produces when we shade above the curve $y = x^2$ and below the curve $2x + 3y = 4$. The shaded portion appears to be identical to the shaded portion in Figure 7.28c.   ***Now try Exercise 19.***

EXAMPLE 5   Solving a System of Inequalities

Solve the system.

$$2x + y \leq 10$$

$$2x + 3y \leq 14$$

$$x \geq 0$$

$$y \geq 0$$

continued

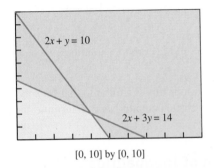

[0, 10] by [0, 10]

FIGURE 7.30 The solution (shaded) of the system in Example 5. The boundary points are included.

SOLUTION The solution is in the first quadrant because $x \geq 0$ and $y \geq 0$, it lies below each of the two lines $2x + y = 10$ and $2x + 3y = 14$, and includes all of its boundary points (Figure 7.30).

Now try Exercise 23.

Linear Programming

Sometimes decision making in management science requires that we find a minimum or a maximum of a linear function

$$f = a_1 x_1 + a_2 x_2 + \cdots + a_n x_n,$$

called an **objective function**, over a set of points. Such a problem is a **linear programming problem**. In two dimensions, the function f takes the form $f = ax + by$ and the set of points is the solution of a system of inequalities, called **constraints**, such as the one in Figure 7.30. The solution of the system of inequalities is the set of **feasible** *xy* points for the optimization problem.

It can be shown that if a linear programming problem has a solution it occurs at one of the **vertex points**, or **corner points**, along the boundary of the region. We use this information in Examples 6 and 7.

EXAMPLE 6 Solving a Linear Programming Problem

Find the maximum and minimum values of the objective function $f = 5x + 8y$, subject to the constraints given by the system of inequalities.

$$2x + y \leq 10$$

$$2x + 3y \leq 14$$

$$x \geq 0$$

$$y \geq 0$$

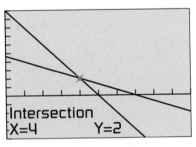

[0, 10] by [−5, 10]

FIGURE 7.31 The lines $2x + 3y = 14$ and $2x + y = 10$ intersect at $(4, 2)$. (Example 6)

SOLUTION The feasible *xy* points are those graphed in Figure 7.30. Figure 7.31 shows that the two lines $2x + 3y = 14$ and $2x + y = 10$ intersect at $(4, 2)$. The corner points are:

$(0, 0)$,

$(0, 14/3)$, the y-intercept of $2x + 3y = 14$,

$(5, 0)$, the x-intercept of $2x + y = 10$, and

$(4, 2)$, the point of intersection of $2x + 3y = 14$ and $2x + y = 10$.

The following table evaluates f at the corner points of the region in Figure 7.31.

(x, y)	$(0, 0)$	$(0, 14/3)$	$(4, 2)$	$(5, 0)$
f	0	112/3	36	25

The maximum value of f is 112/3 at $(0, 14/3)$. The minimum value is 0 at $(0, 0)$.

Now try Exercise 31.

Here is one way to analyze the linear programming problem in Example 6. By assigning positive values to f in $f = 5x + 8y$, we obtain a family of parallel lines whose distance from the origin increases as f increases. (See Exercise 47.) This family of lines sweeps across the region of feasible solutions. Geometrically, we can see that there is a minimum and maximum value for f if the line $f = 5x + 8y$ is to intersect the region of feasible solutions.

EXAMPLE 7   Purchasing Fertilizer

Johnson's Produce is purchasing fertilizer with two nutrients: N (nitrogen) and P (phosphorous). They need at least 180 units of N and 90 units of P. Their supplier has two brands of fertilizer for them to buy. Brand A costs $10 a bag and has 4 units of N and 1 unit of P. Brand B costs $5 a bag and has 1 unit of each nutrient. Johnson's Produce can pay at most $800 for the fertilizer. How many bags of each brand should be purchased to minimize cost?

SOLUTION

Model

$$\text{Let } x = \text{number of bags of Brand A}$$

$$\text{Let } y = \text{number of bags of Brand B}$$

Then C = the total cost = $10x + 5y$ is the objective function to be minimized. The constraints are:

$4x + y \geq 180$	Amount of N is at least 180.
$x + y \geq 90$	Amount of P is at least 90.
$10x + 5y \leq 800$	Total cost to be at most $800.
$x \geq 0, y \geq 0$	

Solve Graphically

The region of feasible xy points is the intersection of the graphs of $4x + y \geq 180$, $x + y \geq 90$, and $10x + 5y \leq 800$ in the first quadrant (Figure 7.32).

The region has three corner points at the points of intersections of the three lines $4x + y = 180$, $x + y = 90$, and $10x + 5y = 800$: (10, 140), (70, 20), and (30, 60). The values of the objective function C at the corner points are:

$$C(10, 140) = 10(10) + 5(140) = 800$$

$$C(70, 20) = 10(70) + 5(20) = 800$$

$$C(30, 60) = 10(30) + 5(60) = 600$$

Interpret

The minimum cost for the fertilizer is $600 when 30 bags of Brand A and 60 bags of Brand B are purchased. For this purchase, Johnson's Produce gets exactly 180 units of nutrient N and 90 units of nutrient P.   *Now try Exercise 37.*

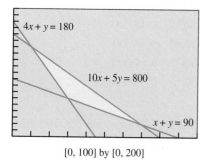

$4x + y = 180$

$10x + 5y = 800$

$x + y = 90$

[0, 100] by [0, 200]

FIGURE 7.32 The feasible region in Example 7.

The region in Example 8 is unbounded. Using the discussion following Example 6, we can see geometrically that the linear programming problem in Example 8 does not have a maximum value but, fortunately, does have a minimum value.

EXAMPLE 8   Minimizing Operating Cost

Gonza Manufacturing has two factories that produce three grades of paper: low grade, medium grade, and high grade. It needs to supply 24 tons of low grade, 6 tons of medium grade, and 30 tons of high grade paper. Factory A produces 8 tons of low grade, 1 ton of medium grade, 2 tons of high grade paper daily, and costs $2000 per day to operate. Factory B produces 2 tons of low grade, 1 ton of medium grade, 8 tons of high grade paper daily, and costs $4000 per day to operate. How many days should each factory operate to fill the orders at minimum cost?

SOLUTION

Model

Let x = the number of days Factory A operates.

Let y = the number of days Factory B operates.

Then C = total operating cost = $2000x + 4000y$ is the objective function to be minimized. The constraints are:

$$8x + 2y \geq 24 \qquad \text{Amount of low grade is at least 24.}$$
$$x + y \geq 6 \qquad \text{Amount of medium grade is at least 6.}$$
$$2x + 8y \geq 30 \qquad \text{Amount of high grade is at least 30.}$$
$$x \geq 0, y \geq 0$$

Solve Graphically

The region of feasible points is the intersection of the graphs of $8x + 2y \geq 24$, $x + y \geq 6$, and $2x + 8y \geq 30$ in the first quadrant (Figure 7.33).

The region has four corner points:

(0, 12), the y-intercept of $8x + 2y = 24$,

(2, 4), the point of intersection of $8x + 2y = 24$ and $x + y = 6$,

(3, 3), the point of intersection of $x + y = 6$ and $2x + 8y = 30$,

(15, 0), the x-intercept of $2x + 8y = 30$.

The values of the objective function C at the corner points are:

$$C(0, 12) = 2000(0) + 4000(12) = 48{,}000$$
$$C(2, 4) = 2000(2) + 4000(4) = 20{,}000$$
$$C(3, 3) = 2000(3) + 4000(3) = 18{,}000$$
$$C(15, 0) = 2000(15) + 4000(0) = 30{,}000$$

Interpret

The minimum operational cost is $18,000 when the two factories are operated for 3 days each. The two factories will produce 30 tons of low grade, 6 tons of medium grade, and 30 tons of high grade. They will have a surplus of 6 tons of low grade paper.

Now try Exercise 39.

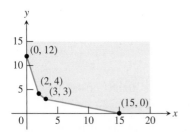

FIGURE 7.33 The graph of the feasible points in Example 8.

QUICK REVIEW 7.5 *(For help, go to Sections P.4 and 7.1.)*

In Exercises 1–4, find the *x*- and *y*-intercepts of the line and draw its graph.

1. $2x - 3y = 6$

2. $5x + 10y = 30$

3. $\dfrac{x}{20} + \dfrac{y}{50} = 1$

4. $\dfrac{x}{30} - \dfrac{y}{20} = 1$

In Exercises 5–9, find the point of intersection of the two lines. (We will use these values in Examples 7 and 8.)

5. $4x + y = 180$ and $x + y = 90$

6. $x + y = 90$ and $10x + 5y = 800$

7. $4x + y = 180$ and $10x + 5y = 800$

8. $8x + 2y = 24$ and $x + y = 6$

9. $x + y = 6$ and $2x + 8y = 30$

10. Solve the system of equations:

$$y = x^2$$
$$2x + 3y = 4$$

SECTION 7.5 EXERCISES

In Exercises 1–6, match the inequality with its graph. Indicate whether the boundary is included in or excluded from the graph. All graphs are drawn in $[-4.7, 4.7]$ by $[-3.1, 3.1]$.

1. $x \le 3$

2. $y > 2$

3. $2x - 5y \ge 2$

4. $y > (1/2)x^2 - 1$

5. $y \ge 2 - x^2$

6. $x^2 + y^2 < 4$

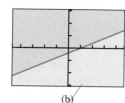

(a)

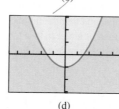

(b)

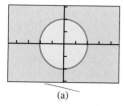

(c)

(d)

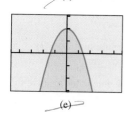

(e)

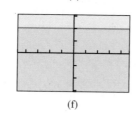

(f)

In Exercises 7–16, graph the inequality. State the boundary of the region.

7. $x \le 4$

8. $y \ge -3$

9. $2x + 5y \le 7$

10. $3x - y > 4$

11. $y < x^2 + 1$

12. $y \ge x^2 - 3$

13. $x^2 + y^2 < 9$

14. $x^2 + y^2 \ge 4$

15. $y \ge \dfrac{e^x + e^{-x}}{2}$

16. $y < \sin x$

In Exercises 17–22, solve the system of inequalities.

17. $5x - 3y > 1$
$3x + 4y \le 18$

18. $4x + 3y \le -6$
$2x - y \le -8$

19. $y \le 2x + 3$
$y \ge x^2 - 2$

20. $x - 3y - 6 < 0$
$y > -x^2 - 2x + 2$

21. $y \ge x^2$
$x^2 + y^2 \le 4$

22. $x^2 + y^2 \le 9$
$y \ge |x|$

In Exercises 23–26, solve the system of inequalities.

23. $2x + y \le 80$
$x + 2y \le 80$
$x \ge 0$
$y \ge 0$

24. $3x + 8y \ge 240$
$9x + 4y \ge 360$
$x \ge 6$
$y \ge 0$

25. $5x + 2y \le 20$
$2x + 3y \le 18$
$x + y \ge 2$
$x \ge 0$
$y \ge 0$

26. $7x + 3y \le 210$
$3x + 7y \le 210$
$x + y \ge 30$

In Exercises 27–30, write a system of inequalities whose solution is the region shaded in the given figure. All boundaries are to be included.

27. Group Activity

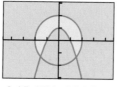

$[-4.7, 4.7]$ by $[-3.1, 3.1]$

28. Group Activity

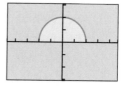

$[-4.7, 4.7]$ by $[-3.1, 3.1]$

29. Group Activity

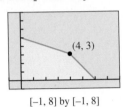

[−1, 8] by [−1, 8]

30. Group Activity

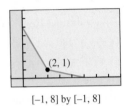

[−1, 8] by [−1, 8]

In Exercises 31–36, find the minimum and maximum, if they exist, of the objective function f, subject to the constraints.

31. Objective function: $f = 4x + 3y$
Constraints:
$$x + y \leq 80$$
$$x - 2y \leq 0$$
$$x \geq 0, y \geq 0$$

32. Objective function: $f = 10x + 11y$
Constraints:
$$x + y \leq 90$$
$$3x - y \geq 0$$
$$x \geq 0, y \geq 0$$

33. Objective function: $f = 7x + 4y$
Constraints:
$$5x + y \geq 60$$
$$x + 6y \geq 60$$
$$4x + 6y \geq 204$$
$$x \geq 0, y \geq 0$$

34. Objective function: $f = 15x + 25y$
Constraints:
$$3x + 4y \geq 60$$
$$x + 8y \geq 40$$
$$11x + 28y \leq 380$$
$$x \geq 0, y \geq 0$$

35. Objective function: $f = 5x + 2y$
Constraints:
$$2x + y \geq 12$$
$$4x + 3y \geq 30$$
$$x + 2y \geq 10$$
$$x \geq 0, y \geq 0$$

36. Objective function: $f = 3x + 5y$
Constraints:
$$3x + 2y \geq 20$$
$$5x + 6y \geq 52$$
$$2x + 7y \geq 30$$
$$x \geq 0, y \geq 0$$

37. Mining Ore Pearson's Metals mines two ores: R and S. The company extracts minerals A and B from each type of ore. It costs $50 per ton to extract 80 lb of A and 160 lb of B from ore R. It costs $60 per ton to extract 140 lb of A and 50 lb of B from ore S. Pearson's must produce at least 4000 lb of A and 3200 lb of B. How much of each ore should be processed to minimize cost? What is the minimum cost?

38. Planning a Diet Paul's diet is to contain at least 24 units of carbohydrates and 16 units of protein. Food substance A costs $1.40 per unit and each unit contains 3 units of carbohydrates and 4 units of protein. Food substance B costs $0.90 per unit and each unit contains 2 units of carbohydrates and 1 unit of protein. How many units of each food substance should be purchased in order to minimize cost? What is the minimum cost?

39. Producing Gasoline Two oil refineries produce three grades of gasoline: A, B, and C. At each refinery, the three grades of gasoline are produced in a single operation in the following proportions: Refinery 1 produces 1 unit of A, 2 units of B, and 1 unit of C; Refinery 2 produces 1 unit of A, 4 units of B, and 4 units of C. For the production of one operation, Refinery 1 charges $300 and Refinery 2 charges $600. A customer needs 100 units of A, 320 units of B, and 200 units of C. How should the orders be placed if the customer is to minimize his cost?

40. Maximizing Profit A manufacturer wants to maximize the profit for two products. Product A yields a profit of $2.25 per unit, and product B yields a profit of $2.00 per unit. Demand information requires that the total number of units produced be no more than 3000 units, and that the number of units of product B produced be greater than or equal to half the number of units of product A produced. How many of each unit should be produced to maximize profit?

Standardized Test Questions

41. True or False The graph of a linear inequality in x and y is a half-line. Justify your answer.

42. True or False The boundary of the solution of $2x - 3y < 5$ is the graph of $3y = 2x - 5$. Justify your answer.

In Exercises 43–46, you may use a graphing calculator to solve the problem.

For Exercises 43–44, use the figure below which shows the graphs of the two lines $3x + 4y = 5$ and $2x - 3y = 4$.

43. Multiple Choice Which of the following represents the solution of the system $3x + 4y \geq 5$
$$2x - 3y \leq 4?$$

(A) Region I plus its boundary

(B) Region I without its boundary

(C) Region II plus its boundary

(D) Region II without its boundary

(E) Region IV plus its boundary

44. Multiple Choice Which of the following represents the solution of the system $3x + 4y < 5$
$$2x - 3y > 4?$$

(A) Region II plus its boundary

(B) Region III plus its boundary

(C) Region III without its boundary

(D) Region IV plus its boundary

(E) Region IV without its boundary

Exercises 45–46 refer to the following linear programming problem:
Objective function: $f = 5x + 10y$
Constraints:
$$2x + y \leq 10$$
$$x + 3y \leq 12$$
$$x \geq 0, y \geq 0$$

45. Multiple Choice Which of the following is not a corner point?

(A) $(0, 0)$ (B) $(5, 0)$

(C) $(0, 4)$ (D) $(3, 4)$

(E) $(3.6, 2.8)$

46. Multiple Choice What is the maximum value of f in the feasible region of the problem?

(A) 0   (B) 25   (C) 40   (D) 46   (E) 55

Explorations

47. Revisiting Example 6 Consider the objective function $f = 5x + 8y$ of Example 6.

(a) Prove that for any two real number values for f, the two lines are parallel.

(b) **Writing to Learn** For $f > 0$, give reasons why the line moves further away from the origin as the value of f increases.

(c) **Writing to Learn** Give a geometric explanation of why the region of Example 6 must contain a minimum and a maximum value for f.

48. Writing to Learn Describe all the possible ways that two distinct parabolas of the form $y = f(x)$ can intersect. Give examples.

Extending the Ideas

49. Implicit Functions The equation
$$\frac{x^2}{9} + \frac{y^2}{4} = 1$$
defines y as two implicit functions of x. Solve for y to find the two functions and draw the graph of the equation.

50. Implicit Functions The equation
$$x^2 - y^2 = 4$$
defines y as two implicit functions of x. Solve for y to find the two functions and draw the graph of the equation.

51. Solve the system of inequalities:
$$\frac{x^2}{9} + \frac{y^2}{4} \leq 1$$
$$y \geq x^2 - 1$$
[*Hint*: See Exercise 49.]

52. Graph the inequality $x^2 - y^2 \leq 4$. [*Hint*: See Exercise 50.]

Math at Work

I got into electrical engineering because I enjoy working on computers. Also, it is a skill one can use to get a good job.

One of the ways I use mathematics in my field is when sending a picture between computers. A picture on a computer is made of pixels, which are tiny little dots of color. There are many pixels in a picture, and therefore, storing the picture as the sum total of its pixels makes for a very large file. This large file would take a very long time to send to another computer. Therefore, a mathematical model is used to represent the picture. The mathematical model can use one symbol to represent many pixels, so that

only that symbol needs to be sent, instead of many pixels. Then, at the other computer, the picture can be translated back into the pixels.

Another way we represent a picture is by textures. Different mathematical models can be used to represent different textures, and then the computer can differentiate between different textures in one picture.

Ngao Mayuma

CHAPTER 7 **Key Ideas**

PROPERTIES, THEOREMS, AND FORMULAS

Matrix Operations   580; 582
Theorem Inverses of $n \times n$ Matrices   586
Properties of Matrices   587
Theorem Invertible Square Linear Systems   601

PROCEDURES

Solving Systems of Equations Algebraically   568, 569, 594, 600
Partial Fraction Decomposition of $\dfrac{f(x)}{d(x)}$ 608

CHAPTER 7 **Review Exercises**

The collection of exercises marked in red could be used as a chapter test.

In Exercises 1 and 2, find **(a)** $A + B$, **(b)** $A - B$, **(c)** $-2A$, and **(d)** $3A - 2B$.

1. $A = \begin{bmatrix} -1 & 3 \\ 4 & 0 \end{bmatrix}$, $B = \begin{bmatrix} 2 & -1 \\ 4 & 3 \end{bmatrix}$

2. $A = \begin{bmatrix} 2 & 3 & -1 & 2 \\ 1 & 4 & -2 & -3 \\ 0 & -3 & 2 & 1 \end{bmatrix}$, $B = \begin{bmatrix} -1 & 2 & 0 & 4 \\ 2 & -1 & 3 & 3 \\ -2 & 4 & 1 & 3 \end{bmatrix}$

In Exercises 3–8, find the products AB and BA, or state that a given product is not possible.

3. $A = \begin{bmatrix} -1 & 4 \\ 0 & 6 \end{bmatrix}$, $B = \begin{bmatrix} 3 & -1 & 5 \\ 0 & -2 & 4 \end{bmatrix}$

4. $A = \begin{bmatrix} -1 & 2 \\ 3 & -1 \\ 4 & 3 \end{bmatrix}$, $B = \begin{bmatrix} -2 & 3 & 1 \\ 2 & 1 & 0 \\ -1 & 2 & -3 \end{bmatrix}$

5. $A = \begin{bmatrix} -1 & 4 \end{bmatrix}$, $B = \begin{bmatrix} 5 & -3 \\ 2 & 1 \end{bmatrix}$

6. $A = \begin{bmatrix} -1 & 1 \\ 0 & 1 \end{bmatrix}$, $B = \begin{bmatrix} 3 & -4 \\ 1 & 2 \\ 3 & 1 \\ 1 & 1 \end{bmatrix}$

7. $A = \begin{bmatrix} 0 & 1 & 0 \\ 1 & 0 & 0 \\ 0 & 0 & 1 \end{bmatrix}$, $B = \begin{bmatrix} 2 & -3 & 4 \\ 1 & 2 & -3 \\ -2 & 1 & -1 \end{bmatrix}$

8. $A = \begin{bmatrix} 0 & 1 & 0 & 0 \\ 1 & 0 & 0 & 0 \\ 0 & 0 & 0 & 1 \\ 0 & 0 & 1 & 0 \end{bmatrix}$, $B = \begin{bmatrix} -2 & 1 & 0 & 1 \\ 3 & 0 & 2 & 1 \\ -1 & 1 & 2 & -1 \\ 3 & -2 & 1 & 0 \end{bmatrix}$

In Exercises 9 and 10, use multiplication to verify that the matrices are inverses.

9. $A = \begin{bmatrix} 1 & -2 & 1 & 1 \\ 1 & -1 & 0 & 3 \\ 1 & -1 & 2 & 2 \\ 2 & -4 & 2 & 3 \end{bmatrix}$, $B = \begin{bmatrix} 8 & 1.5 & 0.5 & -4.5 \\ 2 & 0.5 & 0.5 & -1.5 \\ -1 & -0.5 & 0.5 & 0.5 \\ -2 & 0 & 0 & 1 \end{bmatrix}$

10. $A = \begin{bmatrix} -1 & 1 & 1 \\ 2 & 1 & 0 \\ -1 & 0 & 2 \end{bmatrix}$, $B = \begin{bmatrix} -0.4 & 0.4 & 0.2 \\ 0.8 & 0.2 & -0.4 \\ -0.2 & 0.2 & 0.6 \end{bmatrix}$

In Exercises 11 and 12, find the inverse of the matrix if it has one. If it does, use multiplication to support your result.

11. $\begin{bmatrix} 1 & 2 & 0 & -1 \\ 2 & -1 & 1 & 2 \\ 2 & 0 & 1 & 2 \\ -1 & 1 & 1 & 4 \end{bmatrix}$ **12.** $\begin{bmatrix} -1 & 0 & 1 \\ 2 & -1 & 1 \\ 1 & 1 & 1 \end{bmatrix}$

In Exercises 13 and 14, evaluate the determinant of the matrix.

13. $\begin{bmatrix} 1 & -3 & 2 \\ 2 & 4 & -1 \\ -2 & 0 & 1 \end{bmatrix}$ **14.** $\begin{bmatrix} -2 & 3 & 0 & 1 \\ 3 & 0 & 2 & 0 \\ 5 & 2 & -3 & 4 \\ 1 & -1 & 2 & 3 \end{bmatrix}$

In Exercises 15–18, find the reduced row echelon form of the matrix.

15. $\begin{bmatrix} 1 & 0 & 2 \\ 3 & 1 & 5 \\ 1 & -1 & 3 \end{bmatrix}$ **16.** $\begin{bmatrix} 2 & 1 & 1 & 1 \\ -3 & -1 & -2 & 1 \\ 5 & 2 & 2 & 3 \end{bmatrix}$

17. $\begin{bmatrix} 1 & 2 & 3 & 1 \\ 2 & 3 & 3 & -2 \\ 1 & 2 & 4 & 6 \end{bmatrix}$ **18.** $\begin{bmatrix} 1 & -2 & 0 & 4 \\ -2 & 5 & 3 & -6 \\ 2 & -4 & 1 & 9 \end{bmatrix}$

In Exercises 19–22, state whether the system of equations has a solution. If it does, solve the system.

19. $3x - y = 1$
$x + 2y = 5$

20. $x - 2y = -1$
$-2x + y = 5$

21. $x + 2y = 1$
$4y - 4 = -2x$

22. $x - 2y = 9$
$3y - \dfrac{3}{2}x = -9$

In Exercises 23–28, use Gaussian elimination to solve the system of equations.

23.
$$x + z + w = 2$$
$$x + y + z = 3$$
$$3x + 2y + 3z + w = 8$$

24.
$$x + w = -2$$
$$x + y + z + 2w = -2$$
$$-x - 2y - 2z - 3w = 2$$

25.
$$x + y - 2z = 2$$
$$3x - y + z = 4$$
$$-2x - 2y + 4z = 6$$

26.
$$x + y - 2z = 2$$
$$3x - y + z = 1$$
$$-2x - 2y + 4z = -4$$

27.
$$-x - 6y + 4z - 5w = -13$$
$$2x + y + 3z - w = 4$$
$$2x + 2y + 2z = 6$$
$$-x - 3y + z - 2w = -7$$

28.
$$-x + 2y + 2z - w = -4$$
$$y + z = -1$$
$$-2x + 2y + 2z - 2w = -6$$
$$-x + 3y + 3z - w = -5$$

In Exercises 29–32, solve the system of equations by using inverse matrices.

29.
$$x + 2y + z = -1$$
$$x - 3y + 2z = 1$$
$$2x - 3y + z = 5$$

30.
$$x + 2y - z = -2$$
$$2x - y + z = 1$$
$$x + y - 2z = 3$$

31.
$$2x + y + z - w = 1$$
$$2x - y + z - w = -2$$
$$-x + y - z + w = -3$$
$$x - 2y + z - w = 1$$

32.
$$x - 2y + z - w = 2$$
$$2x + y - z - w = -1$$
$$x - y + 2z - w = -1$$
$$x + 3y - z + w = 4$$

In Exercises 33–36, solve the system of equations by finding the reduced row echelon form of the augmented matrix.

33.
$$x + 2y - 2z + w = 8$$
$$2x + 3y - 3z + 2w = 13$$

34.
$$x + 2y - 2z + w = 8$$
$$2x + 7y - 7z + 2w = 25$$
$$x + 3y - 3z + w = 11$$

35. $x + 2y + 4z + 6w = 6$
$$3x + 4y + 8z + 11w = 11$$
$$2x + 4y + 7z + 11w = 10$$
$$3x + 5y + 10z + 14w = 15$$

36.
$$x + 2z - 2w = 5$$
$$2x + y + 4z - 3w = 7$$
$$4x + y + 7z - 6w = 15$$
$$2x + y + 5z - 4w = 9$$

In Exercises 37 and 38, find the equilibrium point for the demand and supply curves.

37. $p = 100 - x^2$
$$p = 20 + 3x$$

38. $p = 80 - \dfrac{1}{10}x^2$
$$p = 5 + 4x$$

In Exercises 39–44, solve the system of equations graphically.

39. $3x - 2y = 5$
$$2x + y = -2$$

40. $y = x - 1.5$
$$y = 0.5x^2 - 3$$

41. $y = -0.5x^2 + 3$
$$y = 0.5x^2 - 1$$

42. $x^2 + y^2 = 4$
$$y = 2x^2 - 3$$

43. $y = 2\sin x$
$$y = 2x - 3$$

44. $y = \ln 2x$
$$y = 2x^2 - 12x + 15$$

In Exercises 45 and 46, find the coefficients of the function so that its graph goes through the given points.

45. Curve Fitting $f(x) = ax^3 + bx^2 + cx + d$
$$(2, 8), \quad (4, 5), \quad (6, 3), \quad (9, 4)$$

46. Curve Fitting $f(x) = ax^4 + bx^3 + cx^2 + dx + e$
$$(-2, -4), \quad (1, 2), \quad (3, 6), \quad (4, -2), \quad (7, 8)$$

In Exercises 47–52, find the partial fraction decomposition of the rational function.

47. $\dfrac{3x - 2}{x^2 - 3x - 4}$

48. $\dfrac{x - 16}{x^2 + x - 2}$

49. $\dfrac{3x + 5}{x^3 + 4x^2 + 5x + 2}$

50. $\dfrac{3(3 + 2x + x^2)}{x^3 + 3x^2 - 4}$

51. $\dfrac{5x^2 - x - 2}{x^3 + x^2 + x + 1}$

52. $\dfrac{-x^2 - 5x + 2}{x^3 + 2x^2 + 4x + 8}$

In Exercises 53–56, match the function with its graph. Do this without using your grapher.

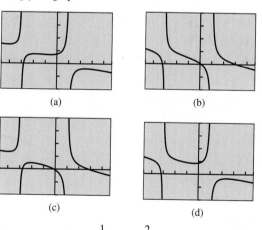

(a)

(b)

(c)

(d)

53. $y = -x + 2 - \dfrac{1}{x + 3} + \dfrac{2}{x - 1}$

54. $y = -x + 2 + \dfrac{1}{x + 3} - \dfrac{2}{x - 1}$

55. $y = -x + 2 + \dfrac{1}{x + 3} + \dfrac{2}{x - 1}$

56. $y = -x + 2 - \dfrac{1}{x + 3} - \dfrac{2}{x - 1}$

In Exercises 57 and 58, graph the inequality.

57. $2x - y \leq 1$ **58.** $x + 3y < 2$

In Exercises 59–64, solve the system of inequalities. Give the coordinates of any corner points.

59. $4x + 9y \geq 360$ **60.** $7x + 10y \leq 70$
$\quad\ 9x + 4y \geq 360$ $\quad\quad\ 2x + y \leq 10$
$\quad\quad\ x + y \leq 90$ $\quad\quad\quad x + y \geq 3$
$\quad\quad\quad\quad\quad\quad\quad\quad\quad\quad x \geq 0$
$\quad\quad\quad\quad\quad\quad\quad\quad\quad\quad y \geq 0$

61. $x - 3y + 6 < 0$
$\quad\quad\quad y > x^2 - 6x + 7$

62. $x + 2y \geq 4$
$\quad\quad\ y \leq 9 - x^2$

63. $x^2 + y^2 \leq 4$
$\quad\quad\ y \geq x^2$

64. $\quad\ y \leq x^2 + 4$
$\quad x^2 + y^2 \geq 4$

In Exercises 65–68, find the minimum and maximum, if they exist, of the objective function *f*, subject to the constraints.

65. Objective function: $f = 7x + 6y$
Constraints:
$$7x + 5y \geq 100$$
$$2x + 5y \geq 50$$
$$x \geq 0, y \geq 0$$

66. Objective function: $f = 11x + 5y$
Constraints:
$$5x + 2y \geq 60$$
$$5x + 8y \geq 120$$
$$x \geq 0, y \geq 0$$

67. Objective function: $f = 3x + 7y$
Constraints:
$$5x + 2y \geq 100$$
$$x + 4y \geq 110$$
$$5x + 11y \leq 460$$
$$x \geq 0, y \geq 0$$

68. Objective function: $f = 9x + 14y$
Constraints:
$$x + y \leq 120$$
$$9x + 2y \geq 240$$
$$3x + 10y \geq 360$$

69. Rotating Coordinate Systems The *xy*-coordinate system is rotated through the angle 45° to obtain the $x'y'$-coordinate system.

(a) If the coordinates of a point in the *xy*-coordinate system are (1, 2), what are the coordinates of the rotated point in the $x'y'$-coordinate system?

(b) If the coordinates of a point in the $x'y'$-coordinate system are (1, 2), what are the coordinates of the point in the *xy*-coordinate system that was rotated to it?

70. Medicare Disbursements Table 7.7 shows the total Medicare Disbursements in billions of dollars for several years. Let $x = 0$ stand for 1990, $x = 1$ for 1991, and so forth.

Table 7.7 Total Medicare Disbursements	
Year	Disbursements (billions)
1990	109.7
1995	180.1
1997	210.3
1998	213.4
1999	212.0
2000	219.3
2001	241.2
2002	256.9

Source: U.S. Centers for Medicare and Medicaid Services, unpublished data, Statistical Abstract of the United States, 2004–2005.

(a) Find a linear regression model and superimpose its graph on a scatter plot of the data.

(b) Find a logistic regression model and superimpose its graph on a scatter plot of the data.

(c) Find when the models in parts (a) and (b) predict the same disbursement amounts.

(d) Writing to Learn Which model appears to be a better fit for the data? Explain.

Which model would you choose to make predictions beyond 2000?

71. Population Table 7.8 gives the population (in thousands) of the states of Hawaii and Idaho for several years. Let $x = 0$ stand for 1980, $x = 1$ for 1981, and so forth.

Table 7.8 Population		
Year	Hawaii (thousands)	Idaho (thousands)
1980	965	944
1990	1108	1007
1995	1197	1177
1998	1215	1252
1999	1210	1276
2000	1212	1294
2001	1225	1321
2002	1241	1343
2003	1258	1366

Source: U.S. Bureau of the Census, Statistical Abstract of the U.S., 1998, 2004–2005.

(a) Find a linear regression model for Hawaii's data and superimpose its graph on a scatter plot of Hawaii's data.

(b) Find a linear regression model for Idaho's data and superimpose its graph on a scatter plot of Idaho's data.

(c) Using the models in parts (a) and (b), when was the population of the two states the same?

72. (a) The 2003 population data for three states is listed below. Use the data in the first table of the Chapter 7 Project to create a 3 × 2 matrix that estimates the number of males and females in each state.

State	Population (millions)
California	35.5
Florida	17.0
Rhode Island	1.1

(b) Write the data from the 2003 Census table below in the form of a 3 × 2 matrix.

State	% Pop under 18 years	% Pop 65 years or older
California	26.5	10.6
Florida	23.1	17.0
Rhode Island	22.8	14.0

(c) Multiply your 3 × 2 matrix in part (b) by the scalar 0.01 to change the values from percentages to decimals.

(d) Use matrix multiplication to multiply the transpose of the matrix from part (c) by the matrix from part (a). What information does the resulting matrix provide?

(e) How many males under age 18 lived in these three states in 2000? How many females age 65 or older lived in these three states?

73. Using Matrices A stockbroker sold a customer 200 shares of stock A, 400 shares of stock B, 600 shares of stock C, and 250 shares of stock D. The price per share of A, B, C, and D are $80, $120, $200, and $300, respectively.

(a) Write a 1 × 4 matrix N representing the number of shares of each stock the customer bought.

(b) Write a 1 × 4 matrix P representing the price per share of each stock.

(c) Write a matrix product that gives the total cost of the stocks that the customer bought.

74. Basketball Attendance At Whetstone High School 452 tickets were sold for the first basketball game. There were two ticket prices: $0.75 for students and $2.00 for nonstudents. How many tickets of each type were sold if the total revenue from the sale of tickets was $429?

75. Truck Deliveries Brock's Discount TV has three types of television sets on sale: a 13-in. portable, a 27-in. remote, and a 50-in. console. They have three types of vehicles to use for delivery: vans, small trucks, and large trucks. The vans can carry 8 portable, 3 remote, and 2 console TVs; the small trucks, 15 portable, 10 remote, and 6 console TVs; and the large trucks, 22 portable, 20 remote, and 5 console TVs. On a given day of the sale they have 115 portable, 85 remote, and 35 console TVs to deliver. How many vehicles of each type are needed to deliver the TVs?

76. Investments Jessica invests $38,000; part at 7.5% simple interest and the remainder at 6% simple interest. If her annual interest income is $2600, how much does she have invested at each rate?

77. Business Loans Thompson's Furniture Store borrowed $650,000 to expand its facilities and extend its product line. Some of the money was borrowed at 4%, some at 6.5%, and the rest at 9%. How much was borrowed at each rate if the annual interest was $46,250 and the amount borrowed at 9% was twice the amount borrowed at 4%?

78. Home Remodeling Sanchez Remodeling has three painters: Sue, Esther, and Murphy. Working together they can paint a large room in 4 hours. Sue and Murphy can paint the same size room in 6 hours. Esther and Murphy can paint the same size room in 7 hours. How long would it take each of them to paint the room alone?

79. Swimming Pool Three pipes, A, B, and C, are connected to a swimming pool. When all three pipes are running, the pool can be filled in 3 hr. When only A and B are running, the pool can be filled in 4 hr. When only B and C are running, the pool can be filled in 3.75 hr. How long would it take each pipe running alone to fill the pool?

80. Writing to Learn If the products AB and BA are defined for the $n \times n$ matrix A, what can you conclude about the order of matrix B? Explain.

81. Writing to Learn If A is an $m \times n$ matrix and B is a $p \times q$ matrix, and if AB is defined, what can you conclude about their orders? Explain.

CHAPTER 7 Project

Analyzing Census Data

The data below was gathered from the U.S. Census Bureau (www.census.gov). Examine the male and female population data from 1990 to 2004.

Population (millions)	Male	Female
1990	121.3	127.5
1995	128.3	134.5
1996	129.5	135.7
1997	130.8	137.0
1998	132.0	138.3
1999	133.3	139.4
2000	138.1	143.4
2001	140.1	145.1
2002	141.5	146.4
2003	143.0	147.8
2004	144.5	149.1

1. Plot the data using 1990 as the year zero. Find a linear regression model for each.

2. What do the slope and y-intercept mean in each equation?

3. What conclusions can you draw? According to these models, will the male population ever become greater than the female population? Was the male population ever greater than the female population? Is this enough data to create a model for a hundred years or more? Explain your answers.

4. Notice that the data above only gave information for a span of 15 years. This is often not enough information to accurately answer the questions asked above. Data over a small period often appears to be linear and can be modeled with a linear equation that works well over that limited domain. The chart below gives more data. Now use this data to plot the number of males versus time and females versus time using 1890 as year zero.

5. Notice that this data does not seem to be linear. Often times, you may remember from Chapter 3, a logistic model is used to model population growth. Find the logistic regression model for each data plot. What is the intersection of the curves and what does it represent? Would any of your responses in question number 3 above change? If so, how? Why?

6. Go to the U.S. Census Bureau web site (www.census.gov). How does your model predict the populations for the current year?

7. Use the census data for 2000. What percentage of the population is male and what percentage is female?

8. Go to the U.S. Census Bureau web site (www.census.gov) and use the most recent data along with the concepts from this chapter to collect and analyze other data.

Population (millions)	Male	Female	Population (millions)	Male	Female
1890	32.2	30.7	1950	75.2	76.1
1900	38.8	37.2	1960	88.3	91.0
1910	47.3	44.6	1970	98.9	104.3
1920	53.9	51.8	1980	110.1	116.5
1930	62.1	60.6	1990	121.3	127.5
1940	66.0	65.6	2000	138.1	143.4

Analytic Geometry in Two and Three Dimensions

The oval-shaped lawn behind the White House in Washington, D.C. is called *the Ellipse*. It has views of the Washington Monument, the Jefferson Memorial, the Department of Commerce, and the Old Post Office Building. The Ellipse is 616 ft long, 528 ft wide, and is in the shape of a conic section. Its shape can be modeled using the methods of this chapter. See page 652.

Chapter 8 Overview

Analytic geometry combines number and form. It is the marriage of algebra and geometry that grew from the works of Frenchmen René Descartes (1596–1650) and Pierre de Fermat (1601–1665). Their achievements allowed geometry problems to be solved algebraically and algebra problems to be solved geometrically—two major themes of this book. Analytic geometry opened the door for Newton and Leibniz to develop calculus.

In Sections 8.1–8.4, we will learn that parabolas, ellipses, and hyperbolas are all conic sections and can all be expressed as second-degree equations. We will investigate their uses, including the reflective properties of parabolas and ellipses and how hyperbolas are used in long-range navigation. In Section 8.5, we will see how parabolas, ellipses, and hyperbolas are unified in the polar-coordinate setting. In Section 8.6, we will move from the two-dimensional plane to revisit the concepts of point, line, midpoint, distance, and vector in three-dimensional space.

8.1
Conic Sections and Parabolas

What you'll learn about
- Conic Sections
- Geometry of a Parabola
- Translations of Parabolas
- Reflective Property of a Parabola

. . . and why
Conic sections are the paths of nature: Any free-moving object in a gravitational field follows the path of a conic section.

Conic Sections

Imagine two nonperpendicular lines intersecting at a point *V*. If we fix one of the lines as an *axis* and rotate the other line (the *generator*) around the axis, then the generator sweeps out a **right circular cone** with **vertex** *V*, as illustrated in Figure 8.1. Notice that *V* divides the cone into two parts called **nappes**, with each nappe of the cone resembling a pointed ice-cream cone.

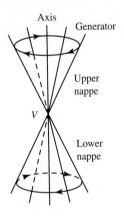

FIGURE 8.1 A right circular cone (of two nappes).

A **conic section** (or **conic**) is a cross section of a cone, in other words, the intersection of a plane with a right circular cone. The three basic conic sections are the *parabola*, the *ellipse*, and the *hyperbola* (Figure 8.2a).

Some atypical conics, known as **degenerate conic sections**, are shown in Figure 8.2b. Because it is atypical and lacks some of the features usually associated with an ellipse, a circle is considered to be a degenerate ellipse. Other degenerate conic sections can be obtained from cross sections of a degenerate cone; such cones occur when the generator and axis of the cone are parallel or perpendicular. (See Exercise 73.)

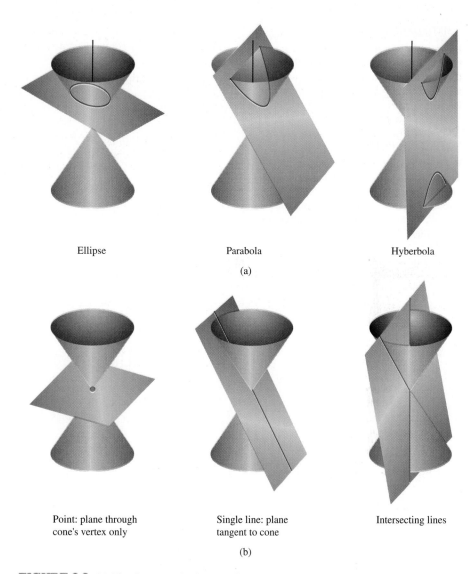

Ellipse   Parabola   Hyberbola

(a)

Point: plane through   Single line: plane   Intersecting lines
cone's vertex only   tangent to cone

(b)

FIGURE 8.2 (a) The three standard types of conic sections and (b) three degenerate conic sections.

The conic sections can be defined algebraically as the graphs of **second-degree (quadratic) equations in two variables**, that is, equations of the form

$$Ax^2 + Bxy + Cy^2 + Dx + Ey + F = 0,$$

where A, B, and C are not all zero.

Geometry of a Parabola

In Section 2.1 we learned that the graph of a quadratic function is an upward or downward opening parabola. We have seen the role of the parabola in free-fall and projectile motion. We now investigate the geometric properties of parabolas.

DEFINITION Parabola

A **parabola** is the set of all points in a plane equidistant from a particular line (the **directrix**) and a particular point (the **focus**) in the plane. (See Figure 8.3.)

The line passing through the focus and perpendicular to the directrix is the **(focal) axis** of the parabola. The axis is the line of symmetry for the parabola. The point where the parabola intersects its axis is the **vertex** of the parabola. The vertex is located midway between the focus and the directrix and is the point of the parabola that is closest to both the focus and the directrix. See Figure 8.3.

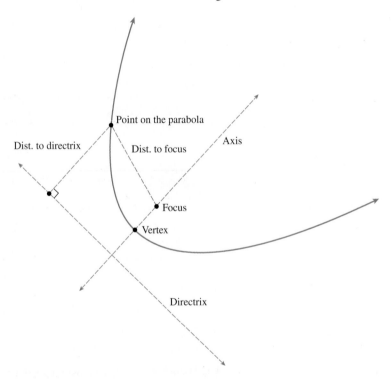

FIGURE 8.3 *Structure of a Parabola.* The distance from each point on the parabola to both the focus and the directrix is the same.

A DEGENERATE PARABOLA

If the focus F lies on the directrix l, the parabola "degenerates" to the line through F perpendicular to l. Henceforth, we will assume F does not lie on l.

LOCUS OF A POINT

Before the word *set* was used in mathematics, the Latin word *locus*, meaning "place," was often used in geometric definitions. The locus of a point was the set of possible places a point could be and still fit the conditions of the definition. Sometimes, conics are still defined in terms of loci.

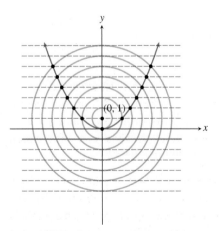

FIGURE 8.4 The geometry of a parabola.

EXPLORATION 1 Understanding the Definition of Parabola

1. Prove that the vertex of the parabola with focus $(0, 1)$ and directrix $y = -1$ is $(0, 0)$. (See Figure 8.4.)

2. Find an equation for the parabola shown in Figure 8.4.

3. Find the coordinates of the points of the parabola that are highlighted in Figure 8.4.

We can generalize the situation in Exploration 1 to show that an equation for the parabola with focus $(0, p)$ and directrix $y = -p$ is $x^2 = 4py$. (See Figure 8.5.)

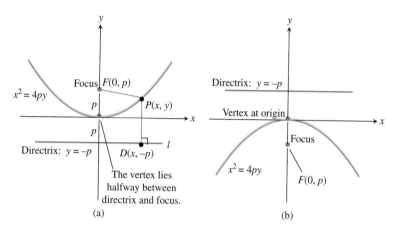

FIGURE 8.5 Graphs of $x^2 = 4py$ with (a) $p > 0$ and (b) $p < 0$.

We must show first that a point $P(x, y)$ that is equidistant from $F(0, p)$ and the line $y = -p$ satisfies the equation $x^2 = 4py$, and then that a point satisfying the equation $x^2 = 4py$ is equidistant from $F(0, p)$ and the line $y = -p$:

Let $P(x, y)$ be equidistant from $F(0, p)$ and the line $y = -p$. Notice that

$$\sqrt{(x - 0)^2 + (y - p)^2} = \text{distance from } P(x, y) \text{ to } F(0, p), \text{ and}$$

$$\sqrt{(x - x)^2 + (y - (-p))^2} = \text{distance from } P(x, y) \text{ to } y = -p.$$

Equating these distances and squaring yields:

$$(x - 0)^2 + (y - p)^2 = (x - x)^2 + (y - (-p))^2$$

$$x^2 + (y - p)^2 = 0 + (y + p)^2 \qquad \text{Simplify.}$$

$$x^2 + y^2 - 2py + p^2 = y^2 + 2py + p^2 \qquad \text{Expand.}$$

$$x^2 = 4py \qquad \text{Combine like terms.}$$

By reversing the above steps, we see that a solution (x, y) of $x^2 = 4py$ is equidistant from $F(0, p)$ and the line $y = -p$.

The equation $x^2 = 4py$ is the **standard form** of the equation of an upward or downward opening parabola with vertex at the origin. If $p > 0$, the parabola opens upward; if $p < 0$, it opens downward. An alternative algebraic form for such a parabola is $y = ax^2$, where $a = 1/(4p)$. So the graph of $x^2 = 4py$ is also the graph of the quadratic function $f(x) = ax^2$.

When the equation of an upward or downward opening parabola is written as $x^2 = 4py$, the value p is interpreted as the **focal length** of the parabola—the *directed* distance from the vertex to the focus of the parabola. A line segment with endpoints on a parabola is a **chord** of the parabola. The value $|4p|$ is the **focal width** of the parabola—the length of the chord through the focus and perpendicular to the axis.

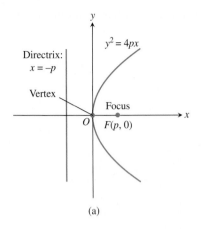

(a)

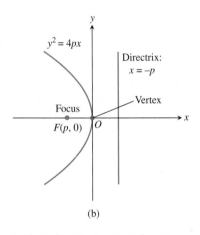

(b)

FIGURE 8.6 Graph of $y^2 = 4px$ with (a) $p > 0$ and (b) $p < 0$.

Parabolas that open to the right or to the left are *inverse relations* of upward or downward opening parabolas. So equations of parabolas with vertex (0, 0) that open to the right or to the left have the standard form $y^2 = 4px$. If $p > 0$, the parabola opens to the right, and if $p < 0$, to the left. (See Figure 8.6.)

Parabolas with Vertex (0, 0)		
• **Standard equation**	$x^2 = 4py$	$y^2 = 4px$
• **Opens**	Upward or downward	To the right or to the left
• **Focus**	$(0, p)$	$(p, 0)$
• **Directrix**	$y = -p$	$x = -p$
• **Axis**	y-axis	x-axis
• **Focal length**	p	p
• **Focal width**	$|4p|$	$|4p|$

See Figures 8.5 and 8.6.

EXAMPLE 1   Finding the Focus, Directrix, and Focal Width

Find the focus, the directrix, and the focal width of the parabola $y = -(1/2)x^2$.

SOLUTION Multiplying both sides of the equation by -2 yields the standard form $x^2 = -2y$. The coefficient of y is $4p = -2$, and $p = -1/2$. So the focus is $(0, p) = (0, -1/2)$. Because $-p = -(-1/2) = 1/2$, the directrix is the line $y = 1/2$. The focal width is $|4p| = |-2| = 2$.

Now try Exercise 1.

EXAMPLE 2   Finding an Equation of a Parabola

Find an equation in standard form for the parabola whose directrix is the line $x = 2$ and whose focus is the point $(-2, 0)$.

SOLUTION Because the directrix is $x = 2$ and the focus is $(-2, 0)$, the focal length is $p = -2$ and the parabola opens to the left. The equation of the parabola in standard form is $y^2 = 4px$, or more specifically, $y^2 = -8x$.

Now try Exercise 15.

Translations of Parabolas

When a parabola with the equation $x^2 = 4py$ or $y^2 = 4px$ is translated horizontally by h units and vertically by k units, the vertex of the parabola moves from (0, 0) to (h, k). (See Figure 8.7.) Such a translation does not change the focal length, the focal width, or the direction the parabola opens.

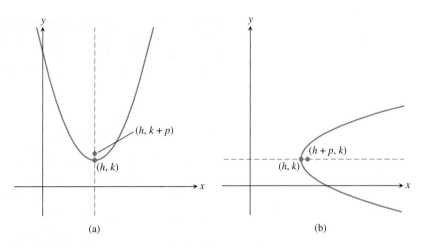

FIGURE 8.7 Parabolas with vertex (h, k) and focus on (a) $x = h$ and (b) $y = k$.

Parabolas with Vertex (h, k)

- **Standard equation** $(x - h)^2 = 4p(y - k)$ $(y - k)^2 = 4p(x - h)$

- **Opens**   Upward or   To the right or
                  downward   to the left

- **Focus** $(h, k + p)$ $(h + p, k)$

- **Directrix** $y = k - p$ $x = h - p$

- **Axis** $x = h$ $y = k$

- **Focal length** p p

- **Focal width** $|4p|$ $|4p|$

See Figure 8.7.

EXAMPLE 3   Finding an Equation of a Parabola

Find the standard form of the equation for the parabola with vertex $(3, 4)$ and focus $(5, 4)$.

SOLUTION The axis of the parabola is the line passing through the vertex $(3, 4)$ and the focus $(5, 4)$. This is the line $y = 4$. So the equation has the form

$$(y - k)^2 = 4p(x - h).$$

Because the vertex $(h, k) = (3, 4)$, $h = 3$ and $k = 4$. The directed distance from the vertex $(3, 4)$ to the focus $(5, 4)$ is $p = 5 - 3 = 2$, so $4p = 8$. Thus the equation we seek is

$$(y - 4)^2 = 8(x - 3).$$

Now try Exercise 21.

When solving a problem like Example 3, it is a good idea to sketch the vertex, the focus, and other features of the parabola as we solve the problem. This makes it easy to see whether the axis of the parabola is horizontal or vertical and the relative positions of its features. Exploration 2 "walks us through" this process.

EXPLORATION 2 Building a Parabola

Carry out the following steps using a sheet of rectangular graph paper.

1. Let the focus F of a parabola be $(2, -2)$ and its directrix be $y = 4$. Draw the x- and y-axes on the graph paper. Then sketch and label the focus and directrix of the parabola.

2. Locate, sketch, and label the axis of the parabola. What is the equation of the axis?

3. Locate and plot the vertex V of the parabola. Label it by name and coordinates.

4. What are the focal length and focal width of the parabola?

5. Use the focal width to locate, plot, and label the endpoints of a chord of the parabola that parallels the directrix.

6. Sketch the parabola.

7. Which direction does it open?

8. What is its equation in standard form?

Sometimes it is best to sketch a parabola by hand, as in Exploration 2; this helps us see the structure and relationships of the parabola and its features. At other times, we may want or need an accurate, active graph. If we wish to graph a parabola using a function grapher, we need to solve the equation of the parabola for y, as illustrated in Example 4.

EXAMPLE 4 Graphing a Parabola

Use a function grapher to graph the parabola $(y - 4)^2 = 8(x - 3)$ of Example 3.

SOLUTION

$$(y - 4)^2 = 8(x - 3)$$

$$y - 4 = \pm\sqrt{8(x - 3)} \qquad \text{Extract square roots.}$$

$$y = 4 \pm \sqrt{8(x - 3)} \qquad \text{Add 4.}$$

Let $Y_1 = 4 + \sqrt{8(x - 3)}$ and $Y_2 = 4 - \sqrt{8(x - 3)}$, and graph the two equations in a window centered at the vertex, as shown in Figure 8.8.

Now try Exercise 45.

CLOSING THE GAP

In Figure 8.8, we centered the graphing window at the vertex (3, 4) of the parabola to ensure that this point would be plotted. This avoids the common grapher error of a gap between the two upper and lower parts of the conic section being plotted.

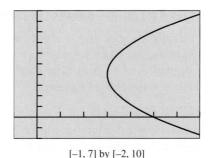

[–1, 7] by [–2, 10]

FIGURE 8.8 The graphs of $Y1 = 4 + \sqrt{8(x-3)}$ and $Y2 = 4 - \sqrt{8(x-3)}$ together form the graph of $(y-4)^2 = 8(x-3)$. (Example 4)

EXAMPLE 5   Using Standard Forms with a Parabola

Prove that the graph of $y^2 - 6x + 2y + 13 = 0$ is a parabola, and find its vertex, focus, and directrix.

SOLUTION Because this equation is quadratic in the variable y, we complete the square with respect to y to obtain a standard form.

$$y^2 - 6x + 2y + 13 = 0$$
$$y^2 + 2y = 6x - 13 \qquad \text{Isolate the } y\text{-terms.}$$
$$y^2 + 2y + 1 = 6x - 13 + 1 \qquad \text{Complete the square.}$$
$$(y + 1)^2 = 6x - 12$$
$$(y + 1)^2 = 6(x - 2)$$

This equation is in the standard form $(y - k)^2 = 4p(x - h)$, where $h = 2$, $k = -1$, and $p = 6/4 = 3/2 = 1.5$. It follows that

- the vertex (h, k) is $(2, -1)$;
- the focus $(h + p, k)$ is $(3.5, -1)$, or $(7/2, -1)$;
- the directrix $x = h - p$ is $x = 0.5$, or $x = 1/2$.

Now try Exercise 49.

Reflective Property of a Parabola

The main applications of parabolas involve their use as reflectors of sound, light, radio waves, and other electromagnetic waves. If we rotate a parabola in three-dimensional space about its axis, the parabola sweeps out a **paraboloid of revolution**. If we place a signal source at the focus of a reflective paraboloid, the signal reflects off the surface in lines parallel to the axis of symmetry, as illustrated in Figure 8.9a. This property is used by flashlights, headlights, searchlights, microwave relays, and satellite up-links.

The principle works for signals traveling in the reverse direction as well; signals arriving parallel to a parabolic reflector's axis are directed toward the reflector's focus. This property is used to intensify signals picked up by radio telescopes and television satellite dishes, to focus arriving light in reflecting telescopes, to concentrate heat in solar ovens, and to magnify sound for sideline microphones at football games. See Figure 8.9b.

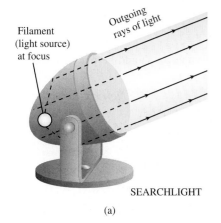

Filament (light source) at focus

Outgoing rays of light

SEARCHLIGHT

(a)

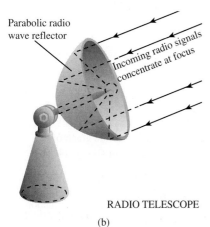

Parabolic radio wave reflector

Incoming radio signals concentrate at focus

RADIO TELESCOPE

(b)

FIGURE 8.9 Examples of parabolic reflectors.

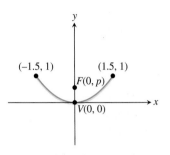

FIGURE 8.10 Cross section of parabolic reflector in Example 6.

EXAMPLE 6   Studying a Parabolic Microphone

On the sidelines of each of its televised football games, the FBTV network uses a parabolic reflector with a microphone at the reflector's focus to capture the conversations among players on the field. If the parabolic reflector is 3 ft across and 1 ft deep, where should the microphone be placed?

SOLUTION

We draw a cross section of the reflector as an upward opening parabola in the Cartesian plane, placing its vertex V at the origin (see Figure 8.10). We let the focus F have coordinates $(0, p)$ to yield the equation

$$x^2 = 4py.$$

Because the reflector is 3 ft across and 1 ft deep, the points $(\pm 1.5, 1)$ must lie on the parabola. The microphone should be placed at the focus, so we need to find the value of p. We do this by substituting the values we found into the equation:

$$x = 4py$$

$$(\pm 1.5)^2 = 4p(1)$$

$$2.25 = 4p$$

$$p = 2.25/4 = 0.5625$$

Because $p = 0.5625$ ft, or 6.75 inches, the microphone should be placed inside the reflector along its axis and 6.75 inches from its vertex.

Now try Exercise 59.

QUICK REVIEW 8.1   *(For help, go to Sections P.2, P.5, and 2.1.)*

In Exercises 1 and 2, find the distance between the given points.

1. $(-1, 3)$ and $(2, 5)$ **2.** $(2, -3)$ and (a, b)

In Exercises 3 and 4, solve for y in terms of x.

3. $2y^2 = 8x$ **4.** $3y^2 = 15x$

In Exercises 5 and 6, complete the square to rewrite the equation in vertex form.

5. $y = -x^2 + 2x - 7$ **6.** $y = 2x^2 + 6x - 5$

In Exercises 7 and 8, find the vertex and axis of the graph of f. Describe how the graph of f can be obtained from the graph of $g(x) = x^2$, and graph f.

7. $f(x) = 3(x - 1)^2 + 5$ **8.** $f(x) = -2x^2 + 12x + 1$

In Exercises 9 and 10, write an equation for the quadratic function whose graph contains the given vertex and point.

9. Vertex $(-1, 3)$, point $(0, 1)$

10. Vertex $(2, -5)$, point $(5, 13)$

SECTION 8.1 EXERCISES

In Exercises 1–6, find the vertex, focus, directrix, and focal width of the parabola.

1. $x^2 = 6y$

2. $y^2 = -8x$

3. $(y - 2)^2 = 4(x + 3)$

4. $(x + 4)^2 = -6(y + 1)$

5. $3x^2 = -4y$

6. $5y^2 = 16x$

In Exercises 7–10, match the graph with its equation.

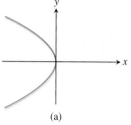

(a)

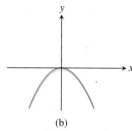

(b)

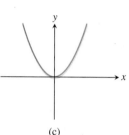

(c)

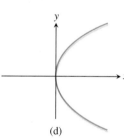

(d)

7. $x^2 = 3y$

8. $x^2 = -4y$

9. $y^2 = -5x$

10. $y^2 = 10x$

In Exercises 11–30, find an equation in standard form for the parabola that satisfies the given conditions.

11. Vertex $(0, 0)$, focus $(-3, 0)$

12. Vertex $(0, 0)$, focus $(0, 2)$ $y^2 = 4px$ $x^2 = 4py$

13. Vertex $(0, 0)$, directrix $y = 4$

14. Vertex $(0, 0)$, directrix $x = -2$

15. Focus $(0, 5)$, directrix $y = -5$

16. Focus $(-4, 0)$, directrix $x = 4$

17. Vertex $(0, 0)$, opens to the right, focal width $= 8$

18. Vertex $(0, 0)$, opens to the left, focal width $=12$

19. Vertex $(0, 0)$, opens downward, focal width $= 6$

20. Vertex $(0, 0)$, opens upward, focal width $= 3$

21. Focus $(-2, -4)$, vertex $(-4, -4)$

22. Focus $(-5, 3)$, vertex $(-5, 6)$

23. Focus $(3, 4)$, directrix $y = 1$

24. Focus $(2, -3)$, directrix $x = 5$

25. Vertex $(4, 3)$, directrix $x = 6$

26. Vertex $(3, 5)$, directrix $y = 7$

27. Vertex $(2, -1)$, opens upward, focal width $= 16$

28. Vertex $(-3, 3)$, opens downward, focal width $= 20$

29. Vertex $(-1, -4)$, opens to the left, focal width $= 10$

30. Vertex $(2, 3)$, opens to the right, focal width $= 5$

In Exercises 31–36, sketch the graph of the parabola by hand.

31. $y^2 = -4x$

32. $x^2 = 8y$

33. $(x + 4)^2 = -12(y + 1)$

34. $(y + 2)^2 = -16(x + 3)$

35. $(y - 1)^2 = 8(x + 3)$

36. $(x - 5)^2 = 20(y + 2)$

In Exercises 37–48, graph the parabola using a function grapher.

37. $y = 4x^2$

38. $y = -\dfrac{1}{6}x^2$

39. $x = -8y^2$

40. $x = 2y^2$

41. $12(y + 1) = (x - 3)^2$

42. $6(y - 3) = (x + 1)^2$

43. $2 - y = 16(x - 3)^2$

44. $(x + 4)^2 = -6(y - 1)$

45. $(y + 3)^2 = 12(x - 2)$

46. $(y - 1)^2 = -4(x + 5)$

47. $(y + 2)^2 = -8(x + 1)$

48. $(y - 6)^2 = 16(x - 4)$

In Exercises 49–52, prove that the graph of the equation is a parabola, and find its vertex, focus, and directrix.

49. $x^2 + 2x - y + 3 = 0$

50. $3x^2 - 6x - 6y + 10 = 0$

51. $y^2 - 4y - 8x + 20 = 0$

52. $y^2 - 2y + 4x - 12 = 0$

In Exercises 53–56, write an equation for the parabola.

53.

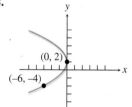

54.

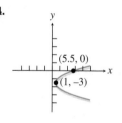

55.

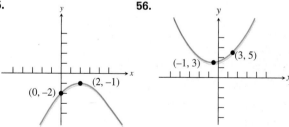

56.

57. **Writing to Learn** Explain why the derivation of $x^2 = 4py$ is valid regardless of whether $p > 0$ or $p < 0$.

58. **Writing to Learn** Prove that an equation for the parabola with focus $(p, 0)$ and directrix $x = -p$ is $y^2 = 4px$.

59. **Designing a Flashlight Mirror** The mirror of a flashlight is a paraboloid of revolution. Its diameter is 6 cm and its depth is 2 cm. How far from the vertex should the filament of the light bulb be placed for the flashlight to have its beam run parallel to the axis of its mirror?

60. **Designing a Satellite Dish** The reflector of a television satellite dish is a paraboloid of revolution with diameter 5 ft and a depth of 2 ft. How far from the vertex should the receiving antenna be placed?

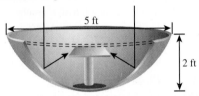

61. **Parabolic Microphones** The Sports Channel uses a parabolic microphone to capture all the sounds from golf tournaments throughout a season. If one of its microphones has a parabolic surface generated by the parabola $x^2 = 10y$, locate the focus (the electronic receiver) of the parabola.

62. **Parabolic Headlights** Stein Glass, Inc., makes parabolic headlights for a variety of automobiles. If one of its headlights has a parabolic surface generated by the parabola $x^2 = 12y$, where should its light bulb be placed?

63. **Group Activity Designing a Suspension Bridge** The main cables of a suspension bridge uniformly distribute the weight of the bridge when in the form of a parabola. The main cables of a particular bridge are attached to towers that are 600 ft apart. The cables are attached to the towers at a height of 110 ft above the roadway and are 10 ft above the roadway at their lowest points. If vertical support cables are at 50-ft intervals along the level roadway, what are the lengths of these vertical cables?

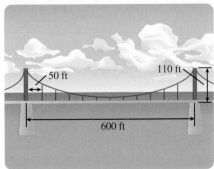

64. **Group Activity Designing a Bridge Arch** Parabolic arches are known to have greater strength than other arches. A bridge with a supporting parabolic arch spans 60 ft with a 30-ft wide road passing underneath the bridge. In order to have a minimum clearance of 16 ft, what is the maximum clearance?

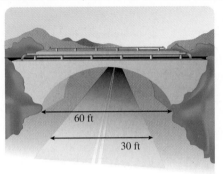

Standardized Test Questions

65. **True or False** Every point on a parabola is the same distance from its focus and its axis. Justify your answer.

66. **True or False** The directrix of a parabola is parallel to the parabola's axis. Justify your answer.

In Exercises 67–70, solve the problem without using a calculator.

67. **Multiple Choice** Which of the following curves is **not** a conic section?

(A) circle

(B) ellipse

(C) hyperbola

(D) oval

(E) parabola

68. **Multiple Choice** Which point do all conics of the form $x^2 = 4py$ have in common?

(A) $(1, 1)$

(B) $(1, 0)$

(C) $(0, 1)$

(D) $(0, 0)$

(E) $(-1, -1)$

69. **Multiple Choice** The focus of $y^2 = 12x$ is

(A) $(3, 3)$.

(B) $(3, 0)$.

(C) $(0, 3)$.

(D) $(0, 0)$.

(E) $(-3, -3)$.

70. Multiple Choice The vertex of $(y - 3)^2 = -8(x + 2)$ is

(A) $(3, -2)$.

(B) $(-3, -2)$.

(C) $(-3, 2)$.

(D) $(-2, 3)$.

(E) $(-2, -3)$.

Explorations

71. Dynamically Constructing a Parabola Use a geometry software package, such as *Cabri Geometry II* ™, *The Geometer's Sketchpad*®, or similar application on a handheld device to construct a parabola geometrically from its definition. (See Figure 8.3.)

(a) Start by placing a line *l* (directrix) and a point *F* (focus) not on the line in the construction window.

(b) Construct a point *A* on the directrix, and then the segment *AF*.

(c) Construct a point *P* where the perpendicular bisector of *AF* meets the line perpendicular to *l* through *A*.

(d) What curve does *P* trace out as *A* moves?

(e) Prove your answer to part (d) is correct.

72. Constructing Points of a Parabola Use a geometry software package, such as *Cabri Geometry II* ™, *The Geometer's Sketchpad*®, or similar application on a handheld device, to construct Figure 8.4, associated with Exploration 1.

(a) Start by placing the coordinate axes in the construction window.

(b) Construct the line $y = -1$ as the directrix and the point $(0, 1)$ as the focus.

(c) Construct the horizontal lines and concentric circles shown in Figure 8.4.

(d) Construct the points where these horizontal lines and concentric circles meet.

(e) Prove these points lie on the parabola with directrix $y = -1$ and focus $(0, 1)$.

73. Degenerate Cones and Degenerate Conics Degenerate cones occur when the generator and axis of the cone are parallel or perpendicular.

(a) Draw a sketch and describe the "cone" obtained when the generator and axis of the cone are parallel.

(b) Draw sketches and name the types of degenerate conics obtained by intersecting the degenerate cone in part (a) with a plane.

(c) Draw a sketch and describe the "cone" obtained when the generator and axis of the cone are perpendicular.

(d) Draw sketches and name the types of degenerate conics obtained by intersecting the degenerate cone in part (c) with a plane.

Extending the Ideas

74. Tangent Lines A **tangent line** of a parabola is a line that intersects but does not cross the parabola. Prove that a line tangent to the parabola $x^2 = 4py$ at the point (a, b) crosses the y-axis at $(0, -b)$.

75. Focal Chords A **focal chord** of a parabola is a chord of the parabola that passes through the focus.

(a) Prove that the x-coordinates of the endpoints of a focal chord of $x^2 = 4py$ are $x = 2p(m \pm \sqrt{m^2 + 1})$, where *m* is the slope of the focal chord.

(b) Using part (a), prove the minimum length of a focal chord is the focal width $|4p|$.

76. Latus Rectum The focal chord of a parabola perpendicular to the axis of the parabola is the **latus rectum**, which is Latin for "right chord." Using the results from Exercises 74 and 75, prove:

(a) For a parabola the two endpoints of the latus rectum and the point of intersection of the axis and directrix are the vertices of an isosceles right triangle.

(b) The legs of this isosceles right triangle are tangent to the parabola.

32. Major axis endpoints are $(-2, -3)$ and $(-2, 7)$, minor axis length 4

33. The foci are $(1, -4)$ and $(5, -4)$; the major axis endpoints are $(0, -4)$ and $(6, -4)$.

34. The foci are $(-2, 1)$ and $(-2, 5)$; the major axis endpoints are $(-2, -1)$ and $(-2, 7)$.

35. The major axis endpoints are $(3, -7)$ and $(3, 3)$; the minor axis length is 6.

36. The major axis endpoints are $(-5, 2)$ and $(3, 2)$; the minor axis length is 6.

In Exercises 37–40, find the center, vertices, and foci of the ellipse.

37. $\dfrac{(x + 1)^2}{25} + \dfrac{(y - 2)^2}{16} = 1$ **38.** $\dfrac{(x - 3)^2}{11} + \dfrac{(y - 5)^2}{7} = 1$

39. $\dfrac{(y + 3)^2}{81} + \dfrac{(x - 7)^2}{64} = 1$ **40.** $\dfrac{(y - 1)^2}{25} + \dfrac{(x + 2)^2}{16} = 1$

In Exercises 41–44, graph the ellipse using a parametric grapher.

41. $\dfrac{y^2}{25} + \dfrac{x^2}{4} = 1$ **42.** $\dfrac{x^2}{30} + \dfrac{y^2}{20} = 1$

43. $\dfrac{(x + 3)^2}{12} + \dfrac{(y - 6)^2}{5} = 1$ **44.** $\dfrac{(y + 1)^2}{15} + \dfrac{(x - 2)^2}{6} = 1$

In Exercises 45–48, prove that the graph of the equation is an ellipse, and find its vertices, foci, and eccentricity.

45. $9x^2 + 4y^2 - 18x + 8y - 23 = 0$

46. $3x^2 + 5y^2 - 12x + 30y + 42 = 0$

47. $9x^2 + 16y^2 + 54x - 32y - 47 = 0$

48. $4x^2 + y^2 - 32x + 16y + 124 = 0$

In Exercises 49 and 50, write an equation for the ellipse.

49.
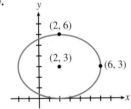

50.
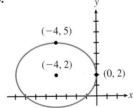

51. Writing to Learn Prove that an equation for the ellipse with center $(0, 0)$, foci $(0, \pm c)$, and semimajor axis $a > c \geq 0$ is $y^2/a^2 + x^2/b^2 = 1$, where $b^2 = a^2 - c^2$.
[*Hint*: Refer to derivation at the beginning of the section.]

52. Writing to Learn Dancing Planets Using the data in Table 8.1, prove that the planet with the most eccentric orbit sometimes is closer to the Sun than the planet with the least eccentric orbit.

Table 8.1 Semimajor Axes and Eccentricities of the Planets

Planet	Semimajor Axis (Gm)	Eccentricity
Mercury	57.9	0.2056
Venus	108.2	0.0068
Earth	149.6	0.0167
Mars	227.9	0.0934
Jupiter	778.3	0.0485
Saturn	1427	0.0560
Uranus	2869	0.0461
Neptune	4497	0.0050
Pluto	5900	0.2484

Source: Shupe, et al., National Geographic Atlas of the World (rev. 6th ed.). Washington, DC: National Geographic Society, 1992, plate 116, and other sources.

53. The Moon's Orbit The Moon's apogee (farthest distance from Earth) is 252,710 miles, and perigee (closest distance to Earth) is 221,463 miles. Assuming the Moon's orbit of Earth is elliptical with Earth at one focus, calculate and interpret a, b, c, and e.

54. Hot Mercury Given that the diameter of the Sun is about 1.392 Gm, how close does Mercury get to the Sun's surface?

55. Saturn Find the perihelion and aphelion distances of Saturn.

56. Venus and Mars Write equations for the orbits of Venus and Mars in the form $x^2/a^2 + y^2/b^2 = 1$.

57. Sungrazers One comet group, known as the sungrazers, passes within a Sun's diameter (1.392 Gm) of the solar surface. What can you conclude about $a - c$ for orbits of the sungrazers?

58. Halley's Comet The orbit of Halley's comet is 36.18 AU long and 9.12 AU wide. What is its eccentricity?

59. Lithotripter For an ellipse that generates the ellipsoid of a lithotripter, the major axis has endpoints $(-8, 0)$ and $(8, 0)$. One endpoint of the minor axis is $(0, 3.5)$. Find the coordinates of the foci.

60. Lithotripter (Refer to Figure 8.21.) A lithotripter's shape is formed by rotating the portion of an ellipse below its minor axis about its major axis. If the length of the major axis is 26 in. and the length of the minor axis is 10 in., where should the shock-wave source and the patient be placed for maximum effect?

Group Activities In Exercises 61 and 62, solve the system of equations algebraically and support your answer graphically.

61. $\dfrac{x^2}{4} + \dfrac{y^2}{9} = 1$ **62.** $\dfrac{x^2}{9} + y^2 = 1$

$x^2 + y^2 = 4$ $x - 3y = -3$

63. Group Activity Consider the system of equations

$$x^2 + 4y^2 = 4$$
$$y = 2x^2 - 3$$

(a) Solve the system graphically.

(b) If you have access to a grapher that also does symbolic algebra, use it to find the exact solutions to the system.

64. Writing to Learn Look up the adjective *eccentric* in a dictionary and read its various definitions. Notice that the word is derived from *ex-centric*, meaning "out-of-center" or "off-center." Explain how this is related to the word's everyday meanings as well as its mathematical meaning for ellipses.

Standardized Test Questions

65. True or False The distance from a focus of an ellipse to the closer vertex is $a(1+e)$, where a is the semimajor axis and e is the eccentricity. Justify your answer.

66. True or False The distance from a focus of an ellipse to either endpoint of the minor axis is half the length of the major axis. Justify your answer.

In Exercises 67–70, you may use a graphing calculator to solve the problem.

67. Multiple Choice One focus of $x^2 + 4y^2 = 4$ is

(A) (4, 0). (B) (2, 0).

(C) $(\sqrt{3}, 0)$. (D) $(\sqrt{2}, 0)$.

(E) (1, 0).

68. Multiple Choice The focal axis of $\dfrac{(x-2)^2}{25} + \dfrac{(y-3)^2}{16} = 1$ is

(A) $y = 1$. (B) $y = 2$.

(C) $y = 3$. (D) $y = 4$.

(E) $y = 5$.

69. Multiple Choice The center of

$$9x^2 + 4y^2 - 72x - 24y + 144 = 0 \text{ is}$$

(A) (4, 2). (B) (4, 3).

(C) (4, 4). (D) (4, 5).

(E) (4, 6).

70. Multiple Choice The perimeter of a triangle with one vertex on the ellipse $x^2/a^2 + y^2/b^2 = 1$ and the other two vertices at the foci of the ellipse would be

(A) $a + b$. (B) $2a + 2b$.

(C) $2a + 2c$. (D) $2b + 2c$.

(E) $a + b + c$.

Explorations

71. Area and Perimeter The area of an ellipse is $A = \pi ab$, but the perimeter cannot be expressed so simply:

$$P \approx \pi(a + b)\left(3 - \frac{\sqrt{(3a + b)(a + 3b)}}{a + b}\right)$$

(a) Prove that, when $a = b = r$, these become the familiar formulas for the area and perimeter (circumference) of a circle.

(b) Find a pair of ellipses such that the one with greater area has smaller perimeter.

72. Writing to Learn Kepler's Laws We have encountered Kepler's First and Third Laws (p. 193). Using a library or the Internet,

(a) Read about Kepler's life, and write in your own words how he came to discover his three laws of planetary motion.

(b) What is Kepler's Second Law? Explain it with both pictures and words.

73. Pendulum Velocity vs. Position As a pendulum swings toward and away from a motion detector, its distance (in meters) from the detector is given by the position function $x(t) = 3 + \cos(2t - 5)$, where t represents time (in seconds). The velocity (in m/sec) of the pendulum is given by $y(t) = -2 \sin(2t - 5)$,

(a) Using parametric mode on your grapher, plot the (x, y) relation for velocity versus position for $0 \le t \le 2\pi$

(b) Write the equation of the resulting conic in standard form, in terms of x and y, and eliminating the parameter t.

74. Pendulum Velocity vs. Position A pendulum that swings toward and away from a motion detector has a distance (in feet) from the detector of $x(t) = 5 + 3 \sin(\pi t + \pi/2)$ and a velocity (in ft/sec) of $y(t) = 3\pi \cos(\pi t + \pi/2)$, where t represents time (in seconds).

(a) Prove that the plot of velocity versus position (distance) is an ellipse.

(b) **Writing to Learn** Describe the motion of the pendulum.

Extending the Ideas

75. Prove that a nondegenerate graph of the equation

$$Ax^2 + Cy^2 + Dx + Ey + F = 0$$

is an ellipse if $AC > 0$.

76. Writing to Learn The graph of the equation

$$\frac{(x-h)^2}{a^2} + \frac{(y-k)^2}{b^2} = 0$$

is considered to be a degenerate ellipse. Describe the graph. How is it like a full-fledged ellipse, and how is it different?

EXAMPLE 4   Locating Key Points of a Hyperbola

Find the center, vertices, and foci of the hyperbola

$$\frac{(x + 2)^2}{9} - \frac{(y - 5)^2}{49} = 1.$$

SOLUTION The center (h, k) is $(-2, 5)$. Because the semitransverse axis $a = \sqrt{9} = 3$, the vertices are

$$(h + a, k) = (-2 + 3, 5) = (1, 5) \quad \text{and}$$

$$(h - a, k) = (-2 - 3, 5) = (-5, 5).$$

Because $c = \sqrt{a^2 + b^2} = \sqrt{9 + 49} = \sqrt{58}$, the foci $(h \pm c, k)$ are $(-2 \pm \sqrt{58}, 5)$, or approximately $(5.62, 5)$ and $(-9.62, 5)$.   ***Now try Exercise 39.***

With the information found about the hyperbola in Example 4 and knowing that its semiconjugate axis $b = \sqrt{49} = 7$, we could easily sketch the hyperbola. Obtaining an accurate graph of the hyperbola using a function grapher is another matter. Often, the best way to graph a hyperbola using a graphing utility is to use parametric equations.

EXPLORATION 1   Graphing a Hyperbola Using Its Parametric Equations

1. Use the Pythagorean trigonometry identity $\sec^2 t - \tan^2 t = 1$ to prove that the parameterization $x = -1 + 3/\cos t$, $y = 1 + 2 \tan t$ $(0 \le t \le 2\pi)$ will produce a graph of the hyperbola $(x + 1)^2/9 - (y - 1)^2/4 = 1$.

2. Using Dot graphing mode, graph $x = -1 + 3/\cos t$, $y = 1 + 2 \tan t$ $(0 \le t \le 2\pi)$ in a square viewing window to support part 1 graphically. Switch to Connected graphing mode, and regraph the equation. What do you observe? Explain.

3. Create parameterizations for the hyperbolas in Examples 1, 2, 3, and 4.

4. Graph each of your parameterizations in part 3 and check the features of the obtained graph to see whether they match the expected geometric features of the hyperbola. If necessary, revise your parameterization and regraph until all features match.

5. Prove that each of your parameterizations is valid.

Eccentricity and Orbits

DEFINITION Eccentricity of a Hyperbola

The **eccentricity** of a hyperbola is

$$e = \frac{c}{a} = \frac{\sqrt{a^2 + b^2}}{a},$$

where a is the semitransverse axis, b is the semiconjugate axis, and c is the distance from the center to either focus.

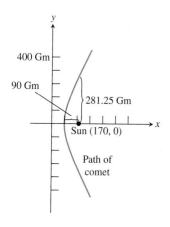

FIGURE 8.28 The graph of one branch of $x^2/6400 - y^2/22{,}500 = 1$. (Example 5)

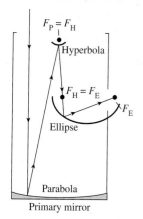

FIGURE 8.29 Cross section of a reflecting telescope.

For a hyperbola, because $c > a$, the eccentricity $e > 1$. In Section 8.2 we learned that the eccentricity of an ellipse satisfies the inequality $0 \le e < 1$ and that, for $e = 0$, the ellipse is a circle. In Section 8.5 we will generalize the concept of eccentricity to all types of conics and learn that the eccentricity of a parabola is $e = 1$.

Kepler's first law of planetary motion says that a planet's orbit is elliptical with the Sun at one focus. Since 1609, astronomers have generalized Kepler's law; the current theory states: A celestial body that travels within the gravitational field of a much more massive body follows a path that closely approximates a conic section that has the more massive body as a focus. Two bodies that do not differ greatly in mass (such as Earth and the Moon, or Pluto and its moon Charon) actually revolve around their balance point, or *barycenter*. In theory, a comet can approach the Sun from interstellar space, make a partial loop about the Sun, and then leave the solar system returning to deep space; such a comet follows a path that is one branch of a hyperbola.

EXAMPLE 5   Analyzing a Comet's Orbit

A comet following a hyperbolic path about the Sun has a perihelion distance of 90 Gm. When the line from the comet to the Sun is perpendicular to the focal axis of the orbit, the comet is 281.25 Gm from the Sun. Calculate a, b, c, and e. What are the coordinates of the center of the Sun if we coordinatize space so that the hyperbola is given by

$$\frac{x^2}{a^2} - \frac{y^2}{b^2} = 1?$$

SOLUTION The perihelion distance is $c - a = 90$. When $x = c$, $y = \pm b^2/a$ (see Exercise 74). So $b^2/a = 281.25$, or $b^2 = 281.25a$. Because $b^2 = c^2 - a^2$, we have the system

$$c - a = 90 \quad \text{and} \quad c^2 - a^2 = 281.25a,$$

which yields the equation:

$$(a + 90)^2 - a^2 = 281.25a$$

$$a^2 + 180a + 8100 - a^2 = 281.25a$$

$$8100 = 101.25a$$

$$a = 80$$

So $a = 80$ Gm, $b = 150$ Gm, $c = 170$ Gm, and $e = 17/8 = 2.125$. If the comet's path is the branch of the hyperbola with positive x-coordinates, then the Sun is at the focus $(c, 0) = (170, 0)$. See Figure 8.28.

Now try Exercise 55.

Reflective Property of a Hyperbola

Like other conics, a hyperbola can be used to make a reflector of sound, light, and other waves. If we rotate a hyperbola in three-dimensional space about its focal axis, the hyperbola sweeps out a **hyperboloid of revolution**. If a signal is directed toward a focus of a reflective hyperboloid, the signal reflects off the hyperbolic surface to the other focus. In Figure 8.29 light reflects off a primary parabolic mirror toward the mirror's focus $F_P = F_H$, which is also the focus of a small hyperbolic mirror. The light is then reflected off the hyperbolic mirror, toward the hyperboloid's other focus $F_H = F_E$, which is also the focus of an elliptical mirror. Finally the light is reflect into the observer's eye, which is at the second focus of the ellipsoid F_E.

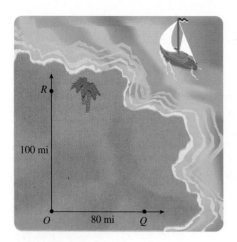

FIGURE 8.30 Strategically located LORAN transmitters O, Q, and R. (Example 6)

Reflecting telescopes date back to the 1600s when Isaac Newton used a primary parabolic mirror in combination with a flat secondary mirror, slanted to reflect the light out the side to the eyepiece. French optician G. Cassegrain was the first to use a hyperbolic secondary mirror, which directed the light through a hole at the vertex of the primary mirror (see Exercise 70). Today, reflecting telescopes such as the Hubble Space Telescope have become quite sophisticated and must have nearly perfect mirrors to focus properly.

Long-Range Navigation

Hyperbolas and radio signals are the basis of the LORAN (long-range navigation) system. Example 6 illustrates this system using the definition of hyperbola and the fact that radio signals travel 980 ft per microsecond (1 microsecond $= 1$ μsec $= 10^{-6}$ sec).

EXAMPLE 6   Using the LORAN System

Radio signals are sent simultaneously from transmitters located at points O, Q, and R (Figure 8.30). R is 100 mi due north of O, and Q is 80 mi due east of O. The LORAN receiver on sloop *Gloria* receives the signal from O 323.27 μsec after the signal from R, and 258.61 μsec after the signal from Q. What is the sloop's bearing and distance from O?

SOLUTION   The *Gloria* is at a point of intersection between two hyperbolas: one with foci O and R, the other with foci O and Q.

The hyperbola with foci $O(0, 0)$ and $R(0, 100)$ has center $(0, 50)$ and transverse axis

$$2a = (323.27 \text{ μsec})(980 \text{ ft/μsec})(1 \text{ mi/5280 ft}) \approx 60 \text{ mi.}$$

Thus $a \approx 30$ and $b = \sqrt{c^2 - a^2} \approx \sqrt{50^2 - 30^2} = 40$, yielding the equation

$$\frac{(y - 50)^2}{30^2} - \frac{x^2}{40^2} = 1.$$

The hyperbola with foci $O(0, 0)$ and $Q(80, 0)$ has center $(40, 0)$ and transverse axis

$$2a = (258.61 \text{ μsec})(980 \text{ ft/μsec})(1 \text{ mi/5280 ft}) \approx 48 \text{ mi.}$$

Thus $a \approx 24$ and $b = \sqrt{c^2 - a^2} \approx \sqrt{40^2 - 24^2} = 32$, yielding the equation

$$\frac{(x - 40)^2}{24^2} - \frac{y^2}{32^2} = 1.$$

The *Gloria* is at point P where upper and right-hand branches of the hyperbolas meet (see Figure 8.31). Using a grapher we find that $P \approx (187.09, 193.49)$. So the bearing from point O is

$$\theta \approx 90° - \tan^{-1}\left(\frac{193.49}{187.09}\right) \approx 44.04°,$$

and the distance from point O is

$$d \approx \sqrt{187.09^2 + 193.49^2} \approx 269.15.$$

So the *Gloria* is about 187.1 mi east and 193.5 mi north of point O on a bearing of roughly 44°, and the sloop is about 269 mi from point O.

Now try Exercise 57.

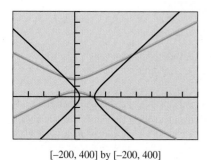

[–200, 400] by [–200, 400]

FIGURE 8.31 Graphs for Example 6.

QUICK REVIEW 8.3 *(For help, go to Sections P.2, P.5, and 7.1.)*

In Exercises 1 and 2, find the distance between the given points.

1. $(4, -3)$ and $(-7, -8)$

2. $(a, -3)$ and (b, c)

In Exercises 3 and 4, solve for y in terms of x.

3. $\dfrac{y^2}{16} - \dfrac{x^2}{9} = 1$ **4.** $\dfrac{x^2}{36} - \dfrac{y^2}{4} = 1$

In Exercises 5–8, solve for x algebraically.

5. $\sqrt{3x + 12} - \sqrt{3x - 8} = 10$

6. $\sqrt{4x + 12} - \sqrt{x + 8} = 1$

7. $\sqrt{6x^2 + 12} - \sqrt{6x^2 + 1} = 1$

8. $\sqrt{2x^2 + 12} - \sqrt{3x^2 + 4} = -8$

In Exercises 9 and 10, solve the system of equations.

9. $c - a = 2$ and $c^2 - a^2 = 16a/3$

10. $c - a = 1$ and $c^2 - a^2 = 25a/12$

SECTION 8.3 EXERCISES

In Exercises 1–6, find the vertices and foci of the hyperbola.

1. $\dfrac{x^2}{16} - \dfrac{y^2}{7} = 1$ **2.** $\dfrac{y^2}{25} - \dfrac{x^2}{21} = 1$

3. $\dfrac{y^2}{36} - \dfrac{x^2}{13} = 1$ **4.** $\dfrac{x^2}{9} - \dfrac{y^2}{16} = 1$

5. $3x^2 - 4y^2 = 12$ **6.** $9x^2 - 4y^2 = 36$

In Exercises 7–10, match the graph with its equation.

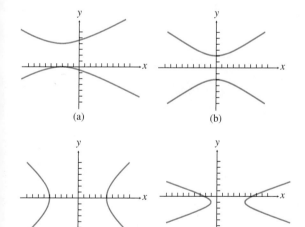

(a)

(b)

(c)

(d)

7. $\dfrac{x^2}{25} - \dfrac{y^2}{16} = 1$ **8.** $\dfrac{y^2}{4} - \dfrac{x^2}{9} = 1$

9. $\dfrac{(y - 2)^2}{4} - \dfrac{(x + 3)^2}{16} = 1$ **10.** $\dfrac{(x - 2)^2}{9} - (y + 1)^2 = 1$

In Exercises 11–16, sketch the graph of the hyperbola by hand.

11. $\dfrac{x^2}{49} - \dfrac{y^2}{25} = 1$ **12.** $\dfrac{y^2}{64} - \dfrac{x^2}{25} = 1$

13. $\dfrac{y^2}{25} - \dfrac{x^2}{16} = 1$ **14.** $\dfrac{x^2}{169} - \dfrac{y^2}{144} = 1$

15. $\dfrac{(x + 3)^2}{16} - \dfrac{(y - 1)^2}{4} = 1$ **16.** $\dfrac{(x - 1)^2}{2} - \dfrac{(y + 3)^2}{4} = 1$

In Exercises 17–22, graph the hyperbola using a function grapher.

17. $\dfrac{x^2}{36} - \dfrac{y^2}{16} = 1$ **18.** $\dfrac{y^2}{64} - \dfrac{x^2}{16} = 1$

19. $\dfrac{x^2}{4} - \dfrac{y^2}{9} = 1$ **20.** $\dfrac{y^2}{16} - \dfrac{x^2}{9} = 1$

21. $\dfrac{x^2}{4} - \dfrac{(y - 3)^2}{5} = 1$ **22.** $\dfrac{(y - 3)^2}{9} - \dfrac{(x + 2)^2}{4} = 1$

In Exercises 23–38, find an equation in standard form for the hyperbola that satisfies the given conditions.

23. Foci $(\pm 3, 0)$, transverse axis length 4

24. Foci $(0, \pm 3)$, transverse axis length 4

25. Foci $(0, \pm 15)$, transverse axis length 8

26. Foci $(\pm 5, 0)$, transverse axis length 3

27. Center at $(0, 0)$, $a = 5$, $e = 2$, horizontal focal axis

28. Center at $(0, 0)$, $a = 4$, $e = 3/2$, vertical focal axis

29. Center at $(0, 0)$, $b = 5$, $e = 13/12$, vertical focal axis

30. Center at $(0, 0)$, $c = 6$, $e = 2$, horizontal focal axis

31. Transverse axis endpoints $(2, 3)$ and $(2, -1)$, conjugate axis length 6

32. Transverse axis endpoints $(5, 3)$ and $(-7, 3)$, conjugate axis length 10

33. Transverse axis endpoints $(-1, 3)$ and $(5, 3)$, slope of one asymptote 4/3

34. Transverse axis endpoints $(-2, -2)$ and $(-2, 7)$, slope of one asymptote 4/3

35. Foci $(-4, 2)$ and $(2, 2)$, transverse axis endpoints $(-3, 2)$ and $(1, 2)$

36. Foci $(-3, -11)$ and $(-3, 0)$, transverse axis endpoints $(-3, -9)$ and $(-3, -2)$

37. Center at $(-3, 6)$, $a = 5$, $e = 2$, vertical focal axis

38. Center at $(1, -4)$, $c = 6$, $e = 2$, horizontal focal axis

In Exercises 39–42, find the center, vertices, and the foci of the hyperbola.

39. $\dfrac{(x+1)^2}{144} - \dfrac{(y-2)^2}{25} = 1$ **40.** $\dfrac{(x+4)^2}{12} - \dfrac{(y+6)^2}{13} = 1$

41. $\dfrac{(y+3)^2}{64} - \dfrac{(x-2)^2}{81} = 1$ **42.** $\dfrac{(y-1)^2}{25} - \dfrac{(x+5)^2}{11} = 1$

In Exercises 43–46, graph the hyperbola using a parametric grapher in Dot graphing mode.

43. $\dfrac{y^2}{25} - \dfrac{x^2}{4} = 1$ **44.** $\dfrac{x^2}{30} - \dfrac{y^2}{20} = 1$

45. $\dfrac{(x+3)^2}{12} - \dfrac{(y-6)^2}{5} = 1$ **46.** $\dfrac{(y+1)^2}{15} - \dfrac{(x-2)^2}{6} = 1$

In Exercises 47–50, graph the hyperbola, and find its vertices, foci, and eccentricity.

47. $4(y-1)^2 - 9(x-3)^2 = 36$

48. $4(x-2)^2 - 9(y+4)^2 = 1$

49. $9x^2 - 4y^2 - 36x + 8y - 4 = 0$

50. $25y^2 - 9x^2 - 50y - 54x - 281 = 0$

In Exercises 51 and 52, write an equation for the hyperbola.

51.

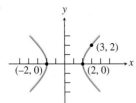

52.

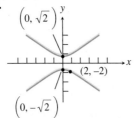

53. Writing to Learn Prove that an equation for the hyperbola with center (0, 0), foci (0, ±c), and semitransverse axis a is $y^2/a^2 - x^2/b^2 = 1$, where $c > a > 0$ and $b^2 = c^2 - a^2$. [*Hint:* Refer to derivation at the beginning of the section.]

54. Degenerate Hyperbolas Graph the degenerate hyperbola.

(a) $\dfrac{x^2}{4} - y^2 = 0$ (b) $\dfrac{y^2}{9} - \dfrac{x^2}{16} = 0$

55. Rogue Comet A comet following a hyperbolic path about the Sun has a perihelion of 120 Gm. When the line from the comet to the Sun is perpendicular to the focal axis of the orbit, the comet is 250 Gm from the Sun. Calculate a, b, c, and e. What are the coordinates of the center of the Sun if the center of the hyperbolic orbit is (0, 0) and the Sun lies on the positive x-axis?

56. Rogue Comet A comet following a hyperbolic path about the Sun has a perihelion of 140 Gm. When the line from the comet to the Sun is perpendicular to the focal axis of the orbit, the comet is 405 Gm from the Sun. Calculate a, b, c, and e. What are the coordinates of the center of the Sun if the center of the hyperbolic orbit is (0, 0) and the Sun lies on the positive x-axis?

57. Long-Range Navigation Three LORAN radio transmitters are positioned as shown in the figure, with R due north of O and Q due east of O. The cruise ship *Princess Ann* receives simultaneous

signals from the three transmitters. The signal from O arrives 323.27 μsec after the signal from R, and 646.53 μsec after the signal from Q. Determine the ship's bearing and distance from point O.

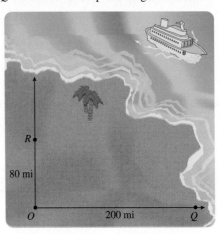

58. Gun Location Observers are located at positions A, B, and C with A due north of B. A cannon is located somewhere in the first quadrant as illustrated in the figure. A hears the sound of the cannon 2 sec before B, and C hears the sound 4 sec before B. Determine the bearing and distance of the cannon from point B. (Assume that sound travels at 1100 ft/sec.)

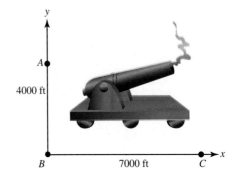

Group Activities In Exercises 59 and 60, solve the system of equations algebraically and support your answer graphically.

59. $\dfrac{x^2}{4} - \dfrac{y^2}{9} = 1$ **60.** $\dfrac{x^2}{4} - y^2 = 1$

$x - \dfrac{2\sqrt{3}}{3}y = -2$ $x^2 + y^2 = 9$

61. Group Activity Consider the system of equations

$$\dfrac{x^2}{4} - \dfrac{y^2}{25} = 1$$

$$\dfrac{x^2}{25} + \dfrac{y^2}{4} = 1$$

(a) Solve the system graphically.

(b) If you have access to a grapher that also does symbolic algebra, use it to find the the exact solutions to the system.

62. Writing to Learn Escape of the Unbound When NASA launches a space probe, the probe reaches a speed sufficient for it to become unbound from Earth and escape along a hyperbolic trajectory. Look up *escape speed* in an astronomy textbook or on the Internet, and write a paragraph in your own words about what you find.

Standardized Test Questions

63. True or False The distance from a focus of a hyperbola to the closer vertex is $a(e - 1)$, where a is the semitransverse axis and e is the eccentricity. Justify your answer.

64. True or False Unlike that of an ellipse, the Pythagorean relation for a hyperbola is the usual $a^2 + b^2 = c^2$. Justify your answer.

In Exercises 65–68, you may use a graphing calculator to solve the problem.

65. Multiple Choice One focus of $x^2 - 4y^2 = 4$ is

(A) $(4, 0)$.

(B) $(\sqrt{5}, 0)$.

(C) $(2, 0)$.

(D) $(\sqrt{3}, 0)$.

(E) $(1, 0)$.

66. Multiple Choice The focal axis of $\dfrac{(x + 5)^2}{9} - \dfrac{(y - 6)^2}{16} = 1$ is

(A) $y = 2$.

(B) $y = 3$.

(C) $y = 4$.

(D) $y = 5$.

(E) $y = 6$.

67. Multiple Choice The center of $4x^2 - 12y^2 - 16x - 72y - 44 = 0$ is

(A) $(2, -2)$.

(B) $(2, -3)$.

(C) $(2, -4)$.

(D) $(2, -5)$.

(E) $(2, -6)$.

68. Multiple Choice The slopes of the asymptotes of the hyperbola $\dfrac{x^2}{4} - \dfrac{y^2}{3} = 1$ are

(A) ± 1.

(B) $\pm 3/2$.

(C) $\pm \sqrt{3}/2$.

(D) $\pm 2/3$.

(E) $\pm 4/3$.

Explorations

69. Constructing Points of a Hyperbola Use a geometry software package, such as *Cabri Geometry II™*, *The Geometer's Sketchpad®*, or a similar application on a hand-held device, to carry out the following construction.

(a) Start by placing the coordinate axes in the construction window.

(b) Construct two points on the *x*-axis at $(\pm 5, 0)$ as the foci.

(c) Construct concentric circles of radii $r = 1, 2, 3, \ldots, 12$ centered at these two foci.

(d) Construct the points where these concentric circles meet and have a difference of radii of $2a = 6$, and overlay the conic that passes through these points if the software has a conic tool.

(e) Find the equation whose graph includes all of these points.

70. Cassegrain Telescope A Cassegrain telescope as described in the section has the dimensions shown in the figure. Find the standard form for the equation of the hyperbola centered at the origin with the focal axis the *x*-axis.

Extending the Ideas

71. Prove that a nondegenerate graph of the equation
$$Ax^2 + Cy^2 + Dx + Ey + F = 0$$
is a hyperbola if $AC < 0$.

72. Writing to Learn The graph of the equation
$$\frac{(x - h)^2}{a^2} - \frac{(y - k)^2}{b^2} = 0$$
is considered to be a degenerate hyperbola. Describe the graph. How is it like a full-fledged hyperbola, and how is it different?

73. Conjugate Hyperbolas The hyperbolas
$$\frac{(x - h)^2}{a^2} - \frac{(y - k)^2}{b^2} = 1 \quad \text{and} \quad \frac{(y - k)^2}{b^2} - \frac{(x - h)^2}{a^2} = 1$$
obtained by switching the order of subtraction in their standard equations are **conjugate hyperbolas**. Prove that these hyperbolas have the same asymptotes and that the conjugate axis of each of these hyperbolas is the transverse axis of the other hyperbola.

74. Focal Width of a Hyperbola Prove that, for the hyperbola
$$\frac{x^2}{a^2} - \frac{y^2}{b^2} = 1,$$
if $x = c$, then $y = \pm b^2/a$. Why is it reasonable to define the **focal width** of such hyperbolas to be $2b^2/a$?

75. Writing to Learn Explain how the standard form equations for the conics are related to
$$Ax^2 + Bxy + Cy^2 + Dx + Ey + F = 0.$$

8.4
Translation and Rotation of Axes

What you'll learn about

- Second-Degree Equations in Two Variables
- Translating Axes versus Translating Graphs
- Rotation of Axes
- Discriminant Test

. . . and why

You will see ellipses, hyperbolas, and parabolas as members of the family of conic sections rather than as separate types of curves.

Second-Degree Equations in Two Variables

In Section 8.1, we began with a unified approach to conic sections, learning that parabolas, ellipses, and hyperbolas are all cross sections of a right circular cone. In Sections 8.1–8.3, we gave separate plane-geometry definitions for parabolas, ellipses, and hyperbolas that led to separate kinds of equations for each type of curve. In this section and the next, we once again consider parabolas, ellipses, and hyperbolas as a unified family of interrelated curves.

In Section 8.1, we claimed that the conic sections can be defined algebraically in the Cartesian plane as the graphs of *second-degree equations in two variables*, that is, equations of the form

$$Ax^2 + Bxy + Cy^2 + Dx + Ey + F = 0,$$

where A, B, and C are not all zero. In this section, we investigate equations of this type, which are really just *quadratic equations in x and y*. Because they are quadratic equations, we can adapt familiar methods to this unfamiliar setting. That is exactly what we do in Examples 1–3.

EXAMPLE 1   Graphing a Second-Degree Equation

Solve for y, and use a function grapher to graph

$$9x^2 + 16y^2 - 18x + 64y - 71 = 0.$$

SOLUTION Rearranging terms yields the equation:

$$16y^2 + 64y + (9x^2 - 18x - 71) = 0.$$

The quadratic formula gives us

$$y = \frac{-64 \pm \sqrt{64^2 - 4(16)(9x^2 - 18x - 71)}}{2(16)}$$

$$= \frac{-8 \pm 3\sqrt{-x^2 + 2x + 15}}{4}$$

$$= -2 \pm \frac{3}{4}\sqrt{-x^2 + 2x + 15}$$

Let

$$Y1 = -2 + 0.75\sqrt{-x^2 + 2x + 15} \text{ and } Y2 = -2 - 0.75\sqrt{-x^2 + 2x + 15},$$

and graph the two equations in the same viewing window, as shown in Figure 8.32. The combined figure appears to be an ellipse.

Now try Exercise 1.

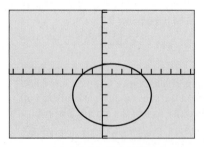

[–9.4, 9.4] by [–6.2, 6.2]

FIGURE 8.32 The graph of $9x^2 + 16y^2 - 18x + 64y - 71 = 0$. (Example 1)

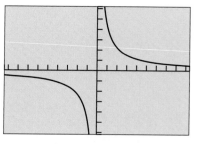

[–9.4, 9.4] by [–6.2, 6.2]

FIGURE 8.33 The graph of $2xy - 9 = 0$. (Example 2)

In the equation in Example 1, there was no Bxy term. None of the examples in Sections 8.1–8.3 included such a *cross-product* term. A cross-product term in the equation causes the graph to tilt relative to the coordinate axes, as illustrated in Examples 2 and 3.

EXAMPLE 2 Graphing a Second-Degree Equation

Solve for y, and use a function grapher to graph

$$2xy - 9 = 0.$$

SOLUTION This equation can be rewritten as $2xy = 9$ or as $y = 9/(2x)$. The graph of this equation is shown in Figure 8.33. It appears to be a hyperbola with a slant focal axis. *Now try Exercise 5.*

EXAMPLE 3 Graphing a Second-Degree Equation

Solve for y, and use a function grapher to graph

$$x^2 + 4xy + 4y^2 - 30x - 90y + 450 = 0.$$

SOLUTION We rearrange the terms as a quadratic equation in y:

$$4y^2 + (4x - 90)y + (x^2 - 30x + 450) = 0.$$

The quadratic formula gives us

$$y = \frac{-(4x - 90) \pm \sqrt{(4x - 90)^2 - 4(4)(x^2 - 30x + 450)}}{2(4)}$$

$$= \frac{45 - 2x \pm \sqrt{225 - 60x}}{4}$$

Let

$$y_1 = \frac{45 - 2x + \sqrt{225 - 60x}}{4} \text{ and } y_2 = \frac{45 - 2x - \sqrt{225 - 60x}}{4},$$

and graph the two equations in the same viewing window, as shown in Figure 8.34a. The combined figure appears to be a parabola, with a slight gap due to grapher failure. The combined graph should connect at a point for which the radicand $225 - 60x = 0$, that is, when $x = 225/60 = 15/4 = 3.75$. Figure 8.34b supports this analysis. *Now try Exercise 9.*

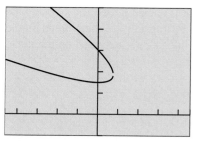

[–23, 23] by [–5, 25]
(a)

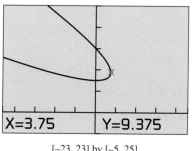

X=3.75   Y=9.375

[–23, 23] by [–5, 25]
(b)

FIGURE 8.34 The graph of $x^2 + 4xy + 4y^2 - 30x - 90y + 450 = 0$ (a) with a gap and (b) with the trace feature activated at the connecting point. (Example 3)

The graphs obtained in Examples 1–3 all *appear* to be conic sections, but how can we be sure? If they are conics, then we probably have classified Examples 1 and 2 correctly, but couldn't the graph in Example 3 (Figure 8.34) be part of an ellipse or one branch of a hyperbola? We now set out to answer these questions and to develop methods for simplifying and classifying second-degree equations in two variables.

Translating Axes versus Translating Graphs

The coordinate axes are often viewed as a permanent fixture of the plane, but this just isn't so. We can shift the position of axes just as we have been shifting the position of graphs since Chapter 1. Such a **translation of axes** produces a new set of axes parallel to the original axes, as shown in Figure 8.35 on the next page.

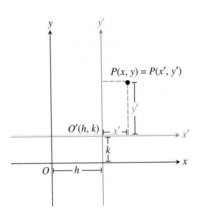

FIGURE 8.35 A translation of Cartesian coordinate axes.

Figure 8.35 shows a plane containing a point P that is named in two ways: using the coordinates (x, y) and the coordinates (x', y'). The coordinates (x, y) are based on the original x- and y-axes and the original origin O, while (x', y') are based on the translated x'- and y'-axes and the corresponding origin O'.

Translation-of-Axes Formulas

The coordinates (x, y) and (x', y') based on parallel sets of axes are related by either of the following **translation formulas**:

$$x = x' + h \quad \text{and} \quad y = y' + k$$

or

$$x' = x - h \quad \text{and} \quad y' = y - k.$$

We use the second pair of translation formulas in Example 4.

EXAMPLE 4   Revisiting Example 1

Prove that $9x^2 + 16y^2 - 18x + 64y - 71 = 0$ is the equation of an ellipse. Translate the coordinate axes so that the origin is at the center of this ellipse.

SOLUTION We complete the square of both x and y:

$$9x^2 - 18x + 16y^2 + 64y = 71$$
$$9(x^2 - 2x + 1) + 16(y^2 + 4y + 4) = 71 + 9(1) + 16(4)$$
$$9(x - 1)^2 + 16(y + 2)^2 = 144$$
$$\frac{(x - 1)^2}{16} + \frac{(y + 2)^2}{9} = 1$$

This is a standard equation of an ellipse. If we let $x' = x - 1$ and $y' = y + 2$, then the equation of the ellipse becomes

$$\frac{(x')^2}{16} + \frac{(y')^2}{9} = 1.$$

Figure 8.36 shows the graph of this final equation in the new $x'y'$ coordinate system, with the original xy-axes overlaid. Compare Figures 8.32 and 8.36.

Now try Exercise 21.

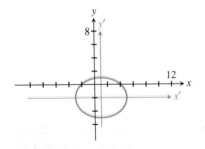

FIGURE 8.36 The graph of $(x')^2/16 + (y')^2/9 = 1$. (Example 4)

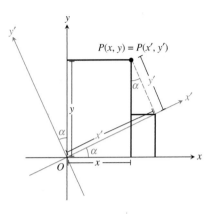

FIGURE 8.37 A rotation of Cartesian coordinate axes.

Rotation of Axes

To show that the equation in Example 2 or 3 is the equation of a conic section, we need to rotate the coordinate axes so that one axis aligns with the (focal) axis of the conic. In such a **rotation of axes**, the origin stays fixed, and we rotate the x- and y-axes through an angle α to obtain the x'- and y'-axes. (See Figure 8.37.)

Figure 8.37 shows a plane containing a point P named in two ways: as (x, y) and as (x', y'). The coordinates (x, y) are based on the original x- and y-axes, while (x', y') are based on the rotated x'- and y'-axes.

Rotation-of-Axes Formulas

The coordinates (x, y) and (x', y') based on rotated sets of axes are related by either of the following **rotation formulas**:

$$x' = x \cos \alpha + y \sin \alpha \quad \text{and} \quad y' = -x \sin \alpha + y \cos \alpha,$$

or

$$x = x' \cos \alpha - y' \sin \alpha \quad \text{and} \quad y = x' \sin \alpha + y' \cos \alpha.$$

where α, $0 < \alpha < \pi/2$, is the *angle of rotation*.

The first pair of equations was established in Example 10 of Section 7.2. The second pair can be derived directly from the geometry of Figure 8.37 (see Exercise 55) and is used in Example 5.

EXAMPLE 5 Revisiting Example 2

Prove that $2xy - 9 = 0$ is the equation of a hyperbola by rotating the coordinate axes through an angle $\alpha = \pi/4$.

SOLUTION Because $\cos(\pi/4) = \sin(\pi/4) = 1/\sqrt{2}$, the rotation equations become

$$x = \frac{x' - y'}{\sqrt{2}} \quad \text{and} \quad y = \frac{x' + y'}{\sqrt{2}}.$$

So by rotating the axes, the equation $2xy - 9 = 0$ becomes

$$2\left(\frac{x' - y'}{\sqrt{2}}\right)\left(\frac{x' + y'}{\sqrt{2}}\right) - 9 = 0$$

$$(x')^2 - (y')^2 - 9 = 0$$

To see that this is the equation of a hyperbola, we put it in standard form:

$$(x')^2 - (y')^2 = 9$$

$$\frac{(x')^2}{9} - \frac{(y')^2}{9} = 1$$

Figure 8.38 shows the graph of the original equation in the original xy system with the $x'y'$-axes overlaid. *Now try Exercise 37.*

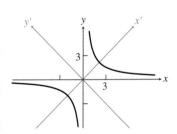

FIGURE 8.38 The graph of $2xy - 9 = 0$. (Example 5)

In Example 5 we converted a second-degree equation in x and y into a second-degree equation in x' and y' using the rotation formulas. By choosing the angle of rotation appropriately, there was no $x'y'$ cross-product term in the final equation, which allowed us to put it in standard form. We now generalize this process.

Coefficients for a Conic in a Rotated System

If we apply the rotation formulas to the general second-degree equation in x and y, we obtain a second-degree equation in x' and y' of the form

$$A'x'^2 + B'x'y' + C'y'^2 + D'x' + E'y' + F' = 0,$$

where the coefficients are

$$A' = A \cos^2 \alpha + B \cos \alpha \sin \alpha + C \sin^2 \alpha$$

$$B' = B \cos 2\alpha + (C - A) \sin 2\alpha$$

$$C' = C \cos^2 \alpha - B \cos \alpha \sin \alpha + A \sin^2 \alpha$$

$$D' = D \cos \alpha + E \sin \alpha$$

$$E' = E \cos \alpha - D \sin \alpha$$

$$F' = F$$

In order to eliminate the cross-product term and thus align the coordinate axes with the focal axis of the conic, we rotate the coordinate axes through an angle α that causes B' to equal 0. Setting $B' = B \cos 2\alpha + (C - A) \sin 2\alpha = 0$ leads to the following useful result.

Angle of Rotation to Eliminate the Cross-Product Term

If $B \neq 0$, an angle of rotation α such that

$$\cot 2\alpha = \frac{A - C}{B} \quad \text{and} \quad 0 < \alpha < \frac{\pi}{2}$$

will eliminate the term $B'x'y'$ from the second-degree equation in the rotated $x'y'$ coordinate system.

EXAMPLE 6   Revisiting Example 3

Prove that $x^2 + 4xy + 4y^2 - 30x - 90y + 450 = 0$ is the equation of a parabola by rotating the coordinate axes through a suitable angle α.

SOLUTION The angle of rotation α must satisfy the equation

$$\cot 2\alpha = \frac{A - C}{B} = \frac{1 - 4}{4} = -\frac{3}{4}.$$

So

$$\cos 2\alpha = -\frac{3}{5},$$

continued

and thus

$$\cos \alpha = \sqrt{\frac{1 + \cos 2\alpha}{2}} = \sqrt{\frac{1 + (-3/5)}{2}} = \frac{1}{\sqrt{5}},$$

$$\sin \alpha = \sqrt{\frac{1 - \cos 2\alpha}{2}} = \sqrt{\frac{1 - (-3/5)}{2}} = \frac{2}{\sqrt{5}}.$$

Therefore the coefficients of the transformed equation are

$$A' = 1 \cdot \frac{1}{5} + 4 \cdot \frac{2}{5} + 4 \cdot \frac{4}{5} = \frac{25}{5} = 5$$

$$B' = 0$$

$$C' = 4 \cdot \frac{1}{5} - 4 \cdot \frac{2}{5} + 1 \cdot \frac{4}{5} = 0$$

$$D' = -30 \cdot \frac{1}{\sqrt{5}} - 90 \cdot \frac{2}{\sqrt{5}} = -\frac{210}{\sqrt{5}} = -42\sqrt{5}$$

$$E' = -90 \cdot \frac{1}{\sqrt{5}} + 30 \cdot \frac{2}{\sqrt{5}} = -\frac{30}{\sqrt{5}} = -6\sqrt{5}$$

$$F' = 450$$

So the equation $x^2 + 4xy + 4y^2 - 30x - 90y + 450 = 0$ becomes

$$5x'^2 - 42\sqrt{5}x' - 6\sqrt{5}y' + 450 = 0.$$

After completing the square of the x-terms, the equation becomes

$$\left(x' - \frac{21}{\sqrt{5}}\right)^2 = \frac{6}{\sqrt{5}}\left(y' - \frac{3\sqrt{5}}{10}\right).$$

If we translate using $h = 21/\sqrt{5}$ and $k = 3\sqrt{5}/10$, then the equation becomes

$$(x'')^2 = \frac{6}{\sqrt{5}}(y''),$$

a standard equation of a parabola.

Figure 8.39 shows the graph of the original equation in the original xy coordinate system, with the $x''y''$-axes overlaid. ***Now try Exercise 39.***

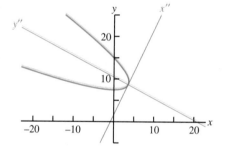

FIGURE 8.39 The graph of $x^2 + 4xy + 4y^2 - 30x - 90y + 450 = 0$. (Example 6)

Discriminant Test

Example 6 demonstrates that the algebra of rotation can get ugly. Fortunately, we can determine which type of conic a second-degree equation represents by looking at the sign of the **discriminant** $B^2 - 4AC$.

Discriminant Test

The second-degree equation $Ax^2 + Bxy + Cy^2 + Dx + Ey + F = 0$ graphs as

- a hyperbola if $B^2 - 4AC > 0$,

- a parabola if $B^2 - 4AC = 0$,

- an ellipse if $B^2 - 4AC < 0$,

except for degenerate cases.

This test hinges on the fact that the discriminant $B^2 - 4AC$ is **invariant under rotation**; in other words, even though A, B, and C do change when we rotate the coordinate axes, the combination $B^2 - 4AC$ maintains its value.

EXAMPLE 7   Revisiting Examples 5 and 6

(a) In Example 5, before the rotation $B^2 - 4AC = (2)^2 - 4(0)(0) = 4$, and after the rotation $B'^2 - 4A'C' = (0)^2 - 4(1)(-1) = 4$. The positive discriminant tells us the conic is a hyperbola.

(b) In Example 6, before the rotation $B^2 - 4AC = (4)^2 - 4(1)(4) = 0$, and after the rotation $B'^2 - 4A'C' = (0)^2 - 4(5)(0) = 0$. The zero discriminant tells us the conic is a parabola.   *Now try Exercise 43.*

Not only is the discriminant $B^2 - 4AC$ invariant under rotation, but also its *sign* is invariant under translation and under algebraic manipulations that preserve the equivalence of the equation, such as multiplying both sides of the equation by a nonzero constant.

The discriminant test can be applied to degenerate conics. Table 8.2 displays the three basic types of conic sections grouped with their associated degenerate conics. Each conic or degenerate conic is shown with a sample equation and the sign of its discriminant.

Table 8.2 Conics and the Equation $Ax^2 + Bxy + Cy^2 + Dx + Ey + F = 0$

Conic	Sample Equation	A	B	C	D	E	F	Sign of Discriminant
Hyperbola	$x^2 - 2y^2 = 1$	1		-2			-1	Positive
Intersecting lines	$x^2 + xy = 0$	1	1					Positive
Parabola	$x^2 = 2y$	1				-2		Zero
Parallel lines	$x^2 = 4$	1					-4	Zero
One line	$y^2 = 0$			1				Zero
No graph	$x^2 = -1$	1					1	Zero
Ellipse	$x^2 + 2y^2 = 1$	1		2			-1	Negative
Circle	$x^2 + y^2 = 9$	1		1			-9	Negative
Point	$x^2 + y^2 = 0$	1		1				Negative
No graph	$x^2 + y^2 = -1$	1		1			1	Negative

QUICK REVIEW 8.4 *(For help, go to Sections 4.7 and 5.4.)*

In Exercises 1–10, assume $0 \le \alpha < \pi/2$.

1. Given that $\cot 2\alpha = 5/12$, find $\cos 2\alpha$.

2. Given that $\cot 2\alpha = 8/15$, find $\cos 2\alpha$.

3. Given that $\cot 2\alpha = 1/\sqrt{3}$, find $\cos 2\alpha$.

4. Given that $\cot 2\alpha = 2/\sqrt{5}$, find $\cos 2\alpha$.

5. Given that $\cot 2\alpha = 0$, find α.

6. Given that $\cot 2\alpha = \sqrt{3}$, find α.

7. Given that $\cot 2\alpha = 3/4$, find $\cos \alpha$.

8. Given that $\cot 2\alpha = 3/\sqrt{7}$, find $\cos \alpha$.

9. Given that $\cot 2\alpha = 5/\sqrt{11}$, find $\sin \alpha$.

10. Given that $\cot 2\alpha = 45/28$, find $\sin \alpha$.

SECTION 8.4 EXERCISES

In Exercises 1–12, solve for y, and use a function grapher to graph the conic.

1. $x^2 + y^2 - 6x + 10y + 18 = 0$

2. $4x^2 + y^2 + 24x - 2y + 21 = 0$

3. $y^2 - 8x - 8y + 8 = 0$

4. $x^2 - 4y^2 + 6x - 40y + 91 = 0$

5. $-4xy + 16 = 0$

6. $2xy + 6 = 0$

7. $xy - y - 8 = 0$

8. $2x^2 - 5xy + y = 0$

9. $2x^2 - xy + 3y^2 - 3x + 4y - 6 = 0$

10. $-x^2 + 3xy + 4y^2 - 5x - 10y - 20 = 0$

11. $2x^2 - 4xy + 8y^2 - 10x + 4y - 13 = 0$

12. $2x^2 - 4xy + 2y^2 - 5x + 6y - 15 = 0$

In Exercises 13–16, write an equation in standard form for the conic shown.

13.

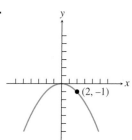

14.

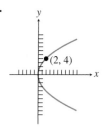

$(2, 4)$

$(2, -1)$

15.

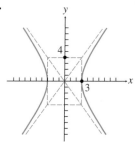

4

3

16.

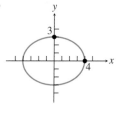

3

4

In Exercises 17–20, using the point $P(x, y)$ and the translation information, find the coordinates of P in the translated $x'y'$ coordinate system.

17. $P(x, y) = (2, 3),\ h = -2,\ k = 4$

18. $P(x, y) = (-2, 5),\ h = -4,\ k = -7$

19. $P(x, y) = (6, -3),\ h = 1,\ k = \sqrt{5}$

20. $P(x, y) = (-5, -4),\ h = \sqrt{2},\ k = -3$

In Exercises 21–30, identify the type of conic, write the equation in standard form, translate the conic to the origin, and sketch it in the translated coordinate system.

21. $4y^2 - 9x^2 - 18x - 8y - 41 = 0$

22. $2x^2 + 3y^2 + 12x - 24y + 60 = 0$

23. $x^2 + 2x - y + 3 = 0$ **24.** $3x^2 - 6x - 6y + 10 = 0$

25. $9x^2 + 4y^2 - 18x + 16y - 11 = 0$

26. $16x^2 - y^2 - 32x - 6y - 57 = 0$

27. $y^2 - 4y - 8x + 20 = 0$ **28.** $2x^2 - 4x + y^2 - 6y = 9$

29. $2x^2 - y^2 + 4x + 6 = 0$ **30.** $y^2 - 2y + 4x - 12 = 0$

31. Writing to Learn Translation Formulas Use the geometric relationships illustrated in Figure 8.35 to explain the translation formulas $x = x' + h$ and $y = y' + k$.

32. Translation Formulas Prove that if $x = x' + h$ and $y = y' + k$, then $x' = x - h$ and $y' = y - k$.

In Exercises 33–36, using the point $P(x, y)$ and the rotation information, find the coordinates of P in the rotated $x'y'$ coordinate system.

33. $P(x, y) = (-2, 5),\ \alpha = \pi/4$

34. $P(x, y) = (6, -3),\ \alpha = \pi/3$

35. $P(x, y) = (-5, -4),\ \cot 2\alpha = -3/5$

36. $P(x, y) = (2, 3),\ \cot 2\alpha = 0$

In Exercises 37–40, identify the type of conic, and rotate the coordinate axes to eliminate the xy term. Write and graph the transformed equation.

37. $xy = 8$

38. $3xy + 15 = 0$

39. $2x^2 + \sqrt{3}xy + y^2 - 10 = 0$

40. $3x^2 + 2\sqrt{3}xy + y^2 - 14 = 0$

In Exercises 41 and 42, identify the type of conic, solve for y, and graph the conic. Approximate the angle of rotation needed to eliminate the xy term.

41. $16x^2 - 20xy + 9y^2 - 40 = 0$

42. $4x^2 - 6xy + 2y^2 - 3x + 10y - 6 = 0$

In Exercises 43–52, use the discriminant $B^2 - 4AC$ to decide whether the equation represents a parabola, an ellipse, or a hyperbola.

43. $x^2 - 4xy + 10y^2 + 2y - 5 = 0$

44. $x^2 - 4xy + 3x + 25y - 6 = 0$

45. $9x^2 - 6xy + y^2 - 7x + 5y = 0$

46. $-xy + 3y^2 - 4x + 2y + 8 = 0$

47. $8x^2 - 4xy + 2y^2 + 6 = 0$

48. $3x^2 - 12xy + 4y^2 + x - 5y - 4 = 0$

49. $x^2 - 3y^2 - y - 22 = 0$

50. $5x^2 + 4xy + 3y^2 + 2x + y = 0$

51. $4x^2 - 2xy + y^2 - 5x + 18 = 0$

52. $6x^2 - 4xy + 9y^2 - 40x + 20y - 56 = 0$

53. Revisiting Example 5 Using the results of Example 5, find the center, vertices, and foci of the hyperbola $2xy - 9 = 0$ in the original coordinate system.

54. Revisiting Examples 3 and 6 Use information from Examples 3 and 6

  (a) to prove that the point $P(x, y) = (3.75, 9.375)$ where the graphs of $Y1 = (45 - 2x + \sqrt{225 - 60x})/4$ and $Y2 = (45 - 2x - \sqrt{225 - 60x})/4$ meet is not the vertex of the parabola,

  (b) to prove that the point $V(x, y) = (3.6, 8.7)$ is the vertex of the parabola.

55. Rotation Formulas Prove $x = x' \cos \alpha - y' \sin \alpha$ and $y = x' \sin \alpha + y' \cos \alpha$ using the geometric relationships illustrated in Figure 8.37.

56. Rotation Formulas Prove that if $x' = x \cos \alpha + y \sin \alpha$ and $y' = -x \sin \alpha + y \cos \alpha$, then $x = x' \cos \alpha - y' \sin \alpha$ and $y = x' \sin \alpha + y' \cos \alpha$.

Standardized Test Questions

57. True or False The graph of the equation $Ax^2 + Cy^2 + Dx + Ey + F = 0$ (A and C not both zero) has a focal axis aligned with the coordinate axes. Justify your answer.

58. True or False The graph of the equation $x^2 + y^2 + Dx + Ey + F = 0$ is a circle or a degenerate circle. Justify your answer.

In Exercises 59–62, solve the problem without using a calculator.

59. Multiple Choice Which of the following is **not** a reason to translate the axes of a conic?

  (A) to simplify its equation

  (B) to eliminate the cross-product term

  (C) to place its center or vertex at the origin

  (D) to make it easier to identify its type

  (E) to make it easier to sketch by hand

60. Multiple Choice Which of the following is **not** a reason to rotate the axes of a conic?

  (A) to simplify its equation

  (B) to eliminate the cross-product term

  (C) to place its center or vertex at the origin

  (D) to make it easier to identify its type

  (E) to make it easier to sketch by hand

61. Multiple Choice The vertices of $9x^2 + 16y^2 - 18x + 64y - 71 = 0$ are

  (A) $(1 \pm 4, -2)$ **(B)** $(1 \pm 3, -2)$

  (C) $(4 \pm 1, 3)$ **(D)** $(4 \pm 2, 3)$

  (E) $(1, -2 \pm 3)$

62. Multiple Choice The asymptotes of the hyperbola $xy = 4$ are

  (A) $y = x, y = -x.$ **(B)** $y = 2x, y = -\dfrac{x}{2}.$

  (C) $y = -2x, y = \dfrac{x}{2}.$ **(D)** $y = 4x, y = -\dfrac{x}{4}.$

  (E) the coordinate axes

Explorations

63. Axes of Oblique Conics The axes of conics that are not aligned with the coordinate axes are often included in the graphs of conics.

  (a) Recreate the graph shown in Figure 8.38 using a function grapher including the x'- and y'-axes. What are the equations of these rotated axes?

  (b) Recreate the graph shown in Figure 8.39 using a function grapher including the x''- and y''-axes. What are the equations of these rotated and translated axes?

64. The Discriminant Determine what happens to the sign of $B^2 - 4AC$ within the equation
$$Ax^2 + Bxy + Cy^2 + Dx + Ey + F = 0$$
when

  (a) the axes are translated h units horizontally and k units vertically,

  (b) both sides of the equation are multiplied by the same nonzero constant k.

Extending the Ideas

65. Group Activity Working together prove that the formulas for the coefficients A', B', C', D', E', and F' in a rotated system given on page 670 are correct.

66. Identifying a Conic Develop a way to decide whether $Ax^2 + Cy^2 + Dx + Ey + F = 0$, with A and C not both 0, represents a parabola, an ellipse, or a hyperbola. Write an example to illustrate each of the three cases.

67. Rotational Invariant Prove that $B'^2 - 4A'C' = B^2 - 4AC$ when the xy coordinate system is rotated through an angle α.

68. Other Rotational Invariants Prove each of the following are invariants under rotation:

  (a) F,   **(b)** $A + C$,   **(c)** $D^2 + E^2$.

69. Degenerate Conics Graph all of the degenerate conics listed in Table 8.2. Recall that *degenerate cones* occur when the generator and axis of the cone are parallel or perpendicular. (See Figure 8.2.) Explain the occurrence of all of the degenerate conics listed on the basis of cross sections of typical or degenerate right circular cones.

8.5
Polar Equations of Conics

. . . and why

You will learn the approach to conics used by astronomers.

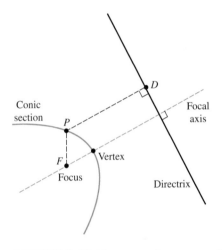

FIGURE 8.40 The geometric structure of a conic section.

Eccentricity Revisited

Eccentricity and polar coordinates provide ways to see once again that parabolas, ellipses, and hyperbolas are a unified family of interrelated curves. We can define these three curves simultaneously by generalizing the focus-directrix definition of parabola given in Section 8.1.

Focus-Directrix Definition Conic Section

A **conic section** is the set of all points in a plane whose distances from a particular point (the **focus**) and a particular line (the **directrix**) in the plane have a constant ratio. (We assume that the focus does not lie on the directrix.)

The line passing through the focus and perpendicular to the directrix is the **(focal) axis** of the conic section. The axis is a line of symmetry for the conic. The point where the conic intersects its axis is a **vertex** of the conic. If P is a point of the conic, F is the focus, and D is the point of the directrix closest to P, then the constant ratio PF/PD is the **eccentricity e** of the conic (see Figure 8.40). A parabola has one focus and one directrix. Ellipses and hyperbolas have two focus-directrix pairs, and either focus-directrix pair can be used with the eccentricity to generate the entire conic section.

Focus-Directrix-Eccentricity Relationship

If P is a point of a conic section, F is the conic's focus, and D is the point of the directrix closest to P, then

$$e = \frac{PF}{PD} \quad \text{and} \quad PF = e \cdot PD,$$

where e is a constant and the eccentricity of the conic. Moreover, the conic is

- a hyperbola if $e > 1$,
- a parabola if $e = 1$,
- an ellipse if $e < 1$.

In this approach to conic sections, the eccentricity e is a strictly positive constant, and there are no circles or other degenerate conics.

REMARKS

- To be consistent with our work on parabolas, we could use 2p for the distance from the focus to the directrix, but following George B. Thomas, Jr. we use k for this distance. This simplifies our polar equations of conics.

- Rather than religiously using polar coordinates and equations, we use a mixture of the polar and Cartesian systems. So, for example, we use $x = k$ for the directrix rather than $r \cos \theta = k$ or $r = k \sec \theta$.

Writing Polar Equations for Conics

Our focus-directrix definition of conics works best in combination with polar coordinates. Recall that in polar coordinates the origin is the *pole* and the x-axis is the *polar axis*. To obtain a polar equation for a conic section, we position the pole at the conic's focus and the polar axis along the focal axis with the directrix to the right of the pole (Figure 8.41). If the distance from the focus to the directrix is k, the Cartesian equation of the directrix is $x = k$. From Figure 8.41, we see that

$$PF = r \quad \text{and} \quad PD = k - r \cos \theta.$$

So the equation $PF = e \cdot PD$ becomes

$$r = e(k - r \cos \theta),$$

which when solved for r is

$$r = \frac{ke}{1 + e \cos \theta}.$$

In Exercise 53, you are asked to show that this equation is still valid if $r < 0$ or $r \cos \theta > k$. This one equation can produce all sizes and shapes of nondegenerate conic sections. Figure 8.42 shows three typical graphs for this equation. In Exploration 1, you will investigate how changing the value of e affects the graph of $r = ke/(1 + e \cos \theta)$.

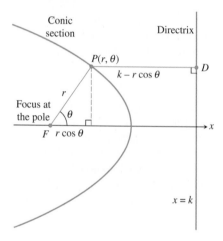

FIGURE 8.41 A conic section in the polar plane.

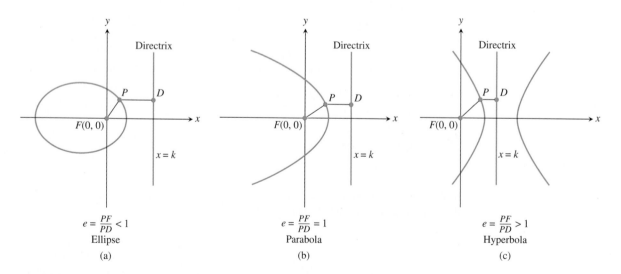

FIGURE 8.42 The three types of conics possible for $r = ke/(1 + e \cos \theta)$.

EXPLORATION 1 Graphing Polar Equations of Conics

Set your grapher to Polar and Dot graphing modes, and to Radian mode. Using $k = 3$, an xy window of $[-12, 24]$ by $[-12, 12]$, θmin $= 0$, θmax $= 2\pi$, and θstep $= \pi/48$, graph

$$r = \frac{ke}{1 + e \cos \theta}$$

for $e = 0.7, 0.8, 1, 1.5, 3$. Identify the type of conic section obtained for each e value. Overlay the five graphs, and explain how changing the value of e affects the graph of $r = ke/(1 + e \cos \theta)$. Explain how the five graphs are similar and how they are different.

Polar Equations for Conics

The four standard orientations of a conic in the polar plane are as follows.

(a) $r = \dfrac{ke}{1 + e \cos \theta}$

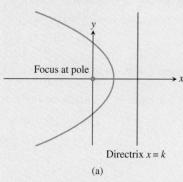

(a)

(b) $r = \dfrac{ke}{1 - e \cos \theta}$

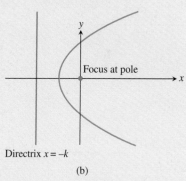

(b)

(c) $r = \dfrac{ke}{1 + e \sin \theta}$

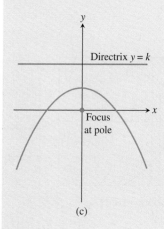

(c)

(d) $r = \dfrac{ke}{1 - e \sin \theta}$

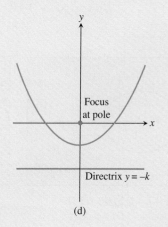

(d)

EXAMPLE 1 Writing and Graphing Polar Equations of Conics

Given that the focus is at the pole, write a polar equation for the specified conic, and graph it.

(a) Eccentricity $e = 3/5$, directrix $x = 2$.

(b) Eccentricity $e = 1$, directrix $x = -2$.

(c) Eccentricity $e = 3/2$, directrix $y = 4$.

SOLUTION

(a) Setting $e = 3/5$ and $k = 2$ in $r = \dfrac{ke}{1 + e \cos \theta}$ yields

$$r = \frac{2(3/5)}{1 + (3/5) \cos \theta}$$

$$= \frac{6}{5 + 3 \cos \theta}.$$

Figure 8.43a shows this ellipse and the given directrix.

(b) Setting $e = 1$ and $k = 2$ in $r = \dfrac{ke}{1 - e \cos \theta}$ yields

$$r = \frac{2}{1 - \cos \theta}.$$

Figure 8.43b shows this parabola and its directrix.

(c) Setting $e = 3/2$ and $k = 4$ in $r = \dfrac{ke}{1 + e \sin \theta}$ yields

$$r = \frac{4(3/2)}{1 + (3/2) \sin \theta}$$

$$= \frac{12}{2 + 3 \sin \theta}.$$

Figure 8.43c shows this hyperbola and the given directrix.

Now try Exercise 1.

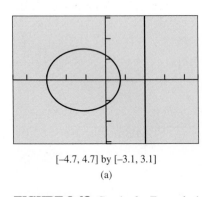

[−4.7, 4.7] by [−3.1, 3.1]

(a)

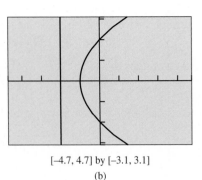

[−4.7, 4.7] by [−3.1, 3.1]

(b)

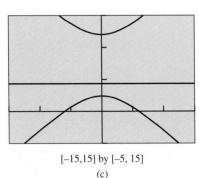

[−15,15] by [−5, 15]

(c)

FIGURE 8.43 Graphs for Example 1.

Analyzing Polar Equations of Conics

The first step in analyzing the polar equations of a conic section is to use the eccentricity to identify which type of conic the equation represents. Then we determine the equation of the directrix.

EXAMPLE 2 Identifying Conics from Their Polar Equations

Determine the eccentricity, the type of conic, and the directrix.

(a) $r = \dfrac{6}{2 + 3 \cos \theta}$ **(b)** $r = \dfrac{6}{4 - 3 \sin \theta}$

SOLUTION

(a) Dividing numerator and denominator by 2 yields $r = 3/(1 + 1.5 \cos \theta)$. So the eccentricity $e = 1.5$, and thus the conic is a hyperbola. The numerator $ke = 3$, so $k = 2$, and thus the equation of the directrix is $x = 2$.

(b) Dividing numerator and denominator by 4 yields $r = 1.5/(1 - 0.75 \sin \theta)$. So the eccentricity $e = 0.75$, and thus the conic is an ellipse. The numerator $ke = 1.5$, so $k = 2$, and thus the equation of the directrix is $y = -2$.

Now try Exercise 7.

All of the geometric properties and features of parabolas, ellipses, and hyperbolas developed in Sections 8.1–8.3 still apply in the polar coordinate setting. In Example 3 we use this prior knowledge.

EXAMPLE 3 Analyzing a Conic

Analyze the conic section given by the equation $r = 16/(5 - 3 \cos \theta)$. Include in the analysis the values of e, a, b, and c.

SOLUTION Dividing numerator and denominator by 5 yields

$$r = \frac{3.2}{1 - 0.6 \cos \theta}.$$

So the eccentricity $e = 0.6$, and thus the conic is an ellipse. Figure 8.44 shows this ellipse. The vertices (endpoints of the major axis) have polar coordinates $(8, 0)$ and $(2, \pi)$. So $2a = 8 + 2 = 10$, and thus $a = 5$.

The vertex $(2, \pi)$ is 2 units to the left of the pole, and the pole is a focus of the ellipse. So $a - c = 2$, and thus $c = 3$. An alternative way to find c is to use the fact that the eccentricity of an ellipse is $e = c/a$, and thus $c = ae = 5 \cdot 0.6 = 3$.

To find b we use the Pythagorean relation of an ellipse:

$$b = \sqrt{a^2 - c^2} = \sqrt{25 - 9} = 4.$$

With all of this information, we can write the Cartesian equation of the ellipse:

$$\frac{(x - 3)^2}{25} + \frac{y^2}{16} = 1.$$

Now try Exercise 31.

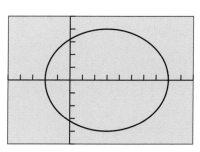

[–5, 10] by [–5, 5]

FIGURE 8.44 Graph of the ellipse $r = 16/(5 - 3 \cos \theta)$. (Example 3)

Orbits Revisited

Polar equations for conics are used extensively in celestial mechanics, the branch of astronomy based on the work of Kepler and others who have studied the motion of celestial bodies. The polar equations of conic sections are well suited to the *two-body problem* of celestial mechanics for several reasons. First, the same equations are used for ellipses, parabolas, and hyperbolas—the paths of one body traveling about another. Second, a focus of the conic is always at the pole. This arrangement has two immediate advantages:

• The pole can be thought of as the center of the larger body, such as the Sun, with the smaller body, such as Earth, following a conic path about the larger body.

• The coordinates given by a polar equation are the distance *r* between the two bodies and the direction θ from the larger body to the smaller body relative to the axis of the conic path of motion.

For these reasons, polar coordinates are preferred over Cartesian coordinates for studying orbital motion.

Table 8.3 Semimajor Axes and Eccentricities of the Planets

Planet	Semimajor Axis (Gm)	Eccentricity
Mercury	57.9	0.2056
Venus	108.2	0.0068
Earth	149.6	0.0167
Mars	227.9	0.0934
Jupiter	778.3	0.0485
Saturn	1427	0.0560
Uranus	2869	0.0461
Neptune	4497	0.0050
Pluto	5900	0.2484

Source: Shupe, et al., National Geographic Atlas of the World (rev. 6th ed.).
Washington, DC: National Geographic Society, 1992, plate 116, and other sources.

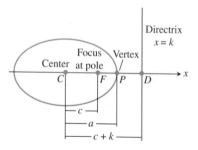

FIGURE 8.45 Geometric relationships within an ellipse.

To use the data in Table 8.3 to create polar equations for the elliptical orbits of the planets, we need to express the equation $r = ke/(1 + e \cos \theta)$ in terms of a and e. We apply the formula $PF = e \cdot PD$ to the ellipse shown in Figure 8.45:

$$e \cdot PD = PF$$

$e(c + k - a) = a - c$	From Figure 8.45
$e(ae + k - a) = a - ae$	Use $e = c/a$.
$ae^2 + ke - ae = a - ae$	Distribute the e.
$ae^2 + ke = a$	Add ae.
$ke = a - ae^2$	Subtract ae^2.
$ke = a(1 - e^2)$	Factor.

So the equation $r = ke/(1 + e \cos \theta)$ can be rewritten as follows:

> **Ellipse with Eccentricity e and Semimajor Axis a**
>
> $$r = \frac{a(1 - e^2)}{1 + e \cos \theta}$$

In this form of the equation, when $e = 0$, the equation reduces to $r = a$, the equation of a circle with radius a.

EXAMPLE 4   Analyzing a Planetary Orbit

Find a polar equation for the orbit of Mercury, and use it to approximate its aphelion (farthest distance from the Sun) and perihelion (closest distance to the Sun).

SOLUTION Setting $e = 0.2056$ and $a = 57.9$ in

$$r = \frac{a(1 - e^2)}{1 + e \cos \theta} \quad \text{yields} \quad r = \frac{57.9(1 - 0.2056^2)}{1 + 0.2056 \cos \theta}.$$

Mercury's aphelion is

$$r = \frac{57.9(1 - 0.2056^2)}{1 - 0.2056} \approx 69.8 \text{ Gm.}$$

Mercury's perihelion is

$$r = \frac{57.9(1 - 0.2056^2)}{1 + 0.2056} \approx 46.0 \text{ Gm.}$$

Now try Exercise 41.

QUICK REVIEW 8.5 *(For help, go to Section 6.4.)*

In Exercises 1 and 2, solve for r.

1. $(3, \theta) = (r, \theta + \pi)$

2. $(-2, \theta) = (r, \theta + \pi)$

In Exercises 3 and 4, solve for θ.

3. $(1.5, \pi/6) = (-1.5, \theta), -2\pi \le \theta \le 2\pi$

4. $(-3, 4\pi/3) = (3, \theta), -2\pi \le \theta \le 2\pi$

In Exercises 5 and 6, find the focus and directrix of the parabola.

5. $x^2 = 16y$

6. $y^2 = -12x$

In Exercises 7–10, find the foci and vertices of the conic.

7. $\dfrac{x^2}{9} + \dfrac{y^2}{4} = 1$

8. $\dfrac{y^2}{25} + \dfrac{x^2}{9} = 1$

9. $\dfrac{x^2}{16} - \dfrac{y^2}{9} = 1$

10. $\dfrac{y^2}{36} - \dfrac{x^2}{4} = 1$

SECTION 8.5 EXERCISES

In Exercises 1–6, find a polar equation for the conic with a focus at the pole and the given eccentricity and directrix. Identify the conic, and graph it.

1. $e = 1$, $x = -2$ **2.** $e = 5/4$, $x = 4$

3. $e = 3/5$, $y = 4$ **4.** $e = 1$, $y = 2$

5. $e = 7/3$, $y = -1$ **6.** $e = 2/3$, $x = -5$

In Exercises 7–14, determine the eccentricity, type of conic, and directrix.

7. $r = \dfrac{2}{1 + \cos \theta}$ **8.** $r = \dfrac{6}{1 + 2 \cos \theta}$

9. $r = \dfrac{5}{2 - 2 \sin \theta}$ **10.** $r = \dfrac{2}{4 - \cos \theta}$

11. $r = \dfrac{20}{6 + 5 \sin \theta}$ **12.** $r = \dfrac{42}{2 - 7 \sin \theta}$

13. $r = \dfrac{6}{5 + 2 \cos \theta}$ **14.** $r = \dfrac{20}{2 + 5 \sin \theta}$

In Exercises 15–20, match the polar equation with its graph, and identify the viewing window.

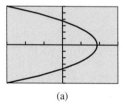

(a)

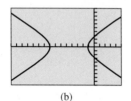

(b)

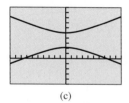

(c)

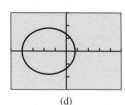

(d)

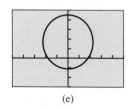

(e)

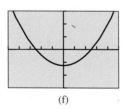

(f)

15. $r = \dfrac{8}{3 - 4 \cos \theta}$ **16.** $r = \dfrac{4}{3 + 2 \cos \theta}$

17. $r = \dfrac{5}{2 - 2 \sin \theta}$ **18.** $r = \dfrac{9}{5 - 3 \sin \theta}$

19. $r = \dfrac{15}{2 + 5 \sin \theta}$ **20.** $r = \dfrac{15}{4 + 4 \cos \theta}$

In Exercises 21–24, find a polar equation for the ellipse with a focus at the pole and the given polar coordinates as the endpoints of its major axis.

21. $(1.5, 0)$ and $(6, \pi)$ **22.** $(1.5, 0)$ and $(1, \pi)$

23. $(1, \pi/2)$ and $(3, 3\pi/2)$ **24.** $(3, \pi/2)$ and $(0.75, -\pi/2)$

In Exercises 25–28, find a polar equation for the hyperbola with a focus at the pole and the given polar coordinates as the endpoints of its transverse axis.

25. $(3, 0)$ and $(-15, \pi)$ **26.** $(-3, 0)$ and $(1.5, \pi)$

27. $\left(2.4, \dfrac{\pi}{2}\right)$ and $\left(-12, \dfrac{3\pi}{2}\right)$ **28.** $\left(-6, \dfrac{\pi}{2}\right)$ and $\left(2, \dfrac{3\pi}{2}\right)$

In Exercises 29 and 30, find a polar equation for the conic with a focus at the pole.

29.

$(3, \pi)$, $(0.75, 0)$

30.

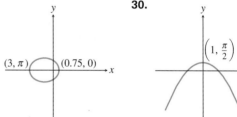

$\left(1, \dfrac{\pi}{2}\right)$

In Exercises 31–36, graph the conic, and find the values of e, a, b, and c.

31. $r = \dfrac{21}{5 - 2 \cos \theta}$ **32.** $r = \dfrac{11}{6 - 5 \sin \theta}$

33. $r = \dfrac{24}{4 + 2 \sin \theta}$ **34.** $r = \dfrac{16}{5 + 3 \cos \theta}$

35. $r = \dfrac{16}{3 + 5 \cos \theta}$ **36.** $r = \dfrac{12}{1 - 5 \sin \theta}$

In Exercises 37 and 38, determine a Cartesian equation for the given polar equation.

37. $r = \dfrac{4}{2 - \sin \theta}$ **38.** $r = \dfrac{6}{1 + 2 \cos \theta}$

In Exercises 39 and 40, use the fact that $k = 2p$ is twice the focal length and half the focal width, to determine a Cartesian equation of the parabola whose polar equation is given.

39. $r = \dfrac{4}{2 - 2 \cos \theta}$ **40.** $r = \dfrac{12}{3 + 3 \cos \theta}$

41. Halley's Comet The orbit of Halley's comet has a semimajor axis of 18.09 AU and an orbital eccentricity of 0.97. Compute its perihelion and aphelion distances.

42. Uranus The orbit of the planet Uranus has a semimajor axis of 19.18 AU and an orbital eccentricity of 0.0461. Compute its perihelion and aphelion distances.

In Exercises 43 and 44, the velocity of an object traveling in a circular orbit of radius r (distance from center of planet in meters) around a planet is given by

$$v = \sqrt{\frac{3.99 \times 10^{14}\,k}{r}} \ \text{m/sec},$$

where k is a constant related to the mass of the planet and the orbiting object.

43. Group Activity Lunar Module A lunar excursion module is in a circular orbit 250 km above the surface of the Moon. Assume that the Moon's radius is 1740 km and that $k = 0.012$. Find the following.

(a) the velocity of the lunar module

(b) the length of time required for the lunar module to circle the moon once

44. Group Activity Mars Satellite A satellite is in a circular orbit 1000 mi above Mars. Assume that the radius of Mars is 2100 mi and that $k = 0.11$. Find the velocity of the satellite.

Standardized Test Questions

45. True or False The equation $r = ke/(1 + e \cos \theta)$ yields no true circles. Justify your answer.

46. True or False The equation $r = a(1 - e^2)/(1 + e \cos \theta)$ yields no true parabolas. Justify your answer.

In Exercises 47–50, solve the problem without using a calculator.

47. Multiple Choice Which ratio of distances is constant for a point on a nondegenerate conic?

(A) distance to center : distance to directrix

(B) distance to focus : distance to vertex

(C) distance to vertex : distance to directrix

(D) distance to focus : distance to directrix

(E) distance to center : distance to vertex

48. Multiple Choice Which type of conic section has an eccentricity greater than one?

(A) an ellipse

(B) a parabola

(C) a hyperbola

(D) two parallel lines

(E) a circle

49. Multiple Choice For a conic expressed by $r = ke/(1 + e \sin \theta)$, which point is located at the pole?

(A) the center

(B) a focus

(C) a vertex

(D) an endpoint of the minor axis

(E) an endpoint of the conjugate axis

50. Multiple Choice Which of the following is **not** a polar equation of a conic?

(A) $r = 1 + 2 \cos \theta$

(B) $r = 1/(1 + \sin \theta)$

(C) $r = 3$

(D) $r = 1/(2 - \cos \theta)$

(E) $r = 1/(1 + 2 \cos \theta)$

Explorations

51. Planetary Orbits Use the polar equation $r = a(1 - e^2)/(1 + e \cos \theta)$ in completing the following activities.

(a) Use the fact that $-1 \le \cos \theta \le 1$ to prove that the perihelion distance of any planet is $a(1 - e)$ and the aphelion distance is $a(1 + e)$.

(b) Use $e = c/a$ to confirm that $a(1 - e) = a - c$ and $a(1 + e) = a + c$.

(c) Use the formulas $a(1 - e)$ and $a(1 + e)$ to compute the perihelion and aphelion distances of each planet listed in Table 8.4.

(d) For which of these planets is the difference between the perihelion and aphelion distance the greatest?

Table 8.4 Semimajor Axes and Eccentricities of the Six Innermost Planets		
Planet	Semimajor Axis (AU)	Eccentricity
Mercury	0.3871	0.206
Venus	0.7233	0.007
Earth	1.0000	0.017
Mars	1.5237	0.093
Jupiter	5.2026	0.048
Saturn	9.5547	0.056

Source: Encrenaz & Bibring. The Solar System (2nd ed.). New York: Springer, p. 5.

52. Using the Astronomer's Equation for Conics Using Dot mode, $a = 2$, an xy window of $[-13, 5]$ by $[-6, 6]$, $\theta\text{min} = 0$, $\theta\text{max} = 2\pi$, and $\theta\text{step} = \pi/48$, graph $r = a(1 - e^2)/(1 + e \cos \theta)$ for $e = 0, 0.3, 0.7, 1.5, 3$. Identify the type of conic section obtained for each e value. What happens when $e = 1$?

Extending the Ideas

53. Revisiting Figure 8.41 In Figure 8.41, if $r < 0$ or $r \cos \theta > k$, then we must use $PD = |k - r \cos \theta|$ and $PF = |r|$. Prove that, even in these cases, the resulting equation is still $r = ke/(1 + e \cos \theta)$.

54. Deriving Other Polar Forms for Conics Using Figure 8.41 as a guide, draw an appropriate diagram for and derive the equation.

(a) $r = \dfrac{ke}{1 - e \cos \theta}$

(b) $r = \dfrac{ke}{1 + e \sin \theta}$

(c) $r = \dfrac{ke}{1 - e \sin \theta}$

55. Revisiting Example 3 Use the formulas $x = r \cos \theta$ and $x^2 + y^2 = r^2$ to transform the polar equation $r = \dfrac{16}{5 - 3 \cos \theta}$ into the Cartesian equation $\dfrac{(x - 3)^2}{25} + \dfrac{y^2}{16} = 1$.

56. Focal Widths Using polar equations, derive formulas for the focal width of an ellipse and the focal width of a hyperbola. Begin by defining focal width for these conics in a manner analogous to the definition of the focal width of a parabola given in Section 8.1.

57. Prove that for a hyperbola the formula $r = ke/(1 - e \cos \theta)$ is equivalent to $r = a(e^2 - 1)/(1 - e \cos \theta)$, where a is the semi-transverse axis of the hyperbola.

58. Connecting Polar to Rectangular Consider the ellipse

$$\frac{x^2}{a^2} + \frac{y^2}{b^2} = 1,$$

where half the length of the major axis is a, and the foci are $(\pm c, 0)$ such that $c^2 = a^2 - b^2$. Let L be the vertical line $x = a^2/c$.

(a) Prove that L is a directrix for the ellipse. [*Hint:* Prove that PF/PD is the constant c/a, where P is a point on the ellipse, and D is the point on L such that PD is perpendicular to L.]

(b) Prove that the eccentricity is $e = c/a$.

(c) Prove that the distance from F to L is $a/e - ea$.

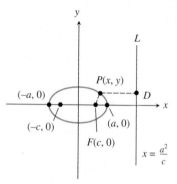

59. Connecting Polar to Rectangular Consider the hyperbola

$$\frac{x^2}{a^2} - \frac{y^2}{b^2} = 1,$$

where half the length of the transverse axis is a, and the foci are $(\pm c, 0)$ such that $c^2 = a^2 + b^2$. Let L be the vertical line $x = a^2/c$.

(a) Prove that L is a directrix for the hyperbola. [*Hint:* Prove that PF/PD is the constant c/a, where P is a point on the hyperbola, and D is the point on L such that PD is perpendicular to L.]

(b) Prove that the eccentricity is $e = c/a$.

(c) Prove that the distance from F to L is $ea - a/e$

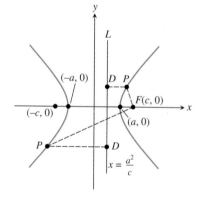

8.6
Three-Dimensional Cartesian Coordinate System

What you'll learn about

- Three-Dimensional Cartesian Coordinates
- Distance and Midpoint Formulas
- Equation of a Sphere
- Planes and Other Surfaces
- Vectors in Space
- Lines in Space

... and why

This is the analytic geometry of our physical world.

Three-Dimensional Cartesian Coordinates

In Sections P.2 and P.4, we studied Cartesian coordinates and the associated basic formulas and equations for the two-dimensional plane; we now extend these ideas to *three-dimensional space*. In the plane, we used two axes and ordered pairs to name points; in space, we use three mutually perpendicular axes and ordered triples of real numbers to name points. See Figure 8.46.

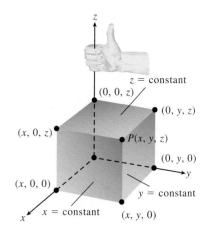

FIGURE 8.46 The point $P(x, y, z)$ in Cartesian space.

Notice that Figure 8.46 exhibits several important features of the *three-dimensional Cartesian coordinate system*:

- The axes are labeled x, y, and z, and these three **coordinate axes** form a **right-handed coordinate frame**: When you hold your right hand with fingers curving from the positive x-axis toward the positive y-axis, your thumb points in the direction of the positive z-axis.

- A point P in space uniquely corresponds to an ordered triple (x, y, z) of real numbers. The numbers x, y, and z are the **Cartesian coordinates of P**.

- Points on the axes have the form $(x, 0, 0)$, $(0, y, 0)$, or $(0, 0, z)$, with $(x, 0, 0)$ on the x-axis, $(0, y, 0)$ on the y-axis, and $(0, 0, z)$ on the z-axis.

In Figure 8.47, the axes are paired to determine the **coordinate planes**:

- The coordinate planes are the **xy-plane**, the **xz-plane**, and the **yz-plane**, and have equations $z = 0$, $y = 0$, and $x = 0$, respectively.

- Points on the coordinate planes have the form $(x, y, 0)$, $(x, 0, z)$, or $(0, y, z)$, with $(x, y, 0)$ on the xy-plane, $(x, 0, z)$ on the xz-plane, and $(0, y, z)$ on the yz-plane.

- The coordinate planes meet at the **origin**, $(0, 0, 0)$.

- The coordinate planes divide space into eight regions called **octants**, with the **first octant** containing all points in space with three positive coordinates.

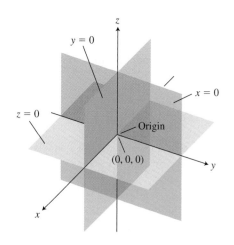

FIGURE 8.47 The coordinate planes divide space into eight octants.

EXAMPLE 1   Locating a Point in Cartesian Space

Draw a sketch that shows the point (2, 3, 5).

SOLUTION To locate the point (2, 3, 5), we first sketch a right-handed three-dimensional coordinate frame. We then draw the planes $x = 2$, $y = 3$, and $z = 5$, which parallel the coordinate planes $x = 0$, $y = 0$, and $z = 0$, respectively. The point (2, 3, 5) lies at the intersection of the planes $x = 2$, $y = 3$, and $z = 5$, as shown in Figure 8.48.   ***Now try Exercise 1.***

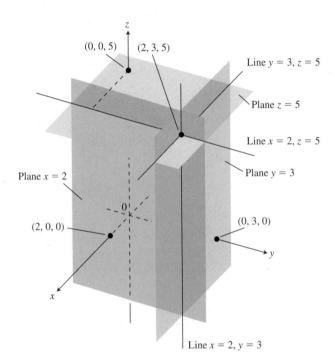

FIGURE 8.48 The planes $x = 2$, $y = 3$, and $z = 5$ determine the point (2, 3, 5). (Example 1)

Distance and Midpoint Formulas

The distance and midpoint formulas for space are natural generalizations of the corresponding formulas for the plane.

Distance Formula (Cartesian Space)

The distance $d(P, Q)$ between the points $P(x_1, y_1, z_1)$ and $Q(x_2, y_2, z_2)$ in space is

$$d(P, Q) = \sqrt{(x_1 - x_2)^2 + (y_1 - y_2)^2 + (z_1 - z_2)^2}.$$

Just as in the plane, the coordinates of the midpoint of a line segment are the averages for the coordinates of the endpoints of the segment.

Midpoint Formula (Cartesian Space)

The midpoint M of the line segment PQ with endpoints $P(x_1, y_1, z_1)$ and $Q(x_2, y_2, z_2)$ is

$$M = \left(\frac{x_1 + x_2}{2}, \frac{y_1 + y_2}{2}, \frac{z_1 + z_2}{2} \right).$$

EXAMPLE 2 Calculating a Distance and Finding a Midpoint

Find the distance between the points $P(-2, 3, 1)$ and $Q(4, -1, 5)$, and find the midpoint of line segment PQ.

SOLUTION The distance is given by

$$d(P, Q) = \sqrt{(-2 - 4)^2 + (3 + 1)^2 + (1 - 5)^2}$$

$$= \sqrt{36 + 16 + 16}$$

$$= \sqrt{68} \approx 8.25$$

The midpoint is

$$M = \left(\frac{-2 + 4}{2}, \frac{3 - 1}{2}, \frac{1 + 5}{2} \right) = (1, 1, 3).$$

Now try Exercises 5 and 9.

Equation of a Sphere

A *sphere* is the three-dimensional analogue of a circle: In space, the set of points that lie a fixed distance from a fixed point is a **sphere**. The fixed distance is the **radius**, and the fixed point is the **center** of the sphere. The point $P(x, y, z)$ is a point of the sphere with center (h, k, l) and radius r if and only if

$$\sqrt{(x - h)^2 + (y - k)^2 + (z - l)^2} = r.$$

Squaring both sides gives the standard equation shown below.

Standard Equation of a Sphere

A point $P(x, y, z)$ is on the sphere with center (h, k, l) and radius r if and only if

$$(x - h)^2 + (y - k)^2 + (z - l)^2 = r^2.$$

Drawing Lesson

How to Draw Three-Dimensional Objects to Look Three-Dimensional

1. Make the angle between the positive *x*-axis and the positive *y*-axis large enough.

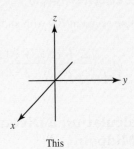

This

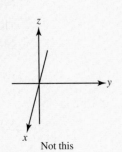

Not this

2. Break lines. When one line passes behind another, break it to show that it doesn't touch and that part of it is hidden.

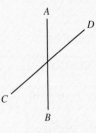

Intersecting

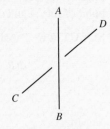

CD behind *AB*

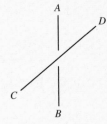

AB behind *CD*

3. Dash or omit hidden portions of lines. Don't let the line touch the boundary of the parallelogram that represents the plane, unless the line lies in the plane.

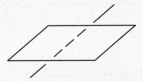

Line below plane

Line above plane

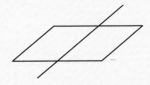

Line *in* plane

4. Spheres: Draw the sphere first (outline and equator); draw axes, if any, later. Use line breaks and dashed lines.

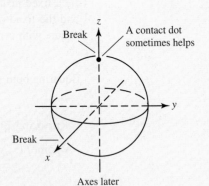

Hidden part dashed

Sphere first

Break

A contact dot sometimes helps

Break

Axes later

EXAMPLE 3   Finding the Standard Equation of a Sphere

The standard equation of the sphere with center $(2, 0, -3)$ and radius 7 is

$$(x - 2)^2 + y^2 + (z + 3)^2 = 49.$$

Now try Exercise 13.

Planes and Other Surfaces

In Section P.4, we learned that every line in the Cartesian plane can be written as a first-degree (linear) equation in two variables; that is, every line can be written as

$$Ax + By + C = 0,$$

where A and B are not both zero. Conversely, every first-degree equation in two variables represents a line in the Cartesian plane.

In an analogous way, every **plane** in Cartesian space can be written as a **first-degree equation in three variables**:

Equation for a Plane in Cartesian Space

Every plane can be written as

$$Ax + By + Cz + D = 0,$$

where A, B, and C are not all zero. Conversely, every first-degree equation in three variables represents a plane in Cartesian space.

EXAMPLE 4   Sketching a Plane in Space

Sketch the graph of $12x + 15y + 20z = 60$.

SOLUTION Because this is a first-degree equation, its graph is a plane. Three points determine a plane. To find three points, we first divide both sides of $12x + 15y + 20z = 60$ by 60:

$$\frac{x}{5} + \frac{y}{4} + \frac{z}{3} = 1.$$

In this form, it is easy to see that the points $(5, 0, 0)$, $(0, 4, 0)$, and $(0, 0, 3)$ satisfy the equation. These are the points where the graph crosses the coordinate axes. Figure 8.49 shows the completed sketch.

Now try Exercise 17.

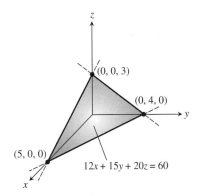

FIGURE 8.49 The intercepts $(5, 0, 0)$, $(0, 4, 0)$, and $(0, 0, 3)$ determine the plane $12x + 15y + 20z = 60$. (Example 4)

Equations in the three variables x, y, and z generally graph as surfaces in three-dimensional space. Just as in the plane, second-degree equations are of particular interest. Recall that second-degree equations in two variables yield conic sections in the Cartesian plane. In space, second-degree equations in *three* variables yield **quadric surfaces**: The paraboloids, ellipsoids, and hyperboloids of revolution that have special reflective properties are all quadric surfaces, as are such exotic-sounding surfaces as hyperbolic paraboloids and elliptic hyperboloids.

Other surfaces of interest include graphs of **functions of two variables**, whose equations have the form $z = f(x, y)$. Some examples are $z = x \ln y$, $z = \sin(xy)$, and $z = \sqrt{1 - x^2 - y^2}$. The last equation graphs as a hemisphere (see Exercise 63). Equations of the form $z = f(x, y)$ can be graphed using some graphing calculators and most computer algebra software. Quadric surfaces and functions of two variables are studied in most university-level calculus course sequences.

Vectors in Space

In space, just as in the plane, the sets of equivalent directed line segments (or arrows) are *vectors*. They are used to represent forces, displacements, and velocities in three dimensions. In space, we use ordered triples to denote vectors:

$$\mathbf{v} = \langle v_1, v_2, v_3 \rangle.$$

The **zero vector** is $\mathbf{0} = \langle 0, 0, 0 \rangle$, and the **standard unit vectors** are $\mathbf{i} = \langle 1, 0, 0 \rangle$, $\mathbf{j} = \langle 0, 1, 0 \rangle$, and $\mathbf{k} = \langle 0, 0, 1 \rangle$. As shown in Figure 8.50, the vector $\mathbf{v}$ can be expressed in terms of these standard unit vectors:

$$\mathbf{v} = \langle v_1, v_2, v_3 \rangle = v_1 \mathbf{i} + v_2 \mathbf{j} + v_3 \mathbf{k}.$$

The vector $\mathbf{v}$ that is represented by the arrow from $P(a, b, c)$ to $Q(x, y, z)$ is

$$\mathbf{v} = \overrightarrow{PQ} = \langle x - a, y - b, z - c \rangle = (x - a)\mathbf{i} + (y - b)\mathbf{j} + (z - c)\mathbf{k}.$$

A vector $\mathbf{v} = \langle v_1, v_2, v_3 \rangle$ can be multiplied by a scalar (real number) c as follows:

$$c\mathbf{v} = c\langle v_1, v_2, v_3 \rangle = \langle cv_1, cv_2, cv_3 \rangle.$$

Many other properties of vectors generalize in a natural way when we move from two to three dimensions:

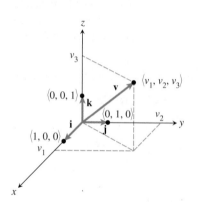

FIGURE 8.50 The vector $\mathbf{v} = \langle v_1, v_2, v_3 \rangle$.

Vector Relationships in Space

For vectors $\mathbf{v} = \langle v_1, v_2, v_3 \rangle$ and $\mathbf{w} = \langle w_1, w_2, w_3 \rangle$,

- **Equality:** $\mathbf{v} = \mathbf{w}$ if and only if $v_1 = w_1$, $v_2 = w_2$, and $v_3 = w_3$
- **Addition:** $\mathbf{v} + \mathbf{w} = \langle v_1 + w_1, v_2 + w_2, v_3 + w_3 \rangle$
- **Subtraction:** $\mathbf{v} - \mathbf{w} = \langle v_1 - w_1, v_2 - w_2, v_3 - w_3 \rangle$
- **Magnitude:** $|\mathbf{v}| = \sqrt{v_1^2 + v_2^2 + v_3^2}$
- **Dot product:** $\mathbf{v} \cdot \mathbf{w} = v_1 w_1 + v_2 w_2 + v_3 w_3$
- **Unit vector:** $\mathbf{u} = \mathbf{v}/|\mathbf{v}|$, $\mathbf{v} \neq \mathbf{0}$, is the unit vector in the direction of $\mathbf{v}$.

EXAMPLE 5 Computing with Vectors

(a) $3\langle -2, 1, 4 \rangle = \langle 3 \cdot -2, 3 \cdot 1, 3 \cdot 4 \rangle = \langle -6, 3, 12 \rangle$

(b) $\langle 0, 6, -7 \rangle + \langle -5, 5, 8 \rangle = \langle 0 - 5, 6 + 5, -7 + 8 \rangle = \langle -5, 11, 1 \rangle$

(c) $\langle 1, -3, 4 \rangle - \langle -2, -4, 5 \rangle = \langle 1 + 2, -3 + 4, 4 - 5 \rangle = \langle 3, 1, -1 \rangle$

(d) $|\langle 2, 0, -6 \rangle| = \sqrt{2^2 + 0^2 + 6^2} = \sqrt{40} \approx 6.32$

(e) $\langle 5, 3, -1 \rangle \cdot \langle -6, 2, 3 \rangle = 5 \cdot (-6) + 3 \cdot 2 + (-1) \cdot 3$

$$= -30 + 6 - 3 = -27$$

Now try Exercises 23–26.

EXAMPLE 6 Using Vectors in Space

A jet airplane just after takeoff is pointed due east. Its air velocity vector makes an angle of 30° with flat ground with an airspeed of 250 mph. If the wind is out of the southeast at 32 mph, calculate a vector that represents the plane's velocity relative to the point of takeoff.

SOLUTION Let **i** point east, **j** point north, and **k** point up. The plane's air velocity is

$$\mathbf{a} = \langle 250 \cos 30°, 0, 250 \sin 30° \rangle \approx \langle 216.506, 0, 125 \rangle,$$

and the wind velocity, which is pointing northwest, is

$$\mathbf{w} = \langle 32 \cos 135°, 32 \sin 135°, 0 \rangle \approx \langle -22.627, 22.627, 0 \rangle.$$

The velocity relative to the ground is $\mathbf{v} = \mathbf{a} + \mathbf{w}$, so

$$\mathbf{v} \approx \langle 216.506, 0, 125 \rangle + \langle -22.627, 22.627, 0 \rangle$$

$$\approx \langle 193.88, 22.63, 125 \rangle$$

$$= 193.88\mathbf{i} + 22.63\mathbf{j} + 125\mathbf{k}$$

Now try Exercise 33.

In Exercise 64, you will be asked to interpret the meaning of the velocity vector obtained in Example 6.

Lines in Space

We have seen that first-degree equations in three variables graph as planes in space. So how do we get lines? There are several ways. First notice that to specify the x-axis, which is a line, we could use the pair of first-degree equations $y = 0$ and $z = 0$. As alternatives to using a pair of Cartesian equations, we can specify any line in space using

• one vector equation, or

• a set of three parametric equations.

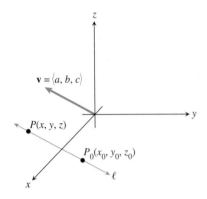

FIGURE 8.51 The line ℓ is parallel to the direction vector $\mathbf{v} = \langle a, b, c \rangle$.

Suppose ℓ is a line through the point $P_0(x_0, y_0, z_0)$ and in the direction of a nonzero vector $\mathbf{v} = \langle a, b, c \rangle$ (Figure 8.51). Then for any point $P(x, y, z)$ on ℓ,

$$\overrightarrow{P_0P} = t\mathbf{v}$$

for some real number t. The vector $\mathbf{v}$ is a **direction vector** for line ℓ. If $\mathbf{r} = \overrightarrow{OP} = \langle x, y, z \rangle$ and $\mathbf{r}_0 = \overrightarrow{OP_0} = \langle x_0, y_0, z_0 \rangle$, then $\mathbf{r} - \mathbf{r}_0 = t\mathbf{v}$. So an equation of the line ℓ is $\mathbf{r} = \mathbf{r}_0 + t\mathbf{v}$.

Equations for a Line in Space

If ℓ is a line through the point $P_0(x_0, y_0, z_0)$ in the direction of a nonzero vector $\mathbf{v} = \langle a, b, c \rangle$, then a point $P(x, y, z)$ is on ℓ if and only if

- **Vector form:** $\mathbf{r} = \mathbf{r}_0 + t\mathbf{v}$, where $\mathbf{r} = \langle x, y, z \rangle$ and $\mathbf{r}_0 = \langle x_0, y_0, z_0 \rangle$; or

- **Parametric form:** $x = x_0 + at$, $y = y_0 + bt$, and $z = z_0 + ct$,

where t is a real number.

EXAMPLE 7   Finding Equations for a Line

The line through $P_0(4, 3, -1)$ with direction vector $\mathbf{v} = \langle -2, 2, 7 \rangle$ can be written

- in vector form as $\mathbf{r} = \langle 4, 3, -1 \rangle + t\langle -2, 2, 7 \rangle$; or

- in parametric form as $x = 4 - 2t$, $y = 3 + 2t$, and $z = -1 + 7t$.

Now try Exercise 35.

EXAMPLE 8   Finding Equations for a Line

Using the standard unit vectors $\mathbf{i}$, $\mathbf{j}$, and $\mathbf{k}$, write a vector equation for the line containing the points $A(3, 0, -2)$ and $B(-1, 2, -5)$, and compare it to the parametric equations for the line.

SOLUTION The line is in the direction of

$$\mathbf{v} = \overrightarrow{AB} = \langle -1 - 3, 2 - 0, -5 + 2 \rangle = \langle -4, 2, -3 \rangle.$$

So using $\mathbf{r}_0 = \overrightarrow{OA}$, the vector equation of the line becomes:

$$\mathbf{r} = \mathbf{r}_0 + t\mathbf{v}$$
$$\langle x, y, z \rangle = \langle 3, 0, -2 \rangle + t\langle -4, 2, -3 \rangle$$
$$\langle x, y, z \rangle = \langle 3 - 4t, 2t, -2 - 3t \rangle$$
$$x\mathbf{i} + y\mathbf{j} + z\mathbf{k} = (3 - 4t)\mathbf{i} + 2t\mathbf{j} + (-2 - 3t)\mathbf{k}$$

The parametric equations are the three component equations

$$x = 3 - 4t, \ y = 2t, \text{ and } z = -2 - 3t.$$

Now try Exercise 41.

QUICK REVIEW 8.6 *(For help, go to Sections 6.1 and 6.3.)*

In Exercises 1–3, let $P(x, y)$ and $Q(2, -3)$ be points in the *xy*-plane.

1. Compute the distance between P and Q.

2. Find the midpoint of the line segment PQ.

3. If P is 5 units from Q, describe the position of P.

In Exercises 4–6, let $\mathbf{v} = \langle -4, 5 \rangle = -4\mathbf{i} + 5\mathbf{j}$ be a vector in the *xy*-plane.

4. Find the magnitude of $\mathbf{v}$.

5. Find a unit vector in the direction of $\mathbf{v}$.

6. Find a vector 7 units long in the direction of $-\mathbf{v}$.

7. Give a geometric description of the graph of $(x + 1)^2 + (y - 5)^2 = 25$ in the *xy*-plane.

8. Give a geometric description of the graph of $x = 2 - t$, $y = -4 + 2t$ in the *xy*-plane.

9. Find the center and radius of the circle $x^2 + y^2 + 2x - 6y + 6 = 0$ in the *xy*-plane.

10. Find a vector from $P(2, 5)$ to $Q(-1, -4)$ in the *xy*-plane.

SECTION 8.6 EXERCISES

In Exercises 1–4, draw a sketch that shows the point.

1. $(3, 4, 2)$ **2.** $(2, -3, 6)$

3. $(1, -2, -4)$ **4.** $(-2, 3, -5)$

In Exercises 5–8, compute the distance between the points.

5. $(-1, 2, 5), (3, -4, 6)$

6. $(2, -1, -8), (6, -3, 4)$

7. $(a, b, c), (1, -3, 2)$

8. $(x, y, z), (p, q, r)$

In Exercises 9–12, find the midpoint of the segment PQ.

9. $P(-1, 2, 5), Q(3, -4, 6)$

10. $P(2, -1, -8), Q(6, -3, 4)$

11. $P(2x, 2y, 2z), Q(-2, 8, 6)$

12. $P(-a, -b, -c), Q(3a, 3b, 3c)$

In Exercises 13–16, write an equation for the sphere with the given point as its center and the given number as its radius.

13. $(5, -1, -2), 8$ **14.** $(-1, 5, 8), \sqrt{5}$

15. $(1, -3, 2), \sqrt{a}, a > 0$ **16.** $(p, q, r), 6$

In Exercises 17–22, sketch a graph of the equation. Label all intercepts.

17. $x + y + 3z = 9$ **18.** $x + y - 2z = 8$

19. $x + z = 3$ **20.** $2y + z = 6$

21. $x - 3y = 6$ **22.** $x = 3$

In Exercises 23–32, evaluate the expression using $\mathbf{r} = \langle 1, 0, -3 \rangle$, $\mathbf{v} = \langle -3, 4, -5 \rangle$, and $\mathbf{w} = \langle 4, -3, 12 \rangle$.

23. $\mathbf{r} + \mathbf{v}$ **24.** $\mathbf{r} - \mathbf{w}$

25. $\mathbf{v} \cdot \mathbf{w}$ **26.** $|\mathbf{w}|$

27. $\mathbf{r} \cdot (\mathbf{v} + \mathbf{w})$ **28.** $(\mathbf{r} \cdot \mathbf{v}) + (\mathbf{r} \cdot \mathbf{w})$

29. $\mathbf{w}/|\mathbf{w}|$ **30.** $\mathbf{i} \cdot \mathbf{r}$

31. $\langle \mathbf{i} \cdot \mathbf{v}, \mathbf{j} \cdot \mathbf{v}, \mathbf{k} \cdot \mathbf{v} \rangle$ **32.** $(\mathbf{r} \cdot \mathbf{v})\mathbf{w}$

In Exercises 33 and 34, let $\mathbf{i}$ point east, $\mathbf{j}$ point north, and $\mathbf{k}$ point up.

33. Three-Dimensional Velocity An airplane just after takeoff is headed west and is climbing at a 20° angle relative to flat ground with an airspeed of 200 mph. If the wind is out of the northeast at 10 mph, calculate a vector $\mathbf{v}$ that represents the plane's velocity relative to the point of takeoff.

34. Three-Dimensional Velocity A rocket soon after takeoff is headed east and is climbing at a 80° angle relative to flat ground with an airspeed of 12,000 mph. If the wind is out of the southwest at 8 mph, calculate a vector $\mathbf{v}$ that represents the rocket's velocity relative to the point of takeoff.

In Exercises 35–38, write the vector and parametric forms of the line through the point P_0 in the direction of $\mathbf{v}$.

35. $P_0(2, -1, 5), \mathbf{v} = \langle 3, 2, -7 \rangle$

36. $P_0(-3, 8, -1), \mathbf{v} = \langle -3, 5, 2 \rangle$

37. $P_0(6, -9, 0), \mathbf{v} = \langle 1, 0, -4 \rangle$

38. $P_0(0, -1, 4), \mathbf{v} = \langle 0, 0, 1 \rangle$

In Exercises 39–48, use the points $A(-1, 2, 4)$, $B(0, 6, -3)$, and $C(2, -4, 1)$.

39. Find the distance from A to the midpoint of BC.

40. Find the vector from A to the midpoint of BC.

41. Write a vector equation of the line through A and B.

42. Write a vector equation of the line through A and the midpoint of BC.

43. Write parametric equations for the line through A and C.

44. Write parametric equations for the line through B and C.

45. Write parametric equations for the line through B and the midpoint of AC.

46. Write parametric equations for the line through C and the midpoint of AB.

47. Is $\triangle ABC$ equilateral, isosceles, or scalene?

48. If M is the midpoint of BC, what is the midpoint of AM?

In Exercises 49–52, **(a)** sketch the line defined by the pair of equations, and **(b) Writing to Learn** give a geometric description of the line, including its direction and its position relative to the coordinate frame.

49. $x = 0$, $y = 0$

50. $x = 0$, $z = 2$

51. $x = -3$, $y = 0$

52. $y = 1$, $z = 3$

53. Write a vector equation for the line through the distinct points $P(x_1, y_1, z_1)$ and $Q(x_2, y_2, z_2)$.

54. Write parametric equations for the line through the distinct points $P(x_1, y_1, z_1)$ and $Q(x_2, y_2, z_2)$.

55. Generalizing the Distance Formula Prove that the distance $d(P, Q)$ between the points $P(x_1, y_1, z_1)$ and $Q(x_2, y_2, z_2)$ in space is $\sqrt{(x_1 - x_2)^2 + (y_1 - y_2)^2 + (z_1 - z_2)^2}$ by using the point $R(x_2, y_2, z_1)$, the two-dimensional distance formula within the plane $z = z_1$, the one-dimensional distance formula within the line $\mathbf{r} = \langle x_2, y_2, t \rangle$, and the Pythagorean theorem. [*Hint:* A sketch may help you visualize the situation.]

56. Generalizing a Property of the Dot Product Prove $\mathbf{u} \cdot \mathbf{u} = |\mathbf{u}|^2$ where $\mathbf{u}$ is a vector in three-dimensional space.

Standardized Test Questions

57. True or False $x^2 + 4y^2 = 1$ represents a surface in space. Justify your answer.

58. True or False The parametric equation $x = 1 + 0t$, $y = 2 - 0t$, $z = -5 + 0t$ represent a line in space. Justify your answer.

In Exercises 59–62, solve the problem without using a calculator.

59. Multiple Choice A first-degree equation in three variables graphs as

(A) a line.

(B) a plane.

(C) a sphere.

(D) a paraboloid.

(E) an ellipsoid.

60. Multiple Choice Which of the following is **not** a quadric surface?

(A) a plane

(B) a sphere

(C) an ellipsoid

(D) an elliptic paraboloid

(E) a hyperbolic paraboloid

61. Multiple Choice If $\mathbf{v}$ and $\mathbf{w}$ are vectors and c is a scalar, which of these is **a scalar**?

(A) $\mathbf{v} + \mathbf{w}$

(B) $\mathbf{v} - \mathbf{w}$

(C) $\mathbf{v} \cdot \mathbf{w}$

(D) $c\mathbf{v}$

(E) $|\mathbf{v}|\mathbf{w}$

62. Multiple Choice The parametric form of the line $\mathbf{r} = \langle 2, -3, 0 \rangle + t \langle 1, 0, -1 \rangle$ is

(A) $x = 2 - 3t$, $y = 0 + 1t$, $z = 0 - 1t$.

(B) $x = 2t$, $y = -3 + 0t$, $z = 0 - 1t$.

(C) $x = 1 + 2t$, $y = 0 - 3t$, $z = -1 + 0t$.

(D) $x = 1 + 2t$, $y = -3$, $z = -1t$.

(E) $x = 2 + t$, $y = -3$, $z = -t$.

Explorations

63. Group Activity Writing to Learn The figure shows a graph of the ellipsoid $x^2/9 + y^2/4 + z^2/16 = 1$ drawn in a *box* using *Mathematica* computer software.

(a) Describe its cross sections in each of the three coordinate planes, that is, for $z = 0$, $y = 0$, and $x = 0$. In your description, include the name of each cross section and its position relative to the coordinate frame.

(b) Explain algebraically why the graph of $z = \sqrt{1 - x^2 - y^2}$ is half of a sphere. What is the equation of the related whole sphere?

(c) By hand, sketch the graph of the hemisphere $z = \sqrt{1 - x^2 - y^2}$. Check your sketch using a 3D grapher if you have access to one.

(d) Explain how the graph of an ellipsoid is related to the graph of a sphere and why a sphere is a *degenerate* ellipsoid.

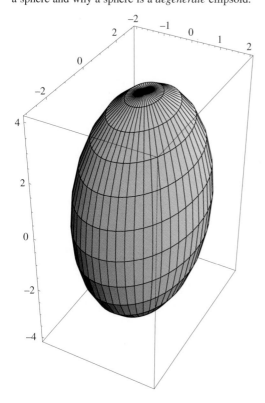

64. Revisiting Example 6 Read Example 6. Then using $\mathbf{v} = 193.88\mathbf{i} + 22.63\mathbf{j} + 125\mathbf{k}$, establish the following:

(a) The plane's compass bearing is 83.34°.

(b) Its speed downrange (that is, ignoring the vertical component) is 195.2 mph.

(c) The plane is climbing at an angle of 32.63°.

(d) The plane's overall speed is 231.8 mph.

Extending the Ideas

The **cross product u × v** of the vectors $\mathbf{u} = u_1\mathbf{i} + u_2\mathbf{j} + u_3\mathbf{k}$ and $\mathbf{v} = v_1\mathbf{i} + v_2\mathbf{j} + v_3\mathbf{k}$ is

$$\mathbf{u} \times \mathbf{v} = \begin{vmatrix} \mathbf{i} & \mathbf{j} & \mathbf{k} \\ u_1 & u_2 & u_3 \\ v_1 & v_2 & v_3 \end{vmatrix}$$

$$= (u_2v_3 - u_3v_2)\mathbf{i} + (u_3v_1 - u_1v_3)\mathbf{j} + (u_1v_2 - u_2v_1)\mathbf{k}.$$

Use this definition in Exercises 65–68.

65. $\langle 1, -2, 3 \rangle \times \langle -2, 1, -1 \rangle$

66. $\langle 4, -1, 2 \rangle \times \langle 1, -3, 2 \rangle$

67. Prove that $\mathbf{i} \times \mathbf{j} = \mathbf{k}$.

68. Assuming the theorem about angles between vectors (Section 6.2) holds for three-dimensional vectors, prove that $\mathbf{u} \times \mathbf{v}$ is perpendicular to both $\mathbf{u}$ and $\mathbf{v}$ if they are nonzero.

CHAPTER 8 **Key Ideas**

PROPERTIES, THEOREMS, AND FORMULAS

PROCEDURES

CHAPTER 8 Review Exercises

The collection of exercises marked in red could be used as a chapter test.

In Exercises 1–4, find the vertex, focus, directrix, and focal width of the parabola, and sketch the graph.

1. $y^2 = 12x$

2. $x^2 = -8y$

3. $(x + 2)^2 = -4(y - 1)$

4. $(y + 2)^2 = 16x$

In Exercises 5–12, identify the type of conic. Find the center, vertices, and foci of the conic, and sketch its graph.

5. $\dfrac{y^2}{8} + \dfrac{x^2}{5} = 1$

6. $\dfrac{y^2}{16} - \dfrac{x^2}{49} = 1$

7. $\dfrac{x^2}{25} - \dfrac{y^2}{36} = 1$

8. $\dfrac{x^2}{49} - \dfrac{y^2}{9} = 1$

9. $\dfrac{(x + 3)^2}{18} - \dfrac{(y - 5)^2}{28} = 1$

10. $\dfrac{(y - 3)^2}{9} - \dfrac{(x - 7)^2}{12} = 1$

11. $\dfrac{(x - 2)^2}{16} + \dfrac{(y + 1)^2}{7} = 1$

12. $\dfrac{y^2}{36} + \dfrac{(x + 6)^2}{20} = 1$

In Exercises 13–20, match the equation with its graph.

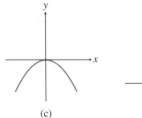

(a)

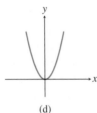

(b)

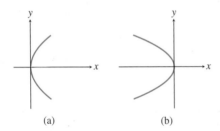

(c)

(d)

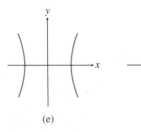

(e)

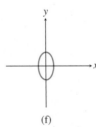

(f)

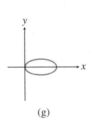

(g)

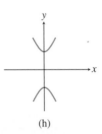

(h)

13. $y^2 = -3x$

14. $\dfrac{(x - 2)^2}{4} + y^2 = 1$

15. $\dfrac{y^2}{5} - x^2 = 1$

16. $\dfrac{x^2}{9} - \dfrac{y^2}{25} = 1$

17. $\dfrac{y^2}{3} + x^2 = 1$

18. $x^2 = y$

19. $x^2 = -4y$

20. $y^2 = 6x$

In Exercises 21–28, identify the conic. Then complete the square to write the conic in standard form, and sketch the graph.

21. $x^2 - 6x - y - 3 = 0$

22. $x^2 + 4x + 3y^2 - 5 = 0$

23. $x^2 - y^2 - 2x + 4y - 6 = 0$

24. $x^2 + 2x + 4y - 7 = 0$

25. $y^2 - 6x - 4y - 13 = 0$

26. $3x^2 - 6x - 4y - 9 = 0$

27. $2x^2 - 3y^2 - 12x - 24y + 60 = 0$

28. $12x^2 - 4y^2 - 72x - 16y + 44 = 0$

29. Prove that the parabola with focus $(0, p)$ and directrix $y = -p$ has the equation $x^2 = 4py$.

30. Prove that the equation $y^2 = 4px$ represents a parabola with focus $(p, 0)$ and directrix $x = -p$.

In Exercises 31–36, identify the conic. Solve the equation for y and graph it.

31. $3x^2 - 8xy + 6y^2 - 5x - 5y + 20 = 0$

32. $10x^2 - 8xy + 6y^2 + 8x - 5y - 30 = 0$

33. $3x^2 - 2xy - 5x + 6y - 10 = 0$

34. $5xy - 6y^2 + 10x - 17y + 20 = 0$

35. $-3x^2 + 7xy - 2y^2 - x + 20y - 15 = 0$

36. $-3x^2 + 7xy - 2y^2 - 2x + 3y - 10 = 0$

In Exercises 37–48, find the equation for the conic in standard form.

37. Parabola: vertex $(0, 0)$, focus $(2, 0)$

38. Parabola: vertex $(0, 0)$, opens downward, focal width $= 12$

39. Parabola: vertex $(-3, 3)$, directrix $y = 0$

40. Parabola: vertex $(1, -2)$, opens to the left, focal length $= 2$

41. Ellipse: center $(0, 0)$, foci $(\pm 12, 0)$, vertices $(\pm 13, 0)$

42. Ellipse: center $(0, 0)$, foci $(0, \pm 2)$, vertices $(0, \pm 6)$

43. Ellipse: center $(0, 2)$, semimajor axis = 3, one focus is $(2, 2)$.

44. Ellipse: center $(-3, -4)$, semimajor axis = 4, one focus is $(0, -4)$.

45. Hyperbola: center $(0, 0)$, foci $(0, \pm 6)$, vertices $(0, \pm 5)$

46. Hyperbola: center $(0, 0)$, vertices $(\pm 2, 0)$, asymptotes $y = \pm 2x$

47. Hyperbola: center $(2, 1)$, vertices $(2 \pm 3, 1)$, one asymptote is $y = (4/3)(x - 2) + 1$

48. Hyperbola: center $(-5, 0)$, one focus is $(-5, 3)$, one vertex is $(-5, 2)$.

In Exercises 49–54, find the equation for the conic in standard form.

49. $x = 5 \cos t, y = 2 \sin t, 0 \le t \le 2\pi$

50. $x = 4 \sin t, y = 6 \cos t, 0 \le t \le 4\pi$

51. $x = -2 + \cos t, y = 4 + \sin t, 2\pi \le t \le 4\pi$

52. $x = 5 + 3 \cos t, y = -3 + 3 \sin t, -2\pi \le t \le 0$

53. $x = 3 \sec t, y = 5 \tan t, 0 \le t \le 2\pi$

54. $x = 4 \sec t, y = 3 \tan t, 0 \le t \le 2\pi$

In Exercises 55–62, identify and graph the conic, and rewrite the equation in Cartesian coordinates.

55. $r = \dfrac{4}{1 + \cos \theta}$

56. $r = \dfrac{5}{1 - \sin \theta}$

57. $r = \dfrac{4}{3 - \cos \theta}$

58. $r = \dfrac{3}{4 + \sin \theta}$

59. $r = \dfrac{35}{2 - 7 \sin \theta}$

60. $r = \dfrac{15}{2 + 5 \cos \theta}$

61. $r = \dfrac{2}{1 + \cos \theta}$

62. $r = \dfrac{4}{4 - 4 \cos \theta}$

In Exercises 63–74, use the points $P(-1, 0, 3)$ and $Q(3, -2, -4)$ and the vectors $\mathbf{v} = \langle -3, 1, -2 \rangle$ and $\mathbf{w} = \langle 3, -4, 0 \rangle$.

63. Compute the distance from P to Q.

64. Find the midpoint of segment PQ.

65. Compute $\mathbf{v} + \mathbf{w}$.

66. Compute $\mathbf{v} - \mathbf{w}$.

67. Compute $\mathbf{v} \cdot \mathbf{w}$.

68. Compute the magnitude of $\mathbf{v}$.

69. Write the unit vector in the direction of $\mathbf{w}$.

70. Compute $(\mathbf{v} \cdot \mathbf{w})(\mathbf{v} + \mathbf{w})$.

71. Write an equation for the sphere centered at P with radius 4.

72. Write parametric equations for the line through P and Q.

73. Write a vector equation for the line through P in the direction of $\mathbf{v}$.

74. Write parametric equations for the line in the direction of $\mathbf{w}$ through the midpoint of PQ.

75. Parabolic Microphones B-Ball Network uses a parabolic microphone to capture all the sounds from the basketball players and coaches during each regular season game. If one of its microphones has a parabolic surface generated by the parabola $18y = x^2$, locate the focus (the electronic receiver) of the parabola.

76. Parabolic Headlights Specific Electric makes parabolic headlights for a variety of automobiles. If one of its headlights has a parabolic surface generated by the parabola $y^2 = 15x$ (see figure), where should its lightbulb be placed?

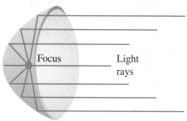

77. Writing to Learn Elliptical Billiard Table Elliptical billiard tables have been constructed with spots marking the foci. Suppose such a table has a major axis of 6 ft and minor axis of 4 ft.

(a) Explain the strategy that a "pool shark" who knows conic geometry would use to hit a blocked spot on this table.

(b) If the table surface is coordinatized so that $(0, 0)$ represents the center of the table and the x-axis is along the focal axis of the ellipse, at which point(s) should the ball be aimed?

78. Weather Satellite The Nimbus weather satellite travels in a north-south circular orbit 500 meters above Earth. Find the following. (Assume Earth's radius is 6380 km.)

(a) The velocity of the satellite using the formula for velocity v given for Exercises 43 and 44 in Section 8.5 with $k = 1$

(b) The time required for Nimbus to circle Earth once

79. Elliptical Orbits The velocity of a body in an elliptical Earth orbit at a distance r (in meters) from the focus (the center of Earth) is

$$v = \sqrt{3.99 \times 10^{14} \left(\frac{2}{r} - \frac{1}{a} \right)} \text{ m/sec,}$$

where a is the semimajor axis of the ellipse. An Earth satellite has a maximum altitude (at *apogee*) of 18,000 km and has a minimum altitude (at *perigee*) of 170 km. Assuming Earth's radius is 6380 km, find the velocity of the satellite at its apogee and perigee.

80. Icarus The asteroid Icarus is about 1 mi wide. It revolves around the Sun once every 409 Earth days and has an orbital eccentricity of 0.83. Use Kepler's first and third laws to determine Icarus's semimajor axis, perihelion distance, and aphelion distance.

CHAPTER 8 Project

Ellipses as Models of Pendulum Motion

As a simple pendulum swings back and forth, a plot of its velocity versus its position is elliptical in nature and can be modeled using a standard form of the equation of an ellipse,

$$\frac{(x-h)^2}{a^2} + \frac{(y-k)^2}{b^2} = 1 \text{ or } \frac{(y-k)^2}{a^2} + \frac{(x-h)^2}{b^2} = 1,$$

where x represents the pendulum's position relative to a fixed point and y represents the pendulum's velocity. In this project, you will use a motion detection device to collect position (distance) and velocity data for a swinging pendulum, then find a mathematical model that describes the pendulum's velocity with respect to position.

COLLECTING THE DATA

Construct a simple pendulum by fastening about 0.5 meter of string to a ball. Set up a Calculator-Based Ranger (CBR) system to collect distance and velocity readings for 4 seconds (enough time to capture at least one complete swing of the pendulum). See the printed or online CBR guidebook for specific setup instructions. Start the pendulum swinging toward and away from the detector, then activate the CBR system. The table below is a sample set of data collected in the manner just described.

EXPLORATIONS

1. If you collected data using a CBR, a plot of distance versus time may be shown on your grapher screen. Go to the plot setup screen and create a scatter plot of velocity versus distance. If you do not have access to a CBR, use the distance and velocity data from the table below to create a scatter plot.

2. Find values for a, b, h, and k so that the equation

$$\frac{(x-h)^2}{a^2} + \frac{(y-k)^2}{b^2} = 1 \text{ or } \frac{(y-k)^2}{a^2} + \frac{(x-h)^2}{b^2} = 1$$

fits the velocity versus position data plot. To graph this model you will have to solve the appropriate equation for y and enter it into the calculator in Y1 and Y2.

3. With respect to the ellipse modeled above, what do the variables a, b, h, and k represent?

4. What are the physical meanings of a, b, h, and k with respect to the motion of the pendulum?

5. Set up plots of distance versus time and velocity versus time. Find models for both of these plots and use them to graph the plot of the ellipse using parametric equations.

Time (sec)	Distance from the CBR (m)	Velocity (m/sec)	Time (sec)	Distance from the CBR (m)	Velocity (m/sec)
0	0.682	−0.3	0.647	0.454	0.279
0.059	0.659	−0.445	0.706	0.476	0.429
0.118	0.629	−0.555	0.765	0.505	0.544
0.176	0.594	−0.621	0.824	0.54	0.616
0.235	0.557	−0.638	0.882	0.576	0.639
0.294	0.521	−0.605	0.941	0.612	0.612
0.353	0.489	−0.523	1	0.645	0.536
0.412	0.463	−0.4	1.059	0.672	0.418
0.471	0.446	−0.246	1.118	0.69	0.266
0.529	0.438	−0.071	1.176	0.699	0.094
0.588	0.442	0.106	1.235	0.698	−0.086

Discrete Mathematics

As the use of cellular telephones, modems, pagers, and fax machines has grown in recent years, the increasing demand for unique telephone numbers has necessitated the creation of new area codes in many areas of the United States. Counting the number of possible telephone numbers in a given area code is a *combinatorial* problem, and such problems are solved using the techniques of *discrete* mathematics. See page 707 for more on the subject of telephone area codes.

Chapter 9 Overview

The branches of mathematics known broadly as algebra, analysis, and geometry come together so beautifully in calculus that it has been difficult over the years to squeeze other mathematics into the curriculum. Consequently, many worthwhile topics like probability and statistics, combinatorics, graph theory, and numerical analysis that could easily be introduced in high school are (for many students) either first seen in college electives or else never seen at all. This situation is gradually changing as the applications of noncalculus mathematics become increasingly more important in the modern, computerized, data-driven workplace. Therefore, besides introducing important topics like sequences and series and the Binomial Theorem, this chapter will touch on some other discrete topics that might prove useful to you in the near future.

9.1
Basic Combinatorics

What you'll learn about

- Discrete Versus Continuous
- The Importance of Counting
- The Multiplication Principle of Counting
- Permutations
- Combinations
- Subsets of an *n*-Set

. . . and why

Counting large sets is easy if you know the correct formula.

Discrete Versus Continuous

A point has no length and no width, and yet intervals on the real line—which are made up of these dimensionless points—have length! This little mystery illustrates the distinction between *continuous* and *discrete* mathematics. Any interval (a, b) contains a **continuum** of real numbers, which is why you can zoom in on an interval forever and there will still be an interval there. Calculus concepts like limits and continuity depend on the mathematics of the continuum. In *discrete* mathematics, we are concerned with properties of numbers and algebraic systems that do not depend on that kind of analysis. Many of them are related to the first kind of mathematics that most of us ever did, namely counting. Counting is what we will do for the rest of this section.

The Importance of Counting

We begin with a relatively simple counting problem.

FIGURE 9.1 A tree diagram for ordering the letters *ABC*. (Example 1)

EXAMPLE 1   Arranging Three Objects in Order

In how many different ways can three distinguishable objects be arranged in order?

SOLUTION It is not difficult to list all the possibilities. If we call the objects *A, B,* and *C,* the different orderings are: *ABC, ACB, BAC, BCA, CAB,* and *CBA.* A good way to visualize the six choices is with a *tree diagram,* as in Figure 9.1. Notice that we have three choices for the first letter. Then, branching off each of those three choices are two choices for the second letter. Finally, branching off each of the $3 \times 2 = 6$ branches formed so far is one choice for the third letter. By beginning at the "root" of the tree, we can proceed to the right along any of the $3 \times 2 \times 1 = 6$ branches and get a different ordering each time. We conclude that there are six ways to arrange three distinguishable objects in order.   *Now try Exercise 3.*

Scientific studies will usually manipulate one or more **explanatory** variables and observe the effect of that manipulation on one or more **response** variables. The key to understanding the significance of the effect is to know what is likely to occur *by chance alone*, and that often depends on counting. For example, Exploration 1 shows a real-world application of Example 1.

EXPLORATION 1 Questionable Product Claims

A salesman for a copying machine company is trying to convince a client to buy his $2000 machine instead of his competitor's $5000 machine. To make his point, he lines up an original document, a copy made by his machine, and a copy made by the more expensive machine on a table and asks 60 office workers to identify which is which. To everyone's surprise, not a single worker identifies all three correctly. The salesman states triumphantly that this proves that all three documents look the same to the naked eye and that therefore the client should buy his company's less expensive machine.

What do you think?

1. Each worker is essentially being asked to put the three documents in the correct order. How many ways can the three documents be ordered?

2. Suppose all three documents really *do* look alike. What fraction of the workers would you expect to put them into the correct order by chance alone?

3. If zero people out of 60 put the documents in the correct order, should we conclude that "all three documents look the same to the naked eye"?

4. Can you suggest a more likely conclusion that we might draw from the results of the salesman's experiment?

The Multiplication Principle of Counting

You would not want to draw the tree diagram for ordering five objects (*ABCDE*), but you should be able to see in your mind that it would have $5 \times 4 \times 3 \times 2 \times 1 = 120$ branches. A tree diagram is a geometric visualization of a fundamental counting principle known as the *Multiplication Principle*.

Multiplication Principle of Counting

If a procedure P has a sequence of stages $S_1, S_2, \ldots, S_n$ and if

$$S_1 \text{ can occur in } r_1 \text{ ways,}$$

$$S_2 \text{ can occur in } r_2 \text{ ways,}$$

$$\vdots$$

$$S_n \text{ can occur in } r_n \text{ ways,}$$

then the number of ways that the procedure P can occur is the product

$$r_1 r_2 \cdots r_n.$$

It is important to be mindful of how the choices at each stage are affected by the choices at preceding stages. For example, when choosing an order for the letters *ABC* we have 3 choices for the first letter, but only 2 choices for the second and 1 for the third.

EXAMPLE 2　Using the Multiplication Principle

The Tennessee license plate shown here consists of three letters of the alphabet followed by three numerical digits (0 through 9). Find the number of different license plates that could be formed

(a) if there is no restriction on the letters or digits that can be used;

(b) if no letter or digit can be repeated.

SOLUTION　Consider each license plate as having six blanks to be filled in: three letters followed by three numerical digits.

(a) If there are no restrictions on letters or digits, then we can fill in the first blank 26 ways, the second blank 26 ways, the third blank 26 ways, the fourth blank 10 ways, the fifth blank 10 ways, and the sixth blank 10 ways. By the Multiplication Principle, we can fill in all six blanks in $26 \times 26 \times 26 \times 10 \times 10 \times 10 = 17{,}576{,}000$ ways. There are 17,576,000 possible license plates with no restrictions on letters or digits.

(b) If no letter or digit can be repeated, then we can fill in the first blank 26 ways, the second blank 25 ways, the third blank 24 ways, the fourth blank 10 ways, the fifth blank 9 ways, and the sixth blank 8 ways. By the Multiplication Principle, we can fill in all six blanks in $26 \times 25 \times 24 \times 10 \times 9 \times 8 = 11{,}232{,}000$ ways. There are 11,232,000 possible license plates with no letters or digits repeated.

Now try Exercise 5.

LICENSE PLATE RESTRICTIONS

Although prohibiting repeated letters and digits as in Example 2 would make no practical sense (why rule out more than 6 million possible plates for no good reason?), states do impose some restrictions on license plates. They rule out certain letter progressions that could be considered obscene or offensive.

Permutations

One important application of the Multiplication Principle of Counting is to count the number of ways that a set of *n* objects (called an ***n*-set**) can be arranged in order. Each such ordering is called a **permutation** of the set. Example 1 showed that there are $3! = 6$ permutations of a 3-set. In fact, if you understood the tree diagram, you can probably guess how many permutations there are of an *n*-set.

FACTORIALS

If *n* is a positive integer, the symbol *n*! (read "*n* factorial") represents the product $n(n-1)(n-2)(n-3) \cdots 2 \cdot 1$. We also define $0! = 1$.

> **Permutations of an *n*-set**
>
> There are *n*! permutations of an *n*-set.

Usually the elements of a set are distinguishable from one another, but we can adjust our counting when they are not, as we see in Example 3.

EXAMPLE 3　Distinguishable Permutations

Count the number of different 9-letter "words" (don't worry about whether they're in the dictionary) that can be formed using the letters in each word.

(a) DRAGONFLY　　**(b)** BUTTERFLY　　**(c)** BUMBLEBEE

continued

SOLUTION

(a) Each permutation of the 9 letters forms a different word. There are $9! = 362{,}880$ such permutations.

(b) There are also 9! permutations of these letters, but a simple permutation of the two T's does not result in a new word. We correct for the overcount by dividing by 2!. There are $\dfrac{9!}{2!} = 181{,}440$ *distinguishable* permutations of the letters in BUTTERFLY.

(c) Again there are 9! permutations, but the three B's are indistinguishable, as are the three E's, so we divide by 3! twice to correct for the overcount. There are $\dfrac{9!}{3!3!} = 10{,}080$ distinguishable permutations of the letters in BUMBLEBEE.

Now try Exercise 9.

Distinguishable Permutations

There are $n!$ distinguishable permutations of an n-set containing n distinguishable objects.

If an n-set contains n_1 objects of a first kind, n_2 objects of a second kind, and so on, with $n_1 + n_2 + \cdots + n_k = n$, then the number of distinguishable permutations of the n-set is

$$\frac{n!}{n_1!n_2!n_3! \cdots n_k!}.$$

In many counting problems, we are interested in using n objects to fill r blanks in order, where $r < n$. These are called **permutations of n objects taken r at a time**. The procedure for counting them is the same; only this time we run out of blanks before we run out of objects.

The first blank can be filled in n ways, the second in $n - 1$ ways, and so on until we come to the rth blank, which can be filled in $n - (r - 1)$ ways. By the Multiplication Principle, we can fill in all r blanks in $n(n - 1)(n - 2) \cdots (n - r + 1)$ ways. This expression can be written in a more compact (but less easily computed) way as $n!/(n - r)!$.

PERMUTATIONS ON A CALCULATOR

Most modern calculators have an $_nP_r$ selection built in. They also compute factorials, but remember that factorials get very large. If you want to count the number of permutations of 90 objects taken 5 at a time, be sure to use the $_nP_r$ function. The expression $90!/85!$ is likely to lead to an overflow error.

Permutation Counting Formula

The number of permutations of n objects taken r at a time is denoted $_nP_r$ and is given by

$$_nP_r = \frac{n!}{(n - r)!} \quad \text{for} \quad 0 \le r \le n.$$

If $r > n$, then $_nP_r = 0$.

Notice that $_nP_n = n!/(n - n)! = n!/0! = n!/1 = n!$, which we have already seen is the number of permutations of a complete set of n objects. This is why we define $0! = 1$.

EXAMPLE 4   Counting Permutations

Evaluate each expression without a calculator.

(a) $_6P_4$ **(b)** $_{11}P_3$ **(c)** $_nP_3$

SOLUTION

(a) By the formula, $_6P_4 = 6!/(6 - 4)! = 6!/2! = (6 \cdot 5 \cdot 4 \cdot 3 \cdot 2!)/2! = 6 \cdot 5 \cdot 4 \cdot 3 = 360$.

(b) Although you could use the formula again, you might prefer to apply the Multiplication Principle directly. We have 11 objects and 3 blanks to fill:

$$_{11}P_3 = 11 \cdot 10 \cdot 9 = 990.$$

(c) This time it is definitely easier to use the Multiplication Principle. We have n objects and 3 blanks to fill; so assuming $n \geq 3$,

$$_nP_3 = n(n - 1)(n - 2).$$

Now try Exercise 15.

EXAMPLE 5   Applying Permutations

Sixteen actors answer a casting call to try out for roles as dwarfs in a production of *Snow White and the Seven Dwarfs*. In how many different ways can the director cast the seven roles?

SOLUTION The 7 different roles can be thought of as 7 blanks to be filled, and we have 16 actors with which to fill them. The director can cast the roles in $_{16}P_7 = 57,657,600$ ways.   *Now try Exercise 12.*

Combinations

When we count permutations of n objects taken r at a time, we consider different orderings of the same r selected objects as being different permutations. In many applications we are only interested in the ways to *select* the r objects, regardless of the order in which we arrange them. These unordered selections are called **combinations of n objects taken r at a time**.

Combination Counting Formula

The number of combinations of n objects taken r at a time is denoted $_nC_r$ and is given by

$$_nC_r = \frac{n!}{r!(n - r)!} \quad \text{for} \quad 0 \leq r \leq n.$$

If $r > n$, then $_nC_r = 0$.

We can verify the $_nC_r$ formula with the Multiplication Principle. Since every permutation can be thought of as an *unordered* selection of r objects *followed* by a particular *ordering* of the objects selected, the Multiplication Principle gives $_nP_r = {_nC_r} \cdot r!$.

Therefore

$$_nC_r = \frac{_nP_r}{r!} = \frac{1}{r!} \cdot \frac{n!}{(n-r)!} = \frac{n!}{r!(n-r)!}.$$

A WORD ON NOTATION

Some textbooks use $P(n, r)$ instead of $_nP_r$ and $C(n, r)$ instead of $_nC_r$. Much more common is the notation $\binom{n}{r}$ for $_nC_r$. Both $\binom{n}{r}$ and $_nC_r$ are often read "n choose r."

EXAMPLE 6   Distinguishing Combinations from Permutations

In each of the following scenarios, tell whether permutations (ordered) or combinations (unordered) are being described.

(a) A president, vice-president, and secretary are chosen from a 25-member garden club.

(b) A cook chooses 5 potatoes from a bag of 12 potatoes to make a potato salad.

(c) A teacher makes a seating chart for 22 students in a classroom with 30 desks.

SOLUTION

(a) Permutations. Order matters because it matters who gets which office.

(b) Combinations. The salad is the same no matter what order the potatoes are chosen.

(c) Permutations. A different ordering of students in the same seats results in a different seating chart.

Notice that once you know what is being counted, getting the correct number is easy with a calculator. The number of possible choices in the scenarios above are: (a) $_{25}P_3 = 13,800$, (b) $_{12}C_5 = 792$, and (c) $_{30}P_{22} \approx 6.5787 \times 10^{27}$.

Now try Exercise 19.

COMBINATIONS ON A CALCULATOR

Most modern calculators have an nCr selection built in. As with permutations, it is better to use the nCr function than to use the formula $\dfrac{n!}{r!(n-r)!}$, as the individual factorials can get too large for the calculator.

EXAMPLE 7   Counting Combinations

In the Miss America pageant, 51 contestants must be narrowed down to 10 finalists who will compete on national television. In how many possible ways can the ten finalists be selected?

SOLUTION   Notice that the *order* of the finalists does not matter at this phase; all that matters is which women are selected. So we count combinations rather than permutations.

$$_{51}C_{10} = \frac{51!}{10! \, 41!} = 12,777,711,870.$$

The 10 finalists can be chosen in 12,777,711,870 ways.

Now try Exercise 27.

EXAMPLE 8   Picking Lottery Numbers

The Georgia Lotto requires winners to pick 6 integers between 1 and 46. The order in which you select them does not matter; indeed, the lottery tickets are always printed with the numbers in ascending order. How many different lottery tickets are possible?

SOLUTION There are $_{46}C_6 = 9{,}366{,}819$ possible lottery tickets of this type. (That's more than enough different tickets for every person in the state of Georgia!)

Now try Exercise 29.

Subsets of an *n*-Set

As a final application of the counting principle, consider the pizza topping problem.

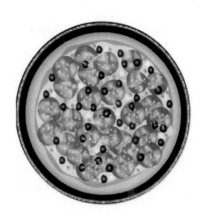

EXAMPLE 9   Selecting Pizza Toppings

Armando's Pizzeria offers patrons any combination of up to 10 different toppings: pepperoni, mushroom, sausage, onion, green pepper, bacon, prosciutto, black olive, green olive, and anchovies. How many different pizzas can be ordered

(a) if we can choose any three toppings?

(b) if we can choose any number of toppings (0 through 10)?

SOLUTION

(a) Order does not matter (for example, the sausage-pepperoni-mushroom pizza is the same as the pepperoni-mushroom-sausage pizza), so the number of possible pizzas is $_{10}C_3 = 120$.

(b) We could add up all the numbers of the form $_{10}C_r$ for $r = 0, 1, \ldots, 10$, but there is an easier way to count the possibilities. Consider the ten options to be lined up as in the statement of the problem. In considering each option, we have two choices: yes or no. (For example, the pepperoni-mushroom-sausage pizza would correspond to the sequence YYYNNNNNNN.) By the Multiplication Principle, the number of such sequences is $2 \cdot 2 \cdot 2 \cdot 2 \cdot 2 \cdot 2 \cdot 2 \cdot 2 \cdot 2 \cdot 2 = 1024$, which is the number of possible pizzas.

Now try Exercise 37.

The solution to Example 9b suggests a general rule that will be our last counting formula of the section.

Formula for Counting Subsets of an *n*-Set

There are 2^n subsets of a set with n objects (including the empty set and the entire set).

EXAMPLE 10   Analyzing an Advertised Claim

A national hamburger chain used to advertise that it fixed its hamburgers "256 ways," since patrons could order whatever toppings they wanted. How many toppings must have been available?

continued

SOLUTION We need to solve the equation $2^n = 256$ for n. We could solve this easily enough by trial and error, but we will solve it with logarithms just to keep the method fresh in our minds.

$$2^n = 256$$

$$\log 2^n = \log 256$$

$$n \log 2 = \log 256$$

$$n = \frac{\log 256}{\log 2}$$

$$n = 8$$

There must have been 8 toppings from which to choose.

Now try Exercise 39.

WHY ARE THERE NOT 1000 POSSIBLE AREA CODES?

While there are 1000 three-digit numbers between 000 and 999, not all of them are available for use as area codes. For example, area codes cannot begin with 0 or 1, and numbers of the form *abb* have been reserved for other purposes.

CHAPTER OPENER PROBLEM (from page 699)

PROBLEM: There are 680 three-digit numbers that are available for use as area codes in North America. As of April 2002, 305 of them were actually being used (*Source: www.nanpa.com*). How many additional three-digit area codes are available for use? Within a given area code, how many unique telephone numbers are theoretically possible?

SOLUTION: There are $680 - 305 = 375$ additional area codes available. Within a given area code, each telephone number has seven digits chosen from the ten digits 0 through 9. Since each digit can theoretically be any of 10 numbers, there are

$$10 \cdot 10 \cdot 10 \cdot 10 \cdot 10 \cdot 10 \cdot 10 = 10^7 = 10,000,000$$

different telephone numbers possible within a given area code.

Putting these two results together, we see that the unused area codes in April 2002 represented an additional 3.75 billion possible telephone numbers!

QUICK REVIEW 9.1

In Exercises 1–10, give the number of objects described. In some cases you might have to do a little research or ask a friend.

1. The number of cards in a standard deck
2. The number of cards of each suit in a standard deck
3. The number of faces on a cubical die
4. The number of possible totals when two dice are rolled
5. The number of vertices of a decagon
6. The number of musicians in a string quartet
7. The number of players on a soccer team
8. The number of prime numbers between 1 and 10, inclusive
9. The number of squares on a chessboard
10. The number of cards in a contract bridge hand

SECTION 9.1 EXERCISES

In Exercises 1–4, count the number of ways that each procedure can be done.

1. Line up three people for a photograph.

2. Prioritize four pending jobs from most to least important.

3. Arrange five books from left to right on a bookshelf.

4. Award ribbons for 1st place to 5th place to the top five dogs in a dog show.

5. Homecoming King and Queen There are four candidates for homecoming queen and three candidates for king. How many king-queen pairs are possible?

6. Possible Routes There are three roads from town *A* to town *B* and four roads from town *B* to town *C*. How many different routes are there from *A* to *C* by way of *B*?

7. Permuting Letters How many 9-letter "words" (not necessarily in any dictionary) can be formed from the letters of the word LOGARITHM? (Curiously, one such arrangement spells another word related to mathematics. Can you name it?)

8. Three-Letter Crossword Entries Excluding J, Q, X, and Z, how many 3-letter crossword puzzle entries can be formed that contain no repeated letters? (It has been conjectured that all of them have appeared in puzzles over the years, sometimes with painfully contrived definitions.)

9. Permuting Letters How many distinguishable 11-letter "words" can be formed using the letters in MISSISSIPPI?

10. Permuting Letters How many distinguishable 11-letter "words" can be formed using the letters in CHATTANOOGA?

11. Electing Officers The 13 members of the East Brainerd Garden Club are electing a President, Vice-President, and Secretary from among their members. How many different ways can this be done?

12. City Government From among 12 projects under consideration, the mayor must put together a prioritized (that is, ordered) list of 6 projects to submit to the city council for funding. How many such lists can be formed?

In Exercises 13–18, evaluate each expression without a calculator. Then check with your calculator to see if your answer is correct.

13. $4!$

14. $(3!)(0!)$

15. $_6P_2$

16. $_9P_2$

17. $_{10}C_7$

18. $_{10}C_3$

In Exercises 19–22, tell whether permutations (ordered) or combinations (unordered) are being described.

19. 13 cards are selected from a deck of 52 to form a bridge hand.

20. 7 digits are selected (without repetition) to form a telephone number.

21. 4 students are selected from the senior class to form a committee to advise the cafeteria director about food.

22. 4 actors are chosen to play the Beatles in a film biography.

23. License Plates How many different license plates begin with two digits, followed by two letters and then three digits if no letters or digits are repeated?

24. License Plates How many different license plates consist of five symbols, either digits or letters?

25. Tumbling Dice Suppose that two dice, one red and one green, are rolled. How many different outcomes are possible for the pair of dice?

26. Coin Toss How many different sequences of heads and tails are there if a coin is tossed 10 times?

27. Forming Committees A 3-woman committee is to be elected from a 25-member sorority. How many different committees can be elected?

28. Straight Poker In the original version of poker known as "straight" poker, a five-card hand is dealt from a standard deck of 52. How many different straight poker hands are possible?

29. Buying Discs Juan has money to buy only three of the 48 compact discs available. How many different sets of discs can he purchase?

30. Coin Toss A coin is tossed 20 times and the heads and tails sequence is recorded. From among all the possible sequences of heads and tails, how many have exactly seven heads?

31. Drawing Cards How many different 13-card hands include the ace and king of spades?

32. Job Interviews The head of the personnel department interviews eight people for three identical openings. How many different groups of three can be employed?

33. Scholarship Nominations Six seniors at Rydell High School meet the qualifications for a competitive honor scholarship at a major university. The university allows the school to nominate up to three candidates, and the school always nominates at least one. How many different choices could the nominating committee make?

34. Pu-pu Platters A Chinese restaurant will make a Pu-pu platter "to order" containing any one, two, or three selections from its appetizer menu. If the menu offers five different appetizers, how many different platters could be made?

trig functions

cos
tan
sin

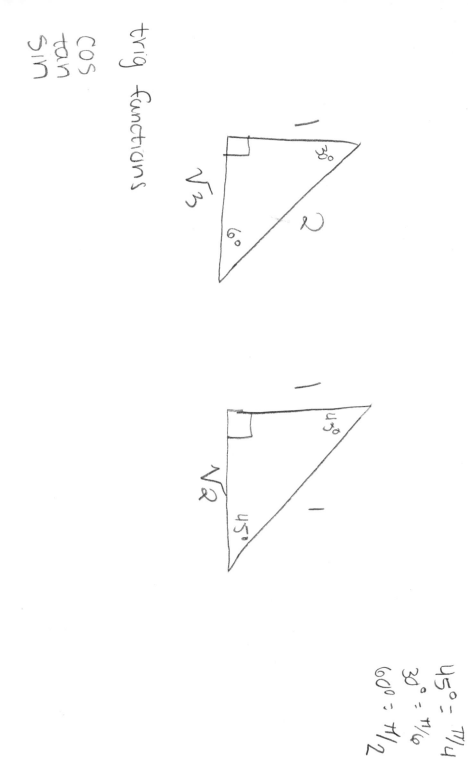

$45° = π/4$
$30° = π/6$
$60° = π/2$

35. Yahtzee In the game of Yahtzee, five dice are tossed simultaneously. How many outcomes can be distinguished if all the dice are different colors?

36. Indiana Jones and the Final Exam Professor Indiana Jones gives his class 20 study questions, from which he will select 8 to be answered on the final exam. How many ways can he select the questions?

37. Salad Bar Mary's lunch always consists of a full plate of salad from Ernestine's salad bar. She always takes equal amounts of each salad she chooses, but she likes to vary her selections. If she can choose from among 9 different salads, how many essentially different lunches can she create?

38. Buying a New Car A new car customer has to choose from among 3 models, each of which comes in 4 exterior colors, 3 interior colors, and with any combination of up to 6 optional accessories. How many essentially different ways can the customer order the car?

39. Pizza Possibilities Luigi sells one size of pizza, but he claims that his selection of toppings allows for "more than 4000 different choices." What is the smallest number of toppings Luigi could offer?

40. Proper Subsets A subset of set A is called *proper* if it is neither the empty set nor the entire set A. How many proper subsets does an n-set have?

41. True-False Tests How many different answer keys are possible for a 10-question True-False test?

42. Multiple-Choice Tests How many different answer keys are possible for a 10-question multiple-choice test in which each question leads to choice a, b, c, d, or e?

Standardized Test Questions

43. True or False If a and b are positive integers such that $a + b = n$, then $\binom{n}{a} = \binom{n}{b}$. Justify your answer.

44. True or False If a, b, and n are integers such that $a < b < n$, then $\binom{n}{a} < \binom{n}{b}$. Justify your answer.

You may use a graphing calculator when evaluating Exercises 45–48.

45. Multiple Choice Lunch at the Gritsy Palace consists of an entrée, two vegetables, and a dessert. If there are four entrées, six vegetables, and six desserts from which to choose, how many essentially different lunches are possible?

(A) 16

(B) 25

(C) 144

(D) 360

(E) 720

46. Multiple Choice How many different ways can the judges choose 5th to 1st places from ten Miss America finalists?

(A) 50

(B) 120

(C) 252

(D) 30,240

(E) 3,628,800

47. Multiple Choice Assuming r and n are positive integers with $r < n$, which of the following numbers does *not* equal 1?

(A) $(n - n)!$

(B) $_nP_n$

(C) $_nC_n$

(D) $\binom{n}{n}$

(E) $\binom{n}{r} \div \binom{n}{n - r}$

48. Multiple Choice An organization is electing 3 new board members by approval voting. Members are given ballots with the names of 5 candidates and are allowed to check off the names of all candidates whom they would approve (which could be none, or even all five). The three candidates with the most checks overall are elected. In how many different ways can a member fill out the ballot?

(A) 10

(B) 20

(C) 32

(D) 125

(E) 243

Explorations

49. Group Activity For each of the following numbers, make up a counting problem that has the number as its answer.

(a) $_{52}C_3$

(b) $_{12}C_3$

(c) $_{25}P_{11}$

(d) 2^5

(e) $3 \cdot 2^{10}$

50. Writing to Learn You have a fresh carton containing one dozen eggs and you need to choose two for breakfast. Give a counting argument based on this scenario to explain why $_{12}C_2 = {}_{12}C_{10}$.

51. Factorial Riddle The number 50! ends in a string of consecutive 0's.

(a) How many 0's are in the string?

(b) How do you know?

52. Group Activity Diagonals of a Regular Polygon In Exploration 1 of Section 1.7, you reasoned from data points and quadratic regression that the number of diagonals of a regular polygon with n vertices was $(n^2 - 3n)/2$.

(a) Explain why the number of segments connecting all pairs of vertices is $_nC_2$.

(b) Use the result from part (a) to prove that the number of diagonals is $(n^2 - 3n)/2$.

Extending the Ideas

53. Writing to Learn Suppose that a chain letter (illegal if money is involved) is sent to five people the first week of the year. Each of these five people sends a copy of the letter to five more people during the second week of the year. Assume that everyone who receives a letter participates. Explain how you know with certainty that someone will receive a second copy of this letter later in the year.

54. A Round Table How many different seating arrangements are possible for 4 people sitting around a round table?

55. Colored Beads Four beads—red, blue, yellow, and green—are arranged on a string to make a simple necklace as shown in the figure. How many arrangements are possible?

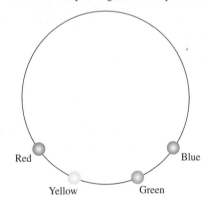

56. Casting a Play A director is casting a play with two female leads and wants to have a chance to audition the actresses two at a time to get a feeling for how well they would work together. His casting director and his administrative assistant both prepare charts to show the amount of time that would be required, depending on the number of actresses who come to the audition. Which time chart is more reasonable, and why?

Number who audition	Time required (minutes)	Number who audition	Time required (minutes)
3	10	3	10
6	45	6	30
9	110	9	60
12	200	12	100
15	320	15	150

57. Bridge Around the World Suppose that a contract bridge hand is dealt somewhere in the world every second. What is the fewest number of years required for every possible bridge hand to be dealt? (See Quick Review Exercise 10.)

58. Basketball Lineups Each NBA basketball team has 12 players on its roster. If each coach chooses 5 starters without regard to position, how many different sets of 10 players can start when two given teams play a game?

9.2
The Binomial Theorem

Powers of Binomials

Many important mathematical discoveries have begun with the study of patterns. In this chapter, we want to introduce an important polynomial theorem called the Binomial Theorem, for which we will set the stage by observing some patterns.

If you expand $(a + b)^n$ for $n = 0, 1, 2, 3, 4$, and 5, here is what you get:

$$(a + b)^0 = 1$$

$$(a + b)^1 = 1a^1b^0 + 1a^0b^1$$

$$(a + b)^2 = 1a^2b^0 + 2a^1b^1 + 1a^0b^2$$

$$(a + b)^3 = 1a^3b^0 + 3a^2b^1 + 3a^1b^2 + 1a^0b^3$$

$$(a + b)^4 = 1a^4b^0 + 4a^3b^1 + 6a^2b^2 + 4a^1b^3 + 1a^0b^4$$

$$(a + b)^5 = 1a^5b^0 + 5a^4b^1 + 10a^3b^2 + 10a^2b^3 + 5a^1b^4 + 1a^0b^5$$

Can you observe the patterns and predict what the expansion of $(a + b)^6$ will look like? You can probably predict the following:

1. The powers of a will decrease from 6 to 0 by 1's.

2. The powers of b will increase from 0 to 6 by 1's.

3. The first two coefficients will be 1 and 6.

4. The last two coefficients will be 6 and 1.

At first you might not see the pattern that would enable you to find the other so-called *binomial coefficients*, but you should see it after the following Exploration.

EXPLORATION 1 **Exploring the Binomial Coefficients**

1. Compute $_3C_0$, $_3C_1$, $_3C_2$, and $_3C_3$. Where can you find these numbers in the binomial expansions above?

2. The expression $4\,_nC_r$ $\{0, 1, 2, 3, 4\}$ tells the calculator to compute $_4C_r$ for each of the numbers $r = 0, 1, 2, 3, 4$ and display them as a list. Where can you find these numbers in the binomial expansions above?

3. Compute $5\,_nC_r$ $\{0, 1, 2, 3, 4, 5\}$. Where can you find these numbers in the binomial expansions above?

By now you are probably ready to conclude that the binomial coefficients in the expansion of $(a + b)^n$ are just the values of $_nC_r$ for $r = 0, 1, 2, 3, 4, \ldots, n$. We also hope you are wondering *why* this is true.

The expansion of

$$(a + b)^n = \underbrace{(a + b)(a + b)(a + b) \cdots (a + b)}_{n \text{ factors}}$$

consists of all possible products that can be formed by taking one letter (either a or b) from each factor $(a + b)$. The number of ways to form the product $a^r b^{n-r}$ is exactly the same as the number of ways to choose r factors to contribute an a, since the rest of the factors will obviously contribute a b. The number of ways to choose r factors from n factors is $_nC_r$.

DEFINITION Binomial Coefficient

The binomial coefficients that appear in the expansion of $(a + b)^n$ are the values of $_nC_r$ for $r = 0, 1, 2, 3, \ldots, n$.

A classical notation for $_nC_r$, especially in the context of binomial coefficients, is $\binom{n}{r}$. Both notations are read "n choose r."

TABLE TRICK

You can also use the table display to show binomial coefficients. For example, let Y1 = 5 $_nC_r$ X, and set TblStart = 0 and $\triangle Tbl$ = 1 to display the binomial coefficients for $(a + b)^5$.

EXAMPLE 1   Using $_nC_r$ to Expand a Binomial

Expand $(a + b)^5$, using a calculator to compute the binomial coefficients.

SOLUTION Enter 5 $_nC_r$ {0, 1, 2, 3, 4, 5} into the calculator to find the binomial coefficients for $n = 5$. The calculator returns the list {1, 5, 10, 10, 5, 1}. Using these coefficients, we construct the expansion:

$$(a + b)^5 = 1a^5 + 5a^4b + 10a^3b^2 + 10a^2b^3 + 5ab^4 + 1b^5.$$

Now try Exercise 3.

Pascal's Triangle

If we eliminate the plus signs and the powers of the variables a and b in the "triangular" array of binomial coefficients with which we began this section, we get:

```
              1
            1   1
          1   2   1
        1   3   3   1
      1   4   6   4   1
    1   5  10  10   5   1
```

THE NAME GAME

The fact that Pascal's triangle was not discovered by Pascal is ironic, but hardly unusual in the annals of mathematics. We mentioned in Chapter 5 that Heron did not discover Heron's formula, and Pythagoras did not even discover the Pythagorean theorem. The history of calculus is filled with similar injustices.

This is called **Pascal's triangle** in honor of Blaise Pascal (1623–1662), who used it in his work but certainly did not discover it. It appeared in 1303 in a Chinese text, the *Precious Mirror*, by Chu Shih-chieh, who referred to it even then as a "diagram of the old method for finding eighth and lower powers."

For convenience, we refer to the top "1" in Pascal's triangle as row 0. That allows us to associate the numbers along row n with the expansion of $(a + b)^n$.

Pascal's triangle is so rich in patterns that people still write about them today. One of the simplest patterns is the one that we use for getting from one row to the next, as in the following example.

EXAMPLE 2   Extending Pascal's Triangle

Show how row 5 of Pascal's triangle can be used to obtain row 6, and use the information to write the expansion of $(x + y)^6$.

SOLUTION The two outer numbers of every row are 1's. Each number between them is the sum of the two numbers immediately above it. So row 6 can be found from row 5 as follows:

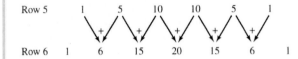

These are the binomial coefficients for $(x + y)^6$, so

$$(x + y)^6 = x^6 + 6x^5y + 15x^4y^2 + 20x^3y^3 + 15x^2y^4 + 6xy^5 + y^6.$$

Now try Exercise 7.

The technique used in Example 1 to extend Pascal's triangle depends on the following recursion formula.

Recursion Formula for Pascal's Triangle

$$\binom{n}{r} = \binom{n - 1}{r - 1} + \binom{n - 1}{r} \text{ or, equivalently, } {}_nC_r = {}_{n-1}C_{r-1} + {}_{n-1}C_r$$

Here's a counting argument to explain why it works. Suppose we are choosing r objects from n objects. As we have seen, this can be done in ${}_nC_r$ ways. Now identify one of the n objects with a special tag. How many ways can we choose r objects if the tagged object is among them? Well, we have $r - 1$ objects yet to be chosen from among the $n - 1$ that are untagged, so ${}_{n-1}C_{r-1}$. How many ways can we choose r objects if the tagged object is *not* among them? This time we must choose all r objects from among the $n - 1$ without tags, so ${}_{n-1}C_r$. Since our selection of r objects must either contain the tagged object or not contain it, ${}_{n-1}C_{r-1} + {}_{n-1}C_r$ counts all the possibilities with no overlap. Therefore, ${}_nC_r = {}_{n-1}C_{r-1} + {}_{n-1}C_r$.

It is not necessary to construct Pascal's triangle to find specific binomial coefficients, since we already have a formula for computing them: ${}_nC_r = \binom{n}{r} = \dfrac{n!}{r!(n - r)!}$. This formula can be used to give an algebraic formula for the recursion formula above, but we will leave that as an exercise for the end of the section.

**THE BINOMIAL THEOREM
IN Σ NOTATION**

For those who are already familiar with summation notation, here is how the Binomial Theorem looks:

$$(a + b)^n = \sum_{r=0}^{n} \binom{n}{r} a^{n-r} b^r.$$

Those who are not familiar with this notation will learn about it in Section 9.4.

EXAMPLE 3   Computing Binomial Coefficients

Find the coefficient of x^{10} in the expansion of $(x + 2)^{15}$.

SOLUTION The only term in the expansion that we need to deal with is $_{15}C_{10} x^{10} 2^5$. This is

$$\frac{15!}{10!5!} \cdot 2^5 \cdot x^{10} = 3003 \cdot 32 \cdot x^{10} = 96{,}096\, x^{10}.$$

The coefficient of x^{10} is 96,096.   *Now try Exercise 15.*

The Binomial Theorem

We now state formally the theorem about expanding powers of binomials, known as the Binomial Theorem. For tradition's sake, we will use the symbol $\binom{n}{r}$ instead of $_nC_r$.

The Binomial Theorem

For any positive integer n,

$$(a + b)^n = \binom{n}{0} a^n + \binom{n}{1} a^{n-1} b + \cdots + \binom{n}{r} a^{n-r} b^r + \cdots + \binom{n}{n} b^n,$$

where

$$\binom{n}{r} = {}_nC_r = \frac{n!}{r!(n-r)!}.$$

EXAMPLE 4   Expanding a Binomial

Expand $(2x - y^2)^4$.

SOLUTION We use the Binomial Theorem to expand $(a + b)^4$ where $a = 2x$ and $b = -y^2$.

$$(a + b)^4 = a^4 + 4a^3 b + 6a^2 b^2 + 4ab^3 + b^4$$
$$(2x - y^2)^4 = (2x)^4 + 4(2x)^3(-y^2) + 6(2x)^2(-y^2)^2$$
$$+ 4(2x)(-y^2)^3 + (-y^2)^4$$
$$= 16x^4 - 32x^3 y^2 + 24x^2 y^4 - 8xy^6 + y^8$$

Now try Exercise 17.

Factorial Identities

Expressions involving factorials combine to give some interesting identities, most of them relying on the basic identities shown below (actually two versions of the same identity).

Basic Factorial Identities

For any integer $n \geq 1$, $n! = n(n-1)!$

For any integer $n \geq 0$, $(n+1)! = (n+1)n!$

> **EXAMPLE 5   Proving an Identity with Factorials**
>
> Prove that $\binom{n+1}{2} - \binom{n}{2} = n$ for all integers $n \geq 2$.
>
> **SOLUTION**
>
> $$\binom{n+1}{2} - \binom{n}{2} = \frac{(n+1)!}{2!(n+1-2)!} - \frac{n!}{2!(n-2)!}$$
>
> $$= \frac{(n+1)(n)(n-1)!}{2(n-1)!} - \frac{n(n-1)(n-2)!}{2(n-2)!}$$
>
> $$= \frac{n^2+n}{2} - \frac{n^2-n}{2}$$
>
> $$= \frac{2n}{2}$$
>
> $$= n$$

Now try Exercise 33.

QUICK REVIEW 9.2   *(Prerequisite skill Section A.2)*

In Exercises 1–10, use the distributive property to expand the binomial.

1. $(x + y)^2$

2. $(a + b)^2$

3. $(5x - y)^2$

4. $(a - 3b)^2$

5. $(3s + 2t)^2$

6. $(3p - 4q)^2$

7. $(u + v)^3$

8. $(b - c)^3$

9. $(2x - 3y)^3$

10. $(4m + 3n)^3$

SECTION 9.2 EXERCISES

In Exercises 1–4, expand the binomial using a calculator to find the binomial coefficients.

1. $(a + b)^4$

2. $(a + b)^6$

3. $(x + y)^7$

4. $(x + y)^{10}$

In Exercises 5–8, expand the binomial using Pascal's triangle to find the coefficients.

5. $(x + y)^3$

6. $(x + y)^5$

7. $(p + q)^8$

8. $(p + q)^9$

In Exercises 9–12, evaluate the expression by hand (using the formula) before checking your answer on a grapher.

9. $\binom{9}{2} = {}_9C_2$

10. $\binom{15}{11}$

11. $\binom{166}{166} = 166C_{166}$

12. $\binom{166}{0}$

In Exercises 13–16, find the coefficient of the given term in the binomial expansion.

13. $x^{11}y^3$ term, $(x + y)^{14}$

14. x^5y^8 term, $(x + y)^{13}$

15. x^4 term, $(x - 2)^{12}$

16. x^7 term, $(x - 3)^{11}$

In Exercises 17–20, use the Binomial Theorem to find a polynomial expansion for the function.

17. $f(x) = (x - 2)^5$

18. $g(x) = (x + 3)^6$

19. $h(x) = (2x - 1)^7$

20. $f(x) = (3x + 4)^5$

In Exercises 21–26, use the Binomial Theorem to expand each expression.

21. $(2x + y)^4$

22. $(2y - 3x)^5$

23. $(\sqrt{x} - \sqrt{y})^6$

24. $(\sqrt{x} + \sqrt{3})^4$

25. $(x^{-2} + 3)^5$

26. $(a - b^{-3})^7$

27. Determine the largest integer n for which your calculator will compute $n!$.

28. Determine the largest integer n for which your calculator will compute $\binom{n}{100}$.

29. Prove that $\binom{n}{1} = \binom{n}{n-1} = n$ for all integers $n \geq 1$.

30. Prove that $\binom{n}{r} = \binom{n}{n-r}$ for all integers $n \geq r \geq 0$.

31. Use the formula $\binom{n}{r} = \dfrac{n!}{r!(n-r)!}$ to prove that

$\binom{n}{r} = \binom{n-1}{r-1} + \binom{n-1}{r}$. (This is the pattern in Pascal's triangle that appears in Example 2.)

32. Find a counterexample to show that each statement is *false*.

(a) $(n + m)! = n! + m!$

(b) $(nm)! = n!m!$

33. Prove that $\binom{n}{2} + \binom{n+1}{2} = n^2$ for all integers $n \geq 2$.

34. Prove that $\binom{n}{n-2} + \binom{n+1}{n-1} = n^2$ for all integers $n \geq 2$.

Standardized Test Questions

35. True or False The coefficients in the polynomial expansion of $(x - y)^{50}$ alternate in sign. Justify your answer.

36. True or False The sum of any row of Pascal's triangle is an even integer. Justify your answer.

You may use a graphing calculator when evaluating Exercises 37–40.

37. Multiple Choice What is the coefficient of x^4 in the expansion of $(2x + 1)^8$?

(A) 16

(B) 256

(C) 1120

(D) 1680

(E) 26,680

38. Multiple Choice Which of the following numbers does *not* appear on row 10 of Pascal's Triangle?

(A) 1

(B) 5

(C) 10

(D) 120

(E) 252

39. Multiple Choice The *sum* of the coefficients of $(3x - 2y)^{10}$ is

(A) 1

(B) 1024

(C) 58,025

(D) 59,049

(E) 9,765,625

40. Multiple Choice $(x + y)^3 + (x - y)^3 =$

(A) 0

(B) $2x^3$

(C) $2x^3 - 2y^3$

(D) $2x^3 + 6xy^2$

(E) $6x^2y + 2y^3$

Explorations

41. Triangular Numbers Numbers of the form $1 + 2 + \cdots + n$ are called **triangular numbers** because they count numbers in triangular arrays, as shown below:

(a) Compute the first 10 triangular numbers.

(b) Where do the triangular numbers appear in Pascal's triangle?

(c) **Writing to Learn** Explain why the diagram below shows that the nth triangular number can be written as $n(n + 1)/2$.

(d) Write the formula in part (c) as a binomial coefficient. (This is why the triangular numbers appear as they do in Pascal's triangle.)

42. Group Activity Exploring Pascal's Triangle Break into groups of two or three. Just by looking at patterns in Pascal's triangle, guess the answers to the following questions. (It is easier to make a conjecture from a pattern than it is to construct a proof!)

(a) What positive integer appears the least number of times?

(b) What number appears the greatest number of times?

(c) Is there any positive integer that does *not* appear in Pascal's triangle?

(d) If you go along any row alternately adding and subtracting the numbers, what is the result?

(e) If p is a prime number, what do all the interior numbers along the pth row have in common?

(f) Which rows have all even interior numbers?

(g) Which rows have all odd numbers?

(h) What other patterns can you find? Share your discoveries with the other groups.

Extending the Ideas

43. Use the Binomial Theorem to prove that the sum of the entries along the nth row of Pascal's triangle is 2^n. That is,

$$\binom{n}{0} + \binom{n}{1} + \binom{n}{2} + \cdots + \binom{n}{n} = 2^n.$$

[*Hint*: Use the Binomial Theorem to expand $(1 + 1)^n$.]

44. Use the Binomial Theorem to prove that the alternating sum along any row of Pascal's triangle is zero. That is,

$$\binom{n}{0} - \binom{n}{1} + \binom{n}{2} - \cdots + (-1)^n \binom{n}{n} = 0.$$

45. Use the Binomial Theorem to prove that

$$\binom{n}{0} + 2\binom{n}{1} + 4\binom{n}{2} + \cdots + 2^n\binom{n}{n} = 3^n.$$

9.3
Probability

What you'll learn about

- Sample Spaces and Probability Functions
- Determining Probabilities
- Venn Diagrams and Tree Diagrams
- Conditional Probability
- Binomial Distributions

. . . and why

Everyone should know how mathematical the "laws of chance" really are.

FIGURE 9.2 A sum of 4 on a roll of two dice. (Example 1d)

Sample Spaces and Probability Functions

Most people have an intuitive sense of probability. Unfortunately, this sense is not often based on a foundation of mathematical principles, so people become victims of scams, misleading statistics, and false advertising. In this lesson, we want to build on your intuitive sense of probability and give it a mathematical foundation.

EXAMPLE 1   Testing Your Intuition About Probability

Find the probability of each of the following events.

(a) Tossing a head on one toss of a fair coin.

(b) Tossing two heads in a row on two tosses of a fair coin.

(c) Drawing a queen from a standard deck of 52 cards.

(d) Rolling a sum of 4 on a single roll of two fair dice.

(e) Guessing all 6 numbers in a state lottery that requires you to pick 6 numbers between 1 and 46, inclusive.

SOLUTION

(a) There are two equally likely outcomes: {T, H}. The probability is 1/2.

(b) There are four equally likely outcomes: {TT, TH, HT, HH}. The probability is 1/4.

(c) There are 52 equally likely outcomes, 4 of which are queens. The probability is 4/52, or 1/13.

(d) By the Multiplication Principle of Counting (Section 9.1), there are $6 \times 6 = 36$ equally likely outcomes. Of these, three {(1, 3), (3, 1), (2, 2)} yield a sum of 4 (Figure 9.2). The probability is 3/36, or 1/12.

(e) There are $_{46}C_6 = 9{,}366{,}819$ equally likely ways that 6 numbers can be chosen from 46 numbers without regard to order. Only one of these choices wins the lottery. The probability is $1/9{,}366{,}819 \approx 0.00000010676$.   ***Now try Exercise 5.***

Notice that in each of these cases we first counted the number of possible outcomes of the experiment in question. The set of all possible outcomes of an experiment is the **sample space** of the experiment. An **event** is a subset of the sample space. Each of our sample spaces consisted of a finite number of **equally likely outcomes**, which enabled us to find the probability of an event by counting.

Probability of an Event (Equally Likely Outcomes)

If E is an event in a finite, nonempty sample space S of equally likely outcomes, then the **probability** of the event E is

$$P(E) = \frac{\text{the number of outcomes in } E}{\text{the number of outcomes in } S}.$$

Probability theory got its start in letters between Blaise Pascal (1623–1662) and Pierre de Fermat (1601–1665) concerning games of chance, but it has come a long way since then. Modern mathematicians like David Blackwell (1919), the first African-American to receive a fellowship to the Institute for Advanced Study at Princeton, have greatly extended both the theory and the applications of probability, especially in the areas of statistics, quantum physics, and information theory. Moreover, the work of John Von Neumann (1903–1957) has led to a separate branch of modern discrete mathematics that really is about games, called Game Theory.

The hypothesis of equally likely outcomes is critical here. Many people guess wrongly on the probability in Example 1d because they figure that there are 11 possible outcomes for the sum on two fair dice: {2, 3, 4, 5, 6, 7, 8, 9, 10, 11, 12} and that 4 is one of them. (That reasoning is correct so far.) The reason that 1/11 is not the probability of rolling a sum of 4 is that all those sums are *not equally likely*.

On the other hand, we can *assign* probabilities to the 11 outcomes in this smaller sample space in a way that is consistent with the number of ways each total can occur. The table below shows a **probability distribution**, in which each outcome is assigned a unique probability by a *probability function*.

Outcome	Probability
2	1/36
3	2/36
4	3/36
5	4/36
6	5/36
7	6/36
8	5/36
9	4/36
10	3/36
11	2/36
12	1/36

We see that the outcomes are not equally likely, but we can find the probabilities of events by adding up the probabilities of the outcomes in the event, as in the following example.

EXAMPLE 2   Rolling the Dice

Find the probability of rolling a sum divisible by 3 on a single roll of two fair dice.

SOLUTION The event E consists of the outcomes {3, 6, 9, 12}. To get the probability of E we add up the probabilities of the outcomes in E (see the table of the probability distribution):

$$P(E) = \frac{2}{36} + \frac{5}{36} + \frac{4}{36} + \frac{1}{36} = \frac{12}{36} = \frac{1}{3}.$$

Now try Exercise 7.

Notice that this method would also have worked just fine with our 36-outcome sample space, in which every outcome has probability 1/36. In general, it is easier to work with sample spaces of equally likely events because it is not necessary to write out the probability distribution. When outcomes do have unequal probabilities, we need to know what probabilities to assign to the outcomes.

Not every function that assigns numbers to outcomes qualifies as a probability function.

EMPTY SET

A set with no elements is the *empty set*, denoted by $\varnothing$.

DEFINITION Probability Function

A **probability function** is a function P that assigns a real number to each outcome in a sample space S subject to the following conditions:

1. $0 \leq P(O) \leq 1$ for every outcome O;

2. the sum of the probabilities of all outcomes in S is 1;

3. $P(\varnothing) = 0$.

The probability of any event can then be defined in terms of the probability function.

Probability of an Event (Outcomes not Equally Likely)

Let S be a finite, nonempty sample space in which every outcome has a probability assigned to it by a probability function P. If E is any event in S, the **probability** of the event E is the sum of the probabilities of all the outcomes contained in E.

EXAMPLE 3   Testing a Probability Function

Is it possible to weight a standard 6-sided die in such a way that the probability of rolling each number n is exactly $1/(n^2 + 1)$?

SOLUTION The probability distribution would look like this:

Outcome	Probability
1	1/2
2	1/5
3	1/10
4	1/17
5	1/26
6	1/37

This is not a valid probability function, because $1/2 + 1/5 + 1/10 + 1/17 + 1/26 + 1/37 \neq 1$.   *Now try Exercise 9a.*

Determining Probabilities

It is not always easy to determine probabilities, but the arithmetic involved is fairly simple. It usually comes down to multiplication, addition, and (most importantly) counting. Here is the strategy we will follow:

Strategy for Determining Probabilities

1. Determine the sample space of all possible outcomes. When possible, choose outcomes that are equally likely.

2. If the sample space has equally likely outcomes, the probability of an event E is determined by counting:

$$P(E) = \frac{\text{the number of outcomes in } E}{\text{the number of outcomes in } S}.$$

3. If the sample space does not have equally likely outcomes, determine the probability function. (This is not always easy to do.) Check to be sure that the conditions of a probability function are satisfied. Then the probability of an event E is determined by adding up the probabilities of all the outcomes contained in E.

EXAMPLE 4 Choosing Chocolates, Sample Space I

Sal opens a box of a dozen chocolate cremes and generously offers two of them to Val. Val likes vanilla cremes the best, but all the chocolates look alike on the outside. If four of the twelve cremes are vanilla, what is the probability that both of Val's picks turn out to be vanilla?

SOLUTION The experiment in question is the selection of two chocolates, without regard to order, from a box of 12. There are $_{12}C_2 = 66$ outcomes of this experiment, and all of them are equally likely. We can therefore determine the probability by counting.

The event E consists of all possible pairs of 2 vanilla cremes that can be chosen, without regard to order, from 4 vanilla cremes available. There are $_4C_2 = 6$ ways to form such pairs.

Therefore, $P(E) = 6/66 = 1/11$.   *Now try Exercise 25.*

Many probability problems require that we think of events happening in succession, often with the occurrence of one event affecting the probability of the occurrence of another event. In these cases, we use a law of probability called the Multiplication Principle of Probability.

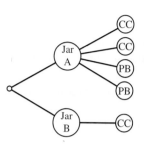

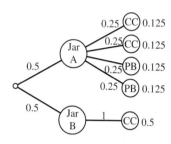

FIGURE 9.5 A tree diagram for Example 7.

FIGURE 9.6 The tree diagram for Example 7 with the probabilities filled in. Notice that the five cookies are not equally likely to be drawn. Notice also that the probabilities of the five cookies do add up to 1.

ORDERED OR

Notice that in
sample space
regarded, whe
have a sample
matters. (For e
distinct outcor
in Example 5 b
ing the probab
draw, second c
the other. In E:
counting unorc

EXAMPLE 7   Using a Tree Diagram

Two identical cookie jars are on a counter. Jar A contains 2 chocolate chip and 2 peanut butter cookies, while jar B contains 1 chocolate chip cookie. We select a cookie at random. What is the probability that it is a chocolate chip cookie?

SOLUTION It is tempting to say 3/5, since there are 5 cookies in all, 3 of which are chocolate chip. Indeed, this would be the answer if all the cookies were in the same jar. However, the fact that they are in different jars means that the 5 cookies are *not equally likely outcomes*. That lone chocolate chip cookie in jar B has a much better chance of being chosen than any of the cookies in jar A. We need to think of this as a two-step experiment: first pick a jar, then pick a cookie from that jar.

Figure 9.5 gives a visualization of the two-step process. In Figure 9.6, we have filled in the probabilities along each branch, first of picking the jar, then of picking the cookie. The probability at the *end* of each branch is obtained by multiplying the probabilities from the root to the branch. (This is the Multiplication Principle.) Notice that the probabilities of the 5 cookies (as predicted) are not equal.

The event "chocolate chip" is a set containing three outcomes. We add their probabilities together to get the correct probability:

$P(\text{chocolate chip}) = 0.125 + 0.125 + 0.5 = 0.75.$ ***Now try Exercise 29.***

Conditional Probability

The probability of drawing a chocolate chip cookie in Example 7 is an example of **conditional probability**, since the "cookie" probability is **dependent** on the "jar" outcome. A convenient symbol to use with conditional probability is $P(A\,|\,B)$, pronounced "P of A given B," meaning "the probability of the event A, given that event B occurs." In the cookie jars of Example 7,

$$P(\text{chocolate chip}\,|\,\text{jar } A) = \frac{2}{4} \quad \text{and} \quad P(\text{chocolate chip}\,|\,\text{jar } B) = 1.$$

(In the tree diagram, these are the probabilities along the *branches* that come out of the two jars, not the probabilities at the *ends* of the branches.)

The Multiplication Principle of Probability can be stated succinctly with this notation as follows:

$$P(A \text{ and } B) = P(A) \cdot P(B\,|\,A).$$

This is how we found the numbers at the ends of the branches in Figure 9.6.

As our final example of a probability problem, we will show how to use this formula in a different but equivalent form, sometimes called the **conditional probability formula**:

Conditional Probability Formula

If the event B depends on the event A, then $P(B \mid A) = \dfrac{P(A \text{ and } B)}{P(A)}.$

EXAMPLE 8 Using the Conditional Probability Formula

Suppose we have drawn a cookie at random from one of the jars described in Example 7. Given that it is chocolate chip, what is the probability that it came from jar *A*?

SOLUTION By the formula,

$$P(\text{jar } A \mid \text{chocolate chip}) = \frac{P(\text{jar } A \text{ and chocolate chip})}{P(\text{chocolate chip})}$$

$$= \frac{(1/2)(2/4)}{0.75} = \frac{0.25}{0.75} = \frac{1}{3}$$

Now try Exercise 31.

EXPLORATION 1 Testing Positive for HIV

As of the year 2003, the probability of an adult in the United States having HIV/AIDS was 0.006 (Source: *2004 CIA World Factbook*). The ELISA test is used to detect the virus antibody in blood. If the antibody is present, the test reports positive with probability 0.997 and negative with probability 0.003. If the antibody is not present, the test reports positive with probability 0.015 and negative with probability 0.985.

1. Draw a tree diagram with branches to nodes "antibody present" and "antibody absent" branching from the root. Fill in the probabilities for North American adults (age 15–49) along the branches. (Note that these two probabilities must add up to 1.)

2. From the node at the end of each of the two branches, draw branches to "positive" and "negative." Fill in the probabilities along the branches.

3. Use the Multiplication Principle to fill in the probabilities at the ends of the four branches. Check to see that they add up to 1.

4. Find the probability of a positive test result. (Note that this event consists of two outcomes).

5. Use the conditional probability formula to find the probability that a person with a positive test result actually *has* the antibody, i.e., *P*(antibody present | positive).

You might be surprised that the answer to part 5 is so low, but it should be compared with the probability of the antibody being present *before* seeing the positive test result, which was 0.006. Nonetheless, that is why a positive ELISA test is followed by further testing before a diagnosis of HIV/AIDS is made. This is the case with many diagnostic tests.

Binomial Distributions

We noted in our "Strategy for Determining Probabilities" that it is not always easy to determine a probability distribution for a sample space with unequal probabilities. An interesting exception for those who have studied the Binomial Theorem (Section 9.2) is the binomial distribution.

EXAMPLE 9　Repeating a Simple Experiment

We roll a fair die four times. Find the probability that we roll:

(a) all 3's.　　　**(b)** no 3's.　　　**(c)** exactly two 3's.

SOLUTION

(a) We have a probability 1/6 of rolling a 3 each time. By the Multiplication Principle, the probability of rolling a 3 all four times is $(1/6)^4 \approx 0.00077$.

(b) There is a probability 5/6 of rolling something other than 3 each time. By the Multiplication Principle, the probability of rolling a non-3 all four times is $(5/6)^4 \approx 0.48225$.

(c) The probability of rolling two 3's followed by two non-3's (again by the Multiplication Principle) is $(1/6)^2(5/6)^2 \approx 0.01929$. However, that is not the only outcome we must consider. In fact, the two 3's could occur *anywhere* among the four rolls, in exactly $\binom{4}{2} = 6$ ways. That gives us 6 outcomes, each with probability $(1/6)^2(5/6)^2$. The probability of the event "exactly two 3's" is therefore $\binom{4}{2}(1/6)^2(5/6)^2 \approx 0.11574$.

> *Now try Exercise 47.*

Did the form of those answers look a little familiar? Watch what they look like when we let $p = 1/6$ and $q = 5/6$:

$$P(\text{four 3's}) = p^4$$

$$P(\text{no 3's}) = q^4$$

$$P(\text{two 3's}) = \binom{4}{2}p^2q^2$$

You can probably recognize these as three of the terms in the expansion of $(p + q)^4$. This is no coincidence. In fact, the terms in the expansion

$$(p + q)^4 = p^4 + 4p^3q^1 + 6p^2q^2 + 4p^1q^3 + q^4$$

give the exact probabilities of 4, 3, 2, 1, and 0 threes (respectively) when we toss a fair die four times! That is why this is called a *binomial probability distribution*. The general theorem follows.

BINOMIAL PROBABILITIES ON A CALCULATOR

Your calculator might be programmed to find values for the binomial probability distribution function (binompdf). The solutions to Example 10 in one calculator syntax, for example, could be obtained by:

(a) binompdf(20, .9, 20) (20 repetitions, 0.9 probability, 20 successes)

(b) binompdf(20, .9, 18) (20 repetitions, 0.9 probability, 18 successes)

(c) 1 − binomcdf(20, .9, 17) (1 minus the *cumulative* probability of 17 or fewer successes)

Check your owner's manual for more information.

THEOREM Binomial Distribution

Suppose an experiment consists of n independent repetitions of an experiment with two outcomes, called "success" and "failure." Let $P(\text{success}) = p$ and $P(\text{failure}) = q$. (Note that $q = 1 - p$.)

Then the terms in the binomial expansion of $(p + q)^n$ give the respective probabilities of exactly $n, n - 1, \ldots, 2, 1, 0$ successes. The distribution is shown below:

Number of successes out of n independent repetitions	Probability
n	p^n
$n - 1$	$\binom{n}{n-1}p^{n-1}q$
$\vdots$	$\vdots$
r	$\binom{n}{r}p^r q^{n-r}$
$\vdots$	$\vdots$
1	$\binom{n}{1}pq^{n-1}$
0	q^n

EXAMPLE 10 Shooting Free Throws

Suppose Michael makes 90% of his free throws. If he shoots 20 free throws, and if his chance of making each one is independent of the other shots (an assumption you might question in a game situation), what is the probability that he makes

(a) all 20?

(b) exactly 18?

(c) at least 18?

SOLUTION We could get the probabilities of all possible outcomes by expanding $(0.9 + 0.1)^{20}$, but that is not necessary in order to answer these three questions. We just need to compute three specific terms.

(a) $P(20 \text{ successes}) = (0.9)^{20} \approx 0.12158$

(b) $P(18 \text{ successes}) = \binom{20}{18}(0.9)^{18}(0.1)^2 \approx 0.28518$

(c) $P(\text{at least 18 successes}) = P(18) + P(19) + P(20)$

$$= \binom{20}{18}(0.9)^{18}(0.1)^2 + \binom{20}{19}(0.9)^{19}(0.1) + (0.9)^{20}$$

$$\approx 0.6769$$

Now try Exercise 49.

QUICK REVIEW 9.3 *(Prerequisite skill Section 9.1)*

In Exercises 1–8, tell how many outcomes are possible for the experiment.

1. A single coin is tossed.
2. A single 6-sided die is rolled.
3. Three different coins are tossed.
4. Three different 6-sided dice are rolled.
5. Five different cards are drawn from a standard deck of 52.
6. Two chips are drawn simultaneously without replacement from a jar containing 10 chips.

7. Five people are lined up for a photograph.
8. Three-digit numbers are formed from the numbers {1, 2, 3, 4, 5} without repetition.

In Exercises 9 and 10, evaluate the expression by pencil and paper. Verify your answer with a calculator.

9. $\dfrac{_5C_3}{_{10}C_3}$

10. $\dfrac{_5C_2}{_{10}C_2}$

SECTION 9.3 EXERCISES

In Exercises 1–8, a red die and a green die have been rolled. What is the probability of the event?

1. The sum is 9.
2. The sum is even.
3. The number on the red die is greater than the number on the green die.
4. The sum is less than 10.
5. Both numbers are odd.
6. Both numbers are even.
7. The sum is prime.
8. The sum is 7 or 11.

9. **Writing to Learn** Alrik's gerbil cage has four compartments, A, B, C, and D. After careful observation, he estimates the proportion of time the gerbil spends in each compartment and constructs the table below.

Compartment	A	B	C	D
Proportion	0.25	0.20	0.35	0.30

(a) Is this a valid probability function? Explain.

(b) Is there a problem with Alrik's reasoning? Explain.

10. (Continuation of Exercise 9) Suppose Alrik determines that his gerbil spends time in the four compartments A, B, C, and D in the ratio 4:3:2:1. What proportions should he fill in the table above? Is this a valid probability function?

The maker of a popular chocolate candy that is covered in a thin colored shell has released information about the overall color proportions in its production of the candy, which is summarized in the following table.

Color	Brown	Red	Yellow	Green	Orange	Tan
Proportion	0.3	0.2	0.2	0.1	0.1	0.1

In Exercises 11–16, a single candy of this type is selected at random from a newly-opened bag. What is the probability that the candy has the given color(s)?

11. Brown or tan
12. Red, green, or orange
13. Red
14. Not red
15. Neither orange nor yellow
16. Neither brown nor tan

A peanut version of the same candy has all the same colors except tan. The proportions of the peanut version are given in the following table.

Color	Brown	Red	Yellow	Green	Orange
Proportion	0.3	0.2	0.2	0.2	0.1

In Exercises 17–22, a candy of this type is selected at random from each of two newly-opened bags. What is the probability that the two candies have the given color(s)?

17. Both are brown.
18. Both are orange.
19. One is red, and the other is green.
20. The first is brown, and the second is yellow.
21. Neither is yellow.
22. The first is not red, and the second is not orange.

Exercises 23–32 concern a version of the card game "bid Euchre" that uses a pack of 24 cards, consisting of ace, king, queen, jack, 10, and 9 in each of the four suits (spades, hearts, diamonds, and clubs). In bid Euchre, a hand consists of 6 cards. Find the probability of each event.

23. **Euchre** A hand is all spades.
24. **Euchre** All six cards are from the same suit.
25. **Euchre** A hand includes all four aces.

26. Euchre A hand includes two jacks of the same color (called the right and left bower).

27. Using Venn Diagrams *A* and *B* are events in a sample space *S* such that $P(A) = 0.6$, $P(B) = 0.5$, and $P(A \text{ and } B) = 0.3$.

 (a) Draw a Venn diagram showing the overlapping sets *A* and *B* and fill in the probabilities of the four regions formed.

 (b) Find the probability that *A* occurs but *B* does not.

 (c) Find the probability that *B* occurs but *A* does not.

 (d) Find the probability that neither *A* nor *B* occurs.

 (e) Are events *A* and *B* independent? (That is, does $P(A|B) = P(A)$?)

28. Using Venn Diagrams *A* and *B* are events in a sample space *S* such that $P(A) = 0.7$, $P(B) = 0.4$, and $P(A \text{ and } B) = 0.2$.

 (a) Draw a Venn diagram showing the overlapping sets *A* and *B* and fill in the probabilities of the four regions formed.

 (b) Find the probability that *A* occurs but *B* does not.

 (c) Find the probability that *B* occurs but *A* does not.

 (d) Find the probability that neither *A* nor *B* occurs.

 (e) Are events *A* and *B* independent? (That is, does $P(A|B) = P(A)$?)

In Exercises 29 and 30, it will help to draw a tree diagram.

29. Piano Lessons If it rains tomorrow, the probability is 0.8 that John will practice his piano lesson. If it does not rain tomorrow, there is only a 0.4 chance that John will practice. Suppose that the chance of rain tomorrow is 60%. What is the probability that John will practice his piano lesson?

30. Predicting Cafeteria Food If the school cafeteria serves meat loaf, there is a 70% chance that they will serve peas. If they do not serve meat loaf, there is a 30% chance that they will serve peas anyway. The students know that meat loaf will be served exactly once during the 5-day week, but they do not know which day. If tomorrow is Monday, what is the probability that

 (a) the cafeteria serves meat loaf?

 (b) the cafeteria serves meat loaf and peas?

 (c) the cafeteria serves peas?

31. Conditional Probability There are two precalculus sections at West High School. Mr. Abel's class has 12 girls and 8 boys, while Mr. Bonitz's class has 10 girls and 15 boys. If a West High precalculus student chosen at random happens to be a girl, what is the probability she is from Mr. Abel's class? [*Hint*: The answer is not 12/22.]

32. Group Activity Conditional Probability Two boxes are on the table. One box contains a normal coin and a two-headed coin; the other box contains three normal coins. A friend reaches into a box, removes a coin, and shows you one side: a head. What is the probability that it came from the box with the two-headed coin?

33. Renting Cars Floppy Jalopy Rent-a-Car has 25 cars available for rental—20 big bombs and 5 midsize cars. If two cars are selected at random, what is the probability that both are big bombs?

34. Defective Calculators Dull Calculators, Inc., knows that a unit coming off an assembly line has a probability of 0.037 of being defective. If four units are selected at random during the course of a workday, what is the probability that none of the units are defective?

35. Causes of Death The government designates a single cause for each death in the United States. The resulting data indicate that 45% of deaths are due to heart and other cardiovascular disease and 22% are due to cancer.

 (a) What is the probability that the death of a randomly selected person will be due to cardiovascular disease or cancer?

 (b) What is the probability that the death will be due to some other cause?

36. Yahtzee In the game of *Yahtzee*, on the first roll five dice are tossed simultaneously. What is the probability of rolling five of a kind (which is Yahtzee!) on the first roll?

37. Writing to Learn Explain why the following statement cannot be true. The probabilities that a computer salesperson will sell zero, one, two, or three computers in any one day are 0.12, 0.45, 0.38, and 0.15, respectively.

38. HIV Testing A particular test for HIV, the virus that causes AIDS, is 0.7% likely to produce a false positive result—a result indicating that the human subject has HIV when in fact the person is not carrying the virus. If 60 individuals who are HIV-negative are tested, what is the probability of obtaining at least one false result?

39. Graduate School Survey The Earmuff Junction College Alumni Office surveys selected members of the class of 2000. Of the 254 who graduated that year, 172 were women, 124 of whom went on to graduate school. Of the male graduates, 58 went on to graduate school. What is the probability of the given event?

 (a) The graduate is a woman.

 (b) The graduate went on to graduate school.

 (c) The graduate was a woman who went on to graduate school.

40. Indiana Jones and the Final Exam Professor Indiana Jones gives his class a list of 20 study questions, from which he will select 8 to be answered on the final exam. If a given student knows how to answer 14 of the questions, what is the probability that the student will be able to answer the given number of questions correctly?

 (a) All 8 questions

 (b) Exactly 5 questions

 (c) At least 6 questions

41. Graduation Requirement To complete the kinesiology requirement at Palpitation Tech you must pass two classes chosen from aerobics, aquatics, defense arts, gymnastics, racket sports, recreational activities, rhythmic activities, soccer, and volleyball. If you decide to choose your two classes at random by drawing two class names from a box, what is the probability that you will take racket sports and rhythmic activities?

42. Writing to Learn During July in Gunnison, Colorado, the probability of at least 1 hour a day of sunshine is 0.78, the probability of at least 30 minutes of rain is 0.44, and the probability that it will be cloudy all day is 0.22. Write a paragraph explaining whether this statement could be true.

In Exercises 43–50, ten dimes dated 1990 through 1999 are tossed. Find the probability of each event.

43. Tossing Ten Dimes Heads on the 1990 dime only

44. Tossing Ten Dimes Heads on the 1991 and 1996 dimes only

45. Tossing Ten Dimes Heads on all 10 dimes

46. Tossing Ten Dimes Heads on all but one dime

47. Tossing Ten Dimes Exactly two heads

48. Tossing Ten Dimes Exactly three heads

49. Tossing Ten Dimes At least one head

50. Tossing Ten Dimes At least two heads

Standardized Test Questions

51. True or False A sample space consists of equally likely events. Justify your answer.

52. True or False The probability of an event can be greater than 1. Justify your answer.

Evaluate Exercises 53–56 without using a calculator.

53. Multiple Choice The probability of rolling a total of 5 on a pair of fair dice is

(A) $\dfrac{1}{4}$.

(B) $\dfrac{1}{5}$.

(C) $\dfrac{1}{6}$.

(D) $\dfrac{1}{9}$.

(E) $\dfrac{1}{11}$.

54. Multiple Choice Which of the following numbers could not be the probability of an event?

(A) 0

(B) 0.95

(C) $\dfrac{\sqrt{3}}{4}$

(D) $\dfrac{3}{\pi}$

(E) $\dfrac{\pi}{2}$

55. Multiple Choice If A and B are independent events, then $P(A \mid B) =$

(A) $P(A)$.

(B) $P(B)$.

(C) $P(B \mid A)$.

(D) $P(A) \cdot P(B)$.

(E) $P(A) + P(B)$.

56. Multiple Choice A fair coin is tossed three times in succession. What is the probability that exactly one of the coins shows heads?

(A) $\dfrac{1}{8}$

(B) $\dfrac{1}{3}$

(C) $\dfrac{3}{8}$

(D) $\dfrac{1}{2}$

(E) $\dfrac{2}{3}$

Explorations

57. Empirical Probability In real applications, it is often necessary to approximate the probabilities of the various outcomes of an experiment by performing the experiment a large number of times and recording the results. Barney's Bread Basket offers five different kinds of bagels. Barney records the sales of the first 500 bagels in a given week in the table shown below:

Type of Bagel	Number Sold
Plain	185
Onion	60
Rye	55
Cinnamon Raisin	125
Sourdough	75

(a) Use the observed sales number to approximate the probability that a random customer buys a plain bagel. Do the same for each other bagel type and make a table showing the approximate probability distribution.

(b) Assuming independence of the events, find the probability that three customers in a row all order plain bagels.

(c) **Writing to Learn** Do you think it is reasonable to assume that the orders of three consecutive customers actually are independent? Explain.

58. Straight Poker In the original version of poker known as "straight" poker, a 5-card hand is dealt from a standard deck of 52 cards. What is the probability of the given event?

(a) A hand will contain at least one king.

(b) A hand will be a "full house" (any three of one kind and a pair of another kind).

59. Married Students Suppose that 23% of all college students are married. Answer the following questions for a random sample of eight college students.

(a) How many would you expect to be married?

(b) Would you regard it as unusual if the sample contained five married students?

(c) What is the probability that five or more of the eight students are married?

60. Group Activity Investigating an Athletic Program A university widely known for its track and field program claims that 75% of its track athletes get degrees. A journalist investigates what happened to the 32 athletes who began the program over a 6-year period that ended 7 years ago. Of these athletes, 17 have graduated and the remaining 15 are no longer attending any college. If the university's claim is true, the number of athletes who graduates among the 32 examined should have been governed by binomial probability with $p = 0.75$.

(a) What is the probability that exactly 17 athletes should have graduated?

(b) What is the probability that 17 or fewer athletes should have graduated?

(c) If you were the journalist, what would you say in your story on the investigation?

Extending the Ideas

61. Expected Value If the outcomes of an experiment are given numerical values (such as the total on a roll of two dice, or the payoff on a lottery ticket), we define the **expected value** to be the sum of all the numerical values times their respective probabilities.

For example, suppose we roll a fair die. If we roll a multiple of 3, we win $3; otherwise we lose $1. The probabilities of the two possible payoffs are shown in the table below:

Value	Probability
+3	2/6
−1	4/6

The expected value is

$3 \times (2/6) + (-1) \times (4/6) = (6/6) - (4/6) = 1/3$.

We interpret this to mean that we would win an average of 1/3 dollar per game in the long run.

(a) A game is called *fair* if the expected value of the payoff is zero. Assuming that we still win $3 for a multiple of 3, what should we pay for any other outcome in order to make the game fair?

(b) Suppose we roll *two* fair dice and look at the total under the original rules. That is, we win $3 for rolling a multiple of 3 and lose $1 otherwise. What is the expected value of this game?

62. Expected Value (Continuation of Exercise 61) Gladys has a personal rule never to enter the lottery (picking 6 numbers from 1 to 46) until the payoff reaches 4 million dollars. When it does reach 4 million, she always buys ten different $1 tickets.

(a) Assume that the payoff for a winning ticket is 4 million dollars. What is the probability that Gladys holds a winning ticket? (Refer to Example 1 of this section for the probability of any ticket winning.)

(b) Fill in the probability distribution for Gladys's possible payoffs in the table below. (Note that we subtract $10 from the $4 million, since Gladys has to pay for her tickets even if she wins.)

Value	Probability
−10	
+3,999,990	

(c) Find the expected value of the game for Gladys.

(d) **Writing to Learn** In terms of the answer in part (b), explain to Gladys the long-term implications of her strategy.

9.4
Sequences

Infinite Sequences

One of the most natural ways to study patterns in mathematics is to look at an ordered progression of numbers, called a **sequence**. Here are some examples of sequences:

1. $5, 10, 15, 20, 25$

2. $2, 4, 8, 16, 32, \ldots, 2^k, \ldots$

3. $\left\{ \dfrac{1}{k} : k = 1, 2, 3, \ldots \right\}$

4. $\{a_1, a_2, a_3, \ldots, a_k, \ldots\}$, which is sometimes abbreviated $\{a_k\}$

The first of these is a **finite sequence**, while the other three are **infinite sequences**. Notice that in (2) and (3) we were able to define a rule that gives the kth number in the sequence (called the **kth term**) as a function of k. In (4) we do not have a rule, but notice how we can use subscript notation (a_k) to identify the kth term of a "general" infinite sequence. In this sense, an infinite sequence can be thought of as a *function* that assigns a unique number (a_k) to each natural number k.

EXAMPLE 1   Defining a Sequence Explicitly

Find the first 6 terms and the 100th term of the sequence $\{a_k\}$ in which $a_k = k^2 - 1$.

SOLUTION   Since we know the kth term *explicitly* as a function of k, we need only to evaluate the function to find the required terms:

$$a_1 = 1^2 - 1 = 0, \quad a_2 = 3, \quad a_3 = 8, \quad a_4 = 15, \quad a_5 = 24, \quad a_6 = 35, \quad \text{and}$$

$$a_{100} = 100^2 - 1 = 9999.$$

Now try Exercise 1.

Explicit formulas are the easiest to work with, but there are other ways to define sequences. For example, we can specify values for the first term (or terms) of a sequence, then define each of the following terms **recursively** by a formula relating it to previous terms. Example 2 shows how this is done.

EXAMPLE 2   Defining a Sequence Recursively

Find the first 6 terms and the 100th term for the sequence defined recursively by the conditions:

$$b_1 = 3$$

$$b_n = b_{n-1} + 2 \quad \text{for all } n > 1.$$

continued

SOLUTION We proceed one term at a time, starting with $b_1 = 3$ and obtaining each succeeding term by adding 2 to the term just before it:

$$b_1 = 3$$
$$b_2 = b_1 + 2 = 5$$
$$b_3 = b_2 + 2 = 7$$

etc.

Eventually it becomes apparent that we are building the sequence of odd natural numbers beginning with 3:

$$\{3, 5, 7, 9, \ldots\}.$$

The 100th term is 99 terms beyond the first, which means that we can get there quickly by adding 99 2's to the number 3:

$$b_{100} = 3 + 99 \times 2 = 201.$$

Now try Exercise 5.

Limits of Infinite Sequences

Just as we were concerned with the end behavior of functions, we will also be concerned with the end behavior of sequences.

DEFINITION Limit of a Sequence

Let $\{a_n\}$ be a sequence of real numbers, and consider $\lim\limits_{n \to \infty} a_n$. If the limit is a finite number L, the sequence **converges** and L is the **limit of the sequence**. If the limit is infinite or nonexistent, the sequence **diverges**.

EXAMPLE 3   Finding Limits of Sequences

Determine whether the sequence converges or diverges. If it converges, give the limit.

(a) $\dfrac{1}{1}, \dfrac{1}{2}, \dfrac{1}{3}, \dfrac{1}{4}, \ldots, \dfrac{1}{n}, \ldots$

(b) $\dfrac{2}{1}, \dfrac{3}{2}, \dfrac{4}{3}, \dfrac{5}{4}, \ldots$

(c) $2, 4, 6, 8, 10, \ldots$

(d) $-1, 1, -1, 1, \ldots, (-1)^n, \ldots$

SOLUTION

(a) $\lim\limits_{n \to \infty} \dfrac{1}{n} = 0$, so the sequence converges to a limit of 0.

(b) Although the nth term is not explicitly given, we can see that $a_n = \dfrac{n+1}{n}$.
$$\lim_{n \to \infty} \frac{n+1}{n} = \lim_{n \to \infty} \left(1 + \frac{1}{n}\right) = 1 + 0 = 1.$$ The sequence converges to a limit of 1.

(c) This time we see that $a_n = 2n$. Since $\lim\limits_{n \to \infty} 2n = \infty$, the sequence diverges.

(d) This sequence oscillates forever between two values and hence has no limit. The sequence diverges.

Now try Exercise 13.

Proof

We can construct the sequence forward by starting with a_1 and *adding d* each time, or we can construct the sequence backward by starting at a_n and *subtracting d* each time. We thus get two expressions for the sum we are looking for:

$$\sum_{k=1}^{n} a_k = a_1 + (a_1 + d) + (a_1 + 2d) + \cdots + (a_1 + (n-1)d)$$

$$\sum_{k=1}^{n} a_k = a_n + (a_n - d) + (a_n - 2d) + \cdots + (a_n - (n-1)d)$$

Summing vertically, we get

$$2\sum_{k=1}^{n} a_k = (a_1 + a_n) + (a_1 + a_n) + \cdots + (a_1 + a_n)$$

$$2\sum_{k=1}^{n} a_k = n(a_1 + a_n)$$

$$\sum_{k=1}^{n} a_k = n\left(\frac{a_1 + a_n}{2}\right)$$

If we substitute $a_1 + (n-1)d$ for a_n, we get the alternate formula

$$\sum_{k=1}^{n} a_k = \frac{n}{2}(2a_1 + (n-1)d).$$

EXAMPLE 1   Summing the Terms of an Arithmetic Sequence

A corner section of a stadium has 8 seats along the front row. Each successive row has two more seats than the row preceding it. If the top row has 24 seats, how many seats are in the entire section?

SOLUTION The numbers of seats in the rows form an arithmetic sequence with

$$a_1 = 8, \quad a_n = 24, \quad \text{and} \quad d = 2.$$

Solving $a_n = a_1 + (n-1)d$, we find that

$$24 = 8 + (n-1)(2)$$
$$16 = (n-1)(2)$$
$$8 = n - 1$$
$$n = 9$$

Applying the Sum of a Finite Arithmetic Sequence Theorem, the total number of seats in the section is $9(8 + 24)/2 = 144$.

We can support this answer numerically by computing the sum on a calculator:

$$\text{sum(seq}(8 + (N-1)2, N, 1, 9) = 144.$$

Now try Exercise 7.

As you might expect, there is also a convenient formula for summing the terms of a finite geometric sequence.

THEOREM Sum of a Finite Geometric Sequence

Let $\{a_1, a_2, a_3, \ldots, a_n\}$ be a finite geometric sequence with common ratio $r \neq 1$.

Then the sum of the terms of the sequence is

$$\sum_{k=1}^{n} a_k = a_1 + a_2 + \cdots + a_n$$

$$= \frac{a_1(1 - r^n)}{1 - r}$$

Proof

Because the sequence is geometric, we have

$$\sum_{k=1}^{n} a_k = a_1 + a_1 \cdot r + a_1 \cdot r^2 + \cdots + a_1 \cdot r^{n-1}.$$

Therefore,

$$r \cdot \sum_{k=1}^{n} a_k = a_1 \cdot r + a_1 \cdot r^2 + \cdots + a_1 \cdot r^{n-1} + a_1 \cdot r^n.$$

If we now *subtract* the lower summation from the one above it, we have (after eliminating a lot of zeros):

$$\left(\sum_{k=1}^{n} a_k \right) - r \cdot \left(\sum_{k=1}^{n} a_k \right) = a_1 - a_1 \cdot r^n$$

$$\left(\sum_{k=1}^{n} a_k \right)(1 - r) = a_1(1 - r^n)$$

$$\sum_{k=1}^{n} a_k = \frac{a_1(1 - r^n)}{1 - r}$$

EXAMPLE 2 Summing the Terms of a Geometric Sequence

Find the sum of the geometric sequence $4, -4/3, 4/9, -4/27, \ldots, 4(-1/3)^{10}$.

SOLUTION We can see that $a_1 = 4$ and $r = -1/3$. The nth term is $4(-1/3)^{10}$, which means that $n = 11$. (Remember that the exponent on the nth term is $n - 1$, not n.) Applying the Sum of a Finite Geometric Sequence Theorem, we find that

$$\sum_{n=1}^{11} 4\left(-\frac{1}{3} \right)^{n-1} = \frac{4(1 - (-1/3)^{11})}{1 - (-1/3)} \approx 3.000016935.$$

We can support this answer by having the calculator do the actual summing:

sum(seq($4(-1/3)$^$(N - 1)$, N, 1, 11) = 3.000016935.   ***Now try Exercise 13.***

9.6
Mathematical Induction

What you'll learn about

- The Tower of Hanoi Problem
- Principle of Mathematical Induction
- Induction and Deduction

. . . and why

The principle of mathematical induction is a valuable technique for proving combinatorial formulas.

FIGURE 9.11 The Tower of Hanoi Game. The object is to move the entire stack of washers to the rightmost peg, one washer at a time, never placing a larger washer on top of a smaller washer.

TOWER OF HANOI HISTORY

The legend notwithstanding, the Tower of Hanoi dates back to 1883, when Édouard Lucas marketed the game as "La Tour de Hanoï," brought back from the Orient by "Professor N. Claus de Siam"—an anagram of "Professor Lucas d'Amiens." The legend appeared shortly thereafter. The game has been a favorite among computer programmers, so a web search on "Tower of Hanoi" will bring up multiple sites that allow you to play it on your home computer.

The Tower of Hanoi Problem

You might be familiar with a game that is played with a stack of round washers of different diameters and a stand with three vertical pegs (Figure 9.11). The game is not difficult to win once you get the hang of it, but it takes a while to move all the washers even when you know what you are doing. A mathematician, presented with this game, wants to figure out the *minimum* number of moves required to win the game—not because of impatience, but because it is an interesting mathematical problem.

In case mathematics is not sufficient motivation to look into the problem, there is a legend attached to the game that provides a sense of urgency. The legend has it that a game of this sort with 64 golden washers was created at the beginning of time. A special order of far eastern monks has been moving the washers at one move per second ever since, always using the minimum number of moves required to win the game. When the final washer is moved, that will be the end of time. The Tower of Hanoi Problem is simply to figure out how much time we have left.

We will solve the problem by proving a general theorem that gives the minimum number of moves for any number of washers. The technique of proof we use is called the principle of mathematical induction, the topic of this section.

THEOREM **The Tower of Hanoi Solution**

The minimum number of moves required to move a stack of n washers in a Tower of Hanoi game is $2^n - 1$.

Proof

(The Anchor) First, we note that the assertion is true when $n = 1$. We can certainly move the one washer to the right peg in (minimally) one move, and $2^1 - 1 = 1$.

(The Inductive Hypothesis) Now let us assume that the assertion holds for $n = k$; that is, the minimum number of moves required to move k washers is $2^k - 1$. (So far the only k we are sure of is 1, but keep reading.)

(The Inductive Step) We next consider the case when $n = k + 1$ washers. To get at the bottom washer, we must first move the entire stack of k washers sitting on top of it. *By the assumption we just made*, this will take a minimum of $2^k - 1$ moves. We can then move the bottom washer to the free peg (1 move). Finally, we must move the stack of k washers back onto the bottom washer—again, *by our assumption*, a minimum of $2^k - 1$ moves. Altogether, moving $k + 1$ washers requires

$$(2^k - 1) + 1 + (2^k - 1) = 2 \cdot 2^k - 1 = 2^{k+1} - 1$$

moves. Since that agrees with the formula in the statement of the proof, we have shown the assertion to be true for $n = k + 1$ washers—under the assumption that it is true for $n = k$.

Remarkably, we are done. Recall that we *did* prove the theorem to be true for $n = 1$. Therefore, by the Inductive Step, it must also be true for $n = 2$. By the Inductive Step again, it must be true for $n = 3$. And so on, for all positive integers n.

If we apply the Tower of Hanoi Solution to the legendary Tower of Hanoi Problem, the monks will need $2^{64} - 1$ seconds to move the 64 golden washers. The largest current conjecture for the age of the universe is something on the order of 20 billion years. If you convert $2^{64} - 1$ seconds to years, you will find that the end of time (at least according to this particular legend) is not exactly imminent. In fact, you might be surprised at how much time is left!

EXPLORATION 1 Winning the Game

One thing that the Tower of Hanoi Solution does not settle is how to get the stack to finish on the rightmost peg rather than the middle peg. Predictably, it depends on where you move the first washer, but it also depends on the height of the stack. Using a website game, or coins of different sizes, or even the real game if you have one, play the game with 1 washer, then 2, then 3, then 4, and so on, keeping track of what your first move must be in order to have the stack wind up on the rightmost peg in $2^n - 1$ moves. What is the general rule for a stack of n washers?

Principle of Mathematical Induction

The proof of the Tower of Hanoi Solution used a general technique known as the principle of mathematical induction. It is a powerful tool for proving all kinds of theorems about positive integers. We *anchor* the proof by establishing the truth of the theorem for 1, then we show the *inductive hypothesis* that "true for k" implies "true for $k + 1$."

Principle of Mathematical Induction

Let P_n be a statement about the integer n. Then P_n is true for all positive integers n provided the following conditions are satisfied:

1. (the anchor) P_1 is true;

2. (the inductive step) if P_k is true, then P_{k+1} is true.

FIGURE 9.12 The Principle of Mathematical Induction visualized by dominoes. The toppling of domino #1 guarantees the toppling of domino n for all positive integers n.

A good way to visualize how the principle works is to imagine an infinite sequence of dominoes stacked upright, each one close enough to its neighbor so that any kth domino, if it falls, will knock over the $(k + 1)$st domino (Figure 9.12). Given that fact (the inductive step), the toppling of domino 1 guarantees the toppling of the entire infinite sequence of dominoes.

Let us use the principle to prove a fact that we already know.

EXAMPLE 1 Using Mathematical Induction

Prove that $1 + 3 + 5 + \cdots + (2n - 1) = n^2$ is true for all positive integers n.

SOLUTION Call the statement P_n. We could verify P_n by using the formula for the sum of an arithmetic sequence, but here is how we prove it by mathematical induction.

continued

(The Anchor) For $n = 1$, the equation reduces to P_1: $1 = 1^2$, which is true.

(The Inductive Hypothesis) Assume that the equation is true for $n = k$. That is, assume

$$P_k: 1 + 3 + \cdots + (2k - 1) = k^2 \text{ is true.}$$

(The Inductive Step) The next term on the left-hand side would be $2(k + 1) - 1$. We add this to both sides of P_k and get

$$1 + 3 + \cdots + (2k - 1) + (2(k + 1) - 1) = k^2 + (2 (k + 1) - 1)$$
$$= k^2 + 2k + 1$$
$$= (k + 1)^2$$

This is exactly the statement P_{k+1}, so the equation is true for $n = k + 1$. Therefore, P_n is true for all positive integers, by mathematical induction.

Now try Exercise 1.

Notice that we did *not* plug in $k + 1$ on both sides of the equation P_n in order to verify the inductive step; if we had done that, there would have been nothing to verify. If you find yourself verifying the inductive step without using the inductive hypothesis, you can assume that you have gone astray.

EXAMPLE 2   Using Mathematical Induction

Prove that $1^2 + 2^2 + 3^2 + \cdots + n^2 = [n(n + 1)(2n + 1)]/6$ is true for all positive integers n.

SOLUTION Let P_n be the statement $1^2 + 2^2 + 3^2 + \cdots + n^2 = [n(n + 1)(2n + 1)]/6$.

(The Anchor) P_1 is true because $1^2 = [1(2)(3)]/6$.

(The Inductive Hypothesis) Assume that P_k is true, so that
$$1^2 + 2^2 + \cdots + k^2 = \frac{k(k + 1)(2k + 1)}{6}.$$

(The Inductive Step) The next term on the left-hand side would be $(k + 1)^2$. We add this to both sides of P_k and get

$$1^2 + 2^2 + \cdots + k^2 + (k + 1)^2 = \frac{k(k + 1)(2k + 1)}{6} + (k + 1)^2$$
$$= \frac{k(k + 1)(2k + 1) + 6(k + 1)^2}{6}$$
$$= \frac{(k + 1)(2k^2 + k + 6k + 6)}{6}$$
$$= \frac{(k + 1)(k + 2)(2k + 3)}{6}$$
$$= \frac{(k + 1)((k + 1) + 1)(2(k + 1) + 1)}{6}$$

This is exactly the statement P_{k+1}, so the equation is true for $n = k + 1$. Therefore, P_n is true for all positive integers, by mathematical induction.

Now try Exercise 13.

REBECCA SKLOOT

Journalist, Teacher, Author of The Immortal Life of Henrietta Lacks

Visit http://rebeccaskloot.com/.

HENRIETTA LACKS
FOUNDATION

Visit http://henriettalacksfoundation.org/.

The Lacks Family

Visit http://www.lacksfamily.net/.

Applications of mathematical induction can be quite different from the first two examples. Here is one involving divisibility.

EXAMPLE 3 Proving Divisibility

Prove that $4^n - 1$ is evenly divisible by 3 for all positive integers n.

SOLUTION Let P_n be the statement that $4^n - 1$ is evenly divisible by 3 for all positive integers n.

(**The Anchor**) P_1 is true because $4^1 - 1 = 3$ is divisible by 3.

(**The Inductive Hypothesis**) Assume that P_k is true, so that $4^k - 1$ is divisible by 3.

(**The Inductive Step**) We need to prove that $4^{k+1} - 1$ is divisible by 3.

Using a little algebra, we see that $4^{k+1} - 1 = 4 \cdot 4^k - 1 = 4(4^k - 1) + 3$.

By the inductive hypothesis, $4^k - 1$ is divisible by 3. Of course, so is 3. Therefore, $4(4^k - 1) + 3$ is a sum of multiples of 3, and hence divisible by 3. This is exactly the statement P_{k+1}, so P_{k+1} is true. Therefore, P_n is true for all positive integers, by mathematical induction. *Now try Exercise 19.*

Induction and Deduction

The words *induction* and *deduction* are usually used to contrast two patterns of logical thought. We reason by **induction** when we use evidence derived from particular examples to draw conclusions about general principles. We reason by **deduction** when we reason from general principles to draw conclusions about specific cases.

When mathematicians prove theorems, they use deduction. In fact, even a "proof by mathematical induction" is a deductive proof, since it consists of applying the general principle to a particular formula. We have been careful to use the term **mathematical induction** in this section to distinguish it from inductive reasoning, which is often good for inspiring conjectures—but not for proving general principles.

Exploration 2 illustrates why mathematicians do not rely on inductive reasoning.

EXPLORATION 2 Is $n^2 + n + 41$ Prime for all n?

1. Plug in the numbers from 1 to 10. Are the results all prime?
2. Repeat for the numbers from 11 to 20.
3. Repeat for the numbers from 21 to 30. (Ready to make your conjecture?)
4. What is the smallest value of n for which $n^2 + n + 41$ is not prime?

THE FOUR-COLOR MAP THEOREM

In 1852, Francis Guthrie conjectured that any map on a flat surface could be colored in at most four colors so that no two bordering regions would share the same color. Mathematicians tried unsuccessfully for almost 150 years to prove (or disprove) the conjecture, until Kenneth Appel and Wolfgang Haken finally proved it in 1976.

There is one situation in which (nonmathematical) induction can constitute a proof. In **enumerative induction**, one reasons from specific cases to the general principle by considering *all possible cases*. This is simple enough when proving a theorem like "All one-digit prime numbers are factors of 210," but it can involve some very elegant mathematics when the number of cases is seemingly infinite. Such was the case in the proof of the Four-Color Map Theorem, in which all possible cases were settled with the help of a clever computer program.

QUICK REVIEW 9.6 *(Prerequisite skill Sections A.2 and 1.2)*

In Exercises 1–3, expand the product.

1. $n(n + 5)$ **2.** $(n + 2)(n - 3)$

3. $k(k + 1)(k + 2)$

In Exercises 4–6, factor the polynomial.

4. $n^2 + 2n - 3$

5. $k^3 + 3k^2 + 3k + 1$

6. $n^3 - 3n^2 + 3n - 1$

In Exercises 7–10, evaluate the function at the given domain values or variable expressions.

7. $f(x) = x + 4;\ f(1), f(t), f(t + 1)$

8. $f(n) = \dfrac{n}{n + 1};\ f(1), f(k), f(k + 1)$

9. $P(n) = \dfrac{2n}{3n + 1};\ P(1), P(k), P(k + 1)$

10. $P(n) = 2n^2 - n - 3;\ P(1), P(k), P(k + 1)$

SECTION 9.6 EXERCISES

In Exercises 1–4, use mathematical induction to prove that the statement holds for all positive integers.

1. $2 + 4 + 6 + \cdots + 2n = n^2 + n$

2. $8 + 10 + 12 + \cdots + (2n + 6) = n^2 + 7n$

3. $6 + 10 + 14 + \cdots + (4n + 2) = n(2n + 4)$

4. $14 + 18 + 22 + \cdots + (4n + 10) = 2n(n + 6)$

In Exercises 5–8, state an explicit rule for the *n*th term of the recursively defined sequence. Then use mathematical induction to prove the rule.

5. $a_n = a_{n-1} + 5, a_1 = 3$ **6.** $a_n = a_{n-1} + 2, a_1 = 7$

7. $a_n = 3a_{n-1}, a_1 = 2$ **8.** $a_n = 5a_{n-1}, a_1 = 3$

In Exercises 9–12, write the statements P_1, P_k, and P_{k+1}. (Do not write a proof.)

9. P_n: $1 + 2 + \cdots + n = \dfrac{n(n + 1)}{2}$

10. P_n: $1^2 + 3^2 + 5^2 + \cdots + (2n - 1)^2 = \dfrac{n(2n - 1)(2n + 1)}{3}$

11. P_n: $\dfrac{1}{1 \cdot 2} + \dfrac{1}{2 \cdot 3} + \cdots + \dfrac{1}{n \cdot (n + 1)} = \dfrac{n}{n + 1}$

12. P_n: $\displaystyle\sum_{k=1}^{n} k^4 = \dfrac{n(n + 1)(2n + 1)(3n^2 + 3n - 1)}{30}$

In Exercises 13–20, use mathematical induction to prove that the statement holds for all positive integers.

13. $1 + 5 + 9 + \cdots + (4n - 3) = n(2n - 1)$

14. $1 + 2 + 2^2 + \cdots + 2^{n-1} = 2^n - 1$

15. $\dfrac{1}{1 \cdot 2} + \dfrac{1}{2 \cdot 3} + \dfrac{1}{3 \cdot 4} + \cdots + \dfrac{1}{n(n + 1)} = \dfrac{n}{n + 1}$

16. $\dfrac{1}{1 \cdot 3} + \dfrac{1}{3 \cdot 5} + \cdots + \dfrac{1}{(2n - 1)(2n + 1)} = \dfrac{n}{2n + 1}$

17. $2^n \geq 2n$ **18.** $3^n \geq 3n$

19. 3 is a factor of $n^3 + 2n$.   **20.** 6 is a factor of $7^n - 1$.

In Exercises 21 and 22, use *mathematical induction* to prove that the statement holds for all positive integers. (We have already seen each proved in another way.)

21. The sum of the first *n* terms of a geometric sequence with first term a_1 and common ratio $r \neq 1$ is $a_1(1 - r^n)/(1 - r)$.

22. The sum of the first *n* terms of an arithmetic sequence with first term a_1 and common difference *d* is

$$S_n = \frac{n}{2}[2a_1 + (n - 1)d].$$

In Exercises 23 and 24, use mathematical induction to prove that the formula holds for all positive integers.

23. Triangular Numbers $\displaystyle\sum_{k=1}^{n} k = \dfrac{n(n + 1)}{2}$

24. Sum of the First *n* Cubes $\displaystyle\sum_{k=1}^{n} k^3 = \dfrac{n^2(n + 1)^2}{4}$

[Note that if you put the results from Exercises 23 and 24 together, you obtain the pleasantly surprising equation

$$1^3 + 2^3 + 3^3 + \cdots + n^3 = (1 + 2 + 3 + \cdots + n)^2.]$$

In Exercises 25–30, use the results of Exercises 21–24 and Example 2 to find the sums.

25. $1 + 2 + 3 + \cdots + 500$ **26.** $1^2 + 2^2 + \cdots + 250^2$

27. $4 + 5 + 6 + \cdots + n$ **28.** $1^3 + 2^3 + 3^3 + \cdots + 75^3$

29. $1 + 2 + 4 + 8 + \cdots + 2^{34}$

30. $1 + 8 + 27 + \cdots + 3375$

In Exercises 31–34, use the results of Exercises 21–24 and Example 2 to find the sum in terms of n.

31. $\displaystyle\sum_{k=1}^{n} (k^2 - 3k + 4)$ **32.** $\displaystyle\sum_{k=1}^{n} (2k^2 + 5k - 2)$

33. $\displaystyle\sum_{k=1}^{n} (k^3 - 1)$ **34.** $\displaystyle\sum_{k=1}^{n} (k^3 + 4k - 5)$

35. Group Activity Here is a proof by mathematical induction that any gathering of n people must all have the same blood type.

(**Anchor**) If there is 1 person in the gathering, everyone in the gathering obviously has the same blood type.

(**Inductive Hypothesis**) Assume that any gathering of k people must all have the same blood type.

(**Inductive Step**) Suppose $k + 1$ people are gathered. Send one of them out of the room. The remaining k people must all have the same blood type (by the inductive hypothesis). Now bring the first person back and send someone else out of the room. You get another gathering of k people, all of whom must have the same blood type. Therefore all $k + 1$ people must have the same blood type, and we are done by mathematical induction.

This result is obviously false, so there must be something wrong with the proof. Explain where the proof goes wrong.

36. Writing to Learn Kitty is having trouble understanding mathematical induction proofs because she does not understand the inductive hypothesis. If we can assume it is true for k, she asks, why can't we assume it is true for n and be done with it? After all, a variable is a variable! Write a response to Kitty to clear up her confusion.

Standardized Test Questions

37. True or False The goal of mathematical induction is to prove that a statement P_n is true for all real numbers n. Justify your answer.

38. True or False If P_n is the statement "$(n + 1)^2 = 4n$," then P_1 is true. Justify your answer.

You may use a graphing calculator when solving Exercises 39–42.

39. Multiple Choice In a proof by mathematical induction that $1 + 2 + 3 + \cdots + n = \dfrac{n(n + 1)}{2}$ for all positive integers n, the inductive hypothesis would be to assume that

(A) $n = 1$.

(B) $k = 1$.

(C) $1 = \dfrac{1(1+1)}{2}$.

(D) $1 + 2 + 3 + \cdots + n = \dfrac{n(n + 1)}{2}$ for all positive integers n.

(E) $1 + 2 + 3 + \cdots + k = \dfrac{k(k + 1)}{2}$ for some positive integer k.

40. Multiple Choice The first step in a proof by mathematical induction is to prove

(A) the anchor.

(B) the inductive hypothesis.

(C) the inductive step.

(D) the inductive principle.

(E) the inductive foundation.

41. Multiple Choice Which of the following could be used to prove that $1 + 3 + 5 + \cdots + (2n - 1) = n^2$ for all positive integers n?

  I. Mathematical induction

  II. The formula for the sum of a finite arithmetic sequence

  III. The formula for the sum of a finite geometric sequence

(A) I only

(B) I and II only

(C) I and III only

(D) II and III only

(E) I, II, and III

42. Multiple Choice Mathematical induction can be used to prove that, for any positive integer n, $\displaystyle\sum_{k=1}^{n} k^3 =$

(A) $\dfrac{n(n + 1)}{2}$

(B) $\dfrac{n^2(n + 1)^2}{2}$

(C) $\dfrac{n^2(n + 1)^2}{4}$

(D) $\dfrac{n^3(n + 1)^3}{2}$

(E) $\dfrac{n^3(n + 1)^3}{8}$

Explorations

43. Use mathematical induction to prove that 2 is a factor of $(n + 1)(n + 2)$ for all positive integers n.

44. Use mathematical induction to prove that 6 is a factor of $n(n + 1)(n + 2)$ for all positive integers n. (You may assume the assertion in Exercise 43 to be true.)

45. Give an alternate proof of the assertion in Exercise 43 based on the fact that $(n + 1)(n + 2)$ is a product of two consecutive integers.

46. Give an alternate proof of the assertion in Exercise 44 based on the fact that $n(n + 1)(n + 2)$ is a product of three consecutive integers.

Extending the Ideas

In Exercises 47 and 48, use mathematical induction to prove that the statement holds for all positive integers.

47. Fibonacci Sequence and Series $F_{n+2} - 1 = \sum_{k=1}^{n} F_k$, where $\{F_n\}$ is the Fibonacci sequence.

48. If $\{a_n\}$ is the sequence $\sqrt{2}, \sqrt{2 + \sqrt{2}}, \sqrt{2 + \sqrt{2 + \sqrt{2}}}, \dots$, then $a_n < 2$.

49. Let a be any integer greater than 1. Use mathematical induction to prove that $a - 1$ divides $a^n - 1$ evenly for all positive integers n.

50. Give an alternate proof of the assertion in Exercise 49 based on the Factor Theorem of Section 2.4.

It is not necessary to anchor a mathematical induction proof at $n = 1$; we might only be interested in the integers greater than or equal to some integer c. In this case, we simply modify the anchor and inductive step as follows:

Anchor: P_c is true.

Inductive Step: If P_k is true for some $k \geq c$, then P_{k+1} is true.

(This is sometimes called the **Extended Principle of Mathematical Induction**.) Use this principle to prove the statements in Exercises 51 and 52.

51. $3n - 4 \geq n$, for all $n \geq 2$ **52.** $2^n \geq n^2$, for all $n \geq 4$

53. Proving the interior angle formula Use extended mathematical induction to prove P_n for $n \geq 3$.

P_n: The sum of the interior angles of an n-sided polygon is $180°(n - 2)$.

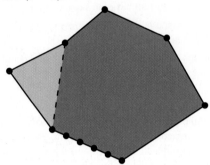

9.7

Statistics and Data (Graphical)

. . . and why

Graphical displays of data are increasingly prevalent in professional and popular media. We all need to understand them.

Statistics

Statistics is a branch of science that draws from both discrete mathematics and the mathematics of the continuum. The aim of statistics is to draw meaning from data and communicate it to others.

The objects described by a set of data are **individuals**, which may be people, animals, or things. The characteristic of the individuals being identified or measured is a **variable**. Variables are either *categorical* or *quantitative*. If the variable identifies each individual as belonging to a distinct class, such as male or female, then the variable is a **categorical variable**. If the variable takes on numerical values for the characteristic being measured, then the variable is a **quantitative variable**.

Examples of quantitative variables are heights of people, and weights of lobsters. You have already seen many tables of quantitative data in this book; indeed, most of the data-based exercises have been solved using techniques that are basic tools for statisticians. So far, however, most of our attention has been restricted to finding models that relate quantitative variables to each other. In the last two sections of this chapter, we will look at some of the other graphical and algebraic tools that can be used to draw meaning from data and communicate it to others.

Displaying Categorical Data

The National Center for Health Statistics reported that the leading causes of death in 2001 were heart disease, cancer, and stroke. Table 9.3 gives more detailed information.

Table 9.3 Leading Causes of Death in the United States in 2001		
Cause of Death	Number of Deaths	Percentage
Heart Disease	700,142	29.0
Cancer	553,768	22.9
Stroke	163,538	6.8
Other	1,018,977	41.3

Source: National Center for Health Statistics, as reported in The World Almanac and Book of Facts 2005.

Because the causes of death are categories, not numbers, "cause of death" is a categorical variable. The numbers of deaths and percentages, while certainly numerical, are not values of a *variable*, because they do not describe the *individuals*. Nonetheless, the numbers can communicate information about the categorical variables by telling us the relative size of the categories in the 2001 population.

We can get that information directly from the numbers, but it is very helpful to see the comparative sizes visually. This is why you will often see categorical data displayed graphically, as a **bar chart** (Figure 9.13a), a **pie chart** (Figure 9.13b), or a **circle graph** (Figure 9.13c). For variety, the popular press also makes use of **picture graphs** suited to the categories being displayed. For example, the bars in Figure 9.13a could be made to look like tombstones of different sizes to emphasize that these are causes of death. In each case, the graph provides a visualization of the relative sizes of the categories, with the pie chart and circle graph providing the added visualization of how the categories are parts of a whole population.

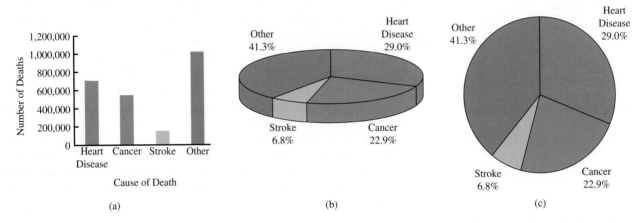

FIGURE 9.13 Causes of death in the United States in 2001 shown in (a) a bar graph, (b) a 3-D pie chart, and (c) a circle graph.

In bar charts of categorical data, the *y*-axis has a numerical scale, but the *x*-axis is labeled by category. The rectangular bars are separated by spaces to show that no continuous numerical scale is implied. (In this respect, a bar graph differs from a *histogram*, to be described later in this section.) A circle graph or a pie chart consists of shaded or colored sections of a circle or "pie." The central angles for the sectors are found by multiplying the percentages by 360°. For example, the angle for the sector representing cancer victims in Figure 9.13c is

$$22.9\% \cdot 360° = 82.4°.$$

It used to require time, skill, and mathematical savvy to draw data displays that were both visually appealing and geometrically accurate, but modern spreadsheet programs have made it possible for anyone with a computer to produce high-quality graphs from tabular data with the click of a button.

Stemplots

A quick way to organize and display a small set of quantitative data is with a **stemplot**, also called a **stem-and-leaf plot**. Each number in the data set is split into a **stem**, consisting of its initial digit or digits, and a **leaf**, which is its final digit.

EXAMPLE 1 Making a Stemplot

Table 9.4 gives the percentage of each state's population that was 65 or older in the last official Census (2000). Make a stemplot for the data.

 Table 9.4 Percentages of State Residents in 2000 Who Were 65 or Older

AL	13.0	HI	13.3	MA	13.5	NM	11.7	SD	14.3
AK	5.7	ID	11.3	MI	12.3	NY	12.9	TN	12.4
AZ	13.0	IL	12.1	MN	12.1	NC	12.0	TX	9.9
AR	14.0	IN	12.4	MS	12.1	ND	14.7	UT	8.5
CA	10.6	IO	14.9	MO	13.5	OH	13.3	VT	12.7
CO	9.7	KS	13.3	MT	13.4	OK	13.2	VA	11.2
CT	13.8	KY	12.5	NE	13.6	OR	12.8	WA	11.2
DE	13.0	LA	11.6	NV	11.0	PA	15.6	WV	15.3
FL	17.6	ME	14.4	NH	12.0	RI	14.5	WI	13.1
GA	9.6	MD	11.3	NJ	13.2	SC	12.1	WY	11.7

Source: U.S. Census Bureau, 2001.

SOLUTION To form the stem-and-leaf plot, we use the whole number part of each number as the stem and the tenths digit as the leaf. We write the stems in order down the first column and, for each number, write the leaf in the appropriate stem row. Then we arrange the leaves in each stem row in ascending order. The final plot looks like this:

Stem	Leaf
5	7
6	
7	
8	5
9	6 7 9
10	6
11	0 2 2 3 3 6 7 7
12	0 0 1 1 1 1 3 4 4 5 7 8 9
13	0 0 0 1 2 2 3 3 3 4 5 5 6 8
14	0 3 4 5 7 9
15	3 6
16	
17	6

Notice that we include the "leafless" stems (6, 7, 16) in our plot, as those empty gaps are significant features of the visualization. For the same reason, be sure that each "leaf" takes up the same space along the stem. A branch with twice as many leaves should appear to be twice as long. ***Now try Exercise 1.***

EXPLORATION 1 Using Information from a Stemplot

By looking at both the stemplot and the table, answer the following questions about the distribution of senior citizens among the 50 states.

1. Judging from the stemplot, what was the approximate *average* national percentage of residents who were 65 or older?

2. In how many states were more than 15% of the residents 65 or older?

3. Which states were in the bottom tenth of all states in this statistic?

4. The numbers 5.7 and 17.6 are so far above or below the other numbers in this stemplot that statisticians would call them *outliers*. Quite often there is some special circumstance that sets outliers apart from the other individuals under study and explains the unusual data. What could explain the two outliers in this stemplot?

Sometimes the data are so tightly clustered that a stemplot has too few stems to give a meaningful visualization of the data. In such cases we can spread the data out by splitting the stems, as in Example 2.

EXAMPLE 2 Making a Split-Stem Stemplot

The per capita federal aid to state and local governments for the top 15 states (in this category) in 2003 are shown in Table 9.5. Make a stemplot that provides a good visualization of the data. What is the average of the 15 numbers? Why is the stemplot a better summary of the data than the average?

Table 9.5 Federal Aid to State and Local Governments Per Capita (2003) in Dollars					
AL	3713	ND	1900	MI	1709
WY	2829	ME	1836	MT	1563
NY	2262	WV	1823	MA	1533
NM	2005	SD	1799	KY	1449
VT	1913	R1	1724	LA	1443

Source: U.S. Census Bureau Public Information Office.

SOLUTION We first round the data to $100 units, which does not affect the visualization. Then, to spread out the data a bit, we *split* each stem, putting leaves 0–4 on the lower stem and leaves 5–9 on the upper stem.

Stem	Leaf
1	4 4
1	5 6 7 7 8 8 8 9 9
2	0 3
2	8
3	
3	7

continued

The average of the 15 numbers is $1,967, but this is misleading. The table shows that eleven of the numbers are lower than that, while only four are higher. It is better to observe that the distribution is clustered fairly tightly around $1,800, with the number for Alabama being a significant outlier on the high end.

Now try Exercise 3.

Sometimes it is easier to compare two sets of data if we have a visualization that allows us to view both stemplots simultaneously. **Back-to-back stem-plots** use the same stems, but leaves from one set of data are added to the left, while leaves from another set are added to the right.

EXAMPLE 3   Making Back-to-Back Stemplots

Mark McGwire and Barry Bonds entered the major leagues in 1986 and had overlapping careers until 2001, the year that McGwire retired. During that period they averaged 36.44 and 35.44 home runs per year, respectively. Compare their annual home run totals with a back-to-back stemplot. Can you tell which player was more consistent as a home run hitter?

SOLUTION We form a back-to-back stemplot with McGwire's totals branching off to the left and Bond's to the right.

Mark McGwire		Barry Bonds
9 9 3	0	
	1	6 9
9 2	2	4 5 5
9 9 3 2 2	3	3 3 4 4 7 7
9 2	4	0 2 6 9
8 2	5	
5	6	
0	7	3

The single-digit home run years for McGwire can be explained by fewer times at bat (late entry into the league in 1986 and injuries in 1993 and 1994). If those years are ignored as anomalies, McGwire's numbers seem to indicate more consistency. Bond's record-setting 73 in 2001 was (and still is) an outlier of such magnitude that it actually inspired more skepticism than admiration among baseball fans.

Now try Exercise 5.

Table 9.6 Major League Home Run Totals for Mark McGwire and Barry Bonds through 2001															
Year	86	87	88	89	90	91	92	93	94	95	96	97	98	99	00	01
McGwire	3	49	32	33	39	22	42	9	9	39	52	58	70	65	32	29
Bonds	16	25	24	19	33	25	34	46	37	33	42	40	37	34	49	73

Source: Major League Baseball Enterprises, 2002.

Frequency Tables

The visual impact of a stemplot comes from the lengths of the various rows of leaves, which is just a way of seeing *how many* leaves branch off each stem. The number of leaves for a particular stem is the **frequency** of observations within each stem interval. Frequencies are often recorded in a **frequency table**. Table 9.7 shows a frequency table for Mark McGwire's yearly home run totals from 1986 to 2001 (see Example 3). The table shows the **frequency distribution**—literally the way that the total frequency of 16 is "distributed" among the various home run intervals. This same information is conveyed visually in a stemplot, but notice that the stemplot has the added numerical advantage of displaying what the numbers in each interval actually are.

Table 9.7 Frequency Table for Mark McGwire's Yearly Home Run Totals, 1986–2001 (Higher frequencies in a table correspond to longer leaf rows in a stemplot. Unlike a stemplot, a frequency Table does not display what the numbers in each interval actually are.)

Home Runs	Frequency	Home Runs	Frequency
0–9	3	40–49	2
10–19	0	50–59	2
20–29	2	60–69	1
30–39	5	70–79	1
		Total	16

Histograms

A **histogram**, closely related to a stemplot, displays the information of a frequency table. Visually, a histogram is to quantitative data what a bar chart is to categorical data. Unlike a bar chart, however, both axes of a histogram have numerical scales, and the rectangular bars on adjacent intervals have no intentional gaps between them.

Figure 9.14 shows a histogram of the information in Table 9.7, where each bar corresponds to an interval in the table and the height of each bar represents the frequency of observations within the interval.

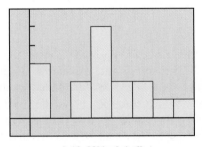

[–10, 80] by [–1, 6]

FIGURE 9.14 A histogram showing the distribution of Mark McGwire's annual home run totals from 1986 to 2001. This is a visualization of the data in Table 9.7.

EXAMPLE 4 Graphing a Histogram on a Calculator

Make a histogram of Hank Aaron's annual home run totals given in Table 9.8, using intervals of width 5.

Table 9.8 Regular Season Home Run Statistics for Hank Aaron

Year	Home Runs	Year	Home Runs	Year	Home Runs
1954	13	1962	45	1970	38
1955	27	1963	44	1971	47
1956	26	1964	24	1972	34
1957	44	1965	32	1973	40
1958	30	1966	44	1974	20
1959	39	1967	39	1975	12
1960	40	1968	29	1976	10
1961	34	1969	44		

Source: The Baseball Encyclopedia (7th ed., 1988, New York: MacMillan) p. 695.

SOLUTION We first make a frequency table for the data, using intervals of width 5. (This is not needed for the calculator to produce the histogram, but we will compare this with the result.)

Home Runs	Frequency	Home Runs	Frequency
10–14	3	30–34	4
15–19	0	35–39	3
20–24	2	40–44	6
25–29	3	45–49	2
		Total	23

To scale the x-axis to be consistent with the intervals of the table, let Xmin = 0, Xmax = 55, and Xscl = 5. Notice that the maximum frequency is 6 years (40–44 home runs), so the y-axis ought to extend at least to 7. Enter the data from Table 9.8 into list L1 and plot a histogram in the window $[0, 55]$ by $[-1, 7]$. (See Figure 9.15a.) Tracing along the histogram should reveal the same frequencies as in the frequency table we made. (See Figure 9.15b.) *Now try Exercise 11.*

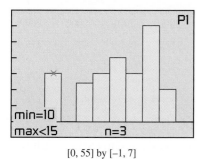

[0, 55] by [−1, 7]

(a)

[0, 55] by [−1, 7]

(b)

FIGURE 9.15 Calculator histograms of Hank Aaron's yearly home run totals. (Example 4)

Time Plots

We have seen in this book many examples of functions in which the input variable is time. It is also quite common to consider quantitative data as a function of time. If we make a scatter plot of the data (y) against the time (x) that it was measured, we can analyze the patterns as the variable changes over time. To help with the visualization, the discrete points from left to right are connected by line segments, just as a grapher would do in connect mode. The resulting **line graph** is called a **time plot**.

Time plots reveal trends in data over time. These plots frequently appear in magazines and newspapers and on the Internet, a typical example being the graph of the historic 1999 climb of the Dow Jones Industrial Average (DJIA) in Figure 9.16.

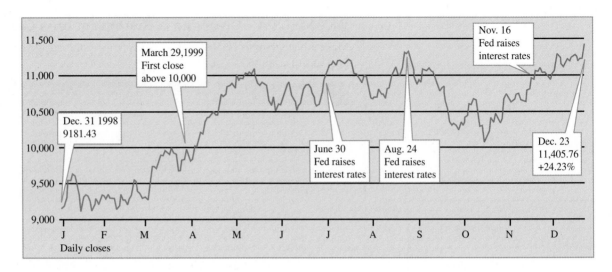

FIGURE 9.16 Time plot of the Dow Jones Industrial Average during the spectacular year 1999. Investors get a good visualization of where the stock market has been, although the trick is to figure out where it is going. *(Source: Quote.com, as reported by the Associated Press in the Chattanooga Times/Free Press.)*

Table 9.9 Millions of Units of CDs Shipped to Retailers Each Year 1991–2003												
Year	1991	1992	1993	1994	1995	1996	1997	1998	1999	2000	2001	2002	2003
CDs	333.3	407.5	495.4	662.1	722.9	778.9	753.1	847.0	938.9	942.5	881.9	803.3	745.9

Source: Recording Industry Association of America, as reported in the World Almanac and Book of Facts 2005.

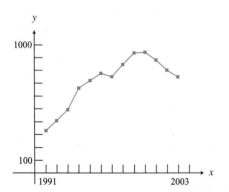

FIGURE 9.17 A time plot of CD volume over the years from 1991 to 2003, as reflected in shipments to retailers. (Example 5)

EXAMPLE 5   Drawing a Time Plot

Table 9.9 gives the number of compact disc (CD) units shipped to retailers each year in the 13-year period from 1991 to 2003. (The numbers are shown in millions, net after returns.) Display the data in a time plot and analyze the 13-year trend.

SOLUTION The horizontal axis represents time (in years) from 1991 to 2003. The vertical axis represents the number of CDs shipped that year, in millions of units. Since the visualization is enhanced by showing both axes in the viewing window, the vertical axis of a time plot is usually translated to cross the x-axis at or near the beginning of the time interval in the data. You can create this effect on your grapher by entering the years as $\{1, 2, 3, \ldots, 13\}$ rather than $\{1991, 1992, 1993, \ldots, 2003\}$. The labeling on the x-axis can then display the years, as shown in Figure 9.17.

The time plot shows that CD shipments increased steadily from 1991 to 1999 with especially good years from 1994 to 1996; then they declined steadily from 2000 to 2003. The music industry knew the reason for the dramatic turnaround—do you?

Now try Exercise 17.

Table 9.10 Millions of Units of Cassettes Shipped to Retailers, 1991–2003												
Year	1991	1992	1993	1994	1995	1996	1997	1998	1999	2000	2001	2002	2003
Cass	360.1	366.4	339.5	345.4	272.6	225.3	172.6	158.5	123.6	76	45	31.1	17.2

Source: Recording Industry Association of America, as reported in the World Almanac and Book of Facts 2005.

EXAMPLE 6   Overlaying Two Time Plots

Table 9.10 gives the number of cassette units shipped to retailers each year in the 13-year period from 1991 to 2003. (The numbers are shown in millions, net after returns.) Compare the cassette trend with the CD trend by overlaying the time plots for the two products.

SOLUTION The two time plots are shown in Figure 9.18. The popularity of cassette tapes declined as more consumers switched to CD (and eventually MP3) technology.

Now try Exercise 19.

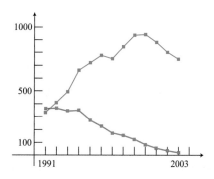

FIGURE 9.18 A time plot comparing cassette shipments and CD shipments for the time period from 1991 to 2003. (Example 6)

QUICK REVIEW 9.7

In Exercises 1–6, solve for the required value.

1. 457 is what percent of 2953?
2. 827 is what percent of 3950?
3. 52° is what percent of 360°?
4. 98° is what percent of 360°?
5. 734 is 42.6% of what number?
6. 5106 is 55.5% of what number?

In Exercises 7–10, round the given value to the nearest whole number in the specified units.

7. $234,598.43 (thousands of dollars)
8. 237,834,289 (millions)
9. 848.36 thousands (millions)
10. 1432 millions (billions)

SECTION 9.7 EXERCISES

Table 9.11 shows the home run statistics for Roger Maris during his major league career.

Table 9.11 Regular Season Home Run Statistics for Roger Maris

Year	Home Runs
1957	14
1958	28
1959	16
1960	39
1961	61
1962	33
1963	23
1964	26
1965	8
1966	13
1967	9
1968	5

Source: The Baseball Encyclopedia, 7th ed., 1988.

1. Make a stemplot of the data in Table 9.11. Are there any outliers?

2. Make a back-to-back stemplot comparing the annual home run production of Roger Maris (Table 9.11) with that of Hank Aaron (Table 9.8 on page 765). Write a brief interpretation of the stemplot.

In Exercises 3–6, construct the indicated stemplot from the data in Table 9.12. Then write a brief interpretation of the stemplot.

Table 9.12 Life Expectancy by Gender for the Nations of South America

Nation	Male	Female
Argentina	72.0	79.7
Bolivia	62.5	67.9
Brazil	67.5	75.6
Chile	73.1	79.8
Colombia	67.6	75.4
Ecuador	73.2	79.0
Guyana	60.1	64.8
Paraguay	72.1	77.3
Peru	67.5	71.0
Suriname	66.8	71.6
Uruguay	72.7	79.2
Venezuela	71.0	77.3

Source: The World Almanac and Book of Facts, 2005.

3. A stemplot showing life expectancies of males in the nations of South America (Round to nearest year and use split stems.)

4. A stemplot showing life expectancies of females in the nations of South America (Round to nearest year and use split stems.)

5. A back-to-back stemplot for life expectancies of males and females in the nations of South America (Round to nearest year and use split stems.)

6. A stemplot showing the difference between female and male life expectancies in the nations of South America (Use unrounded data and do not split stems.)

In Exercises 7 and 8, use the data in Table 9.12 to construct the indicated frequency table, using intervals 60.0–64.9, 65.0–69.9, etc.

7. Life expectancies of males in the nations of South America

8. Life expectancies of females in the nations of South America

In Exercises 9–12, draw a histogram for the given table.

9. The frequency table in Exercise 7

10. The frequency table in Exercise 8

11. Table 9.13 of Willie Mays' annual home run totals, using intervals 1–5, 6–10, 11–15, etc.

12. Table 9.13 of Mickey Mantle's annual home run totals, using intervals 0–4, 5–9, 10–14, etc.

Table 9.13 Regular Season Home Run Statistics for Willie Mays and Mickey Mantle

Year	Mays	Mantle	Year	Mays	Mantle
1951	20	13	1962	38	30
1952	4	23	1963	47	15
1953	41	21	1964	52	35
1954	51	27	1965	37	19
1955	36	37	1966	22	23
1956	35	52	1967	23	22
1957	29	34	1968	13	18
1958	34	42	1969	28	
1959	29	31	1970	18	
1960	40	40	1971	8	
1961	49	54	1972	6	

Source: The Baseball Encyclopedia, 7th ed., 1988.

In Exercises 13–16, make a time plot for the indicated data.

13. Willie Mays's annual home run totals given in Table 9.13

14. Mickey Mantle's annual home run totals given in Table 9.13

15. Mark McGwire's home run totals given in Table 9.6 on page 763

16. Hank Aaron's home run totals given in Table 9.8 on page 765

Table 9.14 shows the total amount of money won (in units of $1000, rounded to the nearest whole number) by the leading money winners in women's (LPGA) and men's (PGA) professional golf for selected years between 1970 and 2003.

Table 9.14 Yearly Winnings (in Thousands of Dollars) of the Top Money Winners in Men's and Women's Golf for Selected Years 1970–2003		
Year	Men (PGA)	Women (LPGA)
1970	157	30
1975	323	95
1980	531	231
1985	542	416
1990	1165	864
1995	1655	667
1996	1780	1002
1997	2067	1237
1998	2591	1093
1999	6617	1592
2000	9188	1877
2001	5688	2106
2002	6913	2864
2003	7574	2030

Source: The World Almanac and Book of Facts 2005.

17. Make a time plot for the men's winnings in Table 9.14. Write a brief interpretation of the time plot.

18. Make a time plot for the women's winnings in Table 9.14. Write a brief interpretation of the time plot.

19. Compare the trends in Table 9.14 by overlaying the time plots. Write a brief interpretation.

20. Writing to Learn (Continuation of Exercise 19) The data in Table 9.14 show that the earnings for the top PGA player rose by a modest 68% in the decade from 1975 to 1985, while the earnings for the top LPGA player rose by a whopping 338%. Although this was, in fact, a strong growth period for women's sports, statisticians would be unlikely to draw any conclusions from a comparison of these two numbers. Use the visualization from the comparative time plot in Exercise 19 to explain why.

In Exercises 21 and 22, compare performances by overlaying time plots.

21. The time plots from Exercises 13 and 14 to compare the performances of Mays and Mantle

22. The time plots from Exercises 15 and 16 to compare the performances of McGwire and Aaron

In Exercises 23 and 24, analyze the data as indicated.

23. The salaries of the workers in one department of the Garcia Brothers Company (given in thousands of dollars) are as follows:

33.5, 35.3, 33.8, 29.3, 36.7, 32.8, 31.7, 36.3, 33.5, 28.2, 34.8, 33.5, 35.3, 29.7, 38.5, 32.7, 34.8, 34.2, 31.6, 35.4

(a) Complete a stemplot for this data set.

(b) Create a frequency table for the data.

(c) Draw a histogram for the data. What viewing window did you use?

(d) Why does a time plot not work well for the data?

24. The average wind speeds for one year at 44 climatic data centers around the United States are as follows:

9.0, 6.9, 9.1, 9.2, 10.2, 12.5, 12.0, 11.2, 12.9, 10.3, 10.6, 10.9, 8.7, 10.3, 11.0, 7.7, 11.4, 7.9, 9.6, 8.0, 10.7, 9.3, 7.9, 6.2, 8.3, 8.9, 9.3, 11.6, 10.6, 9.0, 8.2, 9.4, 10.6, 9.5, 6.3, 9.1, 7.9, 9.7, 8.8, 6.9, 8.7, 9.0, 8.9, 9.3

(a) Complete a stemplot for this data set.

(b) Create a frequency table for the data.

(c) Draw a histogram for the data. What viewing window did you use?

(d) Why does a circle graph not work well for the data?

In Exercises 25 and 26, compare by overlaying time plots for the data in Table 9.15

Table 9.15 Population (in millions) of the Six Most Populous States						
Census	CA	FL	IL	NY	PA	TX
1900	1.5	0.5	4.8	7.3	6.3	3.0
1910	2.4	0.8	5.6	9.1	7.7	3.9
1920	3.4	1.0	6.5	10.4	8.7	4.7
1930	5.7	1.5	7.6	12.6	9.6	5.8
1940	6.9	1.9	7.9	13.5	9.9	6.4
1950	10.6	2.8	8.7	14.8	10.5	7.7
1960	15.7	5.0	10.1	16.8	11.3	9.6
1970	20.0	6.8	11.1	18.2	11.8	11.2
1980	23.7	9.7	11.4	17.6	11.9	14.2
1990	29.8	12.9	11.4	18.0	11.9	17.0
2000	33.9	16.0	12.4	19.0	12.3	20.9

Source: The World Almanac and Book of Facts 2005.

25. The populations of California, New York, and Texas from 1990 through 2000

26. The populations of Florida, Illinois, and Pennsylvania from 1990 through 2000

Standardized Test Questions

27. True or False If there are stems without leaves in the interior of a stemplot, it is best to omit them. Justify your answer.

28. True or False The highest and lowest numbers in a set of data are called outliers. Justify your answer.

Answer Exercises 29–32 without using a calculator.

29. Multiple Choice A time plot is an example of a

(A) histogram.   (B) bar graph.

(C) line graph.   (D) pie chart.

(E) table.

30. Multiple Choice A back-to-back stemplot is particularly useful for

(A) identifying outliers.

(B) comparing two data distributions.

(C) merging two sets of data.

(D) graphing home runs.

(E) distinguishing stems from leaves.

31. Multiple Choice The histogram below would most likely result from which set of data?

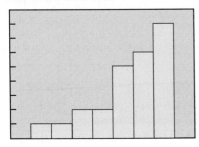

(A) test scores from a fairly easy test

(B) weights of children in a third-grade class

(C) winning soccer scores for a team over the course of a season

(D) ages of all the people visiting the Bronx Zoo at a given point in time

(E) prices of all the desserts on the menu at a certain restaurant

32. Multiple Choice A sector on a pie chart with a central angle of 45° corresponds to what percentage of the data?

(A) 8%   (B) 12.5%

(C) 15%   (D) 25%

(E) 45%

Explorations

33. Group Activity Measure the resting pulse rates (beats per minute) of the members of your class. Make a stemplot for the data. Are there any outliers? Can they be explained?

34. Group Activity Measure the heights (in inches) of the members of your class. Make a back-to-back stemplot to compare the distributions of the male heights and the female heights. Write a brief interpretation of the stemplot.

Extending the Ideas

35. Time Plot of Periodic Data Some data are periodic functions of time. If data vary in an annual cycle, the period is 1 year. Use the information in Table 9.16 to overlay the time plots for the average daily high and low temperatures for Beijing, China.

Table 9.16 Average Daily High and Low Temperatures in °C for Beijing, China		
Month	High	Low
January	2	−9
February	5	−7
March	12	−1
April	20	7
May	27	13
June	31	18
July	32	22
August	31	21
September	27	14
October	21	7
November	10	−1
December	3	−7

Source: National Geographic Atlas of the World (rev. 6th ed., 1992, Washington, D.C.), plate 132.

36. Find a sinusoidal function that models each time plot in Exercise 35. (See Sections 4.4 and 4.8.)

9.8
Statistics and Data (Algebraic)

. . . and why

The language of statistics is
becoming more commonplace
in our everyday world.

Parameters and Statistics

The various numbers that are associated with a data set are called **statistics**. They
serve to describe the individuals from which the data come, so the gathering and processing of such numerical information is often called **descriptive statistics**. You saw
many examples of descriptive statistics in Section 9.7.

The *science* of statistics comes in when we use descriptive statistics (like the results
of a study of 1500 smokers) to make judgments, called *inferences*, about entire *populations* (like all smokers). Statisticians are really interested in the numbers called
parameters that are associated with entire populations. Since it is usually either
impractical or impossible to measure entire populations, statisticians gather statistics
from carefully chosen **samples**, then use the science of **inferential statistics** to
make inferences about the parameters.

EXAMPLE 1   Distinguishing a Parameter from a Statistic

A 1996 study called *Kids These Days: What Americans Really Think About the Next
Generation* reported that 33% of adolescents say there is no adult at home when they
return from school. The report was based on a survey of 600 randomly selected people aged 12 to 17 years old and had a margin of error of ±4% (Source: *Public
Agenda*). Did the survey measure a parameter or a statistic, and what does that
"margin of error" mean?

SOLUTION The survey did not measure all adolescents in the population, so it did
not measure a parameter. They *sampled* 600 adolescents and found a statistic. On the
other hand, the statement "33% of adolescents say there is no adult at home" is making an inference about *all* American adolescents. We should interpret that statement in
terms of the margin of error, as meaning "between 29% and 37% of all American adolescents would say that there is no adult at home when they return from school." In other
words, the statisticians are confident that the parameter is within ±4% of their sample
statistic, even though they only sampled 600 adolescents—a tiny fraction of the ado-
lescent population! The mathematics that gives them that confidence is based on the
laws of probability and is scientifically reliable, but we will not go into it here.

Now try Exercise 1.

Mean, Median, and Mode

If you wanted to study the effect of chicken feed additives on the thickness of egg
shells, you would need to sample many eggs from different hens under various feed-
ing conditions. Suppose you were to gather data from 50 eggs from hens eating feed
A and 50 eggs from hens eating feed B. How would you compare the two? The simplest way would be to find the *average* egg shell thickness for each feed and compare
those two numbers.

The word "average," however, can have several different meanings, all of them some-how *measures of center.*

• If we say, "The average on last week's test was 83.4," we are referring to the **mean**. (This is what most people usually think of as "average.")

• If we say, "The average test score puts you right in the middle of the class," we are referring to the **median**.

• If we say, "The average American student starts college at age 18," we are referring to the **mode**.

We will look at each of these measures separately.

DEFINITION **Mean**

The **mean** of a list of *n* numbers $\{x_1, x_2, \ldots, x_n\}$ is

$$\bar{x} = \frac{x_1 + x_2 + \cdots + x_n}{n} = \frac{1}{n} \sum_{i=1}^{n} x_i.$$

The mean is also called the *arithmetic mean*, *arithmetic average*, or *average value.*

EXAMPLE 2   Computing a Mean

Find the mean annual home run total for Roger Maris's major league career, 1957–1968 (Table 9.11 on page 768).

SOLUTION According to Table 9.11, we are looking for the mean of the following list of 12 numbers: $\{14, 28, 16, 39, 61, 33, 23, 26, 8, 13, 9, 5\}$.

$$\bar{x} = \frac{14 + 28 + 16 + \cdots + 9 + 5}{12} = \frac{275}{12} \approx 22.9$$

Now try Exercise 9.

As common as it is to use the mean as a measure of center, sometimes it can be mis-leading. For example, if you were to find the mean annual salary of a Geography major from the University of North Carolina working in Chicago in 1997, it would probably be a number in the millions of dollars. This is because the group being measured, which is not very large, includes an outlier named Michael Jordan. The mean can be strongly affected by outliers.

We call a statistic **resistant** if it is not strongly affected by outliers (See Exploration 1, Section 9.7). While the mean is not a resistant measure of center, the *median* is.

DEFINITION **Median**

The **median** of a list of *n* numbers $\{x_1, x_2, \ldots, x_n\}$ arranged in order (either ascending or descending) is

• the middle number if *n* is odd, and

• the mean of the two middle numbers if *n* is even.

EXAMPLE 3   Finding a Median

Find the median of Roger Maris's annual home run totals. (See Example 2.)

SOLUTION First, we arrange the list in ascending order: {5, 8, 9, 13, 14, 16, 23, 26, 28, 33, 39, 61}. Since there are 12 numbers, the median is the mean of the 6th and 7th numbers:

$$\frac{16 + 23}{2} = 19.5.$$

Notice that this number is quite a bit smaller than the mean (22.9). The mean is strongly affected by the outlier representing Maris's record-breaking season, while the median is not. The median would still have been 19.5 if he had hit only 41 home runs that season, or, indeed, if he had hit 81.   *Now try Exercise 11.*

Both the mean and the median are important measures of center. A somewhat less important measure of center (but a measure of possible significance statistically) is the *mode*.

À LA MODE

The mode can also be used for categorical variables.

DEFINITION Mode

The **mode** of a list of numbers is the number that appears most frequently in the list.

EXAMPLE 4   Finding a Mode

Find the mode for the annual home run totals of Hank Aaron (Table 9.8 on page 765). Is this number of any significance?

SOLUTION It helps to arrange the list in ascending order: {10, 12, 13, 20, 24, 26, 27, 29, 30, 32, 34, 34, 38, 39, 39, 40, 40, 44, 44, 44, 44, 45, 47}.

Most numbers in the list appear only once. Three numbers appear twice, and the number 44 appears four times. The mode is 44.

It is rather unusual to have that many repeats of a number in a list of this sort. (By comparison, the Maris list has no repeats—and hence no mode—while the lists for Mays, Mantle, McGwire, and Bonds contain no number more than twice.) The mode in this case has special significance only to baseball trivia buffs, who recognize 44 as Aaron's uniform number!

Now try Exercise 17.

Example 4 demonstrates why the mode is less useful as a measure of center. The mode (44) is a long way from the median (34) and the mean (32.83), either of which does a much better job of representing Aaron's annual home run output over the course of his career.

EXAMPLE 5   Using a Frequency Table

A teacher gives a 10-point quiz and records the scores in a frequency table (Table 9.17) as shown below. Find the mode, median, and mean of the data.

Table 9.17 Quiz Scores for Example 5										
Score	10	9	8	7	6	5	4	3	2	1	0
Frequency	2	2	3	8	4	3	3	2	1	1	1

SOLUTION The total of the frequencies is 30, so there are 30 scores.

The *mode* is 7, since that is the score with the highest frequency.

The *median* of 30 numbers will be the mean of the 15th and 16th numbers. The table is already arranged in descending order, so we count the frequencies from left to right until we come to 15. We see that the 15th number is a 7 and the 16th number is a 6. The median, therefore, is 6.5.

To find the *mean*, we multiply each number by its frequency, add the products, and divide the total by 30:

$$\bar{x} = \frac{\begin{bmatrix} 10(2) + 9(2) + 8(3) + 7(8) + 6(4) + 5(3) \\ + 4(3) + 3(2) + 2(1) + 1(1) + 0(1) \end{bmatrix}}{30}$$

$$= 5.9\overline{3}$$

Now try Exercise 19.

The formula for finding the mean of a list of numbers $\{x_1, x_2, \ldots, x_n\}$ with frequencies $\{f_1, f_2, \ldots, f_n\}$ is

$$\bar{x} = \frac{x_1 f_1 + x_2 f_2 + \cdots + x_n f_n}{f_1 + f_2 + \cdots + f_n} = \frac{\sum x_i f_i}{\sum f_i}.$$

This same formula can be used to find a **weighted mean**, in which numbers $\{x_1, x_2, \ldots, x_n\}$ are given **weights** before the mean is computed. The weights act the same way as frequencies.

EXAMPLE 6   Working with a Weighted Mean

At Marty's school, it is an administrative policy that the final exam must count 25% of the semester grade. If Marty has an 88.5 average going into the final exam, what is the minimum exam score needed to earn a 90 for the semester?

continued

SOLUTION The preliminary average (88.5) is given a weight of 0.75 and the final exam (x) is given a weight of 0.25. We will assume that a semester average of 89.5 will be rounded to a 90 on the transcript. Therefore,

$$\frac{88.5(0.75) + x(0.25)}{0.75 + 0.25} = 89.5$$

$$0.25x = 89.5(1) - 88.5(0.75)$$

$$x = 92.5$$

Interpreting the answer, we conclude that Marty needs to make a 93 on the final exam.

Now try Exercise 21.

The Five-Number Summary

Measures of center tell only part of the story of a data set. They do *not* indicate how widely distributed or highly variable the data are. *Measures of spread* do. The simplest and crudest measure of spread is the **range**, which is the difference between the maximum and minimum values in the data set:

$$\text{Range} = \text{maximum} - \text{minimum}.$$

For example, the range of numbers in Roger Maris's annual home run production is $61 - 5 = 56$. Like the mean, the range is a statistic that is strongly influenced by outliers, so it can be misleading. A more resistant (and therefore more useful) measure is the *interquartile range*, which is the range of the middle half of the data.

FINDING QUARTILES

When we are finding the quartiles for a data set with an odd number of values, we do *not* consider the middle value to be included in either the lower or the upper half of the data.

Just as the median separates the data into halves, the **quartiles** separate the data into fourths. The **first quartile** Q_1 is the median of the lower half of the data, the **second quartile** is the median, and the **third quartile** Q_3 is the median of the upper half of the data. The **interquartile range** (*IQR*) measures the spread between the first and third quartiles, comprising the middle half of the data:

$$IQR = Q_3 - Q_1.$$

Taken together, the maximum, the minimum, and the three quartiles give a fairly complete picture of both the center and the spread of a data set.

> **DEFINITION** Five-Number Summary
>
> The **five-number summary** of a data set is the collection
>
> $$\{\text{minimum}, Q_1, \text{median}, Q_3, \text{maximum}\}.$$

EXAMPLE 7   Five-Number Summary and Spread

Find the five-number summaries for the male and female life expectancies in South American nations (Table 9.12 on page 768) and compare the spreads.

COMPUTING STATISTICS ON A CALCULATOR

Modern calculators will usually process lists of data and give statistics like the mean, median, and quartiles with a push of a button. Consult your owner's manual.

SOLUTION Here are the lists in ascending order.

Males:

{59.0, 60.5, 61.5,   66.7, 67.9, 68.5,   69.0, 70.3, 71.4,   71.9, 72.1, 72.6}

continued

Females:

$\{66.2, 66.7, 67.7, \quad 72.8, 74.3, 74.4, \quad 74.6, 76.5, 76.6, \quad 78.8, 79.0, 79.4\}$

We have spaced the lists to show where the quartiles appear. The median of the 12 values is midway between the 6th and 7th values. The first quartile is the median of the lower 6 values (i.e., midway between the 3rd and 4th), and the third quartile is the median of the upper 6 values (i.e., midway between the 9th and 10th).

The five-number summaries are shown below.

Males: $\{59.0, 64.1, 68.75, 71.65, 72.6\}$

Females: $\{66.2, 70.25, 74.5, 77.7, 79.4\}$

The males have a range of $72.6 - 59.0 = 13.6$ and an *IQR* of $71.65 - 64.1 = 7.55$.

The females have a range of $79.4 - 66.2 = 13.2$ and an *IQR* of $77.7 - 70.25 = 7.45$.

Not only do the women live longer, but there is less variability in their life expectancies (as measured by the *IQR*). Male life expectancy is more strongly affected by different political conditions within countries (war, civil strife, crime, etc.).

Now try Exercise 23.

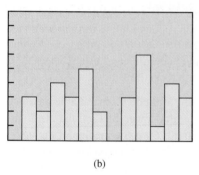

(a)

(b)

FIGURE 9.19 Which data set has more variability? (Exploration 1)

You can learn a lot about the data by considering the *shape* of the distribution, as visualized in a histogram. Try answering the questions in Exploration 1.

EXPLORATION 1 Interpreting Histograms

1. Of the two histograms shown in Figure 9.19, which displays a data set with more variability?

2. Of the three histograms in Figure 9.20, which has a median less than its mean? Which has a median greater than its mean? Which has a median approximately equal to its mean?

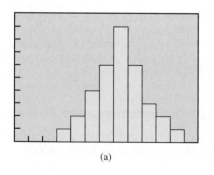

(a)

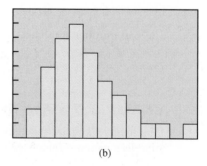

(b)

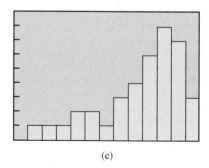

(c)

FIGURE 9.20 Which graph shows a data set in which the mean is less than the median? Greater than the median? Approximately equal to the median? (Exploration 1)

The distribution in Figure 9.20a is **symmetric** because it looks approximately the same when reflected about a vertical line through the median. The distribution in Figure 9.20b is **skewed right** because it has a longer "tail" to the right. The values in the tail will pull nonresistant measures (like the mean) to the *right*, leaving resistant measures (like the median) behind. The distribution in Figure 9.20c is **skewed left**, because nonresistant measures are pulled to the *left*.

You can also see symmetry and skewedness in stemplots, which have the same shapes vertically as histograms have horizontally.

Boxplots

A **boxplot** (sometimes called a **box-and-whisker plot**) is a graph that depicts the five-number summary of a data set. The plot consists of a central rectangle (box) that extends from the first quartile to the third quartile, with a vertical segment marking the median. Line segments (whiskers) extend at the ends of the box to the minimum and maximum values. For example, the five-number summary for the male life expectancies in South American nations (Example 7) was {59.0, 64.1, 68.75, 71.65, 72.6}. The boxplot for the data is shown in Figure 9.21. Notice that the box and the whisker extend further to the left of the median than to the right, suggesting a distribution that is skewed left. (The histogram obtained in Exercise 9 of Section 9.7 confirms that this is the case.)

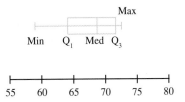

FIGURE 9.21 A boxplot for the five-number summary of the male life expectancies in Example 7. (The features on the box are labeled here for illustrative purposes; it is not necessary in general to label the min, the quartiles, or the max.)

EXAMPLE 8   Comparing Boxplots

Draw boxplots for the male and female data in Example 7 and describe briefly the information displayed in the visualization.

SOLUTION The five-number summaries are:

Males: {59.0, 64.1, 68.75, 71.65, 72.6}

Females: {66.2, 70.25, 74.5, 77.7, 79.4}

The boxplots can be graphed simultaneously (Figure 9.22).

From this graph we can see that the middle half of the female life expectancies are all greater than the median of the male life expectancies. The median life expectancy for the women among South American nations is greater than the maximum for the men.

Now try Exercise 31.

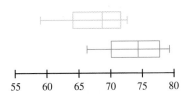

FIGURE 9.22 A single graph showing boxplots for male and female life expectancies in the nations of South America gives a good visualization of the differences in the two data sets. (Example 8)

If we look at the boxplot for Roger Maris's annual home run totals {14, 28, 16, 39, 61, 33, 23, 26, 8, 13, 9, 5}, we see that the whisker on the right is very long (Figure 9.23). This is a visualization of the effect of the outlier (61), which is much larger than the third quartile (30.5).

In fact, a boxplot gives us a convenient way to think of an outlier: a number that makes one of the whiskers noticeably longer than the box. The usual rule of thumb for "noticeably longer" is 1.5 times as long. Since the length of the box is the *IQR*, that leads us to the following numerical check.

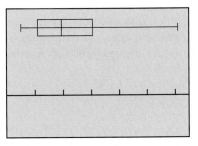

[0, 65] by [−5, 10]

FIGURE 9.23 A boxplot of Roger Maris's annual home run production (Table 9.11 on page 768). The outlier (61) results in a long whisker on the right because the maximum is much larger than Q_3.

A number in a data set can be considered an **outlier** if it is more than $1.5 \times IQR$ below the first quartile or above the third quartile.

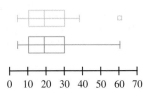

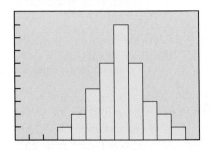

FIGURE 9.24 A modified boxplot and a regular boxplot of Roger Maris's annual home run totals.

EXAMPLE 9   Identifying an Outlier

Is 61 an outlier in Roger Maris's home run data according to the $1.5 \times IQR$ criterion?

SOLUTION Maris's totals, in order: $\{5, 8, 9, 13, 14, 16, 23, 26, 28, 33, 39, 61\}$

His five-number summary: $\{5, 11, 19.5, 30.5, 61\}$

His IQR: $30.5 - 11 = 19.5$

So,
$$Q_3 + 1.5 \times IQR = 30.5 + 1.5 \times 19.5 = 59.75$$

Since $61 > 59.75$, the rule of thumb identifies it as an outlier.

Now try Exercise 39.

By their very nature, outliers can distort the overall picture we get of the data. For that reason, statisticians will frequently look for reasons to omit them from their statistical displays and calculations. (This can, of course, be risky. You want to omit a strange laboratory reading if you suspect equipment error, but you do not want to ignore a potential scientific discovery.) A **modified boxplot** is a compromise visualization that separates outliers as isolated points, extending the whiskers only to the farthest nonoutliers. Figure 9.24 shows a modified boxplot of Roger Maris's home run data, as compared to a regular boxplot.

Variance and Standard Deviation

You might be surprised that the five-number summary and its boxplot graph do not even make reference to the mean, which is a more familiar measure than a median or a quartile. This is because the mean, being a nonresistant measure, is less reliable in the presence of outliers or skewed data.

On the other hand, the mean is an excellent measure of center when outliers and skewedness are not present, which is quite often the case. Indeed, histograms of data from all kinds of real-world sources tend to look something like Figure 9.25, in which frequencies are higher close to the mean and lower as you move away from the mean in either direction. Statisticians call these *normal* distributions. (We will make that term more precise shortly.)

For normally distributed data, the mean is the preferred measure of center. There is also a measure of variability for normal data that is better than the *IQR*, called the *standard deviation*. Like the mean, the standard deviation is strongly affected by outliers and can be misleading if outliers are present.

DEFINITION **Standard Deviation**

The **standard deviation** of the numbers $\{x_1, x_2, \ldots, x_n\}$ is
$$\sigma = \sqrt{\frac{1}{n} \sum_{i=1}^{n} (x_i - \bar{x})^2},$$
where $\bar{x}$ denotes the mean. The **variance** is σ^2, the square of the standard deviation.

FIGURE 9.25 Histograms based on data gathered from real-world sources are often symmetric and higher in the middle, without outliers. Frequency distributions with graphs of this shape are called *normal*.

If we define the "deviation" of a data value to be how much it differs from the mean, then the variance is just the mean of the squared deviations. The standard deviation is the square *root* of the *mean* of the *squared deviations*, which is why it is sometimes called the *root mean square deviation*. The symbol "σ" is a lowercase Greek letter sigma.

Calculating a standard deviation by hand can be tedious, but with modern calculators it is usually only necessary to enter the list of data and push a button. In fact, most calculators give you a choice of two standard deviations, one slightly larger than the other. The larger one (usually called s) is based on the formula

$$s = \sqrt{\frac{1}{n-1} \sum_{i=1}^{n} (x_i - \bar{x})^2}.$$

The difference is that the σ formula is for finding the true parameter, which means that it only applies when $\{x_1, x_2, \ldots, x_n\}$ is the whole **population**. If $\{x_1, x_2, \ldots, x_n\}$ is a *sample* from the population, then the s formula actually gives a better estimate of the parameter than the σ formula does. So use the larger standard deviation when your data come from a sample (which is almost always the case).

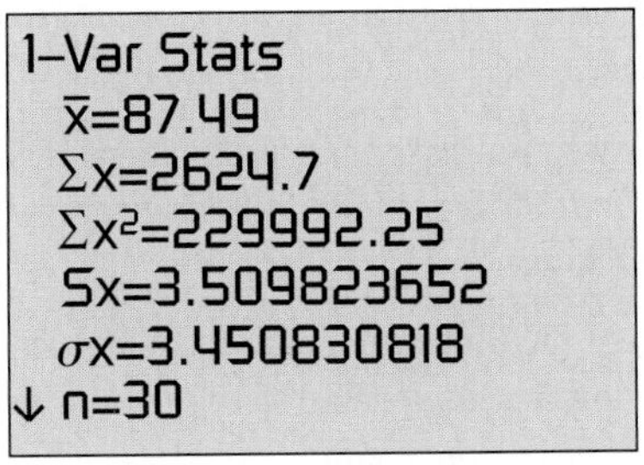

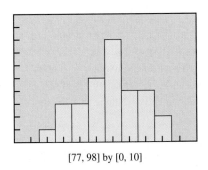

FIGURE 9.26 Single-variable statistics in a typical calculator display. (Example 10)

FIGURE 9.27 The weights of the loon chicks in Example 10 appear to be normal, with no outliers or strong skewedness. We conclude that the mean and standard deviation are appropriate measures of center and variability, respectively.

EXAMPLE 10 Finding Standard Deviation with a Calculator

A researcher measured 30 newly hatched loon chicks and recorded their weights in grams as shown in Table 9.18.

Table 9.18 Weights in Grams of 30 Loon Chicks									
79.5	87.5	88.5	89.2	91.6	84.5	82.1	82.3	85.7	89.8
84.0	84.8	88.2	88.2	82.9	89.8	89.2	94.1	88.0	91.1
91.8	87.0	87.7	88.0	85.4	94.4	91.3	86.4	85.7	86.0

Based on the sample, estimate the mean and standard deviation for the weights of newly hatched loon chicks. Are these measures useful in this case, or should we use the five-number summary?

SOLUTION We enter the list of data into a calculator and choose the command that will produce statistics of a single variable. The output from one such calculator is shown in Figure 9.26.

The mean is $\bar{x} = 87.49$ grams. For standard deviation, we choose $Sx = 3.51$ grams because the calculations are based on a sample of loon chicks, not the entire population of loon chicks.

A histogram (Xscl = 2) in the window [77, 98] by [0, 10] shows that the distribution is normal (as we would expect from nature), containing no outliers or strong skewedness. Therefore, the mean and standard deviation are appropriate measures (Figure 9.27). *Now try Exercise 35.*

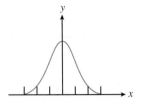

FIGURE 9.28 The graph of $y = e^{-x^2/2}$. This is a Gaussian (or normal) curve.

Normal Distributions

Although we use the word *normal* in many contexts to suggest typical behavior, in the context of statistics and data distributions it is really a technical term. If you graph the function

$$y = e^{-x^2/2}$$

in the window $[-3, 3]$ by $[0, 1]$, you will see what **normal** means mathematically (Figure 9.28).

The shape corresponds to the kind of distribution we have been calling "normal." In fact, this curve, called a **Gaussian curve** or **normal curve** is a *precise mathematical model* for normal behavior. That is where the mean and standard deviation come in.

The standard deviation of the curve in Figure 9.28 is 1. Using calculus, we can find that about 68% of the total area under this curve lies between -1 and 1 (Figure 9.29a). Since any normal distribution has this shape, *about 68% of the data in any normal distribution lie within 1 standard deviation of the mean.*

Similarly, we can find that about 95% of the total area under the Gaussian curve lies between -2 and 2 (Figure 9.29b), implying that *about 95% of the data in any normal distribution lie within 2 standard deviations of the mean.*

Similarly, we can find that about 99.7% (nearly all) of the total area under the Gaussian curve lies between -3 and 3 (Figure 9.29c), implying that *about 99.7% of the data in any normal distribution lie within 3 standard deviations of the mean.*

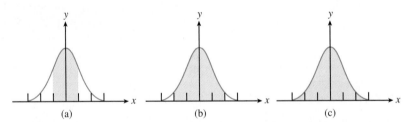

FIGURE 9.29 (a) About 68% of the area under a Gaussian curve lies within 1 unit of the mean. (b) About 95% of the area lies within 2 units of the mean. (c) About 99.7% of the area lies within 3 units of the mean. If we think of the units as standard deviations, this gives us a model for *any* normal distribution.

The 68–95–99.7 Rule

If the data for a population are normally distributed with mean μ and standard deviation σ, then

- approximately 68% of the data lie between $\mu - 1\sigma$ and $\mu + 1\sigma$;

- approximately 95% of the data lie between $\mu - 2\sigma$ and $\mu + 2\sigma$;

- approximately 99.7% of the data lie between $\mu - 3\sigma$ and $\mu + 3\sigma$.

What makes this rule so useful is that normal distributions are common in a wide variety of statistical applications. We close the chapter with a simple application.

EXAMPLE 11   Using the 68–95–99.7 Rule

Based on the research data presented in Example 10, would a loon chick weighing 95 grams be in the top 2.5% of all newly hatched loon chicks?

SOLUTION We assume that the weights of newly hatched loon chicks are normally distributed in the whole population. Since we do not know the mean and standard deviation for the whole population (the parameters μ and σ), we use $\bar{x} = 87.49$ and $Sx = 3.51$ as estimates.

Look at Figure 9.29b. The shaded region contains 95% of the area, so the two identical white regions at either end must each contain 2.5% of the area. That is, to be in the top 2.5%, a loon chick will have to weigh at least 2 standard deviations more than the mean:

$$\bar{x} + 2Sx = 87.49 + 2(3.51) = 94.51 \text{ grams.}$$

Since $95 > 94.51$, a 95-gram loon chick is indeed in the top 2.5%.

Now try Exercise 41.

If you study statistics more deeply someday you will learn that there is more going on in Example 11 than meets the eye. For starters, we need to know that the chicks are really a random sample of all loon chicks (not, for example, from the same geographical area). Also, we lose some accuracy by using a statistic to estimate the true standard deviation, and statisticians have ways of taking that into account.

This section has only offered a brief glimpse into how statisticians use mathematics. If you are interested in learning more, we urge you to find a good statistics textbook and pursue it!

QUICK REVIEW 9.8   *(Prerequisite skill Section 9.5)*

In Exercises 1–6, write the sum in expanded form.

1. $\displaystyle\sum_{i=1}^{7} x_i$

2. $\displaystyle\sum_{i=1}^{5} (x_i - \bar{x})$

3. $\displaystyle\frac{1}{7}\sum_{i=1}^{7} x_i$

4. $\displaystyle\frac{1}{5}\sum_{i=1}^{5} (x_i - \bar{x})$

5. $\displaystyle\frac{1}{5}\sum_{i=1}^{5} (x_i - \bar{x})^2$

6. $\displaystyle\sqrt{\frac{1}{5}\sum_{i=1}^{5} (x_i - \bar{x})^2}$

In Exercises 7–10, write the sum in sigma notation.

7. $x_1f_1 + x_2f_2 + x_3f_3 + \cdots + x_8f_8$

8. $(x_1 - \bar{x})^2 + (x_2 - \bar{x})^2 + \cdots + (x_{10} - \bar{x})^2$

9. $\dfrac{1}{50}[(x_1 - \bar{x})^2 + (x_2 - \bar{x})^2 + \cdots + (x_{50} - \bar{x})^2]$

10. $\sqrt{\dfrac{1}{7}[(x_1 - \bar{x})^2 + (x_2 - \bar{x})^2 + \cdots + (x_7 - \bar{x})^2]}$

In Exercises 63–66, find the sum of the terms of the arithmetic sequence.

63. $-11, -8, -5, -2, 1, 4, 7, 10$

64. $13, 9, 5, 1, -3, -7, -11$

65. $2.5, -0.5, -3.5, \ldots, -75.5$

66. $-5, -3, -1, 1, \ldots, 55$

In Exercises 67–70, find the sum of the terms of the geometric sequence.

67. $4, -2, 1, -\dfrac{1}{2}, \dfrac{1}{4}, -\dfrac{1}{8}$

68. $-3, -1, -\dfrac{1}{3}, -\dfrac{1}{9}, -\dfrac{1}{27}$

69. $2, 6, 18, \ldots, 39{,}366$

70. $1, -2, 4, -8, \ldots, -8192$

In Exercises 71 and 72, find the sum of the first 10 terms of the arithmetic or geometric sequence.

71. $2187, 729, 243, \ldots$ **72.** $94, 91, 88, \ldots$

In Exercises 73 and 74, graph the sequence.

73. $a_n = 1 + \dfrac{(-1)^n}{n}$ **74.** $a_n = 2n^2 - 1$

75. Annuity Mr. Andalib pays $150 at the end of each month into an account that pays 8% interest compounded monthly. At the end of 10 years, the balance in the account, in dollars, is

$$150\left(1 + \frac{0.08}{12}\right)^0 + 150\left(1 + \frac{0.08}{12}\right)^1 + \cdots + 150\left(1 + \frac{0.08}{12}\right)^{119}.$$

Use the formula for the sum of a finite geometric series to find the balance.

76. Annuity What is the minimum monthly payment at month's end that must be made in an account that pays 8% interest compounded monthly if the balance at the end of 10 years is to be at least $30,000?

In Exercises 77–82, determine whether the geometric series converges. If it does, find its sum.

77. $\displaystyle\sum_{j=1}^{\infty} 2\left(\frac{3}{4}\right)^j$ **78.** $\displaystyle\sum_{k=1}^{\infty} 2\left(-\frac{1}{3}\right)^k$

79. $\displaystyle\sum_{j=1}^{\infty} 4\left(-\frac{4}{3}\right)^j$ **80.** $\displaystyle\sum_{k=1}^{\infty} 5\left(\frac{6}{5}\right)^k$

81. $\displaystyle\sum_{k=1}^{\infty} 3(0.5)^k$ **82.** $\displaystyle\sum_{k=1}^{\infty} (1.2)^k$

In Exercises 83–86, write the sum in sigma notation.

83. $-8 - 3 + 2 + \cdots + 92$

84. $4 - 8 + 16 - 32 + \cdots - 2048$

85. $1^2 + 3^2 + 5^2 + \cdots$

86. $1 + \dfrac{1}{2} + \dfrac{1}{2^2} + \dfrac{1}{2^3} + \cdots$

In Exercises 87–90, use summation formulas to evaluate the expression.

87. $\displaystyle\sum_{k=1}^{n} (3k + 1)$ **88.** $\displaystyle\sum_{k=1}^{n} 3k^2$

89. $\displaystyle\sum_{k=1}^{25} (k^2 - 3k + 4)$ **90.** $\displaystyle\sum_{k=1}^{175} (3k^2 - 5k + 1)$

In Exercises 91–94, use mathematical induction to prove that the statement is true for all positive integers n.

91. $1 + 3 + 6 + \cdots + \dfrac{n(n + 1)}{2} = \dfrac{n(n + 1)(n + 2)}{6}$

92. $1 \cdot 2 + 2 \cdot 3 + 3 \cdot 4 + \cdots + n(n + 1) = \dfrac{n(n + 1)(n + 2)}{3}$

93. $2^{n-1} \le n!$

94. $n^3 + 2n$ is divisible by 3.

In Exercises 95–98, construct **(a)** a stemplot, **(b)** a frequency table, and **(c)** a histogram for the indicated data.

95. Real Estate Prices Use intervals of $10,000. The median sales prices (in units of $10,000) for homes in 30 randomly selected metropolitan areas in 2001 were as follows:

10.7, 11.4, 12.7, 11.5, 14.6, 13.6, 9.2, 21.9, 16.1, 12.2, 13.5, 12.6, 12.0, 14.7, 23.4, 12.4, 17.0, 11.7, 11.5, 10.6, 14.1, 15.4, 15.8, 17.6, 14.7, 11.7, 12.7, 9.1, 16.4, 14.8

(Source: *National Association of Realtors,* as reported in *The World Almanac and Book of Facts, 2002.*)

96. Popular Web Sites Use intervals of 5 million. The number of visitors (in units of 10 million) to the top 25 Web sites in 2001 (as measured May 1–31) were as follows:

7.1, 6.1, 5.8, 3.3, 2.9, 2.8, 2.6, 2.0, 2.0, 2.0, 1.9, 1.9, 1.9, 1.5, 1.4, 1.4, 1.3, 1.3, 1.2, 1.2, 1.1, 1.1, 1.1, 1.1, 1.1

(Source: *Media Matrix,* as reported in *The New York Times Almanac, 2002.*)

97. Beatles Songs Use intervals of length 10. The lengths (in seconds) of 24 randomly selected Beatles songs that appeared on singles are as follows, in order of release date:

143, 120, 120, 139, 124, 144, 131, 132, 148, 163, 140, 177, 136, 124, 179, 131, 180, 137, 156, 202, 191, 197, 230, 190

(Source: *Personal collection.*)

98. Passing Yardage In 1995, Warren Moon of the Minnesota Vikings became the first pro quarterback to pass for 60,000 total yards. Use intervals of 1000 yards for Moon's regular season passing yards given in Table 9.22.

Table 9.22 Regular Season Passing Yardage Statistics for Warren Moon

Year	Yards	Year	Yards
1978	1112	1987	2806
1979	2382	1988	2327
1980	3127	1989	3631
1981	3959	1990	4689
1982	5000	1991	4690
1983	5648	1992	2521
1984	3338	1993	3485
1985	2709	1994	4264
1986	3489	1995	4228

Source: The Minnesota Vikings, as reported by Julie Stacey in USA Today on September 25, 1995 and www.nfl.com

In Exercises 99–102, find the five-number summary, the range, the interquartile range, the standard deviation, and the variance σ for the specified data (use σ and σ^2). Identify any outliers.

99. The data in Exercise 95   **100.** The data in Exercise 96

101. The data in Exercise 97   **102.** The data in Exercise 98

In Exercises 103–106, construct **(a)** a boxplot and **(b)** a modified boxplot for the specified data.

103. The data in Exercise 95   **104.** The data in Exercise 96

105. The data in Exercise 97   **106.** The data in Exercise 98

107. Make a back-to-back stemplot of the data in Exercise 97, showing the earlier 12 songs in one plot and the later 12 songs in the other. Write a sentence interpreting the stemplot.

108. Make simultaneous boxplots of the data in Exercise 97, showing the earlier 12 songs in one boxplot and the later 12 songs in the other.

(a) Which set of data has the greater range?

(b) Which set of data has the greater interquartile range?

109. Time Plots Make a time plot for the data in Exercise 97, assuming equal time intervals between songs. Interpret the trend revealed in the time plot.

110. Damped Time Plots Statisticians sometimes use a technique called *damping* to smooth out random fluctuations in a time plot. Find the mean of the first four numbers in Exercise 97, the mean of the next four numbers, and so on. Then graph the six means as a function of time. Is there a clear trend?

111. Find row 9 of Pascal's triangle.

112. Show algebraically that

$$_nP_k \times {}_{n-k}P_j = {}_nP_{k+j}.$$

113. Baseball Bats Suppose that the probability of producing a defective baseball bat is 0.02. Four bats are selected at random. What is the probability that the lot of four bats contains the following?

(a) No defective bats

(b) One defective bat

114. Lightbulbs Suppose that the probability of producing a defective lightbulb is 0.0004. Ten lightbulbs are selected at random. What is the probability that the lot of 10 contains the following?

(a) No defective lightbulbs

(b) Two defective lightbulbs

CHAPTER 9 Project

Analyzing Height Data

The set of data below was gathered from a class of 30 precalculus students. Use this set of data or collect the data for your own class and use it for analysis.

Heights of students in inches				
66	69	72	64	68
70	71	66	65	63
72	59	64	63	66
68	63	64	71	71
69	62	61	67	69
64	73	75	61	70

1. Create a stem-and-leaf plot of the data using split stems. From this data, what is the approximate average height of a student in the class?

2. Create a frequency table for the data using an interval of 2. What information does this give?

3. Create a histogram for the data using an interval of 2. What conclusions can you draw from this representation of the data? Can you estimate the average height for males and the average height for females?

4. Compute the mean, median, and mode for the data set. Discuss whether each is a good measure of the average height of a student in the class. Is each a good predictor for average height of students in other precalculus classes?

5. What can you say about the data if the mean and median values are close?

6. Find the five-number summary for the class heights.

7. Create a boxplot and explain what information it gives about the data set.

8. A new student is now added to the class. He is a 7'2" star basketball player. Add his height to the data set. Recalculate the mean, median, and five-number summary. Create a new boxplot and use your calculator to plot it underneath the boxplot for the original class. How does this new student affect the statistics?

9. Explain why this new student would be considered an outlier and the importance of identifying outliers when calculating statistics and making predictions from them.

10. Suppose now that three additional basketball players transferred into the class. They are 7'0", 6'11" and 6'10". Recalculate the statistics from number 9 and discuss the implications of using these statistics to make predictions for other precalculus classes.

An Introduction to Calculus: Limits, Derivatives, and Integrals

Windmills have long been used to pump water from wells, grind grain, and saw wood. They are more recently being used to produce electricity. The propeller radius of these windmills range from one to one hundred meters, and the power output ranges from a hundred watts to a thousand kilowatts. See page 800 in Section 10.1 for some more information and a question and answer about windmills.

Chapter 10 Overview

By the beginning of the 17th century, algebra and geometry had developed to the point where physical behavior could be modeled both algebraically and graphically, each type of representation providing deeper insights into the other. New discoveries about the solar system had opened up fascinating questions about gravity and its effects on planetary motion, so that finding the mathematical key to studying motion became the scientific quest of the day. The analytic geometry of René Descartes (1596–1650) put the final pieces into place, setting the stage for Isaac Newton (1642–1727) and Gottfried Leibniz (1646–1716) to stand "on the shoulders of giants" and see beyond the algebraic boundaries that had limited their predecessors. With geometry showing them the way, they created the new form of algebra that would come to be known as the calculus.

In this chapter we will look at the two central problems of motion much as Newton and Leibniz did, connecting them to geometric problems involving tangent lines and areas. We will see how the obvious geometric solutions to both problems led to algebraic dilemmas, and how the algebraic dilemmas led to the discovery of calculus. The language of limits, which we have used in this book to describe asymptotes, end behavior, and continuity, will serve us well as we make this transition.

10.1
Limits and Motion: The Tangent Problem

What you'll learn about

- Average Velocity
- Instantaneous Velocity
- Limits Revisited
- The Connection to Tangent Lines
- The Derivative

. . . and why

The derivative allows us to analyze rates of change, which are fundamental to understanding physics, economics, engineering, and even history.

Average Velocity

Average velocity is the change in position divided by the change in time, as in the following familiar-looking example.

EXAMPLE 1 Computing Average Velocity

An automobile travels 120 miles in 2 hours and 30 minutes. What is its average velocity over the entire $2\frac{1}{2}$ hour time interval?

SOLUTION The average velocity is the change in position (120 miles) divided by the change in time (2.5 hours). If we denote position by s and time by t, we have

$$v_{\text{ave}} = \frac{\triangle s}{\triangle t} = \frac{120 \text{ miles}}{2.5 \text{ hours}} = 48 \text{ miles per hour.}$$

Now try Exercise 1.

Notice that the average velocity does not tell us how fast the automobile is traveling at any given moment during the time interval. It could have gone at a constant 48 mph all the way, or it could have sped up, slowed down, or even stopped momentarily several times during the trip. Scientists like Galileo Galilei (1564–1642), who studied motion prior to Newton and Leibniz, were looking for formulas that would give velocity as a *function* of time, that is, formulas that would give *instantaneous* values of v for individual values of t. The step from average to instantaneous sounds simple enough, but there were complications, as we shall see.

Instantaneous Velocity

Galileo experimented with gravity by rolling a ball down an inclined plane and recording its approximate velocity as a function of elapsed time. Here is what he might have asked himself when he began his experiments:

A Velocity Question

A ball rolls a distance of 16 feet in 4 seconds. What is the instantaneous velocity of the ball at a moment of time 3 seconds after it starts to roll?

A DISCLAIMER

Readers who know a little calculus will recognize that the ball in the Velocity Question really does have a nonzero instantaneous velocity after 3 seconds (as we will eventually show). The point of the discussion is to show how difficult it is to demonstrate that fact at a single instant, since both time and the position of the ball appear not to change.

You might want to visualize the ball being frozen at that moment, and then try to determine its velocity. Well, then the ball would have zero velocity, because it is frozen! This approach seems foolish, since, of course, the ball is moving.

Is this a trick question? On the contrary, it is actually quite profound—it is exactly the question that Galileo (among many others) was trying to answer. Notice how easy it is to find the *average* velocity:

$$v_{ave} = \frac{\triangle s}{\triangle t} = \frac{16 \text{ feet}}{4 \text{ seconds}} = 4 \text{ feet per second.}$$

Now, notice how inadequate our algebra becomes when we try to apply the same formula to *instantaneous* velocity:

$$v_{ave} = \frac{\triangle s}{\triangle t} = \frac{0 \text{ feet}}{0 \text{ seconds}},$$

which involves division by 0—and is therefore undefined!

So Galileo did the best he could by making $\triangle t$ as small as experimentally possible, measuring the small values of $\triangle s$, and then finding the quotients. It only *approximated* the instantaneous velocity, but finding the exact value appeared to be algebraically out of the question, since division by zero was impossible.

Limits Revisited

Newton invented "fluxions" and Leibniz invented "differentials" to explain instantaneous rates of change without resorting to zero denominators. Both involved mysterious quantities that could be infinitesimally small without really being zero. (Their 17th-century colleague Bishop Berkeley called them "ghosts of departed quantities" and dismissed them as nonsense.) Though not well understood by many, the strange quantities modeled the behavior of moving bodies so effectively that most scientists were willing to accept them on faith until a better explanation could be developed. That development, which took about a hundred years, led to our modern understanding of limits.

Since you are already familiar with limit notation, we can show you how this works with a simple example.

EXAMPLE 3   Finding the Slope of a Tangent Line

Use limits to find the slope of the tangent line to the graph of $s = t^2$ at the point $(1, 1)$ (Figure 10.2).

SOLUTION This will look a lot like the solution to Example 2.

$$\lim_{t \to 1} \frac{\triangle s}{\triangle t} = \lim_{t \to 1} \frac{t^2 - 1^2}{t - 1}$$

$$= \lim_{t \to 1} \frac{(t + 1)(t - 1)}{t - 1} \qquad \text{Factor the numerator.}$$

$$= \lim_{t \to 1} (t + 1) \cdot \frac{t - 1}{t - 1}$$

$$= \lim_{t \to 1} (t + 1) \qquad \text{Since } t \neq 1, \frac{t - 1}{t - 1} = 1$$

$$= 2$$

Now try Exercise 17(a).

If you compare Example 3 to Example 2 it should be apparent that a method for solving the tangent line problem can be used to solve the instantaneous velocity problem, and vice versa. They are geometric and algebraic versions of the same problem!

THE TANGENT LINE PROBLEM

Although we have focused on Galileo's work with motion problems in order to follow a coherent story, it was Pierre de Fermat (1601–1665) who first developed a "method of tangents" for general curves, recognizing its usefulness for finding relative maxima and minima. Fermat is best remembered for his work in number theory, particularly for Fermat's Last Theorem, which states that there are no positive integers x, y, and z that satisfy the equation $x^n + y^n = z^n$ if n is an integer greater than 2. Fermat wrote in the margin of a textbook, "I have a truly marvelous proof that this margin is too narrow to contain," but if he had one, he apparently never wrote it down. Although mathematicians tried for over 330 years to prove (or disprove) Fermat's Last Theorem, nobody succeeded until Andrew Wiles of Princeton University finally proved it in 1994.

The Derivative

Velocity, the rate of change of position with respect to time, is only one application of the general concept of "rate of change." If $y = f(x)$ is *any* function, we can speak of how y changes as x changes.

DEFINITION Average Rate of Change

If $y = f(x)$, then the **average rate of change** of y with respect to x on the interval $[a, b]$ is

$$\frac{\triangle y}{\triangle x} = \frac{f(b) - f(a)}{b - a}.$$

Geometrically, this is the slope of the **secant line** through $(a, f(a))$ and $(b, f(b))$.

Using limits, we can proceed to a definition of the *instantaneous* rate of change of y with respect to x at the point where $x = a$. This instantaneous rate of change is called the *derivative*.

DIFFERENTIABILITY

We say a function is "differentiable" at *a* if $f'(a)$ exists, because we can find the limit of the "quotient of differences."

DEFINITION Derivative at a Point

The **derivative of the function** *f* **at** $x = a$, denoted by $f'(a)$ and read "*f* prime of *a*" is

$$f'(a) = \lim_{x \to a} \frac{f(x) - f(a)}{x - a},$$

provided the limit exists.

Geometrically, this is the slope of the **tangent line** through $(a, f(a))$.

A more computationally useful formula for the derivative is obtained by letting $x = a + h$ and looking at the limit as *h* approaches 0 (equivalent to letting *x* approach *a*).

DEFINITION Derivative at a Point (easier for computing)

The **derivative of the function** *f* **at** $x = a$, denoted by $f'(a)$ and read "*f* prime of *a*" is

$$f'(a) = \lim_{h \to 0} \frac{f(a + h) - f(a)}{h},$$

provided the limit exists.

The fact that the derivative of a function at a point can be viewed geometrically as the slope of the line tangent to the curve $y = f(x)$ at that point provides us with some insight as to how a derivative might fail to exist. Unless a function has a well-defined "slope" when you zoom in on it at *a*, the derivative at *a* will not exist. For example, Figure 10.3 shows three cases for which $f(0)$ exists but $f'(0)$ does not.

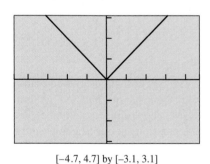

[−4.7, 4.7] by [−3.1, 3.1]

(a)

$f(x) = |x|$ has a graph with no definable slope at $x = 0$.

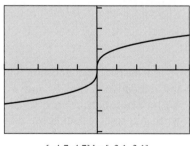

[−4.7, 4.7] by [−3.1, 3.1]

(b)

$f(x) = \sqrt[3]{x}$ has a graph with a vertical tangent line (no slope) at $x = 0$.

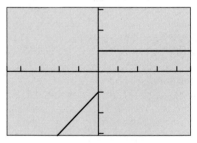

[−4.7, 4.7] by [−3.1, 3.1]

(c)

$$f(x) = \begin{cases} x - 1 & \text{for } x < 0 \\ 1 & \text{for } x \geq 0 \end{cases}$$

has a graph with no definable slope at $x = 0$.

FIGURE 10.3 Three examples of functions defined at $x = 0$ but not differentiable at $x = 0$.

EXAMPLE 4 Finding a Derivative at a Point

Find $f'(4)$ if $f(x) = 2x^2 - 3$.

SOLUTION

$$
\begin{aligned}
f'(4) &= \lim_{h \to 0} \frac{f(4+h) - f(4)}{h} \\[2mm]
&= \lim_{h \to 0} \frac{2(4+h)^2 - 3 - (2 \cdot 4^2 - 3)}{h} \\[2mm]
&= \lim_{h \to 0} \frac{2(16 + 8h + h^2) - 32}{h} \\[2mm]
&= \lim_{h \to 0} \frac{16h + 2h^2}{h} \\[2mm]
&= \lim_{h \to 0} (16 + 2h) \qquad \frac{h}{h} = 1, \text{ since } h \neq 0. \\[2mm]
&= 16
\end{aligned}
$$

Now try Exercise 23.

The derivative can also be thought of as a function of x. Its domain consists of all values in the domain of f for which f is differentiable. The function f' can be defined by adapting the second definition above.

DEFINITION Derivative

If $y = f(x)$, then the **derivative of the function f with respect to x**, is the function f' whose value at x is

$$
f'(x) = \lim_{h \to 0} \frac{f(x+h) - f(x)}{h},
$$

for all values of x where the limit exists.

To emphasize the connection with slope $\triangle y / \triangle x$, Leibniz used the notation dy/dx for the derivative. (The dy and dx were his "ghosts of departed quantities.") This **Leibniz notation** has several advantages over the "prime" notation, as you will learn when you study calculus. We will use both notations in our examples and exercises.

EXAMPLE 5 Finding the Derivative of a Function

(a) Find $f'(x)$ if $f(x) = x^2$.

(b) Find $\dfrac{dy}{dx}$ if $y = \dfrac{1}{x}$.

continued

SOLUTION

(a)
$$f'(x) = \lim_{h \to 0} \frac{f(x+h) - f(x)}{h}$$

$$= \lim_{h \to 0} \frac{(x+h)^2 - x^2}{h}$$

$$= \lim_{h \to 0} \frac{x^2 + 2xh + h^2 - x^2}{h}$$

$$= \lim_{h \to 0} \frac{2xh + h^2}{h}$$

$$= \lim_{h \to 0} (2x + h) \qquad \text{Since } h \neq 0, \frac{h}{h} = 1.$$

$$= 2x$$

So $f'(x) = 2x$.

(b)
$$\frac{dy}{dx} = \lim_{h \to 0} \frac{f(x+h) - f(x)}{h}$$

$$= \lim_{h \to 0} \frac{\dfrac{1}{x+h} - \dfrac{1}{x}}{h}$$

$$= \lim_{h \to 0} \frac{\dfrac{x - (x+h)}{x(x+h)}}{h}$$

$$= \lim_{h \to 0} \frac{-h}{x(x+h)} \cdot \frac{1}{h}$$

$$= \lim_{h \to 0} \frac{-1}{x(x+h)}$$

$$= -\frac{1}{x^2}$$

So $\dfrac{dy}{dx} = -\dfrac{1}{x^2}$.

Now try Exercise 29.

CHAPTER OPENER PROBLEM
(from page 791)

PROBLEM: For an efficient windmill, the power generated in watts is given by the equation

$$P = kr^2v^3$$

where r is the radius of the propeller in meters, v is the wind velocity in meters per second, and k is a constant with units of kg/m^3. The exact value of k depends on various characteristics of the windmill.

Suppose a windmill has a propeller with radius 5 meters and $k = 0.134$ kg/m^3.

(a) Find the function $P(v)$ which gives power as a function of wind velocity.

(b) Find $P'(7)$, the rate of change in power generated with respect to wind velocity, when the wind velocity is 7 meters per second.

SOLUTION:

(a) Since $r = 5$ m and $k = 0.134$ kg/m^2, we have

$$P = kr^2v^3 = (0.134)(5^2)v^3 = 3.35v^3$$

So, $P(v) = 3.35v^3$, where v is in meters per second and P is in watts.

(b) $P'(7) = \lim_{h \to 0} \dfrac{P(7 + h) - P(7)}{h}$

$\qquad = \lim_{h \to 0} \dfrac{3.35(7 + h)^3 - 3.35(7)^3}{h}$

$\qquad = \lim_{h \to 0} \dfrac{3.35[(7^3 + 147h + 21h^2 + h^3) - 7^3]}{h}$

$\qquad = \lim_{h \to 0} \dfrac{3.35(147h + 21h^2 + h^3)}{h}$

$\qquad = \lim_{h \to 0} [3.35(147 + 21h + h^2)]$

$\qquad = 3.35(147)$

$\qquad = 492.45$

The rate of change in power generated is about 492 watts per meter/sec.

QUICK REVIEW 10.1 *(For help, go to Sections P.1 and P.4.)*

In Exercises 1 and 2, find the slope of the line determined by the points.

1. $(-2, 3), (5, -1)$ **2.** $(-3, -1), (3, 3)$

In Exercises 3–6, write an equation for the specified line.

3. Through $(-2, 3)$ with slope $= 3/2$
4. Through $(1, 6)$ and $(4, -1)$
5. Through $(1, 4)$ and parallel to $y = (3/4)x + 2$
6. Through $(1, 4)$ and perpendicular to $y = (3/4)x + 2$

In Exercises 7–10, simplify the expression assuming $h \neq 0$.

7. $\dfrac{(2 + h)^2 - 4}{h}$

8. $\dfrac{(3 + h)^2 + 3 + h - 12}{h}$

9. $\dfrac{1/(2 + h) - 1/2}{h}$

10. $\dfrac{1/(x + h) - 1/x}{h}$

SECTION 10.1 EXERCISES

1. Average Velocity A bicyclist travels 21 miles in 1 hour and 45 minutes. What is her average velocity during the entire 1 3/4 hour time interval?

2. Average Velocity An automobile travels 540 kilometers in 4 hours and 30 minutes. What is its average velocity over the entire 4 1/2 hour time interval?

In Exercises 3–6, the position of an object at time t is given by $s(t)$. Find the instantaneous velocity at the indicated value of t.

3. $s(t) = 3t - 5$ at $t = 4$

4. $s(t) = \dfrac{2}{t + 1}$ at $t = 2$

5. $s(t) = at^2 + 5$ at $t = 2$

6. $s(t) = \sqrt{t + 1}$ at $t = 1$
[*Hint*: "rationalize the numerator."]

In Exercises 7–10, use the graph to estimate the slope of the tangent line, if it exists, to the graph at the given point.

7. $x = 0$

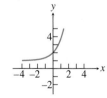

8. $x = 1$

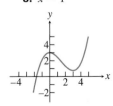

9. $x = 2$

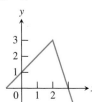

10. $x = 4$

In Exercises 11–14, graph the function in a square viewing window and, without doing any calculations, estimate the derivative of the function at the given point by interpreting it as the tangent line slope, if it exists at the point.

11. $f(x) = x^2 - 2x + 5$ at $x = 3$
12. $f(x) = \dfrac{1}{2}x^2 + 2x - 5$ at $x = 2$
13. $f(x) = x^3 - 6x^2 + 12x - 9$ at $x = 0$
14. $f(x) = 2 \sin x$ at $x = \pi$

15. A Rock Toss A rock is thrown straight up from level ground. The distance (in ft) the ball is above the ground (the position function) is $f(t) = 3 + 48t - 16t^2$ at any time t (in sec). Find

(a) $f'(0)$.

(b) the initial velocity of the rock.

16. Rocket Launch A toy rocket is launched straight up in the air from level ground. The distance (in ft) the rocket is above the ground (the position function) is $f(t) = 170t - 16t^2$ at any time t (in sec). Find

(a) $f'(0)$.

(b) the initial velocity of the rocket.

In Exercises 17–20, use the limit definition to find

(a) the slope of the graph of the function at the indicated point,

(b) an equation of the tangent line at the point.

(c) Sketch a graph of the curve near the point without using your graphing calculator.

17. $f(x) = 2x^2$ at $x = -1$

18. $f(x) = 2x - x^2$ at $x = 2$

19. $f(x) = 2x^2 - 7x + 3$ at $x = 2$

20. $f(x) = \dfrac{1}{x + 2}$ at $x = 1$

In Exercises 21 and 22, estimate the slope of the tangent line to the graph of the function, if it exists, at the indicated points.

21. $f(x) = |x|$ at $x = -2, 2,$ and 0.

22. $f(x) = \tan^{-1}(x + 1)$ at $x = -2, 2,$ and 0.

In Exercises 23–28, find the derivative, if it exists, of the function at the specified point.

23. $f(x) = 1 - x^2$ at $x = 2$

24. $f(x) = 2x + \dfrac{1}{2}x^2$ at $x = 2$

25. $f(x) = 3x^2 + 2$ at $x = -2$

26. $f(x) = x^2 - 3x + 1$ at $x = 1$

27. $f(x) = |x + 2|$ at $x = -2$

28. $f(x) = \dfrac{1}{x + 2}$ at $x = -1$

In Exercises 29–32, find the derivative of f.

29. $f(x) = 2 - 3x$ **30.** $f(x) = 2 - 3x^2$

31. $f(x) = 3x^2 + 2x - 1$ **32.** $f(x) = \dfrac{1}{x - 2}$

33. Average Speed A lead ball is held at water level and dropped from a boat into a lake. The distance the ball falls at 0.1 sec time intervals is given in Table 10.1.

Table 10.1 Distance Data of the Lead Ball

Time (sec)	Distance (ft)
0	0
0.1	0.1
0.2	0.4
0.3	0.8
0.4	1.5
0.5	2.3
0.6	3.2
0.7	4.4
0.8	5.8
0.9	7.3

(a) Compute the average speed from 0.5 to 0.6 seconds and from 0.8 to 0.9 seconds.

(b) Find a quadratic regression model for the distance data and overlay its graph on a scatter plot of the data.

(c) Use the model in part (b) to estimate the depth of the lake if the ball hits the bottom after 2 seconds.

34. Finding Derivatives from Data A ball is dropped from the roof of a two-story building. The distance in feet above ground of the falling ball is given in Table 10.2 where t is in seconds.

Table 10.2 Distance Data of the Ball

Time (sec)	Distance (ft)
0.2	30.00
0.4	28.36
0.6	25.44
0.8	21.24
1.0	15.76
1.2	9.02
1.4	0.95

(a) Use the data to estimate the average velocity of the ball in the interval $0.8 \le t \le 1$.

(b) Find a quadratic regression model s for the data in Table 10.2 and overlay its graph on a scatter plot of the data.

(c) Find the derivative of the regression equation and use it to estimate the velocity of the ball at time $t = 1$.

In Exercises 35–38, complete the following.

(a) Draw a graph of the function.

(b) Find the derivative of the function at the given point if it exists.

(c) Writing to Learn If the derivative does not exist at the point, explain why not.

35. $f(x) = \begin{cases} 4 - x & \text{if } x \le 2 \\ x + 3 & \text{if } x > 2 \end{cases}$ at $x = 2$

36. $f(x) = \begin{cases} 1 + (x - 2)^2 & \text{if } x \le 2 \\ 1 - (x - 2)^2 & \text{if } x > 2 \end{cases}$ at $x = 2$

37. $f(x) = \begin{cases} \dfrac{|x - 2|}{x - 2} & \text{if } x \ne 2 \\ 1 & \text{if } x = 2 \end{cases}$ at $x = 2$

38. $f(x) = \begin{cases} \dfrac{\sin x}{x} & \text{if } x \ne 0 \\ 1 & \text{if } x = 0 \end{cases}$ at $x = 0$

In Exercises 39–42, sketch a possible graph for a function that has the stated properties.

39. The domain of f is $[0, 5]$ and the derivative at $x = 2$ is 3.

40. The domain of f is $[0, 5]$ and the derivative is 0 at both $x = 2$ and $x = 4$.

41. The domain of f is $[0, 5]$ and the derivative at $x = 2$ is undefined.

42. The domain of f is $[0, 5]$, f is nondecreasing on $[0, 5]$, and the derivative at $x = 2$ is 0.

43. Writing to Learn Explain why you can find the derivative of $f(x) = ax + b$ without doing any computations. What is $f'(x)$?

44. Writing to Learn Use the *first* definition of derivative at a point to express the derivative of $f(x) = |x|$ at $x = 0$ as a limit. Then explain why the limit does not exist. (A graph of the quotient for x values near 0 might help.)

Standardized Test Questions

45. True or False When a ball rolls down a ramp, its instantaneous velocity is always zero. Justify your answer.

46. True or False If the derivative of the function f exists at $x = a$, then the derivative is equal to the slope of the tangent line at $x = a$. Justify your answer.

In Exercises 47–50, choose the correct answer. You may use a calculator.

47. Multiple Choice If $f(x) = x^2 + 3x - 4$, find $f'(x)$.

(A) $x^2 + 3$ (B) $x^2 - 4$ (C) $2x - 1$ (D) $2x + 3$
(E) $2x - 3$

48. Multiple Choice If $f(x) = 5x - 3x^2$, find $f'(x)$.

(A) $5 - 6x$ (B) $5 - 3x$ (C) $5x - 6$ (D) $10x - 3$
(E) $5x - 6x^2$

49. Multiple Choice If $f(x) = x^3$, find the derivative of f at $x = 2$.

(A) 3 (B) 6 (C) 12 (D) 18 (E) Does not exist

50. Multiple Choice If $f(x) = \frac{1}{x-3}$, find the derivative of f at $x = 1$.

(A) $-\frac{1}{4}$ (B) $\frac{1}{4}$ (C) $-\frac{1}{2}$ (D) $\frac{1}{2}$ (E) Does not exist

Explorations

Graph each function in Exercises 51–54 and then answer the following questions.

(a) Writing to Learn Does the function have a derivative at $x = 0$? Explain.

(b) Does the function appear to have a tangent line at $x = 0$? If so, what is an equation of the tangent line?

51. $f(x) = |x|$

52. $f(x) = |x^{1/3}|$

53. $f(x) = x^{1/3}$

54. $f(x) = \tan^{-1} x$

55. Free Fall A water balloon dropped from a window will fall a distance of $s = 16t^2$ feet during the first t seconds. Find the balloon's **(a)** average velocity during the first 3 seconds of falling and **(b)** instantaneous velocity at $t = 3$.

56. Free Fall on Another Planet It can be established by experimentation that heavy objects dropped from rest free fall near the surface of another planet according to the formula $y = gt^2$, where y is the distance in meters the object falls in t seconds after being dropped. An object falls from the top of a 125 m spaceship which landed on the surface. It hits the surface in 5 seconds.

(a) Find the value of g.

(b) Find the average speed for the fall of the object.

(c) With what speed did the object hit the surface?

Extending the Ideas

57. Graphing the Derivative The graph of $f(x) = x^2 e^{-x}$ is shown below. Use your knowledge of the geometric interpretation of the derivative to sketch a rough graph of the derivative $y = f'(x)$.

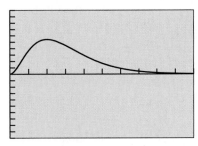

[0, 10] by [–1, 1]

58. Group Activity The graph of $y = f'(x)$ is shown below. Determine a possible graph for the function $y = f(x)$.

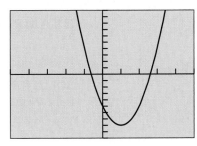

[–5, 5] by [–10, 10]

10.2
Limits and Motion: The Area Problem

Distance from a Constant Velocity

"Distance equals rate times time" is one of the earliest problem-solving formulas that we learn in school mathematics. Given a velocity and a time period, we can use the formula to compute distance traveled—as in the following standard example.

EXAMPLE 1   Computing Distance Traveled

An automobile travels at a constant rate of 48 miles per hour for 2 hours and 30 minutes. How far does the automobile travel?

SOLUTION We apply the formula $d = rt$:

$$d = (48 \text{ mi/hr})(2.5 \text{ hr}) = 120 \text{ miles}.$$

Now try Exercise 1.

The similarity to Example 1 in Section 10.1 is intentional. In fact, if we represent distance traveled (i.e., the change in position) by $\triangle s$ and the time interval by $\triangle t$, the formula becomes

$$\triangle s = (48 \text{ mph}) \, \triangle t,$$

which is equivalent to

$$\frac{\triangle s}{\triangle t} = 48 \text{ mph}.$$

So the two Example 1s are nearly identical—except that Example 1 of Section 10.1 did not make an assumption about constant velocity. What we computed in that instance was the average velocity over the 2.5-hour interval. This suggests that we could have actually solved the following, slightly different, problem to open this section.

EXAMPLE 2   Computing Distance Traveled

An automobile travels at an *average* rate of 48 miles per hour for 2 hours and 30 minutes. How far does the automobile travel?

SOLUTION The distance traveled is $\triangle s$, the time interval has length $\triangle t$, and $\triangle s/\triangle t$ is the average velocity.

Therefore,

$$\triangle s = \frac{\triangle s}{\triangle t} \cdot \triangle t = (48 \text{ mph})(2.5 \text{ hr}) = 120 \text{ miles}.$$

Now try Exercise 5.

So, given average velocity over a time interval, we can easily find distance traveled. But suppose we have a velocity function $v(t)$ that gives instantaneous velocity as a changing function of time. How can we use the instantaneous velocity function to find distance traveled over a time interval? This was the other intriguing problem about instantaneous velocity that puzzled the 17th-century scientists—and once again, algebra was inadequate for solving it, as we shall see.

Distance from a Changing Velocity

When Galileo began his experiments, here's what he might have asked himself about using a changing velocity to find distance:

A Distance Question

Suppose a ball rolls down a ramp and its velocity is always $2t$ feet per second, where t is the number of seconds after it started to roll. How far does the ball travel during the first 3 seconds?

One might be tempted to offer the following "solution":

Velocity times $\triangle t$ gives $\triangle s$. But instantaneous velocity occurs at an instant of time, so $\triangle t = 0$. That means $\triangle s = 0$. So, at any given instant of time, the ball doesn't move. Since any time interval consists of instants of time, the ball never moves at all! (You might well ask: Is this another trick question?)

As was the case with the Velocity Question in Section 10.1, this foolish-looking example conceals a very subtle algebraic dilemma—and, far from being a trick question, it is exactly the question that needed to be answered in order to compute the distance traveled by an object whose velocity varies as a function of time. The scientists who were working on the tangent line problem realized that the distance-traveled problem must be related to it, but, surprisingly, their geometry led them in another direction. The distance traveled problem led them not to tangent lines, but to areas.

Limits at Infinity

Before we see the connection to areas, let us revisit another limit concept that will make instantaneous velocity easier to handle, just as in the last section. We will again be content with an informal definition.

DEFINITION (INFORMAL) Limit at Infinity

When we write "$\lim_{x \to \infty} f(x) = L$," we mean that $f(x)$ gets arbitrarily close to L as x gets arbitrarily large.

EXPLORATION 1 An Infinite Limit

A gallon of water is divided equally and poured into teacups. Find the amount in each teacup and the *total amount* in *all* the teacups if there are

1. 10 teacups

2. 100 teacups

3. 1 billion teacups

4. an infinite number of teacups

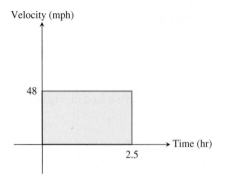

FIGURE 10.4 For constant velocity, the area of the rectangle is the same as the distance traveled, since it represents the product of the same two quantities: (48 mph)(2.5 hr) = 120 miles.

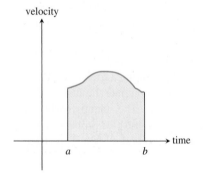

FIGURE 10.5 If the velocity varies over the time interval $[a, b]$, does the shaded region give the distance traveled?

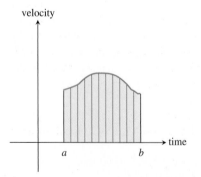

FIGURE 10.6 The region is partitioned into vertical strips. If the strips are narrow enough, they are almost indistinguishable from rectangles. The sum of the areas of these "rectangles" will give the total area and can be interpreted as distance traveled.

The preceding Exploration probably went pretty smoothly until you came to the infinite number of teacups. At that point you were probably pretty comfortable in saying what the *total amount* would be, and probably a little uncomfortable in saying how much would be in each teacup. (Theoretically it would be zero, which is just one reason why the actual experiment cannot be performed.) In the language of limits, the total amount of water in the infinite number of teacups would look like this:

$$\lim_{n\to\infty}\left(n\cdot\frac{1}{n}\right)=\lim_{n\to\infty}\frac{n}{n}=1\text{ gallon}$$

while the total amount in each teacup would look like this:

$$\lim_{n\to\infty}\frac{1}{n}=0\text{ gallons.}$$

Summing up an infinite number of nothings to get something is mysterious enough when we use limits; *without* limits it seems to be an algebraic impossibility. That is the dilemma that faced the 17th-century scientists who were trying to work with instantaneous velocity. Once again, it was geometry that showed the way when the algebra failed.

The Connection to Areas

If we graph the constant velocity $v = 48$ in Example 1 as a function of time t, we notice that the area of the shaded rectangle is the same as the distance traveled (Figure 10.4). This is no mere coincidence, either, as the area of the rectangle and the distance traveled over the time interval are both computed by multiplying the same two quantities:

$$(48\text{ mph})(2.5\text{ hr}) = 120\text{ miles.}$$

Now suppose we graph a velocity function that varies continuously as a function of time (Figure 10.5). Would the area of this irregularly-shaped region still give the total distance traveled over the time interval $[a, b]$?

Newton and Leibniz (and, actually, many others who had considered this question) were convinced that it obviously would, and that is why they were interested in a calculus for finding areas under curves. They imagined the time interval being partitioned into many tiny subintervals, each one so small that the velocity over it would essentially be constant. Geometrically, this was equivalent to slicing the area into narrow strips, each one of which would be nearly indistinguishable from a narrow rectangle (Figure 10.6).

The idea of partitioning irregularly-shaped areas into approximating rectangles was not new. Indeed, Archimedes had used that very method to approximate the area of a circle with remarkable accuracy. However, it was an exercise in patience and perseverance, as Example 3 will show.

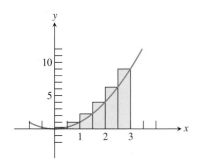

FIGURE 10.7 The area under the graph of $f(x) = x^2$ is approximated by six rectangles, each with base 1/2. The height of each rectangle is the function value at the right-hand endpoint of the subinterval (Example 3).

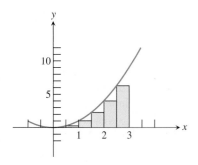

FIGURE 10.8 If we change the rectangles in Figure 10.7 so that their heights are determined by function values at the *left-hand* endpoints, we get an area approximation (6.875 square units) that underestimates the true area.

EXAMPLE 3 Approximating an Area with Rectangles

Use the six rectangles in Figure 10.7 to approximate the area of the region below the graph of $f(x) = x^2$ over the interval $[0, 3]$.

SOLUTION The base of each approximating rectangle is 1/2. The height is determined by the function value at the right-hand endpoint of each subinterval. The areas of the six rectangles and the total area are computed in the table below:

Subinterval	Base of rectangle	Height of rectangle	Area of rectangle
$[0, 1/2]$	1/2	$f(1/2) = (1/2)^2 = 1/4$	$(1/2)(1/4) = 0.125$
$[1/2, 1]$	1/2	$f(1) = (1)^2 = 1$	$(1/2)(1) = 0.500$
$[1, 3/2]$	1/2	$f(3/2) = (3/2)^2 = 9/4$	$(1/2)(9/4) = 1.125$
$[3/2, 2]$	1/2	$f(2) = (2)^2 = 4$	$(1/2)(4) = 2.000$
$[2, 5/2]$	1/2	$f(5/2) = (5/2)^2 = 25/4$	$(1/2)(25/4) = 3.125$
$[5/2, 3]$	1/2	$f(3) = (3)^2 = 9$	$(1/2)(9) = 4.500$
		Total Area:	11.375

The six rectangles give a (rather crude) approximation of 11.375 square units for the area under the curve from 0 to 3. ***Now try Exercise 11.***

Figure 10.7 shows that the *right rectangular approximation method* (RRAM) in Example 4 overestimates the true area. If we were to use the function values at the left-hand endpoints of the subintervals (LRAM), we would obtain a rectangular approximation (6.875 square units) that underestimates the true area (Figure 10.8). The average of the two approximations is 9.125 square units, which is actually a pretty good estimate of the true area of 9 square units. If we were to repeat the process with 20 rectangles, the average would be 9.01125. This method of converging toward an unknown area by refining approximations is tedious, but it works—Archimedes used a variation of it 2200 years ago to estimate the area of a circle, and in the process demonstrated that the ratio of the circumference to the diameter was between 3.140845 and 3.142857.

The calculus step is to move from a finite number of rectangles (yielding an approximate area) to an infinite number of rectangles (yielding an exact area). This brings us to the definite integral.

A sum of the form $\sum_{i=1}^{n} f(x_i)\triangle x$ in which x_1 is in the first subinterval, x_2 is in the second, and so on, is called a **Riemann sum**, in honor of Georg Riemann (1826–1866), who determined the functions for which such sums had limits as $n\rightarrow\infty$.

The Definite Integral

In general, begin with a continuous function $y = f(x)$ over an interval $[a, b]$. Divide $[a, b]$ into n subintervals of length $\triangle x = (b - a)/n$. Choose any value x_1 in the first subinterval, x_2 in the second, and so on. Compute $f(x_1), f(x_2), f(x_3), \ldots, f(x_n)$, multiply each value by $\triangle x$, and sum up the products. In sigma notation, the sum of the products is

$$\sum_{i=1}^{n} f(x_i)\triangle x.$$

The *limit* of this sum as n approaches infinity is the solution to the area problem, and hence the solution to the problem of distance traveled. Indeed, it solves a variety of other problems as well, as you will learn when you study calculus. The limit, if it exists, is called a *definite integral*.

Notice that the notation for the definite integral (another legacy of Leibniz) parallels the sigma notation of the sum for which it is a limit. The "Σ" in the limit becomes a stylized "S," for "sum." The "$\triangle x$" becomes "dx" (as it did in the derivative), and the "$f(x_i)$" becomes simply "$f(x)$" because we are effectively summing up *all* the $f(x)$ values along the interval (times an arbitrarily small change in x), rendering the subscripts unnecessary.

DEFINITION Definite Integral

Let f be a function defined on $[a, b]$ and let $\sum_{i=1}^{n} f(x_i)\triangle x$ be defined as above. The definite integral of f over $[a, b]$, denoted $\int_{a}^{b} f(x)\ dx$, is given by

$$\int_{a}^{b} f(x)dx = \lim_{n\to\infty} \sum_{i=1}^{n} f(x_i)\triangle x,$$

provided the limit exists. If the limit exists, we say f is **integrable** on $[a, b]$.

The solution to Example 4 shows that it can be tedious to approximate a definite integral by working out the sum for a large value of n. One of the crowning achievements of calculus was to demonstrate how the exact value of a definite integral could be obtained without summing up any products at all. You will have to wait until calculus to see how that is done; meanwhile, you will learn in Section 10.4 how to use a calculator to take the tedium out of finding definite integrals by summing.

You can also use the area connection to your advantage, as shown in these next two examples.

EXAMPLE 4   Computing an Integral

Find $\int_{1}^{5} 2x\ dx$.

SOLUTION This will be the area under the line $y = 2x$ over the interval $[1, 5]$. The graph in Figure 10.9 shows that this is the area of a trapezoid.

Using the formula $A = h\left(\dfrac{b_1 + b_2}{2}\right)$, we find that

$$\int_{1}^{5} 2x\ dx = 4\left(\frac{2(1) + 2(5)}{2}\right) = 24$$

Now try Exercise 23.

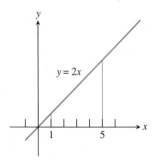

FIGURE 10.9 The area of the trapezoid equals $\int_{1}^{5} 2x\ dx$. (Example 4)

EXAMPLE 5 Computing an Integral

Suppose a ball rolls down a ramp so that its velocity after t seconds is always $2t$ feet per second. How far does it fall during the first 3 seconds?

SOLUTION The distance traveled will be the same as the area under the velocity graph, $v(t) = 2t$, over the interval $[0, 3]$. The graph is shown in Figure 10.10. Since the region is triangular, we can find its area: $A = (1/2)(3)(6) = 9$. The distance traveled in the first 3 seconds, therefore, is $\triangle s = (1/2)(3 \text{ sec})(6 \text{ feet/sec}) = 9$ feet.

Now try Exercise 45.

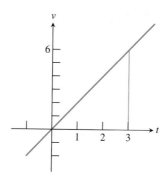

FIGURE 10.10 The area under the velocity graph $v(t) = 2t$, over the interval $[0, 3]$ is the distance traveled by the ball in Example 5 during the first 3 seconds.

QUICK REVIEW 10.2 (For help, go to Sections 1.1 and 9.4.)

In Exercises 1 and 2, list the elements of the sequence.

1. $a_k = \dfrac{1}{2}\left(\dfrac{1}{2}k\right)^2$ for $k = 1, 2, 3, 4, \dots, 9, 10$

2. $a_k = \dfrac{1}{4}\left(2 + \dfrac{1}{4}k\right)^2$ for $k = 1, 2, 3, 4, \dots, 9, 10$

In Exercises 3–6, find the sum.

3. $\displaystyle\sum_{k=1}^{10} \dfrac{1}{2}(k + 1)$ 　　　 **4.** $\displaystyle\sum_{k=1}^{n} (k + 1)$

5. $\displaystyle\sum_{k=1}^{10} \dfrac{1}{2}(k + 1)^2$ 　　　 **6.** $\displaystyle\sum_{k=1}^{n} \dfrac{1}{2}k^2$

7. A truck travels at an average speed of 57 mph for 4 hours. How far does it travel?

8. A pump working at 5 gal/min pumps for 2 hours. How many gallons are pumped?

9. Water flows over a spillway at a steady rate of 200 cubic feet per second. How many cubic feet of water pass over the spillway in 6 hours?

10. A county has a population density of 560 people per square mile in an area of 35,000 square miles. What is the population of the county?

SECTION 10.2 EXERCISES

In Exercises 1–4, explain how to represent the problem situation as an area question and then solve the problem.

1. A train travels at 65 mph for 3 hours. How far does it travel?

2. A pump working at 15 gal/min pumps for one-half hour. How many gallons are pumped?

3. Water flows over a spillway at a steady rate of 150 cubic feet per second. How many cubic feet of water pass over the spillway in one hour?

4. A city has a population density of 650 people per square mile in an area of 20 square miles. What is the population of the city?

5. An airplane travels at an average velocity of 640 kilometers per hour for 3 hours and 24 minutes. How far does the airplane travel?

6. A train travels at an average velocity of 24 miles per hour for 4 hours and 50 minutes. How far does the train travel?

In Exercises 7–10, estimate the area of the region above the x-axis and under the graph of the function from $x = 0$ to $x = 5$.

7.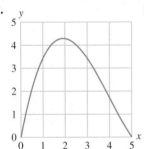

8.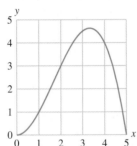

9.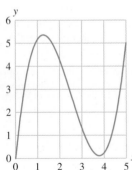

10.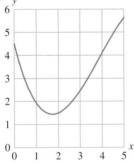

In Exercises 11 and 12, use the 8 rectangles shown to approximate the area of the region below the graph of $f(x) = 10 - x^2$ over the interval $[-1, 3]$.

11. 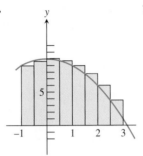   **12.**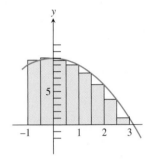

In Exercises 13–16, partition the given interval into the indicated number of subintervals.

13. $[0, 2]$; 4   **14.** $[0, 2]$; 8

15. $[1, 4]$; 6   **16.** $[1, 5]$; 8

In Exercises 17–20, complete the following.

(a) Draw the graph of the function for x in the specified interval. Verify that the function is nonnegative in that interval.

(b) On the graph in part (a), draw and shade the approximating rectangles for the RRAM using the specified partition. Compute the RRAM area estimate without using a calculator.

(c) Repeat part (b) using the LRAM.

(d) Average the RRAM and LRAM approximations from parts (b) and (c) to find an average estimate of the area.

17. $f(x) = x^2$; $[0, 4]$; 4 subintervals

18. $f(x) = x^2 + 2$; $[0, 6]$; 6 subintervals

19. $f(x) = 4x - x^2$; $[0, 4]$; 4 subintervals

20. $f(x) = x^3$; $[0, 3]$; 3 subintervals

In Exercises 21–28, find the definite integral by computing an area. (It may help to look at a graph of the function.)

21. $\int_3^7 5\, dx$ **22.** $\int_{-1}^4 6\, dx$

23. $\int_0^5 3x\, dx$ **24.** $\int_1^7 0.5x\, dx$

25. $\int_1^4 (x + 3)\, dx$ **26.** $\int_1^4 (3x - 2)\, dx$

27. $\int_{-2}^2 \sqrt{4 - x^2}\, dx$ **28.** $\int_0^6 \sqrt{36 - x^2}\, dx$

It can be shown that the area enclosed between the *x*-axis and one arch of the sine curve is 2. Use this fact in Exercises 29–38 to compute the definite integral. (It may help to look at a graph of the function.)

29. $\int_0^\pi \sin x \, dx$

30. $\int_0^\pi (\sin x + 2) \, dx$

31. $\int_2^{\pi+2} \sin (x - 2) \, dx$

32. $\int_{-\pi/2}^{\pi/2} \cos x \, dx$

33. $\int_0^{\pi/2} \sin x \, dx$

34. $\int_0^{\pi/2} \cos x \, dx$

35. $\int_0^\pi 2 \sin x \, dx$ [*Hint*: All the rectangles are twice as tall.]

36. $\int_0^{2\pi} \sin \left(\dfrac{x}{2}\right) dx$ [*Hint*: All the rectangles are twice as wide.]

37. $\int_0^{2\pi} |\sin x| \, dx$

38. $\int_{-\pi}^{3\pi/2} |\cos x| \, dx$

In Exercises 39–42, find the integral assuming that *k* is a number between 0 and 4.

39. $\int_0^4 (kx + 3) \, dx$

40. $\int_0^k (4x + 3) \, dx$

41. $\int_0^4 (3x + k) \, dx$

42. $\int_k^4 (4x + 3) \, dx$

43. Writing to Learn Let $g(x) = -f(x)$ where f has nonnegative function values on an interval $[a, b]$. Explain why the area above the graph of g is the same as the area under the graph of f in the same interval.

44. Writing to Learn Explain how you can find the area under the graph of $f(x) = \sqrt{16 - x^2}$ from $x = 0$ to $x = 4$ by mental computation only.

45. Falling Ball Suppose a ball is dropped from a tower and its velocity after *t* seconds is always 32*t* feet per second. How far does the ball fall during the first 2 seconds?

46. Accelerating Automobile Suppose an automobile accelerates so that its velocity after *t* seconds is always 6*t* feet per second. How far does the car travel in the first 7 seconds?

47. Rock Toss A rock is thrown straight up from level ground. The velocity of the rock at any time *t* (sec) is $v(t) = 48 - 32t$ ft/sec.

(a) Graph the velocity function.

(b) At what time does the rock reach its maximum height?

(c) Find how far the rock has traveled at its maximum height.

48. Rocket Launch A toy rocket is launched straight up from level ground. Its velocity function is $f(t) = 170 - 32t$ feet per second, where *t* is the number of seconds after launch.

(a) Graph the velocity function.

(b) At what time does the rocket reach its maximum height?

(c) Find how far the rocket has traveled at its maximum height.

49. Finding Distance Traveled as Area A ball is pushed off the roof of a three-story building. Table 10.3 gives the velocity (in feet per second) of the falling ball at 0.2-second intervals until it hits the ground 1.4 seconds later.

Table 10.3 Velocity Data of the Ball	
Time	Velocity
0.2	−5.05
0.4	−11.43
0.6	−17.46
0.8	−24.21
1.0	−30.62
1.2	−37.06
1.4	−43.47

(a) Draw a scatter plot of the data.

(b) Find the approximate building height using RRAM areas as in Example 4. Use the fact that if the velocity function is always negative the distance traveled will be the same as if the absolute value of the velocity values were used.

50. Work Work is defined as force times distance. A full water barrel weighing 1250 pounds has a significant leak and must be lifted 35 feet. Table 10.4 displays the weight of the barrel measured after each 5 feet of movement. Find the approximate work in foot-pounds done in lifting the barrel 35 feet.

Table 10.4 Weight of a Leaking Water Barrel	
Distance (ft)	Weight (lb)
0	1250
5	1150
10	1050
15	950
20	850
25	750
30	650

Standardized Test Questions

51. True or False When estimating the area under a curve using LRAM, the accuracy typically improves as the number *n* of subintervals is increased.

52. True or False The statement $\lim\limits_{x \to \infty} f(x) = L$ means that $f(x)$ gets arbitrarily large as *x* gets arbitrarily close to *L*.

It can be shown that the area of the region enclosed by the curve $y = \sqrt{x}$, the x-axis, and the line $x = 9$ is 18. Use this fact in Exercises 53–56 to choose the correct answer. Do not use a calculator.

53. Multiple Choice $\displaystyle\int_0^9 2\sqrt{x}\, dx$

(A) 36   (B) 27   (C) 18   (D) 9   (E) 6

54. Multiple Choice $\displaystyle\int_0^9 \left(\sqrt{x} + 5\right) dx$

(A) 14   (B) 23   (C) 33   (D) 45   (E) 63

55. Multiple Choice $\displaystyle\int_5^{14} \left(\sqrt{x - 5}\right) dx$

(A) 9   (B) 13   (C) 18   (D) 23   (E) 28

56. Multiple Choice $\displaystyle\int_0^3 \sqrt{3x}\, dx$

(A) 54   (B) 18   (C) 9   (D) 6   (E) 3

Explorations

57. Group Activity You may have erroneously assumed that the function f had to be positive in the definition of the definite integral. It is a fact that $\displaystyle\int_0^{2\pi} \sin x\, dx = 0$. Use the definition of the definite integral to explain why this is so. What does this imply about $\displaystyle\int_0^1 (x - 1)\, dx$?

58. Area Under a Discontinuous Function Let

$$f(x) = \begin{cases} 1 & \text{if } x < 2 \\ x & \text{if } x > 2 \end{cases}.$$

(a) Draw a graph of f. Determine its domain and range.

(b) **Writing to Learn** How would you define the area under f from $x = 0$ to $x = 4$? Does it make a difference if the function has no value at $x = 2$?

Extending the Ideas

Group Activity From what you know about definite integrals, decide whether each of the following statements is true or false for integrable functions (in general). Work with your classmates to justify your answers.

59. $\displaystyle\int_a^b f(x)\, dx + \int_a^b g(x)\, dx = \int_a^b (f(x) + g(x))\, dx$

60. $\displaystyle\int_a^b 8 \cdot f(x)\, dx = 8 \cdot \int_a^b f(x)\, dx$

61. $\displaystyle\int_a^b f(x) \cdot g(x)\, dx = \int_a^b f(x)\, dx \cdot \int_a^b g(x)\, dx$

62. $\displaystyle\int_a^c f(x)\, dx + \int_c^b f(x)\, dx = \int_a^b f(x)\, dx$ for $a < c < b$

63. $\displaystyle\int_a^b f(x) = \int_b^a f(x)$

64. $\displaystyle\int_a^a f(x)\, dx = 0$

10.3
More on Limits

A Little History

Progress in mathematics occurs gradually and without much fanfare in the early stages. The fanfare occurs much later, after the discoveries and innovations have been cleaned up and put into perspective. Calculus is certainly a case in point. Most of the ideas in this chapter pre-date Newton and Leibniz. Others were solving calculus problems as far back as Archimedes of Syracuse (ca. 287–212 B.C.), long before calculus was "discovered." What Newton and Leibniz did was to develop the rules of the game, so that derivatives and integrals could be computed algebraically. Most importantly, they developed what has come to be called the Fundamental Theorem of Calculus, which explains the connection between the "tangent line problem" and the "area problem."

But the methods of Newton and Leibniz depended on mysterious "infinitesimal" quantities that were small enough to vanish and yet were not zero. Jean Le Rond d'Alembert (1717–1783) was a strong proponent of replacing infinitesimals with limits (the strategy that would eventually work), but these concepts were not well understood until Karl Weierstrass (1815–1897) and his student Eduard Heine (1821–1881) introduced the formal, unassailable definitions that are used in our higher mathematics courses today. By that time, Newton and Leibniz had been dead for over 150 years.

Defining a Limit Informally

There is nothing difficult about the following limit statements:

$$\lim_{x \to 3} (2x - 1) = 5 \qquad \lim_{x \to \infty} (x^2 + 3) = \infty \qquad \lim_{n \to \infty} \frac{1}{n} = 0$$

That is why we have used limit notation throughout this book. Particularly when electronic graphers are available, analyzing the limiting behavior of functions algebraically, numerically, and graphically can tell us much of what we need to know about the functions.

What *is* difficult is to come up with an air-tight definition of what a limit really is. If it had been easy, it would not have taken 150 years. The subtleties of the "epsilon-delta" definition of Weierstrass and Heine are as beautiful as they are profound, but they are not the stuff of a precalculus course. Therefore, even as we look more closely at limits and their properties in this section, we will continue to refer to our "informal" definition of limit (essentially that of d'Alembert). We repeat it here for ready reference:

DEFINITION (INFORMAL) Limit at *a*

When we write "$\lim_{x \to a} f(x) = L$," we mean that $f(x)$ gets arbitrarily close to L as x gets arbitrarily close (but not equal) to a.

EXPLORATION 1 What's the Limit?

As a class, discuss the following two limit statements until you really understand why they are true. Look at them every way you can. Use your calculators. Do you see how the above definition verifies that they are true? In particular, can you defend your position against the challenges that follow the statements? (This exploration is intended to be free-wheeling and philosophical. You can't *prove* these statements without a stronger definition.)

1. $\lim\limits_{x \to 2} 7x \neq 14.000000000000000001$

Challenges:

- Isn't $7x$ getting "arbitrarily close" to that number as x approaches 2?
- How can you tell that 14 is the limit and 14.000000000000000001 is not?

2. $\lim\limits_{x \to 0} \dfrac{x^2 + 2x}{x} = 2$

Challenges:

- How can the limit be 2 when the quotient isn't even defined at 0?
- Won't there be an asymptote at $x = 0$? The denominator equals 0 there.
- How can you *tell* that 2 is the limit and 1.99999999999999999999 is not?

EXAMPLE 1 Finding a Limit

Find $\lim\limits_{x \to 1} \dfrac{x^3 - 1}{x - 1}$.

Solve Graphically

The graph in Figure 10.11a suggests that the limit exists and is about 3.

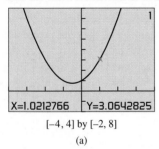

X=1.0212766 Y=3.0642825

[−4, 4] by [−2, 8]

(a)

FIGURE 10.11a A graph of $f(x) = (x^3 - 1)/(x - 1)$.

Solve Numerically

The table also gives compelling evidence that the limit is 3.

X	Y1	
.997	2.991	
.998	2.994	
.999	2.997	
1	ERROR	
1.001	3.003	
1.002	3.006	
1.003	3.009	

Y1 ☰ (X³−1)/(X−1)

(b)

FIGURE 10.11b A table of values for $f(x) = (x^3 - 1)/(x - 1)$.

Solve Algebraically

$$\lim_{x \to 1} \frac{x^3 - 1}{x - 1}$$

$$= \lim_{x \to 1} \frac{(x - 1)(x^2 + x + 1)}{x - 1}$$

$$= \lim_{x \to 1} (x^2 + x + 1)$$

$$= 1 + 1 + 1$$

$$= 3$$

As convincing as the graphical and numerical evidence is, the best evidence is algebraic. The limit is 3.

Now try Exercise 11.

Properties of Limits

When limits exist, there is nothing unusual about the way they interact algebraically with each other. You could easily predict that the following properties would hold. These are all theorems that one could prove with a rigorous definition of limit, but we must state them without proof here.

Properties of Limits

If $\lim\limits_{x \to c} f(x)$ and $\lim\limits_{x \to c} g(x)$ both exist, then

1. **Sum Rule** $\qquad\qquad \lim\limits_{x \to c} (f(x) + g(x)) = \lim\limits_{x \to c} f(x) + \lim\limits_{x \to c} g(x)$

2. **Difference Rule** $\qquad \lim\limits_{x \to c} (f(x) - g(x)) = \lim\limits_{x \to c} f(x) - \lim\limits_{x \to c} g(x)$

3. **Product Rule** $\qquad\quad \lim\limits_{x \to c} (f(x) \cdot g(x)) = \lim\limits_{x \to c} f(x) \cdot \lim\limits_{x \to c} g(x)$

4. **Constant Multiple Rule** $\quad \lim\limits_{x \to c} (k \cdot g(x)) = k \cdot \lim\limits_{x \to c} g(x)$

5. **Quotient Rule** $\qquad\quad \lim\limits_{x \to c} \dfrac{f(x)}{g(x)} = \dfrac{\lim\limits_{x \to c} f(x)}{\lim\limits_{x \to c} g(x)},$

   provided $\lim\limits_{x \to c} g(x) \ne 0$

6. **Power Rule** $\qquad\qquad \lim\limits_{x \to c} (f(x))^n = (\lim\limits_{x \to c} f(x))^n$ for n

   a positive integer

7. **Root Rule** $\qquad\qquad \lim\limits_{x \to c} \sqrt[n]{f(x)} = \sqrt[n]{\lim\limits_{x \to c} f(x)}$ for $n \ge 2$

   a positive integer, provided $\sqrt[n]{\lim\limits_{x \to c} f(x)}$

   and $\lim\limits_{x \to c} \sqrt[n]{f(x)}$ are real numbers.

EXAMPLE 2 Using the Limit Properties

You will learn in Example 10 that $\lim\limits_{x \to 0} \dfrac{\sin x}{x} = 1$. Use this fact, along with the limit properties, to find the following limits:

(a) $\lim\limits_{x \to 0} \dfrac{x + \sin x}{x}$ $\qquad$ **(b)** $\lim\limits_{x \to 0} \dfrac{1 - \cos^2 x}{x^2}$ $\qquad$ **(c)** $\lim\limits_{x \to 0} \dfrac{\sqrt[3]{\sin x}}{\sqrt[3]{x}}$

SOLUTION

(a) $\lim\limits_{x \to 0} \dfrac{x + \sin x}{x} = \lim\limits_{x \to 0} \left(\dfrac{x}{x} + \dfrac{\sin x}{x} \right)$

$\qquad\qquad\qquad = \lim\limits_{x \to 0} \dfrac{x}{x} + \lim\limits_{x \to 0} \dfrac{\sin x}{x}$ $\qquad$ Sum Rule

$\qquad\qquad\qquad = 1 + 1$

$\qquad\qquad\qquad = 2$

continued

(b) $\lim\limits_{x \to 0} \dfrac{1 - \cos^2 x}{x^2} = \lim\limits_{x \to 0} \dfrac{\sin^2 x}{x^2}$ Pythagorean identity

$$= \lim\limits_{x \to 0} \left(\dfrac{\sin x}{x}\right)\left(\dfrac{\sin x}{x}\right)$$

$$= \lim\limits_{x \to 0} \left(\dfrac{\sin x}{x}\right) \cdot \lim\limits_{x \to 0} \left(\dfrac{\sin x}{x}\right)$$ Product Rule

$$= 1 \cdot 1$$

$$= 1$$

(c) $\lim\limits_{x \to 0} \dfrac{\sqrt[3]{\sin x}}{\sqrt[3]{x}} = \lim\limits_{x \to 0} \sqrt[3]{\dfrac{\sin x}{x}}$

$$= \sqrt[3]{\lim\limits_{x \to 0} \dfrac{\sin x}{x}}$$ Root Rule

$$= \sqrt[3]{1}$$

$$= 1$$

Now try Exercise 19.

Limits of Continuous Functions

Recall from Section 1.2 that a function is continuous at a if $\lim\limits_{x \to a} f(x) = f(a)$. This means that the limit (at a) of a function can be found by "plugging in a" provided the function is continuous at a. (The condition of continuity is essential when employing this strategy. For example, plugging in 0 does not work on any of the limits in Example 2.)

EXAMPLE 3   Finding Limits by Substitution

Find the limits.

(a) $\lim\limits_{x \to 0} \dfrac{e^x - \tan x}{\cos^2 x}$ **(b)** $\lim\limits_{n \to 16} \dfrac{\sqrt{n}}{\log_2 n}$

SOLUTION

You might not recognize these functions as being continuous, but you can use the limit properties to write the limits in terms of limits of basic functions.

(a) $\lim\limits_{x \to 0} \dfrac{e^x - \tan x}{\cos^2 x} = \dfrac{\lim\limits_{x \to 0} (e^x - \tan x)}{\lim\limits_{x \to 0} (\cos^2 x)}$ Quotient Rule

$$= \dfrac{\lim\limits_{x \to 0} e^x - \lim\limits_{x \to 0} \tan x}{(\lim\limits_{x \to 0} \cos x)^2}$$ Difference and Power Rules

$$= \dfrac{e^0 - \tan 0}{(\cos 0)^2}$$ Limits of continuous functions

$$= \dfrac{1 - 0}{1}$$

$$= 1$$

continued

(b) $\displaystyle\lim_{n\to16}\frac{\sqrt{n}}{\log_2 n} = \frac{\displaystyle\lim_{n\to16}\sqrt{n}}{\displaystyle\lim_{n\to16}\log_2 n}$ $\qquad$ Quotient Rule

$\qquad\qquad\qquad = \dfrac{\sqrt{16}}{\log_2 16}$ $\qquad$ Limits of continuous functions

$\qquad\qquad\qquad = \dfrac{4}{4}$

$\qquad\qquad\qquad = 1$

Now try Exercise 23.

Example 3 hints at some important properties of continuous functions that follow from the properties of limits. If f and g are both continuous at $x = a$, then so are $f + g$, $f - g$, fg, and f/g (with the assumption that $g(a)$ does not create a zero denominator in the quotient). Also, the nth power and nth root of a function that is continuous at a will also be continuous at a (with the assumption that $\sqrt{f(a)}$ is real).

One-sided and Two-sided Limits

We can see that the limit of the function in Figure 10.11 is 3 whether x approaches 1 from the left or right. Sometimes the values of a function f can approach different values as x approaches a number c from opposite sides. When this happens, the limit of f as x approaches c from the left is the **left-hand limit** of f at c and the limit of f as x approaches c from the right is the **right-hand limit** of f at c. Here is the notation we use:

left-hand: $\qquad \displaystyle\lim_{x\to c^-} f(x)$ $\qquad$ *The limit of f as x approaches c from the left.*

right-hand: $\qquad \displaystyle\lim_{x\to c^+} f(x)$ $\qquad$ *The limit of f as x approaches c from the right.*

EXAMPLE 4   Finding Left- and Right-Hand Limits

Find $\displaystyle\lim_{x\to2^-} f(x)$ and $\displaystyle\lim_{x\to2^+} f(x)$ where $f(x) = \begin{cases} -x^2 + 4x - 1 & \text{if } x \le 2 \\ 2x - 3 & \text{if } x > 2 \end{cases}$.

SOLUTION Figure 10.12 suggests that the left- and right-hand limits of f exist but are not equal. Using algebra we find:

$$\lim_{x\to2^-} f(x) = \lim_{x\to2^-} (-x^2 + 4x - 1) \qquad \text{Definition of } f$$

$$= -2^2 + 4\cdot 2 - 1$$

$$= 3$$

$$\lim_{x\to2^+} f(x) = \lim_{x\to2^+} (2x - 3) \qquad \text{Definition of } f$$

$$= 2\cdot 2 - 3$$

$$= 1$$

You can use trace or tables to support the above results.

Now try Exercise 27, parts (a) and (b).

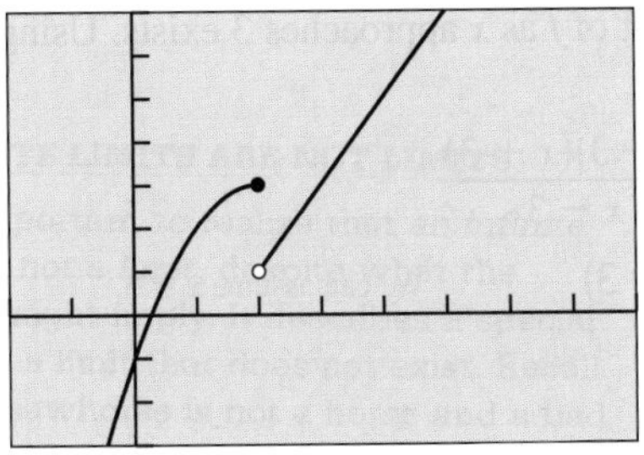

[−2, 8] by [−3, 7]

FIGURE 10.12 A graph of the piecewise-defined function.

$f(x) = \begin{cases} -x^2 + 4x - 1 & x \le 2 \\ 2x - 3 & x > 2 \end{cases}$

(Example 4)

QUICK REVIEW 10.3   *(For help, go to Sections 1.2 and 1.3.)*

In Exercises 1 and 2, find **(a)** $f(-2)$, **(b)** $f(0)$, and **(c)** $f(2)$.

1. $f(x) = \dfrac{2x + 1}{(2x - 4)^2}$ **2.** $f(x) = \dfrac{\sin x}{x}$

In Exercises 3 and 4, find the **(a)** vertical asymptotes and **(b)** horizontal asymptotes of the graph of f, if any.

3. $f(x) = \dfrac{2x^2 + 3}{x^2 - 4}$ **4.** $f(x) = \dfrac{x^3 + 1}{2 - x - x^2}$

In Exercises 5 and 6, the end behavior asymptote of the function f is one of the following. Which one is it?

(a) $y = 2x^2$ **(b)** $y = -2x^2$ **(c)** $y = x^3$ **(d)** $y = -x^3$

5. $f(x) = \dfrac{2x^3 - 3x^2 + 1}{3 - x}$ **6.** $f(x) = \dfrac{x^4 + 2x^2 + x + 1}{x - 3}$

In Exercises 7 and 8, find **(a)** the points of continuity and **(b)** the points of discontinuity of the function.

7. $f(x) = \sqrt{x + 2}$ **8.** $g(x) = \dfrac{2x + 1}{x^2 - 4}$

Exercises 9 and 10 refer to the piecewise-defined function

$$f(x) = \begin{cases} 3x + 1 & \text{if } x \leq 1 \\ 4 - x^2 & \text{if } x > 1 \end{cases}.$$

9. Draw the graph of f.

10. Find the points of continuity and the points of discontinuity of f.

SECTION 10.3 EXERCISES

In Exercises 1–10, find the limit by direct substitution if it exists.

1. $\lim\limits_{x \to -1} x\,(x - 1)^2$ **2.** $\lim\limits_{x \to 3} (x - 1)^{12}$

3. $\lim\limits_{x \to 2} (x^3 - 2x + 3)$ **4.** $\lim\limits_{x \to -2} (x^3 - x + 5)$

5. $\lim\limits_{x \to 2} \sqrt{x + 5}$ **6.** $\lim\limits_{x \to -2} (x - 4)^{2/3}$

7. $\lim\limits_{x \to 0} (e^x \sin x)$ **8.** $\lim\limits_{x \to \pi} \ln\left(\sin \dfrac{x}{2}\right)$

9. $\lim\limits_{x \to a} (x^2 - 2)$ **10.** $\lim\limits_{x \to a} \dfrac{x^2 - 1}{x^2 + 1}$

In Exercises 11–18, **(a)** explain why you cannot use substitution to find the limit and **(b)** find the limit algebraically if it exists.

11. $\lim\limits_{x \to -3} \dfrac{x^2 + 7x + 12}{x^2 - 9}$ **12.** $\lim\limits_{x \to 3} \dfrac{x^2 - 9}{x^2 + 2x - 15}$

13. $\lim\limits_{x \to -1} \dfrac{x^3 + 1}{x + 1}$ **14.** $\lim\limits_{x \to 2} \dfrac{x^3 - 2x^2 + x - 2}{x - 2}$

15. $\lim\limits_{x \to -2} \dfrac{x^2 - 4}{x + 2}$ **16.** $\lim\limits_{x \to -2} \dfrac{|x^2 - 4|}{x + 2}$

17. $\lim\limits_{x \to 0} \sqrt{x - 3}$ **18.** $\lim\limits_{x \to 0} \dfrac{x - 2}{x^2}$

In Exercises 19–22, use the fact that $\lim\limits_{x \to 0} \dfrac{\sin x}{x} = 1$, along with the limit properties, to find the following limits.

19. $\lim\limits_{x \to 0} \dfrac{\sin x}{2x^2 - x}$ **20.** $\lim\limits_{x \to 0} \dfrac{\sin 3x}{x}$

21. $\lim\limits_{x \to 0} \dfrac{\sin^2 x}{x}$ **22.** $\lim\limits_{x \to 0} \dfrac{x + \sin x}{2x}$

In Exercises 23–26, find the limits.

23. $\lim\limits_{x \to 0} \dfrac{e^x - \sqrt{x}}{\log_4(x + 2)}$ **24.** $\lim\limits_{x \to 0} \dfrac{3\sin x - 4\cos x}{5\sin x + \cos x}$

25. $\lim\limits_{x \to \pi/2} \dfrac{\ln(2x)}{\sin^2 x}$ **26.** $\lim\limits_{x \to 27} \dfrac{\sqrt{x + 9}}{\log_3 x}$

In Exercises 27–30, use the given graph to find the limits or to explain why the limits do not exist.

27. (a) $\lim\limits_{x \to 2^-} f(x)$

 (b) $\lim\limits_{x \to 2^+} f(x)$

 (c) $\lim\limits_{x \to 2} f(x)$

28. (a) $\lim\limits_{x \to 3^-} f(x)$

 (b) $\lim\limits_{x \to 3^+} f(x)$

 (c) $\lim\limits_{x \to 3} f(x)$

29. (a) $\lim\limits_{x \to 3^-} f(x)$

 (b) $\lim\limits_{x \to 3^+} f(x)$

 (c) $\lim\limits_{x \to 3} f(x)$

30. (a) $\lim\limits_{x\to 1^-} f(x)$

(b) $\lim\limits_{x\to 1^+} f(x)$

(c) $\lim\limits_{x\to 1} f(x)$

In Exercises 31 and 32, the graph of a function $y = f(x)$ is given. Which of the statements about the function are true and which are false?

31. (a) $\lim\limits_{x\to -1^+} f(x) = 1$

(b) $\lim\limits_{x\to 0^-} f(x) = 0$

(c) $\lim\limits_{x\to 0^-} f(x) = 1$

(d) $\lim\limits_{x\to 0^-} f(x) = \lim\limits_{x\to 0^+} f(x)$

(e) $\lim\limits_{x\to 0} f(x)$ exists   **(f)** $\lim\limits_{x\to 0} f(x) = 0$

(g) $\lim\limits_{x\to 0} f(x) = 1$ **(h)** $\lim\limits_{x\to 1} f(x) = 1$

(i) $\lim\limits_{x\to 1} f(x) = 0$ **(j)** $\lim\limits_{x\to 2^-} f(x) = 2$

32. (a) $\lim\limits_{x\to -1^+} f(x) = 1$

(b) $\lim\limits_{x\to 2} f(x)$ does not exist

(c) $\lim\limits_{x\to 2} f(x) = 2$

(d) $\lim\limits_{x\to 1^-} f(x) = 2$

(e) $\lim\limits_{x\to 1^+} f(x) = 1$

(f) $\lim\limits_{x\to 1} f(x)$ does not exist

(g) $\lim\limits_{x\to 0^+} f(x) = \lim\limits_{x\to 0^-} f(x)$

(h) $\lim\limits_{x\to c} f(x)$ exists for every c in $(-1, 1)$.

(i) $\lim\limits_{x\to c} f(x)$ exists for every c in $(1, 3)$.

In Exercises 33 and 34, use a graph of f to find **(a)** $\lim\limits_{x\to 0^-} f(x)$, **(b)** $\lim\limits_{x\to 0^+} f(x)$, and **(c)** $\lim\limits_{x\to 0} f(x)$ if they exist.

33. $f(x) = (1 + x)^{1/x}$ **34.** $f(x) = (1 + x)^{1/(2x)}$

35. Group Activity Assume that $\lim\limits_{x\to 4} f(x) = -1$ and $\lim\limits_{x\to 4} g(x) = 4$. Find the limit.

(a) $\lim\limits_{x\to 4} (g(x) + 2)$ **(b)** $\lim\limits_{x\to 4} xf(x)$

(c) $\lim\limits_{x\to 4} g^2(x)$ **(d)** $\lim\limits_{x\to 4} \dfrac{g(x)}{f(x) - 1}$

36. Group Activity Assume that $\lim\limits_{x\to a} f(x) = 2$ and $\lim\limits_{x\to a} g(x) = -3$. Find the limit.

(a) $\lim\limits_{x\to a} (f(x) + g(x))$ **(b)** $\lim\limits_{x\to a} (f(x) \cdot g(x))$

(c) $\lim\limits_{x\to a} (3g(x) + 1)$ **(d)** $\lim\limits_{x\to a} \dfrac{f(x)}{g(x)}$

In Exercises 37–40, complete the following for the given piecewise-defined function f.

(a) Draw the graph of f.

(b) Determine $\lim\limits_{x\to a^+} f(x)$ and $\lim\limits_{x\to a^-} f(x)$.

(c) Writing to Learn Does $\lim\limits_{x\to a} f(x)$ exist? If it does, give its value. If it does not exist, give an explanation.

37. $a = 2$, $f(x) = \begin{cases} 2 - x & \text{if } x < 2 \\ 1 & \text{if } x = 2 \\ x^2 - 4 & \text{if } x > 2 \end{cases}$

38. $a = 1$, $f(x) = \begin{cases} 2 - x & \text{if } x < 1 \\ x + 1 & \text{if } x \geq 1 \end{cases}$

39. $a = 0$, $f(x) = \begin{cases} |x - 3| & \text{if } x < 0 \\ x^2 - 2x & \text{if } x \geq 0 \end{cases}$

40. $a = -3$, $f(x) = \begin{cases} 1 - x^2 & \text{if } x \geq -3 \\ 8 - x & \text{if } x < -3 \end{cases}$

In Exercises 41–46, find the limit.

41. $\lim\limits_{x\to 2^+} \text{int}\,(x)$ **42.** $\lim\limits_{x\to 2^-} \text{int}\,(x)$

43. $\lim\limits_{x\to 0.0001} \text{int}\,(x)$ **44.** $\lim\limits_{x\to 5/2^-} \text{int}\,(2x)$

45. $\lim\limits_{x\to -3^+} \dfrac{x + 3}{|x + 3|}$ **46.** $\lim\limits_{x\to 0^-} \dfrac{5x}{|2x|}$

In Exercises 47–54, find **(a)** $\lim\limits_{x\to\infty} y$ and **(b)** $\lim\limits_{x\to -\infty} y$.

47. $y = \dfrac{\cos x}{1 + x}$ **48.** $y = \dfrac{x + \sin x}{x}$

49. $y = 1 + 2^x$ **50.** $y = \dfrac{x}{1 + 2^x}$

51. $y = x + \sin x$ **52.** $y = e^{-x} + \sin x$

53. $y = -e^x \sin x$ **54.** $y = e^{-x} \cos x$

In Exercises 55–60, use graphs and tables to find the limit and identify any vertical asymptotes.

55. $\lim\limits_{x\to 3^-} \dfrac{1}{x - 3}$ **56.** $\lim\limits_{x\to 3^+} \dfrac{x}{x - 3}$

57. $\lim\limits_{x\to -2^+} \dfrac{1}{x + 2}$ **58.** $\lim\limits_{x\to -2^-} \dfrac{x}{x + 2}$

59. $\lim\limits_{x\to 5} \dfrac{1}{(x - 5)^2}$ **60.** $\lim\limits_{x\to 2} \dfrac{1}{x^2 - 4}$

In Exercises 61–64, determine the limit algebraically if possible. Support your answer graphically.

61. $\lim\limits_{x\to 0} \dfrac{(1 + x)^3 - 1}{x}$ **62.** $\lim\limits_{x\to 0} \dfrac{1/(3 + x) - 1/3}{x}$

63. $\lim\limits_{x\to 0} \dfrac{\tan x}{x}$ **64.** $\lim\limits_{x\to 2} \dfrac{x - 4}{x^2 - 4}$

In Exercises 65–72, find the limit.

65. $\lim\limits_{x \to 0} \dfrac{|x|}{x^2}$

66. $\lim\limits_{x \to 0} \dfrac{x^2}{|x|}$

67. $\lim\limits_{x \to 0} \left[x \sin\left(\dfrac{1}{x}\right) \right]$

68. $\lim\limits_{x \to 27} \cos\left(\dfrac{1}{x}\right)$

69. $\lim\limits_{x \to 1} \dfrac{x^2 + 1}{x - 1}$

70. $\lim\limits_{x \to \infty} \dfrac{\ln x^2}{\ln x}$

71. $\lim\limits_{x \to \infty} \dfrac{\ln x}{\ln x^2}$

72. $\lim\limits_{x \to \infty} 3^{-x}$

Standardized Test Questions

73. True or False If $f(x) = \begin{cases} x + 2 & \text{if } x \le 3 \\ 8 - x & \text{if } x > 3 \end{cases}$, then

$\lim\limits_{x \to 3} f(x)$ is undefined. Justify your answer.

74. True or False If $f(x)$ and $g(x)$ are two functions and $\lim\limits_{x \to 0} f(x)$ does not exist, then $\lim\limits_{x \to 0} [f(x) \cdot g(x)]$ cannot exist. Justify your answer.

Multiple Choice In Exercises 75–78, match the function $y = f(x)$ with the table. Do not use a calculator.

X	Y₁	
2.7	-52.3	
2.8	-82.2	
2.9	-172.1	
3	ERROR	
3.1	188.1	
3.2	98.2	
3.3	68.3	
X=2.7		

(A)

X	Y₁	
2.7	3.7	
2.8	3.8	
2.9	3.9	
3	ERROR	
3.1	4.1	
3.2	4.2	
3.3	4.3	
X=2.7		

(B)

X	Y₁	
2.7	23.7	
2.8	33.8	
2.9	63.9	
3	ERROR	
3.1	-55.9	
3.2	-25.8	
3.3	-15.7	
X=2.7		

(C)

X	Y₁	
2.7	24.39	
2.8	25.24	
2.9	26.11	
3	ERROR	
3.1	27.91	
3.2	28.84	
3.3	29.79	
X=2.7		

(D)

X	Y₁	
2.7	3.7	
2.8	3.8	
2.9	3.9	
3	4	
3.1	4.1	
3.2	4.2	
3.3	4.3	
X=2.7		

(E)

75. $y = \dfrac{x^2 - 2x - 3}{x - 3}$

76. $y = \dfrac{x^2 + 2x + 3}{x - 3}$

77. $y = \dfrac{x^2 - 2x - 9}{x - 3}$

78. $y = \dfrac{x^3 - 27}{x - 3}$

Explorations

In Exercises 79–82, complete the following for the given piecewise-defined function f.

(a) Draw the graph of f.

(b) At what points c in the domain of f does $\lim\limits_{x \to c} f(x)$ exist?

(c) At what points c does only the left-hand limit exist?

(d) At what points c does only the right-hand limit exist?

79. $f(x) = \begin{cases} \cos x & \text{if } -\pi \le x < 0 \\ -\cos x & \text{if } 0 \le x \le \pi \end{cases}$

80. $f(x) = \begin{cases} \sin x & \text{if } -\pi \le x < 0 \\ \csc x & \text{if } 0 \le x \le \pi \end{cases}$

81. $f(x) = \begin{cases} \sqrt{1 - x^2} & \text{if } -1 \le x < 0 \\ x & \text{if } 0 \le x < 1 \\ 2 & \text{if } x = 1 \end{cases}$

82. $f(x) = \begin{cases} x^2 & \text{if } -2 \le x < 0 \quad \text{or} \quad 0 < x \le 2 \\ 1 & \text{if } x = 0 \\ 2x & \text{if } x < -2 \quad \text{or} \quad x > 2 \end{cases}$

83. Rabbit Population The population of rabbits over a 2-year period in a certain county is given in Table 10.5.

Table 10.5 Rabbit Population	
Beginning of Month	Number (in thousands)
0	10
2	12
4	14
6	16
8	22
10	30
12	35
14	39
16	44
18	48
20	50
22	51

(a) Draw a scatter plot of the data in Table 10.5.

(b) Find a logistic regression model for the data. Find the limit of that model as time approaches infinity.

(c) What can you conclude about the limit of the rabbit population growth in the county?

(d) Provide a reasonable explanation for the population growth limit.

Group Activity In Exercises 84–87, sketch a graph of a function $y = f(x)$ that satisfies the stated conditions. Include any asymptotes.

84. $\lim\limits_{x \to 0} f(x) = \infty$, $\lim\limits_{x \to \infty} f(x) = \infty$, $\lim\limits_{x \to -\infty} f(x) = 2$

85. $\lim\limits_{x \to 4} f(x) = -\infty$, $\lim\limits_{x \to \infty} f(x) = -\infty$, $\lim\limits_{x \to -\infty} f(x) = 2$

86. $\lim\limits_{x \to \infty} f(x) = 2$, $\lim\limits_{x \to -2^+} f(x) = -\infty$,
$\lim\limits_{x \to -2^-} f(x) = -\infty$, $\lim\limits_{x \to -\infty} f(x) = \infty$

87. $\lim\limits_{x \to 1} f(x) = \infty$, $\lim\limits_{x \to 2^+} f(x) = -\infty$, $\lim\limits_{x \to 2^-} f(x) = -\infty$, $\lim\limits_{x \to -\infty} f(x) = \infty$

Extending the Ideas

88. Properties of Limits Find the limits of f, g, and fg as x approaches c.

(a) $f(x) = \dfrac{2}{x^2}$, $g(x) = x^2$, $c = 0$

(b) $f(x) = \left|\dfrac{1}{x}\right|$, $g(x) = \sqrt[3]{x}$, $c = 0$

(c) $f(x) = \left|\dfrac{3}{x - 1}\right|$, $g(x) = (x - 1)^2$, $c = 1$

(d) $f(x) = \dfrac{1}{(x - 1)^4}$, $g(x) = (x - 1)^2$, $c = 1$

(e) Writing to Learn Suppose that $\lim\limits_{x \to c} f(x) = \infty$ and $\lim\limits_{x \to c} g(x) = 0$. Based on your results in parts (a)–(d), what can you say about $\lim\limits_{x \to a} (f(x) \cdot g(x))$?

89. Limits and the Area of a Circle Consider an n-sided regular polygon made up of n congruent isosceles triangles, each with height h and base b. The figure shows an 8-sided regular polygon.

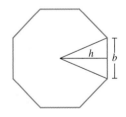

(a) Show that the area of an 8-sided regular polygon is $A = 4hb$ and the area of the n-sided regular polygon is $A = (1/2)nhb$.

(b) Show that the base b of the n-sided regular polygon is $b = 2h \tan (180/n)°$.

(c) Show that the area A of the n-sided regular polygon is $A = nh^2 \tan (180/n)°$.

(d) Let $h = 1$. Construct a table of values for n and A for $n = 4, 8, 16, 100, 500, 1000, 5000, 10000, 100000$. Does A have a limit as $n \to \infty$?

(e) Repeat part (d) with $h = 3$.

(f) Give a convincing argument that $\lim\limits_{n \to \infty} A = \pi h^2$, the area of a circle of radius h.

90. Continuous Extension of a Function Let
$$f(x) = \begin{cases} x^2 - 3x + 3 & \text{if } x \neq 2 \\ a & \text{if } x = 2. \end{cases}$$

(a) Sketch several possible graphs for f.

(b) Find a value for a so that the function is continuous at $x = 2$.

In Exercises 91–93, **(a)** graph the function, **(b)** verify that the function has one removable discontinuity, and **(c)** give a formula for a continuous extension of the function. [*Hint:* See Exercise 90.]

91. $y = \dfrac{2x + 4}{x + 2}$

92. $y = \dfrac{x - 5}{5 - x}$

93. $y = \dfrac{x^3 - 1}{x - 1}$

10.4
Numerical Derivatives and Integrals

Derivatives on a Calculator

As computers and sophisticated calculators have become indispensable tools for modern engineers and mathematicians (and, ultimately, for modern students of mathematics), numerical techniques of differentiation and integration have re-emerged as primary methods of problem-solving. This is no small irony, as it was precisely to avoid the tedious computations inherent in such methods that calculus was invented in the first place. Although nothing can diminish the magnitude of calculus as a significant human achievement, and although nobody can get far in mathematics or science without it, the modern fact is that applying the old-fashioned methods of limiting approximations—with the help of a calculator—is often the most efficient way to solve a calculus problem.

Most graphing calculators have built-in algorithms that will approximate derivatives of functions numerically with good accuracy at most points of their domains. We will use the notation NDER $f(a)$ to denote such a calculator derivative approximation to $f'(a)$.

For small values of h, the regular difference quotient

$$\frac{f(a + h) - f(a)}{h}$$

is often a good approximation of $f'(a)$. However, the same value of h will usually produce a better approximation of $f'(a)$ if we use the **symmetric difference quotient**

$$\frac{f(a + h) - f(a - h)}{2h},$$

as illustrated in Figure 10.22.

Many graphing utilities use the symmetric difference quotient with a default value of $h = 0.001$ for computing NDER $f(a)$. When we refer to the numerical derivative in this book, we will assume that it is the symmetric difference quotient with $h = 0.001$.

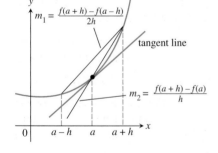

FIGURE 10.22 The symmetric difference quotient (slope m_1) usually gives a better approximation of the derivative for a given value of h than does the regular difference quotient (slope m_2).

DEFINITION Numerical Derivative

In this book, we define the **numerical derivative of f at a** to be

$$\text{NDER } f(a) = \frac{f(a + 0.001) - f(a - 0.001)}{0.002}.$$

Similarly, we define the **numerical derivative of f** to be the function

$$\text{NDER } f(x) = \frac{f(x + 0.001) - f(x - 0.001)}{0.002}$$

EXAMPLE 1   Computing a Numerical Derivative

Let $f(x) = x^3$. Compute NDER $f(2)$ by calculating the symmetric difference quotient with $h = 0.001$. Compare it to the actual value of $f'(x)$.

SOLUTION

$$\text{NDER } f(x) = \frac{f(2 + 0.001) - f(2 - 0.001)}{0.002}$$

$$= \frac{f(2.001) - f(1.999)}{0.002}$$

$$= 12.000001$$

The actual value is

$$f'(2) = \lim_{h \to 0} \frac{(2 + h)^3 - 2^3}{h}$$

$$= \lim_{h \to 0} \frac{8 + 12h + 6h^2 + h^3 - 8}{h}$$

$$= \lim_{h \to 0} (12 + 6h + h^2)$$

$$= 12$$

Now try Exercise 3.

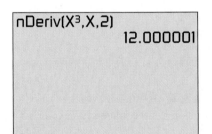

FIGURE 10.23 Applying the numerical derivative command on a graphing calculator. (Example 1)

The numerical derivative in this case is obviously quite accurate. In practice, it is not necessary to key in the symmetric difference quotient, as it is done by the calculator with its built-in algorithm. Figure 10.23 shows the command that would be used on one such calculator to find the numerical derivative in Example 1.

If $f'(a)$ exists, then NDER $f(a)$ usually gives a good approximation to the actual value. On the other hand, the algorithm will sometimes return a value for NDER $f(a)$ when $f'(a)$ does not exist. (See Exercise 51.)

Definite Integrals on a Calculator

Recall from the history of the area problem (Section 10.2) that the strategy of summing up thin rectangles to approximate areas is ancient. The thinner the rectangles, the better the approximation—and, of course, the more tedious the computation. Today, thanks to technology, we can employ the ancient strategy without the tedium.

Many graphing calculators have built-in algorithms to compute definite integrals with great accuracy. We use the notation NINT $(f(x), x, a, b)$ to denote such a calculator approximation to $\int_a^b f(x)\, dx$. Unlike NDER, which uses a fixed value of $\triangle x$, NINT will vary the value of $\triangle x$ until the numerical integral gets close to a limiting value, often resulting in an exact answer (at least to the number of digits in the calculator display). Because the algorithm for NINT finds the definite integral by Riemann sum approximation rather than by calculus, we call it a **numerical integral**.

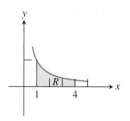

FIGURE 10.24 The graph of $f(x) = 1/x$ with the area under the curve between $x = 1$ and $x = 4$ shaded. (Example 2)

EXAMPLE 2   Finding a Numerical Integral

Use NINT to find the area of the region R enclosed between the x-axis and the graph of $y = 1/x$ from $x = 1$ to $x = 4$.

SOLUTION   The region is shown in Figure 10.24.

The area can be written as the definite integral $\int_1^4 \frac{1}{x}\,dx$, which we find on a graphing calculator: NINT $(1/x, x, 1, 4) = 1.386294361$. The exact answer (as you will learn in a calculus course) is ln 4, which agrees in every displayed digit with the NINT value! Figure 10.25 shows the syntax for numerical integration on one type of calculator.

Now try Exercise 13.

```
fnInt(1/X,X,1,4)

           1.386294361
```

FIGURE 10.25 A numerical integral approximation for $\int_1^4 \frac{1}{x}\,dx$. (Example 2)

EXPLORATION 1   A Do-It-Yourself Numerical Integrator

Recall that a definite integral is the limit at infinity of a Riemann sum—that is, a sum of the form $\sum_{k=1}^{n} f(x_k)\triangle x$. You can use your calculator to evaluate sums of sequences using LIST commands. (It is not as accurate as NINT, and certainly not as easy, but at least you can see the summing that takes place.)

1. The integral in Example 2 can be computed using the command

    sum(seq(1/(1 + K · 3/50) · 3/50, K, 1, 50)).

   This uses 50 RRAM rectangles, each with width $\triangle x = 3/50$. Find the sum on your calculator and compare it to the NINT value.

2. Study the command until you see how it works. Adapt the command to find the RRAM approximation for 100 rectangles and compute it on your calculator. Does the approximation get better?

3. What definite integral is approximated by the command

    sum(seq(sin(0 + K · π/50) · π/50, K, 1, 50))?

   Compute it on your calculator and compare it to the NINT value for the same integral.

4. Write a command that uses 50 RRAM rectangles to approximate $\int_4^9 \sqrt{x}\,dx$. Compute it on your calculator and compare it to the NINT value for the same integral.

Remember that we were originally motivated to find areas because of their connection to the problem of distance traveled. To show just one of the many applications of integration, we use the numerical integral to solve a distance problem in Example 3.

EXAMPLE 3   Finding Distance Traveled

An automobile is driven at a variable rate along a test track for 2 hours so that its velocity at any time t $(0 \le t \le 2)$ is given by $v(t) = 30 + 10 \sin 6t$ miles per hour. How far does the automobile travel during the 2-hour test?

continued

SOLUTION According to the analysis found in Section 10.2, the distance traveled is given by $\int_0^2 (30 + 10 \sin (6t)) \, dt$. We use a calculator to find the numerical integral:

$$\text{NINT} (30 + 10 \sin (6t), t, 0, 2) \approx 60.26.$$

Interpreting the answer, we conclude that the automobile travels 60.26 miles.

Now try Exercise 21.

Computing a Derivative from Data

Sometimes all we are given about a problem situation is a scatter plot obtained from a set of data—a numerical model of the problem. There are two ways to get information about the derivative of the model.

1. To approximate the derivative at a point: Remember that the average rate of change over a small interval, $\triangle y/\triangle x$, approximates the derivative at points in that interval. (Generally, the approximation is better near the middle of the interval than it is near the endpoints.) The average rate of change on an interval between two data points can be computed directly from the data.

2. To approximate the derivative function: Regression techniques can be used to fit a curve to the data, and then NDER can be applied to the regression model to approximate the derivative. Alternatively, the values of $\triangle y/\triangle x$ can be plotted, and then regression techniques can be used to fit a derivative approximation through those points.

EXAMPLE 4 Finding Derivatives from Data

Table 10.6 shows the height (in feet) of a falling ball above ground level as measured by a motion detector at time intervals of 0.04 seconds.

(a) Estimate the instantaneous speed of the ball at $t = 0.2$ seconds.

(b) Draw a scatter plot of the data and use quadratic regression to model the height s of the ball above the ground as a function of t.

(c) Use NDER to approximate $s'(0.2)$ and compare it to the value found in (a).

SOLUTION

(a) Since 0.2 is the midpoint of the time interval [0.16, 0.24], the average rate of change $\triangle s/\triangle t$ on the interval [0.16, 0.24] should give a good approximation to $s'(0.2)$.

$$s'(0.2) \approx \frac{\triangle s}{\triangle t} = \frac{4.30 - 5.45}{0.24 - 0.16} = -14.375.$$

The speed is about 14.375 feet per second at $t = 0.2$.

(b) The scatter plot is shown in Figure 10.26, along with the quadratic regression curve. A graphing calculator gives $s(t) \approx -17.12t^2 - 7.74t + 7.13$ as the equation of the quadratic regression model.

(c) The calculator computes NDER $s(0.2)$ to be -14.588, which agrees quite well with the approximation in (a). In fact, the difference is only 0.213 feet per second, less than 1.5% of the speed of the ball.

Now try Exercise 23.

WHO'S DRIVING???

We will candidly admit that the conditions in Example 3 would be virtually impossible to replicate in a real setting, even if one could imagine a reason for doing so. Mathematics textbooks are filled with such unreal real-world problems, but they do serve a purpose when students are being exposed to new material. Real real-world problems are often either too easy to illustrate the concept or too hard for beginners to solve.

Table 10.6 Falling Ball	
Time (sec)	Position (ft)
0.04	6.80
0.08	6.40
0.12	5.95
0.16	5.45
0.20	4.90
0.24	4.30
0.28	3.60
0.32	2.90
0.36	2.15
0.40	1.30
0.44	0.40

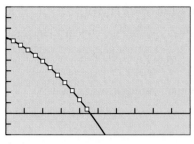

[0, 1] by [–2, 10]

FIGURE 10.26 A scatter plot of the data in Table 10.6 together with its quadratic regression model. (Example 4)

Table 10.7 Change over Intervals from Table 10.6	
Midpoint	$\triangle s/\triangle t$
0.06	−10.00
0.10	−11.25
0.14	−12.50
0.18	−13.75
0.22	−15.00
0.26	−17.50
0.30	−17.50
0.34	−18.75
0.38	−21.25
0.42	−22.50

EXAMPLE 5   Finding Derivatives from Data

This example also uses the falling ball data in Table 10.6.

(a) Compute the average velocity, $\triangle s/\triangle t$, on each subinterval of length 0.04. Make a table showing the midpoints of the subintervals in one column and the values of $\triangle s/\triangle t$ in the second column.

(b) Make a scatter plot showing the numbers in the second column as a function of the numbers in the first column and find a linear regression model to model the data.

(c) Use the linear regression model in (b) to approximate the velocity of the ball at $t = 0.2$ and compare it to the values found in Example 4.

SOLUTION

(a) The first subinterval, which begins at 0.04 and ends at 0.08, has a midpoint of $(0.04 + 0.08)/2 = 0.06$. On that interval, $\triangle s/\triangle t = (6.40 - 6.80)/(0.08 - 0.04) = -10.00$. The rest of the midpoints and values of $\triangle s/\triangle t$ are computed similarly and are shown in Table 10.7.

(b) The scatter plot and the regression line are shown in Figure 10.27. A graphing calculator gives $v(t) = -34.470t - 7.727$ as the regression line.

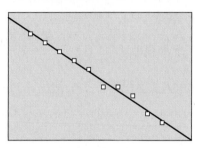

[0, 0.5] by [–25, –7]

FIGURE 10.27 A scatter plot of the data in Table 10.7 together with its linear regression model. (Example 5)

(c) The linear regression model gives $v(0.2) \approx -14.62$, which is close to the values found in Example 4.   *Now try Exercise 25.*

Computing a Definite Integral from Data

If we are given a set of data points, the x-coordinates of the points define subintervals between the smallest and largest x-values in the data. We can form a Riemann sum

$$\sum_{k=1}^{n} f(x_k) \, \triangle x$$

using the lengths of the subintervals for $\triangle x$ and either the left or right endpoints of the intervals as the x_k's. The Riemann sum then approximates the definite integral of the function over the interval.

Example 6 illustrates how this is done.

Table 10.8 Velocity of the Moving Body	
Time (sec)	Velocity (m/sec)
0.00	0.00
0.25	0.28
0.50	0.53
0.75	0.73
1.00	0.90
1.25	1.01
1.50	1.11
1.75	1.18
2.00	1.21
2.25	1.17
2.50	1.13
2.75	1.05
3.00	0.91
3.25	0.72
3.50	0.55
3.75	0.26

EXAMPLE 6   Finding a Definite Integral Using Data

Table 10.8 shows the velocity of a moving body (in meters per second) measured at regular quarter-second intervals. Estimate the distance traveled by the body from $t = 0$ to $t = 3.75$.

SOLUTION Figure 10.28 gives a scatter plot of the velocity data.

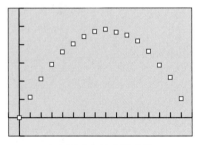

[−0.25, 4] by [−0.25, 1.5]

FIGURE 10.28 A scatter plot of the velocity data in Table 10.8. (Example 6)

The distance traveled is $\int_0^{3.75} v(t)\, dx$, which we approximate with a Riemann sum constructed directly from the data. We sum up 15 products of the form $v(t_k)\, \triangle t$, using the right endpoint for t_k each time. (This is the RRAM approximation in the notation of Section 10.2.) Note that $\triangle t = 0.25$ for every subinterval.

$$\int_0^{3.75} v(t)\, dx = \sum_{k=1}^{15} v(t_k)\, \triangle t$$

$$= 0.25(0.28 + 0.53 + 0.73 + 0.90 + 1.01 + 1.11 \\ + 1.18 + 1.21 + 1.17 + 1.13 + 1.05 + 0.91 + 0.72 \\ + 0.55 + 0.26)$$

$$= 3.185$$

So the distance traveled by the body is about 3.2 meters.

Now try Exercise 27.

QUICK REVIEW 10.4 *(For help, go to Sections P.4 and 1.3.)*

In Exercises 1–6, find $\triangle y / \triangle x$ on the interval $[1, 4]$ under the given conditions.

1. $y = x^2$
2. $y = \sqrt{x}$
3. $y = \log_2 x$
4. $y = 3^x$
5. $y = 2$ when $x = 1$ and $y = 11$ when $x = 4$.
6. The graph of $y = f(x)$ passes through points $(4, 10)$ and $(1, -2)$.

In Exercises 7–10, compute the quotient $(f(1 + h) - f(1 - h))/2h$ with the given f and h.

7. $f(x) = \sin x, h = 0.01$
8. $f(x) = x^4, h = 0.001$
9. $f(x) = \ln x, h = 0.001$
10. $f(x) = e^x, h = 0.0001$

SECTION 10.4 EXERCISES

In Exercises 1–10, use NDER on a calculator to find the numerical derivative of the function at the specified point.

1. $f(x) = 1 - x^2$ at $x = 2$

2. $f(x) = 2x + \dfrac{1}{2}x^2$ at $x = 2$

3. $f(x) = 3x^2 + 2$ at $x = -2$

4. $f(x) = x^2 - 3x + 1$ at $x = 1$

5. $f(x) = |x + 2|$ at $x = -2$

6. $f(x) = \dfrac{1}{x + 2}$ at $x = -1$

7. $f(x) = \ln 2x$ at $x = 1$

8. $f(x) = 2 \ln x$ at $x = 1$

9. $f(x) = 3 \sin x$ at $x = \pi$

10. $f(x) = \sin 3x$ at $x = \pi$

In Exercises 11–20, use NINT on a calculator to find the numerical integral of the function over the specified interval.

11. $f(x) = x^2, [0, 4]$

12. $f(x) = x^2, [-4, 0]$

13. $f(x) = \sin x, [0, \pi]$

14. $f(x) = \sin x, [\pi, 2\pi]$

15. $f(x) = \cos x, [0, \pi]$

16. $f(x) = |\cos x|, [0, \pi]$

17. $f(x) = 1/x, [1, e]$

18. $f(x) = 1/x, [e, 2e]$

19. $f(x) = \dfrac{2}{1 + x^2}, [0, 10^8]$

20. $f(x) = \sec^2 x - \tan^2 x, [0, 10]$

21. Travel Time A truck is driven at a variable rate for 3 hours so that its velocity at any time t ($0 \le t \le 3$) is given by $v(t) = 35 - 12 \cos 4t$ miles per hour. How far does the truck travel during the 3 hours? Round your answer to the nearest hundredth.

22. Travel Time A bicyclist rides for 90 minutes, and her velocity at any time t hours ($0 \le t \le 1.5$) is given by $v(t) = 12 - 8 \sin 5t$ miles per hour. How far does she travel during the 90 minutes? Round your answer to the nearest hundredth.

23. Finding Derivatives from Data A ball is dropped from the roof of a 30-story building. The height in feet above the ground of the falling ball is measured at 1/2-second intervals and recorded in the table.

Time (sec)	Height (ft)
0	500
0.5	495
1.0	485
1.5	465
2.0	435
2.5	400
3.0	355
3.5	305
4.0	245
4.5	175
5.0	100
5.5	15

(a) Use the average velocity on the interval $[1, 2]$ to estimate the velocity of the ball at $t = 1.5$ seconds.

(b) Draw a scatter plot of the data.

(c) Find a quadratic regression model for the data.

(d) Use NDER of the model in part (c) to estimate the velocity of the ball at $t = 1.5$ seconds.

(e) Use the model to estimate how fast the ball is going when it hits the ground.

24. Estimating Average Rate of Change from Data
Table 10.9 gives the U.S. Gross Domestic Product Data in billions of dollars for the years 1990–2003.

Table 10.9 U.S. Gross Domestic Product Data	
Year	Amount (billions of dollars)
1990	5803.1
1995	7397.7
1997	8304.3
1998	8747.0
1999	9268.4
2000	9817.0
2001	10,100.8
2002	10,480.8
2003	10,987.9

Source: Statistical Abstract of the United States: 2004–2005.

(a) Find the average rate of change of the gross domestic product from 1997 to 1998 and then from 2001 to 2002.

(b) Find a quadratic regression model for the data in Table 10.9 and overlay its graph on a scatter plot of the data. Let $x = 0$ stand for 1990, $x = 1$ stand for 1991, and so forth.

(c) Use the model in part (b), and a calculator NDER computation, to estimate the rate of change of the gross domestic product in 1997 and in 2001.

(d) **Writing to Learn** Use the model in part (b) to predict the gross domestic product in 2007. Is this reasonable? Why or why not?

25. **Estimating Velocity** Refer to the data in Exercise 23.

(a) Compute the average velocity, $\triangle y/\triangle x$, on each subinterval of length 0.5. Make a table showing the midpoints of the subintervals in one column and the average velocities in the second column.

(b) Make a scatter plot showing the numbers in the second column as a function of the numbers in the first column and find a linear regression model to model the data.

(c) Use the linear regression model in part (b) to approximate the velocity of the ball at $t = 1.5$ seconds, and compare your result to the value found in Exercise 23(d).

26. **Approximating Rate of Change** Refer to the data in Exercise 24.

(a) Compute the average rate of change, $\triangle y/\triangle x$, on each subinterval. Make a table showing the midpoints of the subintervals in one column and the average rates of change in the second column.

(b) Make a scatter plot showing the numbers in the second column as a function of the numbers in the first column and find a linear regression model to model the data.

(c) Use the linear regression model in part (b) to approximate the rate of change in 1997 and in 2001, and compare your results to the values found in Exercise 24(c).

27. **Estimating Distance** A stone is dropped from a cliff and its velocity (in feet per second) at regular 0.5-second intervals is shown in Table 10.10. Estimate the distance that the stone travels from $t = 0$ to $t = 2.5$.

Table 10.10 Velocity of the Stone	
Time (sec)	Velocity (ft/sec)
0	0
0.5	16
1	32
1.5	48
2	64
2.5	80

28. **Estimating Distance** Table 10.11 shows the velocity of a moving object in meters per second, measured at regular 0.2-second intervals. Estimate the distance traveled by the body from $t = 0$ to $t = 1.6$.

Table 10.11 Velocity of the Object	
Time (sec)	Velocity (m/sec)
0	1.20
0.2	0.98
0.4	0.72
0.6	0.50
0.8	0.34
1.0	0.30
1.2	0.44
1.4	0.79
1.6	1.40

29. **Writing to Learn** Analyze the following program, which produces an LRAM approximation for the function entered in Y1 in the calculator. Then write a short paragraph explaining how it works.

```
PROGRAM:LRAM
:Input "A",A
:Input "B",B
:Input "N",N
:sum(seq(((B−A)/
N)Y₁(A+K((B−A)/N
)),K,0,N−1))→C
:Disp "AREA =",C
```

30. **Writing to Learn** Analyze the following program, which produces an RRAM approximation for the function entered in Y1 in the calculator. Then write a short paragraph explaining how it works.

```
PROGRAM:RRAM
:Input "A",A
:Input "B",B
:Input "N",N
:sum(seq(((B−A)/
N)Y₁(A+K((B−A)/N
)),K,1,N))→C
:Disp "AREA =",C
```

In Exercises 31–42, complete the following for the indicated interval $[a, b]$.

(a) Verify the given function is nonnegative.

(b) Use a calculator to find the LRAM, RRAM, and average approximations for the area under the graph of the function from $x = a$ to $x = b$ with 10, 20, 50, and 100 approximating rectangles. (You may want to use the programs in Exercises 29 and 30.)

(c) Compare the average area estimate in part (b) using 100 approximating rectangles with the calculator NINT area estimate, if your calculator has this feature.

31. $f(x) = x^2 - x + 1;$ $[0, 4]$

32. $f(x) = 2x^2 - 2x + 1;$ $[0, 6]$

33. $f(x) = x^2 - 2x + 1;$ $[0, 4]$

34. $f(x) = x^2 + x + 5;$ $[0, 6]$

35. $f(x) = x^2 + x + 5;$ $[2, 6]$ **36.** $f(x) = x^3 + 1;$ $[1, 5]$

37 $f(x) = \sqrt{x + 2};$ $[0, 4]$ **38.** $f(x) = \sqrt{x - 2};$ $[3, 6]$

39. $f(x) = \cos x;$ $[0, \pi/2]$ **40.** $f(x) = x \cos x;$ $[0, \pi/2]$

41. $f(x) = xe^{-x};$ $[0, 2]$ **42.** $f(x) = \dfrac{1}{x - 2};$ $[3, 5]$

Standardized Test Questions

43. True or False The numerical derivative algorithm NDER always uses the same value of $\triangle x$ (or h) to complete its calculations. Justify your answer.

44. True or False The numerical integral algorithm always uses the same value of $\triangle x$ to complete its calculations. Justify your answer.

In Exercises 45–48, choose the correct answer. Do not use a calculator.

45. Multiple Choice Estimating Area Under a Curve Which of the following will typically produce the most accurate estimate of an area under a curve?

(A) NDER

(B) NINT

(C) LRAM, 10 rectangles

(D) RRAM, 25 rectangles

(E) LRAM, 60 rectangles

46. Multiple Choice Estimating Derivative Values Given a continuous function f, which of the following expressions will typically produce the most accurate estimate of $f'(a)$?

(A) $\dfrac{f(a + 0.05) - f(a - 0.05)}{0.05}$

(B) $\dfrac{f(a + 0.05) - f(a - 0.05)}{0.1}$

(C) $\dfrac{f(a + 0.01) - f(a)}{0.01}$

(D) $\dfrac{f(a + 0.01) - f(a - 0.01)}{0.01}$

(E) $\dfrac{f(a + 0.01) - f(a - 0.01)}{0.02}$

47. Multiple Choice Using a Numerical Integral Which of the following *cannot* be estimated using a numerical integral?

(A) The area under a curve that represents some function $f(x)$

(B) The distance traveled, when the velocity function is known

(C) The instantaneous velocity of an object, when the position function is known

(D) The change of a city's population over a 10-year period, when the rate-of-change function is known

(E) The change of a child's height over a 4-year period, when the rate-of-change function is known

48. Multiple Choice Using a Numerical Derivative Which of the following *cannot* be estimated using a numerical derivative?

(A) The instantaneous velocity of an object, when the position function is known

(B) The slope of a curve that represents some function $g(x)$

(C) The growth rate of a city's population, when the population is known as a function of time

(D) The area under a curve that represents some function $f(x)$

(E) The rate of change of an airplane's altitude, when the altitude is known as a function of time

Explorations

49. Let $f(x) = 2x^2 + 3x + 1$ and $g(x) = x^3 + 1$.

(a) Compute the derivative of f.

(b) Compute the derivative of g.

(c) Using $x = 2$ and $h = 0.001$, compute the standard difference quotient

$$\frac{f(x + h) - f(x)}{h}$$

and the symmetric difference quotient

$$\frac{f(x + h) - f(x - h)}{2h}.$$

(d) Using $x = 2$ and $h = 0.001$, compare the approximations to $f'(2)$ in part (c). Which is the better approximation?

(e) Repeat parts (c) and (d) for g.

50. When Are Derivatives and Areas Equal? Let $f(x) = 1 + e^x$.

(a) Draw a graph of f for $0 \le x \le 1$.

(b) Use NDER on your calculator to compute the derivative of f at 1.

(c) Use NINT on your calculator to compute the area under f from $x = 0$ to $x = 1$ and compare it to the answer in part (b).

(d) Group Activity What do you think the exact answers to parts (b) and (c) are?

51. Calculator Failure Many calculators report that NDER of $f(x) = |x|$ evaluated at $x = 0$ is equal to 0. Explain why this is incorrect. Explain why this error occurs.

52. Grapher Failure Graph the function $f(x) = |x|/x$ in the window $[-5, 5]$ by $[-3, 3]$ and explain why $f'(0)$ does not exist. Find the value of NDER $f(0)$ on the calculator and explain why it gives an incorrect answer.

Extending the Ideas

53. Group Activity Finding Total Area The **total area** bounded by the graph of the function $y = f(x)$ and the x-axis from $x = a$ to $x = b$ is the area below the graph of $y = |f(x)|$ from $x = a$ to $x = b$.

(a) Find the total area bounded by the graph of $f(x) = \sin x$ and the x-axis from $x = 0$ to $x = 2\pi$.

(b) Find the total area bounded by the graph of $f(x) = x^2 - 2x - 3$ and the x-axis from $x = 0$ to $x = 5$.

54. Writing to Learn If a function is *unbounded* in an interval $[a, b]$ it may have *finite* area. Use your knowledge of limits at infinity to explain why this might be the case.

55. Writing to Learn Let f and g be two continuous functions with $f(x) \geq g(x)$ on an interval $[a, b]$. Devise a limit of sums definition of the area of the region between the two curves. Explain how to compute the area if the area under both curves is already known.

56. Area as a Function Consider the function $f(t) = 2t$.

(a) Use NINT on a calculator to compute $A(x)$ where A is the area under the graph of f from $t = 0$ to $t = x$ for $x = 0.25, 0.5, 1,$ 1.5, 2, 2.5, and 3.

(b) Make a table of pairs $(x, A(x))$ for the values of x given in part (a) and plot them using graph paper. Connect the plotted points with a smooth curve.

(c) Use a quadratic regression equation to model the data in part (b) and overlay its graph on a scatter plot of the data.

(d) Make a conjecture about the exact value of $A(x)$ for any x greater than zero.

(e) Find the derivative of the $A(x)$ found in part (d). Record any observations.

57. Area as a Function Consider the function $f(t) = 3t^2$.

(a) Use NINT on a calculator to compute $A(x)$ where A is the area under the graph of f from $t = 0$ to $t = x$ for $x = 0.25, 0.5, 1,$ 1.5, 2, 2.5, and 3.

(b) Make a table of pairs $(x, A(x))$ for the values of x given in part (a) and plot them using graph paper. Connect the plotted points with a smooth curve.

(c) Use a cubic regression equation to model the data in part (b) and overlay its graph on a scatter plot of the data.

(d) Make a conjecture about the exact value of $A(x)$ for any x greater than zero.

(e) Find the derivative of the $A(x)$ found in part (d). Record any observations.

58. Group Activity Based on Exercises 56 and 57, discuss how derivatives (slope functions) and integrals (area functions) may be connected.

CHAPTER 10 **Key Ideas**

PROPERTIES, THEOREMS, AND FORMULAS

Limit at *a* (informal definition)   794

Average Rate of Change   796

Derivative at a Point   797

Derivative at a Point (easier for computing)   797

Derivative   798

Limit at Infinity (informal definition)   805

Definite Integral   808

Properties of Limits   815

Limits at Infinity (informal definition)   819

Symmetric Difference Quotient   826

Numerical Derivative of *f* at *a*   826

Numerical Derivative of *f*   826

Numerical Integral   827

PROCEDURE

Computing a Derivative From Data   829

CHAPTER 10 **Review Exercises**

The collection of exercises marked in red could be used as a chapter test.

In Exercises 1–4, use the graph of the function $y = f(x)$ to find
(a) $\lim_{x \to 1^-} f(x)$ and **(b)** $\lim_{x \to 1} f(x)$.

1.

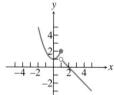

2.

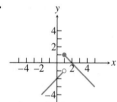

3.

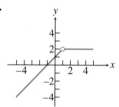

4.

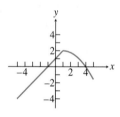

In Exercises 5–10, find the limit at the indicated point, if it exists. Support your answer graphically.

5. $f(x) = \dfrac{x - 1}{x^2 + 1}, x = -1$

6. $f(x) = \dfrac{\sin 5x}{x}, x = 0$

7. $f(x) = \dfrac{x^2 - 3x - 10}{x + 2}, x = -2$

8. $f(x) = |x - 1|, x = 1$

9. $f(x) = 2 \tan^{-1} x, x = 0$

10. $f(x) = \dfrac{2}{1 - 2^x}, x = 0$

In Exercises 11–14, find the limit. Support your answer with an appropriate table.

11. $\lim\limits_{x \to -\infty} \dfrac{-1}{(x + 2)^2}$

12. $\lim\limits_{x \to \infty} \dfrac{x + 5}{x - 3}$

13. $\lim\limits_{x \to \infty} \dfrac{2 - x^2}{x}$

14. $\lim\limits_{x \to -\infty} \dfrac{x^2}{x - 2}$

In Exercises 15–18, find the limit.

15. $\lim\limits_{x \to 2^+} \dfrac{1}{x - 2}$

16. $\lim\limits_{x \to 2^-} \dfrac{1}{x^2 - 4}$

17. $\lim\limits_{x \to 0} \dfrac{1/(2 + x) - 1/2}{x}$

18. $\lim\limits_{x \to 0} \dfrac{(2 + x)^3 - 8}{x}$

In Exercises 19–20, find the vertical and horizontal asymptotes, if any.

19. $f(x) = \dfrac{x - 5}{x^2 + 6x + 5}$

20. $f(x) = \dfrac{x^2 + 1}{2x - 4}$

In Exercises 21–26, find the limit algebraically.

21. $\lim\limits_{x \to 3} \dfrac{x^2 + 2x - 15}{3 - x}$

22. $\lim\limits_{x \to 1} \dfrac{x^2 - 4x + 3}{x - 1}$

23. $\lim\limits_{x \to 0} \dfrac{1/(-3 + x) + 1/3}{x}$

24. $\lim\limits_{x \to 2} \dfrac{\tan (3x - 6)}{x - 2}$

25. $\lim\limits_{x \to 2} \dfrac{x^2 - 5x + 6}{x^2 - 3x + 2}$

26. $\lim\limits_{x \to 3} \dfrac{(x - 3)^2}{x - 3}$

In Exercises 27 and 28, state a formula for the continuous extension of the function. (See Exercise 90, Section 10.3.)

27. $f(x) = \dfrac{x^3 - 1}{x - 1}$

28. $f(x) = \dfrac{x^2 - 6x + 5}{x - 5}$

In Exercises 29 and 30, use the limit definition to find the derivative of the function at the specified point, if it exists. Support your answer numerically with an NDER calculator estimate.

29. $f(x) = 1 - x - 2x^2$ at $x = 2$

30. $f(x) = (x + 3)^2$ at $x = 2$

In Exercises 31 and 32, find **(a)** the average rate of change of the function over the interval $[3, 3.01]$ and **(b)** the instantaneous rate of change at $x = 3$.

31. $f(x) = x^2 + 2x - 3$　　**32.** $f(x) = \dfrac{3}{x + 2}$

In Exercises 33 and 34, find **(a)** the slope and **(b)** an equation of the line tangent to the graph of the function at the indicated point.

33. $f(x) = x^3 - 2x + 1$ at $x = 1$

34. $f(x) = \sqrt{x - 4}$ at $x = 7$

In Exercises 35 and 36, find the derivative of f.

35. $f(x) = 5x^2 + 7x - 1$　　**36.** $f(x) = 2 - 8x + 3x^2$

In Exercises 37 and 38, complete the following for the indicated interval $[a, b]$.

(a) Verify the given function is nonnegative.

(b) Use a calculator to find the LRAM, RRAM, and average approximations for the area under the graph of the function from $x = a$ to $x = b$ with 50 approximating rectangles.

37. $f(x) = (x - 5)^2$; $[0, 4]$

38. $f(x) = 2x^2 - 3x + 1$; $[1, 5]$

39. Gasoline Prices The annual average retail price for unleaded regular gasoline in the United States for the years 1990–2003 is given in Table 10.12.

Table 10.12 U.S. Average Unleaded Regular Gasoline Price	
Year	Price (cents per gallon)
1990	116.4
1991	114.0
1992	112.7
1993	110.8
1994	111.2
1995	114.7
1996	123.1
1997	123.4
1998	105.9
1999	116.5
2000	151.0
2001	146.1
2002	135.8
2003	159.1

Source: Energy Information Administration, U.S. Department of Energy, World Almanac and Book of Facts, 2005.

(a) Draw a scatter plot of the data in Table 10.12. Use $x = 0$ for 1990, $x = 1$ for 1991, and so forth.

(b) Find the average rate of change from 1990 to 1991 and from 1997 to 1998.

(c) From what year to the next consecutive year does the average rate exhibit the greatest increase?

(d) From what year to the next consecutive year does the average rate exhibit the greatest decrease?

(e) Find a linear regression model for the data and overlay its graph on a scatter plot of the data.

(f) Group Activity Find a cubic regression model for the data and overlay its graph on a scatter plot of the data. Discuss pro and con arguments that this is a good model for the gasoline price data. Compare the cubic model with the linear model. Which one does your group think is best? Why?

(g) Writing to Learn Use the cubic regression model found in part (f) and NDER to find the instantaneous rate of change in 1997, 1998, 1999, and 2000. How could the cubic model give some misleading information?

(h) Writing to Learn Use the cubic regression model found in part (f) to predict the average price of a gallon of unleaded regular gasoline in 2007. Do you think this is a reasonable estimate? Give reasons.

40. An Interesting Connection

Let $A(x) = $ NINT $(\cos t, t, 0, x)$

(a) Draw a scatter plot of the pairs $(x, A(x))$ for $x = 0, 0.4, 0.8, 0.12, \ldots, 6.0, 6.4$.

(b) Find a function that seems to model the data in part (a) and overlay its graph on a scatter plot of the data.

(c) Assuming that the function found in part (b) agrees with $A(x)$ for all values of x, what is the derivative of $A(x)$?

(d) Writing to Learn Describe what seems to be true about the derivative of NINT $(f(t), t, 0, x)$.

CHAPTER 10 Project

Estimating Population Growth Rates

Las Vegas, Nevada and its surrounding cities represent one of the fastest-growing areas in the United States. Clark County, the county in which Las Vegas is located, has grown by over one million people in the past three decades. The data in the table below (obtained from the web site http://cber.unlv.edu/pop.html) summarizes the growth of Clark County from 1970 through 2004.

Year	Population
1970	277,230
1980	463,087
1990	770,280
1995	1,055,435
1999	1,327,145
2000	1,394,440
2001	1,485,855
2002	1,549,657
2003	1,620,748
2004	1,715,337

EXPLORATIONS

1. Enter the data in the table above into your graphing calculator or computer. (Let $t = 0$ represent 1970.) Make a scatter plot of the data.

2. Find the average population growth rates for Clark County from 1970 to 2004, from 1980 to 2004, from 1990 to 2004, and from 1995 to 2004.

3. Use your calculator or computer to find an exponential regression equation to model the population data set (see your grapher's guidebook for instructions on how to do this).

4. Use the exponential model you just found in question (3) and your calculator/computer NDER feature to estimate the instantaneous population growth rate in 2004. Which of the average growth rates you found in question (2) most closely matches this instantaneous growth rate? Explain why this makes sense.

5. Use the exponential regression model you found in question (3) to predict the population of Clark County in the years 2010, 2020, and 2030. Compare your predictions with the predictions in the table below, obtained from the web site www.co.clark.nv.us. Which predictions seem more reasonable? Explain.

Year	Population
2010	2,089,102
2020	2,578,221
2030	2,941,398

Source: http://www.co.clark.nv.us/ comprehensive_planning/Advanced/ Demographics/Population_Forecasts/ Pop_Forecast_2002to2035.htm.

Appendixes Overview

This section contains a review of some basic algebraic skills. (You should read Section P.1 before reading this appendix.) Radical and rational expressions are introduced and radical expressions are simplified algebraically. We add, subtract, and multiply polynomials and factor simple polynomials by a variety of techniques. These factoring techniques are used to add, subtract, multiply, and divide fractional expressions.

A.1
Radicals and Rational Exponents

What you'll learn about

- Radicals

- Simplifying Radical Expressions

- Rationalizing the Denominator

- Rational Exponents

. . . and why

You need to review these basic algebraic skills if you don't remember them.

Radicals

If $b^2 = a$, then b is a **square root** of a. For example, both 2 and -2 are square roots of 4 because $2^2 = (-2)^2 = 4$. Similarly, b is a **cube root** of a if $b^3 = a$. For example, 2 is a cube root of 8 because $2^3 = 8$.

DEFINITION Real nth Root of a Real Number

Let n be an integer greater than 1 and a and b real numbers.

1. If $b^n = a$, then b is an **nth root** of a.

2. If a has an nth root, the **principal nth root** of a is the nth root having the same sign as a.

The principal nth root of a is denoted by the **radical expression** $\sqrt[n]{a}$. The positive integer n is the **index** of the radical and a is the **radicand**.

Every real number has exactly one real *nth root* whenever n is odd. For instance, 2 is the only real cube root of 8. When n is even, positive real numbers have two real *nth roots* and negative real numbers have no real *nth roots*. For example, the real fourth roots of 16 are ± 2, and -16 has no real fourth roots. The *principal* fourth root of 16 is 2.

When $n = 2$, special notation is used for roots. We omit the index and write $\sqrt{a}$ instead of $\sqrt[2]{a}$. If a is a positive real number and n a positive even integer, its two nth roots are denoted by $\sqrt[n]{a}$ and $-\sqrt[n]{a}$.

EXAMPLE 1 Finding Principal nth Roots

(a) $\sqrt{36} = 6$ because $6^2 = 36$.

(b) $\sqrt[3]{\dfrac{27}{8}} = \dfrac{3}{2}$ because $\left(\dfrac{3}{2}\right)^3 = \dfrac{27}{8}$.

(c) $\sqrt[3]{-\dfrac{27}{8}} = -\dfrac{3}{2}$ because $\left(-\dfrac{3}{2}\right)^3 = -\dfrac{27}{8}$.

(d) $\sqrt[4]{-625}$ is not a real number because the index 4 is even and the radicand -625 is negative (there is *no* real number whose fourth power is negative).

Now try Exercises 7 and 9.

PRINCIPAL *n*TH ROOTS AND CALCULATORS

Most calculators have a key for the principal *n*th root. Use this feature of your calculator to check the computations in Example 1.

Here are some properties of radicals together with examples that help illustrate their meaning.

Properties of Radicals

Let u and v be real numbers, variables, or algebraic expressions, and m and n be positive integers greater than 1. We assume that all of the roots are real numbers and all of the denominators are not zero.

Property	Example
1. $\sqrt[n]{uv} = \sqrt[n]{u} \cdot \sqrt[n]{v}$	$\sqrt{75} = \sqrt{25 \cdot 3}$
	$= \sqrt{25} \cdot \sqrt{3} = 5\sqrt{3}$
2. $\sqrt[n]{\dfrac{u}{v}} = \dfrac{\sqrt[n]{u}}{\sqrt[n]{v}}$	$\dfrac{\sqrt[4]{96}}{\sqrt[4]{6}} = \sqrt[4]{\dfrac{96}{6}} = \sqrt[4]{16} = 2$
3. $\sqrt[m]{\sqrt[n]{u}} = \sqrt[m \cdot n]{u}$	$\sqrt{\sqrt[3]{7}} = \sqrt[2 \cdot 3]{7} = \sqrt[6]{7}$
4. $\left(\sqrt[n]{u}\right)^n = u$	$\left(\sqrt[4]{5}\right)^4 = 5$
5. $\sqrt[n]{u^m} = \left(\sqrt[n]{u}\right)^m$	$\sqrt[3]{27^2} = \left(\sqrt[3]{27}\right)^2 = 3^2 = 9$
6. $\sqrt[n]{u^n} = \begin{cases} \lvert u \rvert & n \text{ even} \\ u & n \text{ odd} \end{cases}$	$\sqrt{(-6)^2} = \lvert -6 \rvert = 6$
	$\sqrt[3]{(-6)^3} = -6$

CAUTION

Without the restriction that preceded the list, Property 5 would need special attention. For example,

$$\sqrt{(-3)^2} \neq \left(\sqrt{-3}\right)^2$$

because $\sqrt{-3}$ on the right is not a real number.

Simplifying Radical Expressions

Many simplifying techniques for roots of real numbers have been rendered obsolete because of calculators. For example, when determining the decimal form of $1/\sqrt{2}$, it was once very common first to change the fraction so that the radical was in the numerator:

$$\frac{1}{\sqrt{2}} = \frac{1}{\sqrt{2}} \cdot \frac{\sqrt{2}}{\sqrt{2}} = \frac{\sqrt{2}}{2}.$$

Using paper and pencil, it was then easier to divide a decimal approximation for $\sqrt{2}$ by 2 than to divide that decimal into 1. Now either form is quickly computed with a calculator. However, these techniques are still valid for radicals involving algebraic expressions and for numerical computations when you need exact answers. Example 2 illustrates the technique of *removing factors from radicands*.

PROPERTIES OF EXPONENTS

Check the Properties of Exponents on page 8 of Section P.1 to see why

$$16 = 2^4 \text{ and } 9x^4 = (3x^2)^2.$$

EXAMPLE 2   Removing Factors from Radicands

(a) $\sqrt[4]{80} = \sqrt[4]{16 \cdot 5}$ Finding greatest fourth-power factor

$\phantom{\textbf{(a)} \sqrt[4]{80}} = \sqrt[4]{2^4 \cdot 5}$ $16 = 2^4$

$\phantom{\textbf{(a)} \sqrt[4]{80}} = \sqrt[4]{2^4} \cdot \sqrt[4]{5}$ Property 1

$\phantom{\textbf{(a)} \sqrt[4]{80}} = 2\sqrt[4]{5}$ Property 6

(b) $\sqrt{18x^5} = \sqrt{9x^4 \cdot 2x}$ Finding greatest square factor

$\phantom{\textbf{(b)} \sqrt{18x^5}} = \sqrt{(3x^2)^2 \cdot 2x}$ $9x^4 = (3x^2)^2$

$\phantom{\textbf{(b)} \sqrt{18x^5}} = 3x^2\sqrt{2x}$ Properties 1 and 6

continued

(c) $\sqrt[4]{x^4 y^4} = \sqrt[4]{(xy)^4}$ Finding greatest fourth-power factor

      $= |xy|$ Property 6

(d) $\sqrt[3]{-24y^6} = \sqrt[3]{(-2y^2)^3 \cdot 3}$ Finding greatest cube factor

      $= -2y^2 \sqrt[3]{3}$ Properties 1 and 6

Now try Exercises 29 and 33.

Rationalizing the Denominator

The process of rewriting fractions containing radicals so that the denominator is free of radicals is **rationalizing the denominator**. When the denominator has the form $\sqrt[n]{u^k}$, multiplying numerator and denominator by $\sqrt[n]{u^{n-k}}$ and using Property 6 will eliminate the radical from the denominator because

$$\sqrt[n]{u^k} \cdot \sqrt[n]{u^{n-k}} = \sqrt[n]{u^{k+n-k}} = \sqrt[n]{u^n}.$$

Example 3 illustrates the process.

EXAMPLE 3   Rationalizing the Denominator

(a) $\sqrt{\dfrac{2}{3}} = \dfrac{\sqrt{2}}{\sqrt{3}} = \dfrac{\sqrt{2}}{\sqrt{3}} \cdot \dfrac{\sqrt{3}}{\sqrt{3}} = \dfrac{\sqrt{6}}{3}$

(b) $\dfrac{1}{\sqrt[4]{x}} = \dfrac{1}{\sqrt[4]{x}} \cdot \dfrac{\sqrt[4]{x^3}}{\sqrt[4]{x^3}} = \dfrac{\sqrt[4]{x^3}}{\sqrt[4]{x^4}} = \dfrac{\sqrt[4]{x^3}}{|x|}$

(c) $\sqrt[5]{\dfrac{x^2}{y^3}} = \dfrac{\sqrt[5]{x^2}}{\sqrt[5]{y^3}} = \dfrac{\sqrt[5]{x^2}}{\sqrt[5]{y^3}} \cdot \dfrac{\sqrt[5]{y^2}}{\sqrt[5]{y^2}} = \dfrac{\sqrt[5]{x^2 y^2}}{\sqrt[5]{y^5}} = \dfrac{\sqrt[5]{x^2 y^2}}{y}$

Now try Exercise 37.

Rational Exponents

We know how to handle exponential expressions with integer exponents (see Section P.1). For example, $x^3 \cdot x^4 = x^7$, $(x^3)^2 = x^6$, $x^5/x^2 = x^3$, $x^{-2} = 1/x^2$, and so forth. But exponents can also be rational numbers. How should we define $x^{1/2}$? If we assume that the same rules that apply for integer exponents also apply for rational exponents we get a clue. For example, we want

$$x^{1/2} \cdot x^{1/2} = x^1.$$

This equation suggests that $x^{1/2} = \sqrt{x}$. In general, we have the following definition.

DEFINITION Rational Exponents

Let u be a real number, variable, or algebraic expression, and n an integer greater than 1. Then

$$u^{1/n} = \sqrt[n]{u}.$$

If m is a positive integer, m/n is in reduced form, and all roots are real numbers, then

$$u^{m/n} = (u^{1/n})^m = (\sqrt[n]{u})^m \qquad \text{and} \qquad u^{m/n} = (u^m)^{1/n} = \sqrt[n]{u^m}.$$

The numerator of a rational exponent is the *power* to which the base is raised, and the denominator is the *root* to be taken. The fraction m/n needs to be in reduced form because, for instance,

$$u^{2/3} = (\sqrt[3]{u})^2$$

is defined for all real numbers u (every real number has a cube root), but

$$u^{4/6} = (\sqrt[6]{u})^4$$

is defined only for $u \geq 0$ (only nonnegative real numbers have sixth roots).

SIMPLIFYING RADICALS

If you also want the radical form in Example 4d to be simplified, then continue as follows:

$$\frac{1}{\sqrt{z^3}} = \frac{1}{\sqrt{z^3}} \cdot \frac{\sqrt{z}}{\sqrt{z}} = \frac{\sqrt{z}}{z^2}$$

EXAMPLE 4   Converting Radicals to Exponentials and Vice Versa

(a) $\sqrt{(x + y)^3} = (x + y)^{3/2}$

(b) $3x\sqrt[5]{x^2} = 3x \cdot x^{2/5} = 3x^{7/5}$

(c) $x^{2/3}y^{1/3} = (x^2 y)^{1/3} = \sqrt[3]{x^2 y}$

(d) $z^{-3/2} = \dfrac{1}{z^{3/2}} = \dfrac{1}{\sqrt{z^3}}$

Now try Exercises 43 and 47.

An expression involving powers is *simplified* if each factor appears only once and all exponents are positive. Example 5 illustrates.

EXAMPLE 5   Simplifying Exponential Expressions

(a) $(x^2 y^9)^{1/3}(xy^2) = (x^{2/3}y^3)(xy^2) = x^{5/3}y^5$

(b) $\left(\dfrac{3x^{2/3}}{y^{1/2}}\right)\left(\dfrac{2x^{-1/2}}{y^{2/5}}\right) = \dfrac{6x^{1/6}}{y^{9/10}}$

Now try Exercise 61.

Example 6 suggests how to simplify a sum or difference of radicals.

EXAMPLE 6   Combining Radicals

(a) $2\sqrt{80} - \sqrt{125} = 2\sqrt{16 \cdot 5} - \sqrt{25 \cdot 5}$ Find greatest square factors.

$\qquad\qquad = 8\sqrt{5} - 5\sqrt{5}$ Remove factors from radicands.

$\qquad\qquad = 3\sqrt{5}$ Distributive property

(b) $\sqrt{4x^2 y} - \sqrt{y^3} = \sqrt{(2x)^2 y} - \sqrt{y^2 y}$ Find greatest square factors.

$\qquad\qquad = 2|x|\sqrt{y} - |y|\sqrt{y}$ Remove factors from radicands.

$\qquad\qquad = (2|x| - |y|)\sqrt{y}$ Distributive property.

Now try Exercise 71.

Here's a summary of the procedures we use to *simplify expressions* involving radicals.

Simplifying Radical Expressions

1. Remove factors from the radicand (see Example 2).
2. Eliminate radicals from denominators and denominators from radicands (see Example 3).
3. Combine sums and differences of radicals, if possible (see Example 6).

APPENDIX A.1 EXERCISES

In Exercises 1–6, find the indicated real roots.

1. Square roots of 81

2. Fourth roots of 81

3. Cube roots of 64

4. Fifth roots of 243

5. Square roots of 16/9

6. Cube roots of $-27/8$

In Exercises 7–12, evaluate the expression without using a calculator.

7. $\sqrt{144}$

8. $\sqrt{-16}$

9. $\sqrt[3]{-216}$

10. $\sqrt[3]{216}$

11. $\sqrt[3]{-\dfrac{64}{27}}$

12. $\sqrt{\dfrac{64}{25}}$

In Exercises 13–22, use a calculator to evaluate the expression.

13. $\sqrt[4]{256}$

14. $\sqrt[5]{3125}$

15. $\sqrt[3]{15.625}$

16. $\sqrt{12.25}$

17. $81^{3/2}$

18. $16^{5/4}$

19. $32^{-2/5}$

20. $27^{-4/3}$

21. $\left(-\dfrac{1}{8}\right)^{-1/3}$

22. $\left(-\dfrac{125}{64}\right)^{-1/3}$

In Exercises 23–26, use the information from the grapher screens below to evaluate the expression.

```
1.5^3
          3.375
4.41^2
          19.4481
```

```
1.3^2
          1.69
2.1^4
          19.4481
```

23. $\sqrt{1.69}$

24. $\sqrt{19.4481}$

25. $\sqrt[4]{19.4481}$

26. $\sqrt[3]{3.375}$

In Exercises 27–36, simplify by removing factors from the radicand.

27. $\sqrt{288}$

28. $\sqrt[3]{500}$

29. $\sqrt[3]{-250}$

30. $\sqrt[4]{192}$

31. $\sqrt{2x^3y^4}$

32. $\sqrt[3]{-27x^3y^6}$

33. $\sqrt[4]{3x^8y^6}$

34. $\sqrt[3]{8x^6y^4}$

35. $\sqrt[5]{96x^{10}}$

36. $\sqrt{108x^4y^9}$

In Exercises 37–42, rationalize the denominator.

37. $\dfrac{4}{\sqrt[3]{2}}$

38. $\dfrac{1}{\sqrt{5}}$

39. $\dfrac{1}{\sqrt[5]{x^2}}$

40. $\dfrac{2}{\sqrt[4]{y}}$

41. $\sqrt[3]{\dfrac{x^2}{y}}$

42. $\sqrt[5]{\dfrac{a^3}{b^2}}$

In Exercises 43–46, convert to exponential form.

43. $\sqrt[3]{(a+2b)^2}$

44. $\sqrt[5]{x^2y^3}$

45. $2x\sqrt[3]{x^2y}$

46. $xy\sqrt[4]{xy^3}$

In Exercises 47–50, convert to radical form.

47. $a^{3/4}b^{1/4}$

48. $x^{2/3}y^{1/3}$

49. $x^{-5/3}$

50. $(xy)^{-3/4}$

In Exercises 51–56, write using a single radical.

51. $\sqrt{\sqrt{2x}}$

52. $\sqrt{\sqrt[3]{3x^2}}$

53. $\sqrt[4]{\sqrt{xy}}$

54. $\sqrt[3]{\sqrt{ab}}$

55. $\dfrac{\sqrt[5]{a^2}}{\sqrt[3]{a}}$

56. $\sqrt{a}\sqrt[3]{a^2}$

In Exercises 57–64, simplify the exponential expression.

57. $\dfrac{a^{3/5}a^{1/3}}{a^{3/2}}$

58. $(x^2y^4)^{1/2}$

59. $(a^{5/3}b^{3/4})(3a^{1/3}b^{5/4})$

60. $\left(\dfrac{x^{1/2}}{y^{2/3}}\right)^6$

61. $\left(\dfrac{-8x^6}{y^{-3}}\right)^{2/3}$

62. $\dfrac{(p^2q^4)^{1/2}}{(27q^3p^6)^{1/3}}$

63. $\dfrac{(x^9y^6)^{-1/3}}{(x^6y^2)^{-1/2}}$

64. $\left(\dfrac{2x^{1/2}}{y^{2/3}}\right)\left(\dfrac{3x^{-2/3}}{y^{1/2}}\right)$

In Exercises 65–74, simplify the radical expression.

65. $\sqrt{9x^{-6}y^4}$

66. $\sqrt{16y^8z^{-2}}$

67. $\sqrt[4]{\dfrac{3x^8y^2}{8x^2}}$

68. $\sqrt[5]{\dfrac{4x^6y}{9x^3}}$

69. $\sqrt[3]{\dfrac{4x^2}{y^2}}\cdot\sqrt[3]{\dfrac{2x^2}{y}}$

70. $\sqrt[5]{9ab^6}\cdot\sqrt[5]{27a^2b^{-1}}$

71. $3\sqrt{48}-2\sqrt{108}$

72. $2\sqrt{175}-4\sqrt{28}$

73. $\sqrt{x^3}-\sqrt{4xy^2}$

74. $\sqrt{18x^2y}+\sqrt{2y^3}$

In Exercises 75–82, replace $\bigcirc$ with $<$, $=$, or $>$ to make a true statement.

75. $\sqrt{2+6}\,\bigcirc\,\sqrt{2}+\sqrt{6}$

76. $\sqrt{4}+\sqrt{9}\,\bigcirc\,\sqrt{4+9}$

77. $(3^{-2})^{-1/2}\,\bigcirc\,3$

78. $(2^{-3})^{1/3}\,\bigcirc\,2$

79. $\sqrt[4]{(-2)^4}\,\bigcirc\,-2$

80. $\sqrt[3]{(-2)^3}\,\bigcirc\,-2$

81. $2^{2/3}\,\bigcirc\,3^{3/4}$

82. $4^{-2/3}\,\bigcirc\,3^{-3/4}$

83. The time t (in seconds) that it takes for a pendulum to complete one period is approximately $t = 1.1\sqrt{L}$, where L is the length (in feet) of the pendulum. How long is the period of a pendulum of length 10 ft?

84. The time t (in seconds) that it takes for a rock to fall a distance d (in meters) is approximately $t = 0.45\sqrt{d}$. How long does it take for the rock to fall a distance of 200 m?

85. Writing to Learn Explain why $\sqrt[n]{a}$ and a real nth root of a need not have the same value.

A.2
Polynomials and Factoring

Adding, Subtracting, and Multiplying Polynomials

A **polynomial in x** is any expression that can be written in the form

$$a_nx^n + a_{n-1}x^{n-1} + \cdots + a_1x + a_0,$$

where n is a nonnegative integer, and $a_n \neq 0$. The numbers $a_{n-1}, \ldots, a_1, a_0$ are real numbers called **coefficients**. The **degree of the polynomial** is n and the **leading coefficient** is a_n. Polynomials with one, two, or three terms are **monomials**, **binomials**, or **trinomials**, respectively. A polynomial written with powers of x in *descending order* is in **standard form**.

To add or subtract polynomials, we add or subtract *like terms* using the distributive property. Terms of polynomials that have the same variable each raised to the same power are **like terms**.

> ### EXAMPLE 1 Adding and Subtracting Polynomials
> **(a)** $(2x^3 - 3x^2 + 4x - 1) + (x^3 + 2x^2 - 5x + 3)$
>
> **(b)** $(4x^2 + 3x - 4) - (2x^3 + x^2 - x + 2)$
>
> **SOLUTION**
>
> **(a)** We group like terms and then combine them as follows:
>
> $$(2x^3 + x^3) + (-3x^2 + 2x^2) + (4x + (-5x)) + (-1 + 3)$$
> $$= 3x^3 - x^2 - x + 2$$
>
> **(b)** We group like terms and then combine them as follows:
>
> $$(0 - 2x^3) + (4x^2 - x^2) + (3x - (-x)) + (-4 - 2)$$
> $$= -2x^3 + 3x^2 + 4x - 6$$
>
> *Now try Exercises 9 and 11.*

To **expand the product** of two polynomials we use the distributive property. Here is what the procedure looks like when we multiply the binomials $3x + 2$ and $4x - 5$.

$(3x + 2)(4x - 5)$

$$= 3x(4x - 5) + 2(4x - 5) \qquad \text{Distributive property}$$
$$= (3x)(4x) - (3x)(5) + (2)(4x) - (2)(5) \qquad \text{Distributive property}$$
$$= \underbrace{12x^2}_{\substack{\text{Product of} \\ \text{First terms}}} - \underbrace{15x}_{\substack{\text{Product of} \\ \text{Outer terms}}} + \underbrace{8x}_{\substack{\text{Product of} \\ \text{Inner terms}}} - \underbrace{10}_{\substack{\text{Product of} \\ \text{Last terms}}}$$

In the above **FOIL method** for products of binomials, the outer (O) and inner (I) terms are like terms and can be added to give

$$(3x + 2)(4x - 5) = 12x^2 - 7x - 10.$$

Multiplying two polynomials requires multiplying each term of one polynomial by every term of the other polynomial. A convenient way to compute a product is to arrange the polynomials in standard form one on top of another so their terms align vertically, as illustrated in Example 2.

EXAMPLE 2   Multiplying Polynomials in Vertical Form

Write $(x^2 - 4x + 3)(x^2 + 4x + 5)$ in standard form.

SOLUTION

$$
\begin{array}{l}
x^2 - 4x + 3 \\
\underline{x^2 + 4x + 5} \\
x^4 - 4x^3 + 3x^2 \qquad\qquad = x^2(x^2 - 4x + 3) \\
\quad 4x^3 - 16x^2 + 12x \qquad = 4x(x^2 - 4x + 3) \\
\quad 5x^2 - 20x + 15 \quad = 5(x^2 - 4x + 3) \\
\overline{x^4 + 0x^3 - 8x^2 - 8x + 15} \qquad \text{Add.}
\end{array}
$$

Thus,

$$(x^2 - 4x + 3)(x^2 + 4x + 5) = x^4 - 8x^2 - 8x + 15.$$

Now try Exercise 33.

Special Products

Certain products provide patterns that will be useful when we factor polynomials. Here is a list of some special products for binomials.

Special Binomial Products

Let u and v be real numbers, variables, or algebraic expressions.

1. **Product of a sum and a difference:** $(u + v)(u - v) = u^2 - v^2$

2. **Square of a sum:** $(u + v)^2 = u^2 + 2uv + v^2$

3. **Square of a difference:** $(u - v)^2 = u^2 - 2uv + v^2$

4. **Cube of a sum:** $(u + v)^3 = u^3 + 3u^2v + 3uv^2 + v^3$

5. **Cube of a difference:** $(u - v)^3 = u^3 - 3u^2v + 3uv^2 - v^3$

EXAMPLE 3   Using Special Products

Expand the products.

(a) $(3x + 8)(3x - 8) = (3x)^2 - 8^2$ Product of a sum and a difference

$= 9x^2 - 64$ Simplify.

(b) $(5y - 4)^2 = (5y)^2 - 2(5y)(4) + 4^2$ Square of a difference

$= 25y^2 - 40y + 16$ Simplify.

continued

(c) $(2x - 3y)^3 = (2x)^3 - 3(2x)^2(3y)$
$$+ 3(2x)(3y)^2 - (3y)^3 \qquad \text{Cube of a difference}$$
$$= 8x^3 - 36x^2y + 54xy^2 - 27y^3 \qquad \text{Simplify.}$$

Now try Exercises 23, 25, and 27.

Factoring Polynomials Using Special Products

When we write a polynomial as a product of two or more **polynomial factors** we are **factoring a polynomial**. Unless specified otherwise, we factor polynomials into factors of lesser degree and with integer coefficients in this appendix. A polynomial that cannot be factored using integer coefficients is a **prime polynomial**.

A polynomial is **completely factored** if it is written as a product of its prime factors. For example,

$$2x^2 + 7x - 4 = (2x - 1)(x + 4)$$

and

$$x^3 + x^2 + x + 1 = (x + 1)(x^2 + 1)$$

are completely factored (it can be shown that $x^2 + 1$ is prime). However,

$$x^3 - 9x = x(x^2 - 9)$$

is not completely factored because $(x^2 - 9)$ is *not* prime. In fact, $x^2 - 9 = (x - 3)(x + 3)$ and

$$x^3 - 9x = x(x - 3)(x + 3)$$

is completely factored.

The first step in factoring a polynomial is to remove common factors from its terms using the distributive property as illustrated by Example 4.

EXAMPLE 4 Removing Common Factors

(a) $2x^3 + 2x^2 - 6x = 2x(x^2 + x - 3)$ $2x$ is the common factor.

(b) $u^3v + uv^3 = uv(u^2 + v^2)$ uv is the common factor.

Now try Exercise 43.

Recognizing the expanded form of the five special binomial products will help us factor them. The special form that is easiest to identify is the difference of two squares. The two binomial factors have opposite signs:

$$u^2 - v^2 = (u + v)(u - v).$$

Two squares   Square roots

Difference   Opposite signs

EXAMPLE 5   Factoring the Difference of Two Squares

(a) $25x^2 - 36 = (5x)^2 - 6^2$ Difference of two squares

            $= (5x + 6)(5x - 6)$ Factors are prime.

(b) $4x^2 - (y + 3)^2 = (2x)^2 - (y + 3)^2$ Difference of two squares

                  $= [2x + (y + 3)][2x - (y + 3)]$ Factors are prime.

                  $= (2x + y + 3)(2x - y - 3)$ Simplify.

Now try Exercise 45.

A perfect square trinomial is the square of a binomial and has one of the two forms shown here. The first and last terms are squares of u and v, and the middle term is twice the product of u and v. The operation signs before the middle term and in the binomial factor are the same.

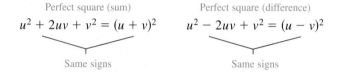

Perfect square (sum)   Perfect square (difference)

$$u^2 + 2uv + v^2 = (u + v)^2 \qquad u^2 - 2uv + v^2 = (u - v)^2$$

         Same signs   Same signs

EXAMPLE 6   Factoring Perfect Square Trinomials

(a) $9x^2 + 6x + 1 = (3x)^2 + 2(3x)(1) + 1^2$ $u = 3x, v = 1$

                $= (3x + 1)^2$

(b) $4x^2 - 12xy + 9y^2 = (2x)^2 - 2(2x)(3y) + (3y)^2$ $u = 2x, v = 3y$

                  $= (2x - 3y)^2$

Now try Exercise 49.

In the sum and difference of two cubes, notice the pattern of the signs.

     Same signs   Same signs

$$u^3 + v^3 = (u + v)(u^2 - uv + v^2) \qquad u^3 - v^3 = (u - v)(u^2 + uv + v^2)$$

         Opposite signs   Opposite signs

EXAMPLE 7   Factoring the Sum and Difference of Two Cubes

(a) $x^3 - 64 = x^3 - 4^3$ Difference of two cubes

            $= (x - 4)(x^2 + 4x + 16)$ Factors are prime.

(b) $8x^3 + 27 = (2x)^3 + 3^3$ Sum of two cubes

            $= (2x + 3)(4x^2 - 6x + 9)$ Factors are prime.

Now try Exercise 55.

Factoring Trinomials

Factoring the trinomial $ax^2 + bx + c$ into a product of binomials with integer coefficients requires factoring the integers a and c.

$$\overset{\text{Factors of } a}{\overbrace{}}$$

$$ax^2 + bx + c = (\square x + \square)(\square x + \square)$$

$$\underset{\text{Factors of } c}{\underbrace{}}$$

Because the number of integer factors of a and c is finite, we can list all possible binomial factors. Then we begin checking each possibility until we find a pair that works. (If no pair works, then the trinomial is prime.) Example 8 illustrates.

EXAMPLE 8   Factoring a Trinomial with Leading Coefficient = 1

Factor $x^2 + 5x - 14$.

SOLUTION The only factor pair of the leading coefficient is 1 and 1. The factor pairs of 14 are 1 and 14, and 2 and 7. Here are the four possible factorizations of the trinomial:

$$(x + 1)(x - 14) \qquad (x - 1)(x + 14)$$
$$(x + 2)(x - 7) \qquad (x - 2)(x + 7)$$

If you check the middle term from each factorization you will find that

$$x^2 + 5x - 14 = (x - 2)(x + 7).$$

Now try Exercise 59.

With practice you will find that it usually is not necessary to list all possible binomial factors. Often you can test the possibilities mentally.

EXAMPLE 9   Factoring a Trinomial with Leading Coefficient ≠ 1

Factor $35x^2 - x - 12$.

SOLUTION The factor pairs of the leading coefficient are 1 and 35, and 5 and 7. The factor pairs of 12 are 1 and 12, 2 and 6, and 3 and 4. The possible factorizations must be of the form

$$(x - *)(35x + ?), \qquad (x + *)(35x - ?),$$
$$(5x - *)(7x + ?), \qquad (5x + *)(7x - ?),$$

where * and ? are one of the factor pairs of 12. Because the two binomial factors have opposite signs, there are 6 possibilities for each of the four forms—a total of 24 possibilities in all. If you try them, mentally and systematically, you should find that

$$35x^2 - x - 12 = (5x - 3)(7x + 4).$$

Now try Exercise 63.

We can extend the technique of Examples 8 and 9 to trinomials in two variables as illustrated in Example 10.

EXAMPLE 10   Factoring Trinomials in *x* and *y*

Factor $3x^2 - 7xy + 2y^2$.

SOLUTION The only way to get $-7xy$ as the middle term is with $3x^2 - 7xy + 2y^2 = (3x - ?y)(x - ?y)$.

The signs in the binomials must be negative because the coefficient of y^2 is positive *and* the coefficient of the middle term is negative. Checking the two possibilities, $(3x - y)(x - 2y)$ and $(3x - 2y)(x - y)$, shows that

$$3x^2 - 7xy + 2y^2 = (3x - y)(x - 2y).$$

Now try Exercise 67.

Factoring by Grouping

Notice that $(a + b)(c + d) = ac + ad + bc + bd$. If a polynomial with four terms is the product of two binomials, we can group terms to factor. There are only three ways to group the terms and two of them work. So, if two of the possibilities fail, then it is not factorable.

EXAMPLE 11   Factoring by Grouping

(a) $3x^3 + x^2 - 6x - 2$

$$= (3x^3 + x^2) - (6x + 2) \qquad \text{Group terms.}$$
$$= x^2(3x + 1) - 2(3x + 1) \qquad \text{Factor each group.}$$
$$= (3x + 1)(x^2 - 2) \qquad \text{Distributive property}$$

(b) $2ac - 2ad + bc - bd$

$$= (2ac - 2ad) + (bc - bd) \qquad \text{Group terms.}$$
$$= 2a(c - d) + b(c - d) \qquad \text{Factor each group.}$$
$$= (c - d)(2a + b) \qquad \text{Distributive property}$$

Now try Exercise 69.

Here is a checklist for factoring polynomials.

Factoring Polynomials

1. Look for common factors.

2. Look for special polynomial forms.

3. Use factor pairs.

4. If there are four terms, try grouping.

APPENDIX A.2 EXERCISES

In Exercises 1–4, write the polynomial in standard form and state its degree.

1. $2x - 1 + 3x^2$

2. $x^2 - 2x - 2x^3 + 1$

3. $1 - x^7$

4. $x^2 - x^4 + x - 3$

In Exercises 5–8, state whether the expression is a polynomial.

5. $x^3 - 2x^2 + x^{-1}$

6. $\dfrac{2x - 4}{x}$

7. $(x^2 + x + 1)^2$

8. $1 - 3x + x^4$

In Exercises 9–18, simplify the expression. Write your answer in standard form.

9. $(x^2 - 3x + 7) + (3x^2 + 5x - 3)$

10. $(-3x^2 - 5) - (x^2 + 7x + 12)$

11. $(4x^3 - x^2 + 3x) - (x^3 + 12x - 3)$

12. $-(y^2 + 2y - 3) + (5y^2 + 3y + 4)$

13. $2x(x^2 - x + 3)$

14. $y^2(2y^2 + 3y - 4)$

15. $-3u(4u - 1)$

16. $-4v(2 - 3v^3)$

17. $(2 - x - 3x^2)(5x)$

18. $(1 - x^2 + x^4)(2x)$

In Exercises 19–40, expand the product. Use vertical alignment in Exercises 33 and 34.

19. $(x - 2)(x + 5)$

20. $(2x + 3)(4x + 1)$

21. $(3x - 5)(x + 2)$

22. $(2x - 3)(2x + 3)$

23. $(3x - y)(3x + y)$

24. $(3 - 5x)^2$

25. $(3x + 4y)^2$

26. $(x - 1)^3$

27. $(2u - v)^3$

28. $(u + 3v)^3$

29. $(2x^3 - 3y)(2x^3 + 3y)$

30. $(5x^3 - 1)^2$

31. $(x^2 - 2x + 3)(x + 4)$

32. $(x^2 + 3x - 2)(x - 3)$

33. $(x^2 + x - 3)(x^2 + x + 1)$

34. $(2x^2 - 3x + 1)(x^2 - x + 2)$

35. $(x - \sqrt{2})(x + \sqrt{2})$

36. $(x^{1/2} - y^{1/2})(x^{1/2} + y^{1/2})$

37. $(\sqrt{u} + \sqrt{v})(\sqrt{u} - \sqrt{v})$

38. $(x^2 - \sqrt{3})(x^2 + \sqrt{3})$

39. $(x - 2)(x^2 + 2x + 4)$

40. $(x + 1)(x^2 - x + 1)$

In Exercises 41–44, factor out the common factor.

41. $5x - 15$

42. $5x^3 - 20x$

43. $yz^3 - 3yz^2 + 2yz$

44. $2x(x + 3) - 5(x + 3)$

In Exercises 45–48, factor the difference of two squares.

45. $z^2 - 49$

46. $9y^2 - 16$

47. $64 - 25y^2$

48. $16 - (x + 2)^2$

In Exercises 49–52, factor the perfect square trinomial.

49. $y^2 + 8y + 16$

50. $36y^2 + 12y + 1$

51. $4z^2 - 4z + 1$

52. $9z^2 - 24z + 16$

In Exercises 53–58, factor the sum or difference of two cubes.

53. $y^3 - 8$

54. $z^3 + 64$

55. $27y^3 - 8$

56. $64z^3 + 27$

57. $1 - x^3$

58. $27 - y^3$

In Exercises 59–68, factor the trinomial.

59. $x^2 + 9x + 14$

60. $y^2 - 11y + 30$

61. $z^2 - 5z - 24$

62. $6t^2 + 5t + 1$

63. $14u^2 - 33u - 5$

64. $10v^2 + 23v + 12$

65. $12x^2 + 11x - 15$

66. $2x^2 - 3xy + y^2$

67. $6x^2 + 11xy - 10y^2$

68. $15x^2 + 29xy - 14y^2$

In Exercises 69–74, factor by grouping.

69. $x^3 - 4x^2 + 5x - 20$

70. $2x^3 - 3x^2 + 2x - 3$

71. $x^6 - 3x^4 + x^2 - 3$

72. $x^6 + 2x^4 + x^2 + 2$

73. $2ac + 6ad - bc - 3bd$

74. $3uw + 12uz - 2vw - 8vz$

In Exercises 75–90, factor completely.

75. $x^3 + x$

76. $4y^3 - 20y^2 + 25y$

77. $18y^3 + 48y^2 + 32y$

78. $2x^3 - 16x^2 + 14x$

79. $16y - y^3$

80. $3x^4 + 24x$

81. $5y + 3y^2 - 2y^3$

82. $z - 8z^4$

83. $2(5x + 1)^2 - 18$

84. $5(2x - 3)^2 - 20$

85. $12x^2 + 22x - 20$

86. $3x^2 + 13xy - 10y^2$

87. $2ac - 2bd + 4ad - bc$

88. $6ac - 2bd + 4bc - 3ad$

89. $x^3 - 3x^2 - 4x + 12$

90. $x^4 - 4x^3 - x^2 + 4x$

91. Writing to Learn Show that the grouping

$$(2ac + bc) - (2ad + bd)$$

leads to the same factorization as in Example 11b. Explain why the third possibility,

$$(2ac - bd) + (-2ad + bc)$$

does not lead to a factorization.

A.3
Fractional Expressions

What you'll learn about

- Domain of an Algebraic Expression
- Reducing Rational Expressions
- Operations with Rational Expressions
- Compound Rational Expressions

. . . and why

You need to review these basic algebraic skills if you don't remember them.

Domain of an Algebraic Expression

A quotient of two algebraic expressions, besides being another algebraic expression, is a **fractional expression**, or simply a fraction. If the quotient can be written as the ratio of two polynomials, the fractional expression is a **rational expression**. Here are examples of each.

$$\frac{x^2 - 5x + 2}{\sqrt{x^2 + 1}} \qquad \frac{2x^3 - x^2 + 1}{5x^2 - x - 3}$$

The one on the left is a fractional expression but not a rational expression. The other is both a fractional expression and a rational expression.

Unlike polynomials, which are defined for all real numbers, some algebraic expressions are not defined for some real numbers. The set of real numbers for which an algebraic expression is defined is the **domain of the algebraic expression**.

EXAMPLE 1   Finding Domains of Algebraic Expressions

(a) $3x^2 - x + 5$ **(b)** $\sqrt{x - 1}$ **(c)** $\dfrac{x}{x - 2}$

SOLUTION

(a) The domain of $3x^2 - x + 5$, like that of any polynomial, is the set of all real numbers.

(b) Because only nonnegative numbers have square roots, $x - 1 \geq 0$, or $x \geq 1$. In interval notation, the domain is $[1, \infty)$.

(c) Because division by zero is undefined, $x - 2 \neq 0$, or $x \neq 2$. The domain is the set of all real numbers except 2.   ***Now try Exercises 11 and 13.***

Reducing Rational Expressions

Let u, v, and z be real numbers, variables, or algebraic expressions. We can write rational expressions in simpler form using

$$\frac{uz}{vz} = \frac{u}{v}$$

provided $z \neq 0$. This requires that we first factor the numerator and denominator into prime factors. When all factors common to numerator and denominator have been removed, the rational expression (or rational number) is in **reduced form**.

EXAMPLE 2   Reducing Rational Expressions

Write $(x^2 - 3x)/(x^2 - 9)$ in reduced form.

SOLUTION

$$\frac{x^2 - 3x}{x^2 - 9} = \frac{x(x - 3)}{(x + 3)(x - 3)} \qquad \text{Factor completely.}$$

$$= \frac{x}{x + 3}, \quad x \neq 3 \qquad \text{Remove common factors.}$$

continued

We include $x \neq 3$ as part of the reduced form because 3 is not in the domain of the original rational expression and thus should not be in the domain of the final rational expression.

Now try Exercise 35.

Two rational expressions are **equivalent** if they have the same domain and have the same value for all numbers in the domain. The reduced form of a rational expression must have the same domain as the original rational expression. This is why we attached the restriction $x \neq 3$ to the reduced form in Example 2.

Operations with Rational Expressions

Two fractions are **equal**, $u/v = z/w$, if and only if $uw = vz$. Here is how we operate with fractions.

Operations with Fractions

Let u, v, w, and z be real numbers, variables, or algebraic expressions. All of the denominators are assumed to be different from zero.

Operation	Example
1. $\dfrac{u}{v} + \dfrac{w}{v} = \dfrac{u + w}{v}$	$\dfrac{2}{3} + \dfrac{5}{3} = \dfrac{2 + 5}{3} = \dfrac{7}{3}$
2. $\dfrac{u}{v} + \dfrac{w}{z} = \dfrac{uz + vw}{vz}$	$\dfrac{2}{3} + \dfrac{4}{5} = \dfrac{2 \cdot 5 + 3 \cdot 4}{3 \cdot 5} = \dfrac{22}{15}$
3. $\dfrac{u}{v} \cdot \dfrac{w}{z} = \dfrac{uw}{vz}$	$\dfrac{2}{3} \cdot \dfrac{4}{5} = \dfrac{2 \cdot 4}{3 \cdot 5} = \dfrac{8}{15}$
4. $\dfrac{u}{v} \div \dfrac{w}{z} = \dfrac{u}{v} \cdot \dfrac{z}{w} = \dfrac{uz}{vw}$	$\dfrac{2}{3} \div \dfrac{4}{5} = \dfrac{2}{3} \cdot \dfrac{5}{4} = \dfrac{10}{12} = \dfrac{5}{6}$

5. For subtraction, replace "$+$" by "$-$" in 1 and 2.

INVERT AND MULTIPLY

The division step shown in 4 is often referred to as invert the divisor (the fraction following the division symbol) and multiply the result times the numerator (the first fraction).

EXAMPLE 3 Multiplying and Dividing Rational Expressions

(a) $\dfrac{2x^2 + 11x - 21}{x^3 + 2x^2 + 4x} \cdot \dfrac{x^3 - 8}{x^2 + 5x - 14}$

$= \dfrac{(2x - 3)(x + 7)}{x(x^2 + 2x + 4)} \cdot \dfrac{(x - 2)(x^2 + 2x + 4)}{(x - 2)(x + 7)}$ Factor completely.

$= \dfrac{2x - 3}{x}, \quad x \neq 2, \quad x \neq -7$ Remove common factors.

(b) $\dfrac{x^3 + 1}{x^2 - x - 2} \div \dfrac{x^2 - x + 1}{x^2 - 4x + 4}$

$= \dfrac{(x^3 + 1)(x^2 - 4x + 4)}{(x^2 - x - 2)(x^2 - x + 1)}$ Invert and multiply.

$= \dfrac{(x + 1)(x^2 - x + 1)(x - 2)^{2^1}}{(x + 1)(x - 2)(x^2 - x + 1)}$ Factor completely.

$= x - 2, \quad x \neq -1, \quad x \neq 2$ Remove common factors.

Now try Exercises 49 and 55.

NOTE ON EXAMPLE

The numerator, $x^2 + 4x - 6$, of the final expression in Example 4 is a prime polynomial. Thus, there are no common factors.

EXAMPLE 4   Adding Rational Expressions

$$\frac{x}{3x - 2} + \frac{3}{x - 5} = \frac{x(x - 5) + 3(3x - 2)}{(3x - 2)(x - 5)} \qquad \text{Definition of addition}$$

$$= \frac{x^2 - 5x + 9x - 6}{(3x - 2)(x - 5)} \qquad \text{Distributive property}$$

$$= \frac{x^2 + 4x - 6}{(3x - 2)(x - 5)} \qquad \text{Combine like terms.}$$

Now try Exercise 59.

If the denominators of fractions have common factors, then it is often more efficient to find the LCD before adding or subtracting the fractions. The **LCD (least common denominator)** is the product of all the prime factors in the denominators, where each factor is raised to the greatest power found in any one denominator for that factor.

EXAMPLE 5   Using the LCD

Write the following expression as a fraction in reduced form.

$$\frac{2}{x^2 - 2x} + \frac{1}{x} - \frac{3}{x^2 - 4}$$

SOLUTION The factored denominators are $x(x - 2)$, x, and $(x - 2)(x + 2)$, respectively. The LCD is $x(x - 2)(x + 2)$.

$$\frac{2}{x^2 - 2x} + \frac{1}{x} - \frac{3}{x^2 - 4}$$

$$= \frac{2}{x(x - 2)} + \frac{1}{x} - \frac{3}{(x - 2)(x + 2)} \qquad \text{Factor.}$$

$$= \frac{2(x + 2)}{x(x - 2)(x + 2)} + \frac{(x - 2)(x + 2)}{x(x - 2)(x + 2)} - \frac{3x}{x(x - 2)(x + 2)} \qquad \text{Equivalent fractions}$$

$$= \frac{2(x + 2) + (x - 2)(x + 2) - 3x}{x(x - 2)(x + 2)} \qquad \text{Combine numerators.}$$

$$= \frac{2x + 4 + x^2 - 4 - 3x}{x(x - 2)(x + 2)} \qquad \text{Expand terms.}$$

$$= \frac{x^2 - x}{x(x - 2)(x + 2)} \qquad \text{Simplify.}$$

$$= \frac{x(x - 1)}{x(x - 2)(x + 2)} \qquad \text{Factor.}$$

$$= \frac{x - 1}{(x - 2)(x + 2)}, \quad x \neq 0 \qquad \text{Reduce.}$$

Now try Exercise 61.

Compound Rational Expressions

Sometimes a complicated algebraic expression needs to be changed to a more familiar form before we can work on it. A **compound fraction** (sometimes called a **complex fraction**), in which the numerators and denominators may themselves contain fractions, is such an example. One way to simplify a compound fraction is to write both the numerator and denominator as single fractions and then invert and multiply. If the fraction then takes the form of a rational expression, then we write the expression in reduced or simplest form.

EXAMPLE 6   Simplifying a Compound Fraction

$$\dfrac{3 - \dfrac{7}{x+2}}{1 - \dfrac{1}{x-3}} = \dfrac{\dfrac{3(x+2)-7}{x+2}}{\dfrac{(x-3)-1}{x-3}} \qquad \text{Combine fractions.}$$

$$= \dfrac{\dfrac{3x-1}{x+2}}{\dfrac{x-4}{x-3}} \qquad \text{Simplify.}$$

$$= \dfrac{(3x-1)(x-3)}{(x+2)(x-4)}, \quad x \ne 3 \qquad \text{Invert and multiply.}$$

Now try Exercise 63.

A second way to simplify a compound fraction is to multiply the numerator and denominator by the LCD of all fractions in the numerator and denominator as illustrated in Example 7.

EXAMPLE 7   Simplifying Another Compound Fraction

Use the LCD to simplify the compound fraction

$$\dfrac{\dfrac{1}{a^2} - \dfrac{1}{b^2}}{\dfrac{1}{a} - \dfrac{1}{b}}.$$

SOLUTION The LCD of the four fractions in the numerator and denominator is a^2b^2.

$$\dfrac{\dfrac{1}{a^2} - \dfrac{1}{b^2}}{\dfrac{1}{a} - \dfrac{1}{b}} = \dfrac{\left(\dfrac{1}{a^2} - \dfrac{1}{b^2}\right)a^2b^2}{\left(\dfrac{1}{a} - \dfrac{1}{b}\right)a^2b^2} \qquad \text{Multiply numerator and}$$

$$= \dfrac{b^2 - a^2}{ab^2 - a^2b} \qquad \text{Simplify.}$$

$$= \dfrac{(b+a)(b-a)}{ab(b-a)} \qquad \text{Factor.}$$

$$= \dfrac{b+a}{ab}, \quad a \ne b \qquad \text{Reduce.}$$

Now try Exercise 69.

APPENDIX A.3 EXERCISES

In Exercises 1–8, rewrite as a single fraction.

1. $\dfrac{5}{9} + \dfrac{10}{9}$ 　　　　　　**2.** $\dfrac{17}{32} - \dfrac{9}{32}$

3. $\dfrac{20}{21} \cdot \dfrac{9}{22}$ 　　　　　　**4.** $\dfrac{33}{25} \cdot \dfrac{20}{77}$

5. $\dfrac{2}{3} \div \dfrac{4}{5}$ 　　　　　　**6.** $\dfrac{9}{4} \div \dfrac{15}{10}$

7. $\dfrac{1}{14} + \dfrac{4}{15} - \dfrac{5}{21}$ 　　**8.** $\dfrac{1}{6} + \dfrac{6}{35} - \dfrac{4}{15}$

In Exercises 9–18, find the domain of the algebraic expression.

9. $5x^2 - 3x - 7$ 　　　　**10.** $2x - 5$

11. $\sqrt{x - 4}$ 　　　　　　**12.** $\dfrac{2}{\sqrt{x + 3}}$

13. $\dfrac{2x + 1}{x^2 + 3x}$ 　　　　　**14.** $\dfrac{x^2 - 2}{x^2 - 4}$

15. $\dfrac{x}{x - 1}, \quad x \ne 2$ 　　**16.** $\dfrac{3x - 1}{x - 2}, \quad x \ne 0$

17. $x^2 + x^{-1}$ 　　　　　**18.** $x(x + 1)^{-2}$

In Exercises 19–26, find the missing numerator or denominator so that the two rational expressions are equal.

19. $\dfrac{2}{3x} = \dfrac{?}{12x^3}$ 　　　**20.** $\dfrac{5}{2y} = \dfrac{15y}{?}$

21. $\dfrac{x - 4}{x} = \dfrac{x^2 - 4x}{?}$ 　　**22.** $\dfrac{x}{x + 2} = \dfrac{?}{x^2 - 4}$

23. $\dfrac{x + 3}{x - 2} = \dfrac{?}{x^2 + 2x - 8}$ 　**24.** $\dfrac{x - 4}{x + 5} = \dfrac{x^2 - x - 12}{?}$

25. $\dfrac{x^2 - 3x}{?} = \dfrac{x - 3}{x^2 + 2x}$ 　**26.** $\dfrac{?}{x^2 - 9} = \dfrac{x^2 + x - 6}{x - 3}$

In Exercises 27–32, consider the original fraction and its reduced form from the specified example. Explain why the given restriction is needed on the reduced form.

27. Example 3a, $x \ne 2, x \ne -7$ 　**28.** Example 3b, $x \ne -1, x \ne 2$

29. Example 4, none 　　　　　**30.** Example 5, $x \ne 0$

31. Example 6, $x \ne 3$ 　　　　**32.** Example 7, $a \ne b$

In Exercises 33–44, write the expression in reduced form.

33. $\dfrac{18x^3}{15x}$ 　　　　　　**34.** $\dfrac{75y^2}{9y^4}$

35. $\dfrac{x^3}{x^2 - 2x}$ 　　　　　**36.** $\dfrac{2y^2 + 6y}{4y + 12}$

37. $\dfrac{z^2 - 3z}{9 - z^2}$ 　　　　　**38.** $\dfrac{x^2 + 6x + 9}{x^2 - x - 12}$

39. $\dfrac{y^2 - y - 30}{y^2 - 3y - 18}$ 　　　**40.** $\dfrac{y^3 + 4y^2 - 21y}{y^2 - 49}$

41. $\dfrac{8z^3 - 1}{2z^2 + 5z - 3}$ 　　　**42.** $\dfrac{2z^3 + 6z^2 + 18z}{z^3 - 27}$

43. $\dfrac{x^3 + 2x^2 - 3x - 6}{x^3 + 2x^2}$ 　**44.** $\dfrac{y^2 + 3y}{y^3 + 3y^2 - 5y - 15}$

In Exercises 45–62, simplify.

45. $\dfrac{3}{x - 1} \cdot \dfrac{x^2 - 1}{9}$ 　　**46.** $\dfrac{x + 3}{7} \cdot \dfrac{14}{2x + 6}$

47. $\dfrac{x + 3}{x - 1} \cdot \dfrac{1 - x}{x^2 - 9}$ 　**48.** $\dfrac{18x^2 - 3x}{3xy} \cdot \dfrac{12y^2}{6x - 1}$

49. $\dfrac{x^3 - 1}{2x^2} \cdot \dfrac{4x}{x^2 + x + 1}$ 　**50.** $\dfrac{y^3 + 2y^2 + 4y}{y^3 + 2y^2} \cdot \dfrac{y^2 - 4}{y^3 - 8}$

51. $\dfrac{2y^2 + 9y - 5}{y^2 - 25} \cdot \dfrac{y - 5}{2y^2 - y}$ 　**52.** $\dfrac{y^2 + 8y + 16}{3y^2 - y - 2} \cdot \dfrac{3y^2 + 2y}{y + 4}$

53. $\dfrac{1}{2x} \div \dfrac{1}{4}$ 　　　　**54.** $\dfrac{4x}{y} \div \dfrac{8y}{x}$

55. $\dfrac{x^2 - 3x}{14y} \div \dfrac{2xy}{3y^2}$ 　　**56.** $\dfrac{7x - 7y}{4y} \div \dfrac{14x - 14y}{3y}$

57. $\dfrac{\dfrac{2x^2 y}{(x - 3)^2}}{\dfrac{8xy}{x - 3}}$ 　　　　**58.** $\dfrac{\dfrac{x^2 - y^2}{2xy}}{\dfrac{y^2 - x^2}{4x^2 y}}$

59. $\dfrac{2x + 1}{x + 5} - \dfrac{3}{x + 5}$ 　　**60.** $\dfrac{3}{x - 2} + \dfrac{x + 1}{x - 2}$

61. $\dfrac{3}{x^2 + 3x} - \dfrac{1}{x} - \dfrac{6}{x^2 - 9}$

62. $\dfrac{5}{x^2 + x - 6} - \dfrac{2}{x - 2} + \dfrac{4}{x^2 - 4}$

In Exercises 63–70, simplify the compound fraction.

63. $\dfrac{\dfrac{x}{y^2} - \dfrac{y}{x^2}}{\dfrac{1}{y^2} - \dfrac{1}{x^2}}$ 　　　**64.** $\dfrac{\dfrac{1}{x} + \dfrac{1}{y}}{\dfrac{1}{x^2} - \dfrac{1}{y^2}}$

65. $\dfrac{2x + \dfrac{13x - 3}{x - 4}}{2x + \dfrac{x + 3}{x - 4}}$ 　　**66.** $\dfrac{2 - \dfrac{13}{x + 5}}{2 + \dfrac{3}{x - 3}}$

67. $\dfrac{\dfrac{1}{(x + h)^2} - \dfrac{1}{x^2}}{h}$ 　　**68.** $\dfrac{\dfrac{x + h}{x + h + 2} - \dfrac{x}{x + 2}}{h}$

69. $\dfrac{\dfrac{b}{a} - \dfrac{a}{b}}{\dfrac{1}{a} - \dfrac{1}{b}}$ 　　　**70.** $\dfrac{\dfrac{1}{a} + \dfrac{1}{b}}{\dfrac{b}{a} - \dfrac{a}{b}}$

In Exercises 71–74, write with positive exponents and simplify.

71. $\left(\dfrac{1}{x} + \dfrac{1}{y}\right)(x + y)^{-1}$ 　**72.** $\dfrac{(x + y)^{-1}}{(x - y)^{-1}}$

73. $x^{-1} + y^{-1}$ 　　　　**74.** $(x^{-1} + y^{-1})^{-1}$

B
Key Formulas

B.1 Formulas from Algebra

Exponents

If all bases are nonzero:

$$u^m u^n = u^{m+n} \qquad \frac{u^m}{u^n} = u^{m-n}$$

$$u^0 = 1 \qquad u^{-n} = \frac{1}{u^n}$$

$$(uv)^m = u^m v^m \qquad (u^m)^n = u^{mn}$$

$$\left(\frac{u}{v}\right)^m = \frac{u^m}{v^m}$$

Radicals and Rational Exponents

If all roots are real numbers:

$$\sqrt[n]{uv} = \sqrt[n]{u} \cdot \sqrt[n]{v} \qquad \sqrt[n]{\frac{u}{v}} = \frac{\sqrt[n]{u}}{\sqrt[n]{v}} \ (v \neq 0)$$

$$\sqrt[m]{\sqrt[n]{u}} = \sqrt[mn]{u} \qquad (\sqrt[n]{u})^n = u$$

$$\sqrt[n]{u^m} = (\sqrt[n]{u})^m \qquad \sqrt[n]{u^n} = \begin{cases} |u| & n \text{ even} \\ u & n \text{ odd} \end{cases}$$

$$u^{1/n} = \sqrt[n]{u} \qquad u^{m/n} = (u^{1/n})^m = (\sqrt[n]{u})^m$$

$$u^{m/n} = (u^m)^{1/n} = \sqrt[n]{u^m}$$

Special Products

$$(u + v)(u - v) = u^2 - v^2$$

$$(u + v)^2 = u^2 + 2uv + v^2$$

$$(u - v)^2 = u^2 - 2uv + v^2$$

$$(u + v)^3 = u^3 + 3u^2v + 3uv^2 + v^3$$

$$(u - v)^3 = u^3 - 3u^2v + 3uv^2 - v^3$$

Factoring Polynomials

$$u^2 - v^2 = (u + v)(u - v)$$

$$u^2 + 2uv + v^2 = (u + v)^2$$

$$u^2 - 2uv + v^2 = (u - v)^2$$

$$u^3 + v^3 = (u + v)(u^2 - uv + v^2)$$

$$u^3 - v^3 = (u - v)(u^2 + uv + v^2)$$

Inequalities

If $u < v$ and $v < w$, then $u < w$.

If $u < v$, then $u + w < v + w$.

If $u < v$ and $c > 0$, then $uc < vc$.

If $u < v$ and $c < 0$, then $uc > vc$.

If $c > 0$, $|u| < c$ is equivalent to $-c < u < c$.

If $c > 0$, $|u| > c$ is equivalent to $u < -c$ or $u > c$.

Quadratic Formula

If $a \neq 0$, the solutions of the equation $ax^2 + bx + c = 0$ are given by

$$x = \frac{-b \pm \sqrt{b^2 - 4ac}}{2a}.$$

Logarithms

If $0 < b \neq 1, 0 < a \neq 1, x, R, S, > 0$

$y = \log_b x$ if and only if $b^y = x$

$$\log_b 1 = 0 \qquad \log_b b = 1$$

$$\log_b b^y = y \qquad b^{\log_b x} = x$$

$$\log_b RS = \log_b R + \log_b S \qquad \log_b \frac{R}{S} = \log_b R - \log_b S$$

$$\log_b R^c = c \log_b R \qquad \log_b x = \frac{\log_a x}{\log_a b}$$

Determinants

$$\begin{vmatrix} a & b \\ c & d \end{vmatrix} = ad - bc$$

Arithmetic Sequences and Series

$$a_n = a_1 + (n - 1)d$$

$$S_n = n\left(\frac{a_1 + a_n}{2}\right) \text{ or } S_n = \frac{n}{2}[2a_1 + (n - 1)d]$$

Geometric Sequences and Series

$$a_n = a_1 \cdot r^{n-1}$$

$$S_n = \frac{a_1(1 - r^n)}{1 - r} \ (r \neq 1)$$

$$S = \frac{a_1}{1 - r} \ (|r| < 1) \text{ infinite geometric series}$$

Factorial

$$n! = n \cdot (n - 1) \cdot (n - 2) \cdot \cdots \cdot 3 \cdot 2 \cdot 1$$

$$n \cdot (n - 1)! = n!, 0! = 1$$

Binomial Coefficient

$$\binom{n}{r} = \frac{n!}{r!(n-r)!} \text{ (integers } n \text{ and } r, n \geq r \geq 0)$$

Binomial Theorem

If n is a positive integer

$$(a+b)^n = \binom{n}{0}a^n + \binom{n}{1}a^{n-1}b + \cdots + \binom{n}{r}a^{n-r}b^r + \cdots + \binom{n}{n}b^n$$

B.2 Formulas from Geometry

Triangle

$h = a \sin \theta$

$\text{Area} = \dfrac{1}{2}bh$

Trapezoid

$\text{Area} = \dfrac{h}{2}(a+b)$

Circle

$\text{Area} = \pi r^2$

$\text{Circumference} = 2\pi r$

Sector of Circle

$\text{Area} = \dfrac{\theta r^2}{2}$ (θ in radians)

$s = r\theta$ (θ in radians)

Right Circular Cone

$\text{Volume} = \dfrac{\pi r^2 h}{3}$

$\text{Lateral surface area} = \pi r \sqrt{r^2 + h^2}$

Right Circular Cylinder

$\text{Volume} = \pi r^2 h$

$\text{Lateral surface area} = 2\pi rh$

Right Triangle

Pythagorean Theorem:

$c^2 = a^2 + b^2$

Parallelogram

$\text{Area} = bh$

Circular Ring

$\text{Area} = \pi(R^2 - r^2)$

Ellipse

$\text{Area} = \pi ab$

Cone

$\text{Volume} = \dfrac{Ah}{3}$ ($A = \text{Area of base}$)

Sphere

$\text{Volume} = \dfrac{4}{3}\pi r^3$

$\text{Surface area} = 4\pi r^2$

B.3 Formulas from Trigonometry

Angular Measure

π radians $= 180°$

So, 1 radian $= \dfrac{180}{\pi}$ degrees,

and 1 degree $= \dfrac{\pi}{180}$ radians.

Reciprocal Identities

$$\sin x = \frac{1}{\csc x} \qquad\qquad \csc x = \frac{1}{\sin x}$$

$$\cos x = \frac{1}{\sec x} \qquad\qquad \sec x = \frac{1}{\cos x}$$

$$\tan x = \frac{1}{\cot x} \qquad\qquad \cot x = \frac{1}{\tan x}$$

Quotient Identities

$$\tan x = \frac{\sin x}{\cos x} \qquad\qquad \cot x = \frac{\cos x}{\sin x}$$

Pythagorean Identities

$$\sin^2 x + \cos^2 x = 1$$

$$1 + \tan^2 x = \sec^2 x$$

$$1 + \cot^2 x = \csc^2 x$$

Odd-Even Identities

$$\sin(-x) = -\sin x \qquad\qquad \csc(-x) = -\csc x$$

$$\cos(-x) = \cos x \qquad\qquad \sec(-x) = \sec x$$

$$\tan(-x) = -\tan x \qquad\qquad \cot(-x) = -\cot x$$

Sum and Difference Identities

$$\sin(u + v) = \sin u \cos v + \cos u \sin v$$

$$\sin(u - v) = \sin u \cos v - \cos u \sin v$$

$$\cos(u + v) = \cos u \cos v - \sin u \sin v$$

$$\cos(u - v) = \cos u \cos v + \sin u \sin v$$

$$\tan(u + v) = \frac{\tan u + \tan v}{1 - \tan u \tan v}$$

$$\tan(u - v) = \frac{\tan u - \tan v}{1 + \tan u \tan v}$$

Cofunction Identities

$$\cos\left(\frac{\pi}{2} - u\right) = \sin u$$

$$\sin\left(\frac{\pi}{2} - u\right) = \cos u$$

$$\tan\left(\frac{\pi}{2} - u\right) = \cot u$$

$$\cot\left(\frac{\pi}{2} - u\right) = \tan u$$

$$\sec\left(\frac{\pi}{2} - u\right) = \csc u$$

$$\csc\left(\frac{\pi}{2} - u\right) = \sec u$$

Double-Angle Identities

$$\sin 2u = 2 \sin u \cos u$$

$$\cos 2u = \cos^2 u - \sin^2 u$$

$$= 2 \cos^2 u - 1$$

$$= 1 - 2 \sin^2 u$$

$$\tan 2u = \frac{2 \tan u}{1 - \tan^2 u}$$

Power-Reducing Identities

$$\sin^2 u = \frac{1 - \cos 2u}{2}$$

$$\cos^2 u = \frac{1 + \cos 2u}{2}$$

$$\tan^2 u = \frac{1 - \cos 2u}{1 + \cos 2u}$$

Half-Angle Identities

$$\sin\frac{u}{2} = \pm\sqrt{\frac{1 - \cos u}{2}}$$

$$\cos\frac{u}{2} = \pm\sqrt{\frac{1 + \cos u}{2}}$$

$$\tan\frac{u}{2} = \pm\sqrt{\frac{1 - \cos u}{1 + \cos u}}$$

$$= \frac{1 - \cos u}{\sin u} = \frac{\sin u}{1 + \cos u}$$

Triangles

Law of sin

$$\frac{\sin A}{a} = \frac{\text{si}}{}$$

Law of cos

$a^2 = b^2 + $

$b^2 = a^2 + $

$c^2 = a^2 + b$

Area:

$$\text{Area} = \frac{1}{2} bc$$

$$= \frac{1}{2} ac$$

$$\text{Area} = \sqrt{s(}$$

where s

Trigonomet

$z = a + bi = $

$= $

De Moivre's

$z^n = [r(\cos \theta$

$= r^n(\cos n$

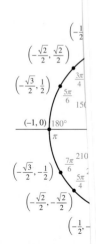

Some statements involve **quantifiers** and are more complicated to negate. Quantifiers include words such as *all, some, every,* and *there exists.*

The quantifiers *all, every,* and *no* refer to each and every element in a set and are **universal quantifiers** . The quantifiers *some* and *there exists at least one* refer to one or more, or possibly all, of the elements in a set. *Some* and *there exists* are called **existential quantifiers** . Examples with universal and existential quantifiers follow:

1. All roses are red. [universal]
2. Every student is important. [universal]
3. For each counting number x, $x + 0 = x$. [universal]
4. Some roses are red. [existential]
5. There exists at least one even counting number less than 3. [existential]
6. There are women who are taller than 200 cm. [existential]

Venn diagrams can be used to picture statements involving quantifiers. For example, Figures C.1a and C.1b picture statements (1) and (4). The x in Figure C.1b is used to show that there must be at least one element of the set of roses that is red.

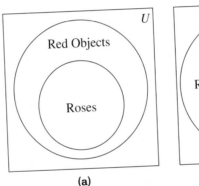
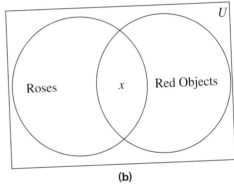

(a)   (b)

FIGURE C.1 **(a)** All roses are red. **(b)** Some roses are red.

Consider the following statement involving the existential quantifier *some*. "Some professors at Paxson University have blue eyes." This means that at least one professor at Paxson University has blue eyes. It does not rule out the possibilities that all the Paxson professors have blue eyes or that some of the Paxson professors do not have blue eyes. Because the negation of a true statement is false, neither "Some professors at Paxson University do not have blue eyes" nor "All professors at Paxson have blue eyes" are negations of the original statement. One possible negation of the original statement is "No professors at Paxson University have blue eyes."

Statement	Negation
Some a are b.	No a is b.
Some a are not b.	All a are b.
All a are b.	Some a are not b.
No a is b.	Some a are b.

EXAMPLE 2   Negation with Quantifiers

Negate each of the following statements:

(a) All students like hamburgers.

(b) Some people like mathematics.

(c) There exists a counting number x such that $3x = 6$.

(d) For all counting numbers x, $3x = 3x$.

SOLUTION

(a) Some students do not like hamburgers.

(b) No people like mathematics.

(c) For all counting numbers x, $3x \neq 6$.

(d) There exists a counting number x such that $3x \neq 3x$.

Now try Exercise 5, parts (e) and (f).

There is a symbolic system defined to help in the study of logic. If p represents a statement, the negation of the statement p is denoted by $\sim p$. **Truth tables** are often used to show all possible true-false patterns for statements. Table C.1 summarizes the truth tables for p and $\sim p$.

Observe that p and $\sim p$ are analogous to sets P and $\overline{P}$. If x is an element of P, then x is not an element of $\overline{P}$.

Table C.1 Negation

p	$\sim p$
T	F
F	T

Compound Statements

From two given statements, it is possible to create a new, **compound statement** by using a connective such as *and*. For example, "It is snowing" and "the ski run is open" together with *and* give "It is snowing and the ski run is open." Other compound statements can be obtained by using the connective *or*. For example, "It is snowing or the ski run is open."

The symbols $\wedge$ and $\vee$ are used to represent the connectives *and* and *or*, respectively. For example, if p represents "It is snowing," and if q represents "The ski run is open," then "It is snowing and the ski run is open" is denoted by $p \wedge q$. Similarly, "It is snowing or the ski run is open" is denoted by $p \vee q$.

The truth value of any compound statement, such as $p \wedge q$, is defined using the truth table of each of the simple statements. Because each of the statements p and q may be either true or false, there are four distinct possibilities for the truth values of p and q, as shown in Table C.2. The compound statement $p \wedge q$, is the **conjunction** of p and q and is defined to be true if, and only if, both p and q are true. Otherwise, it is false.

Table C.2 Conjunction

p	q	$p \wedge q$
T	T	T
T	F	F
F	T	F
F	F	F

The compound statement $p \vee q$—that is, p or q—is a **disjunction**. In everyday language, *or* is not always interpreted in the same way. In logic, we use an *inclusive or*. The statement "I will go to a movie or I will read a book" means that I will either go to a movie, or read a book, or do both. Hence, in logic, p or q, symbolized as $p \vee q$, is defined to be false if both p and q are false and true in all other cases. This is summarized in Table C.3.

Table C.3 Disjunction

p	q	$p \vee q$
T	T	T
T	F	T
F	T	T
F	F	F

EXAMPLE 1   Converse, Inverse, Contrapositive

Write the converse, the inverse, and the contrapositive for each of the following statements:

(a) If $2x = 6$, then $x = 3$.

(b) If I am in San Francisco, then I am in California.

SOLUTION

(a) *Converse:* If $x = 3$, then $2x = 6$.
   Inverse: If $2x \neq 6$, then $x \neq 3$.
   Contrapositive: If $x \neq 3$, then $2x \neq 6$.

(b) *Converse:* If I am in California, then I am in San Francisco.
   Inverse: If I am not in San Francisco, then I am not in California.
   Contrapositive: If I am not in California, then I am not in San Francisco.

Now try Exercise 3, parts (a) and (b).

Table C.7 shows that an implication and its converse do not always have the same truth value. However, an implication and its contrapositive always have the same truth value. Also, the converse and inverse of a conditional statement are logically equivalent.

Table C.7 Converse, Inverse, Contrapositive

p	q	$\sim p$	$\sim q$	Implication $p \rightarrow q$	Converse $q \rightarrow p$	Inverse $\sim p \rightarrow \sim q$	Contra-positive $\sim q \rightarrow \sim p$
T	T	F	F	T	T	T	T
T	F	F	T	F	T	T	F
F	T	T	F	T	F	F	T
F	F	T	T	T	T	T	T

Connecting a statement and its converse with the connective *and* gives $(p \rightarrow q) \wedge (q \rightarrow p)$. This compound statement can be written as $p \leftrightarrow q$ and usually is read "p if and only if q." The statement "p if and only if q" is a **biconditional**. A truth table for $p \leftrightarrow q$ is given in Table C.8. Observe that $p \leftrightarrow q$ is true if and only if both statements are true or both are false.

Table C.8 Biconditional

p	q	$p \rightarrow q$	$q \rightarrow p$	Biconditional $(p \rightarrow q) \wedge (q \rightarrow p)$ or $p \leftrightarrow q$
T	T	T	T	T
T	F	F	T	F
F	T	T	F	F
F	F	T	T	T

EXAMPLE 2 Biconditionals

Given the following statements, classify each of the biconditionals as true or false:

$p: 2 = 2$ $r: 2 = 1$

$q: 2 \neq 1$ $s: 2 + 3 = 1 + 3$

(a) $p \leftrightarrow q$ **(b)** $p \leftrightarrow r$

(c) $s \leftrightarrow q$ **(d)** $r \leftrightarrow s$

SOLUTION

(a) $p \to q$ is true and $q \to p$ is true, so $p \leftrightarrow q$ is true.

(b) $p \to r$ is false and $r \to p$ is true, so $p \leftrightarrow r$ is false.

(c) $s \to q$ is true and $q \to s$ is false, so $s \leftrightarrow q$ is false.

(d) $r \to s$ is true and $s \to r$ is true, so $r \leftrightarrow s$ is true.

Now try Exercise 5, parts (a) and (f).

Now consider the following statement:

It is raining or it is not raining.

This statement, which can be modeled as $p \lor (\sim p)$, is always true, as shown in Table C.9. A statement that is always true is called a **tautology**. One way to make a tautology is to take two logically equivalent statements such as $p \to q$ and $\sim q \to \sim p$ (from Table C.7) and form them into a biconditional as follows:

$$p \to q \leftrightarrow (\sim q \to \sim p)$$

Because $p \to q$ and $\sim q \to \sim p$ have the same truth values, $(p \to q) \leftrightarrow (\sim q \to \sim p)$ is a tautology.

Table C.9 A Tautology

p	$\sim p$	$p \lor (\sim p)$
T	F	T
F	T	T

Valid Reasoning

In problem solving, the reasoning used is said to be **valid** if the conclusion follows unavoidably from the hypotheses. Consider the following example:

Hypotheses: All roses are red.

This flower is a rose.

Conclusion: Therefore, this flower is red.

The statement "All roses are red" can be written as the implication, "If a flower is a rose, then it is red" and pictured with the Venn diagram is Figure C.2a.

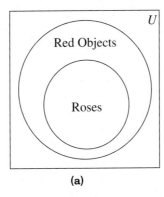
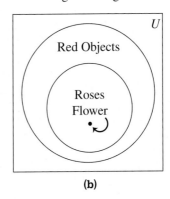

(a)   **(b)**

FIGURE C.2 (a) All roses are red. **(b)** This flower is a rose.

The information "This flower is a rose" implies that this flower must belong to the circle containing roses, as pictured in Figure C.2b. This flower also must belong to the circle containing red objects. Thus the reasoning is valid because it is impossible to draw a picture satisfying the hypotheses and contradicting the conclusion.

Consider the following argument:

Hypotheses: All elementary school teachers are mathematically literate.
               Some mathematically literate people are not children.

Conclusion: Therefore, no elementary school teacher is a child.

Let E be the set of elementary school teachers, M be the set of mathematically literate people, and C be the set of children. Then the statement "All elementary school teachers are mathematically literate" can be pictured as in Figure C.3a. The statement "Some mathematically literate people are not children" can be pictured in several ways. Three of these are illustrated in Figure C.3b–d.

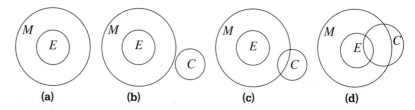

(a)   (b)   (c)   (d)

FIGURE C.3 (a) All elementary school teachers are mathematically literate. **(b)–(d)** Some mathematically literate people are not children.

According to Figure C.3d, it is possible that some elementary school teachers are children, and yet the given statements are satisfied. Therefore, the conclusion that "No elementary school teacher is a child" does not follow from the given hypotheses. Hence, the reasoning is not valid.

If a single picture can be drawn to satisfy the hypotheses of an argument and contradict the conclusion, the argument is not valid. However, to show that an argument is valid, all possible pictures must be considered to show that there are no contradictions. There must be no way to satisfy the hypotheses and contradict the conclusion if the argument is valid.

EXAMPLE 3   Argument Validity

Determine if the following argument is valid:

Hypotheses: In Washington, D.C., all senators wear power ties.
               No one in Washington, D.C., over 6 ft tall wears a power tie.

Conclusion: Persons over 6 ft tall are not senators in Washington, D.C.

SOLUTION

If S represents the set of senators and P represents the set of people who wear power ties, the first hypothesis is pictured as shown in Figure C.4a. If T represents the set of people in Washington, D.C., over 6 ft tall, the second hypothesis is pictured in Figure C.4b. Because people over 6 ft tall are outside the circle representing power tie wearers and senators are inside the circle P, the conclusion is valid and no person over 6 ft tall can be a senator in Washington, D.C.   ***Now try Exercise 14(a).***

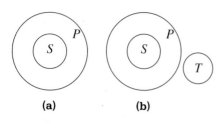

(a)   (b)

FIGURE C.4 (a) In Washington, D.C., all senators wear power ties. **(b)** No one in Washington, D.C., over 6 ft tall wears a power tie.

Glossary

ABSOLUTE MAXIMUM A value $f(c)$ is an absolute maximum value of f if $f(c) \geq f(x)$ for all x in the domain of f, p. 96.

ABSOLUTE MINIMUM A value $f(c)$ is an absolute minimum value of f if $f(c) \leq f(x)$ for all x in the domain of f, p. 96.

ABSOLUTE VALUE OF A COMPLEX NUMBER The absolute value of the complex number $z = a + bi$ is given by $\sqrt{a^2 + b^2}$; also, the length of the segment from the origin to z in the complex plane, p. 551.

ABSOLUTE VALUE OF A REAL NUMBER Denoted by $|a|$, represents the number a or the positive number $-a$ if $a < 0$, p. 14.

ABSOLUTE VALUE OF A VECTOR See *Magnitude of a vector*.

ACUTE ANGLE An angle whose measure is between $0°$ and $90°$, p. 360.

ACUTE TRIANGLE A triangle in which all angles measure less than $90°$, p. 478.

ADDITION PROPERTY OF EQUALITY If $u = v$ and $w = z$, then $u + w = v + z$, p. 24.

ADDITION PROPERTY OF INEQUALITY If $u < v$, then $u + w < v + w$, p. 27.

ADDITIVE IDENTITY FOR THE COMPLEX NUMBERS $0 + 0i$ is the complex number zero, p. 54.

ADDITIVE INVERSE OF A REAL NUMBER The opposite of b, or $-b$, p. 6.

ADDITIVE INVERSE OF A COMPLEX NUMBER The opposite of $a + bi$, or $-a - bi$, p. 54.

ALGEBRAIC EXPRESSION A combination of variables and constants involving addition, subtraction, multiplication, division, powers, and roots, p. 6.

ALGEBRAIC MODEL An equation which relates variable quantities associated with a phenomena being studied, p. 71.

AMBIGUOUS CASE A triangle in which two sides and a nonincluded angle are known, p. 479.

AMPLITUDE See *Sinusoid*.

ANCHOR See *Mathematical induction*.

ANGLE Union of two rays with a common endpoint (the vertex). The beginning ray (the initial side) can be rotated about its endpoint to obtain the final position (the terminal side), p. 370.

A different method for determining if an argument is valid uses **direct reasoning** and a form of argument called the Law of Detachment (or **Modus Ponens**). For example, consider the following true statements:

> If the sun is shining, then we shall take a trip.
> The sun is shining.

Using these two statements, we can conclude that we shall take a trip. In general, the **Law of Detachment** is stated as follows:

> *If a statement in the form "if p, then q" is true, and p is true, then q must also be true.*

The Law of Detachment is sometimes described schematically as follows, where all statements above the horizontal line are true and the statement below the horizontal line is the conclusion.

$$\frac{\begin{array}{c} p \to q \\ p \end{array}}{q}$$

The Law of Detachment follows from the truth table for $p \to q$ given in Table C.6. The only case in which both p and $p \to q$ are true is when q is true (line 1 in the table).

EXAMPLE 4   Applications of the Law of Detachment

Determine if each of the following arguments is valid:

Hypotheses:	If you eat spinach, then you will be strong.
	You eat spinach.
Conclusion:	Therefore, you will be strong.
Hypotheses:	If Claude goes skiing, he will break his leg.
	If Claude breaks his leg, he cannot enter the dance contest.
	Claude goes skiing.
Conclusion:	Therefore, Claude cannot enter the dance contest.

SOLUTION

(a) Using the Law of Detachment, we see that the conclusion is valid.

(b) By using the Law of Detachment twice, we see that the conclusion is valid.

Now try Exercise 14(d).

8. Consider the statement "If every digit of a number is 6, then the number is divisible by 3." Determine whether each of the following is logically equivalent to the statement.

   (a) If every digit of a number is not 6, then the number is not divisible by 3.

   (b) If a number is not divisible by 3, then some digit of the number is not 6.

   (c) If a number is divisible by 3, then every digit of the number is 6.

9. Write a statement logically equivalent to the statement "If a number is a multiple of 8, then it is a multiple of 4."

10. Use truth tables to prove that the following are tautologies:

   (a) $(p \rightarrow q) \rightarrow [(p \wedge r) \rightarrow q]$ Law of Added Hypothesis

   (b) $[(p \rightarrow q) \wedge p] \rightarrow q$ Law of Detachment

   (c) $[(p \rightarrow q) \wedge \sim q] \rightarrow \sim p$ Modus Tollens

   (d) $[(p \rightarrow q) \wedge (q \rightarrow r)] \rightarrow (p \rightarrow r)$ Chain Rule

11. (a) Suppose that $p \rightarrow q$, $q \rightarrow r$, and $r \rightarrow s$ are all true, but s is false. What can you conclude about the truth value of p?

   (b) Suppose that $(p \wedge q) \rightarrow r$ is true, r is false, and q is true. What can you conclude about the truth value of p?

   (c) Suppose that $p \rightarrow q$ is true and $q \rightarrow p$ is false. Can q be true? Why or why not?

12. Translate the following statements into symbolic form. Give the meanings of the symbols that you use.

   (a) If Mary's little lamb follows her to school, then its appearance there will break the rules and Mary will be sent home.

   (b) If it is not the case that Jack is nimble and quick, then Jack will not make it over the candlestick.

   (c) If the apple had not hit Isaac Newton on the head, then the laws of gravity would not have been discovered.

13. For
from

   (a)

   (b)

   (c)

   (d)

14. Inv

   (a)

   (b)

   (c)

   (d)

15. W

   (a)

   (b)

   (c)

   (d)

SYMMETRIC DIFFERENCE QUOTIENT OF f AT a
$\dfrac{f(x + h) - f(x - h)}{2h}$, p. 826.

SYMMETRIC MATRIX A matrix $A = [a_{ij}]$ with the property $a_{ij} = a_{ji}$ for all i and j, p. 591.

SYMMETRIC ABOUT THE ORIGIN A graph in which $(-x, -y)$ is on the graph whenever (x, y) is; or a graph in which $(-r, \theta)$ or $(r, \theta + \pi)$ is on the graph whenever (r, θ) is, p. 98.

SYMMETRIC ABOUT THE x-AXIS A graph in which $(x, -y)$ is on the graph whenever (x, y) is; or a graph in which $(r, -\theta)$ or $(-r, \pi - \theta)$ is on the graph whenever (r, θ) is, p. 97.

SYMMETRIC ABOUT THE y-AXIS A graph in which $(-x, y)$ is on the graph whenever (x, y) is; or a graph in which $(-r, -\theta)$ or $(r, \pi - \theta)$ is on the graph whenever (r, θ) is, p. 97.

SYMMETRIC PROPERTY OF EQUALITY If $a = b$, then $b = a$, p. 24.

SYNTHETIC DIVISION A procedure used to divide a polynomial by a linear factor, $x - a$, p. 217.

SYSTEM A set of equations or inequalities, p. 568.

TANGENT The function $y = \tan x$, p. 396.

TANGENT LINE OF f AT $x = a$ The line through $(a, f(a))$ with slope $f'(a)$ provided $f'(a)$ exists, p. 797.

TERMINAL POINT See *Arrow*.

TERMINAL SIDE OF AN ANGLE See *Angle*.

TERM OF A POLYNOMIAL (FUNCTION) An expression of the form $a_n x^n$ in a polynomial (function), p. 200.

TERMS OF A SEQUENCE The range elements of a sequence, p. 732.

THIRD QUARTILE See *Quartile*.

TIME PLOT A line graph in which time is measured on the horizontal axis, p. 765.

TRANSFORMATION A function that maps real numbers to real numbers, p. 138.

TRANSITIVE PROPERTY If $a = b$ and $b = c$, then $a = c$. Similar properties hold for the inequality symbols $<, \leq, >, \geq$, pp. 24, 27.

TRANSLATION See *Horizontal translation, Vertical translation*.

TRANSPOSE OF A MATRIX The matrix A^T obtained by interchanging the rows and columns of A, p. 583.

TRANSVERSE AXIS The line segment whose endpoints are the vertices of a hyperbola, p. 657.

TREE DIAGRAM A visualization of the *Multiplication Principle of Probability*, p. 723.

TRIANGULAR FORM A special form for a system of linear equations that facilitates finding the solution, p. 594.

TRIANGULAR NUMBER A number that is a sum of the arithmetic series $1 + 2 + 3 + \cdots + n$ for some natural number n, p. 716.

TRICHOTOMY PROPERTY For real numbers a and b, exactly one of the following is true: $a < b$, $a = b$, or $a > b$, p. 4.

TRIGONOMETRIC FORM OF A COMPLEX NUMBER $r(\cos \theta + i \sin \theta)$, p. 551.

UNBOUNDED INTERVAL An interval that extends to $-\infty$ or ∞ (or both), p. 5.

UNION OF TWO SETS A AND B The set of all elements that belong to A or B or both, p. 60.

UNIT CIRCLE A circle with radius 1 centered at the origin, p. 377.

UNIT RATIO See *Conversion factor*.

UNIT VECTOR Vector of length 1, pp. 506.

UNIT VECTOR IN THE DIRECTION OF A VECTOR A unit vector that has the same direction as the given vector, p. 506.

UPPER BOUND FOR f Any number B for which $f(x) \leq B$ for all x in the domain of f, p. 95.

UPPER BOUND FOR REAL ZEROS A number d is an upper bound for the set of real zeros of f if $f(x) \neq 0$ whenever $x > d$, p. 220.

UPPER BOUND TEST FOR REAL ZEROS A test for finding an upper bound for the real zeros of a polynomial, p. 220.

VALUE OF AN ANNUITY $FV = \dfrac{R(1 + i)^n - 1}{i}$, p. 339.

VALUE OF AN INVESTMENT $A = P\left(1 + \dfrac{r}{k}\right)^{kt}$ or $A = Pe^{rt}$, p. 337.

VARIABLE A letter that represents an unspecified number, p. 6.

VARIABLE (IN STATISTICS) A characteristic of individuals that is being identified or measured, p. 759.

VARIANCE The square of the standard deviation, p. 778.

VECTOR An ordered pair $\langle a, b \rangle$ of real numbers in the plane, or an ordered triple $\langle a, b, c \rangle$ of real numbers in space. A vector has both magnitude and direction, p. 502.

VECTOR EQUATION FOR A LINE IN SPACE The line through $P_0(x_0, y_0, z_0)$ in the direction of the nonzero vector $\mathbf{V} = \langle a, b, c \rangle$ has vector equation $\mathbf{r} = \mathbf{r}_0 + t\mathbf{v}$, where $\mathbf{r} = \langle x, y, z \rangle$, p. 692.

VELOCITY A vector that specifies the motion of an object in terms of its speed and direction, p. 508.

VENN DIAGRAM A visualization of the relationships among events within a sample space, p. 723.

VERTEX OF A CONE See *right circular cone.*

VERTEX OF A PARABOLA The point of intersection of a parabola and its line of symmetry, pp. 177, 634.

VERTEX OF AN ANGLE See *Angle.*

VERTEX FORM FOR A QUADRATIC FUNCTION
$f(x) = a(x - h)^2 + k$, p. 178.

VERTICAL ASYMPTOTE The line $x = a$ is a vertical asymptote of the graph of the function f if $\lim_{x \to a^+} f(x) = \pm\infty$ or $\lim_{x \to a^-} f(x) = \pm\infty$, pp. 100, 240.

VERTICAL COMPONENT See *Component form of a vector.*

VERTICAL LINE $x = a$, p. 34.

VERTICAL LINE TEST A test for determining whether a graph is a function, p. 87.

VERTICAL STRETCH OR SHRINK See *Stretch, Shrink.*

VERTICAL TRANSLATION A shift of a graph up or down, p. 138.

VERTICES OF AN ELLIPSE The points where the ellipse intersects its focal axis, p. 644.

VERTICES OF A HYPERBOLA The points where a hyperbola intersects the line containing its foci, p. 659.

VIEWING WINDOW The rectangular portion of the coordinate plane specified by the dimensions [Xmin, Xmax] by [Ymin, Ymax], p. 34.

WEIGHTED MEAN A mean calculated in such a way that some elements of the data set have higher weights (that is, are counted more strongly in determining the mean) than others, p. 774.

WEIGHTS See *Weighted mean.*

WHOLE NUMBERS The numbers 0, 1, 2, 3, …, p. 2.

WINDOW DIMENSIONS The restrictions on x and y that specify a viewing window. See *Viewing window.*

WORK The product of a force applied to an object over a given distance $W = |\mathbf{F}| |\overrightarrow{AB}|$, p. 519.

WRAPPING FUNCTION The function which associated points on the unit circle with points on the real number line, p. 377.

x-AXIS Usually the horizontal coordinate line in a Cartesian coordinate system with positive direction to the right, p. 14.

x-COORDINATE The directed distance from the y-axis (yz-plane) to a point in a plane (space), or the first number in an ordered pair (triple), p. 14.

x-INTERCEPT A point that lies on both the graph and the x-axis, p. 34.

XMAX The x-value of the right side of the viewing window, p. 34.

XMIN The x-value of the left side of the viewing window, p. 34.

XSCL The scale of the tick marks on the x-axis in a viewing window, p. 34.

xy-PLANE The points $(x, y, 0)$ in Cartesian space, p. 685.

xz-PLANE The points $(x, 0, z)$ in Cartesian space, p. 685.

y-AXIS Usually the vertical coordinate line in a Cartesian coordinate system with positive direction up, p. 14.

y-COORDINATE The directed distance from the x-axis (xz-plane) to a point in a plane (space), or the second number in an ordered pair (triple), p. 14.

y-INTERCEPT A point that lies on both the graph and the y-axis, p. 33.

YMAX The y-value of the top of the viewing window, p. 34.

YMIN The y-value of the bottom of the viewing window, p. 34.

YSCL The scale of the tick marks on the y-axis in a viewing window, p. 34.

yz-PLANE The points $(0, y, z)$ in Cartesian space, p. 685.

z-AXIS Usually the third dimension in Cartesian space, p. 685.

z-COORDINATE The directed distance from the xy-plane to a point in space, or the third number in an ordered triple, p. 685.

ZERO FACTOR PROPERTY If $ab = 0$, then either $a = 0$ or $b = 0$, pp. 45, 75.

ZERO FACTORIAL See *n factorial.*

ZERO OF A FUNCTION A value in the domain that makes the function value zero, p. 217.

ZERO MATRIX A matrix consisting entirely of zeros, p. 581.

ZERO VECTOR The vector $\langle 0, 0 \rangle$ or $\langle 0, 0, 0 \rangle$, p. 503.

ZOOM OUT A procedure of a graphing utility used to view more of the coordinate plane (used, for example, to find the end behavior of a function), p. 203.

Selected Answers

Quick Review P.1

1. $\{1, 2, 3, 4, 5, 6\}$ **3.** $\{-3, -2, -1\}$ **5. (a)** 1187.75 **(b)** -4.72 **7.** $-3; 1.375$ **9.** $0, 1, 2, 3, 4, 5, 6$

Exercises P.1

1. -4.625 (terminating)   **3.** $-2.1\overline{6}$ (repeating)   **5.** 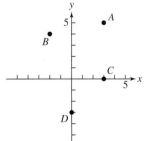 ; All real numbers less than or equal to 2

7. ; All real numbers less than 7   **9.** ; All real numbers less than 0

11. $-1 \le x < 1$ **13.** $-\infty < x < 5$, or $x < 5$ **15.** $-1 < x < 2$ **17.** $(-3, \infty)$ **19.** $(-2, -1)$ **21.** $(-3, 4]$ **23.** The real numbers greater than 4 and less than or equal to 9   **25.** The real numbers greater than or equal to -3, or the real numbers which are at least -3

27. The real numbers greater than -1 **29.** $-3 < x \le 4$; endpoints -3 and 4; bounded; half-open   **31.** $x < 5$; endpoint 5; unbounded; open

33. $x \ge 29$ or $[29, \infty)$; x = Bill's age   **35.** $1.099 \le x \le 1.399$ or $[1.099, 1.399]$; x = dollars per gallon of gasoline   **37.** $ax^2 + ab$

39. $(a + d)x^2$ **41.** $\pi - 6$ **43.** 5 **45. (a)** Associative property of multiplication   **(b)** Commutative property of multiplication

(c) Addition inverse property   **(d)** Addition identity property   **(e)** Distributive property of multiplication over addition   **47.** $\dfrac{x^2}{y^2}$ **49.** $\dfrac{16}{x^4}$

51. $x^4 y^4$ **53.** 3.6930338×10^{10} **55.** $1.93175805 \times 10^{11}$ **57.** 4.839×10^8 **59.** $0.000\,000\,033\,3$ **61.** $5,870,000,000,000$

63. 2.6028×10^{-8} **65. (a)** Because $a^m \ne 0$, $a^m a^0 = a^{m+0} = a^m$ implies that $a^0 = 1$.

(b) Because $a^m \ne 0$, $a^m a^{-m} = a^{m-m} = a^0 = 1$ implies that $a^{-m} = \dfrac{1}{a^m}$.

67. False. For example, the additive inverse of -5 is 5, which is positive.   **69.** E   **71.** B   **73.** $0, 1, 2, 3, 4, 5, 6$

75. $-6, -5, -4, -3, -2, -1, 0, 1, 2, 3, 4, 5, 6$

SECTION P.2

Quick Review P.2

1.
0.5 1 1.5 2 2.5 3

3.
$-5\ -4\ -3\ -2\ -1\ \ 0\ \ 1\ \ 2\ \ 3\ \ 4\ \ 5$

Distance: $\sqrt{7} - \sqrt{2} \approx 1.232$

5.

7. 5.5 **9.** 10

Exercises P.2

1. $A(1, 0)$, $B(2, 4)$, $C(-3, -2)$, $D(0, -2)$ **3. (a)** First quadrant   **(b)** On the y-axis, between quadrants I and II   **(c)** Second quadrant   **(d)** Third quadrant   **5.** 6 **7.** 6 **9.** $4 - \pi$ **11.** 19.9 **13.** 8 **15.** 5 **17.** 7 **19.** Perimeter $= 2\sqrt{41} + \sqrt{82} \approx 21.86$; Area $= 20.5$ **21.** Perimeter $= 2\sqrt{20} + 16 \approx 24.94$; Area $= 32$ **23.** 0.65 **25.** $(2, 6)$ **27.** $\left(-\dfrac{1}{3}, -\dfrac{3}{4}\right)$

29.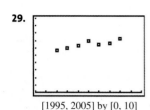

[1995, 2005] by [0, 10]

31.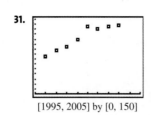

[1995, 2005] by [0, 150]

33.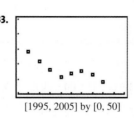

[1995, 2005] by [0, 50]

35. (a) about \$183,000 **(b)** about \$277,000   **37.** The three sides have lengths 5, 5, and $5\sqrt{2}$. Since two sides have the same length, the triangle is isosceles.   **39. (a)** (no answer) **(b)** $8^2 + 5^2 = 64 + 25 = 89 = (\sqrt{89})^2$ **41.** $(x - 1)^2 + (y - 2)^2 = 25$

43. $(x + 1)^2 + (y + 4)^2 = 9$ **45.** center: (3, 1); radius: 6   **47.** center: (0, 0); radius: $\sqrt{5}$ **49.** $|x - 4| = 3$ **51.** $|x - c| < d$

53. 7; 6   **55.** Midpoint is $\left(\dfrac{5}{2}, \dfrac{7}{2}\right)$. Distances from this point to vertices are equal to $\sqrt{18.5}$.   **57.** $x \le -8$ or $x \ge 2$

59. True. $\dfrac{\text{length of } AM}{\text{length of } AB} = \dfrac{1}{2}$ because M is the midpoint of AB. By similar triangles, $\dfrac{\text{length of } AM'}{\text{length of } AC} = \dfrac{\text{length of } AM}{\text{length of } AB} = \dfrac{1}{2}$, so M' is the midpoint of AC.   **61.** C   **63.** E

65. If the legs have lengths a and b, and the hypotenuse is c units long, then without loss of generality, we can assume the vertices are (0, 0), (a, 0), and (0, b). Then the midpoint of the hypotenuse is $\left(\dfrac{a + 0}{2}, \dfrac{b + 0}{2}\right) = \left(\dfrac{a}{2}, \dfrac{b}{2}\right)$. The distance to the other vertices is $\sqrt{\left(\dfrac{a}{2}\right)^2 + \left(\dfrac{b}{2}\right)^2} = \sqrt{\dfrac{a^2}{4} + \dfrac{b^2}{4}} = \dfrac{c}{2} = \dfrac{1}{2}c$.

67. $Q(a, -b)$ **69.** $Q(-a, -b)$

SECTION P.3

Quick Review P.3

1. $4x + 5y + 9$ **3.** $3x + 2y$ **5.** $\dfrac{5}{y}$ **7.** $\dfrac{2x + 1}{x}$ **9.** $\dfrac{11x + 18}{10}$

Exercises P.3

1. (a) and (c)   **3.** (b)   **5.** yes   **7.** no   **9.** no   **11.** $x = 8$ **13.** $t = 4$ **15.** $x = 1$ **17.** $y = -\dfrac{4}{5} = -0.8$

19. $x = \dfrac{7}{4} = 1.75$ **21.** $x = \dfrac{4}{3}$ **23.** $z = \dfrac{8}{19}$ **25.** $x = \dfrac{17}{10} = 1.7$ **27.** $t = \dfrac{31}{9}$ **29. (a)** The figure shows that $x = -2$ is a solution of the equation $2x^2 + x - 6 = 0$. **(b)** The figure shows that $x = \dfrac{3}{2}$ is a solution of the equation $2x^2 + x - 6 = 0$.   **31.** (a)   **33.** (b) and (c)

35. $x < 6$

37. $x \ge -2$

39. $-4 \le x < 3$

41. $x \ge 3$

43. $x \le -\dfrac{19}{5}$ **45.** $-\dfrac{1}{2} \le y \le \dfrac{17}{2}$ **47.** $-\dfrac{5}{2} \le z < \dfrac{3}{2}$ **49.** $x > \dfrac{21}{5}$ **51.** $y < \dfrac{7}{6}$ **53.** $x \le \dfrac{34}{7}$ **55.** $x = 1$ **57.** $x = 3, 4, 5, 6$

59. Multiply both sides of the first equation by 2.   **61. (a)** no **(b)** yes   **63.** False. $-6 < -2$ because -6 lies to the left of -2 on the number line.   **65.** E   **67.** A   **69. (a)** (no answer) **(b)** (no answer) **(c)** 800/801 > 799/800 **(d)** $-103/102 > -102/101$ **(e)** If your calculator returns 0 when you enter $2x + 1 < 4$, you can conclude that the value stored in x is not a solution of the inequality $2x + 1 < 4$.

71. $b_1 = \dfrac{2A}{h} - b_2$ **73.** $F = \dfrac{9}{5}C + 32$

SECTION P.4

Exploration 1

1. The graphs of $y = mx + b$ and $y = mx + c$ have the same slope but different y-intercepts.

3.

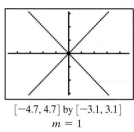

$[-4.7, 4.7]$ by $[-3.1, 3.1]$
$m = 1$

$[-4.7, 4.7]$ by $[-3.1, 3.1]$
$m = 3$

$[-4.7, 4.7]$ by $[-3.1, 3.1]$
$m = 4$

$[-4.7, 4.7]$ by $[-3.1, 3.1]$
$m = 5$

In each case, the two lines appear to be at right angles to one another.

Quick Review P.4

1. $x = -\dfrac{7}{3}$ **3.** $x = 12$ **5.** $y = \dfrac{2}{5}x - \dfrac{21}{5}$ **7.** $y = \dfrac{17}{5}$ **9.** $\dfrac{2}{3}$

Exercises P.4

1. -2 **3.** $\dfrac{4}{7}$ **5.** 8 **7.** $x = 2$ **9.** $y = 16$ **11.** $y - 4 = 2(x - 1)$ **13.** $y + 4 = -2(x - 5)$ **15.** $x - y + 5 = 0$ **17.** $y + 3 = 0$

19. $x - y + 3 = 0$ **21.** $y = -3x + 5$ **23.** $y = -\dfrac{1}{4}x + 4$ **25.** $y = -\dfrac{2}{5}x + \dfrac{12}{5}$

27.

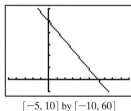

$[-5, 10]$ by $[-10, 60]$

29.

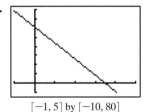

$[-1, 5]$ by $[-10, 80]$

31. (a): the slope is 1.5, compared to 1 in (b).   **33.** $x = 4$; $y = 21$

35. $x = -10$; $y = -7$ **37.** Ymin $= -30$, Ymax $= 30$, Yscl $= 3$

39. Ymin $= -20/3$, Ymax $= 20/3$, Yscl $= 2/3$

41. (a) $y = 3x - 1$ (b) $y = -\dfrac{1}{3}x + \dfrac{7}{3}$ **43.** (a) $y = -\dfrac{2}{3}x + 3$

(b) $y = \dfrac{3}{2}x - \dfrac{7}{2}$ **45.** (a) 3187.5; 42,000   (b) 9.57 years

(c) $3187.5t + 42,000 = 74,000$; $t = 10.04$ (d) 12 years

47. 32,000 ft

49. $m = \dfrac{3}{8} > \dfrac{4}{12}$, so asphalt shingles are acceptable.

51. (a) $y = 0.4x - 793.3$ (b) \$7.5 trillion   (c) \$9.1 trillion   (d)

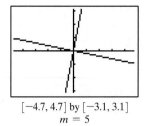

$[1995, 2005]$ by $[5, 10]$

53. (a)

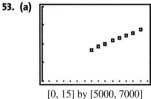

$[0, 15]$ by $[5000, 7000]$

$y = 75x + 5327$ (b)

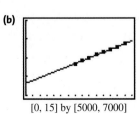

$[0, 15]$ by $[5000, 7000]$

(c) The year 2006 is represented by $x = 16$. So the value of y for $x = 16$ is ≈ 6527 million, a little larger than the U.S. Census Bureau estimate of 6525 million.

55. 9   **57.** $b = 5$; $a = 6$ **59. (a)** No; perpendicular lines have slopes with opposite signs.   **(b)** No; perpendicular lines have slopes with opposite signs.   **61.** False. The slope of a vertical line is undefined. For example, the vertical line through (3, 1) and (3, 6) would have slope $(6 - 1)/(3 - 3) = 5/0$, which is undefined.   **63.** A   **65.** E

61. False. The slope of a vertical line is undefined. For example, the vertical line through (3, 1) and (3, 6) would have slope $(6 - 1)/(3 - 3) = 5/0$, which is undefined.

67. (a)

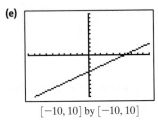

$[-5, 5]$ by $[-5, 5]$

(b)

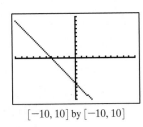

$[-5, 5]$ by $[-5, 5]$

(c)

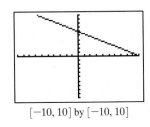

$[-5, 5]$ by $[-5, 5]$

(d) a is the x-intercept and b is the y-intercept when $c = 1$.

(e)

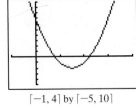

$[-10, 10]$ by $[-10, 10]$

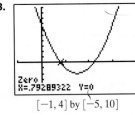

$[-10, 10]$ by $[-10, 10]$

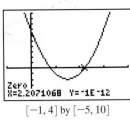

$[-10, 10]$ by $[-10, 10]$

(f) When $c = -1$, a is the opposite of the x-intercept and b is the opposite of the y-intercept.

a is half the x-intercept and b is half the y-intercept when $c = 2$.

69. As in the diagram, we can choose one point to be the origin, and another to be on the x-axis. The midpoints of the sides, starting from the origin and working around counterclockwise in the diagram, are then $A\left(\dfrac{a}{2}, 0\right)$, $B\left(\dfrac{a + b}{2}, \dfrac{c}{2}\right)$, $C\left(\dfrac{b + d}{2}, \dfrac{c + e}{2}\right)$, and $D\left(\dfrac{d}{2}, \dfrac{e}{2}\right)$.

The opposite sides are therefore parallel, since the slopes of the four lines connecting those points are: $m_{AB} = \dfrac{c}{b}$; $m_{BC} = \dfrac{e}{d - a}$;

$m_{CD} = \dfrac{c}{b}$; $m_{DA} = \dfrac{e}{d - a}$.   **71.** A has coordinates $\left(\dfrac{b}{2}, \dfrac{c}{2}\right)$, while B is $\left(\dfrac{a + b}{2}, \dfrac{c}{2}\right)$, so the line

containing A and B is the horizontal line $y = \dfrac{c}{2}$, and the distance from A to B is $\left|\dfrac{a + b}{2} - \dfrac{b}{2}\right| = \dfrac{a}{2}$.

SECTION P.5

Exploration 1

1.

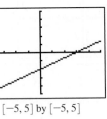

$[-1, 4]$ by $[-5, 10]$

3.

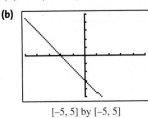

Zero
X=.79289322 Y=0

$[-1, 4]$ by $[-5, 10]$

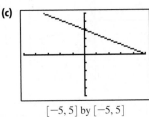

Zero
X=2.2071068 Y=-1E-12

$[-1, 4]$ by $[-5, 10]$

By this method, we have zeros at 0.79 and 2.21.   **5.** The answers in parts 2, 3, and 4 are the same.   **7.** 0.792893; 2.207107

Quick Review P.5

1. $9x^2 - 24x + 16$ **3.** $6x^2 - 7x - 5$ **5.** $(5x - 2)^2$ **7.** $(3x + 1)(x^2 - 5)$ **9.** $\dfrac{(x - 2)(x + 1)}{(2x + 1)(x + 3)}$

Exercises P.5

1. $x = -4$ or $x = 5$ **3.** $x = 0.5$ or $x = 1.5$ **5.** $x = -\dfrac{2}{3}$ or $x = 3$ **7.** $x = \pm\dfrac{5}{2}$ **9.** $x = -4 \pm \sqrt{\dfrac{8}{3}}$ **11.** $y = \pm\sqrt{\dfrac{7}{2}}$

13. $x = -7$ or $x = 1$ **15.** $x = \dfrac{7}{2} - \sqrt{11} \approx 0.18$ or $x = \dfrac{7}{2} + \sqrt{11} \approx 6.82$ **17.** $x = 2$ or $x = 6$

19. $x = -4 - 3\sqrt{2} \approx -8.24$ or $x = -4 + 3\sqrt{2} \approx 0.24$ **21.** $x = -1$ or $x = 4$ **23.** $-\dfrac{5}{2} + \dfrac{\sqrt{73}}{2} \approx 1.77$ or $-\dfrac{5}{2} - \dfrac{\sqrt{73}}{2} \approx -6.77$

25. x-intercept: 3; y-intercept: -2 **27.** x-intercepts: $-2, 0, 2$; y-intercept: 0

29.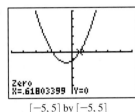
$[-5, 5]$ by $[-5, 5]$
31.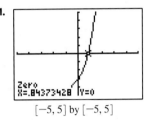
$[-5, 5]$ by $[-5, 5]$
33.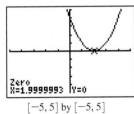
$[-5, 5]$ by $[-5, 5]$

35. $x^2 + 2x - 1 = 0$; $x \approx 0.4$ **37.** 1.62; -0.62 **39.** $t = 6$ or $t = 10$ **41.** $x = 1$ or $x = -6$ **43.** $x = -3$ or $x = 1$

45. (a) $y_1 = 3\sqrt{x + 4}$ (the one that begins on the x-axis) and $y_2 = x^2 - 1$ **(b)** $y = 3\sqrt{x + 4} - x^2 + 1$ **(c)** The x-coordinates of the intersections in the first picture are the same as the x-coordinates where the second graph crosses the x-axis. **47.** $x = -2$ or $x = 1$ **49.** $x = 3$ or $x = -2$

51. $x \approx -4.56$ or $x \approx -0.44$ or $x = 1$ **53.** $x = -2 \pm 2\sqrt{3}$ **55.** $x \approx -2.41$ or $x \approx 2.91$ **57. (a)** There must be 2 distinct real zeros, because $b^2 - 4ac > 0$ implies that $\pm\sqrt{b^2 - 4ac}$ are 2 distinct real numbers. **(b)** There must be 1 real zero, because $b^2 - 4ac = 0$ implies that $\pm\sqrt{b^2 - 4ac} = 0$, so the root must be $x = -\dfrac{b}{2a}$. **(c)** There must be no real zeros, because $b^2 - 4ac < 0$ implies that $\pm\sqrt{b^2 - 4ac}$ are not real numbers. **59.** 80 yd wide; 110 yd long **61.** ≈ 11.98 ft **63.** False. Notice that $2(-3)^2 = 18$, so x could also be -3. **65.** B **67.** E

69. (a) $c = 2$ **(b)** $c = 4$ **(c)** $c = 5$ **(d)** $c = -1$ **(e)** There is no other possible number of solutions of this equation. For any c, the solution involves solving two quadratic equations, each of which can have 0, 1, or 2 solutions. **71.** $2.5 \pm \dfrac{1}{2}\sqrt{13}$, or approximately 0.697 and 4.303

SECTION P.6

Quick Review P.6

1. $x + 9$ **3.** $a + 2d$ **5.** $x^2 - x - 6$ **7.** $x^2 - 2$ **9.** $x^2 - 2x - 1$

Exercises P.6

1. $8 + 2i$ **3.** $13 - 4i$ **5.** $5 - (1 + \sqrt{3})i$ **7.** $-5 + i$ **9.** $7 + 4i$ **11.** $-5 - 14i$ **13.** $-48 - 4i$ **15.** $5 - 10i$ **17.** $4i$

19. $\sqrt{3}i$ **21.** $x = 2, y = 3$ **23.** $x = 1, y = 2$ **25.** $5 + 12i$ **27.** -1 **29.** 13 **31.** 25 **33.** $2/5 - 1/5i$ **35.** $3/5 + 4/5i$

37. $1/2 - 7/2i$ **39.** $7/5 - 1/5i$ **41.** $x = -1 \pm 2i$ **43.** $x = \dfrac{7}{8} \pm \dfrac{\sqrt{15}}{8}i$ **45.** False. Any complex number bi has this property.

47. E **49.** A **51. (a)** $i; -1; -i; 1; i; -1; -i; 1$ **(b)** $-i; -1; i; 1; -i; -1; i; 1$ **(c)** 1 **(d)** (no answer)

53. $(a + bi) - (a - bi) = 2bi$, real part is zero

55. $\overline{(a + bi) \cdot (c + di)} = \overline{(ac - bd) - (ad + bc)i} = (ac - bd) - (ad + bc)i$ and
$\overline{(a + bi)} \cdot \overline{(c + di)} = (a - bi) \cdot (c - di) = (ac - bd) - (ad + bc)i$ are equal

57. $(-i)^2 - i(-i) + 2 = 0$ but $(i)^2 - i(i) + 2 \neq 0$. Because the coefficient of x in $x^2 - ix + 2 = 0$ is not a real number, the complex conjugate, i, of $-i$, need not be a solution.

SECTION P.7

Quick Review P.7

1. $-2 < x < 5$ **3.** $x = 1$ or $x = -5$ **5.** $x(x - 2)(x + 2)$ **7.** $\dfrac{z + 5}{z}$ **9.** $\dfrac{4x^2 - 4x - 1}{(x - 1)(3x - 4)}$

Exercises P.7

1. ├─────────────────────────────┤ $(-\infty, -9] \cup [1, \infty)$ **3.** ├─────────────────────────┤ $(1, 5)$
$\;\;\;-12\,-10\,-8\,-6\,-4\,-2\;\;0\;\;2\;\;4\;\;6\;\;8$ $\;\;\;-2\;-1\;\;0\;\;1\;\;2\;\;3\;\;4\;\;5\;\;6\;\;7\;\;8$

5. ├─────────────────────────┤ $\left(-\dfrac{2}{3}, \dfrac{10}{3}\right)$ **7.** ├─────────────────────────────┤ $(-\infty, -11] \cup [7, \infty)$
$\;\;\;-5\,-4\,-3\,-2\,-1\;\;0\;\;1\;\;2\;\;3\;\;4\;\;5$ $\;\;\;-12\,-10\,-8\,-6\,-4\,-2\;\;0\;\;2\;\;4\;\;6\;\;8$

9. $[-7, -3/2]$ **11.** $(-\infty, -5) \cup \left(\dfrac{3}{2}, \infty\right)$ **13.** $(-\infty, -2) \cup \left(\dfrac{1}{3}, \infty\right)$ **15.** $[-1, 0] \cup [1, \infty)$ **17.** $(-0.24, 4.24)$

19. $\left(-\infty, -\dfrac{1}{2}\right) \cup \left(\dfrac{4}{3}, \infty\right)$ **21.** $(-\infty, -1.41] \cup [0.08, \infty)$ **23.** $\left(-\infty, \dfrac{1}{2}\right) \cup \left(\dfrac{1}{2}, \infty\right)$ **25.** No solution   **27.** $[-2.08, 0.17] \cup [1.19, \infty)$

29. $(1.11, \infty)$ **31. (a)** $x^2 + 1 > 0$ **(b)** $x^2 + 1 < 0$ **(c)** $x^2 \le 0$ **(d)** $(x + 2)(x - 5) \le 0$ **(e)** $(x + 1)(x - 4) > 0$ **(f)** $x(x - 4) \ge 0$

33. (a) $t = 4$ sec up; $t = 12$ sec down   **(b)** when t is in the interval $[4, 12]$ **(c)** When t is in the interval $(0, 4]$ or $[12, 16)$ **35.** Reveals the boundaries of the solution set   **37. (a)** 1 in. $< x < 34$ in.   **(b)** When x is in the interval $(1, 25]$.   **39.** no more than \$100,000   **41.** True. The absolute value of any real number is always greater than or equal to zero.   **43.** D   **45.** D   **47.** $(-5.69, -4.11) \cup (0.61, 2.19)$

CHAPTER P REVIEW EXERCISES

1. Endpoints 0 and 5; bounded   **3.** $2x^2 - 2x$ **5.** v^4 **7.** 3.68×10^9 **9.** 5,000,000,000   **11. (a)** 5.0711×10^{10} **(b)** 4.63×10^9
(c) 5.0×10^8 **(d)** 3.995×10^9 **(e)** 1.4497×10^{10} **13.** 19; 4.5   **15.** $5\sqrt{5}$; $2\sqrt{5}$; $\sqrt{145}$ **17.** $(x - 0)^2 + (y - 0)^2 = 2^2$, or $x^2 + y^2 = 4$
19. Center: $(-5, -4)$; radius: 3   **21. (a)** $\sqrt{20} \approx 4.47$, $\sqrt{80} \approx 8.94$, 10   **(b)** $(\sqrt{20})^2 + (\sqrt{80})^2 = 20 + 80 = 100 = 10^2$ **23.** $a = 7$; $b = 9$

25. $y + 1 = -\dfrac{2}{3}(x - 2)$ **27.** $y = \dfrac{4}{5}x - 4.4$ **29.** $y = 4$ **31.** $y = -\dfrac{2}{5}x - \dfrac{11}{5}$

33. (a)

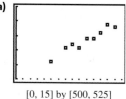

[0, 15] by [500, 525]

(b) $y = 1.6x + 498$ **(c)** 507.6, which is very close to 508.   **(d)** 524

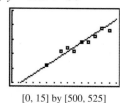

[0, 15] by [500, 525]

35. 2.5   **37.** $x = -3$ **39.** $y = 3$ **41.** $x = 2 - \sqrt{7} \approx -0.65$; $x = 2 + \sqrt{7} \approx 4.65$ **43.** $x = \dfrac{1}{3}$ or $x = -\dfrac{3}{2}$ **45.** $x = \dfrac{7}{2}$ or $x = -4$

47. $x = \dfrac{5}{2}$ **49.** $x = 0$, $x = 3$ **51.** $3 \pm 2i$ **53.** $x = \dfrac{3}{4} - \dfrac{\sqrt{17}}{4} \approx -0.28$ or $x = \dfrac{3}{4} + \dfrac{\sqrt{17}}{4} \approx 1.78$ **55.** $x = 0$ or $x = -\dfrac{2}{3}$ or $x = 7$

57. $x \approx 2.36$ **59.** $(-6, 3]$ ├─────────────┤ $-10\,-8\,-6\,-4\,-2\;\;0\;\;2\;\;4\;\;6\;\;8\;\;10$ **61.** $\left(-\infty, \dfrac{1}{3}\right]$ **63.** $(-\infty, -2] \cup \left[-\dfrac{2}{3}, \infty\right)$

65. $(-\infty, -0.37) \cup (1.37, \infty)$ **67.** $(-\infty, -2.82] \cup [-0.34, 3.15]$ **69.** $(-\infty, -17) \cup (3, \infty)$ **71.** $(-\infty, \infty)$ **73.** $1 + 3i$
75. $7 + 4i$ **77.** 25   **79.** $4i$ **81. (a)** $t \approx 8$ sec up; $t \approx 12$ sec down   **(b)** When t is in the interval $(0, 8]$ or $[12, 20)$ (approximately).
(c) When t is in the interval $[8, 12]$ (approximately).   **83. (a)** When w is in the interval $(0, 18.5]$.   **(b)** When w is in the interval $(22.19, \infty)$ (approximately).

SECTION 1.1

Exploration 1

1. 0.75   **3.** 0.8025   **5.** $124.61

Exploration 2

1. Percentages must be ≤ 100

3. A statistician might look for adverse economic factors in 1990, especially those that would affect people near or below the poverty line.

Quick Review 1.1

1. $(x + 4)(x - 4)$ **3.** $(9y + 2)(9y - 2)$ **5.** $(4h^2 + 9)(2h + 3)(2h - 3)$ **7.** $(x + 4)(x - 1)$ **9.** $(2x - 1)(x - 5)$

Exercises 1.1

1. (d)(q)   **3.** (a)(p)   **5.** (e)(l)   **7.** (g)(t)   **9.** (i)(m)   **11. (a)** Increasing, except for a slight drop from 1999 to 2004.   **(b)** 1974 to 1979

13. Women ($\square$), Men ($+$)

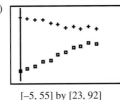

[-5, 55] by [23, 92]

15. Women: $y = 0.582x + 32.3$, men: $y = -0.211x + 83.5$ **17.** 2018, 69.9%

19. (a) and **(b)**

L3
3.7975
4.375
5.5405
5.8986
6.657

21. square stones   **23.** $y = 1.2t^2$

25. The lower line shows the minimum salaries, since they are lower than the average salaries.

27. Year 15. There is a clear drop in the average salary right after the 1994 strike.

29. $\pm\sqrt{\dfrac{13}{3}}$ **31.** $-1; 4$ **33.** $-1.5; 4$ **35.** $-\dfrac{7}{2} \pm \dfrac{\sqrt{105}}{2}$ **37.** 5

39.

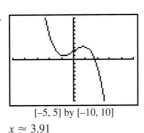

[-10, 10] by [-10, 10]

$x \approx 1.77$

41.

[-10, 10] by [-10, 10]

$x \approx -1.47$

43.

[-5, 5] by [-10, 10]

$x \approx 3.91$

45.

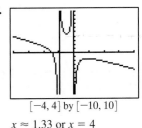

[-4, 4] by [-10, 10]

$x \approx 1.33$ or $x = 4$

47. (a) $46.94   **(b)** 210 mi

49. (a) $y = (x^{200})^{1/200} = x^{200/200} = x^1 = x$ for all $x \geq 0$.   **(b)**

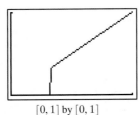

[0, 1] by [0, 1]

(c) Yes   **(d)** For values of x close to 0, x^{200} is so small that the calculator is unable to distinguish it from zero. It returns a value of $0^{1/200} = 0$ rather than x.

51. (a) -3 or 1.1 or 1.15   **(b)** -3

53. Let n be any integer. $n^2 + 2n = n(n + 2)$, which is either the product of two odd integers or the product of two even integers. The product of two odd integers is odd. The product of two even integers is a multiple of 4, since each even integer in the product contributes a factor of 2 to the product. Therefore, $n^2 + 2n$ is either odd or a multiple of 4.

55. False; a product is zero if at least one factor is zero.   **57.** C   **59.** B   **61. (a)** March   **(b)** $120   **(c)** June, after three months of poor performance   **(d)** About $2000   **(e)** After reaching a low in June, the stock climbed back to a price near $140 by December. LaToya's shares had gained $2000 by that point.   **(f)** Any graph that decreases steadily from March to December would favor Ahmad's strategy over LaToya's.

63. (a)

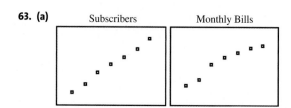

Subscribers   Monthly Bills

[7, 15] by [50, 200]   [7, 15] by [35, 55]

(b) The linear model for subscribers as a function of years after 1990 is $y = 18.53x - 79.04$.

(c) The fit is very good:

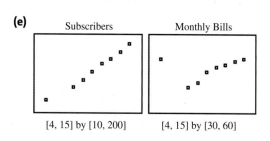

[7, 15] by [50, 200]

(d) The monthly bill scatter plot has an obviously curved shape that could be modeled more effectively by a function with a curved graph. Some possibilities: quadratic (parabola), logarithm, sine, power (e.g., square root), logistic. (We will learn about these curves later in the book.)

(e)

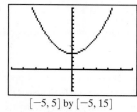

Subscribers   Monthly Bills

[4, 15] by [10, 200]   [4, 15] by [30, 60]

(f) Cellular phone technology was still emerging in 1995, so the growth rate was not as fast, explaining the lower slope on the subscriber scatter plot. The new technology was also more expensive before competition drove prices down, explaining the anomaly on the monthly bill scatter plot.

SECTION 1.2

Exploration 1

1. From left to right, the tables are (c) constant, (b) decreasing, and (a) increasing.

3. positive; negative; 0

Quick Review 1.2

1. $x = \pm 4$ **3.** $x < 10$ **5.** $x = 16$ **7.** $x < 16$ **9.** $x < -2, x \geq 3$

Exercises 1.2

1. Function   **3.** Not a function; y has two values for each positive value of x.   **5.** Yes   **7.** No

9.

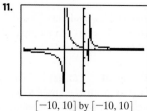

$(-\infty, \infty)$

[−5, 5] by [−5, 15]

11.

[−10, 10] by [−10, 10]

13.

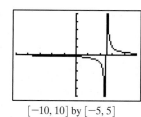

$[-10, 10]$ by $[-5, 5]$

15.

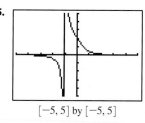

$[-5, 5]$ by $[-5, 5]$

17. $(-\infty, 10]$ **19.** $(-\infty, \infty) \cup [0, \infty)$

21. Yes, non-removable   **23.** Yes, non-removable

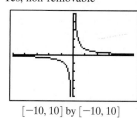

$[-10, 10]$ by $[-10, 10]$

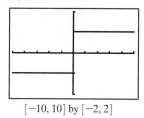

$[-10, 10]$ by $[-2, 2]$

25. Local maxima at $(-1, 4)$ and $(5, 5)$, local minimum at $(2, 2)$. The function increases on $(-\infty, -1]$, decreases on $[-1, 2]$, increases on $[2, 5]$, and decreases on $[5, \infty)$.   **27.** $(-1, 3)$ and $(3, 3)$ are neither, $(1, 5)$ is a local maximum, and $(5, 1)$ is a local minimum. The function increases on $(-\infty, 1]$, decreases on $[1, 5]$, and increases on $[5, \infty)$.

29. Decreasing on $(-\infty, -2]$; increasing on $[-2, \infty)$

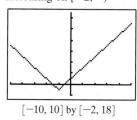

$[-10, 10]$ by $[-2, 18]$

31. Decreasing on $(-\infty, -2]$; constant on $[-2, 1]$; increasing on $[1, \infty)$

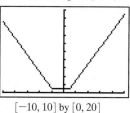
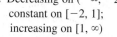

$[-10, 10]$ by $[0, 20]$

33. Increasing on $(-\infty, 1]$; decreasing on $[1, \infty)$

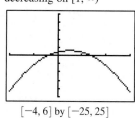

$[-4, 6]$ by $[-25, 25]$

35. Bounded   **37.** Bounded below   **39.** Bounded

41. f has a local minimum of $y = 3.75$ at $x = 0.5$. It has no maximum.

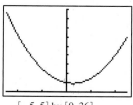

$[-5, 5]$ by $[0, 36]$

43. Local minimum: $y \approx -4.09$ at $x \approx -0.82$.
Local maximum: $y \approx -1.91$ at $x \approx 0.82$.

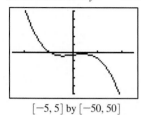

$[-5, 5]$ by $[-50, 50]$

45. Local maximum: $y \approx 9.16$ at $x \approx -3.20$.
Local minimum: $y = 0$ at $x = 0$ and $y = 0$ at $x = -4$.

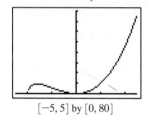

$[-5, 5]$ by $[0, 80]$

47. Even   **49.** Even   **51.** Even   **53.** Odd

55.

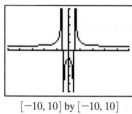

$[-10, 10]$ by $[-10, 10]$

$y = 1; x = 1$

57.

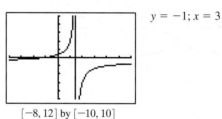

$y = -1; x = 3$

$[-8, 12]$ by $[-10, 10]$

59.

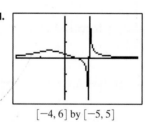

$[-10, 10]$ by $[-10, 10]$

$y = 1; x = 1; x = -1$ **61.**

$y = 0; x = 2$

$[-4, 6]$ by $[-5, 5]$

63. (b)   **65.** (a)

67. (a) $f(x)$ crosses the horizontal asymptote at $(0, 0)$.

(b) $g(x)$ crosses the horizontal asymptote at $(0, 0)$.

(c) $h(x)$ intersects the horizontal asymptote at $(0, 0)$.

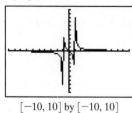

$[-10, 10]$ by $[-10, 10]$

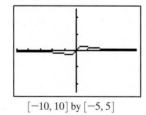

$[-10, 10]$ by $[-5, 5]$

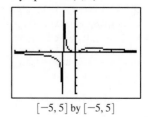

$[-5, 5]$ by $[-5, 5]$

69. (a) The vertical asymptote is $x = 0$, and this function is undefined at $x = 0$ (because a denominator can't be zero).

71. True; this is the definition of the graph of a function.

73. B   **75.** C

(b)

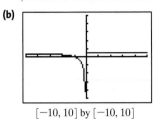

$[-10, 10]$ by $[-10, 10]$

Add the point $(0, 0)$.

(c) Yes

77. (a)

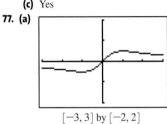

$[-3, 3]$ by $[-2, 2]$

$k = 1$ **(b)** $\dfrac{x}{1 + x^2} < 1 \Leftrightarrow x < 1 + x^2 \Leftrightarrow x^2 - x + 1 > 0$; but the discriminant of $x^2 - x + 1$ is negative (-3), so the graph never crosses the x-axis on the interval $(0, \infty)$.

(c) $k = -1$

(d) $\dfrac{x}{1 + x^2} > -1 \Leftrightarrow x > -1 - x^2 \Leftrightarrow x^2 + x + 1 > 0$; but the discriminant of $x^2 - x + 1$ is negative (-3), so the graph never crosses the x-axis on the interval $(-\infty, 0)$.

79. (a) (b) (c) (d) (e) Answers vary   **81. (a) (b) (c) (d)** Answers vary   **83. (a)** 2. It is in the range.   **(b)** 3. It is not in the range.   **(c)** $h(x)$ is not bounded above.   **(d)** 2. It is in the range.   **(e)** 1. It is in the range.

85. Since f is odd, $f(-x) = -f(x)$ for all x. In particular, $f(-0) = -f(0)$. This is equivalent to saying that $f(0) = -f(0)$, and the only number which equals its opposite is 0. Therefore $f(0) = 0$, which means the graph must pass through the origin.

SECTION 1.3

Exploration 1

1. $f(x) = 1/x$, $f(x) = \ln x$ **3.** $f(x) = 1/x$, $f(x) = e^x$, $f(x) = 1/(1 + e^{-x})$ **5.** No. There is a removable discontinuty at $x = 0$.

Quick Review 1.3

1. 59.34   **3.** $7 - \pi$ **5.** 0   **7.** 3   **9.** -4

Exercises 1.3

1. (e)   **3.** (j)   **5.** (i)   **7.** (k)   **9.** (d)   **11.** (l)   **13.** Ex. 8   **15.** Ex. 7, 8   **17.** Ex. 2, 4, 6, 10, 11, 12   **19.** $y = x$, $y = x^3$, $y = 1/x$, $y = \sin x$ **21.** $y = x^2$, $y = 1/x$, $y = |x|$ **23.** $y = 1/x$, $y = e^x$, $y = 1/(1 + e^{-x})$ **25.** $y = 1/x$, $y = \sin x$, $y = \cos x$, $y = 1/(1 + e^{-x})$ **27.** $y = x$, $y = x^3$, $y = 1/x$, $y = \sin x$ **29.** Domain: all reals; Range: $[-5, \infty)$ **31.** Domain: $(-6, \infty)$; Range: all reals   **33.** Domain: all reals; Range: all integers   **35. (a)** Increasing on $[10, \infty)$ **(b)** Neither   **(c)** Minimum value of 0 at $x = 10$ **(d)** Square root function, shifted 10 units right   **37. (a)** Increasing on $(-\infty, \infty)$ **(b)** Neither   **(c)** None   **(d)** Logistic function, stretched vertically by a factor of 3   **39. (a)** Increasing on $[0, \infty)$; decreasing on $(-\infty, 0]$ **(b)** Even   **(c)** Minimum of -10 at $x = 0$ **(d)** Absolute value function, shifted 10 units down   **41. (a)** Increasing on $[2, \infty)$; decreasing on $(-\infty, 2]$ **(b)** Neither   **(c)** Minimum of 0 at $x = 2$ **(d)** Absolute value function, shifted 2 units right   **43.** $y = 2$, $y = -2$

45.

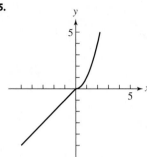

No points of discontinuity

47.

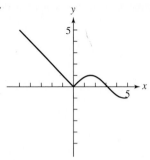

No points of discontinuity

49.

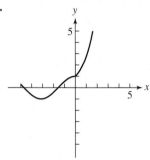

No points of discontinuity

51.

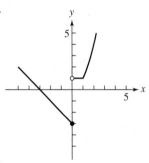

$x = 0$

53. (a)

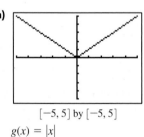

$[-5, 5]$ by $[-5, 5]$

$g(x) = |x|$

(b) $f(x) = \sqrt{x^2} = \sqrt{|x|^2} = |x| = g(x)$

55. (a)

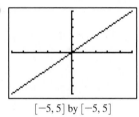

$[-5, 5]$ by $[-5, 5]$

$f(x) = x$

(b) The fact that $\ln(e^x) = x$ shows that the natural logarithm function takes on arbitrarily large values. In particular, it takes on the value L when $x = e^L$.

57. Domain: all real numbers; Range: all integers; Continuity: There is a discontinuity at each integer value of x; Increasing/decreasing behavior: constant on intervals of the form $[k, k + 1)$, where k is an integer; Symmetry: none; Boundedness: not bounded; Local extrema: every non-integer is both a local minimum and local maximum; Horizontal asymptotres: none; Vertical asymptotes: none; End behavior: $\text{int}(x) \to -\infty$ as $x \to -\infty$ and $\text{int}(x) \to \infty$ as $x \to \infty$.
59. True; the asymptotes are $x = 0$ and $x = 1$.   **61.** D   **63.** E **65. (a)** even **(b)** even **(c)** odd
67. (a) Pepperoni count ought to be proportional to the area of the pizza, which is proportional to the square of the radius   **(b)** 0.75   **(c)** Yes, very well.   **(d)** The fact that the pepperoni count fits the expected quadratic model so perfectly suggests that the pizzeria uses such a chart. If repeated observations produced the same results, there would be little doubt.
69. (a) $f(x) = 1/x, f(x) = e^x, f(x) = \ln x, f(x) = \cos x, f(x) = 1/(1 + e^{-x})$ **(b)** $f(x) = x$ **(c)** $f(x) = e^x$ **(d)** $f(x) = \ln x$
(e) The odd functions: $x, x^3, 1/x, \sin x$

SECTION 1.4

Exploration 1

f	g	$f \circ g$
$2x - 3$	$\dfrac{x + 3}{2}$	x
$\lvert 2x + 4 \rvert$	$\dfrac{(x - 2)(x + 2)}{2}$	x^2
$\sqrt{x}$	x^2	$\lvert x \rvert$
x^5	$x^{0.6}$	x^3
$x - 3$	$\ln(e^3 x)$	$\ln x$
$2 \sin x \cos x$	$\dfrac{x}{2}$	$\sin x$
$1 - 2x^2$	$\sin\left(\dfrac{x}{2}\right)$	$\cos x$

Quick Review 1.4

1. $(-\infty, -3) \cup (-3, \infty)$ **3.** $(-\infty, 5]$ **5.** $[1, \infty)$ **7.** $(-\infty, \infty)$ **9.** $(-1, 1)$

Exercises 1.4

1. $(f + g)(x) = 2x - 1 + x^2$; $(f - g)(x) = 2x - 1 - x^2$; $(fg)(x) = (2x - 1)(x^2) = 2x^3 - x^2$. There are no restrictions on any of the domains, so all three domains are $(-\infty, \infty)$.

3. $(f + g)(x) = \sqrt{x} + \sin x$; $(f - g)(x) = \sqrt{x} - \sin x$; $(fg)(x) = \sqrt{x} \sin x$. Domain in each case is $[0, \infty)$.

5. $(f/g)(x) = \dfrac{\sqrt{x + 3}}{x^2}$; $x + 3 \geq 0$ and $x \neq 0$, so the domain is $[-3, 0) \cup (0, \infty)$.

$(g/f)(x) = \dfrac{x^2}{\sqrt{x + 3}}$; $x + 3 > 0$, so the domain is $(-3, \infty)$.

7. $(f/g)(x) = x^2 / \sqrt{1 - x^2}$; $1 - x^2 > 0$, so $x^2 < 1$; the domain is $(-1, 1)$.
  $(g/f)(x) = \sqrt{1 - x^2} / x^2$; $1 - x^2 \geq 0$ and $x \neq 0$; the domain is $[-1, 0) \cup (0, 1]$.

9.

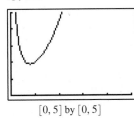

$[0, 5]$ by $[0, 5]$

11. $5; -6$ **13.** $8; 3$ **15.** $f(g(x)) = 3x - 1$; $(-\infty, \infty)$; $g(f(x)) = 3x + 1$; $(-\infty, \infty)$

17. $f(g(x)) = x - 1$; $[-1, \infty)$; $g(f(x)) = \sqrt{x^2 - 1}$; $(-\infty, 1] \cup [1, \infty)$

19. $f(g(x)) = 1 - x^2$; $[-1, 1]$; $g(f(x)) = \sqrt{1 - x^4}$; $[-1, 1]$

21. $f(g(x)) = \dfrac{3x}{2}$; $(-\infty, 0) \cup (0, \infty)$; $g(f(x)) = \dfrac{2x}{3}$; $(-\infty, 0) \cup (0, \infty)$

23. One possibility: $f(x) = \sqrt{x}$ and $g(x) = x^2 - 5x$

25. One possibility: $f(x) = \lvert x \rvert$ and $g(x) = 3x - 2$

27. One possibility: $f(x) = x^5 - 2$ and $g(x) = x - 3$

29. One possibility: $f(x) = \cos x$ and $g(x) = \sqrt{x}$

31. $V = \dfrac{4}{3}\pi r^3 = \dfrac{4}{3}\pi(48 + 0.03t)^3$; 775,734.6 in.^{3} **33.** $t \approx 3.63$ sec   **35.** $(3, -1)$ **37.** $y = \sqrt{25 - x^2}$ and $y = -\sqrt{25 - x^2}$

39. $y = \sqrt{x^2 - 25}$ and $y = -\sqrt{x^2 - 25}$ **41.** $y = 1 - x$ and $y = x - 1$ **43.** $y = x$ and $y = -x$ or $y = |x|$ and $y = -|x|$

45. False; x is not in the domain of $(f/g)(x)$ if $g(x) = 0$.   **47.** C   **49.** E

51.

f	g	D
e^x	$2 \ln x$	$(0, \infty)$
$(x^2 + 2)^2$	$\sqrt{x - 2}$	$[2, \infty)$
$(x^2 - 2)^2$	$\sqrt{2 - x}$	$(-\infty, 2]$
$\dfrac{1}{(x - 1)^2}$	$\dfrac{x + 1}{x}$	$x \neq 0$
$x^2 - 2x + 1$	$x + 1$	$(-\infty, \infty)$
$\left(\dfrac{x + 1}{x}\right)^2$	$\dfrac{1}{x - 1}$	$x \neq 1$

53. (a) $g(x) = 0$ **(b)** $g(x) = 1$ **(c)** $g(x) = x$ **55.**

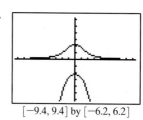

$[-9.4, 9.4]$ by $[-6.2, 6.2]$

$y = \dfrac{-x^2 \pm \sqrt{x^4 + 20}}{2}$

SECTION 1.5

Exploration 1

1. T starts at -4, at the point $(-8, -3)$. It stops at T $= 2$, at the point $(8, 3)$. 61 points are computed.
3. The graph is less smooth because the plotted points are further apart.
5. The grapher skips directly from the point $(0, -1)$ to the point $(0, 1)$, corresponding to the T-values T $= -2$ and T $= 0$. The two points are connected by a straight line, hidden by the Y-axis.   **7.** Leave everything else the same, but change Tmin back to -4 and Tmax to -1.

Quick Review 1.5

1. $y = \dfrac{1}{3}x + 2$ **3.** $y = \pm\sqrt{x - 4}$ **5.** $y = \dfrac{3x + 2}{1 - x}$ **7.** $y = \dfrac{4x + 1}{x - 2}$ **9.** $y = x^2 - 3$, $y \geq -3$, and $x \geq 0$

Exercises 1.5

1. $(6, 9)$ **3.** $(15, 2)$ **5. (a)** $(-6, -10), (-4, -7), (-2, -4), (0, -1), (2, 2), (4, 5), (6, 8)$ **(b)** $1.5x - 1$; It is a function.

(c)

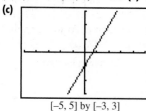

$[-5, 5]$ by $[-3, 3]$

7. (a) $(9, -5), (4, -4), (1, -3), (0, -2), (1, -1),$
$(4, 0), (9, 1)$

(b) $x = (y + 2)^2$; It is not a function.

(c)

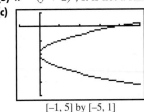

$[-1, 5]$ by $[-5, 1]$

9. (a) No   **(b)** Yes   **11. (a)** Yes   **(b)** Yes   **13.** $f^{-1}(x) = \dfrac{1}{3}x + 2, (-\infty, \infty)$ **15.** $f^{-1}(x) = \dfrac{x + 3}{2 - x}, (-\infty, 2) \cup (2, \infty)$

17. $f^{-1}(x) = x^2 + 3, x \geq 0$

19. $f^{-1}(x) = \sqrt[3]{x}, (-\infty, \infty)$ **21.** $f^{-1}(x) = x^3 - 5, (-\infty, \infty)$

23. One-to-one   **25.** One-to-one

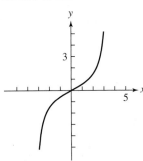

27. $f(g(x)) = 3\left[\dfrac{1}{3}(x + 2)\right] - 2 = x + 2 - 2 = x; g(f(x)) = \dfrac{1}{3}[(3x - 2) + 2] = \dfrac{1}{3}(3x) = x$

29. $f(g(x)) = [(x - 1)^{1/3}]^3 + 1 = (x - 1)^1 + 1 = x - 1 + 1 = x; g(f(x)) = [(x^3 + 1) - 1]^{1/3} = (x^3)^{1/3} = x^1 = x.$

31. $f(g(x)) = \dfrac{\dfrac{1}{x - 1} + 1}{\dfrac{1}{x - 1}} = (x - 1)\left(\dfrac{1}{x - 1} + 1\right) = 1 + x - 1 = x; g(f(x)) = \dfrac{1}{\dfrac{x + 1}{x} - 1} = \left(\dfrac{1}{\dfrac{x + 1}{x} - 1}\right) \cdot \dfrac{x}{x} = \dfrac{x}{x + 1 - x} = \dfrac{x}{1} = x$

33. (a) 108 euros   **(b)** $y = \dfrac{25}{27}x$. This converts euros (x) to dollars (y).   **(c)** \$44.44

35. $y = e^x$ and $y = \ln x$ are inverses. If we restrict the domain of the function $y = x^2$ to the interval $[0, \infty)$, then the restricted function and $y = \sqrt{x}$ are inverses.   **37.** $y = |x|$ **39.** True. All the ordered pairs swap domain and range values.   **41.** E   **43.** C   **45.** (Answers may vary.)
(a) If the graph of f is unbroken, its reflection in the line $y = x$ will be also.   **(b)** Both f and its inverse must be one-to-one in order to be inverse functions.   **(c)** Since f is odd, $(-x, -y)$ is on the graph whenever (x, y) is. This implies that $(-y, -x)$ is on the graph of f^{-1} whenever (x, y) is. That implies that f^{-1} is odd.   **(d)** Let $y = f(x)$. Since the ratio of Δy to Δx is positive, the ratio of Δx to Δy is positive. Any ratio of Δy to Δx on the graph of f^{-1} is the same as some ratio of Δx to Δy on the graph of f, hence positive. This implies that f^{-1} is increasing.   **47. (a)** $y = 0.75x + 31$
(b) $y = \dfrac{4}{3}(x - 31)$. It converts scaled scores to raw scores.   **49. (a)** No   **(b)** No   **(c)** $45°$; yes   **51.** When $k = 1$, the scaling function is linear. Opinions will vary as to which is the best value of k.

SECTION 1.6

Exploration 1

1. They raise or lower the parabola along the y-axis.   **3.** Yes

Exploration 2

1. Graph C. Points with positive y-coordinates remain unchanged, while points with negative y-coordinates are reflected across the x-axis.
3. Graph F. The graph will be a reflection across the y-axis of graph C.

Exploration 3

1.

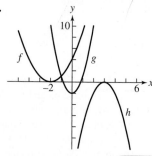

$[-4.7, 4.7]$ by $[-1.1, 5.1]$

The 1.5 and the 2 stretch the graph vertically; the 0.5 and the 0.25 shrink the graph vertically.

Quick Review 1.6

1. $(x + 1)^2$ **3.** $(x + 6)^2$ **5.** $(x - 5/2)^2$ **7.** $x^2 - x + 2$ **9.** $x^3 - 6x + 5$

Exercises 1.6

1. Vertical translation down 3 units   **3.** Horizontal translation left 4 units
5. Horizontal translation to the right 100 units
7. Horizontal translation to the right 1 unit, and Vertical translation up 3 units
9. Reflection across x-axis   **11.** Reflection across y-axis
13. Vertically stretch by 2

15. Horizontally stretch by $\dfrac{1}{0.2} = 5$, or vertically shrink by $0.2^3 = 0.008$

17. Translate right 6 units to get g
19. Translate left 4 units, and reflect across the x-axis to get g
21.

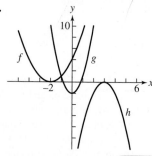

23.

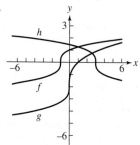

25. $f(x) = \sqrt{x + 5}$ **27.** $f(x) = -\sqrt{x + 2} + 3 = 3 - \sqrt{x + 2}$
29. (a) $-x^3 + 5x^2 + 3x - 2$ **(b)** $-x^3 - 5x^2 + 3x + 2$
31. (a) $y = -f(x) = -(\sqrt[3]{8x}) = -2\sqrt[3]{x}$ **(b)** $y = f(-x) = \sqrt[3]{8(-x)} = -2\sqrt[3]{x}$
33. Let f be an odd function; that is, $f(-x) = -f(x)$ for all x in the domain of f. To reflect the graph of $y = f(x)$ across the y-axis, we make the tranformation $y = f(-x)$. But $f(-x) = -f(x)$ for all x in the domain of f, so this transformation results in $y = -f(x)$.
That is exactly the translation that reflects the graph of f across the x-axis, so the two reflections yield the same graph.

35.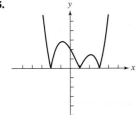

37.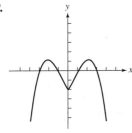

39. (a) $2x^3 - 8x$ **(b)** $27x^3 - 12x$

41. (a) $2x^2 + 2x - 4$ **(b)** $9x^2 + 3x - 2$

43. Starting with $y = x^2$, translate right 3 units, vertically stretch by 2, and translate down 4 units.

45. Starting with $y = x^2$, horizontally shrink by $\frac{1}{3}$ and translate down 4 units.

47. $y = 3(x - 4)^2$ **49.** $y = 2|x + 2| - 4$

51.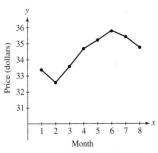

53.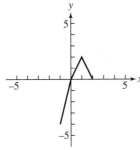

55. Reflections have more effect on points that are farther away from the line of reflection. Translations affect the distance of points from the axes, and hence change the effect of the reflections.

57. First vertically stretch by $\frac{9}{5}$, then translate up 32 units.

59. False; it is translated left. **61.** C **63.** A

65. (a)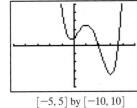

(b) Change the y-value by multiplying by the conversion rate from dollars to yen, a number that changes according to international market conditions. This results in a vertical stretch by the conversion rate.

67. (a) The original graph is on the top; the graph of $y = |f(x)|$ is on the bottom.

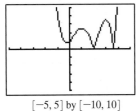

$[-5, 5]$ by $[-10, 10]$

(b) The original graph is on the top; the graph of $y = f(|x|)$ is on the bottom.

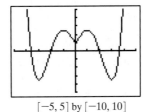

$[-5, 5]$ by $[-10, 10]$

(c)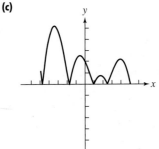

(d)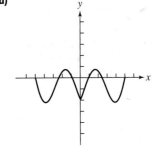

SECTION 1.7

Exploration 1

1.

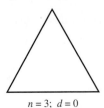
$n = 3; \ d = 0$

$n = 4; \ d = 2$

$n = 5; \ d = 5$

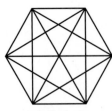
$n = 6; \ d = 9$

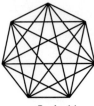
$n = 7; \ d = 14$

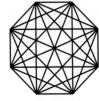
$n = 8; \ d = 20$

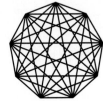
$n = 9; \ d = 27$

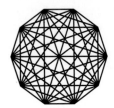
$n = 10; \ d = 35$

2, 5, 9, 14, 20, 27

3. Linear: $r^2 = 0.9758$; power: $r^2 = 0.9903$;
quadratic: $R^2 = 1$; cubic: $R^2 = 1$; quartic: $R^2 = 1$

5. Since the quadratic curve fits the points perfectly, there is
nothing to be gained by adding a cubic term or a quartic term.
The coefficients of these terms in the regressions are zero.

Quick Review 1.7

1. $h = 2(A/b)$ **3.** $h = V/(\pi r^2)$ **5.** $r = \sqrt[3]{\dfrac{3V}{4\pi}}$ **7.** $h = \dfrac{A - 2\pi r^2}{2\pi r} = \dfrac{A}{2\pi r} - r$ **9.** $P = \dfrac{A}{(1 + r/n)^{nt}} = A(1 + r/n)^{-nt}$

Exercises 1.7

1. $3x + 5$ **3.** $0.17x$ **5.** $(x + 12)(x)$ **7.** $1.045x$ **9.** $0.60x$ **11.** Let C be the total cost and n be the number of items
produced; $C = 34{,}500 + 5.75n$.

13. Let R be the revenue and n be the number of items sold: $R = 3.75n$.   **15.** $V = 2\pi r^3$ **17.** $A = a^2\sqrt{15}/4$ **19.** $A = 24r^2$

21. $x + 4x = 620; x = 124; 4x = 496$ **23.** $1.035x = 36{,}432; x = 35{,}200$ **25.** $182 = 52t$, so $t = 3.5$ hr

27. $0.60(33) = 19.8, 0.75(27) = 20.25$; The \$33 shirt is a better bargain, because the sale price is cheaper.   **29.** 15.95%

31. (a) $0.10x + 0.45(100 - x) = 0.25(100)$ **(b)** Use $x \approx 57.14$ gallons of the 10% solution and about 42.86 gal of the 45% solution.

33. (a) $V = x(10 - 2x)(18 - 2x)$ **(b)** $(0,5)$ **(c)** Approx. 2.06 in. by 2.06 in.   **35.** 6 in.   **37.** Approx. 21.36 in.   **39.** Approx. 11.42 mph

41. True; the correlation coefficient is close to 1 if there is a good fit.   **43.** C   **45.** B   **47. (a)** $C = 100{,}000 + 30x$ **(b)** $R = 50x$

(c) $x = 5000$ pairs of shoes   **(d)** The point of intersection corresponds to the break-even point, where $C = R$.

49. (a) $y_1 = u(x) = 125{,}000 + 23x$ **(b)** $y_2 = s(x) = 125{,}000 + 31x$ **(c)** $y_3 = R_u(x) = 56x$ **(d)** $y_4 = R_s(x) = 79x$

(e)

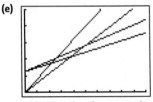
$[0, 10{,}000]$ by $[0, 500{,}000]$

(f) You should recommend stringing the rackets; fewer strung rackets need to
be sold to begin making a profit (since the intersection of y_2 and y_4 occurs
for smaller x than the intersection of y_1 and y_3).

51. (a)

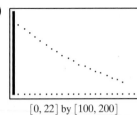

[0, 22] by [100, 200]

(b) List L3 = {112.3, 106.5, 101.5, 96.6, 92.0, 87.2, 83.1, 79.8, 75.0, 71.7, 68, 64.1, 61.5, 58.5, 55.9, 53.0, 50.8, 47.9, 45.2, 43.2}

(c) $y = 118.07 \times 0.951^x$. It fits the data extremely well.

CHAPTER 1 REVIEW EXERCISES

1. d **3.** i **5.** b **7.** g **9.** a **11. (a)** All reals **(b)** All reals **13. (a)** All reals **(b)** $[0, \infty)$

15. (a) All reals **(b)** $[8, \infty)$ **17. (a)** All reals except 0 and 2 **(b)** All reals except 0

19. Continuous **21. (a)** Vertical asymptotes at $x = 0$ and $x = 5$ **(b)** $y = 0$

23. (a) none **(b)** $y = 7$ and $y = -7$ **25.** $(-\infty, \infty)$ **27.** $(-\infty, -1), (-1, 1), (1, \infty)$ **29.** Not bounded **31.** Bounded above

33. (a) none **(b)** -7, at $x = -1$

35. (a) -1, at $x = 0$ **(b)** none **37.** Even **39.** Neither **41.** $(x - 3)/2$ **43.** $2/x$

45.

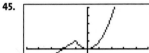

[-5, 5] by [-5, 5]

47.

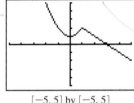

[-5, 5] by [-5, 5]

49.

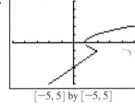

[-5, 5] by [-5, 5]

51.

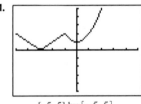

[-5, 5] by [-5, 5]

53. $(f \circ g)(x) = \sqrt{x^2 - 4}; (-\infty, -2] \cup [2, \infty)$ **55.** $(f \circ g)(x) = \sqrt{x}(x^2 - 4); [0, \infty)$

57. $\lim\limits_{x \to \infty} \sqrt{x} = \infty$ **59.** $\pi s^2/2$ **61.** $100\pi h$ **63.** $40 - t/(50\pi)$

65. (a)

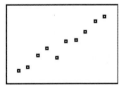

[4, 15] by [940, 1700]

(b) The regression line is $y = 61.133x + 725.333$.

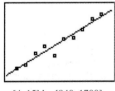

[4, 15] by [940, 1700]

(c) 1,948 (thousands of barrels)

67. (a) $h = 2\sqrt{3 - r^2}$ **(b)** $2\pi r^2\sqrt{3 - r^2}$ **(c)** $[0, \sqrt{3}]$

(d)

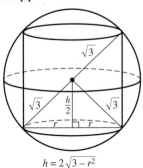

[0, $\sqrt{3}$] by [0, 20]

(e) 12.57 in.^{3}

Chapter 1 Project

1.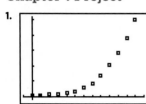

$[-1, 13]$ by $[-100, 2600]$

3. $4690, 7085$ **5.** $y = \dfrac{4914.198}{1 + 269.459e^{-0.468x}}$

SECTION 2.1

Exploration 1

1. $-\$2000$ per year   **3.** $\$50,000; \$18,000$

Quick Review 2.1

1. $y = 8x + 3.6$ **3.** $y = -0.6x + 2.8$ **5.** $x^2 + 6x + 9$ **7.** $3x^2 - 36x + 108$ **9.** $2(x - 1)^2$

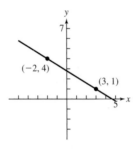

Exercises 2.1

1. Not a polynomial function because of the exponent -5
3. Polynomial of degree 5 with leading coefficient 2
5. Not a polynomial function because of the radical

7. $f(x) = \dfrac{5}{7}x + \dfrac{18}{7}$ **9.** $f(x) = -\dfrac{4}{3}x + \dfrac{2}{3}$

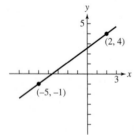   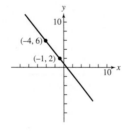

11. $f(x) = -x + 3$

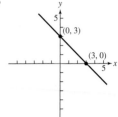

13. (a)　　**15.** (b)　　**17.** (e)

19. Translate the graph of $y = x^2$ 3 units right and the result 2 units down.

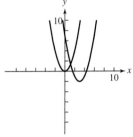

21. Translate the graph of $y = x^2$ 2 units left, vertically shrink the resulting graph by a factor of $\frac{1}{2}$, and translate that graph 3 units down.

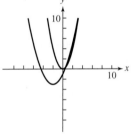

23. Vertex: $(1, 5)$; axis: $x = 1$　　**25.** Vertex: $(1, -7)$; axis: $x = 1$

27. Vertex: $\left(-\frac{5}{6}, -\frac{73}{12}\right)$; axis: $x = -\frac{5}{6}$; $f(x) = 3\left(x + \frac{5}{6}\right)^2 - \frac{73}{12}$

29. Vertex: $(4, 19)$; axis: $x = 4$; $f(x) = -(x - 4)^2 + 19$

31. Vertex: $\left(\frac{3}{5}, \frac{11}{5}\right)$; axis: $x = \frac{3}{5}$; $g(x) = 5\left(x - \frac{3}{5}\right)^2 + \frac{11}{5}$

33. $f(x) = (x - 2)^2 + 2$; Vertex: $(2, 2)$; axis: $x = 2$; opens upward; does not intersect x-axis

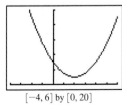

$[-4, 6]$ by $[0, 20]$

35. $f(x) = -(x + 8)^2 + 74$; Vertex: $(-8, 74)$; axis: $x = -8$; opens downward; intersects x-axis at about -16.602 and 0.602, or $(-8 \pm \sqrt{74})$

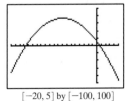

$[-20, 5]$ by $[-100, 100]$

37. $f(x) = 2\left(x + \frac{3}{2}\right)^2 + \frac{5}{2}$;

Vertex: $\left(-\frac{3}{2}, \frac{5}{2}\right)$; axis: $x = -\frac{3}{2}$;

opens upward; does not intersect

x-axis; vertically stretched by 2

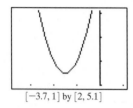

$[-3.7, 1]$ by $[2, 5.1]$

39. $y = 2(x + 1)^2 - 3$ **41.** $y = -2(x - 1)^2 + 11$ **43.** $y = 2(x - 1)^2 + 3$ **45** Strong positive   **47.** Weak positive

49. (a)

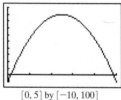

$[15, 45]$ by $[20, 50]$

(b) Strong positive

51 $940

53. (a) $y \approx 0.541x + 4.072$. The slope tells us that hourly compensation for production workers increases about 54¢/yr.   **(b)** About $25.70

55. (a) $[0, 100]$ by $[0, 1000]$ is one possibility.   **(b)** either 107,335 units or 372,665 units

57. 3.5 ft

59. (a) $R(x) = (26{,}000 - 1000x)(0.50 + 0.05x)$ **(b)**

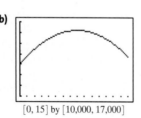

$[0, 15]$ by $[10{,}000, 17{,}000]$

(c) 90 cents per can; $16,200

61. (a) About 215 ft above the field   **(b)** About 6.54 sec   **(c)** About 117 ft/sec downward

63. (a) $h = -16t^2 + 80t - 10$ **(b)** 90 ft, 2.5 sec

$[0, 5]$ by $[-10, 100]$

65. 2006

67. (a)

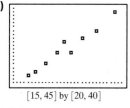

[15, 45] by [20, 40]

(c) On average, the children gained 0.68 pounds per month.

(d)

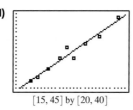

[15, 45] by [20, 40]

(b) $y \approx 0.68x + 9.01$

(e) ≈ 29.41 lb

69. The Identity Function $f(x) = x$

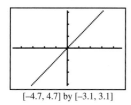

[−4.7, 4.7] by [−3.1, 3.1]

Domain: $(-\infty, \infty)$; Range: $(-\infty, \infty)$; Continuous; Increasing for all x; Symmetric about the origin; Not bounded; No local extrema; No horizontal or vertical asymptotes; End behavior: $\lim_{x \to -\infty} f(x) = -\infty$, $\lim_{x \to \infty} f(x) = \infty$

71. False. The initial value is $f(0) = -3$

73. E

75. B

77. (a) (i), (iii), and (v)   **(b)** (i), (iii), (iv), (v), and (vi) **(c)** (ii) is not a function.

81. (a) The two solutions are $\dfrac{-b + \sqrt{b^2 - 4ac}}{2a}$ and $\dfrac{-b - \sqrt{b^2 - 4ac}}{2a}$; their sum is $2\left(-\dfrac{b}{2a}\right) = -\dfrac{b}{a}$.

(b) The product of the two solutions given above is $\dfrac{b^2 - (b^2 - 4ac)}{4a^2} = \dfrac{c}{a}$.

83. $\left(\dfrac{a + b}{2}, -\dfrac{(a - b)^2}{4}\right)$

85. Suppose $f(x) = mx + b$ with m and b constants and $m \neq 0$. Let x_1 and x_2 be real numbers with $x_1 \neq x_2$. Then the average rate of change of f is

$\dfrac{f(x_2) - f(x_1)}{x_2 - x_1} = \dfrac{(mx_2 + b) - (mx_1 + b)}{x_2 - x_1} = \dfrac{m(x_2 - x_1)}{x_2 - x_1} = m$, a nonzero constant. On the other hand, suppose m and x_1 are constants and $m \neq 0$.

Let x be a real number with $x \neq x_1$ and let f be a function defined on all real numbers such that $\dfrac{f(x) - f(x_1)}{x - x_1} = m$. Then $f(x) - f(x_1) = m(x - x_1)$

and $f(x) = mx + (f(x_1) - mx_1)$. Notice that the expression $f(x_1) - mx_1$ is a constant; call it b. Then $f(x_1) - mx_1 = b$; so, $f(x_1) = mx_1 + b$ and $f(x) = mx + b$ for all $x \neq x_1$. Thus f is a linear function.

SECTION 2.2

Exploration 1

1.

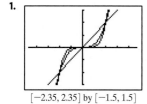

[−2.35, 2.35] by [−1.5, 1.5]   [−5, 5] by [−15, 15]

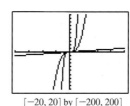

[−20, 20] by [−200, 200]

The pairs (0, 0), (1, 1) and (−1, −1) are common to all three graphs.

Quick Review 2.2

1. $\sqrt[3]{x^2}$ **3.** $1/d^2$ **5.** $1/\sqrt[5]{q^4}$ **7.** $3x^{3/2}$ **9.** $\approx 1.71x^{-4/3}$

Exercises 2.2

1. power $= 5$, constant $= -\dfrac{1}{2}$ **3.** not a power function

5. power $= 1$, constant $= c^2$ **7.** power $= 2$, constant $= \dfrac{g}{2}$

9. power $= -2$, constant $= k$

11. degree $= 0$, coefficient $= -4$ **13.** degree $= 7$, coefficient $= -6$ **15.** degree $= 2$, coefficient $= 4\pi$

17. $A = ks^2$ **19.** $I = V/R$ **21.** $E = mc^2$

23. The weight w of an object varies directly with its mass m, with the constant of variation g.

25. The refractive index n of a medium is inversely proportional to v, the velocity of light in the medium, with constant of variation c, the constant velocity of light in free space.

27. power $= 4$, constant $= 2$; Domain: $(-\infty, \infty)$;
Range: $[0, \infty)$; Continuous; Decreasing on $(-\infty, 0)$.
Increasing on $(0, \infty)$; Even. Symmetric with respect
to y-axis; Bounded below, but not above;
Local minimum at $x = 0$; Asymptotes: none;
End Behavior: $\lim\limits_{x \to -\infty} 2x^4 = \infty$, $\lim\limits_{x \to \infty} 2x^4 = \infty$

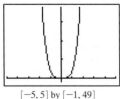

$[-5, 5]$ by $[-1, 49]$

29. power $= \dfrac{1}{4}$, constant $= \dfrac{1}{2}$; Domain: $[0, \infty)$;

Range: $[0, \infty)$; Continuous; Increasing on $[0, \infty)$;
Bounded below; Neither even nor odd;
Local minimum at $(0, 0)$; Asymptotes: none;

End Behavior: $\lim\limits_{x \to \infty} \dfrac{1}{2}\sqrt[4]{x} = \infty$

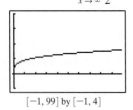

$[-1, 99]$ by $[-1, 4]$

31. shrink vertically by $\dfrac{2}{3}$; f is even.

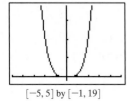

$[-5, 5]$ by $[-1, 19]$

33. stretch vertically by 1.5 and reflect
over the x-axis; f is odd.

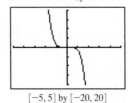

$[-5, 5]$ by $[-20, 20]$

35. shrink vertically by $\frac{1}{4}$; f is even.　　**37.** (g)　　**39.** (d)　　**41.** (h)

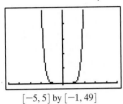

$[-5, 5]$ by $[-1, 49]$

43. $k = 3$, $a = \frac{1}{4}$. f is increasing in Quadrant I. f is undefined for $x < 0$.　　**45.** $k = -2$, $a = \frac{4}{3}$. f is decreasing in Quadrant IV. f is even.

47. $k = \frac{1}{2}$, $a = -3$. f is decreasing in Quadrant I. f is odd.

49. $y = \frac{8}{x^2}$, power $= -2$, constant $= 8$　　**51.** 2.21 L　　**53.** 1.24×10^8 m/sec

55. (a)

(b) $r \approx 231.204 \cdot w^{-0.297}$　　**(c)**

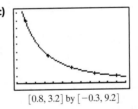

(d) Approximately 37.67 beats/min, which is very close to Clark's observed value

$[-2, 71]$ by $[50, 450]$　　$[-2, 71]$ by $[50, 450]$

57. (a)

(b) $y \approx 7.932 \cdot x^{-1.987}$; yes　　**(c)**

(d) Approximately 2.76 $\frac{\text{W}}{\text{m}^2}$ and 0.697 $\frac{\text{W}}{\text{m}^2}$, respectively

$[0.8, 3.2]$ by $[-0.3, 9.2]$　　$[0.8, 3.2]$ by $[-0.3, 9.2]$

59. False, because $f(-x) = (-x)^{1/3} = -x^{1/3} = -f(x)$. The graph of f is symmetric about the origin.　　**61.** E　　**63.** B

65. (a)

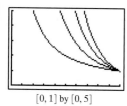
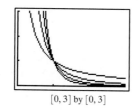
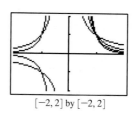

$[0, 1]$ by $[0, 5]$ $[0, 3]$ by $[0, 3]$ $[-2, 2]$ by $[-2, 2]$

The graphs of $f(x) = x^{-1}$ and $h(x) = x^{-3}$ are similar and appear in the 1st and 3rd quadrants only. The graphs of $g(x) = x^{-2}$ and $k(x) = x^{-4}$ are similar and appear in the 1st and 2nd quadrants only. The pair $(1, 1)$ is common to all four functions.

	f	g	h	k
Domain	$x \neq 0$	$x \neq 0$	$x \neq 0$	$x \neq 0$
Range	$y \neq 0$	$y > 0$	$y \neq 0$	$y > 0$
Continuous	yes	yes	yes	yes
Increasing		$(-\infty, 0)$		$(-\infty, 0)$
Decreasing	$(-\infty, 0), (0, \infty)$	$(0, \infty)$	$(-\infty, 0), (0, \infty)$	$(0, \infty)$
Symmetry	w.r.t. origin	w.r.t. y-axis	w.r.t. origin	w.r.t. y-axis
Bounded	not	below	not	below
Extrema	none	none	none	none
Asymptotes	x-axis, y-axis	x-axis, y-axis	x-axis, y-axis	x-axis, y-axis
End Behavior	$\lim\limits_{x \to \pm\infty} f(x) = 0$	$\lim\limits_{x \to \pm\infty} g(x) = 0$	$\lim\limits_{x \to \pm\infty} h(x) = 0$	$\lim\limits_{x \to \pm\infty} k(x) = 0$

(b)

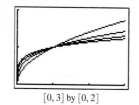
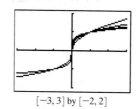

$[0, 1]$ by $[0, 1]$ $[0, 3]$ by $[0, 2]$ $[-3, 3]$ by $[-2, 2]$

The graphs of $f(x) = x^{1/2}$ and $h(x) = x^{1/4}$ are similar and appear in the 1st quadrant only. The graphs of $g(x) = x^{1/3}$ and $k(x) = x^{1/5}$ are similar and appear in the 1st and 3rd quadrants only. The pairs $(0, 0)$, $(1, 1)$ are common to all four functions.

	f	g	h	k
Domain	$[0, \infty)$	$(-\infty, \infty)$	$[0, \infty)$	$(-\infty, \infty)$
Range	$y \geq 0$	$(-\infty, \infty)$	$y \geq 0$	$(-\infty, \infty)$
Continuous	yes	yes	yes	yes
Increasing	$[0, \infty)$	$(-\infty, \infty)$	$[0, \infty)$	$(-\infty, \infty)$
Decreasing				
Symmetry	none	w.r.t. origin	none	w.r.t. origin
Bounded	below	not	below	not
Extrema	min at $(0, 0)$	none	min at $(0, 0)$	none
Asymptotes	none	none	none	none
End behavior	$\lim\limits_{x \to \infty} f(x) = \infty$	$\lim\limits_{x \to \infty} g(x) = \infty$ $\lim\limits_{x \to -\infty} g(x) = -\infty$	$\lim\limits_{x \to \infty} h(x) = \infty$	$\lim\limits_{x \to \infty} k(x) = \infty$ $\lim\limits_{x \to -\infty} k(x) = -\infty$

67. $T \approx a^{1.5}$. Squaring both sides shows that approximately $T^2 = a^3$.   **69.** If $f(x)$ is even, $g(-x) = 1/f(-x) = 1/f(x) = g(x)$. If $f(x)$ is odd, $g(-x) = 1/f(-x) = 1/(-f(x)) = -1/f(x) = -g(x)$. If $g(x) = 1/f(x)$, then $f(x) \cdot g(x) = 1$ and $f(x) = 1/g(x)$. So by the reasoning used above, if $g(x)$ is even, so is $f(x)$, and if $g(x)$ is odd, so is $f(x)$.   **71. (a)** The force F acting on an object varies jointly as the mass m of the object and the acceleration a of the object.   **(b)** The kinetic energy KE of an object varies jointly as the mass m of the object and the square of the velocity v of the object.   **(c)** The force of gravity F acting on two objects varies jointly as their masses m_1 and m_2 and inversely as the square of the distance r between their centers, with the constant of variation G, the universal gravitational constant.

SECTION 2.3

Exploration 1

1. (a) $\infty; -\infty$ **(b)** $-\infty; \infty$ **(c)** $\infty; -\infty$ **(d)** $-\infty; \infty$ **3. (a)** $-\infty; \infty$ **(b)** $-\infty; -\infty$ **(c)** $\infty; \infty$ **(d)** $\infty; -\infty$

Exploration 2

1. $y = 0.0061x^3 + 0.0177x^2 - 0.5007x + 0.9769$

Quick Review 2.3

1. $(x - 4)(x + 3)$ **3.** $(3x - 2)(x - 3)$ **5.** $x(3x - 2)(x - 1)$ **7.** $x = 0, x = 1$ **9.** $x = -6, x = -3, x = 1.5$

Exercises 2.3

1. Shift $y = x^3$ to the right by 3 units, stretch vertically by 2. y-intercept: $(0, -54)$

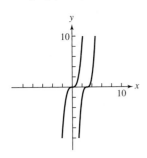

3. Shift $y = x^3$ to the left by 1 unit, vertically shrink by $\dfrac{1}{2}$, reflect over the x-axis, and then vertically shift up 2 units. y-intercept: $\left(0, \dfrac{3}{2}\right)$

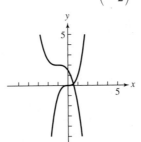

5. Shift $y = x^4$ to the left 2 units, vertically stretch by 2, reflect over the x-axis, and vertically shift down 3 units. y-intercept: $(0, -35)$

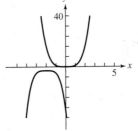

7. local maximum: $\approx (0.79, 1.19)$, zeros: $x = 0$ and $x \approx 1.26$.   **9.** (c)   **11.** (a)

13. One possibility:

$[-100, 100]$ by $[-1000, 1000]$

15. One possibility:

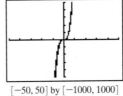

$[-50, 50]$ by $[-1000, 1000]$

17.

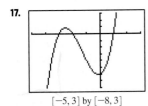

$[-5, 3]$ by $[-8, 3]$

$\lim\limits_{x \to \infty} f(x) = \infty$; $\lim\limits_{x \to -\infty} f(x) = -\infty$

19.

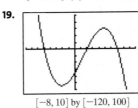

$[-8, 10]$ by $[-120, 100]$

$\lim\limits_{x \to \infty} f(x) = -\infty$; $\lim\limits_{x \to -\infty} f(x) = \infty$

21.

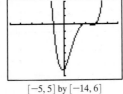

$[-5, 5]$ by $[-14, 6]$

$\lim\limits_{x \to \infty} f(x) = \infty$; $\lim\limits_{x \to -\infty} f(x) = \infty$

23.

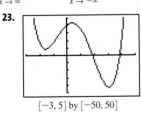

$[-3, 5]$ by $[-50, 50]$

$\lim\limits_{x \to \infty} f(x) = \infty$; $\lim\limits_{x \to -\infty} f(x) = \infty$

25. ∞, ∞ **27.** $-\infty, \infty$

29. (a) There are 3 zeros: they are -2.5, 1, and 1.1.   **31.** (c) There are 3 zeros: approximately -0.273 $\left(\text{actually } -\dfrac{3}{11}\right)$, -0.25, and 1.

33. -4 and 2   **35.** $\dfrac{2}{3}$ and $-\dfrac{1}{3}$ **37.** 0, $-\dfrac{2}{3}$, and 1

39. Degree 3; zeros: $x = 0$ (mult. 1, graph crosses x-axis), $x = 3$ (mult. 2, graph is tangent)

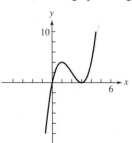

41. Degree 5; zeros: $x = 1$ (mult. 3, graph crosses x-axis), $x = -2$ (mult. 2, graph is tangent)

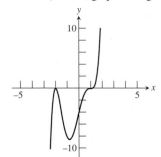

43.

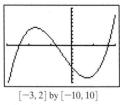

$[-3, 2]$ by $[-10, 10]$
-2.43, -0.74, 1.67

45.

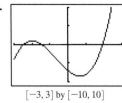

$[-3, 3]$ by $[-10, 10]$
-2.47, -1.46, 1.94

47.

$[-6, 4]$ by $[-100, 20]$
-4.90, -0.45, 1, 1.35

49. 0, -6, and 6   **51.** -5, 1, 11

53. $f(x) = x^3 - 5x^2 - 18x + 72$ **55.** $f(x) = x^3 - 4x^2 - 3x + 12$ **57.** $y = 0.25x^3 - 1.25x^2 - 6.75x + 19.75$

59. $y = -2.21x^4 + 45.75x^3 - 339.79x^2 + 1075.25x - 1231$

61. It follows from the Intermediate Value Theorem.

63. (a)

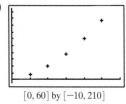

$[0, 60]$ by $[-10, 210]$

(b) $y = 0.051x^2 + 0.97x + 0.26$ **(c)**

$[0, 60]$ by $[-10, 210]$

(d) ≈ 56.39 ft   **(e)** 67.74 mph

65. (a)

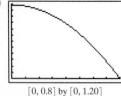

$[0, 0.8]$ by $[0, 1.20]$

(b) 0.3391 cm   **67.** $0 < x \le 0.929$ or $3.644 \le x < 5$

69. True. Because f is continuous and $f(1) = -2$ and $f(2) = 2$, the Intermediate Value Theorem assures us that the graph of f crosses the x-axis between $x = 1$ and $x = 2$.   **71.** C   **73.** B

77. The exact behavior near $x = 1$ is hard to see. A zoomed-in view around the point $(1, 0)$ suggests that the graph just touches the x-axis at 0 without actually crossing it — that is, $(1, 0)$ is a local maximum. One possible window is $[0.9999, 1.0001]$ by $[-1 \times 10^{-7}, 1 \times 10^{-7}]$.

79. A maximum and minimum are not visible in the standard window, but can be seen on the window $[0.2, 0.4]$ by $[5.29, 5.3]$.

81. The graph of $y = 3(x^3 - x)$ increases, then decreases, then increases; the graph of $y = x^3$ only increases. Therefore, this graph can not be obtained from the graph of $y = x^3$ by the transformations studied in Chapter 1 (translations, reflections, and stretching/shrinking). Since the right side includes only these transformations, there can be no solution.

83. (a) Substituting $x = 2$, $y = 7$, we find that $7 = 5(2 - 2) + 7$, so Q is on line L, and also $f(2) = -8 + 8 + 18 - 11 = 7$, so Q is on the graph of $f(x)$.

(b)

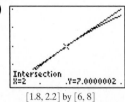

Intersection
X=2   Y=7.0000002
$[1.8, 2.2]$ by $[6, 8]$

(c) The line L also crosses the graph of $f(x)$ at $(-2, -13)$.

85. (a) $\dfrac{8}{D-u} = \dfrac{x}{D}$ and $\dfrac{8}{u} = \dfrac{y}{D}$ imply $D - u = \dfrac{uy}{x}$. $\dfrac{8}{u} = \dfrac{y-8}{D-u}$ implies $D - u = \dfrac{u(y-8)}{8}$. Combining these yields $\dfrac{uy}{x} = \dfrac{u(y-8)}{8}$, which

implies $\dfrac{8}{x} = \dfrac{y-8}{y}$. **(b)** Equation (a) says $\dfrac{8}{x} = 1 - \dfrac{8}{y}$. So, $\dfrac{8}{y} = 1 - \dfrac{8}{x} = \dfrac{x-8}{x}$. Thus $y = \dfrac{8x}{x-8}$. **(c)** By the Pythagorean Theorem,

$y^2 + D^2 = 900$ and $x^2 + D^2 = 400$. Subtracting equal quantities yields $y^2 - x^2 = 500$. So, $500 = \left(\dfrac{8x}{x-8}\right)^2 - x^2$. Thus, $500\,(x-8)^2 = $

$64x^2 - x^2(x-8)^2$, or $500x^2 - 8000x + 32000 = 64x^2 - x^4 + 16x^3 - 64x^2$. This is equivalent to $x^4 - 16x^3 + 500x^2 - 8000x + 32000 = 0$.
(d) Notice that $8 < x < 20$. So, the solution we seek is $x \approx 11.71$, which yields $y \approx 25.24$ and $D \approx 16.21$.

SECTION 2.4

Quick Review 2.4

1. $x^2 - 4x + 7$ **3.** $7x^3 + x^2 - 3$ **5.** $x(x+2)(x-2)$ **7.** $4(x+5)(x-3)$ **9.** $(x+2)(x+1)(x-1)$

Exercises 2.4

1. $f(x) = (x-1)^2 + 2;\ \dfrac{f(x)}{x-1} = x - 1 + \dfrac{2}{x-1}$

3. $f(x) = (x^2 + x + 4)(x+3) - 21;\ \dfrac{f(x)}{x+3} = x^2 + x + 4 - \dfrac{21}{x+3}$

$\dfrac{f(x)}{2x+1} = 2x^2 - 5x + \dfrac{7}{2} - \dfrac{9/2}{2x+1}$ **5.** $f(x) = (x^2 - 4x + 12)(x^2 + 2x - 1) - 32x + 18;$

$\dfrac{f(x)}{x^2 + 2x - 1} = x^2 - 4x + 12 + \dfrac{-32x + 18}{x^2 + 2x - 1}$ **7.** $x^2 - 6x + 9 + \dfrac{-11}{x+1}$ **9.** $9x^2 + 97x + 967 + \dfrac{9670}{x-10}$

11. $-5x^3 - 20x^2 - 80x - 317 + \dfrac{-1269}{4-x}$ **13.** 3   **15.** -43 **17.** 5   **19.** Yes   **21.** No   **23.** Yes

25. $f(x) = (x+3)(x-1)(5x-17)$ **27.** $2x^3 - 6x^2 - 12x + 16$ **29.** $2x^3 - 8x^2 + \dfrac{19}{2}x - 3$ **31.** $f(x) = 3(x+4)(x-3)(x-5)$

33. $\dfrac{\pm 1}{\pm 1,\ \pm 2,\ \pm 3,\ \pm 6};\ 1$ **35.** $\dfrac{\pm 1,\ \pm 3,\ \pm 9}{\pm 1,\ \pm 2};\ \dfrac{3}{2}$ **45.** No zeros outside window

47. There *are* zeros not shown (approx. -11.002 and 12.003)

49. Rational zero: $\dfrac{3}{2}$; irrational zeros: $\pm\sqrt{2}$ **51.** Rational: -3; irrational: $1 \pm \sqrt{3}$

53. Rational: -1 and 4; irrational: $\pm\sqrt{2}$ **55.** Rational: $-\dfrac{1}{2}$ and 4; irrational: none   **57.** \$36.27; 53.7   **59.** -2

61. (b) 2 is a zero of $f(x)$ **(c)** $(x-2)(x^3 + 4x^2 - 3x - 19)$ **(d)** One irrational zero is $x \approx 2.04$.
  (e) $f(x) \approx (x-2)(x-2.04)(x^2 + 6.04x + 9.3116)$
63. False. $(x+2)$ is a factor if and only if $f(-2) = 0$.   **65.** A   **67.** B
69. (d) $x \approx 0.6527$ m
71. (a) Shown is one possible view, on the window $[0, 600]$ by $[0, 500]$.

[0, 600] by [0, 500]

(b) The maximum population, after 300 days, is 460 turkeys.
(c) $P = 0$ when $t \approx 523.22$ — about 523 days after release.
(d) Answers will vary.

73. (a) 0 or 2 positive zeros, 1 negative zero   **(b)** no positive zeros, 1 or 3 negative zeros   **(c)** 1 positive zero, no negative zeros   **(d)** 1 positive zero, 1 negative zero

75. Answers will vary, but should include a diagram of the synthetic division and a summary:

$$4x^3 - 5x^2 + 3x + 1 = \left(x - \frac{1}{2}\right)\left(4x^2 - 3x + \frac{3}{2}\right) + \frac{7}{4}$$
$$= (2x - 1)\left(2x^2 - \frac{3}{2}x + \frac{3}{4}\right) + \frac{7}{4}$$

77. (a)   (b) $-\dfrac{7}{3}, \dfrac{1}{2},$ and 3   **(c)** There are no rational zeros.

79. (a) Approximate zeros: $-3.126, -1.075, 0.910, 2.291.$ **(b)** $f(x) \approx g(x) = (x + 3.126)(x + 1.075)(x - 0.910)(x - 2.291)$
(c) Graphically: graph the original function and the approximate factorization on a variety of windows and observe their similarity. Numerically: Compute $f(c)$ and $g(c)$ for several values of c.

SECTION 2.5

Exploration 1

1. $f(2i) = (2i)^2 - i(2i) + 2 = -4 + 2 + 2 = 0;$ $f(-i) = (-i)^2 - i(-i) + 2 = -1 - 1 + 2 = 0;$ no.
3. The Complex Conjugate Zeros Theorem does not necessarily hold true for a polynomial function with *complex* coefficients.

Quick Review 2.5

1. $1 + 3i$ **3.** $7 + 4i$ **5.** $(2x - 3)(x + 1)$ **7.** $\dfrac{5}{2} \pm \dfrac{\sqrt{19}}{2}i$ **9.** $\pm 1, \pm 2, \pm 1/3, \pm 2/3$

Exercises 2.5

1. $x^2 + 9$; zeros: $\pm 3i$; x-intercepts: none   **3.** $x^4 - 2x^3 + 5x^2 - 8x + 4$; zeros: 1 (mult. 2), $\pm 2i$; x-intercept: $x = 1$ **5.** $x^2 + 1$
7. $x^3 - x^2 + 9x - 9$ **9.** $x^4 - 5x^3 + 7x^2 - 5x + 6$ **11.** $x^3 - 11x^2 + 43x - 65$ **13.** $x^5 + 4x^4 + x^3 - 10x^2 - 4x + 8$
15. $x^4 - 10x^3 + 38x^2 - 64x + 40$ **17.** (b)   **19.** (d)   **21.** 2 complex zeros; none real   **23.** 3 complex zeros; 1 real
25. 4 complex zeros; 2 real

27. Zeros: $x = 1, x = -\dfrac{1}{2} \pm \dfrac{\sqrt{19}}{2}i; f(x) = \dfrac{1}{4}(x - 1)(2x + 1 + \sqrt{19}i)(2x + 1 - \sqrt{19}i)$

29. Zeros: $x = \pm 1, x = -\dfrac{1}{2} \pm \dfrac{\sqrt{23}}{2}i; f(x) = \dfrac{1}{4}(x - 1)(x + 1)(2x + 1 + \sqrt{23}i)(2x + 1 - \sqrt{23}i)$

31. Zeros: $x = -\dfrac{7}{3}, x = \dfrac{3}{2}, x = 1 \pm 2i; f(x) = (3x + 7)(2x - 3)(x - 1 + 2i)(x - 1 - 2i)$

33. Zeros: $x = \pm\sqrt{3}, x = 1 \pm i; f(x) = (x - \sqrt{3})(x + \sqrt{3})(x - 1 + i)(x - 1 - i)$

35. Zeros: $x = \pm\sqrt{2}, x = 3 \pm 2i; f(x) = (x - \sqrt{2})(x + \sqrt{2})(x - 3 + 2i)(x - 3 - 2i)$

37. $(x - 2)(x^2 + x + 1)$ **39.** $(x - 1)(2x^2 + x + 3)$ **41.** $(x - 1)(x + 4)(x^2 + 1)$ **43.** $h \approx 3.776$ ft
45. Yes, $f(x) = (x + 2)^2 = x^3 + 6x^2 + 12x + 8.$ **47.** No, either the degree would be at least 5 or some of the coefficients would be nonreal.
49. $f(x) = -2x^4 + 12x^3 - 20x^2 - 4x + 30$

51. (a) $D \approx -0.0820t^3 + 0.9162t^2 - 2.5126t + 3.3779$

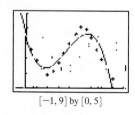

$[-1, 9]$ by $[0, 5]$

(b) Sally walks toward the detector, turns and walks away (or walks backward), then walks toward the detector again.
(c) $t \approx 1.81$ sec $(D \approx 1.35$ m$)$ and $t \approx 5.64$ sec $(D \approx 3.65$ m$)$.

53. False. If $1 - 2i$ is a zero, then $1 + 2i$ must also be a zero. **55.** E **57.** C

59. (a)

Power	Real Part	Imaginary Part
7	8	−8
8	16	0
9	16	16
10	0	32

(b) $(1 + i)^7 = 8 - 8i$
$(1 + i)^8 = 16$
$(1 + i)^9 = 16 + 16i$
$(1 + i)^{10} = 32i$

(c) Reconcile as needed.

61. $f(i) = i^3 - i(i)^2 + 2i(i) + 2 = -i + i - 2 + 2 = 0$ **63.** Synthetic division shows that $f(i) = 0$ (the remainder), and at the same time gives $f(x) \div (x - i) = x^2 + 3x - i = h(x)$, so $f(x) = (x - i)(x^2 + 3x - i)$. **65.** $-4, 2 + 2\sqrt{3}i, 2 - 2\sqrt{3}i$

SECTION 2.6

Exploration 1

1. $g(x) = \dfrac{1}{x - 2}$

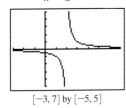

$[-3, 7]$ by $[-5, 5]$

3. $k(x) = \dfrac{3}{x + 4} - 2$

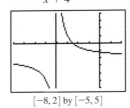

$[-8, 2]$ by $[-5, 5]$

Quick Review 2.6

1. $x = -3, x = \dfrac{1}{2}$ **3.** $x = \pm 2$ **5.** $x = 1$ **7.** $2; 7$ **9.** $3; -5$

Exercises 2.6

1. Domain: all $x \neq -3$; $\lim\limits_{x \to -3^-} f(x) = -\infty$, $\lim\limits_{x \to -3^+} f(x) = \infty$

3. Domain: all $x \neq -2, 2$; $\lim\limits_{x \to -2^-} f(x) = -\infty$, $\lim\limits_{x \to -2^+} f(x) = \infty$, $\lim\limits_{x \to 2^-} f(x) = \infty$, $\lim\limits_{x \to 2^+} f(x) = -\infty$

5. Translate right 3 units.
Asymptotes: $x = 3, y = 0$

7. Translate left 3 units, reflect across x-axis,
vertically stretch by 7, translate up 2 units.
Asymptotes: $x = -3, y = 2$

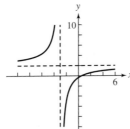

9. Translate left 4 units, vertically stretch by 13, translate down 2 units.
Asymptotes: $x = -4$, $y = -2$

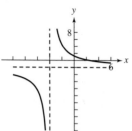

11. ∞ **13.** 0   **15.** ∞ **17.** 5

19. Vertical asymptote: none; Horizontal asymptote: $y = 2$; $\lim\limits_{x \to -\infty} f(x) = \lim\limits_{x \to \infty} f(x) = 2$

21. Vertical asymptotes: $x = 0$, $x = 1$; Horizontal asymptote: $y = 0$; $\lim\limits_{x \to 0^-} f(x) = \infty$, $\lim\limits_{x \to 0^+} f(x) = -\infty$, $\lim\limits_{x \to 1^-} f(x) = -\infty$, $\lim\limits_{x \to 1^+} f(x) = \infty$, $\lim\limits_{x \to -\infty} f(x) = \lim\limits_{x \to \infty} f(x) = 0$

23. Intercepts: $\left(0, \dfrac{2}{3}\right)$ and $(2, 0)$
Asymptotes: $x = -1$, $x = 3$, and $y = 0$

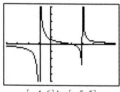

$[-4, 6]$ by $[-5, 5]$

25. No intercepts
Asymptotes: $x = -1$, $x = 0$, $x = 1$, and $y = 0$

$[-4.7, 4.7]$ by $[-10, 10]$

27. Intercepts: $(0, 2)$, $(-1.28, 0)$, and $(0.78, 0)$; Asymptotes: $x = 1$, $x = -1$, and $y = 2$

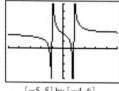

$[-5, 5]$ by $[-4, 6]$

29. Intercept: $\left(0, \dfrac{3}{2}\right)$
Asymptotes: $x = -2$, $y = x - 4$

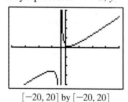

$[-20, 20]$ by $[-20, 20]$

31. (d); Xmin $= -2$, Xmax $= 8$, Xscl $= 1$, and Ymin $= -3$, Ymax $= 3$, Yscl $= 1$

33. (a); Xmin $= -3$, Xmax $= 5$, Xscl $= 1$, and Ymin $= -5$, Ymax $= 10$, Yscl $= 1$

35. (e); Xmin $= -2$, Xmax $= 8$, Xscl $= 1$, and Ymin $= -3$, Ymax $= 3$, Yscl $= 1$

37. Intercept: $\left(0, -\dfrac{2}{3}\right)$; asymptotes: $x = -1$, $x = \dfrac{3}{2}$, $y = 0$; $\lim\limits_{x \to -1^-} f(x) = \infty$, $\lim\limits_{x \to -1^+} f(x) = -\infty$, $\lim\limits_{x \to (3/2)^-} f(x) = -\infty$, $\lim\limits_{x \to (3/2)^+} f(x) = \infty$;

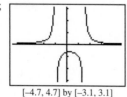

$[-4.7, 4.7]$ by $[-3.1, 3.1]$

Domain: $x \neq -1, \dfrac{3}{2}$; Range: $\left(-\infty, -\dfrac{16}{25}\right] \cup (0, \infty)$; Continuity: all $x \neq -1, \dfrac{3}{2}$; Increasing: $(-\infty, -1)$, $\left(-1, \dfrac{1}{4}\right]$, Decreasing: $\left[\dfrac{1}{4}, \dfrac{3}{2}\right)$, $\left(\dfrac{3}{2}, \infty\right)$; Unbounded; Local Maximum at $\left(\dfrac{1}{4}, -\dfrac{16}{25}\right)$; Horizontal asymptote: $y = 0$; Vertical asymptotes: $x = -1$, $x = \dfrac{3}{2}$; End behavior: $\lim\limits_{x \to -\infty} f(x) = \lim\limits_{x \to \infty} f(x) = 0$

69. (a

(b) N

(c) M

71. (a)

73. Ho

Int

$h(x$

77. The

and

SEC

Quick

1. $2x^2 +$

7. $\dfrac{3 \pm }{4}$

Exerc

1. $x = -$

11. $x = \dfrac{1}{2}$

17. $x = -$

23. $x = 3$

31. (a) Tl

(b) $y = 0$

33. (a) C

35. (a) $P($

37. (a) S

39. (a) $R($

---

39. Intercepts: $\left(0, \dfrac{1}{12}\right)$, $(1, 0)$; asymptotes: $x = -3$, $x = 4$, $y = 0$; $\lim\limits_{x \to -3^-} h(x) = -\infty$, $\lim\limits_{x \to -3^+} h(x) = \infty$, $\lim\limits_{x \to 4^-} h(x) = -\infty$, $\lim\limits_{x \to 4^+} h(x) = \infty$

[−5.875, 5.875] by [−3.1, 3.1]

Domain: $x \neq -3, 4$; Range: $(-\infty, \infty)$;
Continuity: all $x \neq -3, 4$;
Decreasing: $(-\infty, -3)$, $(-3, 4)$, $(4, \infty)$;
No symmetry; Unbounded; No extrema;
Horizontal asymptote: $y = 0$; Vertical asymptotes: $x = -3$, $x = 4$;
End behavior: $\lim\limits_{x \to -\infty} h(x) = \lim\limits_{x \to \infty} h(x) = 0$

41. Intercepts: $(-2, 0)$, $(1, 0)$, $\left(0, \dfrac{2}{9}\right)$; asymptotes: $x = -3$, $x = 3$, $y = 1$; $\lim\limits_{x \to -3^-} f(x) = \infty$, $\lim\limits_{x \to -3^+} f(x) = -\infty$, $\lim\limits_{x \to 3^-} f(x) = -\infty$, $\lim\limits_{x \to 3^+} f(x) = \infty$

[−9.4, 9.4] by [−3, 3]

Domain: $x \neq -3, 3$; Range: $(-\infty, 0.260] \cup (1, \infty)$; Continuity: all $x \neq -3, 3$;
Increasing: $(-\infty, -3)$, $(-3, -0.675)$; Decreasing: $(-0.675, 3)$, $(3, \infty)$;
No symmetry; Unbounded; Local maximum at $(-0.675, 0.260)$;
Horizontal asymptote: $y = 1$; Vertical asymptotes: $x = -3$, $x = 3$;
End behavior: $\lim\limits_{x \to -\infty} f(x) = \lim\limits_{x \to \infty} f(x) = 1$

43. Intercepts: $(-3, 0)$, $(1, 0)$, $\left(0, -\dfrac{3}{2}\right)$; asymptotes: $x = -2$, $y = x$; $\lim\limits_{x \to -2^-} h(x) = \infty$, $\lim\limits_{x \to -2^+} h(x) = -\infty$

[−9.4, 9.4] by [−15, 15]

Domain: $x \neq -2$, Range: $(-\infty, \infty)$;
Continuity: all $x \neq -2$;
Increasing: $(-\infty, -2)$, $(-2, \infty)$;
No symmetry; Unbounded; No extrema;
Horizontal asymptote: none; Vertical asymptote: $x = -2$;
Slant asymptote: $y = x$; End behavior: $\lim\limits_{x \to -\infty} h(x) = -\infty$, $\lim\limits_{x \to \infty} h(x) = \infty$

45. $y = x + 3$

(a)
[−10, 20] by [−10, 30]

(b)
[−40, 40] by [−40, 40]

47. $y = x^2 - 3x + 6$

(a)
[−10, 10] by [−30, 60]

(b)
[−50, 50] by [−1500, 2500]

49. $y = x^3 + 2x^2 + 4x + 6$

(a)
[−5, 5] by [−100, 200]

(b)
[−20, 20] by [−5000, 5000]

43. (a)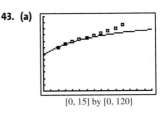

$[0, 15]$ by $[0, 120]$

(b) About 98.3 billion dollars

45. False. An extraneous solution is a solution of the equation cleared of fractions that is *not* a solution of the original equation.

47. D   **49.** E

51. (a) $f(x) = \dfrac{x^2 + 2x}{x^2 + 2x}$ **(b)** $x \neq 0, -2$ **(c)** $f(x) = \begin{cases} 1, & x \neq -2, 0 \\ \text{undefined}, & x = -2 \text{ or } x = 0 \end{cases}$ **(d)** The graph appears to be the horizontal line $y = 1$ with holes at $x = -2$ and $x = 0$.

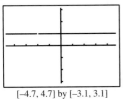

$[-4.7, 4.7]$ by $[-3.1, 3.1]$

53. $x = \dfrac{y}{y - 1}$ **55.** $x = \dfrac{2y - 3}{y - 2}$

SECTION 2.8

Exploration 1

1. (a) $\underset{-3 \qquad\quad 2}{\dfrac{(+)(-)(+)}{\text{Negative}} \bigg| \dfrac{(+)(-)(+)}{\text{Negative}} \bigg| \dfrac{(+)(+)(+)}{\text{Positive}}}\; x$

(b)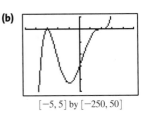

$[-5, 5]$ by $[-250, 50]$

3. (a) $\underset{-4 \qquad\quad 2}{\dfrac{(+)(+)(-)(-)}{\text{Positive}} \bigg| \dfrac{(+)(+)(+)(-)}{\text{Negative}} \bigg| \dfrac{(+)(+)(+)(-)}{\text{Negative}}}\; x$

(b)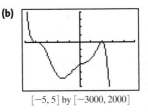

$[-5, 5]$ by $[-3000, 2000]$

Quick Review 2.8

1. $\lim\limits_{x \to \infty} f(x) = \infty;\ \lim\limits_{x \to -\infty} f(x) = -\infty$ **3.** $\lim\limits_{x \to \infty} g(x) = \infty,\ \lim\limits_{x \to -\infty} g(x) = \infty$ **5.** $(x^3 + 5)/x$ **7.** $\dfrac{x^2 - 7x - 2}{2x^2 - 5x - 3}$

9. (a) $\pm 1, \pm\dfrac{1}{2}, \pm 3, \pm\dfrac{3}{2}$ **(b)** $(x + 1)(2x - 3)(x + 1)$

Exercises 2.8

1. (a) $x = -2, -1, 5$ **(b)** $-2 < x < -1$ or $x > 5$ **(c)** $x < -2$ or $-1 < x < 5$

3. (a) $x = -7, -4, 6$ **(b)** $x < -7$ or $-4 < x < 6$ or $x > 6$ **(c)** $-7 < x < -4$

5. (a) $x = 8, -1$ **(b)** $-1 < x < 8$ or $x > 8$ **7.** $(-1, 3) \cup (3, \infty)$ **9.** $(-\infty, -1) \cup (1, 2)$ **11.** $[-2, 1/2] \cup [3, \infty)$

13. $[-1, 0] \cup [2, \infty)$ **15.** $(-1, 3/2) \cup (2, \infty)$ **17.** $[-1.15, \infty)$ **19.** $(3/2, 2)$

21. (a) $(-\infty, \infty)$ **(b)** $(-\infty, \infty)$ **(c)** There are no solutions.   **(d)** There are no solutions.

23. (a) $x \neq \dfrac{4}{3}$ **(b)** $(-\infty, \infty)$ **(c)** There are no solutions.   **(d)** $x = \dfrac{4}{3}$

25. (a) $x = 1$ **(b)** $x = -\dfrac{3}{2}, 4$ **(c)** $-\dfrac{3}{2} < x < 1$ or $x > 4$ **(d)** $x < -\dfrac{3}{2}$, or $1 < x < 4$

27. (a) $x = 0, -3$ **(b)** $x < -3$ **(c)** $x > 0$ **(d)** $-3 < x < 0$ **29. (a)** $x = -5$ **(b)** $x = -\dfrac{1}{2}, x = 1, x < -5$ **(c)** $-5 < x < -\dfrac{1}{2}$ or $x > 1$

81.

(d) $-\dfrac{1}{2} < x < 1$ **31. (a)** $x = 3$ **(b)** $x = 4, x < 3$ **(c)** $3 < x < 4$ or $x > 4$ **(d)** $f(x)$ is never negative.

85.

33. $(-\infty, -2) \cup (1, 2)$ **35.** $[-1, 1]$ **37.** $(-\infty, -4) \cup (3, \infty)$ **39.** $[-1, 0] \cup [1, \infty)$ **41.** $(0, 2) \cup (2, \infty)$ **43.** $\left(-4, \dfrac{1}{2}\right)$

87.

45. $(0, 2)$ **47.** $(-\infty, 0) \cup (\sqrt[3]{2}, \infty)$ **49.** $(-\infty, -1) \cup [1, 3)$ **51.** $[-3, \infty)$ **53.** $[5, \infty)$ **57.** 1 in. $< x <$ 34 in.

59. 0 in. $\leq x \leq 0.69$ in. or 4.20 in. $\leq x \leq 6$ in.

61. (a) $S = 2\pi x^2 + 1000/x$ **(b)** 1.12 cm $\leq x \leq$ 11.37 cm, 1.23 cm $\leq h \leq$ 126.88 cm   **(c)** about 348.73 cm^2

63. (a) $y \approx 993.870x + 19025.768$ **(b)** After 2011   **65.** False, because the factor x^4 does not change sign at $x = 0$.   **67.** C   **69.** D

71. Vertical asymptotes: $x = -1, x = 3$; x-intercepts: $(-2, 0), (1, 0)$; y-intercept: $\left(0, \dfrac{4}{3}\right)$

89.

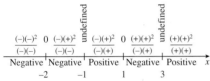

By hand:

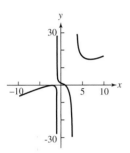

Grapher support:

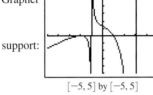

$[-5, 5]$ by $[-5, 5]$ $[0, 10]$ by $[-40, 40]$

91. (

93. (

95. (

73. (a) $|x - 3| < 1/3 \Rightarrow |3x - 9| < 1 \Rightarrow |3x - 5 - 4| < 1 \Rightarrow |f(x) - 4| < 1$.
(b) If x stays within the dashed vertical lines, $f(x)$ will stay within the dashed horizontal lines.
(c) $|x - 3| < 0.01 \Rightarrow |3x - 9| < 0.03 \Rightarrow |3x - 5 - 4| < 0.03 \Rightarrow |f(x) - 4| < 0.03$. The dashed lines would be closer when $x = 3$ and $y = 4$.
75. $0 < a < b \Rightarrow a^2 < ab$ and $ab < b^2$; so, $a^2 < b^2$.

Cha

Answ

1.

CHAPTER 2 REVIEW EXERCISES

1. $y = -x - 5$

$[-15, 5]$ by $[-15, 5]$

3. Starting from $y = x^2$, translate right 2 units and vertically stretch by 3 (either order), then translate up 4 units.

SE

Exp

1. (0,
   $y =$

Exp

1.

5. Vertex: $(-3, 5)$; axis: $x = -3$ **7.** Vertex: $(-4, 1)$; axis: $x = -4$
9. $y = (5/9)(x + 2)^2 - 3$ **11.** $y = \dfrac{1}{2}(x - 3)^2 - 2$

Quic

1. -6

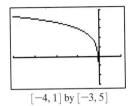

[−5, 5] by [−3, 3]

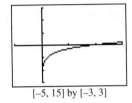

[−4, 1] by [−3, 5]

45. Starting from $y = \ln x$:
reflect across the y-axis
and translate right 2 units.

47. Starting with $y = \log x$:
shift down 1 unit.

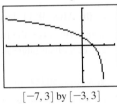

[−7, 3] by [−3, 3]

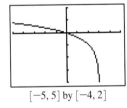

[−5, 15] by [−3, 3]

51. Starting from $y = \log x$:
reflect across the y-axis,
translate right 3 units,
vertically stretch by 2,
translate down 1 unit.

49. Starting from $y = \log x$:
reflect across both axes
and vertically stretch by 2.

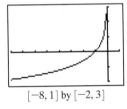

[−8, 1] by [−2, 3]

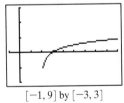

[−5, 5] by [−4, 2]

53. Domain: $(2, \infty)$; Range: $(-\infty, \infty)$; Continuous;
Always increasing; Not symmetric; Not bounded;
No local extrema; Asymptote at $x = 2$; $\displaystyle\lim_{x \to \infty} f(x) = \infty$

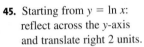

[−1, 9] by [−3, 3]

55.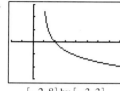

[−2, 8] by [−3, 3]

Domain: $(1, \infty)$; Range: $(-\infty, \infty)$; Continuous;
Always decreasing; Not symmetric; Not bounded;
No local extrema; Asymptote: $x = 1$; $\lim\limits_{x \to \infty} f(x) = -\infty$

57.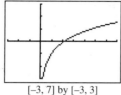

[−3, 7] by [−3, 3]

Domain: $(0, \infty)$; Range: $(-\infty, \infty)$; Continuous;
Always increasing on its domain; Not symmetric;
Not bounded; No local extrema;
Asymptote: $x = 0$; $\lim\limits_{x \to \infty} x = \infty$

59. (a) 10 dB   **(b)** 70 dB   **(c)** 150 dB   **61.** 2023   **63.** True, by definition.   **65.** C   **67.** B

69.

$f(x)$	3^x	$\log_3 x$
Domain	$(-\infty, \infty)$	$(0, \infty)$
Range	$(0, \infty)$	$(-\infty, \infty)$
Intercepts	$(0, 1)$	$(1, 0)$
Asymptotes	$y = 0$	$x = 0$

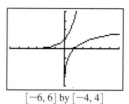

[−6, 6] by [−4, 4]

71. $b = \sqrt[e]{e}$; (e, e) **73.** reflect across the x-axis

SECTION 3.4

Exploration 1

1. $0.90309 = 0.30103 + 0.60206$ **3.** $0.90309 = 3 \times 0.30103$ **5.** 1.20412; 1.50515; 1.80618

Exploration 2

1. false   **3.** true   **5.** false   **7.** false

Quick Review 3.4

1. 2   **3.** -2 **5.** $x^3 y^2$ **7.** $|x|^{3/}|y|$ **9.** $1/(3|u|)$

Exercises 3.4

1. $3 \ln 2 + \ln x$ **3.** $\log 3 - \log x$ **5.** $5 \log_2 y$ **7.** $3 \log x + 2 \log y$ **9.** $2 \ln x - 3 \ln y$ **11.** $\frac{1}{4}\log x - \frac{1}{4}\log y$ **13.** $\log xy$

15. $\ln (y/3)$ **17.** $\log \sqrt[3]{x}$ **19.** $\ln (x^2 y^3)$ **21.** $\log (x^4 y/z^3)$ **23.** 2.8074 **25.** 2.4837 **27.** -3.5850 **29.** $\ln x/\ln 3$

31. $\ln (a + b)/\ln 2$ **33.** $\log x/\log 2$ **35.** $-\log (x + y)/\log 2$

37. Let $x = \log_b R$ and $y = \log_b S$. Then $\dfrac{R}{S} = \dfrac{b^x}{b^y} = b^{x-y}$. So $\log_b\left(\dfrac{R}{S}\right) = x - y = \log_b R - \log_b S$.

39. Starting with $g(x) = \ln x$: vertically shrink by a

factor of $\dfrac{1}{\ln 4} \approx 0.72$.

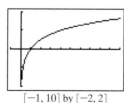

[−1, 10] by [−2, 2]

41. Starting with $g(x) = \ln x$: reflect across the x-axis, then vertically shrink by a factor of $\dfrac{1}{\ln 3} \approx 0.91$.

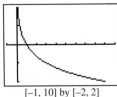

[−1, 10] by [−2, 2]

43. (b): $[-5, 5]$ by $[-3, 3]$, with $\mathrm{Xscl} = 1$ and $\mathrm{Yscl} = 1$

45. (d): $[-2, 8]$ by $[-3, 3]$, with $\mathrm{Xscl} = 1$ and $\mathrm{Yscl} = 1$

47.

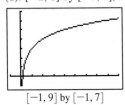

[−1, 9] by [−1, 7]

Domain: $(0, \infty)$; Range: $(-\infty, \infty)$; Continuous;
Always increasing; Asymptote: $x = 0$;
$\displaystyle\lim_{x \to \infty} f(x) = \infty$

49.

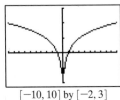

[−10, 10] by [−2, 3]

Domain: $(-\infty, 0) \cup (0, \infty)$; Range: $(-\infty, \infty)$;
Discontinuous at $x = 0$; Decreasing on interval $(-\infty, 0)$;
Increasing on interval $(0, \infty)$; Asymptote: $x = 0$;
$\displaystyle\lim_{x \to \infty} f(x) = \infty$; $\displaystyle\lim_{x \to -\infty} f(x) = \infty$

51. (a) 0 (b) 10 (c) 60 (d) 80 (e) 100 (f) 120   **53.** ≈ 9.6645 lumens   **55.** vertical stretch by a factor of ≈ 0.9102

57. True, by the product rule for logarithms   **59.** B   **61.** A

63. (a) $2.75x^{5.0}$ (b) $49{,}616$

(c)

$\ln(x)$	1.39	1.87	2.14	2.30
$\ln(y)$	7.94	10.37	11.71	12.52

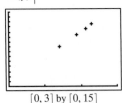

[0, 3] by [0, 15]

(d) $(\ln y) = 5.00 \,(\ln x) + 1.01$

(e) $a \approx 5$, $b \approx 1$ so $f(x) = e^1 x^5 = e x^5 \approx 2.72 x^5$:
The two equations are almost the same.

65. (a)

$\log(w)$	-0.70	-0.52	0.30	0.70	1.48	1.70	1.85
$\log(r)$	2.62	2.48	2.31	2.08	1.93	1.85	1.86

(b) $\log r = (-0.30)\log w + 2.36$

(c)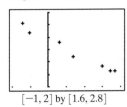

$[-1, 2]$ by $[1.6, 2.8]$

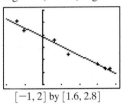

$[-1, 2]$ by $[1.6, 2.8]$

(e) One possible answer: Consider the power function $y = a \cdot x^b$ then: $\log y = \log(a \cdot x^b) = \log a + \log x^b = \log a + b \log x = b(\log x) + \log a$; Which is clearly a linear function of the form $f(t) = mt + c$ where $m = b$, $c = \log a$, $f(t) = \log y$ and $t = \log x$. As a result, there is a linear relationship between $\log y$ and $\log x$.

67. $(6.41, 93.35)$ **69. (a)** Domain of f and g: $(3, \infty)$ **(b)** Domain of f and g: $(5, \infty)$ **(c)** Domain of f: $(-\infty, -3) \cup (-3, \infty)$; Domain of g: $(-3, \infty)$; Answers will vary.

71. $\dfrac{\log x}{\ln x} = \dfrac{\log x}{\log x/\log e} = \log e, \ x > 0, \ x \neq 1$

SECTION 3.5

Exploration 1

1. 1.60206, 2.60206, 3.60206, 4.60206, 5.60206, 6.60206, 7.60206, 8.60206, 9.60206, 10.60206
3. The decimal parts are exactly equal.

Quick Review 3.5

1. $f(g(x)) = e^{2 \ln(x^{1/2})} = e^{\ln x} = x$ and $g(f(x)) = \ln(e^{2x})^{1/2} = \ln(e^x) = x$

3. $f(g(x)) = \dfrac{1}{3}\ln(e^{3x}) = \dfrac{1}{3}(3x) = x$ and $g(f(x)) = e^{3(1/3 \ln x)} = e^{\ln x} = x$ **5.** 7.783×10^8 km

7. 602,000,000,000,000,000,000,000   **9.** 5.766×10^{12}

Exercises 3.5

1. 10   **3.** 12   **5.** -3 **7.** 10,000   **9.** 5.25   **11.** ≈ 24.2151 **13.** ≈ 39.6084 **15.** ≈ -0.4055 **17.** ≈ 4.3956
19. Domain: $(-\infty, -1) \cup (0, \infty)$; graph (e)   **21.** Domain: $(-\infty, -1) \cup (0, \infty)$; graph (d)   **23.** Domain: $(0, \infty)$; graph (a)
25. $x = 1000$ or $x = -1000$ **27.** $\pm\sqrt{10}$ **29.** $x \approx 3.5949$ **31.** $x \approx \pm 2.0634$ **33.** $x \approx -9.3780$ **35.** $x \approx 2.3028$ **37.** 4
39. 3   **41.** 1.5   **43.** 3   **45.** about 20 times greater   **47. (a)** 1.26×10^{-4}; 1.26×10^{-12} **(b)** 10^8 **(c)** 8   **49.** ≈ 28.41 min
51. (a)

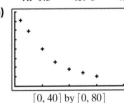

$[0, 40]$ by $[0, 80]$

(b)

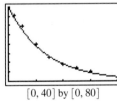

$[0, 40]$ by $[0, 80]$

$T(x) \approx 79.47 \cdot 0.93^x$

(c) 89.47°C

53. (a)

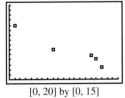

$[0, 20]$ by $[0, 15]$

(b) The scatter plot is better because it accurately represents the times between the measurements. The equal spacing on the bar graph suggests that the measurements were taken at equally spaced intervals, which distorts our perception of how the consumption has changed over time.

55. Logarithmic seems best — the scatterplot of (x, y) looks most logarithmic. (The data can be modeled by $y = 3 + 2 \ln x$.)

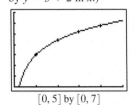

$[0, 5]$ by $[0, 7]$

57. Exponential — the scatterplot of (x, y) is *exactly* exponential. $\left(\text{The data can be modeled by } y = \frac{3}{2} \cdot 2^x.\right)$

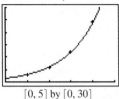

$[0, 5]$ by $[0, 30]$

59. False. The order of magnitude is its *common* logarithm.   **61.** B   **63.** E   **65.** logistic regression
67. (a) As k increases, the bell curve stretches vertically.   **(b)** As c increases, the bell curve compresses horizontally.
69. (a)

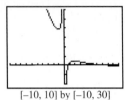

$[-10, 10]$ by $[-10, 30]$

(b) $[0, 10]$ by $[-5, 3]$; 2.3807   **71.** $y = a \ln x + b$, a logarithmic regression

r cannot be negative since it is a distance.
73. $x \approx 1.3066$ **75.** $0 < x < 1.7115$ (approx.)   **77.** $x > 9$

SECTION 3.6

Exploration 1

1.

k	A
10	1104.6
20	1104.9
30	1105
40	1105
50	1105.1
60	1105.1
70	1105.1
80	1105.1
90	1105.1
100	1105.1

A approaches a limit of about 1105.1.

Quick Review 3.6
1. 7   **3.** 1.8125%   **5.** 65%   **7.** 150   **9.** $315

Exercises 3.6
1. $2251.10   **3.** $19,908.59   **5.** $2122.17   **7.** $86,496.26   **9.** $1728.31   **11.** $30,402.43   **13.** $14,755.51   **15.** $70,819.63
17. $43,523.31   **19.** $293.24   **21.** 6.63 years — round to 6 years 9 months   **23.** 13.78 years — round to 13 years 10 months
25. $\approx 10.13\%$ **27.** 7.07%   **29.** 12.14 — round to 12 years 3 months   **31.** 7.7016 years; $48,217.82   **33.** 17.33%; $127,816.26
35. 17.42 — round to 17 years 6 months   **37.** 10.24 — round to 11 years   **39.** 9.93 — round to 10 years   **41.** $\approx 6.14\%$ **43.** $\approx 6.50\%$
45. 5.1% quarterly   **47.** $42,211.46   **49.** $239.41 per month   **51.** $219.51 per month   **53.** $676.56   **55. (a)** 172 months (14 years, 4 months) **(b)** $137,859.60   **57.** One possible answer: The APY is the percentage increase from the initial balance $S(0)$ to the end-of-year balance $S(1)$; specifically, it is $S(1)/S(0) - 1$. Multiplying the initial balance by P results in the end-of-year balance being multiplied by the same

amount, so that the ratio remains unchanged.   **59.** One possible answer: Some of these situations involve counting things (e.g., populations), so that they can only take on whole number values — exponential models which predict, e.g., 439.72 fish, have to be interpreted in light of this fact.; Technically, bacterial growth, radioactive decay, and compounding interest also are "counting problems" — for example, we cannot have fractional bacteria, or fractional molecules of radioactive material, or fractions of pennies. However, because these are generally very large numbers, it is easier to ignore the fractional parts. (This might also apply when one is talking about, e.g., the population of the whole world.); Another distinction: while we often use an exponential model for all these situations, it generally fits better (over long periods of time) for radioactive decay than for most of the others. Rates of growth in populations (esp. human populations) tend to fluctuate more than exponential models suggest. Of course, an exponential model also fits well in compound interest situations where the interest rate is held constant, but there are many cases where interest rates change over time.

61. False. The limit is $A = Pe^{rt} = 100e^{0.05} \approx \$105.13$.   **63.** B   **65.** E   **67.** $364.38   **69. (a)** 8%   **(b)** 12   **(c)** $100

CHAPTER 3 REVIEW EXERCISES

1. $-3\sqrt[3]{4}$ **3.** $3 \cdot 2^{x/2}$ **5.** $f(x) = 2^{-2x} + 3$ — starting from 2^x, horizontally shrink by $\frac{1}{2}$, reflect across y-axis, and translate up 3 units.   **7.** $f(x) = -2^{-3x} - 3$ — starting from 2^x, horizontally shrink by $\frac{1}{3}$, reflect across y-axis, reflect across x-axis, translate down 3 units.

9. Starting from e^x, horizontally shrink by $\frac{1}{2}$, then translate right $\frac{3}{2}$ units — or translate right 3 units, then horizontally shrink by $\frac{1}{2}$.

11. y-intercept: $(0, 12.5)$; Asymptotes: $y = 0$, $y = 20$

13. exp. decay; $\lim\limits_{x \to \infty} f(x) = 2$, $\lim\limits_{x \to -\infty} f(x) = \infty$

15.

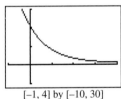

$[-1, 4]$ by $[-10, 30]$

Domain: $(-\infty, \infty)$; Range: $(1, \infty)$; Continuous;
Always decreasing; Not symmetric;
Bounded below by $y = 1$, which is also the only asymptote;
No local extrema; $\lim\limits_{x \to \infty} f(x) = 1$; $\lim\limits_{x \to -\infty} f(x) = \infty$

17.

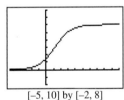

$[-5, 10]$ by $[-2, 8]$

Domain: $(-\infty, \infty)$; Range: $(0, 6)$; Continuous;
Increasing; Symmetric about $(1.20, 3)$;
Bounded by the asymptotes $y = 0$, $y = 6$; No extrema;
$\lim\limits_{x \to \infty} f(x) = 6$; $\lim\limits_{x \to -\infty} f(x) = 0$

19. $f(x) = 24 \cdot 1.053^x$ **21.** $f(x) = 18 \cdot 2^{x/21}$ **23.** $f(x) \approx 30/(1 + 1.5e^{-0.55x})$ **25.** $f(x) \approx \dfrac{20}{1 + 3e^{-0.37x}}$ **27.** 5   **29.** 1/3   **31.** $3^5 = x$

33. $y = xe^2$

35. Translate left 4 units.

37. Translate right 1 unit, reflect across x-axis, and translate up 2 units.

39.

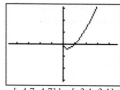

$[-4.7, 4.7]$ by $[-3.1, 3.1]$

Domain: $(0, \infty)$; Range: $\left[-\dfrac{1}{e}, \infty\right) \approx [-0.37, \infty)$;

Continuous; Decreasing on $(0, 0.37]$;

Increasing on $[0.37, \infty)$; Not symmetric;

Bounded below;

Local minimum at $\left(\dfrac{1}{e}, -\dfrac{1}{e}\right)$; $\lim\limits_{x \to \infty} f(x) = \infty$

41.

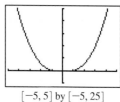

$[-5, 5]$ by $[-5, 25]$

Domain: $(-\infty, 0) \cup (0, \infty)$; Range: $[-0.18, \infty)$;

Discontinuous at $x = 0$;

Decreasing on $(-\infty, -0.61]$, $(0, 0.61]$;

Increasing on $[-0.61, 0)$, $[0.61, \infty)$;

Symmetric across y-axis; Bounded below;

Local minima at $(-0.61, -0.18)$ and $(0.61, -0.18)$;

No asymptotes; $\lim\limits_{x \to \infty} f(x) = \infty$; $\lim\limits_{x \to -\infty} f(x) = \infty$

43. $\log 4 \approx 0.6021$ **45.** ≈ 22.5171 **47.** 0.0000001 **49.** 4 **51.** ≈ 2.1049 **53.** ≈ 99.5112 **55.** $\ln x/\ln 2$ **57.** $\log x/\log 5$

59. (c)   **61.** (b)   **63.** \$515.00   **65.** Pe^{rt} **67.** \$28,794.06   **69.** -0.3054

71. $P(t) \approx 2.0956 \cdot 1.01218^t$, $P(105) \approx 7.5$ million

73. (a) 90 units   **(b)** 32.8722 units   **(c)**

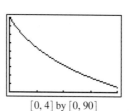

$[0, 4]$ by $[0, 90]$

75. (a) $P(t) = 89,000(0.982)^t$ **(b)** 31.74 years   **77. (a)** $P(t) = 20 \cdot 2^t$ **(b)** $81,920$; 2.3058×10^{19} **(c)** ≈ 8.9658 months

79. (a) $S(t) = S_0 \cdot (1/2)^{t/1.5}$ **(b)** 1,099,500 metric tons   **(c)** $S_0/2$; $S_0/4$ **81.** 6.31   **83.** 11.75 years

85. 137.7940 — about 11 years 6 months   **87.** $\approx 8.57\%$ **89.** ≈ 5.84 lumens   **91.** $\dfrac{1}{10} < b < 10$; $0 < b < \dfrac{1}{10}$ or $b > 10$

93. (a) 16   **(b)** about $11\dfrac{1}{2}$ days   **(c)** 8.7413 — about 8 or 9 days.   **95.** ≈ 41.54 minutes   **97. (a)** 9% **(b)** 4   **(c)** \$100

99. (a) Grace's balance will always remain $1000 since interest is not added to it. Every year she receives 5% of that $1000 in interest; after t years, she has been paid $5t\%$ of the $1000 investment, meaning that altogether she has $1000 + 1000 \cdot 0.05t = 1000(1 + 0.05t)$.

(b)

Years	Not Compounded	Compounded
0	1000.00	1000.00
1	1050.00	1051.27
2	1100.00	1105.17
3	1150.00	1161.83
4	1200.00	1221.40
5	1250.00	1284.03
6	1300.00	1349.86
7	1350.00	1419.07
8	1400.00	1491.82
9	1450.00	1568.31
10	1500.00	1648.72

Chapter 3 Project

Answers are based on the sample data shown in the table.

3.
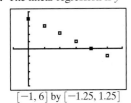
$[-1, 6]$ by $[0, 3]$

5. $y \approx 2.7188 \cdot 0.788^x$

7. A different ball would change the rebound percentage P.

9. $y = He^{\ln(P)x}$ so $y = 2.7188e^{-0.238x}$

11. The linear regression is $y \approx -0.253x + 1.005$. Since $\ln y = (\ln P)x + \ln H$, the slope is $\ln P$ and the y-intercept is $\ln H$.

$[-1, 6]$ by $[-1.25, 1.25]$

SECTION 4.1

Exploration 1

1. $2\pi r$ **3.** No, not quite, since the distance πr would require a piece of thread π times as long, and $\pi > 3$.

Quick Review 4.1

1. 5π in.   **3.** $\dfrac{6}{\pi}$ m   **5. (a)** 47.52 ft   **(b)** 39.77 km   **7.** 88 ft/sec   **9.** 6 mph

Exercises 4.1

1. 23.2°   **3.** 118.7375°   **5.** 21°12′   **7.** 118°19′12″   **9.** $\pi/3$ **11.** $2\pi/3$ **13.** ≈ 1.2518 rad   **15.** ≈ 1.0716 rad   **17.** 30°

19. 18°   **21.** 140°   **23.** $\approx 114.59°$ **25.** 50 in.   **27.** $6/\pi$ ft   **29.** 3 (radians)   **31.** $360/\pi$ cm   **33.** $\theta = \dfrac{9}{11}$ rad and $s_2 = 36$ cm

35. 24 inches   **37.** ≈ 5.4 inches   **39. (a)** 45°   **(b)** 22.5°   **(c)** 247.5°   **41.** ESE is closest at 112.5°.   **43.** ≈ 4.23 statute miles

45. ≈ 387.85 rpm   **47.** $\approx 12,566.37$

49.

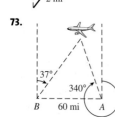

51. ≈ 778 nautical miles   **53. (a)** $16\pi \approx 50.265$ in.   **(b)** $2\pi \approx 6.283$ ft   **55. (a)** 4π rad/sec   **(b)** 28π cm/sec
(c) 7π rad/sec   **57.** True. Horse A travels $2\pi(2r) = 2(2\pi r)$ units of distance in the same amount of time as horse B travels $2\pi r$ units of distance, and so is moving twice as fast.   **59.** C   **61.** B   **63.** 38°02′   **65.** 5°37′
67. 80 naut mi   **69.** 902 naut mi   **71.** The whole circle's area is πr^2; the sector with central angle θ makes up $\dfrac{\theta}{2\pi}$ of that area, or $\dfrac{\theta}{2\pi} \cdot \pi r^2 = \dfrac{1}{2}r^2\theta.$

73.

SECTION 4.2

Exploration 1

1. sin and csc, cos and sec, and tan and cot   **3.** sec θ **5.** sin θ and cos θ

Exploration 2

1. Let $\theta = 60°$. Then $\sin\theta = \dfrac{\sqrt{3}}{2} \approx 0.866$ $\csc\theta = \dfrac{2}{\sqrt{3}} \approx 1.155$

$\cos\theta = \dfrac{1}{2}$ $\sec\theta = 2$

$\tan\theta = \sqrt{3} \approx 1.732$ $\cot\theta = \dfrac{1}{\sqrt{3}} \approx 0.577$

3. The value of a trig function at θ is the same as the value of its co-function at $90° - \theta$.

Quick Review 4.2

1. $5\sqrt{2}$ **3.** 6   **5.** 100.8 in.   **7.** 7.9152 km   **9.** ≈ 1.0101 (no units)

Exercises 4.2

1. $\sin\theta = \dfrac{4}{5}$, $\cos\theta = \dfrac{3}{5}$, $\tan\theta = \dfrac{4}{3}$, $\csc\theta = \dfrac{5}{4}$, $\sec\theta = \dfrac{5}{3}$, $\cot\theta = \dfrac{3}{4}$

3. $\sin\theta = \dfrac{12}{13}$, $\cos\theta = \dfrac{5}{13}$, $\tan\theta = \dfrac{12}{5}$, $\csc\theta = \dfrac{13}{12}$, $\sec\theta = \dfrac{13}{5}$, $\cot\theta = \dfrac{5}{12}$

5. $\sin\theta = \dfrac{7}{\sqrt{170}}$, $\cos\theta = \dfrac{11}{\sqrt{170}}$, $\tan\theta = \dfrac{7}{11}$, $\csc\theta = \dfrac{\sqrt{170}}{7}$, $\sec\theta = \dfrac{\sqrt{170}}{11}$, $\cot\theta = \dfrac{11}{7}$

7. $\sin\theta = \dfrac{\sqrt{57}}{11}$, $\cos\theta = \dfrac{8}{11}$, $\tan\theta = \dfrac{\sqrt{57}}{8}$, $\csc\theta = \dfrac{11}{\sqrt{57}}$, $\sec\theta = \dfrac{11}{8}$, $\cot\theta = \dfrac{8}{\sqrt{57}}$

9. $\cos\theta = \dfrac{2\sqrt{10}}{7}$, $\tan\theta = \dfrac{3}{2\sqrt{10}}$, $\csc\theta = \dfrac{7}{3}$, $\sec\theta = \dfrac{7}{2\sqrt{10}}$, $\cot\theta = \dfrac{2\sqrt{10}}{3}$

11. $\sin\theta = \dfrac{4\sqrt{6}}{11}$, $\tan\theta = \dfrac{4\sqrt{6}}{5}$, $\csc\theta = \dfrac{11}{4\sqrt{6}}$, $\sec\theta = \dfrac{11}{5}$, $\cot\theta = \dfrac{5}{4\sqrt{6}}$

13. $\sin\theta = \dfrac{5}{\sqrt{106}}$, $\cos\theta = \dfrac{9}{\sqrt{106}}$, $\csc\theta = \dfrac{\sqrt{106}}{5}$, $\sec\theta = \dfrac{\sqrt{106}}{9}$, $\cot\theta = \dfrac{9}{5}$

15. $\sin\theta = \dfrac{3}{\sqrt{130}}$, $\cos\theta = \dfrac{11}{\sqrt{130}}$, $\tan\theta = \dfrac{3}{11}$, $\csc\theta = \dfrac{\sqrt{130}}{3}$, $\sec\theta = \dfrac{\sqrt{130}}{11}$, $\cot\theta = \dfrac{11}{3}$

17. $\sin\theta = \dfrac{9}{23}$, $\cos\theta = \dfrac{8\sqrt{7}}{23}$, $\tan\theta = \dfrac{9}{8\sqrt{7}}$, $\sec\theta = \dfrac{23}{8\sqrt{7}}$, $\cot\theta = \dfrac{8\sqrt{7}}{9}$ **19.** $\dfrac{\sqrt{3}}{2}$ **21.** $\sqrt{3}$ **23.** $\dfrac{\sqrt{2}}{2}$ **25.** $\sqrt{2}$

27. $\sqrt{4/3} = 2/\sqrt{3} = 2\sqrt{3}/3$ **29.** 0.961   **31.** 0.943   **33.** 0.268   **35.** 1.524   **37.** 0.810   **39.** 2.414   **41.** $30° = \dfrac{\pi}{6}$

43. $60° = \dfrac{\pi}{3}$ **45.** $60° = \dfrac{\pi}{3}$ **47.** $30° = \dfrac{\pi}{6}$ **49.** $\dfrac{15}{\sin 34°} \approx 26.82$ **51.** $\dfrac{32}{\tan 57°} \approx 20.78$

53. $\dfrac{6}{\sin 35°} \approx 10.46$ **55.** $b \approx 33.79$, $c \approx 35.96$, $\beta = 70°$

57. $b \approx 22.25$, $c \approx 27.16$, $\alpha = 35°$ **59.** As θ gets smaller and smaller, the side opposite θ gets smaller and smaller, so its ratio to the hypotenuse approaches 0 as a limit.   **61.** ≈ 205.26 ft   **63.** ≈ 74.16 ft^2 **65.** ≈ 378.80 ft   **67.** False. This is true only if θ is an acute angle in a right triangle.   **69.** E   **71.** D   **73.** Sine values should be increasing, cosine values should be decreasing, and only tangent values can be greater than 1. Therefore, the first column is tangent, the second column is sine, and the third column is cosine.
75. The distance d_A from A to the mirror is $5\cos 30°$; the distance from B to the mirror is $d_B = d_A - 2$.

Then $PB = \dfrac{d_B}{\cos\beta} = \dfrac{d_A - 2}{\cos 30°} = 5 - \dfrac{2}{\cos 30°} = 5 - \dfrac{4}{\sqrt{3}} \approx 2.69$ m

77. One possible proof: $(\sin\theta)^2 + (\cos\theta)^2 = \left(\dfrac{a}{c}\right)^2 + \left(\dfrac{b}{c}\right)^2 = \dfrac{a^2}{c^2} + \dfrac{b^2}{c^2} = \dfrac{a^2 + b^2}{c^2} = \dfrac{c^2}{c^2} = 1$ (Pythagorean theorem: $a^2 + b^2 = c^2$.)

SECTION 4.3

Exploration 1
1. The side opposite θ in the triangle has length y and the hypotenuse has length r. Therefore $\sin\theta = \dfrac{opp}{hyp} = \dfrac{y}{r}$.   **3.** $\tan\theta = y/x$

Exploration 2
1. The x-coordinates on the unit circle lie between -1 and 1, and $\cos t$ is always an x-coordinate on the unit circle.
3. The points corresponding to t and $-t$ on the number line are wrapped to points above and below the x-axis with the same x-coordinates. Therefore $\cos t$ and $\cos(-t)$ are equal.   **5.** Since 2π is the distance around the unit circle, both t and $t + 2\pi$ get wrapped to the same point.
7. By the observation in (6), $\tan t$ and $\tan(t + \pi)$ are ratios of the form $\dfrac{y}{x}$ and $\dfrac{-y}{-x}$, which are either equal to each other or both undefined.
9. Answers will vary. For example, there are similar statements that can be made about the functions cot, sec, and csc.

Quick Review 4.3
1. $-30°$ **3.** $45°$ **5.** $\sqrt{3}/3$ **7.** $\sqrt{2}$ **9.** $\cos\theta = \dfrac{12}{13}$, $\tan\theta = \dfrac{5}{12}$, $\csc\theta = \dfrac{13}{5}$, $\sec\theta = \dfrac{13}{12}$, $\cot\theta = \dfrac{12}{5}$

Exercises 4.3
1. $450°$ **3.** $\sin\theta = \dfrac{2}{\sqrt{5}}$, $\cos\theta = -\dfrac{1}{\sqrt{5}}$, $\tan\theta = -2$, $\csc\theta = \dfrac{\sqrt{5}}{2}$, $\sec\theta = -\sqrt{5}$, $\cot\theta = -\dfrac{1}{2}$

$\tan\theta = -\dfrac{3}{4}$, $\csc\theta = -\dfrac{5}{3}$, $\sec\theta = \dfrac{5}{4}$, $\cot\theta = -\dfrac{4}{3}$ **5.** $\sin\theta = -\dfrac{1}{\sqrt{2}}$, $\cos\theta = -\dfrac{1}{\sqrt{2}}$, $\tan\theta = 1$, $\csc\theta = -\sqrt{2}$, $\sec\theta = -\sqrt{2}$,

$\cot\theta = 1$

7. $\sin\theta = \dfrac{4}{5}$, $\cos\theta = \dfrac{3}{5}$, $\tan\theta = \dfrac{4}{3}$, $\csc\theta = \dfrac{5}{4}$, $\sec\theta = \dfrac{5}{3}$, $\cot\theta = \dfrac{3}{4}$
$\sec\theta = -1$, $\cot\theta$ undefined   **9.** $\sin\theta = 1$, $\cos\theta = 0$, $\tan\theta$ undefined, $\csc\theta = 1$, $\sec\theta$ undefined, $\cot\theta = 0$

11. $\sin\theta = -\dfrac{2}{\sqrt{29}}$, $\cos\theta = \dfrac{5}{\sqrt{29}}$, $\tan\theta = -\dfrac{2}{5}$, $\csc\theta = -\dfrac{\sqrt{29}}{2}$, $\sec\theta = \dfrac{\sqrt{29}}{5}$, $\cot\theta = -\dfrac{5}{2}$ **13.** $+, +, +$ **15.** $-, -, +$

17. $-$ **19.** $-$ **21.** (a)   **23.** (a)   **25.** $-1/2$ **27.** 2   **29.** $\dfrac{1}{2}$ **31.** 1   **33.** $\dfrac{\sqrt{3}}{2}$ **35.** $-\dfrac{\sqrt{3}}{2}$

37. (a) -1 (b) 0 (c) Undefined

39. (a) 0 (b) -1 (c) 0   **41.** (a) 1 (b) 0 (c) Undefined   **43.** $\sin\theta = \dfrac{\sqrt{5}}{3}$; $\tan\theta = \dfrac{\sqrt{5}}{2}$

45. $\tan\theta = -\dfrac{2}{\sqrt{21}}$; $\sec\theta = \dfrac{5}{\sqrt{21}}$ **47.** $\sec\theta = -\dfrac{5}{4}$; $\csc\theta = \dfrac{5}{3}$ **49.** $1/2$ **51.** 0

53. The calculator's value of the irrational number π is necessarily an approximation. When multiplied by a very large number, the slight error of the original approximation is magnified sufficiently to throw the trigonometric functions off.

55. $\mu = \dfrac{\sin 83°}{\sin 36°} \approx 1.69$ **57.** (a) 0.4 in. (b) ≈ 0.1852 in.

59. The difference in the elevations is 600 ft, so $d = 600/\sin\theta$. Then: (a) ≈ 848.53 ft (b) 600 ft (c) ≈ 933.43 ft   **61.** True. Acute angles determine reference triangles in QI, where cosine is positive, while obtuse angles determine reference triangles in QII, where cosine is negative.
63. E   **65.** A   **67.** $5\pi/6$ **69.** $7\pi/4$ **71.** The two triangles are congruent: both have hypotenuse 1, and the corresponding angles are congruent—the smaller acute angle has measure t in both triangles, and the two acute angles in a right triangle add up to $\pi/2$.   **73.** One possible answer: Starting from the point (a, b) on the unit circle—at an angle of t, so that $\cos t = a$—then measuring a quarter of the way around the circle (which corresponds to adding $\pi/2$ to the angle), we end at $(-b, a)$, so that $\sin(t + \pi/2) = a$. For (a, b) in quadrant I, this is shown in the figure; similar illustrations can be drawn for the other quadrants.
75. Starting from the point (a, b) on the unit circle—at an angle of t, so that $\cos t = a$—then measuring a quarter of the way around the circle (which corresponds to adding $\pi/2$ to the angle), we end at $(-b, a)$, so that $\sin(t + \pi/2) = a$. This holds true when (a, b) is in quadrant II, just as it did for quadrant I.

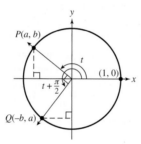

77. $|\theta| < 0.2441$ (approximately)
79. This Taylor polynomial is generally a very good approximation for $\sin\theta$—in fact, the relative error is less than 1% for $|\theta| < 1$ (approx.). It is better for θ close to 0; it is slightly larger than $\sin\theta$ when $\theta < 0$ and slightly smaller when $\theta > 0$.

SECTION 4.4

Exploration 1

1. $\pi/2$ (at the point $(0, 1)$)   **3.** Both graphs cross the x-axis when the y-coordinate on the unit circle is 0.   **5.** The sine function tracks the y-coordinate of the point as it moves around the unit circle. After the point has gone completely around the unit circle (a distance of 2π), the same pattern of y-coordinates starts over again.

Quick Review 4.4

1. In order: $+, +, - -$ **3.** In order: $+, -, +, -$ **5.** $-5\pi/6$ **7.** vertically stretch by 3   **9.** vertically shrink by 0.5

Exercises 4.4

1. Amplitude 2; vertical stretch by a factor of 2
3. Amplitude 4; vertical stretch by a factor of 4, reflection across x-axis   **5.** Amplitude 0.73; vertical shrink by a factor of 0.73
7. Period $\dfrac{2\pi}{3}$; horizontal shrink by a factor of $\dfrac{1}{3}$ **9.** Period $\dfrac{2\pi}{7}$; horizontal shrink by a factor of $\dfrac{1}{7}$, reflection across y-axis   **11.** Period π; horizontal shrink by a factor of $\dfrac{1}{2}$, vertical stretch by a factor of 3
13. Amplitude 3, period 4π,   **15.** Amplitude $\dfrac{3}{2}$, period π,

frequency $\dfrac{1}{4\pi}$

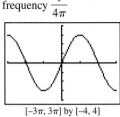

[−3π, 3π] by [−4, 4]

frequency $\dfrac{1}{\pi}$

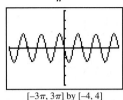

[−3π, 3π] by [−4, 4]

17.

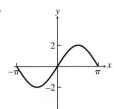

19.

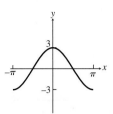

21.

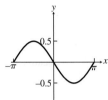

23.

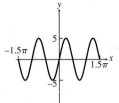

25.

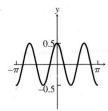

27.

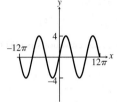

29. Period π; amplitude 1.5; $[-2\pi, 2\pi]$ by $[-2, 2]$

31. Period π; amplitude 3; $[-2\pi, 2\pi]$ by $[-4, 4]$

33. Period 6; amplitude 4; $[-3, 3]$ by $[-5, 5]$

35. Maximum: 2 $\left(\text{at } -\dfrac{3\pi}{2} \text{ and } \dfrac{\pi}{2}\right)$; minimum: $-2 \left(\text{at } -\dfrac{\pi}{2} \text{ and } \dfrac{3\pi}{2}\right)$; zeros: 0, $\pm\pi$, $\pm2\pi$

37. Maximum: 1 (at 0, $\pm\pi$, $\pm2\pi$); minimum: $-1 \left(\text{at } \pm\dfrac{\pi}{2} \text{ and } \pm\dfrac{3\pi}{2}\right)$; zeroes: $\pm\dfrac{\pi}{4}, \pm\dfrac{3\pi}{4}, \pm\dfrac{5\pi}{4}, \pm\dfrac{7\pi}{4}$

39. Maximum: 1 $\left(\text{at } \pm\dfrac{\pi}{2} \text{ and } \pm\dfrac{3\pi}{2}\right)$; minimum: -1 (at 0, $\pm\pi$, $\pm2\pi$); zeros: $\pm\dfrac{\pi}{4}, \pm\dfrac{3\pi}{4}, \pm\dfrac{5\pi}{4}, \pm\dfrac{7\pi}{4}$

41. One possibility is $y = \sin(x + \pi)$.

43. Starting from $y = \sin x$, horizontally shrink by $\dfrac{1}{3}$ and vertically shrink by 0.5.

45. Starting from $y = \cos x$, horizontally stretch by 3, vertically shrink by $\dfrac{2}{3}$, reflect across x-axis.

47. Starting from $y = \cos x$, horizontally shrink by $\dfrac{3}{2\pi}$ and vertically stretch by 3.

49. Starting with y_1, vertically stretch by $\dfrac{5}{3}$.

51. Starting with y_1, horizontally shrink by $\dfrac{1}{2}$.   **53.** (a) and (b)   **55.** (a) and (b)

57. One possibility is $y = 3 \sin 2x$.   **59.** One possibility is $y = 1.5 \sin 12(x - 1)$.

61. Amplitude 2, period 2π, phase shift $\dfrac{\pi}{4}$, vertical translation 1 unit up

63. Amplitude 5, period $\dfrac{2\pi}{3}$, phase shift $\dfrac{\pi}{18}$, vertical translation 0.5 unit up

65. Amplitude 2, period 1, phase shift 0, vertical translation 1 unit up

67. Amplitude $\dfrac{7}{3}$, period 2π, phase shift $-\dfrac{5}{2}$, vertical translation 1 unit down

69. $y = 2 \sin 2x$ ($a = 2, b = 2, h = 0, k = 0$)

71. (a) two   **(b)** (0, 1) and $(2\pi, 1.3^{-2\pi}) \approx (6.28, 0.19)$ **73.** ≈ 15.90 sec   **75. (a)** 1:00 A.M.   **(b)** 8.90 ft; 10.52 ft   **(c)** 4:06 A.M.

77. (a) The maximum d is approximately 21.4 cm.   **(b)** ≈ 0.83 sec   **(c)** $d(t) = -7.1 \cos\left(\dfrac{2\pi x}{0.83}\right) + 14.3$
     The amplitude is 7.1 cm; scatterplot:

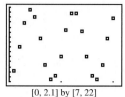

[0, 2.1] by [7, 22]

(d)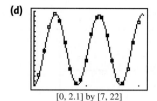

[0, 2.1] by [7, 22]

79. One possible solution is
$$T = 21.5 \cos\left(\frac{\pi}{6}(x - 7)\right) + 57.5.$$

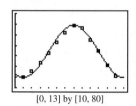

[0, 13] by [10, 80]

81. False. $y = \sin 2x$ is a horizontal *stretch* of $y = \sin 4x$ by a factor of 2, so it has twice the period.   **83.** D   **85.** C

87. (a)

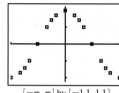

$[-\pi, \pi]$ by $[-1.1, 1.1]$

(b) $0.0246x^4 + 0x^3 - 0.4410x^2 + 0x + 0.9703$

(c) The coefficients are fairly similar.

89. (a) 1/262 sec   **(b)** $f = 262 \dfrac{1}{\sec}$ ("cycles per sec"), or 262 Hertz

91. (a) $a - b$ must equal 1.   **(b)** $a - b$ must equal 2

(c) $a - b$ must equal k

(c)

(c) $a - b$ must equal k.

93. $B = (0, 3); C = \left(\dfrac{3\pi}{4}, 0\right)$ **95.** $B = \left(\dfrac{\pi}{4}, 2\right); C = \left(\dfrac{3\pi}{4}, 0\right)$

97. (a) If b is negative, then $b = -B$, where B is positive. Then $y = a \sin[-B(x - H)] + k = -a \sin[B(x - H)] + k$, since sine is an odd function. We will see in part (d) what to do if the number out front is negative.

(b) A sine graph can be translated a quarter of a period to the left to become a cosine graph of the same sinusoid. Thus
$$y = a \sin\left[b\left((x - h) + \frac{1}{4} \cdot \frac{2\pi}{b}\right)\right] + k = a \sin\left[b\left(x - \left(h - \frac{\pi}{2b}\right)\right)\right] + k \text{ has the same graph as } y = a \cos[b(x - h)] + k.$$
We therefore choose $H = h - \dfrac{\pi}{2b}$.

(c) The angles $\theta + \pi$ and θ determine diametrically opposite points on the unit circle, so they have point symmetry with respect to the origin. The y-coordinates are therefore opposites, so $\sin(\theta + \pi) = -\sin \theta$.

(d) By the identity in (c), $y = a \sin[b(x - h) + \pi] + k = -a \sin[b(x - h)] + k$. We therefore choose $H = h - \dfrac{\pi}{b}$.

(e) Part (b) shows how to convert $y = a \cos[b(x - h)] + k$ to $y = a \sin[b(x - H)] + k$, and parts (a) and (d) show how to ensure that a and b are positive.

SECTION 4.5

Exploration 1

1. The graphs do not seem to intersect.

Quick Review 4.5

1. π **3.** 6π **5.** Zero: 3; asymptote: $x = -4$ **7.** Zero: -1; asymptotes: $x = 2$ and $x = -2$ **9.** even

Exercises 4.5

1. The graph of $y = 2 \csc x$ must be vertically stretched by 2 compared to $y = \csc x$, so $y_1 = 2 \csc x$ and $y_2 = \csc x$.

3. $y_1 = 3 \csc 2x$, $y_2 = \csc x$

5.

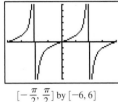

$\left[-\dfrac{\pi}{2}, \dfrac{\pi}{2}\right]$ by $[-6, 6]$

Horizontal shrink of $y = \sec x$ by factor 1/3; asymptotes at odd multiples of $\pi/6$

7.

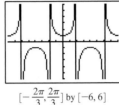

$\left[-\dfrac{2\pi}{3}, \dfrac{2\pi}{3}\right]$ by $[-6, 6]$

Horizontal shrink of $y = \cot x$ by factor 1/2, vertical stretch by factor 2; asymptotes at multiples of $\pi/2$

9.

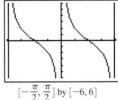

$\left[-\dfrac{\pi}{2}, \dfrac{\pi}{2}\right]$ by $[-6, 6]$

Horizontal shink of $y = \tan x$ by factor 1/2; asymptotes at multiples of $\pi/4$.

11.

$[-4\pi, 4\pi]$ by $[-6, 6]$

Horizontal stretch of $y = \csc x$ by factor 2; asymptotes at multiples of 2π

13. Graph (a); Xmin $= -\pi$ and Xmax $= \pi$

15. Graph (c); Xmin $= -\pi$ and Xmax $= \pi$

17. Domain: All reals except integer multiples of π; Range: $(-\infty, \infty)$; Continuous on its domain; Decreasing on each interval in its domain; symmetry with respect to the origin (odd); Not bounded above or below; No local extrema; No horizontal asymptotes; Vertical asymptotes: $x = k\pi$ for all integers k; End behavior: $\lim\limits_{x \to \infty} \cot x$ and $\lim\limits_{x \to -\infty} \cot x$ do not exist.

19. Domain: All reals except integer multiples of π; Range: $(-\infty, -1] \cup [1, \infty)$; Continuous on its domain; On each interval centered at $x = \dfrac{\pi}{2} + 2k\pi$, where k is an integer, decreasing on the left half of the interval and increasing on the right, for $x = \dfrac{3\pi}{2} + 2k\pi$, increasing on the first half of the interval and decreasing on the second half; Symmetric with respect to the origin (odd); Not bounded above or below; Local minimum 1 at each $x = \dfrac{\pi}{2} + 2k\pi$ and local maximum -1 at each $x = \dfrac{3\pi}{2} + 2k\pi$, where k is an integer; No horizontal asymptotes; Vertical asymptotes: $x = k\pi$ for all integers k; End behavior: $\lim\limits_{x \to \infty} \csc x$ and $\lim\limits_{x \to -\infty} \csc x$ do not exist.

21. Starting with $y = \tan x$, vertically stretch by 3.

23. Starting with $y = \csc x$, vertically stretch by 3.

25. Starting with $y = \cot x$, horizontally stretch by 2, vertically stretch by 3, and reflect across x-axis.

27. Starting with $y = \tan x$, horizontally shrink by $\dfrac{2}{\pi}$, reflect across x-axis, and shift up by 2 units.

29. $\pi/3$ **31.** $5\pi/6$ **33.** $5\pi/2$ **35.** $x \approx 0.92$ **37.** $x \approx 5.25$ **39.** $x \approx 0.52$ or $x \approx 2.62$

41. (a) The reflection of (a, b) across the origin is $(-a, -b)$. **(b)** Definition of tangent **(c)** $\tan t = \dfrac{b}{a} = \dfrac{-b}{-a} = \tan(t - \pi)$

(d) Since points on opposite sides of the unit circle determine the same tangent ratio, $\tan(t \pm \pi) = \tan t$ for all numbers t in the domain. Other points on the unit circle yield triangles with different tangent ratios, so no smaller period is possible. **(e)** The same argument that uses the ratio $\dfrac{b}{a}$ above can be repeated using the ratio $\dfrac{a}{b}$, which is the cotangent ratio.

43. For any x, $\left(\dfrac{1}{f}\right)(x + p) = \dfrac{1}{f(x + p)} = \dfrac{1}{f(x)} = \left(\dfrac{1}{f}\right)(x)$. This is not true for any smaller value of p, since this is the smallest value

that works for f.

45. (a) $d = 350 \sec x$ **(b)** $\approx 16{,}831$ ft   **47.** $\approx \pm 0.905$ **49.** $\approx \pm 1.107$ or $\approx \pm 2.034$

51. False. It is increasing only over intervals on which it is defined, i.e., intervals bounded by consecutive asymptotes.

53. A   **55.** D

57. About $(-0.44, 0) \cup (0.44, \pi)$

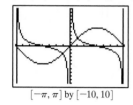

$[-\pi, \pi]$ by $[-10, 10]$

59. $\cot x$ is not defined at 0; the definition of "increasing on (a, b)" requires that the function be defined everywhere in (a, b). Also, choosing $a = -\pi/4$ and $b = \pi/4$, we have $a < b$ but $f(a) = 1 > f(b) = -1$.

61. $\csc x = \sec\left(x - \dfrac{\pi}{2}\right)$

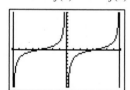

$[-\pi, \pi]$ by $[-10, 10]$

63. $d = \dfrac{30}{\cos x} = 30 \sec x$ **65.** ≈ 0.8952 radians $\approx 51.29°$

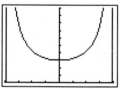

$[-0.5\pi, 0.5\pi]$ by $[0, 100]$

SECTION 4.6

Exploration 1

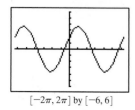

$[-2\pi, 2\pi]$ by $[-6, 6]$

Sinusoid

$[-2\pi, 2\pi]$ by $[-6, 6]$

Sinusoid

$[-2\pi, 2\pi]$ by $[-6, 6]$

Not a Sinusoid

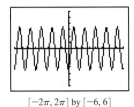

$[-2\pi, 2\pi]$ by $[-6, 6]$

Sinusoid

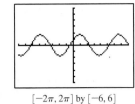

$[-2\pi, 2\pi]$ by $[-6, 6]$

Sinusoid

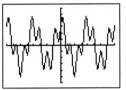

$[-2\pi, 2\pi]$ by $[-6, 6]$

Not a Sinusoid

Quick Review 4.6

1. Domain: $(-\infty, \infty)$; range: $[-3, 3]$ **3.** Domain: $[1, \infty)$; range: $[0, \infty)$

5. Domain: $(-\infty, \infty)$; range: $[-2, \infty)$

7. As $x \to -\infty, f(x) \to \infty$; as $x \to \infty, f(x) \to 0$.

9. $(f \circ g)(x) = x - 4$, domain: $[0, \infty)$; $(g \circ f)(x) = \sqrt{x^2 - 4}$, domain: $(-\infty, -2] \cup [2, \infty)$

Exercises 4.6

1. Periodic.   **3.** Not periodic.   **5.** Not periodic.   **7.** Periodic.

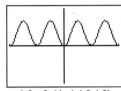

$[-2\pi, 2\pi]$ by $[-1.5, 1.5]$

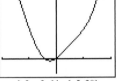

$[-2\pi, 2\pi]$ by $[-5, 20]$

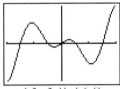

$[-2\pi, 2\pi]$ by $[-6, 6]$

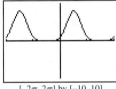

$[-2\pi, 2\pi]$ by $[-10, 10]$

9. Since the period of $\cos x$ is 2π, we have
$\cos^2(x + 2\pi) = (\cos(x + 2\pi))^2 = (\cos x)^2 = \cos^2 x$.
The period is therefore an exact divisor of 2π,
and we see graphically that it is π. A graph for
$-\pi \le x \le \pi$ is shown:

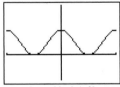

$[-\pi, \pi]$ by $[-1, 2]$

11. Since the period of $\cos x$ is 2π, we have
$\sqrt{\cos^2(x + 2\pi)} = \sqrt{(\cos(x + 2\pi))^2} = \sqrt{(\cos x)^2} = \sqrt{\cos^2 x}$.
The period is therefore an exact divisor of 2π,
and we see graphically that it is π. A graph for
$-\pi \le x \le \pi$ is shown:

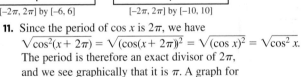

$[-\pi, \pi]$ by $[-1, 2]$

13. Domain: $(-\infty, \infty)$; range: $[0, 1]$;

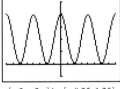

$[-2\pi, 2\pi]$ by $[-0.25, 1.25]$

15. Domain: all $x \ne n\pi$,
n an integer; range: $[0, \infty)$;

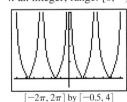

$[-2\pi, 2\pi]$ by $[-0.5, 4]$

17. Domain: all $x \neq \dfrac{\pi}{2} + n\pi$,

n an integer; range: $(-\infty, 0]$;   **19.** $y = 2x - 1$; $y = 2x + 1$ **21.** $y = 1 - 0.3x$; $y = 3 - 0.3x$ **23.**   yes

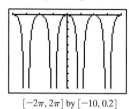

$[-2\pi, 2\pi]$ by $[-10, 0.2]$

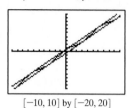

$[-10, 10]$ by $[-20, 20]$

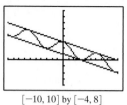

$[-10, 10]$ by $[-4, 8]$

25. Yes   **27.** No   **29.** $a \approx 3.61$, $b = 2$, $h \approx 0.49$

31. $a \approx 2.24$, $b = \pi$, $h \approx 0.35$ **33.** $a \approx 2.24$, $b = 1$, $h \approx -1.11$

35.

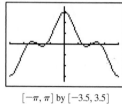

$[-\pi, \pi]$ by $[-3.5, 3.5]$

37.

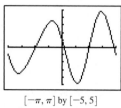

$[-\pi, \pi]$ by $[-5, 5]$

39. (a)   **41.** (c)

43. The damping factor is e^{-x}, and the damping occurs as $x \to \infty$.

45. No damping   **47.** The damping factor is x^3, and the damping occurs as $x \to 0$.

49. f oscillates between 1.2^{-x} and -1.2^{-x}.

As $x \to \infty$, $f(x) \to 0$.

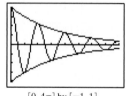

$[0, 4\pi]$ by $[-1, 1]$

51. f oscillates between $\dfrac{1}{x}$ and $-\dfrac{1}{x}$.

As $x \to \infty$, $f(x) \to 0$.

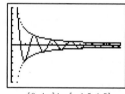

$[0, 4\pi]$ by $[-1.5, 1.5]$

53. 2π

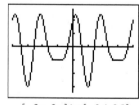

$[-2\pi, 2\pi]$ by $[-3.4, 2.8]$

55. 2π

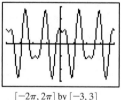

$[-2\pi, 2\pi]$ by $[-3, 3]$

57. Period 2π

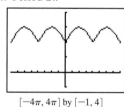

$[-4\pi, 4\pi]$ by $[-1, 4]$

59. Not periodic

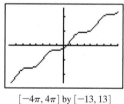

$[-4\pi, 4\pi]$ by $[-13, 13]$

61. Not periodic

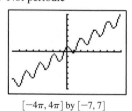

$[-4\pi, 4\pi]$ by $[-7, 7]$

63. Domain: $(-\infty, \infty)$;
range: $(-\infty, \infty)$

65. Domain: $(-\infty, \infty)$;
range: $[1, \infty)$

67. Domain: $\ldots \cup [-2\pi, -\pi] \cup [0, \pi] \cup [2\pi, 3\pi] \cup \ldots$; that is, all x with $2n\pi \leq x \leq (2n + 1)\pi$, n an integer; range: $[0, 1]$

69. Domain: $(-\infty, \infty)$; range: $[0, 1]$

71. (a)

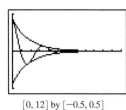

[0, 12] by [−0.5, 0.5]

73. Not periodic **75. (a)**

77. Graph (d), shown on $[-2\pi, 2\pi]$ by $[-4, 4]$

79. Graph (b), shown on $[-2\pi, 2\pi]$ by $[-4, 4]$

(b) For $t > 0.51$ (approximately).

81. False. For example, the function has a relative minimum of 0 at $x = 0$ that is not repeated anywhere else.

83. B **85.** D

87. (a) Answers will vary — for example, on a TI-81: $\dfrac{\pi}{47.5} = 0.0661\ldots \approx 0.07$;

on a TI-82: $\dfrac{\pi}{47} = 0.0668\ldots \approx 0.07$; on a TI-85: $\dfrac{\pi}{63} = 0.0498\ldots \approx 0.05$; on a TI-92: $\dfrac{\pi}{119} = 0.0263\ldots \approx 0.03$.

(b) Period: $p = \pi/125 = 0.0251\ldots$. For any of the TI graphers, there are from 1 to 3 cycles between each pair of pixels; the graphs produced are therefore inaccurate, since so much detail is lost.

89. Domain: $(-\infty, \infty)$; range: $[-1, 1]$;

horizontal asymptote: $y = 1$;

zeros at $\ln\left(\dfrac{\pi}{2} + n\pi\right)$,

n a non-negative integer

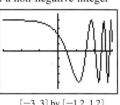

[−3, 3] by [−1.2, 1.2]

91. Domain: $[0, \infty)$; range: $(-\infty, \infty)$;

zeroes at $n\pi$, n a nonnegative integer

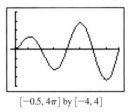

[−0.5, 4π] by [−4, 4]

93. Domain: $(-\infty, 0) \cup (0, \infty)$;

range: approximately $[-0.22, 1)$;

horizontal asymptote: $y = 0$;

zeros at $n\pi$, n a nonzero integer

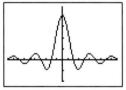

[−5π, 5π] by [−0.5, 1.2]

95. Domain: $(-\infty, 0) \cup (0, \infty)$;

range: approximately $[-0.22, 1)$;

horizontal asymptote: $y = 1$;

zeros at $\dfrac{1}{n\pi}$, n a nonzero integer

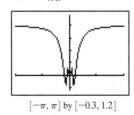

[−π, π] by [−0.3, 1.2]

SECTION 4.7

Exploration 1

1. x **3.** $\sqrt{1 + x^2}$ **5.** $\sqrt{1 + x^2}$

Quick Review 4.7

1. $\sin x$: positive; $\cos x$: positive; $\tan x$: positive **3.** $\sin x$: negative; $\cos x$: negative; $\tan x$: positive **5.** 1/2 **7.** −1/2 **9.** −1/2

Exercises 4.7

1. $\pi/3$ **3.** 0   **5.** $\pi/3$ **7.** $-\pi/4$ **9.** $-\pi/4$ **11.** $\pi/2$ **13.** $21.22°$ **15.** $-85.43°$ **17.** 1.172   **19.** -0.478

21. $\lim\limits_{x \to \infty} \tan^{-1}(x^2) = \dfrac{\pi}{2}$ and $\lim\limits_{x \to -\infty} \tan^{-1}(x^2) = \dfrac{\pi}{2}$ **23.** $\sqrt{3}/2$ **25.** $\pi/4$ **27.** 1/2   **29.** $\pi/6$ **31.** 1/2

33. Domain: $[-1, 1]$; Range: $\left[-\dfrac{\pi}{2}, \dfrac{\pi}{2}\right]$; Continuous; Increasing; Symmetric with respect to the origin (odd); Bounded;

Absolute maximum of $\dfrac{\pi}{2}$, absolute minimum of $-\dfrac{\pi}{2}$; No asymptotes; No end behavior (bounded domain)

35. Domain: $(-\infty, \infty)$; Range: $\left(-\dfrac{\pi}{2}, \dfrac{\pi}{2}\right)$; Continuous; Increasing; Symmetric with respect to the origin (odd); Bounded;

No local extrema; Horizontal asymptotes: $y = \dfrac{\pi}{2}$ and $y = -\dfrac{\pi}{2}$; End behavior: $\lim\limits_{x \to \infty} \tan^{-1} x = \dfrac{\pi}{2}$ and $\lim\limits_{x \to -\infty} \tan^{-1} x = -\dfrac{\pi}{2}$

37. Domain: $\left[-\dfrac{1}{2}, \dfrac{1}{2}\right]$; Range: $\left[-\dfrac{\pi}{2}, \dfrac{\pi}{2}\right]$; Starting from $y = \sin^{-1} x$, horizontally shrink by $\dfrac{1}{2}$.

39. Domain: $(-\infty, \infty)$; Range: $\left(-\dfrac{5\pi}{2}, \dfrac{5\pi}{2}\right)$; Starting from $y = \tan^{-1} x$, horizontally stretch by 2 and vertically stretch by 5 (either order).

41. 1   **43.** $\sin \dfrac{1}{2} \approx 0.479$ **45.** $\dfrac{1}{3}$ **47.** $x/\sqrt{1 + x^2}$ **49.** $x/\sqrt{1 - x^2}$ **51.** $\dfrac{1}{\sqrt{1 + 4x^2}}$

53. (b)

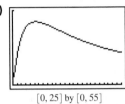

[0, 25] by [0, 55]

(c) 2 or 15 ft.

55. (a) $\theta = \tan^{-1} \dfrac{s}{500}$ **(b)** As s changes from 10 to 20 ft, θ changes from about $1.1458°$ to $2.2906°$—it almost exactly doubles (a 99.92% increase). As s changes from 200 to 210 ft, θ changes from about $21.80°$ to $22.78°$—an increase of less than $1°$, and a very small relative change (only about 4.5%).   **(c)** The x-axis represents the height and the y-axis represents the angle: the angle cannot grow past $90°$ (in fact, it *approaches* but never exactly equals $90°$).

57. False. This is only true for $-1 \le x \le 1$, the domain of $\sin^{-1} x$.   **59.** E   **61.** C

63. The cotangent function restricted to the interval $(0, \pi)$ is one-to-one and has an inverse. The unique angle y between 0 and π (non-inclusive) such that $\cot y = x$ is called the inverse cotangent (or arccotangent) of x, denoted $\cot^{-1} x$ or arccot x. The domain of $y = \cot^{-1} x$ is $(-\infty, \infty)$ and the range is $(0, \pi)$.

65. (a) Domain all reals, range $[-\pi/2, \pi/2]$, period 2π **(b)** Domain all reals, range $[0, \pi]$, period 2π **(c)** Domain all reals except $\pi/2 + n\pi$, n an integer, range $(-\pi/2, \pi/2)$, period π. Discontinuity is not removable.

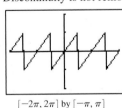

[$-2\pi, 2\pi$] by [$-0.5\pi, 0.5\pi$]   [$-2\pi, 2\pi$] by [$0, \pi$]   [$-2\pi, 2\pi$] by [$-\pi, \pi$]

67. $y = \dfrac{\pi}{2} - \tan^{-1} x$ **69.** $\dfrac{18}{\pi} \tan^{-1} x + 33$

71. (a) $y = \pi/2$ **(b)** $y = \pi/2, y = 3\pi/2$ **(c)** The graph on the left.   **(d)** The graph on the left.

SECTION 4.8

Exploration 1

1. the unit circle

3. Since the grapher is plotting points along the unit circle, it covers the circle at a constant speed. Toward the extremes its motion is mostly vertical, so not much horizontal progress (which is all that we see) occurs. Toward the middle, the motion is mostly horizontal, so it moves faster.

Quick Review 4.8

1. $b = 15 \cot 31° \approx 24.964$, $c = 15 \csc 31° \approx 29.124$

3. $b = 28 \cot 28° - 28 \cot 44° \approx 23.665$, $c = 28 \csc 28° \approx 59.642$, $a = 28 \csc 44° \approx 40.308$ **5.** Complement: $58°$; supplement: $148°$

7. $45°$ **9.** Amplitude: 3; period: π

Exercises 4.8

1. $300\sqrt{3} \approx 519.62$ ft   **3.** $120 \cot 10° \approx 680.55$ ft   **5.** wire length $= 5 \sec 80° \approx 28.79$ ft; tower height $= 5 \tan 80° \approx 28.36$ ft

7. $185 \tan 80°1'12'' \approx 1051$ ft   **9.** $100 \tan 83°12' \approx 839$ ft   **11.** $10 \tan 55° \approx 14.3$ ft   **13.** $4.25 \tan 35° \approx 2.98$ mi

15. $200(\tan 40° - \tan 30°) \approx 52.35$ ft   **17.** distance: $60\sqrt{2} \approx 84.85$ naut mi; bearing is $140°$ **19.** $1097 \cot 19° \approx 3186$ ft

21. $325 \tan 63° \approx 638$ ft   **23.** $36.5 \tan 15° \approx 9.8$ ft   **25.** $\dfrac{550}{\cot 70° - \cot 80°} \approx 2931$ ft

27. (a) 8 cycles/sec **(b)** $d = 6 \cos 16\pi t$ **(c)** about 4.1 in. left of the starting position   **29.** $d = 3 \cos 4\pi t$ cm

31. (a) 25 ft **(b)** 33 ft **(c)** $\pi/10$ radians/sec

33. (a) $\pi/6$ **(b)** $a = (82 - 48)/2 = 17$ and $k = 82 - 17 = 65$

(c) $(3 + 1 = 4)$ **(d)** The fit is very good:

(e) Setting $17 \sin (\pi/6(t - 4)) + 65 = 70$, we get $t = 4.57$ or $t = 9.43$. These represent (approximately) days #139 and #287 of a 365-day year, namely May 19 and October 14.

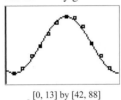

$[0, 13]$ by $[42, 88]$

35. (a) March **(b)** November

37. True. Since the frequency and the period are reciprocals, the higher the frequency, the shorter the period.   **39.** D   **41.** D

43. (a)

(b) The first is the best.   **(c)** About $\dfrac{2464}{2\pi} = \dfrac{1232}{\pi}$

≈ 392 oscillations/sec

$[0, 0.0062]$ by $[-0.5, 1]$

45. $2.5 \cot \dfrac{\pi}{7} \approx 5.2$ cm   **47.** $AC \approx 33.6$ in.; $BD \approx 12.9$ in   **49.** $\tan^{-1} 0.06 \approx 3.4°$

51. (a)

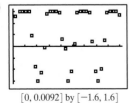

(b) One pretty good match is $y = 1.51971 \sin[2467(t - 0.0002)]$ (that is, $a = 1.51971$, $b = 2467$, $h = 0.0002$). Answers will vary but should be close to these values.

(c) Frequency: about $\dfrac{2467}{2\pi} \approx 393$ Hz; It appears to be a G.

(d) G

$[0, 0.0092]$ by $[-1.6, 1.6]$

CHAPTER 4 REVIEW EXERCISES

1. positive y-axis; $450°$ **3.** QIII; $-3\pi/4$ **5.** QI; $13\pi/30$ **7.** QI; $15°$

9. $270°$ or $\dfrac{3\pi}{2}$ radians   **11.** $30° = \pi/6$ rad   **13.** $120° = 2\pi/3$ rad   **15.** $360° + \tan^{-1}(-2) \approx 296.565° \approx 5.176$ radians

17. $1/2$ **19.** 1 **21.** $1/2$ **23.** 2 **25.** -1 **27.** 0

29. $\sin\left(-\dfrac{\pi}{6}\right) = -\dfrac{1}{2}, \cos\left(-\dfrac{\pi}{6}\right) = \dfrac{\sqrt{3}}{2}, \tan\left(-\dfrac{\pi}{6}\right) = -\dfrac{1}{\sqrt{3}}, \csc\left(-\dfrac{\pi}{6}\right) = -2, \sec\left(-\dfrac{\pi}{6}\right) = \dfrac{2}{\sqrt{3}}, \cot\left(-\dfrac{\pi}{6}\right) = -\sqrt{3}$

31. $\sin(-135°) = -\dfrac{1}{\sqrt{2}}, \cos(-135°) = -\dfrac{1}{\sqrt{2}}, \tan(-135°) = 1, \csc(-135°) = -\sqrt{2}, \sec(-135°) = -\sqrt{2}, \cot(-135°) = 1$

33. $\sin\alpha = \dfrac{5}{13}, \cos\alpha = \dfrac{12}{13}, \tan\alpha = \dfrac{5}{12}, \csc\alpha = \dfrac{13}{5}, \sec\alpha = \dfrac{13}{12}, \cot\alpha = \dfrac{12}{5}$

35. $\sin\theta = \dfrac{15}{17}, \cos\theta = \dfrac{8}{17}, \tan\theta = \dfrac{15}{8}, \csc\theta = \dfrac{17}{15}, \sec\theta = \dfrac{17}{8}, \cot\theta = \dfrac{8}{15}$

37. ≈ 4.075 radians   **39.** $a = 15\sin 35° \approx 8.604, b = 15\cos 35° \approx 12.287, \beta = 55°$

41. $b = 7\tan 48° \approx 7.774, c = \dfrac{7}{\cos 48°} \approx 10.461, \alpha = 42°$

43. $a = 2\sqrt{6} \approx 4.90, \alpha \approx 44.42°, \beta \approx 45.58°$ **45.** QIII   **47.** QII

49. $\sin\theta = \dfrac{2}{\sqrt{5}}, \cos\theta = -\dfrac{1}{\sqrt{5}}, \tan\theta = -2, \csc\theta = \dfrac{\sqrt{5}}{2}, \sec\theta = -\sqrt{5}, \cot\theta = -\dfrac{1}{2}$

51. $\sin\theta = -\dfrac{3}{\sqrt{34}}, \cos\theta = -\dfrac{5}{\sqrt{34}}, \tan\theta = \dfrac{3}{5}, \csc\theta = -\dfrac{\sqrt{34}}{3}, \sec\theta = -\dfrac{\sqrt{34}}{5}, \cot\theta = \dfrac{5}{3}$

53. Starting from $y = \sin x$, translate left π units.

55. Starting from $y = \cos x$, translate left $\dfrac{\pi}{2}$ units, reflect across x-axis, and translate up 4 units.

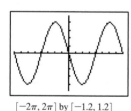

$[-2\pi, 2\pi]$ by $[-1.2, 1.2]$

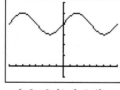

$[-2\pi, 2\pi]$ by $[-1, 6]$

57. Starting from $y = \tan x$, horizontally shrink by $\dfrac{1}{2}$.

59. Starting from $y = \sec x$, horizontally stretch by 2, vertically stretch by 2, and reflect across x-axis (in any order).

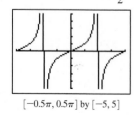

$[-0.5\pi, 0.5\pi]$ by $[-5, 5]$

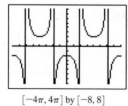

$[-4\pi, 4\pi]$ by $[-8, 8]$

61. Amplitude: 2; period: $\dfrac{2\pi}{3}$; phase shift: 0; domain: $(-\infty, \infty)$; range: $[-2, 2]$

63. Amplitude: 1.5; period: π; phase shift: $\dfrac{\pi}{8}$; domain: $(-\infty, \infty)$; range: $[-1.5, 1.5]$

65. Amplitude: 4; period: π; phase shift: $\dfrac{1}{2}$; domain: $(-\infty, \infty)$; range: $[-4, 4]$

67. $a \approx 4.47$, $b = 1$, and $h \approx 1.11$ **69.** $\approx 49.996° \approx 0.873$ radians   **71.** $45° = \pi/4$ rad

73. Starting from $y = \sin^{-1} x$, horizontally shrink by $\dfrac{1}{3}$. Domain: $\left[-\dfrac{1}{3}, \dfrac{1}{3}\right]$; range: $\left[-\dfrac{\pi}{2}, \dfrac{\pi}{2}\right]$

75. Starting from $y = \sin^{-1} x$, translate right 1 unit, horizontally shrink by $\dfrac{1}{3}$, translate up 2 units. Domain: $\left[0, \dfrac{2}{3}\right]$; range: $\left[2 - \dfrac{\pi}{2}, 2 + \dfrac{\pi}{2}\right]$

77. $5\pi/6$ **79.** $3\pi/4$ **81.** $3\pi/2$ **83.** As $|x| \to \infty$, $\dfrac{\sin x}{x^2} \to 0$.   **85.** 1   **87.** 3/4

89. Periodic; period: π; domain: $x \ne \dfrac{\pi}{2} + n\pi$, n an integer; range: $[1, \infty)$

91. Not periodic; domain: $x \ne \dfrac{\pi}{2} + n\pi$, n an integer; range: $(-\infty, \infty)$ **93.** $4\pi/3$ **95.** $100 \tan 78° \approx 470$ m

97. $150(\cot 18° - \cot 42°) \approx 295$ ft

99.

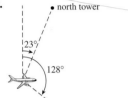

101. $62 \tan 72°24' \approx 195.4$ ft   **103.** $22\pi/15 \approx 4.6$ in.   **105.** **(a)** Day 123 (May 3)

(b) Day 287 (October 14)

Chapter 4 Project

Answers are based on the sample data shown in the table.

1.

3. The constant a represents half the distance the pendulum bob swings as it moves from its highest point to its lowest point; k represents the distance from the detector to the pendulum bob when it is in mid-swing.

5. $y \approx 0.22 \sin (3.87x - 0.16) + 0.71$; Most calculator/computer regression models are expressed in the form $y = a \sin(bx + p) + k$, where $-p/b = h$ in the equation $y = a \sin(b(x - h)) + k$. The latter equation form differs from $y = a \cos(b(x - h)) + k$ only in h.

SECTION 5.1

Exploration 1

1. $\cos \theta = \dfrac{1}{\sec \theta}$, $\sec \theta = \dfrac{1}{\cos \theta}$, and $\tan \theta = \dfrac{\sin \theta}{\cos \theta}$

3. $\csc \theta = \dfrac{1}{\sin \theta}$, $\cot \theta = \dfrac{1}{\tan \theta}$, and $\cot \theta = \dfrac{\cos \theta}{\sin \theta}$

Quick Review 5.1

1. 1.1760 rad $= 67.380°$ **3.** 2.4981 rad $= 143.130°$ **5.** $(a - b)^2$ **7.** $(2x + y)(x - 2y)$ **9.** $\dfrac{y - 2x}{xy}$ **11.** xy

Exercises 5.1

1. $\sin \theta = 3/5$ and $\cos \theta = 4/5$ **3.** $\tan \theta = -\sqrt{15}$ and $\cot \theta = -1/\sqrt{15} = -\sqrt{15}/15$ **5.** 0.45   **7.** −0.73   **9.** $\sin x$ **11.** 1
13. $\tan^2 x$ **15.** $\cos x \sin^2 x$ **17.** −1   **19.** −1   **21.** 1   **23.** $\cos x$ **25.** 2   **27.** $\sec x$ **29.** $\tan x$ **31.** $\tan x$ **33.** $2\csc^2 x$
35. $-\sin x$ **37.** $\cot x$
39. $(\cos x + 1)^2$ **41.** $(1 - \sin x)^2$ **43.** $(2\cos x - 1)(\cos x + 1)$ **45.** $(2 \tan x - 1)^2$ **47.** $1 - \sin x$ **49.** $1 - \cos x$
51. $\pi/6, \pi/2, 5\pi/6, 3\pi/2$ **53.** $0, \pi$ **55.** $\pi/3, 2\pi/3, 4\pi/3, 5\pi/3$

57. $\pm\dfrac{\pi}{3} + 2n\pi, \ n = 0, \pm 1, \pm 2, \ldots$ **59.** $n\pi, n = 0, \pm 1, \pm 2, \ldots$ **61.** $n\pi, n = 0, \pm 1, \pm 2, \ldots$

63. $\{\pm 1.1918 + 2n\pi \mid n = 0, \pm 1, \pm 2, \ldots\}$ **65.** $\{0.3047 + 2n\pi \text{ or } 2.8369 + 2n\pi \mid n = 0, \pm 1, \pm 2, \ldots\}$

67. $\{\pm 0.8861 + n\pi \mid n = 0, \ \pm 1, \pm 2, \ldots\}$ **69.** $|\sin \theta|$ **71.** $3|\tan \theta|$ **73.** $9|\sec \theta|$

75. True. Since secant is an even function, $\sec\left(x - \dfrac{\pi}{2}\right) = \sec\left(\dfrac{\pi}{2} - x\right)$, which equals $\csc x$ by one of the cofunction identities.

77. D   **79.** C

81. $\sin x, \cos x = \pm\sqrt{1 - \sin^2 x}, \tan x = \pm\dfrac{\sin x}{\sqrt{1 - \sin^2 x}}, \csc x = \dfrac{1}{\sin x}, \sec x = \pm\dfrac{1}{\sqrt{1 - \sin^2 x}}, \cot x = \pm\dfrac{\sqrt{1 - \sin^2 x}}{\sin x}$

83. The two functions are parallel to each other,
separated by 1 unit for every x. At any x,
the distance between the two graphs is
$\sin^2 x - (-\cos^2 x) = \sin^2 x + \cos^2 x = 1$.

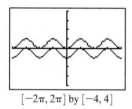

$[-2\pi, 2\pi]$ by $[-4, 4]$

85. (a)

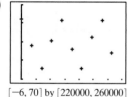

$[-6, 70]$ by $[220000, 260000]$

(b) The equation is $y = 13{,}111 \sin (0.22997 x + 1.571) + 238{,}855$.

$[-6, 70]$ by $[220000, 260000]$

(c) $(2\pi)/0.22998 \approx 27.32$ days. This is the number of days that it takes the moon to make one complete orbit of the Earth (known as the moon's sidereal period).
(d) 225,744 miles
(e) $y = 13{,}111 \cos(-0.22997x) + 238855$, or
$y = 13{,}111 \cos(0.22997x) + 238855$.

87. Factor the left-hand side: $\sin^4 \theta - \cos^4 \theta = (\sin^2 \theta - \cos^2 \theta)(\sin^2 \theta + \cos^2 \theta) = (\sin^2 \theta - \cos^2 \theta) \cdot 1 = \sin^2 \theta - \cos^2 \theta$
89. Use the hint:
$\sin(\pi - x) = \sin(\pi/2 - (x - \pi/2))$
$= \cos(x - \pi/2)$ Cofunction identity
$= \cos(\pi/2 - x)$ Since cos is even
$= \sin x$ Cofunction identity
91. Since A, B, and C are angles of a triangle, $A + B = \pi - C$. So: $\sin(A + B) = \sin(\pi - C) = \sin C$.

SECTION 5.2

Exploration 1

1. The graphs lead us to conclude that this is not an identity. **3.** Yes

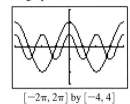

$[-2\pi, 2\pi]$ by $[-4, 4]$

5. No. The graph window cannot show the full graphs, so they could differ outside the viewing window. Also, the function values could be so close that the graphs *appear* to coincide.

Quick Review 5.2

1. $\dfrac{\sin x + \cos x}{\sin x \cos x}$ **3.** $\dfrac{1}{\sin x \cos x}$ **5.** 1 **7.** No. Any negative x.

9. No. Any x for which $\sin x < 0$, e.g., $x = -\pi/2$. **11.** yes

Exercises 5.2

1. One possible proof: $\dfrac{x^3 - x^2}{x} - (x - 1)(x + 1) = \dfrac{x(x^2 - x)}{x} - (x^2 - 1) = x^2 - x - (x^2 - 1) = -x + 1 = 1 - x$

3. One possible proof: $\dfrac{x^2 - 4}{x - 2} - \dfrac{x^2 - 9}{x + 3} = \dfrac{(x + 2)(x - 2)}{x - 2} - \dfrac{(x + 3)(x - 3)}{x + 3} = x + 2 - (x - 3) = 5$ **5.** yes **7.** no **9.** yes

11. $(\cos x)(\tan x + \sin x \cot x) = \cos x \cdot \dfrac{\sin x}{\cos x} + \cos x \sin x \cdot \dfrac{\cos x}{\sin x} = \sin x + \cos^2 x$

13. $(1 - \tan x)^2 = 1 - 2 \tan x + \tan^2 x = (1 + \tan^2 x) - 2 \tan x = \sec^2 x - 2 \tan x$

15. $\dfrac{(1 - \cos u)(1 + \cos u)}{\cos^2 u} = \dfrac{1 - \cos^2 u}{\cos^2 u} = \dfrac{\sin^2 u}{\cos^2 u} = \tan^2 u$ **17.** $\dfrac{\cos^2 x - 1}{\cos x} = \dfrac{-\sin^2 x}{\cos x} = -\dfrac{\sin x}{\cos x} \sin x = -\tan x \sin x$

19. Multiply out the expression on the left side.

21. $(\cos t - \sin t)^2 + (\cos t + \sin t)^2 = \cos^2 t - 2 \cos t \sin t + \sin^2 t + \cos^2 t + 2 \cos t \sin t + \sin^2 t = 2 \cos^2 t + 2 \sin^2 t = 2$

23. $\dfrac{1 + \tan^2 x}{\sin^2 x + \cos^2 x} = \dfrac{\sec^2 x}{1} = \sec^2 x$

25. $\dfrac{\cos \beta}{1 + \sin \beta} = \dfrac{\cos^2 \beta}{\cos \beta(1 + \sin \beta)} = \dfrac{1 - \sin^2 \beta}{\cos \beta(1 + \sin \beta)} = \dfrac{(1 - \sin \beta)(1 + \sin \beta)}{\cos \beta(1 + \sin \beta)} = \dfrac{1 - \sin \beta}{\cos \beta}$

27. $\dfrac{\tan^2 x}{\sec x + 1} = \dfrac{\sec^2 x - 1}{\sec x + 1} = \sec x - 1 = \dfrac{1}{\cos x} - 1 = \dfrac{1 - \cos x}{\cos x}$

29. $\cot^2 x - \cos^2 x = \left(\dfrac{\cos x}{\sin x}\right)^2 - \cos^2 x = \dfrac{\cos^2 x(1 - \sin^2 x)}{\sin^2 x} = \cos^2 x \cdot \dfrac{\cos^2 x}{\sin^2 x} = \cos^2 x \cot^2 x$

31. $\cos^4 x - \sin^4 x = (\cos^2 x + \sin^2 x)(\cos^2 x - \sin^2 x) = 1(\cos^2 x - \sin^2 x) = \cos^2 x - \sin^2 x$

33. $(x \sin \alpha + y \cos \alpha)^2 + (x \cos \alpha - y \sin \alpha)^2 = (x^2 \sin^2 \alpha + 2xy \sin \alpha \cos \alpha + y^2 \cos^2 \alpha) + (x^2 \cos^2 \alpha - 2xy \cos \alpha \sin \alpha + y^2 \sin^2 \alpha) = x^2 \sin^2 \alpha + y^2 \cos^2 \alpha + x^2 \cos^2 \alpha + y^2 \sin^2 \alpha = (x^2 + y^2)(\sin^2 \alpha + \cos^2 \alpha) = x^2 + y^2$

35. $\dfrac{\tan x}{\sec x - 1} = \dfrac{\tan x(\sec x + 1)}{\sec^2 x + 1} = \dfrac{\tan x(\sec x + 1)}{\tan^2 x} = \dfrac{\sec x + 1}{\tan x}$. See also #26.

37. $\dfrac{\sin x - \cos x}{\sin x + \cos x} = \dfrac{(\sin x - \cos x)(\sin x + \cos x)}{(\sin x + \cos x)^2} = \dfrac{\sin^2 x - \cos^2 x}{\sin^2 x + 2 \sin x \cos x + \cos^2 x} = \dfrac{\sin^2 x - (1 - \sin^2 x)}{1 + 2 \sin x \cos x} = \dfrac{2 \sin^2 x - 1}{1 + 2 \sin x \cos x}$

39. $\dfrac{\sin t}{1 - \cos t} + \dfrac{1 + \cos t}{\sin t} = \dfrac{\sin^2 t + (1 + \cos t)(1 - \cos t)}{(\sin t)(1 - \cos t)} = \dfrac{1 - \cos^2 t + 1 - \cos^2 t}{(\sin t)(1 - \cos t)} = \dfrac{2(1 - \cos^2 t)}{(\sin t)(1 - \cos t)} = \dfrac{2(1 + \cos t)}{\sin t}$

41. $\sin^2 x \cos^3 x = \sin^2 x \cos^2 x \cos x = \sin^2 x(1 - \sin^2 x)(\cos x) = (\sin^2 x - \sin^4 x)(\cos x)$

43. $\cos^5 x = \cos^4 x \cos x = (\cos^2 x)^2(\cos x) = (1 - \sin^2 x)^2 (\cos x) = (1 - 2\sin^2 x + \sin^4 x)(\cos x)$

45. $\dfrac{\tan x}{1 - \cot x} + \dfrac{\cot x}{1 - \tan x} = \dfrac{\tan x}{1 - \cot x} \cdot \dfrac{\sin x}{\sin x} + \dfrac{\cot x}{1 - \tan x} \cdot \dfrac{\cos x}{\cos x} = \left(\dfrac{\sin^2 x/\cos x}{\sin x - \cos x} + \dfrac{\cos^2 x/\sin x}{\cos x - \sin x} \right) \dfrac{\sin x \cos x}{\sin x \cos x}$

$= \dfrac{\sin^3 x - \cos^3 x}{\sin x \cos x(\sin x - \cos x)} = \dfrac{\sin^2 x + \sin x \cos x + \cos^2 x}{\sin x \cos x} = \dfrac{1 + \sin x \cos x}{\sin x \cos x} = \dfrac{1}{\sin x \cos x} + 1 = \csc x \sec x + 1$

47. $\dfrac{2\tan x}{1 - \tan^2 x} + \dfrac{1}{2\cos^2 x - 1} = \dfrac{2\tan x}{1 - \tan^2 x} \cdot \dfrac{\cos^2 x}{\cos^2 x} + \dfrac{1}{\cos^2 x - \sin^2 x} = \dfrac{2\sin x \cos x}{\cos^2 x - \sin^2 x} + \dfrac{\cos^2 x + \sin^2 x}{\cos^2 x - \sin^2 x}$

$= \dfrac{2\sin x \cos x + \cos^2 x + \sin^2 x}{(\cos x - \sin x)(\cos x + \sin x)} = \dfrac{(\cos x + \sin x)^2}{(\cos x - \sin x)(\cos x + \sin x)} = \dfrac{\cos x + \sin x}{\cos x - \sin x}$

49. $\cos^3 x = (\cos^2 x)(\cos x) = (1 - \sin^2 x)(\cos x)$

51. $\sin^5 x = (\sin^4 x)(\sin x) = (\sin^2 x)^2(\sin x) = (1 - \cos^2 x)^2(\sin x) = (1 - 2\cos^2 x + \cos^4 x)(\sin x)$ **53.** (d)   **55.** (c)   **57.** (b)

59. True. If x is in the domain of both sides of the equation, then $x \geq 0$. The equation holds for all $x \geq 0$, so it is an identity.

61. E   **63.** B   **65.** $\sin x$ **67.** 1   **69.** 1

71. If A and B are complementary angles, then $\sin^2 A + \sin^2 B = \sin^2 A + \sin^2(\pi/2 - A) = \sin^2 A + \cos^2 A = 1$.

73. Multiply and divide by $1 - \sin t$ under the radical: $\sqrt{\dfrac{1 - \sin t}{1 + \sin t} \cdot \dfrac{1 - \sin t}{1 - \sin t}} = \sqrt{\dfrac{(1 - \sin t)^2}{1 - \sin^2 t}} = \sqrt{\dfrac{(1 - \sin t)^2}{\cos^2 t}} = \dfrac{|1 - \sin t|}{|\cos t|}$ since $\sqrt{a^2} = |a|$.

Now, since $1 - \sin t \geq 0$, we can dispense with the absolute value in the numerator, but it must stay in the denominator.

75. $\sin^6 x + \cos^6 x = (\sin^2 x)^3 + \cos^6 x = (1 - \cos^2 x)^3 + \cos^6 x = (1 - 3\cos^2 x + 3\cos^4 x - \cos^6 x) + \cos^6 x$
$= 1 - 3\cos^2 x(1 - \cos^2 x) = 1 - 3\cos^2 x \sin^2 x$

77. $\ln|\tan x| = \ln\left|\dfrac{\sin x}{\cos x}\right| = \ln|\sin x| - \ln|\cos x|$

79. (a) They are not equal. Shown is the window $[-2\pi, 2\pi]$ by $[-2, 2]$; graphing on nearly any viewing window does not show any apparent difference—but using TRACE, one finds that the y coordinates are not identical. Likewise, a table of values will show slight differences; for example, when $x = 1$, $y_1 = 0.53988$ while $y_2 = 0.54030$.

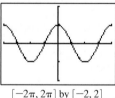

$[-2\pi, 2\pi]$ by $[-2, 2]$

(b) One choice for h is 0.001 (shown). The function y_3 is a combination of three sinusoidal functions $(1000 \sin(x + 0.001)$, $1000 \sin x$, and $\cos x)$, all with period 2π.

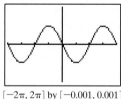

$[-2\pi, 2\pi]$ by $[-0.001, 0.001]$

81. In the decimal window, the x coordinates used to plot the graph on the calculator are (e.g.) 0, 0.1, 0.2, 0.3, etc.—that is, $x = n/10$, where n is an integer. Then $10\pi x = \pi n$, and the sine of integer multiples of π is 0; therefore, $\cos x + \sin 10\pi x = \cos x + \sin \pi n = \cos x + 0 = \cos x$. However, for other choices of x, such as $x = \dfrac{1}{\pi}$, we have $\cos x + \sin 10\pi x = \cos x + \sin 10 \neq \cos x$.

SECTION 5.3

Exploration 1

1. No   **3.** $\tan\left(\dfrac{\pi}{3} + \dfrac{\pi}{3}\right) = -\sqrt{3}$, $\tan\dfrac{\pi}{3} + \tan\dfrac{\pi}{3} = 2\sqrt{3}$. (Many other answers are possible.)

Quick Review 5.3

1. $45° - 30°$ **3.** $210° - 45°$ **5.** $2\pi/3 - \pi/4$ **7.** no   **9.** yes

Exercises 5.3

1. $(\sqrt{6} - \sqrt{2})/4$ **3.** $(\sqrt{6} + \sqrt{2})/4$ **5.** $(\sqrt{2} + \sqrt{6})/4$ **7.** $2 + \sqrt{3}$ **9.** $(\sqrt{2} - \sqrt{6})/4$ **11.** $\sin 25°$ **13.** $\sin 7\pi/10$ **15.** $\tan 66°$

17. $\cos\left(x - \dfrac{\pi}{7}\right)$ **19.** $\sin 2x$ **21.** $\tan(2y + 3x)$ **23.** $\sin\left(x - \dfrac{\pi}{2}\right) = \sin x \cos \dfrac{\pi}{2} - \cos x \sin \dfrac{\pi}{2} = \sin x \cdot 0 - \cos x \cdot 1 = -\cos x$

25. $\cos\left(x - \dfrac{\pi}{2}\right) = \cos x \cos \dfrac{\pi}{2} + \sin x \sin \dfrac{\pi}{2} = \cos x \cdot 0 + \sin x \cdot 1 = \sin x$

27. $\sin\left(x + \dfrac{\pi}{6}\right) = \sin x \cos \dfrac{\pi}{6} + \cos x \sin \dfrac{\pi}{6} = \sin x \cdot \dfrac{\sqrt{3}}{2} + \cos x \cdot \dfrac{1}{2}$

29. $\tan\left(\theta + \dfrac{\pi}{4}\right) = \dfrac{\tan \theta + \tan(\pi/4)}{1 - \tan \theta \tan(\pi/4)} = \dfrac{\tan \theta + 1}{1 - \tan \theta \cdot 1} = \dfrac{1 + \tan \theta}{1 - \tan \theta}$

31. Equations (b) and (f)

33. Equations (d) and (h). **35.** $x = n\pi, n = 0, \pm 1, \pm 2, \ldots$

37. $\sin\left(\dfrac{\pi}{2} - u\right) = \sin \dfrac{\pi}{2} \cos u - \cos \dfrac{\pi}{2} \sin u = 1 \cdot \cos u - 0 \cdot \sin u = \cos u$

39. $\cot\left(\dfrac{\pi}{2} - u\right) = \dfrac{\cos(\pi/2 - u)}{\sin(\pi/2 - u)} = \dfrac{\sin u}{\cos u} = \tan u$, using the first two cofunction identities.

41. $\csc\left(\dfrac{\pi}{2} - u\right) = \dfrac{1}{\sin(\pi/2 - u)} = \dfrac{1}{\cos u} = \sec u$, using the second cofunction identity.

43. $y \approx 5 \sin(x + 0.9273)$ **45.** $y \approx 2.236 \sin(3x + 0.4636)$

47. $\sin(x - y) + \sin(x + y) = (\sin x \cos y - \cos x \sin y) + (\sin x \cos y + \cos x \sin y) = 2 \sin x \cos y$

49. $\cos 3x = \cos[(x + x) + x] = \cos(x + x) \cos x - \sin(x + x) \sin x = (\cos x \cos x - \sin x \sin x) \cos x$
$- (\sin x \cos x + \cos x \sin x) \sin x = \cos^3 x - \sin^2 x \cos x - 2\cos x \sin^2 x = \cos^3 x - 3 \sin^2 x \cos x$

51. $\cos 3x + \cos x = \cos(2x + x) + \cos(2x - x)$; use Exercise 48 with x replaced with $2x$ and y replaced with x.

53. $\tan(x + y) \tan(x - y) = \left(\dfrac{\tan x + \tan y}{1 - \tan x \tan y}\right) \cdot \left(\dfrac{\tan x - \tan y}{1 + \tan x \tan y}\right) = \dfrac{\tan^2 x - \tan^2 y}{1 - \tan^2 x \tan^2 y}$ since both the numerator and denominator

are factored forms for differences of squares.

55. $\dfrac{\sin(x + y)}{\sin(x - y)} = \dfrac{\sin x \cos y + \cos x \sin y}{\sin x \cos y - \cos x \sin y} = \dfrac{\sin x \cos y + \cos x \sin y}{\sin x \cos y - \cos x \sin y} \cdot \dfrac{1/(\cos x \cos y)}{1/(\cos x \cos y)}$

$= \dfrac{(\sin x \cos y)/(\cos x \cos y) + (\cos x \sin y)/(\cos x \cos y)}{(\sin x \cos y)/(\cos x \cos y) - (\cos x \sin y)/(\cos x \cos y)} = \dfrac{(\sin x/\cos x) + (\sin y/\cos y)}{(\sin x/\cos x) - (\sin y/\cos y)} = \dfrac{\tan x + \tan y}{\tan x - \tan y}$

57. False. For example, $\cos 3\pi + \cos 4\pi = 0$, but 3π and 4π are not supplementary.

59. A **61.** B

63. $\tan(u - v) = \dfrac{\sin(u - v)}{\cos(u - v)} = \dfrac{\sin u \cos v - \cos u \sin v}{\cos u \cos v + \sin u \sin v} = \dfrac{\dfrac{\sin u \cos v}{\cos u \cos v} - \dfrac{\cos u \sin v}{\cos u \cos v}}{\dfrac{\cos u \cos v}{\cos u \cos v} + \dfrac{\sin u \sin v}{\cos u \cos v}} = \dfrac{\dfrac{\sin u}{\cos u} - \dfrac{\sin v}{\cos v}}{1 + \dfrac{\sin u \sin v}{\cos u \cos v}} = \dfrac{\tan u - \tan v}{1 + \tan u \tan v}$

65. The identity would involve $\tan\left(\dfrac{3\pi}{2}\right)$, which does not exist. $\tan\left(x - \dfrac{3\pi}{2}\right) = \dfrac{\sin\left(x - \dfrac{3\pi}{2}\right)}{\cos\left(x - \dfrac{3\pi}{2}\right)} = \dfrac{\sin x \cos \dfrac{3\pi}{2} - \cos x \sin \dfrac{3\pi}{2}}{\cos x \cos \dfrac{3\pi}{2} + \sin x \sin \dfrac{3\pi}{2}}$

$= \dfrac{\sin x \cdot 0 - \cos x \cdot (-1)}{\cos x \cdot 0 + \sin x \cdot (-1)} = -\cot x$

67. $\dfrac{\cos(x + h) - \cos x}{h} = \dfrac{\cos x \cos h - \sin x \sin h - \cos x}{h} = \dfrac{\cos x(\cos h - 1) - \sin x \sin h}{h} = \cos x\left(\dfrac{\cos h - 1}{h}\right) - \sin x \dfrac{\sin h}{h}$

69. $\sin(A + B) = \sin(\pi - C) = \sin \pi \cos C - \cos \pi \sin C = 0 \cdot \cos C - (-1)\sin C = \sin C$

71. $\tan A + \tan B + \tan C = \dfrac{\sin A}{\cos A} + \dfrac{\sin B}{\cos B} + \dfrac{\sin C}{\cos C} = \dfrac{\sin A(\cos B \cos C) + \sin B(\cos A \cos C)}{\cos A \cos B \cos C} + \dfrac{\sin C(\cos A \cos B)}{\cos A \cos B \cos C}$

$= \dfrac{\cos C(\sin A \cos B + \cos A \sin B) + \sin C(\cos A \cos B)}{\cos A \cos B \cos C} = \dfrac{\cos C \sin(A + B) + \sin C(\cos(A + B) + \sin A \sin B)}{\cos A \cos B \cos C}$

$= \dfrac{\cos C \sin(\pi - C) + \sin C(\cos(\pi - C) + \sin A \sin B)}{\cos A \cos B \cos C} = \dfrac{\cos C \sin C + \sin C(-\cos C) + \sin C \sin A \sin B}{\cos A \cos B \cos C}$

$= \dfrac{\sin A \sin B \sin C}{\cos A \cos B \cos C} = \tan A \tan B \tan C$

73. This equation is easier to deal with after rewriting it as $\cos 5x \cos 4x + \sin 5x \sin 4x = 0$. The left side of this equation is the expanded form of $\cos(5x - 4x)$, which of course equals $\cos x$; the graph shown is simply $y = \cos x$. The equation $\cos x = 0$ is easily solved on the interval $[-2\pi, 2\pi]$: $x = \pm\dfrac{\pi}{2}$ or $x = \pm\dfrac{3\pi}{2}$. The original graph is so crowded that one cannot see where crossings occur. The window shown is $[-2\pi, 2\pi]$ by $[-1.1, 1.1]$.

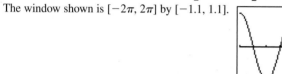

$[-2\pi, 2\pi]$ by $[-1.1, 1.1]$

75. $B = B_{in} + B_{ref} = \dfrac{E_0}{c}\cos\!\left(\omega t - \dfrac{\omega x}{c}\right) + \dfrac{E_0}{c}\cos\!\left(\omega t + \dfrac{\omega x}{c}\right) = \dfrac{E_0}{c}\!\left[\cos\!\left(\omega t - \dfrac{\omega x}{c}\right) + \cos\!\left(\omega t + \dfrac{\omega x}{c}\right)\right]$

$= \dfrac{E_0}{c}\!\left(2\cos \omega t \cos\dfrac{\omega x}{c}\right) = 2\dfrac{E_0}{c}\cos \omega t \cos\dfrac{\omega x}{c}$

The next-to-last step follows by the identity in Exercise 48.

SECTION 5.4

Exploration 1

1. $\sin^2\dfrac{\pi}{8} = \dfrac{1 - \cos(\pi/4)}{2} = \dfrac{1 - (\sqrt{2}/2)}{2} \cdot \dfrac{2}{2} = \dfrac{2 - \sqrt{2}}{4}$ **3.** $\sin^2\dfrac{9\pi}{8} = \dfrac{1 - \cos(9\pi/4)}{2} = \dfrac{1 - (\sqrt{2}/2)}{2} \cdot \dfrac{2}{2} = \dfrac{2 - \sqrt{2}}{4}$

Quick Review 5.4

1. $x = \dfrac{\pi}{4} + n\pi, n = 0, \pm1, \pm2, \ldots$ **3.** $x = \dfrac{\pi}{2} + n\pi, n = 0, \pm1, \pm2, \ldots$

5. $x = -\dfrac{\pi}{4} + n\pi, n = 0, \pm1, \pm2, \ldots$ **7.** $x = \dfrac{\pi}{6} + 2n\pi$ or $x = \dfrac{5\pi}{6} + 2n\pi$ or $x = \pm\dfrac{2\pi}{3} + 2n\pi, n = 0, \pm1, \pm2,$ **9.** 10 1/2 sq units

Exercises 5.4

1. $\cos 2u = \cos(u + u) = \cos u \cos u - \sin u \sin u = \cos^2 u - \sin^2 u$

3. Starting with the result of Exercise 1: $\cos 2u = \cos^2 u - \sin^2 u = (1 - \sin^2 u) - \sin^2 u = 1 - 2\sin^2 u$ **5.** $0, \pi$ **7.** $\dfrac{\pi}{6}, \dfrac{5\pi}{6}, \dfrac{3\pi}{2}$

9. $0, \dfrac{\pi}{4}, \dfrac{3\pi}{4}, \pi, \dfrac{5\pi}{4}, \dfrac{7\pi}{4}$ **11.** $2\sin\theta\cos\theta + \cos\theta$ or $(\cos\theta)(2\sin\theta + 1)$

13. $2\sin\theta\cos\theta + 4\cos^3\theta - 3\cos\theta$ or $2\sin\theta\cos\theta + \cos^3\theta - 3\sin^2\theta\cos\theta$

15. $\sin 4x = \sin 2(2x) = 2\sin 2x \cos 2x$

17. $2\csc 2x = \dfrac{2}{\sin 2x} = \dfrac{2}{2\sin x \cos x} = \dfrac{1}{\sin^2 x} \cdot \dfrac{\sin x}{\cos x} = \csc^2 x \tan x$

19. $\sin 3x = \sin 2x \cos x + \cos 2x \sin x = 2 \sin x \cos^2 x + (2 \cos^2 x - 1) \sin x = (\sin x)(4 \cos^2 x - 1)$

21. $\cos 4x = \cos 2(2x) = 1 - 2 \sin^2 2x = 1 - 2(2 \sin x \cos x)^2 = 1 - 8 \sin^2 x \cos^2 x$

23. $\dfrac{\pi}{3}, \pi, \dfrac{5\pi}{3}$ **25.** $\dfrac{\pi}{4}, \dfrac{\pi}{2}, \dfrac{3\pi}{4}, \dfrac{5\pi}{4}, \dfrac{3\pi}{2}, \dfrac{7\pi}{4}$ **27.** $0, \dfrac{\pi}{3}, \dfrac{\pi}{2}, \dfrac{2\pi}{3}, \pi, \dfrac{4\pi}{3}, \dfrac{3\pi}{2}, \dfrac{5\pi}{3}$ **29.** $\dfrac{\pi}{2}, \dfrac{3\pi}{2}, 0.1\pi, 0.9\pi, 1.3\pi, 1.7\pi$

31. $(1/2)\sqrt{2 - \sqrt{3}}$ **33.** $(1/2)\sqrt{2 - \sqrt{3}}$ **35.** $-2 - \sqrt{3}$

37. (a) Starting from the right side: $\dfrac{1}{2}(1 - \cos 2u) = \dfrac{1}{2}[1 - (1 - 2 \sin^2 u)] = \dfrac{1}{2}(2 \sin^2 u) = \sin^2 u.$

(b) Starting from the right side: $\dfrac{1}{2}(1 + \cos 2u) = \dfrac{1}{2}[1 + (2 \cos^2 u - 1)] = \dfrac{1}{2}(2 \cos^2 u) = \cos^2 u.$

39. $\sin^4 x = (\sin^2 x)^2 = \left[\dfrac{1}{2}(1 - \cos 2x)\right]^2 = \dfrac{1}{4}(1 - 2 \cos 2x + \cos^2 2x) = \dfrac{1}{4}\left[1 - 2 \cos 2x + \dfrac{1}{2}(1 + \cos 4x)\right]$

$= \dfrac{1}{8}(2 - 4 \cos 2x + 1 + \cos 4x) = \dfrac{1}{8}(3 - 4 \cos 2x + \cos 4x)$ **41.** $\sin^3 2x = \sin 2x \sin^2 2x = \sin 2x \cdot \dfrac{1}{2}(1 - \cos 4x) = \dfrac{1}{2}(\sin 2x)(1 - \cos 4x)$

43. $\dfrac{\pi}{3}, \pi, \dfrac{5\pi}{3}$; general solution: $\pm\dfrac{\pi}{3} + 2n\pi$ or $\pi + 2n\pi$, $n = 0, \pm1, \pm2, \ldots$

45. $0, \dfrac{\pi}{2}$; general solution: $2n\pi$ or $\dfrac{\pi}{2} + 2n\pi$, $n = 0, \pm1, \pm2, \ldots$

47. False. For example, $f(x) = 2 \sin x$ has period 2π and $g(x) = \cos x$ has period 2π, but the product $f(x) g(x) = 2 \sin x \cos x$

$= \sin 2x$ has period π. **49.** D **51.** E

53. (a) In the figure, the triangle with side lengths $x/2$ and R is a right triangle, since R is given as the

perpendicular distance. Then the tangent of the angle $\theta/2$ is the ratio "opposite over adjacent": $\tan \dfrac{\theta}{2} = \dfrac{x/2}{R}$. Solving for x gives the

desired equation. The central angle θ is $2\pi/n$ since one full revolution of 2π radians is divided evenly into n sections.

(b) $5.87 \approx 2R \tan \dfrac{\theta}{2}$, where $\theta = \dfrac{2\pi}{11}$, so $R \approx 5.87 / \left(2 \tan \dfrac{\pi}{11}\right) \approx 9.9957$, $R = 10$.

55. $\theta = \dfrac{\pi}{6}$; the maximum value is about 12.99 ft^3.

57. $\csc 2u = \dfrac{1}{\sin 2u} = \dfrac{1}{2 \sin u \cos u} = \dfrac{1}{2} \cdot \dfrac{1}{\sin u} \cdot \dfrac{1}{\cos u} = \dfrac{1}{2} \csc u \sec u$

59. $\sec 2u = \dfrac{1}{\cos 2u} = \dfrac{1}{1 - 2 \sin^2 u} = \left(\dfrac{1}{1 - 2 \sin^2 u}\right)\left(\dfrac{\csc^2 u}{\csc^2 u}\right) = \dfrac{\csc^2 u}{\csc^2 u - 2}$

61. $\sec 2u = \dfrac{1}{\cos 2u} = \dfrac{1}{\cos^2 u - \sin^2 u} = \left(\dfrac{1}{\cos^2 u - \sin^2 u}\right)\left(\dfrac{\sec^2 u \csc^2 u}{\sec^2 u \csc^2 u}\right) = \dfrac{\sec^2 u \csc^2 u}{\csc^2 u - \sec^2 u}$

63. (a)

$[-30, 370]$ by $[-60, 60]$

(b)

$[-30, 370]$ by $[-60, 60]$

This is a fairly good fit, but not really as good as one might expect from data generated by a sinusoidal physical model.

$y = 41.656 \sin(0.015x - 0.825) - 1.473$

(c) The residual list: $\{3.64, 7.56, 3.35, -5.94, -9.35, -3.90, 5.12, 9.43, 3.90, -4.57, -9.72, -3.22\}$

(d)

$[-30, 370]$ by $[-15, 15]$

$y = 8.856 \sin(0.0346x + 0.576) - 0.331$

This is another fairly good fit, which indicates that the residuals are not due to chance. There is a periodic variation that is most probably due to physical causes.

(e) The first regression indicates that the data are periodic and nearly sinusoidal. The second regression indicates that the *variation* of the data around the predicted values is also periodic and nearly sinusoidal. Periodic variation around periodic models is a predictable consequence of bodies orbiting bodies, but ancient astronomers had a difficult time reconciling the data with their simpler models of the universe.

SECTION 5.5

Exploration 1

1. If $BC \leq AB$, the segment will not reach from point B to the dotted line. On the other hand, if $BC > AB$, then a circle of radius BC will intersect the dotted line in a unique point. (Note that the line only extends to the left of point A.)
3. The second point (C_2) is the reflection of the first point (C_1) on the other side of the altitude.
5. If $BC \geq AB$, then BC can only extend to the right of the altitude, thus determining a unique triangle.

Quick Review 5.5

1. bc/d **3.** ad/b **5.** 13.314   **7.** 17.458°   **9.** 224.427°

Exercises 5.5

1. $C = 75°$; $a \approx 4.5$; $c \approx 5.1$ **3.** $B = 45°$; $b \approx 15.8$; $c \approx 12.8$
5. $C = 110°$; $a \approx 12.9$; $c \approx 18.8$
7. $C = 77°$; $a \approx 4.1$; $c \approx 7.3$ **9.** $B \approx 20.1°$; $C \approx 127.9°$; $c \approx 25.3$
11. $C \approx 37.2°$; $A \approx 72.8°$; $a \approx 14.2$ **13.** zero   **15.** two   **17.** two   **19.** $B_1 \approx 72.7°$; $C_1 \approx 43.3°$; $c_1 \approx 12.2$; $B_2 \approx 107.3°$; $C_2 \approx 8.7°$; $c_2 \approx 2.7$ **21.** $A_1 \approx 78.2°$; $B_1 \approx 33.8°$; $b_1 \approx 10.8$; $A_2 \approx 101.8°$; $B_2 \approx 10.2°$; $b_2 \approx 3.4$ **23. (a)** $6.691 < b < 10$ **(b)** $b \approx 6.691$ or $b \geq 10$ **(c)** $b < 6.691$
25. (a) No: this is an SAS case.   **(b)** No: only two pieces of information given   **27.** no triangle is formed   **29.** no triangle is formed
31. $A = 99°$; $a \approx 28.3$; $b \approx 19.1$
33. $A_1 \approx 24.6°$; $B_1 \approx 80.4°$; $a_1 \approx 20.7$; $A_2 \approx 5.4°$; $B_2 \approx 99.6°$; $a_2 \approx 4.7$ **35.** Cannot be solved with Law of Sines (an SAS case).
37. (a) 54.6 ft   **(b)** 51.9 ft   **39.** ≈ 24.9 ft   **41.** 1.9 ft   **43.** ≈ 108.9 ft   **45.** 36.6 mi to A; 28.9 mi to B
47. True. By the Law of Sines, $\dfrac{\sin A}{a} = \dfrac{\sin B}{b}$, which is equivalent to $\dfrac{\sin A}{\sin B} = \dfrac{a}{b}$.
49. C   **51.** A
53. (b) Possible answers: $a = 1$, $b = \sqrt{3}$, $c = 2$ (or any set of three numbers proportional to these).
(c) Any set of three identical numbers.
55. (a) $h = AB \sin A$ **(b)** $BC < AB \sin A$ **(c)** $BC \geq AB$ or $BC = AB \sin A$ **(d)** $AB \sin A < BC < AB$
57. $AC \approx 8.7$ mi; $BC \approx 12.2$ mi; $h \approx 5.2$ mi

SECTION 5.6

Exploration 1

1. 8475.742818 paces^2 **3.** 0.0014714831 square miles
5. The estimate of "a little over an acre" seems questionable, but the roughness of their measurement system does not provide firm evidence that it is incorrect. If Jim and Barbara wish to make an issue of it with the owner, they would be well-advised to get some more reliable data.

Quick Review 5.6

1. $A \approx 53.130°$ **3.** $A \approx 132.844°$
5. (a) $\cos A = \dfrac{x^2 + y^2 - 81}{2xy}$ **(b)** $A = \cos^{-1}\left(\dfrac{x^2 + y^2 - 81}{2xy}\right)$
7. One answer: $(x - 1)(x - 2)$ **9.** One answer $(x - i)(x + i) = x^2 + 1$

Exercises 5.6

1. $A \approx 30.7°$; $C \approx 18.3°$, $b \approx 19.2$ **3.** $A \approx 76.8°$; $B \approx 43.2°$, $C \approx 60°$
5. $B \approx 89.3°$; $C \approx 35.7°$, $a \approx 9.8$
7. $A \approx 28.5°$; $B \approx 56.5°$, $c \approx 25.1$ **9.** No triangles possible   **11.** $A \approx 24.6°$; $B \approx 99.2°$, $C \approx 56.2$
13. $B_1 \approx 72.9°$; $C_1 \approx 65.1°$, $c_1 \approx 9.487$; $B_2 \approx 107.1°$; $C_2 \approx 30.9°$, $c_2 \approx 5.376$
15. no triangle   **17.** ≈ 222.33 ft^2 **19.** ≈ 107.98 cm^2 **21.** ≈ 8.18 **23.** no triangle is formed   **25.** ≈ 216.15 **27.** ≈ 314.05
29. ≈ 1.445 radians   **31.** ≈ 374.1 in.^{2} **33.** ≈ 498.8 in.^{2} **35.** ≈ 130.42 ft

37. (a) ≈ 42.5 ft   **(b)** The home-to-second segment is the hypotenuse of a right triangle, so the distance from the pitcher's rubber to second base is $60\sqrt{2} - 40 \approx 44.9$ ft.   **(c)** $\approx 93.3°$

39. (a) $\tan^{-1}(1/3) \approx 18.4°$ **(b)** ≈ 4.5 ft   **(c)** ≈ 7.6 ft

41. ≈ 12.5 yd   **43.** $\approx 37.9°$

45. True. By the Law of Cosines, $b^2 + c^2 - 2bc \cos A = a^2$, which is a positive number. Since $b^2 + c^2 - 2bc \cos A > 0$, it follows that $b^2 + c^2 > 2bc \cos A$.

47. B   **49.** C   **51.** Area $= (nr^2/2) \sin (360°/n)$

53. (a) Ship A: $\dfrac{30.2 - 15.1}{1 \text{ hr}} = 15.1$ knots; Ship B: $\dfrac{37.2 - 12.4}{2 \text{ hrs}} = 12.4$ knots   **(b)** $35.18°$ **(c)** 34.8 nautical mi

55. 6.9 in.^{2}

CHAPTER 5 REVIEW EXERCISES

1. $\sin 200°$ **3.** 1

5. $\cos 3x = \cos(2x + x) = \cos 2x \cos x - \sin 2x \sin x = (\cos^2 x - \sin^2 x)\cos x - (2 \sin x \cos x)\sin x$

$= \cos^3 x - 3 \sin^2 x \cos x = \cos^3 x - 3(1 - \cos^2 x)\cos x = \cos^3 x - 3 \cos x + 3 \cos^3 x = 4 \cos^3 x - 3 \cos x$

7. $\tan^2 x - \sin^2 x = \sin^2 x \left(\dfrac{1 - \cos^2 x}{\cos^2 x} \right) = \sin^2 x \cdot \dfrac{\sin^2 x}{\cos^2 x} = \sin^2 x \tan^2 x$

9. $\csc x - \cos x \cot x = \dfrac{1}{\sin x} - \cos x \cdot \dfrac{\cos x}{\sin x} = \dfrac{1 - \cos^2 x}{\sin x} = \dfrac{\sin^2 x}{\sin x} = \sin x$

11. $\dfrac{1 + \tan \theta}{1 - \tan \theta} + \dfrac{1 + \cot \theta}{1 - \cot \theta} = \dfrac{(1 + \tan \theta)(1 - \cot \theta) + (1 + \cot \theta)(1 - \tan \theta)}{(1 - \tan \theta)(1 - \cot \theta)}$

$= \dfrac{(1 + \tan \theta - \cot \theta - 1) + (1 + \cot \theta - \tan \theta - 1)}{(1 - \tan \theta)(1 - \cot \theta)} = \dfrac{0}{(1 - \tan \theta)(1 - \cot \theta)} = 0$

13. $\cos^2 \dfrac{t}{2} = \left[\pm \sqrt{\dfrac{1}{2}(1 + \cos t)} \right]^2 = \dfrac{1}{2}(1 + \cos t) = \left(\dfrac{1 + \cos t}{2} \right) \left(\dfrac{\sec t}{\sec t} \right) = \dfrac{1 + \sec t}{2 \sec t}$

15. $\dfrac{\cos \phi}{1 - \tan \phi} + \dfrac{\sin \phi}{1 - \cot \phi} = \left(\dfrac{\cos \phi}{1 - \tan \phi} \right) \left(\dfrac{\cos \phi}{\cos \phi} \right) + \left(\dfrac{\sin \phi}{1 - \cot \phi} \right) \left(\dfrac{\sin \phi}{\sin \phi} \right) = \dfrac{\cos^2 \phi}{\cos \phi - \sin \phi} + \dfrac{\sin^2 \phi}{\sin \phi - \cos \phi}$

$= \dfrac{\cos^2 \phi - \sin^2 \phi}{\cos \phi - \sin \phi} = \cos \phi + \sin \phi$

17. $\sqrt{\dfrac{1 - \cos y}{1 + \cos y}} = \sqrt{\dfrac{(1 - \cos y)^2}{(1 + \cos y)(1 - \cos y)}} = \sqrt{\dfrac{(1 - \cos y)^2}{1 - \cos^2 y}} = \sqrt{\dfrac{(1 - \cos y)^2}{\sin^2 y}} = \dfrac{|1 - \cos y|}{|\sin y|}$

$= \dfrac{1 - \cos y}{|\sin y|}$; since $1 - \cos y \geq 0$, we can drop that absolute value.

19. $\tan\left(u + \dfrac{3\pi}{4} \right) = \dfrac{\tan u + \tan (3\pi/4)}{1 - \tan u \tan (3\pi/4)} = \dfrac{\tan u + (-1)}{1 - \tan u (-1)} = \dfrac{\tan u - 1}{1 + \tan u}$

21. $\tan \dfrac{1}{2}\beta = \dfrac{1 - \cos \beta}{\sin \beta} = \dfrac{1}{\sin \beta} - \dfrac{\cos \beta}{\sin \beta} = \csc \beta - \cot \beta$

23. Yes: $\sec x - \sin x \tan x = \dfrac{1}{\cos x} - \dfrac{\sin^2 x}{\cos x} = \dfrac{1 - \sin^2 x}{\cos x} = \dfrac{\cos^2 x}{\cos x} = \cos x$

25. Many answers are possible, for example, $(\cos x - \sin x)(1 + 4 \sin x \cos x)$.   **27.** Many answers are possible, for example, $1 - 4 \sin^2 x \cos^2 x - 2 \sin x \cos x$.

29. $\dfrac{\pi}{12} + n\pi$ or $\dfrac{5\pi}{12} + n\pi$, $n = 0, \pm 1, \pm 2, \dots$

31. $-\dfrac{\pi}{4} + n\pi$ **33.** $\tan 1$ **35.** $x \approx 1.12$ **37.** $x \approx 1.15$ **39.** $\pi/3, 5\pi/3$

41. $\dfrac{3\pi}{2}$ **43.** No solutions

45. $\left[0, \dfrac{\pi}{6}\right) \cup \left(\dfrac{5\pi}{6}, \dfrac{7\pi}{6}\right) \cup \left(\dfrac{11\pi}{6}, 2\pi\right)$ **47.** $(\pi/3, 5\pi/3)$ **49.** $y \approx 5\sin(3x + 0.9273)$

51. $C = 68°$, $b \approx 3.9$, $c \approx 6.6$ **53.** no triangle is formed   **55.** $C = 72°$; $a \approx 2.9$, $b \approx 5.1$ **57.** $A \approx 44.4°$, $B \approx 78.5°$, $C \approx 57.1°$

59. ≈ 7.5 **61. (a)** $\approx 5.6 < b < 12$ **(b)** $b \approx 5.6$ or $b \geq 12$ **(c)** $b < 5.6$

63. ≈ 0.6 mi   **65.** 1.25 rad

67. (a) $\sin\theta + \dfrac{1}{2}\sin 2\theta$ **(b)** $\theta = 60°$; ≈ 1.30 square units

69. (a) $h = 4000\sec\dfrac{\theta}{2} - 4000$ miles   **(b)** $\approx 35.51°$

71. area of circle $-$ area of hexagon $= 256\pi - 6 \cdot 64\sqrt{3} \approx 139.140$ cm^2

73. $405\pi/24 \approx 53.01$ cm^3

75. (a) By the product-to-sum formula in Exercise 74c, $2\sin\dfrac{u+v}{2}\cos\dfrac{u-v}{2}$

$= 2 \cdot \dfrac{1}{2}\left(\sin\dfrac{u+v+u-v}{2} + \sin\dfrac{u+v-(u-v)}{2}\right) = \sin u + \sin v$

(b) By the product-to-sum formula in Exercise 74c, $2\sin\dfrac{u-v}{2}\cos\dfrac{u+v}{2} = 2 \cdot \dfrac{1}{2}\left(\sin\dfrac{u-v+u+v}{2} + \sin\dfrac{u-v-(u+v)}{2}\right)$

$= \sin u + \sin(-v) = \sin u - \sin v$ **(c)** By the product-to-sum formula in Exercise 74b, $2\cos\dfrac{u+v}{2}\cos\dfrac{u-v}{2}$

$= 2 \cdot \dfrac{1}{2}\left(\cos\dfrac{u+v-(u-v)}{2} + \cos\dfrac{u+v+u-v}{2}\right) = \cos v + \cos u = \cos u + \cos v$ **(d)** By the product-to-sum formula in

Exercise 74a, $-2\sin\dfrac{u+v}{2}\sin\dfrac{u-v}{2} = -2 \cdot \dfrac{1}{2}\left(\cos\dfrac{u+v-(u-v)}{2} - \cos\dfrac{u+v+u-v}{2}\right) = -(\cos v - \cos u) = \cos u - \cos v$

77. (a) Any inscribed angle that intercepts an arc of $180°$ is a right angle.

(b) Two inscribed angles that intercept the same arc are congruent.   **(c)** In right $\triangle A'BC$, $\sin A' = \dfrac{\text{opp}}{\text{hyp}} = \dfrac{a}{d}$.

(d) Because $\angle A'$ and $\angle A$ are congruent, $\dfrac{\sin A}{a} = \dfrac{\sin A'}{a} = \dfrac{a/d}{a} = \dfrac{1}{d}$.   **(e)** Of course. They both equal $\dfrac{\sin A}{a}$ by the Law of Sines.

Chapter 5 Project

1.

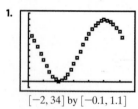

$[-2, 34]$ by $[-0.1, 1.1]$

5. One possible model is $y = 0.493\sin\left(\dfrac{2\pi}{29.36}(x + 11.654)\right) + 0.533$.

SECTION 6.1

Exploration 1

1. $\langle 5, 2\rangle$ **3.** $\langle 6, -7\rangle$

Quick Review 6.1

1. $\dfrac{9\sqrt{3}}{2}$; 4.5   **3.** -5.36; -4.50 **5.** $33.85°$ **7.** $60.95°$ **9.** $180° + \tan^{-1}(5/2) \approx 248.20°$

Exercises 6.1

1. Both vectors represent $\langle 3, -2 \rangle$ by the HMT Rule.   **3.** Both vectors represent $\langle -2, -2 \rangle$ by the HMT Rule.   **5.** $\langle 5, 2 \rangle$; $\sqrt{29}$
7. $\langle -5, 1 \rangle$; $\sqrt{26}$ **9.** $\langle -2, -24 \rangle$; $2\sqrt{145}$ **11.** $\langle -11, -7 \rangle$; $\sqrt{170}$ **13.** $\langle 1, 7 \rangle$ **15.** $\langle -3, 8 \rangle$ **17.** $\langle 4, -9 \rangle$ **19.** $\langle -4, -18 \rangle$
21. $\approx -0.45\mathbf{i} + 0.89\mathbf{j}$ **23.** $\approx -0.45\mathbf{i} - 0.89\mathbf{j}$

25. **(a)** $\left\langle \dfrac{2}{\sqrt{5}}, \dfrac{1}{\sqrt{5}} \right\rangle$ **(b)** $\dfrac{2}{\sqrt{5}}\mathbf{i} + \dfrac{1}{\sqrt{5}}\mathbf{j}$ **27.** **(a)** $\left\langle -\dfrac{4}{\sqrt{41}}, -\dfrac{5}{\sqrt{41}} \right\rangle$ **(b)** $-\dfrac{4}{\sqrt{41}}\mathbf{i} - \dfrac{5}{\sqrt{41}}\mathbf{j}$

29. $\approx \langle 16.31, 7.61 \rangle$ **31.** $\approx \langle -14.52, 44.70 \rangle$ **33.** 5; $\approx 53.13°$ **35.** 5; $\approx 306.87°$ **37.** 7; $135°$ **39.** $\langle \sqrt{2}, -\sqrt{2} \rangle$
41. $\approx \langle -223.99, 480.34 \rangle$ **43.** **(a)** $\approx \langle -111.16, 305.40 \rangle$ **(b)** ≈ 362.85 mph; bearing $\approx 337.84°$ **45.** **(a)** $\approx \langle 3.42, 9.40 \rangle$ **(b)** The horizontal component is the (constant) horizontal speed of the basketball as it travels toward the basket. The vertical component is the vertical velocity of the basketball, affected by both the initial speed and the downward pull of gravity.   **47.** $\approx \langle 2.20, 1.43 \rangle$ **49.** $|\mathbf{F}| \approx 100.33$ lb and $\theta \approx -1.22°$
51. $\approx 342.86°$; ≈ 9.6 mph   **53.** ≈ 13.66 mph; ≈ 7.07 mph   **55.** True. $\mathbf{u}$ and $-\mathbf{u}$ have the same length but opposite directions. Thus, the length of $-\mathbf{u}$ is also 1   **57.** D   **59.** A

SECTION 6.2

Exploration 1

1. $\langle -2 - x, -y \rangle$, $\langle 2 - x, -y \rangle$ **3.** Answers will vary

Quick Review 6.2

1. $\sqrt{13}$ **3.** 1   **5.** $\langle 3, \sqrt{3} \rangle$ **7.** $\langle -1, -\sqrt{3} \rangle$ **9.** $\left\langle \dfrac{4}{\sqrt{13}}, \dfrac{6}{\sqrt{13}} \right\rangle$

Exercises 6.2

1. 72   **3.** -47 **5.** 30   **7.** -14 **9.** 13   **11.** 4   **13.** $\approx 115.6°$ **15.** $\approx 64.65°$ **17.** $165°$ **19.** $135°$ **21.** $\approx 94.86°$

25. $-\dfrac{21}{10}\langle 3, 1 \rangle$; $-\dfrac{21}{10}\langle 3, 1 \rangle + \dfrac{17}{10}\langle -1, 3 \rangle$ **27.** $\dfrac{82}{85}\langle 9, 2 \rangle$; $\dfrac{82}{85}\langle 9, 2 \rangle + \dfrac{29}{85}\langle -2, 9 \rangle$ **29.** $47.73°, 74.74°, 57.53°$ **31.** ≈ -20.78

33. Parallel   **35.** Neither   **37.** Orthogonal   **39.** **(a)** $(4, 0)$ and $(0, -3)$ **(b)** $(4.6, -0.8)$ or $(3.4, 0.8)$ **41.** **(a)** $(7, 0)$ and $(0, -3)$

(b) $\approx (7.39, -0.92)$ or $(6.61, 0.92)$ **43.** $\langle -1, 4 \rangle$ or $\left\langle \dfrac{53}{13}, \dfrac{8}{13} \right\rangle$ **45.** ≈ 138.56 pounds   **47.** **(a)** ≈ 415.82 pounds   **(b)** ≈ 1956.30 pounds

49. 14,300 foot-pounds   **51.** ≈ 21.47 foot-pounds   **53.** ≈ 85.38 foot-pounds   **55.** $100\sqrt{39} \approx 624.5$ foot-pounds

61. False. If one of $\mathbf{u}$ or $\mathbf{v}$ is the zero vector, then $\mathbf{u} \cdot \mathbf{v} = 0$ but $\mathbf{u}$ and $\mathbf{v}$ are not perpendicular.   **63.** D   **65.** A
67. **(a)** $2 \cdot 0 + 5 \cdot 2 = 10$ and $2 \cdot 5 + 5 \cdot 0 = 10$ **(b)** $\dfrac{5}{29}\langle 5, -2 \rangle$; $\dfrac{1}{29}\langle 62, 155 \rangle$ **(c)** $|\mathbf{w}_2| = \dfrac{31\sqrt{29}}{29}$ **(d)** $d = \dfrac{|2x_0 + 5y_0 - 10|}{\sqrt{29}}$
(e) $d = \dfrac{|ax_0 + by_0 - c|}{\sqrt{a^2 + b^2}}$

SECTION 6.3

Exploration 1

1.
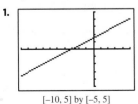
$[-10, 5]$ by $[-5, 5]$

3. $t = 12$ **5.** Tmin ≤ -2 and Tmax ≥ 5.5

Exploration 2

3.

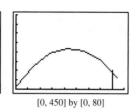 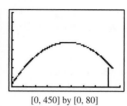 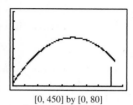

| [0, 450] by [0, 80] | [0, 450] by [0, 80] | [0, 450] by [0, 80] | [0, 450] by [0, 80] |

Quick Review 6.3

1. (a) $\langle -3, -2 \rangle$ **(b)** $\langle 4, 6 \rangle$ **(c)** $\langle 7, 8 \rangle$ **3.** $y + 2 = \dfrac{8}{7}(x + 3)$ or $y - 6 = \dfrac{8}{7}(x - 4)$

5.

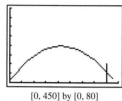

[−3, 7] by [−7, 7]

7. $x^2 + y^2 = 4$ **9.** 20π rad/sec

Exercises 6.3

1. (b) $[-5, 5]$ by $[-5, 5]$ **3. (a)** $[-5, 5]$ by $[-5, 5]$

5. (a)

t	-2	-1	0	1	2
x	0	1	2	3	4
y	$1/2$	-2	und.	4	$5/2$

5. (b)

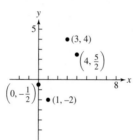

7.

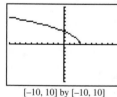

[−10, 10] by [−10, 10]

9.

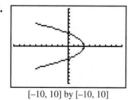

[−10, 10] by [−10, 10]

11. $y = x - 1$: line through $(0, -1)$ and $(1, 0)$ **13.** $y = -2x + 3$, $3 \le x \le 7$: line segment with endpoints $(3, -3)$ and $(7, -11)$

15. $x = (y - 1)^2$: parabola that opens to right with vertex at $(0, 1)$ **17.** $y = x^3 - 2x + 3$: cubic polynomial   **19.** $x = 4 - y^2$ parabola

that opens to left with vertex at $(4, 0)$ **21.** $t = x + 3$, so $y = \dfrac{2}{x + 3}$, on domain: $[-8, -3) \cup (-3, 2]$ **23.** $x^2 + y^2 = 25$, circle of

radius 5 centered at $(0, 0)$ **25.** $x^2 + y^2 = 4$, three-fourths of a circle of radius 2 centered at $(0, 0)$ (not in Quadrant II)

27. $x = 6t - 2$; $y = -3t + 5$ **29.** $x = 3t + 3$, $y = 4 - 7t$, $0 \le t \le 1$ **31.** $x = 5 + 3 \cos t$, $y = 2 + 3 \sin t$, $0 \le t \le 2\pi$ **33.** $0.5 < t < 2$

35. $-3 \le t < -2$ **37. (b)** Ben is ahead by 2 ft.   **39. (a)** $y = -16t^2 + 1000$ **(c)** 744 ft   **41. (a)** $0 < t < \pi/2$ **(b)** $0 < t < \pi$

(c) $\pi/2 < t < 3\pi/2$ **43. (a)** about 2.80 sec   **(b)** ≈ 7.18 ft   **45. (a)** yes   **(b)** 1.59 ft   **47** no   **49.** $y \approx -10.00$ ft/sec or

551.20 ft/sec   **51.** $x = 35 \cos\left(\dfrac{\pi}{6}t\right)$ and $y = 50 + 35 \sin\left(\dfrac{\pi}{6}t\right)$ **53. (a)** When $t = \pi$ (or 3π, or 5π, etc.), $y = 2$. This corresponds to the

highest points on the graph. **(b)** 2π units   **55.** (no answer)   **57.** (no answer)   **59.** True. Both correspond to the rectangular equation

$y = 3x + 4$.

61. A   **63.** D

65. (a) 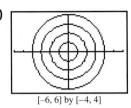 **(b)** a **(c)** 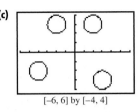 **(d)** $(x - h)^2 + (y - k)^2 = a^2$; circle of radius a centered at (h, k)

(e) $x = 3 \cos t - 1$; $y = 3 \sin t + 4$

$[-6, 6]$ by $[-4, 4]$ $[-6, 6]$ by $[-4, 4]$

67. (a) Jane is travelling in a circle of radius 20 feet and origin $(0, 20)$, which yields $x_1 = 20 \cos(nt)$ and $y_1 = 20 + 20 \sin(nt)$. Since the Ferris wheel is making one revolution (2π) every 12 seconds, $2\pi = 12n$, so $n = \dfrac{2\pi}{12} = \dfrac{\pi}{6}$. Thus, $x_1 = 20 \cos\left(\dfrac{\pi}{6}t\right)$ and $y_1 = 20 + 20 \sin\left(\dfrac{\pi}{6}t\right)$, in radian mode. **(b)** Since the ball was released at 75 ft in the positive x-direction and gravity acts in the negative y-direction at 16 ft/s², we have $x_2 = at + 75$ and $y_2 = -16t^2 + bt$, where a is the initial speed of the ball in the x-direction and b is the initial speed of the ball in the y-direction. The initial velocity vector of the ball is $60 \langle \cos 120°, \sin 120° \rangle = \langle -30, 30\sqrt{3} \rangle$, so $a = -30$ and $b = 30\sqrt{3}$. As a result, $x_2 = -30t + 75$ and $y_2 = -16t^3 + (30\sqrt{3})t$ are the parametric equations for the ball.

(c)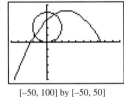

$[-50, 100]$ by $[-50, 50]$

Jane and the ball will be close to each other but not at the exact same point at $t = 2.2$ seconds.

(d) $d(t) = \sqrt{\left(20 \cos\left(\dfrac{\pi}{6}t\right) + 30t - 75\right)^2 + \left(20 + 20 \sin\left(\dfrac{\pi}{6}t\right) + 16t^2 - 30\sqrt{3}t\right)^2}$

(e) The minimum distance occurs at $t \approx 2.2$ seconds, when $d(t) \approx 1.64$ feet. Jane will have a good chance of catching the ball.

69. about 4.11 ft   **71. (a)** (no answer) **(b)** (no answer)   **73.** $t = \dfrac{1}{3}, \dfrac{2}{3}$; $t = \dfrac{1}{4}, \dfrac{1}{2}, \dfrac{3}{4}$

SECTION 6.4

Exploration 1

1. (no answer)   **3.** $(-2, \pi/3), (2, \pi/2), (3, 0), (1, \pi), (4, 3\pi/2)$

Quick Review 6.4

1. (a) II **(b)** III   **3.** $7\pi/4, -9\pi/4$ **5.** $520°, -200°$ **7.** $(x - 3)^2 + y^2 = 4$ **9.** ≈ 11.14

Exercises 6.4

1. $\left(-\dfrac{3}{2}, \dfrac{3\sqrt{3}}{2}\right)$ **3.** $(-1, -\sqrt{3})$ **5. (a)**

θ	$\pi/4$	$\pi/2$	$5\pi/6$	π	$4\pi/3$	2π
r	$3\sqrt{2}/2$	3	$3/2$	0	$-3\sqrt{3}/2$	0

(b)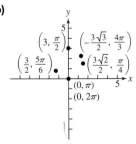

$\left(3, \dfrac{\pi}{2}\right)$ $\left(-\dfrac{3\sqrt{3}}{2}, \dfrac{4\pi}{3}\right)$

$\left(\dfrac{3}{2}, \dfrac{5\pi}{6}\right)$ $\left(\dfrac{3\sqrt{2}}{2}, \dfrac{\pi}{4}\right)$

$(0, \pi)$
$(0, 2\pi)$

7.

$\dfrac{4\pi}{3}$

$\left(3, \dfrac{4\pi}{3}\right)$

9.

$\dfrac{2\pi}{5}$

$\left(-1, \dfrac{2\pi}{5}\right)$

11. $(2, 30°)$
$30°$

13. $120°$

$\left(-2, 120°\right)$

15. $\left(\dfrac{3}{4}, \dfrac{3}{4}\sqrt{3}\right)$ **17.** $(-2.70, 1.30)$ **19.** $(2, 0)$ **21.** $(0, -2)$ **23.** $\left(2, \dfrac{\pi}{6} + 2n\pi\right)$ and $\left(-2, \dfrac{\pi}{6} + (2n + 1)\pi\right)$, n an integer

25. $(1.5, -20° + 360n°)$ and $(-1.5, 160° + 360n°)$, n an integer   **27.** **(a)** $\left(\sqrt{2}, \dfrac{\pi}{4}\right)$ or $\left(-\sqrt{2}, \dfrac{5\pi}{4}\right)$ **(b)** $\left(\sqrt{2}, \dfrac{\pi}{4}\right)$ or $\left(-\sqrt{2}, -\dfrac{3\pi}{4}\right)$

(c) The answers from part (a), and also $\left(\sqrt{2}, \dfrac{9\pi}{4}\right)$ or $\left(-\sqrt{2}, \dfrac{13\pi}{4}\right)$.   **29.** **(a)** $(\sqrt{29}, 1.95)$ or $(-\sqrt{29}, 5.09)$

(b) $(-\sqrt{29}, -1.19)$ or $(\sqrt{29}, 1.95)$ **(c)** The answers from part (a), plus $(\sqrt{29}, 8.23)$ or $(-\sqrt{29}, 11.38)$ **31.** **(b)**   **33.** **(c)**

35. $x = 3$ — a vertical line   **37.** $x^2 + \left(y + \dfrac{3}{2}\right)^2 = \dfrac{9}{4}$ — a circle centered at $\left(0, -\dfrac{3}{2}\right)$ with radius $\dfrac{3}{2}$

39. $x^2 + \left(y - \dfrac{1}{2}\right)^2 = \dfrac{1}{4}$ — a circle centered at $\left(0, \dfrac{1}{2}\right)$ with radius $\dfrac{1}{2}$ **41.** $(x + 2)^2 + (y - 1)^2 = 5$ — a circle centered at $(-2, 1)$ with radius $\sqrt{5}$

43. $r = 2/\cos\theta = 2\sec\theta$

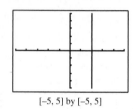

[-5, 5] by [-5, 5]

45. $r = \dfrac{5}{2\cos\theta - 3\sin\theta}$

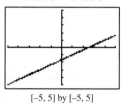

[-5, 5] by [-5, 5]

47. $r^2 - 6r\cos\theta = 0$, so $r = 6\cos\theta$ **49.** $r^2 + 6r\cos\theta + 6r\sin\theta = 0$, so $r = -6\cos\theta - 6\sin\theta$

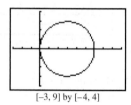

[-3, 9] by [-4, 4]

[-12, 6] by [-9, 3]

51. $2\sqrt{3} \approx 3.46$ mi   **53.** $\left(\dfrac{a}{\sqrt{2}}, \dfrac{\pi}{4}\right), \left(\dfrac{a}{\sqrt{2}}, \dfrac{3\pi}{4}\right), \left(\dfrac{a}{\sqrt{2}}, \dfrac{5\pi}{4}\right)$, and $\left(\dfrac{a}{\sqrt{2}}, \dfrac{7\pi}{4}\right)$

55. False. $(r, \theta) = (r, \theta + 2n\pi)$ for any integer n. These are all distinct polar coordinates.   **57.** C   **59.** A   **61. (a)** If $\theta_1 - \theta_2$ is an odd integer multiple of π, then the distance is $|r_1 + r_2|$. If $\theta_1 - \theta_2$ is an even integer multiple of π, then the distance is $|r_1 - r_2|$.   **(b)**
63. ≈ 6.24 **65.** ≈ 7.43 **67.** $x = f(\theta)\cos(\theta)$, $y = f(\theta)\sin(\theta)$ **69.** $x = 5(\cos\theta)(\sin\theta)$, $y = 5\sin^2\theta$ **71.** $x = 4\cot\theta$, $y = 4$

SECTION 6.5

Quick Review 6.5

1. Minimum: -3 at $x = \left\{\dfrac{\pi}{2}, \dfrac{3\pi}{2}\right\}$; Maximum: 3 at $x = \{0, \pi, 2\pi\}$ **3.** Minimum: 0 at $x = \left\{\dfrac{\pi}{4}, \dfrac{3\pi}{4}, \dfrac{5\pi}{4}, \dfrac{7\pi}{4}\right\}$; Maximum: 2 at $x = \{0, \pi, 2\pi\}$
5. no; no; yes   **7.** $\sin\theta$ **9.** $\cos^2\theta - \sin^2\theta$

Exercises 6.5

1. (a)

θ	0	$\pi/4$	$\pi/2$	$3\pi/4$	π	$5\pi/4$	$3\pi/2$	$7\pi/4$
r	3	0	-3	0	3	0	-3	0

(b)

$\left(0, \left\{\dfrac{\pi}{4}, \dfrac{3\pi}{4}, \dfrac{5\pi}{4}, \dfrac{7\pi}{4}\right\}\right)$

$\left(-3, \dfrac{3\pi}{2}\right)$

$(3, 0)$

$(3, \pi)$

$\left(-3, \dfrac{\pi}{2}\right)$

3. $k = \pi$

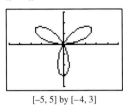

[−5, 5] by [−4, 3]

5. $k = 2\pi$

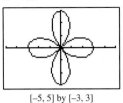

[−5, 5] by [−3, 3]

7. r_3 is graph (b)   **9.** Graph (b) is $r = 2 - 2\cos\theta$ **11.** Graph (a) is $r = 2 - 2\sin\theta$ **13.** Symmetric about the y-axis
15. Symmetric about the x-axis   **17.** All three symmetries   **19.** Symmetric about the y-axis   **21.** Maximum r is 5—when $\theta = 2n\pi$ for any integer n.   **23.** Maximum r is 3 (along with -3)—when $\theta = 2n\pi/3$ for any integer n.

25. Domain: All reals
Range: $r = 3$
Continuous
Symmetric about the x-axis,
   y-axis, and origin
Bounded
Maximum r-value: 3
No asymptotes

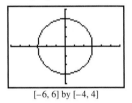

[−6, 6] by [−4, 4]

27. Domain: $\theta = \pi/3$
Range: $(-\infty, \infty)$
Continuous
Symmetric about the origin
Unbounded
Maximum r-value: none
No asymptotes

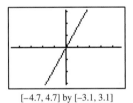

[−4.7, 4.7] by [−3.1, 3.1]

29. Domain: All reals
Range: $[-2, 2]$
Continuous
Symmetric about the y-axis
Bounded
Maximum r-value: 2
No asymptotes

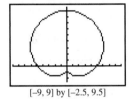

[−3, 3] by [−2, 2]

31. Domain: All reals
Range: $[1, 9]$
Continuous
Symmetric about the y-axis
Bounded
Maximum r-value: 9
No asymptotes

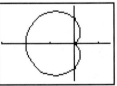

[−9, 9] by [−2.5, 9.5]

33. Domain: All reals
Range: $[0, 8]$
Continuous
Symmetric about the x-axis
Bounded
Maximum r-value: 8
No asymptotes

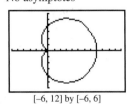

[−6, 12] by [−6, 6]

35. Domain: All reals
Range: $[3, 7]$
Continuous
Symmetric about the x-axis
Bounded
Maximum r-value: 7
No asymptotes

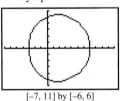

[−7, 11] by [−6, 6]

37. Domain: All reals
Range: $[-3, 7]$
Continuous
Symmetric about the x-axis
Bounded
Maximum r-value: 7
No asymptotes

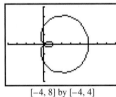

[−4, 8] by [−4, 4]

39. Domain: All reals
Range: $[0, 2]$
Continuous
Symmetric about the x-axis
Bounded
Maximum r-value: 2
No asymptotes

[−3, 1.5] by [−1.5, 1.5]

41. Domain: All reals
Range: $[0, \infty)$
Continuous
No symmetry
Unbounded
Maximum r-value: none
No asymptotes
Graph for $\theta \geq 0$:

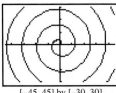

[−45, 45] by [−30, 30]

43. Domain: $\left[0, \dfrac{\pi}{2}\right] \cup \left[\pi, \dfrac{3\pi}{2}\right]$
Range: $[0, 1]$
Continuous on domain
Symmetric about the origin
Bounded
Maximum r-value: 1
No asymptotes

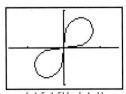

[−1.5, 1.5] by [−1, 1]

45. $\{6, 2, 6, 2\}$ **47.** $\{3, 5, 3, 5, 3, 5, 3, 5, 3, 5\}$ **49.** r_1 and r_2 **51.** r_2 and r_3 **53. (a)** A 4-petal rose curve with 2 short petals of length 1 and 2 long petals of length 3   **(b)** Symmetric about the origin   **(c)** Maximum r-value: 3   **55. (a)** A 6-petal rose curve with 3 short petals of length 2, and 3 long petals of length 4   **(b)** Symmetric about the x-axis   **(c)** Maximum r-value: 4   **57.** (no answer)
59. (no answer)   **61.** False. The spiral $r = \theta$ is unbounded.   **63.** D   **65.** B
67. (e) Domain: All reals
Range: $[-|a|, |a|]$
Continuous
Symmetric about the x-axis
Bounded
Maximum r-value: $|a|$
No asymptotes
69. (a) For r_1: $0 \le \theta \le 4\pi$ (or any interval that   **(b)** r_1: 10 (overlapping) petals;
is 4π units long). For r_2: same answer. r_2: 14 (overlapping) petals
71. Starting with the graph of r_1, if we rotate counterclockwise (centered at the origin) by $\pi/4$ radians (45°),
we get the graph of r_2; rotating r_1 counterclockwise by $\pi/3$ radians (60°) gives the graph of r_3.

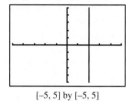 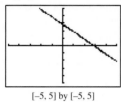 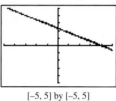

[−5, 5] by [−5, 5]   [−5, 5] by [−5, 5]   [−5, 5] by [−5, 5]

73. (no answer)

SECTION 6.6

Quick Review 6.6

1. $2 + 3i$, $2 - 3i$ **3.** $-4 - 4i$ **5.** $\theta = \dfrac{5\pi}{6}$ **7.** $\dfrac{4\pi}{3}$ **9.** 1

Exercises 6.6

1.

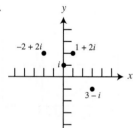

3. $3\left(\cos \dfrac{\pi}{2} + i \sin \dfrac{\pi}{2}\right)$ **5.** $2\sqrt{2}\left(\cos \dfrac{\pi}{4} + i \sin \dfrac{\pi}{4}\right)$ **7.** $4\left(\cos \dfrac{2\pi}{3} + i \sin \dfrac{2\pi}{3}\right)$ **9.** $\approx \sqrt{13}(\cos 0.59 + i \sin 0.59)$

11. $3\left(\cos \dfrac{\pi}{6} + i \sin \dfrac{\pi}{6}\right)$ **13.** $\dfrac{3\sqrt{3}}{2} - \dfrac{3}{2}i$ **15.** $5/2 - (5/2)\sqrt{3}i$ **17.** $\dfrac{\sqrt{6}}{2} - \dfrac{\sqrt{2}}{2}i$ **19.** $14(\cos 155° + i \sin 155°)$

21. $15\left(\cos \dfrac{23\pi}{12} + i \sin \dfrac{23\pi}{12}\right)$ **23.** $\dfrac{2}{3}(\cos 30° - i \sin 30°)$ **25.** $2(\cos \pi + i \sin \pi)$ **27. (a)** $5 + i; \dfrac{1}{2} - \dfrac{5}{2}i$ **(b)** Same as part (a).

29. (a) $18 - 4i; \approx 0.35 + 0.41i$ **(b)** Same as part (a).   **31.** $-\dfrac{\sqrt{2}}{2} + i\dfrac{\sqrt{2}}{2}$ **33.** $4\sqrt{2} + 4\sqrt{2}i$ **35.** $-4 - 4i$ **37.** -8

39. $\dfrac{-1 + \sqrt{3}i}{\sqrt[3]{4}}; \dfrac{-1 - \sqrt{3}i}{\sqrt[3]{4}}; \sqrt[3]{2}$ **41.** $\sqrt[3]{3}\left(\cos \dfrac{4\pi}{9} + i \sin \dfrac{4\pi}{9}\right), \sqrt[3]{3}\left(\cos \dfrac{10\pi}{9} + i \sin \dfrac{10\pi}{9}\right), \sqrt[3]{3}\left(\cos \dfrac{16\pi}{9} + i \sin \dfrac{16\pi}{9}\right)$

43. $\approx \sqrt[3]{5}(\cos 1.79 + i \sin 1.79)$, $\approx \sqrt[3]{5}(\cos 3.88 + i \sin 3.88)$, $\approx \sqrt[3]{5}(\cos 5.97 + i \sin 5.97)$

45. $\cos \dfrac{\pi}{5} + i \sin \dfrac{\pi}{5}$, $\cos \dfrac{3\pi}{5} + i \sin \dfrac{3\pi}{5}$, -1, $\cos \dfrac{7\pi}{5} + i \sin \dfrac{7\pi}{5}$, $\cos \dfrac{9\pi}{5} + i \sin \dfrac{9\pi}{5}$

47. $\sqrt[5]{2}\left(\cos \dfrac{\pi}{30} + i \sin \dfrac{\pi}{30}\right)$, $\sqrt[5]{2}\left(\cos \dfrac{13\pi}{30} + i \sin \dfrac{13\pi}{30}\right)$, $\sqrt[5]{2}\left(\cos \dfrac{5\pi}{6} + i \sin \dfrac{5\pi}{6}\right)$, $\sqrt[5]{2}\left(\cos \dfrac{37\pi}{30} + i \sin \dfrac{37\pi}{30}\right)$, $\sqrt[5]{2}\left(\cos \dfrac{49\pi}{30} + i \sin \dfrac{49\pi}{30}\right)$

49. $\sqrt[5]{2}\left(\cos \dfrac{\pi}{10} + i \sin \dfrac{\pi}{10}\right)$, $\sqrt[5]{2}i$, $\sqrt[5]{2}\left(\cos \dfrac{9\pi}{10} + i \sin \dfrac{9\pi}{10}\right)$, $\sqrt[5]{2}\left(\cos \dfrac{13\pi}{10} + i \sin \dfrac{13\pi}{10}\right)$, $\sqrt[5]{2}\left(\cos \dfrac{17\pi}{10} + i \sin \dfrac{17\pi}{10}\right)$

51. $\sqrt[8]{2}\left(\cos \dfrac{\pi}{16} + i \sin \dfrac{\pi}{16}\right)$, $\sqrt[8]{2}\left(\cos \dfrac{9\pi}{16} + i \sin \dfrac{9\pi}{16}\right)$, $\sqrt[8]{2}\left(\cos \dfrac{17\pi}{16} + i \sin \dfrac{17\pi}{16}\right)$, $\sqrt[8]{2}\left(\cos \dfrac{25\pi}{16} + i \sin \dfrac{25\pi}{16}\right)$

53. $\sqrt{2}\left(\cos \dfrac{\pi}{12} + i \sin \dfrac{\pi}{12}\right)$, $-1 + i$, $\sqrt{2}\left(\cos \dfrac{17\pi}{12} + i \sin \dfrac{17\pi}{12}\right)$

55. $\dfrac{1+i}{\sqrt[6]{4}}$, $\sqrt[6]{2}\left(\cos \dfrac{7\pi}{12} + i \sin \dfrac{7\pi}{12}\right)$, $\sqrt[6]{2}\left(\cos \dfrac{11\pi}{12} + i \sin \dfrac{11\pi}{12}\right)$, $\sqrt[6]{2}\left(\cos \dfrac{5\pi}{4} + i \sin \dfrac{5\pi}{4}\right)$, $\sqrt[6]{2}\left(\cos \dfrac{19\pi}{12} + i \sin \dfrac{19\pi}{12}\right)$,

$\sqrt[6]{2}\left(\cos \dfrac{23\pi}{12} + i \sin \dfrac{23\pi}{12}\right)$

57. 1, $-\dfrac{1}{2} \pm \dfrac{\sqrt{3}}{2} i$ **59.** ± 1, $\dfrac{1}{2} \pm \dfrac{\sqrt{3}}{2}i$, $-\dfrac{1}{2} \pm \dfrac{\sqrt{3}}{2}i$

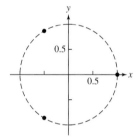   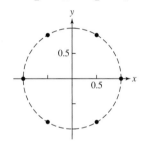

61. -8; -2 and $1 \pm \sqrt{3}i$ **65.** False. For example, the complex number $1 + i$ has infinitely many trigonometric forms. Here are two:

$\sqrt{2}\left(\cos \dfrac{\pi}{4} + i \sin \dfrac{\pi}{4}\right)$, $\sqrt{2}\left(\cos \dfrac{9\pi}{4} + i \sin \dfrac{9\pi}{4}\right)$.   **67.** B   **69.** A

71. (a) (no answer) **(b)** r^2 **(c)** $\cos (2\theta) + i \sin (2\theta)$ **(d)** (no answer)

73. Set the calculator for rounding to 2 decimal places. In part (b), use Degree mode.

(a)    **(b)** 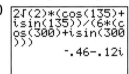   **(c)**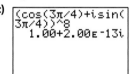

75. $x(t) = (\sqrt{3})^t \cos(0.62t)$ **79.** 1, $-\dfrac{1}{2} + \dfrac{\sqrt{3}}{2} i$, $-\dfrac{1}{2} - \dfrac{\sqrt{3}}{2} i$

$y(t) = (\sqrt{3})^t \sin(0.62t)$

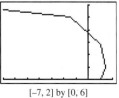   **81.** -1, $\dfrac{1}{2} + \dfrac{\sqrt{3}}{2}i$, $\dfrac{1}{2} - \dfrac{\sqrt{3}}{2}i$

$[-7, 2]$ by $[0, 6]$

83. -1, $\approx 0.81 + 0.59i$, $0.81 - 0.59i$, $-0.31 + 0.95i$, $-0.31 - 0.95i$

CHAPTER 6 REVIEW EXERCISES

CHAPTER 6 REVIEW EXERCISES

1. $\langle -2, -3 \rangle$ **3.** $\sqrt{37}$ **5.** 6   **7.** $\langle 3, 6 \rangle$; $3\sqrt{5}$ **9.** $\langle -8, -3 \rangle$; $\sqrt{73}$ **11. (a)** $\left\langle -\dfrac{2}{\sqrt{5}}, \dfrac{1}{\sqrt{5}} \right\rangle$ **(b)** $\left\langle \dfrac{6}{\sqrt{5}}, -\dfrac{3}{\sqrt{5}} \right\rangle$

13. (a) $\tan^{-1}\!\left(\dfrac{3}{4}\right) \approx 0.64$, $\tan^{-1}\!\left(\dfrac{5}{2}\right) \approx 1.19$ **(b)** ≈ 0.55 **15.** $\approx (-2.27, -1.06)$ **17.** $(\sqrt{2}, -\sqrt{2})$

19. $\left(1, -\dfrac{2\pi}{3} + (2n + 1)\pi\right)$ and $\left(-1, -\dfrac{2\pi}{3} + 2n\pi\right)$, n an integer   **21. (a)** $\left(-\sqrt{13}, \pi + \tan^{-1}\!\left(-\dfrac{3}{2}\right)\right) \approx (-\sqrt{13}, 2.16)$ or

$\left(\sqrt{13}, 2\pi + \tan^{-1}\!\left(-\dfrac{3}{2}\right)\right) \approx (\sqrt{13}, 5.30)$ **(b)** $\left(\sqrt{13}, \tan^{-1}\!\left(-\dfrac{3}{2}\right)\right) \approx (\sqrt{13}, -0.98)$ or $\left(-\sqrt{13}, \pi + \tan^{-1}\!\left(-\dfrac{3}{2}\right)\right) \approx (-\sqrt{13}, 2.16)$

(c) The answers from part (a), and also $\left(-\sqrt{13}, 3\pi + \tan^{-1}\!\left(-\dfrac{3}{2}\right)\right) \approx (-\sqrt{13}, 8.44)$ or $\left(\sqrt{13}, 4\pi + \tan^{-1}\!\left(-\dfrac{3}{2}\right)\right) \approx (\sqrt{13}, 11.58)$

23. (a) $(5, 0)$ or $(-5, \pi)$ or $(5, 2\pi)$ **(b)** $(-5, -\pi)$ or $(5, 0)$ or $(-5, \pi)$ **(c)** The answers from part (a), and also $(-5, 3\pi)$ or $(5, 4\pi)$

25. $y = -\dfrac{3}{5}x + \dfrac{29}{5}$: line through $\left(0, \dfrac{29}{5}\right)$ with slope $m = -\dfrac{3}{5}$ **27.** $x = 2(y + 1)^2 + 3$: parabola that opens to right with vertex at $(3, -1)$

29. $y = \sqrt{x + 1}$: square root function starting at $(-1, 0)$ **31.** $x = 2t + 3$, $y = 3t + 4$ **33.** $a = -3$, $b = 4$, $|z_1| = 5$ **35.** $3\sqrt{3} + 3i$

37. $-1.25 - 1.25\sqrt{3}i$ **39.** $3\sqrt{2}\left(\cos\dfrac{7\pi}{4} + i \sin\dfrac{7\pi}{4}\right)$. Other representations would use angles $\dfrac{7\pi}{4} + 2n\pi$, n an integer.

41. $\approx \sqrt{34}[\cos(5.25) + i \sin(5.25)]$ Other representations would use angles $\approx 5.25 + 2n\pi$, n an integer.

43. $12(\cos 90° + i \sin 90°)$; $\dfrac{3}{4}(\cos 330° + i \sin 330°)$ **45. (a)** $243\left(\cos\dfrac{5\pi}{4} + i \sin\dfrac{5\pi}{4}\right)$ **(b)** $-\dfrac{243\sqrt{2}}{2} - \dfrac{243\sqrt{2}}{2}i$

47. (a) $125(\cos \pi + i \sin \pi)$ **(b)** -125 **49.** $\sqrt[8]{18}\left(\cos\dfrac{\pi}{16} + i \sin\dfrac{\pi}{16}\right)$, $\sqrt[8]{18}\left(\cos\dfrac{9\pi}{16} + i \sin\dfrac{9\pi}{16}\right)$,

$\sqrt[8]{18}\left(\cos\dfrac{17\pi}{16} + i \sin\dfrac{17\pi}{16}\right)$, $\sqrt[8]{18}\left(\cos\dfrac{25\pi}{16} + i \sin\dfrac{25\pi}{16}\right)$

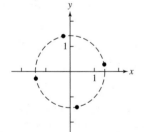

51. $1, \cos\dfrac{2\pi}{5} + i \sin\dfrac{2\pi}{5}, \cos\dfrac{4\pi}{5} + i \sin\dfrac{4\pi}{5}, \cos\dfrac{6\pi}{5} + i \sin\dfrac{6\pi}{5}, \cos\dfrac{8\pi}{5} + i \sin\dfrac{8\pi}{5}$

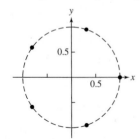

53. (b)   **55.** (a)   **57.** not shown   **59.** (c)   **61.** $x^2 + y^2 = 4$—a circle with center $(0, 0)$ and radius 2

63. $\left(x + \dfrac{3}{2}\right)^2 + (y + 1)^2 = \dfrac{13}{4}$—a circle of radius $\dfrac{\sqrt{13}}{2}$ with center $\left(-\dfrac{3}{2}, -1\right)$

65. $r = -\dfrac{4}{\sin \theta} = -4 \csc \theta$

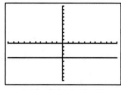

[–10, 10] by [–10, 10]

67. $r = 6 \cos \theta - 2 \sin \theta$

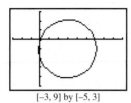

[–3, 9] by [–5, 3]

69.

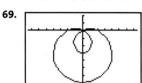

[–7.5, 7.5] by [–8, 2]

Domain: All reals
Range: $[-3, 7]$
Continuous
Symmetric about the y-axis
Bounded
Maximum r-value: 7
No asymptotes

71.

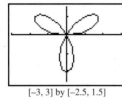

[–3, 3] by [–2.5, 1.5]

Domain: All reals
Range: $[-2, 2]$
Continuous
Symmetric about the y-axis
Bounded
Maximum r-value: 2
No asymptotes

73. **(a)** (no answer) **(b)** (no answer) **(c)** (no answer) **(d)**

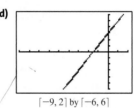

[–9, 2] by [–6, 6]

75. **(a)** $\approx \langle -463.64, 124.23 \rangle$ **(b)** ≈ 508.29 mph; bearing $\approx 283.84°$ **77.** **(a)** ≈ 826.91 pounds **(b)** 2883.79 pounds

79. **(a)** $h = -16t^2 + 245t + 200$ **(b)** Graph and trace: $x = 17$ and $y = -16t^2 + 245t + 200$

(c)

[0, 18] by [0, 1200]

(d) 924 ft **(e)** ≈ 1138 ft; $t \approx 7.66$ **(f)** about 16.09 sec with $0 \le t \le 16.1$ (upper limit may vary) on [0, 18] by [0, 1200]. This graph will appear as a vertical line from about (17, 0) to about (17, 1138). Tracing shows how the arrow begins at a height of 200 ft, rises to over 1000 ft, then falls back to the ground.

81. $x = 40 \sin\left(\dfrac{2\pi}{15}t\right)$, $y = 50 - 40 \cos\left(\dfrac{2\pi}{15}t\right)$

83. **(a)**

[–7.5, 7.5] by [–5, 5]

(b) All 4's should be changed to 5's.
85. $t \approx 1.06$ sec, $x \approx 68.65$ ft **87.** **(a)** ≈ 77.59 ft
(b) ≈ 4.404 sec **89.** ≈ 17.65 ft

Answers are based on the sample data shown in the table.

1.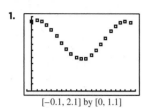

[–0.1, 2.1] by [0, 1.1]

3.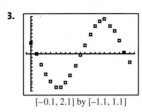

[–0.1, 2.1] by [–1.1, 1.1]

5.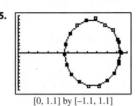

[0, 1.1] by [–1.1, 1.1]

SECTION 7.1

Exploration 1

1.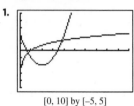

[0, 10] by [–5, 5]

Quick Review 7.1

1. $y = \dfrac{5}{3} - \dfrac{2}{3}x$ **3.** $x = -\dfrac{2}{3}, x = 1$ **5.** $0, 2, -2$ **7.** $y = (-4x + 6)/5$ **9.** $-4x - 6y = -10$

Exercises 7.1

1. (a) No **(b)** Yes **(c)** No   **3.** $(9, -2)$ **5.** $(50/7, -10/7)$ **7.** $(-1/2, 2)$ **9.** No solution   **11.** $(\pm 3, 9)$

13. $(-3/2, 27/2)$ and $(1/3, 2/3)$ **15.** $(0, 0)$ and $(3, 18)$ **17.** $\left(\dfrac{-1 + 3\sqrt{89}}{10}, \dfrac{3 + \sqrt{89}}{10}\right)$ and $\left(\dfrac{-1 - 3\sqrt{89}}{10}, \dfrac{3 - \sqrt{89}}{10}\right)$

19. $(8, -2)$ **21.** $(4, 2)$ **23.** No solution   **25.** Infinitely many solutions   **27.** $(0, 1)$ and $(3, -2)$ **29.** No solution

31. One solution   **33.** Infinitely many solutions   **35.** $\approx (0.69, -0.37)$ **37.** $\approx(-2.32, -3.16), (0.47, -1.77)$ and $(1.85, -1.08)$

39. $(-1.2, 1.6)$ and $(2, 0)$ **41.** $\approx(2.05, 2.19)$ and $(-2.05, 2.19)$ **43.** Demand curve   Supply curve $(3.75, 143.75)$

45. (a) $y = -0.0938x^2 + 15.0510x - 28.2375$ **(b)** $y = \dfrac{353.6473}{(1 + 8.6873e^{-0.1427x})}$ **(c)** Quadratic: 2006, logistic: 2007

(d) The quadratic regression predicts that the expenditures will eventually be zero.

(e) The logistic regression predicts that the expenditures will eventually level off at about 354 billion dollars.

[0, 30] by [–100, 500]

[0, 30] by [–100, 500]

47. (a) $y = 127.6351x + 2587.0010$ **(b)** $y = 31.3732x + 5715.9742$ **(c)** 2012

49. ≈ 5.28 m $\times \approx 94.72$ m

51. current speed ≈ 1.06 mph; rowing speed ≈ 3.56 mph

53. medium: $0.79; large: $0.95

55. $a = 2/3$ and $b = 14/3$ **57. (a)** 300 miles

59. False. A system of two linear equations in two variables has either 0, 1, or infinitely many solutions.

61. C   **63.** D

[–5, 30] by [0, 8000]

[–5, 30] by [0, 8000]

65. (a) $y = (3/2)\sqrt{4 - x^2}, y = -(3/2)\sqrt{4 - x^2}$

(b)

[−4.7, 4.7 by −3.1, 3.1]

$\approx(-1.29, 2.29)$ or $(1.91, -0.91)$

67. $(\pm\sqrt{2/3},\ 10/3)$ **69.** 12.5 units

SECTION 7.2

Exploration 1

1. $A = \begin{bmatrix} 2 & 1 \\ 5 & 4 \end{bmatrix}$; $B = \begin{bmatrix} -1 & 2 \\ 2 & 5 \end{bmatrix}$ **3.** $\begin{bmatrix} 8 & -1 \\ 11 & 2 \end{bmatrix}$

Exploration 2

1. $\det A = -a_{12}a_{21}a_{33} + a_{13}a_{21}a_{32} + a_{11}a_{22}a_{33} - a_{13}a_{22}a_{31} - a_{11}a_{23}a_{32} + a_{12}a_{23}a_{31}$

3. Since each term in the expansion contains an element from each row and each column, at least one factor in each term is a zero. Therefore, the expansion will be the sum of n zero terms, or zero.

Quick Review 7.2

1. $(3, 2)$; $(x, -y)$ **3.** $(-2, 3)$; (y, x) **5.** $(3\cos\theta, 3\sin\theta)$ **7.** $\sin\alpha\cos\beta + \cos\alpha\sin\beta$ **9.** $\cos\alpha\cos\beta - \sin\alpha\sin\beta$

Exercises 7.2

1. 2×3; not square   **3.** 3×2; not square   **5.** 3×1; not square   **7.** 3   **9.** 4

11. **(a)** $\begin{bmatrix} 3 & 0 \\ -3 & 1 \end{bmatrix}$ **(b)** $\begin{bmatrix} 1 & 6 \\ 1 & 9 \end{bmatrix}$ **(c)** $\begin{bmatrix} 6 & 9 \\ -3 & 15 \end{bmatrix}$ **(d)** $\begin{bmatrix} 1 & 15 \\ 4 & 22 \end{bmatrix}$ **13.** **(a)** $\begin{bmatrix} 1 & 1 \\ -2 & 0 \\ -1 & 0 \end{bmatrix}$ **(b)** $\begin{bmatrix} -7 & 1 \\ 2 & -2 \\ 5 & 2 \end{bmatrix}$ **(c)** $\begin{bmatrix} -9 & 3 \\ 0 & -3 \\ 6 & 3 \end{bmatrix}$ **(d)** $\begin{bmatrix} -18 & 2 \\ 6 & -5 \\ 13 & 5 \end{bmatrix}$

15. **(a)** $\begin{bmatrix} -3 \\ 1 \\ 4 \end{bmatrix}$ **(b)** $\begin{bmatrix} -1 \\ 1 \\ -4 \end{bmatrix}$ **(c)** $\begin{bmatrix} -6 \\ 3 \\ 0 \end{bmatrix}$ **(d)** $\begin{bmatrix} -1 \\ 2 \\ -12 \end{bmatrix}$ **17.** **(a)** $\begin{bmatrix} -4 & -18 \\ -11 & -17 \end{bmatrix}$ **(b)** $\begin{bmatrix} 5 & -12 \\ 0 & -26 \end{bmatrix}$ **19.** **(a)** $\begin{bmatrix} 2 & 2 \\ -11 & 12 \end{bmatrix}$ **(b)** $\begin{bmatrix} 4 & 8 & -5 \\ -5 & 4 & -6 \\ -2 & -8 & 6 \end{bmatrix}$

21. **(a)** $\begin{bmatrix} 6 & -7 & -2 \\ 3 & 7 & 3 \\ 8 & -1 & -1 \end{bmatrix}$ **(b)** $\begin{bmatrix} 2 & 1 & 3 \\ 5 & 0 & 0 \\ -18 & -3 & 10 \end{bmatrix}$ **23.** **(a)** $[-8]$ **(b)** $\begin{bmatrix} -10 & 5 & -15 \\ 8 & -4 & 12 \\ 4 & -2 & 6 \end{bmatrix}$ **25.** not possible; $[18\ \ 14]$

27. **(a)** $\begin{bmatrix} -1 & 3 & 4 \\ 2 & 0 & 1 \\ 1 & 2 & 1 \end{bmatrix}$ **(b)** $\begin{bmatrix} 1 & 2 & 1 \\ 1 & 0 & 2 \\ 4 & 3 & -1 \end{bmatrix}$ **29.** $a = 5, b = 2$ **31.** $a = -2, b = 0$ **33.** $AB = BA = I_2$ **35.** $\begin{bmatrix} -1 & 1.5 \\ 1 & -1 \end{bmatrix}$

37. no inverse   **39.** no inverse   **41.** -14 **43.** $\begin{bmatrix} 1 \\ -1/3 \end{bmatrix}$

45. **(a)** The distance from city X to city Y is the same as the distance from city Y to city X.
   (b) Each entry represents the distance from city X to city X.   **47.** **(a)** $[382\ \ 227.50]$ **49.** **(a)** AB^T or BA^T **(b)** $(A - C)B^T$

51. **(a)** $\approx (1.37\ \ 0.37)$ **(b)** $\approx (0.37\ \ 1.37)$ **55.** $A \cdot A^{-1} = I_2$ **57.** $[x\ \ y]\begin{bmatrix} -1 & 0 \\ 0 & 1 \end{bmatrix}$ **59.** $[x\ \ y]\begin{bmatrix} 0 & -1 \\ -1 & 0 \end{bmatrix}$

61. $[x \quad y]\begin{bmatrix} c & 0 \\ 0 & 1 \end{bmatrix}$ **63.** False. It can be negative. For example, the determinant of $A = \begin{bmatrix} 1 & 0 \\ 2 & -1 \end{bmatrix}$ is -1.   **65.** B   **67.** D

71. (a) $A \cdot A^{-1} = A^{-1} \cdot A = I_2$ **(c)** It is the inverse of A.

73. (b) The constant term equals $-\det A$.   **(c)** The coefficient of x^2 is the opposite of the sum of the elements of the main diagonal in A.

SECTION 7.3

Exploration 1

1. $x + y + z$ must equal 60 L.

3. The number of liters of 35% solution must equal twice the number of liters of 55% solution.

5. $\begin{bmatrix} 3.75 \\ 37.5 \\ 18.75 \end{bmatrix}$

Quick Review 7.3

1. 12.8 L   **3.** 38 L   **5.** $(-1, 6)$ **7.** $y = -z + w + 1$ **9.** $\begin{bmatrix} -0.5 & -0.75 \\ 0.5 & 0.25 \end{bmatrix}$

Exercises 7.3

1. $\left(\dfrac{25}{2}, \dfrac{7}{2}, -2\right)$ **3.** $(1, 2, 1)$ **5.** No solution

7. $\left(\dfrac{9}{2}, \dfrac{7}{2}, 4, -\dfrac{15}{2}\right)$ **9.** $\begin{bmatrix} 2 & -6 & 4 \\ 1 & 2 & -3 \\ 0 & -8 & 4 \end{bmatrix}$ **11.** $\begin{bmatrix} 0 & -10 & 10 \\ 1 & 2 & -3 \\ -3 & 1 & -2 \end{bmatrix}$ **13.** R_{12} **15.** $(-3)R_2 + R_3$

For Exercises 17–20, possible answers are given.

17. $\begin{bmatrix} 1 & 3 & -1 \\ 0 & 1 & -1.2 \\ 0 & 0 & 1 \end{bmatrix}$ **19.** $\begin{bmatrix} 1 & 2 & 3 & -4 \\ 0 & 1 & 0 & -0.6 \\ 0 & 0 & 1 & -9.2 \end{bmatrix}$ **21.** $\begin{bmatrix} 1 & 0 & 2 & 1 \\ 0 & 1 & -1 & 2 \\ 0 & 0 & 0 & 0 \end{bmatrix}$ **23.** $\begin{bmatrix} 1 & 0 & -1 & 3 \\ 0 & 1 & 2 & -1 \end{bmatrix}$

25. $\begin{bmatrix} 2 & -3 & 1 & 1 \\ -1 & 1 & -4 & -3 \\ 3 & 0 & -1 & 2 \end{bmatrix}$ **27.** $\begin{bmatrix} 2 & -5 & 1 & -1 & -3 \\ 1 & 0 & -2 & 1 & 4 \\ 0 & 2 & -3 & -1 & 5 \end{bmatrix}$

In Exercises 29–32, the variable names are arbitrary.

29. $\begin{aligned} 3x + 2y &= -1 \\ -4x + 5y &= 2 \end{aligned}$ **31.** $\begin{aligned} 2x + z &= 3 \\ -x + y &= 2 \\ 2y - 3z &= -1 \end{aligned}$

33. $(2, -1, 4);$ $\begin{bmatrix} 1 & -2 & 1 & 8 \\ 0 & 1 & -1 & -5 \\ 0 & 0 & 1 & 4 \end{bmatrix}$ **35.** $(-2, 3, 1)$

37. No solution   **39.** $(2 - z, 1 + z, z)$ **41.** no solution   **43.** $(z + w + 2, 2z - w - 1, z, w)$

45. $\begin{bmatrix} 2 & 5 \\ 1 & -2 \end{bmatrix}\begin{bmatrix} x \\ y \end{bmatrix} = \begin{bmatrix} -3 \\ 1 \end{bmatrix}$ **47.** $3x - y = -1 \quad 2x + 4y = 3$ **49.** $(-2, 3)$ **51.** $(-2, -5, -7)$ **53.** $(-1, 2, -2, 3)$ **55.** $(0, -10, 1)$

57. $(3, 3, -2, 0)$ **59.** $\left(2 - \dfrac{3}{2}z, -\dfrac{1}{2}z - 4, z\right)$ **61.** $(-2w - 1, w + 1, -w, w)$ **63.** $(-w - 2, -z + 0.5, z, w)$ **65.** No solution

67. $f(x) = 2x^2 - 3x - 2$ **69.** $f(x) = (-c - 3)x^2 + x + c$, for any c

71. (a) $y = 2.0734x + 234.0268$ **(b)** $y = 3.5302x + 141.7246$ **(c)** 2043   **73.** 825 children, 410 adults, 165 senior citizens

75. $14,500 CDs, $5500 bonds, $60,000 growth funds

67. $f(x) = 2x^2 - 3x - 2$ **69.** $f(x) = (-c - 3)x^2 + x + c$, for any c

71. (a) $y = 2.0734x + 234.0268$ **(b)** $y = 3.5302x + 141.7246$ **(c)** 2043

73. 825 children, 410 adults, 165 senior citizens

75. \$14,500 CDs, \$5500 bonds, \$60,000 growth funds

77. \$0 CDs, \$38,983.05 bonds, \$11,016.95 growth fund

79. 22 nickels, 35 dimes, and 17 quarters

81. $(16/3, 220/3)$ **85.** False. The determinant of the matrix must be not equal to zero   **87.** D   **89.** D

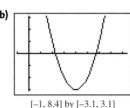

[−3, 30] by [0, 400]   [−3, 30] by [0, 400]

93. (a) $C(x) = x^2 - 8x + 13$ **(b)**   **(c)** $4 \pm \sqrt{3}$ **(d)** $\det A = C(0) = 13$ **(e)** $a_{11} + a_{22} = (4 - \sqrt{3}) + (4 + \sqrt{3}) = 8$

[−1, 8.4] by [−3.1, 3.1]

SECTION 7.4

Exploration 1

1. (a) $3 = A_2$ **(b)** $2 = A_1$

Quick Review 7.4

1. $\dfrac{3x - 5}{x^2 - 4x + 3}$ **3.** $\dfrac{4x^2 + 6x + 1}{x^3 + 2x^2 + x}$ **5.** $3x^2 - 2 + \dfrac{3}{x - 2}$ **7.** $(x + 1)(x - 3)(x^2 + 4)$ **9.** $A = 3, B = -1, C = 1$

Exercises 7.4

1. $\dfrac{A_1}{x} + \dfrac{A_2}{x - 2} + \dfrac{A_3}{x + 2}$ **3.** $\dfrac{A_1}{x} + \dfrac{A_2}{x^2} + \dfrac{A_3}{x^3} + \dfrac{A_4}{x - 1} + \dfrac{A_5}{(x - 1)^2} + \dfrac{B_1 x + C_1}{x^2 + 9}$ **5.** $\dfrac{-3}{x + 4} + \dfrac{4}{x - 2}$ **7.** $\dfrac{3}{x^2 + 1} + \dfrac{2x - 1}{(x^2 + 1)^2}$

9. $\dfrac{1}{x - 2} + \dfrac{2}{(x - 2)^2} + \dfrac{1}{(x - 2)^3}$ **11.** $\dfrac{2}{x + 3} + \dfrac{-1}{(x + 3)^2} + \dfrac{3x - 1}{x^2 + 2} + \dfrac{x + 2}{(x^2 + 2)^2}$ **13.** $\dfrac{1}{x - 5} + \dfrac{-1}{x - 3}$ **15.** $\dfrac{2}{x - 1} + \dfrac{-2}{x + 1}$

17. $\dfrac{1}{2x} + \dfrac{-1/2}{x + 2}$ **19.** $\dfrac{1}{x - 3} + \dfrac{-2}{x + 4}$ **21.** $\dfrac{-2}{x + 3} + \dfrac{5}{2x - 1}$ **23.** $\dfrac{2}{x^2 + 1} + \dfrac{3}{(x^2 + 1)^2}$ **25.** $\dfrac{2}{x} + \dfrac{-1}{x - 1} + \dfrac{2}{(x - 1)^2}$

27. $\dfrac{1}{x} + \dfrac{-1}{x^2} + \dfrac{2}{x - 3}$ **29.** $\dfrac{2x}{x^2 + 2} + \dfrac{-1}{(x^2 + 2)^2}$ **31.** $\dfrac{2}{x - 1} + \dfrac{-x}{x^2 + x + 1}$

33. $2 + \dfrac{x + 5}{x^2 - 1}; \dfrac{3}{x - 1} + \dfrac{-2}{x + 1}$

Graph of $(2x^2 + x + 3)/(x^2 - 1)$:   Graph of $3/(x - 1)$:   Graph of $-2/(x + 1)$:

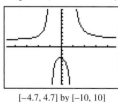   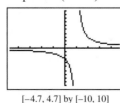   

[−4.7, 4.7] by [−10, 10]   [−4.7, 4.7] by [−10, 10]   [−4.7, 4.7] by [−10, 10]

35. $x - 1 + \dfrac{x - 2}{x^2 + x}; \dfrac{3}{x + 1} + \dfrac{-2}{x}$

Graph of $y = (x^3 - 2)/(x^2 + x)$ Graph of $y = x - 1$:   Graph of $y = 3/(x + 1)$:   Graph of $y = -2/x$:

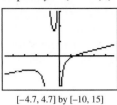   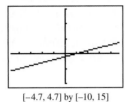   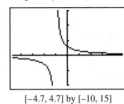   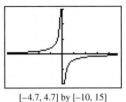

[−4.7, 4.7] by [−10, 15]   [−4.7, 4.7] by [−10, 15]   [−4.7, 4.7] by [−10, 15]   [−4.7, 4.7] by [−10, 15]

37. (c)   **39.** (d)   **41.** (a)   **43.** $\dfrac{-1}{ax} + \dfrac{1}{a(x - a)}$ **45.** $\dfrac{-3}{(b - a)(x - a)} + \dfrac{3}{(b - a)(x - b)}$

47. True. The behavior of f near $x = 3$ is the same as the behavior of $y = \dfrac{1}{x - 3}$ and $\lim\limits_{x \to 3^-} \dfrac{1}{x - 3} = -\infty$.   **49.** E   **51.** B

53. (a) $A = 3$ **(b)** $B = -2, C = 2$ **55.** $b/(x - 1)^2$

SECTION 7.5

Quick Review 7.5

1. $(3, 0); (0, -2)$ **3.** $(20, 0); (0, 50)$ **5.** $(30, 60)$ **7.** $(10, 140)$ **9.** $(3, 3)$

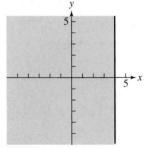   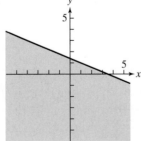

Exercises 7.5

1. Graph (c); boundary included   **3.** Graph (b); boundary included   **5.** Graph (e); boundary included

7.   **9.**   **11.**

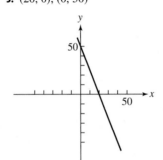   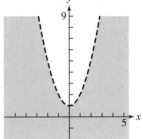

boundary $x = 4$ included   boundary $2x + 5y = 7$ included   boundary $y = x^2 + 1$ excluded

13.

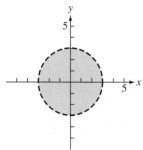

boundary $x^2 + y^2 = 9$ excluded

15.

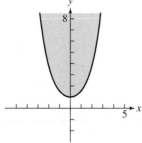

boundary $y = (e^x + e^{-x})/2$ included

17.

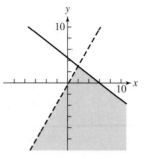

boundary $y = \sin x$ excluded

19.

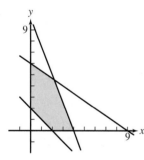

21.

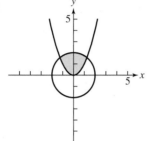

23.

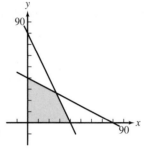

25.

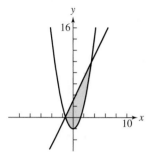

27. $x^2 + y^2 \leq 4$
$y \geq -x^2 + 1$

29. $y \leq -\dfrac{1}{2}x + 5$

$y \leq -\dfrac{3}{2}x + 9$

$x \geq 0$

$y \geq 0$

31. The minimum is 0 at (0, 0); the maximum is 880/3 at (160/3, 80/3).

33. The minimum is 162 at (6, 30); there is no maximum.

35. The minimum is 24 at (0, 12); there is no maximum.

37. ≈ 13.48 tons of ore R and ≈ 20.87 tons of ore S; \$1926.20

39. x operations at Refinery 1 and y operations at Refinery 2 such that $2x + 4y = 320$ with $40 \leq x \leq 120$.

41. False. It is a half-plane.   **43.** A   **45.** D

49. $y_1 = 2\sqrt{1 - \dfrac{x^2}{9}}$; $y_2 = -2\sqrt{1 - \dfrac{x^2}{9}}$

51.

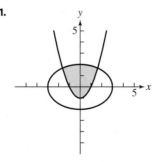

CHAPTER 7 REVIEW EXERCISES

1. (a) $\begin{bmatrix} 1 & 2 \\ 8 & 3 \end{bmatrix}$ **(b)** $\begin{bmatrix} -3 & 4 \\ 0 & -3 \end{bmatrix}$ **(c)** $\begin{bmatrix} 2 & -6 \\ -8 & 0 \end{bmatrix}$ **(d)** $\begin{bmatrix} -7 & 11 \\ 4 & -6 \end{bmatrix}$ **3.** $\begin{bmatrix} -3 & -7 & 11 \\ 0 & -12 & 24 \end{bmatrix}$; not possible **5.** [3 7]; not possible

7. $\begin{bmatrix} 1 & 2 & -3 \\ 2 & -3 & 4 \\ -2 & 1 & -1 \end{bmatrix}; \begin{bmatrix} -3 & 2 & 4 \\ 2 & 1 & -3 \\ 1 & -2 & -1 \end{bmatrix}$ **9.** $AB = BA = I_4$ **11.** $\begin{bmatrix} -2 & -5 & 6 & -1 \\ 0 & -1 & 1 & 0 \\ 10 & 24 & -27 & 4 \\ -3 & -7 & 8 & -1 \end{bmatrix}$ **13.** 20   **15.** $\begin{bmatrix} 1 & 0 & 2 \\ 0 & 1 & -1 \\ 0 & 0 & 0 \end{bmatrix}$

17. $\begin{bmatrix} 1 & 0 & 0 & 8 \\ 0 & 1 & 0 & -11 \\ 0 & 0 & 1 & 5 \end{bmatrix}$ **19.** $(1, 2)$ **21.** no solution   **23.** $(-z - w + 2, w + 1, z, w)$ **25.** No solution

27. $(-2z + w + 1, z - w + 2, z, w)$ **29.** $(9/4, -3/4, -7/4)$ **31.** no solution   **33.** $(-w + 2, z + 3, z, w)$ **35.** $(-2, 1, 3, -1)$

37. Demand curve $\approx (7.57, 42.71)$ Supply curve   **39.** $(x, y) \approx (0.14, -2.29)$ **41.** $(x, y) = (-2, 1)$ or $(x, y) = (2, 1)$

43. $(x, y) \approx (2.27, 1.53)$ **45.** $(a, b, c, d) = (17/840, -33/280, -571/420, 386/35)$ **47.** $\dfrac{1}{x + 1} + \dfrac{2}{x - 4}$

49. $\dfrac{-1}{x + 2} + \dfrac{1}{x + 1} + \dfrac{2}{(x + 1)^2}$ **51.** $\dfrac{2}{x + 1} + \dfrac{3x - 4}{x^2 + 1}$ **53.** (c)   **55.** (b)

57. 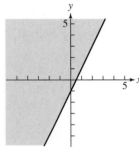   **59.** 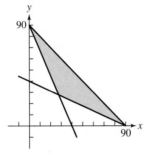   **61.** 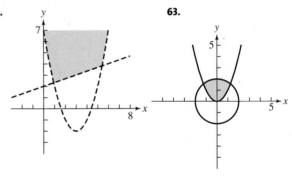   **63.**

Corners at $\approx(-1.25, 1.56)$
and $\approx(1.25, 1.56)$.
Boundaries included.

Corners at $(0, 90)$, $(90, 0)$,
$(360/13, 360/13)$.
Boundaries included.

Corners at $\approx(0.92, 2.31)$
and $\approx(5.41, 3.80)$.
Boundaries included.

65. The minimum is 106 at $(10, 6)$; there is no maximum.   **67.** The minimum is 205 at $(10, 25)$; the maximum is 292 at $(4, 40)$.

69. (a) $\approx(2.12, 0.71)$ **(b)** $\approx(-0.71, 2.12)$ **71. (a)** $y = 12.2614x + 979.5909$ **(b)** $y = 19.8270x + 893.9566$ **(c)** 1991

71. (a) $y = 12.2614x + 979.5909$ **(b)** $y = 19.8270x + 893.9566$

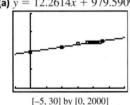   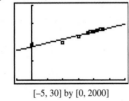

$[-5, 30]$ by $[0, 2000]$ $[-5, 30]$ by $[0, 2000]$

77. \$160,000 at 4%, \$170,000 at 6.5%, \$320,000 at 9%
79. Pipe A: 15 hours; Pipe B: $\approx$5.45 hours; Pipe C: 12 hours
81. n must be equal to p.

73. (a) $N = \begin{bmatrix} 200 & 400 & 600 & 250 \end{bmatrix}$ **(b)** $P = \begin{bmatrix} \$80 & \$120 & \$200 & \$300 \end{bmatrix}$ **(c)** $NP^T = \$259,000$ **75.** Answers will vary

Chapter 7 Project

1. 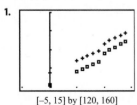   **3.** Yes; no; no   **5.** Males: $y \approx \dfrac{412.574}{1 + 10.956e^{-0.01539x}}$; Females: $y \approx \dfrac{315.829}{1 + 9.031e^{-0.01831x}}$; $(45, 64)$;

This represents the time when the female population became greater than the male population. $(159, 212)$; This represents the time when the male population will again become greater than the female population.

7. Approx. 49.1% male and 50.9% female

$[-5, 15]$ by $[120, 160]$

Males: $y \approx 1.7585x + 119.5765$

Females: $y \approx 1.6173x + 126.4138$

SECTION 8.1

Exploration 1

1. The axis of the parabola with focus (0, 1) and directrix $y = -1$ is the y-axis because it is perpendicular to $y = -1$ and passes through (0, 1). The vertex lies on this axis midway between the directrix and the focus, so the vertex is the point (0, 0).

3. $\{(-2\sqrt{6}, 6), (-2\sqrt{5}, 5), (-4, 4), (-2\sqrt{3}, 3), (-2\sqrt{2}, 2), (-2, 1), (0, 0), (2, 1), (2\sqrt{2}, 2), (2\sqrt{3}, 3), (4, 4), (2\sqrt{5}, 5), (2\sqrt{6}, 6)\}$

Exploration 2

1.
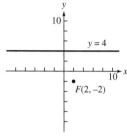

3.
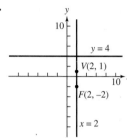

5.
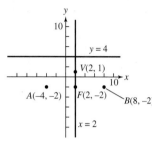

7. downward

Quick Review 8.1

1. $\sqrt{13}$ **3.** $y = \pm 2\sqrt{x}$ **5.** $y + 6 = -(x - 1)^2$

7. Vertex: (1, 5); $f(x)$ can be obtained from $g(x)$ by stretching x^2 by 3, shifting up 5 units, and shifting right 1 unit.

9. $f(x) = -2(x + 1)^2 + 3$

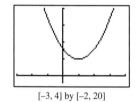

[-3, 4] by [-2, 20]

Exercises 8.1

1. Vertex: (0, 0); Focus: $\left(0, \dfrac{3}{2}\right)$; Directrix: $y = -\dfrac{3}{2}$; Focal width: 6   **3.** Vertex: (-3, 2); Focus: (-2, 2); Directrix: $x = -4$; Focal width: 4

5. Vertex: (0, 0); Focus: $\left(0, -\dfrac{1}{3}\right)$; Directrix: $y = \dfrac{1}{3}$; Focal width: $\dfrac{4}{3}$ **7.** (c)   **9.** (a)   **11.** $y^2 = -12x$ **13.** $x^2 = -16y$ **15.** $x^2 = 20y$

17. $y^2 = 8x$ **19.** $x^2 = -6y$ **21.** $(y + 4)^2 = 8(x + 4)$ **23.** $(x - 3)^2 = 6(y - 5/2)$ **25.** $(y - 3)^2 = -8(x - 4)$

27. $(x - 2)^2 = 16(y + 1)$ **29.** $(y + 4)^2 = -10(x + 1)$ **31.**

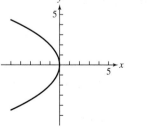

33.

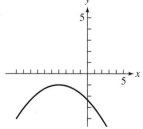

35.

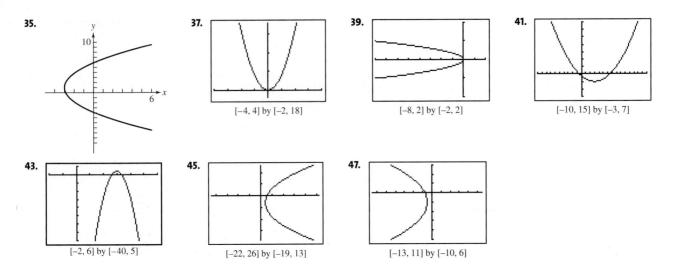

37. [−4, 4] by [−2, 18]   **39.** [−8, 2] by [−2, 2]   **41.** [−10, 15] by [−3, 7]

43. [−2, 6] by [−40, 5]   **45.** [−22, 26] by [−19, 13]   **47.** [−13, 11] by [−10, 6]

49. Completing the square, the equation becomes $(x + 1)^2 = y − 2$, a parabola with vertex $(−1, 2)$, focus $(−1, 9/4)$, and directrix $y = 7/4$.
51. Completing the square, the equation becomes $(y − 2)^2 = 8(x − 2)$, a parabola with vertex $(2, 2)$, focus $(4, 2)$, and directrix $x = 0$.
53. $(y − 2)^2 = −6x$ **55.** $(x − 2)^2 = −4(y + 1)$ **57.** The derivation only requires that p is a fixed real number.   **59.** The filament should be placed 1.125 cm from the vertex along the axis of the mirror.   **61.** The electronic receiver is located 2.5 units from the vertex along the axis of the parabolic microphone.   **63.** Starting at the leftmost tower, the lengths of the cables are: $\approx$ {79.44, 54.44, 35, 21.11, 12.78, 10, 12.78, 21.11, 35, 54.44, 79.44}.   **65.** False. Every point on a parabola is the same distance from its focus and its directrix.   **67.** D
69. B
71. (a)–(c)   **(d)** parabola

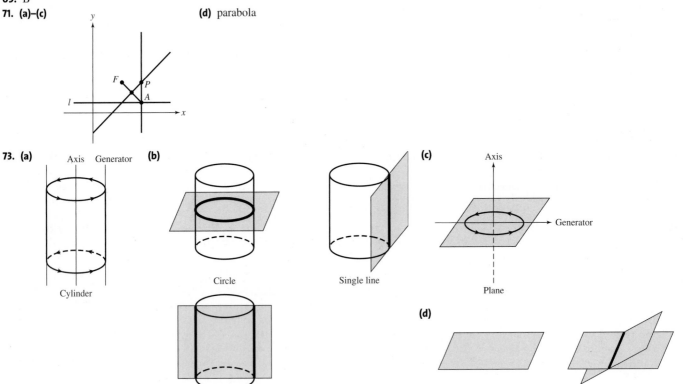

73. (a) Axis Generator Cylinder   **(b)** Circle Single line Two parallel lines   **(c)** Axis Generator Plane   **(d)** Plane Line

SECTION 8.2

Exploration 1

1. $x = -2 + 3\cos t$ and $y = 5 + 7\sin t$; $\cos t = \dfrac{x+2}{3}$ and $\sin t = \dfrac{y-5}{7}$; $\cos^2 t + \sin^2 t = 1$ yields the equation $\dfrac{(x+2)^2}{9} + \dfrac{(y-5)^2}{49} = 1.$

3. Example 1: $x = 3\cos t$ and $y = 2\sin t$
Example 2: $x = 2\cos t$ and $y = \sqrt{13}\sin t$
Example 3: $x = 3 + 5\cos t$ and $y = -1 + 4\sin t$

5. Example 1: $x = 3\cos t$, $y = 2\sin t$; $\cos t = \dfrac{x}{3}$, $\sin t = \dfrac{y}{2}$; $\cos^2 t + \sin^2 t = 1$ yields $\dfrac{x^2}{9} + \dfrac{y^2}{4} = 1$, or $4x^2 + 9y^2 = 36$.

Example 2: $x = 2\cos t$, $y = \sqrt{13}\sin t$; $\cos t = \dfrac{x}{2}$, $\sin t = \dfrac{y}{\sqrt{13}}$; $\sin^2 t + \cos^2 t = 1$ yields $\dfrac{y^2}{13} + \dfrac{x^2}{4} = 1$.

Example 3: $x = 3 + 5\cos t$, $y = -1 + 4\sin t$; $\cos t = \dfrac{x-3}{5}$, $\sin t = \dfrac{y+1}{4}$; $\cos^2 t + \sin^2 t = 1$ yields $\dfrac{(x-3)^2}{25} + \dfrac{(y+1)^2}{16} = 1$.

Exploration 2

3. $a = 8$ cm, $b \approx 7.75$ cm, $c = 2$ cm, $e = 0.25$, $b/a \approx 0.97$; $a = 7$ cm, $b \approx 6.32$ cm, $c = 3$ cm, $e \approx 0.43$, $b/a \approx 0.90$; $a = 6$ cm, $b \approx 4.47$ cm, $c = 4$ cm, $e \approx 0.67$, $b/a \approx 0.75$

5.

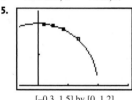

[−0.3, 1.5] by [0, 1.2]

$b/a = \sqrt{1 - e^2}$

Quick Review 8.2

1. $\sqrt{61}$ **3.** $y = \pm\dfrac{3}{2}\sqrt{4 - x^2}$ **5.** $x = 8$ **7.** $x = 2, x = -2$ **9.** $x = \dfrac{3 \pm \sqrt{15}}{2}$

Exercises 8.2

1. Vertices: $(4, 0), (-4, 0)$; Foci: $(3, 0), (-3, 0)$ **3.** Vertices: $(0, 6), (0, -6)$; Foci: $(0, 3), (0, -3)$ **5.** Vertices: $(2, 0), (-2, 0)$; Foci: $(1, 0),$ $(-1, 0)$ **7.** (d)   **9.** (a)   **11.**

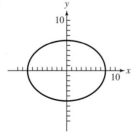

13.

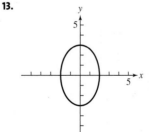

15.

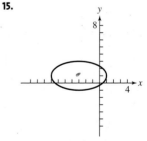

17.

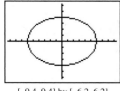

[−9.4, 9.4] by [−6.2, 6.2]

19.

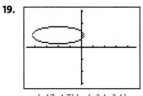

[−17, 4.7] by [−3.1, 3.1]

21. $\dfrac{y^2}{9} + \dfrac{x^2}{4} = 1$ **23.** $x^2/25 + y^2/21 = 1$ **25.** $\dfrac{y^2}{25} + \dfrac{x^2}{16} = 1$ **27.** $\dfrac{y^2}{36} + \dfrac{x^2}{16} = 1$ **29.** $\dfrac{x^2}{25} + \dfrac{y^2}{16} = 1$

31. $\dfrac{(y - 2)^2}{36} + \dfrac{(x - 1)^2}{16} = 1$ **33.** $(x - 3)^2/9 + (y + 4)^2/5 = 1$ **35.** $(y + 2)^2/25 + (x - 3)^2/9 = 1$

37. Center: $(-1, 2)$; Vertices: $(-6, 2)$, $(4, 2)$; Foci: $(-4, 2)$, $(2, 2)$ **39.** Center: $(7, -3)$; Vertices: $(7, 6)$, $(7, -12)$; Foci: $(7, -3 \pm \sqrt{17})$

45. Vertices: $(1, -4)$, $(1, 2)$; Foci: $(1, -1 \pm \sqrt{5})$; Eccentricity: $\dfrac{\sqrt{5}}{3}$

41.

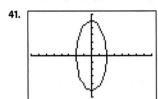

$[-8, 8]$ by $[-6, 6]$

$x = 2 \cos t,\ y = 5 \sin t$

43.

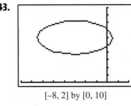

$[-8, 2]$ by $[0, 10]$

$x = 2\sqrt{3} \cos t - 3,$
$y = \sqrt{5} \sin t + 6$

47. Vertices: $(-7, 1)$, $(1, 1)$; Foci: $(-3 \pm \sqrt{7}, 1)$; Eccentricity: $\dfrac{\sqrt{7}}{4}$ Foci: $(4, -8 \pm \sqrt{3})$; Eccentricity: $\dfrac{\sqrt{3}}{2}$ **49.** $\dfrac{(x - 2)^2}{16} + \dfrac{(y - 3)^2}{9} = 1$

53. $a = 237{,}086.5$, $b \approx 236{,}571$, $c = 15{,}623.5$, $e \approx 0.066$ **55.** ≈ 1347 Gm, ≈ 1507 Gm   **57.** $a - c < 1.5(1.392) = 2.088$

59. $(\pm\sqrt{51.75}, 0) \approx (\pm 7.19, 0)$ **61.** $(-2, 0)$, $(2, 0)$

63. (a) Approximate solutions: $(\pm 1.04, -0.86)$, $(\pm 1.37, 0.73)$

(b) $\left(\pm \dfrac{\sqrt{94 - 2\sqrt{161}}}{8}, -\dfrac{1 + \sqrt{161}}{16}\right), \left(\pm \dfrac{\sqrt{94 + 2\sqrt{161}}}{8}, \dfrac{-1 + \sqrt{161}}{16}\right)$

65. False. The distance is $a(1 - e)$.   **67.** C   **69.** B

71. (a) When $a = b = r$, $A = \pi ab = \pi rr = \pi r^2$ and $P \approx \pi(2r) \cdot (3 - \sqrt{(4r)(4r)}/(2r)) = \pi(2r) \cdot (3 - 2) = 2\pi r$.   **(b)** Answers will vary.

73. (a)

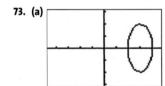

$[-4.7, 4.7]$ by $[-3.1, 3.1]$

(b) $y^2/4 + (x - 3)^2 = 1$

SECTION 8.3

Exploration 1

1. $x = -1 + 3/\cos t = -1 + 3 \sec t$ and $y = 1 + 2 \tan t$; $\sec t = \dfrac{x + 1}{3}$ and $\tan t = \dfrac{y - 1}{2}$; $\sec^2 t - \tan^2 t = 1$ yields the equation

$\dfrac{(x + 1)^2}{9} - \dfrac{(y - 1)^2}{4} = 1$.

3. Example 1: $x = 3/\cos t$, $y = 2 \tan t$; Example 2: $x = 2 \tan t$, $y = \sqrt{5}/\cos t$; Example 3: $x = 3 + 5/\cos t$, $y = -1 + 4 \tan t$; Example 4: $x = -2 + 3/\cos t$, $y = 5 + 7 \tan t$

5. Example 1: $x = 3/\cos t = 3 \sec t$, $y = 2 \tan t$; $\sec t = x/3$, $\tan t = y/2$; $\sec^2 t - \tan^2 t = 1$ yields $x^2/9 - y^2/4 = 1$,

or $4x^2 - 9y^2 = 36$. Example 2: $x = 2 \tan t$, $y = \sqrt{5}/\cos t = \sqrt{5} \sec t$; $\tan t = x/2$, $\sec t = y/\sqrt{5}$; $\sec^2 t - \tan^2 t = 1$

yields $y^2/5 - x^2/4 = 1$. Example 3: $x = 3 + 5/\cos t = 3 + 5 \sec t$, $y = -1 + 4 \tan t$; $\sec t = (x - 3)/5$, $\tan t = (y + 1)/4$;

$\sec^2 t - \tan^2 t = 1$ yields $(x - 3)^2/25 - (y + 1)^2/16 = 1$. Example 4: $x = -2 + 3/\cos t = -2 + 3 \sec t$, $y = 5 + 7 \tan t$;

$\sec t = (x + 2)/3$, $\tan t = (y - 5)/7$; $\sec^2 t - \tan^2 t = 1$ yields $(x + 2)^2/9 - (y - 5)^2/49 = 1$.

Quick Review 8.3

1. $\sqrt{146}$ **3.** $y = \pm\frac{4}{3}\sqrt{9 + x^2}$ **5.** no solution   **7.** $x = 2, x = -2$ **9.** $a = 3, c = 5$

Exercises 8.3

1. Vertices: $(\pm4, 0)$; Foci: $(\pm\sqrt{23}, 0)$ **3.** Vertices: $(0, \pm6)$; Foci: $(0, \pm7)$ **5.** Vertices: $(\pm2, 0)$; Foci: $(\pm\sqrt{7}, 0)$ **7.** (c)   **9.** (a)

11.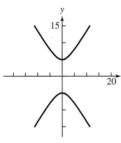

13.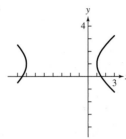

15.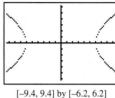

17.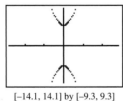
$[-9.4, 9.4]$ by $[-6.2, 6.2]$

19.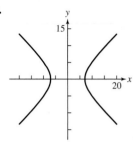
$[-9.4, 9.4]$ by $[-6.2, 6.2]$

21.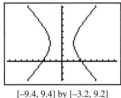
$[-9.4, 9.4]$ by $[-3.2, 9.2]$

23. $x^2/4 - y^2/5 = 1$ **25.** $y^2/16 - x^2/209 = 1$ **27.** $\dfrac{x^2}{25} - \dfrac{y^2}{75} = 1$

29. $\dfrac{y^2}{144} - \dfrac{x^2}{25} = 1$ **31.** $\dfrac{(y - 1)^2}{4} - \dfrac{(x - 2)^2}{9} = 1$

33. $\dfrac{(x - 2)^2}{9} - \dfrac{(y - 3)^2}{16} = 1$ **35.** $\dfrac{(x + 1)^2}{4} - \dfrac{(y - 2)^2}{5} = 1$

37. $\dfrac{(y - 6)^2}{25} - \dfrac{(x + 3)^2}{75} = 1$ **39.** Center: $(-1, 2)$; Vertices: $(11, 2)$, $(-13, 2)$; Foci: $(12, 2)$, $(-14, 2)$ **41.** Center: $(2, -3)$; Vertices: $(2, 5)$, $(2, -11)$; Foci: $(2, -3 \pm \sqrt{145})$

43.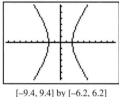
$[-14.1, 14.1]$ by $[-9.3, 9.3]$

45.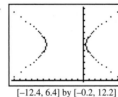
$[-12.4, 6.4]$ by $[-0.2, 12.2]$

47.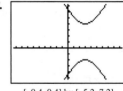
$[-9.4, 9.4]$ by $[-5.2, 7.2]$
Vertices: $(3, -2)$, $(3, 4)$; Foci: $(3, 1 \pm \sqrt{13})$; $e = \dfrac{\sqrt{13}}{3}$

49.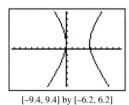
$[-9.4, 9.4]$ by $[-6.2, 6.2]$
Vertices: $(0, 1)$, $(4, 1)$; Foci: $(2 \pm \sqrt{13}, 1)$; $e = \dfrac{\sqrt{13}}{2}$

51. $\dfrac{x^2}{4} - \dfrac{5y^2}{16} = 1$

55. $a = 1440, b = 600, c = 1560, e = 13/12$; The Sun is centered at $(1560, 0)$.

57. A bearing and distance of about $40.29°$ and 1371.11 miles, respectively

59. $(-2, 0)$, $(4, 3\sqrt{3})$

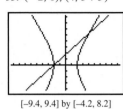

[−9.4, 9.4] by [−4.2, 8.2]

61. (a)

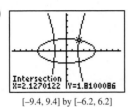

[−9.4, 9.4] by [−6.2, 6.2]

Four solutions:
$(\pm 2.13, \pm 1,81)$

(b) $\left(\pm 10 \sqrt{\dfrac{29}{641}}, \pm 10 \sqrt{\dfrac{21}{641}} \right)$

63. True, because $c - a = ae - a$. **65.** B **67.** B

69. (a–d)

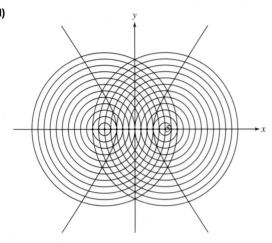

(e) $x^2/9 - y^2/16 = 1$

SECTION 8.4

Quick Review 8.4

1. $\cos 2\alpha = 5/13$ **3.** $\cos 2\alpha = 1/2$ **5.** $\alpha = \pi/4$ **7.** $\cos \alpha = 2/\sqrt{5}$ **9.** $\sin \alpha = 1/\sqrt{12}$

Exercises 8.4

1. $y = -5 \pm \sqrt{-x^2 + 6x + 7}$ **3.** $y = 4 \pm 2\sqrt{2x + 2}$

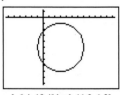

[−6.4, 12.4] by [−11.2, 1.2]

[−19.8, 17.8] by [−8.4, 16.4]

5. $y = 4/x$ **7.** $y = 8/(x - 1)$ **9.** $y = \dfrac{1}{6}(x - 4 \pm \sqrt{-23x^2 + 28x + 88})$

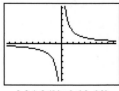

[−9.4, 9.4] by [−6.2, 6.2]

[−10, 12] by [−12, 12]

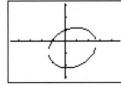

[−4.7, 4.7] by [−3.1, 3.1]

11. $y = \dfrac{1}{4}(x - 1 \pm \sqrt{3(-x^2 + 6x + 9)})$ **13.** $x^2 = -4y$ **15.** $\dfrac{x^2}{9} - \dfrac{y^2}{16} = 1$ **17.** $(x', y') = (4, -1)$ **19.** $(x', y') = (5, -3 - \sqrt{5})$

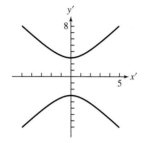

[–2, 8] by [–3, 3]

21. Hyperbola: $\dfrac{(y-1)^2}{9} - \dfrac{(x+1)^2}{4} = 1$; $\dfrac{(y')^2}{9} - \dfrac{(x')^2}{4} = 1$ **23.** Parabola: $(x+1)^2 = y - 2$; $(x')^2 = y'$

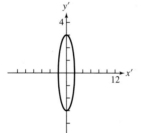

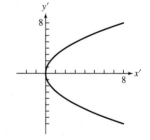

25. Ellipse: $\dfrac{(y+2)^2}{9} + \dfrac{(x-1)^2}{4} = 1$; $\dfrac{(y')^2}{9} + \dfrac{(x')^2}{4} = 1$ **27.** Parabola: $(y-2)^2 = 8(x-2)$; $(y')^2 = 8x'$

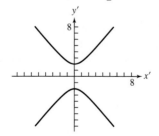

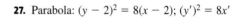

29. Hyperbola: $\dfrac{y^2}{4} - \dfrac{(x+1)^2}{2} = 1$; $\dfrac{(y')^2}{4} - \dfrac{(x')^2}{2} = 1$

31. The horizontal distance from O to P is $x = h + x' = x' + h$, and the vertical distance is $y = k + y' = y' + k$.

33. $(3\sqrt{2}/2, 7\sqrt{2}/2)$ **35.** $\approx (-5.94, 2.38)$

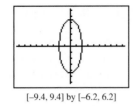

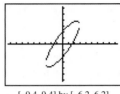

[−9.4, 9.4] by [−6.2, 6.2]   [−9.4, 9.4] by [−6.2, 6.2]   [−9.4, 9.4] by [−6.2, 6.2]

$\alpha \approx 0.954 \approx 54.65°$

43. $-24 < 0$; ellipse   **45.** 0; parabola   **47.** $-48 < 0$; ellipse   **49.** $12 > 0$; hyperbola   **51.** $-12 < 0$; ellipse

53. In the "old" coordinate system, the center is $(0, 0)$, the vertices are $\left(\dfrac{3\sqrt{2}}{2}, \dfrac{3\sqrt{2}}{2} \right)$ and $\left(-\dfrac{3\sqrt{2}}{2}, -\dfrac{3\sqrt{2}}{2} \right)$, and the foci are $(3, 3)$ and $(-3, -3)$.   **57.** True, because there is no xy term.   **59.** B   **61.** A

63. (a) $y = \pm x$ **(b)** $y = 2x + 3/2,\ y = (-1/2)x + 21/2$

69. Intersecting lines:

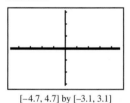

[−4.7, 4.7] by [−3.1, 3.1]

A plane containing the axis of a cone intersects the cone.

Parallel lines:

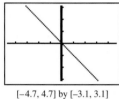

[−4.7, 4.7] by [−3.1, 3.1]

A degenerate cone is created by a generator that is parallel to the axis, producing a cylinder. A plane parallel to a generator of the cylinder intersects the cylinder and its interior.

One line:

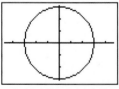

[−4.7, 4.7] by [−3.1, 3.1]

A plane containing a generator of a cone intersects the cone.

No graph:

[−4.7, 4.7] by [−3.1, 3.1]

A plane parallel to a generator of a cylinder fails to intersect the cylinder. Also, a degenerate cone is created by a generator that is perpendicular to the axis, producing a plane. A second plane perpendicular to the axis of this degenerate cone fails to intersect it.

Circle:

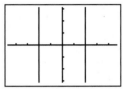

[−4.7, 4.7] by [−3.1, 3.1]

A plane perpendicular to the axis of a cone intersects the cone but not its vertex.

Point:

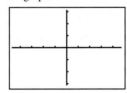

[−4.7, 4.7] by [−3.1, 3.1]

A plane perpendicular to the axis of a cone intersects the vertex of the cone.

SECTION 8.5

Exploration 1

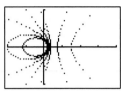

[−12, 24] by [−12, 12]

$e = 0.7$, $e = 0.8$: an ellipse; $e = 1$: a parabola; $e = 1.5$, $e = 3$: a hyperbola
The graphs have a common focus, $(0, 0)$, and a common directrix, the line $x = 3$. As e increases, the graphs move away from the focus and toward the directrix.

Quick Review 8.5

1. $r = -3$ **3.** $\theta = \dfrac{7\pi}{6}$, $\theta = -\dfrac{5\pi}{6}$ **5.** The focus is $(0, 4)$, and the directrix is $y = -4$.   **7.** Foci: $(\pm\sqrt{5}, 0)$; Vertices: $(\pm 3, 0)$

9. Foci: $(\pm 5, 0)$; Vertices: $(\pm 4, 0)$

Exercises 8.5

1. $r = \dfrac{2}{1 - \cos\theta}$; Parabola   **3.** $r = \dfrac{12}{5 + 3\sin\theta}$; Ellipse   **5.** $r = \dfrac{7}{3 - 7\sin\theta}$; Hyperbola

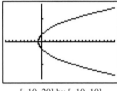

[−10, 20] by [−10, 10]

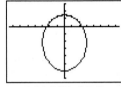

[−7.5, 7.5] by [−7, 3]

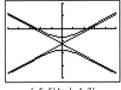

[−5, 5] by [−4, 2]

7. $e = 1$, Parabola; Directrix: $x = 2$ **9.** $e = 1$, Parabola; Directrix: $y = -\dfrac{5}{2} = -2.5$

11. $e = \dfrac{5}{6}$, Ellipse; Directrix: $y = 4$ **13.** $e = \dfrac{2}{5} = 0.4$, Ellipse; Directrix: $x = 3$

15. (b); $[-15, 5]$ by $[-10, 10]$ **17.** (f); $[-5, 5]$ by $[-3, 3]$ **19.** (c); $[-10, 10]$ by $[-5, 10]$

21. $r = \dfrac{12}{5 + 3\cos\theta}$ **23.** $r = \dfrac{3}{2 + \sin\theta}$ **25.** $r = \dfrac{15}{2 + 3\cos\theta}$

27. $r = \dfrac{12}{2 + 3\sin\theta}$ **29.** $r = \dfrac{6}{5 + 3\cos\theta}$

31.

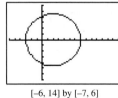

[−6, 14] by [−7, 6]

33.

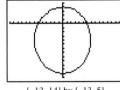

[−13, 14] by [−13, 5]

35.

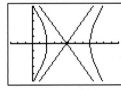

[−3, 12] by [−5, 5]

$e = 0.4$, $a = 5$, $b = \sqrt{21}$, $c = 2$ $e = \dfrac{1}{2}$, $a = 8$, $b = 4\sqrt{3}$, $c = 4$ $e = \dfrac{5}{3}$, $a = 3$, $b = 4$, $c = 5$

37. $\dfrac{9(y - 4/3)^2}{64} + \dfrac{3x^2}{16} = 1$ **39.** $y^2 = 4(x + 1)$

41. Perihelion distance ≈ 0.54 AU; Aphelion distance ≈ 35.64 AU   **43. (a)** $v \approx 1551$ m/sec $= 1.551$ km/sec   **(b)** about 2 hr 14 min

45. True. For a circle, $e = 0$, so the equation is $r = 0$, which graphs as a point.   **47.** D   **49.** B

51. (c)

Planet	Perihelion distance (AU)	Aphelion distance (AU)
Mercury	0.307	0.467
Venus	0.718	0.728
Earth	0.983	1.017
Mars	1.382	1.665
Jupiter	4.953	5.452
Saturn	9.020	10.090

(d) The difference is greatest for Saturn.

55. $5r - 3r \cos \theta = 16 \Rightarrow 5r = 3x + 16$. So, $25r^2 = 25(x^2 + y^2) = (3x + 16)^2$. $25x^2 + 25y^2 = 9x^2 + 96x + 256 \Rightarrow 16x^2 - 96x + 25y^2 = 256$.

Completing the square yields $\dfrac{(x - 3)^2}{25} + \dfrac{y^2}{16} = 1$, the desired result.

SECTION 8.6

Quick Review 8.6

1. $\sqrt{(x - 2)^2 + (y + 3)^2}$ **3.** P lies on the circle of radius 5 centered at $(2, -3)$.

5. $\left\langle \dfrac{-4}{\sqrt{41}}, \dfrac{5}{\sqrt{41}} \right\rangle$ **7.** Circle of radius 5 centered at $(-1, 5)$

9. Center: $(-1, 3)$; Radius: 2

Exercises 8.6

1.

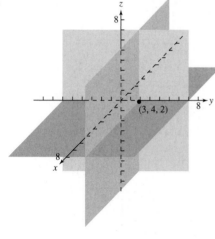

3.

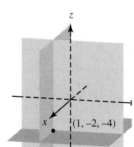

5. $\sqrt{53}$
7. $\sqrt{(a - 1)^2 + (b + 3)^2 + (c - 2)^2}$
9. $(1, -1, 11/2)$
11. $(x - 1, y + 4, z + 3)$
13. $(x - 5)^2 + (y + 1)^2 + (z + 2)^2 = 64$
15. $(x - 1)^2 + (y + 3)^2 + (z - 2)^2 = a$

17.

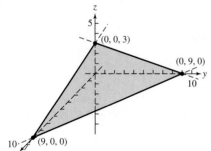

19.

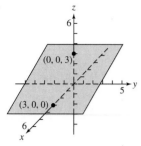

21.

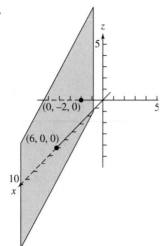

23. $\langle -2, 4, -8 \rangle$

25. -84

27. -20

29. $\left\langle \dfrac{4}{13}, -\dfrac{3}{13}, \dfrac{12}{13} \right\rangle$ **31.** $\langle -3, 4, -5 \rangle$ **33.** $\mathbf{v} = -195.01\mathbf{i} - 7.07\mathbf{j} + 68.40\mathbf{k}$ **35.** $\mathbf{r} = \langle 2, -1, 5 \rangle + t\langle 3, 2, -7 \rangle$; $x = 2 + 3t$, $y = -1 +$ $2t$, $z = 5 - 7t$ **37.** $\mathbf{r} = \langle 6, -9, 0 \rangle + t\langle 1, 0, -4 \rangle$; $x = 6 + t$, $y = -9$, $z = -4t$ **39.** $\sqrt{30}$ **41.** $\mathbf{r} = \langle -1, 2, 4 \rangle + t\langle 1, 4, -7 \rangle$

43. $x = -1 + 3t$, $y = 2 - 6t$, $z = 4 - 3t$ **45.** $x = \dfrac{1}{2}t$, $y = 6 - 7t$, $z = -3 + \dfrac{11}{2}t$ **47.** scalene

49. (a)

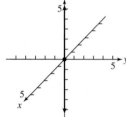

(b) the z-axis; a line through the origin in the direction $\mathbf{k}$

51. (a)

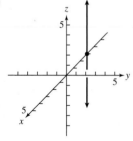

(b) the intersection of the xz plane ($y = 0$) and the plane $x = -3$; a line parallel to the z-axis through $(-3, 0, 0)$

53. $\mathbf{r} = \langle x_1 + (x_2 - x_1)t, y_1 + (y_2 - y_1)t, z_1 + (z_2 - z_1)t \rangle$

57. True. The equation can be viewed as an equation in three variables, where the coefficient of z is zero.

59. B   **61.** C   **65.** $\langle -1, -5, -3 \rangle$

67. $\mathbf{i} \times \mathbf{j} = \langle 1, 0, 0 \rangle \times \langle 0, 1, 0 \rangle = \langle 0 - 0, 0 - 0, 1 - 0 \rangle = \langle 0, 0, 1 \rangle = \mathbf{k}$

CHAPTER 8 REVIEW EXERCISES

1.

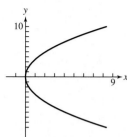

Vertex: (0, 0); Focus: (3, 0);
Directrix: $x = -3$; Focal width: 12

3.

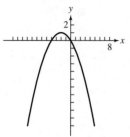

Vertex: $(-2, 1)$; Focus: $(-2, 0)$;
Directrix: $y = 2$; Focal width: 4

5.

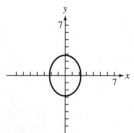

Ellipse; Center: (0, 0);
Vertices: $(0, \pm 2\sqrt{2})$; Foci: $(0, \pm\sqrt{3})$

7.

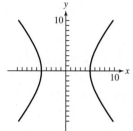

Hyperbola; Center: (0, 0);
Vertices: $(\pm 5, 0)$; Foci: $(\pm\sqrt{61}, 0)$

9.

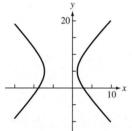

Hyperbola; Center: $(-3, 5)$; Vertices:
$(-3 \pm 3\sqrt{2}, 5)$; Foci: $(-3 \pm \sqrt{46}, 5)$

11.

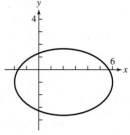

Ellipse; Center: $(2, -1)$; Vertices:
$(6, -1), (-2, -1)$; Foci: $(5, -1), (-1, -1)$

13. (b) **15.** (h) **17.** (f) **19.** (c)

21.

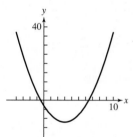

Parabola; $(x - 3)^2 = y + 12$

23.

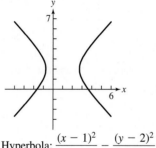

Hyperbola; $\dfrac{(x - 1)^2}{3} - \dfrac{(y - 2)^2}{3} = 1$

25.

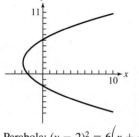

Parabola; $(y - 2)^2 = 6\left(x + \dfrac{17}{6}\right)$

27.

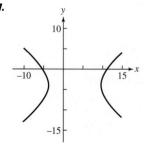

Hyperbola; $\dfrac{(y + 4)^2}{30} - \dfrac{(x - 3)^2}{45} = 1$

29. See proof on page 635.

31.

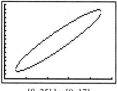

[0, 25] by [0, 17]

Ellipse; $y = \dfrac{1}{12}[8x + 5 \pm \sqrt{-8x^2 + 200x - 455}]$

33.

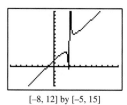

[−8, 12] by [−5, 15]

Hyperbola; $y = \dfrac{3x^2 - 5x - 10}{2x - 6}$

35.

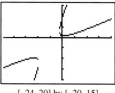

[−24, 20] by [−20, 15]

Hyperbola; $y = \dfrac{1}{4}[7x + 20 \pm \sqrt{25x^2 + 272x + 280}]$

37. $y^2 = 8x$ **39.** $(x + 3)^2 = 12(y - 3)$ **41.** $\dfrac{x^2}{169} + \dfrac{y^2}{25} = 1$ **43.** $x^2/9 + (y - 2)^2/5 = 1$

45. $\dfrac{y^2}{25} - \dfrac{x^2}{11} = 1$ **47.** $\dfrac{(x - 2)^2}{9} - \dfrac{(y - 1)^2}{16} = 1$ **49.** $x^2/25 + y^2/4 = 1$ **51.** $(x + 2)^2 + (y - 4)^2 = 1$ **53.** $x^2/9 - y^2/25 = 1$

55.

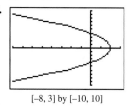

[−8, 3] by [−10, 10]

Parabola; $y^2 = -8(x - 2)$

57.

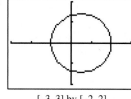

[−3, 3] by [−2, 2]

Ellipse; $\dfrac{4(x - 1/2)}{9} + \dfrac{y^2}{2} = 1$

59.

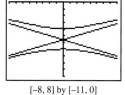

[−8, 8] by [−11, 0]

Hyperbola; $\dfrac{81(y + 49/9)^2}{196} - \dfrac{9x^2}{245} = 1$

61.

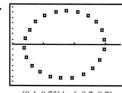

[−20, 4] by [−8, 8]

Parabola; $y^2 = -4(x - 1)$

63. $\sqrt{69}$

65. $\langle 0, -3, -2 \rangle$

67. -13

69. $\langle 3/5, -4/5, 0 \rangle$

71. $(x + 1)^2 + y^2 + (z - 3)^2 = 16$

73. $r = \langle -1, 0, 3 \rangle + t\langle -3, 1, -2 \rangle$

75. $(0, 4.5)$ **79.** At apogee, $v \approx 2633$ m/sec; At perigee, $v \approx 9800$ m/sec

Chapter 8 Project

Answers are based on the sample data provided.

1.

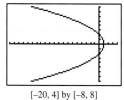

[0.4, 0.75] by [−0.7, 0.7]

3. With respect to the graph of the ellipse, the point (h, k) represents the center of the ellipse.

The value a is the semimajor axis, and b is the semiminor axis.

5. The parametric equations for the sample data set are $x_{1T} \approx 0.131 \sin(4.80T + 2.10) + 0.569$ and $y_{1T} \approx 0.639 \sin(4.80T - 2.65)$

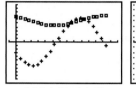
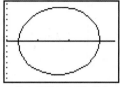

[−0.1, 1.4] by [−1, 1]   [0.4, 0.75] by [−0.7, 0.7]

SECTION 9.1

Exploration 1

1. 6   **3.** No

Quick Review 9.1

1. 52   **3.** 6   **5.** 10   **7.** 11   **9.** 64

Exercises 9.1

1. 6   **3.** 120   **5.** 12   **7.** 362,880 (ALGORITHM)   **9.** 34,650   **11.** 1716   **13.** 24   **15.** 30   **17.** 120   **19.** combinations
21. combinations   **23.** 19,656,000   **25.** 36   **27.** 2300   **29.** 17,296   **31.** 37,353,738,800   **33.** 41   **35.** 7776
37. 511   **39.** 12   **41.** 1024
43. True. Both equal $\dfrac{n!}{a!\,b!}$.   **45.** D   **47.** B
51. (a) 12   **(b)** There are 12 factors of 5 in 50!, one in each of 5, 10, 15, 20, 30, 35, 40, and 45 and two in each of 25 and 50.
        Each factor of 5, when paired with one of the 47 factors of 2, yields a factor of 10 and consequently a 0 at the end of 50!
55. 3   **57.** $\approx 20{,}123$ years

SECTION 9.2

Exploration 1

1. 1, 3, 3, 1; These are (in order) the coefficients in the expansion of $(a + b)^3$.
3. {1 5 10 10 5 1}; These are (in order) the coefficients in the expansion of $(a + b)^5$.

Quick Review 9.2

1. $x^2 + 2xy + y^2$ **3.** $25x^2 - 10xy + y^2$
5. $9s^2 + 12st + 4t^2$ **7.** $u^3 + 3u^2v + 3uv^2 + v^3$
9. $8x^3 - 36x^2y + 54xy^2 - 27y^3$

Exercises 9.2

1. $a^4 + 4a^3b + 6a^2b^2 + 4ab^3 + b^4$ **3.** $x^7 + 7x^6y + 21x^5y^2 + 35x^4y^3 + 35x^3y^4 + 21x^2y^5 + 7xy^6 + y^7$ **5.** $x^3 + 3x^2y + 3xy^2 + y^3$
7. $p^8 + 8p^7q + 28p^6q^2 + 56p^5q^3 + 70p^4q^4 + 56p^3q^5 + 28p^2q^6 + 8pq^7 + q^8$ **9.** 36   **11.** 1   **13.** 364   **15.** 126,720
17. $f(x) = x^5 - 10x^4 + 40x^3 - 80x^2 + 80x - 32$ **19.** $h(x) = 128x^7 - 448x^6 + 672x^5 - 560x^4 + 280x^3 - 84x^2 + 14x - 1$
21. $16x^4 + 32x^3y + 24x^2y^2 + 8xy^3 + y^4$ **23.** $x^3 - 6x^{5/2}y^{1/2} + 15x^2y - 20x^{3/2}y^{3/2} + 15xy^2 - 6x^{1/2}y^{5/2} + y^3$
25. $x^{-10} + 15x^{-8} + 90x^{-6} + 270x^{-4} + 405x^{-2} + 243$
35. True. The signs of the coefficients are determined by the powers of the $(-y)$.
37. C   **39.** A
41. (a) 1, 3, 6, 10, 15, 21, 28, 36, 45, 55   **(b)** They appear diagonally down the triangle, starting with either of the 1's in row 2.   **(c)**   **(d)**
43. $2^n = (1 + 1)^n = \dbinom{n}{0} 1^n 1^0 + \dbinom{n}{1} 1^{n-1} 1^1 + \dbinom{n}{2} 1^{n-2} 1^2 + \cdots + \dbinom{n}{n} 1^0 1^n = \dbinom{n}{0} + \dbinom{n}{1} + \dbinom{n}{2} + \cdots + \dbinom{n}{n}$

SECTION 9.3

Exploration 1

1.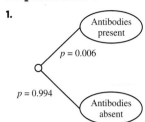

3.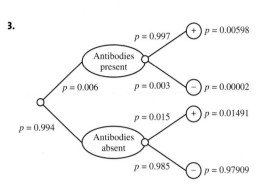

5. ≈ 0.286

Quick Review 9.3

1. 2   **3.** 8   **5.** 2,598,960   **7.** 120   **9.** $\dfrac{1}{12}$

Exercises 9.3

1. 1/9   **3.** 5/12   **5.** 1/4   **7.** 5/12   **9. (a)** No; the numbers do not add up to 1.   **(b)** Yes; assuming the gerbil cannot be in more than one compartment at a time, the proportions cannot sum to more than 1.

11. 0.4   **13.** 0.2   **15.** 0.7   **17.** 0.09   **19.** 0.08   **21.** 0.64   **23.** 1/134,596   **25.** 5/3542

27. (a)

0.3	0.3	0.2
A	B	
		0.2

(b) 0.3   **(c)** 0.2   **(d)** 0.2   **(e)** yes   **29.** 0.64   **31.** 3/5   **33.** 19/30   **35. (a)** 0.67   **(b)** 0.33

39. (a) 86/127   **(b)** 91/127   **(c)** 62/127   **41.** 1/36   **43.** 1/1024   **45.** 1/1024

47. 45/1024   **49.** 1023/1024   **51.** False. A sample space consists of outcomes, which are not necessarily equally likely.

53. D   **55.** A   **57. (a)**

Type of Bagel	Probability
Plain	0.37
Onion	0.12
Rye	0.11
Cinnamon Raisin	0.25
Sourdough	0.15

(b) ≈ 0.051

59. (a) ≈ 2 **(b)** yes   **(c)** $\approx 1.913\%$
61. (a) \$1.50   **(b)** 1/3

SECTION 9.4

Quick Review 9.4

1. 19   **3.** 80   **5.** 10/11   **7.** 2560   **9.** 15

Exercises 9.4

1. $2, \dfrac{3}{2}, \dfrac{4}{3}, \dfrac{5}{4}, \dfrac{6}{5}, \dfrac{7}{6}; \dfrac{101}{100}$ **3.** 0, 6, 24, 60, 120, 210; 999,900   **5.** 8, 4, 0, -4; -20 **7.** 2, 6, 18, 54; 4374   **9.** 2, -1, 1, 0; 3

11. Diverges   **13.** Converges to 0   **15.** Converges to -1 **17.** Converges to 0   **19.** Diverges

21. (a) 4   **(b)** 42   **(c)** $a_1 = 6$ and $a_n = a_{n-1} + 4$ for $n \geq 2$ **(d)** $a_n = 6 + 4(n-1)$

23. (a) 3   **(b)** 22   **(c)** $a_1 = -5$ and $a_n = a_{n-1} + 3$ for $n \geq 2$ **(d)** $a_n = -5 + 3(n-1)$

25. (a) 3   **(b)** 4374   **(c)** $a_1 = 2$ and $a_n = 3a_{n-1}$ for $n \geq 2$ **(d)** $a_n = 2 \cdot 3^{n-1}$

27. (a) -2 **(b)** -128 **(c)** $a_1 = 1$ and $a_n = -2a_{n-1}$ for $n \geq 2$ **(d)** $a_n = (-2)^{n-1}$

29. $a_1 = -20; a_n = a_{n-1} + 4$ for $n \geq 2$ **31.** $a_1 = \pm\dfrac{3}{2}, r = \pm 2$, and $a_n = 3(\pm 2)^{n-2}$

33.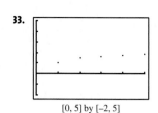

[0, 5] by [−2, 5]

35.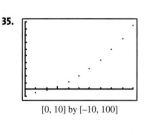

[0, 10] by [−10, 100]

37. 700, 702.3, 704.6, 706.9, . . . , 815, 817.3   **39.** 775   **41.** 9

43. True. The common ratio r must be positive, so the sign of the first term determines the sign of every number in the sequence.

45. A   **47.** E   **49. (b)** 3, 5, 8, 13, 21, 34, 55, 89, 144, 233

51. (a)

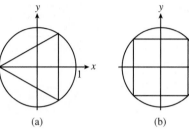

(a)   (b)

(b) $a_n \to 2\pi$

as $n \to \infty$

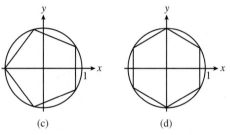

(c)   (d)

55. $a_1 = [1 \quad 1]$, $a_2 = [1 \quad 2]$, $a_3 = [2 \quad 3]$, $a_4 = [3 \quad 5]$, $a_5 = [5 \quad 8]$, $a_6 = [8 \quad 13]$, $a_7 = [13 \quad 21]$. The entries in the terms of this sequence are successive pairs of terms from the Fibonacci sequence.

SECTION 9.5

Exploration 1

1. 45   **3.** 1   **5.** $\dfrac{1}{3}$

Exploration 2

1. $1 + 2 + 3 + \cdots + 99 + 100$ **3.** 101

5. The sum in 4 involves two copies of the same progression, so it doubles the sum of the progression. The answer is 5050.

Quick Review 9.5

1. 22   **3.** 27   **5.** 512   **7.** −40   **9.** 55

Exercises 9.5

1. $\displaystyle\sum_{k=1}^{11} (6k - 13)$ **3.** $\displaystyle\sum_{k=1}^{n+1} k^2$ **5.** $\displaystyle\sum_{k=0}^{\infty} 6(-2)^k$ **7.** 18   **9.** 3240   **11.** 975   **13.** 24,573   **15.** $50.4(1 - 6^{-9}) \approx 50.4$ **17.** 155

19. $\dfrac{8}{3}(1 - 2^{-12}) \approx 2.666$ **21.** −196,495,641   **23. (a)** 0.3, 0.33, 0.333, 0.3333, 0.33333, 0.333333; convergent   **(b)** 1, −1, 2, −2, 3, −3; divergent

25. Yes; 12   **27.** no   **29.** yes; 1   **31.** 707/99   **33.** $-\dfrac{17{,}251}{999}$ **35. (a)** 1.1   **(b)** $20{,}000(1.1)^n$ **(c)** \$370,623.34

37. (a) 120; $1 + 0.07/12$ **(b)** \$20,770.18   **39.** $\approx$ 36 m

41. False. The series might well diverge.

diverge, but $\displaystyle\sum_{n=1}^{8} (n + (-n))$ is constant and converges to 0.

43. A   **45** C   **47. (a)** Heartland: 19,237,759 people; Southeast: 42,614,977 people   **(b)** Heartland: 517,825 mi²; Southeast: 348,999 mi²

(c) Heartland: $\approx$ 37.15 people/mi²; Southeast: $\approx$ 122.11 people/mi²

49.

n	F_n	S_n	$F_{n+2} - 1$
1	1	1	1
2	1	2	2
3	2	4	4
4	3	7	7
5	5	12	12
6	8	20	20
7	13	33	33
8	21	54	54
9	34	88	88

Conjecture: $S_n = F_{n+2} - 1$

SECTION 9.6

Exploration 1

Start with the rightmost peg if n is odd and the middle peg if n is even.

Exploration 2

1. Yes   **3.** Still all prime

Quick Review 9.6

1. $n^2 + 5n$ **3.** $k^3 + 3k^2 + 2k$ **5.** $(k + 1)^3$ **7.** $5; t + 4; t + 5$ **9.** $\dfrac{1}{2}; \dfrac{2k}{3k + 1}; \dfrac{2k + 2}{3k + 4}$

Exercises 9.6

1. P_n: $2 + 4 + 6 + \cdots + 2n = n^2 + n$. P_1 is true: $2(1) = 1^2 + 1$. Now assume P_k is true: $2 + 4 + 6 + \cdots + 2k = k^2 + k$.
Add $2(k + 1)$ to both sides: $2 + 4 + 6 + \cdots + 2k + 2(k + 1) = k^2 + k + 2(k + 1) = k^2 + 3k + 2 = k^2 + 2k + 1 + k + 1 =$
$(k + 1)^2 + (k + 1)$, so P_{k+1} is true. Therefore, P_n is true for all $n \geq 1$.

3. P_n: $6 + 10 + 14 + \cdots + (4n + 2) = n(2n + 4)$. P_1 is true: $4(1) + 2 = 1(2(1) + 4)$.
Now assume P_k is true: $6 + 10 + 14 + \cdots + (4k + 2) = k(2k + 4)$. Add $4(k + 1) + 2 = 4k + 6$ to both sides:
$6 + 10 + 14 + \cdots + (4k + 2) + [4(k + 1) + 2] = k(2k + 4) + 4k + 6 = 2k^2 + 8k + 6 = (k + 1)(2k + 6) = (k + 1)[2(k + 1) + 4]$,
so P_{k+1} is true. Therefore, P_n is true for all $n \geq 1$.

5. P_n: $a_n = 5n - 2$. P_1 is true: $a_1 = 5 \cdot 1 - 2 = 3$. Now assume P_k is true: $a_k = 5k - 2$. To get a_{k+1}, add 5 to a_k;
that is, $a_{k+1} = (5k - 2) + 5 = 5(k + 1) - 2$. This shows that P_{k+1} is true. Therefore, P_n is true for all $n \geq 1$.

7. P_n: $a_n = 2 \cdot 3^{n-1}$. P_1 is true: $a_1 = 2 \cdot 3^{1-1} = 2 \cdot 3^0 = 2$. Now assume P_k is true: $a_k = 2 \cdot 3^{k-1}$. To get a_{k+1}, multiply a_k by 3;
that is, $a_{k+1} = 3 \cdot 2 \cdot 3^{k-1} = 2 \cdot 3k = 2 \cdot 3^{(k+1)-1}$. This shows that P_{k+1} is true. Therefore, P_n is true for all $n \geq 1$.

9. P_1: $1 = \dfrac{1(1 + 1)}{2}$. P_k: $1 + 2 + \cdots + k = \dfrac{k(k + 1)}{2}$. P_{k+1}: $1 + 2 + \cdots + k + (k + 1) = \dfrac{(k + 1)(k + 2)}{2}$.

11. P_1: $\dfrac{1}{1 \cdot 2} = \dfrac{1}{1 + 1}$. P_k: $\dfrac{1}{1 \cdot 2} + \dfrac{1}{2 \cdot 3} + \cdots + \dfrac{1}{k(k + 1)} = \dfrac{k}{k + 1}$.
P_{k+1}: $\dfrac{1}{1 \cdot 2} + \dfrac{1}{2 \cdot 3} + \cdots + \dfrac{1}{k(k + 1)} + \dfrac{1}{(k + 1)(k + 2)} = \dfrac{k + 1}{k + 2}$.

13. P_n: $1 + 5 + 9 + \cdots + (4n - 3) = n(2n - 1)$. P_1 is true: $4(1) - 3 = 1 \cdot (2 \cdot 1 - 1)$.
Now assume P_k is true: $1 + 5 + 9 + \cdots + (4k - 3) = k(2k - 1)$. Add $4(k + 1) - 3 = 4k + 1$ to both sides: $1 + 5 + 9 + \cdots$
$+ (4k - 3) + [4(k + 1) - 3] = k(2k - 1) + 4k + 1 = 2k^2 + 3k + 1 = (k + 1)(2k + 1) = (k + 1)[2(k + 1) - 1]$,
so P_{k+1} is true. Therefore, P_n is true for all $n \geq 1$.

15. P_n: $\dfrac{1}{1 \cdot 2} + \dfrac{1}{2 \cdot 3} + \cdots + \dfrac{1}{n(n+1)} = \dfrac{n}{n+1}$. P_1 is true: $\dfrac{1}{1 \cdot 2} = \dfrac{1}{1+1}$.

Now assume P_k is true: $\dfrac{1}{1 \cdot 2} + \dfrac{1}{2 \cdot 3} + \cdots + \dfrac{1}{k(k+1)} = \dfrac{k}{k+1}$.

Add $\dfrac{1}{(k+1)(k+2)}$ to both sides: $\dfrac{1}{1 \cdot 2} + \dfrac{1}{2 \cdot 3} + \cdots + \dfrac{1}{k(k+1)} + \dfrac{1}{(k+1)(k+2)} = \dfrac{k}{k+1} + \dfrac{1}{(k+1)(k+2)}$

$= \dfrac{k(k+2)+1}{(k+1)(k+2)} = \dfrac{(k+1)(k+1)}{(k+1)(k+2)} = \dfrac{k+1}{k+2} = \dfrac{k+1}{(k+1)+1}$, so P_{k+1} is true. Therefore, P_n is true for all $n \geq 1$.

17. P_n: $2^n \geq 2n$. P_1 is true: $2^1 \geq 2 \cdot 1$ (in fact, they are equal). Now assume P_k is true: $2^k \geq 2k$.
Then $2^{k+1} = 2 \cdot 2^k \geq 2 \cdot 2k = 2 \cdot (k+k) \geq 2(k+1)$, so P_{k+1} is true. Therefore, P_n is true for all $n \geq 1$.

19. P_n: 3 is a factor of $n^3 + 2n$. P_1 is true: 3 is a factor of $1^3 + 2 \cdot 1 = 3$. Now assume P_k is true: 3 is a factor of $k^3 + 2k$.
Then $(k+1)^3 + 2(k+1) = (k^3 + 3k^2 + 3k + 1) + (2k+2) = (k^3 + 2k) + 3(k^2 + k + 1)$. Since 3 is a factor of both terms,
it is a factor of the sum, so P_{k+1} is true. Therefore, P_n is true for all $n \geq 1$.

21. P_n: The sum of the first n terms of a geometric sequence with first term a_1 and common ratio $r \neq 1$ is $\dfrac{a_1(1-r^n)}{1-r}$.

P_1 is true: $a_1 = \dfrac{a_1(1-r^1)}{1-r}$. Now assume P_k is true so that $a_1 + a_1 r + \cdots + a_1 r^{k-1} = \dfrac{a_1(1-r^k)}{(1-r)}$.

Add $a_1 r^k$ to both sides: $a_1 + a_1 r + \cdots + a_1 r^{k-1} + a_1 r^k = \dfrac{a_1(1-r^k)}{(1-r)} + a_1 r^k = \dfrac{a_1(1-r^k) + a_1 r^k(1-r)}{1-r}$

$= \dfrac{a_1 - a_1 r^k + a_1 r^k - a_1 r^{k+1}}{1-r} = \dfrac{a_1 - a_1 r^{k+1}}{1-r}$, so P_{k+1} is true. Therefore, P_n is true for all positive integers n.

23. P_n: $\displaystyle\sum_{k=1}^{n} k = \dfrac{n(n+1)}{2}$. P_1 is true: $\displaystyle\sum_{k=1}^{1} k = 1 = \dfrac{1 \cdot 2}{2}$. Now assume P_k is true: $\displaystyle\sum_{i=1}^{k} i = \dfrac{k(k+1)}{2}$.

Add $(k+1)$ to both sides, and we have $\displaystyle\sum_{i=1}^{k+1} i = \dfrac{k(k+1)}{2} + (k+1) = \dfrac{k(k+1)}{2} + \dfrac{2(k+1)}{2} = \dfrac{(k+1)(k+2)}{2}$

$= \dfrac{(k+1)((k+1)+1)}{2}$, so P_{k+1} is true. Therefore, P_n is true for all $n \geq 1$.

25. 125,250   **27.** $\dfrac{(n-3)(n+4)}{2}$ **29.** $\approx 3.44 \times 10^{10}$ **31.** $\dfrac{n(n^2 - 3n + 8)}{3}$ **33.** $\dfrac{n(n-1)(n^2 + 3n + 4)}{4}$

35. The inductive step does not work for 2 people. Sending them alternately out of the room leaves 1 person (and one blood type) each time, but we cannot conclude that their blood types will match *each other.*

37. False. Mathematical induction is used to show that a statement P_n is true for all positive integers.   **39.** E   **41.** B

43. P_n: 2 is a factor of $(n+1)(n+2)$. P_1 is true because 2 is a factor of $(2)(3)$. Now assume P_k is true so that 2 is a factor of $(k+1)(k+2)$. Then $[(k+1)+1][(k+1)+2] = (k+2)(k+3) = k^2 + 5k + 6 = k^2 + 3k + 2 + 2k + 4$
$= (k+1)(k+2) + 2(k+2)$. Since 2 is a factor of both terms of this sum, it is a factor of the sum, and so P_{k+1} is true. Therefore, P_n is true for all positive integers n.

45. Given any two consecutive integers, one of them must be even. Therefore, their product is even. Since $n+1$ and $n+2$ are consecutive integers, their product is even. Therefore, 2 is a factor of $(n+1)(n+2)$.

47. P_n: $F_{n+2} - 1 = \displaystyle\sum_{k=1}^{n} F_k$. P_1 is true since $F_{1+2} - 1 = F_3 - 1 = 2 - 1 = 1$, which equals $\displaystyle\sum_{k=1}^{1} F_k = 1$.

Now assume that P_k is true: $F_{k+2} - 1 = \displaystyle\sum_{i=1}^{k} F_i$. Then $F_{(k+1)+2} - 1 = F_{k+3} - 1 = F_{k+1} + F_{k+2} - 1$

$= (F_{k+2} - 1) + F_{k+1} = \left(\displaystyle\sum_{i=1}^{k} F_i\right) + F_{k+1} = \displaystyle\sum_{i=1}^{k+1} F_i$, so P_{k+1} is true. Therefore, P_n is true for all $n \geq 1$.

49. P_n: $a - 1$ is a factor of $a^n - 1$. P_1 is true because $a - 1$ is a factor of $a - 1$. Now assume P_k is true so that $a - 1$ is a factor of $a^k - 1$. Then $a^{k+1} - 1 = a \cdot a^k - 1 = a(a^k - 1) + (a - 1)$. Since $a - 1$ is a factor of both terms in the sum, it is a factor of the sum, and so P_{k+1} is true. Therefore, P_n is true for all positive integers n.

51. P_n: $3n - 4 \geq n$ for $n \geq 2$. P_2 is true since $3 \cdot 2 - 4 \geq 2$. Now assume that P_k is true: $3k - 4 \geq k$.
Then $3(k+1) - 4 = 3k + 3 - 4 = (3k - 4) + 3 \geq k + 3 \geq k + 1$, so P_{k+1} is true. Therefore, P_n is true for all $n \geq 2$.

53. Use P_3 as the anchor and obtain the inductive step by representing any n-gon as the union of a triangle and an $(n-1)$-gon.

SECTION 9.7

Exploration 1

1. The average is about 12.8.   **3.** Alaska, Colorado, Georgia, Texas, and Utah

Quick Review 9.7

1. ≈ 15.48%   **3.** ≈ 14.44%   **5.** ≈ 1723   **7.** \$235 thousand   **9.** 1 million

Exercises 9.7

1.
```
0 | 5 8 9
1 | 3 4 6
2 | 3 6 8
3 | 3 9
4 |
5 |
6 | 1
```
61 is an outlier.

3.
```
        Males
6 | 3  0
6 | 7  8  8  8
7 | 1  2  2  3  3  3
7 |
```

5.
```
               Males                    Females
                          3  0 | 6 |
           8  8  8  7  6 | 5 | 8
  3  3  3  2  2  1 | 7 | 1  2
                   | 7 | 5  6  7  7  9  9
                   | 8 | 0  0
```

7.

Life expectancy (years)	Frequency
60.0–64.9	2
65.0–69.9	4
70.0–74.9	6

9.

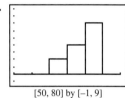

[50, 80] by [−1, 9]

11.

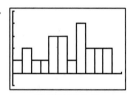

[0, 60] by [−1, 5]

13.

[−1, 25] by [−5, 60]

15.

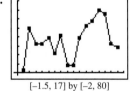

[−1.5, 17] by [−2, 80]

17.

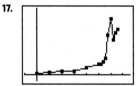

[1965, 2008] by [−1000, 11000]

The top male's earnings appear to be growing exponentially, with unusually high earnings in 1999 and 2000. Since the graph only shows the earnings of the top player (as opposed to a mean or median for all players), it can behave strangely if the top player has a very good year—as Tiger Woods did in 1999 and 2000.

The two home run hitters enjoyed similar success.

19.

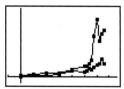

[1965, 2008] by [−1000, 11000]

After approaching parity in 1985, the top PGA player's earnings have grown much faster than the top LPGA player's earnings, even if the unusually good years for Tiger Woods (1999 and 2000) are not considered part of the trend.

21.

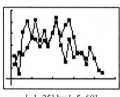

[−1, 25] by [−5, 60]

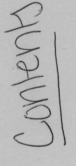

23. (a)

```
28 | 2
29 | 3 7
30 |
31 | 6 7
32 | 7 8
33 | 5 5 5 8
34 | 2 8 8
35 | 3 3 4
36 | 3 7
37 |
38 | 5
```

(b)

Interval	Frequency
25.0–29.9	3
30.0–34.9	11
35.0–39.9	6

(c)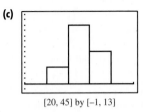

[20, 45] by [−1, 13]

(d) Time is not a variable in the data.

25.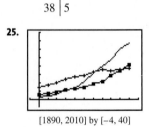

[1890, 2010] by [−4, 40]

——— = CA; + = NY, ■ = TX

27. False. The empty branches are important for visualizing the distribution of the data.

29. C   **31.** A   **35.**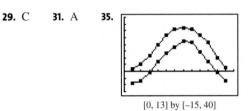

[0, 13] by [−15, 40]

SECTION 9.8

Exploration 1

1. Figure (b)

Quick Review 9.8

1. $x_1 + x_2 + x_3 + x_4 + x_5 + x_6 + x_7$ **3.** $\frac{1}{7}(x_1 + x_2 + x_3 + x_4 + x_5 + x_6 + x_7)$ **5.** $\frac{1}{5}[(x_1 - \bar{x})^2 + (x_2 - \bar{x})^2 + \cdots + (x_5 - \bar{x})^2]$

7. $\sum_{i=1}^{8} x_i f_i$ **9.** $\frac{1}{50} \sum_{i=1}^{50} (x_i - \bar{x})^2$

Exercises 9.8

1. (a) statistic **(b)** parameter **(c)** statistic   **3.** 26.8   **5.** ≈ 60.12 **7.** 3.9 million   **9.** ≈ 15.2 satellites   **11.** 2
13. 30 runs/yr; ≈ 29.8 runs/yr; Mays   **15.** What-Next Fashion
17. median: 87.85; mode: None   **19.** ≈ 3.61 **21. (a)** $\approx 6.42°C$ **(b)** $\approx 6.49°C$ **(c)** The weighted average is the better indicator.
23. Willie Mays: Five-number summary: {4, 20, 31.5, 40, 52}; Range: 48; IQR: 20; No outliers;
Mickey Mantle: Five-number summary: {13, 21, 28.5, 37, 54}; Range: 41; IQR: 16; No outliers
25. {28.2, 31.7, 33.5, 35.3, 38.5}; 10.3; 3.6; No outliers
27. (a)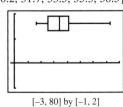

[−3, 80] by [−1, 2]

(b)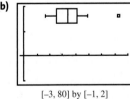

[−3, 80] by [−1, 2]

29. 3/11   **31. (a)** Mays **(b)** Mays   **33.** $\sigma \approx 9.08$; $\sigma^2 = 82.5$ **35.** $\sigma \approx 186.62$; $\sigma^2 \approx 34828.12$ **37.** $\sigma \approx 1.53$; $\sigma^2 \approx 2.34$
39. no   **41. (a)** 68% **(b)** 2.5% **(c)** A parameter   **43.** False. The median is a resistant measure.   **45.** A   **47.** B
49. There are many possible answers; examples are given. **(a)** {2, 2, 2, 3, 6, 8, 20} **(b)** {1, 2, 3, 4, 6, 48, 48} **(c)** {−20, 1, 1, 1, 2, 3, 4, 5, 6}
51. No   **55.** 75.9 years   **57.** 5%

CHAPTER 9 REVIEW EXERCISES

1. 792   **3.** 18,564   **5.** 3,991,680   **7.** 43,670,016   **9.** 14,508,000   **11.** 8,217,822,536   **13.** 26   **15.** 325

17. (a) 5040; Meg Ryan   **(b)** 778,377,600; Britney Spears   **19.** $32x^5 + 80x^4y + 80x^3y^2 + 40x^2y^3 + 10xy^4 + y^5$

21. $243x^{10} + 405x^8y^3 + 270x^6y^6 + 90x^4y^9 + 15x^2y^{12} + y^{15}$

23. $512a^{27} - 2304a^{24}b^2 + 4608a^{21}b^4 - 5376a^{18}b^6 + 4032a^{15}b^8 - 2016a^{12}b^{10} + 672a^9b^{12} - 144a^6b^{14} + 18a^3b^{16} - b^{18}$

25. -1320 **27.** $\{1, 2, 3, 4, 5, 6\}$ **29.** $\{13, 16, 31, 36, 61, 63\}$ **31.** $\{HHH, HHT, HTH, HTT, THH, THT, TTH, TTT\}$

33. $\{HHH, TTT\}$ **35.** 1/64   **37.** 1/4   **39.** 0.25   **41.** 0.24   **43.** 0.64   **45. (a)** 0.5   **(b)** 0.15   **(c)** 0.35   **(d)** ≈ 0.43

47. 0, 1, 2, 3, 4, 5; 39   **49.** $-1, 2, 5, 8, 11, 14; 32$ **51.** $-5, -3.5, -2, -0.5, 1, 2.5; 11.5$ **53.** $-3, 1, -2, -1, -3, -4; -76$

55. Arithmetic with $d = -2.5$; $a_n = 14.5 - 2.5n$ **57.** Geometric with $r = 1.2$; $a_n = 10 \cdot (1.2)^{n-1}$

59. Arithmetic with $d = 4.5$; $a_n = 4.5n - 15.5$ **61.** $a_n = 3(-4)^{n-1}$; $r = -4$ **63.** -4 **65.** -985.5 **67.** 21/8   **69.** 59,048

71. $3280.\overline{4}$ **73.**

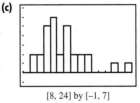

[0, 15] by [0, 2]

75. \$27,441.91   **77.** converges; 6   **79.** diverges   **81.** converges; 3

83. $\displaystyle\sum_{k=1}^{21} (5k - 13)$ **85.** $\displaystyle\sum_{k=0}^{\infty} (2k + 1)^2$ or $\displaystyle\sum_{k=1}^{\infty} (2k - 1)^2$ **87.** $\dfrac{n(3n + 5)}{2}$ **89.** 4650

91. $P_n: 1 + 3 + 6 + \cdots + \dfrac{n(n + 1)}{2} = \dfrac{n(n + 1)(n + 2)}{6}$. P_1 is true: $\dfrac{1(1 + 1)}{2} = \dfrac{1(1 + 1)(1 + 2)}{6}$.

Now assume P_k is true: $1 + 3 + 6 + \cdots + \dfrac{k(k + 1)}{2} = \dfrac{k(k + 1)(k + 2)}{6}$.

Add $\dfrac{(k + 1)(k + 2)}{2}$ to both sides: $1 + 3 + 6 + \cdots + \dfrac{k(k + 1)}{2} + \dfrac{(k + 1)(k + 2)}{2} = \dfrac{k(k + 1)(k + 2)}{6} + \dfrac{(k + 1)(k + 2)}{2}$

$= (k + 1)(k + 2)\left(\dfrac{k}{6} + \dfrac{1}{2}\right) = (k + 1)(k + 2)\left(\dfrac{k + 3}{6}\right) = \dfrac{(k + 1)((k + 1) + 1)((k + 1) + 2)}{6}$, so P_{k+1} is true.

Therefore, P_n is true for all $n \geq 1$.

93. $P_n: 2^{n-1} \leq n!$. P_1 is true: $2^{1-1} \leq 1!$ (they are equal). Now assume P_k is true: $2^{k-1} \leq k!$.

Then $2^{(k+1)-1} = 2 \cdot 2^{k-1} \leq 2 \cdot k! \leq (k + 1)k! = (k + 1)!$, so P_{k+1} is true. Therefore, P_n is true for all $n \geq 1$.

95. (a)

```
 9 | 1 2
10 | 6 7
11 | 4 5 5 7 7
12 | 0 2 4 6 7 7
13 | 5 6
14 | 1 6 7 7 8
15 | 4 8
16 | 1 4
17 | 0 6
18 |
19 |
20 |
21 | 9
22 |
23 | 4
```

(b)

Price	Frequency
90,000– 99,999	2
100,000–109,999	2
110,000–119,999	5
120,000–129,999	6
130,000–139,999	2
140,000–149,999	5
150,000–159,999	2
160,000–169,999	2
170,000–179,999	2
210,000–219,999	1
230,000–239,999	1

(c)

[8, 24] by [−1, 7]

97. (a)

12	0 0 4 4
13	1 1 2 6 7 9
14	0 3 4 8
15	6
16	3
17	7 9
18	0
19	0 1 7
20	2
21	
22	
23	0

(b)

Length (in seconds)	Frequency
120–129	4
130–139	6
140–149	4
150–159	1
160–169	1
170–179	2
180–189	1
190–199	3
200–209	1
210–219	0
220–229	0
230–239	1

(c)

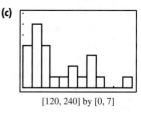

[120, 240] by [0, 7]

99. Five-number summary: {9.1, 11.7, 13.1, 15.4, 23.4}; Range: 14.3; IQR: 3.7; $\sigma \approx 3.19$, $\sigma^2 \approx 10.14$; Outliers: 21.9 and 23.4

101. Five-number summary: {120, 131.5, 143.5, 179.5, 230}; Range: 110; IQR: 48; $\sigma = 29.9$, $\sigma^2 = 891.4$; No outliers

103. (a)

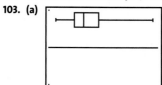

[8, 24] by [−1, 1]

(b)

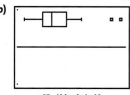

[8, 4] by [−1, 1]

105. (a)

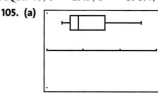

[100, 250] by [−1, 1]

(b)

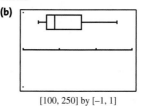

[100, 250] by [−1, 1]

107.

Earlier		Later
4 0 0	12	4
9 2 1	13	1 6 7
8 4 3 0	14	
	15	6
3	16	
7	17	9
	18	0
	19	0 1 7
	20	2
	21	
	22	
	23	0

The songs released in the earlier years tended to be shorter.

109.

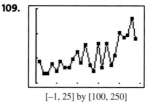

[−1, 25] by [100, 250]

Again, the data demonstrates that songs appearing later tended to be longer.

111. 1 9 36 84 126 126 84 36 9 1

113. (a) ≈ 0.922 **(b)** ≈ 0.075

Chapter 9 Project

Answers are based on the sample data shown in the table.

1.

5	
5	9
6	1 1 2 3 3 3 4 4 4
6	5 6 6 6 7 8 8 9 9 9
7	0 0 1 1 1 2 2 3
7	5

66 or 67 inches

3.

[59, 78] by [−1, 7]

5. The data set is well distributed and probably does not have outliers.

7.

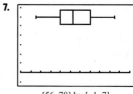

[56, 78] by [−1, 7]

SECTION 10.1

Exploration 1
1. 3   **3.** They are the same.

Quick Review 10.1

1. $-4/7$ **3.** $y = (3/2)x + 6$ **5.** $y - 4 = \dfrac{3}{4}(x - 1)$ **7.** $h + 4$ **9.** $-\dfrac{1}{2(h + 2)}$

Exercises 10.1

1. 12 mi per hour   **3.** 3   **5.** $4a$

7. 1   **9.** no tangent   **11.** 4 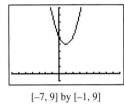   **13.** 12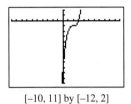

$[-7, 9]$ by $[-1, 9]$ $[-10, 11]$ by $[-12, 2]$

15. (a) 48   **(b)** 48 ft/sec

17. (a) -4 **19. (a)** 1

(b) $y - 2 = -4(x + 1)$ **(b)** $y = x - 5$

(c) 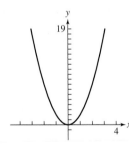   **(c)**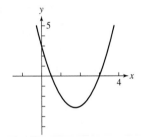

21. -1; 1; none   **23.** -4 **25.** -12 **27.** does not exist   **29.** -3

31. $6x + 2$ **33. (a)** 9 ft/sec; 15 ft/sec   **(b)** $f(x) = 8.94x^2 + 0.05x + 0.01$, $x = $ time in seconds   **(c)** ≈ 35.9 ft

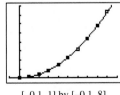

$[-0.1, 1]$ by $[-0.1, 8]$

35. (a)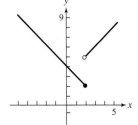

(b) Since the graph of the function does not have a definable slope at $x = 2$, the derivative of f does not exist at $x = 2$.

(c) Derivatives do not exist at points where functions have discontinuities.

37. (a)

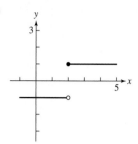

(b) Since the graph of the function does not have a definable slope at $x = 2$, the derivative of f does not exist at $x = 2$.

(c) Derivatives do not exist at points where functions have discontinuities.

39. Possible answer:

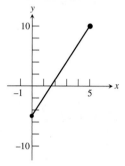

41. Possible answer:

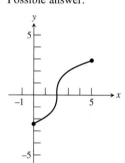

43. The slope of the line is a; $f'(x) = a$

45. False; the instantaneous velocity is a limit of average velocities. It is nonzero when the ball is moving.

47. D   **49.** C

51.

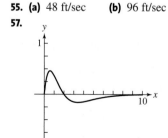

[−4.7, 4.7] by [−3.1, 3.1]

53.

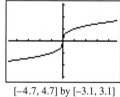

[−4.7, 4.7] by [−3.1, 3.1]

(a) No, there is no derivative because the graph has a corner at $x = 0$.

(b) No

(b) Yes, $x = 0$

(a) No, there is no derivative because the graph has a vertical tangent (no slope) at $x = 0$.

55. (a) 48 ft/sec   **(b)** 96 ft/sec

57.

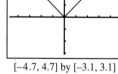

SECTION 10.2

Exploration 1

1. 0.1 gal; 1 gal   **3.** 0.000000001 gal; 1 gal

Quick Review 10.2

1. $\frac{1}{8}, \frac{1}{2}, \frac{9}{8}, 2, \frac{25}{8}, \frac{9}{2}, \frac{49}{8}, 8, \frac{81}{8}, \frac{25}{2}$ **3.** $\frac{65}{2}$ **5.** $\frac{505}{2}$ **7.** 228 miles   **9.** 4,320,000 ft^3

Exercises 10.2

1. 195 mi   **3.** 540,000 ft^3 **5.** 2176 km   **7.** 13; Answers will vary.   **9.** 13; Answers will vary.   **11.** 32.5

13. $\left[0, \frac{1}{2}\right], \left[\frac{1}{2}, 1\right], \left[1, \frac{3}{2}\right], \left[\frac{3}{2}, 2\right]$

15. $\left[1, \frac{3}{2}\right], \left[\frac{3}{2}, 2\right], \left[2, \frac{5}{2}\right], \left[\frac{5}{2}, 3\right], \left[3, \frac{7}{2}\right], \left[\frac{7}{2}, 4\right]$

17. (a)
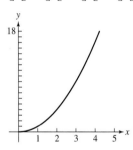
(b)
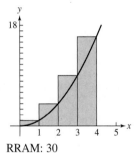
RRAM: 30
(c)
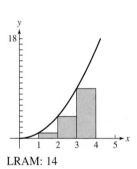
LRAM: 14
(d) Average: 22

19. (a)
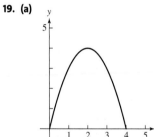
(b)
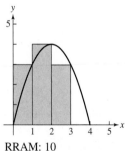
RRAM: 10
(c)
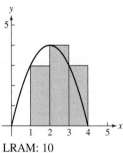
LRAM: 10
(d) Average: 10

21. 20   **23.** 37.5   **25.** 16.5   **27.** 2π **29.** 2   **31.** 2   **33.** 1   **35.** 4   **37.** 4   **39.** $8k + 12$ **41.** $24 + 4k$ **45.** 64 ft

47. (a)
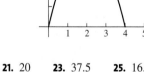
[0, 3] by [0, 50]

49. (a)
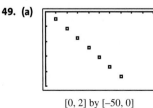
[0, 2] by [−50, 0]

(b) $t = 1.5$ sec   **(c)** 36 ft

(b) 33.86 ft

51. True; the exact area is given by the limit as $n \to \infty$.   **53.** A   **55.** C   **57.** $\int_0^1 (x - 1)\, dx = -\frac{1}{2}$ **59.** true   **61.** false   **63.** false

SECTION 10.3

Exploration 2

1. 50; 0

Quick Review 10.3

1. (a) $-\dfrac{3}{64}$ **(b)** $\dfrac{1}{16}$ **(c)** undefined **3. (a)** $x = -2$ and $x = 2$ **(b)** $y = 2$ **5. (b)** **7. (a)** $[-2, \infty)$ **(b)** None

9.

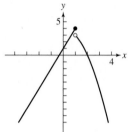

Exercises 10.3

1. -4 **3.** 7 **5.** $\sqrt{7}$ **7.** 0 **9.** $a^2 - 2$ **11. (a)** Division by zero **(b)** $-\dfrac{1}{6}$ **13. (a)** Division by zero **(b)** 3

15. (a) Division by zero **(b)** -4 **17. (a)** The square root of negative numbers is not defined in the real plane. **(b)** The limit does not exist.

19. -1 **21.** 0 **23.** 2 **25.** $\ln \pi$

27. (a) 3 **(b)** 1 **(c)** none **29. (a)** 4 **(b)** 4 **(c)** 4 **31. (a)** true **(b)** true **(c)** false **(d)** False **(e)** false **(f)** false **(g)** false

(h) true **(i)** false **(j)** true **33. (a)** ≈ 2.72 **(b)** ≈ 2.72 **(c)** ≈ 2.72 **35. (a)** 6 **(b)** -4 **(c)** 16 **(d)** -2

37. (a)

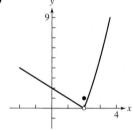

(b) 0; 0
(c) 0

39. (a)

(b) 0; 3
(c) Does not exist; $\lim\limits_{x \to 0^-} f(x) \neq \lim\limits_{x \to 0^+} f(x)$

41. 2 **43.** 0 **45.** 1 **47.** 0; 0 **49.** ∞; 1 **51. (a)** ∞ **(b)** $-\infty$ **53. (a)** Undefined **(b)** 0 **55.** $-\infty$; $x = 3$ **57.** ∞; $x = -2$

59. ∞; $x = 5$ **61.** 3 **63.** 1 **65.** ∞ **67.** 0 **69.** undefined **71.** $\dfrac{1}{2}$ **73.** False; $\lim\limits_{x \to 3} f(x) = 5$ **75.** B **77.** C

79. (a)

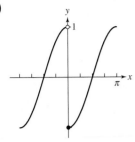

(b) $(-\pi, 0) \cup (0, \pi)$
(c) $x = \pi$
(d) $x = -\pi$

81. (a)

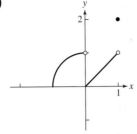

(b) $(-1, 0) \cup (0, 1)$
(c) $x = 1$
(d) $x = -1$

83. (a)

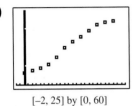

$[-2, 25]$ by $[0, 60]$

(b) $f(x) \approx \dfrac{57.71}{1 + 6.39e^{-0.19x}}$, where $x =$ the number

of months; $\lim\limits_{x \to \infty} f(x) \approx 57.71$

(c) It's about 58,000.

85.

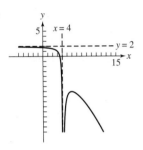

87.

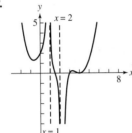

89. (d)

n	A
4	4
8	3.3137
16	3.1826
100	3.1426
500	3.1416
1,000	3.1416
5,000	3.1416
10,000	3.1416
100,000	3.1416

Yes, $A \to \pi$ as $n \to \infty$.

(e)

n	A
4	36
8	29.823
16	28.643
100	28.284
500	28.275
1,000	28.274
5,000	28.274
10,000	28.274
100,000	28.274

As $n \to \infty$, $A \to 9\pi$.

(b) $y = \dfrac{2x + 4}{x + 2} = \dfrac{2(x + 2)}{x + 2} = 2$
(c) $y = 2$

(f) One possible answer:
$$\lim_{n \to \infty} A = \lim_{n \to \infty} nh^2 \tan\left(\frac{180°}{n}\right)$$
$$= h^2 \lim_{n \to \infty} n \tan\left(\frac{180°}{n}\right)$$
$$= h^2\pi = \pi h^2$$

As the number of sides of the polygon increases, the distance between h and the edge of the circle becomes progressively smaller. As $n \to \infty$, $h \to$ radius of the circle.

91. (a)

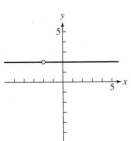

93. (a)

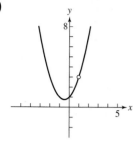

(b) $y = \dfrac{x^3 - 1}{x - 1} = \dfrac{(x - 1)(x^2 + x + 1)}{x - 1}$

$= x^2 + x + 1$

(c) $y = x^2 + x + 1$

SECTION 10.4

Exploration 1

1. 1.364075504   **3.** $\displaystyle\int_0^\pi \sin x\, dx$; sum(seq(sin(0 + K*π/50)*π/50, K, 1, 50)) = 1.999341983; fnInt(sin(X), X, 0, π) = 2

Quick Review 10.4

1. 5   **3.** 2/3   **5.** 3   **7.** ≈ 0.5403   **9.** ≈ 1.000

Exercises 10.4

1. −4   **3.** −12   **5.** 0   **7.** ≈ 1.0000   **9.** ≈ −3.0000   **11.** 64/3   **13.** 2   **15.** ≈ 0   **17.** 1   **19.** ≈ 3.1416   **21.** 106.61 mi

23. (a) −50 ft/sec   **(b)**

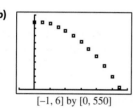

[−1, 6] by [0, 550]

(c) $s(t) = -16.08t^2 + 0.36t + 499.77$
(d) ≈ −47.88 ft/sec
(e) ≈ 179.28 ft/sec

25. (a)

Midpoint	$\Delta s/\Delta t$
0.25	−10
0.75	−20
1.25	−40
1.75	−60
2.25	−70
2.75	−90
3.25	−100
3.75	−120
4.25	−140
4.75	−150
5.25	−170

(b)

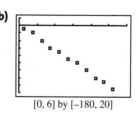

[0, 6] by [−180, 20]

(c) Approx. −47.95 ft/sec; this is close to the results in Exercise 23.

27. 100 ft

31. (b)

N	LRAM	RRAM	Average
10	15.04	19.84	17.44
20	16.16	18.56	17.36
50	16.86	17.82	17.34
100	17.09	17.57	17.33

(c) fnInt gives 17.33; at N_{100}, the average is 17.3344.

33. (b)

N	LRAM	RRAM	Average
10	7.84	11.04	9.44
20	8.56	10.16	9.36
50	9.02	9.66	9.34
100	9.17	9.49	9.33

(c) fnInt gives 9.33; at N_{100}, the average is 9.3344.

35. (b)

N	LRAM	RRAM	Average
10	98.24	112.64	105.44
20	101.76	108.96	105.36
50	103.90	106.78	105.34
100	104.61	106.05	105.33

(c) fnInt gives 105.33; at N_{100}, the average is 105.3344.

37. (b)

N	LRAM	RRAM	Average
10	7.70	8.12	7.91
20	7.81	8.02	7.91
50	7.87	7.95	7.91
100	7.89	7.93	7.91

(c) fnInt gives 7.91, the same result as N_{100}

39. (b)

N	LRAM	RRAM	Average
10	1.08	0.92	1.00
20	1.04	0.96	1.00
50	1.02	0.98	1.00
100	1.01	0.99	1.00

(c) fnInt gives 1, the same result as N_{100}.

41. (b)

N	LRAM	RRAM	Average
10	0.56	0.62	0.59
20	0.58	0.61	0.59
50	0.59	0.60	0.59
100	0.59	0.60	0.59

(c) fnInt $= 0.59$, the same result as N_{100}

43. True; the notation NDER refers to a symmetric difference quotient using $h = 0.001$.   **45.** B   **47.** C

49. (a) $4x + 3$ **(b)** $3x^2$ **(c)** 11.002, 11   **(d)** The symmetric method provides a closer approximation to $f'(2) = 11$.
(e) 12.006001; 12.000001; symmetric

51. The values of $f(0 + h)$ and $f(0 - h)$ are the same.   **53. (a)** 4   **(b)** ≈ 19.67

57. (b)

x	A(x)
0.25	0.0156
0.5	0.125
1	1
1.5	3.375
2	8
2.5	15.625
3	27

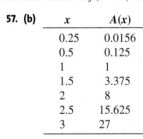

[0, 5] by [−5, 30]

(c) $y \approx x^3$

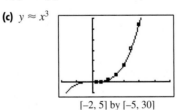

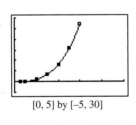

[−2, 5] by [−5, 30]

(d) The exact value of $A(x)$ for any x greater than zero appears to be x^3.
(e) $A'(x) = 3x^2$

CHAPTER 10 REVIEW EXERCISES

1. (a) 2   **(b)** Does not exist   **3. (a)** 2   **(b)** 2

5. −1   **7.** −7   **9.** 0   **11.** 0   **13.** −∞   **15.** ∞

17. $-\dfrac{1}{4}$ **19.** f has vertical asymptotes at $x = -1$ and $x = -5$; f has a horizontal asymptote at $y = 0$ **21.** −8   **23.** $-\dfrac{1}{9}$ **25.** −1

27. $y = \begin{cases} \dfrac{x^3 - 1}{x - 1} & \text{if } x \neq 1 \\ 3 & \text{if } x = 1 \end{cases}$

29. −9   **31. (a)** 8.01   **(b)** 8

33. 1; $y = x - 1$ **35.** $10x + 7$
37. LRAM: 42.2976; RRAM: 40.3776; 41.3376

39. (a)

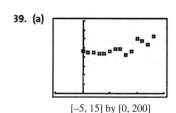

[–5, 15] by [0, 200]

(b) 1990 to 1991: -2.4 cents per year; 1997 to 1998: -17.5 cents per year **(c)** 1999 to 2000
(d) 1997 to 1998 **(e)** $y = 3.0270x + 104.6700$ **(f)** $y = 0.0048x^3 + 0.3659x^2 - 2.4795x +$
116.2006

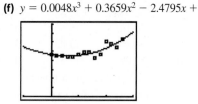

[–5, 15] by [0, 200]

[–5, 15] by [0, 200]

(g) 1997:3.3; 1998:4.3; 1999:5.3; 2000:6.3 **(h)** 203.4 cents per gallon. Could be higher.

Chapter 10 Project

1.

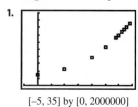

[–5, 35] by [0, 2000000]

3. $y = 271661.8371 \cdot (1.0557797^t)$

5. Regression model predictions: 2,382,109; 4,099,161; 7,053,883. The web site predictions are probably more reasonable, since the scatter plot in Question 1 of this project suggests that growth in recent years has been fairly linear.

APPENDIX A

Appendix A.1

1. 9 or -9 **3.** 4 **5.** $\frac{4}{3}$ or $-\frac{4}{3}$ **7.** 12 **9.** -6 **11.** $-\frac{4}{3}$ **13.** 4 **15.** 2.5 **17.** 729 **19.** 0.25 **21.** -2 **23.** 1.3
25. 2.1 **27.** $12\sqrt{2}$ **29.** $-5\sqrt[3]{2}$ **31.** $|x|y^2\sqrt{2x}$ **33.** $x^2|y|\sqrt[4]{3y^2}$ **35.** $2x^2\sqrt[5]{3}$ **37.** $2\sqrt[3]{4}$ **39.** $\dfrac{\sqrt[5]{x^3}}{x}$ **41.** $\dfrac{\sqrt[3]{x^2y^2}}{y}$

43. $(a + 2b)^{2/3}$ **45.** $2x^{5/3}y^{1/3}$ **47.** $\sqrt[4]{a^3b}$ **49.** $\sqrt[3]{x^{-5}} = 1/\sqrt[3]{x^5}$ **51.** $\sqrt[4]{2x}$ **53.** $\sqrt[8]{xy}$ **55.** $\sqrt[15]{a}$ **57.** $a^{-17/30}$

59. $3a^2b^2$ $(b \geq 0)$ **61.** $4x^4y^2$ **63.** $\dfrac{|x|}{x|y|}$ **65.** $3y^2/|x^3|$ **67.** $|x|\sqrt[4]{6x^2y^2}/2$ **69.** $\dfrac{2x\sqrt[3]{x}}{y}$ **71.** 0 **73.** $(x - 2|y|)\sqrt{x}$

75. $<$ **77.** $=$ **79.** $>$ **81.** $<$ **83.** ≈ 3.48 sec **85.** If n is even, then there are two real nth roots of a (when $a > 0$): $\sqrt[n]{a}$
and $-\sqrt[n]{a}$.

Appendix A.2

1. $3x^2 + 2x - 1$; degree 2 **3.** $-x^7 + 1$; degree 7 **5.** no **7.** yes **9.** $4x^2 + 2x + 4$ **11.** $3x^3 - x^2 - 9x + 3$ **13.** $2x^3 - 2x^2 + 6x$
15. $-12u^2 + 3u$ **17.** $-15x^3 - 5x^2 + 10x$ **19.** $x^2 + 3x - 10$ **21.** $3x^2 + x - 10$ **23.** $9x^2 - y^2$ **25.** $9x^2 + 24xy + 16y^2$
27. $8u^3 - 12u^2v + 6uv^2 - v^3$ **29.** $4x^6 - 9y^2$ **31.** $x^3 + 2x^2 - 5x + 12$ **33.** $x^4 + 2x^3 - x^2 - 2x - 3$ **35.** $x^2 - 2$
37. $u - v$, $u \geq 0$ and $v \geq 0$ **39.** $x^3 - 8$ **41.** $5(x - 3)$ **43.** $yz(z^2 - 3z + 2)$ **45.** $(z + 7)(z - 7)$ **47.** $(8 + 5y)(8 - 5y)$
49. $(y + 4)^2$ **51.** $(2z - 1)^2$ **53.** $(y - 2)(y^2 + 2y + 4)$ **55.** $(3y - 2)(9y^2 + 6y + 4)$ **57.** $(1 - x)(1 + x + x^2)$ **59.** $(x + 2)(x + 7)$ **61.**
$(z - 8)(z + 3)$ **63.** $(2u - 5)(7u + 1)$ **65.** $(3x + 5)(4x - 3)$ **67.** $(2x + 5y)(3x - 2y)$ **69.** $(x - 4)(x^2 + 5)$ **71.** $(x^2 - 3)(x^4 + 1)$
73. $(c + 3d)(2a - b)$ **75.** $x(x^2 + 1)$ **77.** $2y(3y + 4)^2$ **79.** $y(4 + y)(4 - y)$ **81.** $y(1 + y)(5 - 2y)$ **83.** $2(5x + 4)(5x - 2)$
85. $2(2x + 5)(3x - 2)$ **87.** $(2a - b)(c + 2d)$ **89.** $(x - 3)(x + 2)(x - 2)$ **91.** $(2ac + bc) - (2ad + bd) = c(2a + b) - d(2a + b) =$
$(2a + b)(c - d)$; Neither of the groupings $(2ac - bd)$ and $(-2ad + bc)$ have a common factor to remove.

Appendix A.3

1. $\dfrac{5}{3}$ **3.** $\dfrac{30}{77}$ **5.** $\dfrac{5}{6}$ **7.** $\dfrac{1}{10}$ **9.** All real numbers **11.** $x \geq 4$ or $[4, \infty)$ **13.** $x \neq 0$ and $x \neq -3$ **15.** $x \neq 2$ and $x \neq 1$

17. $x \neq 0$ **19.** $8x^2$ **21.** x^2 **23.** $x^2 + 7x + 12$ **25.** $x^3 + 2x^2$ **27.** $(x - 2)(x + 7)$ cancels out during simplification; the restriction indicates that the values 2 and -7 were not valid in the original expression.   **29.** No factors were removed from the expression.   **31.** $(x - 3)$ ends up in the numerator of the simplified expression; the restriction reminds us that it began in the denominator so that 3 is not allowed.

33. $\dfrac{6x^2}{5}, x \neq 0$ **35.** $\dfrac{x^2}{x - 2}, x \neq 0$ **37.** $-\dfrac{z}{z + 3}, z \neq 3$ **39.** $\dfrac{y + 5}{y + 3}, y \neq 6$ **41.** $\dfrac{4z^2 + 2z + 1}{z + 3}, z \neq \dfrac{1}{2}$ **43.** $\dfrac{x^2 - 3}{x^2}, x \neq -2$

45. $\dfrac{x + 1}{3}, x \neq 1$

47. $-\dfrac{1}{x - 3}, x \neq 1$ and $x \neq -3$ **49.** $\dfrac{2(x - 1)}{x}$ **51.** $\dfrac{1}{y}, y \neq 5, y \neq -5$, and $y \neq \dfrac{1}{2}$ **53.** $\dfrac{2}{x}$ **55.** $\dfrac{3(x - 3)}{28}, x \neq 0$ and $y \neq 0$

57. $\dfrac{x}{4(x - 3)}, x \neq 0$ and $y \neq 0$ **59.** $\dfrac{2x - 2}{x + 5}$ **61.** $\dfrac{1}{3 - x}, x \neq 0$ and $x \neq -3$ **63.** $\dfrac{x^2 + xy + y^2}{x + y}, x \neq y, x \neq 0$, and $y \neq 0$

65. $\dfrac{x + 3}{x - 3}, x \neq 4$ and $x \neq \dfrac{1}{2}$ **67.** $-\dfrac{2x + h}{x^2(x + h)^2}, h \neq 0$ **69.** $a + b, a \neq 0, b \neq 0$, and $a \neq b$ **71.** $\dfrac{1}{xy}, x \neq -y$ **73.** $\dfrac{x + y}{xy}$

APPENDIX C

Appendix C.1

1. (a) False statement   **(b)** Not a statement   **(c)** False statement   **(d)** Not a statement   **(e)** Not a statement   **(f)** Not a statement   **(g)** True statement   **(h)** Not a statement   **(i)** Not a statement   **(j)** Not a statement   **3. (a)** There is no natural number x such that $x + 8 = 11$.
(b) There exists a natural number x such that $x + 0 \neq x$.   **(c)** There is no natural x such that $x^2 = 4$.   **(d)** There exists a natural number x such that $x + 1 = x + 2$.   **5. (a)** The book does not have 500 pages.   **(b)** Six is not less than eight.   **(c)** $3 \cdot 5 \neq 15$ **(d)** No people have blond hair.
(e) Some dogs do not have four legs.   **(f)** All cats have nine lives.   **(g)** Some squares are not rectangles.   **(h)** All rectangles are squares.
(i) There exists a natural number x such that $x + 3 \neq 3 + x$.   **(j)** For all natural numbers x, $3 \cdot (x + 2) \neq 12$.   **(k)** Not every counting number is divisible by itself and 1.   **(l)** All natural numbers are divisible by 2.   **(m)** For some natural number x, $5x + 4x \neq 9x$.   **7. (a)** F   **(b)** T   **(c)** T
(d) F   **(e)** F   **(f)** T   **(g)** F   **(h)** F   **(i)** F   **(j)** F   **9. (a)** $R \cup S$ **(b)** $Q \cap \overline{Q}$ **(c)** $\overline{R \cup Q}$ **(d)** $P \cap (R \cup S)$ **11. (a)** The statements $\sim(p \vee q)$ and $\sim p \wedge \sim q$ are equivalent, and the statements $\sim(p \wedge q)$ and $\sim p \vee \sim q$ are equivalent.   **(b)** The corresponding DeMorgan Laws for sets are $\overline{P \cup Q} = \overline{P} \cap \overline{Q}$ and $\overline{P \cap Q} = \overline{P} \cup \overline{Q}$. The analogy comes from letting p mean "x is a member of P" and letting q mean "x is a member of Q." Then, for the first law, $\sim(p \vee q)$ means "x is a member of $\overline{P \cup Q}$," which is equivalent to "x is a member of $\overline{P} \cap \overline{Q}$," which translates into $\sim p \wedge \sim q$.
13 (a) Today is not Wednesday or the month is not June.   **(b)** I did not eat breakfast yesterday, or I did not watch television yesterday.   **(c)** It is not true that both it is raining and it is July.

Appendix C.2

1. (a) $p \rightarrow q$ **(b)** $\sim p \rightarrow q$ **(c)** $p \rightarrow \sim q$ **(d)** $p \rightarrow q$ **(e)** $\sim q \rightarrow \sim p$ **(f)** $q \leftrightarrow p$ **3. (a)** Converse: If you're good in sports, then you eat Meaties; Inverse: If you don't eat Meaties, then you're not good in sports; Contrapositive: If you're not good in sports, then you don't eat Meaties.
(b) Converse: If you don't like mathematics, then you don't like this book; Inverse: If you like this book, then you like mathematics; Contrapositive: If you like mathematics, then you like this book.   **(c)** Converse: If you have cavities, then you don't use Ultra Brush toothpaste; Inverse: If you use Ultra Brush toothpaste, then you don't have cavities; Contrapositive: If you don't have cavities, then you use Ultra Brush toothpaste.   **(d)** Converse: If your grades are high, then you're good at logic; Inverse: If you're not good at logic, then your grades aren't high; Contrapositive: If your grades aren't high, then you're not good at logic.   **5. (a)** T   **(b)** T   **(c)** F   **(d)** F   **(e)** T   **(f)** F   **7.** No
9. If a number is not a multiple of 4, then it is not a multiple of 8.   **11. (a)** p is false.   **(b)** p is false.   **(c)** q can be true, and in fact q true and p false makes $p \rightarrow q$ true and is the only way for $q \rightarrow p$ to be false.   **13. (a)** Helen is poor.   **(b)** Some freshmen are intelligent.   **(c)** If I study for the final, then I will look for a teaching job.   **(d)** There exist triangles that are isosceles.   **15. (a)** If a figure is a square, then it is a rectangle.   **(b)** If a number is an integer, then it is a rational number.   **(c)** If a figure has exactly three sides, then it may be a triangle.   **(d)** If it rains, then it is cloudy.

Applications Index

Index

C

Calculator-Based Laboratory System (CBL™), 247
Calculator Based Ranger (CBR™), 180
Cardioid curve, 545
Cartesian coordinate system, 13, 685
  circles, 18
  conversion with polar, 535
  distance formula, 15–16, 686
  midpoint formula, 17, 687
  plotting data, 13
CASSEGRAIN, G., 662
Cassegrain telescope, 665
Categorical variable, 759
Center
  of circle, 18
  of data, 772
  of ellipse, 644
  of hyperbola, 657
  of sphere, 687
Central angle, 350
Chain rule, 875. See also *Logic*
Change-of-base formula, 313
Characteristic polynomial, 593
Chord of a conic, 635, 657
Circle
  equation of, 18
  parametric equations for, 522
  and radian measure, 377
  segment of, 496
  unit, 377, 380
Circle graph, 760
Circular functions, 378
Closed interval, 5
Coefficient matrix, 597
Coefficient of determination, 158
Coefficient of term, 200
Cofactor, 585
Cofunction identities, 446–447
Column subscript (matrix), 579
Combinations, 704–705
  n objects taken r at a time, 704
Combinatorics, 700
  combinations, 700
  multiplication counting principle, 701
  permutations, 702–704
  tree diagram, 701
Common denominator, least, 854
  compound fraction, 854
  rational expressions, 854–855
Common difference of sequence, 734
Common logarithm, 302
Common ratio of sequence, 735
Commutative property
  of algebraic expressions, 7
  complex numbers, 53

Complement (angle), 446
Completely factored polynomial, 847
Completing the square, 46
Complex conjugate, 55, 230
Complex fraction, 854
Complex number, 53
  absolute value of, 551
  adding and subtracting, 53
  argument of, 551
  conjugates, 55
  and coordinate plane, 53
  exponents of, 54
  modulus of, 551
  multiplying and dividing, 54–55, 552–553
  nth root, 555–556
  and quadratic equations, 56
  and roots, 555–558
  standard form of, 53
  trigonometric form of, 551
  and vectors, 551
  zeros of function, 229, 230
Complex plane, 550
Complx conjugate zero, 230
Component form of vector, 503, 690
Components of a vector, 503
Composition of functions, 118
  absolute value, 142
Compounded annually, 334
Compounded continuously, 336–337
Compound fraction, 854
Compound interest, 334
  value of investment, 337
Compound statement, 865. See also *Logic*
Conclusion, 869. See also *Logic*
Conditional probability, 724
Conditionals (Implications), 869. See also *Logic*
Conic section (conic), 632
  defined as a ratio, 675
  discriminant test, 671
  ellipse, 644
  focus coincident with pole, 676
  hyperbola, 656
  identifying, 672
  parabola, 633
  and polar equations, 676
  rotation of, 670
  and second-degree equation, 633
  standard form of, 636, 645, 656
  and transformations, 636–637, 647, 659
Conjugate, complex, 55
Conjugate axis of hyperbola, 657
Conjugate hyperbolas, 665
Conjunction, 865. See also *Logic*
Constant, 6
  of proportion, 188
  of variation, 188
Constant function on an interval, 92, 93

Constant percentage rate, 290
Constant term of polynomial, 173
Constraints, 620
Continuous at a point, 90
Continuous function, 109
  limit of, 816
Convergent series, 733, 747
Conversion, degree-radian, 353
Conversion factors, 155
Coordinate conversion equations (polar), 535
Coordinate plane
  Cartesian, 13
  complex, 53
  polar, 534
  quadrants of, 14
Coordinate planes (space)
  xy-plane, 685
  xz-plane, 685
  yz-plane, 685
Coordinates of a point, 3, 13, 685
Corner (vertex) point, 620
Correlation coefficient, 158, 174
Cosecant function, 360, 399
  acute angle, 360
  any angle, 373
  graph of, 399
  of special angles, 361
Cosine function, 360, 386
  acute angles, 360
  any angle, 373
  cofunction identities, 487
  of a difference identity, 463–464
  graph of, 386
  harmonic motion, 428
  inverse, 416
  law of cosines, 487–489
  period of, 386, 387
  special angles, 361
  of a sum identity, 464
Cotangent function, 360, 397
  acute angles, 360
  any angle, 373
  graph of, 397, 398
  of special angles, 361
Coterminal angle, 370
Counting. See *Combinatorics*
Counting subsets of an n set, 706
Course, navigational, 351
Cross-product term, 695
Cube
  of difference, 846, 848
  of sum, 846, 848
Cube root, 557, 839
Cubic inequality, 62
Cubic polynomial function, 200
  graphing, 202
  regression, 157

Credits

Photographs